Methods in Enzymology

Volume XLV

PROTEOLYTIC ENZYMES

Part B

METHODS IN ENZYMOLOGY

EDITORS-IN-CHIEF

Sidney P. Colowick   Nathan O. Kaplan

Methods in Enzymology

Volume XLV

Proteolytic Enzymes

Part B

EDITED BY

Laszlo Lorand

DEPARTMENT OF BIOCHEMISTRY AND MOLECULAR BIOLOGY
NORTHWESTERN UNIVERSITY
EVANSTON, ILLINOIS

1976

ACADEMIC PRESS New York San Francisco London
A Subsidiary of Harcourt Brace Jovanovich, Publishers

ACADEMIC PRESS, INC.
111 Fifth Avenue, New York, New York 10003

United Kingdom Edition published by
ACADEMIC PRESS, INC. (LONDON) LTD.
24/28 Oval Road, London NW1

Library of Congress Cataloging in Publication Data

Main entry under title:

Proteolytic enzymes.

    (Methods in enzymology, v. 19, 45   )
    Includes bibliographical references.
    Part B: L. Lorand (volume editor)
    1.   Proteolytic enzymes.   I.   Perlmann, Gertrude E.,
Date   ed.   II.   Lorand, Laszlo,   Date   ed.
III.   Series: Methods in enzymology ; v. 19, etc.
[DNLM: 1.   Peptide hydrolases—Analysis.   W1 ME9615K
v. 19 etc. / QU135 P968]
QP601.M49 vol. 45   [QP609.P78]   574.1'925'08s   75-26936
    [574.1'9256]
ISBN 0-12-181945-0

Table of Contents

Section I. General Aspects

Section II. Blood Clotting Enzymes
A. Enzymes of Blood Coagulation

Section VI. Dipeptidases

Section VII. Endopeptidases

Section VIII. Exopeptidases
A. Aminopeptidases

B. Carboxypeptidases

Section IX. Naturally Occuring Protease Inhibitors
A. Specific Inhibitors of Clotting and Lysis in Blood

B. Inhibitors from Bacteria

C. Inhibitors from Plants

D. Protease Inhibitors from Various Sources

Contributors to Volume XLV

Article numbers are in parentheses following the names of contributors.
Affiliations listed are current.

ANDRANIK BAGDASARIAN (25), *Department of Medicine, University of Pennsylvania School of Medicine, Philadelphia, Pennsylvania*

DANIEL BAGDY (54), *Research Institute for Pharmaceutical Chemistry, Budapest, Hungary*

EVA BARABAS (54), *Research Institute for Pharmaceutical Chemistry, Budapest, Hungary*

G. H. BARLOW (17, 18, 20), *Biochemistry Laboratory, Abbott Laboratories, North Chicago, Illinois*

DENNIS BARRETT (29), *Department of Biological Sciences, University of Denver, Denver, Colorado*

DIANA C. BARTELT (72), *Biology Department, Brookhaven National Laboratory, Upton, New York*

FRANCIS J. BEHAL (41), *Department of Biochemistry, Texas Tech University School of Medicine, Lubbock, Texas*

LASZLO BERESS (79), *Institut für Meereskunde an der Universität Kiel, Kiel, West Germany*

YEHUDITH BIRK (56, 57, 58, 59, 60, 61, 62, 63), *Department of Agricultural Biochemistry, Faculty of Agriculture, The Hebrew University of Jerusalem, Rehovot, Israel*

WILLIAM J. BROCKWAY (21), *The Department of Chemistry, The University of Notre Dame, Notre Dame, Indiana*

EDWARD J. CARROLL, JR. (28), *Department of Zoology, University of Maryland, College Park, Maryland*

FRANCIS J. CASTELLINO (21, 23), *The Department of Chemistry, The University of Notre Dame, Notre Dame, Indiana*

DANA ČECHOVA (71), *Institute of Organic Chemistry and Biochemistry, Czechoslovak Acedemy of Sciences, Flemingovo, Czechoslovakia*

CHARLES G. COCHRANE (7), *Department of Immunopathology, Scripps Clinic and Research Foundation, La Jolla, California*

P. L. COLEMAN (2), *Biology Department, Brookhaven National Laboratory, Upton, New York*

ROBERT W. COLMAN (12, 25), *Department of Medicine, University of Pennsylvania Medical School, Philadelphia, Pennsylvania*

C. G. CURTIS (15), *Department of Biochemistry, University College, Cardiff, Wales, United Kingdom*

PAUL S. DAMUS (53), *Department of Cardiac and Thoracic Surgery, Columbia-Presbyterian Medical Center, New York, New York*

EARL W. DAVIE (7a, 8, 9, 10), *Department of Biochemistry, University of Washington, Seattle, Washington*

THOMAS DIETL (67, 68), *Organische-Chemisches Institut, Lehrstuhl für Organische Chemie und Biochemie, München, West Germany*

GABRIEL R. DRAPEAU (38), *Department of Microbiology, University of Montreal, Montreal, Quebec, Canada*

BEN F. EDWARDS (29), *Department of Biology, Reed College, Portland, Oregon*

FRANZ FIEDLER (24), *Institut für Klinische Chemie und Klinische Biochemie, Universität München, München, West Germany*

EDWIN FINK (73, 74, 75), *Institut für Klinische Chemie und Klinische Biochemie, der Universität München, München, West Germany*

JEAN-MARIE FRERE (51), *Service de Microbiologie, Faculte de Medecine, Université de Liège, Institut de Botanique, Liège, Belgium*

HANS FRITZ (27, 70, 73, 74, 75, 76, 79), *Institut fur Klinische Chemie und Klinische Biochemie der Universität München, München, West Germany*

MARLIES FROHNE (42), *Physiologisch-chemisches Institut der Martin-Luther-Universität, Halle-Wittemberg, Germany*

KAZUO FUJIKAWA (8, 10), *Department of Biochemistry, University of Washington, Seattle, Washington*

BARBARA C. FURIE (16), *Hematology Section, Department of Medicine, Tufts-New England Medical Center and Tufts University Medical School, Boston, Massachusetts*

BRUCE FURIE (16), *Hematology Section, Department of Medicine, Tufts-New England Medical Center and Tufts University School of Medicine, Boston, Massachusetts*

G. MAURICE GAUCHER (34), *Biochemistry Group, Department of Chemistry, The University of Calgary, Calgary, Alberta, Canada*

JEAN-MARIE GHUYSEN (51), *Service de Microbiologie, Faculte de Medecine, Université de Liége, Institut de Botanique, Liége, Belgium*

LASZLO GRAF (54), *Research Institute for Pharmaceutical Chemistry, Budapest, Hungary*

LEWIS J. GREENE (72), *Biology Department, Brookhaven National Laboratory, Upton, New York*

JOHN H. GRIFFIN (7), *Department of Immunopathology, Scripps Clinic and Research Foundation, La Jolla, California*

L. S. HALL (17), *Biochemistry Laboratory, Abbott Laboratories, North Chicago, Illinois*

HORST HANSON (42), *Physiologisch-chemisches Institut der Martin-Luther-Universität, Halle-Wittenberg, Germany*

PETER C. HARPEL (52, 65), *Department of Medicine, Division of Hematology, The New York Hospital-Cornell Medical Center, New York, New York*

RIKIMARU HAYASHI (48), *Research Institute for Food Science, Kyoto University, Uji, Kyoto, Japan*

ROBERT L. HEINRIKSON (64), *Department of Biochemistry, University of Chicago, Chicago, Illinois*

KARL HOCHSTRASSER (76, 77), *Biochemisches Laboratorium der HNO-Klinik, der Universität München, München, West Germany*

THEO HOFMANN (35, 49), *Department of Biochemistry, University of Toronto, Toronto, Ontario, Canada*

SADAAKI IWANAGA (37, 78), *Division of Plasma Proteins, Institute for Protein Research, Osaka University, Suita, Osaka, Japan*

JOLYON JESTY (11), *Department of Medicine—Hematology, State University of New York, Health Sciences Center, Stony Brook, New York*

E. T. KAISER (1), *Departments of Biochemistry and Chemistry, University of Chicago, Chicago, Illinois*

BEATRICE KASSELL (36), *Department of Biochemistry, Medical College of Wisconsin, Milwaukee, Wisconsin*

HISAO KATO (7a), *Department of Biochemistry, University of Washington, Seattle, Washington*

EFRAT KESSLER (45), *Eye and Ear Hospital, University of Pittsburgh School of Medicine, Pittsburgh, Pennsylvania*

FERENC J. KEZDY (1, 64), *Department of Biochemistry, University of Chicago, Chicago, Illinois*

MAMORU KIKUCHI (40), *Noda Institute for Scientific Research, Noda-shi, Chiba-Ken, Japan*

HENRY S. KINGDON (14), *Dental Research Center and the Departments of Medicine and Biochemistry, University of North Carolina School of Medicine, Chapel Hill, North Carolina*

HENNING KLOSTERMEYER (3), *Chemisches Institut der Bundesanstalt für Milchforschung, Kiel, Germany*

TAKEHIKO KOIDE (7a), *Department of Biochemistry, University of Washington, Seattle, Washington*

KAJ KREJCI (70), *Kinderpoliklinik der Universität München, München, West Germany*

H. G. LATHAM, JR. (2), *Biology Department, Brookhaven National Laboratory, Upton, New York*

MARK E. LEGAZ (9), *Department of Bio-*

chemistry, University of Washington, Seattle, Washington

MÉLINA LEYH-BOUILLE (51), Service de Microbiologie, Faculte de Medecine, Université de Liége, Institut de Botanique, Liége, Belgium

GWYNNE H. LITTLE (41), Department of Biochemistry, Texas Tech University School of Medicine, Lubbock, Texas

L. LORAND (4, 15), Department of Biochemistry and Molecular Biology, Northwestern University, Evanston, Illinois

ROGER L. LUNDBLAD (14), Dental Research Center and the Departments of Biochemistry and Pathology, University of North Carolina School of Medicine, Chapel Hill, North Carolina

WERNER MACHLEIDT (79), Institut für Physiologische Chemie und Physikalische Biochemie der Universität München, München, West Germany

STAFFAN MAGNUSSON (54), Department of Molecular Biology, University of Aarhus, Aarhus, Denmark

KENNETH G. MANN (13, 14), Section of Hematology Research, Mayo Clinic and Foundation, Rochester, Minnesota

FRANCIS S. MARKLAND, JR. (19), Department of Biological Chemistry, University of Southern California, School of Medicine, and Cancer Center, Los Angeles, California

TAKASHI MURACHI (39), Department of Clinical Science, Kyoto University, Faculty of Medicine, Sakyoku, Kyoto, Japan

KOZO NARITA (46), Institute for Protein Research, Osaka University, Suita Osaka, Japan

YALE NEMERSON (5, 6, 11), Department of Medicine—Hematology, State University of New York, Health Sciences Center, Stony Brook, New York

MANUEL NIETO (51), Centro de Investigaciones Biologicas, Instituto de Biologia Celular, Madrid, Spain

C. NOLAN (17), Biochemistry Laboratory, Abbott Laboratories, North Chicago, Illinois

GENICHIRO OSHIMA (37), Division of Plasma Proteins, Institute for Protein Research, Osaka University, Suita, Osaka, Japan

ELIZABETH K. PATTERSON (30, 31), The Institute for Cancer Research, Fox Chase Cancer Center, Philadelphia, Pennsylvania

HAROLD R. PERKINS (51), Department of Microbiology, University of Liverpool, Liverpool, United Kingdom

TORBEN ELLEBAEK PETERSON (54), Department of Molecular Biology, University of Aarhus, Aarhus, Denmark

FRANCES ANN PITLICK (5), Department of Internal Medicine, Yale University School of Medicine, New Haven, Connecticut

K. L. POLAKOSKI (26), Department of Obstetrics and Gynecology, Washington University School of Medicine, St. Louis, Missouri

JOHN M. PRESCOTT (32, 33, 44), Department of Biochemistry and Biophysics, Texas A & M University College Station, Texas

MERTON H. PUBOLS (72), Department of Animal Sciences, Washington State University, Pullman, Washington

ROBERT RADCLIFFE (6), Hematology Research Laboratory, Oklahoma Medical Foundation, Oklahoma City, Oklahoma

ERNST H. REIMERDES (3), Chemisches Institut, der Bundesanstalt für Milchforschung, Kiel, Germany

KENNETH C. ROBBINS (22), Michael Reese Blood Center Chicago, Illinois

G. RONCARI (43), Institut fur Molekularbiologie und Biophysik, Eidgenossische Technische Hochschule, Zurich-Honggerberg, Switzerland

ROBERT D. ROSENBERG (53), Department of Medicine, Harvard Medical School, and Beth Israel Hospital, Boston, Massachusetts

KENJI SAKAGUCHI (40), Mitsubishi-Kasei Institute of Life Sciences, Minamicoya, Machiada-shi, Tokyo, Japan

HANS SCHIESSLER (75), Institut für Kli-

nische Chemie und Klinische Biochemie, der Universität München, München, West Germany

WOLF-DIETER SCHLEUNING (27), *Institut für Klinische Chemie und Klinische Biochemie, der Universität München, München, West Germany*

E. N. SHAW (2), *Biology Department, Brookhaven National Laboratory, Upton, New York*

GERALD E. SIEFRING, JR. (21), *The Department of Chemistry, The University of Notre Dame, Notre Dame, Indiana*

JAMES M. SODETZ (21, 23), *The Department of Chemistry, The University of Notre Dame, Notre Dame, Indiana*

WILLIS L. STARNES (41), *Department of Biochemistry, Texas Tech University School of Medicine, Lubbock, Texas*

M. STEINBUCH (66), *Centre National de Transfusion Sanguine, Paris, France*

KENNETH J. STEVENSON (34), *Biochemistry Group, Department of Chemistry, The University of Calgary, Calgary, Alberta, Canada*

K. STOCKER (18), *Pentapharm, Ltd., Basle, Switzerland*

E. STOLL (43), *Institut für Molekularbiologie und Biophysik, Eidgenossische Technische Hochschule, Zurich-Honggerberg, Switzerland*

LOUIS SUMMARIA (22), *Departments of Medicine and Pathology, Pritzker School of Medicine, The University of Chicago, Chicago, Illinois*

TOMOJI SUZUKI (37, 78), *Division of Plasma Proteins, Institute for Protein Research, Osaka University, Suita, Osaka, Japan*

HIDENOBU TAKAHASHI (78), *Division of Plasma Proteins, Institute for Protein Research, Osaka University, Suita, Osaka, Japan*

HARALD TSCHESCHE (67, 68, 69, 74), *Institut für Organische Chemie, Lehrstuhl für Organische Chemie und Biochemie, Technische Universität München, München, West Germany*

SUSUMU TSUNASAWA (46), *Institute for Protein Research, Osaka University, Suita, Osaka, Japan*

HAMAO UMEZAWA (55), *Microbial Chemistry Research Foundation, Institute of Microbial Chemistry, Tokyo, Japan*

FRED W. WAGNER (32), *Department of Biochemistry and Nutrition, University of Nebraska, Lincoln, Nebraska*

PETER H. WARD (36), *Department of Biochemistry, Medical College of Wisconsin, Milwaukee, Wisconsin*

ROBERT M. WEINBERG (12), *Department of Medicine, Harvard Medical School, Boston, Massachusetts*

STELLA H. WILKES (33, 44), *Department of Biochemistry and Biophysics, Texas A & M University, College Station, Texas*

CHRISTINE L. WRIGHT (36), *Department of Biochemistry, Medical College of Wisconsin, Milwaukee, Wisconsin*

GERT WUNDERER (79), *Institut für Klinische Chemie und Klinische Biochemie der Universität München, München, West Germany*

ARIEH YARON (45, 50), *Department of Biophysics, The Weizmann Institute of Science, Rehovot, Israel*

L. J. D. ZANEVELD (26), *Department of Physiology, University of Illinois at the Medical Center, School of Medicine, Chicago, Illinois*

H. ZUBER (43, 47), *Institut für Molekularbiologie und Biophysik, Eidgenossische Technische Hochschule, Zurich-Honggerberg, Switzerland*

GERTRUDE E. PERLMANN

Preface

This volume on "Proteolytic Enzymes," as the previous one (Volume XIX) in the Methods in Enzymology series, was to have been coedited with the late Dr. Gertrude E. Perlmann of The Rockefeller University. Actually, she still participated in much of its organization, diligently sending out letters even from the hospital during the summer of 1974, but, alas, she could not see the task completed. In her passing, I lost a personal friend and an excellent collaborator. My colleagues will surely share my feelings in dedicating this volume to her memory.

Proteolytic enzymes are assuming increasing significance in biological regulation. A published account of a symposium held in the fall of 1974, entitled "Proteases and Biological Control" [(E. Reich, D. B. Rifkin, and E. Shaw, eds.). Cold Spring Harbor, New York, 1975], presents ample evidence that proteolytic enzymes serve as key agents in many intra- and extracellular phenomena. This is a far cry from the days when proteases were considered merely useful tools for protein degradation in nutrition and for protein sequencing. The delicately coordinated systems of the blood coagulation cascade, fibrinolysis, kinin generation, complement activation, fertilization of eggs, cell migration, and mitogenic transformations are but a few of the outstanding examples. In this volume an effort was made to cover rather extensively the enzymes involved in blood coagulation and fibrinolysis and to include the great variety of naturally occurring inhibitors. In general, it is now clear that the field of protease inhibitors is becoming just as exciting as that of the enzymes themselves.

Special thanks are due to Dr. Joyce Bruner-Lorand and to Mrs. Pauline Velasco for their constant help.

Laszlo Lorand

METHODS IN ENZYMOLOGY

EDITED BY

Sidney P. Colowick and Nathan O. Kaplan

VANDERBILT UNIVERSITY
SCHOOL OF MEDICINE
NASHVILLE, TENNESSEE

DEPARTMENT OF CHEMISTRY
UNIVERSITY OF CALIFORNIA
AT SAN DIEGO
LA JOLLA, CALIFORNIA

I. Preparation and Assay of Enzymes
II. Preparation and Assay of Enzymes
III. Preparation and Assay of Substrates
IV. Special Techniques for the Enzymologist
V. Preparation and Assay of Enzymes
VI. Preparation and Assay of Enzymes (*Continued*)
Preparation and Assay of Substrates
Special Techniques
VII. Cumulative Subject Index

METHODS IN ENZYMOLOGY

EDITORS-IN-CHIEF

Sidney P. Colowick   Nathan O. Kaplan

Section I
General Aspects

[1] Active Site Titration of Cysteine Proteases[1]

By F. J. Kézdy and E. T. Kaiser

Cysteine proteases catalyze the hydrolysis of carboxylic acid derivatives through a double-displacement pathway involving the general-acid general-base catalyzed formation and hydrolysis of an acyl-thiol intermediate.[2] The great similarity of this mechanism to that displayed by serine proteases assures that the active site titration methods elaborated for serine proteases[3] are also applicable without much modification to the determination of active site concentrations of sulfhydryl proteases. Indeed, such methods have been used successfully for several plant sulfhydryl proteases, with due allowance for the specificity of the individual enzymes in the choice of the titrating agent.

However, the presence of an active site thiol group as the acyl acceptor confers special properties to the sulfhydryl proteases, as compared to their serine analogs. Among amino acid side chain functions the SH group of cysteine is unique by its reactivity toward electrophilic reagents, oxidizing agents, and heavy metal ions. Since in most cysteine proteases the active-site thiol is the only sulfhydryl group on the protein molecule, the design of titrating agents is greatly facilitated by this unique reactivity.[4] As an example, the use of chromophoric organomercurials could be singled out for the rapid and sensitive quantitation of sulfhydryl groups at specific sites.[5]

The advantages of a single thiol group located in a catalytically favorable position at the active site are somewhat offset, however, by certain unfavorable characteristics inherent to the cysteine proteases. The high intrinsic nucleophilicity of the thiolate ion readily leads to undesired side reactions, such as disulfide exchanges within the protein molecule itself, or the reaction with trace contaminants. Also, free sulfhydryl groups are extremely sensitive toward oxidation by air, and accurate measurements should be carried out in deoxygenated solutions, under nitrogen atmosphere. Most importantly, however, these enzymes are inhibited by minute amounts of heavy metal ions present in virtually all buffers. To counteract these inactivating factors, reproducible kinetic experiments with cysteine proteases require the presence of large amounts

[1] Supported in part by grants GM-15951 and GM-13885 from the National Institutes of Health.
[2] M. L. Bender and L. J. Brubacher, *J. Am. Chem. Soc.* **86**, 5333 (1964).
[3] F. J. Kézdy and E. T. Kaiser, this series Vol. 19 [1].
[4] R. L. Heinrikson and K. J. Kramer, *Prog. Bioorg. Chem.* **3**, 179 (1974).
[5] C. H. McMurray and D. R. Trenton, *Biochem. J.* **115**, 913 (1969).

of protecting agents, such as 1 mM cysteine or dithiothreitol, in conjunction with comparable concentrations of chelating agents, such as EDTA or EGTA. Such conditions preclude the use of thiol-specific reagents unless the reagents incorporate active site-directed binding groups. Finally, the broad pH optimum of most of the sulfhydryl proteases results in significant autoproteolysis upon storage for even short periods. For these reasons, the enzymes are prepared and stored in an inactive form, usually as mercury or as phenylmercury complexes. They are activated then either *in situ* or immediately before use.

Owing to the instability of these enzymes and their possible contamination by enzymatically inactive thiols, the titration of active sites by thiol-specific reagents must be accompanied by measurement of the parallel loss of enzymic activity toward a specific substrate. Optimally, a variety of reagents should be used in conjunction with the "burst titration" by a chromogenic substrate, and all methods should yield consistent results before the data are used to calibrate a convenient rate assay in terms of the molarity of the active sites.

In the following discussion, papain and bromelain will be used as examples to illustrate some of the experimental approaches leading to the establishment of rate assays for cysteine proteases.

The Use of an Irreversible, Covalently Bound Inhibitor: Titration of the Active Site of Papain with α-Bromo-4-hydroxy-3-nitroacetophenone (I)[6-8]

At present the titration procedure to be discussed appears to be the most convenient way of determining the operational normality of the active sites in papain solutions. The reaction that has been utilized is the rapid alkylation of the active site sulfhydryl group of papain by α-bromo-4-hydroxy-3-nitroacetophenone (I). This process gives the irreversibly inactivated species shown as structure (II) in Eq. (1) below. The modification process can be followed by direct spectrophotometric titration, utilizing differences in the ultraviolet absorption spectra of the chromophoric group of the reagent (I) and the modified species (II). Alternatively, rate assays of active enzyme remaining after modification with less than stoichiometric amounts of (I) can be performed employing suitable substrates or sulfhydryl titrants. By linear extrapolation of the rate assay measurements to the point of quantitative inhibition, the equivalence point in the titration can also be established.

[6] R. W. Furlanetto and E. T. Kaiser, *J. Am. Chem. Soc.* **92**, 6980 (1970).
[7] R. W. Furlanetto and E. T. Kaiser, *J. Am. Chem. Soc.* **95**, 6786 (1973).
[8] R. W. Furlanetto, Ph.D. Thesis, University of Chicago, 1972.

(1)

Site of Reaction

The following evidence exists that the modification reaction shown in Eq. (1) occurs at the active site residue Cys_{25}. First, the scheme of Eq. (1) requires that the free sulfhydryl content of papain alkylated by reagent (I) be decreased in proportion to the concentration of the added alkylating agent. To test this, papain purified by affinity chromatography[9] was treated with various concentrations of (I), and the sulfhydryl content of the resultant solutions was measured with the sulfhydryl reagent 5,5'-ditihiobis(2-nitrobenzoic acid).[10] The results obtained clearly showed that the extent of alkylation was parallel to a loss in titratable sulfhydryl groups. Another type of evidence comes from a kinetic study of the alkylation of papain by (I).[7] Specifically, the rate of alkylation of papain by (I) has been measured over the pH range 3 to 10.5. With the exception of the data measured for the more alkaline part of the pH range, it appears that the kinetics of the alkylation reaction can be explained in terms of the attack of the active site sulfhydryl group of papain on (I) assisted by some group in the enzyme acting as a general-base catalyst.[7] At high pH values nucleophilic attack by the thiolate form of Cys_{25} predominates. Thus, the same sulfhydryl group is modified at high pH as in the neutral pH region. However, while the titration of the sulfhydryl group in papain by (I) competes quite effectively in acidic or neutral solution with the reactions of (I) with model sulfhydryl compounds, there is no special selectivity for the sulfhydryl group of papain in very alkaline solutions. Thus, it is not recommended that titrations of the sulfhydryl group in papain be performed with (I) at pH values above pH 8.

[9] S. Blumberg, I. Schechter, and A. Berger, Eur. J. Biochem. 15, 97 (1970).
[10] G. L. Ellman, Arch. Biochem. Biophys. 82, 70 (1959).

Kinetic Considerations

To establish firmly the theoretical basis of an active-site titration procedure, it is necessary to understand the kinetics of the reaction of the titrant with the active site of the enzyme. Some reservations have been expressed concerning the use of irreversible inhibitors as titrating agents, and arguments have been stated in favor of the use of specific substrates instead.[3] The major objections to the use of the irreversible inhibitors are that (1) they do not test the primary chemical reaction catalyzed by the enzyme; (2) they often react with more than one protein functional group; (3) often the reaction does not depend on the specificity or catalytic activity of the enzyme. In the case of compound (I), because of the thorough kinetic investigation of the alkylation of papain with the reagent, only the first objection cannot be eliminated. A major advantage in the use of specific substrates as titrants is that they do allow careful kinetic analysis of the turnover process on the enzyme, thereby providing valuable information concerning the homogeneity of the titrated species.[3,11] In the case of papain an active site titration method based on the use of specific substrate *p*-nitrophenyl *N*-benzyloxycarbonyl L-tyrosinate has been described.[11] However, the titration of the enzyme by this method necessitates a thorough study of the effect of substrate concentration upon the size of the burst observed, and this procedure seems to be considerably less expedient than the methods commonly used for the active site titration of chymotrypsin with substrates, for example. For this reason the direct spectrophotometric titration method described here using reagent (I) is preferred for the determination of the active-site concentrations of papain solutions.

Besides the direct spectrophotometric titration procedure, it was possible to follow the course of the reaction of reagent (I) with papain, using rate assays with *p*-nitrophenyl *N*-benzyloxycarbonyl glycinate (CGN).[12] These assays were carried out in 20 mM phosphate buffer, pH 6.8, 1 mM EDTA, 6.8% acetonitrile with $[\text{CGN}]_0 \gg [\text{E}]_0$. Under these conditions Eq. (2) describes the velocity of the enzyme-catalyzed hydrolysis and $K_{m(\text{app})} \sim 1$ μM. Therefore, when $[\text{CGN}]_0 \sim 0.1$ mM, the initial rate, $v_0 = k_{\text{cat}}[\text{E}]_0$. Since v_0 is directly proportional to $[\text{E}]_0$, it is possible to compare relative enzyme concentrations of two solutions by comparing their initial velocities. By studying the rates of reaction of papain solutions allowed to react with various amounts of reagent (I), it is possible

[11] M. L. Bender, M. L. Begue-Canton, R. L. Blakely, L. J. Brubacher, J. Feder, C. R. Gunter, F. J. Kézdy, J. V. Killheffer, Jr., T. H. Marshall, C. G. Miller, R. W. Roeske, and J. K. Stoops, *J. Am. Chem. Soc.* **88**, 5890 (1966).
[12] J. F. Kirsch and M. Igelstrom, *Biochemistry* **5**, 783 (1966).

to determine the active site concentrations using assays of the remaining activity with the p-nitrophenyl ester. Once the active site concentrations of the papain solutions are determined in this way, then the true value of k_{cat} in Eq. (2) can be calculated, permitting the use of rate assays with CGN for the measurement of absolute concentrations of active papain.

$$v_0 = \frac{k_{cat}[E]_0[CGN]}{K_{m(app)} + [CGN]} \tag{2}$$

Direct Spectrophotometric Titration of the Active Site of Papain

A large change in the extinction coefficient of the o-nitrophenolate group occurs on the reaction of reagent (I) with papain. Although the reaction can be observed at either 262 nm (a drop in ϵ of $6.2 \times 10^3\ M^{-1}$ cm^{-1} at pH 3.5) and or at 323 nm (an increase in ϵ of $6.85 \times 10^3\ M^{-1}$ cm^{-1} at pH 7.0), the longer wavelength was chosen for study because of the large protein absorption below 300 nm. It can be shown that when irreversible enzyme inhibition occurs in the presence of excess inhibitor, this is accompanied by a change in extinction coefficient at 323 nm for which the relationship of Eq. (3) holds. Here $[E]_0$ is the active site concentration, ΔA is the change in absorbance occurring on reaction, $[I]_0$ is the initial inhibitor concentration, and ϵ_E, ϵ_I, and ϵ_{EI} are the extinction coefficients of the free enzyme, inhibitor (reagent I), and inhibited enzyme, respectively.

$$[E]_0 = \frac{\Delta A - \epsilon_I[I]_0}{(\epsilon_{EI} - \epsilon_E - \epsilon_I)} \tag{3}$$

Typically, spectrophotometric titrations of papain solutions using the active-site titrant (I) were performed as follows.[8] The absorbance of a 2-ml solution containing papain in 67 mM phosphate buffer, pH 7.0, 1 mM EDTA was measured in a 1 cm pathlength cell at 323 nm vs air (A_E). A 50-μl aliquot of a stock solution of (I) in acetonitrile was then added with stirring. After 5 min, the absorbance of the solution was again measured at 323 nm (A_{EI}). Finally, the absorbance of the free inhibitor was measured under exactly the same conditions of buffer, pH, wavelength and percent of acetonitrile (A_I). (This reduced an error arising from uncertainty in the concentration of the stock inhibitor solution.) Care was taken to have at least a 1.2-fold excess of reagent (I) present. The final mixture contained 2.42% acetonitrile.

The active site concentration was calculated using Eq. (4) [the operational form of Eq. (3)]. In Eq. (4), A_E, A_{EI}, and A_I are defined as above, $6.85 \times 10^3\ M^{-1}$ cm^{-1} is the value of $(\epsilon_{EI} - \epsilon_E - \epsilon_I)$ determined experimentally and the factor 1.025 corrects for the dilution occurring during the experimental procedure.

$$[E]_0 = \frac{(A_E - A_{EI} - A_I)}{6.85 \times 10^3} \times 1.025 \ (M) \qquad (4)$$

It is clear that to have an excess of inhibitor present either the approximate enzyme concentration must be known or the titration must be repeated using at least two different concentrations of reagent (I). It should be noted that the best results are obtained when enzyme concentrations range from 15 to 30 μM. Below 15 μM the change in absorbance on reaction is small and difficult to measure, and at concentrations above 30 μM the background absorption is large (since excess inhibitor must be used) and it becomes difficult to measure superimposed changes.

The principal limitations of the method described for the titration of papain solutions is that the consumption of considerable amounts of concentrated enzyme solution is required in order to achieve reasonable accuracy. One reagent that has better spectral properties for the titration of the active site of papain, is β-(2-hydroxy-3,5-dinitrophenyl)ethanesulfonic acid sultone.[13] This compound can be used to titrate papain at a wavelength of 400 nm, where the protein absorbance is vanishingly small and the optical change in the titration reaction is more than twice that seen with reagent (I) under optimal conditions. However, the use of the sultone is somewhat limited at the present time owing to the tedious synthetic procedure for its preparation.

Syntheses

Preparation of 4-Hydroxy-3-nitroacetophenone.[8,14] Over a period of 15 min, 10.5 g of 4-hydroxyacetophenone (Aldrich, m.p. 106°–108°, 0.077 mole) was added to 53 ml of fuming nitric acid ($d_{15°}$ 1.49, 79.5 g) while the temperature was maintained at −25° to −30°. The solution was stirred for an additional 60 min at −30° and then poured into 600 g of ice water. The yellowish precipitate was collected by filtration and recrystallized twice from ethanol, yielding 9.6 g (70%) yellow needles, m.p. 133°–134°, lit.[14] m.p. 135°.

Preparation of α-Bromo-4-hydroxy-3-nitroacetophenone.[8,15] To 0.52 g of bromine (0.0053 mole) in 11.0 ml of chloroform at room temperature was added 1.00 g of 4-hydroxy-3-nitroacetophenone (0.0055 mole). The initially red solution turned clear brown, and hydrogen bromide gas

---

[13] P. Campbell and E. T. Kaiser, *J. Am. Chem. Soc.* **95**, 3735 (1973).

[14] P. D. Bartlett and E. N. Trachtenberg, *J. Am. Chem. Soc.* **80**, 5808 (1958).

[15] G. Sipos and R. Szabo, *Acta Phys. Chem.* **7**, 126 (1961).

evolved. The temperature was raised to 40° and maintained there for 15 min; the solution was then extracted once with water and dried over anhydrous magnesium sulfate. The solvent was evaporated under reduced pressure, and the product was recrystallized twice from carbon tetrachloride, yielding 1.02 g (75%) yellow needles, m.p. 91.5°–92.5°, lit.[15] m.p. 93°.

Titration of Bromelain A

The purification of the phenylmercury derivative of bromelain A is described in this volume [64]. For the spectrophotometric titration of the active-site thiol, the enzyme is activated by a reducing agent and then allowed to react with Ellman's reagent[10] in a metal-free buffer. For the spectrophotometric burst titration the specific substrate p-nitrophenyl N^α-benzyloxycarbonyl-L-lysinate (CLN) is used in the presence of 1 mM L-cysteine.

Activation of the Enzyme[16]

In the past, the mercury or organomercurial complexes of cysteine proteases were activated by thiocresol dissolved in toluene, followed by repeated washings with toluene in order to eliminate the excess reagent.[17] The use of immobilized thiol groups on solid supports[18] simplifies considerably the activation process, eliminating at the same time the presence of a water–organic solvent interface, which invariably resulted in some denaturation of the enzyme. In the authors' laboratory the following process was found convenient for the activation of phenylmercury bromelain.

Into a vial containing 1 g of dry Affigel 10 (Bio-Rad Laboratories) one injects with a syringe 25 ml of 0.1 M phosphate buffer, pH 7.0, saturated with L-cystine. The mixture is shaken for 24 hr at 4°. The resulting slurry is used to prepare a 0.8 × 10 cm column. The column is first washed with 500 ml of 0.1 M phosphate buffer, pH 7.0, containing 1 M NaCl, until the eluate is free from N-hydroxysuccinimide, as monitored at 260 mm.

The L-cystine is reduced by passing through the column 200 ml of 0.1 M phosphate buffer, pH 7.0 containing 1 M 2-mercaptoethanol. The column is then washed with approximately 1 liter of 0.1 M phosphate

[16] M. N. Reddy, unpublished observations.
[17] M. Soejima and K. Shimura, *J. Biochem.* (*Tokyo*) 49, 260 (1961).
[18] P. Cuatrecasas and C. B. Anfinsen, this series Vol. 22 [31].

buffer, pH 7.0, until the eluate is free from 2-mercaptoethanol, as measured with Ellman's reagent.

For the activation of the enzyme, 1 ml of 5×10^{-4} M phenylmercury bromelain A is applied to the column at 4° and eluted with 0.1 M phosphate buffer pH 7.0 at a flow-rate of 15 ml/hr. Under these conditions the enzyme is more than 98% activated without any measurable loss of total protein. The column can be reactivated repeatedly by successive washings with 1 M 2-mercaptoethanol and the appropriate buffer.

Sulfhydryl Group Titration[19]

The spectrophotometric titration of the thiol group of bromelain A is carried out in the presence of 60 μM 5,5'-dithiobis(2-nitrobenzoic acid), 10 mM phosphate buffer, pH 7.04, 0.1 M KCl at 25°. Addition of a small aliquot (5–50 μl) of an activated bromelain solution (10–20 mg/ml) to 3 ml of this solution yields a slow, first-order increase of the absorbance at 412 nm. In independent experiments it was shown that under the same conditions L-cysteine reacts 400 times faster with Ellman's reagent, and thus any spectral change due to small residual amounts of cysteine could be corrected for by extrapolating the slow absorbance change to the beginning of the reaction.

The sulfhydryl content of the activated enzyme was determined from the total spectral change, using a $\Delta\epsilon = 13,600$ M^{-1} cm^{-1}. Experiments at several different enzyme concentrations showed that the sulfhydryl titer is linearly proportional to the protein content and the enzymic activity toward CLN. From the amount of protein added, and assuming one thiol group per enzyme molecule, an apparent MW of 36,000 was calculated, indicating that the enzyme preparation was more than 95% pure. Similar results were obtained when the loss of enzymic activity toward CLN was measured upon addition of aliquots of an aqueous solution of p-mercuribenzenesulfonate. The inhibition of the activated enzyme was proportional to the amount of inhibitor added, and total inhibition occurred at stoichiometric equivalence.

Burst Titration[20]

The steady-state kinetics of the bromelain A-catalyzed hydrolysis of CLN obey Eq. (2) with $k_{cat} \simeq 7.5$ sec^{-1} and $K_m = 57$ μM at pH 4.60, 25° and $\mu = 0.1$ M. The rapid initial phase of the reaction is monitored on a Durrum-Gibson stopped-flow spectrophotometer, after mixing equal

[19] R. M. Silverstein and F. J. Kézdy, *Arch. Biochem. Biophys.* **167**, 678 (1975).
[20] R. M. Silverstein, S. H. Perlstein, and F. J. Kézdy, unpublished experiments.

volumes of the substrate dissolved in water (1.6% acetonitrile) and the enzyme in an 0.1 M acetate buffer, pH 4.6, 1 mM L-cysteine. The progress of the reaction is measured at 340 or 370 nm.

The rate of appearance of p-nitrophenol during the first 100 msec of the bromelain-catalyzed hydrolysis of CLN shows a "burst" typical of enzymic reactions involving the rapid formation of an acyl-enzyme intermediate. Analogous with the reaction of bovine trypsin with the same substrate, the reaction of CLN with bromelain is not completely deacylation rate-limiting: the steady-state turnover following the initial rapid "burst" is appreciable. By extrapolating this steady-state straight line to the initial time and by plotting the logarithm of the difference between this line and the experimental curve vs time, one can determine an apparent first-order rate constant, b, for a series of initial substrate concentrations. These constants yield a straight line when plotted as $1/b$ vs $1/S$. From the slope and intercept of this line and the steady-state parameters of the enzymic hydrolysis at the same pH, one can calculate the mechanistically meaningful kinetic parameters defined in Eq. (5).

$$E + S \underset{K_S}{\rightleftharpoons} ES \xrightarrow{k_2} ES' + P_1 \xrightarrow{k_3} E + P_2 \tag{5}$$

E, ES, and ES' represent the free enzyme, the enzyme–substrate complex, and the acyl-enzyme, respectively, and S is the free substrate. P_1 is p-nitrophenol and P_2 is the product carboxylic acid.

It has been shown that the constants, k_2, k_3, and K_S are related to the experimental constants, b, k_{cat}, and K_m, by Eqs. (6)–(8).[3]

$$1/b = 1/k_2 + K_S/k_2 \cdot 1/S \tag{6}$$
$$1/k_{cat} = 1/k_2 + 1/k_3 \tag{7}$$
$$K_m = [k_3/(k_2 + k_3)]K_S \tag{8}$$

The values of the constants calculated from the experimental data are as follows: $k_2 = 140$ sec^{-1}, $k_3 = 7.9$ sec^{-1}, and $K_S = 1.6$ mM. Since equations (6) and (8) allow one to calculate all the parameters, Eq. (7) can be used for an independent verification of the consistency of the constants. In this way, we calculated a value of $k_{cat} = 7 \pm 2$ sec^{-1} in good agreement with the experimentally determined value, 7.5 sec^{-1}, based on enzymic concentrations calculated from sulfhydryl titrations. Thus, our data are fully consistent with the acyl-enzyme hypothesis as shown in Eq. (5).

The extrapolated value of p-nitrophenol liberated in the presteady state (π) is defined for a reaction obeying Eq. (5) by Eq. (9).

$$E_0 = \pi\{[1 + (S/K_m)]/[k_2/(k_2 + k_3)]\}^{1/2} \tag{9}$$

With the use of this equation, the experimental value of π, and the true rate constants, one can calculate the concentration of molarity of active sites in the reaction mixture (E_o). At several different enzyme concentrations, the values thus obtained agreed within $\pm 5\%$ with those obtained by titration of the active site thiol groups, thereby demonstrating that the bromelain A preparation used in these experiments was devoid of impurities.

Conclusions

Because of the high reactivity and selectivity of thiol groups toward function-specific reagents, it is tempting to use a single method such as titration with Ellman's reagent,[10] organomercurials, or alkylating agents. Although, in principle, with pure enzyme solutions the results of such titrations do yield the correct normality for enzyme active sites under appropriate reaction conditions, often competition from contaminating thiol compounds and inactive protein leads to spurious measurements. Burst titrations also have their shortcomings both experimental and theoretical. Therefore, for cysteine proteases even more than for serine proteases, it is imperative that more than one titration method be used in the determination of enzyme active site normality.

[2] Some Sensitive Methods for the Assay of Trypsinlike Enzymes[1]

By P. L. COLEMAN, H. G. LATHAM, JR., and E. N. SHAW

There is a need for more sensitive assays for proteolytic enzymes as increasing numbers are discovered that are present in minute quantities and have biologically important roles. Among these, serine proteases in particular predominate, since they are essential in blood clotting, clot removal, complement action, and fertilization, to name a few areas of current interest. Most of the individual enzymes have a trypsinlike specificity.

Two types of measurements are of value in the assay of proteases: the determination of the concentration of active sites, and the deter-

[1] Research carried out at Brookhaven National Laboratory under the auspices of the U.S. Energy Research and Development Administration and with support from U.S. Public Health Service Grant 17849 from the National Institute of General Medical Sciences.

mination of the rate of hydrolysis of a protein or small-molecular-weight ester or amide.

For the active site titration of many trypsinlike enzymes the nitrophenyl ester of p-guanidinobenzoate (NPGB)[2] has been very useful as a spectroscopic method.[3,4] Its success depends on a rapid acylation of the enzyme with stoichiometric release of nitrophenol, followed by a very slow deacylation owing to the stability of certain p-guanidinobenzoyl enzymes. These conditions are achieved with trypsin, plasmin, and plasma kallikrein, but less satisfactorily with thrombin and factor X_a. The main limitation of NPGB is lack of sensitivity. A limit of detection of about 0.01 absorbance unit is possible (1 nmole) but quantitative measurements with an accuracy of 5% require about an order of magnitude greater.

Elmore and his colleagues[5] have extended the usefulness of the NPGB method by demonstrating the sensitivity of a fluorogenic analog, the methylumbelliferyl ester of p-guanidinobenzoate. The ester is nonfluorescent, but the methylumbelliferone released fluoresces and provides an increase in sensitivity of about two orders of magnitude.

We have been attempting to characterize proteases isolated from a number of sources but not necessarily purified to homogeneity. Frequently, the total amount of enzyme available has been small. We have consequently explored titration by methylumbelliferyl p-guanidinobenzoate as described by Jameson et al.[5] and have developed a rate assay using the methylumbelliferyl ester of N-benzyloxycarbonyl-L-lysine. In addition, we have found the radiometric assays developed by Roffman et al.[6] to be valuable. This chapter is not intended to provide a survey of available methodology, which is evolving rapidly, but to offer a selection of some sensitive methods.

Active Site Titration

Methylumbelliferyl p-Guanidinobenzoate[5]

Substrate Synthesis. p-Guanidinobenzoic acid hydrochloride[3] (1.07 g), 4-methylumbelliferone (0.88 g), and N,N'-dicyclohexylcarbodiimide

[2] Abbreviations used: NPGB, nitrophenyl p-guanidinobenzoate; ZLysMeUmb, N^α-benzyloxycarbonyl-L-lysine methylumbelliferyl ester; ZLys, N^α-benzyloxycarbonyl-L-lysine; TsArgOMe, N^α-toluenesulfonyl-L-arginine methyl ester; BzArgOMe, N^α-benzoyl-L-arginine methyl ester; DNPP, 2,4-dinitrophenyl protamine sulfate.

[3] T. Chase and E. Shaw, this series Vol. 19 [2].

[4] F. J. Kézdy and E. T. Kaiser, this series Vol. 19 [1].

[5] G. W. Jameson, D. V. Roberts, R. W. Adams, W. S. A. Kyle, and D. T. Elmore, *Biochem. J.* **131**, 107 (1973).

[6] S. Roffman, U. Sanocka, and W. Troll, *Anal. Biochem.* **36**, 11 (1970).

(1.08 g) were dissolved in a mixture of pyridine (7.5 ml) and N,N-dimethylformamide (7.5 ml), and the solution was left overnight. N,N'-Dicyclohexylurea was collected and washed with a mixture of pyridine (5 ml) and N,N-dimethylformamide (5 ml). The combined filtrate and washings were evaporated to dryness under reduced pressure, and 0.1 M HCl (25 ml) was added. The slurry was extracted with ethyl acetate (3 × 25 ml), and the phases were separated by centrifugation. The product (59%) was collected by filtration from the aqueous phase, dried under vacuum over P_2O_5, and recrystallized first from acetic acid–diethyl ether,[7] then from 2-methylbutan-2-ol–diethyl ether. It had a melting point of 219°–221°.

Reagents for Fluorimetric Enzyme Titration. Methylumbelliferyl p-guanidinobenzoate, 1 to 2 mM in N,N-dimethylformamide; 4-methylumbelliferone, 1 to 2 × 10^{-4} M in N,N-dimethylformamide; trypsin, 10^{-7} to 10^{-4} M in 1 mM HCl; sodium phosphate buffer, pH 6.0, 0.1 M. The first two solutions are protected from exposure to light with aluminum foil.

Method. The optimal excitation and emission wavelengths for the measurement of methylumbelliferone fluoresence must be determined for the selected combination of buffer and pH. We added 30 μl of the stock methylumbelliferone solution to 3.0 ml of buffer. We scanned the emission spectrum to determine the emission peak, about 450 nm, while maintaining the excitation wavelength constant at the highest absorbing band. Once the emission peak had been determined, we scanned the excitation spectrum at constant emission wavelength. In 0.1 M sodium phosphate buffer, pH 6.0, the excitation and emission wavelengths were 323 and 446 nm, respectively, using a Perkin-Elmer MPF-4 spectrofluorophotometer. The band widths that may be used are variable, and their magnitudes directly affect the magnitudes of the fluorescence measurement; however, there is no need to use wide bands except when approaching the limits of sensitivity.

Unlike absorption spectroscopy, fluorescence yield per mole of fluorophore differs from instrument to instrument and day to day. Therefore, we usually determine a standard curve each time the lamp is fired. It is also wise to check the curve a few times during the initial series of experiments until satisfied that it will remain stable during a working day. We generate a standard curve by addition of 1–5 μl aliquots of methylumbelliferone to 0.4–3.0 ml phosphate buffer (Fig. 1).

Methylumbelliferyl p-guanidinobenzoate, like the nitrophenyl ester, is subject to base-catalyzed hydrolysis, therefore the rate of this reac-

[7] We found crystallization from hot glacial acetic containing a small amount of water to be more useful.

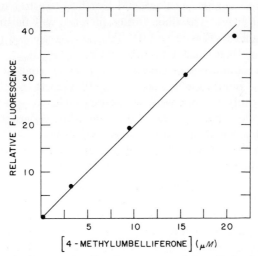

FIG. 1. A standard curve for the fluorescence of 4-methylumbelliferone. Cylindrical quartz cuvettes (500-μl capacity) contained 400 μl of sodium phosphate (0.1 M), pH 6.0. Aliquots (10 and 20 μl) of a standard methylumbelliferone solution (0.126 mM) were added.

tion must be determined in the absence of enzyme. We added 6 μl of methylumbelliferyl p-guanidinobenzoate to 400 μl of buffer and recorded the nonenzymic increase in fluorescence (Fig. 2). Aliquots of trypsin (6 μl) were added until there was no additional increase in fluorescence. The change in fluorescence between the two parallel lines is proportional

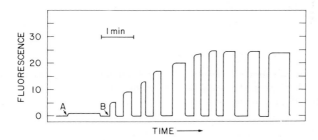

FIG. 2. Recorder tracing of a titration of β-trypsin. The cuvette contains 400 μl of buffer. Methylumbelliferyl p-guanidinobenzoate, 6 μl (1.0 mM), was added at A. Additions of 6-μl aliquots of β-trypsin (about 2 $\times$ 10^{-2} M by weight) began at B. The tracing indicates the fluorescence after the addition of each aliquot. The acylation proceeds too rapidly to be followed at these concentrations. (At the sixth and seventh additions of enzyme there is an indication that the rate of reaction is slightly slower.)

to the amount of enzyme present and can be quantitated by reference to the standard curve. The fluorescence increase will be linear until the end point is reached (Figs. 2 and 3). Titration to the end point is also a more accurate method since it averages pipetting errors.

For some combinations of enzyme and enzyme concentration we observed that the burst might take up to 10 min to reach the rate of endogenous hydrolysis. We were sometimes able to speed the process by addition of the titrant to the enzyme solution. Alternatively, we employed a modification of the titration described below.

To titrate an exceptionally small amount of enzyme or an enzyme that has an inherently slow rate of acylation with methylumbelliferyl p-guanidinobenzoate, Jameson et al.[5] developed a concentrated incubation method. We mix enzyme (5–10 μl), titrant (1–2 μl), and buffer (5–10 μl) in a small total volume (10–20 μl) in a 400-μl cuvette, and allow them to react for 1 min (trypsin), 5–10 min (plasmin), or longer, and terminate the incubation by addition of buffer to a final volume of 400 μl. The difference between the measured fluorescence of the reaction

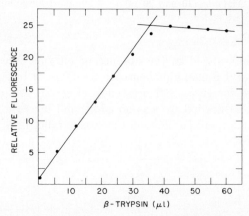

FIG. 3. The titration of β-trypsin by methylumbelliferyl p-guanidinobenzoate. This is a graphical presentation of the data from Fig. 2. The concentration of trypsin was determined in two independent ways from the end-point titration: from the fluorescence yield at the end point, or the known initial concentration of the titrant. The end-point fluorescence is 24.5 units, and the end-point volume of trypsin (the point at which trypsin concentration equals titrant concentration), determined by extrapolation of the two lines, is 35.5 μl. Method 1: From the standard curve (Fig. 1) 24.5 units of fluorescence corresponds to 12.5 μM methylumbelliferone. The stock trypsin concentration = 12.5 μM × [(400 + 6 + 35.5) μl/35.5 μl] = 1.55 × 10⁻⁴ M. Method 2: 1.0 mM × [6 μl/(400 + 6 + 35.5)μl] = 1.36 × 10⁻⁵ M. (adjusted concentration of titrant at end point). The stock concentrations 13.6 μM × [(400 + 6 + 35.5) μl/35.5 μl] = 1.69 × 10⁻⁴ M. Each calculation indicates that there is slightly less active β-trypsin than the weighing suggests.

mixture in the presence and in the absence of enzyme is proportional to the enzyme concentration. This modification requires reproducible pipetting and matched cuvettes. The sensitivity of the titration can be enhanced by diluting the reaction mixture with a buffer in which the fluorescence yield is higher than in pH 6.0 sodium phosphate, e.g., pH 8.0 sodium barbital. We measure the fluorescence immediately upon dilution since the titrant is unstable at the elevated pH. In some cases it is possible with low concentrations of protease that the measurement cannot be made before the base-induced fluorescence masks that from the enzyme.

We have used the concentrated incubation method to titrate 1 pmole of trypsin. Although the Perkin-Elmer fluorometer can give a full-scale recorder deflection with less than 1×10^{-14} mole of methylumbelliferone, the rates of photolytic decomposition of this material and of spontaneous hydrolysis of methylumbelliferyl p-guanidinobenzoate are too great to allow the full range of sensitivity of the instrument to be utilized. To lessen the photolytic effect at low enzyme concentrations, we allow only intermittent light to strike the solution. To decrease the rate of spontaneous hydrolysis we maintain the pH below 7.0 and use sodium phosphate buffer. Sodium cacodylate, Tris-maleate, and citrate-phosphate cause higher rates of spontaneous hydrolysis than sodium phosphate from pH 6.0 to 7.0. Above pH 7.0, this hydrolysis reduces the sensitivity of the measurement. The use of 1 mM HCl as diluent did not sufficiently slow the spontaneous hydrolysis of methylumbelliferone p-guanidinobenzoate. The fluorescence yield of methylumbelliferone also is a function of pH and buffering agent. Sodium phosphate gave the highest fluorescence yield from pH 6.0 to 7.0 and was the only buffer tested in which the yield was not a function of pH over that range. Above pH 7.0 the fluorescence yield tends to increase.

This method has been successfully used by Jameson et al.[5] to titrate α- and β-trypsin, thrombin, and coagulation factor X_a. We have used these procedures to titrate β-trypsin, human plasmin, human plasma kallikrein, and boar acrosin. Subtilisin,[5] chymotrypsin,[5] and bovine spleen cathepsin B gave negative results. Zahnley and Davis[8] reported a technique using NPGB, which allows the determination of both the equilibrium constant of a protease–protease inhibitor complex and the rate at which inhibitor leaves the complex in one experiment. Enzyme and inhibitor are allowed to equilibrate in a cuvette to which NPGB is rapidly mixed. The initial burst of absorbance is equivalent to free enzyme; the rate of increase is equivalent to the rate of release of enzyme

[8] J. C. Zahnley and J. G. Davis, *Biochemistry* **9**, 1428 (1970).

from the complex. Laskowski *et al.*[9] applied this method, using 0.75 nmoles of enzyme and adding methylumbelliferyl *p*-guanidinobenzoate. Sealock and Laskowski[10] have also used this titrant to investigate trypsin-catalyzed peptide bond hydrolysis of trypsin inhibitor.

Jameson *et al.*[5] also described a fluorescent titrant for α-chymotrypsin, 4-methylumbelliferyl *p*-(*N,N,N*-triethylammonium)cinnamate, which reacts with the enzyme in less than 2 min, has no turnover, and is sensitive to 10^{-11} mole of enzyme. Laskowski *et al.*[9] synthesized 4-methylumbelliferyl *p*-ω-dimethylsulfonioacetamido)benzoate, a titrant of chymotrypsin, based on the *p*-nitrophenyl ester described by Wang and Shaw.[11]

Elmore and Smyth[12] have synthesized N^α-methyl-N^α-toluenesulfonyl-L-lysine β-naphthyl ester, which titrates trypsin but not chymotrypsin or thrombin. Whereas this was reported as a spectrophotometric method, it can be easily adapted for fluorometric use.[13]

Other Titration Methods

The active site of serine proteases (esterases) may also be titrated radiometrically using a radiolabeled diisopropylfluorophosphate.[14] As little as 2×10^{-11} mole of enzyme can be detected using the commercially available preparation (New England Nuclear); however, the method is very time consuming and not readily adaptable to routine use.

Rate Assays

Fluorometric Esterase Assays

Enzymic turnover results in amplification, thus rate assays are more sensitive than titrations, which release only a stoichiometric amount of indicator. These assays are particularly valuable in monitoring an enzyme purification.

The methylumbelliferyl ester of *N*-benzyloxycarbonyl-L-lysine (ZLysMeUmb) is a useful fluorogenic substrate for trypsinlike enzymes. It has kinetic properties similar to those of the corresponding nitrophenyl ester with β-trypsin and plasmin (see table), however, the fluorometric assay is at least 200-fold more sensitive than the spectrophotometric.

[9] M. Laskowski, Jr., I. Kato, T. R. Leary, J. Schrode, and R. W. Sealock, *Proteinase Inhibitors, Proc. Int. Res. Conf. (Bayer Symp. V)*, Grosse Ledder, 1973, pp. 597–611. Springer-Verlag, Berlin and New York, 1974.

[10] R. W. Sealock and M. Laskowski, Jr., *Biochemistry* **12**, 3139 (1973).

[11] C.-C. Wang and E. Shaw, *Arch. Biochem. Biophys.* **150**, 259 (1972).

[12] D. T. Elmore and J. J. Smyth, *Biochem. J.* **107**, 97 (1968).

[13] P. H. Bell, C. T. Dziobkowski, and M. E. Englert, *Anal. Biochem.* **61**, 200 (1974).

[14] J. A. Cohen, R. A. Oosterbann, and F. Berends, this series Vol. 11 [81].

COMPARISON OF KINETIC CONSTANTS FOR THE NITROPHENYL AND
METHYLUMBELLIFERYL ESTERS OF CARBOBENZOXYLYSINE[a,b]

Enzyme substrate	K_m ($\times 10^6$ M)	V_{max} (sec^{-1})
β-Trypsin		
Nitrophenyl ester	—	8.3[c]
Methylumbelliferyl ester	7[d]	8.8[d]
Plasmin		
Nitrophenyl ester	22[e]	68[e]
Methylumbelliferyl ester	18[e]	50[e]

[a] J. J. Ruscica and P. L. Coleman, unpublished observations.
[b] N^α-Benzyloxycarbonyl-L-lysine nitrophenyl ester synthesized according to T. Y. Liu, N. Nomura, E. K. Jonsson, and B. G. Wallace, *J. Biol. Chem.* **244**, 5745 (1969).
[c] Sodium maleate, 50 mM, pH 6.0.
[d] Tris-maleate, 50 mM, pH 6.0.
[e] Sodium phosphate, 50 mM, pH 6.0.

Substrate Synthesis. N^α-Benzyloxycarbonyl-N^ε-t-butyloxycarbonyl-L-lysine (Schwarz-Mann) as 50% methylene chloride solution (4 ml) and 468 mg of recrystallized 4-methylumbelliferone are dissolved in 20 ml of a methylene chloride:pyridine solution (1:1, v/v) at 0°. Dicyclohexyl-carbodiimide (522 mg) is added and the mixture is maintained at 0° for 30 min, then held at room temperature for about 2 days. The dicyclo-hexylurea is filtered off, the filtrate is evaporated to dryness *in vacuo* at 40°, and the residue is desiccated over H_2SO_4. The residue is extracted with ethyl acetate (3 × 30 ml), and the combined extracts are washed successively with 5% $NaHCO_3$, cold 0.1 N HCl, and H_2O, then dried over anhydrous magnesium sulfate. The solvent is removed, and the residue (1.42 g) is dissolved in ethyl acetate (20 ml) and treated with gaseous HCl for 2 hr at room temperature. Removal of the solvent and crystallization twice from methanol–ether gave 676 mg, m.p. 158°–162°. Recrystallization from methanol–diethyl ether gave m.p. 162°–164°.

Reagents for Fluorometric Assay. Methylumbelliferone, 5 to 10 × 10^{-5} M in N,N-dimethylformamide (stable for about 1 month at −15°); ZLysMeUmb, 1.0–2.0 mM in N,N-dimethylformamide (prepared fresh daily); plasmin, 6.2 × 10^{-7} M (freshly diluted from a stock solution of 6.2 × 10^{-5} M) in Tris buffer (5 mM), pH 9.0, containing 0.1 M NaCl, 1 mM Na_2EDTA, 2 × 10^{-2} M L-lysine, and 25% (v/v) glycerol; β-trypsin, 2 × 10^{-7} M in 1 mM HCl, Na phosphate buffer, 0.1 M, pH 6.0. The first two solutions are protected from exposure to light with aluminum foil.

Method. Mix 5–10 μl of ZLysMeUmb with 3.0 ml of buffer in the cuvette and record the rate of increase in the fluorescence (same wavelengths as with titration assay) to determine the spontaneous rate of hydrolysis. Enzyme (5–20 μl) is added and a new initial rate of fluorescence change is determined. The difference between the rates is the enzymic rate (Fig. 4). Reference to the standard curve for methylumbelliferone described for the titration assay yields the rate of product formation. We have varied substrate and enzyme concentrations over a wide range to obtain typical saturation kinetics with no apparent problems (see table). Total reaction volume can be decreased to 300 μl with excellent results.

Practical Considerations. The Perkin–Elmer MPF-4 is equipped with a zero suppression dial which allows the electronic substraction of fluorescence without affecting the sensitivity of the photomultiplier. This is a considerable advantage when working with a substrate that has an intrinsic fluorescence or fluorescent contaminant or when assaying small amounts of enzyme for which one is attempting to measure a slow change in fluorescence superimposed on a large "background" fluorescence.

For an instrument without a zero suppression photomultiplier, the limit of sensitivity is determined by the intrinsic fluorescence of the unhydrolyzed compound (plus any already hydrolyzed substrate present before addition of the enzyme). With an instrument capable of zero suppression,

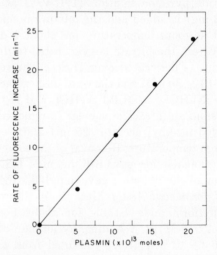

Fig. 4. The rate of fluorescence increase as a function of the amount of plasmin. The photomultiplier sensitivity used was convenient for the range of enzyme concentrations, but not the most sensitive that we employ for the assay; 1×10^{-10} mole methylumbelliferone gave 7.7 units of fluorescence. N^α-Benzyloxycarbonyl-L-lysine methylumbelliferyl ester concentration was 15.6 μM, approximating K_m rather than V_{max} conditions to lessen the rate of spontaneous hydrolysis. Using these conditions, we can assay a sample of plasmin in 1–2 min, or 25–30 per hour.

much of the initial fluorescence of the solution can be subtracted, allowing the use of a photomultiplier setting sensitive to smaller changes in absolute fluorescence. The limit of sensitivity is then governed by the spontaneous rate of hydrolysis of ZLysMeUmb. In this case, the choice of solvent (buffer) becomes very critical. The rates of both spontaneous and enzymic hydrolysis generally increase with increasing pH, and both are affected by variation in the buffering ions, though not necessarily in the same direction. For example, in sodium cacodylate, raising the pH from 6.5 to 7.0 increases the rate of spontaneous hydrolysis but decreases the plasmin-catalyzed rate in contrast to other buffers. Therefore, it is necessary to study a number of buffer systems before using this substrate with a given enzyme.

We can quantitate 1 to 5×10^{-14} mole of trypsin or plasmin with ZLysMeUmb using the Perkin–Elmer MPF-4 fluorometer at the most sensitive useful photomultiplier setting. The absolute limit of detectability will vary slightly from day to day as a function of the lamp intensity and positioning, optical alignments, and small day-to-day changes in the rate of spontaneous hydrolysis of the freshly dissolved ZLysMeUmb solution. Another factor limiting sensitivity is the turnover rate of the enzyme.

Assay of Plasminogen Activator. We devised this assay to quantitate the rate of formation of plasmin, the product of a mouse fibroblast protease, plasminogen activator, acting on plasminogen. The activator is isolated from tissue culture in relatively small amounts. Without further purification, the activity is low. We needed, therefore, an assay which used small amounts of plasminogen activator and which was sensitive to small amounts of plasmin. We can quantitate plasmin generated by an incubation mixture which contains 5 μl of crude, unconcentrated activator and 1 nmole of plasminogen in a total volume of 50 μl. Typically we remove small aliquots (1–4 μl using a Hamilton syringe equipped with a repeating pipetter) from the incubation solution at intervals from 5 min to 4 hr and determine the increase in the rate of methylumbelliferone production with time. The slope is linear and gives an accurate reproducible rate of plasminogen activation.

ZLysMeUmb is also a useful substrate for human plasma kallikrein, boar acrosin, cathepsin B, and human urinary urokinase.

Isotopic TsArgOMe Assay

Roffman *et al.*[6] have developed a sensitive extension of the familiar TAME (N^{α}-toluenesulfonyl-L-arginine methyl ester, TsArgOMe) esterase assay. Previous methods of detecting product formation relied on a change in pH or absorbance and required at least 5 to 20 pmoles of trypsin.[15] The isotopic method depends on the partitioning of tritiated methanol (released by enzymic hydrolysis) and the substrate, TsArgOMe, by their differential solubilities in aqueous and toluene phases of a scintillation counting solution. TsArgOMe is insoluble in toluene, and in the absence of a detergent such as Triton X-100, will not be counted when an aqueous TsArgOMe solution is added to a scintillation counting solution. Methanol is soluble in both phases and can be preferentially extracted into the toluene phase by shaking the vial. This method is sensitive to 4×10^{-14} mole trypsin and needs only small volumes of enzyme.[6]

Substrate Synthesis. The method of synthesis of [methyl-^{3}H]-TsArgOMe according to Roffman *et al.*[6] involving a trans-esterification was used. An initial synthesis used 100 mg of TsArgOMe dissolved in 0.25 ml of methanol (5 mCi, New England Nuclear). HCl was bubbled into the solution for 1–2 min, and the reaction vial was stoppered for 24 hr at room temperature. Solvent was removed under a stream of N_2. The yield was 10% of theoretical, compared with 60% for Roffman *et al.*,[6] and the product had a specific radioactivity of 1.2×10^9 cpm/mole. A subsequent synthesis using 5×10^{-2} Ci [^{3}H]methanol yielded 25% of theoretical incorporation with a specific radioactivity of 37.2×10^9 cpm/mole. A similar synthesis of tritiated N^α-benzoyl-L-arginine methyl ester, [methyl-^{3}H]BzArgOMe, using 5 mCi [^{3}H]MeOH resulted in a specific radioactivity of 8.7×10^8 cpm/mole. Some preparations of TsArgOMe are unstable beyond 6 months, regardless of whether stored as a powder under vacuum or at $-15°$ in 1 mM HCl.

Reagents. [Methyl-^{3}H]TsArgOMe or [methyl-^{3}H]BzArgOMe, 0.1 M in 1 mM HCl; β-trypsin (or other TsArgOMe hydrolase), 10^{-8} to 10^{-6} M in 1 mM HCl; glycine–NaOH buffer, 0.3 M, pH 9.5; scintillation fluor: PPO, 15.2 g; POPOP, 0.19 g; and toluene to 1 liter.

Method. Pipette substrate (1–10 μl), enzyme (1–10 μl), and 10 μl of buffer into a scintillation vial, bring to 30 μl total volume with 1 mM HCl in a scintillation vial, swirl to mix well, and add 10 ml of scintillation fluor. The vials are mechanically shaken for 15–30 min to establish the equilibrium distribution of [methyl-^{3}H]TsArgOMe between the toluene and water phases and counted in a liquid scintillation counter. The samples should be recounted every 20–30 min after a short (15–20 sec) vigorous manual shake. Reproducible methods of mechanical and shaking are obtained after a few attempts. A plot of cpm vs time of incubation yields the rate of TsArgOMe hydrolysis (Fig. 5), which is linear for 4–6 hr at low enzyme concentrations. Since there is also a nonenzymic increase in cpm, an enzyme blank must always be prepared

[15] K. A. Walsh and P. E. Wilcox, this series Vol. 19 [3].

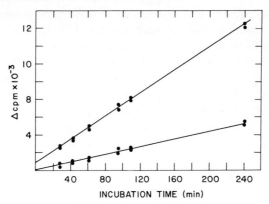

Fig. 5. The appearance of [³H]methanol in the toluene phase in the presence and in the absence of enzyme. The reaction mixtures (in duplicate) contained 6.0 μl of [methyl-³H]N^{α}-toluenesulfonyl-L-arginine methyl ester (0.1 M; 22,300 cpm), 10 μl of glycine buffer (pH 9.0), 4 μl of HCl (1 mM), and either 10 μl of plasminogen activator (in pH 7.4, phosphate buffer, 0.1 M) from mouse 3T3 cells or 10 μl of pH 7.4 phosphate buffer, 0.1 M. The upper line represents the mixture with enzyme; the lower line, without enzyme. The data plotted are the counts per minute (obtained from 0.5-min counts with a Beckman LS-233 liquid scintillation counter) vs the length of time since the reaction was started by addition of substrate. The first count followed 28 min of mechanical shaking. Subsequent counts immediately followed 15–20 sec of vigorous manual shaking.

for all combinations of buffer and substrate. Roffman et al.[6] report about 1% spontaneous hydrolysis per hour, a value we corroborate.

At very low trypsin concentrations, the rate of transport of methanol into the toluene phase is much faster than the rate of TsArgOMe hydrolysis,[6] negating the necessity of the manual shaking before each count. They account for possible pipetting errors by the addition of solid trypsin at the end of the assay.[6] We did not find this necessary to obtain reliable counts.

Since this method requires the insolubility of TsArgOMe in the toluene phase, the assay is subject to one general restriction: nothing can be added to the vial which enhances the solubility of TsArgOMe in the toluene phase. We found spurious results when we attempted to use this assay to analyze the effects of various chloromethyl ketones on a protease from mouse fibroblasts. The inhibitor was known to inactivate the enzyme; however, the apparent rate of TsArgOMe hydrolysis was proportional to the amount of chloromethyl ketone present, even in the absence of enzyme. Other inhibitors did not interfere with the use of this assay.

The sensitivity of this assay is limited by the specific radioactivity of the substrate. The exchange reaction may be performed by using less unlabeled methanol as solvent and using [^{3}H]methanol with a higher total and/or specific radioactivity. Tritiated methanol may be purchased from New England Nuclear, Amersham/Searle, or ICN with specific radioactivities of 40–60, 100–250, and 500–900 Ci/mole, respectively.

This method allows the assay of enzymic activity of solutions whose volumes or concentrations are too small to assay otherwise, and it allows the rapid monitoring of large numbers of samples. We have used it to monitor column fractions for the plasminogen activator of mouse fibroblast (TsArgOMe is also a substrate of this protease) as well as to scan a series of potential inhibitors and to obtain some kinetic properties of the impure activator. Roffman et al.[6] have used the method to assay plasmin and thrombin separately as well as thrombin in the presence of plasmin by the competitive inhibition of plasmin by lysine methyl ester. These authors also synthesized methyl tritiated N^α-acetyl-L-lysine methyl ester for use with this technique.[6]

Other Methods

Hydrolysis rates of small-molecular-weight esters and amides have been determined titrimetrically and spectrophotometrically.[15] Benzoylarginine ethyl ester and TsArgOMe hydrolysis by trypsin also requires at least 8×10^{-11} mole of enzyme. Measured spectrophotometrically, TsArgOMe hydrolysis is sensitive to 2×10^{-11} mole of trypsin.[5] Hydrolysis of N^α-benzyloxycarbonyl-L-lysine nitrophenyl ester requires 5 to 10 pmoles of trypsin.[16]

Additional fluorometric methods have been based on β-naphthylamides and β-naphthyl esters. McDonald et al.[17] described a series of assays for dipeptidyl arylamidases using a number of dipeptidyl β-naphthylamides as substrates. Bell et al.[13] reported that N^α-methyl-N^α-toluenesulfonyl-L-lysine β-naphthyl ester, a titrant for trypsin,[12] is a useful substrate for plasmin. As little as 10^{-15} mole of chymotrypsin can be quantitated by monitoring the hydrolysis of N-benzyloxycarbonyl-L-phenylalanine β-naphthyl ester fluorometrically.[18]

Fluorometric Assay of Proteolytic Activity

Fluorescamine (Fluram, Roche Diagnostics), a nonfluorescent material that reacts with free amino groups and generates fluorescence, is the

[16] P. Coleman and E. Shaw (1974). Unpublished observations.
[17] J. K. McDonald, S. Ellis, and T. J. Reilly, J. Biol. Chem. 241, 1494 (1966).
[18] E. Haas, Y. Elkana, and R. G. Kulka, Anal. Biochem. 40, 218 (1971).

basis of several assays for proteolytic activity. For assay of trypsinlike enzymes, Brown et al.[19] have described the use of the arginine-rich protein, protamine. The N-terminal group was blocked by dinitrophenylation to eliminate background fluorescence. (Succinylation offers the advantage of greater solubility.[20])

Preparation of Dinitrophenylprotamine Sulfate (DNPP). The DNPP was synthesized by adding 1 volume of 1-fluoro-2,4-dinitrobenzene in methyl Cellosolve (0.1 M) to 5 volumes of a 1% solution of protamine sulfate in 0.01 M sodium tetraborate, pH 9.2 at 37°. The resultant yellow precipitate was collected by centrifugation at 1000 g for 10 min, and washed twice with 20 ml of absolute ethanol and twice with 20 ml of acetone. The DNPP was dried at room temperature.

The average molecular weight of DNPP was 4400, as determined by the absorbance of 2,4-dinitrophenylalanine at 250 nm. No fluorescence resulted upon addition of fluorescamine, indicating complete blocking of the amino terminal groups.

Method. The incubation mixture (0.1 ml) used by Brown et al.[19] consisted of equal volumes of enzyme solution in 1 mM HCl and DNPP (10 mg/ml in 0.1 M sodium phosphate buffer, pH 7.5) at 37.5°. Aliquots (10 μl) were removed at 20 min and mixed with 0.6 ml of 0.1 M sodium phosphate, pH 7.0. Addition of fluorescamine (0.2 ml of 0.1 mg/ml dried acetone) with vigorous mixing at room temperature gave the colored product. They determined fluorescence with a Perkin–Elmer MPF-3 fluorometer using 390 nm and 470 nm as excitation and emission wavelengths, respectively.

The assay is sensitive to 2.5×10^{-14} mole of trypsin. The rate of hydrolysis is proportional to enzyme concentration over a 1000-fold range and is linear for 20 min. Thrombin at a concentration as low as 8.3 NIH units/ml is capable of hydrolyzing DNPP. The rate of hydrolysis is proportional to enzyme concentration over at least 5-fold range and is linear for 40 min. This method is capable of detecting small concentrations of arginine-splitting proteases including trypsin, thrombin, and pepsin, whereas chymotrypsin is totally inactive.

Other Methods

Rates of protease activity have been determined using casein as substrate,[21] but the assay is discontinuous, nonlinear with respect to

[19] F. Brown, M. C. Freedman, and W. Troll, *Biochem. Biophys. Res. Commun.* **53**, 75 (1973).
[20] W. Troll (1974). Personal communication.
[21] M. Laskowski, Jr., this series Vol. 2 [2].

protease concentration, and less sensitive than other rate assays, requiring a minimum of 8×10^{-11} mole of trypsin.

Reich and colleagues developed a sensitive method for the assay of the proteolytic activity of minute quantities of plasmin or any enzyme that will activate its zymogen, plasminogen.[22] They label fibrin with ^{125}I, allow a clot to form on a glass plate, then incubate it with either plasmin or plasminogen plus a plasminogen activator. The quantity of solubilized ^{125}I is a measure of the protease activity of plasmin or trypsin.

[22] J. Unkeless, K. Danø, G. M. Kellerman, and E. Reich, *J. Biol. Chem.* **249**, 4295 (1974).

[3] Determination of Proteolytic Activities on Casein Substrates

By ERNST H. REIMERDES and HENNING KLOSTERMEYER

Casein is often used as substrate for the determination of proteolytic activities.[1-3] It is easily available and stable under storage conditions. On account of its complex composition and random structure, this substrate undergoes proteolysis with all the known proteolytic enzymes without any necessity for prior denaturation. The extent of hydrolysis is generally determined, after acid precipitation of the unaltered casein by measurement of the absorption at 280–290 nm in the clear filtrate. This absorption is due to the presence of aromatic amino acids in the acid-soluble casein fragments. There are some problems with the measurement of proteolytic activities with casein.[3,4]

In terms of chemical definition, casein is a product from cow's milk and shows a variable protein composition[5,6] depending on its origin, hygienic qualities, storage, and the procedures for isolation. The best casein samples are composed primarily of the genetic variants of β-casein, α_{s1}-casein and κ-casein. In addition they contain about 10–20% of so-called minor caseins, e.g., γ-casein, R-, S-, TS-, and α_{s2-5}-caseins. Some preparations may be contaminated with albumins and globulins.

[1] W. Rick, *in* "Methoden der enzymatischen Analyse" (H. U. Bergmeyer ed.), 3rd ed., pp. 1046 and 1056. Verlag Chemie, Weinheim, 1974.
[2] R. Ruyssen, *in* "Symposium on Pharmaceutical Enzymes and Their Assay," p. 134. Presse Univ. de France, Paris, 1969.
[3] E. H. Reimerdes and H. Klostermeyer, *Milchwissenschaft* **29**, 517 (1974).
[4] K. Yamauchi and S. Kaminogawa, *Agric. Biol. Chem.* **36**, 249 (1972).
[5] M. P. Thompson, *in* "Milk Proteins Chemistry and Molecular Biology" (H. A. McKenzie, ed.), Vol. 2, p. 117. Academic Press, New York, 1971.
[6] A. G. Mackinlay and R. G. Wake, *in* "Milk Proteins: Chemistry and Molecular Biology" (H. A. McKenzie, ed.), Vol. 2, p. 175. Academic Press, New York, 1971.

After recent sequence analysis of caseins,[7-10] it was possible to study their susceptibilities to proteolysis in greater detail. On examining the hydrolysis of caseins by milk-specific proteolytic enzymes,[3] it was recognized that conflicting results could be obtained by different groups. The major casein components are attacked by enzymes in a rather limited way, and the resulting fragments are only partially acid soluble; not all of these proteolytic products contain aromatic amino acids. Thus, a simple photometric procedure could result in erroneous data.

Acidic proteases such as chymosin specifically hydrolyze the peptide bonds of Phe105–Met106 in κ-casein[11] and Phe23–Phe24 in α-casein,[12] while β-casein is degraded only slowly in a nonspecific manner. Since only κ-casein fragment 106–169 is acid soluble, this product of proteolysis alone would be measured. Even more unfavorable is the situation with trypsin- and plasmin-like enzymes. Only β-caseins are favorable substrates, the Lys28–Lys29-peptide bonds being especially susceptible. The specific proteolytic enzymes of milk, acting during storage of milk, attack the same link and produce varying amounts of γ-caseins corresponding to the 29–209 segment in the β-casein.[7-9] Fragment 1–28 of β-casein does not contain any aromatic amino acids, and therefore no proteolytic reaction would be registered by procedures depending on measuring their content.

It is necessary to search for methods other than those relying on the presence of aromatic amino acids, and a routine procedure is recommended in which the free amino groups are measured by the ninhydrin method[13] after proteolysis.

Assay Procedure

Reagents

I. Substrate solution: 2.5 g casein (Merck 2242 or 2244) or isolated β-casein are carefully triturated with 10 ml of water and 10 ml

[7] J. C. Mercier, F. Grosclaude, and B. Ribadeau-Dumas, *Milchwissenschaft* **27**, (1972).

[8] F. Grosclaude, M. R. Mahé, and B. Ribadeau-Dumas, *Eur. J. Biochem.* **40**, 323 (1973).

[9] J. Jollès, F. Schoentgen, C. Alais, and P. Jollès, *Chimia* **26**, 645 (1972).

[10] J. C. Mercier, C. Brignan, and B. Ribadeau-Dumas, *Eur. J. Biochem.* **35**, 222 (1973).

[11] A. Delfour, J. Jollès, C. Alais, and P. Jollès, *Biochem. Biophys. Res. Commun.* **19**, 452 (1965).

[12] R. D. Hill, E. Lahav, and D. Girol, *J. Dairy Res.* **41**, 147 (1974).

[13] S. Moore and W. H. Stein, *J. Biol. Chem.* **211**, 908 (1970).

of 0.2 N NaOH. After the addition of a further 60 ml of water, the suspension is stirred until complete dissolution of the protein. The pH of the clear solution is adjusted to 7.8 with 0.1 N HCl, the volume is made up to 100 ml with water, and the solution is heated for 15 min at 90°. The substrate as prepared can be used reliably for 24 hr only.

II. Buffer solution: 0.1 M sodium borate, pH 7.8

III. Calcium chloride solution: 2.94 g of $CaCl_2 \cdot 2H_2O$ are dissolved in 900 ml of water, the pH is adjusted to 6.0–6.2, and the volume made up to 1 liter.

IV. Trichloroacetic acid solution, 16%

V. Enzyme solution: an appropriate amount of protein is dissolved in 100 ml of $CaCl_2$ solution (III). This enzyme solution is diluted with buffer such that the final absorption of the ninhydrin color of the proteolysis products should fall between 0.2 and 1.0.

VI. Sodium acetate buffer, 4 N, pH 5.5: 5.44 g of $CH_3COONa \cdot 3H_2O$ are dissolved in distilled water, 100 ml of glacial acetic acid are added, and the volume is made up to 1 liter with the pH adjusted to 5.5.

VII. Ninhydrin solution: 5 g of ninhydrin are dissolved in 188 ml of methyl Cellosolve, and a 62-ml filtered solution of $SnCl_2$ (100 mg) in sodium acetate buffer (VI) is added. This solution has to be stored in a dark bottle in a refrigerator.

Procedure. The reaction mixture consists of 0.25 ml of Substrate solution (I), 0.25 ml of borate buffer (II), and 0.25 ml of enzyme solution (V), it is incubated at 37° for 30 min, then the reaction is terminated by the addition of 0.75 ml of the trichloroacetic acid solution (IV). A blank is obtained by adding trichloroacetic acid to the mixture immediately after the addition of the enzyme solution without any incubation. The solutions are centrifuged for 5 min or filtered to obtain clear supernatants. A 1-ml aliquot of the clear supernatant is mixed with 1 ml of the ninhydrin solution (VII) and 1 ml of acetate buffer (VI) and heated for 10 min at 100° in sealed test tubes. The color is measured at 570 nm. In the 0.2–1 AU range, Beer's law is obeyed.

Section II
Blood Clotting Enzymes

A. ENZYMES OF BLOOD COAGULATION
Articles 4 through 15

B. SNAKE VENOM ENZYMES WITH COAGULATING ACTIVITIES
Articles 16 through 19

[4] Introduction to Clotting and Lysis in Blood Plasma

By L. LORAND

A significant number of contributions in the present volume deals with the enzymes and regulatory proteins involved either as activators or inhibitors of the rather complex events of clotting and lysis in blood plasma. A general synopsis, indicating the interrelationship of components, may be of value for the reader not familiar with this special subject.

The clotting of plasma in vertebrate blood centers around the fibrinogen molecule, a large protein (MW 340,000 in the human species) with a disulfide bridged doublet structure comprising three different constituent chains: $(\alpha\beta\gamma)_2$.[1] Another substrate, the so-called "cold insoluble globulin" (MW 400,000) may also be involved by helping to anchor the fibrin network to the surface of fibroblasts.[2] The clotting of fibrinogen is a *par excellence* example of the enzyme-triggered higher protein associations, and it is controlled in a consecutive manner by two different enzymes: thrombin, a serine-OH protease, and fibrinoligase, a transamidase with a cysteine-SH active center (Fig. 1).[3,4] The explosive cascades of extrinsic and intrinsic initiation systems[5] seemingly evolved for the purpose of generating these two enzymic activities.

Although some of the coagulation components are often referred to by common names, it is also customary to use a numerical designation for the precursor forms of these factors circulating in blood. The Roman numerals represent approximately the order in which their existence was recognized by nomenclature committees. The enzymically active species are denoted by subscripts "a"; other modified forms of the factor may be indicated by apostrophe.

In the intrinsic (endogenous or foreign contact dependent) pathway (Fig. 2) the chain of events leading to coagulation is set in motion merely by the exposure of plasma to nonendothelial surfaces, such as glass *in vitro* or collagen fibers in basement membranes *in vivo*. By contrast, the extrinsic (exogenous or tissue-dependent) pathway (Fig. 3) is brought into action when, as a result of outside injury to the vessel wall, tissue juice becomes mixed with components of the blood plasma.

[1] R. F. Doolittle, *Adv. Protein Chem.* **27**, 1 (1973).

[2] D. F. Mosher, *J. Biol. Chem.* **250**, 6614 (1975).

[3] L. Lorand, *Ann. N.Y. Acad. Sci.* **202**, 6 (1972).

[4] L. Lorand, *in* "Proteases and Biological Control" (E. Reich, D. Rifkin, and E. Shaw, eds). p. 79. Cold Spring Harbor Laboratory, Cold Spring Harbor, New York, 1975.

[5] E. W. Davie and K. Fujikawa, *Annu. Rev. Biochem.* **44**, 799 (1975).

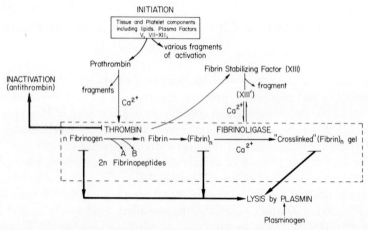

FIG. 1. Outline of the clotting reaction in the plasma of vertebrates. After the initial interaction of tissue, platelet, and plasma factors either by the intrinsic or extrinsic initiation systems (see Figs. 2 and 3), prothrombin is activated to thrombin, a hydrolytic enzyme of great specificity. It brings about the conversion of fibrinogen to fibrin by the consecutive removal of fibrinopeptide fragments (from the N-termini of the α and β chains, respectively) and it also regulates the rate of formation of fibrinoligase, a transamidating enzyme which cross-links fibrin. The pathway of clotting of fibrinogen is depicted within the dashed-line rectangle. Excess thrombin is eliminated by an inactivating mechanism. Fibrinogen–fibrin deposits which may arise in the circulation are removed by plasmin, a trypsinlike enzyme generated from plasminogen by a variety of activators. Heavy arrows outline the "shutoff" mechanisms.

The protein components of the two systems (Fig. 4) include zymogens on one hand (factors VII,[6] XII,[7] XI,[7a] IX,[8] X,[9] II[10,11] and XIII[12]) and regulatory proteins on the other, concerned either with activation ("Tissue factor apoprotein,"[13] factors VIII[14] and V[15]) or inhibition processes ("antithrombin III").[16,16a] In general, the regulatory proteins of activa-

[6] R. Radcliffe and Y. Nemerson, this volume [6].
[7] J. H. Griffin and C. G. Cochrane, this volume [7].
[7a] T. Koide, H. Kato and E. W. Davie, this volume [7a].
[8] K. Fujikawa and E. W. Davie, this volume [8].
[9] K. Fujikawa and E. W. Davie, this volume [10].
[10] K. G. Mann, this volume [13].
[11] R. L. Lundblad, H. S. Kingdon, and K. G. Mann, this volume [14].
[12] C. G. Curtis and L. Lorand, this volume [15].
[13] F. A. Pitlick and Y. Nemerson, this volume [5].
[14] M. E. Legaz and E. W. Davie, this volume [9].
[15] R. W. Colman and R. M. Weinberg, this volume [12].
[16] P. S. Damus and R. D. Rosenberg, this volume [53].
[16a] K. Kurachi, K. Fujikawa, G. Schmer and E. W. Davie, *Biochemistry* **15,** 373 (1976).

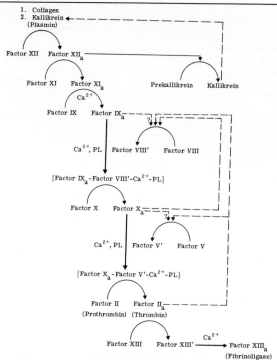

Fig. 2. Known interactions of the enzyme generation cascade in the intrinsic, endogenous, or contact-phase dependent pathway of blood coagulation [modified from E. W. Davie and K. Fujikawa, *Annu. Rev. Biochem.* **44,** 799 (1975)]. After contact activation of factor XII, conversion of this zymogen is greatly enhanced by loops involving kallikrein and plasmin. PL = phospholipid.

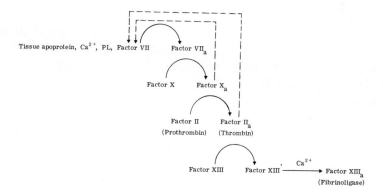

Fig. 3. Enzyme generation cascade on the extrinsic, exogenous, or tissue factor-dependent pathway in blood coagulation. This pathway merges with the intrinsic one, shown in Fig. 2, after the activation of factor X.

Protein Components Involved in Clotting and Lysis in Blood Plasma

		Component of:
A. Zymogens		
1. of "serine" enzymes (proteases)		
Factor XII		intrinsic initiation system
Pre-kallikrein		intrinsic initiation system
Plasminogen		intrinsic initiation system and of fibrinolysis
Factor XI		intrinsic initiation system
Factor IX	vitamin K-dependent synthesis	intrinsic initiation system
Factor VII	vitamin K-dependent synthesis	extrinsic initiation system
Factor X	vitamin K-dependent synthesis	common pathway
Factor II (prothrombin)	vitamin K-dependent synthesis	common pathway
2. of "cysteine" enzyme (transamidase)		
Factor XIII (fibrin stabilizing factor or pro-fibrinoligase)		common pathway
B. Regulatory proteins		
1. activators		
Factor VIII (antihemophilic factor)		intrinsic initiation system
"Tissue-apoprotein"		extrinsic initiation system
Factor V		common pathway
2. inhibitors		
Antithrombin III	binds thrombin, Factors IX_a and X_a	
α_2-Macroglobulin	binds plasmin, kallikrein, thrombin	
CĪ inactivator	binds plasmin, kallikrein	

FIG. 4. Protein components involved in clotting and lysis in blood plasma.

tion require the participation of phospholipids (cephalin, phosphatidyl-ethanolamine, lecithin) and of Ca^{2+} ions.

Coagulation factors, particularly the zymogen forms, seem to be rich in carbohydrate content (as high as 26%, for example, in bovine factor IX[8]), and this may play a role in ensuring their stabilities in circulation. Conversions to the active enzymes occur invariably through processes of limited proteolysis, often resulting in the loss of peptide fragments as well as carbohydrates. Diminution of molecular weights of the parent proteins varies, but fragmentation of prothrombin (factor II) from MW 68,700 to 37,000 appears to be the most extensive.[10,11] It is further the rule that the enzymic forms of the factors comprise two polypeptide chains (usually a light and a heavy one), even when the respective zymogens (e.g., factors IX[8] and II[10,11]) contain single-chain structures. Of all the zymogens, only factor XIII has a heterologous (ab) protomeric subunit composition.[12]

Ordered and controlled reactivity in the hierarchy of the cascade is maintained through the exquisite specificities of individual enzymes. With the sole exception of the very last one (i.e., fibrinoligase or factor XIII_a), all the enzymes belong to the class of serine-OH proteases, which usually show trypsinlike specificities on small substrates, and the ones examined so far display significant sequence homologies with trypsin, suggestive of common ancestry. Although the kinetic constants of thrombin, for example, are indistinguishable from trypsin on small substrates,[17,18] it is well known that thrombin splits only four bonds in fibrinogen[19,20] out of a large number of trypsin-sensitive bonds. The fact that these enzymes comprise two constituent chains may provide a clue as to their restricted specificities. The heavy chain in factors VII_a,[6] IX_a,[8] X_a,[9] and II_a[11] (i.e., thrombin) carries the catalytically essential serine and histidine functionalities. This is equivalent to the single-chain trypsin structure, and, as a first approximation, it may be surmised that in some way the light chains regulate specificities toward protein substrates. Some of the snake venom enzymes[21-23] with fibrinogen clotting activities show an even more restricted specificity than thrombin.

Fibrinoligase (i.e., factor XIII_a), the last enzyme formed in the cascade, is the only one with a cysteine-SH active center,[12,24,25] serving as a rare biochemical example of sulfhydryl enzymes occurring outside of cells. Conversion from its zymogen is also unique in that it includes a hydrolytic feature and a specific Ca^{2+}-dependent heterologous dissociation of subunits:

1. Hydrolysis 2. Dissociation 3. Unmasking

$$(ab) \xrightarrow{\text{thrombin}} (a'b) \xrightarrow[\text{slow}]{\text{high Ca}^{2+}} (a') \xrightarrow[\text{fast}]{\text{low Ca}^{2+}} (a^*)$$

zymogen   enzyme

activation   (b)
peptide   carrier subunit

Unmasking of the active-center cysteine residue on the catalytic subunit (a^*) is directly coupled to the Ca^{2+}-dependent dissociation step.

Four of the zymogens on the cascade [factors VII, IX, X, and II (i.e., prothrombin)] are synthesized in a vitamin-K dependent manner. The vitamin has recently been shown to act as a cofactor in the post-translational carboxylation of these proteins, producing unique γ-car-

[17] F. J. Kézdy, L. Lorand, and K. D. Miller, *Biochemistry* **4**, 2302 (1965).
[18] T. Chase, Jr., and E. Shaw, *Biochemistry* **8**, 2212 (1969).
[19] K. Bailey, F. R. Bettelheim, L. Lorand, and W. R. Middlebrook, *Nature (London)* **167**, 233 (1951).
[20] L. Lorand and W. R. Middlebrook, *Biochem. J.* **52**, 196 (1952).
[21] G. H. Barlow, L. S. Hall, and C. Nolan, this volume [17].
[22] K. Stocker and G. H. Barlow, this volume [18].
[23] F. S. Markland, Jr., this volume [19].
[24] C. G. Curtis, P. Stenberg, C. H. J. Chou, A. Gray, K. L. Brown, and L. Lorand, *Biochem. Biophys. Res. Commun.* **52**, 51 (1973).
[25] C. G. Curtis, K. L. Brown, R. B. Credo, R. A. Domanik, A. Gray, P. Stenberg, and L. Lorand, *Biochemistry* **13**, 3774 (1974).

boxyglutamyl residues.[26,27] These side chains are involved in calcium chelation and, as such, may also be important for binding the phospholipid and the respective regulatory activator proteins to these factors.

Processes of activation are balanced by the removal and neutralization of the active enzymes through a variety of shutoff mechanisms. Hemodilution in rapidly flowing blood must be important; both thrombin and fibrinoligase are bound by fibrin; in addition, normal plasma contains potent protease inhibitors.[16,28,29] Antithrombin III[16] is known to combine with thrombin in a 1:1 stoichiometry and heparin greatly accelerates this reaction; it also inhibits factors IX_a and X_a.[16a] Hirudin[30] provides an interesting biological example of anticoagulant proteins secreted by blood-sucking animals (see also Fritz and Krejci[31]).

Lysis of clots and the organization of thrombi is a complex event which probably involves the release of a number of intracellular enzymes. However, plasma also contains the precursor of a fibrinolytic enzyme, plasmin,[32,33] and this enzyme, too, belongs to the trypsin family of serine-OH proteases. Conversion of plasminogen to plasmin can be brought about either by a direct hydrolytic attack by another protease, urokinase,[34] or indirectly through interaction with streptokinase[35] which, by itself, displays no enzymic activity. It should also be pointed out that factor XII_a, produced at some low rate early in the contact phase of the intrinsic pathway, is known to be able to convert prekallikrein to kallikrein,[36] which, in turn, accelerates the activation of plasminogen. Kallikrein is yet another trypsinlike enzyme.

In summary, the cascade process blood coagulation ensures an orderly reaction pathway for generating the two enzymes: thrombin (factor II_a) and fibrinoligase (factor XIII_a) involved in the clotting of fibrinogen. No

[26] J. Stenflo, P. Fernlund, and P. Roepstorff, in "Proteases and Biological Control" (E. Reich, D. Rifkin, and E. Shaw, eds.), p. 111. Cold Spring Harbor Laboratory. Cold Spring Harbor, New York, 1975.

[27] S. Magnusson. T. E. Petersen, L. Sottrup-Jensen, and H. Claeys, in "Proteases and Biological Control" (E. Reich, D. Rifkin, and E. Shaw, eds.), p. 123. Cold Spring Harbor Laboratory, Cold Spring Harbor, New York, 1975.

[28] P. C. Harpel, this volume [52].

[29] P. C. Harpel, this volume [65].

[30] D. Bagdy, E. Barabás, L. Gráf, T. E. Petersen, and S. Magnusson, this volume [54].

[31] H. Fritz and K. Krejci, this volume [70].

[32] K. C. Robbins and L. Summaria, this volume [22].

[33] F. J. Castellino and J. M. Sodetz, this volume [23].

[34] G. H. Barlow, this volume [20].

[35] F. J. Castellino, J. M. Sodetz, W. J. Brockway, and G. E. Siefring, Jr., this volume [21].

[36] R. W. Colman and A. Bagdasarian, this volume [25].

doubt, there is a significant amplification of signal because each previous trypsinlike enzyme molecule can catalyze the conversion of some hundredfold more of the next zymogen.

[5] Purification and Characterization of Tissue Factor Apoprotein

By FRANCES ANN PITLICK and YALE NEMERSON

Tissue factor is a tissue lipoprotein that initiates the extrinsic pathway of coagulation. The tissue factor (thromboplastic) activity of tissue homogenates is a function of the tissue used and the species from which it was derived.[1] There is general agreement that certain tissues are devoid of this activity. It is of particular interest that erythrocytes, platelets, and leukocytes have no activity in the circulation; with certain stimuli, leukocytes do develop coagulant activity.[2-5] Using previously published procedures,[6] we have prepared tissue factor apoprotein from bovine brain, lung, liver, kidney, and spleen and human placenta. Brain, lung, and placental tissue factors all have similar specific activities; kidney, liver, and spleen tissue factors have relatively low specific activity (5–10% that of brain), yet exhibit similar behavior when examined in aqueous systems (disc gel electrophoresis, for example).

If tissue factor is prepared from saline homogenates of tissue or acetone powders, the coagulant activity is found in a lipoprotein fraction and sediments in the microsomal fraction.[7-11] Both lipid and protein are required for coagulant activity, but the specificity for this activity resides in the protein component.[11-14] After removal of the lipid, the

[1] P. Glas, and T. Astrup, *Am. J. Physiol.* **219**, 1140 (1970).

[2] S. I. Rapaport, and P. F. Hjort, *Thromb. Diath. Haemorrh.* **17**, 222 (1967).

[3] J. Niemetz, and K. Fani, *Nature, (London) New Biol.* **232**, 247 (1971).

[4] R. G. Lerner, R. Goldstein, and G. Cummings, *Proc. Soc. Exp. Biol. Med.* **138**, 145 (1971).

[5] F. R. Rickles, J. A. Hardin, F. A. Pitlick, L. W. Hoyer, and M. E. Conrad, *J. Clin. Invest.* **52**, 1427 (1973).

[6] Y. Nemerson, and F. A. Pitlick, *Biochemistry* **9**, 5100 (1970).

[7] E. Chargaff, D. H. Moore, and A. Bendich, *J. Biol. Chem.* **145**, 593 (1942).

[8] E. R. Hecht, M. H. Cho, and W. H. Seegers, *Am. J. Physiol.* **193**, 584 (1958).

[9] E. Deutsch, K. Irsigler, and H. Lomoschitz, *Thromb. Diath. Haemorrh.* **12**, 12 (1964).

[10] W. J. Williams, *J. Biol. Chem.* **239**, 933 (1964).

[11] M. Hvatum, and H. Prydz, *Biochim. Biophys. Acta* **130**, 92 (1966).

[12] Y. Nemerson, *J. Clin. Invest.* **47**, 72 (1968).

[13] Y. Nemerson, *J. Clin. Invest.* **48**, 322 (1969).

[14] M. Hvatum, and H. Prydz, *Thromb. Diath. Haemorrh.* **21**, 217 (1969).

protein can be solubilized with detergents and purified. While not all phospholipids reconstitute the active complex, certain phospholipids or mixtures of phospholipids can be used interchangeably for reconstitution of the active lipoprotein.

Preparation of Delipidated Tissue Powders

Acetone Dehydration

Bovine lung and other vascular tissues are homogenized in saline to remove serum proteins before acetone extraction; bovine brain is directly extracted with acetone. Fresh or frozen tissue is stripped of membranes and connective tissue, and minced with scissors. The tissue is covered with 0.15 M NaCl and homogenized for 15 sec in a Waring blender at top speed. An equal volume of acetone (certified A.C.S. grade) is then added to the homogenate, and the slurry is filtered under vacuum through Whatman No. 1 paper (dampened with acetone) in a Büchner funnel 26 cm in diameter.

When the flow rate slows appreciably, the slushy homogenate is transferred to a fresh filter apparatus. When filtration is complete, the material is removed from the filter paper, washed with the same amount of acetone, and filtered again. After the acetone has been filtered off, the material is washed with more acetone directly in the Büchner funnel. Washing is continued until the powder no longer looks brown but is white and fluffy and dries rapidly. A vacuum is left on the filtration apparatus for at least another 2 hr (or overnight) until all the material is dry and no longer has any odor of acetone. The powder can then be stored in a freezer.

Tissue for dehydrated brain powders is minced as fine as possible. Enough tissue is added to a large mortar (17 cm diameter) and covered with acetone so that the mortar is no more than half full. The tissue is extracted by grinding with the pestle; the process is repeated until all the brain is extracted. Filtration proceeds as above.

Approximately 15 gallons of acetone are required for extracting a pair of bovine lungs; 7.5–10 gallons are required for extraction of 2500 g of bovine brain; 2500 g of wet tissue yield about 500 g of acetone powder.

Heptane–Butanol Extraction of Acetone Powders

The phospholipid content of the dehydrated tissues is reduced by extensive extraction with a mixture of heptane and butanol. It is most convenient to extract 120 g of acetone powder at a time. Heptane (highest purity, Eastman 2215) and 1-butanol (certified A.C.S.) are mixed in a ratio of 2:1 (v/v). For each gram of acetone powder, 20 ml of heptane–butanol are added; the slurry is stirred for 30 min, then vacuum filtered through a 16.5-cm Büchner funnel with Whatman No. 42 filter paper dampened with heptane–butanol. The extraction with stirring and filtration is repeated at least 4 more times, with care that fine pieces of tissue do not escape into the filtrate. If the fifth filtrate is still cloudy or yellow, extraction and filtration are repeated until the filtrate is clear and colorless.

Vacuum aspiration continues for about 8 hr until no solvent odor remains. The powder is then stored in the freezer. About 70 g of delipidated powder are obtained from 120 g of acetone powder.

Preparation of Tissue Factor Apoprotein

Sodium Deoxycholate Extraction of Delipidated Powders

The delipidated powder is homogenized in 0.5 M NaCl (40 ml per gram of powder) at top speed in a Waring blender for 15 sec. The homogenate is centrifuged for 15 min at 23,000 g using a Beckman ultracentrifuge with a type 15 rotor at 15,000 rpm. Alternatively, the homogenate can be filtered through a 26-cm Büchner funnel using Whatman No. 50 filter paper through which has been poured a slurry consisting of equal amounts of water and Hyflo Super Cel (Fisher Scientific). Enough filter aid is used to form a layer about 2 mm deep. The supernatant (or filtrate) is discarded, and the washed powder is resuspended by homogenization in a minimal volume of 0.25% sodium deoxycholate in water (sodium deoxycholate from Matheson, Coleman and Bell remains in solution better than most other products; sodium deoxycholate solutions that contain precipitate should not be used); 0.25% sodium deoxycholate is then added so that the final proportion is 40 ml per gram of delipidated powder used. The slurry is stirred for 30 min, then centrifuged (or filtered) as above and the supernatant (or filtrate) is retained.

Purification of Tissue Factor Apoprotein with Concanavalin A–Sepharose

With a modification of the procedure described by Allan, Auger, and Crumpton[15] for isolation of concanavalin A (Con A) receptor sites of lymphocyte membranes, we use Con A–Sepharose to remove tissue factor from the sodium deoxycholate extract. After elution with detergent and

[15] D. Allan, J. Auger, and M. J. Crumpton, *Nature (London) New Biol.* **236**, 23 (1972).

α-methyl-D-glucoside, the apoprotein has the same specific activity as earlier preparations which had been through several fractionation procedures.

The sodium deoxycholate extract is passed over a column of Con A–Sepharose (Pharmacia), using 1 ml (bed volume) of Con A–Sepharose for each gram of delipidated powder extracted with sodium deoxycholate. Before use, the column is washed sequentially with 1 column-volume each of 10 mM disodium EDTA, 0.5% sodium deoxycholate, 0.25% sodium deoxycholate–125 mg/ml α-methyl-D-glucoside (Sigma), and, to restore the Con A binding capacity, a buffer consisting of 100 mM sodium acetate, 1.0 M sodium chloride, 10 mM calcium chloride, 10 mM magnesium chloride, pH 6.0, followed by at least 2 column volumes of 50 mM Tris·HCl, pH 7.5. The sodium deoxycholate extract is applied to the column by gravity (1 liter of extract is applied to a column 2.2 × 7 cm at a flow rate of 50–100 ml/hour). Once the sodium deoxycholate extract has passed through the column, the column is washed in sequence with 2 column-volumes each of 50 mM Tris·HCl, pH 7.5, 125 mg/ml α-methyl-D-glucoside, and 50 mM Tris·HCl, pH 7.5. Tissue factor is eluted with 3 column-volumes of 0.1% sodium deoxycholate–125 mg/ml of α-methyl-D-glucoside. The eluate is dialyzed overnight against 50 mM sodium borate, pH 9.0, then against water. The solution is concentrated to at least 1 mg/ml using Diaflo filtration through a PM-10 filter (Amicon, Lexington, Massachusetts). The solution is then mixed with an equal volume of glycerol and stored at −20°.

Data for recovery and purification of tissue factor apoprotein using this technique are presented in Table I.

Characterization of Tissue Factor Apoprotein

In earlier procedures reported from our laboratories,[6] tissue factor apoprotein was fractionated from a sodium deoxycholate extract with

TABLE I

PURIFICATION OF TISSUE FACTOR APOPROTEIN WITH CONCANAVALIN A–SEPHAROSE

	Protein, total (mg)	Activity, total (units)	Activity recovery (%)	Specific activity (units/mg)
Start	1066	650,565		610
Breakthrough and washes	976	284,085	44	
Eluate	21.4	293,055	45	13,442

TABLE II
PARTIAL CHARACTERIZATION OF TISSUE FACTOR APOPROTEIN
ELUTED FROM CONCANAVALIN A–SEPHAROSE[a]

Protein (by the method of Lowry et al.[b])	100 mg
Nucleic acid (from ratio of absorbance at 260 nm to 280 nm[c])	2.5 mg
Phosphorus[d]	
Lipid (extracted into CHCl₃:MeOH; equivalent to 6.7 mg phospholipid)	268 µg
Nonlipid (remaining in aqueous phase; equivalent to 2.2 mg of nucleic acid)	234 µg
Uronic acid (carbazole method[e])	5.6 mg

[a] From F. A. Pitlick, *Biochim. Biophys Acta* to be published.
[b] O. H. Lowry, N. J. Rosebrough, A. L. Farr, and R. J. Randall, *J. Biol. Chem.* **193**, 265 (1951).
[c] S. Chaykin, "Biochemistry Laboratory Techniques." Wiley, New York, 1966.
[d] P. S. Chen, Jr., T. Y. Toribara, and H. Warner, *Anal. Chem.* **28**, 1756 (1956).
[e] T. Bitter and H. M. Muir, *Anal. Biochem.* **4**, 330 (1962).

ammonium sulfate precipitation. Filtration was used to remove material precipitated by 30% saturated ammonium sulfate, and centrifugation (20 min at 23,000 *g*) precipitated the tissue factor activity in 60% saturated ammonium sulfate. The protein solution was then absorbed onto and eluted from DEAE-Sephadex. More recently, we have found that with centrifugation of both ammonium sulfate precipitation steps (1 hr at 23,000 *g* in a Beckman type 15 rotor), the specific activity of the tissue factor is the same as that which has been eluted from DEAE-Sephadex. Consequently, the DEAE-Sephadex adsorption step was discontinued. At this point, the apoprotein was then further purified by gel filtration through Sepharose 6B in 50 m*M* imidazole·HCl–375 m*M* sodium chloride An initial peak from the column is rich in material that absorbs light at 260 nm. The second, much larger, peak contains tissue factor apoprotein; this peak is refiltered through the same column after concentration. The third peak of the initial gel filtration is usually smaller than the tissue factor peak; included in this region are serum protein contaminants, which have been characterized by immunoprecipitin techniques.[16]

When prepared by adsorption to and elution from Con A–Sepharose, bovine brain tissue factor apoprotein has the same specific activity as the material after the final gel filtration using the older procedures. We have partially characterized the composition of the material eluted from Con A–Sepharose, and the results are presented in Table II.

Carbohydrate has previously been reported in tissue factor apoprotein preparations.[6] Using the methods described by Reinhold,[17] we have

[16] F. A. Pitlick, unpublished data.
[17] V. N. Reinhold, this series Vol. 25, p. 244.

TABLE III
CARBOHYDRATE COMPONENTS OF TISSUE FACTOR APOPROTEIN
(mg/100 mg PROTEIN)[a]

Component	Gel filtered		Concanavalin A–Sepharose	
Fucose	1.06	1.10	3.70	2.04
Mannose	0.42	1.12	3.78	6.42
Galactose	4.54	2.30	5.80	4.60
Glucose	2.54	1.80	4.38	2.60
Galactosamine	2.78	2.46	3.18	1.18
Glucosamine	4.58	4.14	6.98	7.82
Sialic acid	5.64	1.74	6.78	6.48
	21.56	14.66	34.60	31.14

[a] From F. A. Pitlick, *Biochim. Biophys.* Acta, to be published.

determined the carbohydrate composition for apoprotein prepared both by gel filtration and by the Con A–Sepharose procedure. As can be seen in Table III, tissue factor apoprotein contains significant amounts of carbohydrate; that prepared by the Con A–Sepharose method is enriched in carbohydrate over the material prepared by older techniques. The carbohydrate composition is in keeping with the membrane localization of the activity.

Consonant with the carbohydrate composition of tissue factor apoprotein and its affinity for Con A–Sepharose, Con A itself inhibits tissue factor activity.[18] When apoprotein is prepared as described above and relipidated so that the final protein concentration is 0.78 μg/ml, Con A, at a concentration of 10 μg/ml, inhibits all the coagulant activity without affecting factors VII and X in the tissue factor assay. α-Methyl-D-glucoside blocks the inhibition when added with the Con A; when added after Con A inhibition has occurred, α-methyl-D-glucoside restores at least 50% of the tissue factor activity.

Diisopropyl phosphorofluoridate, soybean trypsin inhibitor, TPCK, TLCK, o-phenanthroline, dithiothreitol, pCMB, and iodoacetic acid do not inhibit tissue factor activity more than 20% in adequately buffered solutions.[19,20] Glutaraldehyde, on the other hand, completely destroys tissue factor activity (2% glutaraldehyde for 5 min).[16] Tissue factor activity is also sensitive to heat treatment, but some activity may remain even after prolonged heating (1 or 2 hr at 60° or 100°).[8,11,16]

[18] F. A. Pitlick, *J. Clin. Invest.* **55**, 175 (1975).
[19] F. A. Pitlick, Y. Nemerson, A. J. Gottlieb, R. G. Gordon, and W. J. Williams, *Biochemistry* **10**, 2650 (1971).
[20] Y. Nemerson, and M. P. Esnouf, *Proc. Nat. Acad. Sci. U.S.A.* **70**, 310 (1973).

When examined in aqueous systems without detergent, tissue factor apoprotein appears to be electrophoretically homogeneous. Using electrophoretic systems containing sodium dodecyl sulfate (SDS), tissue factor apoprotein is heterogeneous. All bands that stain with Coomassie blue also stain with Schiff reagent after periodic acid oxidation. There is no change in the number of bands or in their mobility after reduction. In our hands, tissue factor activity is quite sensitive to treatment with detergents other than sodium deoxycholate. For example, SDS, even at concentrations less than 0.1%, irreversibly inactivates tissue factor apoprotein.[16]

Preparation of Tissue Factor by Other Techniques

Purification of microsomes with tissue factor activity has been described by Williams[10,21]; isolation of homogeneous tissue factor apoprotein from microsomal preparations has been reported by Bjørklid et al.[22] and by Liu and McCoy.[23]

Earlier work by Hvatum and Prydz[11] suggests that 5-fold purification is obtained with sodium deoxycholate extraction of human brain microsomes; gel filtration of this extract in the same detergent results a 3- to 4-fold increase of specific activity. Final purification is achieved by preparative disc gel electrophoresis in SDS.[22]

Liu and McCoy[33] use similar procedures to isolate the apoprotein from microsomes. By including an ethanol extraction of phospholipids from the microsomes, these investigators achieve a homogeneous protein after a second gel filtration in detergent.

The apoprotein prepared from microsomes requires recombination with phospholipids for full activity. This activity has been determined in a recalcified clotting time assay[22] or by the direct activation of factor X[23]; presumably, there is a trace of factor VII in the factor X reagent for the latter assay to work. At this time, purification and recovery data are insufficient for a comparison of these preparations with the apoprotein described above.

Tissue Factor Assay

When prepared as described above, tissue factor apoprotein is virtually devoid of coagulant activity. Activity is restored by recombination with certain phospholipids and assayed by measuring the amount of acti-

[21] W. J. Williams, J. Biol. Chem. 241, 1840 (1966).
[22] E. Bjørklid, E. Storm, and H. Prydz, Biochem. Biophys. Res. Commun. 55, 969 (1973).
[23] D. T. H. Liu, and L. E. McCoy, Thromb. Res. 7, 199 (1975).

vated factor X formed in a mixture of tissue factor, calcium, and excess amounts of factors VII and X.

Preparation of Mixed Brain Phospholipids

After extraction with heptane–butanol, tissue factor activity of tissue powders is remarkably decreased. Recombination with phospholipids restores coagulant activity to delipidated powders or apoprotein preparations. For routine assays of coagulant activity, recombination with mixed phospholipids from bovine brain is convenient. These phospholipids are derived as a by-product of preparation of delipidated brain powders with a slight modification of the procedures described above.

The first heptane–butanol filtrate is concentrated and dried by rotary evaporation. The lipid is scraped from the walls of the flask, dissolved in benzene–methanol (2:1) (50 ml/per gram of lipid), and clarified by filtration through Whatman No. 1 paper. Aliquots are removed for phosphorus determinations, and the lipid is stored with nitrogen at —80°. The phospholipid concentration of the benzene–methanol solution is estimated from the phosphorous content,[24] assuming the average phosphorus content of phospholipids to be 4%. With care to avoid evaporation, 5- to 10-mg aliquots of phospholipid solution are then dispensed into screw-top glass vials with Teflon-lined caps. The air space over the lipid is filled with prepurified nitrogen before the vials are capped and stored at —80°. Most of the lipid solution is stored in bulk, with 1 to 2 months' supply stored in aliquots. One gram of brain acetone powder yields about 250 mg of phospholipid.

Relipidation of Tissue Factor Apoprotein

To relipidate tissue factor apoprotein, a micellar solution of phospholipid is prepared in sodium deoxycholate and mixed with a solution of tissue factor, and the sodium deoxycholate is removed by dialysis; simple dilution of the relipidated mixture is inadequate to assure full reactivation.

For this procedure, a measured amount of the mixed brain phospholipids is transferred to the glass vessel of a tissue homogenizer (equipped with Teflon pestle with plain tip) and dried under a stream of nitrogen. Enough 0.25% sodium deoxycholate is added to provide a convenient lipid concentration (usually 0.5–1 mg/ml, depending upon the protein concentration of the solution to be relipidated) and the lipid is brought into suspension with homogenization. The solution is then transferred to a small (20- or 30-ml) beaker and sonicated for 15 sec in melting ice.

[24] P. S. Chen, Jr., T. Y. Toribara, and H. Warner, *Anal. Chem.* **28**, 1756 (1956).

The protein concentration of the tissue factor apoprotein solution is determined by the method of Lowry et al.[25] Bovine serum albumin, diluted in the same buffer as the tissue factor apoprotein, is used as a protein standard. The protein concentration of the albumin is determined from the absorbance at 280 nm ($E_{280}^{1\%} = 7.00$). To an aliquot of protein (0.5 ml or 1 ml is a convenient volume) is added enough lipid so that there are 5 mg of lipid per milligram of protein; sufficient 0.25% sodium deoxycholate is added so that the volume of sodium deoxycholate and lipid suspension together is the same as the volume of protein solution. This mixture is then diluted 2-fold with the dialysis buffer, 50 mM imidazole·HCl–100 mM sodium chloride, pH 7.0, and dialyzed overnight against 1 liter of this buffer. Thus, the protein is diluted 4-fold in this procedure.

Preparations of Factors VII and X for Tissue Factor Assay

Factors VII and X are prepared from bovine plasma as described in this volume [6] and [11]. Factor VII is converted to its most active form, the two-chain molecule, by the action of endogenous factor X_a (which accounts for about 0.05% of the protein in this preparation) promoted by addition of calcium and phospholipid. Factor VII (2 mg/ml) is incubated at 37° for about 3 hr with cephalin (125 μg of phospholipid/ml) and 3 mM calcium chloride in 50 mM Tris·HCl–100 mM sodium chloride, pH 7.5. When the apparent factor VII activity no longer increases, the reaction is stopped by the addition of trisodium citrate (12 mM final concentration); phospholipid is removed by centrifugation (100,000 g for 1 hr). Factor VII is stored at −20° in 50% glycerol at a concentration of 0.1 mg/ml (600,000 units/ml). Daily, a fresh aliquot is diluted in 50 mM imidazole·HCl–100 mM sodium chloride, pH 7.0 for tissue factor assay. When prepared this way, a dilution of 1:1000 of the factor VII solution is used.

Factor X is prepared as described in this volume [10]. It is stored in 50% glycerol–50 mM Tris·HCl–100 mM sodium chloride, pH 7.5 at −20°, and aliquots are diluted daily in 50 mM imidazole·HCl–100 mM sodium chloride, pH 7.0, for tissue factor assay. The concentration of the diluted factor X used in the assay is 100 μg/ml. When prepared and stored in this manner, the factor VII and X reagents are stable for several months.

It is also possible to use a crude preparation of factors VII and X prepared from bovine serum. For this procedure, 7.6% sodium citrate, (1

[25] O. H. Lowry, N. J. Rosebrough, A. L. Farr, and R. J. Randall, J. Biol. Chem. 193, 265 (1951).

ml/20 ml serum) and 15% barium chloride (1 ml/10 ml serum) are added to the serum; the slurry is stirred for 20 min, then centrifuged at 23,000 g for 20 min. The precipitate is washed twice with chilled 15% barium chloride and once with cold distilled water. The precipitate is then resuspended in distilled water, using 75.5 ml of water per liter of serum used. Solid ammonium sulfate (20 g per liter of serum) is then added to this slurry to extract the desired coagulation factors. After centrifugation, the same amount of ammonium sulfate is added to the supernatant. After 20 min at 4°, the precipitate is collected by centrifugation, resuspended in a minimal amount of water, and dialyzed overnight against 50 mM imidazole·HCl–100 mM sodium chloride, pH 7.0. Aliquots (0.5 or 1.0 ml) are stored at −80°. The evening before assay, an aliquot is diluted appropriately (see below) and dialyzed against 50 mM imidazole·HCl–100 mM sodium chloride, pH 7.0.

The eluate is tested for suitability as a substrate by comparison of standard curves using a series of dilutions of EDTA bovine brain (see below) tissue factor standard with several dilutions of eluate (1:5, 1:10, 1:20). When the concentration of EDTA bovine brain tissue factor standard is plotted against the final plasma clotting time (see assay below) using log–log coordinates, a straight line should be obtained. Furthermore, the substitution of buffer for EDTA bovine brain tissue factor should result in a clotting time longer than 40 sec. If this buffer blank is faster than 40 sec, it is likely that the original batch of serum contained significant quantities of activated factor X; in this event, eluate must be prepared from a different batch of serum.

Preparation of EDTA-Washed Bovine Brain for Tissue Factor Standard

Bovine brain acetone powder is washed with 20 volumes of 2% disodium EDTA, pH 6.0. After stirring at room temperature for 30 min, the slurry is centrifuged for 2 hr at 37,000 g. The precipitate is resuspended in the same amount of disodium EDTA and centrifuged for 20 min at 37,000 g. The process is repeated for a total of 12 washes. The precipitate is then washed three times in a similar manner with 50 mM imidazole·HCl–100 mM sodium chloride, pH 7.0, resuspended in this buffer (20 ml per gram of acetone powder initially used) and stored in 0.5-ml aliquots at −20°. The day of assay, an aliquot is diluted with 9.5 ml of 50 mM imidazole·HCl–100 mM sodium chloride, pH 7.0, homogenized in a homogenizer with Teflon pestle, transferred to a small beaker, and sonicated for 15 sec in melting ice.

There is variability from one preparation of EDTA brain tissue factor standard to another, so that when a new batch is prepared, it should be titrated on several different days against the batch in use. The newer batch is then diluted appropriately to produce the same standard curve. As used in this laboratory, 100 units of EDTA bovine brain tissue factor standard per milliliter contains about 500 μg of protein and 500 μg of phospholipid per milliliter.

Tissue Factor Assay

The tissue factor assay measures the amount of factor X activated in 1 min by the tissue factor–factor VII complex in the presence of calcium. The amount of factor VII and X, or the dilution of eluate to prepare, is determined by two criteria: to use amounts which are saturating so that clotting times are unaffected by an increase in either factor, and to produce an EDTA bovine brain tissue factor standard dilution curve with linear points using log–log coordinates. The barium citrate eluate is generally diluted 10- or 20-fold; with purified reagents, factor VII is used at a concentration of 600 units/ml and factor X at a concentration of 100 μg/ml.

For assay of coagulant activity, 20 μl of tissue factor, 10 μl of factor VII, 10 μl of factor X (or 20 μl of barium citrate eluate replacing factors VII and X), and 20 μl of 25 mM calcium chloride are incubated for exactly 1 min after the addition of calcium chloride. Then, 100 μl of 25 mM calcium chloride and 100 μl of citrated beef plasma with cephalin are added simultaneously, and the clotting time is determined. Cephalin is prepared from human brain by the method of Bell and Alton.[26] The optimal amount of cephalin to add to the substrate plasma is determined by measuring the partial thromboplastin time of plasma containing different amounts of the phospholipid suspension.[27] The optimal amount should be between 100 μg and 1 mg per milliliter of citrated beef plasma.

Using the assay procedure described above, a typical standard curve using dilutions of EDTA bovine brain tissue factor standard gives the clotting times seen in Table IV. The clotting time for 100 units of EDTA bovine brain tissue factor standard per milliliter is about 15 sec using a Coag-a-mate (Alpha-Medics) and between 20 and 25 sec when judged by eye.

Tissue factor does clot whole plasma in a recalcified clotting time; however, such an assay requires material that is at least 10-fold more concentrated than that used for the two-stage assay. The recalcified clotting time has been an insensitive technique to use for assessing tissue factor activity in cell suspensions, using a cell count of 10^6 cells/ml; the

---

[26] W. N. Bell, and H. G. Alton, *Nature* (*London*) **174**, 880 (1974).
[27] R. P. Proctor, and S. I. Rapaport, *Am. J. Clin. Pathol.* **36**, 212 (1961).

TABLE IV
DILUTION CURVE EDTA BOVINE BRAIN
TISSUE FACTOR STANDARD[a]

TF units	Clotting time (sec)
133	13.8
67	17.3
33	20.6
16	29.5
8	34.5

[a] EDTA-washed bovine brain tissue factor standard was assayed in the 2-stage assay described above using purified factors VII and X and using a Coag-a-mate to determine the clotting time.

two-stage assay readily detects coagulant activity at this cell count. Furthermore, with a very shallow slope for a dilution curve of EDTA bovine brain tissue factor standard, the recalcified clotting time is insensitive to small changes in activity. In addition, we find that a dilution curve of purified relipidated tissue factor apoprotein has a slope different from the standard. Thus, a true assessment of purification is difficult, since each point extrapolates to a different value. While it is not clear at this time why preparations of different purity produce different slopes in the recalcified clotting time, the mechanisms may very well involve the feedback stimulation and inactivation loops discussed in this volume [6] and [11].

Williams has described a two-stage assay (his simplified assay) using a serum eluate from barium sulfate.[10] The slope of his dilution curves with this assay are quite similar to that presented in Table IV.

Acknowledgments

This work was supported in part by grants from the American Heart Association (70-740) and National Institutes of Health (NHLI 16126). We would like to acknowledge the technical assistance of Lionel Clyne, Rita Sznycer-Laszuk, and Audrey Lee.

[6] Bovine Factor VII

By ROBERT RADCLIFFE and YALE NEMERSON

Assay Methods

Principle

Factor VII in conjunction with tissue factor and calcium ions initiates coagulation by the "extrinsic" pathway, and may be assayed by its ability to correct the clotting time of factor VII-deficient plasma upon addition of tissue factor and calcium ions. Plasma congenitally deficient in factor VII is rare, but normal bovine plasma may be selectively rendered deficient in factor VII by the method of Nemerson and Clyne.[1] The latter method derives from the fact that unactivated bovine factor VII is rapidly and irreversibly inhibited by diisopropyl phosphorofluoridate (DFP), in contrast to factor X and prothrombin.[2] Since human factor VII is not readily inhibited by DFP,[2] human plasma cannot be used for the preparation of factor VII-deficient plasma.

Reagents

Thromboplastin prepared from acetone-dehydrated bovine brain powder by the method of Quick[3]

$CaCl_2 \cdot 2 H_2O$, 25 mM, in H_2O

Normal citrated plasma (bovine or human), prepared by collecting blood into $\frac{1}{10}$ volume of 0.12 M trisodium citrate followed by centrifugation at 1500 g for 20 min at 4°. The plasma is decanted and frozen in aliquots at −20°.

Factor VII-deficient bovine plasma, prepared from normal citrated bovine plasma or reconstituted lyophilized citrated bovine plasma (Pentex, Kankakee, Illinois) by the method of Nemerson and Clyne.[1] One gram of Celite is suspended in 100 ml of plasma, and 0.1 ml of DFP is added slowly with stirring. The covered suspension is stirred at 20° for 1 hr. The Celite is removed by centrifugation, and the plasma is frozen in aliquots at −20°. A blank time of over 2 min should be obtained upon addition of thromboplastin and calcium

[1] Y. Nemerson and L. P. Clyne, *J. Lab. Clin. Med.* **83**, 301 (1974).

[2] Y. Nemerson and M. P. Esnouf, *Proc. Natl. Acad. Sci. U.S.A.* **70**, 310 (1972).

[3] A. J. Quick, "Hemorrhagic Diseases," p. 376. Lea & Febiger, Philadelphia, 1959.

ions at 37°, in contrast to a clotting time of 13 sec for normal citrated bovine plasms under these conditions.[4]

Procedure

To 0.1 ml of factor VII-deficient plasma add 0.1 ml of thromboplastin and 0.1 ml of a proper dilution of the test substance in 0.1 M NaCl, 0.05 M Tris·HCl, pH 7.5. Incubate this mixture for 20 sec at 37°. Add 0.1 ml of $CaCl_2$ solution with mixing and start a stopwatch. The test substance should be diluted to yield a clotting time in the range of 30–50 sec. The test reagents are stable at room temperature for several hours.

Calibration

The assay is calibrated with serial dilutions of normal citrated plasma from the species of interest. A straight line is obtained when the clotting times are plotted against concentration of factor VII on log–log graph paper. One milliliter of normal plasma is defined to contain 100 units of factor VII. A 10-fold dilution of human or bovine plasma should yield a clotting time of approximately 20–30 sec or 30–40 sec, respectively.[1] A separate calibration curve must be prepared for each new batch of factor VII-deficient plasma or thromboplastin.

Purification Procedure

Principle

Two major problems are encountered in the purification of bovine factor VII. First, the concentration of factor VII in bovine plasma is approximately 1 mg/liter,[5] in contrast to the much higher concentrations of the other vitamin K-dependent factors from which it must be separated (factor X, 10 mg/liter; factor IX, 5 mg/liter; prothrombin, 100 mg/liter). Second, factor VII is extremely labile to proteolytic cleavage by trace amounts of thrombin, factor X_a, and probably other enzymes generated from plasma.[5] The former problem is overcome by use of affinity column of benzamidine coupled to Sepharose 4B, which binds factor VII tightly but factors II, IX, and X only weakly.[4] The problem of lability is overcome by inclusion of high levels of the proteolytic inhibitor benzamidine[6] in all working solutions. If benzamidine is omitted during purification, a degraded form of factor VII is obtained.[4,5]

---

[4] J. Jesty and Y. Nemerson *J. Biol. Chem.* **249**, 509 (1974).
[5] R. Radcliffe and Y. Nemerson, *J. Biol. Chem.* **250**, 388 (1975).
[6] M. Mares-Guia and E. Shaw, *J. Biol. Chem.* **240**, 1570 (1965).

Procedure

Collection of Blood and Separation of Plasma. One hundred liters of bovine blood are collected into $\frac{1}{10}$ volume of 0.12 M trisodium citrate containing 50 mM benzamidine. Plasma is separated from red cells by passage through a Sharples continuous-flow centrifuge.

Adsorption of Factors II, VII, IX and X to Barium Citrate. A $\frac{1}{12}$ volume of 1 M BaCl$_2$ is added to the citrated plasma and stirred for 30 min at 20°. The barium citrate precipitate with adsorbed protein is collected in the bowl of a continuous-flow centrifuge and washed out with 10 liters of ice-cold 5 mM BaCl$_2$ containing 5 mM benzamidine. The suspension is centrifuged in 4 $\times$ 1-liter buckets at 1500 g for 15 min, and the supernatant is discarded. The pooled precipitate is suspended in 6 liters of 35% saturated (NH$_4$)$_2$SO$_4$, 25 mM benzamidine, pH 7.2, and stirred for 20 min at 4° to elute adsorbed protein. The barium citrate is collected by centrifugation at 1500 g for 10 min and discarded. The supernatant is made to 65% saturation in (NH$_4$)$_2$SO$_4$ and 50 mM in benzamidine, and the protein precipitate is collected by centrifugation at 2500 g for 30 min.

A procedure employing adsorption of plasma with BaSO$_4$ has also been used but is not advised since a severalfold lower yield of factor VII is obtained.[5]

Chromatography on DEAE-Sephadex A50. Two separate batches of DEAE-Sephadex (10 g and 30 g) are swollen for 1 hr in 1 M NaCl and washed extensively with water. The 10-g batch is then equilibrated with 20 mM Tris·HCl, pH 7.5, 25 mM benzamidine. The 30-g batch is equilibrated with 0.1 M NaCl, 0.1 M Tris·HCl, pH 7.5, 25 mM benzamidine.

The (NH$_4$)$_2$SO$_4$ precipitate from the barium citrate eluate is dissolved in 10 liters of 20 mM Tris·HCl, pH 7.5, 25 mM benzamidine. Add the 10-g batch of DEAE-Sephadex suspended in 1 liter of the same buffer, and stir for 20 min at 20°. Filter the suspension through a sintered-glass filter and wash with 3 $\times$ 500 ml of 20 mM Tris·HCl, pH 7.5, 25 mM benzamidine (stirring in each wash for several minutes), 3 $\times$ 500 ml of 0.01 M NaCl, 0.1 M Tris·HCl, pH 7.5, 25 mM benzamidine, and 3 $\times$ 500 ml of 0.12 M NaCl, 0.1 M Tris·HCl, pH 7.5, 25 mM benzamidine. Each wash should be assayed for factor VII as a precaution. The washed DEAE-Sephadex paste is applied to the top of a column (5 cm $\times$ 50 cm) packed with the 30-g batch of DEAE-Sephadex equilibrated in 0.1 M NaCl, 0.1 M Tris·HCl, pH 7.5, 25 mM benzamidine. Chromatography is performed at 4° at a flow rate of 140 ml/hr using a linear 4.5-liter gradient, from 0.12 M NaCl to 0.45 M NaCl, in

0.1 M Tris·HCl, pH 7.5, 25 mM benzamidine. Prothrombin elutes early, with factors VII and IX eluting in the trailing edge of the peak.[2,4,5] Factor X elutes later in a well separated peak. Aliquots must be diluted 20-fold to read the absorbance profile since benzamidine absorbs strongly at 280 nm.

Chromatography and Rechromatography on Benzamidine Coupled to Sepharose. The affinity matrix is prepared by coupling *p*-aminobenzamidine to Sepharose 4B with a nine-carbon spacer group by the method of Cuatrecasas[7] as described by Jesty and Nemerson.[4]

The factor VII pool from the DEAE-Sephadex column is concentrated by ultrafiltration (Amicon; PM-10 membrane) to 100 ml and pumped at a flow rate of 60 ml/hr at 4° onto a column (2.5 cm × 35 cm) of Sepharose-benzamidine equilibrated with 0.1 M NaCl, 50 mM Tris·HCl, pH 7.5, 10 mM benzamidine. Prothrombin and factor IX are eluted with 1 liter of 0.8 M NaCl, 50 mM Tris·HCl, pH 7.5, 10 mM benzamidine at the same flow rate. Factor VII is eluted with a 2-liter linear gradient from 0.2 M guanidine·HCl to 0.85 M guanidine·HCl in 0.2 M NaCl, 50 mM Tris·HCl, pH 7.5, 10 mM benzamidine. The fractions containing factor VII are pooled, concentrated to 20 ml, diluted with 40 ml of 50 mM Tris·HCl, pH 7.5, 10 mM benzamidine, and pumped onto a second column of Sepharose-benzamidine (1.3 cm × 20 cm) at a flow rate of 20 ml/hr. The column is washed with 200 ml of 0.3 M guanidine·HCl, 0.2 M NaCl, 50 mM Tris·HCl, pH 7.5, 10 mM benzamidine, and factor VII eluted with a linear 400-ml gradient from 0.3 M guanidine·HCl to 0.85 M guanidine·HCl in 0.2 M NaCl, 50 mM Tris·HCl, pH 7.5, 10 mM benzamidine. The affinity columns are regenerated by washing with 4 M guanidine·HCl.

Gel Filtration on Sephadex G-100. The factor VII pool from the second affinity column, which contains approximately 50% factor VII (w/w) and 50% contaminant (w/w), mostly prothrombin, is concentrated to 2 ml and applied to a column (1.7 cm × 100 cm) of Sephadex G-100 equilibrated with 0.2 M NaCl, 50 mM Tris·HCl, pH 7.5, 10 mM benzamidine, and chromatography performed with the same buffer at 4°. Two protein peaks are obtained, the first consisting largely of prothrombin and the second being factor VII which gives a single band by several electrophoretic procedures.[4,5] Approximately 10–15 mg of factor VII are obtained from 100 liters of bovine blood (Table I).[5] The factor VII pool is stored frozen at −20° with 10 mM benzamidine present. Aliquots are freed of benzamidine on a column of Sephadex G-25 immediately before use.

[7] P. Cuatrecasas, *J. Biol. Chem.* **245**, 3059 (1970).

TABLE I
PURIFICATION OF FACTOR VII[a]

Step	Protein[b] (mg)	Activity (units × 10⁻⁶)	Yield (%)	Specific activity[b] (units/mg)	Purification (−fold)
Plasma	3.6×10^6	5.0	100	1.4	1
Barium citrate eluate	2.8×10^4	2.0	40	72	52
DEAE pool	3.0×10^3	2.8	56	950	680
Benzamidine pool I	75	2.8	55	37,000	27,000
Benzamidine pool II	21	3.2	63	151,000	107,000
G-100 pool	9	2.4	48	270,000	193,000

[a] Fractionation of barium citrate eluate from 50 liters of bovine plasma.
[b] Calculated from absorbance at 280 nm assuming $A^{1\%}_{1cm} = 10$.

Properties

Chemical Composition. Bovine factor VII is a glycoprotein containing a single peptide chain with a molecular weight of 53,000.[5] The amino acid and carbohydrate composition is shown in Table II.[8] An amino-terminal sequence of Ala-Asx-Gly-Phe-Leu- was obtained by the method of Weiner *et al.,*[9] indicating homology with the amino-terminal sequences of the other vitamin K-dependent factors.[8,10] Digestion with carboxypeptidase A released a valine residue from the carboxyl-terminal end of the peptide chain.[8]

Stability. Bovine factor VII is stable for several hours at neutral pH in the temperature range of 0°–37°, provided the protein concentration is above approximately 20 µg/ml. Activity decays rapidly at lower concentrations unless carrier protein is present (1 mg of ovalbumin or serum albumin per milliliter).[5] Likewise concentrated solutions may be frozen and thawed, but dilute solutions will lose most of their activity.

Proteolytic Activators. Single-chain factor VII is hydrolyzed to a two-chain form (29,500 daltons and 23,500 daltons) by factor X_a and thrombin resulting in an increase in factor VII activity of at least 85-fold.[5] The two chains are disulfide linked, and no activation peptide appears to be released. This fact may explain why the double-chain and single-chain forms of factor VII copurify through the entire above pro-

[8] R. Radcliffe and Y. Nemerson *Fed. Proc., Fed. Am. Soc. Exp. Biol.* **34,** 259 (1975).
[9] A. M. Weiner, T. Platt, and K. Weber, *J. Biol. Chem.* **247,** 3242 (1972).
[10] K. Fujikawa, M. H. Coan, D. L. Enfield, K. Titani, L. H. Ericsson, and E. W. Davie, *Proc. Natl. Acad. Sci. U.S.A.* **71,** 427 (1974).

TABLE II
COMPOSITION OF FACTOR VII AND FACTOR VIIa

Component	Factor VII (moles/ 53,000)	Factor VIIa heavy chain (moles/ 29,500)	Factor VIIa light chain (moles/ 23,500)	Heavy and light chains (moles/ 53,000)
Aspartic acid	28.2	15.2	15.0	30.2
Threonine	15.3	10.2	4.5	14.7
Serine	21.4	13.4	13.0	26.4
Glutamic acid	52.3	25.9	27.0	52.9
Proline	24.6	13.1	8.9	22.0
Glycine	43.2	31.4	20.5	51.9
Alanine	27.5	17.7	9.4	27.1
Half-cystine[a]	23.8	8.1	14.5	22.6
Valine	25.0	18.5	6.7	25.2
Methionine	3.2	1.9	0.6	2.5
Isoleucine	6.6	5.1	4.2	9.3
Leucine	33.7	20.5	10.1	30.6
Tyrosine	7.1	4.0	4.5	8.5
Phenylalanine	13.6	7.5	5.8	13.3
Histidine	10.0	5.8	4.0	9.8
Lysine	9.0	7.1	4.3	11.4
Arginine	26.1	13.2	9.1	22.3
Galactose	4.8[b]			
Mannose	3.2[b]			
N-Acetylgalactosamine	0[b,c]	0[c]	0[c]	0
N-Acetylglucosamine	6.6[b,c]	2.8[c]	3.1[c]	5.9
N-Acetylneuraminic acid	5.6[b]			
Protein	91.7%[d]			
Carbohydrate	8.3%			

[a] Determined as cysteic acid.
[b] Determined by gas-liquid chromatography.
[c] Determined on the amino acid analyzer.
[d] A tryptophan content of 10 moles/53,000 daltons was assumed for purposes of calculation.

cedure.[4,5] The light peptide chain, from the amino-terminal portion of the parent molecule, has an amino terminal sequence of Ala-Asx-Gly-Phe-Leu- and a carboxyl-terminal arginine residue.[8] The heavy chain has a carboxyl-terminal valine residue and an amino-terminal sequence of Ile-Val-Gly-Gly-,[8] showing homology with the heavy chains of factors X_a, IX_a, thrombin, and other serine proteases.[11] After activation of

[11] K. Fujikawa, M. E. Legaz, H. Kato, and E. W. Davie, *Biochemistry* **13**, 4508 (1974).

factor VII by factor X_a or thrombin, a second slower cleavage of the heavy chain to two fragments of 17,000 daltons and 12,500 daltons occurs, resulting in destruction of factor VII activity.[5] These fragments display amino-terminal sequences of Ile-Val-Gly-Gly- and Gly-Val-Thr-Ala, respectively,[8] indicating that the smaller fragment derives from the carboxyl-terminal portion of the parent molecule. This fragment incorporates ^{32}P-labeled DFP,[5] suggesting that this is the location of an active-site serine residue.

The rate of activation of factor VII by factor X_a is enhanced approximately 500-fold by the presence of calcium ions and phospholipids.[5] No cofactor effect is observed for the activation by thrombin. Data have been presented suggesting that factor VII can be activated by kallikrein,[12] insolubilized trypsin,[13] ficin, and papain.[14] It has been observed in this laboratory that Sepharose-trypsin is not a suitable reagent for generation of double-chain factor VII, since only partial activation occurs before activity-destroying cleavages ensue.

Inhibitors. Factor VII is irreversibly inhibited by DFP and phenylmethane sulfonyl fluoride and reversibly inhibited by benzamidine and *p*-aminobenzamidine.[4] Factor VII displays most unusual behavior toward DFP in that both single-chain and double-chain forms incorporate ^{32}P-labeled DFP at the same rate and to the same extent.[5] In contrast, most proenzymes incorporate DFP at a rate orders of magnitude slower than their activated enzyme forms.[15] Even degraded, inactive three-chain factor VII incorporates DFP rapidly.[5]

Specificity. Factor X is the only identified substrate for factor VII. The activation of factor X involves cleavage of an Arg–Ile bond.[16,17] This reaction is enhanced over 2000-fold by the presence of tissue factor and calcium ions. It is clear that two-chain factor VII can catalyze this cleavage rapidly; 25 ng of partially activated factor VII activated 200 μg of factor X to 50% of maximum in 4 min at 37° in the presence of calcium ions and tissue factor.[4] However, it is difficult to establish whether single-chain factor VII possesses coagulant activity. When single-chain material is assayed for activity, a positive finding is obtained indicating 1–2% of the activity that is observed after activation to the two-chain form.[5] However, this positive assay may reflect activa-

[12] H. Gjønnaess, *Thromb. Diath. Haemorrh.* **29**, 633 (1972).
[13] A. Rimon, B. Alexander, and E. Katchalski, *Biochemistry* **5**, 792 (1966).
[14] B. Alexander, L. Pechet and A. Kliman, *Circulation* **26**, 596 (1962).
[15] P. H. Morgan, N. C. Robinson, K. A. Walsh, and H. Neurath, *Proc. Natl. Acad. Sci. U.S.A.* **69**, 3312 (1972).
[16] K. Fujikawa, M. E. Legaz, and E. W. Davie, *Biochemistry* **11**, 4892 (1972).
[17] J. Jesty and Y. Nemerson, this volume [11].

tion of 1–2% of the single-chain factor VII during the assay, rather than inherent single-chain activity. A trace level of double-chain factor VII could generate enough factor X_a to feed back and activate single-chain material, since the activation of factor VII by factor X_a in the presence of phospholipids and calcium ions is rapid.[5] Possible activation by other enzymes generated during the assay, such as intrinsic pathway enzymes, must also be considered.

[7] Human Factor XII (Hageman Factor)

By John H. Griffin and Charles G. Cochrane

Factor XII is an enzyme that circulates in zymogen form in blood and is capable, upon activation, of initiating the clotting, fibrinolytic, and kinin-generating systems.[1–6] Initially described by Ratnoff,[7] factor XII-deficient persons are asymptomatic. In addition to factor XII, the activation in plasma of clotting, fibrinolysis, and kinin generation by negatively charged foreign surfaces also appears to require prekallikrein[4] and a high-molecular-weight form of kininogen.[8] The precise description of the mechanism(s) of activation of the Hageman factor pathways has been limited by the difficulty of isolating satisfactory amounts of the factor XII zymogen and of stabilizing preparations of factor XII in zymogen form. Factor XII in zymogen form has been isolated from human and rabbit blood,[9–11] and structural changes accompanying the

[1] J. Margolis, *Ann. N.Y. Acad. Sci.* **104**, 133 (1963).
[2] O. D. Ratnoff, *Prog. Hematol.* **5**, 204 (1965).
[3] C. G. Cochrane, S. Revak, B. S. Aikin, and K. D. Wuepper, *in* "Inflammation: Mechanisms and Control" (I. H. Lepow and P. A. Ward, eds.), p. 119. Academic Press, New York, 1972.
[4] K. D. Wuepper, *in* "Inflammation: Mechanisms and Control" (I. H. Lepow and P. A. Ward, eds.), p. 93. Academic Press, New York, 1972.
[5] E. H. Magoon, J. Spragg, and K. F. Austen, *Adv. Biosci.* **12**, 225 (1974).
[6] C. G. Cochrane, S. D. Revak, K. D. Wuepper, A. Johnston, and D. C. Morrison, *Adv. Biosci.* **12**, 237 (1974).
[7] O. D. Ratnoff and J. E. Colopy, *J. Clin. Invest.* **34**, 602 (1955).
[8] K. D. Wuepper, D. R. Miller, and M. Lacombe, *Fed. Proc. Fed. Am. Soc. Exp. Biol.* **34**, 859 (1975).
[9] C. G. Cochrane and K. D. Wuepper, *J. Exp. Med.* **134**, 986 (1971).
[10] S. D. Revak, C. G. Cochrane, A. R. Johnston, and T. E. Hugli, *J. Clin. Invest.* **54**, 619 (1974).
[11] H. Saito, O. D. Ratnoff, and V. H. Donaldson, *Circ. Res.* **34**, 641 (1974).

activation of human and rabbit factor XII have been reported.[10,12-14] Nonetheless, variability in reproducing the isolation[9,10] of factor XII and in stabilizing the isolated zymogen[15] led us to develop more reliable methods[16] of isolating and stabilizing factor XII which are described here.

In all manipulations involving factor XII, exposure to glass must be avoided by using plasticware or freshly siliconized glassware (Siliclad, Clay-Adams, Inc.) since factor XII sticks very readily to glass.

Assay Methods

Factor XII activity can be conveniently assayed for its ability to activate isolated plasma prekallikrein or to clot factor XII-deficient plasma, as described below. Less conveniently, it can be assayed for its ability to activate fibrinolysis.[17] Highly purified factor XII can hydrolyze, albeit inefficiently, N-Ac-Gly-Lys-OEt.[18]

Prekallikrein Activation Assay

After activation of factor XII by kaolin or by trypsin, factor XII$_a$ is allowed to convert the zymogen prekallikrein to its active form, kallikrein, which is then conveniently assayed spectrophotometrically for its ability to hydrolyze Bz-Arg-OEt.

Reagents

Tris-buffered saline (TBS): 0.01 M Tris-Cl, 0.15 M NaCl, pH 7.4
Kaolin (Fisher Scientific Co.): acid-washed, 5 mg/ml in TBS
Bz-Arg-OEt (BAEE): (Mann Research): 1 mM in 0.1 M Tris-Cl, 0.05 M NaCl, pH 8.0 (prepared daily and kept at 4° until needed)
Trypsin ("Trypsin-TPCK", Worthington Biochemical Corp.): Stock solution, 1 mg/ml, in 1 mM acetic acid kept at 4° and diluted 1:100 into trypsin assay buffer immediately before each assay.
Trypsin assay buffer: 0.1 M sodium phosphate, 0.05 M NaCl, pH 7.6

[12] C. G. Cochrane, S. D. Revak, and K. D. Wuepper, *J. Exp. Med.* **138**, 1564 (1973).
[13] A. P. Kaplan and K. F. Austen, *J. Exp. Med.* **133**, 696 (1971).
[14] C. R. McMillin, H. Saito, O. D. Ratnoff, and A. C. Walton, *J. Clin. Invest.* **54**, 1312 (1974).
[15] H. Z. Movat and A. H. Ozge-Anwar, *J. Lab. Clin. Med.* **84**, 861 (1974).
[16] J. H. Griffin and C. G. Cochrane, manuscript in preparation.
[17] A. P. Kaplan and K. F. Austen, *J. Exp. Med.* **136**, 1378 (1972).
[18] R. J. Ulevitch, D. Letchford, and C. G. Cochrane. *Thromb. Diath. Haemorrh.* **31**, 30 (1974).

Ovomucoid trypsin inhibitor (Worthington Biochemical Corp.): 2 mg/ml in TBS (solution stable at 4° at least 1 month)

Prekallikrein: Substrate rabbit prekallikrein is partially purified from citrated rabbit plasma as described elsewhere.[19] Briefly, the γ-globulin fraction obtained from a DEAE-Sephadex column is further fractionated on a CM-Sephadex column with elution effected by a NaCl gradient. The pool of prekallikrein activity obtained from the CM-Sephadex column is dialyzed against PBS (0.01 M Na phosphate, 0.15 M NaCl, pH 7.0). The concentration of the pool of prekallikrein is adjusted so that, after total conversion of prekallikrein to kallikrein, a 100-μl aliquot will hydrolyze 0.5 μmole of BAEE per minute in the assay described below. Aliquots of the substrate prekallikrein are stored at $-20°$ in plastic tubes. Aliquot tubes are thawed only once and, thereafter, stored at 4°. Disposable plastic pipettes are used to handle the prekallikrein solution.

Procedure. Factor XII is activated (1) on the negatively charged surface of kaolin or (2) proteolytically by trypsin. (1) Kaolin activation can be useful during early stages of purification because factor XII stays bound to kaolin and plasma protease inhibitors can be washed away so that they do not interfere with prekallikrein activation or with BAEE hydrolysis. Kaolin activation is performed by incubating at 37° for 15 min a 5- to 100-μl aliquot of the test solution containing 0.03–0.5 μg of factor XII with 5.0–500 μg of kaolin in a final volume of 100–200 μl in a 12 × 75 mm plastic test tube. The optimal amount of kaolin must be empirically established for each test solution, and it depends on the amount of interfering protein that can bind to kaolin and thereby compete for the factor XII activation sites. Excess kaolin must be avoided since it binds and inhibits kallikrein at a later stage in this coupled assay. After this incubation, the plastic test tube containing the activation mixture is centrifuged 3 min at 2000 rpm at ambient temperature to form a kaolin pellet, which is washed two times with 200 μl of TBS at 20°. The supernatants are discarded and factor XII$_a$ is bound to the kaolin pellet. (2) Trypsin activation is accomplished by adding a 5–50-μl aliquot of the test solution containing 0.03–0.5 μg of factor XII to a 12 × 75 mm plastic test tube containing 100 μl of trypsin assay buffer and 20 μl of trypsin (0.2 μg). This activation mixture is incubated at 37° for 15 min, then 100 μl of ovomucoid trypsin inhibitor (200 μg) is added to the tube for a further 10 min at 37°. Ovomucoid trypsin inhibitor inhibits trypsin, but not factor XII$_a$ or kallikrein.

[19] K. D. Wuepper and C. G. Cochrane, *J. Exp. Med.* **135**, 1 (1972).

The conversion of the substrate prekallikrein to kallikrein by factor XII_a is accomplished by adding 100 μl of the rabbit prekallikrein reagent solution to the test tube containing activated factor XII and incubating the reaction mixture at 37° for 10–30 min, depending on the amount of factor XII_a present. Then the hydrolysis of BAEE by kallikrein is determined by adding 1.5 ml of 1 mM BAEE to the test tube containing the reaction mixture. After an incubation at 37° if 5–30 min depending on the amount of kallikrein generated, the absorbance increase at 253 nm due to the hydrolysis of BAEE[20] is measured in cuvettes with a 0.5-cm pathlength. It is necessary to have control reaction mixtures lacking factor XII or factor XII_a or prekallikrein and also to use in the reference cuvette an appropriately diluted BAEE solution since 1 mM BAEE has A_{253} of ~1.2 per 0.5 cm. The hydrolysis of BAEE by kallikrein is approximately linear for an absorbance change between 0 and 0.2. Total hydrolysis of 1 mM BAEE gives ΔA_{253} of 0.44 for a 0.5-cm cuvette.

Kallikrein can also be assayed spectrophotometrically based on its hydrolytic activity against the tripeptide substrate, Bz-Pro-Phe-Arg-paranitroanilide (Pentapharm, A. G., Basel).[20a]

Clotting Assay

After activation of factor XII by kaolin, factor XII_a activity is assayed for its ability to clot factor XII-deficient human plasma.

Reagents

TBS
Kaolin (Fisher Scientific): acid washed, 10 mg/ml in TBS
$CaCl_2$, 0.050 M, 37°
Rabbit brain cephalin (Sigma)
Citrated factor XII-deficient and normal human plasmas
Bovine serum albumin (Sigma), 1.0 mg/ml in TBS
All reagents are at 4° except the $CaCl_2$.

Procedure. The following are added in order to a disposable 10 × 75 mm glass tube: 0.1 ml of kaolin, 0.1 ml of cephalin, 0.1 ml of test sample, and 0.1 ml of factor XII-deficient plasma. The tube is mixed gently and placed in a 37° H_2O bath for 8.0 min. Then 75 μl of $CaCl_2$ is added, at which point a stop watch is started. The tube contents are rapidly mixed, and the tube is left standing in the bath until the watch reads 30 sec. The tube is then held in the bath and tilted continually

---

[20] D. C. Morrison and C. G. Cochrane, *J. Exp. Med.* **140**, 797 (1974).
[20a] B. Svenson, *in* "Chemistry and Biology of the Kallikrein-Kinin System in Health and Disease," U.S. Govt. Printing Office, Washington, D.C., in press.

until clotting occurs. A firm clot should appear suddenly, at which point the stop watch is stopped. The reference normal human plasma should be diluted with bovine serum albumin and assayed at dilutions varying from $\frac{1}{20}$ to $\frac{1}{400}$, giving clotting times between 50 and 140 sec. Various dilutions of the sample factor XII solution with bovine serum albumin are made and assayed. To determine the clotting activity of the unknown factor XII solution, the data are plotted on log–log paper showing clotting time versus dilution. By interpolation the clotting activity of the factor XII sample is determined.

Unit of Activity. One unit of factor XII clotting activity is arbitrarily defined as that amount found in 1 ml of a standard pool of normal citrated human plasma. Highly purified human factor XII exhibits specific clotting activities of 59 units/mg [14] to 80 units/mg. [16]

Purification Procedure

Freshly prepared citrated plasma is the starting material for factor XII isolation from human blood. Blood is collected from the antecubital vein of normal healthy volunteers into 50-ml plastic centrifuge tubes containing one-sixth volume acid–citrate–dextrose anticoagulant (ACD: 1 liter contains 13.6 g of citric acid, 25 g of sodium citrate, 20 g of dextrose). The blood is centrifuged at 3000 rpm for 20 min at 20°, and the plasma obtained is then centrifuged again at 5000 rpm for 40 min at 20°. The following is a convenient scheme for isolating factor XII from 500 ml of starting ACD-plasma which contains ca. 12 mg of factor XII.[16]

All steps are carried out in plasticware or siliconized glassware and at 4° (except steps 1 and 2). All buffers contain 1 mM Na$_4$EDTA, 50 μg of hexadimethrine bromide (Polybrene, Aldrich) per milliliter, 1 mM benzamidine·HCl (Aldrich), and 0.02% NaN$_3$. Buffers are prepared fresh for each step to minimize bacterial contamination, since bacterial endotoxin is a potent activator of factor XII.[21] Dialysis tubing and all containers are prerinsed with a 2 g/liter Polybrene solution and then rinsed with H$_2$O before contacting the solution containing factor XII.

Step 1. Heat Treatment. ACD plasma, 500 ml, in a plastic centrifuge bottle is rapidly brought to 56° in a 75° H$_2$O bath with stirring of the plasma by a siliconized thermometer and then placed in a 56° H$_2$O bath for 30 min. The visible precipitate which includes denatured fibrinogen and prekallikrein is removed by centrifugation at 2000 g for 30 min. This defibrination step can be omitted without affecting the final yield, purity, or stability of factor XII.

[21] K. Weber and M. Osborn, *J. Biol. Chem.* **244**, 4406 (1969).

Step 2. DEAE-Sephadex Chromatography. The supernatant is dialyzed against the starting buffer, 40 mM Tris, 10 mM succinic acid, pH 8.4, for chromatography on DEAE-Sephadex A-50 (Pharmacia). One hundred grams of resin is allowed to settle in 3.8 liters of starting buffer in a 4-liter beaker and allowed to swell overnight at 56°. The resin is poured into a 10×40 cm siliconized glass column and washed with 4 liters of starting buffer. The 500-ml sample is applied at 700–900 ml/hr, and 4 liters of starting buffer washes the column at 800 ml/hr. Factor XII is then eluted from the column using a linear gradient formed by 6 liters of starting buffer in the proximal chamber and 6 liters of 0.3 M Tris, 0.12 M succinic acid, 0.3 M NaCl, pH 7.8, in the distal chamber. A flow rate of 600–900 ml/hr is observed for a hydrostatic pressure of 50–80 cm; 45-ml fractions can be collected in 18×250 mm siliconized glass tubes. The fractions are assayed for factor XII activity and for protein by the Lowry method. Factor XII elutes at pH 8.0–8.1 at a buffer conductivity of 8–13 millimho. Fractions containing factor XII activity are found in the ascending portion of the albumin peak, and they are pooled to give 2 liters. This entire step can be performed at 20° and it preferably should be if high molecular weight kininogen is to be purified from the same column.

Pools of high molecular weight kininogen, plasminogen, factor XI, prekallikrein, and plasminogen proactivator have been obtained from this column for further purification. The latter three proteins elute in the γ-fraction which does not bind to the column. Plasminogen elutes just prior to factor XII while high molecular weight kininogen elutes much later at a conductivity of 20–23 millho.

Step 3. Lysine–Sepharose Affinity Chromatography. The 2-liter pool of factor XII is passed through a lysine–Sepharose affinity column in order to remove plasminogen since plasmin is a potent proteolytic activator of factor XII. Lysine–Sepharose is prepared as described elsewhere,[16] and a 200-ml volume of resin is poured into a 6.5×30 cm plastic column and equilibrated with TBS buffer. The 2-liter pool is passed through the lysine–Sepharose column at 300 ml/hr. The 2-liter pool is then brought to 60% saturation with $(NH_4)_2SO_4$ in order to precipitate and to concentrate factor XII. Three liters of saturated $(NH_4)_2SO_4$ (767 g/liter) at 4° are added dropwise over 5 hr to the stirred 2-liter pool of factor XII in a 10-liter plastic bucket. Stirring continues for 45 min before centrifugation at 2500 rpm for 1 hr at 4° in 1-liter plastic bottles to collect the precipitate, which is redissolved in 250 ml of the starting buffer for the next DEAE-Sephadex column.

Step 4. Second DEAE-Sephadex Chromatography. A descending pH gradient is used to elute factor XII from DEAE-Sephadex. Seven grams of DEAE-Sephadex A-50 are allowed to swell at 56° overnight in the starting buffer, 0.03 M imidazole, 0.001 M succinic acid, pH 8.1, and poured into a 2.6 × 35 cm siliconized glass column. The resin is washed with 450 ml of starting buffer before the 250-ml sample which has been dialyzed against starting buffer is applied. The column is again washed with 450 ml of starting buffer. Gradient elution is effected with 600 ml of starting buffer in the stirred proximal chamber and 600 ml of 0.12 M imidazole, 0.045 M succinic acid, pH 6.6, in the distal chamber. The flow rate is controlled at 56 ml/hr by a peristaltic pump, and 8-ml fractions are collected. Factor XII is eluted before the bulk of the protein, and it is found in fractions between pH 7.2 and 6.9. The pool of factor XII activity (about 250 ml) is brought to 60% saturation with $(NH_4)_2SO_4$ at 4°, as outlined above, and the precipitate containing factor XII is dissolved in 30 ml of the starting buffer for the following SP-Sephadex column.

Step 5. SP-Sephadex Chromatograph. An increasing salt gradient at constant pH is employed to elute factor XII from SP-Sephadex. Two grams of SP-Sephadex C-50 are allowed to swell at 56° overnight in the starting buffer, 0.10 M acetate, 0.15 M NaCl, pH 5.3, and sufficient resin to give a bed height of 9 cm is poured into a siliconized glass column, 2.6 cm in diameter. The resin is washed with 100 ml of starting buffer, and the 30-ml sample which has been dialyzed against starting buffer is applied, at which point collection of 6-ml fractions at a flow rate of 56 ml/hr is begun. A 100-ml wash with starting buffer is made. Gradient elution is effected with 150 ml of starting buffer in the stirred proximal chamber and 150 ml of 0.1 M acetate 0.5 M in NaCl, pH 5.3, in the distal chamber. As seen in Fig. 1, the bulk of the protein does not bind to the column, whereas factor XII is bound and elutes between 0.24 M and 0.28 M NaCl.

The pool of factor XII obtained from this column contains 2.4–3.0 mg of protein which is greater than 95% factor XII, corresponding to a yield of 20–25% of the factor XII present in the initial 500 ml of citrated plasma. The clotting activity of factor XII purified with this procedure is 80 ± 11 units/mg, and less than 2% of factor XII is present as factor XII$_a$.

The procedure described above for isolation of human factor XII has been used to isolate rabbit factor XII in similar yield and purity.

Properties

Stability. Maintenance of factor XII in zymogen form poses a difficult problem, and it has often been observed that factor XII can be

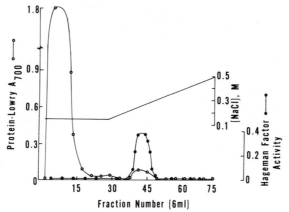

FIG. 1. Elution of human factor XII (Hageman factor) from a sulfopropyl-Sephadex column.

converted to factor XII_a over a period of several weeks even when stored frozen at $-70°$. Nonetheless, factor XII can be successfully and totally preserved in zymogen form for longer than 9 months when kept at 4° in solution in 6 mM acetate, pH 5.0, 0.5 mM EDTA, 0.15 M NaCl.

Purity. The chromatographed preparations of human factor XII described above are found to be greater than 95% homogeneous on SDS polyacrylamide gels[21] and on alkaline polyacrylamide gels.[22]

Physical and Chemical Properties. The electrophoretic mobility of human factor XII is that of a β-globulin and an isoelectric point of 6.3 has been reported.[23] The sedimentation coefficient determined in sucrose density gradients is 4.5 S at 20°.[10] A molecular weight of 76,000 is observed in sodium dodecyl sulfate (SDS) polyacrylamide gel experiments,[10] but less reliable values determined by exclusion chromatography on Sephadex G-200 have ranged from 100,000 to 120,000.[9,23]

Enzymic activation of factor XII in solution results in the appearance of polypeptide fragments of 38,000, 28,000, and about 10,000 MW,[10] and the 28,000 MW fragment contains the enzymic active site while the 38,000 and 10,000 MW fragments contain high affinity binding sites for negatively charged surfaces.[24] Enzymic activation by kallikrein of factor XII bound to negative surfaces either in plasma or in purified systems results in a single cleavage yielding 48,000 and 28,000 MW fragments.[25]

The amino acid content of human factor XII has been determined in

[22] B. J. Davis, *Ann. N.Y. Acad. Sci.* **121**, 404 (1964).
[23] J. Spragg, A. P. Kaplan, and K. F. Austen, *Ann. N.Y. Acad. Sci.* **209**, 372 (1973).
[24] S. D. Revak and C. G. Cochrane, *J. Clin. Invest.* in press.
[25] J. H. Griffin and C. G. Cochrane, *Proc. Nat. Acad. Sci. U.S.A.* in press.

AMINO ACID COMPOSITION OF FACTOR XII

Amino acid[a]	(1) Human[b]	(2) Human[c]	(3) Human[d]	(4) Rabbit[d]
Lysine	6.45	4.2	3.73	3.55
Histidine	3.57	4.5	4.73	5.33
Arginine	5.52	5.9	6.91	6.54
Aspartic acid and asparagine	8.22	6.3	6.56	7.09
Threonine	5.58	5.8	5.98	5.45
Serine	9.73	8.5	6.43	7.16
Glutamic acid and glutamine	12.02	11.3	12.31	11.31
Proline	5.81	8.9	9.94	10.00
Glycine	9.71	10.5	9.02	8.91
Alanine	7.10	9.2	9.33	9.30
Cysteine	3.56	4.2	(3.07)	(2.54)
Valine	5.11	5.4	5.97	5.60
Methionine	0.283	0.1 ± 0.1	0.210	0.223
Isoleucine	2.77	1.4	1.48	1.80
Leucine	8.34	7.6	9.46	9.85
Tyrosine	2.71	2.8	2.43	2.31
Phenylalanine	3.54	2.9	2.60	3.02

[a] Tryptophan was not reported. Cysteine values in columns (3) and (4) are minimal values; those in column (1) were determined as CM-Cys.

[b] S. D. Revak, C. G. Cochrane, A. R. Johnston, and T. E. Hugli, *J. Clin. Invest.* **54**, 619 (1974).

[c] C. R. McMillin, H. Saito, O. D. Ratnoff, and A. C. Walton, *J. Clin. Invest.* **54**, 1312 (1974).

[d] J. H. Griffin and C. G. Cochrane, manuscript in preparation.

several laboratories on different preparations, and the data are shown in the table. The amino acid content of rabbit factor XII is also reported in the table and it shows a remarkable similarity to our more recent data for human factor XII.[16]

The circular dichroism spectrum of human factor XII has been published,[14] and the data suggest that the secondary structure of most of factor XII is nonperiodic.

Recent studies[25] have led to the suggestion that contact activation of factor XII can be entirely explained in terms of interactions between factor XII, prekallikrein, high molecular weight kininogen, and an activating surface. The hypothesis has been advanced[25] that factor XII and high molecular weight kininogen form a complex on the surface which alters the structure of factor XII such that it can be rapidly converted to factor XII_a by a single proteolytic cleavage by kallikrein. Then factor XII_a can reciprocally activate more prekallikrein to kallikrein as well as activate factor XI and plasminogen proactivator leading to activation of the various factor XII-dependent pathways, i.e., kinin-generation, intrinsic coagulation, and fibrinolysis.

Immunochemical Properties. Specific goat precipitating antibodies have been prepared against human factor XII.[10] The absorption of the antiserum with human factor XII-deficient plasma was notably useful in eliminating antibodies directed against trace contaminants of factor XII preparations. The specific antiserum against factor XII is routinely used to determine concentrations of factor XII in plasma samples and in other solutions according to the radial immunodiffusion method of Mancini.[26] The average concentration of factor XII in human plasma has been determined by this method to be 29 μg/ml, i.e., 24 μg/ml in citrated normal human plasma.[10]

Inhibitors. Human factor XII is inhibited by C1 esterase inhibitor[24,27] and by lima bean trypsin inhibitor, but not by soybean or ovomucoid trypsin inhibitor.

Acknowledgments

The studies reported from the authors' laboratory were supported by U.S. Public Health Service Grants NIH AI 07007 and NHLI Grant 16411-01. The skilled technical assistance of Greg Beretta, Alice Kleiss, and Susan Revak is gratefully acknowledged.

[26] G. A. Mancini, A. O. Carbonaro, and J. F. Heremans, *Immunochemistry* **2**, 235 (1965).
[27] A. B. Schreiber, A. P. Kaplan, and K. F. Austen. *J. Clin. Invest.* **52**, 1402 (1973).

[7a] Bovine Factor XI (Plasma Thromboplastin Antecedent)

By Takehiko Koide, Hisao Kato, and Earl W. Davie

Factor XI is a plasma glycoprotein that participates in the early stage of intrinsic blood coagulation.[1] It is absent, or present as an inert protein, in plasma of individuals with a congenital bleeding disorder called plasma thromboplastin antecedent deficiency.[2,3] In normal plasma, factor XI is present in a precursor form and is converted to an active form (factor XI$_a$) by another coagulation factor(s). However, the physiological

[1] E. W. Davie and K. Fujikawa, *Annu. Rev. Biochem.* **44**, 799 (1975).
[2] R. L. Rosenthal, O. H. Dreskin, and N. Rosenthal, *Proc. Soc. Exp. Biol. Med.* **82**, 171 (1953).
[3] C. D. Forbes and O. D. Ratnoff, *J. Lab. Clin. Med.* **79**, 113 (1972).

mechanism for the conversion of factor XI to factor XI$_a$ is not known at the present time.

The role of factor XI$_a$ in blood coagulation has recently been shown by Fujikawa et al.[4] Factor XI$_a$ is a serine protease; it converts factor IX to factor IX$_a$ in the presence of calcium ions. The activation reaction occurs in a two-step reaction. In the first step, an internal peptide bond in factor IX is cleaved, leading to the formation of an intermediate containing two polypeptide chains held together by a disulfide bond(s). This intermediate, which has no enzymic activity, is then converted to factor IX$_a$ by factor XI$_a$. This second step releases an activation peptide from the amino-terminal end of the heavy chain of the factor IX intermediate.

The purification of factor XI has been difficult by conventional methods of protein fractionation since its concentration in plasma is low and it is also very sensitive to protease degradation. In the isolation method described below, heparin–agarose column chromatography and benzamidine–agarose column chromatography play an important role in the isolation procedure. Also, each purification step is performed in the presence of several protease inhibitors, which makes it possible to isolate a highly purified preparation in a precursor form.

Reagents

Factor XI-Deficient Plasma. Bovine factor XI-deficient plasma was very kindly provided by Dr. G. Kociba of Ohio State University, Columbus, Ohio. The deficient plasma was stored in small aliquots at −70° until used.

Heparin–Agarose. Heparin–agarose was prepared as described under factor IX. The heparin-agarose column was used only once and then regenerated as follows: 1 liter of heparin–agarose was stirred with 0.1 N NaOH for 20 min at 4° and rapidly filtered by suction. The gel was then extensively washed with cold water. After packing the heparin–agarose into the column, it was washed with 4 liters of 0.1 M Tris base containing 4.0 M NaCl (no pH adjustment), and then washed and equilibrated with 10 liters of 0.02 M Tris-HCl, pH 7.2, containing 0.05 M NaCl. (Polybrene should not be added to this equilibration buffer.) Heparin–agarose employed for the second heparin–agarose column chromatography was used only after repeated washing with the buffer containing 4 M NaCl followed by equilibration with the starting buffer.

Benzamidine–Agarose. Benzamidine–agarose was prepared according to the method of Schmer[5] using ε-aminocaproic acid as an arm between

[4] K. Fujikawa, M. E. Legaz, H. Kato, and E. W. Davie, *Biochemistry* 13, 4508 (1974).
[5] G. Schmer, *Hoppe-Seyler's Z. Physiol. Chem.* 353, 810 (1972).

benzamidine and the agarose beads. ε-Aminocaproic acid was coupled to agarose A-15m (100–200 mesh) by the same procedure described for the binding of heparin to agarose. Fifty milliliters of ε-aminocaproyl–agarose previously washed with distilled water and 0.1 M 2-(morpholino)ethane-sulfonic acid buffer, pH 4.75, were mixed with 50 ml of 0.1 M 2-(morpholino)ethane sulfonic acid buffer, pH 4.75, containing cyclohexyl morpholinoethyl carbodiimide (100 mg/ml). The mixture was then gently stirred at room temperature for 30 min. One gram of p-aminobenzamidine was added and the mixture was allowed to react for 5 hr. The gel was then washed extensively with 0.05 M imidazole–HCl buffer, pH 6.3, containing 0.05 M NaCl, and stored at 4° in the same buffer containing 0.02% sodium azide.

Phospholipid Suspension. A stock suspension was made as follows: 0.2 g of Centrolex-P was mixed with 0.5 ml of methanol and 9.5 ml of 0.02 M Tris-HCl, pH 7.4, containing 0.1 M NaCl using a motor-driven Potter-Elvehjem homogenizer. The stock suspension was stored in aliquots of 0.5 ml at —20° and diluted with 4.5 ml of 0.02 M Tris-HCl, pH 7.4, containing 0.1 M NaCl immediately before use. The diluted suspension is stable for several hours in an ice bath.

Kaolin Suspension. Fifty milligrams of acid-washed kaolin (Fischer Scientific Co.) was suspended in 10 ml of 0.02 M Tris-HCl, pH 7.4, containing 0.1 M NaCl. The kaolin suspension was carefully mixed each time before use.

Assay Procedures

Factor XI activity was routinely measured by the one-stage kaolin-activated partial thromboplastin time using bovine factor XI-deficient plasma as a substrate. Samples to be assayed for factor XI activity were diluted with 0.02 M Tris-HCl, pH 7.4, containing 0.1 M NaCl. The dilution was sufficient to give a clotting time between 40 and 100 sec. A 0.1-ml aliquot of the diluted sample was incubated at 37° for 10 min with 0.05 ml of bovine factor XI-deficient plasma and 0.1 ml of 0.5% kaolin suspension. After a 10-min preincubation, 0.1 ml of 0.2% phospholipid suspension and 0.1 ml of 0.05 M CaCl$_2$ solution were added to the mixture. An electric timer was activated upon addition of calcium, and the time required for clot formation was recorded. Clot formation was followed by the tilting method. The amount of factor XI activity was estimated from a standard calibration curve made by a serial dilution of normal bovine plasma. The log of the clotting time was plotted against the log of the concentration of platelet-poor normal bovine plasma. One unit of factor XI is defined as that amount of activity present in 1.0 ml of normal bovine plasma. Specific activity is expressed as units per absorbance at 280 nm. For the assay of factor XI_a, the kaolin suspension was replaced by 0.02 M Tris-HCl, pH 7.4, containing 0.1 M NaCl, and the sample mixture was recalcified without preincubation.

Glass tubes (10 × 75 mm disposable culture tubes) were siliconized as described in the factor VIII section and used in the clotting assay.

Purification Steps

Throughout the purification steps, plastic tubes and columns were used, and pipettes and glassware were used after siliconizing as described under factor VIII.

Step 1. Preparation of Plasma. Bovine blood was collected in 10-liter buckets, each containing 1 liter of anticoagulant solution [13.4 g of sodium oxalate, 100 mg of heparin (17,000 units), and 100 mg of crude soybean trypsin inhibitor]. The plasma was isolated at room temperature with a continuous-flow separator (De Laval Model BLE 519). Subsequent steps were all performed at 4°.

Step 2. Barium Sulfate Adsorption and Ammonium Sulfate Precipitation. Barium sulfate adsorption of the plasma was carried out to remove mainly prothrombin, factor VII, factor IX, and factor X prior to ammonium sulfate precipitation. The plasma (20 liters) was mixed and stirred with barium sulfate (20 g/liter) for 30 min, and the slurry was centrifuged for 20 min at 7800 g in a Sorvall RC 3 centrifuge. The barium sulfate supernatant was then brought to 20% saturation by the slow addition of solid ammonium sulfate in the presence of 0.1 mM EDTA. After stirring for 15 min, the precipitate was removed by centrifugation at 7800 g for 15 min. The supernatant was then brought to 50% saturation with solid ammonium sulfate, and after stirring for 30 min the suspension was centrifuged at 7800 g for 40 min. The precipitate was redissolved in 7 liters of cold water containing polybrene (50 mg/liter), DFP (0.2 mM), and soybean trypsin inhibitor[6] (100 mg), and the solution was dialyzed overnight against 100 liters of cold distilled water. The following morning, the dialysis bags were transferred to an 80-liter tank containing 0.02 M Tris-HCl, pH 7.2, and 0.05 M NaCl, and dialysis was continued for an additional 7–8 hr.

Step 3. First Heparin–Agarose Column Chromatography. After dialysis against 0.02 M Tris-HCl, pH 7.2, containing 0.05 M NaCl, the conductivity of the sample solution was adjusted to 7 mmho[7] at 4° by the addi-

---

[6] Kunitz inhibitor.
[7] The conductivity was measured at 4°–5°.

tion of cold distilled water (approximately 2000 ml). DFP (final concentration of 0.2 mM) and polybrene (final concentration of 50 mg/liter) were added to the sample, which was then applied to a heparin–agarose column (8 × 20 cm). This column was previously equilibrated with 0.02 M Tris-HCl, pH 7.2, containing 0.05 M NaCl. After application of the sample,

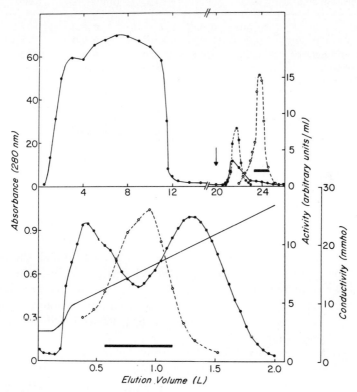

Fig. 1. Elution pattern for bovine factor XI from the first heparin–agarose column and the CM-Sephadex column. *Top panel:* The protein was eluted from the heparin–agarose column (8 × 20 cm) with a linear gradient formed with 3 liters of 0.15 M NaCl in 0.02 M Tris-HCl buffer, pH 7.2, and 3 liters of 0.7 M NaCl in 0.02 M Tris-HCl buffer, pH 7.2, as described under Purification Steps. Fractions (200 ml) were collected at a flow rate of 300 ml/hr. Polybrene (12.5 mg) and soybean trypsin inhibitor (10 mg) were added to each fraction bottle prior to collection of the eluate. Factor XI activity was determined by the one-stage method described under Assay Procedures. ●——●, Absorbance at 280 nm; ○-----○, factor XI activity; ●-----●, factor XII activity. The arrow shows where the gradient was started. *Bottom panel:* The protein was eluted from the CM-Sephadex column (4.5 × 15 cm) with a linear gradient formed with 1 liter of 0.1 M NaCl in 0.02 M phosphate buffer, pH 6.8, and 1 liter of 0.45 M NaCl in 0.02 M phosphate buffer, pH 6.8, as described under Purification Steps. Fractions (9.5 ml) were collected at a flow rate of 95 ml/hr. ●——●, absorbance at 280 nm; ○-----○, factor XI activity.

TABLE I
PURIFICATION OF BOVINE FACTOR XI

Purification step	Volume (ml)	Total activity (units)	Total protein (A_{280})	Specific activity (units/A_{280})	Purification (fold)	Recovery (%)
Plasma	20,000	20,000	1.4×10^6	0.014	1	100
(NH$_4$)$_2$SO$_4$, (20–50%)	9,000	15,800	6.8×10^5	0.023	1.64	79
1st Heparin-agarose	1,400	13,400	2,060	6.50	464	67
CM-Sephadex	580	11,000	380	28.9	2,060	55
2nd Heparin-agarose	125	4,000	23.5	170	12,100	20
Benzamidine-agarose	50	2,790	10.4[a]	268[b]	19,100	14

[a] This value corresponds to 8.3 mg of factor XI.

[b] This value corresponds to 336 units per milligram of protein.

the column was washed with 10 liters of 0.02 M Tris-HCl, pH 7.2, containing 0.05 M NaCl, and 0.2 mM DFP.[8] Factor XI was then eluted at a flow rate of 300 ml/hr with a linear gradient formed with 3 liters of 0.02 M Tris-HCl, pH 7.2, containing 0.15 M NaCl and 3 liters of 0.02 M Tris-HCl, pH 7.2, containing 0.7 M NaCl. Each solution also contained 0.2 mM DFP and 50 mg of polybrene per liter (Fig. 1, top). More than 95% of the contaminating protein passed through the column before the gradient was started. Factor XI was eluted just after factor XII at a conductivity reading of 29–30 mmho. Factor XII was eluted at a conductivity reading of 16–17.5 mmho without appreciable trailing into the factor XI peak. Factor XI was purified 300–600-fold at this stage over the starting plasma (Table I).

Fractions containing factor XI activity were pooled (shown by the solid bar), and benzamidine (final concentration of 1 mM) and DFP (final concentration of 0.2 mM) were added to the pooled fractions, which were dialyzed overnight against 20 liters of 0.02 M phosphate, pH 6.8, containing 0.05 M NaCl. Polybrene (50 mg/liter) and DFP (final concentration of 0.2 mM) were then added to the dialyzed sample.

Step 4. CM-Sephadex Column Chromatography. CM-Sephadex C-50 (200 ml of settled volume) was equilibrated with 0.02 M phosphate buffer, pH 6.8, containing 0.05 M NaCl, and added to the dialyzed sample from the heparin–agarose column. The suspension was stirred slowly for 2 hr and allowed to settle; the supernatant was removed by siphon. The CM-Sephadex was then poured into a 4.5 × 15 cm column, and the column

[8] If only factor XI is being isolated, the NaCl concentration is increased to 0.15 M in the wash buffer. This salt concentration elutes factor XII from the column.

was washed with 500 ml of 0.02 M phosphate buffer, pH 6.8, containing 0.05 M NaCl and polybrene (50 mg/liter).

A linear gradient was then formed with 1 liter of 0.02 M phosphate buffer, pH 6.8, containing 0.1 M NaCl and 1 liter of 0.02 M phosphate buffer, pH 6.8, containing 0.45 M NaCl. Both solutions also contained DFP (0.2 mM) and polybrene (50 mg/liter). About half of the protein was absorbed to the CM-Sephadex before packing the sample into the column. After the gradient was started, factor XI was eluted at a salt concentration with a conductivity of 13–16 mmho (Fig. 1, bottom). The flow rate was 95 ml/hr. Fractions containing factor XI (shown by the bar) were combined.

Step 5. Second Heparin–Agarose Column Chromatography. The pooled fractions from the CM-Sephadex column were adjusted to pH 6.4 with 6 N HCl. Polybrene (50 mg/liter) and DFP (final concentration 0.2 mM) were added and the sample was applied to a second heparin–agarose column (3.2 × 12 cm) previously equilibrated with 0.02 M phosphate buffer, pH 6.4, containing 0.1 M NaCl. The column was then washed with 200 ml of 0.02 M phosphate buffer, pH 6.4, containing 0.20 M NaCl and 0.2 mM DFP. Factor XI was eluted from the column at a flow rate of 80 ml/hr. A linear gradient was formed with 300 ml of 0.02 M phosphate buffer, pH 6.4, containing 0.25 M NaCl, and 300 ml of 0.02 M phosphate buffer, pH 6.4, containing 0.6 M NaCl. Both solutions also contained 0.2 mM DFP and polybrene (50 mg/liter). Factor XI was eluted at a salt concentration with a conductivity of 23–24 mmho (Fig. 2, top). Fractions containing factor XI were combined, and DFP was added to a final concentration of 0.2 mM.

Step 6. Benzamidine–Agarose Column Chromatography. The fractions from the second heparin–agarose column were dialyzed overnight against 2 liters of 0.05 M imidazole-HCl buffer, pH 6.3, containing 0.05 M NaCl), and applied to a 2.3 × 13 cm benzamidine–agarose column (Fig. 2, bottom). This column was previously equilibrated with 0.05 M imidazole-HCl buffer, pH 6.3, containing 0.05 M NaCl. The column was washed with 200 ml of 0.05 M imidazole-HCl buffer, pH 6.3, containing 0.1 M NaCl, 0.01 M guanidine-HCl, and 0.2 mM DFP. Factor XI was then eluted with 200 ml of 0.05 M imidazole-HCl, pH 6.3, containing 0.1 M NaCl, 0.15 M guanidine-HCl, and 0.2 mM DFP. The flow rate was 80 ml/hr. Fractions containing factor XI (shown by the bar) were pooled and DFP was added to a final concentration of 0.2 mM. The sample was then dialyzed overnight against 2 liters of 0.2 M NH$_4$HCO$_3$ or 0.001 N HCl. Salt-free protein was obtained by lyophilization.

Bovine factor XI was purified about 19,000-fold by this method with

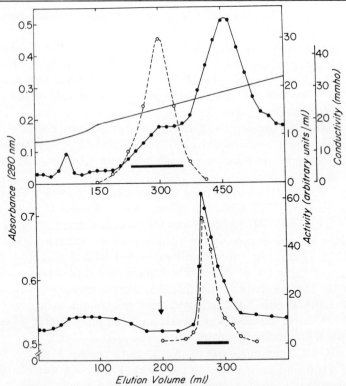

Fig. 2. Elution pattern for bovine factor XI from the second heparin–agarose column and the benzamidine–agarose column. *Top panel:* The protein was eluted from the heparin–agarose column (3.2 × 12 cm) with a linear gradient formed with 300 ml of 0.25 M NaCl in 0.02 M phosphate buffer, pH 6.4, and 300 ml of 0.6 M NaCl in 0.02 M phosphate buffer, pH 6.4, as described under Purification Steps. Fractions (4 ml) were collected at a flow rate of 80 ml/hr. ●——●, Absorbance at 280 nm; ○ - - - - ○, factor XI activity. *Bottom panel:* The benzamidine–agarose column (2.3 × 13 cm) was washed with 200 ml of 0.05 M imidazole–HCl buffer, pH 6.3, containing 0.1 M NaCl, and 0.2 mM DFP. Factor XI was then eluted with 200 ml of 0.05 M imidazole–HCl buffer, pH 6.3, containing 0.1 M NaCl, 0.15 M guanidine–HCl, and 0.2 mM DFP (shown by the arrow). The eluate was collected in 5-ml fractions at a flow rate of 80 ml/hr. ●——●, absorbance at 280 nm; ○ - - - - ○, factor XI activity.

an overall yield of 14%. The isolation procedure takes about 9 days. A summary of the purification is shown in Table I.

The purified preparation was homogeneous when examined by sodium dodecyl sulfate–polyacrylamide gel electrophoresis, immunoelectrophoresis, and sedimentation equilibrium. Factor XI is stable for many months when kept frozen at −20° in the presence of 0.2 mM DFP. The protein retains nearly full activity after dialysis against 0.1 N HCOOH or 0.001 N HCl and after lyophilization.

General Properties of Factor XI

Bovine factor XI is a glycoprotein with a molecular weight of about 130,000. It is composed of two identical or very similar polypeptide chains with a molecular weight of 55,000, and these chains are held together by a disulfide bond(s). The protein contains approximately 11% carbohydrate including hexose, hexosamine, and neuraminic acid. The amino acid and carbohydrate compositions of factor XI are shown in Table II.

The extinction coefficient of bovine factor XI is 12.6 as determined by differential refractometry, and the partial specific volume calculated from the amino acid content and corrected for carbohydrate is 0.707 ml/g.

TABLE II
AMINO ACID AND CARBOHYDRATE
COMPOSITIONS OF BOVINE FACTOR XI

Component	Residues/132,000
Amino acid	
Lysine	59.2
Histidine	41.4
Arginine	43.5
Aspartic acid	88.3
Threonine	80.6
Serine	85.1
Glutamic acid	131.8
Proline	56.0
Glycine	73.1
Alanine	46.9
Half-cystine	45.0
Valine	49.8
Methionine	12.4
Isoleucine	53.4
Leucine	76.7
Tyrosine	31.3
Phenylalanine	44.3
Tryptophan	22.0
Carbohydrate	
Hexose	39.6
Hexosamine	29.4
Neuraminic acid	4.3
Protein (%)	88.9
Carbohydrate (%)	11.1

[8] Bovine Factor IX (Christmas Factor)

By KAZUO FUJIKAWA and EARL W. DAVIE

Factor IX is a plasma glycoprotein that participates in the middle phase of intrinsic blood coagulation. It is inactive or absent in the plasma of individuals with a congenital bleeding disorder known as Christmas disease or hemophilia B. Factor IX exists in normal plasma as a zymogen form and is converted to an active form (factor IX_a) by another coagulation factor, factor XI_a.[1] Factor IX_a is a serine protease, which then converts factor X to factor X_a in the presence of factor VIII, phospholipid, and calcium ions.[2]

Vitamin K is required for the biosynthesis of factor IX as well as factor VII, factor X, and prothrombin. These four coagulation factors also have similar physical-chemical characteristics. For instance, all four are readily absorbed by barium sulfate, barium citrate, and aluminum hydroxide and are eluted with citrate buffer. Recent studies have shown considerable amino acid sequence homology in the amino-terminal portions of the precursor and active enzymes and in the region near the reactive serine residues.[3,4] This suggests that these vitamin-K-dependent proteins have evolved from a common ancestral gene.

The purification of factor IX from bovine and human plasma has been difficult because of its low level in plasma and its high sensitivity to protease degradation. Accordingly, it is highly desirable to carry out the isolation of this protein in the presence of several protease inhibitors. Factor IX often elutes very close to prothrombin, factor X, and factor VII during chromatography with various anionic resins such as DEAE-cellulose and DEAE-Sephadex. Factor IX, however, can be extensively purified on a column of heparin–agarose, and this procedure yields a highly purified bovine factor IX preparation which is homogeneous by several different criteria.[5] A homogenous human factor IX preparation

[1] K. Fujikawa, M. E. Legaz, H. Kato, and E. W. Davie, *Biochemistry* 13, 4508 (1974).

[2] K. Fujikawa, M. H. Coan, M. E. Legaz, and E. W. Davie, *Biochemistry* 13, 5290 (1974).

[3] K. Fujikawa, M. H. Coan, D. L. Enfield, K. Titani, L. H. Ericsson, and E. W. Davie, *Proc. Natl. Acad. Sci. U.S.A.* 71, 427 (1974).

[4] D. L. Enfield, L. H. Ericsson, K. Fujikawa, K. Titani, K. A. Walsh, and H. Neurath, *FEBS Lett.* 47, 132 (1974).

[5] K. Fujikawa, A. R. Thompson, M. E. Legaz, R. G. Meyer, and E. W. Davie, *Biochemistry* 12, 4938 (1973).

has also been described employing a preparative disc gel electrophoresis procedure as a final purification step.[6]

Reagents

Heparin-agarose: Heparin is coupled covalently to agarose beads by the method of Cuatrecasas.[7] A commercial preparation of agarose A-15m (100–200 mesh) or Sepharose 4B is washed with water to remove the contaminating salt before activation. Cyanogen bromide (10 g) is dissolved in 12 ml of dimethylforamide and added immediately with vigorous stirring to a suspension of 50 ml of decanted agarose in an equal volume of cold water. The pH should be kept at 11 by continuous addition of $6N$ NaOH, and the temperature is maintained between 15° and 18° by adding crushed ice. After 15 min of activation, the agarose suspension is poured onto a filter funnel and washed with a large excess of cold 0.1 M NaHCO$_3$ solution. The washed agarose is then mixed with 1 g of heparin (160 units/mg) in 50 ml of 0.1 M NaHCO$_3$ and slowly stirred overnight at 4°. The heparin–agarose is then washed with 0.05 M imidazole-HCl buffer, pH 6.0, and stored in 1 mM sodium azide at 4°.

Solutions: (I) A stock solution of 0.5 M sodium citrate buffer is made by dissolving 588 g of trisodium citrate in 4 liters of water, and the pH is adjusted to 7.0 with 6 N HCl. Citrate buffer of desired molarity is made by diluting the stock solution. Readjustment of pH is not required after dilution. (II) A stock solution of 0.5 M benzamidine-HCl is prepared by dissolving 78.3 g of benzamidine-HCl in 1 liter of water, and the pH is adjusted to 7.0 with 6 N HCl. The solution is stored at 4°. A 0.05 M sodium acetate solution is prepared by dissolving 136.1 g of sodium acetate in 20 liters of water, and the solution is stored at 4°. A stock solution of 1 M diisopropyl fluorophosphate (DFP) is prepared by diluting 1 ml of pure DFP to 5.5 ml with anhydrous isopropanol. This solution is stored in small aliquots at −20°.

DEAE-Sephadex: DEAE-Sephadex (100 g) is prewashed according to the manufacturer's instruction. After a final washing with 0.1 M NaOH, the Sephadex is suspended in 10 liters of 0.05 M citric acid, and the pH is adjusted to 7.0 with 6 N HCl. The DEAE-Sephadex is stored at 4°. After use, the heparin-agarose is regenerated by washing the column with 100 ml of 0.1 M Tris base con-

[6] B. Østerud and R. Flengsrud, *Biochem. J.* **145**, 469 (1975).
[7] P. Cuatrecasas, *J. Biol. Chem.* **245**, 3059 (1970).

taining 2 M NaCl followed by 200 ml of 0.05 M imidazole-HCl buffer, pH 6.0, containing 0.025 M CaCl$_2$.

Phospholipid suspension: A stock solution is made from one of two phospholipid sources and stored in small aliquots at −20°. (A) One vial of rabbit brain cephalin (Sigma) is suspended in 10 ml of 0.15 M NaCl according to the manufacturer's instructions, or (B) 0.2 g of Centrolex-P is mixed with 0.5 ml methanol and 9.5 ml of 0.01 M Tris-HCl buffer, pH 7.4, in 0.15 M NaCl by using a motor-driven Potter-Elvehjem homogenizer. These stock suspensions are diluted with 9 volumes of 0.15 M NaCl before use. The diluted phospholipid suspension is stable for several hours in an ice bath.

Kaolin suspension: Two hundred milligrams of acid-washed kaolin (Fischer Scientific Co.) are suspended in 10 ml of 0.1 M Tris-HCl buffer, pH 7.4. This suspension should be shaken each time before pipetting.

Factor IX-deficient plasma: Blood is collected from a patient with severe Christmas disease in one-tenth volume of 0.1 M sodium oxalate or 0.1 M trisodium citrate. Plasma is obtained by centrifugation of the blood at 12,000 g for 15 min and is stored in small aliquots at −60°.

Assay Procedures

One-Stage Method. Factor IX activity is determined by the kaolin-partial thromboplastin time using human factor IX-deficient plasma as a substrate. Kaolin suspension (0.1 ml) and phospholipid (0.1 ml) are incubated at 37° for 10 min with 0.1 ml of deficient plasma and 0.1 ml of the test sample. One-tenth milliliter of 25 mM CaCl$_2$ solution is then added to the mixture, and the clotting time is recorded with continuous tilting of the tubes at 37°. The amount of factor IX activity is estimated from a standard calibration curve made by a serial dilution of normal bovine plasma. The log of factor IX concentration is plotted against the log of the clotting time. For the assay of factor IX$_a$, the kaolin suspension is replaced by 0.15 M NaCl and preincubation is not necessary.

Two-Stage Method. For an accurate determination of factor IX activity, the two-stage method is preferable. A highly purified preparation of factor XI$_a$, however, is required for this assay.[8] Factor IX ($\sim$500

[8] H. Kato, K. Fujikawa, and M. E. Legaz, *Fed. Proc., Fed. Am. Soc. Exp. Biol.* **33,** 1505 (1974).

μg of protein) is activated at 37° for 30 min by purified factor XI_a (30 μg) in 1.0 ml of 0.05 M Tris-HCl buffer, pH 8.0, containing 5 mM $CaCl_2$. The reaction is terminated by the addition of 0.1 ml of 0.1 M EDTA (pH 7.0), and an aliquot of the reaction mixture is diluted with 0.15 M NaCl to an appropriate concentration of factor IX_a activity. A 0.1 ml of the diluted sample is then preincubated at 37° for 30 sec with 0.1 ml of phospholipid suspension and 0.1 ml of deficient plasma. For most routine assays, normal human or bovine plasma is substituted for factor IX-deficient plasma. One-tenth milliliter of 25 mM $CaCl_2$ solution is then added, and the clotting time is determined employing duplicate samples. The activity of factor IX is again calculated from a standard curve employing normal plasma or the highly purified preparation of factor IX.

One unit of factor IX is defined as that amount of activity present in 1.0 ml of normal bovine plasma. Specific activity is expressed as units per milligram of protein.

Protein is determined by the biuret method[9] using crystalline bovine serum albumin as a standard. The spectrophotometric method for protein concentration is employed for homogeneous preparations of factor IX using an $A_{280}^{1\%}$ of 12.0.

Purification Steps

1. Preparation of Plasma. Bovine blood (100 liters) is collected at a slaughterhouse in ten 10-liter buckets which contain 1 liter of anti-coagulant solution, and plasma is prepared as described under factor VIII isolation.

2. Barium Sulfate Adsorption and Elution. Plasma is stirred for 30 min with barium sulfate (40 g/liter) which adsorbs factor IX along with factor VII, factor X, and prothrombin. Barium sulfate is collected by centrifugation at 7000 g for 10 min. The supernatant can be employed for the isolation of factor VIII (see this volume [9]). The barium sulfate precipitate is washed four times with 3 liters of cold sodium acetate solution. After each washing, the barium sulfate precipitate is removed from the centrifuge cups and mixed with sodium acetate solution in an 8-liter Waring blender. The suspension is homogenized at the highest speed for 1 min followed by centrifugation at 4500 g for 10 min. After the final washing, factor IX is eluted from the $BaSO_4$ by mixing the precipitate with 4 liters of 0.2 M sodium citrate buffer, pH 7.0, containing 1 mM benzamidine, 0.1 mM DFP (add 0.4 ml of 1 M DFP solu-

---

[9] A. G. Gornall, C. S. Bardawill, and M. M. David, *J. Biol. Chem.* 177, 751 (1949).

tion to 4 liters of citrate buffer), and 100 mg of soybean trypsin inhibitor. The solution is stirred for 30 min at 4°. Factor X, factor VII, and prothrombin are also eluted from the barium sulfate along with factor IX. The eluate obtained by centrifugation is added with EDTA to a final concentration of 5×10^{-4} M and dialyzed overnight at 4° against 40 liters of cold water.

3. *Batch Adsorption and Elution on DEAE-Sephadex.* The small particulate matter remaining in the dialysis bags is removed by centrifugation at 7000 g for 15 min. DEAE-Sephadex (500 ml, settled volume), previously equilibrated with sodium citrate buffer, is added to the dialyzate and stirred for 20 min. The solution is allowed to settle for about 15 min and the supernatant (DEAE supernatant) is separated by decantation. The DEAE-Sephadex is then poured onto an empty column (7.5 cm $\times$ 25 cm), and the column is washed with 1 liter of 0.1 M sodium citrate buffer containing 1 mM benzamidine and then with 1 liter of 0.2 M sodium citrate buffer containing 1 mM benzamidine. The eluate of the 0.1 M sodium citrate and the first 600 ml of the 0.2 M sodium citrate eluate are combined with the DEAE supernatant. (For the simultaneous purification of both factor IX and factor X, only the first 250 ml of 0.2 M sodium citrate eluate is taken for factor IX; the next 1500 ml of the eluate are employed for the factor X preparation. This decreases the yield of factor IX, but gives a good yield of factor X). To the combined fraction (about 6.5 liters), DEAE-Sephadex (1 liter of settled volume previously equilibrated with sodium citrate) is added and stirred for 20 min. After the DEAE-Sephadex is settled, the supernatant is decanted and discarded, and the DEAE-Sephadex is transferred to an empty column (7.5 cm $\times$ 50 cm) as described above. After the column is washed with 2 liters of 0.08 M sodium citrate buffer containing 1 mM benzamidine, factor IX is eluted with 0.2 M sodium citrate buffer containing 1 mM benzamidine. The first 450 ml of eluate is discarded. Optical density is then determined on the next fractions, and those with an OD greater than 1.0 are combined. This usually amounts to 800–1000 ml. Soybean trypsin inhibitor (100 mg) and DFP (final concentration of 0.1 mM) are added and the solution is diluted with an equal volume of cold water (batchwise DEAE-Sephadex fraction).

4. *DEAE-Sephadex A-50 Column Chromatography.* DEAE-Sephadex equilibrated with 0.05 M sodium citrate buffer, pH 7.0, is poured in a 5×50 cm column to a height of 25 cm, and the DEAE-Sephadex fraction containing the factor IX is applied to the column. The column is then washed with 500 ml of 0.05 M sodium citrate buffer containing 1 mM benzamidine, and factor IX is eluted with a linear gradient composed of 2 liters of 0.075 M sodium citrate buffer and 2 liters of 0.12 M

sodium citrate buffer. Factor IX activity elutes at the trailing edge of the major protein peak, which is primarily prothrombin (Fig. 1A). The fractions containing factor IX activity (as shown by the solid bar) are combined and dialyzed overnight against 40 liters of 0.05 M imidazole-HCl buffer, pH 6.0 (DEAE-Sephadex column fraction).

  5. Heparin–Agarose Column Chromatography. CaCl$_2$ (1M) is added to the dialyzed sample of factor IX to a final concentration of 2.5 mM, and the solution is applied to a column of heparin agarose (2 $\times$ 10 cm) which was previously equilibrated with 0.05 M imidazole-HCl buffer, pH 6.0,

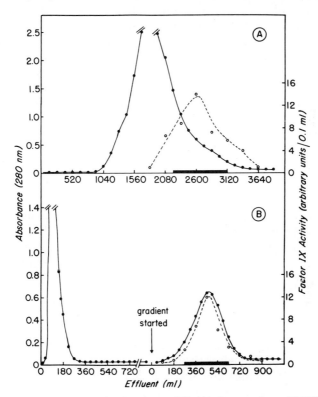

Fig. 1. Elution patterns for bovine factor IX. (A) Pattern from DEAE-Sephadex column. Protein was eluted from the column (5 $\times$ 20 cm) with a linear gradient formed by 2 liters each of 0.075 and 0.12 M sodium citrate buffer, as described under purification steps. Fractions (27 ml) were collected at a flow rate of 140 ml/hr. Factor IX activity was determined by the one-stage method described under assay procedures. (B) Pattern from a heparin–agarose column. Protein was eluted from the column (2 $\times$ 12 cm) with a linear gradient consisting of 500 ml each of 0.25 and 0.7 M NaCl in 0.05 M imidazole (pH 6.0) as described under purification steps. Fractions (9 ml) were collected at a flow rate of 300 ml/hr. (A) and (B): ●——●, absorbance at 280 nm; ○------○, clotting activity.

TABLE I
PURIFICATION OF BOVINE FACTOR IX

Purification step	Volume (ml)	Total protein (mg)	Total activity[a] (units)	Specific activity (units/ mg)	Re- cov- ery (%)	Puri- fica- tion factor
Plasma	45×10^3	3.5×10^{6b}	4.5×10^4	0.013	100	1
BaSO$_4$ eluate	4.9×10^3	11.2×10^{3b}	3.8×10^4	3.39	84.4	260
Batchwise DEAE-Sephadex	700	2.45×10^{3b}	4.6×10^4	18.8	102.2	1,450
DEAE-Sephadex column	770	308^b	4.2×10^4	136	93.3	10,500
Heparin–agarose column	375	117^c	3.4×10^4	291	75.5	22,400

[a] Activity of factor IX was assayed by the two-stage method described under assay procedures. Before assay, each sample (1.0 ml) was diluted 10-fold with 0.05 M imidazole-HCl buffer (pH 6.0) containing 2.5 mM CaCl$_2$ and passed through a heparin–agarose column (0.5 $\times$ 3 cm). The column was washed with 20 ml of 0.25 M NaCl in the same buffer, and factor IX was eluted with the same buffer containing 0.6 M NaCl.
[b] Protein concentration was determined by the biuret method.
[c] Protein concentration was determined by absorbance at 280 nm.

containing 2.5 mM CaCl$_2$. The column is washed with 2 liters of 0.05 M imidazole-HCl buffer, pH 6.0, containing 0.25 M NaCl, 2.5 mM CaCl$_2$ and 1 mM benzamidine-HCl. This step elutes prothrombin.[10] Factor IX is then eluted with a linear gradient formed by 0.5 liter of 0.25 M NaCl and 0.5 liter of 0.75 M NaCl, each containing 0.05 M imidazole-HCl buffer, 2.5 mM CaCl$_2$, and 1 mM benzamidine-HCl (Fig. 1B). The fractions containing factor IX activity are pooled, and EDTA is added to a final concentration of 5 mM. The solution is then concentrated to about 50 ml by Diaflo ultrafiltration (using a PM-10 filter membrane), and the sample is passed through a Sephadex G-25 column (3.5 $\times$ 30 cm) with 0.02 M Tris-HCl buffer, pH 8.0, containing 0.2 M NaCl. The fatcor IX is stable for many months in 50% glycerol at $-20°$. Bovine factor IX is purified about 22,000-fold by this method with an overall yield of 75%. The isolation procedure takes about 1 week. A summary of the purification is shown in Table I. (Eighty nanograms of factor IX$_a$ gives a

[10] With some preparations of heparin-agarose, factor IX was eluted with 0.25 M NaCl containing 2.5 mM CaCl$_2$ and 1 mM benzamidine-HCl. In this situation, prothrombin was eluted with 0.1 M NaCl containing 2.5 mM CaCl$_2$ and 1 mM benzamidine-HCl. Thus, some adjustment of salt concentration may be necessary with different preparations of heparin-agarose.

clotting time of 75 sec with human factor IX-deficient plasma in the one-stage assay described above.)

General Properties of Factor IX

Bovine factor IX is a single-chain polypeptide with a molecular weight of $55,400 \pm 1300$ as determined by sedimentation equilibrium. The purified preparation is homogeneous by sodium dodecyl sulfate (SDS)-polyacrylamide and disc gel electrophoresis, pH 8.9, and immuno-electrophoresis. Factor IX contains 74% protein and 26% carbohydrate.

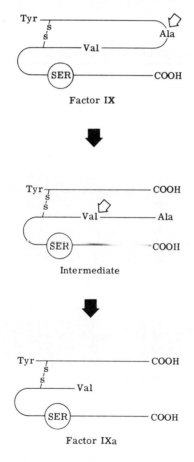

FIG. 2. Mechanism of activation of bovine factor IX by factor XI$_a$. The active center SER is circled. Modified from K. Fujikawa, M. E. Legaz, H. Kato, and E. W. Davie, *Biochemistry* **13**, 4508 (1974).

The carbohydrate includes 10.6% hexose, 6.5% hexosamine, and 8.7% sialic acid.[5] The amino-terminal sequence of factor IX is Tyr-Asn-Ser-Gly-Lys-Leu-Glx-Glx-Phe-Val-Arg-Gly-Asn-Leu-.[3] Eight of these fourteen amino acid residues are homologous with prothrombin and the light chain of factor X. The molecular weight of human factor IX was found to be 70,000 by SDS gel electrophoresis, including about 16% carbohydrate.[6] The molecular weights of the human and bovine preparations are probably the same, however, since bovine factor IX also migrates on SDS polyacrylamide gels with an apparent molecular weight of 68,000.

Factor IX is converted to factor IX_a in the presence of factor XI_a and Ca^{2+}.[1] This reaction occurs in two steps (Fig. 2). In the first step,

TABLE II
AMINO ACID AND CARBOHYDRATE COMPOSITION OF FACTOR IX^a

Component	Factor IX (residues/55,400)	Factor IXa (residues/44,000)	Activation peptide (residues/9000)
Amino acid			
Lysine	27.6	30.5	0.6
Histidine	8.1	8.7	0.2
Arginine	17.0	15.9	0.6
Aspartic acid	36.3	33.2	3.6
Threonine	20.1	20.9	1.6
Serine	29.0	24.6	3.1
Glutamic acid	46.6	45.1	3.8
Proline	13.1	10.7	0.2
Glycine	29.7	32.8	0.7
Alanine	18.8	18.9	1.0
Half-cystine	17.2	21.7	0.3
Valine	24.9	23.6	0.7
Methionine	2.6	2.8	0.6
Isoleucine	19.3	18.9	1.4
Leucine	18.7	19.4	4.0
Tyrosine	16.9	15.3	0.6
Phenylalanine	15.2	13.3	1.3
Tryptophan	11.4		
Carbohydrate			
Hexose	32.6	15.4	12.2
Hexosamine	16.4	7.3	11.8
Neuraminic acid	15.6	1.5	3.7
Protein (%)	74.2	89	30
Carbohydrate (%)	25.8	11	70

[a] From K. Fujikawa, M. E. Legaz, H. Kato, and E. W. Davie, *Biochemistry* **13**, 4508 (1974).

an internal peptide bond in factor IX is cleaved leading to the formation of an intermediate containing two polypeptide chains held together by a disulfide bond(s). In the second step of the activation reaction, an activation peptide (MW 9000) is split from the amino-terminal end of the heavy chain, giving rise to factor IX_a. Factor IX_a is composed of two polypeptide chains containing amino-terminal tyrosine and valine residues. The heavy chain contains the reactive serine. The amino acid and carbohydrate compositions for factor IX, factor IX_a, and the activation peptide are shown in Table II.

[9] Bovine Factor VIII (Antihemophilic Factor)

By MARK E. LEGAZ AND EARL W. DAVIE

The middle phase of intrinsic blood coagulation involves the interaction of factor IX_a (activated Christmas factor), factor VIII (antihemophilic factor), calcium ions, and phospholipid. Incubation of these four components leads to the generation of a strong activator of factor X (Stuart factor), as shown in the following equation:

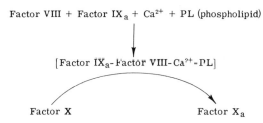

Factor VIII + Factor IX_a + Ca^{2+} + PL (phospholipid)

$[$Factor IX_a-Factor VIII-Ca^{2+}-PL$]$

Factor X   Factor X_a

In this reaction, factor IX_a is the enzyme responsible for the hydrolysis of a specific peptide bond in the heavy chain of factor X. This cleavage occurs between Arg_{51} and Ile_{52}, giving rise to factor $X_{a\alpha}$ and an activation peptide.[1-3] Factor VIII accelerates this reaction several thousandfold. Thus, its role in coagulation appears to be that of a regulatory protein.

[1] K. Fujikawa, M. E. Legaz, and E. W. Davie, *Biochemistry* 11 4892 (1972).
[2] K. Fujikawa, M. H. Coan, M. E. Legaz, and E. W. Davie, *Biochemistry* 13, 5290 (1974).
[3] K. Fujikawa, K. Titani, and E. W. Davie, *Proc. Natl. Acad. Sci. U.S.A.* 72, 3359 (1975).

During the past several years, a number of methods have been described for the purification of milligram quantities of factor VIII.[4-11] Several important steps have greatly aided in the isolation of this trace protein from plasma. These include cryoprecipitation,[12,13] gel filtration,[14] and column chromatography on tricalcium citrate–cellulose.[15-17]

The procedure described in the present article involves a cryoprecipitation step, barium sulfate adsorption of contaminants, glycine precipitation, tricalcium citrate–cellulose chromatography, polyethylene glycol precipitation, and agarose gel filtration. Protease inhibitors are also added throughout the isolation procedure to limit activation and destruction of the molecule.

Reagents

Sodium oxalate
Sodium heparin
Crude soybean trypsin inhibitor
Ammonium sulfate
Barium sulfate (X-ray grade)
Potassium phosphate (monobasic)
Absolute ethanol
Imidazole

[4] E. Bidwell, *Br. J. Haematol.* **1**, 35 (1955).
[5] E. Bidwell, *Br. J. Haematol.* **1**, 386 (1955).
[6] J. P. Hurt, R. H. Wagner, and K. M. Brinkhous, *Thromb. Diath. Haemorrh.* **15**, 327 (1966).
[7] A. J. Johnson, J. Newman, M. B. Howell, and S. Puszkin, *Thromb. Diath. Haemorrh.* **26**, 377 (1967).
[8] S. L. Marchesi, N. R. Shulman, and H. R. Gralnick, *J. Clin. Invest.* **51**, 2151 (1972).
[9] L. Kass, O. D. Ratnoff, and M. A. Leon, *J. Clin. Invest.* **48**, 351 (1969).
[10] G. A. Shapiro, J. C. Andersen, S. V. Pizzo, and P. A. McKee, *J. Clin. Invest.* **52**, 2198 (1973).
[11] P. A. McKee, J. C. Andersen, and M. E. Switzer, *Ann. N.Y. Acad. Sci.* **240**, 8 (1975).
[12] J. G. Pool, E. J. Hershgold, and A. R. Pappenhagen, *Nature (London)* **203**, 312 (1964).
[13] J. G. Pool and A. E. Shannon, *N. Engl. J. Med.* **273**, 1443 (1965).
[14] E. J. Hershgold, A. M. Davison, and M. E. Janszen, *J. Lab. Clin. Med.* **77**, 185 (1971).
[15] G. Schmer, E. P. Kirby, D. C. Teller, and E. W. Davie, *J. Biol. Chem.* **247**, 2512 (1972).
[16] M. E. Legaz, G. Schmer, R. B. Counts, and E. W. Davie, *J. Biol. Chem.* **248**, 3946 (1973).
[17] M. E. Legaz, M. J. Weinstein, C. M. Heldebrant, and E. W. Davie, *Ann. N.Y. Acad. Sci.* **240**, 43 (1975).

ϵ-Aminocaproic acid (ϵAcA)
Sodium citrate
Aluminum hydroxide suspension (Amphogel without flavor)
Trizma base (Sigma)
Succinic acid
Glycine (Sigma)
Tricalcium citrate (Matheson, Coleman and Bell)
Cellulose CF11 powder
Tetrasodium (ethylenedinitrilo)tetraacetate
Sodium azide
Sodium chloride
Agarose (Bio-Rad, A-15m)
β-Alanine
Kaolin (Fischer acid washed)
Centrolex-P (Central Soya)
Dri film silicone (SC87) (General Electric Co.)

Assay

Factor VIII activity was routinely measured employing a one-stage kaolin-activated partial thromboplastin test. Platelet-poor bovine plasma collected in 4% sodium citrate was obtained by centrifugation of the blood for 30 min at 5000 rpm in a Sorvall RC-3 centrifuge. Human factor VIII-deficient plasma was obtained from patients with factor VIII levels of less than 5% and was prepared identically as the bovine plasma. The plasma fractions were stored in small aliquots at $-70°$ until used.

The assay of factor VIII involved a 10-min preincubation of 0.1 ml factor VIII-deficient plasma with 0.1 ml of a solution containing 0.2% Centrolex-P and 0.4% kaolin suspended in 0.05 M Tris-HCl, pH 8.0, containing 0.003 M sodium azide. The phospholipid was prepared by dissolving 0.2 g of Centrolex-P in 0.5 ml of anhydrous ether prior to suspension in 100 ml of Tris-HCl, pH 8.0. The phospholipid was frozen at $-20°$ in 2-ml aliquots and was stable for at least 6 months. The kaolin-phospholipid mixture was incubated for 10 min prior to use, and this solution was stable on ice for 24 hr. After the 10-min incubation of the factor VIII-deficient plasma, phospholipid, and kaolin, a 0.1-ml sample of factor VIII was added followed by immediate recalcification with 0.1 ml of 0.033 M CaCl$_2$. A timer was activated upon addition of calcium, and the time required for clot formation by the tilt method was recorded. The assay was performed in a 37° water bath in siliconized glass tubes to prevent surface adsorption of clotting factors.

The siliconizing solution was prepared by mixing 75 ml of Dri film and 1 liter of petroleum ether. The glass tubes were dipped in the siliconizing solution for 15 sec and then rinsed thoroughly with distilled water. The siliconizing solution can be reused extensively.

Factor VIII activity in a given sample was calculated by comparison with a standard reference curve in which the log of the clotting time is plotted against the log of the concentration of platelet-poor normal bovine plasma. One unit of factor VIII activity was arbitrarily chosen to be that amount of activity present in 1 ml of freshly pooled normal bovine plasma. A standard curve should be prepared every day.

Isolation Procedure

Preparation of Plasma. Bovine blood was collected in 10-liter polyethylene buckets containing 1 liter of anticoagulant. The anticoagulant solution was composed of sodium oxalate, 1.4 g/liter; heparin, 20,000 units/liter; and crude soybean trypsin inhibitor, 100 mg/liter. To ensure thorough mixing of the blood and anticoagulant, the blood was immediately transferred into a 20-liter polyethylene carboy. Plasma was obtained by centrifugation of the blood at 725 rpm in a De Laval blood separator with an output of approximately 1 liter per minute. $BaSO_4$ (40 g/liter) was then added to the plasma to adsorb prothrombin and factors VII, IX, and X. After 30 min of slow stirring, the $BaSO_4$ was removed by centrifugation at 5000 rpm for 10 min at 4° in a Sorvall RC-3 centrifuge. The supernatant was then frozen at −20° in 3-liter polyethylene bottles.

Cryoprecipitate. The supernatant was thawed slowly at room temperature (∼18 hr) until approximately 5% still remained as ice. The pH was then lowered to 6.3 by the addition of a saturated solution of KH_2PO_4. Absolute ethanol (−70°) was added with stirring until a final concentration of 5% was reached. The cryoprecipitate was then recovered by centrifugation at 2500 g for 10 min. The combined protein pellets were washed for an additional 10 min at 4° with 2 liters of 0.02 M imidazole-HCl buffer, pH 6.8, containing 0.5% ε-amino-n-caproic acid (εAcA). The washed protein was dissolved at room temperature in 1 liter of 0.05 M citrate, pH 6.8, containing 0.5% εAcA and to this was added 1/10 volume of aluminum hydroxide with stirring. The solution was mixed for an additional 15 min, and the aluminum hydroxide was removed by centrifugation at 6000 g for 10 min at 15°. The solution was then mixed with an equal volume of 0.5 M Tris-succinate buffer, pH 6.8 (0.5 M succinate titrated to pH 6.8 with Tris base). Glycine, which had been finely powdered in a mortar and pestle, was then slowly added with stirring over a period of about 20 min. The final concentration of glycine was 2.1 M. The solution was allowed to stand for an additional 5 min at 15° to separate factor VIII from a major portion of the cold insoluble globulin contaminant which was soluble under these conditions. The precipitated protein was recovered by centrifugation at 2500 g for 10 min. The flocculent white protein pellet was then squeezed into small pieces by hand employing a disposable rubber glove. The precipitate was added to 1 liter of 0.05 M sodium citrate, pH 6.8, containing 0.5% ϵAcA and slowly stirred. The protein was fully dissolved after 3–4 hr.

Tricalcium Citrate–Cellulose Chromatography. Tricalcium citrate (125 g) and cellulose CF11 powder (125 g) were mixed and washed with 2 liters of distilled water. The water was removed by vacuum filtration, and the solid pellet was washed with 1 liter of M Na citrate, 0.5% ϵAcA, pH 6.8, followed by vacuum filtration to damp dryness. At this time, the factor VIII concentrate was mixed thoroughly with the tricalcium citrate–cellulose powder for 30 min. Citrate buffer was then added to the tricalcium citrate–cellulose suspension ($\sim$500 ml) and the nonadsorbed proteins were removed by centrifugation at room temperature at 6000 g for 30 sec in a Sorvall RC-3 centrifuge. The 1-liter centrifuge bottles containing the tricalcium citrate–cellulose pellets were brought to three-fourths volume with citrate buffer, and after resuspension of the pellets, each was spun as described above. This process was repeated once again, followed by mixing the tricalcium citrate–cellulose pellet with 500 ml of citrate buffer and pouring the slurry directly into a 7.5 × 30 cm polypropylene column. The column was previously packed with 15 g of coarse cellulose powder in order to increase the column flow rate. The column was then allowed to settle and washed with additional citrate buffer ($\sim$2 liters) at a flow rate of 25–30 ml/min until the absorbance of the eluent was less than 0.03 at 280 nm. Factor VIII was eluted from the tricalcium citrate–cellulose column with 0.2 M Tris-HCl, 0.1 M tetrasodium(ethylenedinitrilo)tetraacetate, pH 8.5, containing 0.5% ϵAcA. After 400 ml of buffer had passed through the column (10–15 ml/min), 15-ml fractions were collected at 1-min intervals for 30–40 min. Those fractions containing the majority of the factor VIII activity (150–250 ml) were pooled and the factor VIII precipitated by the slow addition of solid ammonium sulfate to a final concentration of 40%. After 30 min in an ice bucket, the solution was centrifuged in a Sorvall RC-3 for 15 min at 3500 rpm at 4°. The precipitated protein was dissolved in 12–15 ml of 0.05 M Tris-HCl, pH 7.2, containing 0.05 M β-alanine, 0.5% ϵAcA, and 0.003 M sodium azide.

Agarose A15m Gel Filtration. The final purification of bovine factor VIII was accomplished by agarose gel filtration. The protein eluted from two tricalcium citrate–cellulose columns (equivalent to 16 liters of starting plasma) was applied to a 2.5 × 110 cm column of 4% agarose (Bio-Gel A15m, 100–200 mesh) previously equilibrated with 0.05 M Tris-HCl, pH 7.2, containing 0.05 M β-alanine, 0.5% ϵAcA, and 0.003 M sodium azide. The protein was eluted from the column at approximately 15 ml/hr, and factor VIII was recovered in the void volume. It precedes a trace of cold insoluble globulin, which was followed closely by a large fibrinogen peak. The fractions containing factor VIII were assayed for homogeneity by sodium dodecyl sulfate (SDS) agarose polyacrylamide gel electrophoresis.[17] Only those fractions which yield a single band upon reduction were combined and frozen at −20°. The factor VIII coagulant activity is stable for a year or more at −20°.

A purification chart for a typical preparation of bovine factor VIII from 16 liters of plasma is shown in the table. Approximately 21 mg of

PURIFICATION OF BOVINE FACTOR VIII (ANTIHEMOPHILIC FACTOR) FROM PLASMA (16 LITERS)

Purification step	Protein concentration (mg/ml)	Volume (mg)	Total protein (mg)	Specific activity (units/mg)	Total units	Yield (%)	Purification
Plasma	70	16,000	120,000	0.014	16,000	100	1
BaSO₄ supernatant	61	14,100	860,100	0.018	14,480	91	1.3
Ethanol precipitation	12	4,000	48,000	0.24	11,680	73	17
Al(OH)₃ supernatant	10	4,370	43,700	0.24	10,400	65	17
Glycine precipitation	11.7	1,400	16,380	0.53	8,640	54	38
Calcium citrate–cellulose chromatography	0.7	180	126	40	5,020	31	2860
Agarose A-15m chromatography	0.42	50	21	95	2,000	12.5	6800

factor VIII are recovered with a 7000-fold purification and an overall recovery of about 12%. The protein isolated by this procedure was homogeneous when examined by SDS polyacrylamide gel electrophoresis, zone electrophoresis, and immunoelectrophoresis.

Properties of Bovine Factor VIII

Factor VIII is a large glycoprotein with a molecular weight greater than one million.[7-9] Factor VIII isolated by the present procedure has both coagulant activity and platelet aggregating activity (von Willebrand factor). The coagulant activity is increased 50- to 100-fold in the presence of proteases such as thrombin, factor X_a, and trypsin. Whether the coagulant activity and platelet aggregating activity reside in the same molecule has not been established. Thus, it is possible that the protein isolated from the gel filtration step is primarily von Willebrand factor associated with a trace amount of factor VIII.

[10] Bovine Factor X (Stuart Factor)

By KAZUO FUJIKAWA and EARL W. DAVIE

Factor X is a glycoprotein which plays a central role in blood coagulation.[1] It is present in a precursor form in normal plasma, but is inactive in patients with Stuart disease. Factor X is activated by two physiological pathways called the intrinsic and extrinsic systems. In the intrinsic coagulation system, factor X is activated by factor IX_a in the presence of factor VIII, phospholipid, and calcium ions. In the extrinsic system, it is activated by tissue factor and another plasma protein called factor VII. Factor X is also activated by other proteolytic enzymes, such as pancreatic trypsin and a protease from Russell's viper venom. Factor X_a is a serine protease, which in turn activates prothrombin in the presence of factor V, calcium ions, and phospholipid. Recently, other important functions of factor X_a have been reported. Factor X_a activates factor VII, and this reaction may play an important role in the extrinsic coagulation pathway.[2] Factor X_a also modifies factor VIII and potentiates its effect in the intrinsic pathway.[3]

Bovine factor X has been isolated in two different forms, factor X_1 and factor X_2, which are separated by column chromatography on DEAE-Sephadex.[4,5] These two proteins have the same specific activity

[1] E. W. Davie and K. Fujikawa, *Annu. Rev. Biochem.* **44**, 99 (1975).
[2]. R. Radcliffe and Y. Nemerson, *J. Biol. Chem.* **250**, 388 (1975).
[3] E. W. Davie, K. Fujikawa, M. E. Legaz, and H. Kato, *Cold Spring Harbor Conf. Cell Proliferation,* **2**, 65 (1975).
[4] K. Fujikawa, M. W. Legaz, and E. W. Davie, *Biochemistry* **11**, 4892 (1972).
[5] C. M. Jackson, *Biochemistry* **11**, 4873 (1972).

and amino acid content. They also have the same migration on sodium dodecyl sulfate (SDS) or disc gel electrophoresis. The two proteins may differ in their carbohydrate content,[5] although this has not been confirmed.[4]

Factor X was first isolated by Esnouf and Williams as the protein substrate for the Russell's viper venom protease.[6] Thereafter, more efficient methods have been developed by several different groups of investigators in which the yield and quality of this protein have been improved.[4,5,7-9] The use of protease inhibitors during the purification steps has been of considerable importance for the isolation of undegraded protein in the precursor form.

Thus far, a homogeneous preparation of human factor X has not been described.

Reagents

Factor X-deficient plasma: Normal bovine blood is collected in one-tenth volume of 0.1 M sodium oxalate and plasma is obtained by centrifugation at 7000 g for 15 min. The plasma is then filtered through a Seitz filter to remove factor X,[10] and the deficient plasma is stored in small aliquots at −60°. If available, congenital factor X-deficient plasma can also be used.

Russell's viper venom—phospholipid mixture: One-half milligram of crude Russell's viper venom (Ross Allen Reptile Institute, Silver Springs, Florida, or Sigma Chemical Co., St. Louis, Missouri) is mixed with 10 ml of the stock phospholipid suspension described in the previous section under factor IX isolation. This mixture is stored in small aliquots (0.5 ml) at −20° and diluted 10-fold with 0.15 M NaCl before use. This solution is stable for several hours at 4°.

Michaelis buffer. Michaelis buffer is made as follows: 4.9 g of sodium acetate, 7.2 g of sodium barbital, 8.4 g of sodium chloride, and 100 mg of bovine serum albumin are dissolved in 1 liter of distilled water, and the pH is adjusted to 7.4 with 1 N HCl. This buffer is stored at 4°.

Protein is determined by the biuret method[11] using crystalline bovine serum albumin as a standard or the spectrophotometric method is em-

[6] M. P. Esnouf and W. J. Williams, *Biochem. J.* **84**, 62 (1962).
[7] C. M. Jackson and D. J. Hanahan, *Biochemistry* **7**, 4506 (1968).
[8] J. Jesty and M. P. Esnouf, *Biochem. J.* **131**, 791 (1973).
[9] S. P. Bajaj and K. G. Mann, *J. Biol. Chem.* **248**, 7729 (1973).
[10] F. Bachmann, F. Duckert, and F. Koller, *Thromb. Diath. Haemorrh.* **2**, 24 (1958).
[11] A. G. Gornall, C. S. Bardawill, and M. M. David, *J. Biol. Chem.* **177**, 751 (1949).

ployed for the homogeneous preparations using an extinction coefficient of 14.5.

Assay Procedure

A test sample is diluted with Michaelis buffer to give a clotting time between 20 and 60 sec. A 0.1-ml aliquot of the diluted sample is incubated at 37° for 30 sec with 0.1 ml of the diluted Russell's viper venom–phospholipid mixture and 0.1 ml of deficient plasma. A 0.1-ml solution of 0.025 M $CaCl_2$ is then added to the mixture and the clotting time is recorded. A standard activity curve is made by diluting normal plasma with Michaelis buffer at 1:10, 1:20, 1:40, 1:80, and 1:160. Duplicate assays are recommended for each sample. For the assay of factor X_a, Russell's viper venom is deleted from the Russell's viper venom–phospholipid mixture.

Purification Procedure

Reagents. All stock solutions and the preparation of DEAE-Sephadex are the same as those previously described in this volume in Chapters on factor VIII [9] and factor IX [8].

Purification Steps

The purification steps up to the batchwise DEAE-Sephadex fraction are the same as those described in this volume [8]. Fifteen hundred milliliters of 0.2 M sodium citrate eluate from the DEAE-Sephadex is diluted with an equal volume of cold water and diisopropyl phosphorofluoridate (DFP) is added to give a final concentration of 0.1 mM. This solution is then applied to a column of DEAE-Sephadex (4.0 × 20 cm) which was previously equilibrated with 0.05 M sodium citrate buffer, pH 7.0, containing 1 mM benzamidine-HCl. After application of the sample, the protein is eluted by a linear gradient composed of 3 liters of 0.1 M sodium citrate buffer and 3 liters of 0.2 M sodium citrate, both containing 1 mM benzamidine-HCl. The first 1.5 liters of the eluate are discarded; the remaining eluate is collected in 25-ml fractions. When the fractions (dilution 1:100 with Michaelis buffer) are assayed, factor X activity is found in two peaks, which parallel two protein peaks eluted between 0.12 and 0.13 M sodium citrate buffer. The fractions of factor X_1 and factor X_2 are pooled, and DFP is added to a final concentration of 0.1 mM. The solution is then concentrated to about 50 ml by

Diaflo ultrafiltration using a PM-10 filter membrane. An equal volume of glycerol is added to the concentrated sample, and the solution is stored at −20°.

Factor X activity is stable for a number of months under these conditions. About 100 mg of factor X_1 and factor X_2 are obtained from 45 liters of plasma with a recovery of 45%. Before use, the stock sample is dialyzed against an appropriate buffer such as 0.02 M Tris-HCl, pH 7.2, containing 0.2 M NaCl. NH_4HCO_3, 0.1 M, is used for the preparation of salt-free preparations. One unit of factor X is defined as that amount of activity present in 1.0 ml of normal human plasma, and specific activity is expressed as units per milligram of protein. A typical purification is shown in Table I. (One to two nanograms of factor X_a gives a clotting time of about 30 sec with bovine factor X-deficient plasma in the one-stage assay described above.)

Other General Properties of Factor X

Bovine factor X has a molecular weight of 55,100. It is composed of a heavy and a light chain, and these two chains are held together by a disulfide bond(s). A single-chain factor X has also been reported, but

TABLE I
PURIFICATION OF BOVINE FACTOR X[4][a]

Purification step	Volume (ml)	Total protein (mg)	Total activity (units)	Specific activity (units/ mg)	Recovery (%)	Purification factor
Plasma	44,000	3.32×10^{6}[b]	32,600[c]	0.0088	100	1
BaSO₄ eluate	5,000	1.69×10^{4}[b]	27,400[d]	1.6	84	190
Batchwise DEAE-Sephadex	1,680	722	23,400	32	71	3,700
DEAE Sephadex column	1,000	163[e]	15,900	98	49	10,100

[a] From K. Fujikawa, M. W. Legaz, and E. W. Davie, *Biochemistry* **11**, 4892 (1972).

[b] Protein concentration was determined by the biuret method.

[c] The original factor X activity was determined on a small sample collected in the absence of heparin, benzamidine, and soybean trypsin inhibitor.

[d] This sample was passed through a DEAE-Sephadex column (1.0 × 1.0 cm) which was equilibrated with 0.2 M sodium citrate to remove heparin.

[e] Protein concentration was determined by absorbance at 280 nm.

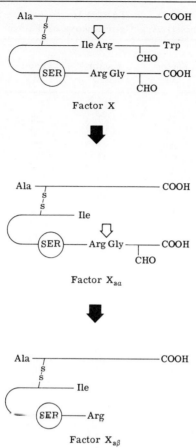

FIG. 1. Mechanism of activation of bovine factor X. The active center Ser-233 is circled. CHO refers to carbohydrate. Modified from K. Fujikawa, M. E. Legaz, and E. W. Davie, *Biochemistry* **13**, 4508 (1974) and K. Fujikawa, K. Titani, and E. W. Davie, *Proc. Natl. Acad. Sci. U.S.A.* **72**, 3359 (1975).

this preparation has not been well characterized.[12] Recently, the total amino acid sequences of the heavy and light chains of bovine factor X_1 have been reported by Enfield *et al.*[13] and Titani *et al.*[14] These studies have shown a great deal of homology between factor X, prothrombin, and factor IX. These three proteins, in addition to factor VII, require vitamin K for their biosynthesis.

[12] P. Mattock and M. P. Esnouf, *Nature (London) New Biol.* **242**, 90 (1973).
[13] D. L. Enfield, L. H. Ericsson, K. A. Walsh, H. Neurath, and K. Titani, *Proc. Natl. Acad. Sci. U.S.A.* **72**, 16 (1975).
[14] K. Titani, K. Fujikawa, D. L. Enfield, L. H. Ericsson, K. A. Walsh, and H. Neurath, *Proc. Natl. Acad. Sci. U.S.A.* **72**, 3082 (1975).

TABLE II
AMINO ACID AND CARBOHYDRATE COMPOSITIONS OF FACTOR X_1 AND FACTOR X_{1a} [a]

Component	Factor X_1	Factor $X_{a\alpha}$	Factor $X_{a\beta}$	Activation peptide	Degradation peptide
Amino acid[b]					
Lysine	23	23	23	0	0
Histidine	12	10	9	2	1
Arginine	25	23	23	2	0
Aspartic acid	27	21	21	6	0
Threonine	31	28	26	3	2
Serine	31	24	23	7	1
Glutamic acid	45	39	38	6	1
Proline	18	15	10	3	5
Glycine	41	38	37	3	1
Alanine	31	26	24	5	2
Half-cystine	24	24	24	0	0
Valine	26	24	23	2	1
Methionine	5	5	5	0	0
Isoleucine	12	11	11	1	0
Leucine	30	23	21	7	2
Tyrosine	10	9	9	1	0
Phenylalanine	21	21	21	0	0
Tryptophan	7	6	5	1	1
Asparagine	14	13	13	1	0
Glutamine	14	13	13	1	0
Molecular weight (protein)[c]	49,801	44,368	42,625	5,452	1,761
Carbohydrate					
Hexose	10	2–3	0	9	2–3
Hexosamine	10	1	0	7	1
Neuraminic acid	7	1	0	5	1
Carbohydrate (%)	9.6	2.1	0	42.6	34.8
Protein (%)[d]	90.4	97.9	100	57.4	65.2
Molecular weight (glycoprotein)[d]	55,070	45,310	42,630	9,490	2,700

[a] From K. Fujikawa, K. Titani, and E. W. Davie, *Proc. Natl. Acad. Sci. U.S.A.* **72,** 3359 (1975).

[b] Calculated from the amino acid sequence.

[c] Ten glutamic acid residues were assumed to be γ-carboxyglutamic acid residues [D. L. Enfield, L. H. Ericsson, K. A. Walsh, H. Neurath, and K. Titani, *Proc. Natl. Acad. Sci. U.S.A.* **72,** 16 (1975)].

[d] Calculated on the basis of three residues of hexose in factor $X_{a\alpha}$ and the degradation peptide. Some revision will be necessary for these values, however, since the carbohydrate analyses have shown some variation. Consequently, a summation of the hexose, hexosamine, and neuraminic acid in factor $X_{a\alpha}$ and the activation peptide does not equal factor X_1.

During the activation of factor X by the intrinsic and extrinsic pathways, a specific peptide bond is hydrolyzed in the amino-terminal region of the heavy chain, as illustrated in Fig. 1.[15-17] This cleavage occurs between Arg_{51} and Ile_{52}, giving rise to factor $X_{a\alpha}$ (MW 45,300) and an activation peptide (MW 9500).

Factor $X_{a\alpha}$ is then converted to factor $X_{a\beta}$ (MW 42,600) by hydrolysis of a second specific peptide bond in the carboxyl-terminal region of the heavy chain. This cleavage occurs between Arg_{290} and Gly_{291}, giving rise to a degradation glycopeptide (MW 2700). Factor $X_{a\alpha}$ and factor $X_{a\beta}$ have equivalent coagulant activity. Thus, the critical event in the activation reaction is the liberation of a new amino-terminal isoleucine in the first step, and this residue probably forms an internal ion pair with Asp_{232}, which is adjacent to the active center Ser_{233}. This probably leads to the charge-relay network analogous to that found in the pancreatic proteases.[18-21]

The amino acid and carbohydrate compositions for factor X_1, factor $X_{a\alpha}$, and factor $X_{a\beta}$ are shown in Table II.[17]

[15] K. Fujikawa, M. E. Legaz, and E. W. Davie, *Biochemistry* 11, 4892 (1972).
[16] K. Fujikawa, M. E. Legaz, and E. W. Davie, *Biochemistry* 13, 4508 (1974).
[17] K. Fujikawa, K. Titani, and E. W. Davie, *Proc. Natl. Acad. Sci. U.S.A.* 72, 3359 (1975).
[18] B. W. Matthews, P. B. Sigler, R. Henderson, and D. M. Blow, *Nature (London)* 214, 652 (1967).
[19] P. B. Sigler, D. M. Blow, B. W. Matthews, and R. Henderson, *J. Mol. Biol.* 35, 143 (1968).
[20] D. M Blow, J. J. Birktoft, and B. S. Hartley, *Nature (London)* 221, 337 (1969).
[21] D. M. Shotton and H. C. Watson, *Nature (London)* 225, 811 (1970).

[11] The Activation of Bovine Coagulation Factor X[1]

By Jolyon Jesty and Yale Nemerson

Factor X (Stuart–Prower factor) is one of the four vitamin K-dependent clotting factors; the others are prothrombin, factor VII, and factor IX. The activation of factor X is the point at which the intrinsic and extrinsic pathways of coagulation converge, the activators being, respectively, a complex of factor IX_a + factor VIII + phospholipid + calcium ions,[2,3] and a complex of factor VII + tissue factor (a lipoprotein)

[1] Supported in part by Grant HL 16126 from National Institutes of Health, U.S. Public Health Service.
[2] C. Hougie, K. W. E. Denson, and R. Biggs, *Thromb. Diath. Haemorrh.* 18, 211 (1967).
[3] P. G. Barton, *Nature (London)* 215, 1508 (1967).

$+$ calcium ions.[4,5] Factor X can also be specifically activated in the presence of Ca^{2+} by a fraction from Russell's viper venom (RVV),[6-9] and it is largely owing to this that the present knowledge of factor X and its activation is so comprehensive.

Activated factor X participates in the coagulation sequence by activating (in the presence of a cofactor, factor V, phospholipid, and Ca^{2+}) prothrombin to thrombin, which then cleaves fibrinogen and factor XIII, leading to clot formation and covalent cross-linking.

The activation of factor X is known to be proteolytic, an NH_2-terminal glycopeptide being released from the heavy chain of factor X to give a new NH_2-terminal sequence in factor X_a, Ile-Val-Gly-Gly.[9,10] This cleavage is probably identical in all mechanisms of activation; Radcliffe and Barton used RVV, the intrinsic and extrinsic activating complexes, and trypsin, and found isoleucine as one of the NH_2-terminal residues in all the isolated forms of factor X_a (the other terminal residue, alanine, is on the light chain).[11]

In activations of factor X in the presence of lipid—for instance, in activations by the tissue factor–factor VII complex (TF-VII) or by RVV in the presence of lipid—another proteolytic cleavage occurs to release a COOH-terminal glycopeptide from the heavy chain of either factor X or X_a; this is a result of the action of factor X_a.[10,12]

The activation of factor X results in the appearance of esterase activity and the ability to incorporate, and be inhibited by, diisopropyl phosphorofluoridate (DFP). This inhibitor reacts with an active serine residue in the now-familiar sequence Gly-Asp-Ser-Gly-Gly-Pro.[13,14] However, the rate of the inactivation of factor X_a by DFP is considerably lower than that observed with many other serine esterases.

Factor X, the zymogen, is a glycoprotein of about 55,000 monomeric molecular weight that tends to aggregate, especially in the presence of

[4] Y. Nemerson, *Biochemistry* **5**, 601 (1966).

[5] W. J. Williams and D. G. Norris, *J. Biol. Chem.* **241**, 1847 (1966).

[6] Abbreviations used are: RVV, Russell's viper venom; TF-VII, the tissue factor–factor VII complex; DFP, diisopropylphosphorofluoridate; TBS, Tris-buffered saline (0.1 M NaCl–50 mM Tris-Cl, pH 7.5) ; SDS, sodium dodecyl sulfate.

[7] R. G. Macfarlane, *Br. J. Haematol.* **7**, 496 (1961).

[8] M. P. Esnouf and W. J. Williams, *Biochem. J.* **84**, 62 (1962).

[9] K. Fujikawa, M. E. Legaz, and E. W. Davie, *Biochemistry* **11**, 4892 (1972).

[10] J. Jesty, A. K. Spencer, Y. Nakashima, Y. Nemerson, and W. Konigsberg *J. Biol. Chem.* **250**, 4497 (1975).

[11] R. D. Radcliffe and P. G. Barton, *J. Biol. Chem.* **248**, 6788 (1973).

[12] J. Jesty, A. K. Spencer, and Y. Nemerson, *J. Biol. Chem.* **249**, 5614 (1974).

[13] J. E. Leveson and M. P. Esnouf, *Br. J. Haematol.* **17**, 173 (1969).

[14] K. Titani, M. A. Hermodson, K. Fujikawa, L. H. Ericsson, K. Walsh, H. Neurath, and E. W. Davie, *Biochemistry* **11**, 4899 (1972).

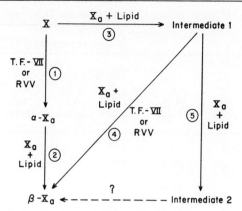

Fig. 1. Pathways of activation of factor X. Reproduced from J. Jesty, A. K. Spencer, Y. Nakashima, Y. Nemerson, and W. Konigsberg, *J. Biol. Chem.* **250**, 4497 (1975).

Ca^{2+}.[15-17] It contains about 11% carbohydrate and consists of two poly-peptide chains linked by disulfide bonds; the heavy chain contains all the carbohydrate.[16,18] About 70% of the carbohydrate is contained in the NH_2-terminal peptide lost on activation,[9] and the remaining carbohy-drate is apparently all contained in the COOH-terminal peptide lost through the action of factor X_a; the final product, β-X_a, contains no carbohydrate.[11]

The two cleavages, of the NH_2- and COOH-termini of the heavy chain of factor X, can occur in either order, resulting in alternative pathways of activation in mixtures that contain lipid.[17] These pathways are summarized in Fig. 1. As the pathway via I_1 is initiated by the action of the product, factor X_a, it occurs to a greater extent in slower activations.[12] In acti-vations with high concentrations of TF-VII, the major pathway is via α-X_a, but the rate of conversion of α-X_a to β-X_a (by the loss of the COOH-terminal peptide) is sufficient to prohibit the isolation of α-X_a from such mixtures free of β-X_a.[12] However, it has been shown that the NH_2-terminal sequence of β-X_a produced in this way is identical with that of α-X_a produced by RVV, and that the same NH_2-terminal pep-tide is produced in both cases.[10] The COOH-terminal peptide produced is identical with that we obtain by incubation of factor X with factor X_a to produce I_1. As shown in Fig. 1, I_1 can be activated directly to β-X_a by the release of the same NH_2-terminal peptide as is released in the

[15] C. M. Jackson and D. J. Hanahan, *Biochemistry* **7**, 4506 (1968).
[16] K. Fujikawa, M. E. Legaz, and E. W. Davie, *Biochemistry* **11**, 4882 (1972).
[17] M. P. Esnouf, P. H. Lloyd, and J. Jesty, *Biochem. J.* **131**, 781 (1973).
[18] C. M. Jackson, *Biochemistry* **11**, 4873 (1972).

AMINO ACID AND CARBOHYDRATE COMPOSITIONS OF DERIVATIVES OF FACTOR X

				Chain or peptide[a]				
Component	MW^b:	X L.C. 17,000	X H.C. 37,000	I_1 H.C. 34,000	α-X_a H.C. 30,000	β-X_a H.C. 27,000	N-terminal peptide 7600	C-terminal peptide 2700
Amino acid[c]								
Lys		7.4	15.7	15.8	15.3	14.9	0.2	0[d]
His		3.3	8.8	7.8	6.9	5.8	1.9	1
Arg		8.2	17.8	17.1	15.5	14.6	2.0	0
Asx		15.2	25.7	25.6	19.2	20.6	7.2	0
Thr		16.5	24.1	22.6	20.2	18.0	2.9	2
Ser		11.1	18.9	19.4	13.3	13.3	6.8	1
Glx		28.1	33.4	32.5	26.3	25.1	7.2	1
Pro		2.4	15.4	10.9	12.6	8.4	3.0	5
Gly		15.3	27.3	26.8	23.7	22.9	3.6	1
Ala		6.5	24.6	22.8	19.5	17.1	5.0	2
Cys/2[e]		15.9	9.4	7.7	9.3	8.6	0	0
Val		5.4	20.0	19.7	18.4	16.9	1.9	1
Met		0	4.7	5.3	4.9	4.9	0	0
Ile		2.1	9.0	9.2	8.0	8.2	0.9	0
Leu		7.7	23.1	20.9	16.3	14.4	6.8	2
Tyr		2.9	6.9	6.4	5.9	6.3	1.0	0
Phe		8.1	12.4	13.1	12.3	11.4	0.2	0
Trp[f]		1.5	7.5	6.5	5.2	4.2	1.4	1
Carbohydrate[c]								
Hexose		0	8.1[g]	ND[h]	3.2	0	4.1[g]	1.6
Hexosamine		0	6.3	—	0.7	0	3.0	0.9
Sialic acid		0	5.3	—	1.4	0	2.4	1.3

[a] Abbreviations used: L.C., light chain; H.C., heavy chain;

[b] The molecular weights of factor X L.C. and H.C. are from C. M. Jackson [*Biochemistry* **11**, 4873 (1972)]. That of the NH_2-terminal peptide is the sum of the polypeptide weight (calculated by adjustment to nearest-integer values for each residue) and the carbohydrate weight. The polypeptide molecular weight of the COOH-terminal peptide is from the amino acid sequence; the carbohydrate weight was added to this.

[c] Amino acid and carbohydrate contents are expressed as moles of residue per mole of chain or peptide.

[d] Calculated from the amino acid sequence.

[e] Determined as the S-carboxymethyl derivative.

[f] The tryptophan contents of factor X L.C. and H.C., α-X_a H.C., and the NH_2-terminal peptide are from K. Fujikawa, M. E. Legaz, and E. W. Davie [*Biochemistry* **11**, 4882 and 4892 (1972)]. The tryptophan contents of I_1 H.C. and β-X_a H.C. are by subtraction of the single Trp residue lost in the COOH-terminal peptide from factor H.C. and α-X_a H.C., respectively.

[g] From Fujikawa *et al.*[f] The figures for the NH_2-terminal peptide are corrected for the lower molecular weight that we estimate.[b]

[h] ND, not determined.

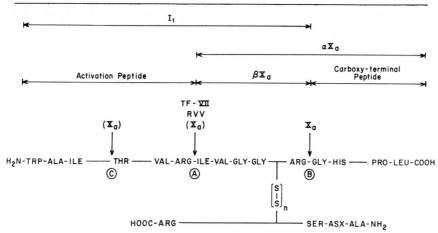

FIG. 2. Sites of cleavage during activation of factor X. Cleavage at A releases the activation peptide [reactions (1) and (4), Fig. 1] and results in the appearance of factor X_a activity. Cleavage at B releases the COOH-terminal peptide [reactions (2) and (3), Fig. 1]. Cleavage at C, in addition to B, occurs very slowly in the presence of factor X_a and results in the appearance of I_2 [reaction (5), Fig. 1]. Reproduced from J. Jesty, A. K. Spencer, Y. Nakashima, Y. Nemerson, and W. Konigsberg, J. Biol. Chem. 250, 4497 (1975).

conversion of α-X_a to β-X_a; and if RVV is used in this activation, lipid is not necessary.[10,12] β-X_a and the two peptides produced in its formation by either pathway account entirely for the amino acid and carbohydrate contents, and the NH_2- and COOH-termini, of factor X (table, Fig. 2).[10]

Fujikawa et al.[19] showed considerable homology in the amino acid sequences of the NH_2-terminal regions of prothrombin, factor IX, and the light chain of factor X; less comprehensive results on single-chain factor VII indicate that this too has a similar NH_2-terminal sequence.[20] Thus it is very likely that factor X is derived from a single-chain precursor; Mattock and Esnouf[21] isolated a possible candidate from bovine plasma, but could not confirm that factor X_a was generated by RVV.

The heavy chains of thrombin, factor X_a, two-chain factor VII, and factor IX_a are also clearly homologous with the pancreatic serine esterases in the NH_2-terminal region.[14,20,22] And third, there is clear homology in the active-site regions of thrombin, factor X_a, the pancreatic enzymes, and what is probably the active-site region of factor IX_a, although in this last case the fragment was obtained from factor IX.[14,22]

[19] K. Fujikawa, M. H. Coan, D. L. Enfield, K. Titani, L. H. Ericsson, and E. W. Davie, Proc. Natl. Acad. Sci. U.S.A. 71, 427 (1974).
[20] R. D. Radcliffe and Y. Nemerson, this volume [6].
[21] P. Mattock and M. P. Esnouf, Nature (London), New Biol. 242, 90 (1973).
[22] D. L. Enfield, L. H. Ericsson, K. Fujikawa, K. Titani, K. A. Walsh, and H. Neurath, FEBS Lett. 47, 132 (1974).

The pancreatic enzymes and the B chain thrombin also show considerable homology in the immediate region of their COOH termini, which apparently remain intact during activation of the respective zymogens.[23-25] In contrast, the COOH terminus of factor X, which is cleaved by the action of factor X_a, shows no homology with these other COOH-terminal sequences. Its amino acid sequence is Gly-His-Ser-Glu-Ala-Pro-Ala-Thr-Trp-Thr(CHO)-Val-Pro-Pro-Pro-Leu-Pro-Leu, CHO indicating the carbohydrate moiety.[10] Not only is the composition of the terminal hexapeptide somewhat unusual, but it also seems that the carbohydrate moiety is attached to the peptide through a linkage that is rare in plasma glycoproteins, N-acetylgalactosaminyl-O-threonine.[10,26]

Assay of Factor X

Principle. Factor X is activated specifically by a component of Russell's viper venom.[7] In single-stage assays for factor X, the sample is added to bovine plasma deficient in factor X; the venom is then added, and the assay is started by the addition of Ca^{2+}. The fact that factor X is always present at less than 10% of its normal value makes it the limiting factor, and the clotting time is a function of its concentration. This single-stage assay is a function not only of the concentration, but also the "activity," of factor X, i.e., the rate at which it is activated by RVV. However, this distinction is not usually a problem in practice.

Procedure. Factor X-deficient plasma can be prepared by filtration of plasma through a Seitz filter,[27] or through charcoal,[28] but both these methods are prone to problems. So unless, there is a need for very large amounts of deficient plasma, it is simpler to buy it (Thame Diagnostics, Thame, Oxon, England; Sigma Chemical Co., St. Louis, Missouri). Before use, the plasma is reconstituted with lipid by adding 1/100 volume of "cephalin" to give a final phospholipid concentration of about 0.1 mg/ml.[29] A stock solution of Russell's viper venom is prepared in the following way: 2 mg of dried venom (Burroughs Wellcome, Tuckahoe, New York; Sigma Chemical Co.) is dissolved in 5 ml of 0.1 M NaCl–50 mM Tris-Cl pH 7.5 (TBS). Glycerol, 5 ml, is then added, and the solu-

[23] S. Magnusson, *in* "The Enzymes" (P. D. Boyer, ed.), 3rd ed., Vol. 3, p. 277. Academic Press, New York, 1971.
[24] B. S. Hartley and D. M. Shotton, *in* "The Enzymes" (P. D. Boyer, ed.), 3rd ed., Vol. 3, p. 323. Academic Press, New York, 1971.
[25] J. Stenflo, *J. Biol. Chem.* **247**, 8167 (1972).
[26] R. G. Spiro, *Annu. Rev. Biochem.* **39**, 599 (1970).
[27] F. Bachmann, F. Duckert, and F. Koller, *Thromb. Diath. Haemorrh.* **2**, 24 (1958).
[28] K. W. E. Denson, *Acta Haematol.* **25**, 105 (1961).
[29] W. N. Bell and H. G. Alton, *Nature (London)* **174**, 880 (1954).

tion is mixed well. This stock solution is stable for at least 4 months at —20°. For use in the assay, 0.2 ml of this stock is diluted with 1.8 ml of TBS; the dilute solution is referred to as RVV/10.

Normal citrated bovine plasma (blood is collected into a $\frac{1}{10}$ volume 0.12 M trisodium citrate, and the red cells are centrifuged down) is defined as containing 100 units of factor X per milliliter, and is used to calibrate the assay. Dilutions of normal plasma (e.g., $\frac{1}{10}$, $\frac{1}{20}$, $\frac{1}{50}$, $\frac{1}{100}$, and $\frac{1}{200}$) are made in TBS and assayed in duplicate at 37°: to a glass tube add 0.1-ml sample, then 0.1 ml of factor X-deficient plasma, and then 0.1 ml RVV/10. Mix, and add 0.1 ml of 25 mM CaCl$_2$, and record the clotting time from this addition. A plot of log (clotting time) against log [factor X] should be linear over this range. To assay an unknown, the sample is diluted sufficiently for the clotting time to fall on the standard line and then assayed in the same way.

The Assay of Factor X$_a$

Principle. Factor X$_a$ forms the prothrombin-converting complex with factor V, phospholipid, and Ca^{2+}, the activity of this complex on prothrombin being a function of the factor X$_a$ concentration. The very low levels of factor X$_a$ that are assayed mean that it is the limiting factor in the generation of thrombin in the assay.

Procedure. There is no standard preparation of factor X$_a$ available, so the assay is calibrated with dilutions of purified factor X$_a$, from 1 to 20 ng/ml. The extremely good stability of factor X$_a$ in 50% glycerol-TBS at —20° permits the use of a stock solution of known purity for years. The concentration of pure factor X$_a$ can be estimated from its extinction; $A_{280}^{1\%} = 9.4$.[30] A suitable dilution of the sample in 0.02% ovalbumin–TBS (0.1 ml) is added to 0.1 ml of 25 mM CaCl$_2$ in a tube at 37°. Factor X-deficient plasma is then added immediately, and the clotting time is recorded from this addition. A plot of log (clotting time) against log [factor X$_a$] should be linear over the range indicated.

The addition of plasma last to the assay mixture gives more reproducible results than the more usual methods where CaCl$_2$ is added last. This may be a result of the inactivation of factor X$_a$ by plasma inhibitors before the addition of CaCl$_2$ in these methods.

The Isolation of Factor X

Factor X, like the other vitamin K-dependent clotting factors, prothrombin, factor VII, and factor IX, can be adsorbed on a variety of salts of divalent metals. The majority of methods for factor X purifica-

---

[30] J. Jesty and M. P. Esnouf, *Biochem. J.* **131**, 791 (1973).

tion use adsorption of factor X from bovine plasma on $BaSO_4$, and subsequent elution with citrate, as the first step in purification.[8,16,17,31-33] It is, however, necessary to use small amounts of $BaSO_4$ in order to avoid contamination with high-molecular-weight material, and this has the disadvantage that the associated adsorption of prothrombin, factor VII, and factor IX is low. Furthermore, $BaSO_4$ with protein adsorbed is very difficult to handle in the subsequent washes and elutions. Barium citrate is more easily handled and results in higher yields of all four factors.

The procedure we prefer involves adsorption on barium citrate, followed by elution and precipitation with ammonium sulfate.[34] This part of the process is carried out for us on 50 liters of plasma by the New England Enzyme Center, and is described by Radcliffe and Nemerson.[20]

The subsequent chromatography of the barium salt eluate on an anion-exchanger is common to all published methods of purification of factor X, and is usually followed by rechromatography on DEAE-Sephadex A50 to obtain pure factor X.[16,17,31-33] In most cases factor X is eluted in the last chromatography as a double peak of almost constant specific activity. So far there are no very significant differences reported between the two forms of factor X; Jackson[18] and Fujikawa et al.[16] show only small differences in amino acid and carbohydrate compositions and molecular weights within the limits of error. Therefore in practice the two forms are generally pooled together.

The first chromatography of the barium citrate eluate is described by Radcliffe and Nemerson.[20] The presence of benzamidine-HCl during this procedure is necessary for the prevention of activation of single-chain factor VII by contaminating factor X_a. Factor X_a also degrades factor X to a species we call I_1, and it is probably this that was observed by Jackson and Hanahan in their preparations of factor X[15]; the presence of the inhibitor completely prevents this. Factor X can be detected only by assay in the fractions of the first chromatography, owing to the high absorbance of the benzamidine-HCl. The rechromatography of factor X is carried out as follows.

Procedure. The fractions containing factor X are pooled (about 3×10^6 units are obtained from 50 liters of starting plasma). Such pools, in 25 mM benzamidine-HCl, can be stored at $-20°$ for some weeks before rechromatography. The pool is concentrated by ultrafiltration (Amicon Corporation, PM-10 membrane) to about 100 ml and then dialyzed against 0.08 M sodium citrate pH 6.5. The solution is then applied at 35

[31] D. Papahadjopoulos, E. T. Yin, and D. J. Hanahan, *Biochemistry* 3, 1931 (1964).
[32] C. M. Jackson, T. F. Johnson, and D. J. Hanahan, *Biochemistry* 7, 4492 (1968).
[33] S. P. Bajaj and K. G. Mann, *J. Biol. Chem.* 248, 7729 (1973).
[34] D. L. Aronson and D. Ménaché, *Biochemistry* 5, 2635 (1966).

ml/hr to a column of DEAE-Sephadex A50 (2.5 × 30 cm) previously equilibrated in the same buffer. The chromatography is done at 4°. After application of the sample, the protein is eluted with a 1-liter gradient, from 0.13 to 0.26 M sodium citrate pH 6.5, at the same flow rate. The fractions are assayed and their extinctions measured at 280 nm. Fractions of specific activity more than 8000 units/mg ($A_{280}^{1\%} = 9.6^{17}$) are pooled, concentrated by ultrafiltration to about 20 ml, and finally dialyzed against 50% glycerol–TBS before storage at −20°. Such solutions are stable for years.

By disc electrophoresis and SDS-gel electrophoresis factor X produced by this method is homogeneous and free of prothrombin, factor IX, and factor VII by bioassay. Such preparations do, however, contain trace amounts of factor X_a—typically about 0.01–0.1%. This level can be reduced by treatment with 20 mM DFP at pH 7.5 for 2 hr at room temperature before storage, but factor X_a activity can never be completely removed by this method. It is contamination with factor X_a that is probably responsible for the generation of factor X_a activity during storage, but such generation can be detected only by assay.

Preparation of β-X_a

The routine preparation of factor X_a that we use involves activation with the coagulant fraction of RVV in the absence of lipid, followed by chromatography of the activated mixture on DEAE-Sephadex to separate the RVV fraction from factor X_a.[30] Although α-X_a is formed initially by RVV in the absence of lipid, over the course of purification this is converted autocatalytically to β-X_a.

Procedure. Factor X, 60 mg in 120 ml of 0.15 M NaCl–50 mM Tris-Cl, pH 7.5, is incubated with 0.1 absorbance unit of the coagulant fraction of RVV[12] in the presence of 5 mM CaCl$_2$ for 30 min at 37°. The reaction is stopped by the addition of trisodium citrate to a concentration of 6 mM, and the mixture is applied directly at 20 ml/hr to a column (1.5 × 16 cm) of DEAE-Sephadex A50 previously equilibrated in 0.15 M NaCl-50 mM Tris-Cl, pH 7.5. The protein is eluted at the same flow rate with a linear 800-ml gradient, 0.2–0.55 M NaCl, in 50 mM Tris-Cl pH 7.5. A breakthrough peak consists largely of the COOH-terminal peptide, which may be isolated by gel filtration on Sephadex G-50, as described for the preparation of I$_1$. The venom coagulant fraction is eluted at about 0.25 M NaCl, close to a small peak that may represent a small proportion of the activation peptide. The factor X is eluted last. The pool of active material is concentrated by ultrafiltration (Amicon Corp., PM-10) and dialyzed against 50% glycerol–TBS before storage at $-20°$. The amount of $\beta\text{-X}_a$ obtained is about 70% of the theoretical yield.

Preparation of $\beta\text{-X}_a$ by Activation with TF-VII

In order to obtain activation of factor X by the tissue factor-factor VII complex, sufficient of the complex must be used to attain activation in less than 15 min. This is a result of the inactivation of factor VII by factor X_a. Under these conditions the major pathway of activation is via $\alpha\text{-X}_a$.[12] Continued incubation of the mixture ensures the complete conversion of $\alpha\text{-X}_a$ to $\beta\text{-X}_a$.

Preparation of the TF-VII Complex. Tissue factor apoprotein is reconstituted with lipid as described in this volume,[35] and dialyzed against 0.1 M NaCl–50 mM Tris-Cl, pH 7.5. The lipoprotein is then centrifuged down at 100,000 g for 1 hr at $4°$, and resuspended in the same buffer to give a final protein concentration of 1 mg/ml. Two-chain factor VII is added to the suspension to a final concentration of 1000 units/ml, and the solution is made 5 mM in $CaCl_2$. After incubation for 10 min at $37°$, the mixture is kept on ice before use; it is stable for 1–2 hr in this form, but slowly loses activity owing to inactivation by endogeneous factor X_a in the factor VII.

Factor X (15 mg in 30 ml of 0.1 M NaCl–50 mM Tris-Cl, pH 7.5) is incubated with factor VII (as the TF-VII complex) at a final concentration of 150 units/ml in the presence of 5 mM $CaCl_2$ at $37°$ for 30 min. The reaction is then stopped by the addition of trisodium citrate to a concentration of 12 mM, and the mixture is then centrifuged for 1 hr at $4°$ at 100,000 g to remove the lipid. The resulting solution containing the reaction products is then chromatographed on a column of DEAE-Sephadex A50 exactly as described for the preparation of $\beta\text{-X}_a$ by activation of factor X with RVV.

It should be noted that $\beta\text{-X}_a$ produced by the method described is identical in amino acid composition and NH_2-terminal and COOH-terminal sequences as $\beta\text{-X}_a$ produced by the simpler method involving activation by RVV in the absence of lipid.

Preparation of I_1 and the COOH-Terminal Peptide

Procedure. Factor X (30 mg in 60 ml of 25 mM Tris-Cl, pH 7.5) is incubated for 110 min at $22°$ with 1.5 mg of $\beta\text{-X}_a$ in the presence of an equimolar dispersion of phosphatidylcholine (Sigma Chemical Co.) and

---

[35] F. A. Pitlick and Y. Nemerson, this volume [5].

phosphatidylserine (ICN Pharmaceuticals) to a final phospholipid concentration of 0.1 mg/ml, and 5 mM CaCl$_2$. The reaction is stopped by the addition of trisodium citrate to a concentration of 12 mM. The mixture is then centrifuged at 100,000 g for 1 hr at 4° to remove the phospholipid. Factor X$_a$ is removed from the mixture by stirring with 2 ml of Sepharose 4B to which soybean trypsin inhibitor has been coupled[36] (substitution is about 5 mg of inhibitor per milliliter of gel). After 20 min at room temperature, the inhibitor-Sepharose is removed by filtration and the mixture is lyophilized. The dried material is taken up in 5 ml of 1 M guanidine-HCl and chromatographed on a column of Sephadex G-50 (2 × 90 cm) equilibrated in 0.2 M NH$_4$HCO$_3$ or other suitable buffer. The gel filtration is done at 4° at a flow rate of 10 ml/hr, and clearly separates I$_1$ from the COOH-terminal peptide. The I$_1$ pool is concentrated by ultrafiltration to about 5 ml, dialyzed against 50% glycerol-TBS, and stored at −20°. The COOH-terminal peptide is stable in water at −20°.

Proteolytic Action of Factor X$_a$

Until recently it was generally a tacit assumption that the only major role of factor X$_a$ is the activation of prothrombin. Recent results, however, show that this is not the case; factor X$_a$ can also cleave at substantial rates two bonds each in factor VII and factor X as well as the two bonds it cleaves in prothrombin. All the cleavages by factor X$_a$ so far studied are accelerated at least 100-fold by phospholipid and calcium ions, and in the case of prothrombin activation a cofactor, factor V, is also involved.

In the activation of prothrombin, factor X$_a$ cleaves first an Arg-Thr bond to form P$_3$ (I$_2$), the inactive immediate precursor of thrombin.[37,38] P$_3$ is then cleaved at an Arg-Ile bond to form the two-chain enzyme.[23,37] The cleavage of prothrombin to form an alternative intermediate, P$_2$ (I$_1$), is not an action of factor X$_a$, and is solely a result of thrombin action.[39] Both cleavages of prothrombin by factor X$_a$ are lipid-dependent and require the lipid-binding sites of prothrombin either intact, in the case of P$_3$ formation, or, in the case of P$_3$ conversion to thrombin, in the form of a complex between P$_3$ and its activation peptide, F1-2.[40]

[36] P. Cuatrecasas, *J. Biol. Chem.* **245**, 3059 (1970).
[37] C. M. Heldebrant, R. J. Butkowski, S. P. Bajaj, and K. G. Mann, *J. Biol. Chem.* **248**, 7149 (1973).
[38] J. Reuterby, D. A. Walz, L. E. McCoy, and W. H. Seegers, *Thrombosis Res.* **4**, 885 (1974).
[39] W. Kisiel and D. J. Hanahan, *Biochem. Biophys. Res. Commun.* **59**, 570 (1974).
[40] C. T. Esmon, W. G. Owen, and C. M. Jackson, *J. Biol. Chem.* **249**, 8045 (1974).

Factor X_a cleaves factor VII at two sites in lipid- and Ca^{2+}-dependent reactions. The first is the rapid cleavage of an Arg-Ile bond, which corresponds with activation and the formation of 2-chain factor VII.[20] The second is the slower inactivation, by the cleavage of a Y-Gly bond, where the nature of Y is not yet known, but is presumed to be Arg. This cleavage occurs in the COOH-terminal region of the heavy chain of the activated material, releasing a 12,000-dalton peptide that contains the active serine residue.[20]

Factor X_a cleaves factor X at three sites in the heavy chain, but the third cleavage [the formation of I_2; reaction (5), Fig. 1] is relatively insignificant even in very slow activations of factor X. The most rapid cleavage is that of an Arg-Gly bond in the COOH-terminal region of the heavy chain of either factor X or α-X_a as discussed in this chapter. A slower cleavage is that of the Arg-Ile bond that is cleaved by other activators to release the normal NH_2-terminal activation peptide, and results in the appearance of factor X_a activity.

So far all the protein substrates that are known for factor X_a are vitamin K-dependent clotting factors, except for one very poor substrate, chymotrypsinogen.[41] And it appears that the lipid-binding sites of the substrate must be either intact or in the form of a complex with the substrate. Both Stenflo et al.[42] and Nelsestuen et al.[43] have demonstrated the existence of γ-carboxylglutamic acid residues in Ca^{2+}-binding peptides from the NH_2-terminal region of prothrombin, and the latter group has preliminary evidence for the existence of "extra" negative charges in a homologous peptide from factor X,[44] whose amino acid composition corresponds closely with residues 5 to 43 of the light chain.[22] In support of the idea that the light chain of factors X and X_a contains the Ca^{2+}-binding sites of the molecules is the identical sequence of a tetrapeptide in the NH_2-terminal regions of prothrombin and the light chain of factor X, Leu-Glu-Glu-Val (residues 6 to 9 in prothrombin, and 5 to 8 in the light chain of factor X).[19] Stenflo et al.[42] showed that both the glutamic residues of this peptide in normal prothrombin and γ-carboxylated, whereas in dicoumarol-induced prothrombin they are not. Thus it seems very likely that factor X contains such residues in the NH_2-terminal region of the light chain, and that these are the Ca^{2+}-binding sites of factors X and X_a.

[41] J. H. Milstone and V. K. Milstone, Proc. Soc. Exp. Biol. Med. 117, 290 (1964).
[42] J. Stenflo, P. Fernlund, W. Egan, and P. Roepstorff, Proc. Natl. Acad. Sci. U.S.A. 71, 2730 (1974).
[43] G. L. Nelsestuen, T. H. Zytkovicz, and J. B. Howard, J. Biol. Chem. 249, 6347 (1974).
[44] J. B. Howard and G. L. Nelsestuen, Fed. Proc., Fed. Am. Soc. Exp. Biol. 33, 1473 (1974).

Of the cleavages by factor X_a described (including that of chymotrypsinogen), four are at Arg-Ile bonds and correspond with the activation of zymogens, while two are at Arg-Gly bonds, and two are at Arg-Thr bonds. Thus although factor X_a is apparently completely specific for arginyl bonds, the distal residue does not in itself appear to specify the site of cleavage.

Acknowledgment

We thank Dr. C. M. Jackson for providing copies of papers before publication.

[12] Factor V

By ROBERT W. COLMAN and ROBERT M. WEINBERG

Bovine Factor V

Assay Method

Principle. Factor V^1 is a plasma protein that is necessary for the optimal rate of conversion of prothrombin to thrombin in both the intrinsic and extrinsic system of blood coagulation. The enzyme responsible for the hydrolytic cleavage of prothrombin to thrombin is activated factor X (X_a),[2] which, in addition to factor V, requires phospholipid and ionic calcium.

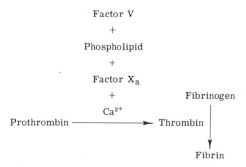

Thus, to assay factor V, one must measure the rate of activation of prothrombin to thrombin. Thrombin activity can be measured by its

[1] A. G. Ware, R. C. Murphy, and W. H. Seegers, *Science* **106**, 618 (1947).

[2] J. H. Milstone, *Fed. Proc., Fed. Am. Soc. Exp. Biol.* **23**, 742 (1964).

ability to clot fibrinogen. There are essentially two different assay systems which have been employed for the measurement of factor V.

The one-stage method[3] is far simpler to perform and is employed routinely in clinical measurements of factor V activity. Human oxalated plasma is incubated at 37° to selectively inactivate factor V. This "aged" human plasma serves as a source of factor X, factor VII, prothrombin, and fibrinogen. A mixture of tissue thromboplastin, factor VII, and calcium converts factor X to factor X_a. The tissue thromboplastin also serves as a source of phospholipid, which is necessary for factor X_a to convert prothrombin to thrombin. The amount of thrombin formed is judged by the clotting time of fibrogen. The rate of conversion of prothrombin to thrombin is measurable but slow in the absence of factor V. A diluted sample of plasma (at least 10-fold to minimize the contribution of other clotting proteins) or a purified preparation can then be assayed for factor V activity. The log of clotting time is inversely proportional to the log of factor V activity. Difficulty in interpretation may arise because of the positive feedback by thrombin through its known ability to increase factor V activity,[1,4-7] and it is also impossible to separately appraise the effects of factor X_a, calcium, prothrombin, and phospholipid concentrations on the action of factor V. Therefore, a two-stage assay was developed using highly purified reagents.[8] Phospholipid, activated factor X, calcium ion, and factor V are preincubated to form a "prothrombinase." After addition of prothrombin, sequential aliquots are removed and tested for thrombin activity using fibrogen as a substrate. At a fixed concentration of factor X_a, prothrombin, and phospholipid, a linear relationship between the log of the initial rate of thrombin generation and the log of factor V concentration is found.

One-Stage Method

Reagents

Factor V-deficient plasma[3]: Nine volumes of human blood (900 ml) are collected into polypropylene bottles containing 1 volume (100 ml) of 2.5% potassium oxalate. The anticoagulated blood is cen-

[3] M. L. Lewis and A G. Ware, *Proc. Soc. Exp. Biol. Med.* **84**, 640 (1953).

[4] P. F. Hjort, *Scand. J. Clin. Lab. Invest.*, Suppl. **27**, 1 (1957).

[5] D. G. Therriault, J. L. Gray, and H. Jensen, *Proc. Soc. Exp. Biol. Med.* **95**, 207 (1957).

[6] S. I. Rapaport, S. Schiffman, M. J. Patch, and S. B. Ames, *Blood* **21**, 221 (1963).

[7] R. W. Colman, *Biochemistry* **8**, 1438 (1969).

[8] C. L. Kandall, T. K. Akinbami, and R. W. Colman, *Br. J. Haematol.* **23**, 655 (1972).

trifuged at 4000 g for 15 min at 4°, and the supernatant plasma is again centrifuged at 9000 g for 30 min at 4°. The resulting plasma, about 500 ml, is dialyzed for 4 hr against 30 liters of 0.05 M NaCl in 0.1 M potassium oxalate. After dialysis, the plasma is adjusted to pH 8.0 with 0.5 N NaOH or 0.5 N HCl if necessary. The dialyzed plasma is then incubated at 37° until the one-stage prothrombin time[9] is greater than 40 sec. The plasma is then adjusted to pH 7.7–7.8 and kept at 37° until the one-stage prothrombin time is greater than 90 sec (usually 36–48 hr). The aged plasma is adjusted to pH 7.4 and stored in 2.5-ml portions at −20°.

Tissue thromboplastin[4]: The blood vessels and membranes are carefully removed from the cerebral hemispheres of human brains obtained less than 12 hr after death. The tissue is then rinsed several times with 0.15 M NaCl. The gray matter of cerebral hemispheres is cut into pieces of about 0.5 cubic inch, homogenized in a Waring blender at low speed for two periods of 5 sec with three volumes of acetone. The brain residues are collected by filtering through a Büchner funnel covered with Whatman filter paper No. 114 and resuspended in five volumes of acetone. The tissue is stirred for 5 min and homogenized again in Waring blender as before at low speed for two periods of 5 sec. The brain residue is collected and, while still on the Büchner funnel, washed with acetone until the filtrate is free of yellow color. The brain residue is spread on the nonabsorbant paper, dried overnight in the dark, ground into powder, and stored in a vacuum desiccator at 4°. The thromboplastin solution may be prepared by extracting 5 g of brain powder with 100 ml of 0.15 M NaCl at 50° for 15 min with occasional stirring. After centrifugation at 1000 rpm for 20 min at 4°, the activity of the supernatant is assessed by the one-stage prothrombin assay: 0.1 ml of the test material is mixed with 0.1 ml of 0.04 M CaCl$_2$ and 0.1 ml of fresh human plasma containing 0.012 M sodium citrate. All clotting times are recorded with an automatic clot timer (Fibrometer). For the thromboplastin solution they should be in the range of 12–13.9 sec, and the material can be stored frozen at −50° for 6 months without loss of potency.

Procedure. The solution to be tested for factor V activity is first diluted 10-fold with 0.02 M Veronal-HCl buffer containing 0.15 M NaCl and 0.1 M Na oxalate at pH 7.4 (VBOS). One-tenth milliliter of the

[9] A. J. Quick, *Am. J. Cli. Pathol.* **15**, 560 (1945).

diluted sample is mixed with 0.1 ml of factor V-depleted plasma; then 0.1 ml of thromboplastin and 0.1 ml of 0.04 M CaCl$_2$ are added to start the reaction. The time required for clot formation is measured, and a standard curve is constructed by assaying in duplicates and in four dilutions (1:10, 1:40, 1:200, 1:1000) 20 normal human plasma samples. The log of the relative concentration of factor V is then plotted against the log of the mean of the 20 clotting times for each dilution. One unit of factor V is defined as the amount contained in 1 ml of average normal plasma.

Two-Stage Method[8]

Reagents

Buffer: 0.02 M Veronal-HCl containing 0.15 M NaCl at pH 7.4 (VBIS)

Phospholipid: Inosithin (mixed soybean phospholipids, Associated Concentrates, New York, New York) is homogenized in the VBIS at a concentration of 2 g/100 ml.

Factor X$_a$: purified by the method of Yin and Wessler[10] from crude bovine thrombin (Parke-Davis) to yield a single band on disc electrophoresis.[11,12] One unit of factor X$_a$ is defined as the activity that would evolve from 1 ml of normal human plasma when the factor X is fully activated by Russell's viper venom.[10]

Prothrombin, purified by the method of Tishkoff et al.[13] One unit of prothrombin is defined as the activity found in 1 ml of normal human plasma.

Factor V, purified by the method of Colman[7] as described below and assayed by the one-stage method as described above.

Bovine fibrinogen (95% clottable; Gallard Schlessinger Chemical Co.) is dissolved in 0.04 M Veronal-HCL (pH 7.4) buffer containing 0.3 M NaCl at 37° to give a stock solution of 12 mg/ml ($E_{280\,nm}^{1\%} = 15.1$), which is then distributed in aliquots for storage at −50°. Before use, the fibrinogen solution is diluted to 3 mg/ml by mixing 1 part of the stock solution with 1 part of water and 2 parts of 0.02 M Veronal-HCl buffer (pH 7.4) containing 0.15 M NaCl.

Bovine thrombin: NIH standard B3 (23.7 units/mg). A standard thrombin curve is constructed by adding 0.2 ml of thrombin in

[10] E. T. Yin and S. Wessler, *J. Biol. Chem.* 243, 112 (1968).
[11] L. Ornstein, *Ann. N.Y. Acad. Sci.* 121, 321 (1964).
[12] B. J. Davis, *Ann. N.Y. Acad. Sci.* 121, 404 (1964).
[13] G. H. Tishkoff, L. C. Williams, and D. M. Brown, *J. Biol. Chem.* 243, 4151 (1968).

VBIS (at concentrations of 0.1–2.0 NIH units/ml) to 0.2 ml of fibrinogen solution (3 mg/ml) at 37°. The reciprocal of the clotting time is plotted against the thrombin concentration; a straight line is obtained in the concentration ranges specified.

Procedure (Fig. 1). The assay is performed at 17° by incubating inosithin (200 μg), factor X_a (0.013 unit), factor V (0.01–0.1 unit), and $CaCl_2$ (0.01 mM) made up to 1 ml with VBIS. After 2 min, prothrombin (0.075 unit in 1.0 ml) is added; then aliquots of 0.1 ml are removed each minute for 10 min and diluted 10-fold with VBIS. The thrombin generated is measured by adding 0.2 ml of fibrinogen solution at 37° and recording the clotting time. Thrombin activity is read from the previously prepared standard curve. A calibration curve is constructed by assaying different concentrations of factor V and plotting the log of the initial rate of thrombin generation against the log of factor V concentration (Fig. 1 inset).

Purification Procedure

Principle. The modern purifications of bovine factor V begins by removing the prothrombin complex (factors II, VII, IX, and X) by adsorption of oxalated plasma with $BaSO_4$. While this does not result in a net purification, it removes the precursors of the two enzymes, which

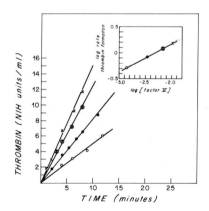

Fig. 1. Two stage assay factor V. Effect of factor V concentration on the intrinsic two-stage assay. The thrombin-generating mixture consisted of factor X_a (0.24 unit/ml), factor II (0.05 units/ml), inositin (200 μg/ml), $CaCl_2$ (0.01 M) and factor V (○, 0.0034; ●, 0.0054; ⊗, 0.0063; ×, 0.0011 unit/ml). A double logarithmic plot of the rate of thrombin formation against factor V concentration is shown in the inset. Reproduced from Fig. 4 of C. L. Kandall, T. K. Akinbami, and R. W. Colman, *Br. J. Haematol.* **23**, 655 (1962).

in their active forms interact with factor V, i.e., factor X_a and thrombin. The next useful step takes advantage of the low isoelectric point ($pI = 4.65$) of factor V^{14} which may be the cause of its tight associations to anionic exchange resins, such as TEAE-cellulose[15,16] or QAE-cellulose.[17] Ammonium sulfate precipitation is also useful but may give rise to changes in the molecular size of factor V.[14,18] Cellulose phosphate chromatography with stepwise elution[15,16] yields considerable purification, but the factor tends to associate after cellulose phosphate adsorption. Therefore, a final step, of gel filtration on 4% agarose, was added to the procedures of Barton and Hanahan[16] and Esnouf and Jobin[15]; this removes any dissociated or aggregated factor V and yields the monomeric (and possibly the dimeric) form.[19]

Procedure. The purification is carried out at room temperature unless otherwise specified. Bovine blood is collected into a plastic bucket containing one-tenth volume of 0.1 M Na-oxalate. The plasma, obtained after centrifugation at 1200 g for 30 min is adsorbed with $BaSO_4$ (100 g per liter of plasma) by stirring constantly for 30 min. The supernatant is collected after centrifugation at 1200 g for 15 min at 4° and stored frozen at −20°. Prior to use, the $BaSO_4$-adsorbed plasma is thawed at 37° and diluted with an equal volume of water. The pH of the diluted $BaSO_4$-adsorbed plasma is adjusted to 7.0 with 1 N HCl, and TEAE-cellulose (Cellex T, Bio-Rad) is added so that each 100 ml of the undiluted $BaSO_4$-adsorbed plasma is mixed with 3.0 g of dry resin and stirred for 20 min. The TEAE-cellulose is collected by filtration through Büchner funnel, washed 4 times by suspending it in a volume corresponding to the original volume of plasma of 0.04 M potassium phosphate buffer at pH 7.0, containing 1 mM NaCN, and stirred for 20 min. After filtration, the TEAE-cellulose is eluted by stirring the resin with one-fourth volume of 0.25 M potassium phosphate buffer (pH 7.0), containing 1 mM NaCN, for 30 min. After removal of TEAE-cellulose by filtration, solid $(NH_4)_2SO_4$ is added to 70% saturation (480 g/liter) and equilibrated for 15 min. After centrifugation (6000 g, 20 min) the precipitate is redissolved in one-tenth volume of 0.04 M potassium phosphate buffer (pH 7.0), containing 1 mM NaCN, and dialyzed against 1 volume of the same as at 4° for 18 hr. The dialyzed solution is centrifuged for 30 min, and

[14] A. C. Greenquist, R. M. Weinberg, A. L. Kuo, and R. W. Colman, *Eur. J. Biochem.* **58**, 213 (1975).
[15] M. P. Esnouf and F. Jobin, *Biochem. J.* **102**, 660 (1967).
[16] P. G. Barton and D. J. Hanahan, *Biochim. Biophys. Acta* **133**, 506 (1967).
[17] F. A. Dombrose, T. Yasui, Z. Roubal, A. Roubal, and W. H. Seegers, *Prep. Biochem.* **2**, 381 (1972).
[18] W. C. Day and P. G. Barton, *Biochim. Biophys. Acta* **261**, 457 (1972).
[19] G. Philip, J. Moran, and R. W. Colman, *Biochemistry* **9**, 2212 (1970).

the supernatant is applied to a 13×5 cm cellulose phosphate column previously equilibrated with 0.04 M potassium phosphate buffer (pH 7.0) containing 1 M NaCN. Approximately 125 ml of wet resin are used for each liter of plasma. The column is washed with the 0.04 M phosphate buffer and then with 0.15 M potassium phosphate buffer (pH 7.0) with 1 mM NaCN until the absorbance of the eluent is less than 0.03. Factor V activity is eluted with 0.25 M potassium phosphate buffer (pH 7.0) containing 1 mM NaCN. The fractions are pooled and concentrated using an Amicon ultrafiltration apparatus with an XM-50 membrane at 4°. The concentrated factor V (about 20 mg) is applied to a 2.4×52 cm Sepharose 4B column which has been equilibrated with 0.04 M potassium phosphate buffer (pH 7.0) with 1 mM NaCN at 4° and is eluted with the same. The predominant activity peak has an approximate molecular weight of 300,000.[19] The fractions are pooled, concentrated by ultrafiltration with an XM-50 membrane, and stored at −50°.

A typical purification is summarized in Table I.

Properties

Stability. The material after cellulose phosphate chromatography and gel filtration on 4% agarose is stable for months when stored at −50° in 25% glycerol. A small loss of activity occurs upon freezing, but thereafter no significant decrease is noted for 6 more months. The original bovine plasma and the $(NH_4)_2SO_4$ precipitates are similarly stable upon freezing but, paradoxically, the TEAE-cellulose eluate seems to lose up

TABLE I
PURIFICATION OF BOVINE PLASMA FACTOR V

Fraction	Total protein (mg)	Total units	Specific activity (units/mg)	Purification factor	Recovery (%)
Plasma	148,500	2600	0.0175	1	100
TEAE-cellulose eluate	1,052	2040	1.94	111	78.5
$(NH_4)_2SO_4$ precipitate	695	2222	3.20	183	85.5
Cellulose phosphate eluate[a]	18.5	900	48.5	2760	34.5
4% agarose gel filtration	16.7	835	50.0	2840	32.0

[a] R. W. Colman, *Biochemistry* **8**, 1438 (1969).

to 50% of its activity on freezing.[20] Esnouf and Jobin found that their cellulose phosphate eluate, when frozen at −20° without glycerol, was completely inactivated.[15]

The biological activity of bovine factor V in plasma is labile at room temperature and above. Human factor V is even more unstable, a fact that is utilized for preparing the factor V depleted substrate for assay purposes. Concentrated cellulose phosphate eluates show greater stability.[7]

The stability of factor V is markedly affected by its ionic environment. This fact has been attributed to dissociation of the molecule by chaotropic anions,[19] since inactivation follows the order $SCN^- > ClO_4^- > NO_3^- > Cl^- \gg SO_4^{2-}$. Chelating agents such as EDTA and citrate rapidly inactivate factor V by specifically binding to an essential calcium in the molecule.[21] Calcium, magnesium, and strontium in fairly high concentrations have been found to markedly stabilize factor V preparations and partially restore the activity of an inactivated preparation.[20,22]

Purity. The purified material is essentially free of all other clotting factors[21] and contains no plasminogen, plasmin, or antithrombin.[8]

Purified preparations are homogeneous by immunoelectrophoresis and ultracentrifugation.[7,15] Although Barton and Hanahan[16] found a single band on cellulose acetate electrophoresis, more sensitive techniques have revealed heterogeneity in these preparations.[7,23] Several steps in the purification procedure seem to produce aggregation of factor V,[18] but all the peaks eluted from Sepharose gel columns have factor V activity.[14]

Physical Properties. Bovine factor V migrates as an α-globulin[14] on immunoelectrophoresis. The molecular weight of the material purified as above is 300,000 as determined by gel filtration on Sepharose 4B column.[19] This is in agreement with most reported values, including determinations by sedimentation equilibrium of 290,000[15]; gel filtration on G-200 Sephadex 350,000[16]; gel filtration on 4% agarose 380,000,[23] although both higher and lower values have also been published.[24,25] Estimates of the Stokes radius have varied between 6.65[23] and 8.5[19] and the partial specific volume was estimated as 0.73 mg/g.[15] The extinction coefficient $E_{280mn}^{1\%}$ has been measured as 11.9[8] and 13.5.[23] The frictional

[20] B. Blombäck and M. Blombäck, *Nature (London)* **198**, 886 (1963).

[21] A. C. Greenquist and R. W. Colman, *Blood* **46**, 769 (1975).

[22] H. J. Weiss, *Thromb. Diath. Haemorrh.* **14**, 32 (1965).

[23] F. A. Dombrose and W. H. Seegers, *Thromb. Diath. Haemorrh., Suppl.* **57**, 241 (1974).

[24] D. Papahadjopoulos, C. Hougie, and D. J. Hanahan, *Biochemistry* **3**, 264 (1964).

[25] Q. Z. Hussain and T. F. Newcomb, *Ann. Biochem. Exp. Med.* **23**, 569 (1963).

TABLE II
AMINO ACID COMPOSITION OF BOVINE FACTOR V

Amino acid	Esnouf and Jobin[a] (moles/10^5 g)	Dombrose and Seegers[b] (moles/10^5 g)
Asp	60.8	57.9
Thr	41.4	45.6
Ser	54.3	64.7
Glu	75.4	63.1
Pro	46.3	40.9
Gly	47.6	40.0
Ala	36.0	48.2
Val	37.1	30.4
Cys[d]	19.2	8.7
Met	8.2	7.2
Ile	26.3	30.5
Leu	41.6	46.4
Tyr	23.2	14.4
Phe	22.2	18.9
Lys	c	30.1
His	25.2	13.2
Arg	21.1	24.1
Trp	11.6	17.5

[a] M. P. Esnouf and F. Jobin, *Biochem. J.* **102**, 660 (1967).
[b] F. A. Dombrose and W. H. Seegers, *Thromb. Diath. Haemorrh., Suppl.* **57**, 241 (1974).
[c] Not recorded by authors.
[d] As cysteic acid.

ratio is 1.42,[23] and the pI = 4.65.[14] Dombrose and Seegers have reported two amino-terminal amino acids, glycine and asparagine.[23]

The amino acid analyses performed on the preparations of Esnouf and Jobin and of Dombrose and Seegers are summarized in Table II. There is reasonable agreement except for the half-cystine, lysine, and histidine.

The carbohydrate contents of preparations varies from lot to lot of starting plasma but generally is between 10 and 20%.[26,27] The variation is apparently due to sialic acid ranging from 2 to 12%. Other carbohydrates include mannose (2.5–3.7%), galactose (2.2–3.2%), and N-acetylglucosamine (2.4–2.5).[28,29] Traces of glucose (0.29%) and N-acetyl-

[26] C. L. Kandall, R. Rosenberg, and R. W. Colman, *Eur. J. Biochem.* **58**, 203 (1975).
[27] J. G. Gumprecht and R. W. Colman, *Arch. Biochem. Biophys.* **196**, 278 (1975).
[28] P. G. Barton, C. M. Jackson, and D. J. Hanahan, *Nature (London)* **214**, 923 (1967).
[29] S. Saraswathi and R. W. Colman, *J. Biol. Chem.* **250**, 8111 (1975).

galactosamine (0.57%) have also been found.[27] About 8.7% of neutral sugars are present.[6] Data given below indicate that sialic acid is terminal and galactose is penultimate. No appreciable amount of phospholipid is present.[23,30]

Interaction of Factor V with X_a. The formation of thrombin from prothrombin *in vivo* is effected by a complex consisting of factor X_a, factor V, phospholipid, and calcium. As first proposed by Milstone,[2] factor X_a is responsible for the proteolytic cleavage.[28] However, the conversion rate is several orders of magnitude slower than it would be in the presence of factor V, lipid, and calcium. No enzymic activity has been found in any of the factor V preparations, and the nature and site of its procoagulant activity remain unknown. In the presence of phospholipid and calcium, but in the absence of prothrombin, factor V increases the esterase activity of factor X_a more than 3-fold.[31] Marciniak[32] demonstrated a change in the catalytic activity of X_a after adding factor V, calcium, and phospholipid.

Interaction of Factor V with Prothrombin. Effects other than on the enzymic activity of factor X_a are possible and may include binding of the "prothrombinase" complex to the prothrombin molecule,[33] and/or alteration of the tertiary structure of prothrombin to render it more susceptible to proteolysis by factor X_a. Factor V binds to prothrombin immobilized on Sepharose,[34] but only when factor V is activated with thrombin.

Interaction of Factor V with Phospholipids. Phospholipid requirement depends on the test employed.[35,36] In an assay[8] employing purified factor V, X_a, prothrombin, and fibrinogen, various phospholipids exhibited a maximum as a function of concentration, and at 250 μM their order of effectiveness was phosphatidylethanolamine > phosphatidylinositol > cardiolipin > phosphatidylcholine.[30] Phosphatidylserine is inert. No single fatty acid was consistently present in the active lipids or absent from the less active and inactive lipids. Unsaturation was necessary but not sufficient; thus, saturated phosphatidylethanolamine is in-

[30] C. L. Kandall, S. B. Shohet, T. K. Akinbami, and R. W. Colman, *Thromb. Diath. Haemorrh.* **34**, 256 (1975).

[31] R. W. Colman, *Br. J. Haematol.* **19**, 675 (1970).

[32] E. Marciniak, *Br. J. Haematol.* **24**, 391 (1973).

[33] C. M. Jackson, W. G. Owen, S. N. Gitel, and C. T. Esmon, *Thromb. Diath. Haemorrh., Suppl.* **57**, 273 (1974).

[34] C. T. Esmon, W. G. Owen, D. L. Diuquid, and C. M. Jackson, *Biochim. Biophys. Acta* **310**, 289 (1973).

[35] A. J. Marcus, *Adv. Lipid Res.* **4**, 1 (1966).

[36] P. G. Barton, "Structural and Functional Aspects of Lipoprotein in Living Systems" (E. Tria and A. M. Scanu, eds.), p. 465. Academic Press, New York, 1969.

active except at very high concentrations, and plant phosphatidylinositol, which is almost fully saturated, is also inert. Effects from mixtures of phospholipids were not necessarily additive; for example, saturated phosphatidylethanolamine with no activity of its own enhanced the activity of phosphatidylethanolamine.[30] Using the technique of gel filtration[3,10] or precipitation, [2,30] several investigators have shown that factor V forms a complex with phospholipid in the absence of calcium. On sucrose density ultracentrifugation, the active lipids such as phosphatidylethanolamine form complexes with factor V of lighter buoyant density than that of the native molecule.[37] These complexes (binding constant 5×10^6 M^{-1}) substitute for both factor V and phospholipid in purified coagulant assays.

Interaction of Factor V with Calcium. Calcium is required for the complex formation between phospholipids and prothrombin[38] and between phospholipids and factor X_a,[24] but is not necessary for the interaction of factor V with phospholipid. Nonetheless, factor V in plasma[22] or in purified preparations[21] is very labile in the presence of chelating agents, especially EDTA. Bovine factor V was shown to contain 1 g-atom of calcium per 300,000 daltons,[21] which cannot be removed by EDTA under nondenaturing conditions. Calcium or phospholipid protect against inhibition by EDTA, suggesting that the calcium in factor V is one of the sites of phospholipid binding.

Interaction of Factor V with Thrombin. Thrombin increases the activity of factor V 5-fold, and this activation is blocked by the thrombin inhibitor hirudin.[7] The rate and extent of activation are proportional to thrombin concentration at least in the range of 0.1–1.0 NIH unit/ml, and both are inversely related to factor V concentration. This finding can be explained by either substrate or product inhibition, and in fact both occur.[39] When factor V is exposed to thrombin (factor V_t) it cannot be further activated by thrombin. Factor V_t inhibits the degree of thrombin activation of unaltered factor V, and the extent of inhibition is proportional to the ratio of factor V_t to unactivated factor V. Factor V_t apparently does not inhibit thrombin directly, since hydrolysis of fibrinogen or of toluenesulfonyl-L-arginine methyl ester is not affected. These results suggest that the inhibition is due to interaction of the factor V derivative with unaltered factor V. Smaller forms of factor V may also occur *in vivo* as a result of proteolysis.[26] After thrombin treatment, some authors have been able to demonstrate an apparent change in

[37] C. L. Kandall, S. B. Shohet, T. K. Akinbami, and R. W. Colman, *Thromb. Diath. Haemorrh.* **34**, 371 (1975).
[38] F. Jobin and M. P. Esnouf, *Biochem. J.* **102**, 666 (1967).
[39] R. W. Colman, J. Moran, and G. Philip, *J. Biol. Chem.* **245**, 5941 (1970).

molecular weight of factor V by gel filtration or by sucrose gradient centrifugation,[16,24] but others have not.[39] As factor V dissociates under mild conditions, the changes may not relate to actual proteolysis by thrombin.[19]

Investigation of the subunit polypeptide structure of factor V has shed some light on the interaction of thrombin and factor V. Purified bovine factor V, subjected to acrylamide disc gel electrophoresis in the presence of sodium dodecyl sulfate (SDS) after vigorous denaturation and reduction, in agreement with the finding of two N-terminal amino acids,[23] reveals two major bands: termed H-chain and L-chain, with chain weights of 125,000 and 70,000, respectively.[26] In addition, there is a series of minor but constant bands of apparently high molecular weights (300,000–1,000,000).

Thrombin breaks the H-chain, producing a peptide of 80,000 daltons called the H-fragment. The remainder of the H-chain is difficult to identify by Coomassie blue staining of the gels but is clearly visible with the periodic acid–Schiff reagent. No detectable change occurs in the L-chain.

Interactions of Factor V with Other Proteolytic Enzymes. In addition to thrombin, three other enzymes can also increase the activity of factor V; namely, papain,[40] Russell's viper venom,[40-42] and plasmin.[40] Plasmin and papain both produce only transient increase inactivity at low concentrations, which are rapidly obscured by further proteolytic degradation of the factor V substrate molecule, with loss of all activity[43] while a purified enzyme from Russell's viper venom produces a stable more active derivative of factor V which may have a lower molecular weight.[42] All three enzymes appear to split the same H-chain[43] as thrombin.[30] It has been suggested that this reaction results in a change in molecular size.[44]

Human plasma kallikrein and *Brothrops jararaca* and *Agkistrodon rhodostoma* venoms do not affect factor V activity.[40] Trypsin seems only to destroy factor V activity,[40] although the possibility of transient activation cannot be ruled out.

Effect of Carbohydrate on Factor V Activity. The carbohydrate side chains may play a role in the coagulant activity or stability of the mole-

[40] R. W. Colman, *Biochemistry* 8, 1445 (1969).
[41] S. Schiffman, I. Theodore, and S. I. Rapaport, *Biochemistry* 8, 1397 (1969).
[42] D. J. Hanahan, M. R. Rolfs, and W. C. Day, *Biochim. Biophys. Acta* 286, 205 (1972).
[43] R. M. Weinberg, A. L. Kuo, and R. W. Colman, *Fed. Proc., Fed. Am. Soc. Exp. Biol.* 33, 225 (Abst No. 123) (1974).
[44] P. A. Owren, *Scand. J. Clin. Lab. Invest.* 1, 131 (1949).

cule, and the function of the carbohydrate residues in bovine factor V were studied through enzymic modifications of sialic acid[27] and galactose.[29] Hydrolysis of the sialic acid residues with neuraminidase (free of proteolytic activity) enhances factor V activity by about 50%.[27] Asialo-factor V is 3.8 times more stable than intact factor V, as determined from the first-order decay constant at 37°. Asialo-factor V is still susceptible to thrombin, but its activation is biphasic and somewhat slower than in unmodified factor V. After neuraminidase treatment factor V forms an inactive complex with phosphatidylethanolamine, in contrast to intact factor V, where the complex retains coagulant activity. In contrast to its behavior after neuraminidase, the oxidation of galactose residues with galactose oxidase, or removal of galactose from asialo-factor V with β-galactosidase, results in complete loss of activity,[29] because of inability to bind phospholipids. Since phospholipid protects against oxidation of galactose residues, it appears that these residues, in addition to endogenous calcium,[29] serve as one of the critical binding sites for phospholipids. The loss of coagulant activity after galactose oxidation can be partially reversed by reduction with sodium borohydride.

Immunochemical Studies on Factor V. Antibodies have been prepared in rabbits to monomeric factor V (molecular weight 300,000)[14] giving two arcs in immunodiffusion, one against the light chain and the other against the heavy chains. The antibody neutralizes bovine factor V activity and, on immunoelectrophoresis, detects factor V as two closely associated arcs in the α-globulin region, Antibodies were also prepared against a high-molecular-weight species containing aggregates of the light chain. This antiserum also exhibited a time and concentration-dependent inactivation of factor V in bovine plasma as well as purified preparations. A single arc, equivalent to the light chain observed on double diffusion, fails to form after denaturation with 5 M guanidine hydrochloride and reduction with dithiothreitol, but not with either alone. There is an identity reaction with the light-chain antigenic components of goat and sheep plasma, but no cross reaction with monkey or human plasma.

Human Factor V

Assay

The one-stage assay[1] or minor variations of it as described for bovine factor V has been used in most studies.

Purification Procedure

Until recently the extreme lability of human plasma factor V has prevented development of an adequate procedure. Lewis and Ware[1] described a 35-fold purification from citrated plasma (0.02 M sodium citrate) with a 12% yield. After $BaCl_2$ adsorption followed by $BaSO_4$ addition to remove the prothrombin complex, a low-ionic-strength precipitation at pH 5.5 gave a 55% yield. Further fractionation with NH_4SO_4 and dialysis against sodium citrate increased stability but decreased yield. Blombäck and Blombäck[20] fractionated Cohn fraction IV-1 with glycine followed by gel filtration on Sephadex G-200, and obtained a 60-fold purification. The preparation required either Ca^{2+} or Mn^{2+} for stability. Giddings[45] described a 300-fold purification using alcohol fractionation, Rivanol precipitation, and agarose gel filtration. Preparations with higher specific activity could be prepared only from serum, and probably represent activation of human factor V by thrombin (see below) rather than purification.

Rosenberg et al.[46] recently prepared the first highly purified factor V. The first step is similar to that of Lewis and Ware[1] involving barium citrate adsorption and isoelectric fractionation, giving a 20-fold purification and a 47% recovery. The following step involves adsorption onto hydroxyapatite and elution with 0.25 M potassium phosphate, and results in additional 10-fold purification and a 40% yield of the preceding step. Concentration and fractionation with 10.5% polyethylene glycol results in a 4-fold purification with no loss of activity, and an additional 3-fold purification with a 60% yield was achieved with DEAE-chromatography upon elution with 0.11 M NaCl. After rapid dialysis against 45% sucrose, the solution was stored at −70°. An overall purification of about 1500-fold was achieved. The yield was 11% and the specific activity was 59 units/mg.

Properties

Stability. Factor V has been called labile factor because of its characteristic instability. In plasma collected in 0.02 M oxalate, 50% of factor V activity is lost at 23° in 2 hr.[47] Even at 4° factor V declines to 40% in acid citrate dextrose solution A and to 31% in citrate phos-

---

[45] J. C. Giddings, *Thromb. Diath. Haemorrh.* **31**, 457 (1974).
[46] J. S. Rosenberg, D. L. Beeler, and R. D. Rosenberg, *J. Biol. Chem.* **250**, 1607 (1975).
[47] S. G. Sandler, C. E. Rath, and A. Ruder, *Ann. Int. Med.* **79**, 485 (1973).

phate dextrose anticoagulant.[48] This instability is enhanced during purification, as shown by Rosenberg.[46] The isoelectric precipitate is stable
at 2° for a maximum of 5 hr, and the hydroxyapatite eluate for only 2
hr at 2°. Thus these steps must be completed rapidly. The final product
is stable at 2° for over 48 hr and at −70° for several months.

Purity and Physical Properties. Only the preparation of Rosenberg[46]
has been characterized to some extent. Isoelectric focusing yields a p*I*
of 5.1 and a single species of protein. Since human factor V is unstable
if maintained for 30 min below pH 7.0, only 5% of the activity was
obtained after elution. Injection of the preparation into rabbits yielded
a single arc against the isoelectric precipitate and precipitated 70% of
the factor V activity of human plasma.

A molecular weight in excess of 300,000 was determined by gel filtration on a Sephadex G-200 column[20] and also be agarose gel filtration.[45]

Synthesis and Destruction. Factor V appears to be synthesized in the
liver.[49] Clinical studies of patients with liver disease who became factor V
deficient[44,46,50] are concordant with the finding of positive immunofluorescence of bovine hepatocytes.[49]

Plasma levels of factor V in man range from 0.5 to 1.0 mg/100 ml.[51]
Congenital deficiency of factor V, first described in 1947 by Owren,[52] is
a moderately severe bleeding disorder characterized by skin hemor-
rhages, nosebleeds, and bleeding from the oral cavity particularly following tooth extraction or tonsillectomy. It is a rare disorder with an
estimated incidence of less than 1 in 10^6 and appears to be inherited
as an autosomal dominant[53] with incomplete expressivity or as an autosomal recessive trait.[54] Immunological studies[55,56] suggest a failure of
synthesis rather than synthesis of an abnormal protein.

The half-life of factor V in man has been determined by plasma
infusion studies in congenitally deficient patients, and values range from
12[57] to 36 hr,[58] with a median of 20 hr.[59]

[48] O. Weisert and M. Jeremic, *Vox Sang.* **24**, 126 (1973).
[49] M. I. Barnhart, J. Ferar, and N. Aoki, *Fed. Proc.*, **22**, 164 (Abst. No. 19) (1963).
[50] R. B. Finkbinder, J. J. McGovern, R. Goldstein, and J. P. Bunker, *Am. J. Med.* **26**, 199 (1959).
[51] E. W. Davie and E. P. Kirby, *Curr. Top. Cell. Regul.* **7**, 53 (1973).
[52] P. A. Owren, *Lancet* **1**, 446 (1947).
[53] C. A. Owen, Jr., and T. Cooper, *Arch. Intern. Med.* **95**, 194 (1955).
[54] C. S. Kingsley, *Quart. J. Med.* **23**, 323 (1954).
[55] D. I. Feinstein, S. I. Rapaport, W. G. McGehee, and M. J. Patch, *J. Clin. Invest.* **49**, 1578 (1970).
[56] J. H. Ferguson, C. L. Johnston, Jr., and D. A. Howell, *Blood* **13**, 382 (1958).
[57] C. F. Borchgrevink and P. A. Owren, *Acta Med. Scand.* **170**, 743 (1961).
[58] W. P. Webster, H. R. Roberts, and G. D. Penick, *Am. J. Med. Sci.* **248**, 194 (1964).
[59] B. Bush and H. Ellis, *Thromb. Diath. Haemorrh.* **14**, 74 (1965).

The catabolic fate of factor V is uncertain. Both enzymic digestion[60] and physical denaturation[61] have been proposed. Thrombin diminishes stability,[48] and plasmin destroys human factor V activity.

Circulating anticoagulants directed against human factor V are rare but have occurred both spontaneously[60,62-65] and after multiple transfusions to patients with congenital deficiency of factor V.[66] Neutralizing antibodies (IgG or IgM) have appeared after surgery,[55,56,63] in association with penicillin[62,64,65] or streptomycin treatment.[56,62,63]

Platelet Factor V

Human platelets contain an accelerator of prothrombin conversion similar to human plasma factor V, which, however, is more stable to heat inactivation[67] or storage,[68,69] and which on transfusion to patients with congenital factor V deficiency corrects the coagulation defect[70] at least as well as plasma.[71] Platelets from a patient with congenital factor V deficiency contained no factor V but acquired it after incubation in normal plasma.[72-74] Platelets which lost factor V activity upon storage regained it after incubation in normal plasma.[75] Conversely, normal platelets lost factor V activity when incubated in factor V-deficient plasma.[76] Factor V is firmly bound to platelets[77] and is not removed by extensive washing.[78,79] The actual role in hemostasis of platelet factor V is not clear.

[60] C. F. Borchgrevink and P. A. Owren, *Acta Med. Scand.* **170**, 375 (1961).
[61] S. Leiken and S. P. Bessman, *Blood* **11**, 916 (1956).
[62] V. Lopez, R. Pflugshaupt, and R. Bütler, *Acta Haematol.* **40**, 275 (1968).
[63] D. A. Handley and B. M. Duncan, *Pathology* **1**, 265 (1969).
[64] I. M. Nilsson, U. Hedner, M. Ekberg, and T. Denneberg, *Acta Med. Scand.* **195**, 73 (1974).
[65] C. A. Onuora, J. Lindenbaum, and H. L. Nossel, *Am. J. Med. Sci.* **265**, 407 (1973).
[66] J. C. Fratantoni, M. Hilgartner, and R. L. Nachman, *Blood* **39**, 751 (1972).
[67] A. G. Ware, J. L. Fahey, and W. H. Seegers, *Am. J. Physiol.* **154**, 140 (1948).
[68] E. Deutsch, S. A. Johnson, and W. H. Seegers, *Circ. Res.* **3**, 110 (1955).
[69] E. F. Lüscher, *Engeb. Physiol.* **50**, 2 (1959).
[70] C. F. Borchgrevink and P. A. Owren, *Acta Med. Scand.* **170**, 375 (1961).
[71] P. A. Owren, *Proc. Int. Union Physiol. Sci. Int. Cong. 22nd*, p. 233 (1962).
[72] P. Hjort, S. I. Rapaport, and P. A. Owren, *Blood* **10**, 1139 (1955).
[73] R. H. Seibert, A. Margolius, and O. D. Ratnoff, *J. Lab. Clin. Med.* **52**, 449 (1958).
[74] E. Deutsch and Z. Wein, *Wein. Z. Inn. Med. Ihre Grenzgeb.* **36**, 355 (1955).
[75] A. Doczy and A. Fejer, *Thromb. Diath. Haemorrh.* **9**, 300 (1963).
[76] G. Raccuglia, *Proc. Soc. Exp. Biol. Med.* **104**, 309 (1960).
[77] S. A. Johnson, W. M. Smathers, and C. L. Schneider, *Am. J. Physiol.* **170**, 631 (1952).
[78] P. G. Iatridis and J. H. Ferguson, *Thromb. Diath. Haemorrh.* **13**, 114 (1965).
[79] Y. Bounameaux, *Rev. Fr. Etud. Clin. Biol.* **2**, 52 (1957).

[13] Prothrombin

By Kenneth G. Mann

Prothrombin is the coagulation proenzyme present in highest concentration in blood (0.07–0.1 mg/ml) and was recognized very early[1] as a prime contributor to the blood coagulation process. The pioneering work of Smith *et al.*[2] and Seegers[3-5] provided most of our present knowledge regarding the requirements for the maximum stimulation of thrombin production. The physiologically significant activator of prothrombin is thought to be a complex of two proteins—factor X_a (a proteolytic enzyme) and factor V (a cofactor protein)—phospholipid, and calcium.[6,7] In addition, the proenzyme can be activated to its product by specific snake venoms[8,9] and by trypsin.[10] During activation of the molecule to its product thrombin, there occurs a complex series of events that results in the deletion of about 42% of the proenzyme as "pro" fragments.[11] Figure 1 presents a conceptual image of the prothrombinase activator of prothrombin. One of the most intriguing and significant features of the prothrombinase catalyst is that the lipid provides a surface upon which the catalyst, factor X_a, and its activators, form the complex and perform their functions. Since plasma is an aqueous medium, the lipid matrix forms a separate phase from the surrounding medium, and in effect the activation of the prothrombin molecule takes place on an oil–water interface. In order to be utilized effectively, prothrombin must also bind to the interface; thus, prothrombin must be partitioned for its product, thrombin, to be generated at a rapid rate.

Prothrombin is synthesized in the liver,[12] and is one of the vitamin

---

[1] P. Morawitz, *Ergeb. Physiol.* **4**, 304 (1905).

[2] E. D. Warner, K. M. Brinkhous, and H. P. Smith, *Am. J. Physiol.* **125**, 296 (1939).

[3] A. G. Ware and W. H. Seegers, *Am. J. Clin. Pathol.* **19**, 471 (1949).

[4] W. H. Seegers, "Blood Clotting Enzymology." Academic Press, New York, 1967.

[5] W. H. Seegers, "Prothrombin." Harvard Univ. Press, Cambridge, Massachusetts, 1962.

[6] D. Papahadjopoulos and D. J. Hanahan. *Biochim. Biophys. Acta* **90**, 436 (1964).

[7] E. R. Cole, J. L. Koppel, and J. H. Olwin, *Thromb. Diath. Haemorrh.* **14**, 431 (1965).

[8] H. Pirkle, M. McIntosh, I. Theodore *et al.*, *Thromb. Res.* **1**, 559 (1972).

[9] K. W. E. Denson, R. Borrett, and R. Biggs, *Br. J. Haematol.* **21**, 219 (1971).

[10] F. Jobin and M. P. Esnouf, *Nature (London)* **211**, 873 (1966).

[11] K. G. Mann, C. M. Heldebrant, D. N. Fass *et al.*, *Thromb. Diath. Haemorrh.*, Suppl. **57**, 179 (1974).

[12] M. I. Barnhart, *Am. J. Physiol.* **199**, 360 (1960).

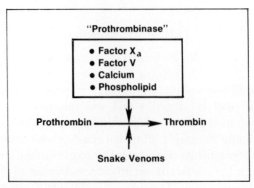

Fɪɢ. 1. Schematic representation of the "prothrombinase" activation of prothrombin.

K-dependent blood coagulation factors.[13] Prothrombin synthesis in the absence of vitamin K or in the presence of vitamin K antagonists (in humans and cattle) results in the formation of an incomplete molecule,[14,15] which lacks certain calcium-binding sites[15,16] that are vitamin K dependent. The incomplete protein is secreted into the plasma at normal levels,[17,18] but is not subject to activation to thrombin by the physiological activator, presumably because the incomplete molecule which lacks the vitamin K-dependent calcium-binding sites is not capable of binding to the factor X_a, factor V, calcium, and phospholipid activator. The abnormal molecule produced in vitamin K deficiency states, however, is still susceptible to activation by snake venom activators.

Assay

Two types of assays have been proposed to be used to evaluate the plasma level of prothrombin: the one-stage assay[19] and the two-stage assay.[3] In both assay systems, the process of prothrombin activation is initiated by adding a source of tissue factor, phospholipid, and calcium to the plasma sample. The tissue factor serves to activate factor VII in

[13] C. A. Owen, Jr., *Mayo Clin. Proc.* **49**, 912 (1974).

[14] J. Stenflo and P.-O. Ganrot, *J. Biol. Chem.* **247**, 8160 (1972).

[15] G. L. Nelsestuen and J. W. Suttie, *J. Biol. Chem.* **247**, 8176 (1972).

[16] J. Stenflo and P.-O. Ganrot, *Biochem. Biophys. Res. Commun.* **50**, 98, 1973.

[17] P.-O. Ganrot and J. E. Niléhn, *Scand. J. Clin. Lab. Invest.* **22**, 23 (1968).

[18] F. Josso, J. M. Lavergne, M. Gouault, O. Prou-Wartell, and J. P. Soulier, *Thromb. Diath. Haemorrh.* **20**, 88 (1968).

[19] A. J. Quick, M. Stanley-Brown, and F. W. Bancroft, *Am. J. Med. Sci.* **190**, 501 (1935).

the plasma, which subsequently activates the endogenous factor X to factor X_a. The factor X_a produced combines with calcium, phospholipid, and factor V to form the catalyst that converts prothrombin to thrombin. In the one-stage assay, the amount of time required to produce a sufficient amount of thrombin to clot the fibrinogen present in the sample is recorded as a function of prothrombin concentration; in the two-stage type of assay, the thrombin produced is evaluated by a separate analysis, and thus both the rate and the amount of thrombin produced from the sample are recorded. In effect, a one-stage assay measures only the rate at which a given amount of thrombin is produced, and two-stage assays measure the quantity of product enzyme that can be produced when the zymogen is subjected to a quasi-physiological activation system. Thus, only the two-stage assay provides an adequate measure of the quantity of functional zymogen.

The procedure our laboratory uses to assay prothrombin is basically a modification of the original two-stage procedure of Ware and Seegers.[3] Since purified prothrombin will not have present in it endogenous factor VII or factor X,[20] defibrinated plasma must be readded to the assay system as a source of these two factors. In addition, the assay of the product (thrombin) is accomplished by a modified NIH thrombin assay.[21] This provides for the evaluation of the concentration of the product directly in terms of the thrombin produced on a molar basis.

First Stage of Assay

Reagents

Defibrinated plasma (factor VII and factor X source): 0.5 ml of citrated bovine plasma, 0.4 ml of 0.9% sodium chloride, and 0.1 ml of a 100 unit/ml solution of Parke-Davis Topical Thrombin in 50% glycerol are mixed, then incubated at 37° for 30 min. Prior to use, the plasma is expressed from the clot.

Prothrombin to be measured: Protein concentration is evaluated spectrophotometrically using the $E_{280}^{1\%}$ for bovine prothrombin 14.4 reported by Cox and Hanahan.[22] Protein concentrations are corrected for Rayleigh scattering using the equation:

$$A_{280} = A_{280\,observed} - 1.706 A_{320\,observed}$$

Factor V: 0.1 ml of AC-globulin (Difco), a barium-adsorbed serum preparation, is added to 15 ml of 0.9% sodium chloride. This

---

[20] S. S. Shapiro and D. F. Waugh, *Thromb. Diath. Haemorrh.* **16**, 469 (1966).
[21] R. L. Lundblad, H. S. Kingdon, and K. G. Mann, this volume [14].

solution is discarded after the completion of the assays. The
freshly prepared solution is stored at 0°.

Tissue factor: Thromboplastin (Difco two-stage reagent) is prepared
by reconstitution of dry reagent with 0.9% sodium chloride. Sim-
plastin (General Diagnostics) may be substituted.

Assay Procedure. Two timers are required for the assay; one to measure the time of activation and one for the thrombin assay. To perform
the assay, prepare an appropriate amount of NIH thrombin assay mixture[21] and the following solutions: solution A: 0.1 ml of defibrinated
plasma and 2.4 ml of factor V reagent; solution B: 0.25 ml solution A
and 0.75 ml of reconstituted thromboplastin (two-stage reagent) and
incubate at 37° for 10 min; solution C: the prothrombin sample appropriately diluted with factor V reagent to a final concentration of between
0.25 and 2.0 mg/ml; solution D: mix 0.25 ml of solution C and 0.75 ml
solution B, start the activation timer, and incubate the reaction mixture
at 37°.

Second Stage (Thrombin Assay)

The reagents and procedures used for the modified NIH thrombin
assay are detailed in this volume in Chapter [14] on thrombin. This
assay is used for the second stage of the prothrombin assay.

Remove aliquots for the NIH thrombin assay at various time intervals (~30 sec) and determine the thrombin activity present in the original sample at each time. Plot the thrombin activity generated in the first
stage vs time. The point of maximum thrombin activity which can be
generated from the prothrombin sample is the prothrombin activity of
that sample. As an example, if a 10-μl sample of the activation mixture
removed at the point of maximum activity has a clot time of 15 sec (1
unit/ml) in the NIH thrombin assay, and the original sample was diluted
1:3 (to prepare solution C) then the activity is

$$1 \text{ unit/ml} \times 310/10 \times 4/1 \times 3/1 = 372 \text{ units/ml}$$

Where 1 unit/ml is the thrombin activity (15 sec clot time in the thrombin assay) of the sample, determined from the clot time by reference to
the standard plot,[21] 310/10 is the dilution of the 10-μl sample in the NIH
thrombin assay, 4/1 is the dilution of solution C into the activation mixture (solution D), and 3/1 is the dilution of the original sample. This
hypothetical sample, then, has an activity of 372 NIH thrombin units/ml
in the modified two-stage assay.

Bovine thrombin has a maximal specific activity of about 2700–3000

NIH units per milligram of protein; thus the theoretical specific activity for pure prothrombin is between 1500 and 1700 NIH thrombin units per milligram of prothrombin. In real assay conditions, however, this value will only be approached, since side reactions and degradation of the product eliminate the possibility of converting all the prothrombin present in the assay to thrombin. Operationally homogeneous bovine prothrombin will have a specific activity of about 1200 NIH thrombin units per milligram of prothrombin in the assay.

Purification

Purification of bovine prothrombin has been reported by a variety of laboratories.[22-30] In general, all purification schemes reported take advantage of a peculiar affinity of the vitamin K-dependent proenzymes for insoluble barium and magnesium salts. It should be noted here that those procedures involving insoluble barium and magnesium salts are suitable only for the normal vitamin K-dependent factors, as the proteins produced in the absence of vitamin K or in the presence of vitamin K antagonists are not adsorbable onto barium salts.

In our early studies, our laboratory used a modification of the method of Ingwall and Scheraga[23] for the purification of prothrombin. In later studies, a procedure was developed that would provide for the isolation of both prothrombin and factor X from the same plasma sample, and this procedure will be detailed here.

Isolation of Prothrombin from Bovine Plasma

The isolation procedure for prothrombin and factor X is routinely carried out with 10–20 liters of plasma.[30] The purification cited for illustrative purposes was carried out with 14 liters of plasma. All steps were performed at 4° unless specified. Figure 2 shows the flow diagram of the steps involved in the purification.

[22] A. C. Cox and D. J. Hanahan. *Biochim. Biophys. Acta* **207**, 49 (1970).
[23] J. S. Ingwall and H. A. Scheraga, *Biochemistry* **8**, 1960 (1969).
[24] G. H. Tishkoff, L. C. Williamson, and D. M. Brown, *J. Biol. Chem.* **243**, 4151 (1968)
[25] G. F. Lanchantin, J. A. Friedmann, and D. W. Hart, *J. Biol. Chem.* **240**, 3276 (1965).
[26] L. E. McCoy and W. H. Seegers, *Thromb. Res.* **1**, 461 (1972).
[27] S. Magnusson, this series Vol. 19, p. 1957.
[28] K. D. Miller, *J. Biol. Chem.* **231**, 987 (1958).
[29] H. C. Moore, D. P. Malhotra, S. Bakerman and J. R. Carter, *Biochim. Biophys. Acta* **111**, 174 (1965).
[30] S. P. Bajaj and K. G. Mann, *J. Biol. Chem.* **248**, 7729 (1973).

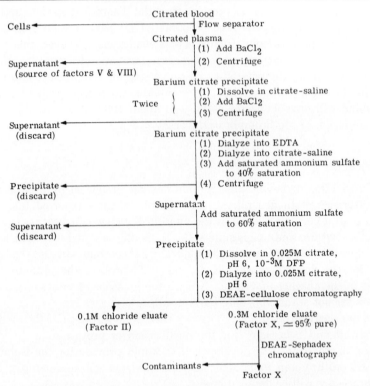

FIG. 2. Flow diagram for prothrombin and factor X isolation. From S. P. Bajaj and K. G. Mann, *J. Biol. Chem.* **248**, 7729 (1973).

Step 1. Collection of Blood. The technology employed in the first two steps is that of Moore *et al.*[29] Slaughterhouse blood was collected in 2.85% trisodium citrate (blood, 8 parts: anticoagulant, 1 part), and the cells and plasma were separated by the means of a continuous-flow separator (DeLaval No. 518).[31] This step was performed at room temperature.

Step 2. Barium Citrate Adsorption and Elution. The vitamin K-dependent factors were initially adsorbed onto barium citrate. The plasma was gently stirred and 1 M $BaCl_2$ (80 ml per liter of plasma, or 1120 ml of $BaCl_2$ per 14 liters of plasma) was added dropwise. Stirring was continued for 10 min after all the $BaCl_2$ had been added. The suspension was then centrifuged for 30 min at 3600 g. We have used a Lourdes continuous-flow centrifuge (BETA-FUGE Model A-2, rotor 1010) for large preparations with good results. The supernatant from the barium citrate

[31] This is an antique hand-driven cream separator. These machines are relatively plentiful in the Midwest dairy regions. Other cream separators or continuous-flow centrifuges may be substituted.

adsorption may be used as a source of factors V and VIII. The barium citrate precipitate obtained was suspended in citrate-saline (9% NaCl and 0.2 M trisodium citrate,[32] diluted 1:9 with distilled H_2O) by means of a Waring blender at a low speed. The total volume of diluted citrate-saline used was one-third of the volume of starting plasma (4670 ml for 14 liters of plasma). The resuspended protein was reprecipitated by addition of the same volume of 1 M $BaCl_2$ used in the initial precipitation step. Stirring was continued for 10 min after the addition of $BaCl_2$, and the suspension was covered and allowed to stand for 1 hr. (The suspension may be allowed to stand overnight, if desired.) The suspension was then centrifuged for 30 min at 3600 g, and the supernatant was discarded. The precipitate was resuspended in diluted citrate-saline as before, and the precipitation was repeated with same volume of 1 M $BaCl_2$. The barium citrate precipitate obtained at this stage may be stored indefinitely in the frozen state ($-20°$).

The barium citrate precipitate was suspended in cold 0.2 M EDTA, pH 7.4 (120 ml per liter of plasma). A Waring blender at low speed was used to obtain a homogeneous suspension. The suspension was dialyzed for 40 min vs 10 volumes of a mixture of 0.2 M EDTA, pH 7.4 (1 part), stock citrate–saline (0.2 M trisodium citrate, 0.9% NaCl, 1 part), and distilled water (8 parts). Dialysis was continued vs a similar volume of citrate–saline (1 part stock citrate–saline and 9 parts distilled water) for an additional 3 hr with the dialyzate changed every 30 min and then continued overnight. The dialysis bags were mixed gently at each dialyzate change to ensure complete mixing.

The dialysis step discussed above may be replaced by the following procedure, which is somewhat simpler and eliminates the need of dialyzing large quantities of particulate material. The barium citrate precipitate is suspended in 0.2 M EDTA pH 7.4 (120 ml of 0.2 M EDTA per liter of starting plasma). After suspension of the material, an equal volume of 0.026 M trisodium citrate, 0.9% NaCl is added. Soybean trypsin inhibitor is then added to a final concentration equivalent to 3.5 mg per liter of starting plasma. After this addition, the suspension is made to 10% of saturation with cold saturated $(NH_4)_2SO_4$ (pH 7.0), and the suspension is allowed to stir overnight in the cold. After a minimum of 12 hr resuspension in this solvent system, the mixture is centrifuged at 6000 g for 15 min. Regardless of whether the resuspension of the barium citrate precipitate is carried out by the dialysis procedure alluded to previously or the direct redissolution procedure involving ammonium sulfate, step 3 is unchanged.

---

[32] Referred to as stock citrate-saline.

Step 3. Fractionation with Ammonium Sulfate. The viscous opaque suspension (EDTA eluate) obtained in step 2 was brought to 40% saturation with respect to ammonium sulfate. A saturated solution of ammonium sulfate, pH 7.0 (at 4°, pH 7 adjusted with concentrated NH₄OH), was added dropwise to the slowly stirring EDTA eluate. Stirring was continued for 15 min after all the ammonium sulfate solution had been added. The opaque suspension was centrifuged at 3600 g for 30 min and the precipitate was discarded. The supernatant was then brought to 60% saturation with respect to ammonium sulfate. Stirring was continued for an additional 10 min. Then the suspension was allowed to stand for 20 min, after which it was centrifuged at 3600 g for 30 min. The supernatant was discarded.

Step 4. DEAE-Cellulose Chromatography. The precipitate obtained in step 3 was dissolved in a minimum volume of 0.025 M (sodium) citrate buffer, pH 6. The protein solution was made 1 mM in diisopropyl phosphorofluoridate (DFP) by the addition of 1 M DFP in isopropanol.[33] The DFP treated protein was dialyzed vs 2 liters of 0.025 M (sodium) citrate buffer, pH 6. The dialysis was continued overnight with two changes.

Invariably after dialysis the solution had a small amount of insoluble protein; this was removed by centrifugation at 3600 g for 20 min. The supernatant was then applied to a DEAE-cellulose column (45 × 4 cm) equilibrated with the same buffer as used for the dialysis. The column was then washed with 0.025 M (sodium) citrate buffer, pH 6.0, until the absorbance of the effluent at 280 nm was less than 0.02. The buffer was then changed to 0.025 M (sodium) citrate, 0.1 M NaCl, pH 6. Prothrombin was eluted with this buffer. After the elution of prothrombin, the column was washed with the same buffer until the effluent had an absorbance of less than 0.01. At this point, a linear gradient of sodium chloride was applied to the column to elute the factor X. The gradient was formed by the presence of 1400 ml of 0.025 M (sodium) citrate, 0.5 M NaCl, pH 6, in the reservior and 1400 ml of 0.025 M (sodium) citrate, 0.1 M NaCl, pH 6, in the mixing chamber. The flow rate of the column was usually 80 ml/hr.

A typical separation of bovine prothrombin and factor X effected by the DEAE-cellulose column is shown in Fig. 3. After application of the prothrombin-factor X containing sample to the column, washing the column with 0.025 M (sodium) citrate buffer, pH 6, results in the elution of peak A. Sodium dodecyl sulfate gel electrophoresis of the nonreduced and reduced samples from this peak revealed a major protein component (about 85%, by visual examination) composed of a single chain and

---

[33] Soybean trypsin inhibitor (0.1 mg/ml) and benzamidine HCl (0.01 M) may also be added in addition to the diisopropyl phosphorofluoridate.

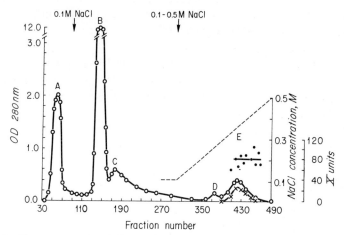

Fig. 3. Chromatograph of the 60% (NH₄)₂SO₄ precipitate on DEAE-cellulose. Absorbance at 280 nm, and factor X activity is plotted vs fraction number. From S. P. Bajaj and K. G. Mann, *J. Biol. Chem.* **248**, 7729 (1973).

having an apparent molecular weight of 55,000 ± 5000 (reduced sample). The ionic strength of the eluting buffer was then increased by the addition of NaCl to 0.1 M and two partially resolved peaks (peaks B and C) were eluted. Both peaks contained prothrombin of identical specific activities (1300 ± 100 NIH thrombin units per milligram of protein). This elution pattern was characteristic of every preparation. However, occasionally the prothrombin peak C gave lower specific activity (about 1000 NIH thrombin units per milligram of protein). When proteins from peaks B and C were rechromatographed separately on a DEAE-cellulose column under the conditions described above, two partially resolved peaks were again eluted. This suggests that the partial resolution obtained is an artifact. These two prothrombin peaks, B and C, do not contain detectable levels of thrombin or factor VII, IX, or X. Application of the gradient results in the elution of peaks D and E. Peak E contains factor X. Electrophoretic analysis of the factor X obtained at this stage indicates that it is about 80% homogeneous. The factor X obtained may be purified to homogeneity by DEAE-Sephadex chromatography.[30] The yield of prothrombin is about 50%, based on the concentration of the protein in plasma.

Concentration and Storage of the Products

The prothrombin and factor X purified by this procedure are concentrated by the addition of solid ammonium sulfate to 80% saturation to the pooled fractions. The resulting precipitate is dissolved in 50% glycerol water (v/v) and stored at —20°. Both proteins are stable for at least a year under these storage conditions.

The prothrombin and factor X isolated by these procedures are electrophoretically and immunologically homogeneous, and effectively devoid of contaminating factor activities when assayed for presumed contamination by either prothrombin or factor X.

Prothrombin from Other Sources

The procedure described here for bovine prothrombin has also been used for the isolation of equine prothrombin without modification of the procedure, and also for the isolation of human and canine prothrombin with slight modifications.[34]

Human prothrombin has been isolated both from fresh human plasma anticoagulated with acid–citrate–dextrose A and from American Red Cross lyophilized factor IX concentrates. Isolations of human prothrombin from commercial prothrombin concentrates have not been successful in our hands, owing to the presence of activated material in these concentrates.[35] In the case of isolation from fresh human plasma anticoagulated with acid–citrate–dextrose A, the initial step is to adjust the pH of the starting plasma from its initial pH 6.6 to pH 8.6 with dilute sodium hydroxide. Succeeding steps of the purification procedure for human prothrombin are the same as for bovine prothrombin with the exception that the behavior of human prothrombin on the DEAE-column is somewhat different than from that of the bovine material. Human prothrombin is eluted from the DEAE-column by elution with 0.15 M sodium chloride present in the buffer, rather than with 0.1 M sodium chloride, as is the case with bovine prothrombin. Figure 4 presents a typical elution profile from DEAE-cellulose of a human prothrombin preparation when the starting material was fresh human plasma. The product obtained in human prothrombin isolation is electrophoretically homogeneous; however, unlike the bovine preparations it is contaminated with a small amount of factor X and factor VII (as detected by activity measurements).

Canine prothrombin isolation is carried out in exactly the same way as the bovine material (including anticoagulation) with the exception of the chromatographic steps. Like human prothrombin, canine prothrombin is eluted at a slightly higher ionic strength from the DEAE-cellulose column, being eluted at an ionic strength of 0.15 M sodium chloride. The

[34] R. J. Butkowski, M.S. Thesis, Univ. of Minnesota, Minneapolis, 1974.
[35] P. M. Blatt, R. L. Lundblad, H. S. Kingdon, G. McLean, and H. R. Roberts, *Ann. Intern. Med.* **81**, 766 (1974).

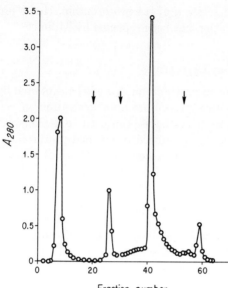

Fɪɢ. 4. Chromatograph of human prothrombin (60% $(NH_4)_2SO_4$ precipitate) on DEAE-cellulose. After the initial wash peak (eluted with 0.025 M sodium citrate pH 6.0), the buffer was changed to 0.025 M sodium citrate, 0.1 M NaCl, pH 6.0 (fraction 20). The buffer is then changed to 0.025 M Na citrate, 0.15 M NaCl, pH 6.0, to elute the prothrombin (fraction 40–50).

canine product is free of factor X, and in fact the canine factor X can be obtained by further elution of this column with the identical gradient used for the bovine factor X isolation. Canine prothrombin as isolated exists as two components that differ in molecular weight by about 2000. The amino acids deleted in the smaller canine prothrombin are from the carboxyl terminal.[34]

The isolation of human prothrombin has been reported by Lanchantin *et al.*[36] and by Kisiel and Hanahan.[37] The latter authors made use of preparative electrophoresis to remove the contaminating vitamin K-dependent factors from human prothrombin. Moore *et al*[29] and Zubairov *et al.*[38] have reported the isolation of canine prothrombin, and Li and Olson[39]

[36] G. F. Lanchantin, J. A. Friedmann, J. DeGroot, and J. W. Mehl, *J. Biol. Chem.* **238**, 238 (1963).
[37] W. Kisiel and D. J. Hanahan, *Biochim. Biophys. Acta* **304**, 703 (1973).
[38] D. M. Zubairov, V. N. Timerbaev, and V. M. Menshov, *Biochemistry (USSR)* **33**, 5 (1968).
[39] L. F. Li and R. E. Olson, *J. Biol. Chem.* **242**, 5611 (1967).

have reported the isolation of rat prothrombin. Horse prothrombin isolation and crystallization has been reported by Miller.[40]

Useful Prothrombin Derivatives

Two derivatives of prothrombin have been exploited by this laboratory to explore the process of prothrombin activation in complex biological systems. These derivatives have been used to demonstrate that the same four fragments that are produced in purified prothrombin activation systems are also produced in a complex biological system such as serum thromboplastin.[41,42]

Fluorescein Isothiocyanate Labeling[43]

Prothrombin may be labeled by a modification of the method of Rinderknecht.[43] The protein sample is brought to pH 8 by the addition of solid sodium bicarbonate and a small aliquot of fluorescein isothiocyanate, 10% on Celite (available from CalBiochem) is added. The reaction mixture is stirred for 10–15 min, and glycine is added to destroy the excess reagent. The protein sample is then centrifuged to remove the Celite, and freed from excess dye by gel filtration on a Bio-Gel P2 column, or by precipitation with ammonium sulfate at 80% saturation.

[^3H-*sialyl*]*Prothrombin*

Prothrombin[42] may be labeled by the procedure of Van Lenten and Ashwell.[44] Prothrombin to be labeled is dialyzed into 0.1 M sodium acetate, 0.15 M sodium chloride, pH 5.6, at 4°. A 9- to 10-fold molar excess of 12 mM sodium metaperiodate with respect to the sialic acid is added to the sample in an ice bath. The reaction is allowed to proceed for 10 min at 0°. At that time, a 100-fold molar excess of ethylene glycol is added to stop the reaction. The reaction mixture is then dialyzed against 0.05 M sodium phosphate, 0.15 M sodium chloride, pH 7.4, for 12 hr. After dialysis, an equimolar quantity (relative to prothrombin) of tritiated sodium borohydride (7.2 Ci/mmole) is added in 0.2 ml of 0.01 M sodium hydroxide. This solution is warmed to room temperature, and the reaction is allowed to proceed for 30 min. In order to ensure complete

[40] K. D. Miller, this series Vol. 19, p. 140.
[41] D. N. Fass and K. G. Mann, *J. Biol. Chem.* **248**, 3280 (1973).
[42] R. J. Butkowski, S. P. Bajaj, and K. G. Mann, *J. Biol. Chem.* **249**, 6562 (1974).
[43] H. Rinderknecht, *Experientia* **16**, 430 (1960).
[44] L. Van Lenten and G. Ashwell, *J. Biol. Chem.* **246**, 1889 (1970).

reduction, a 3-fold molar excess of unlabeled sodium borohydride is then added, and the reaction is continued for an additional 30 min. Excess reagent is removed by dialysis against 0.1 M sodium acetate, 0.15 M sodium chloride, pH 5.6. The labeled prothrombin is recovered by precipitation with ammonium sulfate at 80% saturation and stored at −20° in 50% glycerol. The radiolabeled preparations possess 70–80% of the biological activity of the nonlabeled control. Roughly 90% of the label will be contained within the modified sialic acid, which has been identified as the 7-carbon analog of N-acetylneuraminic acid: [7-³H]5-acetimido-3,5-dideoxy-L-arabino-2-heptulosonic acid.[42]

Congenitally Abnormal Prothrombins

Five abnormal prothrombins have been described: prothrombin Cardeza,[45] prothrombin Barcelona,[46] prothrombin Brussels,[47] prothrombin San Juan,[48] and prothrombin Padua.[49] All five of the reported prothrombin variants appear to be different, the only common feature being a large discrepancy between the biological activity produced on prothrombin activation and the quantity of prothrombin zymogen as identified by immunochemical techniques. In the case of prothrombin Cardeza, an abnormal prothrombin fragment is produced during prothrombin activation without thrombin production. Prothrombin Padua and prothrombin San Juan also generate abnormal prothrombin activation fragments during activation. In the case of prothrombin Barcelona, the defect appears to be in the ability of the variant molecule to be activated by the prothrombinase complex rapidly, and in the case of prothrombin Brussels, a competitive inhibitor to normal prothrombin activation is produced.

Physical and Chemical Properties

Physical and chemical properties of prothrombin as isolated from all species are very similar with the exception of prothrombin isolated from the rat.[39] Values for the molecular weight of bovine and human prothrombin vary to some extent, partly owing to different values used for partial specific volume calculations (0.683[38] to 0.721[23] ml/g). Most values for the molecular weight, however, hover about numbers between 69,000 and

[45] S. S. Shapiro, J. Martinez, and R. R. Holburn, *J. Clin. Invest.* **48**, 2251 (1969).
[46] F. Josso, J. Monasterio De Sanchez, J. M. Lavergne, D. Ménaché, and J. P. Soulier, *Blood* **38**, 9 (1971).
[47] M. J. P. Kahn and A. Goverts, *Thromb. Res.* **5**, 141 (1974).
[48] S. S. Shapiro, N. Maldonald, J. Fradesa, and S. McCord, *J. Clin. Invest.* **33**, 6 (1974).
[49] G. Girolami, A. Bareggi, A. Brunetti, and A. Stecche, *J. Lab. Clin. Med.* **84**, 654 (1974).

74,000.[22,23,50,51] In contrast, rat prothrombin has a molecular weight of 90,000.[52] The sedimentation coefficient of prothrombin has been reported from a high of 5.3 S[53] to a low of 4.6 S.[25] Evaluation of the sedimentation coefficient is made somewhat complex because the protein polymerizes in solvents of low ionic strength.[22] In addition, the choice of solvent in these sedimentation studies also complicates the picture, since it is clear that the chromatographic behavior of prothrombin in citrate and phosphate buffers of equivalent ionic strength is different.[22,30] Similarly, the question of the symmetry of the molecule in solution is complex because of its tendency to aggregate. Cox and Hanahan[22] calculated a frictional ratio of 1.45 from both sedimentation and diffusion data that would indicate the molecule to be assymetrical, whereas Ingwall and Scheraga[23] obtained a value for the intrinsic viscosity of prothrombin that would indicate the molecule to be symmetrical (3.4 ml/g).

A comparison of the amino acid compositions of bovine and human prothrombin obtained from several laboratories is presented in Table I. In general, there is reasonable agreement on the compositions of bovine and human prothrombin, and in addition, the compositions of these two proteins are not unexpectedly similar.

Prothrombin is a glycoprotein which most likely contains between 8 and 10% carbohydrate. Table II provides a comparison of the carbohydrate data obtained in a variety of laboratories. This carbohydrate appears to be distributed in three carbohydrate side chains.[42,54]

Prothrombin Activation Component Nomenclature

Both human and bovine prothrombin activations have been studied by a variety of laboratories,[50-52,54a-60a] and a relatively consistent picture

[50] C. M. Heldebrant, R. J. Butkowski, S. P. Bajaj, and K. G. Mann, *J. Biol. Chem.* **248**, 7149 (1973).
[51] W. G. Owen, C. T. Esmon, and C. M. Jackson, *J. Biol. Chem.* **249**, 594 (1974).
[52] J. J. Morrissey and R. E. Olson, *Fed. Proc., Fed. Am. Soc. Exp. Biol.* **32**, 3 (abstr.) 317 (1973).
[53] W. H. Seegers, E. Marciniak, R. K. Kipfer, and K. Yasunaga, *Arch. Biochem. Biophys.* **121**, 372 (1967).
[54] B. G. Hudson, C. M. Heldebrant, and K. G. Mann, *Thromb. Res.* **6**, 215 (1975).
[54a] G. F. Lanchantin, J. A. Friedmann, and D. W. Hart *J. Biol. Chem.* **243**, 476 (1968).
[55] D. L. Aronson and D. Ménaché, *Biochemistry* **5**, 2635 (1966).
[56] K. S. Stenn, E. R. Blout, *Biochemistry* **11**, 4502 (1972).
[57] H. Pirkle and I. Theodore, *Thromb. Res.* **5**, 511 (1974).
[58] W. Kisiel and D. J. Hanahan, *Biochim. Biophys. Acta* **329**, 221 (1973).
[59] K. G. Mann, C. M. Heldebrant, and D. N. Fass, *Fed. Proc., Fed. Am. Soc. Exp. Biol.* **30**, (abstr.) 539 (1971).
[60] K. G. Mann, C. M. Heldebrant, and D. N. Fass, *J. Biol. Chem.* **246**, 6106 (1971).
[60a] M. R. Downing, R. J. Butkowski, M. Clark, and K. G. Mann, *J. Biol. Chem.* **250**, 8897 (1975).

TABLE I
PROTHROMBIN AMINO ACID COMPOSITION

	Bovine prothrombin				Human prothrombin		
Amino acid	a	b	c	d	e	f	g
Aspartic acid	59	64	68	62	59	56	52
Threonine	27	32	32	29	32	34	39
Serine	32	43	41	41	36	36	38
Glutamic acid	71	78	84	74	77	73	70
Proline	35	38	37	32	32	31	32
Cysteine	17	19	20	20	21	12	16
Glycine	46	56	57	48	46	48	44
Alanine	33	38	38	34	32	38	35
Valine	35	40	38	32	32	32	30
Methionine	5	7	5	5	8	7	7
Isoleucine	18	21	24	17	21	21	22
Leucine	46	48	44	45	41	41	39
Tyrosine	19	22	18	18	18	19	21
Phenylalanine	17	22	22	19	18	24	21
Lysine	30	31	34	31	32	29	27
Histidine	9	12	10	8	11	10	9
Arginine	42	38	41	43	31	38	36
Tryptophan	12	13	18	19	11	13	21
% Carbohydrate	8.2%	(8.2%)	2.3%	9.6%	(8.2%)	(8.2%)	(8.2%)

[a] This laboratory. Based on C. M. Heldebrant, R. J. Butkowski, S. P. Bajaj, and
K. G. Mann, J. Biol. Chem **248**, 7149 (1973), corrected to 8.2% carbohydrate.
[b] W. H. Seegers, E. Marciniak, R. W. Kipfer, and K. Yasunaga, Arch. Biochem.
Biophys. **121**, 372 (1967), adjusted to a molecular weight of 70,000 and 8.2%
carbohydrate.
[c] J. S. Ingwall and H. A. Scheraga, Biochemistry **8**, 1960 (1969), molecular weight
74,000.
[d] W. G. Owen, C. T. Esmon, and C. M. Jackson, J. Biol. Chem. **249**, 594 (1974),
molecular weight 74,000.
[e] This laboratory. Based on M. R. Downing, R. J. Butowski, Clark, M. and K.
G. Mann, J. Biol. Chem. **250**, 8897 (1975), adjusted to 8.2% carbohydrate.
[f] W. Kisiel and D. J. Hanahan, Biochem. Biophys. Acta **304**, 703 (1973), adjusted,
to 8.2% carbohydrate.
[g] G. F. Lanchantin, J. A. Friedmann, J. DeGroot, and J. W. Mehl, J. Biol. Chem.
238, 238 (1963), adjusted to 8.2% carbohydrate.

of the different fragments produced during activation of the prothrombin
molecule has been accumulated. In contrast, however, the nomenclature
for prothrombin fragments has not been consistent. This is partly owing
to the fact that only in a few cases have physical and chemical studies
been carried out to the degree that direct assignments of particular com-

TABLE II
CARBOHYDRATE COMPOSITION OF BOVINE PROTHROMBIN IN
TERMS OF WEIGHT PERCENT

Component	a	b	c	d	e
Neutral sugar	2.9	3.6	3.0	3.8	4.5
Glucosamine	2.5	2.5	1.8	3.3	5.5
Galactosamine	0.1	0	0.2	—	0
Sialic acid	2.7	3.2	4.2	2.8	4.7
Total carbohydrate	8.2	9.3	9.2	9.9	15.0

[a] This laboratory: B. G. Hudson, C. M. Heldebrant, and K. G. Mann, *Thromb. Res.* **6**, 215 (1975).
[b] W. G. Owen, C. T. Esmon, and C. M. Jackson, *J. Biol. Chem.* **249**, 594 (1974).
[c] S. Magnusson, *Ark Kemi* **23**, 285 (1965).
[d] G. H. Tishkoff, L. C. Williamson, and D. M. Brown, *J. Biol. Chem.* **243**, 4151 (1968).
[e] G. L. Nelsestuen and J. W. Suttie, *J. Biol. Chem.* **249**, 6096 (1972).

ponents can be made. Our laboratory was one of the first[59,60] to apply the sodium dodecyl sulfate (SDS) electrophoretic technique[61] to study the changes that occur during the activation of prothrombin. This technique has provided a primary tool for many investigators for the identifications of prothrombin activation components. Thus, coidentities in component nomenclature can frequently be arrived at by inspection of the SDS gel electrophoretic data. Similarly, sequence data, when in agreement with published observations, have been used to establish the identities. Figure 5 provides an illustration of the fragmentation pattern of human prothrombin when it is subjected to factor X_a in the absence of thrombin inhibitors. It can be seen that four components are produced during the activation process when the analysis is conducted by SDS gel electrophoresis. The nomenclature of these different components as devised by different laboratories is presented in Table III.

In order to overcome the obvious difficulties of multiple nomenclatures for the same components of prothrombin, the question of prothrombin activation component nomenclature was placed under the scrutiny of a Task Force on Blood Clotting Zymogens and Zymogen Intermediates of the International Society on Thrombosis and Hemostasis, and the nomenclature question was examined at the Fifth Congress of this Society, held in Paris in July, 1975. As a consequence of this meeting, a tentative

[61] K. Weber and M. Osborn, *J. Biol. Chem.* **244**, 4406 (1969).

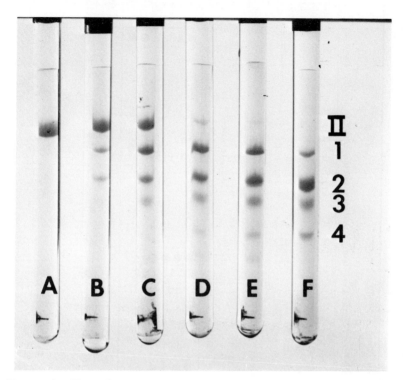

Fɪɢ. 5. An illustration of the (human) prothrombin activation process when evaluated by the sodium dodecyl sulfate electrophoretic technique. The vertical column (II, 1, 2, 3, 4) refers to the activation component nomenclature previously used by this laboratory, and the letter designations on the gel correspond to the time of activation. Current nomenclature (which is used in subsequent sections of the text), refers to these components as prothrombin, prethrombin 1, prethrombin 2, prothrombin fragment 1, and prothrombin fragment 2, respectively. Reproduced from R. J. Butkowski, M. S. Thesis, Univ. of Minnesota, Minneapolis, 1974.

standardized nomenclature system was devised that met with the approval of nearly all investigators who have published in the prothrombin field. This nomenclature system is also presented in Table III; it is the nomenclature system used throughout this manuscript in identifying the products of prothrombin activation.

While the four components identified are those produced on prothrombin activation by factor X_a in any activation system in which a thrombin inhibitor is not present, the addition of thrombin inhibitors to the activation system results in the generation of a new fragment. The

TABLE III

Prothrombin Activation Component Nomenclature

Currently accepted component nomenclature	Component nomenclature previously used in this laboratory	Equivalent component nomenclature by other laboratories			
Prothrombin	Prothrombin[a]	Prothrombin[b,c]	P1[d]	Prothrombin[e]	Prothrombin[f]
Prethrombin 1	Intermediate 1	Intermediate 1	P2	Neoprothrombin-S	Prethrombin
Prethrombin 2	Intermediate 2	Intermediate 2	P3	Neoprothrombin-T	Prethrombin E
Prothrombin fragment 1	Intermediate 3	Fragment 1	Fa	A fragment	PR fragment
Prothrombin fragment 2	Intermediate 4	Fragment 2	Fb	S fragment	O fragment

Basis for Component Identity Conclusion

Laboratory	Physical and chemical properties	Activation studies	Sodium dodecyl sulfate gel electrophoresis	NH2-terminal sequence	Model
a	●	●	●	●	●
b	●	●	●		●
c		●	●		
d		●	●		
e				●	●
f				●[g]	●

[a] C. M. Heldebrant, R. J. Butowski, S. P. Bajaj, and K. G. Mann, *J. Biol. Chem.* **248**, 7149 (1973); C. M. Heldebrant, C. Noyes, H. S. Kingdon, and J. C. Mann, *Biochem. Biophys. Res. Commun.* **54**, 155 (1973).

[b] W. G. Owen, C. T. Esmon, and C. M. Jackson, *J. Biol. Chem.* **249**, 594 (1974); C. T. Esmon, W. G. Owen, and C. M. Jackson; *J. Biol. Chem.* **249**, 606 (1974).

[c] W. Kisiel and D. J. Hanahan, *Biochim. Biophys. Acta* **329**, 221 (1973); W. Kisiel and D. J. Hanahan, *Biochem. Biophys. Res. Commun.* **59**, 570 (1974).

[d] K. S. Stenn and E. R. Blout, *Biochemistry* **11**, 4502 (1972).

[e] S. Magnusson, L. Sottrup-Jensen, T.-E. Petersen, and H. Claeys, "Boerhaave Course on Synthesis of Prothrombin and Related Coagulation Factors," Leiden, MP4, 1974; *Abstr. Cold Spring Harbor Meet. Proteases Biol. Control*, Sept. 10, 1974, p. 7.

[f] J. Reuterby, D. A. Walz, L. E. McCoy, and W. H. Seegers, *Thromb. Res.* **4**, 885 (1974).

[g] Of the O fragment.

work of Stenn and Blout,[56] and Esmon *et al.*,[62] and Kisiel and Hanahan[63] have provided evidence that when thrombin inhibitors are present, a new fragment is produced which has been designated F_x[56] and fragment $1\cdot2$[57,63] by these investigators. This fragment has also been produced in our laboratory using DFP and hirudin as thrombin inhibitors; using our nomenclature system it would be denoted intermediate 3-4.[60a]

Using the recently devised nomenclature system for prothrombin activation components from the Vth Congress of the International Society of Hemostasis and Thrombosis, this activation component would be designated prothrombin fragment $1\cdot2$.

Prothrombin Activation Components

Bovine Prethrombin 1 and Prothrombin Fragment 1

Prethrombin 1 and prothrombin fragment 1 are the sole products of bovine thrombin action on bovine prothrombin. Incubation of bovine prothrombin with a small amount of thrombin results only in the production of these two components (however, see section on Human Prothrombin). In a typical preparation of these two fragments, 200 mg of prothrombin in 30 ml was incubated with 200 units of thrombin in 0.144 M NaCl, 0.0168 M imidazole, pH 7.4. After 3 hr incubation, the reaction was terminated by the addition of DFP to a final concentration of 1 mM. The sample was then dialyzed vs 0.15 M sodium chloride, 0.02 M Tris, pH 7.4, and applied to a 2.5 × 50 cm DEAE-cellulose column equilibrated with the same buffer. Figure 6 presents the elution profile obtained for this chromatographic procedure. After application of the sample, the column was washed with starting buffer, which resulted in the elution of a large, protein containing peak. When no further protein was eluted with this buffer, the column was developed by means of a linear gradient formed by the presence of 500 ml of the initial buffer in the mixing flask, and 500 ml of 0.5 M sodium chloride, 0.6 M Tris, pH 7.4, in the reservoir. The first peak eluted from the column (fractions 10 to 40) corresponds to prethrombin 1. Two partially resolved components are eluted by the gradient. The shoulder corresponding to fractions 60–70 represents a small amount of prothrombin not converted to prethrombin 1 and prothrombin fragment 1, and the major peak (fractions 71 to 90) corresponds to prothrombin fragment 1. The prethrombin 1 and prothrombin fragment 1 protein pools obtained from the chromatographic

[62] C. T. Esmon, W. G. Owen, and C. M. Jackson, *J. Biol. Chem.* **249**, 606 (1974).
[63] W. Kisiel and D. J. Hanahan, *Biochem. Biophys. Res. Commun.* **59**, 570 (1974).

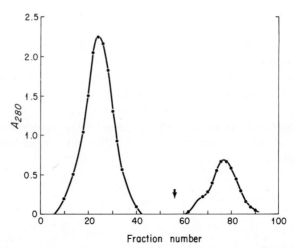

FIG. 6. Chromatograph of bovine prethrombin 1 and prothrombin fragment 1 on DEAE-cellulose. Absorbance at 280 nm is plotted vs fraction number. From C. M. Heldebrant, R. J. Butkowski, S. P. Bajaj, and K. G. Mann, *J. Biol. Chem.* **248**, 7149 (1973).

column were concentrated by precipitation with solid ammonium sulfate to 80% saturation. The individual proteins were then redissolved in 50% glycerol and stored at —20°. These two activation components are stable indefinitely when stored under these conditions.

Human Prethrombin 1 and Prothrombin Fragment 1

Human prethrombin 1 and prothrombin fragment 1 are not the sole products of thrombin treatment of human prothrombin. Prolonged treatment of human prothrombin with thrombin results in the slow production of two additional fragments which are similar when analyzed by electrophoresis in SDS to prethrombin 2 and prothrombin fragment 2. Therefore, it is essential that the reaction that is carried out with human prothrombin not be prolonged beyond 3 hr at a final concentration of 1 unit of thrombin per milligram of prothrombin. The separation of the human fragments, however, is entirely analogous to that used for bovine prothrombin fragment 1 and prethrombin 1.

Bovine Prethrombin 2 and Prothrombin Fragment 2

Treatment of bovine prethrombin 1 with factor X_a results in the production initially of prothrombin fragment 2 and prethrombin 2, and subsequently the production of α-thrombin from the prethrombin 2. This second step (prethrombin 2 → αIIa) proceeds at a rate which is roughly equivalent to the rate of production of prethrombin 2 from prethrombin 1 when factor X_a alone is used as a catalyst. In the presence of the complete prothrombinase catalyst (factor X_a–factor V–calcium and phospholipid) the rate of prethrombin 2 conversion to α-thrombin is faster than the rate of prethrombin 2 production from prethrombin 1. In contrast, the activation of prethrombin 1 with factor X_a in the presence of 25% sodium citrate results in a rapid production of prethrombin 2 and prothrombin fragment 2, and a slow subsequent production of α-thrombin. This alteration in the rates of bond cleavage proves useful in the isolation of prethrombin 2 and prothrombin fragment 2. By using factor X_a in 25% sodium citrate as the activating system for prethrombin 1, one can obtain prethrombin 2 and prothrombin fragment 2 nearly quantitatively without the complication of contamination by thrombin and thrombin degradation products.

To prepare prethrombin 2 and prothrombin fragment 2, prethrombin 1 (final concentration 1–10 mg/ml) in 25% trisodium citrate (w/v) is incubated with factor X_a[64] (final concentration 5–10 units/ml). During incubation, the production of thrombin by this system is evaluated by means of the standard NIH assay.[21] When 10 μl of the reaction mixture will give a clotting time of approximately 15 sec in the NIH thrombin assay (about 30 min), DFP is added to the reaction mixture to a final concentration of 10^{-2} M. The sample is then dialyzed into 0.15 M sodium chloride, 0.02 M Tris, pH 7.4, in the cold and subjected to chromatography on DEAE-cellulose and equilibrated with the same buffer. Figure 7 shows the chromatographic separation of prethrombin 2 and prothrombin fragment 2. Prethrombin 2 is eluted in the wash peak on the column, while prothrombin fragment 2 is subsequently eluted with 0.5 M sodium chloride, 0.066 M Tris, pH 7.4. Frequently because of the noncovalent association of prethrombin 2 and prothrombin fragment 2, prethrombin 2 obtained by this chromatographic procedure in the wash peak will be contaminated with prothrombin fragment 2. The second chromatographic separation is then required. The wash peak can be rechromatographed on sulfopropyl Sephadex C50-120. This chromatographic separation is essentially identical to the purification procedure used for α-thrombin. The wash peak is dissolved in or dialyzed vs 0.025 M sodium phosphate, pH 6.5, and applied to a 2.5 × 50 cm column of

[64] We have used factor X_a (a) activated from factor X with trypsin insolubilized on polyacrylamide[30]; (b) activated with insolubilized Russell's viper venom; (c) insolubilized factor X_a prepared by coupling factor X to Sepharose and activating the insoluble zymogen and Russell's viper venom.

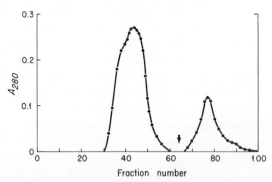

FIG. 7. Chromatograph of bovine prethrombin 2 and prothrombin fragment 2 on DEAE-cellulose. Absorbance at 280 nm is plotted vs fraction number. From C. M. Heldebrant, R. J. Butkowski, S. P. Bajaj, and K. G. Mann, *J. Biol. Chem.* **248**, 7149 (1973).

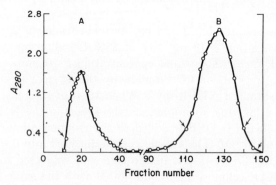

FIG. 8. Rechromatography of bovine prethrombin 2 and prothrombin fragment 2 on sulfopropyl-Sephadex C50-120. Absorbance at 280 nm is plotted against fraction number. Peak A corresponds to prothrombin fragment 2 and is eluted by the wash buffer, 0.025 M sodium phosphate, pH 6.5; peak B corresponds to prethrombin 2 and was eluted with 0.25 M sodium phosphate pH 6.5. From R. J. Butkowski, M. S. Thesis, Univ. of Minnesota, Minneapolis, 1974.

SPC50-120 equilibrated with the same buffer. Prothrombin fragment 2 is eluted in the wash peak, while prethrombin 2 is eluted with 0.25 M sodium phosphate pH 6.5. This separation is presented in Fig. 8.

Human Prothrombin Fragment 2 and Prethrombin 2

Human prethrombin 2 and prothrombin fragment 2 can be prepared in the same way as the bovine activation components with one primary exception: human thrombin, when it is produced, cleaves at the amino terminal of human prethrombin 2, deleting the first 13 residues.[60a] Thus, precautions must be taken to block the production of thrombin during the preparation of this component. We have included DFP (10^{-3} M) or hirudin at a concentration of 100 μg/ml to block the back reaction of thrombin on human prethrombin 2.

If the human prethrombin 1 conversion is carried out so that thrombin activity is generated, the product isolated will not be prethrombin 2, but rather prethrombin 2', from which the first 13 residues have been deleted. This component is designated in the current nomenclature system as prethrombin $2'_{(des\ R1-R13)}$.

Human prothrombin fragment 1·2 can be prepared from prothrombin analogously with human prothrombin, using hirudin (2-fold molar excess) as the thrombin inhibitor. The solvent in this case is 0.02 M Tris, 0.15 M NaCl, pH 7.4.

Immunological Properties of Prothrombin Activation Components

The first successful antibody preparations prepared against prothrombin activation components were reported by Shapiro.[65] We have recently prepared antibodies to prothrombin and the activation fragments, as well as thrombin, by emulsifying the antigen in Freund's adjuvant containing killed tubercule bacilli. Final concentrations of the antigen in each case was 0.1 mg/ml and of the bacilli was 5 mg/ml. We have used rabbits, chickens, and burros to produce the antisera, and the animals in each case received weekly subcutaneous injections of 0.2 ml of these preparations. The antibodies have been evaluated by immunodiffusion[66,67] as well as hemagglutination inhibition, and more recently by radioimmunoassays for prethrombin 1 and prothrombin fragment 1. In our hands,[66] the appropriate immunological recognition between the various prothrombin fragments has been observed in all assays. Antisera prepared against prethrombin 1 recognizes prothrombin, prethrombin 1, prothrombin fragment 2, and prethrombin 2, and antisera directed against prothrombin fragment 1 recognize only prothrombin fragment 1 and prothrombin. Similarly, antisera prepared against prothrombin fragment 2 recognize prethrombin 1, prothrombin, and prothrombin fragment 2, but not prethrombin 2 nor prothrombin fragment 1. Antisera prepared against prethrombin 2 recognize prothrombin and prethrombin 1, but not prothrombin fragment 2 or prothrombin fragment 1. A recent brief communication by Hewett-Emmett et al.[68] has reported immunological

[65] S. Shapiro, Science 162, 127 (1968).
[66] A. H. Auernheimer and F. O. Atchley, Am. J. Clin. Pathol. 38, 548 (1962).
[67] C. Taswell, F. C. McDuffie, and K. G. Mann. Immunochemistry 12, 339 (1975).
[68] D. Hewett-Emmett, L. E. McCoy, H. I. Hassoura, J. Reuterby, D. A. Walz, and W. H. Seegers, Thromb. Res. 5, 421 (1974).

TABLE IV
Physical Properties of Bovine Prothrombin and Its Activation Components

	Prothrombin	Prethrombin 1	Prethrombin 2	Prothrombin fragment 1	Prothrombin fragment 2
			Molecular weights		
a	70,000	51,000[e]	41,000	23,000	13,000
b	74,000	49,700[f]	37,000	24,000	12,900
			Partial specific volumes		
a	0.708	0.719	0.728	0.682	0.704
b	0.711	0.720	—	0.692	0.705
			Extinction coefficient ($E_{280}^{1\%}$)		
a	14.4[c]	16.4	19.5[d]	10.5	12.5
b	14.4	16.1	15.8	10.1	13.8

[a] This laboratory: C. M. Heldebrant, R. J. Butkowski, S. P. Bajaj, and K. G. Mann, *J. Biol. Chem.* **248**, 7149 (1973).
[b] W. G. Owen, C. T. Esmon, and C. M. Jackson, *J. Biol. Chem.* **249**, 594 (1974).
[c] A. C. Cox and D. J. Hanahan, *Biochim. Biophys. Acta* **207**, 49 (1970).
[d] D. J. Winzor and H. A. Scheraga, *J. Phys. Chem.* **68**, 338 (1964).
[e] 60,300–65,000 on sodium dodecyl sulfate electrophoresis.
[f] 61,000 on sodium dodecyl sulfate electrophoresis.

cross-reactivity between prothrombin fragment 1 and prothrombin fragment 2. However, no primary data were presented.

Physical and Chemical Properties of Bovine Prothrombin Activation Components

The physical properties of the bovine prothrombin activation components in terms of molecular weight, partial specific volume, and extinction coefficients are presented in Table IV. Remarkably good agreement is obtained by the two laboratories[50,51] that have provided properties for these fragments. Table V presents the amino acid compositions of bovine prethrombin 1 and prothrombin fragment 1, and preliminary composition data for human prethrombin fragment 1, and Table VI presents similar data for bovine prethrombin 2 (α-thrombin) and prothrombin fragment 2, and human prethrombin 2 and prothrombin fragment 2.

Quantitative amino-terminal analysis[50] and amino terminal sequence data[69] from this laboratory initially allowed the orientation of the bovine prothrombin activation components within the prothrombin primary

[69] C. M. Heldebrant, C. Noyes, H. S. Kingdon, and K. G. Mann, *Biochem. Biophys. Res. Commun.* **54**, 155 (1974).

TABLE V
PRETHROMBIN 1 AND PROTHROMBIN FRAGMENT 1 AMINO ACID COMPOSITION

| Amino acid | Prethrombin 1 | | | | Prothrombin fragment 1 | | |
| | Bovine | | | Human | Bovine | | Human |
	a	b	c	d	a	b	d
Aspartic acid	46	46	48	47	16	16	16
Threonine	17	17	21	18	9	10	15
Serine	22	27	26	24	10	14	11
Glutamic acid	48	49	53	58	24	23	25
Proline	24	22	28	20	10	10	9
Cysteine	9	12	11	14	10	9	7
Glycine	35	38	38	39	12	12	12
Alanine	24	24	26	25	11	10	13
Valine	25	23	28	25	10	9	9
Methionine	4	5	5	7	1	1	2
Isoleucine	14	14	15	17	4	4	4
Leucine	35	36	36	34	11	10	9
Tyrosine	15	14	17	15	6	4	5
Phenylalanine	15	16	17	14	5	4	5
Lysine	22	25	26	26	6	5	6
Histidine	5	6	8	8	2	2	3
Arginine	29	29	30	28	12	14	12
Tryptophan	8	15	13	9	4	5	2
% Carbohydrate	3.75	4.5	(3.75)	(3.75)	16.6	20	(16.6)

[a] This laboratory: C. M. Heldebrant, R. J. Butkowski, S. P. Bajaj, and K. G. Mann, *J. Biol. Chem.* **248**, 7149 (1973).
[b] W. G. Owen, C. T. Esmon, and C. M. Jackson, *J. Biol. Chem.* **249**, 594 (1974).
[c] W. H. Seegers, E. Marciniak, R. K. Kipfer, and K. Yasunaga, *Arch. Biochem. Biophys.* **121**, 372 (1967). Calculated assuming 3.75 % carbohydrate and a molecular weight of 51,000.
[d] This laboratory: M. R. Downing, R. J. Butkowski, M. Clark and K. G. Mann, *J. Biol. Chem.* **250**, 8897 (1975). Calculated assuming the same carbohydrate composition as the bovine material.

structure. Prothrombin fragment 1 corresponds to the amino terminal segment of prothrombin, while prethrombin 1 represents the carboxyl-terminal portion of the prothrombin molecule. Prothrombin fragment 2 represents the amino terminal end of prethrombin 1, while prethrombin 2 represents the carboxyl-terminal end of prethrombin 1. Prethrombin 2 amino-terminal sequence corresponds to the amino terminal sequence of bovine α-thrombin A-chain as presented by Magnusson.[70] The amino-terminal sequences of prothrombin, thrombin, and the activation com-

[70] S. Magnusson, *Thromb. Diath. Haemorrh.*, Suppl. **38**, 97 (1970).

TABLE VI
PRETHROMBIN 2 (α-THROMBIN) AND PROTHROMBIN
FRAGMENT 2 AMINO ACID COMPOSITIONS

| Amino acid | Prethrombin 2 (α-thrombin) | | | Prothrombin fragment 2 | | | |
| | Bovine | | Human | Bovine | | | Human |
	a	b	c	a	b	d	c
Aspartic acid	35(30)	32(30)	34	18	18	17	14
Threonine	16(14)	13(13)	15	4	5	5	4
Serine	17(14)	15(18)	19	8	10	9	9
Glutamic acid	40(40)	32(34)	41	17	15	14	18
Proline	20(19)	16(15)	18	10	10	8	6
Cysteine	6(6)	—(8)	9	3	5	4	5
Glycine	30(27)	26(25)	29	12	12	10	15
Alanine	17(15)	15(15)	17	10	10	10	10
Valine	25(21)	18(20)	18	5	5	4	9
Methionine	5(5)	4(4)	7	0	0	0	0
Isoleucine	15(13)	12(14)	17	1	1	1	1
Leucine	32(28)	27(26)	29	9	9	9	9
Tyrosine	13(11)	9(10)	12	3	4	4	3
Phenylalanine	15(16)	13(12)	11	3	3	3	3
Lysine	25(22)	22(23)	23	3	2	2	4
Histidine	7(11)	6(7)	6	1	0	0	2
Arginine	25(20)	21(20)	25	7	8	8	5
Tryptophan	8(8)	—(11)	8	—	4	2	2
% Carbohydrate	4.67%	—(6.3)	—	0	0	0	0

[a] This laboratory: C. M. Heldebrant, R. J. Butkowski, S. P. Bajaj, and K. G. Mann, *J. Biol. Chem.* **248**, 7149 (1973).
[b] W. G. Owen, C. T. Esmon, and C. M. Jackson, *J. Biol. Chem.* **249**, 594 (1974).
[c] This laboratory: M. R. Downing, R. J. Butkowski, M. Clark and K. G. Mann, *J. Biol. Chem.* **250**, 8897 (1975).
[d] Calculated from the sequence data of J. Reuterby, D. A. Walz, L. E. McCoy, and W. H. Seegers, *Thromb. Res.* **4**, 885 (1974).

ponents are presented in Table VII. A tentative complete sequence for prothrombin fragment 2 has been reported by Reuterby et al.[71] and a tentative complete sequence of prothrombin has been reported by Magnusson.[72,73]

[71] J. Reuterby, D. A. Walz, L. E. McCoy, and W. H. Seegers, *Thromb. Res.* **4**, 885 (1974).
[72] S. Magnusson, L. Sottrup-Jensen, T.-E. Peterson, and H. Claeys, "Boerhaave Course on Synthesis of Prothrombin and Related Coagulation Factors," Leiden MP4, 1974.
[73] S. Magnusson, L. Sottrup-Jensen, T.-E. Petersen, and H. Claeys, *Abstr. Cold Spring Harbor Meet. Proteases Biol. Control,* Sept. 10, 1974, p. 7.

TABLE VII
NH₂-TERMINAL SEQUENCE

Prothrombin and prothrombin fragment 1

Bovine[a,b,c]
Ala Asn Lys Gly Phe Leu *Glu* Glu** Val Arg Lys Gly Asn Leu Glu Arg *Glu** Cys Leu
Human[d,e]
Ala Asn Thr Phe Leu Glu Glu Val Arg Lys Gly Asn Leu Glu Arg Glu . . .
(Glu* = γ-carboxyglutamic acid)

Prethrombin 1 and prothrombin fragment 2

Bovine[a,b,f]
Ser Gly Gly Ser Thr Thr Ser Gln Ser Pro Leu Leu Glu Thr Cys Val Pro Asp Arg
Human[d]
Ser Glu Gly Ser Ser Val Asn Leu Ser Pro Pro Leu Glu Gln Cys Val Pro Asp Arg

Prethrombin 2 and α-thrombin A chain

Bovine[a,b,g]
Thr Ser Glu Asn His Phe Glu Pro Phe Phe Asn Glu Lys Thr Phe Gly Ala Gly Glu
Human[d,h,i]
Thr Ala Thr Ser Glu Tyr Gln Thr Phe Phe Asn Pro Arg ↑ Thr Phe Gly Ser Gly Glu

j

[a] This laboratory: C. M. Heldebrant, C. Noyes, H. S. Kingdon, and K. G. Mann, *Biochem. Biophys. Res. Commun.* **54**, 155 (1973).

[b] Magnusson *et al.* [S. Magnusson, L. Sottrup-Jensen, T. E. Petersen, and H. Claeys, "Boerhaave Course on Synthesis of Prothrombin and Related Coagulation Factors, Leiden, MP4, 1974; *Abstr. Cold Spring Harbor Meet. Proteases Biol. Control*, Sept. 10, 1974, p. 7; S. Magnusson, L. Sottrup-Jensen, and T. E. Petersen, *FEBS Lett.* **44**, 189 (1974)] identified Glu₇, Glu₈, Glu₁₇ as γ-carboxyglutamic acid.

[c] Stenflo identified Glu₇ Glu₈ as γ-carboxyglutamic acid: J. Stenflo, *J. Biol. Chem.* **249**, 5527 (1974).

[d] This laboratory: M. R. Downing, R. J. Butkowski, M. Clark and K. G. Mann, *J. Biol. Chem.* **250**, 8897 (1975).

[e] H. Pirkle and I. Theodore, *Thromb. Res.* **2**, 461 (1973): residues 1–5.

[f] J. Reuterby, D. A. Walz, L. E. McCoy, and W. H. Seegers, *Thromb. Res.* **4**, 885 (1974).

[g] S. Magnusson; this series, Vol. 19, Page 157.

[h] A. R. Thompson, L. H. Ericsson, and D. L. Enfield, *Circulation* **50**, 292 (abstr.), 1119 (1974): human thrombin A-chain.

[i] D. A. Walz and W. H. Seegers, *Biochem. Biophys. Res. Commun.* **60**, 717 (1974): human thrombin A-chain.

[j] Point of thrombin cleavage in human intermediate 2. M. R. Downing, R. J. Butkowski, M. Clark, and K. G. Mann, *J. Biol. Chem.* **250**, 8897 (1975).

It has been mentioned previously that human prothrombin cleavage by thrombin produces, in addition to prethrombin 1 and prothrombin fragment 1 (the initial products), two additional products that are electrophoretically similar to prethrombin 2 and prothrombin fragment 2. These products arise as a result of the cleavage of the prethrombin 1 produced by thrombin at a position 13 residues removed from the amino-terminal of the factor X_a-produced human prethrombin 2. Studies by Thompson et al.[74] and by Walz and Seegers[75] on human thrombin A-chain indicate that the amino terminal sequence of human thrombin A-chain is homologous with bovine thrombin beginning with residue 14. Human prethrombin 2', as isolated in this laboratory from factor X_a activation of prethrombin 1 in the absence of thrombin inhibitors has the same sequence as that reported for human α-thrombin A-chain. However, if human prethrombin 1 is isolated from short-term thrombin treatment of human prothrombin and the resulting prethrombin 1 is subsequently factor X_a activated to prethrombin 2 and prothrombin fragment 2 in the presence of DFP or hirudin, a new fragment is isolated which begins with residues homologous to the bovine prethrombin 2 and α-thrombin A-chain. These data suggest that there are two sites of cleavage of human prothrombin by thrombin: one position analogous to the Arg-Ser bond in bovine prothrombin, and another, cleavage 13 residues removed from the amino terminal of the factor X_a, produced human prethrombin 2.

A summary of the sites of cleavage of human and bovine prothrombin during activation in a noninhibited system are presented in Fig. 9. The cleavage (b') indicated by the dotted line occurs only in human prothrombin. The carboxyl terminals of all the bonds cleaved are arginine,[11,72,73,76] and recent reports by Magnusson et al.[72,73] indicate that cleavages b and c occur after the sequence Ile-Gly-Glu-Arg.

Studies by Stenn and Blout[56] suggested that the fragments produced from prothrombin during activation by factor X_a or by thrombin were different. These authors proposed that bovine prothrombin can be activated by two pathways; one pathway, catalyzed by factor X_a, took place in two steps, while the second pathway was initiated by thrombin and took place in three steps. This proposal was confirmed by Esmon et al.[62] who, repeating the studies of Stenn and Blout,[56] showed that, in the presence of DFP factor X_a produced fragments consistent with cleavage only at positions b and c. These authors were able to show the isolation of the entire "pro" end of the molecule (prothrombin fragment 1 cova-

[74] A. R. Thompson, L. H. Ericsson, and D. L. Enfield, *Circulation* **50**, 292 (abstr) 1119 (1974).
[75] D. A. Walz and W. H. Seegers, *Biochem. Biophys. Res. Commun.* **60**, 717 (1974).
[76] H. Pirkle and I. Theodore, *Thromb. Res.* **5**, 511 (1974).

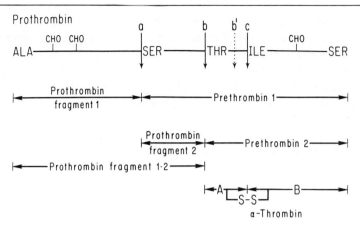

Fig. 9. Schematic representation of the orientation of prothrombin activation components within the prothrombin molecule. CHO represents a carbohydrate side chain. The letter designations a, b, and c represent the sites of cleavage of bovine and human prothrombin. An additional cleavage, b', occurs in human prothrombin. Currently accepted nomenclature for the pieces produced during prothrombin activation are also designated in the figure.

lently linked to prothrombin fragment 2) and prethrombin 2 (α-thrombin) as the products of the reaction. This study has recently been extended to human prothrombin by Kisiel and Hanahan,[63] who provided electrophoretic data which indicate that the same process occurs in human prothrombin activation. The studies with hirudin and human prothrombin have been confirmed by this laboratory.

Calcium and Vitamin K

The vitamin K-dependent blood coagulation factors (factor IX, factor VII, factor X, and prothrombin) all required calcium and phospholipid surfaces for either their activation to enzymes or their subsequent enzymic functions. With respect to prothrombin activation, Papahadjahopoulos and Hanahan[6] demonstrated that calcium was essential for maintaining the integrity of the prothrombin-activating complex. These authors showed that factor V and phospholipid form a complex in the absence of calcium, but that the binding of factor X_a to the complex is dependent upon the presence of calcium. Cole and associates[7] demonstrated that the complexing of factor X_a to phospholipid could occur in the absence of factor V, but not in the absence of calcium.

Ganrot and Niléhn[17] and Josso et al.[18] in 1968 demonstrated by immunochemical techniques that individuals treated with the vitamin K

antagonist drugs (coumarins) had in their plasma an abnormal prothrombin fraction. Suttie *et al.*[77] have isolated a prothrombin liver precursor in the warfarin-treated rat. These studies were extended by Stenflo and Ganrot[14] and Nelsestuen and Suttie,[15] who demonstrated that
prothrombin binds calcium and that the abnormal prothrombin antigen,
which is induced with vitamin K antagonists, does not bind calcium.
Equilibrium dialysis studies conducted by Stenflo and Ganrot[16] and in
our laboratory indicate that prothrombin has 10 or 11 calcium binding
sites. Work in our laboratory[78-80] indicates that these sites are distributed in the two components that are not on the pathway to thrombin,
namely, prothrombin fragment 1 and prothrombin fragment 2. Prothrombin fragment 1 has five or six sites, with a log K association of 3.7, and
prethrombin 1 has four or five sites that are weaker, having a log K association of about 2.5. The calcium-binding sites in prethrombin 1 are
contained in the amino-terminal prothrombin fragment 2 segment of
the molecule. Prethrombin 2, the carboxyl-terminal segment of the prothrombin molecule, and the immediate precursor of thrombin has no
affinity for calcium. The question of cooperativity between these sites in
prothrombin, as well as the precise number of sites, cannot be adequately
determined at the present time because of the relatively low affinities of
these molecules for calcium.

The nature of the vitamin K-dependent alteration in the prothrombin
molecule that confers upon it the tight calcium-binding sites has recently
been elucidated by a number of laboratories. Stenflo and associates[81,82]
and Nelsestuen and co-workers[83-85] have reported the isolation of peptides which contain the strong calcium-binding sites in prothrombin. The
peptide isolated by Stenflo contained the Glu-Glu sequence in position
7 and 8 of the amino terminal of bovine prothrombin. Both Nelsestuen
and Stenflo *et al.* have identified a unique amino acid present in pro-

[77] J. W. Suttie, G. A. Grant, C. T. Esmon, and D. V. Shah, *Mayo Clin. Proc.* **49**, 933 (1974).
[78] S. P. Bajaj, R. J. Butkowski, and K. G. Mann, *Fed. Proc. Fed. Am. Soc. Exp. Biol.* **33**, (abstr.) 1473, (1974).
[79] K. G. Mann and D. N. Fass, *Mayo Clin. Proc.* **49**, 929 (1974).
[80] S. P. Bajaj, R. J. Butkowski, and K. G. Mann, *J. Biol. Chem.* **250**, 2150 (1975).
[81] J. Stenflo, *J. Biol. Chem.* **249**, 5527 (1974).
[82] J. Stenflo, P. Fernlund, W. Egan, *et al., Proc. Natl. Acad. Sci. U.S.A.* **71**, 2730 (1974).
[83] J. B. Howard and G. L. Nelsestuen, *Fed. Proc., Fed. Am. Soc. Exp. Biol.* **33**, (abstr.) 1473 (1974).
[84] G. L. Nelsestuen, T. Zytkovicz, and J. B. Howard, *J. Biol. Chem.* **249**, 6347 (1974).
[85] G. L. Nelsestuen, T. H. Zytkovicz, and J. B. Bryant, *Mayo Clin. Proc.* **49**, 941 (1974).

thrombin as a result of vitamin K action. This unique amino acid corresponds to a γ-carboxyglutamic acid. This unusual residue has also recently been identified by Magnusson et al.,[86] who identified it in 11 of the first 42 positions in the bovine prothrombin amino-terminal sequence, namely positions 7, 8, 15, 17, 20, 21, 26, 27, 30, 33. The human prothrombin sequence reported by our laboratory shows a Glu-Glu sequence at positions 6 and 7 homologous to the γ-COOH Glu, γ-COOH Glu sequence at positions 7 and 8 in the bovine prothrombin sequence identified by both Stenflo[82] and Magnusson.[86]

The significance of the calcium-binding sites in prothrombin can be seen in studies of the kinetics of activation of prothrombin and the two thrombin precursors, prethrombin 1 and prethrombin 2. The addition of calcium to the factor X_a-catalyzed activation of prothrombin increases the rate of prothrombin activation by about 40-fold. The addition of phospholipid to this activation mixture increases the rate by another 2.5-fold, or about 100 times the rate of prothrombin activation in the presence of factor X_a alone. In contrast, the activation of prethrombin 1 and prethrombin 2 in the presence of calcium and/or calcium and phospholipid results in no enhancement of the rate. Thus, the deletion of the prothrombin fragment 1 segment of the molecule eliminates the calcium accelerating effect, and subsequently that of phospholipid in enhancing the rate of prothrombin activation. Calcium-dependent phospholipid binding to prothrombin and its fragments has been reported by Gitel et al.[87] These authors showed that prothrombin fragment 1 is the calcium-dependent phospholipid-binding segment of the prothrombin molecule.

The significance of the weak calcium-binding sites present in the prothrombin fragment 2 segment of the prothrombin molecule can be seen in studies of complete prothrombinase activation of prethrombin 1 and prethrombin 2. Although the addition of calcium and phospholipid to factor X_a as the catalyst for prethrombin 1 activation has no effect on the rate of thrombin production from prethrombin 1, factor V addition results in a dramatic increase in the thrombin production from prethrombin 1.[79,80,88] The stimulatory influence of factor V on the rate of prethrombin 1 activation is not observed in the absence of calcium.[79,80] Similarly, the rate of activation of prethrombin 2 is not enhanced by the addition of calcium, calcium–phospholipid, or calcium–phospholipid–factor V.

[86] S. Magnusson, L. Sottrup-Jensen, and T.-E. Petersen, FEBS Lett. 44, 189 (1974).
[87] S. N. Gitel, W. G. Owen, C. T. Esmon, et al., Proc. Natl. Acad. Sci. U.S.A. 70, 1344 (1973).
[88] C. M. Jackson, W. G. Owen, S. N. Gitel, and C. T. Esmon, Thromb. Diath. Haemorrh., Suppl. 57, 273 (1974).

Thus, the sensitivity to factor V acceleration is lost when the prothrombin fragment 2 segment of the molecule, which contains the weak calcium-binding sites, is deleted.

When unactivated factor V is added, the acceleration of prethrombin 1 activation is preceded by a definite lag. When activated factor V (factor V_a) is used, the lag is substantially decreased, or nonexistent. These observation suggest that activated factor V (factor V_a) can accelerate the activation of prethrombin 1. Esmon et al.[89] have provided evidence that factor V_a, but not factor V, binds to prothrombin in the presence of calcium. Prothrombin fragment 2 noncovalently associates with both bovine α-thrombin[90-92] and with prethrombin 2.[92] The addition of prothrombin fragment 2 to the factor X_a–factor V–calcium–phospholipid activation of prethrombin 2 restores the factor V_a acceleration of thrombin production. Jackson et al.[93] have recently reported that the addition of the whole "pro" piece, prothrombin fragment 1·2 to prethrombin 2 in the presence of factor X_a, factor V_a, calcium and phospholipid results in a rate of prothrombin production faster than the rate of prothrombin conversion to thrombin.

Two conclusions can be drawn from the studies of the calcium binding and kinetics of activation of prothrombin and the thrombin-producing intermediates: (1) the prothrombin fragment 1 segment of the prothrombin molecule which contains the strong calcium binding site is responsible for the calcium-dependent phospholipid binding of prothrombin to the prothrombinase complex; (2) the prothrombin fragment 2 segment of the molecule, which contains the weak calcium binding sites, is implicated in the factor V_a dependent association of the prothrombin molecule with the prothrombinase complex.

Prothrombin Activation

An overall picture of prothrombin activation is beginning to emerge which is consistent with the data provided by all laboratories. An illustration of the hypothetical pathway for prothrombin generation from prothrombin is presented in Fig. 10. Presumably, the initial event in

[89] C. T. Esmon, W. G. Owen, D. L. Duiguid, and C. M. Jackson, Biochim. Biophys. Acta 310, 289 (1973).
[90] C. M. Heldebrant and K. G. Mann, J. Biol. Chem. 248, 3642 (1973).
[91] K. G. Mann, S. P. Bajaj, C. M. Heldebrant et al., Ser. Haematol. 6, 479 (1973).
[92] K. H. Myrmel, R. L. Lundblad, and K. G. Mann, Circulation 50, 292 (abstr.) 118 (1974).
[93] C. M. Jackson, C. T. Esmon, and W. G. Owen, Abstr. Cold Spring Harbor Meet. Proteases Biol. Control, Sept. 10, 1974, p. 7.

HUMAN II ACTIVATION

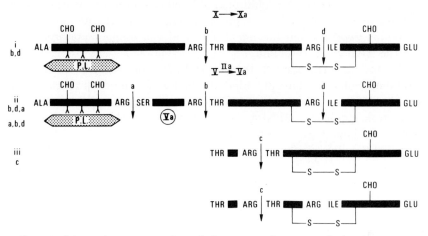

Fig. 10. Schematic representation of the events that occur during human prothrombin activation. For the nomenclature of prothrombin pieces, see Fig. 9. The letters a, b, c, and d represent cleavage sites in the human prothrombin molecule that occur during activation; the orders of these cleavages are presented by the roman numerals i, ii, and iii. P. L. designates phospholipid, and the λ on the prothrombin fragment 1 segment of the molecule represents the γ-carboxyglutamic acids.

prothrombin activation is the formation of factor X_a by either the intrinsic or extrinsic pathway. The factor X_a produced forms a complex with calcium, phospholipid, and probably with unactivated factor V. The binding of the factor V to the complex would be most likely by virtue of lipid–protein interaction rather than direct interaction of prothrombin with factor V, since the work of Esmon et al.[89] indicates that unactivated factor V does not bind to prothrombin. In the initial event of activation, probably only factor X_a, calcium, and phospholipid serve as the catalyst, and bind the prothrombin through the prothrombin fragment 1 segment of the molecule, which contains the vitamin K-dependent calcium-binding sites. Cleavage of the prothrombin by the complex at position b (see Fig. 10) produces the prothrombin fragment 1·2 plus prethrombin 2. Prethrombin 2 is subsequently cleaved at position d to produce α-thrombin. Once α-thrombin is produced, it can serve two functions. It can activate factor V to factor V_a, which results in factor V becoming a calcium-dependent prothrombin and prethrombin 1 binding protein. At this point, the rate of prothrombin activation would be greatly accelerated, and would probably proceed by the two routes originally suggested by Stenn and Blout,[56] one route being catalyzed only by factor X_a cleavage at positions b and d, the other route being catalyzed by a combination of thrombin cleavage at position a and factor X_a cleavage at positions b and d. An additional side effect of thrombin production, at least in the case of human prothrombin, is the cleavage of a small 13-residue peptide from the amino-terminal end of both prethrombin 2 and the A-chain of thrombin.

Prothrombin fragment 2, the factor V_a binding segment of the prothrombin molecule, not only binds to factor V_a and to prethrombin 2, but also binds α-thrombin itself. Therefore, it is conceivable that in the final prothrombinase activator formed, thrombin itself may be a constituent part, bound to the complex noncovalently through prothrombin fragment 2, which binds to the complex via factor V_a and calcium.

[14] Thrombin

By ROGER L. LUNDBLAD, HENRY S. KINGDON, and KENNETH G. MANN

Thrombin is the enzyme involved in the final step in the coagulation of mammalian blood. This step involves the conversion of fibrinogen to fibrin through the cleavage of four arginyl–glycyl peptide bonds.[1] Thrombin therefore is a highly specific proteolytic enzyme. Thrombin also possesses the ability to hydrolyze a variety of ester and amide substrates, but, at least in the case of ester substrates, these synthetic substrates do not effectively measure changes in biological (fibrinogen-clotting) activity.[2,3] In addition, thrombin is not specific for either ester or amide substrates. Therefore it is preferable to use fibrinogen clotting as the specific assay for thrombin.

Assay

The procedure used in our laboratory is a modification of the NIH thrombin assay.[4] The principal difference between this assay and the standard NIH assay procedure is that all reagents are pre-prepared separately in large quantities, and then aliquoted into approximately 2-ml volumes into polyethylene vials (flip-top polyethylene vials Model B, 2-dram capacity, obtainable from Laboratory Supplies Company, Inc.,

---

[1] K. Laki and J. A. Gladner, *Physiol. Rev.* **44,** 127 (1964).
[2] G. F. Lanchantin, C. A. Presant, D. W. Hart, and J. A. Friedmann, *Thromb. Diath. Haemorrh.* **14,** 159 (1965).
[3] R. L. Lundblad, K. G. Mann, and J. H. Harrison, *Biochemistry* **12,** 409 (1973).
[4] "Minimum Requirement for Dried Thrombin," 2nd Revision, Division of Biologic Standards, National Institutes of Health, Bethesda, Maryland, 1946.

29 Jeffrey Lane, Hicksville, New York 11801) and the biological reagents are stored at —20°. Prior to performing an assay, the appropriate number of vials is thawed at 37° in a water bath, and the assay mixture is prepared by mixing all constituent parts save the thrombin. This complete assay mixture (0.3 ml) is placed in the assay tube. In contrast, the NIH assay procedure makes use of the immediate addition of each reagent separately into the assay tube, with introduction of fibrinogen to initiate the reaction. The procedure we have used initiates the reaction by the addition of thrombin, and has several advantages: (1) the assay reagents are mixed in bulk and aliquoted into each assay tube, thus a large number of assay tubes can be prepared rapidly; (2) the initiation of the reaction is made by virtue of the addition of the thrombin to be analyzed to the assay reagent mixture. Since the assay can be performed only over a limited range of clot times (14–25 sec), the amount of thrombin added can be rapidly adjusted to get the assay within useful range. We have performed the assay by adding 1–20 μl of the thrombin to be assayed to the 0.3 ml of assay mixture. Thus the final assay mixture would range from 301 to 320 μl of complete system. This minor dilution variation does not provide a substantial source of error in the thrombin determination. (3) The most precise thrombin assays can be performed if the clotting time measured is between 15 and 20 sec. Thus, once an initial assay is performed, the volume of thrombin added can be rapidly adjusted to provide an assay within this range for a second assay. (4) Since there is a problem of thrombin absorption to glass, the dilution of a thrombin sample prior to assay can provide a substantial source of error if the amount of time the diluted thrombin is allowed to stand in a tube is varied. This procedure, by making use of the addition of the stock thrombin solution directly to the assay mixture, precludes the problem of multiple thrombin dilutions.

It is advisable to acquire a fairly large stock of the standard fibrinogen to be used before setting up procedures for routine assay of thrombin since the fibrinogen tends to be a larger source of variation than any other component in the assay mixture. We have been using the same lot of fibrinogen during the past four years, and our standard assay curve has not varied at all during this time. Figure 1 presents a thrombin assay calibration curve in terms of a direct plot[5] of thrombin NIH units and clotting time. The open circles represent points taken when the curve was prepared, in the fall of 1970. The filled triangles represent points taken in the spring of 1974.[4]

[5] The data may also be expressed in terms of $1/T$ (in terms of seconds) vs NIH thrombin units, or as log NIH thrombin units vs log time. In both cases, a straight-line relationship will be obtained over the useful assay range.

Reagents

Acacia solution: a 15% (w/v) solution of calcium-free gum acacia. The solution is stored frozen in plastic vials. Gum acacia (Scharr and Co., Chicago, Illinois) may be prepared free of calcium by two methods. The calcium content of commercial preparation is approximately 300 μg/ml. (a) A 5% solution of commercial gum acacia is prepared and an excess of potassium oxalate is added. The solution is allowed to stand overnight, then heated (50°) and centrifuged. The supernatant is dialyzed against cold distilled water, and the purified acacia is recovered by lyophilization. (b) A 50% solution of gum acacia is prepared, and 50 g of Dowex 50 WX-12 (500–100 mesh Na⁺ form) resin per 100 ml of acacia solution are added. The solution is stirred several times, the resin is removed by centrifugation, and the purified acacia is recovered from the supernatant by lyophilization. The calcium content of a 15% solution of such preparations is less than 8 μg/ml.

NIH imidazole buffer, pH 7.25, 17.2 g of imidazole are dissolved in 900 ml of 0.1 N HCl and diluted to 1 liter with distilled water; the pH of the solution is then adjusted to 7.25.

Calcium chloride solution: 1% calcium chloride in 0.9% sodium chloride

Fibrinogen solution: 3 g of fibrinogen (Sigma Chemical Co., St. Louis, Missouri, F-4000, fraction 1, type 1) are dissolved in 100 ml of water; the pH is adjusted to 7.2 with 0.5 M dibasic sodium phosphate. The volume is then brought to 150 ml with distilled water. The 2% solution is stored in plastic vials at −20°, and is thawed (37°) immediately prior to use.[6]

Procedure. The assay mixture is composed of 1 part NIH imidazole buffer, 1 part calcium chloride solution, 2 parts acacia solution, 3 parts fibrinogen solution, and 8 parts 0.9% sodium chloride. The fibrinogen should be added last. The mixture is centrifuged to remove insoluble material (a small amount will always be present), and the supernatant is poured into an 18 × 150 test tube. The assay mixture may be stored for several hours at room temperature.

To perform the test, 0.3 ml of the assay mixture is placed in a 10 × 75 or 13 × 75 culture tube; 1–20 μl of the thrombin solution, or an appropriate dilution thereof, are added and the clot timer is started. The tube

---

[6] Some lots of fibrinogen will not go back into solution very well after being frozen. If this is the case, the fibrinogen solution will have to be prepared fresh each day, or another lot selected.

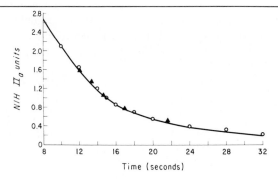

FIG. 1. Thrombin assay calibration curve; clotting time is plotted against NIH thrombin units in the assay. ○, determined in the fall of 1970; ▲, determined in the spring of 1974. The same lot of fibrinogen was used for all assays.

is gently agitated periodically, the end point is the first appearance of fibrin threads. The amount of the thrombin solution added, the dilution of that solution, if any, and the clot time are recorded.

One NIH thrombin unit is defined as that amount of thrombin which will clot 1.0 ml of a standardized fibrinogen solution in 15 ± 0.5 sec at $28 \pm 1.0°$. Our assay is standardized against NIH standard thrombin (lot B-3, 23.7 NIH units/mg),[4] and a standard curve is generated employing known dilutions of this standard (see Fig. 1).

The clotting time is converted into units by means of a direct or (Fig. 1) double-logarithmic plot of clotting time vs dilution of a standard thrombin preparation. The fibrinogen-clotting activity of thrombin is then given in NIH units. Other unitage has been used by several investigators.[7–9]

A number of investigators have recommended the use of additives to the fibrinogen solution, such as acacia[10] or polyethylene glycol,[11] to obtain a sharper end point and more reproducible results. It should be noted that the standard curve obtained for a given lot of fibrinogen remains the same for the period of use of that lot of fibrinogen.[12]

Waugh and colleagues have reported extensive studies on the development of a standard clotting system with respect to variations due to

[7] E. D. Warner, K. M. Brinkhous, and H. P. Smith, *Am. J. Physiol.* **114**, 667 (1936).
[8] W. H. Seegers and H. P. Smith, *Am. J. Physiol.* **137**, 338 (1942).
[9] K. D. Miller and W. H. Copeland, *Exp. Mol. Pathol.* **4**, 431 (1965).
[10] H. P. Smith, E. D. Warner, and K. M. Brinkhous, *J. Exp. Med.* **66**, 801 (1937).
[11] J. W. Fenton, II and M. J. Fasco, *Thromb. Res.* **4**, 809 (1974).
[12] D. J. Baughman, D. F. Waugh, and C. Juvkam-Wold, *Thromb. Diath. Haemorrh.* **20**, 476 (1968).

thrombin, fibrinogen, and routine preparations of reagents utilized in the assay and the clotting assay.[12,13] The influence of fibrinogen and fibrin degradation products on the thrombin clotting time of whole plasma have also been reported.[14,15]

Other Substrates for Thrombin. It is also possible to use small synthetic esters and anilides as substrates for thrombin. These have, for the most part, been recently evaluated by Magnusson.[16] Toluenesulfonyl-L-arginine methyl ester is chronologically the oldest[17] and most extensively used of these substrates. The methodology described for the spectrophotometric assay of trypsin with this substrate[18] is applicable to thrombin. It should be recognized that thrombin is at least 4-fold less active than trypsin on a weight basis toward this substrate.

There had been some question regarding the ability of thrombin to hydrolyze anilide substrates. A careful study of the hydrolysis of several peptide anilide substrates has recently appeared.[19] Furthermore, these anilide substrates appear to be more indicative of the biological activity of thrombin[20] (see below) and have been utilized to measure plasma prothrombin.[21]

Peptide substrates for bovine thrombin have also been the subject of recent study. Scheraga and co-workers have reported studies[22-25] on a series of peptides designed after the structure of the segment of the A(α)-chain of bovine fibrinogen containing the arginyl–glycyl bond hydrolyzed by thrombin in conversion of fibrinogen to fibrin. Lawson and co-workers have reported similar studies[26] using human fibrinogen A(α)-chain as

[13] D. F. Waugh, D. J. Baughman, and C. Juvkam-Wold, *Thromb. Diath. Haemorrh.* **20**, 497 (1968).

[14] H. Arnesen and H. C. Godal, *Scand. J. Haematol.* **10**, 232 (1973).

[15] H. Arnesen, *Scand. J. Haematol.* **10**, 291 (1973).

[16] S. Magnusson, *in* "The Enzymes" (P. D. Boyer, ed.), 3rd ed., Vol. 3, p. 277. Academic Press, New York, 1971.

[17] S. Sherry and W. Troll, *J. Biol. Chem.* **208**, 95 (1954).

[18] K. Walsh, this series Vol. 19, p. 41.

[19] L. Svendsen, B. Blombäck, M. Blombäck, and P. Olsson, *Thromb. Res.* **1**, 267 (1972).

[20] R. L. Lundblad, manuscript in preparation.

[21] K. Bergstrom and M. Blombäck, *Thromb. Res.* **4**, 719 (1974).

[22] R. H. Andreatta, R. K. H. Liem, and H. A. Scheraga, *Proc. Natl. Acad. Sci. U.S.A.* **68**, 253 (1974).

[23] R. K. H. Liem, R. H. Andreatta, and H. A. Scheraga, *Arch. Biochem. Biophys.* **147**, 201 (1971).

[24] R. K. H. Liem and H. A. Scheraga, *Arch. Biochem. Biophys.* **158**, 387 (1973).

[25] R. K. H. Liem and H. A. Scheraga, *Arch. Biochem. Biophys.* **160**, 333 (1974).

[26] W. B. Lawson, A. P. Lobo, and S. M. Yu, *Thromb. Diath. Haemorrh.*, Suppl. **57**, 141 (1974).

model. The use of protamine as a substrate for thrombin has also been reported.[27]

Preparation of Thrombin

Bovine Thrombin. It is clear that, dependent upon the nature of the crude starting material,[28] it is possible to isolate several species of bovine thrombin that differ significantly in both their covalent structure and catalytic activity. These are more extensively discussed in a separate section.

Rasmussen[29] was the first to describe the purification of thrombin by ion-exchange chromatography. The procedure used utilized IRC-50 anion-exchange resin and a stepwise elution with phosphate buffers. The chromatographic purification of thrombin on DEAE-Sephadex has been reported by Strassle.[30] This procedure yielded material with a specific activity of 256 NIH units/mg by weight. Magnusson[31] extended Rasmussen's earlier observations. By starting with thrombin prepared from purified prothrombin, activated preparations were obtained with a specific activity of 2100 NIH units/mg by weight. Baughman and Waugh[32] used a combination of DEAE-cellulose chromatography, phosphocellulose chromatography, and gel filtration to obtain a preparation with a specific activity of 2300 NIH units/mg. Procedures for the separation of factor X_a (activated Stuart factor) from thrombin using chromatography on DEAE-cellulose have also been described. These preparations have specific activities in range of 200–400 NIH units per milligram of protein. Rosenberg and Waugh[33] enlarged earlier observations on the purification of thrombin and presented evidence for the existence of multiple forms of bovine thrombin. The work suggested the existence of six forms of thrombin, but Rosenberg and Waugh were only able to isolate two forms by gradient elution from phosphocellulose. Batt and co-workers[34] used a combination of ion-exchange chromatography (Bio-Rex 70), gel filtra-

---

[27] F. Brown, M. L. Freedman, and W. Troll, *Biochem. Biophys. Res. Commun.* **53**, 75 (1973).

[28] The most common starting material used for the preparation of bovine thrombin has been Parke-Davis topical thrombin.

[29] P. S. Rasmussen, *Biochim. Biophys. Acta* **16**, 157 (1955).

[30] R. Strassle, *Biochim. Biophys. Acta* **73**, 462 (1963).

[31] S. Magnusson, *Ark. Kemi* **24**, 349 (1965).

[32] D. J. Baughman and D. F. Waugh, *J. Biol. Chem.* **242**, 5252 (1967).

[33] R. D. Rosenberg and D. F. Waugh, *J. Biol. Chem.* **245**, 5049 (1970).

[34] C. W. Batt, J. W. Mikulka, K. G. Mann, C. L. Gurracino, R. J. Altiere, R. L. Graham, J. D. Quigley, J. W. Wolf, and C. W. Zafonte, *J. Biol. Chem.* **245**, 4857 (1970).

TABLE I
PURIFICATION OF BOVINE THROMBIN

Starting material	Media utilized	Specific activity (NIH/ units/mg)	Subtype	References[a]
Commercial topical	IRC-50	300–1800	α, β	29
Commercial topical	DEAE-Sephadex A-25	256	α, β[1]	30
Thromboplastin-activated purified prothrombin	IRC-50	1800–2300	α	31
Commercial topical	1. DEAE-cellulose 2. Phosphocellulose 3. Sephadex G-100	2010–2500	α	32
Commercial topical	DEAE-cellulose	250	α, β	b,c
Commercial topical	Bio-Rex-70	1400	α, β, γ	34
Commercial topical	Sulfoethyl or sulfopropyl Sephadex C-50	200–2200	α, β	35
Commercial topical	Bio-Rex-70	2100–2500	α	41
Citrate-activated purified prothrombin	IRC-50	4200	α (3–7 S)	40
3.7 S from IRC-50	IRC-50	8230	? (3.2 S)	40
Partially purified commercial thrombin	Affinity; p-chlorobenzylamino-ε-aminocaproylagarose	2500	α	42
Commercial thrombin	m-Aminobenzylamido-ε-aminocaproylagarose	N.D.	N.A.	44

[a] Numbers refer to text footnotes.
[b] D. M. Kerwin and J. H. Milstone, *Thromb. Diath. Haemorrh.* **17**, 247 (1967).
[c] E. T. Yin and S. Wessler, *J. Biol. Chem.* **243**, 112 (1968).

tion, and final chromatography on TEAE-cellulose to obtain material with a specific activity of 1400 NIH per milligram of protein. It is clear that this preparation also contains multiple thrombin species.[35] A rapid procedure for the purification of bovine thrombin using stepwise chromatography on sulfoethyl[36]- or sulfopropyl-Sephadex[3] has been developed in this laboratory. A column (2.5 × 28 cm) of either sulfoethyl or sulfopropyl-Sephadex C-50 is prepared and equilibrated with 0.025 M sodium phosphate, pH 6.5, under ambient conditions. The thrombin sample (Parke-Davis Topical Thrombin) to be purified, containing 10,000–300,000 units of thrombin, is taken into the initial solvent (0.025 M sodium phosphate, pH 6.5) by dialysis. The sample is applied to the column, and the

[35] K. G. Mann and C. W. Batt, *J. Biol. Chem.* **244**, 6555 (1969).
[36] R. L. Lundblad, *Biochemistry* **10**, 2501 (1971).

column is subsequently developed with the initial solvent. After emergence of the void volume material, which contains factor X_a and, on occasion, a small amount of fibrinogen-clotting activity, the developing solvent is changed to 0.1 M sodium phosphate, pH 6.5, to elute a second protein fraction and subsequently to 0.25 M sodium phosphate, pH 6.5, which elutes the thrombin as shown in Fig. 2. This material is functionally

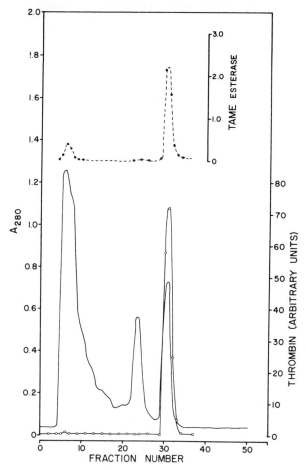

Fig. 2. Chromatographic fractionation of crude bovine thrombin on sulfoethyl-Sephadex C-50. The column was equilibrated with 0.025 M sodium phosphate, pH 6.5, and initially developed with the same solvent. A stepwise elution schedule utilizing 0.1 M sodium phosphate, pH 6.5 (at fraction 15), and subsequently 0.25 M sodium phosphate, pH 6.5 (at fraction 27), was used to finally develop the column. The thrombin was eluted with the final phosphate buffer. TAME, N^α-p-toluenesulfonyl-L-arginine methyl ester. Reprinted with permission from R. Lundblad (1971). *Biochemistry* **10**, 2501. Copyright by the American Chemical Society.

homogeneous; that is, it does not contain measurable levels of other co-agulation factors or plasmin. The specific activity of this material varies from 200 to 2300 NIH units/mg. This variation is due primarily to the respective amounts of α- and β-thrombin present in such a preparation. Preparations high in α-thrombin have high specific activity toward the natural substrate. The relative prportions of α- and β-thrombin are a property of the starting material used for this preparation. By going to a gradient elution[37],[38] instead of a stepwise development, it is possible to separate the various forms of thrombin. An illustration of the separation obtainable using this chromatographic mechanism and these solvents is shown in Fig. 3. (It is frequently necessary to centrifuge the crude material prior to chromatography to remove insoluble material.[3] Such a step affects neither the protein concentration nor the biological activity, and hence probably represents the removal of inorganic contaminants.)

Other chromatographic systems that separate the various forms of thrombin have also been reported. Seegers and co-workers[39] isolated a thrombin species of high specific activity by a second chromatographic fractionation of purified bovine thrombin on IRC-50. These same workers prepared an esterase thrombin by autolysis of purified thrombin followed by gel filtration.[40] Glover and Shaw[41] were able to separate α- and β-thrombin by chromatography on Bio-Rex 70 utilizing a stepwise elution program.

Affinity chromatography has proved to be useful for the purification of thrombin. Thompson and Davie[42] obtained thrombin of high purity by affinity chromatography on p-chlorobenzylamino-ε-aminocaproyl agarose. These investigators found it necessary to subject crude thrombin to a chromatographic purification prior to affinity chromatography to obtain satisfactory purity and activity yield. Schmer[43] also combined classical ion-exchange with affinity chromatography to purify thrombin. p-Amino-benzamidine was utilized as the insolubilized ligand for this latter study. A more recent study[44] has systematically examined various factors affecting the chromatographic behavior of thrombin. In this study on af-

[37] K. G. Mann, C. M. Heldebrant, and D. N. Fass. *J. Biol. Chem.* **246**, 5994 (1971).
[38] R. L. Lundblad, L. C. Uhteg, C. N. Vogel, H. S. Kingdon, and K. G. Mann, *Biochem. Biophys. Res. Commun.* **66**, 482 (1975).
[39] W. H. Seegers, L. McCoy, R. K. Kipfer, and G. Murano, *Arch. Biochem. Biophys.* **128**, 194 (1968).
[40] W. H. Seegers, D. A. Walz, J. Reuterby, and L. E. McCoy, *Thromb. Res.* **4**, 829 (1974).
[41] G. Glover and E. Shaw, *J. Biol. Chem.* **246**, 4594 (1971).
[42] A. R. Thompson and E. W. Davie, *Biochim. Biophys. Acta* **250**, 210 (1971).
[43] G. Schmer, *Hoppe-Seyler's Z. Physiol. Chem.* **353**, 810 (1972).
[44] H. F. Hixon, Jr., and A. H. Nishikawa, *Arch. Biochem. Biophys.* **154**, 501 (1973).

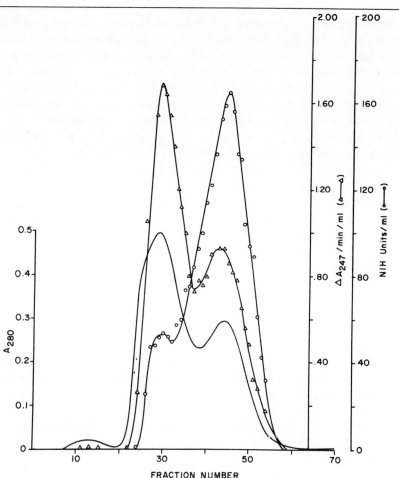

FIG. 3. The chromatographic fractionation of partially purified bovine thrombin on sulfopropyl-Sephadex C-50. The sample was obtained as described under Fig. 1, dialyzed against 0.05 M sodium phosphate, pH 6.5, and applied to the column which had been previously equilibrated with the same solvent. After emergence of the void volume material, the column was developed successively with 0.15 M sodium phosphate, pH 6.5, and a linear gradient to a limit buffer of 0.45 M sodium phosphate, pH 6.5. From R. L. Lundblad *et al.* (1975). *Bichem. Biophys. Res. Commun.* **66**, 482.

finity chromatography, *m*-aminobenzamidine-agarose was found to be very effective in the affinity chromatography of thrombin.

Human Thrombin. There is considerably less literature on the purification of human thrombin. Miller and Copeland[9] prepared crude thrombin by the thromboplastin activation of prothrombin prepared from human plasma Cohn fraction III. Subsequent chromatography on Mallinckrodt CG-50 (Na⁺) and gel filtration yielded thrombin of high purity. This procedure has been subsequently adopted for the large-scale preparation of purified human thrombin by Fenton and colleagues.[45] This material has a specific activity of 3500 NIH Units per milligram of protein. The isolation of multiple molecular forms of human thrombin has been achieved by chromatography of thrombin preparations obtained by the activation of purified human prothrombin either by Taipan snake venom (*Oxyuranus scutellatus scutellatus*)[46] or by citrate activation.[47]

The method developed by this laboratory for bovine α-thrombin is well suited to the purification of the human enzyme.[48,49] The starting material in this case was the prothrombin present in a dry human factor IX complex (American Red Cross, lot No. 5). The Red Cross preparation (0.2 g) was dissolved in 10 ml of 0.0168 M imidazole 0.144 M NaCl, pH 7.4, and activated by the addition of 1 ml of thromboplastin (Difco Two-Stage reagent), 1 ml of Difco Ac-globulin, and 4 ml of 0.25 M $CaCl_2$. After the yield of thrombin became constant, the mixture was dialyzed against 0.025 M sodium phosphate, pH 6.5, centrifuged, and applied to a 2.5 × 50 cm sulfopropyl-Sephadex column, such as is used in the bovine procedure. Figure 4 illustrates the chromatographic behavior of human thrombin in the system. The elution pattern observed is essentially identical to that observed for bovine thrombin. The specific activity of the pooled preparation was 2700 NIH thrombin units per milligram of protein, and electrophoretic analysis revealed the presence of only α-thrombin when analysis was performed with disulfide bonds intact or reduced.

Other Species. The purification of both equine,[50] and porcine[51] thrombin have been reported. In the case of the porcine enzyme, it was possible to isolate multiple molecular species with varying specific activity.

Active Site Titrations of Thrombin

The last decade has seen the increased use of active site titrants to measure proteolytic enzyme concentration. The first-active site titrants

[45] J. W. Fenton, II, W. P. Campbell, J. C. Harrington, and K. D. Miller, *Biochim. Biophys. Acta* **229**, 26 (1971).
[46] G. F. Lanchantin, J. A. Friedmani, and D. W. Hart, *J. Biol. Chem.* **248**, 5956 (1973).
[47] J. J. Gorman and P. A. Castaldi, *Thromb. Res.* **4**, 653 (1974).
[48] K. H. Myrmel and K. G. Mann, unpublished observations, 1974.
[49] K. G. Mann, this volume [13].
[50] Y. Inada, A. Matsushima, I. Kotaku, S. A. Hosain, and K. Shibata, *J. Biochem.* (*Tokyo*) **68**, 193 (1970).
[51] A. Rossi and J. R. Giglio, *Can. J. Biochem.* **52**, 336 (1974).

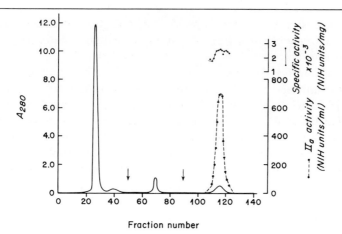

Fig. 4. Purification of human thrombin on sulfopropyl-Sephadex C-50 (120). After the elution of nonthrombin protein, fractions 20–50, with 0.025 M sodium phosphate, pH 6.5, the column was eluted with 0.1 M sodium phosphate, pH 6.5; this resulted in the elution of an additional protein component (fractions 60–80). Human thrombin (fractions 105–125) was eluted with 0.25 M sodium phosphate, pH 6.5.

used on thrombin were N-benzyloxycarbonyl-L-tyrosine p-nitrophenyl ester and N-benzyloxycarbonyl-L-lysine p-nitrophenyl ester.[52] Subsequently, p-nitrophenyl-p'-guanidionebenzoate (p-NPGB) was developed as a titrant for trypsin[53] and subsequently applied to thrombin.[54] A larger number of similar compounds have been recently evaluated.[55] Most investigators use p-NPGB in work on thrombin, using the conditions originally described.[54] It should be recognized that this assay is not specific, and also measures plasmin,[54] trypsin[54] and factor X_a.[56] More recent work suggests that it may be possible to devise "active-site" reagents that would distinguish these latter enzymes.[57]

Factors Enhancing the Activity of Thrombin

In 1965, it was suggested that a peptide or peptides formed during the activation of prothrombin subsequently bound to thrombin and influenced the catalytic activity.[58] Subsequently it was found that intermediate 4

[52] F. J. Kézdy, L. Lorand, and K. D. Miller, *Biochemistry* 4, 2302 (1965).
[53] T. Chase, Jr., and E. Shaw, *Biochem. Biophys. Res. Commun.* 29, 508 (1967).
[54] T. Chase, Jr., and E. Shaw, *Biochemistry* 8, 2212 (1969).
[55] M. J. Casco and J. W. Fenton, II, *Arch. Biochem. Biophys.* 159, 802 (1973).
[56] R. L. Smith, *J. Biol. Chem.* 248, 2418 (1973).
[57] S. C. Wong and E. Shaw. *Arch. Biochem. Biophys.* 161, 536 (1974).
[58] R. H. Landaburu, O. Abdala, and J. Morrone, *Science* 148, 380 (1965).

(fragment 2; 0 fragment) from prothrombin activation could enhance the esterase activity[40,59] of α-thrombin with very slight inhibition of fibrinogen-clotting activity. This phenomenon has been recently investigated in greater detail.[60] The enhancement of esterase activity observed on the binding of intermediate 4 to α-thrombin is due to an increase in the apparent V_{max} while the apparent K_m remained unchanged. It was also noted that the rate of inactivation of thrombin by 1-chloro-3-tosylamido-2-amino-2-heptanone was significantly increased in the presence of intermediate 4. In this respect, it should be noted that cholates have been observed to increase the rate of hydrolysis of certain ester substrates by thrombin[61,62] and subsequently[63] the rate of inactivation of thrombin by 1-chloro-3-tosylamido-2-amino-2-heptanone. With respect to the observation on the interaction of intermediate 4 with thrombin, the binding of "trypsin-binding macroglobulin" to thrombin results in the inactivation of fibrinogen-clotting activity with no apparent change in esterase activity.[64] It has been observed[65,66] that compounds, such as methylguanidine, which potentiate the activity of trypsin with nonspecific substrates[67] and inactivators[68] also increase the rate of thrombin-catalyzed hydrolysis of nonspecific substrates, such as p-nitrophenyl acetate.[69]

The Specificity of Thrombin on Protein and Peptide Substrates

Most proteolytic enzymes which have been extensively studied have a large number of polypeptide substrates and well defined sites of cleavage. Only recently have polypeptide substrates, other than fibrinogen, been described for thrombin. It is abundantly clear that thrombin is far more specific than trypsin with regard to polypeptide substrates even though both enzymes catalyze the hydrolysis of similar low molecular weight ester and amide substrates.

Table II describes the well-defined sites of polypeptide cleavage by

[59] C. M. Heldebrant and K. G. Mann, *J. Biol. Chem.* **248**, 3642 (1973).
[60] K. H. Myrmel, R. L. Lundblad, and K. G. Mann, manuscript in preparation.
[61] J. B. Baird, E. F. Curragh, and D. T. Elmore, *Biochem. J.* **96**, 733 (1965).
[62] T. Exner and J. L. Koppel, *Biochim. Biophys. Acta* **321**, 303 (1973).
[63] T. Exner and J. L. Koppel, *Biochim. Biophys. Acta* **329**, 233 (1973).
[64] G. F. Lanchantin, M. L. Plesset, J. A. Friedmann, and D. W. Hart. *Proc. Soc. Exp. Biol. Med.* **121**, 444 (1966).
[65] L. Lorand and J. L. G. Nilsson, *In* "Drug Design" (E. J. Ariens, ed.), Vol. 3, p. 415. Academic Press, New York, 1972.
[66] R. L. Lundblad, unpublished observations, 1974.
[67] T. Inagami and S. S. York, *Biochemistry* **7**, 4045 (1968).
[68] T. Inagami and T. Murachi, *J. Biol. Chem.* **239**, 1395 (1964).
[69] R. L. Lundblad, *Thromb. Diath. Haemorrh.* **30**, 248 (1973).

TABLE II
Thrombic Cleavage Sites in Macromolecular Substrates

Protein	Cleavage site
Human fibrinogen A(α) chain [a]	↓ Gly-Gly-Gly-Val-Arg-Gly-Pro-Arg-Val-Gln-
Human fibrinogen B(β) [b]	↓ Phe-Phe-Ser-Ala-Arg-Gly-His-Arg-Pro-Leu-Asp
Human fibrinogen A(α) [a]	↓ Val-Arg-Gly-Pro-Arg-Val-Val-Glu-Arg-His
Porcine secretin [c]	↓ Leu-Ser-Arg-Leu-Arg-Asp-Ser-Ala-Arg-Leu
Porcine cholecystokinin pancreozymin [d]	↓ Ala-Pro-Ser-Gly-Arg-Val-Ser-Met-Ile-Lys
Chymotrypsinogen [e]	↓ Ser-Gly-Leu-Ser-Arg-Ile-Val-Asn-Gly-Glu
Human chorionic gonadotropin [f]	↓ Pro-Leu-Arg-Pro-Arg-Cys-Arg-Pro-Ile
	↓ Val-Cys-Asn-Tyr-Arg-Asp-Val-Arg-Phe
	↓ Pro-Gly-Cys-Pro-Arg-Gly-Val-Asn-Pro
	↓ Cys-Ala-Leu-Cys-Arg-Arg-Ser-Thr-Thr
Egg white lysozyme [g]	↓ Gln-Ala-Trp-Ile-Arg-Gly-Cys-Arg-Leu ↓

[a] B. Blombäck, *Br. J. Haematol.* **17**, 145 (1969).
[b] S. Iwanaga, P. Wallén, N.Y. Grandahl, A. Henschen, and B. Blömback, *Eur. J. Biochem.* **8**, 189 (1964).
[c] V. Mutt, S. Magnusson, J. E. Jorpes, and E. Dahl, *Biochemistry* **4**, 2358 (1965).
[d] V. Mutt and J. E. Jorpes, *Eur. J. Biochem.* **6**, 156 (1968).
[e] A. Engel and B. Alexander, *Biochemistry* **5**, 3590 (1966).
[f] F. J. Morgan, S. Birken, and R. E. Canfield, *J. Biol. Chem.* **250**, 5247 (1975).
[g] R. L. Lundblad and J. H. Harrison, unpublished observations, 1969.

thrombin. Other protein sites have been suggested,[70] but data are not available to support these suggestions. In addition, it is clear that thrombin induces changes in several other proteins involved in blood coagula-

[70] D. A. Walz, W. H. Seegers, J. Reuterby, and L. E. McCoy, *Thromb. Res.* **4**, 713 (1974).

tion. Specifically, factor V and factor VIII are modified by thrombin, resulting in first enhanced and then diminished biological activity.[71-75] In addition, the conversion of prothrombin to intermediate 1 is catalyzed by thrombin.[76] In work on platelets, it has been suggested that both platelet myosin and actin can be cleaved by thrombin.[77] Evidence has been presented which suggests the digestion of fibrin by purified thrombin.[78] In these latter cases, no information is available regarding the cleavage site(s).

It is striking that although thrombin will catalyze the hydrolysis of lysine esters and amides, the cleavage of a lysine peptide bond has not been described. Thus, it would appear that thrombin is relatively specific for arginine in protein substrates. Furthermore, only specific arginyl bonds can be hydrolyzed. It has, therefore, been possible to use thrombin as a reagent to obtain highly selective cleavage in proteins useful for studies on the primary structure of proteins.

Multiple Forms of Thrombin

It is clear that multiple forms of human, bovine, and porcine thrombin exist. The relative proportion of the different thrombin species in a given purified enzyme preparation does not significantly affect specific activity toward the ester substrates, but does markedly affect the specific activity toward fibrinogen. The evidence for multiple forms of thrombin in each species will be discussed starting with observations made on proteins of bovine origin.

The existence of multiple species of bovine thrombin possessing different specific activities toward fibrinogen and ester substrates could be implied from early studies on the citrate-activation of partially purified bovine prothrombin preparations.[79] These early experiments showed

[71] S. I. Rapaport, S. Schiffman, M. J. Paker, and S. B. Ames, *Blood* **21**, 221 (1963).

[72] R. Biggs, R. G. Macfarlane, K. W. E. Denson, and B. J. Ash. *Br. J. Haemat.* **11**, 276 (1965).

[73] H. J. Weiss and S. Kochwa. *Br. J. Haematol.* **18**, 89 (1970).

[74] C. Schmer, E. P. Kirby, D. C. Teller, and E. W. Davie. *J. Biol. Chem.* **247**, 2512 (1972).

[75] R. W. Colman, *Biochemistry* **8**, 1438 (1969).

[76] K. G. Mann, C. M. Heldebrant, D. N. Fass, S. P. Bajaj, and R. J. Butkowski, *Thromb. Diath. Haemorrh.,* Suppl. 57, 179 (1974).

[77] L. Muszbek and K. Laki, *Proc. Natl. Acad. Sci. U.S.A.* **71**, 2208 (1974).

[78] D. C. Triantaphyllopoulos and S. Chandra, *Biochim. Biophys. Acta* **328**, 229 (1973).

[79] W. H. Seegers, "Prothrombin," Chapter 16. Harvard Univ. Press, Cambridge, Massachusetts, 1962.

an initial increase of fibrinogen-clotting and esterase activity during pro-thrombin activation, and the subsequent decreases of fibrinogen-clotting activity with no significant change in esterase activity. Subsequent in-vestigators[80] not only confirmed these early observations, but showed that these changes were due to the eventual formation of a species having high esterase activity and low fibrinogen-clotting activity. Mann and Batt[35] described the presence of multiple molecular weight species in purified species of bovine thrombin. Seegers et al.[39] have described the purification of two forms of bovine thrombin, one of which has a molecular weight of 33,000 (3.7 S thrombin) and the other 22,500 (3.2 S thrombin). The 3.2 S thrombin was reported to have approximately twice the specific activity of the 3.7 S species in the clotting of fibrinogen. Another group of investi-gators[33,81] have identified six species of bovine thrombin, which varied considerably in specific activity against fibrinogen. Isoelectric focusing studies conducted with purified α- and β-thrombin[82] indicate that bovine α-thrombin can be resolved into four isoelectric subspecies, while bovine β-thrombin gave rise to three isoelectric subspecies.

In our laboratory, we have been able to separate three forms of bovine thrombin evolved during the citrate activation of prothrombin. These various species can be separated by cation exchange chromatography as previously described. Table III shows the relative specific activities of these various species toward several substrates.

The various forms of bovine thrombin are referred to as α-thrombin, β-thrombin, and so forth. This nomenclature is based on their order of formation from intermediate 2 in prothrombin activation, and is also a reflection of the strength of their adsorption to the sulfopropyl-Sephadex column described herein (e.g., β-thrombin elutes prior to α-thrombin) (see Fig. 3). Tentative structures for bovine α- and β-thrombin are given in Fig. 5. β-thrombin, as shown is formed from α-thrombin by extensive degradation of the A chain and excision of a small, and yet unknown, fragment containing carbohydrate, from the B chain. This results in the three-chain structure shown. γ-Thrombin is thought to arise from further degradation of the B2 chain. There is no evidence to suggest that any of these less active species are formed directly from intermediate 2, but rather they arise from the degradation of α-thrombin by an enzyme or enzymes as yet unknown. It should also be noted that the isolation of a thrombin that possesses no fibrinogen-clotting activity has been reported.[40]

[80] G. F. Lanchantin, J. A. Friedmann, and D. W. Hart, J. Biol. Chem. 242, 2491 (1967).
[81] D. F. Waugh and R. D. Rosenberg, Thromb. Diath. Haemorrh. 27, 183 (1972).
[82] D. N. Fass, C. M. Heldebrant, R. J. Butkowski, and S. J. Bajaj, Fed. Proc., Fed. Am. Soc. Exp. Biol. 32, (abstr.) 318 (1973).

TABLE III
ACTION OF THROMBINS ON VARIOUS SUBSTRATES

Thrombin species	NIH U/mg	TosArg-OMe[a] (μmoles/min/mg)	TosArg-OMe (μmoles/min/μmole)	TosArg-OMe (K_m, obs)	BzArg-Nan[b] (μmoles/min/mg)	p-NPB[c] (μmoles/min/mg)	BzArg-Nan (K_m, obs)
α-Thrombin (39,000)[d]	2500	37.8	1361	1.7×10^{-4}	0.18	0.08	$1.43 \times 10^{-4}\,M$
β-Thrombin (28,000)[d]	134	62.9	1760	3.0×10^{-4}	0.06	0.08	$8.4 \times 10^{-4}\,M$
γ-Thrombin (28,000)[d]	109	47.5	1330	—	0.06	0.04	—

[a] N-α-p-toluenesulfonyl-L-arginine methyl ester.
[b] N-α-Benzoyl-L-arginine-p-nitroanilide.
[c] p-Nitrophenylbutyrate.
[d] Molecular weight of thrombin species [K. G. Mann and C. W. Batt, *J. Biol. Chem.* **244**, 6555 (1969)].

The formation of this component (thrombin E) is thought to occur via autolysis of thrombin with the removal of the amino-terminal portion of the B chain (B1 chain), which would include the active site histidine residue (His 43). This observation is at variance with the observation from this laboratory, as well as with earlier results from other laboratories.[41]

The existence of multiple forms of human thrombin could also be inferred from early studies[83] on the citrate activation of partially purified human prothrombin. Subsequent studies[80] on the chromatographic fractionation of such reaction mixtures revealed the stepwise formation of thrombin species with decreasing fibrinogen-clotting activity, but unchanged esterase activity. Multiple forms of human thrombin have also been separated by isoelectric focusing.[47] At least three species were identified in this manner. These different species appear to have different specific activities toward fibrinogen, but it was not possible to show the existence of a thrombin possessing high esterase activity and low clotting activity. This is comparable to earlier observations[46] on the activation of human prothrombin with Taipan Snake Venom and subsequent "autoly-

[83] G. F. Lanchantin, J. A. Friedmann, and D. W. Hart, *J. Biol. Chem.* **240**, 3276 (1965).

sis" of the thrombin to form a species of lower molecular weight with "enhanced" fibrinogen-clotting activity and unchanged esterase activity. This, then, is somewhat similar to the observation made for bovine 3.7 S and 3.2 S thrombin described previously.[40]

There is also a report[51] describing the existence of multiple forms of porcine thrombin obtained by citrate activation of partially purified porcine prothrombin. Three species, obtained by chromatography on Amberlite IRC-50, possessed varying amounts of fibrinogen-clotting and esterase activity.

Inhibitors

Thrombin is inactivated by both diisopropylphosphorofluoridate[84,85] or phenylmethylsulfonyl fluoride.[36,86] In the former case, it has been established that a serine residue is modified.[87] The location of this modification on the polypeptide chains of thrombin has been identified.[88] In the case of the inactivation with phenylmethylsulfonyl fluoride, it is presumed that the same serine residue is modified, as has been shown for trypsin and chymotrypsin.[89] Saran has also been shown to inactivate thrombin[90] presumably by the same mechanism as that shown for diisopropylphosphorofluoridate (DFP). Geratz[91] has shown that thrombin is irreversibly inactivatel by the "active-site directed" inhibitor, p-m-(m-fluorosulfonylphenylureido)phenoxyethoxybenzamidine. The site of incorporation of this compound is not known, but it is likely that it is the same serine residue modified by the reagents described above.

The modification of histidine in thrombin has also been shown to result in the inactivation of catalytic activity. The "active-site directed" inhibitor, 1-chloro-3-tosylamido-2-amino-2-heptanone (toluenesulfonyllysyl-chloromethyl ketone), originally designed for trypsin,[92] has been shown to react with a specific histidine (His 43) in the B chain of bovine thrombin.[41]

The importance of tyrosine in the function of thrombin can be sug-

[84] J. A. Gladner and K. Laki, *Arch. Biochem. Biophys.* **62**, 501 (1956).
[85] K. D. Miller and H. Van Vunakis, *J. Biol. Chem.* **223**, 227 (1956).
[86] W. H. Seegers, D. Heene, E. Marciniak, N. Ivanovic, and M. J. Caldwell, *Life Sci.* **4**, 425 (1965).
[87] J. A. Gladner and K. Laki, *J. Am. Chem. Soc.* **80**, 1263 (1958).
[88] K. G. Mann, R. Yip, C. M. Heldebrant, and D. N. Fass, *J. Biol. Chem.* **248**, 1868 (1973).
[89] A. M. Gold, *Biochemistry* **4**, 897 (1965).
[90] A. R. Thompson, *Biochim. Biophys. Acta* **198**, 392 (1970).
[91] J. D. Geratz, *FEBS Lett.* **20**, 294 (1972).
[92] E. Shaw, M. Mares-Guia, and W. Cohen, *Biochemistry* **4**, 2219 (1965).

gested from early work[93] on the reaction of tetranitromethane with this enzyme. Subsequent studies with acetic anhydride[94] also implicated tyrosine. More recent work with tetranitromethane[95] and N-acetylimidazole[3] showed that the reaction of either of these reagents with purified bovine thrombin results in the modification of 4–5 moles of tyrosine per mole of thrombin. This modification was associated with the loss of fibrinogen-clotting activity, but no change in esterase activity.

It has also been suggested that the modification of tryptophanyl residues in thrombin results in the loss of enzymic activity.[96,97] Modification of the amino-terminal isoleucine in the B-chain of bovine thrombin[98] by nitrous acid also results in the inhibition of enzyme actvity.

Proflavine also is an inhibitor of thrombin[99] and has been useful in the study of the binding properties of this enzyme.[99,100]

Thrombin had been reported to be sensitive to soybean trypsin inhibitor[101] but there is evidence to suggest that this is not the case.[102] Concanavalin A has been reported to inhibit thrombin,[103] presumably through binding to the carbohydrate moiety of this glycoprotein. The carbohydrate moiety of thrombin can, however, be removed without any deleterious effect on enzymic activity.[104] Heparin[105] and the combination of antithrombin and heparin[106] have been shown to inhibit thrombin. The inhibition of thrombin by a protein factor from liver via binding to the enzymes–substrate complex of thrombin and fibrinogen has also been reported.[107]

One of the most selective natural inhibitors of thrombin is the protein hirudin,[108] which is present in leech saliva. This material is not inhibitory

[93] T. Astrup, *Acta Chem. Scand.* **1**, 744 (1947).
[94] R. H. Landaburu and W. H. Seegers, *Can. J. Biochem. Physiol.* **37**, 1361 (1959).
[95] R. L. Lundblad and J. H. Harrison, *Biochem. Biophys. Res. Commun.* **45**, 1344 (1971).
[96] T. M. Chulkova and V. N. Orekhovich, *Biokhimia* **33** (Part 2), 1000 (1968).
[97] I. Kotaku, A. Matsushima, M. Bando, and Y. Inada, *Biochim. Biophys. Acta* **214**, 400 (1970).
[98] S. Magnusson and T. Hofmann, *Can. J. Biochem.* **48**, 432 (1970).
[99] K. A. Kohler and S. Magnusson, *Arch. Biochem. Biophys.* **160**, 175 (1974).
[100] E. H. H. Li, C. Orton, and R. D. Feinman, *Biochemistry* **13**, 5012 (1974).
[101] G. F. Lanchantin, J. A. Friedmann, and D. W. Hart, *J. Biol. Chem.* **244**, 865 (1969).
[102] R. E. Feeney, G. E. Means, and J. C. Bigler, *J. Biol. Chem.* **244**, 1957 (1969).
[103] S. Karpatkin and M. Karpatkin, *Biochem. Biophys. Res. Commun.* **57**, 1111 (1974).
[104] K. Skaug and T. B. Christensen, *Biochim. Biophys. Acta* **230**, 627 (1971).
[105] F. A. Pitlick, R. L. Lundblad, and E. W. Davie, *J. Biomed. Mater. Res.* **3**, 95 (1969).
[106] R. D. Rosenberg and P. S. Damus, *J. Biol. Chem.* **248**, 6490 (1973).
[107] R. Flensgrud, B. Østerud, and H. Prydz, *Biochem. J.* **129**, 83 (1972).
[108] F. Markwardt, this series Vol. 19, p. 924.

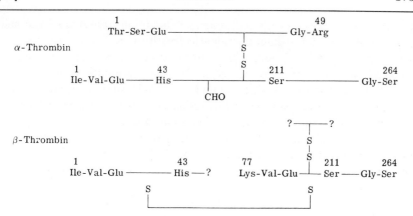

FIG. 5. The tentative structures for α- and β-thrombin.

toward factor X_a esterase activity at concentrations even as high as 1 mg/ml.[109]

Thrombin—Miscellaneous

The concentration of purified thrombin from dilute solutions with preservation of biological activity is a problem frequently encountered when working with this enzyme. In general, it is possible to lyophilize crude preparations of thrombin with no loss of activity, but the lyophilization of purified thrombin is generally associated with unacceptable losses of activity.[31] Precipitation with acetone[31] or ammonium sulfate[39] has been used to concentrate thrombin. The use of ammonium sulfate has been reported to result in the loss of activity with the formation of inactive aggregates.[32] Adsorption to phosphocellulose and subsequent elution with high salt has also been used to concentrate thrombin.[41]

The irreversible adsorption of thrombin to gel filtration matrix material has been reported.[36] Successful use of this material requires a concentrated sample of protein at high ionic strength.

Primary Structure of Thrombin

Bovine Thrombin. The tentative structures of bovine α- and β-thrombins are shown in Fig. 5. A tentative primary structure for bovine α-thrombin has been reported.[110,111] It is likely that this is largely correct.

[109] R. J. Butkowski and K. G. Mann, unpublished observations, 1974.
[110] B. S. Hartley, *Philos. Trans. R. Soc. London B* **257,** 77 (1970).
[111] S. Magnusson, *in* "The Enzymes (P. D. Boyer, ed.) 3rd ed., Vol. 3, p. 277. Academic Press, New York, 1971.

The A chain consists of 49 amino acids. The amino-terminal amino acid
sequence of this chain should be identical with the amino-terminal amino
acid sequences of intermediate 2.[49] This appears to be the case, as recently
reported, with the possible exception of several amide assignments.[112]
The B-chain structure is still incomplete.[113] β-Thrombin probably has a
three-chain structure, as shown in Fig. 5. This structure is supported by
the finding of the same number of cysteine residues in β-thrombin as
compared to α-thrombin, as well as three polypeptide chains upon gel
filtration after performic acid oxidation.[38] The A_1 chain is apparently
quite small, as it is obscured by the urea in the gel filtration system described above after reduction and carboxymethylation in 8.0 M urea, or
on gel filtration in 6 M guanidinium chloride. It is also too small to be
seen on gel electrophoresis.[37] The structure of this fragment is still unknown. The B_1 fragment has been shown to have an amino terminal
amino acid sequence identical with the amino terminal end of the B
chain[38] and also contains the histidine (His 43), which is part of the active site[41] but does not contain carbohydrate.[88] The B_2 chain begins at
residue 77 (lysine)[38,41] and apparently continues through the remainder
of the B-chain sequence. This chain contains the seryl residue (Ser 211),
which reacts with DFP.[88] Thus it would appear that β-thrombin differs
from α-thrombin by loss of most of the A chain and of a portion of the
B chain which contains the carbohydrate moiety of α-thrombin. Much
less information is available on the structure of γ-thrombin. Current evidence[38,85] would suggest that γ-thrombin is derived from β-thrombin by
a single peptide bond cleavage in the B_2-chain in the vicinity of residue
117. It should be noted that a form of thrombin similar to β-thrombin
(thrombin E) has been isolated[40]; it has been suggested that the B_1 chain
is missing in this and the A chain is intact.

Human Thrombin. Human α-thrombin has the same basic two-chain
structure as that shown in the bovine system.[18] The amino-terminal amino
acid sequences[114] of the human β-chain appears to be very similar to that
previously described for the bovine B-chain. The A-chain of human
thrombin contains 36 residues instead of the 49 residues[114,115] in the bovine
A chain. The difference is in the amino-terminal portion, where 13 residues are deleted.

[112] C. M. Heldebrant, C. Noyes, H. S. Kingdon, and K. G. Mann, *Biochem. Biophys.
Res. Commun.* **54**, 155 (1973).
[113] S. Magnusson, L. Sottrup-Jensen, T.-E. Petersen, P. Klemmensen and E. Kouba,
Thromb. Diath. Haemorrh., Suppl. 57, 153 (1974).
[114] A. R. Thompson, L. H. Ericsson, and D. L. Enfield, *Circulation* **50**, 292 (1974).
[115] D. A. Walz and W. H. Seegers, *Biochem. Biophys. Res. Commun.* **60**, 717 (1974).

[15] Fibrin-Stabilizing Factor (Factor XIII)

By C. G. CURTIS and L. LORAND

Fibrin-stabilizing factor (FSF, factor XIII) is the inactive precursor in blood plasma for the enzyme fibrinoligase (activated FSF, FXIII$_a$). This transamidase performs the essential hemostatic function of catalyzing the formation of intermolecular γ-glutamyl-ε-lysine bridges between fibrin molecules, which greatly augments the physical stiffness of the clot structure and its resistance to lysis.[1] The enzyme may also be involved in strengthening the attachment of the fibrin network to the disrupted platelet membranes and to the surface of fibroblasts, the latter being conceivably important in wound healing. It is suggested that fibroblasts are covered by "cold-insoluble globulin" and that, in the presence of the transamidase, fibrin cross-links to this protein.[2,2a]

During coagulation, the fibrin-stabilizing factor zymogen is converted to the active enzyme in two distinct consecutive steps.[3,4] The first requires only thrombin, and the second involves a specific interaction of the thrombin-modified zymogen with calcium ions only:

$$\text{Zymogen} \xrightarrow{\text{thrombin}} \text{zymogen}' \xrightarrow{\text{Ca}^{2+}} \text{enzyme}$$

In addition to the zymogen occurring in plasma where it has an (ab) heterologous protomeric structure, it is also found in platelets where, by contrast, it does not contain the b subunit.[5,6]

Purification of Plasma Factor XIII

The original method for obtaining factor XIII from citrated outdated blood bank human plasma (or oxalated bovine plasma) by a salt-gradient DEAE-cellulose chromatographic procedure, as described by Lorand and

[1] See L. Lorand, *Ann. N.Y. Acad. Sci.* **202**, 6 (1972).
[2] D. F. Mosher, *J. Biol. Chem.* **250**, 6614 (1975).
[2a] D. F. Mosher, *J. Biol. Chem.* **251**, 1639 (1976).
[3] C. G. Curtis, P. Stenberg, C.-H. J. Chou, A. Gray, K. L. Brown, and L. Lorand, *Biochem. Biophys. Res. Commun.* **52**, 51 (1973).
[4] R. D. Cooke and J. J. Holbrook, *Biochem. J.* **141**, 71 (1974).
[5] See P. A. McKee, M. L. Schwartz, S. V. Pizzo, and R. L. Hill, *Ann. N.Y. Acad. Sci.* **202**, 127 (1972).
[6] See H. Bohn, *Ann. N.Y. Acad. Sci.* **202**, 256 (1972).

Gotoh,[7] has been only slightly modified.[8] All solutions now contain 1 mM EDTA, and, after DEAE-cellulose chromatography, the material purified from 5–10 liters of plasma may be applied to a 2.5 × 90 cm Sepharose 6-B column equilibrated with 50 mM Tris chloride, pH 7.5, containing 1mM EDTA. Elution is carried out with the same buffer at a flow rate of 20 ml/hr at 0°, collecting 5-ml fractions. The active peak emerging at V_e/V_o ~1.4 is concentrated by precipitation with 40% ammonium sulfate; it is then dissolved and dialyzed against 50 mM Tris-chloride–1 mM EDTA, pH 7.5, and stored at 4°. Zymogen activity appears to be stable for several months. A minor peak (as monitored at 280 nm) emerging from the Sepharose column at V_e/V_o ~1.8 contains noncatalytic carrier b subunit.

For purification of platelet zymogen Ref. 9 may be consulted.

Assays for Measuring Fibrin-Stabilizing Factor Activity

To meet various requirements, a number of quantitative rate assays were developed for measuring the enzyme activity generated after treatment of the zymogen with thrombin and calcium ions. By omitting thrombin, these tests may be readily adapted also for investigating other endo-γ-glutamyl-ε-lysine transferases.[10] These enzymes, exemplified by guinea pig liver transglutaminase,[11] are widely distributed in many tissues.[10,12]

Assay methods fall into two categories. Procedures A to C utilize the incorporation of labeled synthetic amine substrates into a protein acceptor, such as casein, and have been used extensively both in conjunction with the purified material and also for assays in human plasma. In addition, there are methods (procedures D and E), based entirely on synthetic substrates, which have been employed so far mainly for kinetic and mechanistic studies with the highly purified enzyme.

The method based on the incorporation of dansylcadaverine into casein by measuring the protein-bound fluorescence after interruption of

---

[7] L. Lorand and T. Gotoh, this series Vol. 19, p. 770.

[8] C. G. Curtis, K. L. Brown, R. B. Credo, R. A. Domanik, A. Gray, P. Stenberg, and L. Lorand, *Biochemistry* **13**, 3774 (1974).

[9] M. L. Schwartz, S. V. Pizzo, R. L. Hill, and P. A. McKee, *J. Biol. Chem.* **248**, 1395 (1973).

[10] L. Lorand and P. Stenberg, "Endo-γ-glutamine-ε-lysine Transferases. Enzymes Which Cross-Link Proteins," Handbook of Biochemistry and Molecular Biology (G. D. Fasman, ed.), Vol. II, p. 669. Chemical Rubber Co., Cleveland, 1976.

[11] D. D. Clarke, M. J. Mycek, A. Neidle, and H. Waelsch, *Arch. Biochem. Biophys.* **79**, 338 (1959).

[12] S. I. Chung, *Ann. N.Y. Acad. Sci.* **202**, 240 (1972).

the enzymic reaction with trichloroacetic acid (procedure A) was already described in detail in an earlier volume of this series[7] and will only be mentioned here. This procedure has been widely used in clinical research, for the genetic appraisal of hereditary factor XIII deficiency[13]; for evaluating transfusions administered to the deficient individual[14]; for identifying circulating acquired inhibitors in hemorrhagic disorders of fibrin stabilization[15,16] etc.[17] The covalent enzymic incorporation of dansylcadaverine into casein and some other proteins is accompanied by an increase in quantum yield and a shift in the wavelength of the emitted fluorescent light and may be measured in a continuous manner (procedure B).[18] A "filter paper assay" was developed[19] for measuring the enzymic incorporation of isotopically labeled amine substrates (e.g., ^{14}C or ^{3}H labeled putrescine, histamine glycine ethyl ester, dansylcadaverine) into protein acceptors, such as native or modified casein (procedure C).

Water-soluble and relatively stable thiol esters, typified by β-phenylpropionylthiocholine, were found to be particularly useful for fully synthetic substrate assays with fibrinoligase.[20-22] The enzyme catalyzes the hydrolysis as well as the aminolysis of these esters (Fig. 1) and methods have been developed for monitoring both reactions in a continuous manner. Formation of the thiol product (P_1) may be followed colorimetrically through on-line reaction with Nbs_2 (procedure D).[21] The water-insoluble amide coupling product (P_3) formed, for example, in the reaction between β-phenylpropionylthiocholine and dansylcadaverine, is measured by the continuous change of fluorescence in a heptane extracting phase (procedure E).[22]

[13] L. Lorand, T. Urayama, A. C. Atencio, and D. Y. Y. Hsia, Am. J. Human Genet. 22, 89 (1970).
[14] L. Lorand, T. Urayama, J. deKiewiet, and H. A. Nossel, J. Clin. Invest. 48, 1054 (1969).
[15] L. Lorand, N. Maldonado, J. Fradera, A. C. Atencio, B. Robertson, and T. Urayama, Br. J. Haematol. 23, 17 (1972).
[16] R. Rosenberg, R. W. Colman, and L. Lorand, Br. J. Haematol. 26, 271 (1974).
[17] P. Henriksson, U. Hedner, I. M. Nilsson, J. Boehm, B. Robertson, and L. Lorand, Pediatr. Res. 8, 789 (1974).
[18] L. Lorand, O. M. Lockridge, L. K. Campbell, R. Myhrman, and J. Bruner-Lorand, Anal. Biochem. 44, 221 (1971).
[19] L. Lorand, L. K. Campbell-Wilkes, and L. Cooperstein, Anal. Biochem. 50, 623 (1972).
[20] L. Lorand, C.-H. J. Chou, and I. Simpson, Proc. Natl. Acad. Sci. U.S.A. 69, 2645 (1972).
[21] C. G. Curtis, P. Stenberg, K. L. Brown, A. Baron, K. Chen, A. Gray, I. Simpson, and L. Lorand, Biochemistry 13, 3257 (1974).
[22] P. Stenberg, C. G. Curtis, D. Wing, Y. S. Tong, R. B. Credo, A. Gray, and L. Lorand, Biochem. J. 147, 155 (1975).

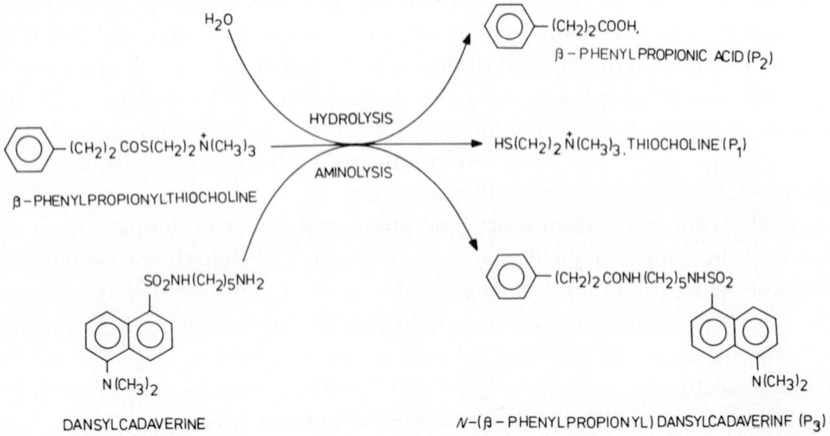

FIG. 1. An outline of the reaction of β-phenylpropionylthiocholine iodide with dansylcadaverine. Formation of thiocholine (P_1) is measured by reaction with 5,5'-dithiobis(2-nitrobenzoic acid) [C. G. Curtis, P. Stenberg, K. L. Brown, A. Baron, K. Chen, A. Gray, I. Simpson, and L. Lorand, *Biochemistry* 13, 3257 (1974)], and the water-insoluble coupling product (P_3) is quantitated by fluorescence measurements after extraction into a heptane phase [P. Stenberg, C. G. Curtis, D. Wing, Y. S. Tong, R. B. Credo, A. Gray, and L. Lorand, *Biochem. J.* 147, 155 (1975); L. Lorand, A. J. Gray, K. Brown, R. B. Credo, C. G. Curtis, R. A. Domanik, and P. Stenberg, *Biochem. Biophys, Res. Commun.* 56, 914 (1974)].

Specific examples are given below for the four more recent procedures (i.e., B to E). When working with plasma rather than with purified factor XIII, it is necessary to desensitize or remove the intrinsic fibrinogen prior to the addition of thrombin. The desensitization procedure which involves mild heat treatment of plasma (56°, 4 min) in the presence of glycerol or ethyleneglycol[14,19] appears to be simple and yields rather reproducible results. In methods B and C, commercially available bovine thrombin is specified as a convenience for activating factor XIII. In our laboratory, however, we prefer to use purified bovine or human thrombin.[23] Both α- and γ-thrombin were found to activate factor XIII equally well.[24]

Procedure A

This procedure is based on incorporation of dansylcadaverine into casein by measuring protein-bound fluorescence.[7]

[23] See R. L. Lundblad, H. S. Kingdon, and K. G. Mann, this volume [14].
[24] R. B. Credo, M. J. Fasco, and J. W. Fenton, II, *Fed. Proc., Fed. Amer. Soc. Exp. Biol.* 35, 648 (Abst. 2441) (1976).

Procedure B. Continuous Rate Assay for Measuring the Enzymic Incorporation of a Fluorescent Amine into a Protein Acceptor[18]

Reagents

Aqueous ethyleneglycol monomethyl ether, 20% (Cellosolve, Matheson).

Dithiothreitol 0.2 M, (Calbiochem) dissolved in 50% glycerol

Thrombin (Parke-Davis Bovine Thrombin Topical). A vial (10,000 NIH units) is dissolved in 20 ml of 25 mM Tris-HCl, pH 7.5, containing 25% glycerol and stored frozen. For the assay the stock thrombin is diluted with an equal volume of 50 mM Tris-HCl buffer containing 40 mM CaCl$_2$.

Borate/potassium chloride 0.1 M, buffer, pH 9.0, containing 1.5% Cellosolve

CaCl$_2$, 0.4 M, dissolved in 50 mM Tris-HCL, pH 7.5

Dansylcadaverine[25] 4 mM, dissolved in 50 mM Tris-HCl, pH 7.5

Acetylated casein, 5%, dissolved in 0.1 M borate/potassium chloride buffer, pH 9.0. Acetylation of Hammarsten casein (Nutritional Biochemicals) was performed as recommended by Fraenkel-Conrat.[26] The modified protein was dialyzed against water and lyophilized.

Procedure. Citrated, frozen plasma is thawed at 37°, an aliquot (0.2 ml) is mixed with 0.2 ml of ethylene glycol monomethyl ether in a 10 × 75 mm glass tube, which is then immersed in a 56° bath (Temp-Blok, Lab-line) for 4 min. After cooling the "desensitized" plasma to room temperature, the following solutions are added: 50 μl of dithiothreitol, 100 μl of diluted thrombin, 1.2 ml of borate/potassium chloride buffer, 50 μl of CaCl$_2$, and 10 μl of dansylcadaverine. The tube is placed in a fluorometer thermostatted at 37°, and the base line is recorded using an excitation wavelength of 360 nm and an emission wavelength of 500 or 546 nm. The enzymic reaction is initiated by the addition of 200 μl of acetylated casein solution.

The linear rate of increase in fluorescence intensity as well as a good signal-to-noise ratio, permits evaluation of factor XIII potency in plasma within a few minutes. As alternatives to plasma, the activity of purified factor XIII preparations and other transamidases may also be monitored by this method. Various types of caseins, as well as succinylated β-lacto-

---

[25] L. Lorand, N. G. Rule, H. H. Ong, R. Furlanetto, A. Jacobsen, J. Downey, N. Oner, and J. Bruner-Lorand, *Biochemistry* **7**, 1214 (1968).

[26] H. Fraenkel-Conrat, this series Vol. 4, p. 251.

globulin and succinylated lysozyme[27] can be used as protein acceptors, although neither of the latter two support the reaction in their native form. An adaptation of this technique utilizes dephosphorylated-acetylated β-casein as the amine acceptor.[28] This substrate has the added advantage that it does not bind calcium ions, which are essential for the activation of transamidating enzymes.

Procedure C. A "Filter Paper Assay" for Measuring the Enzymic Incorporation of Radioactively Labeled Amines into Proteins[19]

Reagents

Bovine Thrombin Topical (Parke-Davis) is used to activate plasma factor XIII: 10,000 NIH units are suspended in 20 ml of 25 mM Tris-HCl, pH 7.5, containing 25% glycerol. This solution is stored in small aliquots at $-20°$.

N,N-Dimethylcasein, 2%, in 50 mM Tris-HCl, pH 7.5. This material is prepared from Hammersten casein (Schwarz-Mann). Alkylation of the free amino groups of lysine[29] prevents casein from acting as an amine donor and cross-linking itself.

1,4-[^{14}C]Putrescine dihydrochloride (usually about 60 mCi/mmole) is obtained from Amersham-Searle Corporation. The dry contents of a single ampoule (250 μCi) are dissolved in 2.5 ml of 10 mM unlabeled putrescine in 50 mM Tris-HCl, pH 7.5, and stored at $-20°$.

Dithiothreitol (DTT), 0.2 M, dissolved in 50 mM Tris-HCl, pH 7.5, and stored at $-20°$

CaCl$_2$, 40 mM, dissolved in 50 mM Tris-HCl, pH 7.5, and stored at $-20°$

Aqueous glycerol, 50%

Trichloroacetic acid (TCA), 5 and 10%

Ethanol/acetone mixture, 1:1 by volume

Procedure. Samples (20 μl) of citrated plasma are dispensed in 5 × 25 mm Pyrex tubes (Kontes) or 2-ml conical dispo-beakers (Scientific Products) to which 5 μl of 50% aqueous glycerol are added. The tubes are gently mixed using a Vortex Genie mixer, and heated at 56° for 4 min in a Lab-line Temp Blok unit. The metal heating block is then placed on ice until the temperature is lowered to approximately 45°, after which it is transferred to another heating unit set at 37°. After equilibra-

[27] I. M. Klotz, this series Vol. 11, p. 576.
[28] R. D. Cooke and J. J. Holbrook, *Biochem. J.* **141**, 71 (1974).
[29] Y. Lin, G. E. Means, and R. E. Feeney, *J. Biol. Chem.* **244**, 789 (1969).

tion, 5 μl of DTT, 5μl of CaCl$_2$, and 5 μl of thrombin solution are added with mixing between each addition. In general, 30 min is allowed for thrombin activation, after which 5–10 μl of ^{14}C-labeled putrescine and 20 μl of N,N-dimethylcasein solution are added and mixed.

After a desired reaction time (usually 30 min) 5–10-μl aliquots of each reaction mixture are spotted on precut 1 cm^2 pieces of Whatman 3 MM filter paper. The papers are immediately dropped into a copper wire basket submerged in a beaker of cold 10% TCA (about 5 ml of TCA per filter paper), which is continuously stirred with a magnetic bar beneath the basket, and the filter papers are washed for 20 min (after the addition of the last paper). The basket is then transferred to a 5% solution of cold TCA. The washing procedure is continued as follows: 5 min each with three changes of cold 5% TCA, 5 min with ethanol/acetone (50:50), and 5 min with acetone. At least 5 ml of wash solution is used for each filter paper. Finally, the filter papers are dried in air, or at ~60°, for 20–30 min, and radioactivity is measured in 10 ml of toluene-based scintillation fluid (containing 3 g of 2, 5-diphenoxazole and 0.3 g of 1,4-bis[2-(5-phenyloxazoyl)]benzene per liter).

Appropriate controls in which CaCl$_2$ is replaced by 5 mM EDTA are included. Blank filter papers to which no radioactivity is applied, and another group of papers to which known amounts of 1,4-[^{14}C]putrescine are applied, are also carried through the wash procedure. These blanks usually range between 20 and 30 cpm above background.

Under the conditions described above, using 20 μl of plasma, the protein-bound isotope remaining on the filter papers ranges between 2500 and 6000 cpm/30 min (i.e., less than 20% of the total radioactivity applied to each test filter paper).

Procedure D. Continuous Rate Assay for Measuring the Production of Free Thiol in the Enzymic Hydrolysis and Aminolysis of Thiol Esters[21]

In the hydrolysis and aminolysis of thiol esters, formation of the thiol product is assayed with Ellman's reagent,[30] [5, 5′-dithiobis(2-nitrobenzoic acid), or Nbs$_2$]. Production of the yellow anion of 3-mercapto-2-nitrobenzoic acid ($\epsilon_{412\ nm} = 1.36 \times 10^4\ M^{-1}cm^{-1}$) is monitored continuously at 412 nm using a double-beam recording spectrophotometer (e.g., Cary Model 16). All measurements are made at 25° in mixtures of 1.0 ml containing 50 mM Tris-acetate buffer, pH 7.5. In general, the direct recording of thiol production in enzymic reactions with Nbs$_2$ is possible only if the enzyme itself (e.g., cholinesterase[31]) is insensitive to this reagent.

[30] G. L. Ellman, *Arch. Biochem. Biophys.* **82**, 70 (1959).
[31] G. L. Ellman, K. D. Courtney, V. A. Andres, Jr., and R. M. Featherstone, *Biochem. Pharmacol.* **7**, 88 (1961).

With fibrinoligase, inclusion of Nbs$_2$ in the reaction mixtures is tolerated because the enzyme is protected against inactivation by the thiol ester substrate.

Reagents. These are dissolved in 50 mM Tris-acetate buffer of pH 7.5 and include

Nbs$_2$, 0.5 mM
CaCl$_2$, 0.05–2.0 M
EDTA, 0.1 M

The optimum concentrations of the different thiolesters and amines will vary for each substrate.[21]

Proteolytic activation of the purified zymogen is achieved by mixing 1 mg of this protein with 3.75 NIH units of purified thrombin[23] in 1.0 ml of 50 mM Tris-acetate buffer, pH 7.5, for 20 min at 25°. In the absence of calcium ions, this thrombin-modified zymogen may be kept on ice for several hours.

In a typical rate assay, the calcium-dependent hydrolytic reactions catalyzed by fibrinoligase are measured as shown in the accompanying tabulation.

Reagent	Reference cuvette	Test cuvette
Nbs$_2$	20 µl	20 µl
CaCl$_2$	—	50 µl
EDTA	50 µl	—
Thiol ester	100 µl	100 µl
Tris-acetate buffer	780 µl	780 µl
Thrombin-modified factor XIII	50 µl	50 µl
Total volume	1.0 ml	1.0 ml

Before adding the thrombin-modified factor XIII (to give a final concentration of approximately 0.2 μM, as determined by active-site titration, see below), the cuvettes are balanced for transmitted light. The contents in each cuvette are mixed and the Ca^{2+} dependent release of thiol is monitored at 412 nm.

The protocol for aminolysis is similar, except for the inclusion of the amine substrate, which is added to both cuvettes immediately after the thiol ester. The final volume of the reaction mixture is maintained at 1.0 ml by appropriate adjustment of the buffer. For measuring nonenzymic hydrolytic rates, Nbs$_2$ is added to both cuvettes, but the thiol ester is omitted from the reference cuvette and is substituted by an equal volume of Tris-acetate buffer.

Procedure E. Continuous Rate Assay for Measuring the Production of Water-Insoluble Fluorescent Amide in the Enzymic Acyl Transfer Reaction between β-Phenylpropionylthiocholine and Dansylcadaverine[22,32]

Amide formation in the enzymic reaction between β-phenylpropionyl-thiocholine and the fluorescent dansylcadaverine can be followed directly by monitoring the continuous extraction of the water-insoluble coupling product into *n*-heptane, layered above the enzymic mixture.

Reagents. All reagents are dissolved in 50 mM Tris-HCl and adjusted to pH 7.5.

β-Phenylpropionylthiocholine iodide, 10 mM ($\epsilon_{225 \text{ nm}} = 1.9 \times 10^4$ $M^{-1}\text{cm}^{-1}$)

Dansylcadaverine, 1 mM ($\epsilon_{327 \text{ nm}} = 4.67 \times 10^4$ $M^{-1}\text{cm}^{-1}$)

CaCl$_2$, 0.5 M

Tris-HCl buffer, 50 mM, pH 7.5

n-Heptane, analytical grade (Fisher H-340)

Procedure. Activation of the zymogen with purified thrombin is carried out as described in procedure D.

In this system, the two water-soluble substrates, β-phenylpropionyl-thiocholine iodide (20 μl) and dansylcadaverine (20 μl), together with the thrombin-activated zymogen (20 μl) and Tris-HCl buffer (20 μl), are added to a circular cuvette (10 mm diameter). *n*-Heptane (2.0 ml) is then layered over the aqueous mixture. In general the enzymic reaction is initiated by the addition of Ca^{2+} ions (20 μl) through the heptane phase directly into the aqueous phase. This procedure makes it possible to record any nonspecific, Ca^{2+}-independent amide formation although their rates are negligible with respect to the enzyme-catalyzed velocities.

Extraction (at 25°) of the water-insoluble amide coupling product, *N*-(β-phenylpropionyl)dansylcadaverine [C$_6$H$_5$(CH$_2$)$_2$CONH(CH$_2$)$_5$NHS-O$_2$C$_{10}$H$_6$N(CH$_3$)$_2$], into the heptane phase is facilitated by stirring the aqueous–heptane interphase with a magnetic bar at 500 rpm. The fluorescence is recorded (excitation wavelength, 340 nm, emission wavelength, 460 nm) in a Ratio Spectrophotofluorometer, and the cuvette is adjusted so that only the fluorescence in the heptane phase is recorded.

The relative fluorescence readings may be converted into actual quantities of the amide coupling product (RCONHR′) by using [methyl-^{14}C]-dansylcadaverine of known specific activity. As the reaction progresses, samples (50 μl) are withdrawn from the *n*-heptane phase at various fluorescence readings and assayed for total radioactivity. In this way, it is possible to construct an accurate calibration curve for the fluorescent

[32] L. Lorand, A. J. Gray, K. Brown, R. B. Credo, C. G. Curtis, R. A. Domanik, and P. Stenberg, *Biochem. Biophys. Res. Commun.* **56**, 914 (1974).

amide coupling product. Using this system, the $K_{m(app)}$ for β-phenyl-propionylthiocholine iodide was found to be 0.4 mM; $K_{m(app)}$ for dansyl-cadaverine = 1.4 mM; k_{cat} = 2.1 sec⁻¹.

A Specific Activity Stain for Detecting the Fibrin-Stabilizing Factor Zymogen (FXIII), Thrombin-Modified Zymogen, and Fibrinoligase (FXIIIₐ) on Disc-Gel Electrophoresis

The ability of fibrinoligase (factor XIIIₐ) to incorporate dansylca-daverine into N,N-dimethylcasein in a calcium-dependent manner has been utilized in the development of a specific fluorescent activity stain-ing procedure for locating the enzyme after disc-gel electrophoresis.[33]

Reagents

"Activity-Stain" solution: 50 mM Tris-acetate buffer, pH 7.5 con-taining 1% N,N-dimethylcasein, 2 mM dansylcadaverine, 5 mM CaCl₂, and 5 mM dithiothreitol. When testing for the zymogen, thrombin (10 U/ml) must also be included.
Aqueous TCA, 10%

Aliquots (approximately 30 μg) of fibrin-stabilizing factor (factor XIII), with or without thrombin activation, are subjected to gel electrophoresis[34] using 7% acrylamide and a buffer system of pH 7.8 containing N-tris(hydroxymethyl)methyl-2-aminoethanesulfonic acid (TES), bis(2-hydroxyethyl)iminotris(hydroxymethyl)methane(Bis-Tris), and cacodylic acid at an ionic strength of 0.01. In order to locate enzy-mic activity, the gels are immersed immediately after electrophoresis in the "Activity-Stain" solution. After 6 hrs at 24°, the gels are fixed in 10% trichloroacetic acid to remove free dansylcadaverine. To observe the casein-bound amine, the gels are suspended in 50 mM Tris-acetate, pH 7.5, and illuminated with UV light (Blak Ray UV122).

This nondenaturing technique is suitable for demonstrating changes in the quaternary structure of fibrin-stabilizing factor (Fig. 2).[1,32]

A "Filter Paper Method" for Titrating the Active Center Sulfhydryl Group of Fibrinoligase.[3,8]

Fibrinoligase is a thiol enzyme with its active center cysteine residue located in the a' subunit.[3,8,35] The amino acid sequence around the reactive

[33] L. Lorand, C. G. Curtis, R. A. Domanik, A. J. Gray, J. Bruner-Lorand, R. Myhrman, and L. K. Campbell-Wilkes, *Fed. Proc., Fed. Am. Soc. Exp. Biol.* **32**, 274 Abstr. 338 (1973).

[34] D. Rodbard and A. Chrambach, *Anal. Biochem.* **40**, 95 (1971).

[35] J. J. Holbrook, R. D. Cooke, and I. B. Kingston, *Biochem. J.* **135**, 901 (1973).

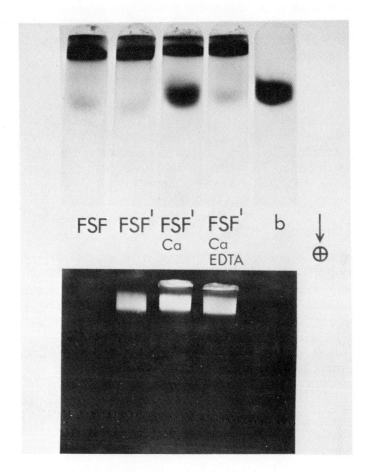

FIG. 2. The reversible, calcium-dependent dissociation of the subunit structure of the thrombin-modified zymogen (FSF′), as seen by disc-gel electrophoresis (pH 7.8) under nondenaturing conditions [L. Lorand, A. J. Gray, K. Brown, R. B. Credo, C. G. Curtis, R. A. Domanik, and P. Stenberg, *Biochem. Biophys. Res. Commun.* **56**, 914 (1974)]. *Top:* Coomassie blue protein stain. *Bottom:* Transamidase specific fluorescent activity stain, comprising dansylcadaverine, *N,N*-dimethylcasein, and calcium chloride. The original zymogen, before exposure to thrombin, is denoted as FSF. The isolated "b" subunit (right-hand gels) is shown for comparison.

cysteine in the human plasma protein is -Gly-Gln-Cys-Trp-,[35] which is identical to that found for guinea pig liver transglutaminase.[36] Activation of the zymogen by thrombin, in itself, is not sufficient to generate transamidase activity.[8] Unmasking of the functional active center requires a

[36] J. E. Folk and P. W. Cole, *J. Biol. Chem.* **241**, 3238 (1966).

specific interaction of the thrombin-modified zymogen with Ca^{2+} ions and general ionic strength also plays a role.[8] The concentration of potential enzymic sites (E_o) in purified preparations of fibrinoligase can be measured by direct titration with 1-[^{14}C]iodoacetamide.

Reagents

Purified thrombin[23] (approximately 800 NIH units/ml)

1-[^{14}C]Iodoacetamide (Radiochemical Centre, Amersham, Bucks). A stock solution of 1-[^{14}C]iodoacetamide (0.758 mM) is prepared by dissolving 1 vial (50 μCi; 33 mCi/mmole) in 2 ml of water and stored frozen. For titration, the stock 1-[^{14}C]iodoacetamide is diluted with 50 mM Tris-HCl buffer, pH 7.5, to give a concentration range between 1 and 50 μM

CaCl$_2$, 0.5 M dissolved in 50 mM Tris-HCl, pH 7.5

Tris-HCl, 50 mM, pH 7.5

In a typical titration, the limited proteolytic activation[37-39] of the isolated factor XIII zymogen is carried out by allowing 1.8 mg of this protein to react with 7 NIH units of thrombin in 0.5-ml solutions of 50 mM Tris-chloride (or acetate) buffer of pH 7.5. After an activation period of 20 min at 25°, aliquots (10 μl) corresponding to approximately 35 μg of protein are taken for alkylation with 1-[^{14}C]iodoacetamide. Labeling of fibrinoligase is carried out over a range of 1-[^{14}C]iodoacetamide concentrations (0–10 μM) by mixing 10 μl of the enzyme, 10 μl of CaCl$_2$, 10 μl of 1-[^{14}C]iodoacetamide, and 20 μl of 50 mM Tris-chloride buffer. Alkylation is initiated by the addition of CaCl$_2$ and allowed to proceed for 20 min at 25°. Aliquots (10 μl) are then tested for incorporation of the isotope and for residual enzymic activity.

In order to measure the protein-bound radioactivity, 10-μl samples are applied to Whatman 3 MM filter paper (1 cm^2) and the papers processed as described for the "filter-paper assay" (C). Quantitation of radioactivity associated with the dried filter papers is determined by comparison with "standards," prepared by the addition of known amounts of 1-[^{14}C]iodoacetamide to untreated Whatman 3 MM papers (1 cm^2). Controls for the washing procedure are obtained by applying samples of 1-[^{14}C]iodoacetamide to untreated filter papers followed by washing as described above.

Residual enzymatic activity is assayed by the continuous fluorescence

---

[37] L. Lorand and K. Konishi, *Arch. Biochem. Biophys.* **105**, 58 (1964).

[38] L. Lorand, J. Downey, T. Gotoh, A. Jacobsen, and S. Tokura, *Biochem. Biophys. Res. Commun.* **31**, 222 (1968).

[39] M. L. Schwartz, S. V. Pizzo, R. L. Hill, and P. A. McKee, *J. Biol. Chem.* **246**, 5851 (1971).

technique described in procedure (E). The inhibition of fibrinoligase activity by 1-[^{14}C]iodoacetamide is accompanied by a proportional uptake of the isotope. By appropriate selection of 1-[^{14}C]iodoacetamide concentration, the fibrinoligase activity can be totally abolished, and at this point, the extent of isotope incorporation approaches a maximum value which is taken as a measure of the titratable active sites of the enzyme. The functional active site concentration of the enzyme may be calculated by using an $E^{1\%}_{280\ nm} = 13.8$[9] and assuming one equivalent of active site per 160,000 g of the (ab) protein. On this basis, the functional purity of different preparations of human fibrinoligase prepared in our laboratories has ranged from 41 to 85%.

Properties

As isolated from plasma, the zymogen form of factor XIII has a heterologous (ab) protomeric structure. Individually, the a and b subunits have molecular weights of 75,000 and 88,000, respectively,[9] which associate into oligomeric assemblies with molecular weights in the 160,000[38] and 320,000[9] ranges.

In a single limited proteolytic attack, thrombin appears to modify only the a subunit, removing a 37 amino acid-containing peptide fragment from its N-terminus[40] (Fig. 3). Thrombin activation alone, however, does not generate transamidase activity, and the modified zymogen still retains its heterologous (a'b) association.[3,8,32] Addition of calcium ions to the thrombin-modified zymogen is followed by a dissociation of the subunits, the unmasking of the active center thiol in the a' subunit and the simultaneous generation of transamidase activity.[8,28] The calcium-dependent steps appear to be reversible and may be summarized as follows:

$$
\begin{array}{ccccc}
\text{1. Hydrolysis} & & \text{2. Dissociation} & & \text{3. Unmasking} \\[4pt]
\text{(ab)} & \xrightarrow{\text{thrombin}} & \text{(a'b)} & \xrightarrow[\text{slow}]{\text{high Ca}^{2+}} & \text{(a')} \xrightarrow[\text{fast}]{\text{low Ca}^{2+}} \text{(a*)} \\
\text{zymogen} & \searrow & & \searrow & \text{enzyme} \\
& \text{activation} & & \text{(b)} & \\
& \text{peptide} & & \text{carrier subunit} &
\end{array}
$$

The a* subunit represents the enzymically active protein (with its unmasked cysteine; a'-SH), which itself may aggregate without much change in catalytic activity, while the b subunit is a catalytically inert glycoprotein functioning as a carrier for the a and a' subunits. Exposure of the reactive sulfhydryl to iodoacetamide titration can be achieved with other group 2a cations with an apparent order of efficacy of $Ca^{2+} > Sr^{2+} > Ba^{2+} > Mg^{2+}$.[8]

[40] T. Takagi and R. F. Doolittle, *Biochemistry* **13**, 750 (1974).

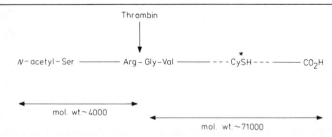

FIG. 3. Limited proteolysis of the "a" subunit of human fibrin-stabilizing factor by thrombin, according to T. Takagi and R. F. Doolittle [*Biochemistry* **13**, 750 (1974)]. The asterisk marks the active center cysteine residue, shown by C. G. Curtis, P. Stenberg, C.-H. J. Chou, A. Gray, K. L. Brown, and L. Lorand, [*Biochem. Biophys. Res Commun.* **52**, 51 (1973)] to become unmasked on this subunit specifically on addition of calcium ions. The N-terminal sequence (Takagi and Doolittle) is

```
      1                 5                   10  11                 15
N-Ac-Ser-Glu-Thr-Ser-Arg-Thr-Ala-Phe-Gly Gly-Arg-Arg-Ala-Val-Pro-Pro-Asn-
         20                  25            28      30
      Asn-Ser-Asn-Ala-Ala-Glu-Asp-Asp-Leu-Pro-Thr-Val-Glu-Leu-Gln-Gly-----
                                    35          ↓
                             Val-Pro-Arg Gly-Val-Asx-Leu-Glx-Glx-----
```

At low calcium levels, the progression curves obtained by continuous assay procedures are characterized by a lag phase subsequent to the addition of the thrombin-modified zymogen before the steady-state rate is achieved.[4,21,22,32] This lag phase is attributed to the dissociation of the b subunit and is abolished at high calcium concentrations. Of the two calcium-dependent steps, dissociation is the slower one[32] and requires a higher concentration of Ca^{2+} [22] Zn^{2+}-ions compete against Ca^{2+} in inhibiting enzyme activity (in step 3 of "Unmasking") with a K_i for $Zn^{2+} \approx 6 \times 10^{-6}$ M, but Zn^{2+} has no effect on the Ca^{2+}-dependent heterologous dissociation (i.e., step 2) of the subunits.[40a]

Inactivation of thrombin-activated platelet factor XIII (like the plasma enzyme) by iodoacetamide, occurs only in the presence of calcium and is a result of selective alkylation of a single cysteine thiol group.[41] In contrast to the plasma enzyme, the progression curves of transamidation between β-phenylpropionylthiocholine and dansylcadaverine catalyzed by thrombin-activated platelet factor XIII show no lag phase prior to steady state even at low calcium concentrations.[22,32] Moreover, the calcium concentration required for optimum activity of the platelet enzyme using this assay procedure is considerably less (<2 mM) than for the plasma enzyme. The platelet enzyme is composed of a single type of subunit, which appears to correspond to the a subunit of the plasma

[40a] R. B. Credo, P. Stenberg, Y. Tong, and L. Lorand, *Fed. Proc., Fed. Am. Soc. Exp. Biol.* **35**, 1631 Abstr. 1390 (1976).

[41] S. I. Chung, M. S. Lewis, and J. E. Folk, *J. Biol. Chem.* **249**, 940 (1974).

enzyme.[9,41] Addition of purified b subunit to the thrombin-modified platelet zymogen induces an appreciable lag phase. This kinetic difference disappears if the b subunit is added to the platelet enzyme itself, i.e., after the addition of calcium ions.[32] Collectively the kinetic data, together with data obtained by ultracentrifugation, gel filtration, and chemical cross-linking,[41] support the view[9] that the platelet zymogen and b subunit can form a hybrid association similar to that found with plasma factor XIII. The platelet zymogen has been reported to have a molecular weight of 160,000 with an a_2 structure.[39] Its activation, in terms of the changes on the protomeric unit may be given as:

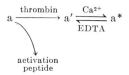

Unmasking of the active center of the thrombin-modified platelet zymogen requires relatively low Ca^{2+} concentration[22] and occurs within seconds.[32]

[16] Coagulant Protein of Russell's Viper Venom

By BARBARA C. FURIE and BRUCE FURIE

Factor X, a plasma zymogen that participates in an intermediate phase of blood coagulation, may be activated physiologically by complexes of factor VIII, activated factor IX, Ca(II), and phospholipid[1] or factor VII, tissue factor, and Ca(II).[2] Alternatively a protease in Russell's viper venom, termed the coagulant protein, can activate factor X in the presence of Ca(II) through a proteolytic mechanism[3-6]:

$$
\begin{array}{ccc}
\left\{ \begin{array}{l} \text{Factor VIII} \\ \text{Factor IX}_a \\ \text{Ca(II)} \\ \text{Phospholipid} \end{array} \right. &
\left\{ \begin{array}{l} \text{Tissue factor} \\ \text{Factor VII} \\ \text{Ca(II)} \end{array} \right. &
\left\{ \begin{array}{l} \text{Coagulant protein of} \\ \text{Russell's viper venom} \\ \text{Ca(II)} \end{array} \right.
\end{array}
$$

Factor X ⎯⎯⎯⎯⎯⎯↓⎯⎯↓⎯⎯⎯⎯⎯⎯⎯↓⎯→ Activated factor X

[1] R. G. Macfarlane, R. Biggs, B. J. Ash, and K. W. E. Denson, *Br. J. Haematol.* **10,** 530, (1964).

[2] C. Hougie, *Proc. Soc. Exp. Biol. Med.* **101,** 132 (1959).

[3] W. J. Williams and M. P. Esnouf, *Biochem. J.* **84,** 52 (1962).

[4] M. P. Esnouf and W. J. Williams, *Biochem. J.* **84,** 62 (1962).

[5] K. Fujikawa, M. E. Legaz, and E. W. Davie, *Biochemistry* **11,** 4892 (1972).

[6] B. C. Furie, B. Furie, A. J. Gottlieb, and W. J. Williams, *Biochim. Biophys. Acta* **365,** 121 (1974).

Historically, the venom coagulant protein has been important in the elucidation of the activation mechanism of blood coagulation. In efforts to develop a therapy for the clotting defect in hemophilia, R. G. Macfarlane examined a number of snake venoms for their ability to accelerate blood coagulation. One venom, that from Russell's viper, clotted hemophilic or normal blood in 17 sec at a 1:10,000 dilution.[7] This reaction was metal dependent and shown to have a requirement for Ca(II). Factor X was subsequently shown to be the substrate of the venom, as demonstrated using an immunological approach employing rabbit antivenom immune serum.[8] It was not until the work of Williams and Esnouf,[3,4] in which the interaction of the coagulant protein of Rusell's viper venom and bovine factor X were analyzed in a chemically defined system, that the activation mechanism of factor X was approached at the molecular level and shown to be proteolytic.

Assay Methods

Principle. The coagulant protein of Russell's viper venom enzymically converts factor X to its activated form.[9] Since activated factor X is an enzyme, the assay of the venom coagulant protein is directly coupled to the assay of activated factor X. Three methods have been employed and are reviewed in detail. The first and second are biological assays that correlate the coagulant protein concentration to the acceleration of blood clotting of relipidated, recalcified platelet-poor human plasma or the rate of activation of purified factor X, respectively. In the absence of a synthetic substrate for the coagulant protein, a third method employs coupled enzyme reactions and monitors the rate of generation of the esterase activity of activated factor X.

A. One-Stage Method[3]

Reagents

CaCl$_2$, 25 mM in 10 mM Tris-HCl, pH 7.4
Citrated human plasma is obtained from a minimum of three normal donors. Whole blood (50 ml) is removed by venipuncture into a plastic syringe and anticoagulated by the addition of 1 ml of 19%

---

[7] R. G. Macfarlane, and B. Barnett, *Lancet* 1, 985 (1934).
[8] R. G. Macfarlane, *Br. J. Haematol.* 7, 496 (1961).
[9] D. Kosow, B. C. Furie, and H. A. Forasteri, *Thromb. Res.* 4, 219 (1974).

sodium citrate. The red cells and platelets are removed by centrifugation, and the plasma samples are pooled and stored at −15°

Phospholipids: Partial thromboplastin (Hyland Laboratories) diluted as per instructions from manufacturer

Procedure. The reaction mixture, consisting of 0.1 ml of enzyme, 0.1 ml of pooled plasma, and 0.1 ml of phospholipid, is incubated at 37° for 1 min in 12×75-mm glass tubes. The reaction is initiated with the addition of 0.1 ml of 25 mM CaCl$_2$ in 10 mM Tris-HCl, pH 7.4 at 37°, and the clotting time is determined. In general, the coagulant protein activity is rate limiting in this system when the clotting time is between 10 and 19 sec; these clotting times are usually reproducible to ±0.5 sec. A dilution of the coagulant protein solution is usually chosen which yields clotting times between 10 and 14 sec. In general, assays are usually performed in duplicate or triplicate.

A plot of the logarithm of the clotting time (seconds) vs the logarithm of the coagulant protein concentration is linear. Clotting activity (units) is expressed relative to the units of activity obtained from standard curves employing crude Russell's viper venom where 0.1 ml of venom yields a clotting time between 10 and 14 sec, and may be defined arbitrarily as containing 100 units of activity. Specific activity is expressed as units of clotting activity per milligram of protein, and is calculated relative to the activity of crude venom.

B. *Two-Stage Method*[4]

Reagents

CaCl$_2$, 25 mM in 10 mM Tris-HCl, pH 7.4
Bovine factor X (see chapter [10], this volume)
Citrated human plasma, platelet poor (see method A, above)
Phospholipid: partial thromboplastin (Hyland Laboratories)

Procedure. Factor X, the venom coagulant protein, and calcium chloride (final concentration 8 mM) are incubated at 37°. The protein concentrations are varied as required so that the concentration of the venom protein is limiting; i.e., the rate of factor X activation is related to the venom protein concentration. After an appropriate time interval (e.g., 1 min), 0.1 ml of the reaction mixture is added to 0.4 ml of 0.01 M Tris-HCl, pH 7.4, 0.1 M NaCl. The diluted reaction mixture (0.1 ml) and 0.1 ml of 25 mM CaCl$_2$ are added simultaneously to a solution containing 0.1 ml of plasma and 0.1 ml of phospholipid at 37°. The clotting time is measured, and the coagulant activity of activated factor X is determined using a standard curve constructed by plotting the logarithm of the clotting time (seconds) against the logarithm of the concentration of activated factor X. The coagulant protein activity may be related to the rate of development of activated factor X.

Alternatively, the generation of activated factor X may be monitored by observation of the development of a protein band associated with activated factor X upon dodecyl sulfate polyacrylamide gel electrophoresis of the above activation mixture.[5] Aliquots are removed from the reaction mixture at various time intervals, and the reaction is terminated by the addition of EDTA to 10 mM. These aliquots are then processed in the usual manner in preparation for gel electrophoresis.

C. Continuous Spectrophotometric Assay[9]

A continuous spectrophotometric assay is based on the following coupled enzymic reactions:

$$
\begin{array}{c}
\text{coagulant protein} \\
\text{Ca(II)} \\
\downarrow \\
\text{Factor X} \xrightarrow{\hspace{3cm}} \text{activated factor X} \\
\downarrow \\
\text{N-carbobenzoxy-L-tyrosine p-nitrophenyl ester} \rightarrow \text{N-carbobenzoxy-L-tyrosine} \\
\text{$+$ nitrophenol}
\end{array}
$$

In the reaction mixture, the coagulant protein is added to a system containing Ca(II), factor X, and the N-CBZ-L-Tyr-p-nitrophenyl ester. As activated factor X is generated by proteolytic cleavage of factor X by the venom coagulant protein, the rate of hydrolysis of N-CBZ-L-Tyr-p-nitrophenyl ester by activated factor X increases exponentially and may be monitored at 400 nm by the absorption of free nitrophenol (Fig. 1). Specifically, the rate of hydrolysis of N-CBZ-L-Tyr-p-nitrophenyl ester (and the rate of generation of nitrophenol) in a system in which activated factor X is developing at a constant rate should conform to the equation for rectilinear motion with constant acceleration: $X = V_0 t + \frac{1}{2} a t^2$ where V_0 is the velocity at zero time and is equivalent to the endogenous rate of hydrolysis, t is time, a is acceleration, and X is the absorbance of the product. From this equation, subtraction of $V_0 t$ from X yields X_{corr}. If X_{corr} is plotted against t^2, a straight line is obtained whose slope equals $a/2$. Thus a is proportional to the velocity of the reaction of the activation of factor X by the coagulant protein. Using this continuous spectrophotometric assay, the kinetic parameters describing the interaction of coagulant protein and factor X may be determined.

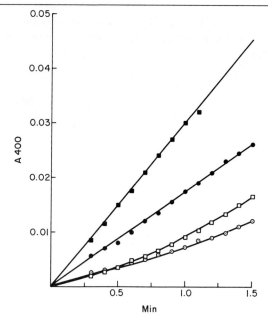

FIG. 1. Development of activated factor X esterolytic activity as a function of time during activation of factor X by the coagulant protein (○, □). These results are compared to factor X which has been completely activated prior to the introduction of the ester substrate (●, ■). Protein concentrations include coagulant protein 24 ng/ml; factor X, 0.015 μM (○, ●); factor X, 0.03 μM (□, ■). Other conditions as per text. From D. Kosow, B. C. Furie, and H. A. Forasteri, *Thromb. Res.* **4**, 219 (1974).

Reagents

   Bovine factor X: see Chapter [10], this volume
   N-CBZ-L-Tyr-p-nitrophenyl ester, 2.5 mM in acetone
   CaCl₂, 0.1 M
   Methanol, 0.2 M
   Tris-HCl, 0.5 M, pH 7.5

Procedure. The reaction mixture, contained in a 1-ml cuvette maintained at 25°, is composed of 0.1 ml of 0.1 M CaCl₂, 0.1 ml of 0.2 M methanol, 0.2 ml of 0.5 M Tris-HCl, pH 7.5, 0.1 ml of coagulant protein, and water to a total volume of 1 ml. N-CBZ-L-Tyr-p-nitrophenyl ester (10 μl of 2.5 mM) is added to the cuvette, and a base line is established at 400 nm. The reaction is initiated with the addition of factor X to a final concentration that is saturating with respect to the coagulant protein concentration. The generation of nitrophenol is monitored at 400 nm over the course of about 2 min. Concentrations of coagulant protein, factor X, and N-CBZ-L-Tyr-p-nitrophenyl ester must be appropriately chosen so that the rate of factor X activation is constant (i.e., the coagulant protein remains saturated with its substrate, factor X) and activated factor X hydrolyzes N-CBZ-L-Tyr-p-nitrophenyl ester maximally (i.e., activated factor X remains saturated with its substrate, N-CBZ-L-Tyr-p-nitrophenyl ester) over the course of the experiment. The coagulant protein activity may be obtained graphically by calculation of a in plots of X_{corr} versus t^2.

Comments

The one-stage assay for the coagulant protein is highly sensitive but is limited to a narrow range of coagulant protein concentrations between 2 to 20 ng per assay. By the design of the assay, great precision is difficult; furthermore, kinetic experiments are precluded because of the inability to vary the substrate concentration. Nonetheless, the sensitivity and rapidity of the assay has made it useful in the estimation of coagulant protein activity. The two-stage assay permits measurement of coagulant protein activity in a chemically defined system with highly purified reagents. However, it has not been possible to standardize coagulant protein activity in terms of the number of micromoles of factor X converted to activated factor X per unit time.

The continuous spectrophotometric assay yields quantitative data of reasonable precision, but is restricted to highly purified systems. N-CBZ-L-Tyr-p-nitrophenyl ester is a substrate for many esterases, and the contamination of a coagulant protein preparation with these esterases complicates or precludes the use of this method. Furthermore, the poor solubility characteristics of this substrate require the addition of methanol to the reaction mixture.

Recently, assays based upon changes in the ultraviolet absorption spectrum of factor X which are associated with activation have shown promise for the development of a direct spectrophotometric assay of the venom coagulant protein.[10] The activation of factor X is associated with a decrease in absorption in the aromatic region, and may be measured by monitoring the change in absorption at 285 nm or 292.5 nm. The coagulant protein activity may be expressed in terms of $\Delta OD_{285 \text{ nm}}$ of factor X per minute.

Purification

Classical Procedure

The coagulant protein of Russell's viper venom may be purified by chromatography using gradient elution from DEAE-cellulose and gel fil-

[10] B. Furie and B. C. Furie, *Fed. Proc., Fed. Am. Soc. Exp. Biol.*, in press.

tration on Sephadex G-200[6] using a modification and extension of an earlier procedure.[3] Crude RVV (200 mg) is dissolved in 20 ml of 0.01 M Tris-PO$_4$, pH 8.5, and dialyzed against 50 volumes of 0.01 M Tris-PO$_4$, pH 8.5, for 24 hr. The solution is clarified by centrifugation and the supernatant (22 ml) is placed on a 2.8 × 30 cm column of DEAE-cellulose equilibrated in 0.01 M Tris-PO$_4$, pH 8.5. The chromatogram is developed by stepwise elution with 0.01 M Tris-PO$_4$, pH 8.5; 0.01 M Tris-PO$_4$, pH 7.2; 0.01 M Tris-PO$_4$, pH 6.0. Fractions are collected at a flow rate of 27 ml/hr.

Coagulant protein activity may be eluted from the DEAE-cellulose column by using a linear 2000-ml NaCl gradient from 0.01 M Tris-PO$_4$ to 0.5 M NaCl–0.01 M Tris-PO$_4$ at pH 6.0. Fractions of high specific coagulant activity are combined and concentrated using a Diaflo ultra-filtration apparatus with a UM-10 membrane. Coagulant protein activity is recovered in about 5% yield with a 7-fold increase in specific activity. The coagulant protein is then dialyzed against 100 volumes of 0.01 M Tris-HCl–0.1 M NaCl, pH 7.4, for 24 hr.

Gel filtration is performed in 2.5 × 100 cm columns of Sephadex G-200 equilibrated with 0.01 M Tris-HCl–0.1 M NaCl, pH 7.4. A single symmetrical protein peak is eluted which contains the coagulant protein activity and is resolved from a contaminating peptidase activity. All of the above procedures are performed at 4°. The coagulant protein yields a single band on sodium dodecyl sulfate (SDS) and disc gel electrophoresis. The activity, purified 7.7-fold from crude venom in 5% yield, is stable at −15° for several months.

Affinity Chromatography

The absence of synthetic ligands which bind tightly and specifically to highly specific proteases, such as the coagulation proteins, has precluded using affinity methods for the purification of these and related proteases. For these reasons, we have employed trivalent lanthanide ions as substitutes for Ca(II) to facilitate the binding of the venom protein of Russell's viper to factor X but to inhibit metal-dependent activation.[11] Because of the general applicability of this approach to other Ca(II)-dependent proteases, the rationale for the use of lanthanide ions as it has been applied to this model system, the interaction of coagulant protein and factor X, is described in detail.

Trivalent lanthanide ions have ionic radii that are similar to Ca(II), e.g., Gd(III), 0.938 Å; Ca(II), 0.940 Å. While substitution of lanthanide ions for Ca(II) may lead to the retention of catalytic activity (e.g.,

[11] B. C. Furie and B. Furie *J. Biol. Chem.* **250**, 601 (1975).

amylase,[12] prothrombin[13]) lanthanide ions competitively inhibit Ca(II)-dependent catalysis in staphylococcal nuclease[14] and the activation of factor X by the coagulant protein of Russell's viper venom.[11] The ability of lanthanide ions to participate in nonproductive enzyme–substrate complex formation has led to a strategy for the affinity purification of proteins involved in Ca(II)-dependent protein–protein interactions using lanthanide ions.[10] The use of lanthanide ions to study Ca(II)-binding proteins is illustrated in our previous studies described below on staphylococcal nuclease and factor X.

Staphylococcal nuclease, a Ca(II)-dependent endonuclease, is an ideal model of the Ca(II)-binding protein.[15] From X-ray crystallographic studies,[16] it would appear that the Ca(II)-binding site of nuclease involves Asp 19, Asp 21, Asp 40, and Glu 43, whose carboxylate groups extend into a planar array with the Ca(II) at the center. The lanthanide ions appear to bind to these same ligands. Studies of the inhibition of Ca(II)-dependent nuclease activity by lanthanide ions produced data consistent with a kinetic model of competitive inhibition. Furthermore, lanthanide ions may be substituted for Ca(II) in the formation of a nuclease–substrate analog–metal ternary complex.[17] The substitution of lanthanide ions for Ca(II) to facilitate nuclease binding to affinity chromatography columns of thymidine diphosphate (a substrate analog) covalently linked to agarose[14] suggested that this method might be extended to study metal-dependent protein–protein interaction in general and those involved in blood coagulation in particular.

We have studied the substitution of trivalent lanthanide ions for Ca(II) in the Ca(II)-dependent activation of bovine factor X by the coagulant protein of Russell's viper venom.[11] Factor X contains two high-affinity metal-binding sites which bind Gd(III), Sm(III), and Yb(III) with a K_d of about 0.4 μM. Dy(III), Yb(III), Tb(III), Gd(III), EU(III), La(III), or Nd(III) do not substitute for Ca(II) in the activation of factor X by the venom coagulant protein at pH 6.8. Kinetic data derived at various Ca(II) and lanthanide ion concentrations are consistent with models of competitive inhibition of Ca(II) by Nd(III) and

[12] D. W. Darnell and E. R. Birnbaum, *Biochemistry* **12**, 3489 (1973).

[13] B. C. Furie, K. G. Mann, and B. Furie, *J. Biol. Chem.* **251**, in press.

[14] B. Furie, A. Eastlake, A. N. Schechter, and C. B. Anfinsen, *J. Biol. Chem.* **248**, 5821 (1973).

[15] C. B. Anfinsen, P. Cuatrecasas, and H. Taniuchi, *in* "The Enzymes" (P. D. Boyer, ed.), 3rd ed. Vol. 4, pp. 177–204. Academic Press, New York, 1971.

[16] A. Arnone, C. J. Bier, F. A. Cotton, V. W. Day, E. E. Hazen, Jr., D. Richardson, J. S. Richardson, and A. Yonath, *J. Biol. Chem.* **246**, 2302 (1971).

[17] B. Furie, J. H. Griffin, R. J. Feldmann, E. A. Sokoloski, and A. N. Schechter, *Proc. Natl. Acad. Sci. U.S.A.* **71**, 2833 (1974).

yield a K_i of 1–4 μM. While lanthanide ions cannot substitute for Ca(II) in the activation of factor X by the venom protein, they do promote protein complex formation of factor X–metal–venom coagulant protein. This property of lanthanide ion interaction with factor X and the venom coagulant protein has facilitated the purification of the coagulant protein from crude venom using affinity chromatography. Using a column containing factor X covalently bound to agarose and equilibrated in 10 mM Nd(III), Tb(III), Gd(III), or La(III), the coagulant protein in crude venom is bound to the insoluble matrix and may be eluted using 10 mM EDTA (Fig. 2). The coagulant protein has been purified 10- to 15-fold

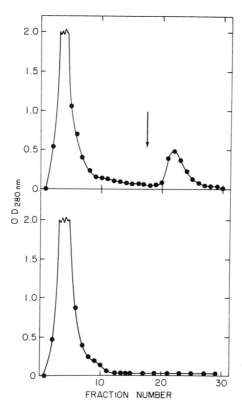

Fig. 2. Affinity purification of the venom coagulant protein on Sepharose-factor X. Upper: A column of Sepharose-factor X was equilibrated at 4° with 10 mM NdCl$_3$, 0.5 M NaCl, 25 mM imidazole, pH 6.8; the crude venom-Nd(III) solution containing 20 mg of protein in 1 ml of buffer was applied and developed with about 15 ml of buffer; bound protein, containing coagulant activity, was eluted using 10 mM EDTA, 0.5 M NaCl, 25 mM imidazole, pH 6.8 (arrow). Lower: Identical experiment as described above except that NdCl$_3$ was deleted from the initial equilibration solution. From B. C. Furie and B. Furie, *J. Biol. Chem.* **250**, 601 (1975).

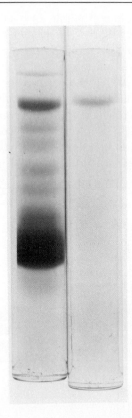

Fig. 3. Sodium dodecyl sulfate gel electrophoresis of venom coagulant protein prepared by affinity chromatography. Left: Crude Russell's viper venom (200 μg). Right: Affinity-purified coagulant protein (21 μg). Gels were stained with Coomassie Blue. From B. C. Furie and B. Furie, *J. Biol. Chem.* **250**, 601 (1975).

in 40% yield from crude venom by this method. The purified protein migrates as a single band on gel electrophoresis in SDS (Fig. 3).

Reagents

Factor X. Bovine factor X (2.5 mg/ml) prepared in 20 mM potassium phosphate–0.13 M NaCl at pH 6.8. See Chapter [10], this volume.

Cyanogen bromide

Sepharose 4B (Pharmacia) washed with distilled water

Ammonium acetate, 0.1 M, at pH 7.1

NaCl, 1 M

Guanidine HCl, 6M

Imidazole, 25 mM–0.15 M NaCl–10 mM NdCl$_3$, pH 6.8
Imidazole, 25 mM–0.15 M NaCl–10 mM EDTA, pH 6.8
Crude Russell's viper venom, 20 mg/ml, in 25 mM imidazole–0.5 M
    NaCl, pH 6.8
NdCl$_3$, 0.1 M, in HCl
Imidazole, 25 mM–0.5 M NaCl–10 mM NdCl$_3$, pH 6.8
Imidazole, 25 mM–0.5 M NaCl–10 mM EDTA, pH 6.8

Procedure. Activated agarose is prepared with 300 mg of cyanogen bromide and 2 ml of washed Sepharose 4B. Three milliliters of factor X (2.5 mg/ml) in 20 mM potassium phosphate–0.13 M NaCl at pH 6.8, are added to the activated Sepharose suspension, and the mixture is stirred for 18 hr at 4 °. The Sepharose–factor X conjugate is then washed sequentially in a sintered-glass funnel with 10 ml each of 0.1 M ammonium acetate at pH 7.1; 1 M NaCl; 6 M guanidine HCl; 25 mM imidazole–0.15 M NaCl–10 mM NdCl$_3$ at pH 6.8; and 25 mM imidazole–0.15 M NaCl–10 mM EDTA adjusted to pH 6.8.

A 0.7 × 3 cm column of Sepharose–factor X is equilibrated at 4° with 25 mM imidazole–0.5 M NaCl–10 mM NdCl$_3$ at pH 6.8. Crude Russell's viper venom, 20 mg, dissolved in 1 ml of 25 mM imidazole–0.5 M NaCl at pH 6.8, is dialyzed for 2 hr at 4° against 25 mM imidazole, 0.5 M NaCl, pH 6.8. The sample is then made 10 mM with respect to NdCl$_3$ by addition of 0.1 M NdCl$_3$ in HCl and the pH readjusted to 6.8. The sample is incubated at 4° for 2 hr. The precipitate which forms is removed by centrifugation and the supernatant is placed onto the column. Most of the crude venom protein does not adhere to the Sepharose–factor X column; the bound protein, about 5% of the total protein, is eluted with 10 mM EDTA–25 mM imidazole–0.5 M NaCl, at pH 6.8 (Fig. 2). This protein fraction shows an average 10- to 15-fold increase in specific activity of the coagulant protein and appears to be electrophoretically homogeneous (Fig. 3). About half of the original coagulant protein activity is associated with the bound-protein fraction. None of the crude venom protein is bound to the Sepharose–factor X column in the absence of metal ions. The Sepharose–factor X column may be recycled into starting buffer and used repeatedly.

Comments

Systematic study of the optimal conditions for affinity chromatography has suggested the importance of certain parameters. Chromatography at 23° yielded material with lower specific activity than when chromatography was performed at 4°. Nonspecific protein binding could be minimized by the use of buffers containing 0.5 M NaCl; the occasional removal of noncovalently bound protein from the Sepharose–factor X column with 6 M guanidine HCl after multiple exposures of the column to crude protein preparations similarly reduced nonspecific protein binding.

The application of this method to the purification of enzymes involved in Ca(II)-dependent protein–protein interactions requires that (1) lanthanide ions competitively inhibit catalysis during enzyme–protein substrate interaction; (2) one protein of the protein pair must be available in purified or at least partially purified form for coupling to Sepharose; (3) the association constant of the protein complex must be sufficiently high at a pH of 7.0 or lower to permit tight binding (the lanthanide ions form insoluble hydroxides at alkaline pH); (4) the protein to be purified must not be precipitated by the lanthanide ion concentrations employed.

Properties

Substrate Specificity. The venom coagulant protein is a highly specific endopeptidase whose only known substrate is factor X. The activation of factor X involves proteolytic cleavage of an Arg-Ile bond on the heavy chain of factor X which yields a small activation fragment with a molecular weight of 11,000 and a large activation fragment (containing the active site serine) with a molecular weight of 44,000.[5,6,18] These fragments appear to form a noncovalent protein complex.[6] The activation of factor X has a absolute Ca(II) requirement.

Although all preparations of venom protein contain N^{α}-toluenesulfonyl-L-arginine methyl ester (TAME) esterase activity, whether or not the venom coagulant protein hydrolyzes TAME remains unclear. Previous experiments have indicated that factor X is a competitive inhibitor of the hydrolysis of TAME by the venom protein[3]; to the contrary, diisopropylphosphofluoridate has been demonstrated to inhibit TAME esterase activity without inhibiting the coagulant activity.[19]

The venom coagulant protein does not hydrolyze L-Phe-L-Phe, L-Phe-Gly, N-CBZ-Gly-L-Phe, N-CBZ-Gly-L-PheNH$_2$, Gly-L-Phe-L-Phe, N-CBZ-Glu-L-Tyr, benzoyl-L-ArgNH$_2$, L-Leu-Gly, Gly-L-Phe, Gly-L-PheNH$_2$, hippuryl-L-Phe, or the B chain of bovine insulin.[20]

Enzyme Characteristics. The venom coagulant protein, monitored by

[18] R. D. Radcliffe and P. G. Barton, *J. Biol. Chem.* **247**, 7735 (1972).
[19] C. M. Jackson, J. G. Gordon, and D. Hanahan, *Biochim. Biophys. Acta* **252**, 255 (1971).
[20] B. Furie, A. J. Gottlieb, and W. J. Williams, *Fed. Proc., Fed. Am. Soc. Exp. Biol.* **29**, 709 (1970).

the two-stage clotting assay, demonstrates maximal activity between pH
7.0 and pH 9.0. The pH optimum is at 7.7.[4] At pH 6.0, the activity is
reduced by 75%.[4] The activation of bovine factor X by the venom coagulant protein has an absolute requirement for Ca(II). The rate of
activation is maximal at concentrations of 7 mM Ca(II).[4]

   Enzyme Kinetics. Although the details of the interaction of the venom
coagulant protein and factor X have not been studied, analysis of the
coagulant protein activity using the continuous spectrophotometric assay
(see method C) monitoring the generation of activated factor X
has yielded limited kinetic data.[9] The activation of factor X by the
venom protein is an enzymic reaction consistent with Michaelis–Menten
kinetics where the coagulant protein is the enzyme and factor X is the
substrate. Specifically, the K_m for this reaction at 25° is 0.25 μM (Fig. 4).

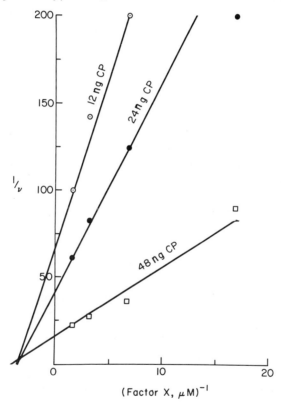

Fig. 4. Lineweaver–Burk plot of the kinetics of activation of bovine factor X by
the venom coagulant protein. Velocity is given as change in the absorption at 400
nm/min². Coagulant protein (CP) concentration as indicated; other conditions as
described in text (Method C). From D. Kosow, B. C. Furie, and H. A. Forasteri,
Thromb. Res. **4,** 219 (1974).

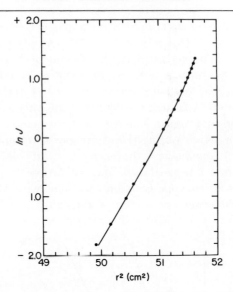

Fig. 5. Plot of ln J versus r^2 for coagulant protein in 0.01 M Tris-HCl, pH 7.4, 0.1 M NaCl. Equilibrium was attained after 40 hr at 16,000 rpm and 4°; 30-mm cell.

Physical Properties. The molecular weight of the venom coagulant protein prepared by either classical chromatographic methods or affinity chromatography has been estimated by SDS polyacrylamide gel electrophoresis using the method of Weber and Osborn.[21] In the presence or absence of β-mercaptoethanol, the electrophoretic migration of the coagulant protein preparations correlates with a molecular weight of about 60,000.[11,20]

The following physical data may be summarized[3]: partial specific volume, $\bar{V}$, 0.72 ml/g; $E^{1\%}_{280\,nm}$, 13.4; diffusion coefficient, D_{20}, 4.66 at pH 6.45; sedimentation velocity, s_{20}, 5.8 at pH 5.8. Our own sedimentation velocity experiments at pH 7.4 have demonstrated a symmetrical peak with an $s_{20,w}$ of 5.52 and 5.76 in separate experiments. Using the determined D_{20} and s_{20} values, Williams and Esnouf estimated a molecular weight of 105,000 for the venom protein.[3]

High-speed sedimentation equilibrium centrifugation of the coagulant protein consistently demonstrated a curvilinear relationship in the plot of the natural logarithm of the concentration, expressed as interference fringes (J) vs the radial distance from the axis of rotation squared (cm²) (Fig. 5).[20] A straight-line fit yields an apparent molecular weight of 124,000 for concentrations from 0.09 to 0.27 mg/ml.

The apparent physical heterogeneity displayed by the coagulant pro-

[21] K. Weber and M. J. Osborn, *J. Biol. Chem.* **244**, 4406 (1969).

tein suggested that these data are consistent with a concentration-dependent self-associating system. The monomer molecular weight of 60,000 determined by SDS gel electrophoresis and an average molecular weight of 124,000 for the protein determined by sedimentation equilibrium centrifugation in Tris buffer at pH 7.4 suggests that, at the concentration studied, the venom protein exists as a dimer.

Metal-Binding Properties. The activation of factor X by the venom coagulant protein has an absolute requirement for Ca(II) and maximal enzymic activity is observed at 7 mM Ca(II). Steady-state rate dialysis experiments employing radioactive lanthanide ions as probes of Ca(II)-binding sites on metal-binding proteins indicate that lanthanide ions do not bind significantly to the venom coagulant protein ($K_d > 1$ mM) but bind tightly to the venom substrate, factor X.[11]

[17] Ancrod,[1] the Coagulating Enzyme from Malayan Pit Viper (*Agkistrodon rhodostoma*) Venom

By C. Nolan, L. S. Hall, and G. H. Barlow

Ancrod is a thrombinlike enzyme in that it splits fibrinopeptide A from fibrinogen, in whole plasma or in purified form, producing a clot of modified fibrin.[2,3] Unlike thrombin, it does not cleave fibrinopeptide B from fibrinogen and does not activate factor XIII, the zymogen of the fibrin cross-linking enzyme, and thus, in the absence of other activators, ancrod-produced clots are not cross-linked.[4,5]

Methods of Assay

Coagulant Activity

The thrombinlike (coagulant) activity of ancrod is conveniently assayed by adding different dilutions of enzyme to a standard control blood plasma or a standard fibrinogen solution and comparing the clotting

[1] Ancrod is the generic name assigned to the coagulant fraction from *A. rhodostoma* venom by the World Health Organization.
[2] W. H. Holleman and L. J. Coen, *Biochim. Biophys. Acta* **200**, 587 (1970).
[3] M. R. Ewart, M. W. C. Hatton, J. M. Basford, and K. S. Dodge, *Biochem. J.* **118**, 603 (1970).
[4] G. H. Barlow, W. H. Holleman, and L. Lorand, *Res. Commun. Chem. Pathol. Pharm.* **1**, 39 (1970).
[5] K. H. Chen, I. Simpson, J. Bruner-Lorand, and L. Lorand, *Fed. Proc., Fed. Am. Soc. Exp. Biol.* **31**, 217 (abstr.) (1972).

times with those obtained with different dilutions of an ancrod or thrombin standard. An assay procedure using purified fibrinogen as substrate has been described[6,7]; a procedure using control plasma is outlined below.

Reagents

Control plasma: Fibritol oxalate (lyophilized normal coagulation control plasma), BBL No. 40778, reconstituted with water

Ancrod standard or NIH thrombin standard[8]

Sample diluent: 0.1% human serum albumin in 0.15 M NaCl containing 0.3% chlorobutanol

Procedure.[9] The control plasma (0.2 ml) and ancrod standard or unknown diluted with sample diluent to contain approximately 3–9 ancrod units/ml are preincubated briefly at 37° until temperature equilibrated. Preincubation of the plasma should not exceed 25 min. A 0.2-ml aliquot of enzyme is then mixed with the plasma (0.2 ml), and the time required for the clot to form is recorded. The assay tube is maintained at 37° with occasional shaking until the clot forms. Clot formation can be determined visually or automatically with a clot timer. Duplicate determinations are performed on each of three or more different dilutions of enzyme. The data can be plotted as the clotting time versus the reciprocal of the activity in units per milliliter, and the activity of the unknown estimated from the standard curve, which is linear in the range of 2–10 units/ml. Under the above conditions, an enzyme solution containing 9 ancrod units/ml (1.8 units per assay tube) produces a clot in about 20 sec.

Activity Units. One ancrod unit (coagulant activity) is equivalent to 0.33 CTA thrombin units.[9]

Esterolytic Activity

Purified ancrod can also be assayed spectrophotometrically or potentiometrically using various α-N-substituted arginine esters as substrates. Other esterases in the crude venom interfere with these assays. Spectrophotometric assays using the ethyl,[6,10,11] methyl,[11] and cyclo-

[6] M. P. Esnouf, this series, Vol. 19, p. 715.

[7] A detailed procedure for the assay of thrombin using purified fibrinogen as substrate has been described by Baughman, this series Vol. 19, p. 145.

[8] Ancrod standard is available from the National Institute for Biological Standards, Holly Hill, London NW36RB, England. NIH thrombin standard is available from the Bureau of Biologics, National Institutes of Health, Bethesda, Maryland 20014.

[9] G. H. Barlow and E. M. Devine, *Thromb. Res.* **5**, 695 (1974).

[10] M. W. C. Hatton, *Biochem. J.* **131**, 799 (1973).

[11] J. P. Collins and J. G. Jones, *Eur. J. Biochem.* **26**, 510 (1972).

hexyl[11] esters of α-N-benzoyl-L-arginine and toluenesulfonyl-L-arginine methyl ester (TAME)[6] have been described. An assay procedure using TAME,[12] adapted from one for trypsin,[13] is described below.

Reagents

TAME-HCl, 2.08 mM (39.4 mg/50 ml) in water (stable for at least 1 week at 4°).

Tris-HCl, 0.1 M, pH 8.0; 12.1 g of tris(hydroxymethyl)aminomethane (Tris) in 900 ml of water adjusted to pH 8.0 with HCl and diluted to 1 liter)

Procedure. At the time of use, a buffered 1.04 mM TAME solution is prepared by mixing equal volumes of the stock solution and the pH 8.0 Tris buffer. For the assay, 3 ml of buffered TAME solution is pipetted into each of two or more cuvettes with 1-cm light paths. To one cuvette (the reference cuvette) add 0.1 ml of water and mix; place the cuvettes in the spectrophotometer and allow them to equilibrate to the temperature of the compartment, regulated at 25°. Balance the reference cuvette at or near zero absorbance at 247 nm using the narrowest slit-width possible. At zero time add 0.1 ml of ancrod solution containing about 2.5–6 μg of enzyme to the sample cuvette, mix, and determine $\Delta A_{247 \text{ nm}}$/min, which is linear up to a ΔA of about 0.2. With spectrophotometers without recorders, the average $\Delta A_{247 \text{ nm}}$/min is determined from 5 absorbance readings taken at 30- or 60-sec intervals, depending on the rate.

Activity Units. One TAME unit is defined as the amount of enzyme that will catalyze the hydrolysis of 1 μmole of TAME per minute under the above conditions. If, for example, in the presence of 4.5 μg of enzyme per cuvette the $\Delta A_{247 \text{ nm}}$/min is 0.0200, then, based on a value of 0.409 cm^{-1} mM^{-1} for the absorbance change accompanying the hydrolysis of 1 μmole of substrate per milliliter of assay solution,[13] the specific activity of the enzyme preparation is

$$\frac{(0.0200 \text{ cm}^{-1} \cdot \text{min}^{-1})(3.1 \text{ ml})}{(0.409 \text{ cm}^{-1} \mu\text{mole}^{-1} \cdot \text{ml})(0.0045 \text{ mg enzyme})}$$

$$= 33.7 \text{ } \mu\text{moles/min/mg of enzyme}$$

$$= 33.7 \text{ TAME units/mg of enzyme}$$

Purification

A two-step chromatographic procedure for purifying ancrod from venom has been described.[6,14] In the first step the crude venom is chromatographed on triethylaminoethyl (TEAE)-cellulose, and in the second step

[12] C. Nolan and L. S. Hall, unpublished observations.
[13] K. A. Walsh, this series Vol. 19, p. 41.
[14] M. P. Esnouf and G. W. Tunnah, *Br. J. Haematol.* **13**, 581 (1967).

the coagulant fraction obtained is rechromatographed on Sephadex G-100. Preparations of greater purity are obtained by first chromatographing the TEAE-cellulose column fraction on IRC-50 and then on Sephadex G-100.[10,11] A more recent method[12] which employs affinity chromatography with agmatine (4-aminobutylguanidine) covalently linked to agarose beads ("agmatine–agarose") as the affinity column support, followed by gel filtration on Sephadex G-100, is described below.[15] Recoveries of approximately 90% are obtained.

Materials and Reagents

Agmatine–agarose: Agmatine was covalently linked by its amino group to agarose beads (Sepharose 4B, Pharmacia) essentially by the method of Cuatrecasas et al.[16] For analysis of the agmatine content, 4–5 ml (packed volume) of agmatine–agarose was further washed on a sintered-glass funnel successively with 50 ml each of water, 1 N acetic acid, water, and acetone, and dried over P_2O_5 under reduced pressure. A 20.0-mg portion of the dry sample was hydrolyzed in 5 ml of constant-boiling HCl (5.7 N) in a sealed Pyrex glass tube for 20 hr at 105°–110°, and the hydrolyzate was analyzed on an amino acid analyzer with agmatine sulfate as a calibration standard. Values of 0.10–0.12 μmole of agmatine per milligram dry weight were obtained.

Buffer I: 0.25 M NaCl in 0.01 M Tris-HCl, pH 8.1, containing 0.1% chlorobutanol by weight as a preservative (14.6 g of NaCl, 0.83 ml of 12 N HCl, 1.0 g of chlorobutanol, and water to 1 liter; the solution is adjusted to pH 8.1 with 2 M Tris before final adjustment of the volume.)

Buffer II: 0.15 M guanidine hydrochloride in 0.01 M Tris-HCl, pH 8.1, containing 0.1% chlorobutanol by weight (14.3 g of guanidine-HCl, 0.83 ml of 12 N HCl, 1.0 g of chlorobutanol, and water to 1 liter; adjust pH to 8.1 with 2 M Tris before final volume adjustment.)

Buffer III: 0.20 M NaCl in 0.02 M sodium citrate, pH 6.8, containing 0.1% chlorobutanol (11.7 g NaCl, 5.8 g trisodium citrate dihydrate, 1.0 g chlorobutanol and water to 1 liter; adjust pH to 6.8 with citric acid before final volume adjustment.)

Affinity Chromatography Step. A 1.9 × 14 cm column of agmatine–agarose is poured in a 1% solution of chlorobutanol in distilled water and carefully equilibrated with buffer I at room temperature. One (1.0)

[15] U.S. Patent 3,879,369.
[16] P. Cuatrecasas, M. Wilchek, and C. B. Anfinsen, *Proc. Natl. Acad. Sci. U.S.A.* **61**, 636 (1968).

gram of lyophilized venom is dissolved in 50 ml of buffer I, giving a clear, yellow solution. The venom solution is applied to the affinity column at a rate of 80–85 ml/hr, and the column is eluted at the same rate with buffer I by gravity flow at room temperature. The effluent is collected in fractions of 10–11 ml, and the absorbance of the effluent at 280 nm is monitored. Elution with buffer I is continued until a large "breakthrough" peak has eluted, and the column is then eluted with approximately 260 ml of buffer II. To determine the point at which elution with buffer II is begun, two criteria are used: elution with buffer I is continued until (a) the $A_{280 nm}$ of the effluent decreases to <0.04 following elution of the breakthrough peak, and (b) no esterolytic activity with benzyloxycarbonyl-L-phenylalanine-p-nitrophenyl ester as substrate can be detected in the effluent. A venom component with this type of esterolytic activity elutes mainly on the descending limb of the breakthrough peak, but detectable levels are present in the effluent until approximately 300–320 ml of effluent have been collected. The coagulant fraction elutes immediately in buffer II, the peak concentration eluting approximately one bed volume after elution with buffer II is begun. The $A_{280 nm}$ of the column effluent returns to <0.03 after elution of the coagulant peak. The coagulant fraction is pooled and concentrated by ultrafiltration. A small amount of coagulant activity that elutes in buffer I just ahead of the coagulant peak is not included in the pooled fraction.

The column can be used repeatedly and is regenerated by washing with 3–4 bed volumes of 0.5 N acetic acid and then immediately reequilibrating in buffer I. The column should be stored at 4° when not in use.

Ultrafiltration Step The coagulant fraction from the affinity column is concentrated to a volume of approximately 10 ml by ultrafiltration under nitrogen pressure at 3°–5° in an Amicon Diaflo apparatus with an Amicon UM-10 membrane.

Gel-Filtration Step. Sephadex G-100 is packed in a 2.5 × 92 cm column and equilibrated with buffer III at room temperature. The concentrated coagulant fraction is warmed to room temperature and applied to the column, and the column is eluted with buffer III at a rate of 50 ml/hr by gravity flow. The effluent is collected in fractions of 5 ml at room temperature and monitored for 280 nm-absorbing material. The coagulant fraction elutes as a symmetrical peak with the peak concentration emerging at 1.46 ± 0.03 column void volumes ($V_e/V_o = 1.46 \pm 0.03$) ; V_e/V_o is greater in very dilute buffers. This coagulant peak was the only $A_{280 nm}$-absorbing material detected except for a very minor amount that elutes just ahead of it. This minor fraction had coagulant activity with a specific activity in the clotting assay approximately equal to that of the main peak and is presumed to have been an aggregated form of ancrod.

It is present at a level of only about 3% of the main coagulant fraction and is not pooled with the main fraction.

Comments. Agmatine is a competitive inhibitor of the coagulant enzyme activity[12,17] and, when bound to the agarose support, permits specific binding of the enzyme. Since the guanidino moiety of agmatine is positively charged at the pH at which the affinity chromatography is performed (pH 8.1), the agmatine-agarose column also functions as an anion exchanger, and thus the enzyme, which is anionic at pH 8.1, is also nonspecifically bound to the column ionically, as are other anionic venom components. In the chromatography of the venom on agmatine–agarose, the column is first eluted with buffered 0.25 M NaCl (buffer I), of relatively high ionic strength, to wash unbound and nonspecifically bound venom components from the column. The enzyme is then eluted with a specific eluting agent, buffered 0.15 M guanidine hydrochloride (buffer II), with a lower ionic strength and the same pH. The guanidinium ion of this buffer, like agmatine, is a competitive inhibitor of the enzyme[10] and competes with the bound agmatine for binding at the active site of the enzyme. Use of the specific eluting agent permits a reduction in the concentration of the chloride ion, the counterion in the ion-exchange process, thus retarding the elution of any possible nonspecifically adsorbed components remaining on the column during the elution of the enzyme. Although some tightly bound protein remains on the column after elution of the enzyme fraction, it is not eluted with several additional bed volumes of the buffered 0.15 guanidine hydrochloride or even with 0.25 M guanidine hydrochloride, and rather extreme conditions, such as 0.5 N acetic acid, are required to elute it.

Little or no additional purification was achieved by gel filtration on Sephadex G-100; this step serves primarily to equilibrate the enzyme preparation with the desired buffer.

Properties

Specificity. Ancrod has a rather high degree of specificity for arginine esters and peptide bonds involving the carboxyl group of arginyl residues. Its proteolytic activity is very limited. Preparations of the enzyme without measurable proteolytic activity toward native casein[10,18] and denatured hemoglobin[10] have been reported. It splits fibrinopeptide A from fibrinogen α(A) chains by hydrolysis of the Arg-Gly bond split by thrombin, but, unlike thrombin, does not cleave the corresponding Arg-Gly bond in the β(B) chains.[2,3] Different preparations of the enzyme

[17] N. G. Rule and L. Lorand, *Biochim. Biophys. Acta* **81**, 130 (1964).
[18] K. S. Soh and K. E. Chan, *Toxicon* **12**, 151 (1974).

have been found to cleave the α chains at additional sites at a slower rate, leaving the β(B) and γ chains apparently intact even after prolonged incubation.[19-21] Ancrod has been reported to split an Arg-His bond (residues 23–24) as well as the Arg-Gly bond (residues 16–17) referred to above, but not an Arg-Val bond (residues 19–20) in an isolated N-terminal segment of the α(A) chains of human fibrinogen; no additional bonds were cleaved when a fragment, the "disulfide knot," containing longer N-terminal segments of the α(A), β(B), and γ chains was incubated with the enzyme.[22]

In contrast to some preparations of lower purity, sufficiently purified ancrod does not activate factor XIII[4,5,20] and the enzyme is without apparent effect on other clotting factors[23] and has no direct effect on plasminogen.[6,20,24]

Esters of α-N-substituted L-arginine are substrates of ancrod, as noted above, whereas various amides of arginine and esters of lysine tested are hydrolyzed relatively very slowly or not at all.[6,10,11,14,24] The initial ancrod preparations of Esnouf and Tunnah[14] were reported to hydrolyze α-N-benzoyl-L-arginine ethyl ester (BAEE), TAME, and benzyloxycarbonyl-L-phenylalanine p-nitrophenyl ester (ZPNE) at the rate of 103, 46, and 5 μmoles/min/mg, respectively, at pH 8 and 37°. The enzyme has since been reported to be without activity toward ZPNE[10] and another aromatic amino acid ester, α-N-acetyl-L-tyrosine ethyl ester[18] (see under heading *Purity*, below).

Inhibitors. Ancrod is inhibited by diisopropylphosphorofluoridate,[11,14] p-toluenesulfonyl fluoride[14] and phenylmethanesulfonyl fluoride,[10,11] indicative of an active-site serine; by the active-site titrant p-nitrophenyl-p-guanidinobenzoate,[11] and by α-N-(p-nitrobenzyloxycarbonyl)-L-arginyl chloromethyl ketone, which reacts with an active histidine in the enzyme.[25] An analogous compound, α-N-(p-toluenesulfonyl)-L-lysyl chloromethyl ketone (TLCK), with a lysine rather than an arginine side chain, reportedly does not inhibit the enzyme.[25]

Agmatine,[12] arginine, and guanidinium ions[10] are competitive inhibitors of the enzyme. Its coagulant activity is inhibited competitively by various arginine esters and, to a lesser extent, by lysine esters.[24]

[19] P. Mattock and M. P. Esnouf, *Nature (London) New Biol.* **233**, 277 (1971).
[20] S. V. Pizzo, M. L. Schwartz, R. L. Hill, and P. A. McKee, *J. Clin. Invest.* **51**, 2841 (1972).
[21] W. Edgar and C. R. M. Prentice, *Thromb. Res.* **2**, 85 (1973).
[22] B. Hessel and M. Blombäck, *FEBS Lett.* **18**, 318 (1971).
[23] W. R. Bell, G. Bolton, and W. R. Pitney, *Br. J. Haematol.* **15**, 589 (1968); H. C. Kwaan, G. H. Barlow, and N. Suwanwela, *Thromb. Res.* **2**, 123 (1973).
[24] T. Exner and J. L. Koppel, *Biochim. Biophys. Acta* **258**, 825 (1972).
[25] J. P. Collins and J. G. Jones, *Eur. J. Biochem.* **42**, 81 (1974).

Normal serum contains at least two inhibitors of the enzyme; one of these has been identified as α_2-macroglobulin and another, tentatively, as antithrombin III.[26] The rate of inactivation of the enzyme by normal serum *in vitro* is relatively slow, being only about 10% of the rate at which thrombin is inactivated.[26]

Purity. Ancrod prepared by the affinity-chromatography procedure described gives a single protein band on disc gel electrophoresis in the presence of sodium dodecyl sulfate (SDS) and a single precipitin line on immunodiffusion against *A. rhodostoma* antivenin and is also homogeneous on the basis of sedimentation and gel filtration properties.

Specific activities of approximately 2000 (1950–2200) ancrod units and 32–35 TAME units per milligram of enzyme are obtained under the assay conditions described. (For the specific activities, enzyme concentrations were based on absorbance at 280 nm and an $E_{280\,nm}^{1\%}$ of 9.85.)

In contrast to some preparations of lower purity, the affinity-column preparations do not activate factor XIII nor hydrolyze casein to a measureable extent. All of four preparations studied as well as each of several electrophoretic forms of the enzyme tested (see *Physical Properties*, below) hydrolyze ZPNE, but not BAEE, at a relatively slow rate (0.9–1.2 μmoles/min/mg at pH 8.5 and 25°), giving TAME-to-ZPNE rate ratios of about 30:1 as compared to a ratio of about 9:1 initially reported by Esnouf and Tunnah.[14] Although an ancrod preparation which reportedly does not hydrolyze ZPNE has been described,[10] the ability of ancrod, itself, to hydrolyze this compound is suggested by these data and may be expected, since ZPNE is known to be a substrate for other enzymes with trypsinlike activity.

Stability. Dilute solutions of the enzyme (100 ancrod units/ml, or approximately 50 μg/ml) at pH 5.0 to 6.8 have remained stable for months when stored at 5°.

Physical Properties. Ancrod prepared by the described procedure has a minimum of eight electrophoretic forms of the enzyme separable by disc gel electrophoresis (in the absence of SDS) and by isoelectric focusing, with isoelectric points ranging from pH 4.2 to 6.2.

The measured partial specific volume, $\bar{V}$, is 0.69, which is in agreement with that previously reported[14] and consistent with the high carbohydrate content of the enzyme (see the table). The $\bar{V}$ value calculated from the composition in the table is 0.697.

The sedimentation coefficient, $s_{20,w}$, for the enzyme is 3.57 ± 0.18 when measured in 0.1 M sodium phosphate buffer, pH 6.8, containing 0.1 M NaCl and 1% chlorobutanol, and the molecular weight determined by sedimentation equilibrium in the same buffer is $35,400 \pm 1400$ based

[26] W. R. Pitney and E. Regoeczi, *Br. J. Haematol.* **19,** 67 (1970).

AMINO ACID AND CARBOHYDRATE COMPOSITION OF ANCROD[a]

Residue	Percent[b] (by weight)
Lys	3.04
His	2.50
Arg	7.83
Asx	9.20
Thr	2.01
Ser	2.77
Glx	4.85
Pro	3.50
Gly	2.50
Ala	1.89
Cys$_{1/2}$	2.91
Val	3.21
Met	1.76
Ile	4.24
Leu	4.16
Tyr	2.32
Phe	2.54
Trp	2.77
Total amino acid content	64.00
Glucosamine[c]	10.57
Hexose (galactose and mannose)	12.81
Fucose	1.96
Sialic acid[d]	10.69
Total carbohydrate content	36.03

[a] From C. Nolan and L. S. Hall, unpublished observations.

[b] The values are the percentage of the total residue weight recovered. The total residue weight recovered accounted for 91.0% of the dry weight of the glycoprotein.

[c] Calculated as N-acetylglucosamine residues.

[d] Calculated as N-acetylneuraminic acid residues.

on a $\bar{V}$ of 0.69. The plot of log c versus r^2 was linear for the entire length of the cell, indicating homogeneity. The $E_{280\,nm}^{1\%}$ for the enzyme is 9.85.

Chemical Composition. Ancrod is a glycoprotein with a high carbohydrate content. The amino acid and carbohydrate compositions of a pool of two preparations isolated by the affinity procedure are given in the table. The value for sialic acid is an average value since the sialic acid content varies from one electrophoretic form of the enzyme to the next within the mixture analyzed.

[18] The Coagulant Enzyme from *Bothrops atrox* Venom (Batroxobin)[1]

By K. STOCKER and G. H. BARLOW

Bothrops atrox (Linnaeus) is a pit viper which is widespread, and comprises several varieties, in South and Central America. The Linnean species *B. atrox* has been proposed by Hoge[2] to be subdivided, according to geographical origin, morphological criteria, and immunological properties of their venom, into the five independent species *B. asper, B. atrox, B. marajoensis, B. moojeni,* and *B. pradoi.* Further subclassifications of *Bothrops atrox* have been proposed by several authors; however, a definitive zoological classification is still lacking. A common property of all *Bothrops atrox* venoms examined in our laboratory is their ability to convert fibrinogen into fibrin, owing to the presence of a thrombinlike enzyme (batroxobin).

Assay Methods

Assay on Human Plasma

Reagents

Citrated human plasma prepared by adding 9 parts freshly drawn venous blood to 1 part sodium citrate solution, 3.8%, and subsequent centrifugation during 30 min at 3000 rpm. A pool obtained from 10 healthy donors is subdivided into 1-ml portions and stored in the frozen state, below —20°. Fresh or frozen plasma may be substituted by commercially available citrated normal human plasma (Citrol, Dade, Miami, Florida).

Diluent for batroxobin contains sodium chloride, 0.9% chlorobutol, 0.3%, and partially hydrolyzed gelatin 0.02% in distilled water, the pH is adjusted to 6.0. Partial hydrolysis of gelatin is carried out by heating an aqueous solution of gelatin, 2%, pH 3.0, for 30 min at 120° in an autoclave.

Method. Citrated human plasma, 0.3 ml, is preincubated for 2 min at 37°, in a polystyrene tube 9 × 80 mm, 0.1 ml of batroxobin dilution is added and the time from the enzyme addition to clot formation is re-

---

[1] Batroxobin is the generic name for the coagulant fraction from *B. atrox* venom by the World Health Organization.

[2] A. R. Hoge, *Mem. Inst. Butantan Sao Paulo* **32,** 109 (1965).

corded. Two batroxobin units represent that amount of enzyme, contained in 0.1 ml, which coagulates 0.3 ml of citrated human plasma in 19 ± 0.2 sec. A standard curve is obtained by plotting the log of the clotting time versus the log of the enzyme concentration. The potency of an unknown batroxobin sample can be determined from the graph.

Assay on Bovine Fibrinogen

Reagents

Bovine fibrinogen, clottability 97–100%, was obtained from Imco, Stockholm. The contents of 1 ampoule is dissolved in 2 ml of distilled water and furnishes a fibrinogen solution, 0.4%, containing 0.15 M Tris-buffer, pH 7.4.

Tris-buffer, 0.15 M pH 7.4: Solution A is made by dissolving 1.211 g of Tris and 0.681 g of imidazole in 200 ml of 0.1 N HCl. Solution B is composed of 1.211 g of Tris, 0.681 g of imidazole, and 0.585 g of NaCl in 300 ml of distilled water. The final solution is prepared by mixing solutions A and B in that proportion to make a final pH of 7.4.

Method. Fibrinogen solution, 0.2 ml, preincubated for 2 min at 37°, is mixed with 0.2 ml of batroxobin dilution (in Tris buffer, pH 7.4), and the time from the enzyme addition to the clot formation is recorded. A log–log plot is constructed, and unknown activities are calculated from the curve.

Quantitative Estimation of Batroxobin in Serum (Micromethod)

Principle. Whereas thrombin is adsorbed and thereby inactivated by fibrin, the activity of batroxobin is not affected by fibrin. If batroxobin-containing serum, e.g., serum of a patient undergoing defibrinogenating therapy with Defibrase, is overlayered by a fibrinogen solution, a clot starts to grow at the interface and reaches a length that is proportional to the batroxobin concentration in the serum.

Reagents

Fibrinogen solution, 0.2%, prepared by dissolving the content of one ampoule of bovine fibrinogen, Imco, Stockholm, in 4 ml of propylene glycol 2.5%, in distilled water

Heparinized serum: normal human serum or patient serum, containing 50 units of heparin per milliliter.

Batroxobin dilutions are prepared by diluting Defibrase, 20 batroxobin units (BU) ml (Pentapharm Ltd., CH-4002, Basle, Switzerland) with heparinized normal serum.

Method. Batroxobin dilution in heparinized serum, 0.1 ml, is carefully filled into glass tubes, 4 × 60 mm, by means of a syringe, and subsequently overlayered with 0.3 ml of fibrinogen solution, 0.2%. The tubes are kept in a humid atmosphere, at 30°, for 15 hr (overnight) in order to allow clot propagation (Fig. 1). The length of each clot is measured by means of a micrometer device and recorded on a semilogarithmic scale as a function of the batroxobin concentration in the serum. From this standard curve the batroxobin concentration in a patient serum may be estimated.

Assay on Bz-L-*Phe*-L-*Val*-L-*Arg*-pNA

Principle. Hydrolysis of benzoylphenylalanylvalylarginine *p*-nitroanilide by batroxobin liberates a quantity of *p*-nitroaniline proportional to the enzyme activity. The liberation of *p*-nitroaniline is measured by following the change in absorbance at 405 nm.

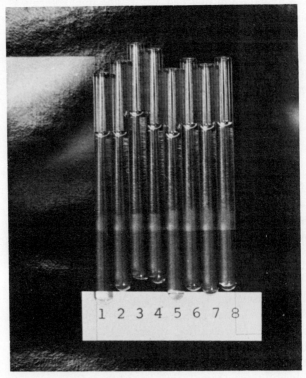

Fig. 1. State of clots induced by overlayering batroxobin dilutions in normal human serum, with bovine fibrinogen, 0.4%. Batroxobin concentrations per milliliter of serum: 1 = 0.0005 BU, 2 = 0.001 BU, 3 = 0.002 BU, 4 = 0.004 BU, 5 = 0.008 BU, 6 = 0.016 BU, 7 = 0.032 BU, 8 = 0.064 BU. BU = batroxobin units.

Reagents

Bz-Phe-Val-Arg-pNA·HCl, substrate S-2160, is commercially available through AB Bofors, Nobel Division, Peptide Research, S-431, ss Molndal, Sweden. A 1 mM solution is prepared by dissolving 0.68 g of substrate in 1 ml of distilled water.

Tris-imidazole buffer, 0.15 M, pH 8.2, prepared as follows: solution A contain 1.211 g Tris and 0.681 g of imidazole in 200 ml of HCl 0.1 M; solution B is composed of 1.211 g of Tris, 0.681 g of imidazole, and 0.585 g of NaCl in 300 ml of distilled water. Solutions A and B are mixed in such a proportion so as to make a final pH of 8.2.

Method. Buffer, 2.00 ml, and batroxobin solution, 0.25 ml, are preincubated for 2 min at 37°, in a 1-cm cuvette; 0.25 ml of substrate solution is added, and adsorbance is recorded during 2 min.

Calculation. Abs./min × 1000 = mU/ml (1 mU of batroxobin converts 1 μmole of substrate per minute).

Qualitative Detection of Batroxobin in Immunopherograms[3]

Principle. Immunopherograms of either crude *B. atrox* venoms or of batroxobin preparations are overlaid by a fibrinogen–agarose gel. Fibrin is formed at the site of the thrombinlike enzyme in the pherogram and allows the identity of the corresponding precipitin line and the estimation of electrophoretic mobility of batroxobin.

Reagents

Barbital buffer, pH 8.6, containing 3.7 g of barbital and 20.6 g of barbital sodium in 2 liters of distilled water

Staining solution containing 500 mg of Amidoschwarz (Merck, Darmstadt, GFR) in a mixture composed of 50 ml of acetic acid and 500 ml of ethanol

Destaining solution, containing 10% acetic acid in ethanol

Agarose plates, 26 × 76 mm, prepared from a solution of 200 mg of agarose in 20 ml of barbital buffer, pH 8.6

Fibrinogen–agarose plates, 26 × 76 mm, prepared as follows: 100 mg of agarose are dissolved in 10 ml of distilled water at 95° and cooled to 40°. 10 ml of bovine fibrinogen, 0.4% (Imco, Stockholm, Sweden) are heated to 40°. Both solutions are mixed and immediately transferred, in 2-ml portions, onto the surface of the glass plates and allowed to gel.

[3] K. Stocker, W. Christ, and P. Leloup, *Toxicon* **12**, 415 (1974).

Polyvalent anti-Bothrops serum (horse), commercially available through Inst. Butantan, Sao Paulo, Brazil or through Lab. Behrens, Caracas, Venezuela.

Method. Immunoelectrophoresis is carried out by conventional techniques in 1% agarose, pH 8.6, 40 V/cm, with 5 μl of samples containing 1–10 mg of protein per milliliter. Immunodiffusion is performed in humid atmosphere for 18 hr after filling the central trough with 20 μl of antiserum.

Intergel-reaction with fibrinogen–agarose is performed by covering the pherogram after immunodiffusion with a fibrinogen–agarose plate and allowing them to react until a distinct turbidity caused by fibrin appears (5–15 min at room temperature). The superposed fibrinogen–agarose gel is cautiously removed and discarded. Excess of fibrinogen, serum and venom proteins is eliminated by washing the pherograms for 6 hr in 0.15 M NaCl. The moist gels are covered with humidified filter paper and dried in an air stream. The filter paper is removed from the gel surface after a brief dip in distilled water and the pherograms are kept for 5 min in the staining solution and subsequently destained by three treatments with destaining solution, washed in 10% acetic acid, and finally rinsed in distilled water. The gels are allowed to dry at room temperature. By this procedure, precipitin lines are revealed as in normal immunoelectrophoresis and thrombinlike activity appears in addition as blue "shadow" of stained fibrin (Fig. 2).

Purification

The purification of batroxobin is achieved by chromatography of the crude *B. atrox* venom on DEAE-Sephadex and subsequent rechromatography of the coagulant fraction on Sephadex G-100.

Chromatography on DEAE-Sephadex. DEAE-Sephadex A-50 (AB Pharmacia, Uppsala. Sweden) is equilibrated with Tris-phosphate buffer, pH 6.0, 0.07 M, and packed into a column, 60 × 600 mm, equipped for ascending chromatography. One gram of *B. atrox* venom is dissolved in 20 ml of Tris-phosphate buffer, pH 6.0, 0.07 M, and applied to the column. The elution is carried out using the same Tris-phosphate buffer; optical density is recorded at 280 nm and fractions of 15 ml are collected. The batroxobin-containing fractions are identified by determination of the clotting time of 0.2 ml of bovine fibrinogen 0.4% after addition of 0.2 ml of eluate. The batroxobin-containing fractions are pooled together, and the buffer is removed by ultrafiltration through a UM-2 Diaflo membrane (Amicon Corp., Lexington, Massachusetts), and several washings with

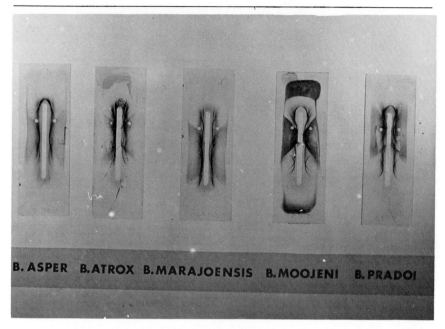

B. ASPER   B.ATROX   B.MARAJOENSIS   B.MOOJENI   B.PRADOI

FIG. 2. Immunopherograms with localized batroxobin activity of the crude venoms of five zoological varieties of *Bothrops atrox.*

ammonium formate solution, 1%. The Tris-phosphate free eluate is finally concentrated on the ultrafilter to one-tenth of its initial volume and freeze dried.

Chromatography on Sephadex G-100. One hundred milligrams of freeze-dried, desalted batroxobin containing eluate from the chromatography on DEAE-Sephadex are dissolved in 0.5 ml of distilled water and applied to a column, 30×100 mm, of Sephadex G-100, previously equilibrated with the solvent composed of tertiary butanol 10% and ammonium formate 1% in distilled water, pH 4.0 (adjusted with NH_4OH). Descending chromatography is carried out using the above solvent system; absorbancy of the eluate is recorded at 280 nm, and fractions of 5 ml are collected. The batroxobin-containing fractions are identified by measuring the clotting time of 0.2 ml of bovine fibrinogen, 0.4%, after addition of 0.2 ml of eluate, previously diluted in the ratio 1:5 with ammonium bicarbonate solution, 1%.

The pooled active eluate is freeze-dried. Almost salt-free, electrophoretically homogeneous batroxobin is obtained. Ammonium formate may be completely removed by dialysis of batroxobin against 0.1 *M* acetic acid and subsequent lyophilization.

TABLE I
SPECIFIC ACTIVITY OF BATROXOBIN ISOLATED FROM VENOM
OF THREE DIFFERENT VARIETIES OF *Bothrops atrox*

Variety	Batroxobin units/mg
B. asper	2000
B. marajoensis	1900
B. moojeni	500

Properties

Purity. Batroxobin isolated following the above method from the venom of one single variety of *Bothrops atrox* shows one precipitin line on immunoelectrophoresis against polyvalent anti-*Bothrops* serum, it also gives one single band in electrophoresis on polyacrylamide gel, 7.5%, pH 2.5. Electrophoresis of the previously reduced and alkylated enzyme in sodium dodecyl sulfate (SDS) containing polyacrylamide, 7%, shows also one single band after staining with either Coomassie blue (peptide moiety) or Schiff's reagent (carbohydrate moiety). The specific activity of batroxobin deriving from the venom of various subspecies of *Bothrops atrox* shows substantial differences (Table I).

Physical Properties. As shown in Fig. 2, the thrombinlike enzyme of different varieties of *Bothrops atrox* migrate with different electrophoretic mobility in agarose, 1%, pH 8.6. Similar differences are also observed in electrophoresis in polyacrylamide gel, even in the presence of SDS, which indicate differences in molecular weight. Differences in the molecular weights of batroxobin originating from *B. marajoensis* and *B. moojeni* venom have been confirmed by ultracentrifugation (Table II).

TABLE II
MOLECULAR WEIGHT OF BATROXOBIN ISOLATED FROM DIFFERENT
ZOOLOGICAL VARIETIES OF *Bothrops atrox*

Batroxobin from	Molecular weight	
	SDS[a] electrophoresis	Ultra-centrifugation
B. asper	32,000	—
B. marajoensis	41,500	43,000
B. moojeni	36,000	37,000

[a] SDS, sodium dodecyl sulfate.

The isoelectric point of batroxobin from *B. moojeni* venom is 6.6, as determined by isoelectric focusing. $OD_{1\,cm}^{1\%}$ of batroxobin from *B. moojeni* venom was found to be 10.5.

Chemical Properties. Reduced and alkylated batroxobin shows in SDS electrophoresis, after staining with either Coomassie blue or Schiff's reagent, one single band that characterizes batroxobin as a single-chain glycopeptide. The neutral carbohydrate content of batroxobin from *B. marajoensis* venom is 10.2% whereas batroxobin from *B. moojeni* venom contains only 5.8% of neutral sugar. Extensive treatment of batroxobin with α-neuraminidase reduces the specific activity of the enzyme from *B. marajoensis* by 52% and of the enzyme from *B. moojeni* venom by 39%. The electrophoretic mobility in polyacrylamide gel, 7.5%, pH 2.5, of neuraminidase-treated batroxobin is significantly increased, whereas the mobility in immunoelectrophoresis in agarose 1%, pH 8.6 is decreased.

The NH_2-terminal amino acid residue of batroxobin from either *B. marajoensis* or *B. moojeni* venom is valine.

Immunological Properties. Batroxobin is antigenic, and specific antibody preparations may be obtained from animals immunized by periodical intramuscular injections. The specific rabbit antisera against batroxobin prepared from (a) *B. marajoensis* venom or (b) *B. moojeni* venom are able to neutralize the clotting activity as well as the amidolytic activity on Bz-Phe-Val-Arg-*p*NA of the crude venoms from *B. asper,* *B. atrox columbiensis,* *B. atrox venezolensis,* *B. marajoensis,* *B. moojeni,* and *B. pradoi.* The clotting activity of *Agkistrodon rhodostoma* venom, however, is not neutralized by those antibody preparations, a fact that confirms previous immunodiffusion experiments.[4,5]

Stability. Batroxobin isolated from either *B. moojeni* or *B. marajoensis* venom remains stable in aqueous solutions in the wide pH range of 2.5–9 for several hours at 20°. A solution of batroxobin in glycerol may be heated for 1 hr at 100° without significant loss of activity. Dilute solutions (20 BU/ml) in physiological saline, pH 6, containing 0.02% of gelatin and 0.3% of chlorobutol have remained stable for more than a year at +4°. Repeated freezing and thawing does not affect the activity of batroxobin solutions.

Specificity. Batroxobin converts fibrinogen into fibrin by specific cleavage of those Arg-Gly bonds in the α-chain, which leads to release of

[4] G. H. Barlow, L. J. Lewis, R. Finlay, D. Martin, and K. Stocker, *Thromb. Res.* 2, 17 (1973).
[5] K. Stocker and G. H. Barlow *in* "Defibrinierung mit thrombinähnlichen Schlangengiftenzymen" (M. Martin and W. Schoop, eds.), pp. 45–63. Huber, Bern, 1974.

fibrinopeptide A, whereas the β-chain of fibrinogen remains unaffected.[6,7] Prolonged incubation of fibrin with batroxobin causes, just as does that of thrombin, release of the tripeptide Gly-Pro-Arg.[8] Batroxobin neither exerts a direct fibrinolytic activity nor activates plasminogen.

Batroxobin acts with a species-dependent preference on fibrinogen of different mammals[9]; hence, coagulation of rabbit fibrinogen takes about 10 times longer than the coagulation of human fibrinogen after additions of equal amounts of batroxobin. This phenomenon might be explained by structural differences of rabbit and human fibrinogen as reflected in the sequence analysis of fibrinopeptide A from different mammals.[10]

In contrast to thrombin, batroxobin does not activate human factor VIII[11] and neither induces nor affects the platelet's aggregation and release reactions.[12] Therefore, it forms a nonretracting clot with platelet-rich human plasma.[13] Batroxobin isolated from the venoms of B. asper and B. moojeni, like thrombin, activates factor XIII, whereas batroxobin from B. marajoensis venom does not activate this factor.[14] Thus, batroxobin from B. asper and B. moojeni venom, in the presence of Ca^{2+}, forms with prothrombin-free human plasma, a clot that remains insoluble in 5 M urea or 1% monochloroacetic acid, whereas batroxobin from B. marajoensis forms a clot that is rapidly dissolved in these agents.

Batroxobin is able to hydrolyze arginine esters such as N^{α}-benzoyl-L-arginine ethyl ester or toluenesulfonyl-L-arginine methyl ester,[15] and it furthermore digests the synthetic fibrinopeptide A analogs Bz-Phe-Val-Arg-pNA and dansyl-Gly-Gly-Val-Arg-GlyOME.[16] Batroxobin originating from the different varieties of Bothrops atrox acts with a different velocity on Bz-Phe-Val-Arg-pNA (Table III).

Activators and Inactivators. Batroxobin-induced coagulation of fibrinogen is accelerated in the presence of imidazole or phenol. Since

[6] B. Blombäck, M. Blombäck, and I. M. Nilsson, *Thromb. Diath. Haemorrh.* **1**, 76 (1957).

[7] K. Stocker and P. W. Straub, *Thromb. Diath. Haemorrh.* **24**, 248 (1970).

[8] B. Hessel and M. Blombäck, *FEBS Lett.* **18**, 318 (1971).

[9] K. Wik, O. Tangen, and F. McKenzie, *Br. J. Haematol.* **23**, 37 (1972).

[10] K. Laki, *Fibrinogen* **1968**, 9 (1968).

[11] S. Lopaciuk and Z. S. Latallo, *Abstr. Int. Congr. Thromb. Haematol, 4th* (1973).

[12] S., Niewiarowski, J. St. Stewart, N. Nath, A. Tai Sha, and G. E. Liebermann, *Am. J. Physiol.* in press

[13] G. DeGaetano, R. Franco, M. B. Donati, A. Bonaccorsi, and S. Garattini, *Thromb. Res.* **4**, 189 (1974).

[14] J. McDonagh and R. P. McDonagh, *Abstr. Int. Congr. Thromb. Haematol., 4th* (1973).

[15] J. Soria, C. Soria, J. Yver, and M. Samama, *Coagulation* **2**, 173 (1969).

[16] K. H. Chen, I. Simpson, J. Bruner-Lorand, and L. Lorand *Abstr. Meet. Fed. Am. Soc. Exp. Biol.* (1972).

TABLE III
ACTION OF BATROXOBIN ORIGINATING FROM *Bothrops marajoensis*
AND *B. moojeni* ON Bz-PHE-VAL-ARG-*p*NA

Batroxobin from	K (M)	V_{max} (moles/min)	
		Per mU	Per BU[a]
B. marajoensis	1.61×10^{-4}	2	2.38×10^{-3}
B. moojeni	2.86×10^{-4}	3.9	0.59×10^{-3}

[a] BU = batroxobin unit.

hydrolysis of Bz-Phe-Val-Arg-*p*NA is also potentiated by imidazole but not by phenol, imidazole may be considered as an activator, whereas phenol seems to catalyze fibrin polymerization. Batroxobin is inactivated by 2.5 mM diisopropyl fluorophosphate, pH 8. In contrast to thrombin, it is not inactivated by incubation for 15 hr at 20° in iodoacetamide, 0.1%, pH 7.4, and, whereas thrombin looses its clotting activity after 1 hr of incubation in SDS 0.01%, pH 7, at 20°, this treatment does not affect the activity of batroxobin.

Batroxobin is inactivated neither by thrombin inhibitors, such as heparin, heparinoids, or hirudin, nor by proteinase inhibitors, such as aprotinin, soybean inhibitor, nor by plasmin inhibitors such as ϵ-aminocaproic acid or tranexamic acid. Batroxobin is bound to α_2-macroglobulin and looses its clotting activity, whereas its amidolytic activity on the small synthetic substrate Bz-Phe-Val-Arg-pNA remains almost unaffected.[17]

[17] N. Egberg, *Thromb. Res.* **4**, 35 (1974).

[19] Crotalase

By FRANCIS S. MARKLAND, JR.

Crotalase is a thrombinlike enzyme isolated from the venom of the eastern diamondback rattlesnake (*Crotalus adamanteus*).[1] In 1886, Mitchell and Reichert[2] demonstrated that animal blood failed to coagulate after treatment with *C. adamanteus* venom. They carried out extensive studies on this venom and showed that the globulin (protein)

[1] F. S. Markland and P. S. Damus, *J. Biol. Chem.* **264**, 6460 (1971).
[2] S. W. Mitchell and E. T. Reichert, *Smithsonian Contrib. Knowl.* **26**, 1 (1886).

fraction was responsible for the defect in coagulation of animal blood. In 1937, Eagle[3] reported that the venom of *C. adamanteus* coagulated plasma by acting directly on fibrinogen and was not dependent on calcium for this action. More recently, Weiss *et al.*[4] reported that the crude venom acts like thrombin in converting fibrinogen to fibrin and had no effect on the activities of factors II, VII, or X *in vitro*. Crotalase was purified from *C. adamanteus* venom,[1] and *in vivo* and *in vitro* studies were carried out in animals and on human plasma, respectively.[5] Important differences in specificity between thrombin and purified crotalase have been described[5] and will be discussed later in this chapter. Other enzymes with properties similar to those of crotalase have been isolated from venoms of different members of the Crotalidae family and are described in this volume: ancrod[6] from the Malayan pit viper (*Agkistrodon rhodostoma*) ; and reptilase[7] from the American lance-headed vipers (*Bothrops jararaca* and *B. atrox*).

Method of Isolation

Crude lyophilized *C. adamanteus* venom is available from several reliable suppliers, including Ross Allen Venom Laboratory (Silver Springs, Florida 32688) and Miami Serpentarium Laboratories (Miami, Florida 33156). Obtaining venom from local sources should be discouraged unless the supplier is a competent herpetologist, pools venom from both juvenile and adult snakes, observes an appropriate milking pattern, and processes his product under standard conditions.

Step 1. Gel Filtration of Crude Venom on Sephadex G-100. Sephadex G-100 (Pharmacia Fine Chemicals, Piscataway, New Jersey) is suspended in 0.10 M sodium chloride for 24–48 hr at room temperature and then equilibrated with 0.04 M sodium acetate, 0.10 M sodium chloride, pH 6.0 (or 0.04 M Tris, 0.10 M sodium chloride, pH 7.1). After packing, the column (5.0 by 90.0 cm) is washed for at least 24 hr with the equilibrating buffer before the sample is applied. Better separation can be achieved if a somewhat longer column (3.8 by 150 cm) is employed. Chromatography and all subsequent steps in purification are performed in the cold room at 3°–5°.

The crude venom (6.5 g) is dissolved in 35 ml of cold 0.15 M sodium chloride (or 0.04 M Tris, 0.10 M sodium chloride, pH 7.1) and the yellow

[3] H. Eagle, *J. Exp. Med.* **65,** 613 (1937).
[4] H. J. Weiss, S. Allan, E. Davidson, and S. Kochwa, *Am. J. Med.* **47,** 625 (1969).
[5] P. S. Damus, F. S. Markland, T. M. Davidson, and J. D. Shanley, *J. Lab. Clin. Med.* **79,** 906 (1972).
[6] G. H. Barlow, L. S. Hall, and C. Nolan, this volume [17].
[7] K. Stocker and G. H. Barlow, this volume [18].

slightly turbid solution is applied to the Sephadex column. Elution is performed with the sodium acetate, sodium chloride buffer (or the Tris, sodium chloride buffer) at a flow rate of 30–55 ml/hr (upward flow). The column as are all other chromatography columns, is monitored by following the absorbance at 280 nm and the fibrinogen clotting activity. Crotalase is eluted at a volume of about 1250 ml, well separated from the first protein peak (which contains the yellow L-amino acid oxidase) and slightly ahead of a large second peak. Active clotting fractions are pooled and concentrated by adding solid ammonium sulfate to 50% saturation (calculated at 25°). The suspension is stirred slowly at 3° for 1 hr and the precipitate is collected by centrifugation at 27,000 g for 20 min at 5°. The precipitate is dissolved in a minimal volume (35–50 ml) of 0.005 M sodium acetate, pH 6.0, containing 0.5 mM EDTA. Concentration can also be performed using the Amicon ultrafiltration apparatus with the type UM-2 membrane (Amicon Corporation, Lexington, Massachusetts). In some preparations, slight increases in activity were observed in the early steps of purification, possibly due either to the removal of an inhibitor of crotalase or to the presence in the crude fractions of a fibrinolytic or fibrinogenolytic enzyme which produced abnormally low crotalase activity in the clotting assay.

Step 2. Chromatography of Crude Crotalase on DEAE-Cellulose. DEAE-cellulose (Whatman, DE-52) is prepared for chromatography by suspension in the starting buffer (5 mM sodium acetate pH 7.0), decanting several times to remove fines, and degassing. The column (1.8 by 55 cm) is packed by gravity and washed with the starting buffer for 24 hr before the sample is applied.

The concentrated fraction from step 1 is dialyzed against starting buffer (several changes) for 24 hr at 3°. The white precipitate that forms (containing some of the coagulant activity) is removed by centrifugation at 25,000 g, and the supernatant fraction is applied to the DEAE-cellulose column. To avoid excessive losses due to precipitation of clotting activity during dialysis, it is advisable to limit the time of dialysis to 24 hr or less; it is also possible to dialyze against sodium acetate buffer of slightly higher ionic strength. The column is initially eluted with the starting buffer at a flow rate of about 25 ml/hr. A linear gradient is initiated at 400 ml, employing 425 ml of the starting buffer in the mixing chamber and 425 ml of 0.10 M sodium acetate, pH 7.0, in the reservoir. A second gradient is initiated at 1250 ml, employing 400 ml of 0.10 M sodium acetate buffer in the mixing chamber and 400 ml of 0.70 M sodium acetate, pH 7.0, in the reservoir. Crotalase activity elutes shortly after the second gradient is started. Peak tubes of clotting activity are pooled and concentrated to about 10 ml by ultrafiltration with the Amicon apparatus.

Step 3. Chromatography of Crotalase on Hydroxyapatite. Hydroxy-apatite (Bio-Gel HT, Bio-Rad Laboratories, Richmond, California) is prepared for chromatography by washing with 0.5 mM potassium phosphate buffer (containing 0.5 mM EDTA), pH 7.0. The column (1.8 × 50 cm) is packed by gravity and washed for 24 hrs with the potassium phosphate buffer at a flow rate of 20 ml/hr.

The concentrated solution from step 2 is dialyzed for 16 hr against potassium phosphate buffer and applied to the hydroxyapatite column. After initial elution with 480 ml of the dilute potassium phosphate buffer (20 ml/hr), a linear gradient is initiated, employing 300 ml of the starting buffer in the mixing chamber and 300 ml of 0.10 M potassium phosphate buffer, pH 7.0, in the reservoir. At 1080 ml a second gradient is started, employing 300 ml of the 0.10 M potassium phosphate buffer in the mixing chamber and 300 ml of 1.0 M potassium phosphate, pH 7.0, in the reservoir. All buffers contain 0.5 mM EDTA. Clotting activity elutes at the end of the first gradient or at the start of the second gradient. Peak clotting fractions are pooled and concentrated by ultrafiltration to about 6 ml, and a small amount of high-molecular-weight protein is removed by passing the fraction through a 3.8 × 52 cm column of Sephadex G-100 in 0.2 M sodium acetate, 0.10 M sodium chloride, pH 7.0. Active clotting fractions are pooled and concentrated by ultrafiltration to about 10 ml.

Step 4. Rechromatography of Crotalase on DEAE-Cellulose. DEAE-cellulose is prepared in 5 mM sodium acetate, pH 7.0, and a 1.8 by 60 cm column is packed as previously described. The concentrated solution from step 3 is dialyzed for 16 hr against dilute sodium acetate buffer (5 mM, pH 7.0) and applied to the column. Elution is achieved by two successive linear gradients operating at a flow rate of 20 ml/hr. The first gradient utilizes 500 ml of starting buffer in the mixing chamber and 500 ml of 0.05 M sodium acetate, pH 7.0, in the reservoir. The second gradient utilizes 500 ml of 0.05 M sodium acetate, pH 7.0, in the mixing chamber and 500 ml of 0.50 M sodium acetate, pH 7.0, in the reservoir. Crotalase is eluted midway through the second gradient. Although the protein peak has a slight shoulder, the leading edge when pooled yields crotalase in a homogeneous state. The trailing edge contains crotalase of a high degree of purity but containing a minor contaminant that is removed by chromatography on SE-Sephadex (Fig. 1). Pooled crotalase fractions are concentrated by ultrafiltration to about 5 mg/ml and stored either in the freezer at −20° or in the refrigerator at 3°–5° using 0.1–0.2 M sodium acetate or Tris-chloride buffer, pH 7.0–7.5.

Table I summarizes the purification steps and shows that the specific clotting activity increases by about 75-fold with a recovery of greater than 50% of the activity. Significant losses occur by selection of only the

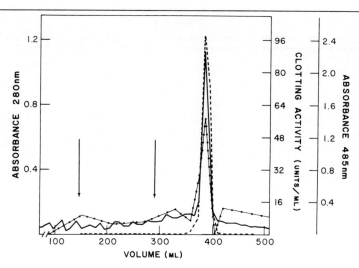

Fig. 1. SE-Sephadex chromatography of crotalase after DEAE-cellulose chromatography (step 4, Table I). The protein was dialyzed against 0.005 M sodium acetate, pH 5.0, and applied to the SE-Sephadex column (0.9 by 53 cm) which had previously been equilibrated with this buffer. Initially, elution was with the equilibrating buffer. Two successive linear gradients were initiated at 140 ml and 290 ml at a flow rate of 9.0 ml/hr. ——, absorbance at 280 nm; - - -, clotting activity; mixing chamber and 75 ml of 0.10 M sodium acetate, pH 5.0, in the reservoir. The second used 100 ml of the 0.10 M buffer in the mixing chamber and 100 ml of 0.70 M sodium acetate, pH 5.0, in the reservoir. Fractions of 3.0 ml were collected at a flow rate of 9.0 ml/hr. ——, absorbance at 280 nm; - - -, clotting activity; ●——●, absorbance at 485 nm after carbohydrate analysis by the phenol–sulfuric acid method as described by C. H. W. Hirs, this series, Vol. 11 [45]. Pooled fractions were dialyzed immediately against 0.10 M Tris buffer, pH 7.4, to avoid loss in activity from prolonged exposure to acid pH. From F. S. Markland and P. S. Damus, *J. Biol. Chem.* **246**, 6460 (1971); reproduced with permission of the *Journal of Biological Chemistry*.

most pure fractions in the final chromatography step. However, if the less pure fraction is additionally purified on SE-Sephadex, the yield is improved significantly. The highest specific activity achieved is about 220 clotting units per 280 nm absorbance unit.

Criteria of Purity. Sodium dodecyl sulfate (SDS) acrylamide gel electrophoresis[8] of highly purified crotalase gives a single band after staining with Coomassie brilliant blue. Acrylamide gel electrophoresis at pH 8.9 by the method of Davis[9] also produces a single band after staining with Amidoschwarz. Sedimentation equilibrium analysis of crotalase

[8] K. Weber and M. Osborn, *J. Biol. Chem.* **244**, 4406 (1969).
[9] B. J. Davis, *Ann. N.Y. Acad. Sci.* **121**, 404 (1964).

TABLE I
PURIFICATION OF CROTALASE FROM 6.5 g OF *Crotalus adamanteus* VENOM[a]

Step	Fraction description	Volume (ml/fraction)	Total protein (A_{280} units)	Total clotting activity (NIH units)	Specific activity (NIH units/A_{280} units)	Cumulative yield (%)
	Crude venom	35	6500[b]	19,750	3.0	100
1	Gel filtration, Sephadex G-100	57	782	26,900	34	136[c]
2	Chromatography, DEAE-cellulose	13.6	266	29,900	112	151[c]
3	Chromatography, hydroxyapatite	10.8	99	13,200	133	67
4	Rechromatography, DEAE-cellulose	6.9	52	11,600	222	59

[a] Adapted from F. S. Markland and P. S. Damus, *J. Biol. Chem.* **246**, 6460 (1971).
[b] Dry weight of crude venom.
[c] Increase in activity is probably due to removal of clotting inhibitor.

by the method of Yphantis[10] indicates a homogeneous component, as evidenced by the straight-line plots of log C versus X^2 (where C is protein concentration expressed in fringes, and X is distance from the axis of rotation to the center of the fringe pattern in centimeters). Finally, chromatography of the less pure fraction from the DEAE-cellulose column (step 4, Table I) on SE-Sephadex gives one major symmetrical peak with constant specific activity throughout the peak tubes (Fig. 1). As seen in the figure there is not only coelution of protein and clotting activity but also of carbohydrate (analyzed by the method of Hirs[11]), indicating that crotalase is a glycoprotein.

Methods of Assay

Principle. Two general types of assay are available for crotalase based on (1) the specific coagulant activity of crotalase on fibrinogen and (2) the esterolytic activity of crotalase toward synthetic basic amino acid ester substrates. In general, the first method is used to monitor crotalase activity during purification; the second method, to evaluate kinetic studies with the highly purified enzyme.

[10] D. A. Yphantis, *Biochemistry* **3**, 297 (1964).
[11] C. H. W. Hirs, this series Vol. 11 [45].

Fibrinogen Clotting Assay

The assay used is basically that described for thrombin[12] but with conditions modified as described below.

Reagents

Fibrinogen (CalBiochem, Human Fraction I, B grade). This is dissolved in 0.04 M Tris-chloride, 0.14 M sodium chloride, pH 7.35, at 37°. Fresh solutions containing 10 mg dry weight per milliliter are made daily and kept at room temperature during use.

Standard thrombin, 21.7 NIH units/mg (obtainable by written request to the National Institute of Health, Biologic Controls, Bethesda, Maryland). This is dissolved in 0.04 M Veronal, 0.14 M sodium chloride, pH 7.35, in a plastic vessel to give 10 units of thrombin per milliliter. Dilutions are made in 0.04 M Tris-chloride, 0.14 M NaCl, pH 7.35; clotting times are determined immediately after dilution.

Procedure. A thrombin standard curve is obtained by plotting the average of duplicate clotting times determined for each of a series of dilutions of standard thrombin against thrombin activity on a log–log scale. Clotting times are obtained after preincubating 0.4 ml of the fibrinogen solution for 3 min at 37° and adding at zero time 0.1 ml of the appropriate thrombin dilution. The plot gives a straight line from 15 sec to 200 sec. The activities of unknown crotalase solutions are determined in duplicate by adding 0.1 ml of an appropriate dilution of the venom to 0.4 ml of fibrinogen preincubated at 37° for 3 min as above. The time from addition of the venom to formation of the clot (with intermittent tilting of the clotting tubes) is recorded. Plastic clotting tubes (12 × 75 mm) are used throughout. By measuring the clotting activity from the thrombin standard curve, crotalase activity can be expressed in NIH clotting units per absorbance unit at 280 nm. Highly purified crotalase has a specific activity of about 220 NIH units per 280 nm unit. A more detailed description of the clotting assay for thrombin has been given.[13]

Esterolytic Assay

The method used is essentially that described by Walsh and Wilcox[14] for the potentiometric assay of serine proteases employing α-N-benzoyl-L-arginine ethyl ester (BAEE).

---

[12] S. Magnusson, this series Vol. 19 [9a].
[13] D. J. Baughman, this series Vol. 19 [9].
[14] K. A. Walsh and P. E. Wilcox, this series Vol. 19 [3].

Reagents

Stock BAEE. A solution of 0.1 M BAEE is prepared in water (pH is not adjusted). This stock solution can be stored in the freezer and is stable for at least a month.

KCl, 1.0 M

Procedure. The reaction is performed in a jacketed vessel in the pH stat at 40°. The reaction mixture consists of 2.5 ml of stock BAEE (final concentration 0.025 M), 1.0 ml of KCl (final concentration 0.1 M), and 6.5 ml of water. The pH is adjusted manually to 8.0 using 0.2 N NaOH and is maintained constant (after addition of crotalase) by continuous automatic addition of 0.02 N NaOH. The reaction is initiated by addition of up to 100 μl of enzyme solution, containing 1–20 μg of crotalase. The slope of the line relating NaOH consumption and time of the reaction gives a direct indication of the micromoles of BAEE utilized per minute. In all assays prior to addition of crotalase, a 5-min period is allowed to measure the spontaneous rate of hydrolysis of BAEE at pH 8.0. The slope of this curve is subtracted from the slope of the enzymic progress curve to obtain the corrected enzymic hydrolysis rate, which is used to calculate enzyme activity. Specific activity is expressed as the number of micromoles of BAEE hydrolyzed per minute per milligram of protein. One unit of enzyme activity is defined as the amount that will hydrolyze 1 μmole of BAEE per minute at 40°. The specific activity of highly purified crotalase is about 160 units/mg.

Properties

Stability. Crotalase has been stored at 5° for more than 6 months in 0.1 M sodium acetate buffer, pH 7.0, 7.5 mg/ml, with the loss of less than 10% of its clotting activity. Storage for 2 years under these conditions resulted in the loss of less than 20–40% of the clotting activity. When stored at −20° in 0.04 M Tris, 0.10 M NaCl, pH 7.4, 4.6 mg/ml, the enzyme is stable for at least one year without measurable loss in either clotting activity or active site concentration (as determined by titration with p-nitrophenyl-p'-guanidino benzoate[15]). The enzyme is stable when stored at neutral or slightly alkaline pH, but at acid pH (pH less than 5.0), there is a time-dependent loss in esterase activity which increases at more acidic pH. At low pH it appears that the loss of esterase activity is irreversible (presumably the same is true for clotting activity).

[15] K. H. Chen, Ph.D. Dissertation, Northwestern University, Evanston, Illinois, 1974.

Crotalase retains almost 100% esterase activity after incubation in 8.0 M urea for 2 hr, when assayed in the absence of urea.

Physical and Chemical Properties. Sedimentation equilibrium analysis of crotalase by the method of Yphantis[10] in 0.1 M Tris, pH 7.6, gave a molecular weight of 33,700. After denaturation of the enzyme by dialysis against 6 M guanidine–HCl containing 0.5% β-mercaptoethanol, pH 7.5, sedimentation equilibrium gave a molecular weight of 34,200. These results indicate that the protein is a monomer containing a single polypeptide chain.

The partial specific volume used for the calculation of the molecular weight is 0.724 ml/g and was calculated from the amino acid and carbohydrate composition by the method of Schachman[16] using a value of 0.67 ml/g as the partial specific volume for each of the carbohydrate residues.[17]

The molecular weight was also estimated by disc gel electrophoresis in the presence of SDS[8] and found to be 31,700. From amino acid and carbohydrate analysis a molecular weight of 33,160 was calculated. An average value of 32,700 was used for all calculations involving molecular weight.

The amino acid composition of crotalase is given in Table II, and for comparison the compositions of thrombin,[18] ancrod,[19] and the thrombinlike enzyme from *Crotalus horridus horridus* venom[20] are listed. There is some similarity in the compositions of crotalase, thrombin, and ancrod.

Carbohydrate analysis by the method of Kim *et al.*[21] gave the following composition: 1 fucose, 2 mannose, 3 galactose, and 5 glucosamine. The carbohydrate content contributes a molecular weight of 1780 toward the overall total of 33,160 (from amino acid and carbohydrate analysis), or 5.4% carbohydrate by weight.

The ultraviolet absorption spectrum of crotalase is typical for a protein and shows a maximum at 280 nm and a minimum at 250 nm, with a ratio of absorbance at 280–260 nm of 1.54 at pH 7.4. There is no visible absorption. The extinction coefficient ($E_{1\,cm}^{1\%}$) at 280 nm is 14.8.

pH Dependence. Crotalase exhibits a sigmoid pH activity profile for the hydrolysis of BAEE with maximum activity above pH 8. The data can be fitted with the calculated titration curve for a group with a pK'

[16] H. K. Schachman, this series Vol. 4 [2].
[17] J. R. Kimmel, H. Markowitz, and D. M. Brown, *J. Biol. Chem.* **234**, 46 (1959).
[18] W. H. Seegers, J. Reuterby, G. Murano, L. E. McCoy, and B. B. L. Agrawal, *Thromb. Diath. Haemorrh.*, Suppl. **47**, 325 (1971).
[19] M. P. Esnouf and G. W. Tunnah, *Br. J. Haematol.* **13**, 581 (1967).
[20] C. A. Bonilla and D. J. MacCarter, *Circulation* **48** (abstr.), 77 (1973).
[21] J. H. Kim, B. Shome, T. Liao, and J. G. Pierce, *Anal. Biochem.* **20**, 258 (1967).

TABLE II
Amino Acid Compositions of Crotalase, Thrombin, Ancrod, and
Thrombinlike Enzyme from Venom of Crotalus horridus horridus

Amino acid	Crotalase	Thrombin[a]	Ancrod[b]	C. h. horridus[c]
Lysine	11	18	10	7
Histidine	9	5	7	4
Arginine	12	17	18	4
Aspartic acid	31	28	30	22
Threonine	14	13	7	7
Serine	16	14	9	12
Glutamic acid	23	29	14	15
Proline	22	14	12	5
Glycine	20	21	16	10
Alanine	11	12	11	8
Half-cystine	14	6	13	5
Valine	17	16	8	9
Methionine	2	4	5	1
Isoleucine	18	11	17	5
Leucine	21	24	14	8
Tyrosine	7	10	6	6
Phenylalanine	13	10	6	6
Tryptophan	6	6	-	5
	267	258	203	139

[a] W. H. Seegers, J. Reuterby, G. Murano, L. E. McCoy and B. B. L. Agrawal, Thromb. Diath. Haemorrh., Suppl. 47, 325 (1971).

[b] Recalculated from M. P. Esnouf and G. W. Tunnah, Br. J. Haematol. 13, 581 (1967).

[c] C. A. Bonilla and D. J. MacCarter, Circulation 48 (abstr.), 77 (1973).

of 6.15, implying that protonation of this group results in inactivation of crotalase. This suggests, as with other serine esterases, that a nonionized histidine residue is involved in the catalytic activity of the enzyme.

Kinetic Properties. The Michaelis–Menten constant (K_m) and maximum velocity (V_{max}) determined at pH 8.0 and 40° are 0.205 mM and 130 μmoles/min per milligram of protein, respectively, for hydrolysis of BAEE. Independent measurements at 25° gave values of 0.034 mM for K_m and 38 sec^{-1} for the catalytic rate constant (k_{cat}).[15] Kinetic analysis of the hydrolysis of toluenesulfonyl-L-arginine methyl ester (TAME) and dansyl-L-arginine methyl ester by crotalase at pH 7.5, 25° and ionic strength 0.13 gave values of 0.294 mM and 0.24 mM, respectively, for K_m and 121 sec^{-1} and 120 sec^{-1}, respectively, for k_{cat}.[15] Table III summarizes the kinetic data for the hydrolysis of ester substrates by crotalase.

TABLE III
HYDROLYSIS OF ESTER SUBSTRATES BY CROTALASE[a]

Substrates	pH	$K_{m(app)}$ $(M \times 10^4)$	k_{cat} (sec^{-1})	$k_{cat}/K_{m(app)}$ $(M^{-1}sec^{-1} \times 10^{-3})$
Benzoylarginine ethyl ester	8	0.34	38	1120
Toluenesulfonylarginine methyl ester	7.5	3	121	412
Dansyl arginine methyl ester	7.5	2.4	120	500

[a] Data from K. H. Chen, Ph.D. Dissertation, Northwestern University, Evanston, Illinois, 1974.

Substrate Specificity. Crotalase exhibits no activity with amino acid amides, and only basic amino acid esters serve as substrates. Thus, hydrolysis of BAEE, N-benzoyl-L-lysine methyl ester, N-benzoyl-L-histidine methyl ester, TAME, and N-dansyl-L-arginine methyl ester was observed using either titrimetric, spectrophotometric, or chromatographic assay. No hydrolytic activity was observed with N-acetyl-L-tyrosine ethyl ester, N-benzoylglycine ethyl ester, or N-acetyl-L-tryptophan ethyl ester.

Crotalase also hydrolyzes the p-nitrophenyl esters of carboenzoxy derivatives of some aromatic and aliphatic amino acids.

Crotalase exhibits peptidase activity on certain synthetic peptides containing an Arg-Gly bond. Thus, in the dipeptide dansyl-Arg-Gly-methyl ester the Arg-Gly bond is cleaved, with no detectable hydrolysis of the ester bond.[15] Similarly, with the hexapeptide dansyl-Gly-Gly-Gly-Val-Arg-Gly-methyl ester, the Arg-Gly bond is cleaved, with no ester hydrolysis.[15] Table IV summarizes the kinetic properties of oligopeptide hydrolysis.

In contrast to thrombin, which cleaves both the A (AP and AY) and B peptides from fibrinogen, crotalase cleaves only the A (AP and AY) peptides.[22] This is assumed to lead to the production of a fibrin polymer of the type $[(\alpha,\beta(B),\gamma)_2]_n$, the same as that formed, presumably, by both ancrod[6] and reptilase.[7] Furthermore, on 24–48 hr of incubation with human fibrinogen, crotalase slowly degrades the $\beta(B)$ chain, producing smaller fragments, at least one of which is still bound within the clot as evidenced by SDS acrylamide gel electrophoresis[8] of the solubilized clot components. There was little or no apparent effect on the α or γ chains of fibrin.[22] These findings may explain the weak direct fibrinolytic effect

[22] F. S. Markland, unpublished observations, 1973.

TABLE IV
HYDROLYSIS OF OLIGOPEPTIDE SUBSTRATES BY CROTALASE[a]

	pH	$K_{m(app)}$ $(M \times 10^4)$	k_{cat} (sec^{-1})	$k_{cat}/K_{m(app)}$ $(M^{-1}sec^{-1} \times 10^{-3})$
Dansyl-Arg-Gly-methyl ester	7.5	53.7	0.025	0.0047
Dansyl-Gly-Gly-Gly-Val-Arg-Gly- methyl ester	7.5	11.8	1.27	1.08

[a] Data from K. H. Chen, Ph.D. Dissertation, Northwestern University, Evanston, Illinois, 1974.

of crotalase.[5] The long-term cleavage of the β(B) chain of fibrinogen by crotalase provides a marked contrast to the specificity of ancrod, which was shown on long-term incubation with human fibrinogen to degrade only the α(A) chain.[22,23]

Another difference in specificity between crotalase and ancrod has also been observed with prothrombin as substrate. It is known that, by cleavage of a single bond, thrombin releases a large amino-terminal fragment from prothrombin but without generation of thrombin activity.[24,25] Similarly, it has been found that crotalase in the presence of phospholipid and calcium cleaves the thrombin-sensitive bond in prothrombin and possibly one other activation bond without generation of thrombin activity. In the absence of either calcium or phospholipid, however, there is no cleavage by crotalase.[26] Ancrod apparently does not cleave the thrombin-sensitive bond or any other bonds in prothrombin.[26]

Crotalase does not appear to activate any of the proteins of the extrinsic blood clotting system, nor does it dramatically alter the clotting activity of factors II, V, VII, VIII, IX, X, XI, or XII as shown by *in vitro* experiments with human plasma.[5] Furthermore, crotalase does not activate factor XIII and has no aggregating effect on washed platelets.[5] These findings further distinguish crotalase from thrombin, which is known to activate clotting factors V, VIII, and XIII, as well as aggregate platelets.[12]

Although crotalase cleaves the Arg-Gly bond in the α(A) chain of

[23] W. Edgar and C. R. M. Prentice, *Thromb. Res.* **2**, 85 (1973).
[24] W. G. Owen, C. T. Esmon, and C. M. Jackson, *J. Biol. Chem.* **249**, 594 (1974).
[25] C. T. Esmon, W. G. Owen, and C. M. Jackson, *J. Biol. Chem.* **249**, 606 (1974).
[26] I. Theodore, F. S. Markland, and H. Pirkle, *Fed. Proc., Fed. Am. Soc. Exp. Biol.* (*Abstr.*) **34**, 860 (1975).

fibrinogen and in oligopeptide substrates, this sequence alone is not sufficient to dictate specificity. This was shown by the inability of crotalase to cleave the Arg-Gly bond in reduced, carboxymethylated bovine insulin B chain. Furthermore, none of the three Arg-X bonds (where X is not glycine) in the β chain of human hemoglobin were cleaved by crotalase.[22]

Inactivators. Heparin, at concentrations of up to 100 units/ml, does not inhibit the clotting activity of crotalase when added to plasma, although thrombin is completely inactivated under these conditions.

Crotalase is a serine esterase like thrombin and trypsin and is, therefore, inhibited by diisopropyl fluorophosphate (DFP). Both esterase and clotting activities are inhibited simultaneously by DFP, showing that the reactive serine residue is involved not only in the esterase, but also in the clotting, activity.

As expected for a serine esterase of the trypsin type,[27] crotalase is inactivated by the chloromethyl ketone of toluenesulfonyl-L-lysine (TLCK) but not the chloromethyl ketone of tosyl-L-phenylalanine. TLCK causes 65–70% loss of activity within 25 hr with simultaneous loss of both clotting and esterase activity. Since crotalase has no reactive cysteine residues, TLCK probably reacts with a histidine residue. This assumption is further supported by the pH activity profile of crotalase, which suggests that protonation of a group with a pK' of about 6.15 (histidine) leads to inactivation of the enzyme.

As noted in the amino acid composition (Table II), crotalase has a high content of half-cystine. Since titration with 5,5'-dithiobis(2-nitrobenzoic acid) (DTNB),[28] either in the presence or in the absence of denaturing agents, indicated the complete absence of sulfhydryl groups, the 14 half-cystine residues must be present as seven disulfide bridges. That at least one of these is critical to the structure or activity of crotalase was demonstrated by the almost complete loss of esterase activity following cleavage of the disulfide bonds by β-mercaptoethanol under nondenaturing conditions.

Tetranitromethane is a highly specific nitrating agent for tyrosine residues in proteins that lack sulfhydryl groups. Nitration of crotalase with this agent produces substantial loss in clotting activity with only partial loss of esterase activity. This may be explained by assuming that the nitrated residues are in the fibrinogen binding site, and that this site is separate from the active site and not involved in ester substrate binding. Further, it would appear that inhibition of clotting activity is due to

---

[27] E. Shaw, M. Mares-Guia, and W. Cohen, *Biochemistry* 4, 2219 (1965).
[28] G. L. Ellman, *Arch. Biochem. Biophys.* 82, 70 (1959).

tyrosine nitration since crotalase has no free sulfhydryl groups. Interestingly, a similar pattern of inhibition was observed with thrombin after nitration[29] or acetylation[30] of four or five tyrosyl residues.

Distribution. Crotalase has been isolated from *Crotalus adamanteus* venom by several different techniques over a period of several years, and there were never different molecular weight forms observed. However, in one purification procedure two chromatographically distinct forms were observed, but with identical physicochemical and enzymic properties. The differences in chromatographic properties were attributed to differences in carbohydrate content of the two species although this was not confirmed by carbohydrate analysis. Probably many other species of the genus *Crotalus* contain venom enzymes similar in nature to crotalase, although to date none have been described. However, Bonilla and Mac-Carter[20] have purified a thrombinlike component from the venom of the timber rattlesnake (*Crotalus horridus horridus*), which has a molecular weight of 19,500 and a distinct amino acid composition and appears to be much more acidic in its elution properties from DEAE-cellulose and more stable to acid treatment than crotalase. This suggests that there may be two distinct molecular forms of thrombinlike enzymes in *Crotalus* venoms, although the possibility may exist that the *C. h. horridus* venom enzyme is a low-activity, acid-degraded form of a crotalase-like enzyme. By comparison of clotting activities on fibrinogen, it would appear that the enzyme from *C. h. horridus*[31] has only 1/15 to 1/30 the activity of crotalase.

[29] R. L. Lundblad and J. H. Harrison, *Biochem. Biophys. Res. Commun.* **45**, 1344 (1971).

[30] R. L. Lundblad, J. H. Harrison, and K. G. Mann, *Biochemistry* **12**, 409 (1973).

[31] C. A. Bonilla, personal communication, 1974.

Section III

Enzymes of Clot Lysis

[20] Urinary and Kidney Cell Plasminogen Activator (Urokinase)

By G. H. Barlow

The presence in urine of an activator substance capable of effecting the transformation of plasminogen to plasmin was first described by Williams[1] and in the following year by Astrup and Sterndorff[2] and Sobel et al.[3] The latter group assigned the name urokinase to this activator. Bernik and Kwaan[4] have shown that the activator produced in culture by embryonic kidney cells was immunologically identical to the urinary activator. Barlow and Lazer[5] have studied other biochemical and physical parameters of the two activators and were unable to show any differences.

Assay Methods

Clot Lysis (Method of Plough[6])

Reagents

Phosphate buffer, 0.1 M, pH 7.2
Fibrinogen (bovine fraction 1), 5 mg/ml
Thrombin, 100 NIH units/ml
Small glass balls, about 7 mm in diameter and about 500 mg in weight

Procedure. In two series of test tubes, set in an ice bath, place 0.025, 0.050, 0.100, 0.150, and 0.200 ml of the unknown solution and the standard solution. The enzyme solutions are made up to contain on the order of 500 CTA units/ml. Each tube is brought to a final volume of 0.6 ml with phosphate buffer. To each tube add 0.1 ml of thrombin solution followed by 0.1 of fibrinogen solution. Invert each tube several times and place in a 37° water bath. A stopwatch is started immediately on addition of the fibrinogen solution. As soon as the fibrin clot is formed introduce on the top of each tube a glass ball, taking care to roll it down the side of the tube so that it lies gently on the clot surface. The time required for the ball to reach the bottom of the tube is taken as the lysis time.

[1] J. R. B. Williams, *Br. J. Exp. Pathol.* 32, 530 (1951).
[2] T. Astrup and I. Sterndorff, *Proc. Soc. Exp. Biol. Med.* 81, 675 (1952).
[3] G. W. Sobel, S. R. Mohler, N. W. Jones, A. B. C. Dowdy, and M. M. Guest, *Am. J. Physiol.* 171, 768 (1952).
[4] M. Bernik and H. C. Kwaan, *J. Lab. Clin. Med.* 70, 650 (1967).
[5] G. H. Barlow and L. S. Lazer, *Thromb. Res.*
[6] J. Plough and N. O. Kjeldgaard, *Biochim. Biophys. Acta* 24, 278 (1957).

A standard curve is made by plotting log CTA units on the abscissa and log lysis time, in minutes, as the ordinate. A straight line is obtained. The activity of the unknown solution is found from the standard curve.

Esterolytic Method

Reagents

Phosphate (0.06 *M*)–NaCl (0.09 *M*) buffer, pH 7.5
Perchloric acid, 0.75 *M*
Methanol-spectroanalyzed, 0.1 *M*
Potassium permanganate, 2%
Sodium sulfite, 10%
Sulfuric acid, 67% (v/v)
N-α-Acetyl-L-lysine methyl ester hydrochloride (ALME) (Cyclo Chemical Corporation, Los Angeles, California), 0.24%
Chromotropic acid reagents: 200 mg of 4,5-dihydroxy-2,7-naphthalenedisulfonic acid disodium (Eastman Kodak Co., Rochester, New York) dissolved in 10 ml of distilled water and then mixed with 90 ml of sulfuric acid reagent.

Procedure. A standard methanol curve is made by diluting the methanol stock with phosphate buffer to give samples in the range 0.25–1.5 μmoles/ml. A 1-ml aliquot of each dilution is transferred to a tube containing 0.5 ml of 0.75 *M* perchloric acid; a blank is prepared in the same manner, using phosphate buffer.

The contents of each tube are mixed, and 1 ml from each is transferred to a test tube suitable for boiling. To each tube is added 0.1 ml of KMnO$_4$ close to the fluid surface. The contents are mixed well, and exactly 1 min is allowed for the oxidation of methanol to HCHO: this is followed by the addition of 0.1 ml of sodium sulfite to each tube, which is immediately shaken vigorously to ensure thorough mixing and decolorizing of the solution. Four milliliters of chromotropic acid reagent are added to each tube, and the contents are mixed well, preferably on a mechanical mixer because of the corrosive and viscous nature of the acid. The tubes, covered with marbles or small beakers, are placed in a boiling water bath for 15 min, then cooled to room temperature; the absorbance is read in a spectrophotometer at 580 nm.

With the standard urokinase, a dilution series is made containing 200–1000 CTA units/ml. An 0.2-ml aliquot of each dilution is mixed with 2 ml of ALME, and the mixtures are incubated at 37°. After 2 and 32 min of incubation, respectively, a 1-ml aliquot is removed and added to

0.5 ml of 0.75 M perchloric acid and kept cold until all tubes are ready for centrifugation. From each supernatant, 1-ml aliquots are transferred to boiling tubes and treated in the same manner as described above for methanol. The absorbance of the 32-min sample is read using the corresponding 2-min sample as a blank. An unknown urokinase preparation is assayed in the manner described above, and the activity is found by direct extrapolation to the control curve and multiplication of the dilution factor.

Definition of Unit. The unit of urokinase activity, referred to as the CTA unit, was adapted in 1964 by the Committee on Thrombolytic Agents, National Heart Institute. This unit is based on the activity of a working standard urokinase preparation,[7] which was independently assayed in several laboratories. The standard urokinase preparation was also assayed by its ability to split the synthetic substate N-acetyl-L-lysine methyl ester (ALME). One unit of urokinase activity releases 5×10^{-4} micromoles/CTA unit per minute at 37°.[8]

Another common unit that is employed is the Plough unit. The relationship between the two units is that 1 CTA unit is equal to 0.7 Plough unit. More recently the unit of choice has become the international unit. This unit being approximately equal to the CTA unit.

Purification

The method described is appropriate for purification from either human urine or the culture harvest media.

The material to be purified is concentrated on an Amicon membrane UM-10, (Amicon, Corporation, Lexington, Massachusetts) to a concentration of approximately 10,000 CTA units/ml.

The concentrate is adjusted to pH 6.7 with H_3PO_4 and passed over a CG-50 ion-exchange column (200–400 mesh), the volume of resin being about 1 ml per 200,000 CTA units of activity. The flow rate averages 45 ml/hr. The column is washed with the equilibrating buffer (0.02 M Na^+ at pH 6.7 with H_3PO_4) and eluted with a 0 to 1 M NaCl gradient in the buffer. The column is monitored by absorbance at 280 nm and by

---

[7] The working standard urokinase is available from WHO International Laboratory for Biological Standards, National Institute for Medical Research, London, England.

[8] This value appears in an article by A. J. Johnson, D. L. Kline, and N. Alkjaersig, *Thromb. Diath. Haemorrh.* **21**, 259 (1969). However, other values ranging from 6.9 to 8.8 $\times 10^{-4}$ have been reported. It is essential that a CTA unit be defined in each laboratory in terms of the standard preparation.

fibrinolytic activity. A pool is made of all tubes with a specific activity of 7500 units/A_{280} or better.

The CG-50 eluate is concentrated and dialyzed to a concentration of about 150,000 CTA units/ml and to a conductivity of 3 mmho. The pH is adjusted to pH 7.0 and applied to a CM-50 Sephadex column that has been equilibrated to 0.1 M Tris-acetic acid at pH 6.5 at room temperature. The capacity of CM-50 Sephadex is about 6 million units per gram dry weight.

The urokinase is eluted with a linear gradient of NaCl from a conductance of 5 mmho in a Tris-acetic acid buffer at pH 9.1. The major urokinase activity elutes as the second protein peak and has a specific activity greater than 60,000 CTA units/mg.

The urokinase from the CM-50 Sephadex is concentrated and dialyzed to 0.15 M sodium chloride containing 0.1% EDTA and gel filtered over a Sephadex G-75 column.

Recently an affinity column method using the competitive inhibitor α-benzylsulfonyl-p-aminophenyalanine has been described by Maciag et al.[9] This method shows promise for the development of a much simplified method for the isolation of the enzyme.

Properties

Activators and Inhibitors. Urokinase reacts as a serine protease and has been shown to be inhibited by diisopropyl fluorophosphate (DFP) and by 4-nitrobenzyl-4-guanidinobenzoate, but not by N-α-tosyl-L-lysine chloromethyl ketone (TLCK).[10]

Nilsson et al.[11] have reported that in certain pathological conditions human plasma contains an inhibitor that reaches levels ten to twenty times the normal level.

Synthetic ϵ-aminocaproic acid inhibits the action of urokinase on plasminogen[12] as well as its action on synthetic ester substrates.[13]

Specificity. Urokinase resembles trypsin and plasmin in its activity against synthetic substrates, although there are differences in the relative rates for the three enzymes, depending on the substrate. However, urokinase is much more selective in its action on proteins, plasminogen being the only substrate against which it has been shown to have any activity.

[9] T. Maciag, M. K. Weibel, and E. K. Pye, this series Vol. 34, [53].
[10] H. Landmann and F. Markwardt, *Experientia* 26, 145 (1970).
[11] I. M. Nilsson, H. Krork, N. H. Sternby, E. Soederberg, and H. Soderstrom, *Acta Med. Scand.* 169, 323 (1961).
[12] N. Alkjaersig, A. P. Fletcher, and S. Sherry, *J. Biol. Chem.* 234, 832 (1959).
[13] L. Lorand and E. V. Condit, *Biochemistry* 4, 265 (1965).

TABLE I
PHYSICAL CONSTANTS ON UROKINASE

Constant	S-1	S-2
Sedimentation coefficients $s_{20,w}$ (S)	2.7	3.3
Molecular weight, sedimentation equilibrium	31,300	54,700
$E_{1\,cm}^{1\%}$ 280 nm, pH 6.5	13.2	13.6
Specific activity (CTA units[a]/mg)	218,000	93,500

[a] Unit of urokinase activity adopted by the NIH Committee on Thrombolytic Agents.

Robbins et al.[14] have shown that the activation of plasminogen to plasmin by urokinase is due to the cleavage of a single arginyl-valine bond.

Physical Properties. Two molecular forms of urokinase have been described by White et al.,[15] and some of the characteristics are shown in Table I. The origin of the two forms is not established, but Lesuk et al.[16] have shown that enzymes such as trypsin can effectively reduce the size of urokinase without loss of activity. They theorized that the smaller size is derived from the larger one by proteolytic activity. However, Burges et al.[17] have shown by chromatography in calibrated Sephadex columns that the activator activity of urine and of crude concentrates from urine shows a molecular weight of $34,500 \pm 2000$, which agrees with type S-1 in Table I.

Table II shows the amino acid composition for the two forms. The enzyme is moderately stable, showing no appreciable loss in activity over years in lyophilized form or over months in sterile solutions at 1 mg/ml or more at refrigerator temperatures. Stability is decreased at salt concentrations below 0.03 M sodium chloride, and precipitation with loss in activity occurs at very low salt concentrations. In diluting to the levels of activity measured in the fibrinolytic assay, it is advisable to add a protein such as human serum albumin fraction V or gelatin to prevent surface denaturation.

Distribution. Activator activity is found in most tissues and in plasma. Bernik et al.[18] have shown that urinary, tissue culture, blood vessel, and

[14] K. C. Robbins, L. Summaria, B. Hsieh, and K. J. Shah, *J. Biol. Chem.* **242**, 2333 (1967).

[15] W. R. White, G. H. Barlow, and M. M. Mozer, *Biochemistry* **5**, 2160 (1966).

[16] A. Lesuk, J. H. Terminiello, J. H. Traver, and J. H. Groff, *Thromb. Diath. Haemorrh.* **18**, 293 (1967).

[17] R. A. Burges, K. W. Brammer, and J. D. Coombes, *Nature (London)* **208**, 894 (1965).

[18] M. B. Bernik, W. F. White, E. P. Oller, and H. C. Kwaan, *J. Lab Clin. Med.* **84**, 546 (1974).

TABLE II
AMINO ACID COMPOSITION OF UROKINASE: AMINO ACID RESIDUES
(g/16 g OF PROTEIN N)[a]

Residue	S-1	S-2
Lysine	7.04 ± 0.23	7.27 ± 0.15
Histidine	4.44 ± 0.34	4.94 ± 0.20
Arginine	7.04 ± 0.25	6.31 ± 0.45
Aspartic acid	6.96 ± 0.26	8.52 ± 0.08
Threonine	6.08 ± 0.14	5.87 ± 0.27
Serine	6.01 ± 0.08	5.56 ± 0.12
Glutamic acid	11.56 ± 0.40	10.42 ± 0.28
Proline	5.04 ± 0.19	5.10 ± 0.23
Glycine	4.04 ± 0.16	4.29 ± 0.07
Alanine	2.27 ± 0.13	2.63 ± 0.08
Half-cystine	3.10 ± 0.14	3.68 ± 0.42
Valine	3.42 ± 0.24	4.21 ± 0.06
Methionine	1.85 ± 0.09	1.85 ± 0.09
Isoleucine	5.58 ± 0.09	4.25 ± 0.04
Leucine	7.35 ± 0.24	6.95 ± 0.05
Tyrosine	6.62 ± 0.37	5.96 ± 0.16
Phenylalanine	4.32 ± 0.60	3.86 ± 0.42
Tryptophan[b]	2.91	3.27
Totals	94.63	94.94

[a] Average or extrapolated values from 24–72 hr hydrolyzates.
[b] Estimated on intact protein.

human heart activator are immunologically identical. However, Thorsen and Astrup[19] have shown that tissue activator is not identical to urokinase.

[19] S. Thorsen and T. Astrup, *Proc. Soc. Exp. Biol. Med.* **130**, 811 (1969).

[21] Streptokinase

By FRANCIS J. CASTELLINO, JAMES M. SODETZ, WILLIAM J. BROCKWAY, and GERALD E. SIEFRING, JR.

Streptokinase[1] is an exocellular protein produced by several strains of streptococci, and it functions in the species-specific conversion of plasminogen to plasmin. Up to now, the only substrates known for

[1] L. R. Christensen, *J. Clin. Invest.* **28**, 163 (1949).

streptokinase are purified plasminogens and plasmins from human,[2] monkey,[3,4] baboon,[3] chimpanzee,[3] cat,[4] dog,[4] and rabbit[5,6] plasmas. Several studies[3,4,7] have been performed in an attempt to evaluate the comparative potency of streptokinase on many of these species of plasminogen. It now appears clear that streptokinase in itself does not possess the inherent ability to catalyze the proteolytic cleavages necessary to convert plasminogen to plasmin, since a synthetic substrate for streptokinase has yet to be discovered. It is believed, based on some recent studies,[6,8-11] that a complex of streptokinase and plasmin or a complex of streptokinase and plasminogen, which has evolved an active site in the plasminogen moiety,[10] are the major activators of plasminogen. Plasmin, in itself, is incapable of readily converting the proenzyme plasminogen into the active enzyme, plasmin.

Assay

Principle. Several different assay methods have been described in the literature. All assays are based on observations demonstrating that when catalytic levels of streptokinase are added to human plasminogen, a given level of activator (streptokinase-plasmin) will be rapidly formed, equivalent to the amount of streptokinase added. This activator will then convert the remaining plasminogen to plasmin. The amount of plasmin formed in a given amount of time after addition of a fixed level of streptokinase is monitored by assaying the action of plasmin on substrates such as fibrin, casein, or synthetic esters.

The usual assay for streptokinase is the clot lysis assay, originally described by Christensen.[1] Fibrin clots are rapidly formed in the presence of fixed levels of streptokinase and plasminogen and the time required for clot dissolution is measured. The amount of streptokinase that will lyse the standard clot in 10 min is arbitrarily defined as 1 unit of streptokinase. An international standard for streptokinase, based on this assay,

[2] H. Milstone, *J. Immunol.* **42,** 109 (1941).
[3] P. A. McKee, W. B. Lemmon, and J. W. Hampton, *Thromb. Diath. Haemorrh.* **26,** 512 (1971).
[4] R. J. Wulf and E. T. Mertz, *Can. J. Biochem.* **47,** 927 (1969).
[5] J. M. Sodetz, W. J. Brockway, and F. J. Castellino, *Biochemistry* **11,** 4451 (1972).
[6] L. A. Schick and F. J. Castellino, *Biochemistry* **12,** 4315 (1973).
[7] L. Summaria, L. Arzadon, P. Bernabe, and K. C. Robbins, *J. Biol. Chem.* **249,** 4760 (1974).
[8] D. K. McClintock and P. H. Bell, *Biochem. Biophys. Res. Commun.* **43,** 694 (1971).
[9] K. N. N. Reddy and G. Markus, *J. Biol. Chem.* **247,** 1683 (1972).
[10] L. A. Schick and F. J. Castellino, *Biochem. Biophys. Res. Commun.* **57,** 47 (1974).
[11] K. N. N. Reddy and G. Markus, *J. Biol. Chem.* **249,** 4851 (1974).

was established by the World Health Organization (WHO) Committee on Biological Standardization. As an alternative assay, the standard can be used for comparison with a given preparation of streptokinase to establish the activity of streptokinase, by means of any assay described. However, the standard is now over a decade old and its viability has undoubtedly deteriorated.

In order to rigorously define a unit of streptokinase we utilize a variation of the esterolytic assay.[12,13] In this assay, we use a known amount of highly purified and well characterized human plasminogen, which is activated by measured catalytic levels of streptokinase for a fixed time. The amount of plasmin formed is monitored by its ability to hydrolyze N-α-toluenesulfonyl-L-arginine methyl ester (TAME).

Reagents

Human plasminogen: This is prepared in a single step by affinity chromatography of fresh diisopropyl fluorophosphate (DFP)-treated human plasma. The chapter on rabbit plasminogen and plasmin in this volume [23] describes the preparation of the resin and the method of application of sample. For purposes of this assay, we remove the plasminogen from the column by elution with 0.05 M ϵ-aminocaproic acid, instead of the ϵ-aminocaproic acid gradient as described for the rabbit system. The plasminogen should possess an amino-terminal glutamic acid. This preparation is dissolved to a concentration of approximately 10 mg/ml in 0.05 M Tris-HCl–0.1 M L-lysine, pH 7.5 ($\epsilon_{1cm}^{1\%}$ at 280 nm = 17.1).[14]

Streptokinase: A stock solution of approximately 0.6–0.8 mg/ml is prepared in 0.001 M Tris-HCl, pH = 7.5 ($\epsilon_{1\ cm}^{1\%}$ at 280 = 9.49).[12]

WHO standard streptokinase: If desired, a standard solution is prepared by dissolving the contents of the vial in 0.4 ml of H_2O and adjusting the pH to 7.5 with NaOH. This can be utilized to convert the units of the esterolytic assay to units of the Christensen clot lysis assay.

Plasmin substrate: A solution of 0.2 M TAME is prepared in H_2O and adjusted to pH 7.5.

Activation and assay solution: A solution of 0.1 M NaCl–0.1 M L-lysine, pH 7.3, at 25° is utilized for this purpose.

Method. Before beginning, a series of streptokinase solutions are prepared by diluting the stock solution, over the range of 1:30 to 1:80, with 0.001 M Tris-HCl, pH 7.5. For the WHO standard, a similar series

[12] F. B. Taylor and J. Botts, *Biochemistry* **7**, 232 (1968).
[13] P. R. Roberts, *J. Biol. Chem.* **235**, 2262 (1960).
[14] K. C. Robbins and L. Summaria, this series Vol. 19, p. 184.

of dilutions are prepared in the range of 1:4 to 1:20. These solutions are kept in an ice bath.

All assays are performed by the potentiometric assay for plasmin, which are described in this volume, [23]. A volume of 1 ml of the NaCl–lysine solution is added to the pH-stat vessel, which is temperature controlled at 25°. Human plasminogen (0.015 ml–0.150 mg) is then added, followed by 0.020 ml of the desired dilution of streptokinase. The pH is adjusted and maintained at 7.5 by automatic titration with 0.014 M NaOH in the pH-stat vessel. After exactly 5 min, 0.100 ml of the stock TAME solution is added and the reaction rate is followed for 1–2 min at pH 7.5 on the recorder. The readings from the recorder are converted to micromoles of TAME cleaved min^{-1}.

Definition of Unit and Specific Activity. To define a unit of streptokinase activity by this method, one must do so indirectly in terms of the amount of plasmin generated from plasminogen under the arbitrarily defined conditions described above. Obviously, if the initial concentration of plasminogen is significantly varied from that suggested, a different specific activity will result. However, we feel that the assay described here (or slight variations according to the particular requirements of the investigator) is very suitable to monitor changes in activity of streptokinase as a result of various modifications. Also, it should be pointed out that, for practical reasons, the level of TAME used is not saturating with respect to plasmin. Thus, another variable is introduced into the assay. Despite this, the assay is reproducible and very sensitive and should satisfy the needs of the investigator.

With these observations in mind, we will define 1 unit of streptokinase activity as the number of micromoles of streptokinase required to generate sufficient plasmin from 0.13 mg of plasminogen per milliliter, to hydrolyze 1 μmole min^{-1} of TAME under the conditions described above. From plots of the rate of TAME hydrolysis vs the amount of streptokinase added, as shown in Fig. 1, we can obtain a specific activity of 25,000 for purified and apparently homogeneous streptokinase. Therefore, 1 unit of streptokinase activity is equivalent to 4.0×10^{-5} μmole of streptokinase. Curves such as those in Fig. 1 are obtained for each new batch of streptokinase and are reproducible within ±10%. The assay is linear to approximately 1 μg of added streptokinase. Although we have found that variations of 5–10% in the initial concentration of plasminogen do not affect the rate of plasmin formation over this range of streptokinase concentrations, strict attention should be given to plasminogen preparations in different laboratories in order to ensure the reproducibility of the data. Plasminogen, prepared as described, is a high-quality preparation and is very reproducible in terms of its own properties.

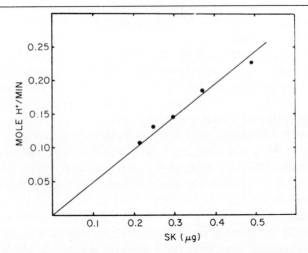

Fɪɢ. 1. Rate of N-α-toluenesulfonyl-ʟ-arginine methyl ester (TAME) hydrolysis by plasmin formed as a result of varying the amounts of streptokinase (SK) added to a constant amount of human plasminogen. Purified Kabikinase was utilized in this experiment and diluted as described in Methods. Activation and assay solutions and conditions are as in Methods.

Purification of Streptokinase

The method of choice for purification of streptokinase depends upon the quality of the starting material. The most suitable starting material available commercially is Kabikinase, which can be obtained from AB Kabi. This material has, as its major protein impurity, plasma albumin, which has been added by the manufacturers as a stabilizer. We use one of the two following methods of removing the albumin from commercial Kabikinase.

Sepharose 4B-Blue Dextran Chromatography

An affinity chromatography column is prepared to remove plasma albumin from Kabikinase preparations. Approximately 75 ml of washed, settled Sepharose 4B is activated with 18.6 g of cyanogen bromide suspended in 75 ml of H_2O. The pH is constantly maintained at 11.0 by repeated addition of 4 N NaOH. After stabilization of the pH, the resin is rapidly washed on a funnel with cold 0.1 N NaHCO$_3$, pH 9.5. A solution of blue dextran containing 0.75 g of blue dextran in 75 ml of 0.1 N NaHCO$_3$ is then added to the resin and allowed to stir gently overnight at 4°. Prior to use, the resin must be extensively washed with

10 M urea, followed by H_2O and finally 0.05 M Tris-HCl–0.5 N NaCl, pH 8.0.

Approximately 1.2 g of Kabikinase are dissolved in 20 ml of cold H_2O, exhaustively dialyzed against H_2O in the cold, and finally lyophilized. The powder is then dissolved in 10 ml of cold H_2O. A quantity of 2.77 g of $(NH_4)_2SO_4$ is slowly added to the protein solution, over a period of 1 hr, with stirring at 4°. The suspension is allowed to remain at 4° for an additional hour, prior to centrifugation. The precipitate is washed twice with 3 ml of 50% saturated $(NH_4)_2SO_4$ in the cold. The precipitate is then dissolved in 5 ml of cold H_2O and dialyzed against H_2O to remove $(NH_4)_2SO_4$. The solution is finally dialyzed against 0.05 M Tris-HCl–0.5 M NaCl at 4°.

The above solution is then passed over a 1 cm $\times$ 10 cm of Sepharose 4B-Blue Dextran at 4°. All material is collected by means of a fraction collector. After application of the sample, additional solvent is passed through the column until no further protein material is eluted. All tubes containing protein which emerged from the column are pooled, dialyzed against H_2O, and lyophilized. The powder contains high-purity streptokinase.

Isoelectric Focusing Procedures

An alternative procedure for preparing high-purity streptokinase from Kabikinase is by a combination of salt cuts and preparative isoelectric focusing.[15] Two salt-cut streptokinase is prepared as described above, repeated twice. This material is finally purified by isoelectric focusing on an LKB 110 ml column at 4°. The anode solution consists of 0.4 ml of 18.1 M H_3PO_4, 26 ml of H_2O, and 26 g of sucrose. The cathode solution consists of 0.5 g of NaOH dissolved in 50 ml of H_2O. After addition of the anode solution, a linear ampholine gradient is layered with the aid of a peristaltic pump. The starting solvent contains 21 g of sucrose dissolved in 28.7 ml of H_2O to which is added 2.82 ml of pH 3–6 ampholines in 40% sucrose (LKB). The limit solvent contains 2.75 g of sucrose in 51.1 ml of H_2O, to which is added 1.15 ml of the same ampholine solution. After 40 ml of the gradient emerges, approximately 15–20 mg of partially purified streptokinase are added directly to the starting solvent. The entire gradient is then allowed to form. At this point a few extra milliliters of the limit solution can be added, followed by the cathode solution. The run is started and allowed to remain for 68 hr. The current is continually adjusted during the run such that the product of (kilovolts) $\times$ (milliamps) never exceeds 3. At the

[15] W. J. Brockway and F. J. Castellino, *Biochemistry* **13**, 2063 (1974).

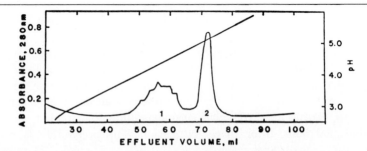

Fig. 2. Isoelectric-focusing profile of partially purified streptokinase. The material was focused for 68 hr at 4° on a pH 3–6 gradient. The absorbance of material eluted from the column and the experimental pH gradient is shown on the graph. Fractions of 1 ml were collected. Peak 2 contained all the streptokinase activity.

conclusion of the experiment, fractions of 1 ml are collected and analyzed for protein content and pH. A typical isoelectric focusing profile for streptokinase is shown in Fig. 2. The peak containing streptokinase is pooled, extensively dialyzed against H_2O, and lyophilized. The powder contains high-purity streptokinase.

Either method of purification results in streptokinase of high specific activity in high yield. Starting with 1.2 g of Kabikinase, we obtain approximately 30–50 mg of streptokinase (the variation is due to the different streptokinase contents of different lots of Kabikinase). This is approximately 60–70% of the theoretical yield. The material is homogeneous on several gel electrophoresis systems, as shown in Fig. 3.

An additional method for preparing a respectable quality streptokinase in a rapid manner is by affinity chromatography on insolubilized DIP-plasmin. Human plasminogen (4.0 ml of 7.5 mg/ml) in 0.05 M Tris-HCl–0.1 M L-lysine, pH 8.0, is activated by 0.9 ml of urokinase (4420 Plough units/ml) in the same buffer for 20 min at room temperature. This solution is then added to 40 ml of a solution containing 0.05 M Tris-HCl–0.1 M L-lysine–0.02 M diisopropyl fluorophosphate (DFP), pH 8.0. The mixture is allowed to remain overnight at 4°. At this point, the solution is dialyzed against H_2O in the cold and lyophilized. Approximately 30 mg of DIP-plasmin are added to 10 ml of 0.1 N sodium citrate, pH 6.2 at 4°. A few crystals of N^α-acetyl-L-lysine are added to enhance solubility. This solution is added to 20 ml of cyanogen bromide-activated Sepharose 4B (see above), which was previously washed with the same citrate buffer, and stirred overnight at 4°. The resin is then filtered and extensively washed with 0.3 M sodium phosphate, pH 8.0, at room temperature. Approximately 95% of the DIP-plasmin is coupled. The resin is then packed in a 0.9 × 11 cm column prior to use.

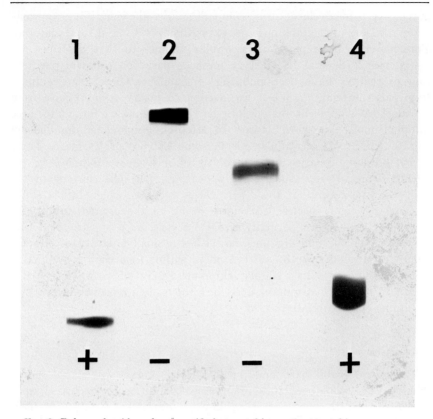

Fig. 3. Polyacrylamide gels of purified streptokinase. 1, streptokinase on an analytical pH 9.5 gel; 2, streptokinase on an analytical pH 3.2 in 6 M urea gel, 3, streptokinase on an analytical pH 4.3 gel; and 4, streptokinase on an analytical sodium dodecyl sulfate–mercaptoethanol polyacrylamide gel. These gels are shown only for evaluation of the material, not for mobility comparisons under different conditions.

Dialyzed and lyophilized Kabikinase (containing 2–5 mg of streptokinase) is placed over the column and then washed with 0.3 M phosphate to remove nonadsorbed protein. Upon elution with 3 M guanidine hydrochloride–0.1 M phosphate, pH 8.0, at room temperature, a single peak is eluted. The material is dialyzed against H_2O and lyophilized. Gel electrophoresis experiments demonstrates that the material contains only streptokinase, but assays indicate a loss of approximately 30% in specific activity.

Other commercially available starting materials are suitable for the preparation of streptokinase. Varidase (Lederle Laboratories) can be purchased through local drug companies in vials containing approximately 0.2 mg of streptokinase per approximately 4 mg of total material. This material is prepared in a similar manner to that described by Christensen.[16] Several satisfactory methods based on classical purification procedures have been published for purification of streptokinase from this material.[12,17] When we attempt to purify streptokinase from Varidase by affinity chromatography on DIP-plasmin, we find up to three different molecular weight forms of streptokinase on sodium dodecyl sulfate (SDS) gel electrophoresis in some batches of Varidase. These forms presumably arise by proteolysis of native streptokinase in the Varidase preparations since their appearance parallels the age of the Varidase preparation. If fresh Varidase is used, we can obtain a single molecular weight species. Therefore, owing to the unpredictability of our final product, we normally avoid Varidase as a starting material. However, for the investigator who wishes a small quantity of streptokinase in a good state of purity and is willing to accept a product of lower specific activity, affinity chromatography of readily available Varidase on DIP-plasmin, as described above, is a reasonable means of obtaining this material.

Properties

Stability. Kabikinase and purified streptokinase are stable indefinitely in the lyophilized state or in aqueous solution in the frozen state. When thawed, the material is stable over several hours in an ice bath, and perhaps significantly longer. The material should not remain for an extended period in solution at room temperature. In our hands, Varidase is not as stable as Kabikinase in the lyophilized state. We notice a deterioration of streptokinase into streptokinase fragments upon prolonged storage. However, the particular streptokinase fragments formed possess good activity, and the overall specific activity of the contents of the vials remains relatively high for a long period. According to the manufacturer, Varidase is stable for 2 weeks after reconstitution of the contents of the vial, if kept at 4°, and for 24 hr if kept at room temperature.

Purity. The streptokinase obtained by the above procedures is homogeneous when examined by polyacrylamide gel electrophoresis in several systems (see Fig. 3) and a single peak is obtained in preparative isoelectric focusing experiments.[15] Single bands are noted in sedimentation

[16] L. R. Christensen, *J. Gen. Physiol.* **30**, 465 (1947).
[17] P. K. Siiteri and R. Douglas, U.S. Patent No. 3,226,304 (1965).

TABLE I

PHYSICAL PROPERTIES OF STREPTOKINASE[a]

Parameter	Value obtained
Molecular weight	$45,000^b$; $50,000^c$; $47,000^d$; $45,000^e$
$s^{\circ}_{20,w}$ (native) (12,15,17)	3.0–3.1 S
$D^{\circ}_{20,w}$ (native) (17)	6.2×10^{-7}
pI	4.7 (18), 5.2 (15)
Electrophoretic mobility (17)	4.19×10^{-5} cm² sec⁻¹ volt⁻¹
Specific viscosity	0.1 dl/g (1 % solution)^f
$\bar{v}$	0.738^g

[a] Numbers in parentheses refer to text footnotes.
[b] Value obtained in native buffers by high speed sedimentation equilibrium (15).
[c] Value obtained in native buffers by approach to equilibrium and calibrated gel filtration techniques (12).
[d] Value obtained in native buffers by sedimentation and diffusion analysis (17).
[e] Value obtained by calibrated sodium dodecyl sulfate–mercaptoethanol polyacrylamide gel analysis (15).
[f] Value obtained in native buffers. The units were not given in the original manuscript. The authors have assumed the units indicated, since the protein appears to be a typical globular protein.
[g] Obtained from density measurements in native buffers (12).

velocity[15,17] and diffusion[17] analysis. Further, a unique molecular weight species is obtained in sedimentation equilibrium determinations.[12,15,18]

Physical Properties. Several groups have investigated the physical properties of native streptokinase, and these are summarized in Table I. In addition, it is well established that streptokinase is a protein consisting of a single subunit with very little carbohydrate[12] and no lipid.[12] Analysis of optical rotation[12] and circular dichroism[15] data suggests a helical content of 10–12%.

Chemical Properties. The amino-terminal amino acid obtained for streptokinase is isoleucine in quantities of 0.92 mole per mole of protein.[15] The carboxyl-terminal residue is lysine.[15,19] The amino-terminal amino acid sequence of streptokinase is NH₂-Ile·Ala·Gly·Pro·Glu·Trp·Leu·Leu·Asp·Arg·Pro·Ser-.[15] A total of 5 cyanogen bromide fragments of molecular weights 16,520; 13,250; 8950; 5180; and 2570 have been obtained for streptokinase.[19] Amino acid compositions obtained for native streptokinase in different laboratories are given in Table II. Good agreement between laboratories can be noted in most respects with regard to these measurements.

[18] E. C. DeRenzo, P. K. Siiteri, B. L. Hutching, and P. H. Bell, *J. Biol. Chem.* **242**, 533 (1967).
[19] F. J. Morgan and A. Henschen, *Biochim. Biophys. Acta* **181**, 93 (1969).

TABLE II
AMINO ACID COMPOSITION OF STREPTOKINASE

Amino acid	Content in streptokinase[a] (moles/mole)				
	Ref. 12[b]	Ref. 15[c]	Ref. 17[d]	Ref. 18[e]	Ref. 19[f]
Aspartic acid	67	57	67	68	67
Serine	27	26	29	30	26
Threonine	22	24	21	24	28
Glutamic acid	52	42	45	46	45
Proline	19	21	22	20	21
Glycine	25	20	21	21	20
Alanine	24	21	24	23	22
Half-cystine	0	0	1	—	0
Valine	26	20	23	23	23
Methionine	1	4	4	3	4
Isoleucine	23	20	23	22	22
Leucine	33	35	40	40	39
Tyrosine	8	20	21	20	21
Phenylalanine	12	14	14	15	15
Histidine	6	9	9	9	9
Lysine	30	29	32	33	34
Arginine	18	18	20	21	19
Tryptophan	—	2	1	1	1
Amides	—	34	52	—	48

[a] For references cited, see text footnotes.
[b] Based on a molecular weight of 50,000 for streptokinase.
[c] Based on a molecular weight of 45,000 for streptokinase.
[d] Based on a molecular weight of 47,000 for streptokinase.
[e] Based on a molecular weight of 47,750 for streptokinase.
[f] Based on a molecular weight of 48,000 for streptokinase.

Mechanism of Activation of Human Plasminogen by Streptokinase.
It has been shown that streptokinase can interact with human plasmin, forming a 1:1 complex.[6,20,21] Compared to plasmin, this complex possesses full capacity for hydrolysis of synthetic esters, such as TAME, but greatly decreased activity, toward protein substrates, such as casein. Further, although plasmin in itself does not readily activate plasminogen, the streptokinase–plasmin complex is a potent human plasminogen activator.[6,9] Thus, this complex is called a plasminogen activator complex. The action of this complex on plasminogen is presumably based upon specific interactions between the streptokinase present in the complex and free plasminogen, resulting in an orientation of plasminogen into the

[20] J. Zybler, W. F. Blatt, and H. Jensen, *Proc. Soc. Exp. Biol. Med.* **102**, 755 (1959).
[21] D. L. Kline and J. B. Fishman, *J. Biol. Chem.* **236**, 2807 (1961).

active site of the activator complex.[22] It has been clearly shown that the active site of the activator complex resides in the plasmin moiety.[6,23,24] It has also been demonstrated that the streptokinase–plasmin activator complex can form when the inactive zymogen, plasminogen, free of plasmin, is used as the starting material.[8,9] This fact led to the proposal[8,9] that there is an initial interaction of plasminogen and streptokinase, resulting in the generation of a proteolytic active site. This site converted the plasminogen within the complex to plasmin. In this manner, the streptokinase–human plasmin activator complex can be generated in the absence of preformed plasmin. This interaction of streptokinase and human plasminogen has been proved by Schick and Castellino,[6] and these same authors have directly demonstrated the generation of an active site in the plasminogen moiety of this complex.[10] In addition, these investigators have shown that the streptokinase–plasminogen complex containing an active site, although very short-lived, can also serve as an activator of plasminogen. Similar observations have been forwarded by Reddy and Markus.[9] Further, it has been shown that the active site in the streptokinase–plasminogen complex is capable of binding pancreatic trypsin inhibitor[25] and capable of hydrolyzing TAME, but not casein.[11] Kinetic comparisons of plasmin, streptokinase–plasmin, and streptokinase–plasminogen are given in Table III.

In addition to the conversion of plasminogen to plasmin within the streptokinase–plasminogen complex, it has been found that streptokinase undergoes a progressive degradation in the activation mixtures. When incubated with human plasminogen, the 45,000–47,000 molecular weight native streptokinase undergoes rapid degradation to a 37,000 molecular weight species.[15] Other studies have reported the transient existence of a 43,000 molecular weight species.[12] We have extensively purified and characterized the 37,000 molecular weight species of streptokinase and find that it arises entirely from degradation at the amino terminus of native streptokinase.[15] Whether the catalyst for this degradation is streptokinase–plasminogen, streptokinase–plasmin, or small amounts of free plasmin is currently unknown. Further, slow degradation of streptokinase to molecular weight forms of 30,000 and lower have been observed.[6,7] The degradation of streptokinase is much more rapid in activation mixtures of plasminogen of other species, such as rabbit.[6,7] The ratios of the var-

[22] F. F. Buck and E. Boggiano, *J. Biol. Chem.* **246**, 2091 (1971).
[23] E. C. DeRenzo, B. Boggiano, W. F. Barg, and F. F. Buck, *J. Biol. Chem.* **242**, 2426 (1967).
[24] L. Summaria, C.-M. Ling, W. R. Groskopf, and K. C. Robbins, *J. Biol. Chem.* **243**, 144 (1968).
[25] K. N. N. Reddy and G. Markus, *Biochem. Biophys. Res. Commun.* **51**, 672 (1973).

TABLE III
COMPARISON OF SOME PROPERTIES OF PLASMIN AND ITS
STREPTOKINASE COMPLEXES[a]

| Activity or inhibition | Plasmin | Relative activities[b] | |
		Streptokinase–plasmin	Streptokinase–plasminogen
Plasminogen activator	0	1	3.5
Fibrinolytic	1	0.2	0
Caseinolytic	1	0.2	0
TAME	1	1	0.5
ALME[c]	1	2.5	0.5
GdnBzoNph[d]	1 Eq	1 Eq	1 Eq
Pancreatic trypsin inhibitor			
a. Azocasein assay	Stoichiometric	Stoichiometric	Undetected owing to lack of activity
b. ALME assay	Stoichiometric	Stoichiometric	Not stoichiometric
Soybean trypsin inhibitor			
a. Azocasein assay	Stoichiometric	Not stoichiometric	Undetected owing to lack of activity
b. ALME assay	Stoichiometric	Very weak	None

[a] Taken from K. N. N. Reddy and G. Markus, *J. Biol. Chem.* **249**, 4851 (1974).
[b] A value of 1 is assigned to plasmin, when present.
[c] N-α-Acetyl-L-lysine methyl ester.
[d] p-Nitrophenyl-p'-guanidinobenzoate.

ious forms of degraded streptokinase present in activation mixtures of several plasminogen species at various times have been reported.[7] We feel that the instability of streptokinase in complexes containing plasminogen or plasmin from species of animals in which the fibrinolytic system exhibits a weak sensitivity or insensitivity to activation by streptokinase[4] can in part be explained by the relative ability of the activator complex to degrade streptokinase rather than activate plasminogen.

Based upon results of the studies cited above, a mechanism for the activation of human plasminogen to plasmin by streptokinase can be forwarded, as described below:

$$SK + HPg \rightleftharpoons SK{\cdot}HPg \rightarrow (SK{\cdot}HPg') \rightarrow SK^*{\cdot}Pm$$
$$HPg \cdot \xrightarrow{\tfrac{(SK{\cdot}HPg')}{SK^*{\cdot}Pm}} HPm$$

with SK + Pm branch above.

where SK is streptokinase; HPg is human plasminogen; (SK·HPg') and

SK·HPg are complexes with and without an active site, respectively; HPm refers to human plasmin; and SK*·HPm refers to the streptokinase–human plasmin complex with a proteolytically altered SK moiety. In this diagram, the SK* possesses a molecular weight of 37,000. As described above, further degradation of SK* is possible. Catalytic levels of streptokinase can thus form catalytic levels of the two activator complexes (SK·Pg′) and SK*·Pm. These complexes, in the second step of the mechanism, will convert the remaining HPg to HPm. A slight variation of the mechanism has been proposed for the conversion of rabbit plasminogen (RPg) to plasmin (RPm), which takes into account the relative instabilities of the (SK·RPg′) and SK*·RPm complexes. Presumably, subtle variations need to be proposed for the plasminogen of other species, depending on their relative sensitivities to activation by streptokinase.

[22] Plasminogen and Plasmin[1]

By KENNETH C. ROBBINS and LOUIS SUMMARIA

Specificity and Assay Methods

Plasmin is a serine protease with trypsinlike specificity.[1a] It cleaves, or hydrolyzes, proteins and peptides at arginyl and lysyl peptide bonds,[2-4] and basic amino acid esters,[4-6] and amides.[7] Plasmin appears to have a preference for lysyl bonds in both protein and ester substrates.[4] Assay methods for plasminogen and plasmin have been devised using both casein (caseinolytic) and fibrin (fibrinolytic) substrates[8]; in the

[1] Work supported in part by the United States Public Health Service grant HL04366 from the National Heart and Lung Institute.

[1a] W. R. Groskopf, B. Hsieh, L. Summaria, and K. C. Robbins, *J. Biol. Chem.* **244**, 359 (1969).

[2] P. Wallén and S. Iwanaga, *Biochim. Biophys. Acta* **154**, 414 (1968).

[3] W. R. Groskopf, B. Hsieh, L. Summaria, and K. C. Robbins, *Biochim. Biophys. Acta* **168**, 376 (1968).

[4] M. J. Weinstein and R. F. Doolittle, *Biochim. Biophys. Acta* **258**, 577 (1972).

[5] S. Sherry, N. Alkjaersig, and A. P. Fletcher, *Thromb. Diath. Haemorrh.* **16**, 18 (1966).

[6] K. C. Robbins, L. Summaria, D. Elwyn, and G. H. Barlow, *J. Biol. Chem.* **240**, 541 (1965).

[7] U. Christensen and S. Müllertz, *Biochim. Biophys. Acta* **334**, 187 (1974).

[8] A. J. Johnson, D. L. Kline, and N. Alkjaersig, *Thromb. Diath. Haemorrh.* **21**, 259 (1969).

caseinolytic assay, the release of trichloroacetic acid, or perchloric acid, soluble peptides are measured, and in the fibrinolytic assay, the time of dissolution of the fibrin clot is determined. A variety of different basic amino acid ester substrates have been used for determining enzyme activity by rate assays (esterolytic).[5-9] Specific active-site titration methods have been developed to determine the amount of enzyme in solution.[7-10] Specific antibody[11] and protein inhibitor assays[7] have also been described. The enzymic activities of the equimolar plasminogen–streptokinase and plasmin–streptokinase complexes can also be measured on the protein, peptide, and synthetic ester substrates.[11a] The activator activity of these complexes can be determined by suitable methods.[11a,12] A method for the determination of human plasminogen and plasmin based on the interaction between the zymogen, or enzyme, and streptokinase has been described.[13] Human plasmin reference preparations are being developed.

Nomenclature of Plasminogen and Plasmin Forms

Human plasminogen when isolated from plasma, or serum, or enriched-plasma fractions, was found to be a single-chain monomeric protein[14] with multiple molecular isoelectric forms which have been isolated by isoelectric focusing methods.[15-18] Those forms with isoelectric points between pH 6.2 and pH 6.6 (4 major forms) could be prepared from plasma, serum, and plasma fraction III[16-19] and had an NH₂-terminal glutamic acid residue,[16,18-22] whereas those forms with isoelectric points between pH 7.2 and pH 8.3 (5 major forms) could be prepared from

[9] R. M. Silverstein, *Thromb. Res.* **3**, 729 (1973).

[10] T. Chase, Jr., and E. Shaw, *Biochemistry* **8**, 2212 (1969).

[11] S. F. Rabiner, I. D. Goldfine, A. Hart, L. Summaria, and K. C. Robbins, *J. Lab. Clin. Med.* **74**, 265 (1969).

[11a] K. N. N. Reddy and G. Markus, *J. Biol. Chem.* **249**, 4851 (1974).

[12] C.-M. Ling, L. Summaria, and K. C. Robbins, *J. Biol. Chem.* **242**, 1419 (1967).

[13] J. Gajewski and G. Markus, *Thromb. Diath. Haemorrh.* **20**, 548 (1968).

[14] G. H. Barlow, L. Summaria, and K. C. Robbins, *J. Biol. Chem.* **244**, 1138 (1969).

[15] L. Summaria, L. Arzadon, P. Bernabe, and K. C. Robbins, *J. Biol. Chem.* **247**, 4691 (1972).

[16] P. Wallén and B. Wiman, *Biochim. Biophys. Acta* **257**, 122 (1972).

[17] K. C. Robbins and L. Summaria, *Ann. N.Y. Acad. Sci.* **209**, 397 (1973).

[18] L. Summaria, L. Arzadon, P. Bernabe, K. C. Robbins, and G. H. Barlow, *J. Biol. Chem.* **248**, 2984 (1973).

[19] D. Collen, E. B. Ong, and A. J. Johnson, *Fed. Proc., Fed. Am. Soc. Exp. Biol.* **31**, 229 (1972).

[20] P. Wallén and B. Wiman, *Biochim. Biophys. Acta* **221**, 20 (1970).

[21] E. E. Rickli and P. A. Cuendet, *Biochim. Biophys. Acta* **250**, 447 (1971).

[22] H. Claeys and J. Vermylen, *Biochim. Biophys. Acta* **342**, 351 (1974).

plasma fraction $III_{2,3}$[15,17,18] and had an NH_2-terminal lysine residue.[18] A pH 6.7 isoelectric form has also been isolated from both Fractions III and $III_{2,3}$ with an NH_2-terminal lysine residue.[18] It is proposed to define these two major groups of human zymogen forms as Glu-plasminogen and Lys-plasminogen, respectively. Lys-plasminogen is probably a partially degraded form of Glu-plasminogen.[23,24] Plasmin, a two-chain molecule, derived from both groups of forms has an NH_2-terminal lysine residue (from heavy (A) chain in the NH_2-terminal portion of the molecule) and an NH_2-terminal valine residue [from light (B) chain in the COOH-terminal portion of the molecule][25,26]; it is proposed to define the enzyme as Lys-plasmin. The enzyme is also a mixture of multiple molecular isoelectric forms with isoelectric points between pH 7.4 and pH 8.5 (3–5 major forms).[15,17]

Rabbit plasminogen is a Glu-zymogen which can be separated from plasma or serum by affinity chromatography methods into two overlapping groups of isoelectric forms by appropriate gradient elution techniques with ε-aminocaproic acid from L-lysine-substituted Sepharose.[26-28] The two groups of forms with isoelectric points between pH 6.2 and pH 7.8, and between 7.0 and pH 8.7, respectively, are both Glu-zymogens.[26,28] Cat and bovine plasminogens were both found to contain only Asp forms whereas dog plasminogen was found to contain only blocked X forms.[26] The cat, rabbit, and bovine multiple enzyme forms were found to be Lys forms whereas the dog multiple enzyme forms were found to be Arg forms.[26]

Method of Isolation of Plasminogen

Mammalian plasminogens can be isolated from plasma, serum, and zymogen enriched-protein fractions derived from plasma and serum by affinity chromatography methods using L-lysine-substituted Sepharose.[15,18,26-31]

[23] K. C. Robbins, I. G. Boreisha, L. Arzadon, L. Summaria, and G. H. Barlow, *J. Biol. Chem.* **250**, 4044 (1975).

[24] K. C. Robbins and L. Summaria, unpublished results.

[25] K. C. Robbins, L. Summaria, B. Hsieh, and R. J. Shah, *J. Biol. Chem.* **242**, 2333 (1967).

[26] K. C. Robbins, P. Bernabe, L. Arzadon, and L. Summaria, *J. Biol. Chem.* **248**, 7242 (1973).

[27] L. Summaria, L. Arzadon, P. Bernabe, and K. C. Robbins, *J. Biol. Chem.* **248**, 6522 (1973).

[28] J. M. Sodetz, W. J. Brockway, and F. J. Castellino, *Biochemistry* **11**, 4451 (1972).

[29] D. G. Deutsch and E. T. Mertz, *Science* **170**, 1095 (1970).

[30] T. H. Liu and E. T. Mertz, *Can. J. Biochem.* **49**, 1055 (1971).

[31] M. W. C. Hatton and E. Regoeczi, *Biochim. Biophys. Acta* **359**, 55 (1974).

Preparation of L-*Lysine-Substituted Sepharose*[29]

Sepharose 4B, 600 ml, was washed with 10 liters of ice-cold water by repeated centrifugations at 4000 rpm at 2°. Sixty grams of cyanogen bromide were dissolved in 1200 ml of water at room temperature. The cyanogen bromide solution was added to the 600 ml of Sepharose in a 3-liter beaker at room temperature. The pH was monitored continuously and maintained between pH 10.5 and 11.0 by the constant addition of 4 *M* sodium hydroxide. The reaction is usually complete within 30 min. The pH at the end of the reaction was 10.5. The cyanogen bromide-activated Sepharose was pulled dry on a Büchner funnel, and was washed on the funnel with 6 liters of 0.1 *M* sodium bicarbonate at pH 9.0. After washing, the Sepharose was hydrated to 800 ml with 0.1 *M* sodium bicarbonate, pH 9.0. Ten grams of L-lysine monohydrochloride dissolved in 40 ml of 0.1 *M* sodium bicarbonate was added to this suspension and mixed in a cold room for 48 hr. The L-lysine-substituted Sepharose was washed with 3 liters of 0.1 *M* sodium bicarbonate, followed by equilibration with 0.1 *M* sodium phosphate buffer, pH 7.4. The L-lysine-substituted Sepharose was packed into a 9.0 × 30.0 cm column and washed with several liters of 0.1 *M* sodium phosphate buffer pH 7.4.

Preparation of Human Plasminogen from Plasma Fraction $III_{2,3}$[15,18]

Human plasma fraction $III_{2,3}$ was obtained as a frozen paste and stored at −25° until used. One hundred grams of the paste was thawed and extracted with 1 liter of 0.05 *M* Tris–0.02 *M* lysine–0.1 *M* NaCl buffer, pH 9.0, for 18 hr in a cold room. The suspension was then centrifuged at 4000 rpm for 1 hr at 2°, and the precipitate was discarded. The supernatant was passed through 600 ml of L-lysine-substituted Sepharose in the 9.0 × 30.0 cm. column at 2°, and then washed through with ice-cold 0.1 *M* phosphate buffer, pH 7.4, until the absorbance of the effluent at 280 nm was less than 0.1. The plasminogen was then eluted with 1 liter of 0.2 *M* ε-aminocaproic acid dissolved in ice-cold 0.1 *M* phosphate buffer, pH 7.4. The zymogen was eluted in a symmetrical elution profile. A small portion of the ascending and descending slopes of the peak was discarded, and the protein in the central portion of the peak, representing 85% of the eluted protein, was precipitated by adding 3.1 g of ammonium sulfate per milliliter of solution. After 18 hr in a cold room, the suspension was centrifuged at 4000 rpm at 2° for 1 hr. The plasminogen precipitate was dissolved in ice-cold 0.05 *M* Tris-0.02 *M* lysine–0.01 *M* NaCl buffer, pH 9.0, at a concentration of approximately 20 mg/ml, in an ice bath. This solution was clarified at 3000 rpm for 1 hr at 2°. The plasminogen solution was then adjusted to 10^{-2} *M* diisopropyl phosphofluoridate (DFP)[32] with a 1 M DFP solution, incubated for 30 min in an ice bath, then clarified by centrifugation at 16,000 rpm for 1 hr at 2°. The plasminogen solution was stored at −25°. Reaffinity chromatography can be used to further purify the zymogen if the specific activity is lower than expected.

Plasminogen was further purified by chromatography on a DEAE-Sephadex column, 6 × 23 cm, equilibrated, and eluted with 0.05 M Tris–0.02 M lysine–0.1 M NaCl buffer, pH 9.0, at 2°. The fractions eluted from the DEAE-Sephadex column were pooled, precipitated with 3.1 g of ammonium sulfate per milliliter of solution, dissolved in the same way as the plasminogen fractions obtained from the L-lysine-substituted Sepharose column, except that no DFP was added, and stored at −25°. The potential specific activity of this plasminogen preparation was 22–26 casein units (26–31 CTA units[8]) per milligram of protein. The yield from this two-step procedure is 175–225 mg of plasminogen per 100 g of fraction $III_{2,3}$ paste. Gel filtration through Sephadex G-150 may also be used as an additional step, if desired.

Preparation of Human Plasminogen from Plasma Fraction III[15,18]

Human plasma fraction III was obtained as a frozen paste and was stored at −25° until used. One hundred grams of paste were thawed and extracted with 1 liter of 0.05 M Tris–0.02 M lysine–0.1 M NaCl buffer, pH 9.0, for 1 hr in a cold room. The suspension was then centrifuged at 4000 rpm for 1 hr at 2°, and the precipitate was discarded. The supernatant was then either frozen and stored at −25°, or was purified, without freezing, by the same two-step method used to purify fraction $III_{2,3}$. The potential specific activity of the plasminogen obtained from the fraction III extract, whether frozen or not frozen, was 20–24 casein units (24–30 CTA units) per milligram of protein. The yield was 70–90 mg of plasminogen per 100 g of fraction III paste.

Preparation of Human Plasminogen from Plasma or Serum[15,18]

One major modification in the affinity chromatography procedure was made when either plasma, or serum, was used as the plasminogen source. One liter of ice-cold plasma, or serum, was added to the L-lysine-substituted Sepharose column at 2° and the column was washed with 0.1 M phosphate buffer, pH 7.4, until absorbance of the effluent at 280 nm was less than 0.1. The column coolant was then raised to 16° and

---

[32] The abbreviations used are: DFP, diisopropyl phosphorofluoridate; TLCK, L-1-chloro-3-tosylamido-7-amino-2-heptanone.

2 liters of 0.3 M phosphate buffer, pH 7.4, was used to wash and elute an absorbed protein shown not to be plasminogen. The 0.3 M phosphate buffer was then washed out of the column with 0.1 M phosphate buffer, pH 7.4, as the temperature of the column coolant was gradually lowered to 2°. The plasminogen was then eluted at 2° with 0.2 M ϵ-aminocaproic acid in 0.1 M phosphate buffer, pH 7.4. The procedure thereafter was identical to that used to purify plasminogen from fraction $III_{2,3}$. The specific activity of the plasminogen was 20–24 casein units (24–30 CTA units) per milligram of protein. The yield was 70–90 mg of plasminogen per liter of plasma, or serum.

Preparation of the Equimolar Plasminogen–Streptokinase Complex[12,33,34]

To prepare the equimolar complex of human, cat, dog, and rabbit plasminogen with streptokinase, 35 μl of the plasminogen solution, containing 20 mg of protein per milliliter in the pH 9.0 Tris–lysine–NaCl buffer was added to 50 μl of highly purified streptokinase (Kabikinase at a concentration of 8.0 mg/ml in 0.067 M phosphate buffer, pH 7.4) in an ice bath. The equimolar complex (at a concentration of 13 mg/ml) was mixed in the ice bath and then incubated at 25° for a selected time interval, from 0 to 40 min. When samples were removed, they were placed in an ice bath and immediately adjusted to 10^{-2} M DFP with a 1.0 M DFP solution. They were left in the ice bath for 30 min and were then frozen at −25°. The time selected for incubation at 25° depended upon whether one desired a plasminogen–streptokinase complex, a mixture of plasminogen–streptokinase and plasmin–streptokinase complexes, or a plasmin–streptokinase complex. Samples could be analyzed in an appropriate acrylamide gel-dodecyl sulfate electrophoretic system for composition.[34]

Preparation of Human Plasmin[35]

Three milliliters of human plasminogen solution in the pH 9.0 Tris–lysine–NaCl buffer, containing 20 mg of protein per milliliter, are adjusted to 25% glycerol by adding 1 ml of 99.5% glycerol (synthetic) in an ice bath. Urokinase, 110 μl, is added from a stock urokinase solution containing 13,000 CTA units per milliliter so that 1 CTA unit of uro-

[33] L. Summaria, K. C. Robbins, and G. H. Barlow, *J. Biol. Chem.* **246**, 2136 (1971).
[34] L. Summaria, L. Arzadon, P. Bernabe, and K. C. Robbins, *J. Biol. Chem.* **249**, 4760 (1974).
[35] K. C. Robbins and L. Summaria, this series Vol. 19, p. 184.

kinase is added per casein unit of plasminogen. The activation mixture was incubated at 25° for approximately 20 hr. It is then assayed for plasmin activity to determine whether complete activation has occurred. The specific activity is usually the same as that of the starting plasminogen solution. The plasmin is either stored in the 25% glycerol activating medium at −25°, or allowed to react with specific inhibitors.

Inhibition and Radiolabeling of Active-Center Serine and Histidine Residues of Plasmin with DFP and TLCK[1,32,36]

A 1.0 M solution of DFP, or radiolabeled DFP (^{14}C or ^{32}P) is added to the plasmin–glycerol activation mixture to a final concentration of 10^{-2} M DFP to react with the active center serine residue. The mixture is incubated for 1 hr in an ice bath; it is then assayed to determine the absence of enzyme activity. The DFP-inactivated plasmin (DIP-plasmin) solution is adjusted to pH 3.0 with 1 M HCl and dialyzed extensively against 10^{-3} M HCl in the cold room. After dialysis, an aliquot of the radiolabeled DIP-plasmin is counted in a Packard Tri-Carb liquid scintillation spectrometer, Model 3003, in Insta Gel to determine incorporation into the active center serine residue of plasmin.

To inhibit or radiolabel the active center histidine residue with TLCK, or radiolabeled TLCK (^{14}C or ^{3}H), the activation mixture is adjusted to pH 7.4 with 1 N HCl in an ice bath. TLCK, or radiolabeled TLCK, is dissolved in a volume of an 0.05 M Tris–0.02 M lysine–0.1 M NaCl–25% glycerol buffer, pH 7.4, so that when it is added to the plasmin solution, the final concentration of TLCK will be 0.005 M. The reaction is carried out at 25° for 3 hr; the mixture is then assayed to determine the absence of enzyme activity. The TLCK-plasmin solution is adjusted to pH 3.0 with 1 N HCl and dialyzed extensively against 10^{-3} M HCl in the cold room. After dialysis, an aliquot is counted to determine the molar incorporation into the active center of plasmin.

Activation of Human Plasminogen with Urokinase in the Presence of Trasylol[37]

Two milliliters of human plasminogen solution in the pH 9.0 Tris–lysine–NaCl buffer, containing 20 mg of protein per milliliter, is mixed with 1.3 ml of 99.5% glycerol and 2.0 ml of Trasylol (FBA Pharma-

[36] L. Summaria, B. Hsieh, W. R. Groskopf, K. C. Robbins, and G. H. Barlow, *J. Biol. Chem.* **242**, 5046 (1967).
[37] L. Summaria, L. Arzadon, P. Bernabe, and K. C. Robbins, *J. Biol. Chem.* **250**, 3988 (1975).

ceuticals), in an ice bath. The ratio of Trasylol to plasminogen is 1000 KIU units of Trasylol per milligram of plasminogen protein; the molar ratio is approximately 2:1. To this mixture is added 1800 CTA units of urokinase. The activation mixture is incubated for 18 hr at 25°. The plasmin–Trasylol mixtures are adjusted to pH 3.0 with 1 N HCl, dialyzed against 10^{-3} M HCl in a cold room and lyophilized. The plasmin–Trasylol complexes may also be dissociated to yield active plasmin by dissolving the complex in a 0.5 M Tris–0.02 M lysine–0.1 M NaCl buffer, pH 9.0, adding solid urea to 8 M, or solid guanidine-HCl to 6 M, and assaying. Radiolabeled DFP can be incorporated into the active center of plasmin by adjusting the plasmin–Trasylol activation mixture to 10^{-2} M DFP prior to the addition of solid urea to 8 M or solid guanidine-HCl to 6 M, followed by incubating the mixture for 24 hr in a cold room. The preparation is then dialyzed at pH 3.0 in a cold room, counted to determine the molar incorporation of DFP, and lyophilized.

Isoelectric Focusing[15,17,18]

Separations by isoelectric focusing methods were carried out using the LKB Model 8100 jacketed column (110 ml or 440 ml capacity), maintained at either 2° or 10°. The bottom electrode solution in all separations contained 60% w/v sucrose and 1% phosphoric acid, and the top electrode solution contained 2% ethylenediamine. A stabilizing linear sucrose gradient (45% to 1%) was formed continuously by the use of a two-chamber gradient mixer (LKB 8121). Carrier ampholytes (Ampholines) were obtained as water solutions with a solids content of 40% w/w, with various pH ranges. One chamber of the gradient mixer was filled (55 ml or 220 ml) with a solution consisting of 45% sucrose, 0.3 M ϵ-aminocaproic acid, and Ampholines at a final concentration of 2%. The second chamber contained only 0.3 M ϵ-aminocaproic acid and 2% Ampholines. For some separations, the ϵ-aminocaproic acid was replaced by 7 M urea, or water. The isoelectric gradient was established by applying a potential of 300 V for 24 hr. Plasminogen and plasmin samples were either dissolved in, or dialyzed against, 0.3 M ϵ-aminocaproic acid. Plasmin-derived heavy (A) and light (B) chains were dissolved in 7 M urea. The sucrose concentration of the sample was adjusted to 30% by adding sucrose crystals, and the sample was introduced into the column by layering it at its isodense position.

The voltage was increased stepwise to between 750 and 1000 V. Isoelectric focusing was carried out at a maximum voltage for 72–96 hr. Fractions of 2.0 ml were collected by gravity flow at a regulated flow rate of 1 ml/min. The protein concentration of each fraction was determined

TABLE I

Isoelectric Forms of Human Plasminogen, Plasmin, and
Plasmin-Derived Heavy (A) and Light (B) Chains

Plasminogen[a] pI	Plasmin[b] pI	Heavy (A) chain[b] pI	Light (B) chain[b] pI
6.2	7.2	4.9	5.8
6.3	7.4		5.9
6.4	7.7		6.0
6.6	7.9		
6.7	8.1		
7.2	8.2		
7.5	8.5		
7.8			
8.1			
8.3			
8.5			

[a] L. Summaria, L. Arzadon, P. Bernabe, and K. C. Robbins, *J. Biol. Chem.* **247**, 4691 (1972); K. C. Robbins and L. Summaria, *Ann. N.Y. Acad. Sci.* **209**, 397 (1973); L. Summaria, L. Arzadon, P. Bernabe, K. C. Robbins, and G. H. Barlow, *J. Biol. Chem.* **248**, 2984 (1973).

[b] L. Summaria, L. Arzadon, P. Bernabe, and K. C. Robbins, *J. Biol. Chem.* **247**, 4691 (1972); K. C. Robbins and L. Summaria, *Ann. N.Y. Acad Sci.* **209**, 397 (1973).

by absorbance measurements at 280 nm; the pH of each fraction was measured at room temperature. The zymogen, or enzyme, activity was determined on fractions exhaustively dialyzed against 10^{-3} M HCl in a cold room. The fractions were then lyophilized. Radioactivity measurements were made on the lyophilized fractions dissolved in 10^{-3} M HCl. The results of experiments with plasminogen, plasmin, and plasmin-derived heavy (A) and light (B) chains are summarized in Table I.

Hydrodynamic Properties[14,23,38]

The hydrodynamic properties of human plasminogen and plasmin are summarized in Table II. The Glu- and Lys-zymogen forms appear to be similar in molecular weights by sedimentation equilibrium methods, in both standard buffers and dodecyl sulfate solutions. The Yphantis method apparently gives values slightly different than those obtained with absorption optics.[38] The sedimentation coefficients of the Glu and Lys forms were different. In the sedimentation velocity runs, the zymogens gave a single symmetrical boundary. Certain amino acids, e.g., L-lysine and ɛ-aminocaproic acid, bind to the zymogens and alter the

[38] I. Sjöholm, B. Wiman, and P. Wallén, *Eur. J. Biochem.* 39 471 (1973).

TABLE II

HYDRODYNAMIC PROPERTIES OF HUMAN PLASMINOGEN AND PLASMIN

	Sedimentation coefficient ($s_{20,w}$)			Partial specific volume ($\bar{v}$)			Molecular weight			Frictional coefficient (f/f_0)		
	1[a]	2[b]	3[c]	1[a]	2[b]	3[c]	1[a]	2[b]	3[c]	1[a]	2[b]	3[c]
Glu-plasminogen	—	5.0	5.1	—	0.709	0.706	—	83,800	92,000	—	1.54	1.50
Lys-plasminogen	4.2	4.4	4.8	0.715	0.714	0.709	81,000	82,400	90,000	1.80	1.63	1.56
Plasmin	3.9	4.3	4.3	0.715	0.714	0.713	75,400	76,500	81,000	1.50	1.64	1.55

[a] G. H. Barlow, L. Summaria, and K. C. Robbins, *J. Biol. Chem.* **244**, 1138 (1969).

[b] K. C. Robbins, I. G. Boreisha, L. Arzadon, L. Summaria, and G. H. Barlow, *J. Biol. Chem.* **250**, 4044 (1975).

[c] I. Sjöholm, B. Wiman, and P. Wallén, *Eur. J. Biochem.* **39**, 471 (1973).

sedimentation coefficients.[39] The circular dichroism spectra of the Glu
and Lys forms are almost identical and indicate that they contain a
large proportion of random structure, about 80%, and about 20% β-
structure with no detectable amounts of α-helix.[38] L-Lysine and ϵ-amino-
caproic acid also influence the circular dichroism spectra of the zymogens
and the enzyme[38] and the rotational diffusion of the zymogen.[40] The
frictional coefficients of the zymogens and the enzyme are similar.

Amino Acid Compositions[6,16,23,24,28,41,42]

The amino acid compositions of various human and rabbit plasminogen
preparations are summarized in Table III. Since the calculations of the
amino acid composition data were not made in the same manner, it is
difficult to compare the data except to note the similarities and differences
in the compositions. The amino acid compositions of human plasmin and
plasmin-derived heavy (A) and light (B) chains are summarized in Table
IV.

Plasmin Chains[1,25,27,37,41]

Cleavage of a single disulfide bond in human Lys-plasmin results in
the formation of two chains that can be readily separated electrophoretically.[25] Complete reduction and alkylation of human Lys-plasmin gives
two S-carboxymethyl chains, which can be readily separated from each
other by dialysis against dilute ammonium bicarbonate solutions into a
soluble heavy (A) chain derivative and an insoluble light (B) chain
derivative.[1] The molecular weight of the isolated heavy (A) chain derivative was determined to be 48,800 by sedimentation equilibrium analysis, and its sedimentation coefficient was determined to be 2.8 S.[41] The
NH$_2$-terminal amino acid was found to be lysine and the COOH-terminal
amino acid was found to be arginine.[25] The heavy (A) chain derivative
has a single isoelectric form with an isoelectric point of pH 4.9.[15] The
molecular weight of the isolated light (B) chain derivative was determined to be 25,700 by sedimentation equilibrium analysis and its sedimentation coefficient was determined to be 1.4 S.[36] The NH$_2$-terminal
residue was determined to be valine and the COOH-terminal residue
was determined to be asparagine.[25] The light (B) chain derivative has

[39] W. J. Brockway and F. J. Castellino, *Arch. Biochem. Biophys.* **151**, 194 (1972).
[40] F. J. Castellino, W. J. Brockway, J. K. Thomas, H.-T. Liao, and A. B. Rawitch, *Biochemistry* **12**, 2787 (1973).
[41] L. Summaria, K. C. Robbins, and G. H. Barlow, *J. Biol. Chem.* **246**, 2143 (1971).
[42] D. K. McClintock, M. E. Englert, C. Dziobkowski, E. H. Snedeker, and P. H. Bell, *Biochemistry* **13**, 5334 (1974).

TABLE III

AMINO ACID COMPOSITION OF HUMAN AND RABBIT PLASMINOGENS

Amino acid	Human					Rabbit	
	Glu-forms[a]	Lys-forms[a]	Lys-forms[b]	Glu-forms[c]	Glu-forms[d]	Glu-forms (1)[e]	Glu-forms (2)[e]
Lysine	40	39	41	50	46	53	50
Histidine	21	21	21	22	22	21	20
Arginine	38	36	37	39	40	55	54
Aspartic acid	72	67	71	76	72	80	78
Threonine	60	57	54	57	65	58	57
Serine	57	57	44	51	56	58	58
Glutamic acid	81	69	69	92	88	84	83
Proline	68	66	65	73	54	58	58
Glycine	60	59	56	58	59	56	57
Alanine	38	33	32	38	37	49	50
Half-cystine	41	44	38	38	40	48	49
Valine	44	41	43	44	41	32	34
Methionine	8	8	9	8	10	9	10
Isoleucine	19	16	19	23	18	22	23
Leucine	40	40	40	43	41	40	41
Tyrosine	27	29	29	28	28	30	32
Phenylalanine	19	16	17	19	18	18	19
Tryptophan	16	18	20	21	22	19	19
	749	707	705	780	757	790	792

[a] K. C. Robbins, I. G. Boreisha, L. Arzadon, L. Summaria, and G. H. Barlow, *J. Biol. Chem.* **250**, 4044 (1975).

[b] K. C. Robbins, L. Summaria, D. Elwyn, and G. H. Barlow, *J. Biol. Chem.* **240**, 541 (1965).

[c] P. Wallén and B. Wiman, *Biochem. Biophys. Acta* **257**, 122 (1972) (pH 6.23 isoelectric form).

[d] D. K. McClintock, M. E. Englert, C. Dziobkowski, E. H. Snedeker, and P. H. Bell, *Biochemistry* **13**, 5334 (1974).

[e] J. M. Sodetz, W. J. Brockway, and F. J. Castellino, *Biochemistry* **11**, 4451 (1972).

TABLE IV
AMINO ACID COMPOSITION OF HUMAN PLASMIN AND PLASMIN-DERIVED
HEAVY (A) AND LIGHT (B) CHAINS

Amino acid	Plasmin[a]	Heavy (A) chain[b]	Light (B) chain[a]
Lysine	38	26	11
Histidine	21	12	7
Arginine	36	22	12
Aspartic acid	60	46	16
Threonine	50	41	15
Serine	40	28	17
Glutamic acid	59	35	24
Proline	61	43	17
Glycine	53	29	24
Alanine	29	16	13
Half-cystine	37	34	12
Valine	36	15	20
Methionine	8	6	2
Isoleucine	15	6	9
Leucine	36	15	20
Tyrosine	26	17	6
Phenylalanine	14	7	7
Tryptophan	16	12	6
	635	410	238

[a] K. C. Robbins and L. Summaria, unpublished results.
[b] L. Summaria, K. C. Robbins, and G. H. Barlow, *J. Biol. Chem.* **246**, 2143 (1971).

three isoelectric forms of pH values 5.8, 5.9, and 6.0.[15] The amino acid composition of Lys-plasmin, and the plasmin-derived heavy (A) and light (B) chains are summarized in Table IV. Heavy (A) and light (B) chain derivatives can be prepared from various mammalian plasmins[27] and plasmin–Trasylol complexes,[37] which have been reduced and alkylated, by the dialysis method (plasmin) or by a combination of the dialysis method and gel filtration through Sephadex G-150 columns equilibrated and eluted with 6 *M* guanidine HCl at pH 3.5 (plasmin–Trasylol complexes).

Equimolar Plasminogen and Plasmin–Streptokinase Complexes[12,33,34,43–45]

Equimolar plasminogen–streptokinase and plasmin–streptokinase complexes have been prepared from a number of mammalian plasmino-

[43] D. K. McClintock and P. H. Bell, *Biochem. Biophys. Res. Commun.* **43**, 694 (1971).
[44] K. N. N. Reddy and G. Markus, *J. Biol. Chem.* **247**, 1683 (1972).
[45] L. A. Schick and F. J. Castellino, *Biochem. Biophys. Res. Commun.* **57**, 47 (1974).

gens.[12,33,34,43-45] An active center is developed in human plasminogen upon complex formation with streptokinase before conversion to plasmin–streptokinase.[43-45] These complexes, particularly the plasmin complexes, probably function as activators of those mammalian species of plasminogen that are not directly activable by streptokinase.[12,24,46] The human plasmin–streptokinase complex is an excellent activator of all mammalian plasminogens, particularly bovine and other non-streptokinase-activable plasminogens.[46] The cat and dog plasmin–streptokinase complexes are qualitatively similar to the human complex in being able to activate bovine plasminogen, but quantitatively, they are poor activators.[24] Cat and dog plasminogens are activable by streptokinase, but they, particularly the dog zymogen, require unusually high concentrations of the activator for complete activation.[34] Rabbit plasminogen forms equimolar complexes,[34,45] but it is difficult, at times impossible, to activate this zymogen.[34]

The human equimolar complex has been prepared and isolated from a mixture of the two components.[12,34] The activator complex contains a single DFP-sensitive residue in the active center which is probably the active site serine of the plasmin moiety.[47,48] Complete loss in both activator and enzymic activity occurs when the activator complex is treated with DFP. Dissociation of the complex either in 8 M urea, or at pH 3.0, gives an active plasmin moiety, but the streptokinase has been fragmented, and there is no activator activity in any of the fragments.[34] Several different streptokinase fragments have been isolated from human complexes[33,49]; a large fragment has been isolated that will form an activator complex with the zymogen.[49] Each fragment in each mammalian complex appears to be identical, suggesting that the fragmentation of streptokinase in each of the mammalian complexes occurs by the same mechanism, but the rate of fragmentation differs in each species.[34]

Active-Center Serine and Histidine Residues[1,36,50]

A single DFP-sensitive serine residue[36] and a single TLCK-sensitive histidine residue[1,50] have been found in the active center of human plasmin, and both residues are located on the light (B) chain of the

[46] R. J. Wulf and E. T. Mertz, *Can. J. Biochem.* 47, 927 (1969).
[47] L. Summaria, C.-M. Ling, W. R. Groskopf, and K. C. Robbins, *J. Biol. Chem.* 243, 144 (1968).
[48] F. F. Buck, B. C. W. Hummel, and E. C. De Renzo, *J. Biol. Chem.* 243, 3648 (1968).
[49] W. J. Brockway and F. J. Castellino, *Biochemistry* 13, 2063 (1974).
[50] K. C. Robbins, P. Bernabe, L. Arzadon, and L. Summaria, *J. Biol. Chem.* 248, 1631 (1973).

Species	Sequences
	1 10
Man	Glu -Pro- Leu-Asp-Asp-Tyr- Val-Asn- X-Gln-Gly-Ala-
Cat	Asp-Pro- Leu-Asp-Asp- Tyr- Val-Asn- X-Gln-Gly-Ala-
Rabbit	Glu -Pro- Leu-Asp-Asp- Tyr- Val-Asn- X-Gln-Gly-Ala-
Ox	Asp- Leu- Leu-Asp-Asp- Tyr- Val-

FIG. 1. NH$_2$-terminal sequences of mammalian plasminogens.

enzyme.[1,36] The partial amino acid sequence of a 31-residue tryptic peptide containing the active-center serine residue was determined to be Val-Glx- (Ser-Thr, Glx)-Leu-(Gly, Ala)-His-Leu-Ala-Cys-Asn-(Thr, Gly,   Gly)-Ser-Cys-Gln-Gly-Asp-Ser*-Gly-Gly-Pro-Leu-Val-Cys-Phe-Glu-Lys-.[51] This sequence is homologous with the active-center serine sequences of plasmin-derived heavy (A) chains from these same mammalian species are summarized in Fig. 2; and the NH$_2$-terminal 21-residue (I) obtained from the light (B) chain and containing the active center histidine residue was determined to be His-Phe-Cys-Gly-Gly-Thr-Leu-Ile-Ser-Pro-Glu-Trp-Val-Leu-Ser-Ala-Ala-His*-Cys-Leu-.[50] This sequence contains the "histidine loop" of human plasmin, and it is homologous to the "histidine loop" sequences of other serine proteases.[52]

Species	Sequences
Man	Lys-Val-Tyr- Leu- Leu-
Cat	Lys - Ile- Tyr- Leu-Val -
Dog	Arg- Ile -Tyr- Leu-Gly -
Rabbit	Lys-Val-Tyr- Leu-Gly -
Ox	Lys -Ile - Tyr- Leu-Val -

FIG. 2. NH$_2$-terminal sequences of plasmin-derived heavy (A) chains from mammalian plasminogens.

NH$_2$-Terminal Sequences[26]

The NH$_2$-terminal 12-residue sequences of human, cat, rabbit, and bovine (7-residue) plasminogens are summarized in Fig. 1. The NH$_2$-terminus of dog plasminogen is blocked. The NH$_2$-terminal 5-residue sequences of plasmin-derived heavy (A) chains from these same mammalian species are summarized in Fig. 2; and the NH$_2$-terminal 21-residue sequences of plasmin-derived light (B) chains from these mammalian species are summarized in Fig. 3. The NH$_2$-terminal 69-residue sequence of human Glu-plasminogen has been determined[53] (Fig. 4). The amino

[51] W. R. Groskopf, L. Summaria, and K. C. Robbins, *J. Biol. Chem.* **244**, 3590 (1969).
[52] B. S. Hartley, *Philos. Trans. R. Soc. London Ser. Biol. Sci.* **257**, 77 (1970).
[53] B. Wiman, *Eur. J. Biochem.* **39**, 1 (1973).

Species		Sequences	
	1	10	20
Man	Val-Val-Gly-Gly-Cys-Val-Ala-His-Pro-His-Ser-Trp-Pro-Try-Gln-Val-Val-Leu-Leu-Arg-Arg-		
Cat	Val-Val-Gly-Gly-Cys-Val-Ala-His-Pro-His-Ser-Trp-Pro-Trp-Gln-Val-Val-Leu-Leu-Lys-Lys-		
Dog	Val-Val-Gly-Gly-Cys-Val-Ala-Lys-Pro-His-Ser-Trp-Pro-Trp-Gln-Ile-Ile-Leu-Leu-Arg-Arg-		
Rabbit	Val-Val-Gly-Gly-Cys-Val-Ala-Lys-Pro-Lys-Ser-Trp-Pro-Trp-Gln-Ile-Ile-Leu-Leu-Lys-Arg-		
Ox	Ile-Val-Gly-Gly-Cys-Val-Ala-Lys-Pro-Lys-Ser-Trp-Pro-Trp-Gln-Val-Ser-Leu-Leu-Lys-Lys-		

FIG. 3. NH_2-terminal sequences of plasmin-derived light (B) chains from mammalian plasminogens.

acid sequence from Met_{69} to the Lys_{78}-Val-Tyr-Leu-sequence has been reported to be Met_{69}-Ser-Asp-Val-Val-Leu-Phe-Glu-Lys_{77}-.[54] Derived plasminogens and heavy (A) chains with other NH_2-terminal sequences and residues have been reported.[55–57]

1   10   20
Glu-Pro-Leu-Asp-Asp-Tyr-Val-Asn-Thr-Gln-Gly-Ala-Ser-Leu-Phe-Ser-Val-Thr-Lys-Lys-
Gln-Leu-Gly-Ala-Gly-Ser-Ile-Glu-Glu-Cys-Ala-Gln-Ala-Lys-Cys-Glu-Glu-Asp-Glu-Glu-
Phe-Thr-Cys-Arg-Ala-Phe-Gln-Tyr-His-Ser-Lys-Glu-Gln-Glu-Cys-Val-Ile-Met-Ala-Glu-
Asn-Arg-Lys-Ser-Ser-Ile-Ile-Arg-Met-

FIG. 4. NH_2-terminal 69-residue sequence of human plasminogen.

Mechanism of Activation[42,53,55,57,58]

The mechanism of activation of plasminogen to plasmin involves two major specific bond cleavages. In the human system, the two principal events are the cleavage of a sensitive arginyl-valine peptide bond in the COOH-terminal portion of Glu-plasminogen to give the two-chain plasmin molecule, followed by the cleavage of a lysyl–lysine peptide bond in the NH_2-terminal portion of the molecule. The latter event results in the release of a peptide, or peptides. The cleavage of a single disulfide bond in the two-chain Lys-plasmin molecule results in two chains.[25] This mechanism of activation appears to be the same mechanism for a number of mammalian plasminogen systems studied.[26] Additional peptide bond cleavages can be found during activation of several mammalian zymogens that appear to be due to autolytic processes.[26] This mechanism of activation also occurs in the conversion of the equimolar plasminogen–streptokinase complexes to plasmin-streptokinase complexes,[34] but in the conver-

[54] B. Wiman and P. Wallén, Eur. J. Biochem. 50, 489 (1975).
[55] B. Wiman and P. Wallén, Eur. J. Biochem. 36, 25 (1973).
[56] E. E. Rickli and W. I. Otavsky, Biochim. Biophys. Acta 295, 381 (1973).
[57] J. M. Sodetz, W. J. Brockway, K. G. Mann, and F. J. Castellino, Biochem. Biophys. Res. Commun. 60, 729 (1974).
[58] P. J. Walther, H. M. Steinman, R. L. Hill, and P. A. McKee, J. Biol. Chem. 249, 1173 (1974).

sion of plasminogen to plasmin in the presence of the plasmin inhibitor Trasylol, only a single peptide bond is cleaved, namely, the arginyl–valine bond in the COOH-terminal portion of the molecule, to give enzymes with heavy (A) chains with the same NH_2-terminal sequences as those found in the parent zymogens[37,57] (Fig. 1). Single peptide bond cleavage at the arginyl–valine bond also appears to occur in the equimolar complexes.[34] A "preactivation peptide" has been described,[58] and other specific peptide bond cleavage sites have been reported.[55–57]

[23] Rabbit Plasminogen and Plasmin Isozymes

By Francis J. Castellino and James M. Sodetz

Assay Method

Assays of plasminogen first require its conversion to the enzyme plasmin by common activators. The enzyme is then assayed on the basis of its esterolytic activity toward N-α-toluenesulfonyl-L-arginine methyl ester (TosArgOMe, TAME) utilizing a recording pH stat to titrate the amount of acid, $N\alpha$-tosyl-L-arginine, liberated with time. If this equipment is not available, a very sensitive colorimetric method for analysis of the concentration of TosArgOMe remaining after incubation of plasmin with this substrate for a given period of time can be used. This method is based upon the quantitative reaction of esters, such as TosArgOMe, with hydroxylamine, yielding the hydroxamic acid derivative. Upon addition of ferric chloride, a ferric ion–hydroxamic acid complex is formed, possessing spectral properties different from ferric chloride. Other assay methods for plasmin are available such as the casein[1] or azocasein[2] methods and the fibrin plate assay.[3] These latter methods will not be discussed in this chapter.

Colorimetric Method

Reagents

Tris buffer, 0.1 M Tris-HCl, pH 8.2
Tris–lysine buffer, 0.05 M Tris-HCl–0.05 M L-lysine, pH 8.0

[1] K. C. Robbins and L. Summaria, this series Vol. 19, p. 184 (1970).
[2] Our procedure for this method can be found in L. A. Schick and F. J. Castellino, *Biochemistry* **12**, 4315 (1973).
[3] T. Astrup and S. Müllertz, *Arch. Biochem. Biophys.* **40**, 346 (1952).

Veronal buffer, 0.05 M, pH 8.3

Stock plasminogen solution, 7–8 mg/ml in Tris–lysine buffer

Stock urokinase solution, 6000 Plough units/ml in Tris–lysine buffer

Stock p-nitrophenyl-p'-guanidinobenzoate (NphBzoGdn) solution, 0.04 M in dimethylformamide stored at 4° is stable for at least 2 months.

Stock N^{α}-tosyl-L-arginine methyl ester (TosArgOMe), 0.1 M in Tris buffer

NaOH, 3.5 N

Hydroxylamine hydrochloride, 2 M

Ferric chloride, 0.11 M in 0.04 M HCl. This solution is filtered over Celite and adjusted to a final pH of 1.2 ± 0.1.

Trichloroacetic acid solution, 200 ml of 1.9 M HCl plus 6 g of trichloroacetic acid

Procedure. In order to accurately determine the specific activity of plasminogen, the protein must first be activated to plasmin. This can be accomplished by mixing 0.13 ml of rabbit plasminogen with 0.02 ml of urokinase and incubating at 22°. After 25 min, the temperature is lowered to 4° in order to inhibit autodigestion of the plasmin. The concentration of active plasmin is determined by active site titration with NphBzoGdn.[4] The titration procedure we use is only slightly different than that described by Chase and Shaw.[4] Our titration consisted of adding 0.04 ml of plasminogen activation mixture to 1.0 ml of NphBzoGdn which has been diluted 50 μM with Veronal buffer. The p-nitrophenolate released is quantitated by measuring the absorbance at 410 nm using a full scale of 0.1 absorbance unit on the recorder. We have successfully carried out these measurements on either a double-beam Cary 15 or a single-beam Gilford 2000 spectrophotometer. The concentration of active plasmin in the stock plasmin solution is calculated using a molar extinction coefficient of 16,595 for p-nitrophenolate[5] at pH 8.3 and a molecular weight of plasmin of 86,000. We routinely obtain 80–85% active plasmin per mole of initial plasminogen under these conditions. If the activated material is retained in the ice bath, no change in the reactivity with NphBzoGdn is noted over a time period of 2 hr.

In order to analyze the kinetics of plasmin using TosArgOMe as the substrate, the stock plasmin solution is diluted with Tris buffer to a final active plasmin concentration of 60–70 μg/ml. The stock TosArgOMe solution is diluted with Tris buffer such that a series of solutions with a concentration range of TosArgOMe of 0.02 M to 0.1 M are obtained. The

[4] T. Chase and E. Shaw, this series Vol. 19, p. 20 (1970).

[5] T. Chase and E. Shaw, *Biochem. Biophys. Res. Commun.* **29**, 508 (1967).

assay is initiated by addition of 0.050 ml of the substrate solution to 0.2 ml of the enzyme. The solutions are allowed to incubate for 10 min at 22°. Controls are simultaneously run at each substrate concentration by substitution of Tris buffer for plasmin. At the conclusion of the incubation time, the concentration of remaining TosArgOMe in each tube is determined in a similar manner to the Roberts[6] modification of the Hestrin method.[7] The enzymic reaction is immediately quenched by addition of 0.20 ml of assay mixture to a solution (prepared immediately prior to use) of 0.20 ml of NaOH and 0.20 ml of hydroxylamine-HCl. A spectrophotometer reference solution is also prepared at this time by substitution of 0.20 ml of Tris buffer for the assay mixture. Hydroxamic acid formation is complete after 30 min at room temperature. At this time, 0.40 ml of trichloroacetic acid solution is added, followed by 4.0 ml of ferric chloride. The samples are deaerated at water-aspirator vacuum conditions with simultaneous vortexing. The absorbance at 525 nm is determined against the reference blank. The amount of TosArgOMe remaining in each assay tube is determined by comparison of the data with a standard curve previously determined in the same fashion for TosArgOMe. Nonenzymic hydrolysis of TosArgOMe is negligible under these conditions but can be evaluated by the control experiments described above.

Potentiometric Method

Reagents

NaCl, 0.1 M, pH 8.0
TosArgOMe stock solution, 0.15 M in 0.1 M NaCl, pH 8.0
Plasmin: A stock solution prepared and activated as described in the preceding section. The concentration of active plasmin is determined as previously described. The final plasmin concentration used is preferably in the range of 3–5 mg/ml.

Procedure. Assays are performed in a pH stat maintained at a constant temperature of 22°. The cell should be thermostatted and capable of continual stirring. Our equipment is purchased from Radiometer and consists of a PHM 26 pH meter, type TTT11 Titrator, TTA-31 Titration Assembly, ABU-11 Auto Burette and a REC-51-S Servograph Recorder. Constant temperature is maintained with a Haake Model FE circulating

[6] P. S. Roberts, *J. Biol. Chem.* **232**, 285 (1958).
[7] S. Hestrin, *J. Biol. Chem.* **180**, 249 (1949).

water bath. The syringe is filled with carefully standardized NaOH at concentrations of approximately 0.02 M.

The substrate is first diluted with the NaCl solution to a range of values of 0–0.15 M. One milliliter of substrate is placed in the pH-Stat vessel and allowed to temperature equilibrate. The pH is then adjusted to pH 8 (if necessary) by automatic addition of base from the pH-Stat. Plasmin is added in a 0.020-ml aliquot such that approximately 45–60 μg of active plasmin is present per milliliter in the reaction vessel. The reaction rate is followed for at least 2 min on the recorder. The results are converted to micromoles of TosArgOMe hydrolyzed min^{-1} mg^{-1} of active plasmin.

Definition of a Unit and Specific Activity. One unit of plasmin is defined as the amount of active plasmin which is required to hydrolyze 1 μmole of TosArgOMe per minute. Extrapolation of reaction rate vs substrate concentration data, as determined above, to infinite substrate concentration, results in a V_{max} of rabbit plasmin of 14.0 μmoles of TosArgOMe cleaved min^{-1} mg^{-1} of active plasmin at 22°. Therefore, 1 TosArgOMe unit of rabbit plasmin activity refers to 0.071 μg of active plasmin. For rapid screening of plasmin specific activity, enzyme rate determinations on a single substrate concentration of 0.15 M TosArgOMe is appropriate for V_{max} analysis. However, the pH-Stat should be used here since the colorimetric method is a subtractive method, and only small changes between initial and final TosArgOMe concentrations are noted owing to the high initial levels of the substrate.

Purification Procedure

Plasminogen

Rabbit plasminogen is best prepared in a single step from whole fresh-frozen rabbit plasma by affinity chromatography. The following method describes the fashion in which we prepare and utilize the affinity resin. Variation of this recommended procedure will lead to resin with slightly different properties than what we describe and will require minor adjustments of the procedure. The basis of the method discussed here is a variation of the original Deutsch and Mertz procedure.[8]

The Sepharose 4B–L-lysine affinity chromatography columns are prepared by adding 10 g of cyanogen bromide in 100 ml of water to 100 ml (settled) of previously washed Sepharose 4B (Pharmacia) at room temperature. The resin is continually stirred and maintained at

[8] D. Deutsch and E. T. Mertz, *Science* **170**, 1095 (1970).

pH 11 by repeated addition of 4 N NaOH. When base uptake ceases, the resin is rapidly washed under suction with 2 liters of cold 0.1 M NaHCO$_3$. A solution of 10 g of L-lysine in 100 ml of 0.1 M phosphate, final pH 9.2, is then added to the resin. The suspension is stirred at 5° overnight. After this, the resin is extensively washed with 0.3 M phosphate, pH 8.0 and packed in a column to a final dimension of 2.5 cm × 11 cm. All subsequent steps are performed at room temperature.

To approximately 500 ml of citrated plasma, diisopropyl fluorophosphate (DFP) is added to a final concentration of 0.001 M. The plasma is then passed over the column at a rapid rate of flow (>100 ml/hr). The column is then washed with 0.3 M phosphate, pH 8.0, until the absorbance of material eluted from the column drops to essentially zero. At this point, the column is developed with a linear gradient, consisting of 250 ml of 0.1 M phosphate, pH 8.0, as the starting solvent and 250 ml of 0.01 M phosphate–0.025 M ε-aminocaproic acid, pH 8.0, as the limit solvent. The flow rate of the column is adjusted to 25 ml/hr, and approximately 3-ml fractions are collected. In this manner, two major fractions containing highly purified plasminogen are obtained, as shown in Fig. 1. Each plasminogen possesses approximately the same specific activity, toward TosArgOMe, after conversion to plasmin by urokinase. The total yield of plasminogen from plasma is approximately 90%.

Each major affinity chromatography resolved form of rabbit plas-

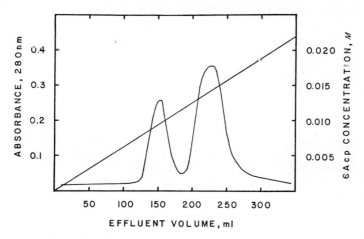

FIG. 1. Elution profile of rabbit plasminogen produced by gradient elution from Sepharose 4B-L-lysine affinity columns. The absorbance at 280 nm and the concentration of 6-aminohexanoic acid (ε-Acp), as a function of the effluent volume are presented on the graph. Taken from J. M. Sodetz, W. J. Brockway, and F. J. Castellino, *Biochemistry* **11**, 4451 (1972). Reprinted with permission. Copyright by the American Chemical Society.

minogen can be resolved into 5 subforms by isoelectric focusing techniques.[9] The purification procedure is herein described. A LKB 110-ml column as well as LKB Ampholines are utilized. The column is maintained at 4° by means of an external circulating temperature-control system. The electrode solutions are 0.4 ml of H_3PO_4, 26 ml of H_2O, and 26 g of sucrose for the anode, and 0.5 g of NaOH in 50 ml of H_2O for the cathode. After the anode solution has been added, a linear sucrose gradient is prepared. The starting solvent consists of 2.82 ml of pH 3–10 Ampholines in 40% sucrose, 28.7 ml of H_2O, and 21.0 g of sucrose. The limit solvent consists of 1.15 ml of Ampholines (pH 3–10) in 40% sucrose, 51.1 ml of H_2O, and 2.75 g of sucrose. The gradient is layered directly above the anode solution with the aid of a peristaltic pump. Each gradient chamber is immersed in ice prior to engaging the pump. After approximately 30 ml of the gradient has emerged, approximately 15–20 mg of lyophilized plasminogen (affinity fraction 1 or 2) are dissolved in the dense solution. The remainder of the gradient is allowed to form. At this point a couple of extra milliliters of limit solvent are added, followed by the cathode solution. Current is then applied, and the voltage is continually adjusted during the run such that the product of (kilovolts) × (milliamperes) never exceeds 3. Run times of 60 hr are common in the procedure. At the conclusion of the run, the column is drained in 1-ml fractions. Each tube is monitored for absorbance at 280 nm and pH at 22°. The protein profile obtained is shown in Fig. 2. The appropriate tubes are pooled, and each pool can be passed over the initial affinity chromatography resin, packed in separate Pasteur pipettes. Excess Ampholines and sucrose are easily and rapidly removed from the plasminogens. Each plasminogen subform could then be removed from the column by the addition of 0.05 M ε-aminocaproic acid to a suitable buffer and further dialyzed.

Plasmin

There are several approaches that can be taken in the preparation of rabbit plasmin, free of activator. The first method to be described is activation of rabbit plasminogen with an insolubilized activator, urokinase. Urokinase (6600 Plough units) is dissolved in 1 ml of 0.1 M $NaHCO_3$ and added to 1 ml of Sepharose 4B which has been activated with cyanogen bromide, as described above. The suspension is stirred overnight at 4°. The urokinase-coupled resin is then washed with large volumes of 0.1 M $NaHCO_3$, prior to use.

[9] J. M. Sodetz, W. J. Brockway, and F. J. Castellino, *Biochemistry* 11, 4451 (1972).

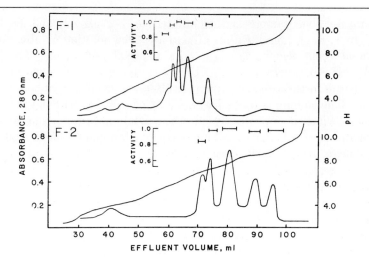

FIG. 2. Isoelectric focusing profiles of rabbit plasminogen. Top: Rabbit fraction 1 focused at 4° on a pH 3–10 gradient. The absorbance at 280 nm, the experimental pH gradient, and activities of each pool after dialysis, concentration, and activation to plasmin by urokinase are indicated on the graph. Specific activities are indicated on the whole pool on the inset, relative to the central fraction having a specific activity of 1.0, by a bar above each fraction. The V_{max} of the central fraction was 13.4 ± 0.4 μmoles of Tos-Arg-OMe (N-α-toluenesulfonyl-L-arginine methyl ester; TAME) cleaved min^{-1} mg^{-1} of plasminogen originally added prior to activation. Bottom: As for top except that rabbit fraction 2 was employed. The V_{max} of the central fraction was 13.7 ± 0.4 μmoles of Tos-Arg-OMe cleaved min^{-1} mg^{-1} of plasminogen originally added prior to activation. Taken from J. M. Sodetz, W. J. Brockway, and F. J. Castellino, *Biochemistry* 11, 4451 (1972). Reprinted with permission. Copyright by the American Chemical Society.

Urokinase-free plasmin is prepared by activation of 1 ml of rabbit plasminogen (7–10 mg/ml in 0.1 M Tris-HCl–0.1 M lysine, pH 8.0) with 1 ml of insolubilized urokinase for 70 min at 30° with frequent mixing. It is advisable to monitor generation of plasmin activity by the potentiometric assay in order to directly determine the point at which to terminate the activation by centrifugation of the resin. The final plasmin solution should be Millipore-filtered to remove the final traces of resin. The centrifuged resin can be reused after extensive washing. Storage of the resin in 50% glycerol in the cold inhibits loss of activity of the resin.

An alternative way to prepare high-quality DFP-inactivated plasmin (DIP-plasmin), free of urokinase, is as follows. Rabbit plasminogen (7–8 mg/ml in 0.05 M Tris-HCl–0.1 M L-lysine) is activated with urokinase (900 Plough units/ml), in the same buffer, for 25 min at room temperature. The sample is then diluted to 1–2 mg/ml, and DFP is added to a final concentration of 0.01 M. The solution is allowed to remain at room temperature for 1 hr and dialyzed against Tris-HCl (0.1 M, pH 8.0) overnight at 4°. Urokinase is removed by then passing the preparation through the Sepharose 4B–L-lysine column at room temperature. The column is washed with 0.3 M phosphate, pH 8.0, to remove urokinase, which is only weakly adsorbed. DIP-plasmin is removed only upon addition of a buffer consisting of 0.1 M phosphate–0.05 M ε-aminocaproic acid, pH 8.0. DIP-plasmin is dialyzed against H_2O or a suitable buffer and lyophilized.

This procedure can also be used to prepare active plasmin. However, autodigestion of plasmin will occur owing to the length of time involved in the procedure, and preparations of variable quality will result. Slight variations of the procedure, such as dialyzing plasmin against dilute acid solutions instead of H_2O or buffers, may result in preparations of higher quality. However, we suggest this latter procedure for DIP-plasmin and the former procedure for active plasmin.

Properties of Rabbit Plasminogen

Storage. All plasminogen forms and subforms are best stored at −18° as powders obtained by lyophilization from H_2O. This material maintains its chemical and physical integrity for at least 6 months.

Purity. At pH 4.3 or at pH 3.2 in 6 M urea, the two rabbit plasminogen major affinity chromatography-resolved forms exhibit single bands on polyacrylamide gels. However, there is a significant mobility difference noted between the two forms. Single bands with near-identical mobilities are observed in sodium dodecyl sulfate (SDS)–mercaptoethanol gel systems At pH 9.5, in native buffers, 5 well resolved bands are obtained for each major form, owing to resolution of the subforms at this pH. There is staggered mobility overlap of the subforms when the two plasminogen major forms are compared at this pH. The subforms, when isolated by preparative isoelectric focusing procedures, retain their distinct mobilities at pH 9.5 and, as expected, migrate in an identical manner to the original major form on pH 4.3 and 3.2 in 6 M urea gel electrophoretic systems. When examined by ultracentrifugal sedimentation velocity methods, each major form or isolated subform behaves as a single component with $s_{20°,w}$ values of 5.6–5.8 S in 0.1 M Tris-HCl, pH 8.0.

Physical Properties. Sedimentation equilibrium analysis on each rabbit plasminogen major form or isolated subform indicates molecular weights of 89,000–94,000. The lower molecular weight range applies to the second plasminogen form, resolved by affinity chromatography, and

TABLE I
PHYSICAL PROPERTIES OF RABBIT PLASMINOGEN

Parameter	F-1[a]	F-1 (pI 6.56)[b]	F-1 (pI 7.78)[b]	F-2[a]	F-2 (pI 7.18)[b]	F-2 (pI 8.42)[b]
Molecular weight[c]	92,000	94,000	93,000	89,000	90,000	89,000
$s_{20°,w}$[d] (S)	5.7	5.6	5.7	5.7	5.7	5.6
$\bar{v}$[e]	0.715	—	—	0.714	—	—
f/f_{min} (translational)[f]	1.38	—	—	1.38	—	—
ρ_h (nsec)[f]	284	—	—	280	—	—
ρ_h/ρ_o (rotational)[f]	3.73	—	—	3.77	—	—

[a] Refers to the first (F-1) and second (F-2) forms resolved from affinity chromatography columns.

[b] Refers to the indicated pI subforms resolved from either F-1 or F-2.

[c] Determined by high-speed sedimentation equilibrium at 20°.

[d] Value in 0.1 M Tris-HCl, pH 8.0.

[e] Determined by amino acid analysis.

[f] Values in 0.1 M phosphate, pH 8.0 at 25°. F. J. Castellino, W. J. Brockway, J. K. Thomas, H.-T. Liao, and A. B. Rawitch, *Biochemistry* **12**, 2787 (1973).

its isolated subforms. The higher molecular weight range applies in a corresponding manner to the first affinity chromatography-resolved plasminogen form and its subforms. Isoelectric points have been determined for the two major plasminogen forms. Affinity chromatography form 1 consists of 5 subforms with the following pI values[9] at 22°: 6.20, 6.56, 6.85, 7.24, and 7.78. Affinity chromatography form 2 also consists of 5 subforms with the following pI values[9] at 22°: 6.95, 7.18, 7.89, 8.24, 8.74. Other physical properties of the two major forms and selected subforms are summarized in Table I.

Chemical Properties. The amino-terminal residue of all native rabbit plasminogen forms and subforms is glutamic acid.[10] The amino-terminal amino acid sequence of all native rabbit plasminogen forms and subforms[10] is NH$_2$-Glu·Pro·Leu·Asp·Asp·Tyr·Val·Asn·Thr·Glu·Gly·Ala. The amino acid compositions of the major affinity chromatography resolved plasminogen forms and selected subforms, prepared by isoelectric focusing procedures, are listed in Table II. No significant differences are noted in this analysis for any of the proteins. Carbohydrate analysis reveals that plasminogen affinity chromatography form 1 (F-1) contains 1.5–1.7% neutral carbohydrate and 3 moles/mole of sialic acid. On the other hand, affinity chromatography form 2 (F-2) contains 0.6–0.8%

[10] F. J. Castellino, G. E. Siefring, Jr., J. M. Sodetz, and R. K. Bretthauer, *Biochem. Biophys. Res. Commun.* **53**, 845 (1973).

TABLE II
AMINO ACID COMPOSITION OF RABBIT PLASMINOGEN[a,b]

Amino acid	F-1	F-1 (pI 6.85)	F-1 (pI 7.78)	F-2	F-2 (pI 7.18)	F-2 (pI 8.42)
Lysine	47	48	47	47	47	47
Histidine	18	18	18	18	18	18
Arginine	51	52	52	53	53	53
Aspartic acid	82	83	82	83	83	83
Threonine	56	56	57	54	55	56
Serine	61	62	60	60	61	59
Glutamic acid	87	88	87	87	88	88
Proline	69	67	67	70	68	70
Glycine	59	60	59	58	59	58
Alanine	47	48	48	48	49	48
Valine	35	35	34	34	35	34
Methionine	9	10	10	10	11	11
Isoleucine	23	23	23	23	23	23
Leucine	42	41	41	42	42	42
Tyrosine	37	36	35	36	36	36
Phenylalanine	21	19	25	18	19	19
Tryptophan	19	19	19	19	19	19
Cysteic acid	49	50	49	50	50	50

[a] Abbreviations are as in Table I.
[b] Values are expressed as moles per mole of protein.

neutral carbohydrate and 2 moles/mole of sialic acid.[10] The extinction coefficient ($\epsilon_{1\,cm}^{1\%}$) has been reported[11] for rabbit plasminogen at 280 nm to be 18.1.

Mechanism of Activation of Rabbit Plasminogen to Plasmin. The activation of single-chain rabbit plasminogen to plasmin, by urokinase, has been shown to involve cleavage of two peptide bonds in the plasminogen molecule.[12] The first bond cleaved by urokinase is in the interior of the molecule and results in a plasmin molecule containing two chains. A heavy chain, derived from the original amino terminus of plasminogen and possessing a molecular weight of 66,000–69,000 is disulfide linked to a light chain of molecular weight 24,000–26,000. Plasmin autocatalytically removes an amino-terminal peptide of molecular weight 6000–8000 from the plasmin heavy chain. This peptide is not covalently linked by disulfide bonds to the remainder of the molecule. Alternatively, the initial plasmin formed is capable of first removing the same amino-

[11] L. Summaria, L. Arzadon, P. Bernabe, and K. C. Robbins, *J. Biol. Chem.* **248,** 6522 (1973).
[12] J. M. Sodetz and F. J. Castellino, *J. Biol. Chem.* **250,** 3041 (1975).

TABLE III

AMINO-TERMINAL AMINO ACID SEQUENCES OF PLASMINOGEN AND PLASMIN

Sequence No.	Residue found in			
	Plasminogen	Plasmin (P)[a]	Plasmin (H)[a]	Plasmin (L)[a]
1	Glu	Glu	Met	Val
2	Pro	Pro	Tyr	Val
3	Leu	Leu	Leu	Gly
4	Asp	Asp	X	Gly
5	Asp	Asp	Glu	X
6	Tyr	Tyr	X	X

[a] Refers to the peptide released (P) and the plasmin heavy (H) and light (L) chains.

terminal peptide from the original plasminogen prior to the urokinase-induced cleavage of the internal bond, leading to final plasmin formation. Activation by either mechanism or a combination of both, leads to the same final plasmin. Amino-terminal amino acid sequences of the plasmin chains are shown in Table III. This activation mechanism and the amino-terminal sequences presented in Table III are similar for each rabbit plasminogen form and subform.

Rabbit plasminogen is also weakly activable by streptokinase.[9]

Ligand Binding. Rabbit plasminogen binds small molecules such as L-lysine, ϵ-aminocaproic acid, and t-4-aminomethylcyclohexane-1-carboxylic acid with resulting gross alterations in the properties of plasminogen. Upon 1:1 complex formation of these small molecules with plasminogen, $s_{20,w}^0$ values of 4.6 S are obtained. This is to be contrasted with the value of approximately 5.6 S for the unliganded protein. The f/f_{min} of 1.38 for native plasminogen is increased to 1.64 upon complex formation with these small molecules, whereas the rotational ρ_h/ρ_0 value decreases from 3.73 to 2.28 upon this same complex formation.[13] These data indicate that compounds of the ϵ-aminocaproic acid class induce a conformational alteration in plasminogen resulting in a more asymmetric and internally more flexible protein molecule.[13] Concentration midpoints for the ϵ-aminocaproic acid induced conformational alteration of 1.89 mM for rabbit plasminogen F-1 and 3.12 mM for F-2 have been reported.[9] The subforms of F-1 and F-2 behave in an analogous manner to the whole F-1 and F-2 in regard to this property. Regarding the activation of plasminogen by urokinase, the above ligands greatly accelerate the rate

[13] F. J. Castellino, W. J. Brockway, J. K. Thomas, H.-T. Liao, and A. B. Rawitch, *Biochemistry* **12**, 2787 (1973).

of activation at concentrations at which the conformational alteration is induced.[12]

Properties of Rabbit Plasmin

Storage. Although we prefer to prepare rabbit plasmin by activation of the zymogen immediately prior to use, the enzyme is relatively stable to storage. After activation of a concentrated solution of plasminogen, glycerol is added to approximately 30–40%. This solution is stable in the cold (−18°) for at least 2 weeks and perhaps significantly longer. Alternatively, after activation, the pH can be lowered to 3.0, lending stability to the enzyme in the cold (4°). Also, the enzyme can be stored for a moderate time frozen in aliquots taken from the activation mixture. These frozen aliquots should not be thawed more than once.

Purity. Similar, but not identical, electrophoretic properties are noted for DFP-inactivated plasmin and that described for plasminogen in a previous section on pH 4.3 gels, pH 3.2 in 6 M urea gels, and pH 9.5 gels. Significantly, when examined in SDS-mercaptoethanol gel systems, two peptide bands, corresponding to the plasmin light and heavy chains, are observed. Sedimentation velocity analysis of DIP-plasmin and plasmin inactivated with pancreatic trypsin inhibitor demonstrates single peaks in the ultracentrifuge with $s_{20°,w}$ values of 4.5–4.7 S.

Physical Properties. Molecular weights of DIP-inactivated rabbit plasmin and its corresponding isolated heavy and light chains, for F-1 and F-2 plasmin, have been determined by high speed analytical ultracentrifugation. This data are shown in Table IV. Molecular weight

TABLE IV
MOLECULAR WEIGHT ANALYSIS OF RABBIT PLASMIN AND ITS
COMPONENT POLYPEPTIDE CHAINS

Component	Molecular weight obtained for	
	F-1[a]	F-2[a]
DIP-plasmin[b]	83,000–86,000	82,000–84,000
Heavy chain[c]	60,000–63,000	59,000–62,000
Light chain[c]	24,000–26,000	24,000–26,000
Peptide[c]	6,000–8,000	6,000–8,000

[a] Represents plasmin obtained from the first (F-1) and second (F-2) form of plasminogen eluted from the affinity-chromatography columns after activation of plasminogen with urokinase.
[b] Diisopropyl fluorophosphate-inactivated plasmin.
[c] These were reduced and alkylated samples in 6 M guanidine hydrochloride.

analysis of the amino-terminal peptide released during activation has been performed on calibrated SDS-mercaptoethanol gels, and the value obtained is also listed in Table IV. The slight differences in molecular weight between the F-1 and F-2 DIP-plasmins and the F-1 and F-2 plasmin heavy chains are due primarily to small differences in carbohydrate content. The $s_{20°,w}$ value for plasmin is 4.6 S and the f/f_{min} value is 1.56.

Chemical Properties. The amino-terminal amino acid of rabbit plasmin is methionine for the heavy chain and valine for the light chain. Amino acid compositions for the plasmin heavy (H) and light (L) chains as well as the peptide (P) released during activation are listed in Table V. Amino-terminal amino acid sequences of the component rabbit plasmin chains have been listed in Table III.

Kinetic Properties. Plasmin is a proteolytic enzyme with a similar substrate specificity to that of trypsin. Plasmin is inhibited by reagents such as DFP and phenylmethylsulfonyl fluoride (PMSF), indicating that a serine is present at the active site of rabbit plasmin. Summaria

TABLE V

AMINO ACID COMPOSITIONS OF THE RABBIT PLASMIN
COMPONENT PEPTIDE CHAINS[a,b]

Amino acid	F-1 H	F-1 L	F-1 P	F-2 H	F-2 L	F-2 P
Lysine	32	10	3	31	10	4
Histidine	13	5	2	12	4	2
Arginine	34	12	3	33	12	3
Aspartic acid	60	21	6	61	19	6
Threonine	39	14	5	39	14	4
Serine	41	17	6	37	16	5
Glutamic acid	54	22	7	53	20	8
Proline	59	12	6	60	13	4
Glycine	35	22	10	37	22	6
Alanine	23	16	4	24	16	5
Valine	11	18	3	13	19	5
Methionine	8	1	1	8	2	1
Isoleucine	11	10	2	12	10	3
Leucine	18	20	4	19	19	5
Tyrosine	26	8	2	27	8	2
Phenylalanine	9	4	2	11	5	2
Tryptophan	18	7	0	18	7	0
Cysteic acid	36	9	2	35	10	2

[a] Abbreviations are as in Table I.
[b] Values are expressed as moles per mole.

et al.[11] have demonstrated that rabbit plasmin also incorporates N^α-toluenesulfonyl-L-lysine chloromethyl ketone, demonstrating the presence of a histidine residue at the active site. Also, these investigators have established that the plasmin active site is located in the light chain of the enzyme.

We have studied steady-state and presteady-state kinetic constants of various substrates with rabbit plasmin. Steady-state constants obtained for TosArgOMe indicate a K_m value of 5.3 ± 1.5 mM for rabbit plasmin F-1 and 7.0 ± 1.5 mM for rabbit plasmin F-2 at 22°. V_{max} values of 14.0 ± 1.3 TosArgOMe units and 13.3 ± 1.3 units at 22° are obtained per milligram of active plasmin F-1 and F-2, respectively. Benzamidine has been found to be a competitive plasmin inhibitor with K_{Diss} (22°) values of 0.77 ± 0.05 mM and 1.33 ± 0.05 mM for rabbit plasmin F-1 and F-2, respectively. In addition, ϵ-aminocaproic acid is a very weak competitive plasmin inhibitor. Proteins such as pancreatic trypsin inhibitor (Kunitz)[14] and soybean trypsin inhibitor[15] are very potent plasmin inhibitors. Pre-steady-state kinetic parameters K_S and k_2 have been determined, at 22°, for rabbit plasmin and p-nitrophenyl-p'-guanidino-benzoate (see Sodetz and Castellino[16] for a definition of the constants). The binding constant, K_S, for this substrate is 5.2 ± 1.0 μM for plasmin F-1 and 11.3 μM for plasmin F-2. Acylation rate constants of 0.125 sec^{-1} have been obtained for both F-1 and F-2 rabbit plasmin with this substrate. Thus, there may be minor differences in the kinetic properties of plasmin F-1 and F-2.

In addition to the weak binding site for ϵ-aminocaproic acid at the active site of plasmin, which is responsible for the competitive inhibition properties of this ligand, there is at least one additional binding site on plasmin for this amino acid. This site appears to be the same as that found on the initial plasminogen and appears to have been conserved upon conversion of plasminogen to plasmin. The K_{Diss} for the binding of ϵ-aminocaproic acid to this additional binding site is approximately the same as for this same site on the original plasminogen molecule,[17] indicating that it is the stronger of the two binding sites on plasmin. However, the final plasmin does not undergo a measurable conformational alteration, as does plasminogen, upon interaction of plasmin with ϵ-aminocaproic acid.

[14] M. Kunitz and J. H. Northrop, *J. Gen. Physiol.* **19**, 991 (1936).
[15] M. Kunitz, *J. Gen. Physiol.* **29**, 149 (1946).
[16] J. M. Sodetz and F. J. Castellino, *Biochemistry* **11**, 3167 (1972).
[17] G. E. Siefring, Jr., and F. J. Castellino, *J. Appl. Physiol.* **38**, 114 (1975).

Section IV

Kallikreins

[24] Pig Pancreatic Kallikreins A and B[1]

By FRANZ FIEDLER

Similarly to other serine proteases of the pancreas, kallikrein also occurs in this gland as an inactive proenzyme.[2] Enzymically active kallikrein is generally isolated from autolyzed pancreas. Habermann[3] was the first to observe that the enzyme thus obtained can be separated by electrophoresis at pH 8–9 into two components called kallikreins A and B. By a procedure initially devised by Schmidt-Kastner, two components evidently corresponding to kallikreins A and B can be isolated also by chromatography of the enzyme on DEAE-cellulose or DEAE-Sephadex.[4,5]

Even when appearing electrophoretically homogeneous in the pH 8–9 range, kallikreins A and B both display microheterogeneity on electrophoresis around pH 4. This microheterogeneity is abolished by treatment with neuraminidase, which removes sialic acid bound to kallikrein.[6,7] Treatment with neuraminidase has been included in the present isolation procedure to obtain electrophoretically homogeneous enzyme for kinetic studies. The separation of kallikreins A and B on DEAE-cellulose works equally well with enzyme containing sialic acid.[4]

As a first step, chromatography on hydroxyapatite, which has been proved very useful for purifying kallikrein,[6] was employed. The procedure has been modified by employing gradient, rather than stepwise, elution, as has also been done by Moriya et al.[8]

Purification Procedure

The present procedure for the isolation of neuraminidase-treated kallikreins A and B starts from prepurified preparations of kallikrein from

[1] This investigation was supported by Deutsche Forschungsgemeinschaft, Sonderforschungsbereich 51.

[2] E. K. Frey, H. Kraut, and E. Werle. "Das Kallikrein-Kinin-System und seine Inhibitoren." Enke, Stuttgart, 1968.

[3] E. Habermann, Hoppe-Seyler's Z. Physiol Chem. 328, 15 (1962).

[4] F. Fiedler and E. Werle, Hoppe-Seyler's Z. Physiol. Chem. 348, 1087 (1967).

[5] C. Kutzbach and G. Schmidt-Kastner, Hoppe-Seyler's Z. Physiol. Chem. 353, 1099 (1972).

[6] H. Fritz, I. Eckert, and E. Werle, Hoppe-Seyler's Z. Physiol. Chem. 348, 1120 (1967).

[7] F. Fiedler, C. Hirschauer, and E. Werle, Hoppe-Seyler's Z. Physiol. Chem. 351, 225 (1970).

[8] H. Moriya, A. Kato, and H. Fukushima, Biochem. Pharmacol. 18, 549 (1969).

partial autolyzates of porcine pancreas obtained from Farbenfabriken Bayer AG, Werk Elberfeld. All operations were routinely performed around 5° unless stated otherwise.

Chromatography on Hydroxyapatite-Cellulose. A kallikrein preparation containing about 40,000 U of the enzyme (about 30 U per milligram of protein) is dissolved in 50 ml of water and dialyzed overnight against 20 liters of water. This solution is applied to a 5.2 × 20 cm column prepared by rapid pouring (to avoid segregation of the material) of a suspension of calcium hydroxyapatite cellulose MN 2200 Apatit-50 (Macherey, Nagel & Co, Düren) in 0.01 M sodium phosphate buffer, pH 7.0. Elution is achieved with a convex gradient, sodium phosphate buffer pH 7.0, 0.2 M in phosphate being fed into a constant-volume mixing vessel containing 5 liters of a 0.01 M sodium phosphate buffer pH 7.0. The eluate is collected in 17-ml fractions at a rate of 50–80 ml/hr (Fig. 1A). Fractions comprising about two-thirds of the kallikrein activity applied to the column with the highest specific activities and an absorbance ratio $A_{280\,nm}/A_{260\,nm} > 1$ are combined, dialyzed against water, and lyophilized.

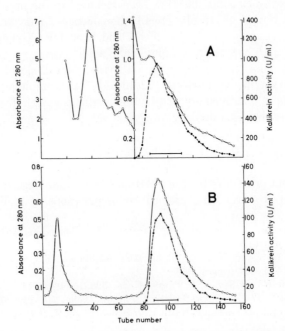

Fig. 1. Chromatography of pig pancreatic kallikrein on hydroxyapatite-cellulose. (A) First chromatography of a preparation of 29 U per milligram of protein. (B) Rechromatography of pooled peak fractions (109 U/mg). The bar indicates fractions of sufficient specific activity to be pooled for the next purification step. O——O, Absorbance at 280 nm; ●---●, enzyme activity.

This chromatographic step was usually repeated once more (Fig. 1B).

Treatment with Neuraminidase. The procedure of Fritz et al.[6] is followed with minor modifications. One hundred milligrams of a kallikrein preparation of about 230 U per milligram of protein as recovered after the second hydroxyapatite step (or more recently as obtained from Farbenfabriken Bayer) is dissolved in 2 ml of a buffer adjusted to pH 6.0 with acetic acid and containing 0.15 M sodium acetate, 0.23 M NaCl, 10 mM $CaCl_2$, and 1.5 mM ethylenediamine tetraacetate. An equal volume of a solution of α-neuraminidase from *Vibrio cholerae* (Serva, Heidelberg; dissolved according to the label in 0.05 M acetate buffer, pH 5.5, and liberating 0.11 μmole of N-acetylneuraminic acid from human glycoprotein at 37° per milliliter of the enzyme per minute) is added, and the mixture is incubated at 37° for 22 hr. It is then dialyzed against water.

Chromatography on DEAE-Cellulose. The dialyzed solution is adjusted to contain 0.1 M ammonium acetate, pH 6.7, in about 40 ml and is applied to a 1.1 × 50 cm column of DEAE-cellulose (Serva DEAE-SS, 0.60 mEq/g) equilibrated with the same ammonium acetate buffer. The column is eluted at a rate of 6 ml/hr (attained with a hydrostatic pressure of about 1.5 m of water) using a convex gradient generated by feeding 0.8 M ammonium acetate, pH 6.7, into a constant-volume mixing vessel containing 220 ml of 0.1 M ammonium acetate, pH 6.7. Fractions of 6 ml are collected. Kallikrein appears as a double peak of the A and B forms, respectivitely, well separated from neuraminidase (Fig. 2A). Sometimes the double peak can be clearly recognized only from activity measurements, but not from absorbance. Fractions of kallikreins A and B are pooled as indicated, dialyzed against distilled water, and rechromatographed separately on DEAE-cellulose exactly as before (Fig. 2, B and C). After dialysis, kallikreins A and B are lyophilized and stored at −20°.

Comments on the Purification Procedure

Table I summarizes a purification of kallikreins A and B. Material of less than satisfactory specific activity is recovered by dialysis and lyophilization and subjected to the same or a preceding step of the isolation procedure. As total recoveries in column eluates amounted to 90% or more of the enzyme activity applied, the purification procedure is on the whole much more economical than might appear from the values quoted in Table I.

Repetition of the DEAE-cellulose step is necessary for a complete separation of kallikreins A and B (Fig. 2, B and C). The reproducibility of the elution volumes on rechromatography of kallikreins A and B

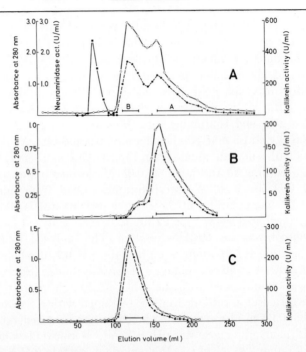

FIG. 2. Chromatography of neuraminidase-treated pig pancreatic kallikrein on DEAE-cellulose. (A) Separation of kallikreins A and B after treatment with neuraminidase. (B) Rechromatography of kallikrein A. (C) Rechromatography of kallikrein B. The columns were prepared at the same time with equal amounts of the ion exchanger and were operated under identical conditions. ◯——◯, Absorbance at 280 nm; ●---●, kallikrein activity; -■-■- neuraminidase activity.

TABLE I

SUMMARY OF PURIFICATION PROCEDURE FOR PIG PANCREATIC KALLIKREINS A AND B

Fraction	Total activity (U)	Specific activity (U/mg protein)	Recovery, main fractions (%)	
Starting material	86,000	29	(100)	
First hydroxyapatite-cellulose	54,000	109	63	
Second hydroxyapatite-cellulose	26,700	224	31	
First DEAE-cellulose				
Kallikrein A	6,800	245	7, 8	
				18, 6
Kallikrein B	9,300	270	10, 8	
Second DEAE-cellulose				
Kallikrein A	4,660	249	5, 4	
				14, 3
Kallikrein B	7,640	281	8, 9	

demonstrates that the separation of the enzyme into two forms is not a chromatographic artifact. In view of the stability of kallikrein it might be possible to carry out the isolation at room temperature.

Other Purification Procedures

The only isolation procedure described for porcine kallikreins A and B starting from pancreatic glands is that of Habermann,[3] which requires preparative column electrophoresis. Applying the conversion factor from biological to esterolytic kallikrein units determined by Arens as quoted by Kutzbach and Schmidt-Kastner,[5] the specific activity of these preparations appears to have been only 80 U/mg.

There are a number of alternative isolation procedures for pig pancreatic kallikreins A and B from industrially processed material. Takami[9] isolated the enzyme in a purity of up to 195 ± 22 U/mg (as calculated from biological units), using chromatography on a DEAE ion exchanger with an ammonium acetate gradient for the separation of kallikreins A and B. Kutzbach and Schmidt-Kastner[5] obtained neuraminidase-treated kallikrein B with a specific activity of 241 U per milligram of protein (if the carbohydrate content as discussed in this chapter is considered). They also reported crystallization of the enzyme. Recently, Zuber and Sache[10] described the isolation of two kallikreins called d_1 and d_2 (and minor amounts of a third form, d_3) by chromatography on DEAE-cellulose with a NaCl gradient. Forms d_1 and d_2 may possibly be identical with kallikreins B and A, respectively. Working with a pH-stat assay system similar to that employed for work reported in this chapter, these authors stated specific activities of up to 134 U per milligram of protein for their preparations.

For isolating kallikreins A and B from porcine pancreatic glands, it might be advisable to follow the purification procedure of Moriya and co-workers (as described in detail in a previous volume of this series[11]) up to the acrinol precipitation step and to carry on from there with one or possibly two chromatographic steps on hydroxyapatite. Although Moriya *et al.* obtained only one kallikrein with their purification method,[8] kallikrein isolated by acrinol precipitation has been resolved into kallikreins A and B.[9]

Properties

Specific Activity and Stability. The mean specific activity ($\pm$SEM) of 9 preparations of neuraminidase-treated pig pancreatic kallikrein B

[9] T. Takami, *J. Jpn. Biochem. Soc.* (*Seikagaku Zasshi*) **41**, 777 (1969).
[10] M. Zuber and E. Sache, *Biochemistry* **13**, 3098 (1974).
[11] M. E. Webster and E. S. Prado, this series Vol. 19, p. 681.

was 282 ± 6 U/mg protein. Though kallikrein A of the same degree of purity has occasionally been obtained, the above isolation procedure, for unknown reasons, usually resulted in preparations of only about 230 U per milligram of protein. Kallikrein A preparations often did not redissolve completely in water after lyophilization, a circumstance never observed with kallikrein B, which can be lyophilized from water solutions without appreciable loss of activity.

Criteria of Purity. The specific activity of kallikreins A and B isolated by the present procedure could not be further increased by rechromatography of the preparations. Each enzyme migrates as a single band during electrophoresis on cellulose acetate foil at pH 8.6, behaving as described by Habermann.[3] Single bands are also obtained on disc electrophoresis,[5] under conditions where protein contaminants amounting to less than 5% should have been easily detected. Neuraminidase-treated kallikrein B of similar specific activity shows a single symmetrical peak on ultracentrifugation.[5] Quantitative end-group studies and amino acid analyses also indicate a high degree of purity.[12]

Composition and Structure. Pig pancreatic kallikreins A and B have an identical amino acid composition (Table II). Both forms of the enzyme contain N-terminal isoleucine and alanine and C-terminal Leu-Ser and proline.[12] The two chains thus indicated can be separated by gel filtration after reduction and aminoethylation or carboxymethylation. Carbohydrate is located on both chains of kallikrein B. Kallikreins A and B differ only in that the B form contains roughly double the amount of carbohydrate (Table III). The half equivalent of fucose residue determined in kallikrein A is indicative of heterogeneity in the carbohydrate content of the enzyme. The existence of pig pancreatic kallikreins in two forms could conceivably be explained in terms of degradation occurring in the course of isolation, especially during autolysis of the pancreas. However, the finding that two prekallikreins (kallikreinogens) are observed also in carefully processed homogenates of porcine pancreas[4,7] would seem to argue against this interpretation.

Kallikreins A and B appear to contain the active site sequence Asp-Ser-Gly[13] characteristic of mammalian serine proteinases. Preliminary data on partial amino acid sequences of kallikrein B are given in Table IV. A high degree of homology with other members of the serine proteinase family is already evident.

Physicochemical Data. Molecular weights and related data for pig pancreatic kallikrein are summarized in Table V. As glycoproteins tend

---

[12] F. Fiedler, C. Hirschauer, and E. Werle, *Hoppe-Seyler's Z. Physiol. Chem.* **356**, 1879 (1975).

[13] F. Fiedler, B. Müller, and E. Werle, *Hoppe-Seyler's Z. Physiol. Chem.* **352**, 1463 (1971).

TABLE II
AMINO ACID COMPOSITION (RESIDUES PER MOLE) OF
PIG PANCREATIC KALLIKREINS A AND B[a]

Asp	27
Thr	14
Ser	14
Glu	23
Pro	16
Gly	22
Ala	13
Cys/2	10
Val	10
Met	4
Ile	12
Leu	19
Tyr	7
Phe	10
Lys	10
His	8
Arg	3
Trp	7
	229

[a] From F. Fiedler, C. Hirschauer, and E. Werle, *Hoppe-Seyler's Z. Physiol. Chem.* **356**, 1879 (1975).

TABLE III
AVERAGE CARBOHYDRATE COMPOSITION (RESIDUES PER MOLE) OF
NEURAMINIDASE-TREATED PIG PANCREATIC KALLIKREINS A AND B[a]

	Kallikrein A	Kallikrein B
Mannose	3.3	5.7
Galactose	0.8	1.5
Fucose	0.5	1.1
Glucosamine	3.9	9.5
Total carbohydrate (by weight)	5.6%	11.5%

[a] F. Fiedler, C. Hirschauer, and E. Werle, *Hoppe-Seyler's Z. Physiol. Chem.* **356**, 1879 (1975).

to give higher than correct molecular weights on gel filtration,[14] results obtained by this method (also diffusion constants and molecular weights calculated from a combination of ultracentrifugation and gel filtration data[10]) must be considered as tentative. Values of 0.734 to 0.75 have been used for the partial specific volume $\bar{v}$ of kallikrein. A recent determination,[5] however, gave 0.722 ml/g, in excellent agreement with values of

[14] P. Andrews, *Methods Biochem. Anal.* **18**, 1 (1970).

TABLE IV
PARTIAL AMINO ACID SEQUENCES OF THE TWO CHAINS OF PIG PANCREATIC KALLIKREIN B[a] AS COMPARED TO PORCINE TRYPSIN[b]

```
                         7                                                          20
Kallikrein, A-chain    Ile - Ile - Gly -Gly -Arg-Glu - Cys-Glu - Lys-Asn-Ser -His -Pro-Trp-Gln - Val -Ala - Ile - Tyr-His -
(N-terminus)
Trypsin                Ile - Val -Gly -Gly -Tyr-Thr-Cys-Ala -Ala -Asn-Ser - Val/Ile -Pro-Tyr-Gln - Val -Ser - Leu-Asn-Ser -

                         30                                          40
                       -Tyr-Ser -Ser -Phe-Gln - Cys-Gly - Gly - Val - Leu-Val -Asn-Pro-Lys-Trp-Val - Leu-Thr-Ala -Ala -
                       -Gly -Ser -His - Phe ——— Cys-Gly - Gly - Ser - Leu- Ile - Asn-Ser -Gln -Trp-Val - Val -Ser -Ala -Ala -

                         82                                90                                          100
Kallikrein, B-chain    Ala -Asp-Gly - Lys-Asp-Tyr-Ser -His-Asp-Leu-Met-Leu- Leu-Arg-Leu-Gln -Ser -Pro-Ala -Lys-
(N-terminus)
Trypsin                -Phe-Asn-Gly -Asn-Thr-Leu-Asp-Asn-Asp - Ile -Met-Leu- Ile - Lys-Leu-Ser -Ser -Pro-Ala -Thr-

                                              110
                       - Ile - Thr-Asp-Ala -Val - Lys-Val - Leu-Glu - Leu-Pro-Thr-Gln -
                       - Leu-Asn-Ser -Arg-Val -Ala -Thr-Val -Ser - Leu-Pro-Arg-Ser -

                         166                170                              180
Kallikrein, B-chain    -Met-Leu-Cys-Ala -Gly -Tyr-Leu-Pro-Gly -Gly - Lys-Asp-Thr-Cys-Met-
Trypsin                -Met - Ile -Cys-Val -Gly -Phe-Leu-Glu -Gly -Gly - Lys-Asp-Ser -Cys-Gln -
```

[a] H. Tschesche, W. Ehret, C. Kutzbach, G. Schmidt-Kastner, and F. Fiedler, unpublished results, 1974.

[b] M. A. Hermodson, L. H. Ericsson, H. Neurath, and K. A. Walsh, *Biochemistry* 12, 3146 (1973).

0.723 and 0.721 calculated from the composition.[10,15] Recalculated molecular weights of kallikrein based on this figure are also given in Table V. The most reliable values for the molecular weight of kallikrein appear to fall below an upper limit of 30,000.

An isoelectric point $pI = 4.05$ has been determined by the electrofocusing of neuraminidase-treated kallikrein A as well as B.[7] The second small peak at pH 3.90 (shown in Fig. 7 of Fiedler *et al.*[7]) disappears on more extensive treatment with neuraminidase. Kallikreins A and B which have not been treated with neuraminidase give a number of species (in decreasing amounts) with isoelectric points of pI 4.05, 3.95, 3.85, 3.71, and 3.64.[7] Corresponding profiles with peaks at pI 4.11, 3.97, 3.92, 3.82, and 3.75 have been observed with kallikrein d_1, and at pI 4.11, 4.01, and 3.93 with kallikrein d_2.[10]

The UV spectra of unresolved kallikrein[6] and of sialidase-treated kallikrein B[5] give an absorption maximum at 281 nm and the $A_{280\text{ nm}}/A_{260\text{ nm}}$ ratio of 1.74.[5] The $A_{280\text{ nm}}^{0.1\%}$, 1 cm has been determined for neuraminidase-treated kallikrein B at pH 7.0 as 1.66 ($\pm5\%$) on a glycoprotein basis[5] (recalculated as 1.88 $\pm$ 0.09 on a protein basis using the composition given in Tables II and III) and as 2.00 $\pm$ 0.05 on a protein basis in

[15] F. Fiedler, B. Müller, and E. Werle, *FEBS Lett.* 24, 41 (1972).

TABLE V
MOLECULAR WEIGHT AND RELATED DATA FOR PIG PANCREATIC KALLIKREIN[a]

Method	Sedimentation constant (S)	Diffusion constant (cm² sec⁻¹)	Molecular weight	Remarks	References
Gel filtration			35,000 (A,B)		b
			34,000 (d₁)		c
			34,500 (d₁)		d
			31,000 (d₂)		d
Ultracentrifugation	2.73	7.6×10^{-7} (d₁)	23,000–24,000	$\bar{v}$ assumed as 0.734	e
	2.85	8.1×10^{-7} (d₂)	33,000	$\bar{v}$ assumed as 0.75	f
			(29,700)	(recalculated with $\bar{v}$ = 0.722)	
	2.56	9.8×10^{-7}	25,200		g
			29,000	$\bar{v}$ assumed as 0.749	h
			(26,100)	(recalculated with $\bar{v}$ = 0.721)	i
			(26,200)	(recalculated with $\bar{v}$ = 0.722)	j
	2.83 (d₁,d₂)	8.39×10^{-7}	32,725 (d₁)	Calculated with Stokes radius from gel filtration	d
			30,643 (d₂)		d
Surface pressure			30,000		d
Light scattering			33,500 (d₁,d₂)		f
Active-site titration			25,700 ± 920		d
Amino acid composition			32,566 ± 1000 (d₁)		i
			33,280 ± 1200 (d₂)		d
Amino acid and carbohydrate composition			26,800 (A)	Without sialic acid	k
			28,600 (B)		k

[a] A, B, d₁, and d₂ refer to the corresponding kallikreins, other data to unresolved pig pancreatic kallikrein.
[b] H. Fritz, I. Trautschold, and E. Werle, Hoppe-Seyler's Z. Physiol. Chem. 342, 253 (1965).
[c] T. Takami, J. Jpn. Biochem. Soc. (Seikagaku Zasshi) 41, 777 (1969).
[d] M. Zuber and E. Sache, Biochemistry 13, 3098 (1974).
[e] E. Habermann, Hoppe-Seyler's Z. Physiol. Chem. 328, 15 (1962).
[f] H. Moriya, J. Pharmacol. Soc. Jpn. (Yakugaku Zasshi) 79, 1453 (1959).
[g] H. Moriya, A. Kato, and H. Fukushima, Biochem. Pharmacol. 18, 549 (1969).
[h] H. Fritz, I. Eckert, and E. Werle, Hoppe-Seyler's Z. Physiol. Chem. 348, 1120 (1967).
[i] F. Fiedler, B. Müller, and E. Werle, FEBS Lett. 24, 41 (1972).
[j] C. Kutzbach and G. Schmidt-Kastner, Hoppe-Seyler's Z. Physiol. Chem. 353, 1099 (1972).
[k] F. Fiedler, C. Hirschauer, and E. Werle, Hoppe-Seyler's Z. Physiol. Chem. 356, 1879 (1975).

TABLE VI

ACYLATION CONSTANTS k_{ac} AND DEACYLATION CONSTANTS k_{deac} FOR THE REACTION OF ACTIVE-SITE ACYLATING REAGENTS WITH KALLIKREINS A AND B AT 25° AND SPECTRAL DATA OF ACYL KALLIKREINS

Reagents	pH	Constant	Kallikrein A	Kallikrein B	Reference
Diisopropyl phosphofluoridate (1 mM)	7.2	k_{ac}	$(9.4 \pm 2.0) \times 10^{-3}$ min^{-1}	$(8.6 \pm 2.2) \times 10^{-3}$ min^{-1}	[a]
p-Nitrophenyl p'-guanidinobenzoate	8.3	k_{deac}	1.76×10^{-2} sec^{-1}	1.76×10^{-2} sec^{-1}	[b]
Cinnamoyl imidazole	5.2	k_{ac}	$130 \pm 8\ M^{-1}$ sec^{-1}	$116 \pm 2\ M^{-1}$ sec^{-1}	[c]
			$119 \pm 6\ M^{-1}$ sec^{-1}		[c]
	8.8	k_{dea}	3.4×10^{-3} sec^{-1}	3.5×10^{-3} sec^{-1}	[c]
Indoleacryloyl imidazole	5.2	k_{ac}	$260\ M^{-1}$ sec^{-1}	$230\ M^{-1}$ sec^{-1}	[c]
	8.8	k_{deac}	1.2×10^{-3} sec^{-1}	1.1×10^{-3} sec^{-1}	[c]

Difference spectra	λ_{max} (nm)	ϵ_{max} ($\times 10^{-4}$ M^{-1} cm^{-1})	λ_{max} (nm)	ϵ_{max} ($\times 10^{-4}$ M^{-1} cm^{-1})	Reference
Cinnamoyl kallikrein	298	1.80	298	1.80	[c]
Indoleacryloyl kallikrein	353	2.24	353	2.17	[c]

[a] F. Fiedler, B. Müller, and E. Werle, *Hoppe-Seyler's Z. Physiol. Chem.* **351**, 1002 (1970).
[b] F. Fiedler, B. Müller, and E. Werle, *FEBS Lett.* **24**, 41 (1972).
[c] F. Fiedler, B. Müller, and E. Werle, *FEBS Lett.* **22**, 1 (1972).

TABLE VII

KINETIC CONSTANTS FOR THE HYDROLYSIS OF α-N-ACYLATED AMINO ACID
ESTERS BY NEURAMINIDASE-TREATED KALLIKREIN B[a]

Amino acid	Ester	K_m (mM)	k_{cat} (sec^{-1})	k_{cat}/K_m (mM^{-1}sec^{-1})
N-α-Toluenesulfonyl-L-arginine	Methyl	0.031 ± 0.008	2.7 ± 0.33	150
N-α-Benzoyl-L-arginine	Ethyl	0.104 ± 0.008	123 ± 8	1180
N-α-Benzoyl-L-arginine	Methyl	0.125 ± 0.020	154 ± 13	1230
N-α-Benzoyl-L-lysine	Methyl	2.00 ± 0.28	15.5 ± 1.4	7.75
N-α-Benzoyl-L-ornithine	Methyl	2.44 ± 0.30	1.80 ± 0.25	0.74
N-α-Benzoyl-L-citrulline	Methyl	12.3 ± 1.4	8.64 ± 0.81	0.70
N-α-Acetyl-L-tyrosine	Ethyl	50 ± 14	11.2 ± 2.2	0.22

[a] pH 8.0, 25°. F. Fiedler, G. Leysath, and E. Werle, *Eur. J. Biochem.* **36,** 152 (1973).

water of pH about 6, valid also for kallikrein A.[12] The circular dichroism spectrum shows maxima at 203, 235, and 282 nm,[5] suggesting that kallikrein has a predominantly random configuration.

Enzymic Properties. Only work relating to isolated kallikreins A and B will be considered here. Table VI gives a summary of rate constants obtained for the two enzymes with active-site acylating agents. Within the limits of experimental precision, kallikreins A and B behave identically. Cinnamoyl and indoleacryloyl kallikreins A and B also show identical difference spectra (Table VI), which resemble most closely those of the corresponding acyl trypsins.[16] A molar turnover (with 10 mM N^α-benzoyl-L-arginine ethyl ester, pH 8.0, 25°) of 119 sec^{-1} was obtained for kallikrein A and 120 ± 1.5 sec^{-1} for kallikrein B.[15] Identical rates were observed for the hydrolysis of N-α-acetyl-L-tyrosine ethyl-ester by kallikreins A and B[17] and also for N-α-benzoyl-L-arginine ethyl-ester and N-α-toluenesulfonyl-L-arginine methyl ester hydrolysis, as described by Takami.[9]

Kinetic constants for the hydrolysis of a number of esters of N-α-acylated amino acids by kallikrein B are presented in Table VII. Arginine esters are the best substrates; lysine esters are attacked at a much slower rate, and other amino acids are very poor substrates. Hydrolytic rates of 0.052 μmoles/mg per minute were found[9] for N-α-benzoyl-L-arginine amide (9 mM) and 0.129 μmole/mg per minute for N-α-benzoyl-DL-arginine p-nitranilide (0.9 mM, pH 8.0, 40°) for kal-

[16] F. Fiedler, B. Müller, and E. Werle, *FEBS Lett.* **22,** 1 (1972).
[17] F. Fiedler, G. Leysath, and E. Werle, *Eur. J. Biochem.* **36,** 152 (1973).

likrein A as well as B, about $\frac{1}{50}$ of the activity of trypsin. Kallikreins A and/or B have been reported[9] to hydrolyze casein, clupein, poly-Arg, poly-Lys, and poly-Lys-Leu while poly-Orn was not affected. On the other hand, kallikrein hydrolyzes casein at a rate of at most $\frac{1}{200}$ of that of trypsin.[3] After partial inactivation with diisopropyl phosphorofluoridate (as a precaution for inactivating other serine proteases, if present), kallikrein B hydrolyzes arginyl peptides effectively in Phe-Arg-X sequences analogous to the C-terminal portion of the kinin molecule.[18] The specific cleavage of a single arginyl bond by kallikrein B has been exploited in sequence analysis.[19]

Assay Methods

If the isolation procedure is applied to prepurified enzyme preparations as in the present work, kallikrein activity can be measured throughout by the hydrolysis of N-α-benzoyl-L-arginine ethyl ester.[20] A Technicon Autoanalyzer was adapted for the determination of kallikrein activity in column eluates. For determination of kallikrein activity in samples of highly purified enzyme, a pH-stat method will be described.

Kallikrein Assay with the Technicon Autoanalyzer[4]

Principle. Buffer, enzyme sample, and N-α-benzoyl-L-arginine ethyl ester are mixed; on passage through the reaction coil of the Autoanalyzer, ethanol is produced and diffuses into a pyrophosphate buffer stream. The ethanol will reduce a stoichiometric amount of NAD in the presence of alcohol dehydrogenase. NADH is determined by absorption at 340 nm and the system is calibrated with a series of standard kallikrein solutions.

Reagents

Tris: 0.05 M Tris-HCl buffer, pH 7.8, containing 0.2 ml of Triton X-100 (Serva) per liter

BAEE: 680 mg of N-α-benzoyl-L-arginine ethyl ester hydrochloride in 100 ml of water (final concentration in test, about 4 mM)

Pyrophosphate: 30.4 g of $Na_4P_2O_7 \cdot 10\ H_2O$, 8.2 g of semicarbazide hydrochloride, 1.8 g of glycine, 0.2 ml of Triton X-100, and 20–24 ml of 2 N NaOH (to pH 8.7) per liter

---

[18] E. Werle, F. Fiedler, and H. Fritz, *in* "Pharmacology and the Future of Man" (*Proc. 5th Int. Congr. Pharmacol.*, San Francisco 1972), Vol. 5, p. 284. Karger, Basel, 1973.

[19] V. Mutt and S. I. Said, *Eur. J. Biochem.* **42**, 581 (1974).

[20] I. Trautschold and E. Werle, *Hoppe-Seyler's Z. Physiol. Chem.* **325**, 48 (1961).

NAD: 150 mg NAD (free acid) in 100 ml of water (final concentration in test, about 0.2 mM)

ADH: 15 mg (3750 U) of alcohol dehydrogenase from yeast (Boehringer) in 100 ml of water (this solution is kept in an ice bath; final concentration in test, about 0.015 mg/ml)

Kallikrein standards: A stock solution of pig pancreatic kallikrein (a crude preparation may be used), the activity of which has been determined by the pH-stat assay, is diluted with 0.05 M Tris-HCl buffer, pH 7.8 (without Triton), to obtain 5-ml aliquots with kallikrein activities of 0.05, 0.1, 0.25, 0.5, 0.75, 1, 1.5, 2, and 3 U/ml.

Kallikrein samples: Aliquots of column eluates are diluted with 0.05 M Tris-HCl buffer, pH 7.8 (without Triton), to obtain 1-ml volumes containing enzyme activities in the 0.05–3 U range (0.5–1.5 U is preferred).

Procedure. An Autoanalyzer manifold is assembled according to the flow diagram in Fig. 3, and instructions of the manufacturer. A base line is established by pumping of reagents, then the turntable of the sampler is loaded with 1-ml cups containing the kallikrein standard solutions, followed by the samples. Samples are fed into the system at a rate of 60 per

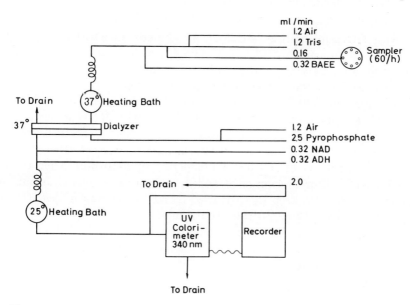

Fig. 3. Flow diagram of the determination of kallikrein on the Technicon Autoanalyzer. BAEE, N-α-benzoyl-L-arginine ethyl ester; ADH, alcohol dehydrogenase. Adapted from F. Fiedler and E. Werle, *Hoppe-Seyler's Z. Physiol. Chem.* **348,** 1087 (1967).

hour. Activity of the samples is evaluated by means of a calibration chart prepared from the recorder tracings for the kallikrein standards.

Kallikrein Assay in a Recording pH-Stat[12,13]

Principle. Enzymic hydrolysis of N-α-benzoyl-L-arginine ethyl ester liberates a carboxyl group that becomes essentially fully dissociated at the pH of assay. The pH is held constant by addition of stoichiometric amounts of alkali (0.1 N NaOH), and the time-dependence of alkali consumption is recorded. Total reaction volume is 20 ml, which minimizes changes in substrate concentration by hydrolysis and by dilution during the titration procedure. In order to prevent possible inhibition of the reaction by heavy-metal ions,[21] 0.1 mM thioglycolic acid is added.

Reagents

NaCl, 0.1 M, prepared in CO_2-free water
Thioglycolic acid, 0.01 M, in 0.1 M NaCl (prepared fresh each day)
N-α-Benzoyl-L-arginine ethyl ester hydrochloride, 0.05 M, in 0.1 M
    NaCl (can be stored in the refrigerator for about 2 weeks)
NaOH, 0.1 M
Kallikrein sample: 10–20 U/ml in water

Procedure. The assay is performed in a Radiometer Autotitrator TTT 1c with Titrigraph SBR 2c, a 0.25-ml Autoburette filled with 0.1 N NaOH, and a titration assembly TTA 3. The double-walled, well-stirred reaction vessel of 50-ml capacity is thermostatted at 25.0°, and a stream of nitrogen is passed over the surface of the reaction mixture.

The reaction vessel contains 15.4 ml of 0.1 M NaCl, 4 ml of 0.05 M N-α-benzoyl-L-arginine ethyl ester, and 0.2 ml of 0.01 M thioglycolic acid, with the pH adjusted to 8.0 by means of 0.1 M NaOH (about 0.3 ml are necessary). After refilling the burette, a base line (most often showing negligible drift) is established for a period of 5–10 min. Then 0.1 ml (1–2 U) of kallikrein solution is added, and recording of the amount of alkali necessary to keep the pH at 8.0 is continued (paper speed 1 cm/min; proportional band, 0.3) for about 6 min.

Units. One unit (U) of pig pancreatic kallikrein is defined as the amount of enzyme that hydrolyzes 1 μmole of N-α-benzoyl-L-arginine ethyl ester per minute under the conditions of the pH-Stat assay.

Other Assay Methods

If an Autoanalyzer is not available, an alternative recommended procedure for monitoring column eluates is the spectrophotometric assay,

[21] F. Fiedler and E. Werle, *Eur. J. Biochem.* **7**, 27 (1968).

also based on the principle of release of ethanol from the arginine ester, determined by means of the alcohol dehydrogenase-linked assay.[6,20] Determination of kallikrein activity in crude pancreatic extracts necessitates the use of biological assays which are less affected by trypsin, as the kallikrein-caused changes of blood pressure in dog Hg.[2]

[25] Human Kallikrein and Prekallikrein

By ROBERT W. COLMAN and ANDRANIK BAGDASARIAN

Human Kallikrein

Kallikreins (EC 3.4.21.8) are proteases that specifically liberate kinins from plasma α_2-globulin substrates known as kininogens. In plasma, kallikrein exists in a precursor form known as prekallikrein. Kallikreins can be divided in two classes depending on the source from which the enzymes are isolated. Tissue or glandular kallikreins have been dealt with in an earlier article in this series.[1] This review will center on human plasma prekallikrein and kallikrein. A scheme of activation of prekallikrein and diagram of kallikrein actions is presented below.

(a) Activation of prekallikrein

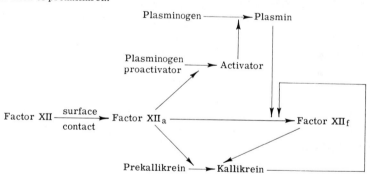

(b) Kallikrein action

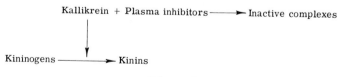

Scheme 1

[1] M. E. Webster and E. S. Prado, this series Vol. 19, p. 681.

Assay Methods

The enzymic activity of kallikrein can be measured by two methods. The rate of release of kinins as determined by bioassay or radioimmunoassay can be quantified. Alternatively, the ability of kallikrein to hydrolyze synthetic basic amino acid esters can be assessed.

Kinin-Releasing Assay

Bioassay. This method utilizes the smooth muscle contractions elicited by bradykinin, which is released by the action of kallikrein on kininogen. Two muscle sources, rat uterus or guinea pig ileum, have been successfully employed for this purpose. While rat uterus is more sensitive to bradykinin, guinea pig ileum has proved to be the more reliable organ for this assay. The bioassay utilizing guinea pig ileum has been described in detail in a previous article.[1]

Radioimmunoassay. One radioimmunoassay method[2] utilized antibody to bradykinin by injecting rabbits with bradykinin coupled to rabbit serum albumin and detected nanogram quantities of bradykinin. Antibody induced by injecting bradykinin conjugated to poly-L-lysine via toluene diisocyanate[3] or by injecting bradykinin coupled to ovalbumin via diisocyanate[4] have also been studied. The latter antibody is appropriately sensitive toward bradykinin. The antiserum is incubated with $[^{125}I]Tyr_8$ bradykinin and known amounts of native bradykinin or test samples. The percent inhibition of binding of the radioactive bradykinin by native bradykinin is then calculated and the concentration of bradykinin formed in the test samples is determined from a standard calibration curve.[5] One unit of kallikrein is defined as the amount of enzyme which, when incubated with 1 ml of heated human plasma containing 3 mM o-phenanthroline for 5 min at 25°, will release 10 ng of bradykinin.

Esterolytic Assay

Principle. Kallikreins are capable of hydrolyzing various synthetic substrates, primarily substituted arginine esters and, less effectively, lysine esters. Arginine esterase activity evolved in plasma following factor

---

[2] T. L. Goodfriend, L. Levine, and D. G. Fasman, *Science* **144**, 1344 (1964).

[3] J. Spragg, K. F. Austen, and E. Haber, *J. Immunol.* **96**, 865 (1966).

[4] R. C. Talamo, E. Haber, and K. F. Austen, *J. Lab. Clin. Med.* **74**, 816 (1969).

[5] R. C. Talamo, K. F. Austen, and E. Haber, *in* "Methods in Investigative and Diagnostic Endocrinology" (S. A. Berson and R. S. Yalow, eds.), Vol. 2B, Chapter 12, Section 3.2. Am. Elsevier, New York, 1973.

XII activation by kaolin was shown to be mainly due to kallikrein.[6] The products after hydrolysis (corresponding amino acid and alcohol) can be quantitated either directly using the spectrophotometer and following the changes in the absorbance at specific wavelength, e.g., 253 nm when substrate is N-α-benzoyl-L-arginine ethyl ester (BAEE)[7] or by titrating the H^+ ion released. The alcohol formed can be measured by its ability to reduce NAD in the presence of alcohol dehydrogenase.[7,8] A colorimetric method for determination of esterolytic activity of kallikrein is available.[9] In this method methanol released from N-α-toluenesulfonyl-L-arginine methyl ester (TAME) is quantitated by the modification of the method of Siegelman et al.[10]

Reagents

Phosphate buffer (sodium), 0.1 M, pH 7.65, containing 0.15 M NaCl
TAME (Sigma Chemical Company, St. Louis, Missouri) was dissolved (0.055 M) fresh in the above phosphate buffer and kept on ice. The final volume of assay mixture is 2.2 ml
Potassium permanganate (2% solution) stored in a brown bottle
Sodium sulfite, 10% solution made up fresh daily
Chromotropic acid solution: 4,5-dihydroxy-2,7-naphthalenedisulfonic acid, disodium salt (Eastman Organic Chemicals, Rochester, New York). Four hundred miligrams were dissolved in 20 ml of distilled water. To this solution 180 ml of 65% sulfuric acid was added. The 65% sulfuric acid is made by slowly adding 600 ml of ice cold sulfuric acid (analytical grade) to 300 ml of cold distilled water in a flask placed in ice water
Trichloroacetic acid (TCA), 15% solution

Procedure. The enzyme solution (0.2 ml) is added to 2.0 ml of TAME solution in ice and immediately transferred to a 37° water bath. After 1 min and 31 min, 1-ml samples are removed and directly placed in tubes containing 0.5 ml of 15% TCA. The solution is centrifuged at 1000 g for 5 min at 25°; 0.5 ml of the clear supernatant is placed into 0.1 ml of $KMnO_4$ and rotated to remove stray drops off the sides of the tubes. The excess $KMnO_4$ is reduced by the addition of 0.1 ml of sodium sulfite, and the colorless mixture is added to 4 ml of the chromotropic acid

[6] R. W. Colman, L. Mattler, and S. Sherry, *J. Clin. Invest.* **48,** 11 (1969).
[7] I. Trautschold and E. Werle, *in* "Methoden der enzymatischen Analyse" (H. U. Bergmeyer, ed.), 2nd ed., p. 880. Verlag Chemie, Weinheim, 1969.
[8] I. Trautschold and E. Werle, *Hoppe-Seyler's Z. Physiol. Chem.* **325,** 48 (1961).
[9] R. W. Colman, L. Mattler, and S. Sherry, *J. Clin. Invest.* **48,** 23 (1969).
[10] A. M. Siegelman, A. S. Carlson, and T. Robertson, *Arch. Biochem. Biophys.* **97,** 159 (1962).

solution. The tubes are then covered by a marble and placed in a boiling water bath for 15 min. The absorbance of the purple solution is determined at 580 nm in a spectrophotometer using the 1-min samples as blank and the 31-min samples as test. The absorbance values are linear from 0.05 to 0.90. The purple color is stable at 4° for 1 week.

Standardization and Definition of Unit. The assay is standardized with 1 ml of methanol (0.771 g), diluted in 100 ml of distilled water to make the stock solution of 0.24 mM. Different dilutions are made in buffer to construct a standard curve. A typical conversion is $A \times 293 = $ μmoles of TAME hydrolyzed per milliliter of sample per hour.

Purification

Several methods are available for the purification of kallikrein from human and animal plasma.[6,11,12] Because of the close resemblance of the properties of kallikrein to other plasma proteins, its purification has presented a challenging problem to the investigators.[13,14] Because of limited space, we shall describe in detail only the method used in our laboratory for the purification of plasma kallikrein.

Reagents

Normal human citrated plasma. Nine volumes of blood were drawn by clean venipuncture into 1 volume of 3.8% sodium citrate (Na$_3$C$_6$H$_5$O$_7$·H$_2$O) in polycarbonate tubes at 4°. The blood is centrifuged at 2500 g for 15 min at 4° to remove the cells, and the plasma is removed with plastic syringe or pipette. The plasma can be used fresh or stored at —50°

Ethanol, 53%

Acetate buffer, 4.8 M, pH 4.0

NaCl 0.06 M

Sodium phosphate, 0.01 M, buffer, pH 6.0

Sodium phosphate, 0.067 M, pH 8.0, containing 0.37 M NaCl

Sodium phosphate, 0.005 M, buffer, pH 8.0

Sodium phosphate, 0.005 M, pH 8.0, containing 0.7 M NaCl

Sodium phosphate, 0.016 M, buffer, pH 8.0, containing 0.13 M NaCl

Carboxymethyl (CM) Sephadex C-50 medium, capacity 4.5 mEq/g (Pharmacia). The resin is swollen in the appropriate buffer for

[11] H. Fritz, G. Wunderer, and B. Dittmann, *Hoppe-Seyler's Z. Physiol. Chem.* **353**, 893 (1972).

[12] M. Yano, S. Nagasawa, and T. Suzuki, *J. Biochem.* **67**, 713 (1970).

[13] A. P. Kaplan and K. F. Austen, *J. Exp. Med.* **136**, 1378 (1972).

[14] P. C. Harpel, *J. Clin. Invest.* **50**, 2084 (1971).

at least 72 hr with frequent change of buffer to facilitate equilibration and to remove the fines. The resin is packed by gravity according to the instruction of the manufacturers. Diethylaminoethyl cellulose (DE-52) is obtained preswollen from Whatman Company. To equilibrate the resin with 0.005 M phosphate pH 8.0 it is first suspended in 0.005 M NaH_2PO_4 with several changes of the monosodium phosphate salt until the pH of the suspension is about 8.0. The resin is then resuspended in the 0.005 M phosphate buffer. Sephadex G-200, particle size is swollen as recommended by the manufacturers and equilibrated with 0.016 M phosphate buffer pH 8.0 containing 0.13 M NaCl

Procedure

Alcohol Fractionation

All operations up to column chromatography are carried out either in metal or plastic containers. In a 2–3 liter capacity metal beaker precooled in a —5° water–ethylene glycol bath place 1000 ml of normal citrated plasma and cool to about 0°. To it add dropwise 893 ml of 53% ethanol precooled in Dry Ice–acetone mixture to below —10°. The mixture is continuously stirred, the temperature being held between —1° and +1°. After addition of 100 ml of the ethanol to plasma, let the mixture temperature equilibrate to —5°. Continue stirring the mixture (containing 25% ethanol) at —5° for an additional 10 min and centrifuge the mixture at 0°, 2000 g for 15 min in plastic bottles with caps. Collect the supernatant into a plastic volumetric cylinder and measure its volume. Add 311 ml of ice-cold distilled water per 1000 ml of supernatant at —5°. Adjust the pH of the solution to 5.2 ± 0.1 with acetate buffer. An additional amount of water should be added to bring up the volume of the buffer and water to 78 ml per 1000 ml of original supernatant. Stir mixture for 20 min and centrifuge to obtain a slimy yellow precipitate, which is Cohn fraction IV-1.[15] Discard supernatant (fraction IV-4). Collect the precipitate into plastic graduated cylinders and add 0.06 M NaCl to five times the volume of the precipitate. Centrifuge at 0° for 15 min at 2000 g. Collect the supernatant (dark lime-green color), and discard the bright yellow precipitated lipoproteins. Adjust the pH of supernatant to 4.8 with a few drops of acetate buffer. At —5° add 39.4 ml of 53% ethanol (precooled to below —10°) per 100 ml of supernatant dropwise with constant stirring to

[15] E. J. Cohn, L. E. Strong, W. L. Hughes, Jr., D. J. Mulford, J. N. Ashworth, M. Melin, and H. L. Taylor, *J. Am. Chem. Soc.* 68, 459 (1946).

obtain 15% alcohol concentration in the solution. Centrifuge mixture; collect greenish blue precipitate and dissolve it in 100 ml of 0.1 M phosphate buffer, pH 6.0. This sample is referred to as the ceruloplasmin-containing fraction.

Column Chromatography

First CM Sephadex. The ceruloplasmin-rich solution is then applied to a 5 × 30 cm column of CM-Sephadex preequilibrated with 0.01 M sodium phosphate buffer, pH 6.0. After application of the sample, the column is washed with starting buffer until all the nonabsorbed proteins are eluted. Then a linear gradient of pH and salt is used. The mixing chamber contains 1000 ml of starting buffer, and the second chamber contains 1000 ml of 0.067 M phosphate, pH 8.0, in 0.51 M NaCl. Fractions (17 ml) are collected overnight at 150 ml/hr flow rate and 0.1–0.2 ml of each fraction is tested for arginine esterase activity. The esterase activity peak elutes after the major protein peak. Fractions containing esterase activity with a specific activity of greater than 0.4 μmole of TAME hydrolyzed per minute per milligram are pooled and concentrated in an Amicon concentrator (Amicon Corporation, Lexington, Massachusetts) using an XM-50 membrane to about 20–25 ml and dialyzed against 0.005 M phosphate, pH 8.0, with several changes. This ceruloplasmin precipitate contains mainly prekallikrein. After CM-Sephadex chromatography, most of the prekallikrein is activated to kallikrein.

DE-52 Chromatography. A 1 × 24 cm column is packed with pre-equilibrated DE-52. The pooled concentrate is applied to the column, and the column is washed with the starting buffer (0.005 M phosphate, pH 8.0). The 3.0-ml fractions collected contains most of the kallikrein activity. Prekallikrein activators or kallikrein II[7] can be eluted from the DE column by salt gradient if desired and further purified.[16]

Second CM-Sephadex Chromatography. The first esterase peak is pooled, concentrated in Amicon concentrator with an XM-50 membrane, and dialyzed against the starting buffer (0.01 M phosphate, pH 6.0). The sample is applied into a column 2.8 × 24 cm packed with CM-Sephadex preequilibrated with the same buffer. A linear salt and pH gradient was used; the first chamber contains 400 ml of the starting buffer and the second chamber contains 400 ml of the limit buffer 0.016 M phosphate, pH 8.0, in 0.13 M NaCl. The flow rate is 60 ml/hr. A major protein peak is evolved with a trailing tail. The major TAME hydrolyzing activity is eluted beyond the major protein peak. Fractions

---

[16] A. Bagdasarian, R. C. Talamo, and R. W. Colman, *J. Biol. Chem.* **248**, 3456 (1973).

with specific activity higher than 5 are pooled, concentrated, and dialyzed against 0.016 M phosphate, pH 8.0, containing 0.13 M NaCl. This esterase activity could not be dissociated throughout the purification from the bradykinin-releasing activity and has been characterized as the major kallikrein.[6] The ratio of TAME hydrolyzing activity to that of other synthetic methyl esters, e.g., N-α-benzoyl-L-arginine (BAME), N-acetyl-L-arginine (AAME), p-toluenesulfonyl-L-lysine (TLME), and N-acetyl-L-lysine (ALME), are significantly different from those obtained by other plasma enzymes, such as thrombin, plasmin, or C$\bar{\text{I}}$ esterase. Moreover, the soybean trypsin inhibitor, which inhibits the esterase activity, also prevents formation of bradykinin by the sample. The purified kallikrein was shown to have no detectable concentrations of coagulation factors, thrombin, plasminogen, or C$\bar{\text{I}}$ esterase. No β_2-glycoprotein (type I) could be detected by immunodiffusion techniques. The preparation was found to contain a substantial degree of contamination with γ-globulin (IgG). In order to remove the IgG the following step was employed.

Sephadex G-200 Gel Filtration. The concentrated solution is applied onto a 2×90 cm column packed with preswollen and equilibrated Sephadex G-200 in a volume of sample not exceeding 2% of the column volume. One-milliliter fractions are collected with a flow rate of 15 ml/hr (Fig. 1). Kallikrein peak is eluted after the major protein peak containing IgG as determined by Partigen plates (Behring Diagnostics, Somerville, New Jersey). Fractions containing kallikrein have low protein content and upon concentration of the pool a considerable loss of activity was found. To prevent activity loss we have been able to stabilize the enzyme by the addition of bovine serum albumin (1 mg/ml) directly to the collecting tubes. Thus, a specific activity of 30 was obtained in the concentrated pool containing the bovine serum albumin. The purification results are summarized in Table I. Purified kallikrein run in acrylamide gel shows a single major protein band (Fig. 2) with esterase activity, kinin release and kallikrein antigen activity corresponding to the protein. Purified kallikrein examined by fibrin plates contains no plasminogen activator or plasmin activities.

Properties

Stability. The presence of IgG substantially stabilizes the enzyme. Kallikrein solutions of greater than 0.5 mg/ml have retained full activity after storage at room temperature overnight and can be kept in the cold room for several days without significant loss of activity. Frozen samples ($-50°$) retain their activity for several months. Solutions of

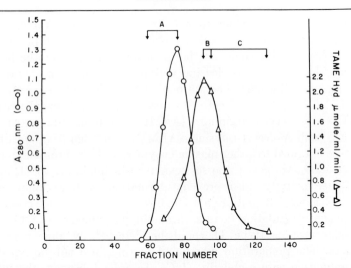

Fig. 1. Gel filtration of partially purified kallikrein. Starting material was 32.4 mg of the second carboxymethyl Sephadex chromatography of kallikrein. The specific activity of the kallikrein was 9.3 μmoles of N-α-toluenesulfonyl-L-arginine methyl ester (TAME) hydrolyzed per milliliter per minute. A 2.0 × 97 cm column was packed with Sephadex G-200 and equilibrated with 0.016 M sodium phosphate, pH 8.0, in 0.13 M NaCl. Fractions of 1.3 ml were collected. The flow rate was 15 ml/hr at 4° in a downward direction. Circles denote the absorption at 280 nm, and the triangles represent the esterase activity. Pools A, B, and C include the fractions denoted in the graph. Reproduced with permission from the publisher from A. Bagdasarian *et al.* (1974). *J. Clin. Invest.* **54**, 1444.

TABLE I

SUMMARY OF PURIFICATION OF KALLIKREIN FROM HUMAN PLASMA

	Total enzyme units (μmoles of TAME[a] hydrolyzed/min)	Protein specific activity yield		
		mg	μmoles of TAME hydrolyzed/min/mg	%
Frozen plasma[b]	1600	60,500	0.027	100
Cohn fraction IV₁	560	7,777	0.072	35
Saline supernatant	528	2,227	0.237	33
Crude ceruloplasmin	608	1,876	0.324	38
CM-Sephadex No. 1	400	624	0.644	25
DE-52	184	132	1.393	11.5
CM Sephadex No. 2	121	33.5	3.61 (8.0)[c]	7.6
Sephadex G-200	69	6.9	10.0 (40.0)[c]	4.3

[a] N-α-Toluenesulfonyl-L-arginine methyl ester.
[b] One liter.
[c] Peak tubes directly off column.

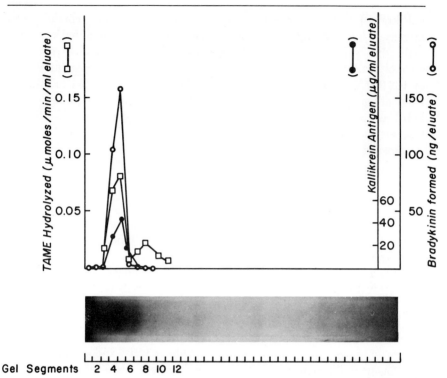

Gel Segments   2 4 6 8 10 12

FIG. 2. Disc gel electrophoresis of the G-200 Sephadex fraction of kallikrein peak pool C. Electrophoresis was performed with 100 μg of protein. The N-α-toluene-sulfonyl-L-arginine methyl ester (TAME) esterase activity eluted from each 6.2-mm segment is plotted. Reproduced with permission from the publisher from A. Bagdasarian *et al.* (1974). *J. Clin. Invest.* **54**, 1444.

kallikrein can be frozen and thawed several times and retain activity.

Optimum pH. For the hydrolysis of TAME, the optimum pH is 7.65. In plasma, kallikrein generates bradykinin at a similar pH. Bovine plasma kallikrein exhibits an optimum pH of 8.8.

Physical Properties

Amino Acid Composition. The relatively low yield of plasma kallikrein has discouraged the investigators from initiating such studies.

Molecular Weight. Using sedimentation equilibrium a molecular weight of 99,800 was obtained[6] for the major plasma kallikrein (K-I). Kallikrein II revealed a molecular weight of 163,000. A monomer–dimer relationship has been suggested based on the following observation. If kallikrein I is gel filtered through Sephadex G-200 under conditions of increased ionic strength, all the activity emerges as one peak with a molecular weight of 160,000–180,000. Investigators who have not used alcohol fractionation have not found kallikrein II. A molecular weight of 108,000 was found for human plasma kallikrein.[17] Plasma kallikreins from other species exhibit similar molecular weights to human plasma kallikrein. The apparent molecular weight of bovine plasma kallikrein has been estimated at 95,000.[12]

Isoelectric Point (pI). Dealt with under prekallikrein properties.

Substrates

Natural Substrates. Plasma kallikrein acting on a kininogen substrate from plasma releases a nonapeptide bradykinin.[18] At least two kininogens exist in human plasma[19] of molecular weights of about 180,000 and 80,000. High-molecular-weight kininogen comprising 20% of the total is a good substrate for plasma kallikrein ($K_m = 0.4$ mg/ml), but low-molecular-weight kininogen (80%) is not ($K_m = 8.3$ mg/ml).[20] High-molecular-weight kininogen has recently[21,22] been shown to be essential for factor XII-catalyzed reactions, since patients lacking it show defective prekallikrein activation, fibrinolysis, and coagulation.

Plasma kallikrein releases several other peptides, aside from bradykinin, from kininogen substrates with molecular weights of 4600, 8000, 18,000 and 51,000.[23] A 4600 MW peptide rich in histidine and lysine can inhibit activation of factor XII, suggesting a negative feedback control of kallikrein–kinin system. Purified human kallikrein was demonstrated to hydrolyze factor XII to 30,000 dalton fragments.[24] These prealbumin fragments are potent activators of prekallikrein and appear to be identical to prekallikrein activators (PKA).[25] Similar findings were

[17] K. F. Austen, J. Spragg, A. D. Schreiber, and A. P. Kaplan, *Int. Congr. Pharmacol.,* 5th, p. 182, San Francisco (1972).
[18] E. Mellanby, *J. Physiol. (London)* **109**, 488 (1949).
[19] S. Jacobsen, *Br. J. Pharmacol.* **26**, 403 (1966).
[20] J. V. Pierce and J. A. Guimaraes, *Int. Conf. Chem. Biol. Kallikrein-Kinin System Health Dis.,* Reston, Virginia (1974).
[21] R. W. Colman, A. Bagdasarian, R. C. Talamo, C. F. Scott, M. Seavey, J. A. Guimaraes, J. V. Pierce, and A. P. Kaplan, *J. Clin. Invest.* **56**, 1650 (1975).
[22] K. D. Wuepper, D. R. Miller, and M. J. Lacombe, *J. Clin. Invest.* **56**, 1663 (1975).
[23] S. Iwanaga, M. Komiya, Y. N. Han, T. Suzuki, S. Oh-Ishi, and M. Katori. *Int. Conf. Chem. Biol. Kallikrein-Kinin System Health Dis.,* Reston, Virginia, (1974).
[24] A. Bagdasarian, B. Lahiri, and R. W. Colman, *J. Biol. Chem.* **248**, 7742 (1973).
[25] A. P. Kaplan and K. F. Austen, *J. Immunol.* **105**, 802 (1970).

TABLE II
PROPERTIES OF KALLIKREIN

Properties[a]	Source	Value	Reference
Isoelectric point	Human	7.7–9.4	69, 72
Sedimentation constant	Human	5.7	7
Molecular weight			
Sedimentation equilibrium	Human	99,800	7
Gel filtration	Bovine	95,000	15
Gel filtration	Human	108,000	20
K_m: TAME	Human	$1.36 \times 10^{-2}\,M$	32, 7
BAEE	Human	$1.36 \times 10^{-4}\,M$	33
BAEE	Bovine	$9.62 \times 10^{-4}\,M$	15
TAME	Bovine	$1.56 \times 10^{-5}\,M$	15
TLME	Bovine	$6.02 \times 10^{-3}\,M$	15
Kininogen I	Human	0.4 mg/ml	23
Kininogen II	Human	8.3 mg/ml	23
V_m: TAME	Human	$0.1\,M$	7

TAME, N-α-toluenesulfonyl-L-arginine methyl ester; BAEE, N-α-benzoyl-L-arginine ethyl ester; TLME, p-toluenesulfonyl-L-lysine methyl ester.

also reported by other workers.[26,27] Such action of kallikrein can be regarded as an amplification of the kinin generation (positive feedback). Unlike plasmin, plasma kallikrein does not hydrolyze casein or fibrin or coagulation factor V[28] and factor XIII.[29]

Synthetic Substrates. Plasma kallikrein like other serine esterases, e.g., trypsin, plasmin, thrombin, are capable of hydrolyzing arginine and, to a lesser degree, lysine esters, such as: TAME, BAME, AAME, TLME, and ALME[10] and benzoyl arginine ethyl ester (BAEE).[8] Substituted p-nitroanilides although good substrates for trypsin[30,31] are not hydrolyzed significantly by kallikreins. K_m for TAME hydrolysis were determined to be 13.6 mM.[32] K_m values have also been determined on bovine kallikrein for TAME, 1.56 mM; for TLME, 0.602 mM; for BAEE, 96.5 μM.[12] Properties of plasma kallikrein are summarized in Table II.

[26] C. G. Cochrane, S. D. Revak, and K. D. Wuepper, *J. Exp. Med.* **138**, 1564 (1973).
[27] S. D. Revak, C. G. Cochrane, A. R. Johnston, and T. E. Hugli, *J. Clin. Invest.* **54**, 619 (1974).
[28] R. W. Colman, *Biochemistry* **8**, 1445 (1969).
[29] A. Bagdasarian, R. W. Colman, and L. Lorand, unpublished data, 1974.
[30] B. F. Erlanger, N. Kokowsky, and W. Cohen, *Arch. Biochem. Biophys.* **95**, 271 (1961).
[31] L. Svendsen, B. Blombäck, M. Blombäck, and P. I. Olsson, *Thromb. Res.* **1**, 267 (1972).
[32] R. W. Colman, J. W. Mason, and S. Sherry, *Ann. Intern. Med.* **71**, 763 (1969).

Inhibitors

Natural Inhibitors. Plasma contains two major inhibitors of kallikrein. C$\bar{1}$ esterase inhibitor (C$\bar{1}$ INH) inhibits both the proteolytic and esterolytic activities of plasma kallikrein[33,34] in a stoichiometric manner[35] with the formation of a stable bimolecular complex.[36,37] The proteolytic activity of kallikrein is also inhibited by α_2-macroglobulin (α_2M).[38,39] In this process the intact subunit of α_2M is cleaved to a 85,000 MW fragment.[40] In addition to C$\bar{1}$ INH and α_2M, antithrombin III (AT-III) has been shown to inhibit kallikrein, and the inhibition is markedly increased in the presence of heparin[41] with the formation of a stable stoichometric complex between kallikrein and AT-III.[42] Trasylol, the protease inhibitor of bovine lung, is an effective inhibitor of pancreatic and submandibular kallikrein[43] but a weaker inhibitor of plasma and urinary kallikrein.[9] Among plant inhibitors, soybean trypsin inhibitor at concentrations as little as 20 μg/ml causes 95% inhibition of kallikrein whereas lima bean trypsin inhibitor at 200 μg/ml has no effect.[16]

Synthetic Inhibitors. Kallikrein like other serine proteases is completely inhibited by 0.1 mM diisopropyl fluorophosphate[9] and diphenylcarbamyl fluoride (6 mM) caused 79% inhibition. p-Carboethoxyphenyl ϵ-guanidine caproate is a good inhibitor of plasma kallikrein esterase activity.[44] Toluenesulfonyl lysyl chloromethyl ketone (3 mM) has no effect on kallikrein.[9]

[33] N. S. Landerman, M. E. Webster, E. L. Becker, and H. D. Ratcliffe. *J. Allergy* **33**, 330 (1962).

[34] L. J. Kagen, *Br. J. Exp. Pathol.* **45**, 604 (1964).

[35] I. Gigli, J. W. Mason, R. W. Colman, and K. F. Austen, *J. Immunol.* **104**, 581 (1970).

[36] A. Bagdasarian, B. Lahiri, R. C. Talamo, P. Wong, and R. W. Colman, *J. Clin. Invest.* **54**, 1444 (1974).

[37] P. C. Harpel, *Int. Conf. Chem. Biol. Kallikrein-Kinin System Health Dis., Reston, Virginia, 1974.*

[38] P. C. Harpel, *J. Exp. Med.* **132**, 329 (1970).

[39] D. J. McConnell, *J. Clin. Invest.* **51**, 1611 (1970).

[40] P. C. Harpel, *J. Exp. Med.* **138**, 508 (1973).

[41] B. Lahiri, R. Rosenberg, R. C. Talamo, B. Mitchell, A. Bagdasarian, and R. W. Colman, *Fed. Proc., Fed. Am. Soc. Exp. Biol.* **33**, 642, Abst. No. 2434, (1974).

[42] B. Lahiri, A. Bagdasarian, B. Mitchell, R. C. Talamo, R. W. Colman, R. D. Rosenberg, unpublished data, 1975.

[43] E. Werle, *in* "Kininogenases-Kallikrein" (1st Symposium on Physiological Properties and Pharmacological Rationale) (G. L. Haberland and J. W. Rohen, eds.), pp. 7–22. Schattauer, Stuttgart, (1973).

[44] M. Muramatu and S. Fujii, *Biochim. Biophys. Acta* **268**, 221 (1972).

Plasma Prekallikrein

Prekallikrein is the precursor protein from which kallikrein is released. That such an inactive precursor exists in plasma is evident from the fact that kininogen, the substrate of kallikrein, exists in high concentrations in plasma whereas only minute quantities of free kinins can be detected. Once kallikrein is activated in plasma, it converts a large fraction of the total kininogen to kinin in a matter of minutes.

Assay Methods

Two types of assays are available for determination of prekallikrein in plasma. The first involves the conversion of prekallikrein to kallikrein and the assay of kallikrein by either the release of kinin from kininogen or the ability to hydrolyze synthetic N-substituted arginine esters. The assays for kallikrein once formed are detailed in the preceding section. The methods of activation of prekallikrein to kallikrein will be delineated as part of the assay procedure. The second method is a direct immunochemical assay of the amount of prekallikrein protein present.

Activation of Prekallikrein to Kallikrein

Principle. Prekallikrein is converted to kallikrein by the action of activated Hageman factor (factor XII_a)[45,46] or its proteolytic fragments (XII_f) large activator and PKA.[24,47,48] The active enzymes can be directly added to plasma or the endogenous factor XII present in plasma can be activated by exposure to a surface such as kaolin. Although factor XII_a can trigger the coagulation and fibrinolytic pathways as well as the kinin-forming reactions, at short time intervals, kallikrein is the only major protease released[9,46] in plasma. Although trypsin can also hydrolyze prekallikrein to kallikrein[49] its use is confined to purified preparations of prekallikrein. In plasma, high concentrations are necessary to overcome its inhibition by α_1AT; only then will it activate other arginine esterases in plasma. Conversely kaolin, which depends on the presence of endogenous factor XII, cannot be used to activate purified prekallikrein preparations but is convenient for plasma. Factors XII_a

[45] J. Margolis, *J. Physiol. (London)* **151**, 238 (1960).
[46] G. J. D. Girey, R. C. Talamo, and R. W. Colman, *J. Lab. Clin. Med.* **80**, 496 (1972).
[47] A. P. Kaplan and K. F. Austen, *J. Exp. Med.* **133**, 696 (1971).
[48] C. G. Cochrane and K. D. Wuepper, *J. Exp. Med.* **134**, 986 (1971).
[49] M. E. Webster and O. D. Ratnoff, *Nature (London)* **192**, 180 (1961).

and XII$_f$ are effective in both plasma[50] and/or partially purified pre-kallikrein.[25]

Reagents

Normal human plasma, prepared as discussed previously and stored in polypropylene tubes at −70°. The plasma should be frozen in aliquots and thawed only once

Sodium phosphate buffer, 0.1 *M* pH 7.65, containing 0.15 *M* NaCl made up in deionized distilled water; used where buffer is designated unless otherwise stated

Kaolin (obtained from Fisher Scientific Co., Fairlawn, New Jersey), used in a stock suspension in buffer of 10 mg/ml. This suspension must be agitated with a Vortex mixer immediately prior to use

Hageman factor (factor XII) purified according to the procedure of Cochrane *et al.*[48] and further eluted off the acrylamide gels[24]

Activators of prekallikrein: These enzymes were obtained in a partially purified state as a by-product of kallikrein preparation and further purified as described earlier.[24] Large activator was a single band on analytical polyacrylamide electrophoresis, and PKA was comprised of several bands in the prealbumin region, which all had activator activity. Activator units are the amount of activator that will liberate from 1 ml of plasma a quantity of kallikrein that can hydrolyze 1 μmole of TAME per minute at 37°[50]

Kallikrein substrates: TAME (0.05 *M* in buffer) or kininogen as described under kallikrein assays

Procedure. Two methods may be used. The first, utilizing measurements of arginine esterase activity, was carried out by the incubation of 0.10 ml of normal plasma with an equal volume of 10 mg of kaolin per milliliter[32] in 0.1 *M* phosphate buffer containing 0.15 *M* NaCl at pH 7.65 or by the addition of excess PKA, large activator, or factor XII$_a$[51] (1 activator unit/ml). After addition of the activator, the mixture was incubated for exactly 60 sec at 25°, and the reaction was then stopped by the addition of ice-cold TAME. The amount of TAME hydrolyzed was then determined as described under kallikrein. The second, employing kinin release, was performed by the incubation of 0.20 ml of normal plasma with an equal volume of 10 mg/ml kaolin for exactly 60 sec at 25°. Similar studies were carried out by addition of 1 acti-

[50] A. Bagdasarian, R. C. Talamo, and R. W. Colman, *Immunol. Commun.* 2, 93 (1973).

[51] B. Pitt, J. Mason, C. R. Conti, and R. W. Colman, *Trans. Assoc. Am. Physicians* 82, 98 (1969).

vator units of PKA per milliliter, large activator, or factor XII_a instead of kaolin, to an incubation mixture of plasma containing 3 mM phenanthroline. The reaction was terminated by the addition of 0.05 ml of 20% trichloroacetic acid per milliliter of the mixture. The level of kinin formed was then determined by the radioimmunoassay[5] as described under kallikrein. The concentration of prekallikrein in plasma can be expressed as the concentration of kallikrein formed, since with these concentrations of activators the reaction is complete. In plasma the normal concentration of prekallikrein as determined in 36 normal subjects is 1.65 μmoles of TAME hydrolyzed per minute per milliliter of plasma.[9] The values in terms of kininogen have not been established, but 2000 ng/ml is released by kallikrein formed by activation of plasma with kaolin or factor XII_a and its derivatives.

Variations of Prekallikrein in Disease

Using the kaolin-activated arginine esterase method, elevated levels of kallikrein have been found in patients treated with estrogen or oral contraceptives,[32] decreased levels in alcoholic cirrhosis,[32,52] and absent prekallikrein in Fletcher factor deficiency.[53] These changes have been attributed to altered synthetic rate. In all these conditions, the level of kallikrein inhibitors,[32] notably C1 esterase,[35] is normal. In addition, various diseases are associated with activation of prekallikrein *in vivo* leading to a decrease in both prekallikrein as it is converted to the active enzyme and kallikrein inhibitors as they complex with kallikrein. Such conditions include carcinoid syndrome,[32,54] dumping syndrome[54,55], hypotensive septicemia,[56] myocardial ischemia,[51] renal allograft rejection,[57] type IIa hyperlipoproteinemia,[58] and during hemodialysis.[59] In the case of the first three of these conditions elevated kinin levels have been demonstrated in some of the patients tested.

[52] P. Wong, R. W. Colman, R. C. Talamo, and B. M. Babior, *Ann. Intern. Med.* **77**, 205 (1972).

[53] K. D. Wuepper, *J. Exp. Med.* **138**, 1345 (1973).

[54] R. W. Colman, P. Y. Wong, and R. C. Talamo, *Int. Conf. Chem. Biol. Kallikrein-Kinin System Health Dis.*, Reston, Virginia, (1974).

[55] P. Y. Wong, R. C. Talamo, B. M. Babior, G. G. Raymond, and R. W. Colman, *Ann. Intern. Med.* **80**, 577 (1974).

[56] J. W. Mason, U. Kleeberg, P. Dolan, and R. W. Colman, *Ann. Intern. Med.* **73**, 545 (1970).

[57] R. W. Colman, G. J. Girey, E. G. Galvanek, and G. Busch, *in* "Coagulation Problems in Transplanted Organs" (K. von Kaulla, ed.), p. 87. Thomas, Springfield, Illinois, 1972.

[58] A. C. Carvalho, R. S. Lees, F. Voliusky, and R. W. Colman, *Blood* **40**, 937 (1972).

[59] E. N. Wardle and D. A. Piercy, *J. Clin. Pathol.* **25**, 1045 (1972).

Immunochemical Assay for Prekallikrein

Principle. When purified kallikrein is injected into a rabbit two precipitin arcs are developed when the antiserum is analyzed by immunodiffusion.[36] One of these arcs is due to interaction with prekallikrein and one to IgG (<8% of total protein) present in the best available preparations of kallikrein. The antiserum was adsorbed with IgG or with Fletcher trait (prekallikrein)-deficient plasma, which lacks both functional and immunochemical prekallikrein.[53] The remaining antibody was shown to be monospecific and directed against kallikrein and prekallikrein as judged by (a) precipitin reaction with purified kallikrein, (b) a failure of the IgG-adsorbed antibody to form an immunoprecipitate with Fletcher factor-deficient plasma,[60] (c) inhibition of the esterolytic and proteolytic activity of kallikrein by the adsorbed antibody,[36] and (d) inhibition of the activation of prekallikrein by activated factor XII or its fragments as measured by esterase or kinin formation. The adsorbed antibody has been used in a radial immunodiffusion assay.

Reagents

Plasma, prepared as in the functional assays of prekallikrein just described

Antibody: Purified kallikrein obtained after gel filtration (Fig. 1, pool C) is mixed with an equal volume of complete Freund's adjuvant and injected in the toe pads of three young Australian white rabbits. Each rabbit should receive 150 μg in a volume of 0.4 ml. After 2 weeks, the rabbits should receive as a booster dose the same amount of kallikrein subcutaneously injected in the neck area. After a week blood samples are drawn from the rabbits and incubated at 37° for 3 hr and at 4° for 18 hr before separating the serum by centrifugation at 2500 g. Complement was inactivated by heating the serum at 56° for 30 min. Additional booster doses were given at 7- to 10-day intervals as required

Two different approaches can be employed to absorb the anti-IgG usually found after injection of kallikrein into rabbits. In one, 0.05 ml of Fletcher factor-deficient human plasma was added to 5 ml of antisera, and the mixture was incubated at 37° for 1 hr and then left at 4° for 3 hr. The heavy precipitate formed was removed by centrifugation at 3000 g at 4°. Addition of 0.05 ml of plasma should fail to produce further precipitation. If a precipitate results continue adsorption until

[60] W. E. Hathaway, L. P. Belhasen, and H. S. Hathaway, *Blood* **26**, 521 (1965).

the last aliquot produces no precipitate. In the second, 0.1 ml purified IgG containing 600 μg of protein, obtained from the gel filtration of kallikrein sample on Sephadex G-200 (Fig. 1, pool A) should be added to 2 ml antisera. Incubation and centrifugation was carried out as above. The antiserum must be tested to determine that it is monospecific by both immunodiffusion and immunoelectrophoresis.

Double diffusion[61] was performed using prepared plates (Pentex, Kankakee, Illinois) containing 0.1 M borate in 0.15 M of NaCl, pH 8.1, in 1% agarose. Twenty microliters of plasma was applied in a hole and antiserum in the center well. Incubation was carried on for 18–36 hr, and one line should result.

Immunoelectrophoresis[62] was carried out in the universal electrophoresis cell, Buchler Instruments. On each microscope slide (2.5 $\times$ 7.5 cm), 3.6 ml of 1% agar gel in 0.04 M barbital buffer at pH 8.6 were placed and allowed to stand at room temperature followed by overnight storing at 4°. After wells were bored in the gels, small volumes of plasma were applied with a plastic microsyringe. A single arc should result after 18–36 hr.

Procedure. The general procedure was that of Mancini.[63] Diffusion plates were made with 6 volumes of 1.5% agarose and 1 volume of absorbed antisera and 1 volume of barbital buffer pH 7.6 containing Merthiolate blue to give a final concentration of 0.01%. On each diffusion dish, 3.6 ml of the mixture were placed and let stand overnight in a moist chamber at 4°. Wells 2.25 mm in diameter were then bored in the gel. Various volumes of plasma or prekallikrein samples from 5–15 μl were placed in the wells, and the reaction was recorded at 18 hr.

A linear relationship between purified kallikrein concentration and the square of the diameter of the precipitin rings were found (Fig. 3). This standard curve was used to estimate the levels of prekallikrein in the normal and patient plasmas. With an antibody adsorbed with IgG, Fletcher factor-deficient plasma demonstrated no detectable kallikrein by radial immunodiffusion.

Variations in Normal Individuals and in Disease. Prekallikrein levels in plasma were 103 $\pm$ 12.7 μg/ml immunochemically.[36] Patients on estrogen therapy had elevated levels, and patients with alcoholic cirrhosis had values about 50% of normal. However in severe cirrhotics the concentration of antigen were less reduced than the functional level, suggesting the synthesis of some nonfunctional prekallikrein (Fig. 4).

[61] O. Ouchterlony, *Acta Pathol. Microbiol. Scand.* **32**, 231 (1953).
[62] J. J. Scheidegger, *Int. Arch. Allergy Appl. Immunonol.* **7**, 103 (1955).
[63] G. Mancini, A. O. Carbonaro, and J. F. Heremans, *Immunochemistry* **2**, 235 (1965).

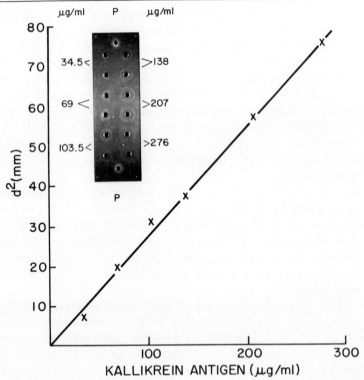

Fig. 3. Radial immunodiffusion assay of prekallikrein. The inset shows the actual rings of the radial immunodiffusion. The plate was prepared as described. The antibody was adsorbed with IgG. Kallikrein at a concentration of 150 μg/ml was used. The concentration of kallikrein used in duplicate is indicated near the inset. The two wells marked "P" contain 10 μl of two normal plasma samples. The kallikrein antigen concentrations are based on a volume of 0.01 ml of sample. Reproduced with permission from the publisher from A. Bagdasarian *et al.* (1974). *J. Clin. Invest.* **54**, 1444.

Purification

No completely acceptable procedure for the purification of human prekallikrein has been described. Chromatography of human plasma in the presence of hexadimethrine (a factor XII inhibitor), successive fractionation on QAE Sephadex, sulfopropyl Sephadex and G-200 Sephadex result in preparation free from IgG and plasminogen preactivator but still containing β_2-glycoprotein.[47] A similar procedure[64] involves preliminary fractionation with saturated $(NH_4)_2SO_4$ (25–55% fraction), chromatography on DEAE-Sephadex 50 and CM-Sephadex. The

[64] K. D. Wuepper and C. G. Cochrane, *J. Exp. Med.* **135**, 1 (1972).

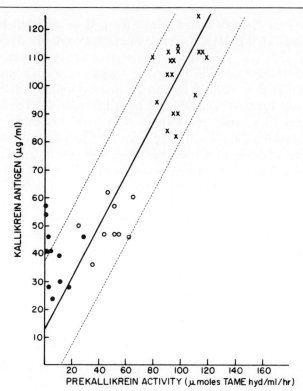

Fig. 4. The relationship of prekallikrein activity to prekallikrein antigen. The solid line is the regression line for normal individuals ($\times$). The equation for the line is $y = 12.6 + 0.93x$. The dashed lines are 95% confidence limits. The equation for the regression line of the patients with mild cirrhosis ($\bigcirc$) is $y = 13.4 + 0.73x$ and for severe cirrhosis ($\bullet$) $y = 35.4 + 0.1x$. Reproduced with permission from the publisher from A. Bagdasarian et al. (1974). J. Clin. Invest. 54, 1444.

prekallikrein is then separated from IgG by Pevikon block electrophoresis and from β_2-glycoprotein by sucrose density gradient.[12] Alternatively, β_2-glycoprotein can be removed by an immunoabsorbant. Neither procedure is well suited for the purification of large enough amounts of prekallikrein for adequate characterization. Rabbit[64] and bovine[65] prekallikrein have been purified with greater success.

Properties

Stability. Formal studies are not available on human prekallikrein, but data have been accumulated for bovine prekallikrein.[65] Prekallikrein

[65] H. Takahashi, S. Nagasawa, and T. Suzuki, J. Biochem. 71, 471 (1972).

loses activity on dialysis against NaCl (0.01 M or less) but is stable at concentrations of 0.1 M NaCl or greater. Between pH 6 and 9, prekallikrein is stable at 51° but loses activity at 61° or greater.

Physical Properties. The molecular weight of human prekallikrein has been reported as 107,000 by sodium dodecyl sulfate (SDS) electrophoresis[64] and 127,000[13] and 98,000[30] by gel filtration.[29] The sedimentation constant in sucrose gradients is 5.2.[64] The isoelectric points reported on human prekallikrein vary from 7.7[64] to 8.3 to 9.4[66] in contrast to rabbit prekallikrein of 5.9 and bovine of 6.98. No amino acid analyses are available. Rabbit prekallikrein stains with PAS and is probably a glycoprotein, but no data are available for human prekallikrein.

Changes on Activation. No change in molecular weight by gel filtration is evident after activation of human prekallikrein with factor XII_a or XII_f. A small change in sedimentation constant from 5.2 to 5.05 occurs.[64] On regular disc or SDS disc electrophoresis in the absence of reducing agents, no change in molecular weight occurs. In contrast, rabbit prekallikrein on activation exhibits an increase in sedimentation constant from 4.2 to 4.5 (sucrose density gradient ultracentrifugation), an increase in the diffusion constant (gel filtration) from 4.46×10^{-7} cm^2/sec to 4.59, and a decrease in molecular weight from 100,000 to 88,000 (SDS electrophoresis). A fragment of 0.95 S and molecular weight 11,000 results. The isoelectric point of rabbit prekallikrein (5.9) shifts to 5.4 on activation.

[66] K. Laake and A. M. Venneröd, *Thromb. Res.* 4, 285 (1974).

Section V

Proteases from Gametes and Developing Embryos

[26] Proacrosin

By K. L. POLAKOSKI and L. J. D. ZANEVELD

Acrosin (EC 3.4.21.10) is located in or on the mammalian sperm acrosome, a lysosomelike structure located on the anterior portion of the sperm head. Acrosin is utilized by the spermatozoa for penetrating the zona pellucida of the ovum, which is a prerequisite for fertilization. This enzyme is present in all ten mammalian species studied to date including man.[1] At ejaculation, spermatozoa mix with seminal plasma and acquire low-molecular-weight proteinase inhibitors,[2,3] which prevent fertilization.[4,5] These inhibitors are removed or inactivated during the sperm's migration through the female reproductive tract (the process of sperm activation is called capacitation).

Boar acrosin has been purified to homogeneity as judged by multiple criteria including the demonstration of a single protein band on sodium dodecyl sulfate (SDS) gel electrophoresis[6,7] Acrosin is a serine endopeptidase which cleaves the carboxyl bond of arginyl and lysyl derivatives with a strong preference for arginine.[8] The acrosin of several species is partially present in a zymogen form called proacrosin.[9] Proacrosin has been partially purified from rabbit testicular tissue,[10,11] rabbit epididymal sperm,[12] and boar ejaculated sperm[9] and has been demon-

---

[1] For a recent review on sperm proteolytic enzymes, the reader is referred to L. J. D. Zaneveld, K. L. Polakoski, and G. F. B. Schumacher, *Cold Spring Harbor Symp. Proteinase Biol. Control*, p. 683 (1975).

[2] L. J. D. Zaneveld, P. N. Srivastava, and W. L. Williams, *J. Reprod. Fertil.* **20**, 337 (1969).

[3] L. J. D. Zaneveld, P. N. Srivastiva, and W. L. Williams, *Proc. Soc. Exp. Biol. Med.* **133**, 1172 (1970).

[4] L. J. D. Zaneveld, K. L. Polakoski, R. T. Robertson, and W. L. Williams, *Proteinase Inhibitors, Proc. Int. Res. Conf. 1st*, Munich, 1970, p. 236 (1971).

[5] L. J. D. Zaneveld, R. T. Robertson, M. Kessler, and W. L. Williams, *J. Reprod. Fertil.* **25**, 387 (1971).

[6] K. L. Polakoski, R. A. McRorie, and W. L. Williams, *J. Biol. Chem.* **248**, 8178 (1973).

[7] K. L. Polakoski and W. L. Williams, *Proteinase Inhibitors, Proc. Int. Res. Conf. 2nd (Boyer Symp. V) Grosse Ledder, 1973, p.* 128 (1974).

[8] K. L. Polakoski and R. A. McRorie, *J. Biol. Chem.* **248**, 8183 (1973).

[9] K. L. Polákoski, *Fed. Proc., Fed. Am. Soc. Exp. Biol.* **33**, 1308 (1974).

[10] S. Meizel, *J. Reprod. Fertil.* **31**, 459 (1972).

[11] S. Meizel and Y. H. J. Huang-Yang, *Biochem. Biophys. Res. Commun.* **53**, 1145 (1973).

[12] S. Miezel and Y. H. J. Huang-Yang, *Anat. Rec.* **175**, 387 (1974).

strated to be present in hamster epididymal sperm[13] and in human ejaculated sperm.[14] This report discusses the assay methods, purification procedures, and properties of boar proacrosin.

Assay Methods

Proacrosin is assayed by determining the amount of acrosin activity produced from the zymogen after activation. Acrosin hydrolyzes protein, amide and ester substrates containing arginyl bonds,[8] Acrosin hydrolysis of benzoyl-L-arginine ethyl ester (BAEE) is conveniently measured by following the initial increase in absorbance at 253 nm.[15] The substrate mixture consists of 0.05 M Tris-HCl containing 0.05 M calcium chloride and 0.5 mM BAEE at pH 8.0. A molar absorption difference of 1150 M^{-1} cm^{-1} [16] is used to convert the change in optical density to micromoles of BAEE hydrolyzed. The protein concentration is estimated by assuming that 1 mg of protein per milliliter exhibits an optical density (OD) of 1.0 at 280 nm in a 1-cm light path. Specific activity is defined as the micromoles of BAEE hydrolyzed per minute per milligram of protein.

Activation of Proacrosin in Crude Preparation

Pure proacrosin is autoactivated by maintaining the zymogen in solution at pH 8.0. This is not possible in the crude preparations because of the presence of a slight excess of proteinase inhibitors.[17] Such preparations are therefore incubated at pH 5.3 where zymogen activation occurs. Preliminary experimentation indicates the activation results from a previously undefined arginyl hydrolase which is active at pH 5.3 and is removed during the purification procedures. The presence of acrosin inhibitors also necessitates that the solutions be preincubated at pH 3.0 or below before addition to the BAEE. This ensures that the enzyme inhibitor complex is completely dissociated so that the total acrosin activity, not just the free enzyme, is measured.

The initial sperm extract is diluted to approximately 2 mg/ml with 0.001 M HCl and adjusted with 1.0 M HCl to a final pH of 2.8. An aliquot of the extract is assayed with BAEE at pH 8.0, which gives the

---

[13] S. K. Mukerji and S. Meizel, personal communications.
[14] K. L. Polakoski, unpublished data.
[15] G. W. Schwert and Y. Takenaka, *Biochim. Biophys. Acta* **16**, 570 (1955).
[16] J. R. Whitaker and M. L. Bender, *J. Am. Chem. Soc.* **87**, 2728 (1965).
[17] H. Fritz, W.-D. Schleuning, and W.-B. Schill, *Proteinase Inhibitors, Proc. Int. Res. Conf. 2nd (Boyer Symp. V) Grosse Ledder, 1973*, p. 118 (1974).

initial amount of active acrosin present in the preparation. To the remaining sample an equal volume of 0.1 M sodium acetate buffer containing 0.1 M calcium chloride at pH 5.3 is added. The incubation mixture is adjusted to 0.5% toluene or 0.5 mM potassium cyanide for retarding bacterial growth. The solution is subsequently incubated at 40° for 24 hr, at which point all the zymogen is normally activated. The sample is adjusted to pH 2.8 with 1 M HCl, and an aliquot is assayed at pH 8.0 with BAEE. By subtracting initial acrosin activity from the acrosin activity obtained after activation, the amount of acrosin that had previously been in the proacrosin form is determined. By this procedure, 55–95% of the total acrosin activity that is extractable from the acrosome was estimated to be proacrosin.

Activation of Purified Proacrosin

An aliquot is adjusted to pH 8.0 by the addition of an equal volume of buffer containing 0.1 M Tris-HCl, 0.1 M Ca^{2+}, and 1 mg of bovine serum albumin (BSA) per milliliter at pH 8.0. The sample is incubated at room temperature, and aliquots are removed for assay at various time intervals. The enzyme activation follows a typical S-shaped zymogen autoactivation curve.[18] Highly purified acrosin undergoes rapid autoproteolysis at pH 8.0. This autolysis is partially retarded by either calcium ions ar BSA and is almost completely retarded when both are present.

Purification Procedure

Step 1. The sperm-rich fraction of a fresh boar ejaculate (150–450 ml) is filtered through cheese cloth to remove large particulate matter. Twenty-milliliter aliquots are suspended over 25 ml of 1 M sucrose at pH 5.5 in a 50-ml centrifuge tube and centrifuged at 2500 g for 30 min. The supernatant is discarded, and the spermatozoa are recovered from the precipitate.[19] The spermatozoa are suspended in 10% glycerol and adjusted to a pH of 2.8 with 1 M HCl. During overnight incubation at 4°, the acrosomes separate from the spermatozoa, the acrosomal membranes become porous, and the enzymes are released.[20] The incubation

---

[18] B. Kassell and J. Kay, *Science* **180**, 1022 (1973).

[19] This washing procedure removes more than 99% of the seminal plasma and recovers at least 90% of the sperm: D. Keller and K. L. Polakoski, unpublished experiments.

[20] E. Fink, H. Schiessler, M. Arnhold, and H. Fritz, *Hoppe-Seyler's Z. Physiol. Chem.* **353**, 1633 (1972).

mixture is centrifuged at 27,000 g for 30 min, and the pellet is discarded. Proacrosin activity in the supernatant is measured under the conditions outlined for crude preparations. Unless otherwise stated, the following procedures are performed in plastic vials between 0° and 4°, at or below pH 3.0.

Step 2. The crude sperm acrosomal extract obtained in step 1 is diluted to an optical density of 1.0 with 1 mM HCl. Solid ammonium sulfate (Schwarz-Mann, ultra pure) is slowly added with stirring at 4° to a final ammonium sulfate concentration of 0.15% (7.5 g/100 ml). The pH is immediately readjusted to 3.0 with 1 M HCl, and the mixture is allowed to settle for at least 1 hr at 4°. The solution is centrifuged at 27,000 g, and the precipitate is discarded. The volume of the supernatant solution is measured. Solid ammonium sulfate is added to a final concentration of 0.35% (11.1 g/100 ml), and the pH is readjusted to 3.0. After 4–6 hr, the solution is centrifuged at 27,000 g and the supernatant is discarded. The precipitate containing proacrosin is suspended in a minimal amount of 0.001 M HCl and dialyzed against 2.1 of 0.1 M NaCl at pH 3.0 for 12 hr.

Step 3. The proacrosin containing fraction obtained in step 2 is applied to a standardized Sephadex G-200 column (2.5 × 65 cm) equilibrated in 0.1 M NaCl at pH 3.0. Fractions, 3.5 ml, are collected at a flow rate of 20 ml/hr. The eluted protein is monitored at OD 280, and the acrosin and proacrosin are assayed with BAEE as described previously. The fractions containing proacrosin emerge after the void volume peak. The samples with an activity greater than 50 units/mg are pooled and dialyzed against 2 liters of 0.05 M sodium acetate at pH 4.0 for 12 hr.

Step 4. The dialyzed protein solution is applied to a column (2.5 × 5 cm) of CM-52 cellulose that has been washed and equilibrated in 0.05 M sodium acetate, pH 4.0. The column is eluted with this same buffer until the OD at 280 nm of the eluent returns to the base line. A linear gradient of 0 to 0.3 M NaCl in 500 ml of 0.05 M sodium acetate is then applied to the column. The flow rate is 20 ml/hr. Three-milliliter fractions are collected, and the protein concentration is monitored by absorbance at 280 nm. The acrosin and proacrosin content are determined as described in the previous section. Proacrosin is eluted from the column between 0.05 and 0.15 M NaCl. The samples with activity greater than 235 units/mg are pooled and dialyzed against 2 liters of 0.001 M HCl, pH 3.0. A convenient method for concentrating the sample is to place the dialysis bag in a glass tray at 4° and to layer dry Sephadex G-200 upon the outside of the bag. This procedure has been previously described for purified acrosin.[6]

The results of the purification procedure are presented in the table.

PURIFICATION OF PROACROSIN

Step	Total protein (mg)	Total activity (units)	Specific activity (units/ mg)	Purifi- cation	Yield (%)
Sperm acrosomal extracts	69.5	201.0	2.9	0	100
Ammonium sulfate fractionation, 15–35%	22.2	129.5	5.8	2	64
Sephadex G-200 chromatography	1.73	110.9	64.2	22	55
CM-cellulose chromatography	0.315	77.3	245.3	85	38

Properties of Proacrosin

The above purification procedure results in a 38% yield of proacrosin with an activity between 235 and 274 units/mg, which compares favorably to the activity of 271 units/mg for homogeneous acrosin. The molecular weight is approximately 62,000, as estimated by gel filtration. The active site directed inhibitor tosyllysylchloromethyl ketone (TLCK) inhibits acrosin but not proacrosin after incubation for 1 hr at pH 8.0 and subsequent extensive dialysis at pH 3.0. Zymogen activation is apparently the result of partial proteolysis and is catalyzed by trypsin or pure acrosin at an enzyme-to-zymogen ratio of 1:20. Chymotrypsin, even at an enyzme-to-zymogen ratio of 1:4, will not activate proacrosin. Calcium is not required for the autoactivation of proacrosin. In fact, the activation is more rapid when calcium is not present.

Proacrosin present in the crude extracts is stable for at least one year below −20° and can be repeatedly frozen and thawed without noticeable loss in activity. The highly purified preparations of proacrosin are stable for several weeks at pH 3.0 at 4°, but substantial loss (25–40%) in activity occurs when the samples are frozen in a Dry Ice acetone bath and thawed at 23°.

Acknowledgments

The authors appreciate the generous support of a grant from the Rockefeller Foundation. The authors are indebted to Drs. Ronald Strickler and David Keller for advice and assistance in the preparation of this manuscript. We would also like to thank Miss Faye Zvibleman for typing the manuscript.

[27] Sperm Acrosin[1]

By WOLF-DIETER SCHLEUNING and HANS FRITZ

First indications of proteolytic activity associated with spermatozoa were reported by Yamane in 1930, who demonstrated that the effect of rabbit, rat, and horse spermatozoa on the second maturation division of ova was similar to that of pancreatin.[1a] The ability of sperm extracts to disperse the investment layers of eggs was shown by the same author.[2] The enzymic nature of the dissolution of the zona pellucida has been subsequently established by many authors.[3-7] In their *in vitro* studies they used mainly trypsin, but also chymotrypsin, papain, ficin, and Pronase for the removal of the zona pellucida from eggs of various mammalian species. In all cases investigated the zona was dissolved by trypsin. The discovery of a trypsinlike enzyme in spermatozoa[8-10] led to the conclusion that this enzyme might be responsible for the penetration ability of the spermatozoa through the zona layer.[11-13]

There is definite evidence that the trypsinlike sperm proteinase is located in the acrosome.[10,11,14-17] Following a suggestion of Zaneveld *et al.*,[18] it was named acrosin (EC 3.4.21.10).

Several methods have been described for the removal of acrosomes

[1] Work supported by Sonderforschungsberich 51 Munich and WHO, Geneva (grant No. 2873).

[1a] J. Yamane, *Cytologia* **1**, 394 (1930).

[2] J. Yamane, *Cytologia* **6**, 293 (1935).

[3] A. W. H. Braden, *Aust. J. Sci.* **5**, 460 (1952).

[4] C. R. Austin, *J. R. Microsc. Soc.* **75**, 141 (1956).

[5] M. C. Chang and D. M. Hunt, *Exp. Cell Res.* **11**, 497 (1956).

[6] R. B. L. Gwatkin, *J. Reprod. Fertil.* **7**, 99 (1968).

[7] Y. Toyoda and M. C. Chang, *Nature (London)* **220**, 589 (1968).

[8] L. M. Buruiana, *Naturwissenschaften* **43**, 523 (1956).

[9] M. Waldschmidt, B. Hoffman, and H. Karg, *Zuchthygiene* **1**, 15 (1966).

[10] P. N. Srivastava, C. E. Adams, and E. F. Hartree, *J. Reprod. Fertil.* **10**, 61 (1965).

[11] R. Stambaugh and J. Buckley, *Science* **161**, 585 (1968).

[12] K. L. Polakoski, L. J. D. Zaneveld, and W. L. Williams, *Biol. Reprod.* **6**, 23 (1972).

[13] L. J. D. Zaneveld, K. L. Polakoski, and W. L. Williams, *Biol. Reprod.* **6**, 30 (1972).

[14] R. Stambaugh and J. Buckley, *J. Reprod. Fertil.* **19**, 423 (1969).

[15] P. Gaddum and R. J. Blandau, *Science* **170**, 749 (1970).

[16] C. R. Brown and E. F. Hartree, *J. Reprod. Fertil.* **36**, 195 (1974).

[17] W.-B. Schill and H. H. Wolff, *Naturwissenschaften* **61**, 172 (1974).

[18] L. J. D. Zaneveld, K. L. Polakoski, R. T. Robertson, and W. L. Williams, *Proteinase Inhibitors, Proc. Int. Res. Conf., 1st,* Munich, 1970, p. 236. de Gruyter, Berlin, 1971.

from spermatozoa[12,19-22] and for the partial purification of acrosins from various species.[12,20,23-27] Starting from acrosomal extracts, thus far only small amounts of electrophoretically homogeneous acrosin preparations have been obtained from rabbit,[12,13] boar,[28,29] ram,[30] bull[31] and human[32,33] spermatozoa. Reviews on this subject have been recently published.[34-37] The purification method described below is generally applicable to the isolation of acid-stable acrosins and yields in a few simple steps the highly purified proteinase. Using this procedure boar,[28] human[32] and hamster[38] acrosin have thus far been purified on an analytical and preparative scale.

Assay Methods

Estimation of Proteolytic Activity

Principle. Formation of trichloroacetic acid-soluble peptides by enzymic degradation of azocasein is followed spectrophotometrically at 366 nm.[39,40]

[19] E. F. Hartree and P. N. Srivastava, *J. Reprod. Fertil.* **9**, 47 (1965).
[20] E. Fink, H. Schiessler, M. Arnhold, and H. Fritz, *Hoppe-Seyler's Z. Physiol. Chem.* **353**, 1633 (1972).
[21] B. W. Morton, *J. Reprod. Fertil.* **15**, 113 (1968).
[22] H. Pedersen, *J. Reprod. Fertil.* **31**, 99 (1972).
[23] S. Multamäki and M. Niemi, *J. Reprod. Fertil.* **17**, 43 (1972).
[24] J. J. L. Ho and S. Meizel, *J. Reprod. Fertil.* **23**, 177 (1970).
[25] L. J. D. Zaneveld, B. M. Dragoje, and G. F. B. Schumacher, *Science* **177**, 702 (1972).
[26] D. L. Garner, *Biol. Reprod.* **9**, 71 (1973).
[27] H. Fritz, B. Förg-Brey, E. Fink, M. Meier, H. Schiessler, and C. Schirren, *Hoppe-Seyler's Z. Physiol. Chem.* **353**, 1943 (1972).
[28] W.-D. Schleuning, H. Schiessler, and H. Fritz, *Hoppe-Seyler's Z. Physiol. Chem.* **354**, 550 (1973).
[28a] W.-D. Schleuning and H. Fritz, *Hoppe-Seyler's Z. Physiol. Chem.* **355**, 125 (1974).
[29] K. L. Polakoski, R. A. McRorie, and W. L. Williams, *J. Biol. Chem.* **248**, 8178 (1973).
[30] D. Morton, personal communication (1975).
[31] D. L. Garner and R. F. Cullison, *J. Chromatogr.* **92**, 445 (1974).
[32] W.-D. Schleuning, R. Hell, and H. Fritz, unpublished results (1975).
[33] E. Gilboa, Y. Elkana, and M. Rigbi, *Eur. J. Biochem.* **39**, 85 (1973).
[34] H. Fritz, H. Schiessler, and W.-D. Schleuning, *Adv. Biosci.* **10**, 271 (1973).
[35] R. A. McRorie and W. L. Williams, *Annu. Rev. Biochem.* **43**, 777 (1974).
[36] L. J. D. Zaneveld, K. L. Polakoski, and G. F. B. Schumacher, *in* "Proteases and Biological Control" (E. Reich, D. B. Rifkin, and E. Shaw, eds.), p. 683. Cold Spring Harbor Laboratory, Cold Spring Harbor, New York (1975).
[37] H. Fritz, W.-D. Schleuning, W.-B. Schill, H. Schiessler, V. Wendt, and G. Winkler, *in* "Proteases and Biological Control" (E. Reich, D. B. Rifkin, and E. Shaw, eds.), p. 715. Cold Spring Harbor Laboratory, Cold Spring Harbor, New York (1975).
[38] E. Fink, R. B. L. Gwatkin, and H. Fritz, unpublished results (1975).
[39] H. Fritz, I. Trautschold, and E. Werle, *in* "Methoden der Enzymatischen Analyse" (H. U. Bergmeyer, ed.), 2nd ed., p. 1021. Verlag Chemie, Weinheim, 1970.
[40] M. Kunitz, *J. Gen. Physiol.* **30**, 291 (1947).

Reagents

Phosphate buffer: 0.1 M sodium phosphate, pH 7.6

Azocasein solution, 2%: 2 g of azocasein dissolved in 500 ml of phosphate buffer under heating to 80° and stirring

Acrosin solution: an aliquot of the lyophilized sample dissolved in 1 mM HCl

Stability of the Solutions. The phosphate buffer is stable for 3 days and the acrosin solution for 1 week at 4°. The azocasein solution has to be prepared daily.

Procedure. Acrosin is adsorbed and inactivated on glass and quartz surfaces.[41] Therefore, plastic material or siliconized glass and quartz ware should be used exclusively. The buffer and substrate solution are brought to 30° prior to assay.

Acrosin solution, 0.1–0.2 ml, is pipetted in plastic tubes; 0.1–0.8 ml of buffer is added, yielding a final volume of 1 ml. This mixture is incubated at 30° for 5 min. Then 2 ml of the azocasein solution are added, the sample is mixed and incubated again for 10 min at 30°. The enzymic reaction is stopped by addition of 2.0 ml 5% (w/v) trichloroacetic acid. After mixing, the sample is allowed to stand at room temperature for 30 min. The precipitate formed is removed by centrifugation. The absorption of the supernatant is read at 366 nm against the solution of a blank sample in which the trichloroacetic acid was added prior to incubation at 30°.

One enzyme unit is defined as the increase in absorbance, ΔA, of 1.0/min under the conditions employed[40]; the specific activity, as the number of units per milligram of protein.

Assay of Esterase Activity

Principle. The hydrolysis of N^α-benzoyl-L-arginine ethyl ester (BAEE) is followed photometrically by measuring the increase in absorbance at 253 nm.[42] A considerably higher sensitivity is attained if the liberated ethanol is estimated photometrically at 366 nm with the aid of alcohol-dehydrogenase, NAD, and semicarbazide.[43] Both methods are described below. Alternatively, the hydrolysis of BAEE may be followed titrimetrically using a pH stat.[44]

---

[41] H. Schiessler, H. Fritz, M. Arnhold, E. Fink, and H. Tschesche, *Hoppe-Seyler's Z. Physiol. Chem.* **353**, 1638 (1972).

[42] G. W. Schwert and Y. Takenaka, *Biochim. Biophys. Acta* **16**, 570 (1955).

[43] I. Trautschold and E. Werle, *Hoppe-Seyler's Z. Physiol. Chem.* **325**, 48 (1961).

[44] K. A. Walsh and P. E. Wilcox, this series Vol. 19, p. 37.

Direct BAEE Assay (253 nm)

Reagents

Tris buffer: 0.1 M Tris-HCl, pH 8.0
Substrate solution: 6 mM BAEE-HCl in distilled water
Acrosin solution: 0.5–10 μg acrosin dissolved in 1 ml of 1 mM HCl

Stability of the Solutions. Stability of Tris buffer and acrosin solution lasts 1 week, and of substrate solution, 2 days at 4°.

Procedure. Plastic ware or siliconized glass and quartz cuvettes have to be used (cf. the foregoing assay procedure). Tris buffer and substrate solution are brought to 25° prior to assay.

Tris buffer (2.3–2.45 ml), acrosin solution (0.05–0.2 ml), and the substrate solution (0.5 ml) are rapidly mixed in a temperature-controlled (25°) 3-ml cuvette of 1 cm light path. The rise in absorbance at 253 nm is read every minute over a period of 10 min against a blank sample containing only Tris buffer and the substrate solution. If the change in absorbance exceeds 0.01 OD/min, the estimation should be repeated applying smaller amounts of acrosin.

One acrosin unit (U) corresponds to the hydrolysis of 1 μmole of substrate per minute, i.e., an increase in absorbance of $\Delta A = 0.385$/min. The specific activity is expressed in units per milligram of protein.

Combined BAEE/ADH Assay (366 nm)

Reagents

Pyrophosphate buffer: 15.2 g of $Na_4P_2O_7$, 4.1 g of semicarbazide, and 0.9 g of glycine are dissolved in 450 ml of distilled water. The pH is adjusted to 8.7 by addition of 2 N NaOH, and the volume brought to 500 ml with distilled water.
NAD solution: 60 mg of NAD dissolved in 3 ml of distilled water
ADH solution: 100 mg of crystallized ADH suspended in 3.4 ml of distilled water
BAEE solution: 6 mM in distilled water
Acrosin solution: 0.5–10 μg acrosin dissolved in 10 ml 0.001 M HCl

Stability of the Solutions. Pyrophosphate buffer and NAD solution are stable for 1 week at 4°, ADH suspension for 1 month at 4°, and BAEE solution for 2 days at 4°.

Procedure. The mixture of 2.3 ml of pyrophosphate buffer, 0.1 ml of NAD solution, 0.02 ml of ADH suspension, and 0.5 ml of BAEE solution is incubated at 25° for 5 min in a 3-ml cuvette (plastic or siliconized glass or quartz) of 1-cm light path. Subsequently, the acrosin solution

(0.1 ml) is admixed and the increase of absorbance at 366 nm is followed for 10 min. The optical density is read against a blank sample containing all reagents except acrosin.

One acrosin unit (U) corresponds to an increase in absorbance of $\Delta A = 1.1/\text{min}$. If the change in absorbance exceeds 0.01 OD/min the estimation should be repeated applying smaller amounts of acrosin.

Assay of Amidase Activity

Principle. The rate of hydrolysis of the trypsin substrates N^{α}-benzoyl-DL-arginine p-nitroanilide[20,45] (DL-BAPA), or preferably N^{α}-benzoyl-L-arginine p-nitroanilide (L-BAPA) or N^{α}-benzoyl-DL-lysine p-nitroanilide[41] (BLNA) is followed photometrically at 405 nm.

Reagents

Triethanolamine (TRA)-HCl buffer: 0.2 M TRA-HCl, pH 7.8
L-BAPA solution: 1.15 mM in distilled water
DL-BLNA solution: 6.0 mM in distilled water
Acrosin solution: 7–15 μg of acrosin in 1 ml of 1 mM HCl

Stability of the Solutions. At 4° TRA buffer is stable for 1 week, substrate solutions for several weeks, and acrosin solution for 1 week.

Procedure. TRA buffer and the substrate solution are brought to 25° prior to assay. TRA buffer (1.8–1.99 ml), the acrosin solution (0.2–0.01 ml), and the substrate solution (1.0 ml) are mixed in a temperature-controlled (25°) 3-ml cuvette (plastic, siliconized glass, or quartz) of 1-cm light path. The increase in absorbance at 405 nm is read every minute for 10 min.

One acrosin unit (U) corresponds to an increase in absorbance of $\Delta A = 3.3/\text{min}$. In the range of $\Delta A = 0.002–0.060/\text{min}$ using L-BAPA and of $\Delta A = 0.002–0.080/\text{min}$ using DL-BLNA, the observed change in absorbance is proportional to the amount of acrosin employed.

Active-Site Titration Using p-Nitrophenyl p'-Guanidinobenzoate (NPGB)

Principle. NPGB is used as an "initial burst" reagent for the titration of trypsin, plasmin, and thrombin.[46] Like trypsin and other related enzymes, acrosin is rapidly and stoichiometrically p-guanidinobenzoylated by NPGB, and this reaction is accompanied by a burstlike liberation of p-nitrophenol. The deacylation reaction proceeds very slowly. Hence, the

---

[45] B. F. Erlanger, N. Kokowski, and W. Cohen, *Arch. Biochem. Biophys.* **95**, 271 (1961).

[46] T. Chase, Jr., and E. Shaw, *Biochem. Biophys. Res. Commun.* **29**, 508 (1967).

number of active acrosin molecules in a given solution can be measured photometrically at 402 nm by the amount of p-nitrophenol liberated in the burst reaction.

Reagents

Veronal buffer: 0.1 M sodium Veronal, pH 8.3

NPGB solution: 0.05 M (16.8 mg/ml) NPGB-HCl in dimethyl-formamide, diluted with four volumes of acetonitrile: 0.01 M NPGB

Acrosin solution: 0.2–2 mg of acrosin dissolved in 1 ml of Veronal buffer

Stability of the Solutions. At 4°, Veronal buffer is stable for 1 week and NPGB solution for 2 weeks. The acrosin solution has to be freshly prepared prior to the assay.

Procedure. The acrosin solution, 1.0 ml, is pipetted into a siliconized 1-ml quartz cuvette of 1-cm light path. Five microliters of the NPGB solution are rapidly admixed, and the absorbance at 402 nm is read for 10 min in 30-sec intervals. A blank sample containing Veronal buffer instead of the acrosin solution serves as the reference.

The very slow postburst liberation of p-nitrophenol is extrapolated back to time zero to give a measure for the initial burst. The increase in absorbance during the initial burst multiplied by 5.46×10^{-5} yields the molarity of liberated p-nitrophenol and therefore the molarity of active acrosin.

Comparison and Evaluation of the Assay Methods

Determination of acrosin activity in sperm extracts is complicated by the presence of strong competitive acrosin inhibitors.[20,27,41,47] In order to get reliable results, the acidic or acidified (pH 2.0–2.7) sperm extract, in which the acrosin inhibitor complex is dissociated, has to be directly applied to the substrate-containing test cuvette in the combined BAEE/ADH assay.[20] The effective concentration of acrosin in this test system is low enough ($\sim$0.3 nM) so that complex formation in the presence of the substrate is not observed during the assay period.[20] With this method a clear correlation was found between sperm count and acrosin activity.[48]

Most of the activity and kinetic data published previously were obtained with DL-BAPA as boar acrosin substrate. However, L-BAPA

[47] L. J. D. Zaneveld, P. N. Srivastava, and W. L. Williams, *Proc. Soc. Exp. Biol. Med.* 133, 1172 (1970).

[48] W.-B. Schill, *Arch. Dermatol. Forsch.* 248, 257 (1973).

now commercially available should be used in preference because D-BAPA acts as a competitive acrosin inhibitor.[49] Nevertheless, under the conditions described above, activities estimated using DL-BAPA (1 mg) are not significantly different from values obtained using 0.5 mg of L-BAPA. Owing to the higher acrosin concentrations required in the DL- and L-BAPA assay, acrosin activities cannot be estimated in the presence of inhibitors.

DL-BLNA is a suitable substrate for acrosin estimation if sufficient amounts of enzyme (without contamination by an inhibitor) are available. In this case the hydrolysis rate is constant over a long period (up to 45 min); proportionality between the amount of enzyme applied and the activity measured is observed over a wide range (1–35 mU), and inactivation effects due to glass adsorption are minimized. The use of this substrate is also recommended if inhibitors with relatively low affinities to acrosin are to be estimated.

The number of active molecules present in a given acrosin preparation may be precisely measured by active site titration with NPGB. The greatest disadvantage, high consumption of enzyme, might be overcome in the future by using fluorescent probes.[50]

Purification Procedure

Step 1. Collection of Semen. Boar semen is collected using an artificial vagina and filtered through gauze. Human semen is obtained from healthy donors. Epididymal spermatozoa, e.g., from hamsters, are collected by rinsing the epididymis with physiological saline.

The semen samples are centrifuged at 600 g for 10 min immediately after collection or liquefaction of the semen samples. The sperm pellet is resuspended twice in saline and centrifuged again to remove substances that are adsorbed unspecifically. Afterward the pellet is frozen at −196° and stored at −40°.

Step 2. Extraction of Spermatozoa. The frozen sperm pellet is thawed at room temperature and suspended in twice its volume of 2% (v/v) acetic acid at 4°. The pH of the suspension is adjusted to 2.5 (boar, human) or 2.7 (hamster, bull) by dropwise addition of 2 N HCl. After gently stirring for 20 min at 4°, the suspension is centrifuged at 4000 g, 4°. This extraction procedure is repeated five times. The combined extracts are centrifuged at 20,000 g for 90 min, 4°. The supernatant is

[49] W.-D. Schleuning, R. Hell, and H. Fritz, unpublished results (1974).
[50] D. V. Roberts, R. W. Adams, D. T. Elmore, G. W. Jameson, and W. S. A. Kyle, *Biochem. J.* **123**, 41 (1971).

concentrated by ultrafiltration to 10% of its original volume in an Amicon cell supplied with an UM-10 membrane, 4°.

Step 3. Separation of Acrosin Inhibitors. The acidic acrosin solution is applied to a Sephadex G-75 column equilibrated and developed with aqueous acetic acid (2%, v/v), pH 2.7, 4°. A 130 × 5 cm column which is developed at a rate of 30 ml/hr, 7.5 ml/tube is suitable for fractionation of 15–20 ml of concentrated acrosin solution. A complete separation of the acrosin fraction (tubes 60–80) from the inhibitor fraction(s) is thus achieved. Whereas only one inhibitor fraction is obtained on fractionating boar sperm extracts, two trypsin inhibitor fractions are separated if acidic human or bull sperm extracts are applied.

One-tenth milliliter of 1 M triethanolamine/HCl buffer, pH 7.8, is added per 1 ml of the acidic acrosin solution obtained by gel filtration. This mixture is adjusted to pH 7.8 with 2 N NaOH. Protein that precipitates between pH 6 and 7 is separated by centrifugation. A 1.0 M NaCl concentration is reached by addition of the calculated amount of sodium chloride crystals.

Step 4. Affinity Chromatography. We used either p-aminobenzamidine succinyl aminohexyl cellulose or p-aminobenzamidine CH-Sepharose 4B as water-insoluble affinity adsorbents. The cellulose adsorbent was purchased from E. Merck, Darmstadt, article No. 10 778. The Sepharose adsorbent was prepared by coupling p-aminobenzamidine (Cyclochemical) to CH-Sepharose 4B (Pharmacia) by the aid of N-cyclohexyl-3-(2-morpholinoethyl)-carbodiimide metho-p-toluenesulfonate (EGA Chemie KG, Heidenheim, Germany): 5 g of CH-Sepharose 4B are swollen in 0.1 M 2-(n-morpholino)ethane sulfonic acid buffer, pH 4.75 (MES-buffer); excess solution is discarded. A solution of 5 g of the carbodiimide derivative in 50 ml of the MES-buffer is added to the moistened gel. The mixture is gently shaken for 30 min at 25°. Subsequently, 1 g of crystalline p-aminobenzamidine-HCl is added and the mixture shaken for 5 hr. The affinity adsorbent thus obtained is extensively washed with the MES-buffer until p-aminobenzamidine not covalently bound is completely removed.

The Sepharose affinity adsorbent is filled into a cooled (4°) plastic column (20 × 2 cm) and equilibrated with 0.05 M TRA/HCl, 0.5 M NaCl, pH 7.8 (TRA/HCl/NaCl-buffer) at a rate of 12 ml/hr. The affinity gel is loaded with acrosin by connecting the column to a cooled (2°) plastic vessel containing the acrosin solution pH 7.8 from step 3. Acrosin, 10–250 U (substrate: L-BAPA), may be applied to the column. The affinity gel is washed subsequently with the equilibration buffer until the transmission of the eluate at 253 nm remains constant. In order to remove nonvolatile salts, the column is now equilibrated with 0.1 M

TABLE I
ISOLATION OF BOAR ACROSIN

Procedure	Step	Specific activity (U^a/mg)	Yield (%)	Purification factor
Acidic extraction (H$^+$)	1	0.05–0.10^b	—	—
Gel filtration (H$^+$)	2	0.5 –1.8	—	5–18
Neutralization, centrifugation	2a	0.6 –2.2	95–100	1– 2
Affinity chromatography	3	14.5	80– 90	7–24
Desalting	4	14.5	95–100	1
Lyophilization	4a	12.8–13.8 (BAEE:165)	92– 95	

^{a} Substrate: N^α-benzoyl-DL-arginine p-nitroanilide.
^{b} Approximate values (inhibitors are present!).

NH$_4$HCO$_3$, pH 7.8. Dissociation of the acrosin–inhibitor complex, and thus elution of acrosin from the column, is achieved with 0.01 M ammonium formate, pH 3.0. Acrosin appears in the eluate near pH 3.5 (10- to 30-fold concentrated compared with the solution applied to the column).

The procedure used for the isolation of boar acrosin employing the cellulose-affinity adsorbent is described in detail by Schleuning et al.[28] In principle, the same degree of purity of acrosin is reached by both methods.

Step 5. Stabilization and Desalting. The acrosin fraction eluted from the Sepharose affinity column in ammonium formate buffer is mixed with sucrose to a final concentration of 1.5% (w/v), frozen in a siliconized glass flask and lyophilized.

The acrosin fraction eluted from the cellulose affinity column has to be desalted by gel chromatography. We employed a 70 × 2 cm Merckogel PGM-2000 column equilibrated and developed with 0.1% (v/v) acetic acid at a rate of 30 ml/hr, 7.5 ml/tube, 4°. The salt free acrosin fraction thus obtained is mixed with sucrose to a final concentration of 1.5%, frozen in a siliconized glass flask, and lyophilized. A summary of the purification procedure is given in Table I.

Properties

Purity. Boar acrosin purified by the described procedure shows a single band in polyacrylamide gel electrophoresis at pH 4.6 employing a 10% gel (system No. 8 of Maurer).[51] After reduction of the disulfide

[51] H. R. Maurer, "Disc Electrophoresis and Related Techniques of Polyacrylamide Gel Electrophoresis," p. 46. de Gruyter, Berlin, 1971.

bridges with 2-mercaptoethanol, usually one band is visible on the gel after SDS-polyacrylamide electrophoresis performed according to Weber *et al.*[52] Some preparations, however, showed one or two additional bands in various amounts, indicating the presence of enzymically modified acrosin molecules in the preparation applied. This interpretation is supported by the results of the end-group analysis. Only one single *N*-terminal amino acid residue (alanine) was detectable for the major component always present, whereas besides alanine also valine and methionine were found in acrosin preparations containing three bands in SDS gel electrophoresis.[49]

Human acrosin isolated by the same procedure was separated into one to four (depending on the activation state) gelatin-digesting fractions in acrylamide gel electrophoresis at pH 4.6. In SDS-polyacrylamide gel electrophoresis different molecular weights were found for these acrosin fractions. The high-molecular-weight form could be converted into more low molecular weight forms by the action of trypsin, kallikrein, and acrosin itself indicating the possibility of self-activation or protease-induced activation of a zymogen precursor. Details of these recent findings will be published elsewhere.[53]

Small amounts of acrosin were isolated by the same procedure from epididymal hamster spermatozoa.[38]

Stability. Acrosins from various species (boar,[28,29] human,[32,33,54] hamster,[38] bull,[23,31] and rabbit[12]) investigated so far are stable in acidic solutions near pH 3 (boar and human acrosin up to pH 2.3) at 4° for several weeks. In solutions at pH 6.5 to 8, boar and human acrosin is stable at 4° for at least 24 hr. The rapid initial loss of acrosin activity in the neutral pH region reported by other authors[11-13] may have several explanations: the presence of inhibitors, adsorption to glass surfaces, self-digestion of acrosin molecules which had been enzymically modified (and thus labilized) during detergent extraction. Acrosin is irreversibly inactivated in solutions of pH higher than 10.

As already stated above, contact of acrosin solutions with glass should be strictly avoided after separation of the inhibitors (step 3). The loss of activity is especially high, if diluted solutions of highly purified acrosin come into contact with glass or quartz surfaces.

During lyophilization of acrosin in the presence of sucrose in siliconized glass flasks a loss of 10–15% of the activity applied is observed. The lyophilized material showed no detectable loss of activity during several months at 4°.

[52] K. Weber, J. R. Pringle, and M. Osborn, this series Vol. 26, p. 3.
[53] W.-D. Schleuning, R. Hell, and H. Fritz, unpublished results (1974).
[54] F. N. Syner, personal communication (1974).

Molecular Weight. For the main component of boar acrosin, a molecular weight of approximately 38,000 was estimated by SDS gel electrophoresis.[28a] Some preparations contained additional fractions with molecular weights of 37,000 and 34,000, or even lower. This additional fractions were mainly obtained from stored ejaculates, thus indicating the occurrence of partially degraded acrosin molecules. Enzymic degradation of acrosin may be caused by other sperm proteinases.

Polakoski *et al.*[29] used a combination of detergent extraction near neutral pH, ion-exchange chromatography, and gel chromatography for the purification of boar acrosin. They found a molecular weight of about 30,000 for their preparation by SDS gel electrophoresis. This considerable difference from the value reported by us, as well as the pronounced instability of their acrosin preparation, might be a consequence of a more extensive enzymic degradation of acrosin during detergent extraction (90 min, 37°).

Garner and Cullison[31] found molecular weights of 44,000, 37,000, and 34,000 for the multiple forms of bull acrosin purified by affinity chromatography employing benzoylarginyl-glycyl-glycyl-tyrosyl-agarose. A molecular weight of about 30,000 was reported for rabbit acrosin.[13]

The human acrosin preparation isolated by the described procedure contained four gelatin-digesting fractions in variable amounts, the ratio of these fractions obviously depending on the actual activation state of acrosin. A molecular weight of about 67,000 was estimated by SDS gel electrophoresis for component I, which represents perhaps the zymogen precursor. The molecular weight of fraction IV, the main product of the activation process, seems to be only slightly different from the value found for boar acrosin (38,000).[32]

Kinetic Data. K_m values reported by various authors employing diverse substrates and acrosin preparations of different degree of purity are given in Table II. Multiple forms of acrosin and the variable activation state of acrosin might be responsible for the considerable differences observed. Therefore, only acrosin preparations of defined purity should be employed for investigations concerned with kinetic measurements.

Acrosin is inhibited not only by inhibitors occurring in male genital tract secretions, but also by nearly all trypsin inhibitors of protein nature found thus far in animals, including humans and plants. Strong competitive acrosin inhibitors of low molecular weight are the microbial peptides leupeptin and antipain.[55] The bimolecular velocity constants for the inhibition of boar acrosin by diisopropyl fluorophosphate and

[55] H. Fritz, B. Förg-Brey, and H. Umezawa, *Hoppe-Seyler's Z. Physiol. Chem.* **354**, 1299 (1973).

TABLE II
MICHAELIS CONSTANTS OF ACROSINS[a]

Substrate	Rabbit	Boar	Human
L-BAEE	$5.2 \times 10^{-6} \ M^b$	$4.8 \times 10^{-5} \ M^c$	$2.8 \times 10^{-5} \ M^f$
		$2.7 \times 10^{-4} \ M^d$	
DL-BAPA	$10.9 \times 10^{-3} \ M^b$	$3.9 \times 10^{-4} \ M^e$	
DL-BLNA		$6.5 \times 10^{-4} \ M^e$	

[a] The considerable differences observed may be due to different activation stages of the acrosins employed. Data obtained with N^α-benzoyl-DL-arginine p-nitroanilide (DL-BAPA) as substrate are questionable because D-BAPA acts as a potent inhibitor of acrosin. L-BAEE, N^α-benzoyl-L-arginine ethyl ester; DL-BLNA, N^α-benzoyl-DL-lysine p-nitroanilide.
[b] R. Stambaugh and J. Buckley, *Biochim. Biophys. Acta* **284**, 473 (1972).
[c] W.-D. Schleuning and H. Fritz, *Hoppe-Seyler's Z. Physiol. Chem.* **355**, 125 (1973).
[d] K. L. Polakoski and R. A. McRorie, *J. Biol. Chem.* **248**, 8183 (1973).
[e] H. Schiessler, H. Fritz, M. Arnhold, E. Fink, and H. Tschesche, *Hoppe-Seyler's Z. Physiol. Chem.* **353**, 1638 (1972).
[f] E. Gilboa, Y. Elkana, and M. Rigbi, *Eur. J. Biochem.* **39**, 85 (1973).

TABLE III
AMINO ACID COMPOSITION OF BOAR ACROSIN[a]

Amino acid	Residues/molecule (approximate values)
Asp	29
Thr	19
Ser	23
Glu	30
Pro	35
Gly	41
Ala	22
Cys	6
Val	23
Met	3
Ile	19
Leu	25
Tyr	11
Phe	9
Lys	19
His	6
Arg	22
Trp	ND[b]

[a] Boar acrosin was purified by affinity chromatography as described in the text and rechromatographed on Sephadex G-100. A 20-hr hydrolyzate was analyzed.
[b] ND, not determined.

N^a-toluene sulfonyl-L-lysyl chloromethyl ketone are comparable to those found for the inhibition of trypsin.[41]

Amino Acid Composition. Preliminary values obtained in the amino acid analysis of boar acrosin are given in Table III. A similar high content of proline was also found in the light chain of plasmin. In contrast to the results published recently by Stambaugh and Smith,[56] no similarity to the amino acid composition of trypsin is to be seen.

Both boar and human acrosin are glycoproteins and, therefore, strongly bound to concanavalin-A Sepharose.[28a,32] On the basis of the amount of glucosamine found in amino acid analysis, the carbohydrate portion of the boar acrosin molecule seems to be relatively low (below 10%).

Enzymic Splitting Specificity. The splitting specificity of the acrosins isolated thus far is very similar to that of trypsin. Synthetic ester and amide derivatives of arginine and lysine are, therefore, suitable substrates for acrosin. Like trypsin, acrosin is a potent kininogenase.[57] Employing equimolar amounts of bovine trypsin and boar acrosin for the digestion of the oxidized β-chain of insulin and of reduced carboxymethylated ribonuclease the same number of arginyl and lysyl bonds was hydrolyzed with similar rates by both enzymes.[58] Hence, the strong preference of acrosin for the hydrolysis of arginyl bonds found by other authors[29] could not be confirmed. The striking similarity of acrosin and trypsin in the splitting specificity and the affinity to natural and synthetic inhibitors suggests that the enzymic reaction mechanisms, including the shape of the active sites of both enzymes, are nearly identical.

Note Added in Proof. Further characteristics of sperm acrosins have recently been published.[59-63]

[56] R. Stambaugh and M. Smith, *Science* **186**, 745 (1974).
[57] S. Palm and H. Fritz, unpublished results (1975).
[58] H. Schiessler, W.-D. Schleuning, and H. Fritz, unpublished results (1974).
[59] H. Scheissler, W.-D. Schleuning, and H. Fritz, *Hoppe-Seyler's Z. Physiol. Chem.* **356**, 1931 (1975).
[60] W.-D. Schleuning, R. Hell, H. Schiessler, and H. Fritz, *Hoppe-Seyler's Z. Physiol. Chem.* **356**, 1915 (1975).
[61] W.-D. Schleuning, H. J. Kolb, R. Hell, and H. Fritz, *Hoppe-Seyler's Z. Physiol. Chem.* **356**, 1923 (1975).
[62] C. R. Brown, Z. Andani, and E. F. Hartree, *Biochem. J.* **149**, 133 (1975).
[63] C. R. Brown, Z. Andani, and E. F. Hartree, *Biochem. J.* **149**, 147 (1975).

[28] Cortical Granule Proteases from Sea Urchin Eggs

By Edward J. Carroll, Jr.

Sea urchin eggs contain protease activity in the cortical granules, a set of secretory organelles adjacent to the plasma membrane.[1,2] After the fusion of cortical granule and plasma membranes at fertilization, the contents of the cortical granules are extruded into the space between the plasma membrane and the overlying vitelline layer. The vitelline layer is semipermeable, as part of the protease activity is found in the supernatant seawater. This protease activity consists of at least two enzymes which have two separable functions *in vivo*.[3] One enzyme, termed sperm receptor hydrolase, modifies the sperm-binding properties of eggs and is important in removing supernumerary sperm-binding sites and spermatozoa as a part of the block against polyspermy. Another enzyme, vitelline delaminase, functions in the first step in elevation of the extracellular vitelline layer by cleaving vitelline layer–plasma membrane attachments. Actual elevation of this layer requires additional cortical granule components.[4]

Assay Methods

The protease activity of the sea urchin egg cortical granule exudate can be determined qualitatively using two bioassays or quantitatively using a radioactive protein or synthetic ester substrate assay.

Bioassays

As noted, two biological activities attributed to the egg cortical granule exudate relate to detachment of the vitelline layer (vitelline delaminase) and destruction of sperm receptors (sperm receptor hydrolase).

Vitelline Delaminase

Principle. Eggs incubated in solutions containing vitelline delaminase activity will elevate a fertilization membrane in the presence of soybean trypsin inhibitor when activated with butyric acid or the divalent cation

[1] V. D. Vacquier, D. Epel, and L. A. Douglas, *Nature (London)* **237**, 34 (1972).
[2] H. Schuel, W. L. Wilson, R. S. Bressler, J. W. Kelly, and J. R. Wilson, *Dev. Biol.* **29**, 307 (1973).
[3] E. J. Carroll, Jr., and D. Epel, *Dev. Biol.* **44**, 22 (1975).
[4] E. J. Carroll, Jr., and D. Epel, *Exp. Cell Res.* **90**, 429 (1975).

ionophore A23187.[4] Control eggs incubated in the absence of this activity exhibit a fertilization membrane that is "blebbed" or has a rosette appearance. Thus this protease activity of the cortical granule exudate functions in modifying connections in the fertilizing egg cell between the plasma membrane and vitelline layer such that normal elevation of the vitelline layer can occur. The ionophore procedure of egg activation is detailed here, as it is the easiest, and most reproducible, method; the butyric acid method of egg activation has been described.[5]

Reagents

Ionophore A23187 (kindly provided by Dr. R. Hamill of the E. Lilly Co., Indianapolis): 5 mM in dimethyl sulfoxide. Always store in a dark, stoppered bottle covered with foil and refrigerate when not in use.

Soybean trypsin inhibitor solution: 3.7 mg/ml in seawater (Sigma, chromatographically purified)

Procedure. A stock egg suspension is prepared containing approximately 500 cells/ml. For the assay, 200 μl of this suspension are pipetted (wide-bore pipette) into 27-mm Syracuse watch glasses (A. H. Thomas Co.) together with 50 μl of seawater or test solution (in seawater). The watch glasses are placed in a covered petri dish in a water bath at 16°. After incubation for 15–60 min, 25 μl of soybean trypsin inhibitor solution and 5 μl of ionophore solution are added with rapid mixing. The cells are then examined at ×400 magnification and scored for the presence of elevated fertilization membranes.

Sperm Receptor Hydrolase

Principle. After incubation of unfertilized eggs in solutions containing this activity, they will not bind sperm, nor will they fertilize. This latter result forms the basis for a bioassay of sperm receptor hydrolase activity.[6]

Reagents. No special reagents are required.

Procedure. A stock egg suspension containing approximately 500 cells/ml is prepared. For each assay, 200 μl of this suspension are pipetted (wide-bore pipette) into 27-mm Syracuse watch glasses. Then 50 μl of test solution (in seawater) or seawater are added; watch glasses are placed in a covered petri plate and then placed in a water bath at 16°. After incubation for 15–60 min, the eggs are inseminated with 50 μl of a sperm suspension prepared by diluting "dry" semen to an absorbance

[5] J. Bryan, *J. Cell Biol.* **45**, 606 (1970).
[6] V. D. Vacquier, M. J. Tegner, and D. Epel, *Exp. Cell Res.* **80**, 111 (1973).

at 340 nm of 1.0. This sperm suspension contains approximately 1.2×10^8 cells/ml.[7] To score percentage of fertilization, the eggs are examined at 20 min for the presence of hyaline layers and elevated fertilization membranes and again at 90 min after insemination for cleavage.

Chemical Assays

Protease Activity

Principle. The extent of proteolysis of radiolabeled sea urchin blastula protein is monitored by determination of the radioactivity solubilized into trichloroacetic acid (TCA).[1]

Reagents

Buffer: 0.1 M Tris-HCl, pH 7.8

Substrate. A 2-liter 24-hr culture (16°–18°) of fertilized *Strongylocentrotus purpuratus* or *Lytechinus pictus* embryos (0.15%, v/v) is reduced to 1 liter. For each 100 ml of this concentrated culture, add 300 μl of a 100 μCi/ml solution of [^{3}H]L-valine. The culture is further incubated for 1–4 hr. The embryos are pelleted by gentle hand centrifugation, and the supernatant seawater is removed by aspiration. The embryos are then extracted in 10% TCA. The resulting precipitate is collected by centrifugation and washed 2 times in 10% TCA by resuspension and centrifugation. A third washing is done in 10% TCA at 90° for 20 min. The TCA-insoluble fraction is then extracted 4 times with chloroform/methanol (1:3 v/v) followed by 3 washes in 95% ethanol. The precipitate fraction is then dissolved in a minimal volume of 0.05 N NaOH and dialyzed against 2 liters of 0.1 M Tris-HCl, pH 7.8, overnight at 4°. The insoluble protein is removed by centrifugation at 20,000 g for 30 min. Typically, specific radioactivities are on the order of 2×10^7 dpm/mg protein. Approximately 15 mg of labeled protein per 100 ml of 0.15% culture are typically obtained using this procedure. An unlabeled preparation of sea urchin embryo protein is also processed to serve as an unlabeled diluent of the radioactive protein.

Carrier protein solution: 50 mg of bovine serum albumin per milliliter in water

Trichloroacetic acid (TCA), 10% (v/v)

Procedure. Dilute the ^{3}H-labeled protein to 2.0 mg/ml in 0.1 M Tris-HCl, pH 7.8. Add 50 μl of enzyme solution to 100 μl of ^{3}H-labeled pro-

[7] V. D. Vacquier and J. E. Payne, *Exp. Cell Res.* **82**, 227 (1973).

tein and 50 μl of 0.1 M Tris-HCl, pH 7.8, in a 12-ml conical centrifuge tube at 25°. The assays are incubated for up to 2 hr, at which time 100 μl of bovine serum albumin carrier protein solution is added with mixing. One ml of ice cold 10% TCA is immediately added with mixing, and the tubes are chilled on an ice bath for 30 min. A zero time blank is prepared by adding the albumin and TCA before enzyme. The tubes are then centrifuged for 5 min at top speed in a clinical centrifuge; 0.8 ml of the supernatant solution is removed and mixed in a scintillation vial with 10 ml of Aquasol (New England Nuclear Corp.) and counted. The limit of linearity of release of TCA-soluble ^{3}H-labeled peptides and amino acid is a function of enzyme concentration and time. Under these conditions, 50 μl of undiluted crude cortical granule exudate gives a linear release of ^{3}H-labeled products for at least 2 hr.

Esterase Activity

Principle. The hydrolysis of the synthetic ester α-N-benzoyl-L-arginine ethyl ester (BAEE) is accompanied by an increase in absorbance at 253 nm.[8]

Reagents

Buffer: 0.1 M Tris-HCl, pH 7.8
Substrate: 10 mM BAEE in water

Procedure. Pipette 1.2 ml of buffer into a 3-ml cuvette. Add 0.7 ml of enzyme solution and initiate the reaction with 100 μl of substrate. Monitor the change in absorbance for approximately 10 min and calculate the rate of absorbance change per minute (corrected for nonenzymic hydrolysis) and divide by 1.15 to convert change in absorbance per minute to micromoles per minute (1 unit). Specific activity is expressed as units per milligram.

Preparation and Purification

Shedding of Gametes. Mature *Strongylocentrotus purpuratus* can be obtained through Pacific Bio-Marine, Venice, California. The shedding of gametes is induced by the intracoelomic injection of 0.5 M KCl. Alternatively, Aristotle's lantern is removed with forceps, the body fluids are drained, and a small amount of 0.5 M KCl is poured into the test. In either case the females are then allowed to shed by inverting the animals

---

[8] G. W. Schwert and Y. Takenaka, *Biochim. Biophys. Acta* **16**, 570 (1955).

and allowing the eggs to fall into beakers filled with seawater. The males are shed "dry," i.e., the test is wrapped with a tissue and the animal is inverted over a dry beaker and allowed to shed at 4°.

Preparation of Cortical Granule Exudate. The eggs are collected and sieved through No. 180 Nitex mesh to remove debris. The eggs are coated with two extracellular layers, which are removed prior to fertilization to facilitate collection of the cortical granule contents. The jelly coat layer which is outermost, is removed by adjustment of the pH of the egg suspension to 5.0 with 0.1–1.0 N HCl. After 3 min the pH is readjusted to 8.0 with unneutralized 1.0 M Tris. The eggs are washed several times by aspiration of the supernatant seawater and resuspension in fresh seawater.

The vitelline layer is an extracellular protein/glycoprotein coat immediately exterior to the egg plasma membrane. Modification of the vitelline layer is effected by brief treatment of eggs with dithiothreitol.[9] The dejellied, washed eggs are adjusted to a 50% suspension (v/v) and mixed with an equal volume of 0.02 M dithiothreitol, pH 9.1. After a 3-min incubation, the eggs are transferred to a 20-fold excess of normal seawater and washed at least three times.[10]

The dejellied, dithiothreitol-treated eggs are adjusted to a 50% (v/v) suspension in seawater containing 10 mM Tris-HCl, pH 8.0, and mixed with an equal volume of sperm suspension for fertilization. The sperm suspension is prepared by adding 6 ml of "dry" sperm up to 100 ml with seawater containing 10 mM Tris-HCl, pH 8.0. Three minutes after insemination, the eggs are removed by gentle hand centrifugation. All further manipulations are performed at 0° 4°. The sperm and insoluble components of the exudate are removed by centrifugation at 20,000 g for 20 min. The supernatant solution is referred to as the crude cortical granule exudate.

Isoelectric Precipitation. The protease activity contained in the exudate is quantitatively precipitated by the gradual addition of 0.1 volume 1.0 M sodium acetate (pH 4.0) with stirring.[3] The preparation is incubated for 1 hr to overnight. The precipitate is harvested by centrifugation at 20,000 g for 30 min, then washed twice by resuspension and homogenization in 0.1 M sodium acetate pH 4.0 (Dounce apparatus) using one-tenth of the original volume of seawater. The pH 4.0-insoluble fraction of the exudate can also be washed by cycles of dissolution at pH 8.0 and reprecipitation at pH 4.0, but this can result in a complete loss of biological activity, as will be detailed later.

[9] D. Epel, A. Weaver, and D. Mazia, *Exp. Cell Res.* **71**, 69 (1970).
[10] V. D. Vacquier, M. J. Tegner, and D. Epel, *Nature (London)* **240**, 352 (1972).

Affinity Chromatography. An effective affinity adsorbent can easily be prepared using the competitive trypsin inhibitor *p*-aminobenzamidine as a ligand.[3] CH-Separose-4B (Pharmacia) is hydrated and washed in 0.5 M NaCl. The gel is adjusted to 10% (v/v), and 40 mg of *p*-amino-benzamidine·2HCl is added for each gram of dry gel. The pH is adjusted to 4.9 with 1.0 N NaOH and the coupling reaction is initiated by the addition of 1-cyclohexyl-3-(2-morpholinoethyl)carbodiimide metho-*p*-toluene-sulfonate (425 mg per gram of dry gel). The pH of the gel suspension is maintained at 4.6–4.9 for 6 hr with very gentle stirring; the gel is then washed with 0.1 M Tris-HCl–0.4 M NaCl, pH 8.0. The capacity of this resin for bovine trypsin is approximately 7 mg per milliliter of resin.

A 2.2×4 cm column of *p*-aminobenzamidine-Sepharose is poured and equilibrated with 0.1 M Tris-HCl–0.4 M NaCl, pH 8.0, by washing the resin bed with several volumes of buffer.

A portion of a protease suspension in pH 4.0 sodium acetate is centrifuged at 20,000 g for 20 min to collect the precipitate. Resuspend the precipitate to a protein concentration of 2.6 mg/ml in water using a Dounce homogenizer. When the precipitate is finely suspended, rapidly add 0.1 volume of 1.0 M Tris-HCl–4.0 M NaCl, pH 8.0, and centrifuge the preparation at 20,000 g for 20 min. Immediately load 15 ml of the enzyme solution onto the column at a flow rate of 60 ml/hr. Wash the column until the absorbance at 280 nm is less than 0.01 when compared to water. Elution of the bound protein is accomplished by applying a 0.1 M NH$_4$OH–0.4 M NaCl (pH $\sim$ 11.6) step to the column. Fractions are collected in test tubes containing sufficient 1.0 M sodium acetate, pH 4.0, to buffer the fraction at that pH. Under these conditions 75% of the input esterase activity and protein is bound to the column. Complete binding of all the protein and esterase activity in the pH 4.0 precipitate fraction can be obtained if the amount of resin is increased or the amount of protein applied is decreased.

As shown in Fig. 1, the protein fraction eluting with 0.1 M NH$_4$OH–0.4 M NaCl does not elute coincidentally with the esterase activity, suggesting heterogeneity in this fraction. Polyacrylamide gel electrofocusing of the pooled fraction (indicated by the bar in Fig. 1) showed the presence of two proteins of isoelectric points 4.7 and 4.9 (Fig. 2).

Complete separation of these two enzymes by affinity chromatography has not yet been successful. The enzymes bind tightly to the resin and seem to elute in an all-or-none fashion. The degree of resolution of the enzymes eluted with a pH change (0.1 M NH$_4$OH–0.4 M NaCl, pH 11.6) is a function of flow rate and column dimension; maximum separation was obtained with a flow rate of 60 ml/hr and a column length:diameter

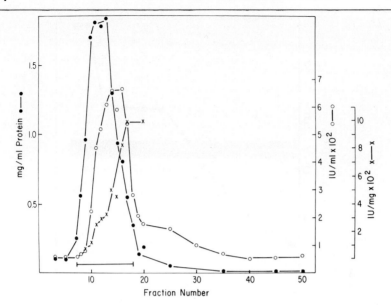

FIG. 1. Chromatography of the pH 4.0 precipitate fraction of the cortical granule exudate on *p*-aminobenzamidine Sepharose. The protein fraction eluting with 0.1 *M* NH₄OH–0.4 *M* NaCl is shown. See text for details. ○——○, IU/ml; ●——●, mg/ml protein; ×——×, IU/mg.

ratio of 20:1. Although complete separation of the proteases was not achieved using affinity chromatographic techniques, use can be made of the widely differing specific activities across the elution profile (Fig. 1). Collection of the leading and trailing edges yields partial separation into a low specific activity and a high specific activity fraction. Fractions collected in this manner yield biologically monospecific enzymes. The low specific activity fractions contain vitelline delaminase activity, and the sperm receptor hydrolase activity is contained in the high specific activity fractions.

Isoelectric Focusing. Preliminary resolution of the two proteases present in cortical granule exudate preparations purified by isoelectric precipitation and affinity chromatography has been obtained using electrofocusing procedures in 4.0 *M* urea.[11] The urea is necessary to preclude precipitation of the enzymes at their isoelectric point.

SOLUTIONS

Lower electrode solution: 1.85% ethylenediamine containing 60% sucrose and 4 *M* urea

[11] E. J. Carroll, Jr., unpublished observations (1975).

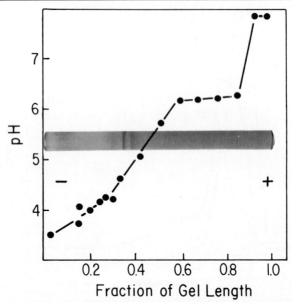

FIG. 2. Polyacrylamide gel electrofocusing of the affinity-purified pooled fraction indicated by the bar in Fig. 1. Details of the conditions for electrofocusing, gel slicing, and pH determination were those of E. J. Carroll, Jr., and D. Epel, *Dev. Biol.* **44**, 22 (1975).

>   Sample solution: 1% pH 3–10 Ampholines containing 55% sucrose
>   and 4 *M* urea
>   Dense solution: 1% pH 3–5 Ampholines containing 50% sucrose
>   and 4 *M* urea
>   Light solution: 1% pH 3–5 Ampholines containing 10% sucrose
>   and 4 *M* urea
>   Upper electrode solution: 1% phosphoric acid containing 4 *M* urea

The LKB model 8101 electrofocusing apparatus is connected to a circulating water bath thermostatted at 0°–4°. The lower electrode solution is pumped into the apparatus. The sample solution is prepared containing approximately 9 mg of protein in a final volume of 5 ml. After the sample solution is loaded into the column, the 10 to 50% linear sucrose gradient is constructed using the light and dense solutions and is pumped into the column. After the loading of the upper electrode solution, the current is applied. Start the electrofocusing with approximately 650 V at 3 mA and gradually increase (over 12 hr) the voltage until 950 V is reached. After approximately 48 hr, the column is fractionated. As shown in Fig. 3, two peaks of protein and esterase activity are well resolved.

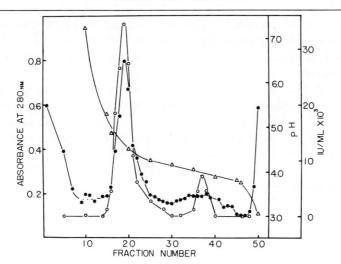

F_IG. 3. Preparative electrofocusing of affinity-purified cortical granule proteases in 4 M urea. ○——○, IU/ml; ●——●, absorbance at 280 nm; △——△, pH.

Properties

Stability. The esterase activity in crude cortical granule exudate and purified fractions is stable for weeks at 0°, pH 4.0 (0.1 M sodium acetate) or for several months (6–8) at pH 4.0, −20° (0.05 M sodium acetate–50% glycerol). The prefered method of storage is the latter; the preparation is thus stored as a precipitate in the unfrozen state. Preparations stored frozen as a precipitate at pH 4.0 do not readily redissolve, and considerable material is denatured. The precipitate can simply be harvested from the glycerol by centrifugation and then dissolved in the desired buffer. When crude enzyme preparations are stored at 0°, pH 8.4, for several days, the specific activity will increase 8- to 10-fold.[3,12] Concentrated and purified fractions will exhibit this dramatic increase in specific activity over a time period of minutes to hours depending on the protein concentration.[11] These "activated" preparations (nothing is implied or intended regarding mechanism in the use of this term) have a complete loss of normal biological activity and apparently digest the vitelline layer.[3]

Purity and Physical Properties. Solutions of sea urchin egg cortical granule protease purified by isoelectric precipitation contain two proteins, one intensely staining and one faintly staining with Coomassie brilliant blue.[3] These proteins are partially resolved by use of the affinity

[12] V. D. Vacquier, unpublished observations.

chromatographic procedure.[3] Complete separation of the two enzymes has been recently attained with preparative electrofocusing in the presence of 4 M urea. Further biological and biochemical characterization of the resolved proteins is currently in progress.

Limited data on the physical properties of the mixture of proteases are relevant to this discussion. The method of Martin and Ames was used to determine the molecular weights of the enzymes in a mixture. A single symmetrical peak of enzymic activity was detected, which corresponded to a molecular weight of 47,000.[3] The presence of only one peak suggested that both vitelline delaminase and sperm receptor hydrolase have the same molecular weight.

Unexpectedly, protease preparations displayed a dispersion of molecular weights when chromatographed on Sepharose 4B. The approximate molecular weights ranged between 10^5 and 10^6.[3] The heterogeneity probably results from aggregation, since rechromatography of the excluded, high-molecular-weight fraction resulted in the reappearance of a fraction of lower molecular weight. The same elution patterns were obtained both at pH 5.6 and 8.0. The observed molecular weight may well be a function of protein concentration.

Activators and Inhibitors. Vacquier has reported that the esterase activity of the crude cortical granule exudate prepared by parthenogenetic activation of sea urchin eggs in 0.5 M KCl–0.004 M EGTA pH 8.0 is activated 10-fold when 0.1 M CaCl$_2$ is added to divalent free assay medium. The rates with divalent free and 0.1 M MgCl$_2$-supplemented assay medium were the same. A 3-fold activation of the divalent free rate was obtained in 0.1 M SrCl$_2$.[13]

Soybean trypsin inhibitor is an effective inhibitor of the crude cortical granule protease activity and of purified fractions. Toluenesulfonyl (tosyl) lysine chloromethyl ketone inhibits the esterase activity of the mixture of enzymes (10 mM, 2 hrs, 4°, 56% inhibition). Tosyl phenylalanine chloromethyl ketone and phenylmethylsulfonyl fluoride are without effect (0.1 mM and 1.0 mM, respectively, 2 hrs, 4°). With the mixture of proteases, the K_i for p-aminobenzamidine (BAEE as substrate) is 1.6×10^{-4} M.

The alkaloid nicotine is an effective inhibitor of the esterase activity of the crude cortical granule exudate; protease activity does not seem to be affected. The vitelline delaminase activity is also unaffected, but sperm receptor hydrolase activity is inhibited by nicotine.[14] Thus it

[13] V. D. Vacquier, *Exp. Cell Res.* **90**, 454 (1975).
[14] E. J. Carroll, Jr., *Amer. Zool.* **15**, 780 (1975).

appears that nicotine is a selective inhibitor of the sperm receptor hydrolase enzyme. These results are significant in that nicotine is a classical polyspermy-including drug,[15] and its mechanism of action is now explained. Additional studies with these inhibitors using the separated proteases is in progress.

Specificity. The cortical granule proteases hydrolyze the synthetic esters BAEE and α-N-tosyl arginine methyl ester. Thus, minimally the enzymes hydrolyze N-blocked esters of lysine and arginine.

Kinetics. The K_m of the crude cortical granule exudate for BAEE is 0.7 mM, and the pH optimum is approximately 8.0.

Distribution. The esterase–protease and vitelline delaminase–sperm receptor hydrolase activities have also been found in *Lytechinus pictus* eggs.[16]

Addendum

Fodor *et al.*[17] have recently reported the isolation of a protease from *Strongylocentrotus purpuratus* cortical granule exudate and unfertilized eggs. The protease was solubilized from both starting materials using 2 M KCl–10% glycerol–1% butanol–50 mM EDTA–0.2 M Tris at pH 8.0 as an extraction medium. Homogeneous enzyme, similar to bovine trypsin in both susceptibility to inhibitors (diisopropyl phosphofluoridate and soybean trypsin inhibitor) and molecular weight (22,500), was obtained following purification of cortical granule exudate and unfertilized egg extracts on columns of immobilized soybean trypsin inhibitor.

Fodor and co-workers did not perform sperm receptor hydrolase or vitelline delaminase assays; however, as they state, their homogeneous enzyme may have been separated from factors which might modulate the functions of the enzymes in fertilization as reported by Carroll and Epel.[18]

Acknowledgments

I thank Professor David Epel for his hospitality and encouragement. The technical assistance of Elizabeth Baker is appreciated. This work was supported by research grants from the National Science Foundation and the Population Council. E. J. Carroll, Jr. is a postdoctoral fellow of the Population Council.

[15] F. J. Longo and E. Anderson, *J. Cell Biol.* **46**, 308 (1970).
[16] E. J. Carroll, Jr. and V. D. Vacquier, unpublished observations (1973).
[17] E. J. Fodor, H Ako, and K. A. Walsh, *Biochemistry* **14**, 4923 (1975).
[18] E. J. Carroll, Jr. and D. Epel, *Dev. Biol.* **44**, 22 (1975).

[29] Hatching Enzyme of the Sea Urchin
Strongylocentrotus purpuratus

By DENNIS BARRETT and BEN F. EDWARDS

During the first few minutes after fertilization, the sea urchin egg constructs a tough protective extracellular coat, the fertilization envelope (the fertilization membrane or chorion in older literature). Some 6–20 hr later, depending on species and temperature, the embryo, now a blastula, grows cilia, hatches out of the fertilization envelope, and swims away. Ishida proposed in 1936[1] that this hatching is accomplished by means of a "specific proteoclastic ferment," which he named hatching enzyme. Continuing the work of the Japanese school, Sugawara[2] thoroughly characterized a hatching protease, and Yasumasu[3] reported purifying the enzyme to crystallinity. Purifying hatching enzyme from the Eastern Pacific urchin *Strongylocentrotus purpuratus* has proved to be a greater challenge.[4,5]

Recent interests in hatching enzyme are 3-fold. In its rare position as an identifiable protein elaborated before gastrula, it stands as a model for the regulation of macromolecular syntheses in early development.[6–10] Second, recovering nucleic acids or even subcellular organelles from homogenates of prehatching sea urchin embryos is made difficult by the shear force required to break the fertilization envelope; the method of choice for removing it with minimal perturbation of the embyro may be to dissolve it from the outside with a preparation of crude hatching enzyme.[11] Finally, the natural occurrence of hatching enzyme as a protease which must attack native rather than denatured protein, together with a preliminary determination that it specifically attacks glutamyl and aspartyl bonds, suggests that its mode of attack will hold considerable interest for protein chemists.

[1] J. Ishida, *Annot. Zool. Jpn.* **15**, 453 (1936).
[2] H. Sugawara, *J. Fac. Sci. Imp. Univ. Tokyo Sect. 4* **6**, 109 (1943).
[3] I. Yasumasu, *Sci. Pap. Coll. Gen. Educ. Univ. Tokyo* **11**, 275 (1961).
[4] D. Barrett, B. F. Edwards, D. B. Wood, and D. J. Lane, *Arch. Biochem. Biophys.* **143**, 261 (1971).
[5] B. F. Edwards, W. R. Allen, and D. Barrett, in preparation.
[6] I. Yasumasu, *Sci. Pap., Coll. Gen. Educ. Univ. Tokyo* **13**, 241 (1963).
[7] D. Barrett, *Am. Zool.* **8**, 816 (1968).
[8] D. Barrett and G. M. Angelo, *Exp. Cell Res.* **57**, 159 (1969).
[9] S. S. Deeb, *J. Exp. Zool.* **181**, 79 (1972).
[10] B. F. Edwards and D. Barrett, *Am. Zool.* **13**, 1316 (1973).
[11] Paul M. Hoque and D. Barrett, unpublished observations, 1971.

Assay Methods

Timing of Fertilization Envelope Dissolution

Principle. Directly detecting the hydrolysis of peptide bonds in the enzyme's natural, insoluble, substrate is not easy. Yasumasu[12] recognized that the fertilization envelope could nonetheless offer a quantitative assay if the parameter observed was *time* required to reach a sharply defined end point, the dissolving of the envelope. The reciprocal of time required for a sample to hatch 50% of the embryos provided a useful measure, linearly related to enzyme concentration; the technique is made more convenient and reproducible[8] by fixing the embryos before use. In Deeb's laboratory *Arbacia lixula* envelopes are allowed to form in 2-mercaptoethanol, so that they are weaker and more quickly digested.[12a]

Reagents

Buffer: 10 mM glycylglycine, or 4-(2-hydroxyethyl)-1-piperazine-propanesulfonic acid (EPPS) or N-2-hydroxyethylpiperazine-N'-2-ethanesulfonic acid (HEPES) with 10 mM CaCl₂, 175 μg of penicillin and 175 μg of streptomycin per milliliter, adjusted with NaOH to pH 8.2

Enzyme samples: Crude enzyme containing up to 1 M salt can be used. However, after the first purification step (urea treatment), envelope-dissolving activity becomes sensitive to salt, so that salt concentration must be reduced by dilution or dialysis to less than 0.1 M.

Fixed embryos: Normal embryos (see below) are allowed to settle at the 4- to 8-cell stage, so that a concentrated suspension can be drawn into a disposable pipette and squirted rapidly (or dripped with constant stirring) into at least 50 volumes of 95% ethanol at −20°. Stored in the freezer, the fixed embryos are stable for years. Shortly before each assay, a small portion is rehydrated by squirting rapidly into at least 10 volumes of ice-cold buffer in a conical centrifuge tube. When the embryos have settled, supernatant is drawn off and they are suspended again, in sufficient buffer to bring the concentration to about 2×10^4 embryos per milliliter.

Procedure. Three 70-μl reaction mixtures are run on each flat microscope slide, previously treated with silicone (e.g., Siliclad, Clay and Adams). Each spot comprises 50 μl of enzyme (including buffer to dilute, inhibitors, etc.) and 20 μl of embryo suspension (agitated immediately

[12] I. Yasumasu, *Univ. Tokyo J. Fac. Sci. Sect. IV* **9,** 39 (1960).
[12a] B. Zaynun, M.Sc. Thesis, American University of Beirut, Lebanon, 1973.

before pipetting to assure uniformity), mixed with a clean dissecting needle or the like. The time of mixing is recorded, and incubation is in a moist chamber (e.g., a plastic box with a moistened paper towel), preferably in an incubator at 35°. It is convenient to run 30 assays or so together, with starting times staggered as necessary.

At approximately 0.5, 1, 2, 4, 8, and 16 hr, each slide is observed briefly by compound microscope with 100-fold magnification. About 100 embryos in each reaction mixture are scored as hatched or unhatched, time of observation is recorded, the slide is tapped to resuspend the embryos (which tend to settle to the center) and is replaced to continue the incubation.

The percent of embryos hatched is plotted against time for each observation. The points usually fall on a sigmoid curve, with fair linearity between 20% and 80%, and the time for 50% hatching is interpolated.

Unit and Limits. The hatching unit is strictly empirical, the reciprocal of the time in hours for 50% hatching (t_{50}^{-1}). Thus its value varies with species, egg batch, and the observer's criterion for envelope dissolution. But duplicate assays done by one observer using one batch of embryos (a single spawning will provide for 10^5 assays) usually agree within 10% and afford a linear response up to 1 unit. Microbial growth may interfere after a day or two of incubation, but usually 0.03 unit can be reliably determined.

Since relatively few bonds (fewer than 5%) need be broken to dissolve the fertilization envelope, the assay is quite sensitive. One hatching unit, measured at 35° on *Strongylocentrotus purpuratus* envelopes, corresponds roughly to 30 mU as defined on a molecular basis below.

Digestion of Modified Casein

Principle. Several modifications of the traditional determination by ninhydrin of the amino groups released by proteolysis have been of value. The extant free amino groups of the substrate are blocked by reductive dimethylation for a substantial reduction of background. And the new amino groups liberated by enzyme are detected by reagents that yield adducts of high extinction coefficient (TNBS, detection by spectrophotometry) or high fluorescent intensity (fluorescamine, detection by spectrofluorometry).

Reagents

Buffer: 10 mM EPPS or HEPES, with 10 mM CaCl$_2$, adjusted with NaOH to pH 8.2. If reaction times longer than 2 hr are to be used, to detect low activities, 50 μg/ml of penicillin may be added.

These buffers, synthesized by Good *et al.*,[13] are close to their pK values and replace primary amines, which interfere, and borate and phosphate, which complex the necessary calcium.

N,N-Dimethylated casein: The reductive methylation of free amino terminals and ϵ-amino groups of lysines is accomplished by the method of Lin *et al.*,[14] modified only slightly in the proportions of reactants. In 150 ml of 0.1 M sodium borate, pH 9.0, 0.5 g casein (Hammarsten quality) is dissolved by heating. After cooling the solution to 0° in an ice bath, 150 mg of sodium borohydride are added with rapid stirring. With a drop of 2-octanol to retard foaming, and with constant stirring, 30 μl of 37% formaldehyde are added each minute for 40 min to a total of 1.2 ml. The protein is acidified to pH 6 with 50% acetic acid, dialyzed against distilled water and lyophilized to dry.

Substrate solution: 1.11 mg/ml of modified casein dissolved in buffer with 70 mM NaCl, by warming to 50°

Enzyme samples: Salt concentration in the enzyme samples is not critical, but low background in primary amines is, so that crude samples usually need to be dialyzed against buffer, in the cold.

EGTA   [ethyleneglycol-bis(β-aminoethyl   ether)-N,N'-tetraacetic acid], 250 mM

NaHCO$_3$, 4 g/100 ml

TNBS (trinitrobenzenesulfonic acid), 1 mg/ml

SDS (sodium dodecyl sulfate), 2 g/100 ml

HCl, 1 M

Fluorescamine {4-phenylspiro[furan-2(3H),1'-phthalan]-3,3'-dione} (Hoffmann-La Roche, Inc., Nutley, New Jersey 07110), 300 μg/ml in reagent-grade dioxane. Solution is stored in freezer, and make up at least weekly; background rises with age.

Chromogenic Procedure. Each tube receives 0.45 ml of substrate solution and is equilibrated briefly in a 35° bath; 0.05 ml of enzyme, diluted with buffer or other constituents as appropriate, is added to start the reaction. With each run is included a substrate blank with buffer instead of enzyme. Incubation is for 60 min, or longer if low activity demands. The reaction is stopped after incubation, and a duplicate enzyme blank is stopped before incubation, by adding 0.10 ml of EGTA reagent, bringing sample to 0.60 ml.

To develop color, each tube receives 0.30 ml of NaHCO$_3$ solution and 0.40 ml of TNBS solution, and incubates in the dark at 50° for 30 min.

[13] N. E. Good, G. D. Winget, W. Winter, J. N. Connolly, S. Izawa, and R. M. M. Singh, *Biochemistry* **5**, 467 (1966).

[14] Y. Lin, G. E. Means, and R. E. Feeney, *J. Biol. Chem.* **244**, 789 (1969).

This reaction is stopped by adding 0.25 ml of SDS solution and 0.25 ml of HCl. The tube is read within the hour at 340 nm against a substrate blank, and the difference is calculated between each incubated sample and the corresponding enzyme blank, stopped at time zero with EGTA. Standards of 1–50 nmoles of amino acid are suitable.

Fluorogenic Procedure. The above protocol is scaled down: 100 μl of substrate solution is incubated with 100 μl of enzyme, in an 8-ml tube tightly capped with Parafilm, at 35°. To stop duplicate reaction mixtures before and after incubation, 1.8 ml of buffer at room temperature is added and then, holding the tube on a vortex mixer, 0.5 ml of fluorescamine solution. Reaction is complete, and excess reagent degraded, within a minute; the fluorophor product is stable for several hours, so that blanks can be read at once, or left to be read with the active samples.[15] Relative fluorescence intensity is read at $\lambda_{excitation}$ = 390 nm, $\lambda_{emission}$ = 470 nm, with standards of 1–100 nmoles of peptide.

Unit and Limits. The unit (U) is the amount of enzyme that releases 1 μmole of amino terminals per minute from dimethylated casein at 35°, pH 8.2.

The rate of hydrolysis is constant with time through at least 16 hr, and proportional to the concentration of enzyme, crude or pure, until about 2% of the casein bonds have been dissolved. Thus in the chromogenic procedure linearity extends to 70 nmoles of amino terminal released; in the usual 60-min hydrolysis, activities up to 70 nmoles/60 min or 1.1 mU may be determined directly, but higher activities require extrapolation or dilution. Seventy nanomoles corresponds to ΔA_{340} = 0.40 (over a background for TNBS and substrate of about 0.5).

In the fluorogenic procedure, linearity of attack extends to 16 nmoles, so that in a 30-min assay up to 0.5 mU can be determined directly. Sixteen nanomoles corresponds to a relative fluorescence intensity of about 0.2, over a background of fluorescamine (negligible) and substrate of about 0.1. In the fluorometric assay, where background due to substrate is not limiting, the concentration of dimethylated casein could be raised within the limit of its solubility, to extend the linear range.

Comparison of Assays

Available instrumentation is likely to dictate the choice of an assay method. Photometry is satisfactory in all respects; fluorometry is somewhat more sensitive, requires smaller enzyme samples, and is quicker and easier to perform. Observing fertilization envelopes dissolve is time-

[15] S. Udenfriend, S. Stein, P. Böhlen, W. Dairman, W. Leimgruber, and M. Weigele, *Science* **178**, 871 (1972).

consuming, less reproducible, and not directly quantifiable in molecular terms; yet it is surprisingly sensitive, and under primitive conditions it may be the only possibility.

Two other reasons for choosing the hatching assay should be considered. First, the enzyme's specificity for its natural substrate may be invoked, and the envelope selected in order to discriminate against other proteases that might contaminate. We[16] do not confirm older reports[17] that the fertilization envelope is invulnerable to other proteases. Rather, common ones like trypsin, papain, Pronase, and pepsin all do attack envelopes, but with specific activities two to three orders of magnitude lower than that of hatching enzyme. The consideration is minor, for if precautions are taken against growth of microbes and cytolysis of embryos, the proteolytic contaminant in crude enzyme is negligible.

Second, hatching enzymes vary sufficiently among species (see p. 371) that fertilization envelopes distinguish among them. Since the differences have not yet been chemically characterized, the species-specificity can be exploited experimentally only in an envelope-dissolving assay.

Ambitions for an assay with a substrate of restricted specificity have so far been frustrated. Copolymers of glutamic acid have not proved to be good substrates. Dipeptides like Z-Glu-Tyr and Z-Glu-Phe are fair substrates[18] (attacked at about one-fifth the rate of glutamyl bonds in casein), but they fail to stabilize hatching enzyme, which when pure degrades rapidly at assay temperatures, in contrast to protein substrates which render the enzyme stable for many hours.

Preparation and Purification

Principles

Sea urchin embryos have a high metabolic rate and are denser than sea water. To assure a supply of oxygen and dilution of waste products sufficient for normal development, they are cultured at a low concentration, with constant gentle stirring in a shallow vessel, so that they maintain a uniform suspension and allow the oxygen content of the water to replenish by diffusion.

Once the enzyme has brought about hatching, the embryos swim away, and enzyme is liberated to the medium. It is collected in the supernatant above the sedimented embryos and fertilization envelope fragments. Thus the more concentrated the suspension of hatching embryos,

---

[16] Donald J. Klingborg and D. Barrett, unpublished observations, 1969.
[17] E. B. Harvey, "The American *Arbacia* and Other Sea Urchins," p. 180. Princeton Univ. Press, Princeton, New Jersey, 1956.
[18] D. Barrett, unpublished observations, 1973.

the smaller the volume of crude product to be handled. A satisfactory compromise between the need for dilution and the convenience of concentration is as follows: embryos are cultured at 0.5% to 1.0% (volume of packed jelly-free eggs to volume of seawater) until hatching, then concentrated to 10% when hatching has begun.

The crude supernatant is >99% contaminant, and this is thought to be mainly solubilized fragments of fertilization envelope. The enzyme adheres to these so tightly that the enzyme appears heterogeneous, and a clean separation is impossible by gel exclusion chromatography, ion exchange chromatography, gel electrophoresis, salt precipitation, or density-gradient fractionation.[4] A limited treatment with chaotrope to dissociate hydrogen bonds renders the enzyme tractable and separable from the contaminant by standard methods.[5,19]

Therefore an alternative strategy is to remove the nascent fertilization envelope before it has hardened and to culture the embryos nude until they simply secrete the enzyme into sea water. This method is reported to be useful for *Arbacia lixula*,[20] but in our hands it produces crude enzyme of no higher specific activity. Since the contaminant here probably arises by lysis of embryos (at envelope removal, or in later culture), it is probably more heterogeneous chemically and harder to remove, and may also include intracellular proteases, which are a particularly unwelcome contaminant.

Preparation of Crude Enzyme

Shedding Gametes and Fertilizing. Procedures in this section are based on the gametes from 100 urchins of the species *Strongylocentrotus purpuratus*. Yields from other species will vary greatly, and these procedures may require qualitative as well as quantitative modifications. We have found it economical to run batches of about six times this quantity (see "Yield" below.) Urchins are collected at peak season (for *S. purpuratus* January to February in northern California, varying along the coast[21]).

Thirty liters of seawater are run through an ultrafilter of 0.5-μm pore diameter (e.g., Millipore Filter Corp., No. HA) to use for all procedures, after discarding the first run-through, which contains toxic detergents.[22]

The skin between the Aristotle's lantern (teeth) and the test is severed with one point of a forceps. The lantern is grasped by the skin and torn

[19] Peter J. Locatelli and D. Barrett, unpublished observations, 1970.

[20] N. P. Kettaneh, M.Sc. Thesis, American Univ. of Beirut, Lebanon, 1974.

[21] R. A. Boolootian, *in* "Physiology of Echinodermata" (R. A. Boolootian, ed.), p. 561. Wiley (Interscience), New York, 1966.

[22] R. D. Cahn, *Science* **155**, 195 (1967).

out. The animal is inverted to pour out most of the coelomic fluid. About 1 ml of 0.5 M KCl is pipetted into the exposed coelom, and the urchin is rotated to assure that the irritant reaches all five gonads.[23] An alternative method, which is slightly less effective, but which the urchin may survive, requires injecting the KCl into the coelom through the peristomial membrane with a 23-gauge needle.

Within a minute gametes flow from each of the gonadopores on the aboral surface, and the individual is sexed by the gametes' color. Females are inverted over a full beaker of seawater to collect eggs; males are inverted, with a collar of tissue paper to intercept water and leaking coelomic fluid, over a syracuse or petri dish. Disposable plastic beakers and dishes, rinsed after use with distilled water (but never detergent), are convenient.

When shedding is essentially complete (30 min), sperm are stored tightly covered at 4°. Eggs are washed through bolting silk of 60-μm mesh and pooled in a 4-liter beaker. After the eggs have settled (25 min), supernatant is drawn off. They are resuspended and settled three or four times, or treated briefly with seawater brought to pH 5 with HCl, in order to remove the jelly coat (the process is followed by staining a drop of eggs with 0.1% toluidine blue in seawater).

It is important to determine the volume of eggs. A random sample from the beaker is centrifuged in a graduated conical tube, by five turns on a hand centrifuge or 30 sec at low speed in a clinical centrifuge swinging-bucket head. The total volume of eggs in suspension is extrapolated from the volume of packed eggs in the sample. Roughly 2×10^6 dejellied eggs occupy 1 ml after gentle centrifugation.

The batch of eggs is inseminated immediately after mixing a few eggs with a drop of sperm suspension on a slide to ascertain that they are fertilizable. Ten drops of sperm, "dry" as collected, are suspended in 100 ml of seawater by vigorously pipetting up and down for a minute. Ten milliliters of sperm suspension is stirred into the eggs in 1 liter of sea water. (If any jelly remains, it is particularly important to suspend the eggs afresh directly before inseminating, since water in which they have sat contains dissolved jelly components that intercept sperm.) A sample is observed at the microscope (100 $\times$) for the elevation of the fertilization envelope which signals fertilization. Stirring or other agitation is avoided between 1.5 and 5 min after insemination, when the envelope weakens, and becomes particularly susceptible to tearing off.[24,25] The 10^4 dilution

[23] M. Kumé and K. Dan, "Invertebrate Zoology," p. 15. National Library of Medicine, Washington, D.C., 1968.
[24] B. Markman, *Acta Zool.* 39, 103 (1958).
[25] D. Barrett, D. J. Klingborg, and S. K. Sholes, in preparation.

of sperm usually suffices, but later additions of three and ten times this quantity of sperm may sometimes increase the percentage fertilized, which should exceed 95. Eggs are settled twice in 2 liters of fresh seawater to dilute sperm away.

Incubating Embryos and Concentrating to Hatch. The embryos are incubated at $\leq 18°$, stirring, as a suspension of 0.5% (calculated on the packed, dejellied, unfertilized eggs) if space permits, or up to 1%, which may retard development slightly. Higher concentrations will seriously retard or arrest the embryos. For adequate oxygenation a straight-sided vessel is used, with water not more than 30 cm deep. A shallow cylindrical plastic laundry tub equipped with a plastic bar 32 cm long, turned at 70 rpm by a 20 W stirring motor, will accommodate 13 liters of suspension, with up to 130 ml of packed eggs (a maximum to expect from 50 females). The addition of 50 μg/ml each of penicillin and streptomycin (diluted from a 1000-fold-concentrated stock solution, which can be stored frozen) effectively prohibits bacterial growth.[26] As hatching time nears, samples are inspected at 15 to 30 min intervals. As soon as 5% of the embryos have hatched, the stirrer is withdrawn, the embryos settle, and as much as possible of the medium is poured or aspirated off. Embryos are resuspended as a 10% suspension in fresh seawater with penicillin and streptomycin, and stirred at $\leq 18°$, in as shallow a layer as practicable, to agitate vigorously without foaming.

In an hour or two, when 95% of the embryos with fertilization envelopes have hatched, stirring is stopped and the culture is chilled on ice, where it may be left overnight if necessary.

After centrifugation at 5000 g for 10 min, the supernatant is rapidly poured off, for the blastulae will swim upward despite the cold. Supernatant may be filtered through Whatman No. 1, if needed, to be sure that it is clear of *all* embryos. The crude supernatant can now be stored in the freezer for years and can withstand several freeze-thaw cycles, but it is easily inactivated by lyophilization. Assays at this stage may err on the high side.[27]

Concentrating the Crude Enzyme. The crude enzyme preparation is further concentrated 10- to 15-fold by ultrafiltration, under 50 psi of

---

[26] W. R. Allen, Ph.D. Thesis, University of California, Davis, 1970.

[27] Fresh crude enzyme contains fragments of fertilization envelope, soluble, but of high molecular weight.[4] Consistently the release of NH_2-terminals with casein substrate by crude enzyme shows biphasic kinetics, with an early burst of reaction of variable duration (typically 1 hr), followed by a 5-fold lower rate which continues for many hours. Since the early phase is not seen with purified preparations, we presume that it represents preferential attack on residual natural substrate. Thus the apparent rate of casein hydrolysis may be inflated in comparison to determinations where casein is the only substrate.

nitrogen pressure, with an Amicon Corp. (Lexington, Massachusetts) filter No. UM-10. This system nominally retains proteins larger than 10,000 daltons, and it retains all the hatching enzyme (bound to heavy particles) but passes salts and smaller peptides. Should the crude enzyme be allowed to dry down on the filter, it can usually be eluted by buffer without substantial loss.

Yield. The expected yield at this stage is about 7.7 pg of pure enzyme per embryo. Thus the batch preparation given above would net about 2 mg; our routine run of six such batches provides 12 mg, which will net 3 mg of enzyme after purification.

Dialysis for Use on Living Embryos. The enzyme at this stage may be used to hatch younger embryos preparatory to isolating polysomes or other fragile structures from them. For this application, the enzyme should first be dialyzed against sea water, to remove peptides and other small molecules that otherwise inhibit normal development,[28] but should not be further diluted. Embryos suspended in it at 15° will lose their fertilization envelopes in about 2 hr, without suffering developmental damage.[11] They should be promptly removed from the enzyme as soon as hatched.

Purification[5]

Step 1. Urea Treatment. Dry urea is added to crude concentrated enzyme on ice, with stirring, to bring it to 6 M. Gentle stirring is continued for 0.5 hr to dissolve urea completely, and incubation is continued for 5.5 hr at 22° without stirring. The chaotrope is then diluted away by dialyzing in the cold against 10 mM EPPS (pH 8.2), 10 mM CaCl$_2$, 70 mM NaCl. If the sac is dialyzed against 25 volumes of fresh buffer every 6 hr, three or four changes will suffice to bring the refractive index of dialyzate down to that of the buffer, $n = 1.3342$.

Typically in the crude concentrate, some enzyme is free and will elute from DEAE in peak I (Fig. 1A), but much of the enzyme is bound to contaminant and will elute in peaks II–IV. An optimal exposure to urea serves to break the attachments (presumably hydrogen bonds) between the enzyme and contaminants, and so to change the contaminant (and perhaps the enzyme) that when urea is removed, the attachments do not reform. Thus in Fig. 1B, after urea treatment, most of the enzyme activity resides in peak I, while most of the protein has moved away from peak I into peaks II, III, and IV.

Step 1 is the only unpredictable one in the procedure. Unfortunately exposure to urea past the optimum results in irreversible denaturation of

[28] D. Barrett, *Am. Zool.* 14, 1252 (1974).

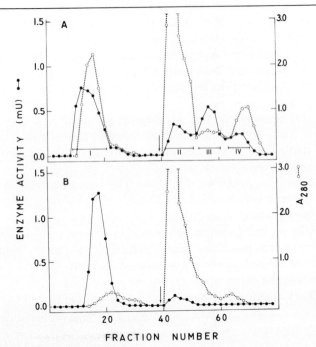

Fig. 1. Chromatography of hatching enzyme on DEAE-cellulose. Sample in 25 ml of 10 mM 4-(2-hydroxyethyl)-1-piperazinepropanesulfonic acid (pH 8.2), 10 mM CaCl$_2$, 70 mM NaCl, was applied to a 50-ml column of Whatman DE-52 in the cold. Elution was by the same buffer up to fraction 30, then by a linear gradient of 0.07 to 1.20 M NaCl in the same EPPS-Ca. Arrow marks the emergence of the salt gradient, determined by refractometry. Fractions were 3 ml; enzyme activity was assayed in 50 μl samples by the casein–trinitrobenzenesulfonic acid method given in the text. The fractions included within peaks I, II, III, and IV are indicated by bars at the base of the peaks in A. Samples: (A) 25 ml of crude concentrate before step 1; (B) 16 ml of enzyme after step 1 (urea-treated and dialyzed). Adapted from B. F. Edwards, Ph.D. thesis, University of California, Davis, 1976.

the enzyme, and the optimal time varies somewhat with the history of the batch, in ways that we have not brought under control. It is advisable, therefore, to pool as large a batch of crude enzyme as possible, and try small samples with 4, 6, and 8 hr of exposure to urea, to allow interpolation of the maximum time which will stop short of decreasing enzyme activity.

Step 2. DEAE-Cellulose Chromatography. A 50-ml column is packed with DEAE-cellulose equilibrated with 10 mM EPPS (pH 8.2), 10 mM CaCl$_2$, 70 mM NaCl, in the cold. Sample dialyzed against the same buffer is applied in 25 ml or less and eluted with 100 ml of the same. (Note that to work up the whole batch obtained from 100 ml of packed eggs will require repeated runs or scaling the procedure up, at this step only.) Unless there is an analytical interest in peaks II–IV with contaminant-bound enzyme (as obtained in Fig. 1), there is no need for a salt gradient, and the column may be regenerated directly, by running through it first 2 M NaCl in the same EPPS-Ca buffer, until A_{280} of effluent returns to base line; then the starting buffer, until refractive index returns to 1.3342. Fractions of 3 ml are convenient, and assay of alternate ones for enzyme activity and protein (by A_{280}) usually allows judicious selection of the peak fractions.

The table demonstrates that this step effects the greatest purification, and the resulting enzyme may be pure enough for many purposes, especially if early fractions are selected and yield is sacrificed.

Critical in this step is the salt concentration of the buffer, which is usually adjusted by refractometry. Lower ionic strength will spread the peak; higher will sharpen the peak, but bring more contaminant off early.

Step 3. Ammonium Sulfate Precipitation. Ammonium sulfate is powdered by mortar and pestle and added slowly to the pooled peak fractions from step 2, stirring on ice, to 2.5 M, with additions of 0.1 M NaOH to maintain pH at 8.5. Precipitate formed after storage at 4° for 6–8 hr is centrifuged out by 15,000 g for 30 min. Ammonium sulfate is added to bring the supernatant to 3.1 M; storage and centrifugation are repeated to obtain the enzyme, as a precipitate, and the tube is carefully dried of excess ammonium sulfate solution. Since the precipitate is stable, this is

PURIFICATION OF HATCHING ENZYME[a]

Step	Volume (ml)	Activity[b] (mU)	Protein[c] (mg)	Specific activity (mU/mg)	Yield (%)	Purification factor
Hatching supernatant	242.0	17.9	74.6	0.24	100	1
Ultrafilter concentrate	17.5	17.2	74.6	0.23	98	1
1. Urea-treated dialyzate	20.1	16.4	69.8	0.23	90	1
2. DEAE peak I	28.0	17.7	0.59	24.9	81	104
3. Ammonium sulfate precipitate	6.0	12.0	0.20	60.1	66	250
4. Sephadex peak	14.0	4.13	0.053	78.0	24	330

[a] From B. F. Edwards, W. R. Allen, and D. Barrett, manuscript in preparation.
[b] Assayed by casein digestion and trinitrobenzenesulfonic acid reaction.
[c] Assayed by method of O. H. Lowry, N. J. Rosebrough, A. L. Farr, and R. J. Randall, *J. Biol. Chem.* **193**, 265 (1951).

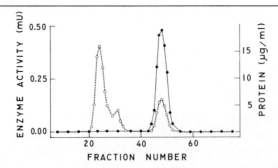

FIG. 2. Gel exclusion chromatography of hatching enzyme on Sephadex G-100 (step 4 of purification procedure). Ammonium sulfate precipitate was applied in 4 ml of 10 mM 4-(2-hydroxyethyl)-1-piperazinepropanesulfonic acid (pH 8.2), 10 mM CaCl$_2$, 150 mM NaCl, 10% (v/v) glycerol, to a column 2.5 × 36 cm in the cold. Elution was at 24 ml/hr; 2-ml fractions were collected and 50-μl samples each assayed for enzyme activity by the casein-trinitrobenzenesulfonic acid method, and for protein [O. H. Lowry, N. J. Rosebrough, A. L. Farr, and R. J. Randall, *J. Biol. Chem.* 193, 265 (1951)]. ●——●, enzyme activity; ○·····○, protein concentration Adapted from B. F. Edwards, Ph.D. thesis, University of California, Davis, 1976.

a good point at which to interrupt the procedure. Ammonium sulfate concentrations may need adjustment with different protein concentrations.

Step 4. Sephadex Gel Exclusion Chromatography. The ammonium sulfate precipitate is dissolved immediately before use in 10 mM EPPS (pH 8.2), 10 mM CaCl$_2$, 150 mM NaCl, 10% (v/v) glycerol. Sample is added to a column of Sephadex G-100, equilibrated in the cold against the same buffer, in about 2% of column volume, and fractions are eluted with the same. Figure 2 provides a convenient scale for running the step.

Enzyme is assayed as usual, but protein is now so dilute that direct spectrophotometry of samples at 280 nm is inadequate for its precise measurement, and a chemical method must be used.

The resulting samples may be concentrated for storage by precipitation with saturated ammonium sulfate.

Properties

Stability.[5,29] While crude enzyme is stable to repeated freeze-thaw cycles, pure enzyme, without substrate, is exceedingly unstable. Similarly, the temperature required to half-inactivate pure enzyme in a given time is 10° lower than that for crude enzyme; and thermal inactivation at 37° follows apparent first-order kinetics, with $t_{1/2} = 0.32$ hr for pure enzyme, 32 hr for crude enzyme.

[20] B. F. Edwards, Ph.D. Thesis, Univ. of California, Davis, 1976.

The addition of 10% glycerol (v/v) dramatically improves the stability to either freezing or heating, increasing the half-life of pure enzyme at 37°, for example, to 12 hr. As little as 0.1 mg/ml of protein, serum albumin or casein, is similarly protective.

The enzyme is unstable to acid, losing more than 95% of activity within 5 min at pH 3.

Purity. In the final purification step, gel exclusion chromatography on Sephadex G–100, specific activities are constant over the seven fractions of the symmetric peak (Fig. 2). A sample from this pool is precipitated over a narrow range of ammonium sulfate concentrations; on SDS–polyacrylamide gel electrophoresis after dansylation,[29] it gives a single fluorescent band, when sufficiently overloaded that a discrete contaminant of 1% would be detectable.

Despite these evidences of homogeneity, it is not excluded, in the light of the above discussion of stability, that some inactivated enzyme accompanies the active enzyme. Therefore the specific activity, 78 mU/mg, although it has been regularly observed at the end of the procedure, may be an underestimate.

Optimal Temperature, pH. Activity of either crude[26] or pure[29] enzyme shows a temperature optimum at 42°, and a pH optimum designated as 8.2, the midpoint of a broad plateau from 7.4 to 8.8.[26,29]

Kinetic Parameters. With pure hatching enzyme, and dimethylated casein as substrate at 35°, Edwards observed $k_{cat} = 0.038$ sec^{-1}, specific activity = 78 mU/mg, and $K_m = 0.93$ mg/ml.[29] If one assumes attack specifically on glutamyl and aspartyl bonds, the K_m corresponds to 0.84 mM, which falls between values similarly calculated[26] for trypsin and chymotrypsin working with their respective specificities on the same substrate.

Molecular Weight. Determinations of molecular weight by two approximate methods are in good agreement. Exclusion chromatography on Sephadex G-100, using globular protein standards, gave MW 30,000 ± 2500. SDS-polyacrylamide gel electrophoresis of the dansylated derivative, with dansylated protein standards, gave MW 28,500.[5,29]

Dependence on Metals. Hatching enzyme is adapted for activity in seawater (since the fertilization envelope poses no osmotic barrier to small ions), with roughly 10 mM Ca^{2+} and 52 mM Mg^{2+}. Dialysis of enzyme against artificial sea water lacking Ca^{2+} and Mg^{2+}, or rapid removal of the divalent cations by chelators, reduces activity to zero.[26,29] Adding back Ca^{2+} restores activity, except for a variable irreversible loss (between 0 and 20% with pure enzyme.) The kinetics of both inactivating enzyme by the chelator EGTA and activating it with Ca^{2+} are rapid, extrapolating back to zero lag period.[29] The enzyme shows a broad optimum of 10–25 mM Ca^{2+} for casein digestion,[29] but a sharp optimum near 10 mM for digestion of fertilization envelopes.[30] Of the other alkaline earth metals, Sr^{2+} can fully substitute for Ca^{2+}, Mg^{2+} is half as good, and Ba^{2+} cannot be adequately tested.[26]

Present evidence cannot establish whether Ca^{2+} participates in events at the active site. Vallee and Wacker have made the distinction between tightly bound ions in "metalloenzymes," and loosely bound ions in "metal–protein complexes."[31] While the firm binding of a metal ion usually correlates with its function in the active site, loosely bound Ca^{2+} is more likely to be implicated in stabilizing the protein conformation. The well-studied neutral proteases employ a single Zn^{2+} at the active site, while Ca^{2+} stabilizes the active conformation[32,33]; in α-amylase, however, Ca^{2+} occupies the active site.[34] Since activity is lost on dialysis, hatching enzyme seems to fit in the category of metal–protein complexes. Seeking quantitative data, Allen removed Ca^{2+} to inactivate, and determined the association constant for adding back the ion to reactivate. For the reaction:

$$\text{Inactive enzyme} + \text{Me}^{2+} \rightleftharpoons \text{active enzyme}$$

K_a approximates 5×10^3 for either Ca^{2+} or Sr^{2+},[26] in contrast to typical values of 10^{10} to 10^{30} for metalloenzymes.[31] But the data fit only for activation by binding a single ion per enzyme molecule, not by binding two or more.[26] It seems premature at present to exclude the possibility that divalent cation functions at the hatching enzyme catalytic site.

The transition and group Ib and IIb elements of the same period as calcium, ionized as divalent cations, are potent inhibitors of hatching enzyme, even with equimolar Ca^{2+} present: Cr^{2+}, Fe^{2+}, Co^{2+}, Cu^{2+}, and Zn^{2+} reduce activity essentially to zero; Mn^{2+} is somewhat less effective.[26]

Inhibitors. Inhibition by the chelators EDTA and EGTA is complete at equivalence and is largely reversible, as discussed above. The enzyme is not inhibited by thiol reagents 2-mercaptoethanol, p-hydroxymercuribenzoate, and N-ethylmaleimide, nor by reagents specific for serine proteases, diisopropyl phosphofluoridate (DFP), phenylmethylsulfonyl fluoride, or soybean trypsin inhibitor or toluenesulfonyl lysine chloromethyl ketone.[7,26]

[30] D. Barrett, unpublished observations, 1968.
[31] B. L. Vallee and W. E. C. Wacker, *in* "The Proteins" (H. Neurath, ed.), 2nd ed., Vol. 5, p. 25. Academic Press, New York, 1970.
[32] H. Matsubara and J. Feder, *in* "The Enzymes" (P. D. Boyer, ed.) 3rd ed., Vol. 3, p. 721. Academic Press, New York, 1971.
[33] M. L. Bade and J. J. Shoukimas, *J. Insect Physiol.* **20**, 281 (1974).
[34] B. L. Vallee, E. A. Stein, W. N. Sumerwell, and E. H. Fischer, *J. Biol. Chem.* **234**, 2901 (1959).

In support of the specificity to be proposed below, Z-Glu (N^α-benzyl-oxycarbonyl-L-glutamate) has been found to inhibit hatching enzymes from *S. purpuratus*[35] and *Arbacia punctulata;* in the latter case the mechanism is competitive and K_I is 2.4 mM.[36]

Strong inhibitory effects of metallic ions have been mentioned above.

Substrate Specificity. Considerable evidence, although much of it is preliminary or indirect, indicates that hatching enzyme is an endopeptidase specific for peptide bonds involving the carboxyl of a dicarboxylic amino acid.

A rigorous demonstration that the enzyme is an *endo*peptidase is lacking, but the assumption is supported by the datum that fertilization envelopes are reduced to soluble fragments by the hydrolysis by hatching enzyme of fewer than 4% of their peptide bonds.[37]

Evidence for the attack of acidic aminoacyl peptide bonds is of three kinds: COOH-terminal analysis of hatching enzyme digests; testing of dipeptides as substrates; and correlation of the extent of a protein's hydrolysis with its content of Glu and Asp.

After hydrazinolysis of a hatching enzyme digest of hemoglobin, only Glu and Asp pass through an Amberlite CG-50 column.[26] This finding positively implicates Glx- and Asx- bonds as sites of enzyme attack, but does not distinguish between the acid and the amide, nor does it exclude Lys- and Arg- bonds, which would not be detected by the procedure.

Dipeptides further restrict the choices for COOH-terminal specificity. Hatching enzyme, soluble[29] or conjugated to Sepharose,[38] does cleave the peptide bonds of Z-Glu-Tyr, Z-Glu-Phe, and more slowly, Z-Glu-Leu. Significantly, it does not cleave Z-Gln-Val-O-Me.[18] It also fails to cleave Tos-Arg-Val, Bz-Arg-naphthylamide, Z-Tyr-Gly-NH₂, and the esters commonly used to detect tryptic and chymotryptic specificities.[7]

Standing out against these indications of attack to the right of an acidic amino acid is Allen's observation[26] that the enzyme attacks only one bond in bovine vasopressin, at a residue identified by hydrazinolysis as Glx. Allen interpreted that the Gln⁴-bond was susceptible, while the AsN⁵-bond was not. An alternative explanation now seems reasonable—that long storage of the vasopressin had allowed spontaneous deamidation of residue 4 from Gln to Glu; AsN residues in protein are far more stable.[39]

At various times the enzyme has been used on a variety of common

[35] Ellen Kraig and D. Barrett, unpublished observations, 1975.
[36] G. W. Lopez and D. Barrett, *Biol. Bull.* **147**, 489 (1974).
[37] D. Barrett and Daniel Mazia, unpublished observations, 1964.
[38] Robert W. Mommsen and D. Barrett, unpublished observations, 1973.
[39] B. Blombäck, this series Vol. 11, p. 398.

purified globular and fibrous proteins, all of which were vigorously attacked except for salmine,[26] which is an exceptional protein in completely lacking the dicarboxylic amino acids.[40] The enzyme attacks copoly-Glu,Ala,Tyr (6:3:1) and more slowly copoly-Glu,Tyr (9:1),[18] though at rather low rates, perhaps because of steric effects[41]

The stoichiometries of insulin and casein hydrolysis by hatching enzyme offer some further corroboration of the restricted specificity we propose. When the hydrolysis of oxidized insulin (which has 4 Glu, no Asp, and 5 amide residues) had proceeded to completion, between 3 and 4 NH_2-terminals had been released.[42] Similarly in casein (which contains roughly[43] 18 residue % Glx, 6.5% Asx, and 12.3% Glu + Asp),[44] hatching enzyme had essentially finished cleaving at about 12% of bonds.[26] In each case reaction did cease because of exhaustion of susceptible bonds, since enzyme was shown still to be active at the end.

Two experiments afford dramatic demonstrations of how well the hatching enzyme is adapted to its natural biological substrate. First, papain, trypsin, and hatching enzyme have similar k_{cat} values on casein substrate, but when the three are set to dissolve fertilization envelopes from without, hatching enzyme works 100 times faster than the other two.[16] Second, when in the latter assay hatching enzyme is working on 200 embryos, other proteins can be added to compete for the enzyme. But with 500 ng of fertilization envelopes, it requires more than 500 μg of cytochrome c, albumin, etc., to effect detectable competition.[16]

What underlies this complementarity between the enzyme and its normal substrate? A first hypothesis requires configurational specificity and is prompted by the consideration that most proteases preferentially attack denatured proteins, while hatching enzyme is adapted to dissolve native, undenatured fertilization envelopes. The experiment cited first above fits particularly well, inasmuch as hatching enzyme has the advantage over the common proteases on native envelopes, but shows no advantage on casein, which has a sufficiently disordered configuration to act in solution like a denatured protein.[45] Allen provides further suggestive evidence in a comparison of the enzyme's attack on casein versus its attack on tryptic and chymotryptic peptides prepared from casein. Its considerably slower attack on the peptides[26] (which differ from casein

[40] T. Ando and S. Watanabe, *Int. J. Pro. Res.* **1**, 221 (1969).

[41] B. Pullman and A. Pullman, *Adv. Protein Chem.* **28**, 348 (1974).

[42] David P. Falconer and D. Barrett, unpublished observations, 1973.

[43] Commercial casein is a nonspecified mixture of several molecules, α-, β-, γ-, and κ-caseins, but their compositions vary little with respect to acidic residues.

[44] Calculated from the amino acid composition of whole bovine casein: T. L. McMeekin and B. D. Polis, *Adv. Protein Chem.* **5**, 201 (1951).

[45] H. A. McKenzie, *Adv. Protein Chem.* **22**, 55 (1967).

only in the loss of bonds which hatching enzyme could not cleave) suggests that size or conformation must be decisive.

A second hypothesis—that specific amino acid sequences in the fertilization envelope near to Glu residues increase their digestibility—has not yet been tested. The overall amino acid composition of fertilization envelopes[18] gives no lead, being unremarkable except for a paucity of hydrophobic residues.

Experimental evidence does not yet warrant the verification or exclusion of either hypothesis.

Similarities to Other Enzymes. Hatching enzyme bears comparison with: pepsin, EC 3.4.23.1,[46] which hydrolyzes glutamyl and aspartyl bonds, but not at a pH where the side chains are ionized; the staphylococcal protease,[47] EC 3.4.21, which attacks glutamyl and aspartyl bonds at neutral pH, but is DFP-sensitive and EDTA-insensitive; and the class of neutral, chelator-sensitive proteases, EC 3.4.24,[46] which have similar metal dependencies, but different substrate specificities. Hatching enzyme seems best classified at present in EC 3.4.24.

Hatching Enzymes of other Species

As cortical granules which precurse fertilization envelopes vary widely in morphology from species to species,[48] and the envelopes themselves vary markedly, at least in thickness and elevation from the egg, so it might be expected that hatching enzymes too will be variable, their evolution being constrained mainly by the requirement that substrate and enzyme coevolve to remain complementary. This general consideration suggests a prudent pessimism in generalizing from the methods and properties here presented for *S. purpuratus* to other species. The few available data reflect on two points: (1) Are other hatching enzymes also tightly bound to fragments of envelope (so that dissociation by urea must precede purification)? and (2) Are their substrate specificities comparable to that proposed for *S. purpuratus* enzyme?

Enzyme Association with Contaminants. Enzyme is found in the supernatant above hatched *S. purpuratus* embryos with about 300 times its weight of soluble envelope fragments, in an association so tenacious that 6 M urea is required to release it. Crude enzyme from *Arbacia punctulata*[49] and *A. lixula*[20] is associated in the same way, and it has

[46] Commission on Biochemical Nomenclature, "Enzyme Nomenclature." Elsevier, Amsterdam, 1973.

[47] G. Drapeau, Y. Boily, and J. Houmard, *J. Biol. Chem.* 247, 6720 (1972).

[48] J. Runnström, *Adv. Morphol.* 5, 222 (1966).

[49] B. F. Edwards and D. Barrett, *Biol. Bull.* 143, 459 (1972).

been speculated[49] that if the enzyme did not adhere to the envelope in some fashion, it should be able to escape the confines of the envelope as soon as it digested small gaps, without ever hatching the embryo.

Nonetheless Yasumasu has reported[3] that *Anthocidaris crassispina* enzyme is accompanied in the crude hatching supernatant by only a 10-fold contaminant, and that is easily separated from it by simple ammonium sulfate precipitation. Comparison of data for *Anthocidaris* and *Strongylocentrotus* purifications indicates that the amount of enzyme secreted by blastulae of the two species is approximately the same. The *Anthocidaris* enzyme appears further to differ most significantly from *Strongylocentrotus* enzyme in lacking a requirement for calcium, since it can be assayed in 25 mM phosphate.

Substrate Specificity. Species specificity in hatching enzymes has been found wherever it has been sought. Whiteley[50] considers the barrier between *S. purpuratus* (sea urchin) and *Dendraster excentricus* (sand dollar) absolute: neither enzyme can digest the other envelope.

More usually, at least among the urchins, enzymes do digest envelopes across species lines. The null hypothesis, that there is no species specificity, predicts that within a set of species tested the envelopes will differ in thickness, and the enzymes in strength, but the hierarchy of enzyme strengths determined on one species' envelopes will hold for each of the other species' envelopes too. Results have departed widely from this expectation in two sets of experiments. No consistent pattern can be discerned at all, in either the group *Arbacia punctulata, Lytechinus variegatus,* and *S. purpuratus,*[51] or the group *A. punctulata, Echinarachnius parma* (sand dollar), *Lytechinus pictus,* and *S. purpuratus,*[52] except that an enzyme usually attacks best the fertilization envelopes of its own species.

The difference between enzymes from congeneric *S. purpuratus* and *S. franciscanus* is small, but has been magnified to a useful tool[8] by assaying in the presence of Mn^{2+}, to which they are differentially susceptible.

The fact that *Arbacia* and *Strongylocentrotus*, in different suborders, cross-digest rather poorly,[51,52] taken with the likelihood that the enzymes share the specificity for Glu and Asp bonds (since both are inhibited by Z-Glu[35,36]), leads to the pleasing speculation that the chemistry of the susceptible peptide bond may have been widely conserved in evolution, while the configuration of surrounding protein has diverged and accounts for species differences.

[50] Arthur H. Whiteley, personal communication, 1970.
[51] Robert W. Mommsen and B. F. Edwards, unpublished observations, 1972.
[52] Gerald M. Edelman, Jeffrey Olliffe, and D. Barrett, unpublished observations, 1974.

On the other hand, Kettaneh, working with *Arbacia lixula* enzyme, has found attack of a variety of unblocked Gly and Ala dipeptides.[20] This pattern would seem to indicate a different specificity altogether, although reciprocal experiments pitting *A. lixula* enzyme on Glu peptides and *S. purpuratus* enzyme on Ala bonds, have not actually been performed. And Vacquier has proposed[53] that *Echinometra vanbrunti* may use a glucanase as a hatching enzyme, although he has not tested the species for a hatching protease.

In short, with the fragmentary evidence at hand one can perceive the outlines of a large family of echinoid hatching enzymes, but how much unity and how much diversity the group can offer the protease chemist is not yet clear.

[53] V. D. Vacquier, *Exp. Cell Res.* **93**, 202 (1975).

Section VI

Dipeptidases

[30] A Dipeptidase from *Escherichia coli* B[1]

By ELIZABETH K. PATTERSON

Dipeptidases hydrolyze only α-L-dipeptides with free amino and carboxyl groups. The exception observed is exceedingly slow hydrolysis of dipeptides with methyl-substituted amino groups. The products of the reaction are the constituent amino acids of the dipeptides. Dipeptides with amino groups blocked by a benzoyl or carbobenzoxy group, dipeptide or amino acid amides, dipeptides with the amino group in the β-position, and tripeptides, are not hydrolyzed.[2]

Since the dipeptidases are Zn-metalloenzymes[3] and easily inhibited by divalent metal ions, precautions to avoid metal ion contamination must be taken. Scrupulously clean rooms and glassware, water of greater than 10^6 ohms resistance, and chemicals of the highest purity must be used.

Assay Method A

Principle. The amino acids resulting from the hydrolyses of dipeptides are estimated by a photometric ninhydrin method that is a modification[2,3] of that of Matheson and Tattrie.[4] The ninhydrin-cyanide reagent gives quantitative color yields that are an order of magnitude higher with most of the constituent amino acids than those produced by the dipeptides. The exception is N-terminal glycyl dipeptides where the difference is too small for accurate determination. This assay method is used to follow the enzyme activity during dipeptidase purification[2] with the substrate Ala-Gly (50 mM).

Reagents

Potassium phosphate buffer, 0.02 M, pH 8.3.

L-Alanylglycine, 0.1 M. The peptide is dissolved in the buffer and brought to pH 8.3 by addition of 0.1 M NaOH. This substrate solution is diluted 1:1 with the enzyme solution in 0.02 M potassium phosphate, pH 8.2, resulting in a 50 mM substrate concentration.

[1] This work was supported by grants from the U.S. Public Health Service and The National Institutes of Health (CA-03102, CA-06927, and RR-05539), and by an appropriation from the Commonwealth of Pennsylvania.
[2] E. K. Patterson, J. S. Gatmaitan, and S. Hayman, *Biochemistry* **12**, 3701 (1973).
[3] S. Hayman, J. S. Gatmaitan, and E. K. Patterson, *Biochemistry* **13**, 4486 (1974).
[4] A. T. Matheson and B. L. Tattrie, *Can. J. Biochem.* **42**, 95 (1964).

L-Alanine, 0.1 *M*, and glycine, 0.1 *M*, are made up with the same concentrations of potassium phosphate buffer and NaOH used for the dipeptide solution.

HCl, 0.1 *N* in 95% alcohol

Sodium citrate buffer, 0.2 *M*, pH 4.7

Ninhydrin–potassium cyanide–methyl Cellosolve reagent. A stock solution containing 2.5 g of ninhydrin (Pierce Chemical Co.) dissolved in 50 ml of methyl Cellosolve (100% ethylene glycol monomethyl ether, Fisher certified reagent kept in the presence of stainless steel wire), and KCN, 0.01 *M*, is made up monthly. The reagent is prepared by first diluting 2 ml of 0.01 *M* aqueous KCN solution to 100 ml with methyl Cellosolve and then adding to it 20 ml of the stock ninhydrin solution. A red color develops which then fades to a yellow color. Storage of this reagent overnight at 4° ensures low blank readings. The reagent is stable for about 2 weeks when stored at 4°.

Procedure. The enzyme solutions are diluted with sufficient buffer so that 50% hydrolysis is not exceeded. At early stages of purification when the protein concentration is above 3 mg/ml, the buffer used is 0.02 *M* potassium phosphate, 0.25 *M* sucrose, pH 8.2. After step 2 (see below) to ensure stability at 40° with lower protein concentrations, the buffer is 2 mg/ml bovine serum albumin (BSA) in 0.02 *M* potassium phosphate, pH 8.2. The total volume of the reaction mixture is 14 μl. Small (250 × 4 mm i.d.) vessels[5] are used. Seven microliters of diluted enzyme solution are added to the bottom of the tubes by using Carlsberg[5] micropipettes (D. B. Micropipettes, Skovskellet 21, Holte, Denmark), which were calibrated by being weighed full and then blown out. The tubes are capped with black rubber "hats" (F1443 Pederson, from A. H. Thomas Co.) and warmed for 1 min in a 40° water bath, then 7 μl of substrate solution is added and stirred by tapping. The following controls are run in parallel with the reactions: (1) phosphate–sucrose or phosphate–BSA instead of enzyme solution, and (2) enzyme solution alone with substrate added after stopping the reaction. After 5 min at 40°, the reaction is terminated by addition of 7 μl of 0.1 *N* HCl in 95% alcohol. Six-microliter aliquots of the above mixtures are transferred to test tubes (100 × 10 mm i.d.), and in the following order are added 1 ml of H_2O, 0.5 ml of citrate buffer, and 1 ml of ninhydrin reagent. After thorough mixing by a vortex mixer, the tubes are capped with metal culture tube caps and heated in a boiling water bath for exactly 7.5 min. After cooling in an ice bath, 2.5 ml of 60% alcohol are added, the contents of the tubes mixed by inverting

[5] K. Linderstrøm-Lang and H. Holter, *C. R. Trav. Lab. Carlsberg* **19**, 1 (1933).

(with Parafilm against the tops) and the absorbance read in a spectro-photometer at 570 nm. A standard curve is obtained by making mixtures of 0.1 M alanine and 0.1 M glycine and 0.1 M alanylglycine to simulate various degrees of hydrolysis. Linearity was obtained with the enzymic reaction when the extent of hydrolysis did not exceed 50%.

Assay Method B

Principle. The assay method of Schmitt and Siebert[6] (1961) depends on the difference in absorbance in the low ultraviolet between dipeptides and their constituent amino acids. This method can be used with low concentrations of purified enzymes when the ultraviolet absorbancy is sufficiently small not to interfere with the absorbancy of the peptide. This method is used for kinetic studies in which the concentration of a variety of dipeptide substrates was varied from 50 down to 0.8 mM.

Reagents

Potassium phosphate 0.030 M, sodium borate 0.035 M, 0.25 M sucrose buffer, pH 8.3 [the enzyme solution contains 0.25 M sucrose for stability to freezing ($-30°$)]

Peptide made up 0.053 M in above buffer and adjusted to pH 8.3 with 0.1 N NaOH

Constituent amino acids, 0.053 M, in the above buffer and NaOH concentration

Procedure.[2] The reaction is carried out in silica micro cells (2 × 10 mm i.d.) in a total volume of 155 μl. The Gilford spectrophotometer is equipped with thermospacers on either side of the cell compartment through which 30° water is circulated; 150 μl of substrate solution of the required concentration in phosphate–borate–sucrose buffer is added to a cuvette, which is then warmed to 30° for 5 min. Then 5 μl of the enzyme dilution in the same buffer is rapidly (15 sec) stirred in by means of the 5-μl pipette, and the reaction is followed at either 230 or 235 nm by means of a recorder. An expanded scale of 0–0.3 is usually employed. The decrease in absorbance observed is frequently preceded by a lag before linearity occurs. Only the linear portion of the record is employed for determination of rates. Changes in absorbance per minute are converted to rates of hydrolysis by use of factors (Table I) calculated for each substrate from the measured differences in absorbance between the dipeptide and its constituent amino acids at the given wavelength and temperature. In comparing the maximum velocities of hy-

[6] A. Schmitt and G. Siebert, *Biochem. Z.* **334**, 96 (1961).

TABLE I
STANDARDIZATION FACTORS FOR PEPTIDE HYDROLYSIS

Peptide	Moles/ $A_{230\ nm}{}^{a}$ ($\times 10^3$)	Moles/ $A_{235\ nm}$ ($\times 10^3$)	Peptide	Moles/ $A_{230\ nm}$ ($\times 10^3$)	Moles/ $A_{235\ nm}$ ($\times 10^3$)
Gly-Gly	6.60	19.7	CH₃Gly-Ala	2.66	
-Ala	4.38	11.7	Ala-Gly	5.12	14.7
-Ser	3.66	11.2	-Ala		8.97
-Thr	3.80		-Abu	3.15	8.58
-Abu[b]	3.15		-Val	2.24	7.02
-Val	3.00	10.1	-Nva	2.58	7.50
-Leu	3.73	11.1	-Leu	2.71	7.75
-Nle	3.63	10.9	-Nle	2.67	6.96
-Ile	2.29	7.56	-Ile	2.12	6.14
-Phe	1.78	5.73	-Phe	1.63	5.31
-Met		5.04	-Lys	3.17	8.10
-Lys	3.75	10.2	-Arg		1.54
-Asn	3.26	7.98	Val-Gly		10.9
-Asp	4.17	11.4	-Ala	2.63	7.11
-Pro	0.64	1.94	-Val	1.89	5.85
Abu-Ala	3.26	8.76	-Leu	2.85	7.65
-Ile	1.47	5.22	-Ile	1.60	4.83
Ile-Ala	2.81	8.31	Leu-Gly	5.04	11.6
-Val	1.66	5.04	-Ala	3.11	7.59
-Leu	1.77	4.90	-Val	1.99	5.40
-Ile	1.56	4.53	-Leu	2.35	5.82
Met-Gly	1.67	5.22	-Phe	1.38	4.17
-Ala	1.66	4.59	Nle-Gly	4.74	14.6
-Leu	1.50	3.95	-Ala	3.12	
Phe-Gly	3.14	5.73	Pro-Gly	4.86	13.8
-Ala	1.15	3.66	Lys-Gly	3.30	7.44

[a] Moles hydrolyzed per liter per absorbance unit. The absorbances of the peptides and amino acids are read at the given wavelengths, and from these values are subtracted the absorbances of the buffer. Then the ratio is obtained between the given molarity and the difference between the values obtained for the peptide and constituent amino acids.

[b] α-NH₂-butyric acid.

drolysis of various substrates, the rate of hydrolysis of a 12.5 mM Ala-Gly standard is measured routinely in order to allow for differences in enzymic activity in separate experiments. Maximum velocities are then calculated relative to that obtained with Ala-Gly as substrate.

Definition of Unit of Enzyme Activity. One unit of enzyme activity is defined as that amount required to hydrolyze 1 μmole of Ala-Gly per minute at 40°. Specific activity (S.A.) is enzyme units per milligram of

TABLE II

PURIFICATION OF A DIPEPTIDASE FROM *Escherichia coli* B[a,b]

Fraction	Protein (mg)	Units	Specific activity (units/mg)	Yield (%)
Streptomycin supernatant	5180	15,400	3.0	—
$(NH_4)_2SO_4$, 30–50%	540	11,300	21	73
G-150 filtrate	225	9,090	40	59
DEAE-cellulose	31	5,700	184	37
DEAE-cellulose[c]	2.0	350	175	—
Hydroxylapatite[d]	0.24	183	760	—

[a] Reprinted with permission from E. K. Patterson, J. S. Gatmaitan, S. Hayman, *Biochemistry* **12**, 3701 (1973). Copyright by the American Chemical Society.

[b] The numbers given refer to the material used as a sample for the next step in purification.

[c] Only one-tenth of the material (one fraction) from the peak of enzyme activity off DEAE was used for the hydroxyapatite step.

[d] The values given refer to the sum of four fractions at the dipeptidase peak.

protein. Protein is assayed by a modified[7] method of Nayyar and Glick[8] in which protein, but not nucleic acid, is precipitated by the dye sulfobromophthalein. BSA is used as a standard. This method does not depend on the presence of aromatic amino acids in the protein. Molecular activities $[V_M/[E] (sec^{-1})]$, moles substrate hydrolyzed at 40° per second per mole enzyme were calculated from molecular weights and specific activities at 40°.

Purification Procedure (Table II)

Step 1. Ammonium Sulfate Precipitation. Escherichia coli strain B is grown in a medium containing 1.1% K_2HPO_4, 0.85% KH_2PO_4, 0.6% Difco yeast extract, and 1% glucose. The growth and harvesting of the cells is carried out by the Grain Processing Corporation, Muscatine, Iowa. The method of Richardson *et al.*[9] is used for preparation of the extract of the late-log phase cells in glycyl-glycine buffer 0.05 *M*, pH 7.0. The temperature is kept below 12°. Streptomycin precipitation[9] is carried out by diluting the extract with equal parts of Tris buffer, 0.05 *M*, pH

[7] S. Hayman and E. K. Patterson, *J. Biol. Chem.* **246**, 660 (1971).

[8] S. N. Nayyar and D. Glick, *J. Histochem. Cytochem.* **2**, 282 (1954).

[9] C. C. Richardson, C. L. Schildkraut, H. V. Aposhian, and A. Kornberg, *J. Biol. Chem.* **239**, 222 (1964).

7.5, 0.001 M EDTA,[10] and slowly adding about $\frac{1}{14}$ its volume of freshly prepared 5% streptocycin sulfate solution. After centrifugation at 15,000 g for 10 min, the supernatant fluid is collected, precipitated at approximately 50% ammonium sulfate saturation, and frozen. This fraction constitutes the starting material (Table II) and, in our case, is a by-product of purification of DNA polymerase by Drs. Lawrence Loeb and John Slater of this Institute, who kindly contributed the material.

All the following operations are carried out in a cold ($2°$–$8°$) box in a dust-free room. The frozen material (103 g) is thawed (80 ml) and diluted with 200 ml of cold ($0°$–$4°$) 0.02 M potassium phosphate buffer, pH 8.3. Cold saturated $(NH_4)_2SO_4$ (Schwarz-Mann Ultra-pure) (120 ml) is stirred in to make a 30% solution. After 15 min stirring, centrifugation at 16,200 rpm for 20 min is carried out in a high-speed head in a PR2 refrigerated centrifuge at $2°$. To the supernatant solution (370 ml) is added 148 ml of saturated $(NH_4)_2SO_4$ to give 50% saturation. After stirring for 0.5 hr, followed by centrifugation at 16,200 rpm for 0.5 hr, the precipitate is dissolved in 0.02 M KPO$_4$ buffer and dialyzed overnight against the same buffer with three changes of buffer.

Step 2. Sephadex G-150 Gel Filtration. Twenty grams of the 44–53 μm fraction (obtained by sieving the dry powder) of Pharmacia Sephadex G-150 is swollen for 5 hr at $70 \pm 5°$ in 0.02 M potassium phosphate buffer, pH 8.3. After cooling and decanting 3 times, the Sephadex slurry is poured into a Pharmacia column measuring 2.5×100 cm and packed at a hydrostatic head of 10–15 cm. The column is washed overnight with phosphate buffer at a constant pressure obtained with a Sigmamotor peristaltic pump (at 8%). The flow rate is about 6 ml/hr. After removing excess Sephadex and inserting the Pharmacia sample applicator, the column (2.5×96.5 cm) is further washed until equilibrated. One-half the volume from step 1 constitutes the sample (20.5 ml containing 278 mg protein and 5770 enzyme units); this is carefully added to the bottom of the applicator by means of a Pasteur pipette with a bent, fire-polished tip and allowed to slowly enter the column with a head of 3–5 cm. When the sample is just through the applicator net, the sides of the applicator are washed down with 1–2 ml of buffer. After this has entered the Sephadex, buffer is added to fill the column completely and the top cover is screwed on. Pumping is again carried out at 8% (6 ml/hr) and fractions of 3–4 ml are collected by means of a timer and fraction collector. Solid sucrose sufficient to give 0.25 M solutions is added to the tubes prior to collection. The effluent is passed through a Gilford spectrophotometer

[10] Since the dipeptidase is a Zn-metalloenzyme, it is preferable to avoid exposure to EDTA.

at 280 nm which records on an Esterline-Angus meter. Fractions are collected only after the meter records outflow of protein. The enzyme recovery is usually 100%. Fractions should be checked for bacterial contamination; if positive, the solutions should be put through Millipore filters.

Step 3. DEAE-Cellulose Chromatography. To prepare the DEAE, 25 g of DE-32 (Whatman) are stirred gently with 375 ml of 0.5 N HCl and allowed to stand for 30 min at room temperature ($22 \pm 2°$). After filtration with suction through a Büchner funnel with sintered glass, the material is transferred to a beaker and washed with water by gentle stirring. The filtration and washing are repeated until the pH of the effluent is 3.7. Then 375 ml of 0.5 N NaOH is gently stirred into the DEAE and the slurry allowed to stand for 30 min. After filtration, this process is repeated and then the material is washed with water until pH 7 is reached. The DEAE may be stored at 4° at this point. For degassing the preparation, after filtration, the DEAE is suspended in a mixture of 100 ml of 1 M potassium phosphate, pH 8.2, and 125 ml of 1 N HCl, pH 4, in a filter flask. The flask is stoppered, evacuated with an aspirator while the DEAE is stirred with a magnetic stirrer until all bubbling has stopped. After filtration, the DEAE is suspended in 1 M potassium phosphate, pH 8.2, and filtered; this process is repeated until the effluent pH is 8.2. Then the material is suspended in 3 M NaCl, 0.02 M potassium phosphate, 0.5 mM $MgCl_2$, pH 7.0. The process of suspension and filtration is repeated until the pH of the effluent is 7.1. The DEAE is then washed with buffer, 0.02 M potassium phosphate, 0.01 M NaCl, 0.5 mM $MgCl_2$, pH 8.2, and filtered; the process is repeated until pH 8.2 is reached. To remove fine particles, the slurry (550 ml) is transferred to a 1-liter cylinder, and 200 ml of the above buffer is stirred into it. After standing for 45 min, the top portion is gently poured off, leaving 225 ml in the cylinder.

The resulting slurry of DEAE is packed in a 2.5×30 cm Pharmacia column using a Sigmamotor pump at a setting of 80% for 2 hr. Then the DEAE in the column (2.5×22 cm) is washed with buffer overnight with the pump at 10%. The sample (S.A. 90–100) consists of the pooled contents of tubes from the peak of enzyme activity of two Sephadex G-150 filtrations (step 2). In the purification given in Table II, 33,000 units of enzyme and 360 mg of protein in 76 ml were applied to the column with the use of a Pharmacia sample applicator, and the sample was introduced into the DEAE with the pump at 35%. Fractions of 1 ml were collected in tubes containing solid sucrose. The enzyme was eluted after the breakthrough peak with an exponential gradient of NaCl (0.01–1.0 M using a 250-ml mixing flask) in phosphate-$MgCl_2$ buffer. The enzyme recovery in this case was 96%.

Step 4. Hydroxylapatite Chromatography. Although this step of the purification outlined in Table II was carried out on a small scale with the material from only one fraction resulting from the DEAE chromatography, this method can be scaled up to accommodate far larger samples. Bio-Gen HTP hydroxylapatite, 0.5 g, is suspended in 100 ml of 0.01 M sodium phosphate, 0.5 mM MgCl$_2$, pH 7.6, in a cylinder. After settling for 5 min, the supernatant is decanted, and the process is repeated. The slurry is then packed in a Pharmacia column (0.9 × 15 cm) giving a height of 1 cm. The sample containing 350 enzyme units, 2.0 mg of protein in 1.5 ml is concentrated, and the buffer is exchanged with the buffer used for the hydroxylapatite by the use of a Schleicher and Schuell conical concentrator in an ice bath. The final volume of the sample is 1.1 ml; this includes a buffer rinse of the cone. The sample is carefully added directly onto the surface of the hydroxylapatite; when it is all in the column, buffer is added and the pump is run at 20%. Fractions of 0.7 ml are collected into tubes containing solid sucrose and when the breakthrough peak is off, an exponential gradient from 0.02 to 0.5 M potassium phosphate, pH 7.6, containing 0.5 mM MgCl$_2$ in a 125-ml mixing flask is used to elute further material. The bulk (about 90%) of the enzyme comes off just after the breakthrough protein. The total recovery is 88%.

The total of these steps (Table II) results in a 250-fold purification from the streptomycin supernatant and 350-fold from the cell extract (specific activity 2.3).

Properties

Purity and Physical Properties. The purified dipeptidase exhibits a single band after sodium dodecyl sulfate-acrylamide gel electrophoresis. Densitometry indicates at least 90% purity. The band corresponds to a molecular weight of about 53,000. The molecular weight obtained by means of Sephadex G-150 gel filtration along with standard proteins is 100,000 ± 5000. Thus, the enzyme may be composed of two identical subunits. The fact that the enzyme contains 2 moles of Zn per mole of enzyme reinforces this assumption.

Stability. The enzyme is stable to freezing at −30° in the presence of buffer (pH 7.5–8.5) and 0.25 M sucrose, and stable to dialysis (18 hr) at 10° against 0.02 M phosphate buffer (pH 7.5 or 8.3). When in the presence of 2 mg/ml bovine serum albumin in phosphate buffer, pH 8.2, the *E. coli* dipeptidase is stable at 0° for 24 hr. In phosphate–borate–0.25 M sucrose buffer of pH's from 6.3 to 9.0, the enzyme (0.37 mg/ml) retained 100% activity for 6 hr at 0°. At 40°, the temperature used for ninhydrin assay of the dipeptidase, it is necessary to dilute the enzyme in BSA to maintain stability for 5–10 min. At 30°, the temperature used for kinetic studies, the enzyme in phosphate-sucrose buffer is stable for the period of the reactions (2–30 min) but decreases to 60% of the initial activity in 24 hr.

Effect of Thiols and SH Reagents. Mercaptoethanol, dithiothreitol, and cysteine (1.0 mM) inhibit the dipeptidase roughly parallel to the metal-chelating ability of these reducing substances.[6] Iodoacetate (1.0 mM) causes a time-dependent inhibition; prior incubation (25°) of the enzyme with iodoacetate results in 20% inhibition after 30 min and 60% after 120 min of preincubation. The reaction is not reversed by cysteine and may be caused by oxidation of a group in the enzyme other than cysteine.

Specificity. The *E. coli* B dipeptidase, in the absence of added metal ions, has been found[2] to hydrolyze at least 36 α-L-dipeptides at molecular activities (V_M/[E] sec^{-1}) varying from 0.1 to 16 $\times$ 10^3. Substrates cleaved most rapidly contain long (Met-) or flat (Phe-) N-terminal R-groups and small (-Gly or -Ala) or positively charged (-Lys) C-terminal groups. Peptides not hydrolyzed include peptides with branched C-terminal groups (-Val, -Ile, and -Leu in some cases) or negatively charged N-terminal groups (Asp-).

Effect of Added Metal Ions. The addition of Mn^{2+} or Co^{2+} to the enzyme changes the substrate specificity.[3] When substrates are cleaved rapidly by the dipeptidase, addition of these metals inhibits the reaction. The more slowly a substrate is hydrolyzed in the absence of metals, the greater the activating effect of Co^{2+} and Mn^{2+} on the reaction. Substrates, such as Ala-Ile, Asp-Gly, and D-Leu-Gly, that are not detectably cleaved without metal addition, are hydrolyzed at appreciable rates when Co^{2+} is present. The activation by Mn^{2+} is accompanied by an increase in K_m, that of Co^{2+} with no change in K_m. Mn^{2+} prevents and Co^{2+} increases the inhibition by high concentrations of substrate. Kinetic data suggest the existence of two binding sites on the dipeptidase for Co^{2+} and two for Mn^{2+}, at least one of which is distinct from a Co^{2+} site. Because Zn^{2+} is lost on long-term incubation of the enzyme with Co^{2+} or Mn^{2+}, the common site probably is that usually occupied by Zn^{2+}.

pH Dependence.[2] With Ala-Gly as substrate, plots of V_M versus pH in the range of pH 6.0 to 9.5 give a bell-shaped curve with a peak at pH 8.0. Plots of K_m versus pH are flat in the region of pH 7.5–8.5, ascending above and below this pH region. From these data, the acid and basic pKs of the enzyme are calculated to be 8.0 and 8.2, and those of the enzyme–substrate complex, 6.7 and 9.1.

Inhibitors.[2] The dipeptidase is inhibited by metal chelators in a manner typical of metalloenzymes. With 0.05 μM enzyme, addition of 50 μM o-phenanthroline results in 40% inhibition in 45 sec. The isomeric non-chelator, m-phenanthroline, has no effect. The inhibition of Ala-Gly hydrolysis by EDTA, however, is time-dependent and logarithmic after a sharp initial drop in activity. Instantaneous full recovery of activity can be achieved by addition of either Zn^{2+} or Mn^{2+} in concentrations in slight excess over the EDTA concentration.

Other inhibitors of the enzymic hydrolysis of L-dipeptides are the D-forms of the dipeptides. D-Leu-Gly and Gly-D-Leu are linear competitive inhibitors of the hydrolysis of L-Ala-Gly with K_i's of 0.35 and 9.0 mM, respectively. The amino acid products of the reaction are also inhibitors, the K_i for alanine being 5.9 mM.

Metal ions other than Mn^{2+} and Co^{2+} also inhibit the enzymic hydrolysis of Ala-Gly (50 mM). The inhibition varies according to the metal ion used and the concentrations (10^{-5} to 10^{-3} M). The inhibitions found for various buffers (HEPES, Tris, n-ethylmorpholine) may be caused by metal ion contamination of the buffers.

Distribution. Dipeptidases are widely distributed, being found in almost all living cells.[11] Rapidly growing or protein-synthesizing cells contain this enzyme in highest specific activity.[11,12] In general, these enzymes are localized in the soluble cell fraction, although in the kidney[13] they are also found in a particulate fraction.

[11] K. V. Linderstrøm-Lang, "Proteins and Enzymes," Lane Medical Lectures, Vol. 6. Stanford Univ. Press, Stanford, California, 1952.
[12] E. K. Patterson, M. E. Dackerman, and J. Schultz, *J. Gen. Physiol.* **32**, 623 (1949).
[13] B. J. Campbell, Y.-C. Linn, R. V. Davis, and E. Ballew, *Biochim. Biophys. Acta* **118**, 371 (1966).

[31] A Dipeptidase from Ehrlich Lettré Mouse Ascites Tumor Cells[1]

By Elizabeth K. Patterson

This dipeptidase has many of the characteristics of the dipeptidase purified from *Escherichia coli* B (see chapter [30], this volume), but it is far more labile and therefore extra precautions as to cleanliness, temperature, and especially pH must be taken.

[1] This work was supported by grants from the U.S. Public Health Service and The National Institutes of Health (CA-03102, CA-06927, and RR-05539) and by an appropriation from the Commonwealth of Pennsylvania.

PURIFICATION OF ASCITES TUMOR DIPEPTIDASE[a,b]

Preparation	Volume (ml)	Total protein (mg)	Total activity (units)	Specific activity[c] (units/mg)	Recovery (%)
Soluble fraction	414	10,519	36,816	3.5 (2.9–4.3)	
(NH₄)₂SO₄, 50–75%	46	3,245	37,204	11.5 (9.4–13.0)	101 (87–118)
Sephadex G-150	76	360	33,083	91.9 (54–92)	90 (48–90)
DEAE-cellulose					
Fractions 90–91	2.8	1.94	5,089	2623 (900–2635)	14 (3–14)
89–94	8.8	8.07	14,253	1766	39

[a] Reproduced with permission from the American Society of Biological Chemists, Inc., S. Hayman and E. K. Patterson, *J. Biol. Chem.* **246**, 660 (1971).

[b] Substrate Ala-Gly 50 mM, 40°, pH 8.2.

[c] The results in this table represent the best purification achieved. The figures in parentheses give the ranges obtained in all purifications. The numbers given refer to the material used as a sample for the next step in purification.

Assay Methods

The identical methods, A and B, described in the previous paper, were used. Method A was used to follow the purification with the substrate, alanylglycine (50 mM), and method B was utilized for kinetic studies with a variety of substrates of varying concentrations. Definitions of unit of enzyme activity, specific activity and molecular activity are also the same.

Purification Procedure[2] (see table)

Step 1. Isolation of the Soluble Fraction of the Cells. Growth and Washing of Cells. The Ehrlich–Lettré hyperdiploid mouse ascites carcinoma is maintained by weekly serial intraperitoneal transplantation of 0.2–0.3 ml of the tumor (depending on the thickness of the ascites) into adult 2.5- to 3-month-old) female mice of the ICR albino randomly bred strain. After 7–8 days of growth, when the highest dipeptidase specific activity is found in the cells,[3] the ascites from about 100 mice is aspirated by sterile syringe and collected in 40-ml heavy-duty centrifuge tubes in an ice bath. The ascites serum, which contains an inhibitor of dipeptidase,[3] is removed from the cells by centrifugation in a PR2 refrigerated (2°) centrifuge for 20 min at 2000 rpm. The cells are then

[2] S. Hayman and E. K. Patterson, *J. Biol. Chem.* **246**, 660 (1971).
[3] A. Malmgren, B. Sylvén, and L. Révész, *Br. J. Cancer* **9**, 473 (1955).

washed in 20 ml of a buffered cold (2°) isotonic saline solution (0.9% NaCl-KCl, Na:K, 10:1, 0.01 M NaHCO$_3$). The pelleted, washed ascites cells are then subjected to a 2-min stirring with 10 ml of cold (2°) water to cytolyze any contaminating red blood cells (all ascites of deep pink or red color are not used) followed by stirring with an equal volume of double-strength saline-bicarbonate solution. After centrifugation (300 rpm) and repetition of this treatment, the cell pellets are light cream in color.

HOMOGENIZATION OF ASCITES CELLS.[2] The packed, washed cells (170–210 ml) are suspended in about one-sixth their volume of 20 mM potassium phosphate, pH 8.2, and transferred to an iced Servall Omnimixer and homogenized at full speed (14,000 rpm) for 5 min. After addition of sufficient 2 M sucrose in phosphate buffer to give 0.25 M sucrose, homogenization is again performed for 2 min. Attempts to break the cells initially in sucrose, or by a variety of other means, failed to break more than a small fraction of the cells. Methods that result in dilute homogenates cannot be used since a concentrated protein solution (20 mg/ml) is necessary for good enzyme stability.

PREPARATION OF THE SOLUBLE FRACTION.[2] The homogenate is centrifuged at 17,500 rpm (20,000 g) in a Spinco Model L centrifuge at 4° for 20 min to pellet the nuclei and mitochondria. The supernatant fluid is then centrifuged at 40,000 rpm (105,000 g) for 2 hr. The top lipid layer is carefully pipetted off, and then the clear soluble fraction, leaving behind the microsomal pellet and fluffy layer. The dipeptidase is localized entirely in the soluble fraction. The yield from 100 mice is 100–150 ml, and this is stored frozen at −30°.

Step 2. Ammonium Sulfate Fractionation.[2] Approximately 400 ml of the soluble fraction is fractionated between 50% and 75% saturated ammonium sulfate. The method is the same as that used for *E. coli* B with the exception that solid ammonium sulfate is added for the 75% precipitation. The resultant precipitate is dissolved in a minimal volume (about 50 ml) of 0.02 M potassium phosphate, pH 8.3, and dialyzed overnight against the same buffer with three buffer changes. The solution has a reddish tinge caused by hemoglobin. The yield in this step is often greater than 100%, probably because of removal of an inhibitor (see the table).

Step 3. Sephadex G-150 Filtration.[2] Preparation of the Sephadex, pouring and washing of the column with 0.02 M potassium phosphate buffer, pH 8.3, are identical to the procedures used in the previous paper. Only about one-third (13–15 ml) of the ammonium sulfate-precipitated material is applied to the column (2.5 × 90 cm). The high protein content (70 mg/ml) of the sample may cause some shrinkage of the column, but this artifact does not appear to impair the resolution of the protein peaks. Flow rate is adjusted to 6 ml/hr by means of a Sigmamotor peristaltic pump set at 8–9%. This slow rate results in far better resolution of the protein peaks than use of faster rates. The effluent is monitored by a Gilson 280 nm meter, and recording is carried out by an Esterline-Angus recorder. When the first protein starts coming off, fractions of 2.5 ml are collected. Sufficient solid sucrose is placed in the empty tubes to give 0.25 M solutions. Leucine aminopeptidase, tripeptidase, prolidase, and hemoglobin are cleanly separated from the dipeptidase which hydrolyzes alanylglycine.[2] The yield of the latter enzyme (see the table) averages about 100%.

 Step 4. DEAE-Cellulose. Methods for preparing the DEAE (35 g of Whatman DE-32) and pouring and washing the column (2.5 × 24 cm) are given in the preceding paper [30] on purification of *E. coli* B dipeptidase. The buffer contains 0.02 M potassium phosphate, 0.01 M NaCl, and 0.5 mM $MgCl_2$, pH 8.0. Fractions from the dipeptidase peak of three G-150 runs are pooled (76 ml), and sufficient NaCl and $MgCl_2$ are added to give 0.01 M NaCl and 0.5 mM $MgCl_2$. After addition of the sample (33,000 enzyme units, 360 mg of protein), the procedure was the same as that described in the previous paper on *E. coli* B dipeptidase purification. When the protein starts to come off, fraction of 1.5 ml are collected in tubes to which sufficient solid sucrose was added to give 0.25 M solutions. When the breakthrough protein is off, elution of enzyme and other proteins is carried out by a linear chloride gradient. This step gives a 20-fold purification (see the table) of the enzyme primarily because the inert protein is not adsorbed. Enzyme recovery in this step varies from 68 to 105%. Although the purification given in the table resulted in the highest specific activity (2600 in each of two tubes) attained for this dipeptidase, the protein peak is not entirely superimposed by the enzyme peak. At least part of the contaminating protein was found to be mouse serum albumin by use of the Ouchterlony[4] immunological technique. A further purification step is useful for removing this albumin.

 Step 5. Hydroxyapatite Chromatography. Preparation, washing, and elution of Bio-Rad HTP (4 g) is carried out as described in the previous paper [30] on the *E. coli* dipeptidase. In a typical run, pooled samples from a DEAE purification are concentrated 2-fold, and the buffer is changed to 0.01 M sodium phosphate, 0.5 mM $MgCl_2$, pH 7.5. The sample contained 8500 units, 7.2 mg of protein in 7.5 ml. The tumor enzyme adsorbs to the column (1 × 12 cm) and is eluted by an exponential gradient (0.02–0.5 M potassium phosphate, pH 7.6, using a 125-ml mixing

[4] E. Weiler, E. W. Melletz, and E. Breuninger-Peck, *Proc. Natl. Acad. Sci. U.S.A.* **54**, 1310 (1965).

flask). The protein peak now follows the enzyme peak to a greater extent than with DEAE, giving specific activities over 2000 in each of four tubes. The mouse serum albumin peak precedes and overlaps the first part of the dipeptidase peak. Recovery of enzyme is about 70%.

Properties

Stability of the Enzyme.[2] This dipeptidase is stable only between pH 7.5 and 8.5. At specific activities of 3–600, the enzyme is stable for months when frozen at −30° in 0.02 M potassium phosphate, 0.25 M sucrose, pH 8.2. The sucrose is necessary for stability. At specific activities over 600, however, there is a slow loss in enzyme activity. In the lower specific activity range, stability to dialysis against buffer at 5° is excellent, but at higher specific activities, loss of activity is observed. Buffer can be changed without loss of enzyme by concentrating and dialyzing the dipeptidase simultaneously. The highly purified dipeptidase is stable for 6 hr at 0° but loses activity overnight. Addition of protein (2 mg/ml bovine serum albumin) stabilizes the enzyme for assay of 5 min incubation at 40°. No protein addition is necessary for 30° assay.

Purity of the Dipeptidase. A concentrated preparation of specific activity 1360 was subjected to acrylamide gel electrophoresis,[5] which resulted in one major and three minor (faint) bands. When preparations of higher specific activity, i.e., 2200, were run under the same conditions, a greater number of bands appeared (5 to 6), but it was found that this preparation had retained only 45% of its original activity. Another preparation (S.A. 2500, 0.3 mg of protein per milliliter) was subjected to sodium dodecyl sulfate (SDS) gel electrophoresis[6] after being frozen for 2 months. Four bands were visible, and densitometry showed the major band to be 38% of the protein. Therefore, until fresh preparations are concentrated and examined before freezing has induced denaturation, it is impossible to estimate dipeptidase purity. The fact that this tumor enzyme has a molecular activity toward its best substrate (Ala-Ile, see below) of 2.3×10^4 compared to the 90% pure bacterial dipeptidase value of 1.6×10^4 should imply a highly purified enzyme.

Molecular Weight.[2] As estimated from Sephadex G-150 gel filtration[2] along with standard proteins, the molecular weight of the tumor enzyme is 85,000 ± 5000. However, a molecular weight of 57,000 ± 1000 was estimated from SDS gel electrophoresis. This latter figure may not represent that of a true subunit, but rather that of some denaturation product. This problem is currently under investigation.

[5] R. J. Davis, *Ann. N.Y. Acad. Sci.* **121**, 404 (1964).
[6] K. Weber and M. Osborn, *J. Biol. Chem.* **244**, 4406 (1969).

Zn Content. The tumor dipeptidase is a metalloenzyme possibly containing 1 mole of Zn per mole of enzyme of MW 85,000.[2] This statement assumes homogeneous enzyme and a true molecular weight. If this turns out to be correct, the tumor dipeptidase would differ from the *E. coli* B dipeptidase (MW 100,000) which contains two subunits (MW 54,000) and two Zn per mole of enzyme.

Specificity.[7] The tumor dipeptidase without metal addition hydrolyzes at least 46 L-α-dipeptides. Maximum velocities are highest with substrates containing uncharged bulky C-terminal (-Ile) and small, uncharged N-terminal (Gly or Ala) side chains. Not detectably hydrolyzed are Lys-Gly, Asp-Gly, and Gly-Asp.

Effect of Added Metal Ions.[7a] Addition of Co^{2+} to reaction mixtures has an appreciable effect on the rates of hydrolysis and K_m of only N-terminal Gly-dipeptides. For instance, the rate of hydrolysis of Gly-Ala is increased 4-fold by 1 mM Co^{2+} and that of CH_3Gly-Ala possibly 200-fold. Dipeptides barely hydrolyzed, such as Gly-Asp, are split at a measurable rate $[V_M/[E] (\text{sec}^{-1}) (\times 10^{-3})]$ of 0.18. The hydrolysis of Gly-Gly is activated 5-fold, and that of Pro-Gly 4-fold. Addition of Mn^{2+} accelerates the hydrolysis of only Pro-Gly (10-fold) and Gly-Lys and Gly-Ser (2-fold) of 15 dipeptides tested. Kinetic affinity experiments imply that Co^{2+} binds (K_A, 1.4 μM) to one site on the enzyme which is presumably the Zn^{2+} site since Zn^{2+} is lost when the dipeptidase is subjected to prolonged exposure to high concentrations of Co^{2+}. On the other hand, kinetic binding experiments with Mn^{2+} indicate that two sites of differing affinities (K_A 0.24 and 17 μM) are involved. Since Zn^{2+} is not removed from the enzyme on incubation with Mn^{2+}, it cannot be assumed that the tight Mn^{2+} site is that of the native Zn^{2+}.

Since in the past[8] dipeptidase have been differentiated according to the substrate hydrolyzed and the metal activator (e.g., glycylglycine dipeptidase which is Co^{2+} stimulated), it appears from the results with added metal ions, both with the tumor and *E. coli* B dipeptidase, that fewer dipeptidases exist in cells than was formerly assumed. When the purified ascites tumor dipeptidase is subjected to 30° for 24 hr, the activities toward Ala-Gly, Gly-Gly-Co^{2+}, and Pro-Gly-Mn^{2+}, decay at the same rate, suggesting that a single enzyme hydrolyzes the three substrates.[2] Also, the ratio of these three activities remains constant during purification. With the *E. coli* B dipeptidase, parallel rates of decay of these activities and ratios during purification are also the case.

[7] E. K. Patterson, J. S. Gatmaitan, and S. Hayman, *Biochemistry* **12**, 3701 (1973).
[7a] E. K. Patterson, J. S. Gatmaitan, and S. Hayman, *Biochemistry* **14**, 4261 (1975).
[8] E. L. Smith, *in* "The Enzymes" (P. D. Boyer, H. Lardy, and K. Myrbäck, eds.), 2nd ed., Vol. 4, p. 1. Academic Press, New York, 1960).

pH Dependence.[7] From plots of maximum velocity (V_{M}) of Ala-Gly hydrolysis versus pH from 6.0–9.5, it is seen that the pH optimum of the tumor enzyme is 8.3. When K_m is plotted against pH, the plot is flat between pH 7.0 and 8.5, rising at pHs above and below these values. From these data, it can be calculated that the acid and basic pK's of the enzyme are 7.3 and 8.8, and the pK's of the enzyme–substrate complex, 7.2 and 9.4, respectively.

Inhibitors. Since the dipeptidase is a Zn-metalloenzyme, it is inhibited by metal-chelators.[2] *o*-Phenanthroline (but not its nonchelating isomer, *m*-phenanthroline) causes a sharp initial drop in enzyme activity followed by a gradual decline. In contrast, EDTA inhibition after an initial lag is first-order throughout the period of observation (7 hr). Of the metals (Zn^{2+}, Mn^{2+}, Co^{2+}, Mg^{2+}) tested, only Zn^{2+} was effective in reactivating the enzyme in the presence of EDTA, and this reactivation was only to 15–20% of the initial activity. The activity toward Ala-Gly is proportional to the Zn^{2+} content of the enzyme. If, however, Zn^{2+} is removed by treating the dipeptidase with Co^{2+} followed by Sephadex G-25 filtration in the presence of Co^{2+}, 55% of the original activity toward Ala-Gly is recovered when Zn^{2+} is added back to the preparations. It appears that the dipeptidase is less subject to denaturation by Zn^{2+} removal if Co^{2+} immediately replaces the lost Zn^{2+}. Other metal chelators such as dithiothreitol and mercaptoethanol inhibit the dipeptidase roughly in proportion to their ability to bind with Zn^{2+}.

Addition of metal ions, other than Co^{2+} and Mn^{2+}, inhibit the hydrolysis of Ala-Gly (50 mM) by the dipeptidase to varying degrees[2] according to the concentrations (10^{-7} to 10^{-2} M) employed. It is notable that Zn^{2+} at levels of 10^{-5} M can inhibit the enzyme from 50 to 100% depending on enzyme concentration. Inhibition by buffers (Tris, Veronal, triethanolanime, *n*-ethylmorpholine, HEPES) may be attributed to metal ion contamination.

Peptides containing D-amino acids are dipeptidase inhibitors.[7] D-Leu-Gly and Gly-D-Leu are linear competitive inhibitors of L-Ala-Gly hydrolysis with K_{I}'s of 2.5 and 9.0 mM, respectively. L-Alanine, a product of the reaction, is a poor competitive inhibitor with a K_{I} of 71 mM.

Distribution. As stated in the previous paper, dipeptidase are found in all living cells but in highest activity in growing or protein-synthesizing cells. Some dipeptidase that have been purified by others are the following: A "glycyl-L-leucine" dipeptidase has been purified[9] about 100-fold from pig intestinal mucosa and gives three bands on acrylamide gel. The authors suggest that this indicates a transformation between

---

[9] O. Norén, H. Sjoström, and L. Josephson, *Acta Chem. Scand.* **25**, 1913 (1971).

different forms of the enzyme during the electrophoresis. From monkey small intestine, another dipeptidase showing highest activity toward glycyl-L-leucine has been purified[10] about 200-fold. This enzyme is extremely unstable, gives one major and three minor protein bands on acrylamide gel electrophoresis and has many properties similar to the ascites tumor dipeptidase. From plant material, *Cucurbita maxima* cotyledons, a dipeptidase has been purified[11] 2000-fold. This enzyme has greatest activity toward L-leucylglycine. It is a stable dipeptidase which on acrylamide gel gives one band.

[10] M. Das and A. N. Radhakrishnan, *Biochem. J.* **128**, 463 (1972).
[11] F. M. Ashton and W. J. Johnson, *Phytochemistry* **6**, 1215 (1967).

Section VII

Endopeptidases

[32] *Leucostoma* Peptidase A

By JOHN M. PRESCOTT and FRED W. WAGNER

Leucostoma peptidase A is the most abundant protease in the venom of the western cottonmouth moccasin (*Agkistrodon piscivorus leucostoma*) from which it can be isolated in good yield and a high degree of homogeneity.[1,2] This endopeptidase possesses some characteristics similar to those of bacterial neutral proteases, inasmuch as it is inhibited by metal complexing reagents, is a metalloprotein that contains calcium and zinc, and has a substrate specificity dictated primarily by the identity of the residue contributing the amino group to the susceptible peptide bond.

Assay Procedures

Leucostoma peptidase A readily digests casein and denatured hemoglobin, either of which can be used as a substrate to assay the proteolytic activity of fractions obtained during enzyme isolation. Details for the assay with hemoglobin substrate are given below. In addition, it is also advisable to assay for esterase and arylamidase activities at each stage of the procedure in order to monitor the elimination of enzymes that accompany *leucostoma* peptidase A through part of the purification procedure.

Hemoglobin Assay for Proteolytic Activity. The procedure is essentially the same as that of Anson,[3] and involves determining the increase in trichloroacetic acid-soluble products that result from digestion of urea-denatured hemoglobin. The substrate is prepared by suspending 5 g of lyophilized, dialyzed hemoglobin (Worthington) in 80 ml of water with 80 g of urea. After the mixture has been incubated at 37° for 1 hr, 125 ml of 0.133 *M* phosphate buffer (pH 8.5) and 10 g of urea are added. The mixture is permitted to stand overnight and the pH is checked to ensure that it is 8.5. Assays are run by adding 1 ml of suitably diluted enzyme solution to 5 ml of the substrate and incubating the mixture at 37° for 10 min. The addition of 10 ml of 5% trichloroacetic acid serves to stop the reaction and to precipitate undigested protein, which is removed by filtration through Whatman No. 3 paper. The absorbance of

[1] F. W. Wagner, A. M. Spiekerman, and J. M. Prescott, *J. Biol. Chem.* **243**, 4486 (1968).
[2] A. M. Spiekerman, K. K. Fredericks, F. W. Wagner, and J. M. Prescott, *Biochim. Biophys. Acta* **293**, 464 (1973).
[3] M. L. Anson, *J. Gen. Physiol.* **22**, 79 (1938).

the filtrate at 280 nm is then determined against a blank made by adding trichloroacetic acid to the substrate prior to addition of the enzyme. One unit is defined as that quantity of enzyme that produces an increase of 1.0 in absorbance in 10 min.

Arylamidase Assay. Although the venom of *A. p. leucostoma* does not contain true aminopeptidase activity, it catalyzes the hydrolysis of L-leucyl-β-naphthylamide[4] and thus contains one or more enzymes that can be classified for operational convenience as arylamidases, the activities of which are assayed by the aminopeptidase assay procedure of Goldbarg and Rutenburg.[5] The reagents required are

L-Leucyl-β-naphthylamide-HCl, (Leu-NA) 0.2 mg/ml in 0.1 M citrate buffer, pH 6.7
Trichloroacetic acid, 40% (w/w)
$NaNO_2$, 0.1%
Ammonium sulfamate, 0.5%
1-Naphthylethylenediamine dihydrochloride (NEDD), 0.05% in 95% ethanol

Assays are run by incubating 0.5 ml of enzyme solution with 0.5 ml of Leu-NA substrate for 30 min, and the reaction is terminated by the addition of 0.5 ml of 40% trichloroacetic acid. To the mixture is added 1 ml of the $NaNO_2$ solution and the reaction is allowed to proceed for 3 min to convert the enzymically liberated β-naphthylamine to its diazonium salt. Excess nitrous acid is neutralized by the addition of 1 ml of 0.5% ammonium sulfamate, and after 2 min the diazonium salt is coupled to NEDD by the addition of 2 ml of the NEDD reagent. Color development is complete in 20 min, after which absorbance is measured at 560 nm. One unit of arylamidase activity is defined as the amount of enzyme required to liberate, in 30 min, sufficient β-naphthylamine to yield a diazotized product having an absorbance of 1.0 at 560 nm.

Esterase Assay. The spectrophotometric assay of Schwert and Takenaka[6] is used to measure the increase in absorbance at 253 nm when N-benzoyl-L-arginine ethyl ester (BAEE) is hydrolyzed. The substrate consists of 0.1 mM BAEE in 5 mM Tris-HCl buffer, pH 7.6, equilibrated at 25°. Cuvettes containing 2.9 ml of substrate are placed in a recording spectrophotometer, and the reaction is started by thoroughly mixing 0.1 ml of enzyme with the substrate. The increase in absorbance is recorded for 2–3 min while the temperature in the cell compartment is maintained

[4] F. W. Wagner and J. M. Prescott, *Comp. Biochem. Physiol.* **17**, 191 (1966).
[5] J. A. Goldbarg and A. M. Rutenburg, *Cancer* **11**, 283 (1958).
[6] G. W. Schwert and Y. Takenaka, *Biochim. Biophys. Acta* **16**, 570 (1955).

at 25°. The initial velocity of the reaction is determined from the recorder tracing, and 1 unit of esterase activity is defined as the amount of enzyme that causes an increase of 1.0 in absorbance in 1 min at 25°.

Enzyme Isolation

All chromatographic procedures are carried out at 4°–5° in order to minimize loss of enzymic activity, and it is convenient, but by no means necessary, to have continuous monitoring of the column effluents at 280 nm.

Step 1. Preparation of Venom for Chromatography. Lyophilized venom of *A. p. leucostoma* is purchased from Miami Serpentarium, Miami, Florida, and stored at −10° until used. Normally, 1–5 g of the dried venom are used in an isolation experiment; the volumes suggested below are suitable for a sample of about 2 g. The venom is dissolved in 20–50 ml of 5 mM Tris-HCl buffer, pH 8.5, and dialyzed against this buffer overnight. The somewhat turbid solution is clarified by centrifugation at 5400 g for 5–10 min and the clear supernatant fluid is stored at 4° until used.

Step 2. Chromatography on DEAE-Sephadex A-50. DEAE-Sephadex A-50 is suspended in distilled water and allowed to stand for several hours, after which the fine particles are decanted. The ion-exchange material is suspended in 0.5 N HCl, filtered immediately, resuspended in distilled water, then washed several times in distilled water. The DEAE-Sephadex is next suspended in 0.5 N NaOH, allowed to stand for 15–20 min then filtered and again suspended in 0.5 N NaOH for a few minutes. After filtration, the resin is washed in distilled water until the washings are neutral to pH paper. The hydroxide form of DEAE-Sephadex thus prepared is suspended in 5 mM Tris-HCl buffer, pH 8.5, and allowed to equilibrate before being poured in a 2.5 × 50 cm column which is throughly washed with the buffer at 5°. The sample from step 1 is applied and the column is washed with the same buffer at a rate of 60 ml/hr until the unadsorbed material emerges. The chromatogram is then developed by the application of a linear gradient from 0 to 0.1 M NaCl, with the effluent being collected in 10 ml fractions at a flow rate of 50–60 ml/hr. Typically, the first cylinder of the gradient-producing apparatus contains 1 liter of the 5 mM Tris-HCl buffer, pH 8.5, and the second contains an equal amount of the same buffer that is 0.1 M in NaCl. For experiments in which less than 1 g of venom is used as the starting material, the total volume of the gradient is reduced to 600 ml and for samples over 5 g it should be increased correspondingly. In some experiments we have completed the chromatogram by washing the column sequentially with 0.2 M and with 1.0 M NaCl as shown in Fig. 1 to ensure elution of all proteolytic activity, but this step is not essential inasmuch as *leucostoma* peptidase A is eluted at about 0.05 M NaCl. This fraction always contains the greatest amount of activity toward hemoglobin substrate, although a small amount of proteolytic activity emerges in the breakthrough fraction, and some lots of venom contain a third chromatographically distinct proteolytic peak.

Step 3. Rechromatography on DEAE-Sephadex A-50. The protease from step 2 (fraction I in Fig. 1) is dialyzed against 5 mM borate–NaOH buffer, pH 9.2, and applied to a 2.5 × 50 cm column of DEAE-Sephadex A-50 that has been equilibrated with this buffer. The chromatogram is developed with approximately 700 ml of a linear gradient of increasing NaCl concentration as in step 2 (0–0.1 M NaCl), with a flow rate of 60 ml/hr. Fractions of 10 ml are collected in tubes containing 1 ml of 0.1 M Na_2HPO_4, pH 8.0, to lower the pH of the eluate. Each tube is read at 280 nm, and every fifth fraction is assayed for the three types of enzymic activity. The proteolytic activity emerges in a prominent peak that is only partially separated from smaller protein fractions that contain esterase and arylamidase activities.

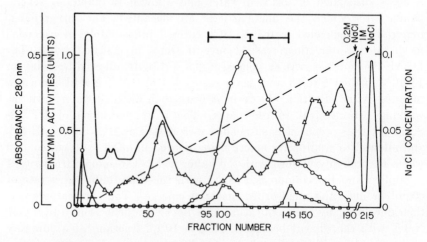

Fig. 1. Chromatography of crude, dialyzed *Agkistrodon piscivorus leucostoma* venom on DEAE-Sephadex A-50 in 5 mM Tris-HCl, pH 8.5. A 2-g sample of venom was dialyzed, then applied to a 2.5 × 50 cm column and eluted at approximately 50 ml/hr into fractions containing 12 ml. ——, absorbance at 280 nm; ○, proteolytic activity toward denatured hemoglobin; △, esterase activity toward *N*-benzoyl-L-arginine ethyl ester; □, arylamidase activity toward Leu-NA; - - -, NaCl gradient. Fraction I (tubes 95–145) was selected for further purification. From F. W. Wagner, A. M. Spiekerman and J. M. Prescott, *J. Biol. Chem.* **243**, 4486 (1968).

Step 4. Gel Filtration. The active material from step 3 is concentrated to approximately 5–10 ml by ultrafiltration with an Amicon UM-05 membrane under a nitrogen atmosphere of 50 psi. The concentrated sample is applied to a 2.5 × 50 cm column of Sephadex G-75 previously equilibrated with 5 mM Tris-HCl buffer, pH 8.5, that is 0.1 M in NaCl, and eluted at a flow rate of 15–20 ml/hr until a total volume of 150–250 ml has been collected in fractions of 3–5 ml. The exact volume at which the chromatogram is considered to be fully developed should be ascertained by monitoring the column effluent at 280 nm, either manually or by a flow analyzer. The emergence of the proteolytic activity frequently is preceded by two relatively small components possessing esterase, but no protease, activity. The third and largest fraction is highly proteolytic but also contains some BAEE activity. In some experiments, this fraction may also be contaminated by arylamidase, and, if so, rechromatography under the same conditions is advisable.

Step 5. Rechromatography on DEAE-Sephadex. The protease fraction from step 4 is dialyzed for 18 hr against 5 mM borate–NaOH buffer, pH 9.6, and applied to a 2.5 × 50 cm column of DEAE-Sephadex A-50, which has been equilibrated with the same buffer. The protease is eluted with 800–850 ml of a linear gradient of NaCl (20 mM to 0.5 M) in the same buffer, with fractions of 6 ml collected at a rate of 15 ml/hr in tubes containing 1 ml of 0.1 M Tris-HCl buffer, pH 7.8. Three components are separated by this procedure; the second and most prominent of these is protease, uncontaminated by esterase; the latter enzyme emerges in the third peak. Those fractions that show protease activity but no esterase are combined to form the final preparation. A summary of a purification experiment is shown in the table.

Properties

Purity. Preparations of *leucostoma* peptidase A, purified as described above, have been found to be homogeneous by moving boundary electrophoresis at several pH values and by polyacrylamide gel electrophoresis at pH 9.5. Moreover, enzymic assays for activity toward BAEE and Leu-NA show the absence of detectable amounts of esterase and arylamidase, respectively. The absence of esterase activity is an important criterion for purity because of the confusion generated by early workers who used unfractionated venoms as a source of enzyme in characterizing proteolytic activity. They found that not only were hemoglobin and casein rapidly hydrolyzed, but a number of venoms also cleaved such trypsin and chymotrypsin substrates as BAEE and N-acetyl-L-tyrosine ethyl ester, respectively. By analogy with the well known animal

SUMMARY OF THE PURIFICATION OF *Leucostoma* PEPTIDASE A [a]

Step	Treatment	Total protein (mg)	Total enzyme units [b]			Specific activity			% Recovery [c] of protease
			Protease	Esterase	Aryl-amidase	Protease	Esterase	Aryl-amidase	
1.	Dialysis	2568	3186	5100	3870	1.24	1.99	1.51	100
2.	Chromatography (DEAE-Sephadex A-50, pH 8.5)	668	2022	528	429	3.03	0.79	0.64	64
3.	Rechromatography (DEAE-Sephadex A-50, pH 9.2)	389	1702	229	132	4.38	0.59	0.34	53
4.	Gel filtration (Sephadex G-75)	211	1260	23	0	5.97	0.11	0.0	40
5.	Rechromatography (DEAE-Sephadex A-50, pH 9.6)	128	984	0	0	7.69	0.00	0.0	31

[a] Reproduced from A. M. Spiekerman, K. K. Fredericks, F. W. Wagner, and J. M. Prescott, *Biochim. Biophys. Acta* **293**, 464 (1973), with permission.
[b] See text for definition of units.
[c] Represents the cumulative recovery in the isolation procedure.

enzymes, but without experimental evidence, some of these workers assumed that the proteolytic and esterolytic activities resulted from a single enzyme; this assumption led to the view that snake venoms contain "trypsin-like" and "chymotrypsin-like" proteases. Subsequently, experiments with carefully isolated proteases, including *leucostoma* peptidase A, showed the enzymes that are responsible for most of the proteolytic activity toward hemoglobin and casein have no esterase activity and, similarly, purified enzymes that cleave BAEE do not hydrolyze hemoglobin or casein.[7] Therefore, one of the criteria for purity of *leucostoma* peptidase A is the absence of activity toward BAEE.[8]

Stability. The protease is stable for several weeks either frozen or at 4°; however, prolonged storage accompanied by freezing and thawing results in some autolysis that is manifest by the appearance of one or two new bands on gel electrophoresis, even when no appreciable loss in activity is evident.

Physical Properties. Sedimentation velocity studies showed $s^0_{20,w}$ to be 2.87 S for the venom protease and the value for $D_{20,w}$ has been found to be 10.66×10^{-7} cm^2 sec^{-1}. From the amino acid content, the partial specific volume was calculated to be 0.711 cm^3/g. Substitution of these values into the Svedberg equation gave a molecular weight of 22,500, and that obtained from the approach to equilibrium method of Archibald was $22,400 \pm 1250$. *Leucostoma* peptidase A has an isoelectric point of 6.5 at an ionic strength of 0.1. The amino acid content is half-Cys$_6$, Asp$_{29}$, Thr$_{15}$, Ser$_{18}$, Glu$_{19}$ Pro$_6$, Gly$_{12}$, Ala$_{11}$, Val$_{15}$, Met$_5$, Ile$_{10}$, Leu$_{19}$, Tyr$_5$, Phe$_6$, Lys$_{10}$, His$_9$, Arg$_6$, and Trp$_4$.[6,7]

Metal Content. Analyses of purified *leucostoma* peptidase A by atomic absorption spectrophotometry reveal the presence of both Ca^{2+} and Zn^{2+} ions in stoichiometric quantities and indicate that the enzyme is a metalloprotein containing 2 g-atoms of calcium and 1 g-atom of zinc per mole. An essential role for metal ions in catalysis is indicated by the fact that both EDTA and *o*-phenanthroline inhibit proteolysis. Removal of Zn^{2+} by *o*-phenanthroline in the presence of excess Ca^{2+} inhibits the action of *leucostoma* peptidase A, which can then be reactivated by the addition of Zn^{2+}. Thus, zinc is the catalytically essential metal, and it is likely that the role of calcium is to stabilize the enzyme.

pH Optimum. *Leucostoma* peptidase A exhibits maximal activity toward hemoglobin substrate at pH 8.5.

Substrate Specificity. As is the case for some other purified proteases from snake venoms, *leucostoma* peptidase A shows little activity toward

[7] P. M. Toom, T. N. Solie, and A. T. Tu, *J. Biol. Chem.* **245**, 2549 (1970).
[8] Venom enzymes that hydrolyze BAEE may be specialized proteases of some type, even though they show no activity toward common protein substrates.

small synthetic substrates. It has no activity against any exopeptidase substrate tested, and experiments with a large number of NH_2-substituted and doubly substituted dipeptides revealed that only a few were susceptible and these were hydrolyzed slowly.[2] However, the protease readily cleaved the B-chain of oxidized insulin at positions Phe_1-Val_2, His_5-Leu_6, His_{10}-Leu_{11}, Ala_{14}-Leu_{15}, Gly_{20}-Glu_{21}, Gly_{23}-Phe_{24}, Phe_{24}-Phe_{25}.[2] The substrate specificity is obviously dictated by the identity of the residue that contributes the NH_2 group to the bond, and a preference is shown for substrates having a hydrophobic residue in that position. It is likely that some minimal substrate size is required for significant rates of hydrolysis, as substrate size has been shown to be important in the action of other purified venom proteases.[9,10]

[9] T. Takahashi and A. Ohsaka, *Biochim. Biophys. Acta* **198**, 293 (1970).
[10] K. Mella, M. Volz, and G. Pfleiderer, *Anal. Biochem.* **21**, 219 (1967).

[33] *Aeromonas* Neutral Protease

By STELLA H. WILKES and JOHN M. PRESCOTT

Culture filtrates of *Aeromonas proteolytica* possess unusually high activity toward casein and hemoglobin substrates,[1,2] surpassing the levels produced by several other microbial species known to yield large quantities of extracellular proteolytic enzymes.[3] Fractionation of culture filtrates has revealed that this activity is due to two endopeptidases, one of which is the subject of this chapter.

Enzyme Assays

Previous work on *Aeromonas* neutral protease in this laboratory has been done with an assay based on the procedure of Anson,[4] using denatured hemoglobin as the substrate. It is frequently desirable, however, to be able to assay proteases by monitoring the cleavage of a single peptide bond of known identity. Details are given below, therefore, for assaying the enzyme both with hemoglobin and with a doubly blocked

[1] J. M. Prescott and C. R. Willms, *Proc. Soc. Exp. Biol. Med.* **103**, 410 (1960).
[2] S. H. Wilkes, B. B. Mukherjee, F. W. Wagner, and J. M. Prescott, *Proc. Soc. Exp. Biol. Med.* **131**, 382 (1969).
[3] L. Keay, M. H. Moseley, R. G. Anderson, R. J. O'Connor, and B. S. Wildi, *Biotechnol. Bioeng. Symp.* **3**, 63 (1972).
[4] M. L. Anson, *J. Gen. Physiol.* **22**, 79 (1938).

dipeptide substrate. For a qualitative assay to locate the active fractions obtained during enzyme purification, the gelatin-digesting test described in Chapter [44] of this volume is convenient. If the neutral protease is to be used in sequence or specificity studies, it is also necessary to assay for the presence of aminopeptidase in fractions obtained during the isolation of the protease and in the final product. This assay is done with L-leucyl-*p*-nitroanilide as the substrate, and details of the procedure are given in Chapter [44].

Assay with Furylacryloyl Peptides. Feder[5] described an assay for neutral proteases based on the decrease in absorbance at 345 nm that accompanies hydrolysis of the peptide bond in furylacryloyl (FA) peptides. *Aeromonas* neutral protease attacks a number of these doubly blocked peptides, including FA-Gly-Leu-NH$_2$, a substrate that is rather widely used to measure neutral protease activity. We have modified the procedure of Feder[5] for use with the *Aeromonas* enzyme as follows. A 10 mM solution of FA-Gly-Leu-NH$_2$ in 20 mM Tris is made by dissolving 30.68 mg of the substrate in 0.1 ml of spectrographically pure dimethylformamide and adding approximately 7 ml of H$_2$O, followed by 1.0 ml of 0.2 M Tris (pH 7.5). After final adjustment of the pH to 7.5, the volume is brought to 10 ml. A solution of 1 mM FA-Gly-Leu-NH$_2$ is used for following activity in the fractions obtained during enzyme isolation; it is prepared by a 10-fold dilution of the 10 mM stock solution with 20 mM Tris buffer, pH 7.5. The reaction is initiated by adding 20 μl of a solution containing approximately 0.5 μg of enzyme to 1 ml of 1 mM FA-Gly-Leu-NH$_2$ (0.870 A) in a 1.5-ml cuvette of 1-cm light path, maintained at 25°. Under these conditions the recorder tracing is linear for approximately 3 min, during which time less than 10% of the substrate is hydrolyzed; the initial hydrolytic velocity is determined from this portion of the tracing. Complete hydrolysis of a 1 mM solution results in an absorbance change of 0.345 at 345 nm and it is thus convenient to use a scale expansion of 0.2 A for this assay. One unit is defined as the amount of enzyme that gives an initial hydrolytic velocity of 1 μmole of FA-Gly-Leu-NH$_2$ per minute. Purified preparations have a specific activity of 40–50 units/mg.

Inasmuch as the K_m value of FA-Gly-Leu-NH$_2$ for *Aeromonas* neutral protease is 3.5 mM, first-order kinetics apply for the above reaction as they do in reactions involving other neutral proteases and similar substrates.[6] In order to more nearly saturate the enzyme, however, higher concentrations of FA-Gly-Leu-NH$_2$ may be employed by using cuvettes of 1-mm light path. For example, we have used the 10 mM stock solution

---

[5] J. Feder, *Biochem. Biophys. Res. Commun.* 32, 326 (1968).

[6] J. Feder and J. M. Schuck, *Biochemistry* 9, 2784 (1970).

of FA-Gly-Leu-NH$_2$ described above to follow the activity of preparations during isolation and have found a specific activity of 200–250 units/mg for the final product. Although the higher substrate concentration gives more reliable values with crude preparations, in general the use of the 1 mM substrate for routine assays is more convenient and economical.

Aeromonas neutral protease may be assayed by the same general procedure using any of several other FA dipeptide amides. Of the substrates tested, FA-Ala-Phe-NH$_2$ is hydrolyzed the most rapidly and has a K_m value one-tenth that of FA-Gly-Leu-NH$_2$. It is relatively insoluble, but can be dissolved in dimethylformamide, then diluted, by the general procedure described above, to yield a stock solution that is 10 mM.

Hemoglobin Assay. This assay is based on the well known principle of determining the increase in trichloroacetic acid-soluble products of the digestion of a protein by an endopeptidase, as described by Anson.[4] It is performed in this laboratory as follows. Hemoglobin substrate is prepared by incubating 5 g of lyophilized hemoglobin (Worthington Biochemical Co.) with 80 g of urea and 80 ml of H$_2$O at 37° for 1 hr with stirring. To this suspension are added 125 ml of phosphate buffer (pH 8.0, 0.133 M) and 10 g of urea; the mixture is allowed to stand overnight, then is adjusted to pH 8.0. One milliliter of enzyme solution (diluted with 10 mM Tricine buffer, pH 8.0) is pipetted into a test tube containing 5 ml of hemoglobin substrate solution that has been equilibrated at 37°; the two solutions are mixed thoroughly and incubated at 37°. After 5 min, the reaction is terminated by the addition of 10 ml of 5% trichloroacetic acid (w/v), and the tube is permitted to stand for 15 min to ensure complete precipitation. The precipitated protein is removed by filtration through Whatman No. 3 paper, and the absorbance of the filtrate at 280 nm is determined. Blanks are prepared for each enzyme dilution by adding the trichloroacetic acid to the substrate before the enzyme is added. One unit is the amount of enzyme required to produce a filtrate having an absorbance of 1.0 in the 5-min assay described.

Enzyme Production and Isolation

Enzyme Production

Medium. Because the aminopeptidase produced by *A. proteolytica* has charge and size properties very similar to those of the neutral protease, quantitative removal of the animopeptidase from preparations of the latter is difficult. When grown in a medium containing enzymically digested soybean protein, however, *A. proteolytica* characteristically produces substantially less aminopeptidase than it does in media containing peptone or enzymically digested casein as the carbon-nitrogen energy source (Table I). Although total proteolytic activity in soybean protein medium is lower than in the casein medium, the high ratio of endopeptidase to aminopeptidase is desirable, as there is much less of the latter enzyme to remove. Various lots of soybean protein differ markedly in the amounts of aminopeptidase they produce, however, and samples of several production batches of soybean protein (C-1 assay protein, Archer-Daniels-Midland Company) should be obtained, hydrolyzed in small quantities and tested for the ability to produce high endopeptidase: aminopeptidase ratios. The reason for the differences in aminopeptidase production on different lots and different types of protein is not clear, but advantage may nevertheless be taken of this effect, and it is well worthwhile to find a protein source that gives low aminopeptidase production.

TABLE I

PRODUCTION OF ENDOPEPTIDASE AND AMINOPEPTIDASE ACTIVITIES BY
Aeromonas proteolytica GROWN ON DIFFERENT
NITROGEN-CARBON ENERGY SOURCES[a,b]

Organic component of medium	Growth[c]	Endo-peptidase (units/ml)[d]	Amino-peptidase (units/ml)[e]	Ratio endo-peptidase: amino-peptidase
Peptone	3.25	7.2	0.05	144
Soybean protein (acid hydrolyzed)	2.80	1.0	0.02	50
Soybean protein (undigested suspension)	3.30	27.8	0.14	199
Soybean protein (enzymically hydrolyzed)	3.35	13.8	0.02	690
Casein (enzymically hydrolyzed)	5.04	40.3	1.56	26

[a] Reproduced from T. B. Griffin and J. M. Prescott, *J. Biol. Chem.* **245**, 1348 (1970) with the permission of the American Society of Biological Chemists, Inc.

[b] All media contained Seven Seas Marine Mix at a concentration of 26.4 g/liter and differed only with respect to the identity of their organic components, which were present at a concentration of 20 g/liter. All cultures were grown in New Brunswick fermentors for a period of 20–22 hr.

[c] Absorbance units/ml, measured at 660 nm.

[d] Hemoglobin units, as defined in the text.

[e] Assayed at 37° by the procedure of J. A. Goldbarg and A. M. Rutenburg, *Cancer* **11**, 283 (1958). One unit is the quantity of enzyme that catalyzes the hydrolysis of L-leucyl-β-naphthylamide at a rate of 1 μmole/min.

For large production runs, hydrolyzates are prepared by suspending 600 g of soybean protein in 6 liters of 0.2 N sodium hydroxide, adjusting the pH to 8.0, then adding 5.56 g of pancreatin (NBC). The mixture is incubated under toluene at 37° for 48 hr with stirring. Two liters of soybean protein hydrolyzate are measured into each of three New Brunswick fermentor jars (Model FS314; 14-liter capacity), 264 g of Seven Seas Marine Mix (Utility Chemical Co., Patterson, New Jersey) are added, and the total volume is adjusted to 10 liters. The assembled fermentors are autoclaved for 90 min at 121° and, in the interest of safety, are left undisturbed until they have cooled. A separately sterilized solution of 1 g of K_2HPO_4 in 10 ml of water is added to each fermentor; after thorough mixing, 10 ml of sterile 0.1 M $ZnCl_2$ solution is added to each container.

Culture and Inoculum. *Aeromonas proteolytica* was isolated by Merkel and Traganza[7] from the intestine of the marine wood borer *Limnoria tripunctata;* its characteristics were described by Merkel *et al.*[8] The organism has been cultured in this laboratory since 1957 and is available from the American Type Culture Collection (No. 15338). It is maintained by monthly transfers of slant cultures on medium consisting of Difco Bacto-Peptone (1.0 g), agar (4.0 g), $FeSO_4 \cdot 7H_2O$ (10 mg), K_2HPO_4 (10 mg), and 200 ml of seawater (artificial or natural), the pH being adjusted to 7.2 before sterilization. Inoculum medium consists of Seven Seas Marine Mix (20 g), Difco Bacto-Peptone (5 g), and water (500 ml); the pH is adjusted to 7.2, the medium is distributed 30 ml per 125-ml Erlenmeyer flask and autoclaved at 121° for 15 min. Loop transfers are made from a slant culture, and the inoculum for large-scale production experiments is grown at room temperature for 8–12 hr with vigorous agitation on a rotary shaker. Three 30-ml inoculum flasks are used for each 10 liters of production medium.

Growth. The three 10-liter cultures of *A. proteolytica* in a typical production run are grown at 26° with filtered air delivered to each fermentor at a rate of 10 liters/min. The medium is also agitated rapidly (400 rpm), as heavy growth and enzyme production are dependent upon thorough aeration. Frothing is controlled by the addition of General Electric Antifoam 60, diluted with an equal volume of water. In order to minimize precipitation of the silicone during sterlizing, it is autoclaved at 121° for only 10 min and then cooled with rapid shaking. Even with these precautions some precipitate forms in the emulsion, but with care

[7] J. R. Merkel and E. D. Traganza, *Bacteriol. Proc.* p. 53 (1958).
[8] J. R. Merkel, E. D. Traganza, B. B. Murkherjee, T. B. Griffin, and J. M. Prescott, *J. Bacteriol.* **87**, 1227 (1964).

it can be used in the automatically controlled antifoam pump on the New Brunswick fermentors without clogging the tubing. Beginning 10–12 hr after inoculation, protease activity is assayed at hourly intervals with the hemoglobin substrate. Maximal production of endopeptidase occurs later in the growth cycle of *A. proteolytica* than does the peak of aminopeptidase activity; therefore, cultures grown for isolation of the neutral protease are terminated at 20–24 hr, at which time the culture medium should contain approximately 8–15 hemoglobin units/ml. Cells are removed by centrifugation in a Sharples Super Centrifuge at 50,000 *g*, and the supernatant fluid is filtered first through a 0.65 μm filter with a fiberglass prefilter, then through a 0.45 μm filter, using a large (293 mm) Millipore filter apparatus.[9] The clear culture fluid is the source of the enzyme.

Enzyme Isolation

An earlier publication from this laboratory[10] described a procedure for isolating *Aeromonas* neutral protease in homogeneous form, and more recently, Pollard[11] isolated the enzyme by a procedure that involved a heat treatment at 60° followed by several steps of gel filtration and ion-exchange chromatography. Preparations thus obtained possessed physical properties identical to those of the enzyme isolated by Griffin and Prescott.[10] Recent experiments in this laboratory have shown that chromatography on Brushite ($CaHPO_4 \cdot 2H_2O$) is more effective than either ion-exchange or gel-filtration chromatography in removing the residual aminopeptidase from preparations of *Aeromonas* neutral protease.[12] Incorporation of this step, along with a heat treatment, has greatly facilitated isolation of the enzyme, and the new procedure described below achieves higher yields with fewer fractionation steps than the procedures previously devised. The new procedure has been used for isolating the neutral protease from culture filtrates produced from both the soybean and casein media, although use of the latter source results in a substantial sacrifice in yield if complete removal of aminopeptidase is desired.

---

[9] It has been the practice in this laboratory to include the filtration step in order to effect the complete removal of cells and debris. It can be omitted, however, if care is taken to ensure complete clarification during the centrifugation.

[10] T. B. Griffin and J. M. Prescott, *J. Biol. Chem.* **245**, 1348 (1970).

[11] D. R. Pollard, Ph.D. Dissertation, Texas A&M University, 1970.

[12] We are indebted to Dr. Bernard A. Link for originating the experiments involving Brushite chromatography and for demonstrating the feasibility of this step in removing the residual aminopeptidase.

Step 1. Concentration and Heat Treatment. Cultural fluid (1360 ml)[13] from *A. proteolytica* grown on soybean protein medium is concentrated 10-fold by rotary evaporation in small portions, the temperature is maintained at 35°–40° to minimize any denaturation by the high salt concentration in this medium during evaporation. After concentration of the culture filtrate, the salt is removed by dialysis; first, against running tap water for 6 hr, then against 4 liters of 10 mM Tricine buffer (pH 8.0) for 12 hr. The dialysis bags should be only partially filled, as the volume of the solution doubles during this procedure and yields a final volume of concentrated, dialyzed culture filtrate approximately one-fifth to one-fourth that of the starting material for step 1.

Incubation of the dialyzed crude neutral protease at 60° for 20–30 min, followed by dialysis, results in the removal of a second endopeptidase that is evident in gel electrophoretograms of the unfractionated culture filtrate. The heat treatment is carried out by dividing the concentrated, dialyzed culture filtrate (total volume, 340 ml) into 10-ml portions that are placed in test tubes and incubated at 37° for 15 min, then at 60° for 25 min. The tubes are removed from the water bath and immediately chilled in an ice bath. The heat-treated sample is dialyzed in 10 mM Tricine buffer, pH 8.0, for 12 hr and is concentrated by ultrafiltration (Amicon membrane UM-10) to a volume of about 10 ml.

Step 2. Gel Filtration. A 2.5 × 85 cm column of Sephadex G-100 is prepared for gravity-flow elution and equilibrated with 10 mM Tricine buffer, pH 8.0, that has been rendered 50 μM with respect to spectrographically pure $Zn(NO_3)_2$. The concentrated sample is centrifuged for 15 min at 12,000 g prior to application to the column, and the chromatogram is developed by elution with the Tricine-Zn^{2+} buffer at a flow rate of 25 ml/hr. The absorbance at 280 nm is determined for each of the 5-ml fractions collected, usually with the aid of a Gilson transferator; plots of the data reveal the presence of three absorbance peaks, as is shown in Fig. 1. The first peak contains an opalescent material and is devoid of protease activity. The fractions underlying the second peak are tested qualitatively for proteolytic activity by the gelatin assay, and those that show activity are assayed quantitatively using either FA-Gly-Leu-NH$_2$ or hemoglobin as the substrate. Most of the amber pigment in the sample is eluted in the third peak, but some pigment overlaps into the tubes in the trailing edge of the protease peak. Elimination of the pigment is best accomplished by visual observation and rejection of those fractions that show its presence. The specific activity of the prepa-

---

[13] Normally, 20–25 liters of culture filtrate result from a production run that begins with 30 liters. This can then be worked up in smaller lots as desired. The example given herein consisted of one such portion of a large production experiment.

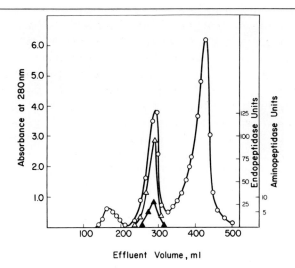

FIG. 1. Chromatography on Sephadex G-100. The culture filtrate had been concentrated, dialyzed, treated at 60° for 25 min, then dialyzed again before being applied to a 2.5 × 85 cm column of Sephadex G-100 equilibrated with 10 mM Tricine buffer, pH 8.0, that was 50 μM with respect to spectrographically pure Zn(NO₃)₂. The same buffer was used to elute the sample at a flow rate of 25 ml/hr. ○, Absorbance at 280 nm; △, neutral protease activity toward FA-Gly-Leu-NH₂; ▲, aminopeptidase activity.

ration at this point indicates a substantial degree of purification, but the preparation is not free of the contaminating aminopeptidase which, at this stage, makes up about 2% of the total protein.

Step 3. Chromatography on Brushite. A 225-ml suspension of Brushite (CaHPO₄·2H₂O; purchased from Bio-Rad) is washed four times with 250 ml of 10 mM Tricine buffer (pH 8.0) that is 1 M in NaCl. After each washing the supernatant liquid is removed by decanting, and after the final washing, the absorbent is equilibrated with 10 mM Tricine buffer (pH 8.0; no added NaCl) for 4 hr. The slurry is poured into a 2.5 × 40 cm column and washed with this buffer for 14 hr at a flow rate of 15 ml/hr. The sample is applied to the surface of the adsorbent and washed in with three 5-ml portions of starting buffer. Elution is accomplished by a linear gradient of NaCl consisting of 300 ml of 10 mM Tricine (pH 8.0) in the first chamber and an equal volume of the same buffer, rendered 0.6 M in NaCl, as the limit buffer; fractions of 5 ml are collected. The absorbance at 280 nm is determined for all fractions and those that give a positive reaction with the gelatin-digesting test are quantitatively assayed for protease activity with either FA-Gly-Leu-NH₂ or hemoglobin and for aminopeptidase activity with leucyl-*p*-nitroanilide.

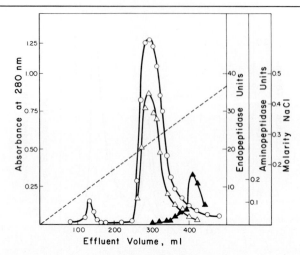

Fig. 2. Chromatography on Brushite. The active peak from Fig. 1 was concentrated to 11.1 ml, applied to a 2.5 × 40 cm column of Brushite and eluted with a linear gradient of NaCl in 10 mM Tricine buffer, pH 8.0. ○, absorbance at 280 nm; △, neutral protease activity toward FA-Gly-Leu-NH₂; ▲, aminopeptidase activity toward L-leucyl-p-nitroanilide; ---, NaCl concentration.

Figure 2 shows the separation achieved by this step. Fractions that reveal the presence of aminopeptidase are rejected, and the remaining fractions under the endopeptidase peak are combined to yield the final product. Recovery of the neutral protease is about 35% for this step after all the fractions that contain aminopeptidase are excluded. An isolation experiment is summarized in Table II.

Properties

Purity. The most difficult contaminants to remove from preparations of *Aeromonas* neutral protease are the aminopeptidase and the amber pigments that occur in culture filtrates of *A. proteolytica.* The endopeptidase can be freed from these and other contaminants, however, by selecting a nitrogen source that produces low amounts of the aminopeptidase and by careful attention to the separations achieved in the various steps. Preparations made by the procedure of Griffin and Prescott[10] and of Pollard[11] have shown a single component when tested by a variety of procedures, including sedimentation velocity, moving-boundary electrophoresis at several pH values, and immunoelectrophoresis. Enzyme prepared by the simplified procedure described above has shown a single band on polyacrylamide gel electrophoresis at pH 8.5, and tests for aminopeptidase were negative.

TABLE II

ISOLATION OF *Aeromonas* NEUTRAL PROTEASE

Step and treatment	Volume (ml)	Total protein[a] (mg)	Endopeptidase				Recovery[d] (%)	Amino-peptidase (units)[e]
			Hb[b] assay		FAGLA[c] assay			
			Total units	Specific activity	Total units	Specific activity		
Culture filtrate	1,360	3,550	10,880	3	36,200	10	100	135
1. Concentration, heat, dialysis	6.4	184	7,670	42	26,000	141	72	215
2. Sephadex G-100	11.1	80	6,180	77	22,170	277	61	206
3. Brushite, fraction 1	70	31	2,400	77	7,850	253	22	0

[a] Based on the Lowry reaction, using purified *Aeromonas* neutral protease as the standard.
[b] Hemoglobin.
[c] Furylacryloylglycyl-L-leucinamide. These assays were performed with 10 m*M* substrate.
[d] Based on the FAGLA assay.
[e] Assayed with L-leucyl-*p*-nitroanilide substrate.

Physical Properties. *Aeromonas* neutral protease has a sedimentation coefficient, $s_{20,w}^0$, of 3.50 S and a diffusion constant, $D_{20,w}^0$, of 8.52×10^{-7} cm^2 sec^{-1}; molecular weight calculations from both sedimentation equilibrium and sedimentation velocity–diffusion experiments yielded a value of 34,800.[10,11] Inasmuch as this value agrees closely with that found by sedimentation equilibrium in the presence of 5 *M* guanidine-HCl, the enzyme evidently consists of a single peptide. *Aeromonas* neutral protease is isoelectric at approximately pH 3.5. The amino acid content is Half-Cys$_4$, Asp$_{52}$, Thr$_{20}$, Ser$_{27}$, Glu$_{19}$, Pro$_{12}$, Gly$_{34}$, Ala$_{25}$, Val$_{24}$, Met$_8$, Ile$_6$, Leu$_{11}$, Tyr$_{22}$, Phe$_{18}$, Lys$_{12}$, His$_6$, Arg$_8$, Trp$_5$.[10,11] The enzyme exhibits a typical protein absorbance with a maximum at 278.5 nm; at this wavelength the molar extinction coefficient is 55,750.[11] The protein content of pigment-free preparations is estimated from absorbance measurements using the value $E_{1\,cm}^{1\%} = 16.0$.

Metal Content. The *Aeromonas* neutral protease, isolated in buffers with or without added Zn^{2+}, contains 1 g-atom of zinc per mole; some preparations have also shown the presence of 1–2 g-atoms of calcium per mole, whereas other active preparations have contained smaller, nonstoichiometric quantities of calcium. Atomic absorption spectrophotometry showed the final preparation from the experiment summarized in Table II to contain 1 g-atom of zinc and 1.7 g-atoms of calcium after 48 hr of dialysis. The presence of Zn^{2+}, but not Ca^{2+}, is essential for activity, inasmuch as dialysis of purified protease against 5 m*M* *o*-phenanthroline in 10 m*M* Tricine containing 10 m*M* Ca^{2+} both inactivates and depletes Zn^{2+}; activity is restored by the addition of Zn^{2+}. It seems likely that Ca^{2+} plays a structural role in *Aeromonas* neutral protease, as it does in thermolysin. Preparations may be rendered inactive also by dialysis against EDTA, and activity is diminished by incubation of the enzyme with dithiothreitol, cysteine, reduced glutathione, and β-mercaptoethanol, presumably because of the ability of these reagents to form complexes with metals.

Catalytic Properties. In concentrations as low as 0.1 μg/ml, *Aeromonas* neutral protease displays activity toward casein, gelatin, and urea-denatured hemoglobin. Specificity studies have indicated a preference for bonds of the type X-Phe, X-Leu, and X-Tyr to which a hydrophobic residue contributes the amino group, and whereas it is essential that the NH$_2$-terminal position of dipeptide substrates be blocked, the enzyme hydrolyzes substrates with free carboxyl groups (e.g., Z-Gly-Phe). Keay *et al.*[3] found the rate of hydrolysis of the X-Phe bond to be 40-fold greater than that for the X-Leu bond in doubly blocked dipeptides and showed that this difference in rate is much more pronounced for *Aeromonas* neutral protease than for any other neutral protease they tested. The use of cuvettes of 1-mm light path has permitted the testing of FA-Gly-Leu-NH$_2$ concentrations over the range of 1.5–14 mM and from these experiments good Lineweaver–Burk plots were obtained. The value of k_{cat}/K_m for the purified enzyme is 42×10^3 sec^{-1} M^{-1}. The *Aeromonas* enzyme, in addition to cleaving peptide bonds, shows esterase activity toward hippurylphenyllactic acid, but no amidase activity has been observed.

Stability. The enzyme is most active at pH 8.0 and retains 90% of its original activity after being stored at pH 7–11 at 10° for more than a week. Enzyme solutions are routinely stored frozen in low concentrations of Tricine or Tris buffer, and may be kept at −20° for several years without loss of activity, whereas lyophilization and reconstitution results in approximately a 20% loss in activity.

Acknowledgments

The research described herein was supported in part by Grant A-003 from the Robert A. Welch Foundation, Houston, Texas. Mrs. Mary E. Bayliss gave generously of her time and skills in helping to develop the enzyme isolation procedure described above.

[34] Thermomycolin

By G. MAURICE GAUCHER and KENNETH J. STEVENSON

The thermophilic fungus *Malbranchea pulchella* var. *sulfurea* produces a single extracellular, thermostable, "serine" protease, which was originally called thermomycolase and is now called thermomycolin (EC 3.4.21.–). While many microbial proteases have been studied in detail, only a few proteases produced by thermophilic fungi have been examined. Unlike their mesophilic counterparts which grow optimally at ~28°, thermophilic fungi prefer temperatures of 40°–55°. Initial observations of growth on casein-agar, led to a screening study[1] of the proteases produced by a selection of such fungi and eventually to a detailed study of thermomycolin.[2-6] Although subject to stringent control via catabolite

[1] P. S. Ong and G. M. Gaucher, *Can. J. Microbiol.* **19**, 129 (1973).

[2] P. S. Ong and G. M. Gaucher, *in* "Fermentation Technology Today," Proceedings of the IVth International Fermentation Symposium, Kyoto, Japan (G. Terui, ed.), p. 271. Soc. Ferment. Technol., Osaka, Japan, 1972.

[3] G. Voordouw, G. M. Gaucher, and R. S. Roche, *Biochem. Biophys. Res. Commun.* **58**, 8 (1974).

[4] G. Voordouw, G. M. Gaucher, and R. S. Roche, *Can. J. Biochem.* **52**, 981 (1974).

[5] P. S. Ong and G. M. Gaucher, *Can. J. Microbiol.* **22**, 165 (1976).

[6] K. J. Stevenson and G. M. Gaucher, *Biochem. J.* **151**, 527 (1975).

repression,[2,5] quantities of protease are readily produced by submerged cultures (8–150 liters) using a relatively inexpensive growth medium of 1–2% casein and salts. A simple purification gives a high yield of homogeneous protease which can be stored as a stable ammonium sulfate precipitate. Stoichiometric inhibition with DFP[7] allows physiochemical studies of this small, globular protease to be made in the absence of significant autolysis. Thermomycolin is particularly thermostable in the presence of Ca^{2+}, is an endoproteinase with a general specificity for apolar residues, and is a member of the subtilisin family of alkaline serine proteases.

Source of Enzyme

The occurrence and taxonomy of the thermophilic fungus *M. pulchella* (Emerson No. 27) have been described,[8] as has the maintenance of stock cultures.[5] A 5-ml spore suspension (400 ppm of Aerosol OT detergent) prepared from one mature (10–14 days at 45°) yeast-glucose agar (5 g of yeast extract, Difco; 10 g of glucose; 20 g of bacto-agar, Difco; and 1 liter of distilled water) surface culture (8-dram vial) provides the inoculum for a primary shake-flask culture (400 ml of a medium, pH 7.4, containing 3 g of yeast extract; 1 g of glucose; 10 g of casein, purified, high nitrogen, Nutritional Biochemicals; 2 g of $NaNO_3$; 1 g of KH_2PO_4; 0.5 g of KCl; 0.5 g of $MgSO_4 \cdot 7H_2O$; 0.01 g of $FeSO_4 \cdot 7H_2O$, and 0.3 mg of $ZnSO_4 \cdot 7H_2O$ per liter of distilled water in a 2-liter Erlenmeyer flask) grown at 45° for 40 hr on a gyrotory shaker (Nw Brunswick, Model G-10; 330 rpm; 1-inch stroke). The vegetative cells from one such shake culture were collected and washed by suction filtration through Whatman No. 4 paper, resuspended in 150 ml of sterile distilled water, blended for 30 sec at maximum speed in the 400-ml chamber of a Sorvall Omni-mixer and used to inoculate 3 secondary shake cultures (400 ml medium in a 2-liter Erlenmeyer flask), or one fermentor culture (see Fig. 1). Because of the deleterious effects of carbohydrate and amino acid nutrients on protease production,[2,5] fermentor cultures are best grown on a simple 2% casein and salts (see above) medium. In order to minimize foaming

---

[7] Abbreviations used are: DFP, diisopropyl fluorophosphate; DIP, diisopropyl-phosphoryl; Ac-Ala₃-OMe, *N*-acetyl-L-alanyl-L-alanyl-L-alanine methyl ester; Cbz-Gly-ONp, *N*-benzyloxycarbonyl-glycine-*p*-nitrophenyl ester; Boc, *t*-butoxy-carbonyl; PBA, 4-phenylbutylamine; Cbz-Phe-CH₂Cl, L-1-benzyloxycarbonyl-amido-2-phenylethyl chloromethyl ketone; Cys(O₃H), cysteic acid; EGTA, ethyl-eneglycol-bis(2-aminoethylether)-*N*,*N*,*N*,'*N'*-tetraacetic acid.

[8] D. G. Cooney and R. Emerson, "Thermophilic Fungi." Freeman, San Francisco, California, 1964.

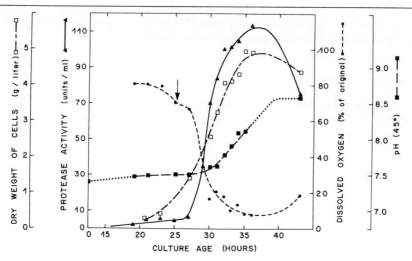

Fig. 1. Growth characteristics of a fermentor culture of *Malbranchea pulchella* var. *sulfurea*. A 14-liter New Brunswick Microferm fermentor [geometry as previously described by P. S. Ong and G. M. Gaucher, *Can. J. Microbiol.* **22**, 165 (1976)] containing 8 liters of medium (1.5% casein and salts) was maintained at 45°, with initial aeration and stirring rates of 1 liter/min and 200 rpm changed (↓) to 1 liter/min and 300 rpm at 25.3 hr. Antifoam (1:10 dilution of Dow-Corning Antifoam C emulsion in water) was added automatically as needed. Arbitrary enzyme units (pH 8.0; Cbz-gly-*p*-nitrophenyl ester substrate) were used as described under Assay of Enzyme. Assuming a culture yield of 110 arbitrary pH 8.0 units/ml and the purification yield and specific activity in Table I, a fermentor culture would yield ~320 mg of pure protease.

and the resultant loss of inoculum, and to optimize growth and protease production, initial areation and stirring rates of 1 liter/min and 200 rpm should be increased to 4 liter/min and 300 rpm only at the beginning of the linear growth phase (i.e., ~25 hr and ~5 units of protease per milliliter). This regime minimizes foaming and prevents the dissolved oxygen content from dropping below 40% of the original value. The typical growth and protease production characteristics given in Fig. 1 indicate that a smaller increase in aeration at ~25 hr results in a greater decrease in dissolved oxygen content. The yield of protease is also somewhat lower. In addition, initial casein concentrations of less than 2% result in a more rapid loss of protease due to autolysis, and agitation in the presence of antifoam can denature *pure* protease.[5] The culture is best harvested at the end of the growth phase (~36 hr) when it exhibits a pH slightly above 8, a more colored yet transparent appearance due to pigment accumulation and significant casein hydrolysis, and a maximum yield of protease (~100–140 arbitrary units/ml). The filtrate obtained by suction filtration may be stored at 4° for about 2 weeks without significant loss of protease and is used as the starting material for the purification.

Assay of Enzyme

Principle. The most convenient assay for thermomycolin is based upon the spectrophotometric determination of the *p*-nitrophenol produced by the hydrolysis of a synthetic ester substrate.

Reagents

Buffer: 10 mM Tris/HCl (pH 8.0) containing 10 mM CaCl$_2$, equilibrated at 30°

Substrate: 1.5 mM Cbz-glycine *p*-nitrophenyl ester (Sigma Chemical Co.) in acetonitrile (spectroquality, Matheson, Coleman, and Bell)

Enzyme: a solution diluted with buffer to yield a ΔOD/min of 0.1–0.2, equilibrated at 30°

Procedure. Diluted enzyme (2.8 ml) and a buffer blank (3.0 ml) are placed in sample and reference cuvettes (1-cm pathlength) in a 30° thermostatted cell compartment of a Beckman DB-G recording spectrophotometer set at 400 nm and 5 inches/min. The enzyme reaction is initiated by adding 0.2 ml of substrate to the sample cuvette with stirring. Final concentrations of substrate and acetonitrile are 0.1 mM and 6.7%, respectively. Nonenzymic hydrolysis of the substrate is measured by adding 0.2 ml of substrate to 2.8 ml of buffer and triplicate determinations are averaged and subtracted from the above measured rates. The error in the assay is $\leq \pm 3.0\%$.

Units. Arbitrary units are defined as the amount of enzyme required to yield 1.0 OD/min at 400 nm, pH 8.0, and 30°. Similar arbitrary units measured at pH 7.4 have also been used,[5] but since the *p*-nitrophenol (ϵ_{400} of anion = 18,300 M^{-1} cm^{-1} at 25°) has a pK_a of 7.15, more protease is required per unit increase in OD[9] and potential errors due to slight variations in assay pH are greater. Assays at 347.5 nm, the isobestic point for ionized and nonionized *p*-nitrophenol, are also less sensitive since ϵ_{400} equals $\sim 2 \epsilon_{347.5}$. A standard unit defined as the amount of enzyme required to yield 1 μmole of *p*-nitrophenol per minute at 400 nm, pH 8.0, and 30° can also be calculated from the total change in OD observed after completion of the reaction.[4,10]

Other Assays. Except for Cbz-L-Ala-*p*-nitrophenyl ester which is hydrolyzed at twice the rate of the glycine ester, no other such esters are

---

[9] One arbitrary pH 7.4 unit yields approximately 2 arbitrary pH 8.0 units.
[10] J. F. Kirsch and M. Ingelström, *Biochemistry* **5**, 783 (1966).

suitable because of their insolubility in water and the inhibitory effect of solvents such as acetonitrile.[5] A titrimetric assay using the methyl ester, N-acetyl-$(L$-Ala$)_3$-OMe (Sigma Chemical Co.) has been successfully used[6] according to a published procedure.[11] Less convenient assays using the protein substrates casein,[1] or Congo red-elastin[6,12] have also been used successfully.

Purification of Enzyme

Filtered culture medium (pH $\sim$8) is concentrated to $\frac{1}{4}$ volume by vacuum evaporation (Buchler Flash-Evaporator) at 45°. Concentration by ammonium sulfate precipitation ($\geq$70% saturation at 23°) is a useful alternative for very large volumes but leads to poorer yields (i.e., $\sim$60–80%). After standing overnight at 4° the precipitate is removed by centrifugation and discarded. The concentrate, after being dialyzed twice over $\sim$48 hr at 4° (20 volumes of 10 mM Tris-maleate buffer, pH 6.0, 10 mM in CaCl$_2$), is clarified by centrifugation and concentrated a second time to $\sim$$\frac{1}{5}$ volume as before. After removing the precipitate, this concentrate is dialyzed twice over 12–48 hr as above, but at pH 6.5, and any precipitate is again removed. The dark pigment is removed by a batch adsorption step. Dry DEAE-Sephadex A-25 (1 g/100 ml) is added to the dialyzed concentrate (pH 6.5–7.5) and the resulting slurry is stirred for at least 24 hr at 4° and then centrifuged. A further reduction in color can be achieved if necessary by a second batch adsorption at pH 7.4, followed by passage through a DEAE-Sephadex column equilibrated with 10 mM Tris-CaCl$_2$ (pH 7.5). The light yellow concentrate can now be stored indefinitely at 4°. In the last step the decolorized concentrate is applied to a Sepharose-4-phenylbutylamine[13] column (Pharmacia K25/45) equilibrated with 10 mM Tris-maleate buffer (pH 6.5, 10 mM CaCl$_2$). The retarded protease is eluted (2.75 ml/min) from this column using the above equilibration buffer. An alternative procedure carried out at 4° yields equally pure protease in $\sim$$\frac{1}{10}$ the volume and about the same yield.[2] After applying the concentrated protease to the above column at pH 7.5 (10 mM Tris-CaCl$_2$), the column is washed with 500 ml of the pH 7.5 buffer, and the protease is eluted with a pH 3.7 buffer (10 mM acetate-CaCl$_2$). Elution at a lower pH will, however, cause irreversible acid denaturation. A typical purification (Table I) yields homogeneous protease as determined by polyacrylamide disc gel electrophoresis and isoelectric focusing.[5] A preparation of pure thermomycolin is very sensitive to auto-

[11] A. Gertler and T. Hofmann, *Can. J. Biochem.* **48**, 384 (1970).
[12] A. Gertler and Y. Birk, *Eur. J. Biochem.* **12**, 170 (1970).
[13] K. J. Stevenson and A. Landman, *Can. J. Biochem.* **49**, 119 (1971).

TABLE I

A SUMMARY OF THE PURIFICATION OF THERMOMYCOLIN[a,b]

Procedure	Volume (ml)	Activity (units/ml)	Total activity (units)	Protein conc. (mg/ml)[c]	Specific activity (units/mg)	Overall Yield (%)	Overall Purification factor
1. Filtered culture medium (pH ~8)	4195	53.4	227,770	6.92	7.72	100.0	1.0
2. Concentration (vac., 45°), centrifugation	850	261.0	221,850	31.68	8.24	97.4	1.07
3. Dialysis (pH 6.0, 4°, 48 hr) centrifugation	1500	146.5	219,750	2.50	58.60	96.5	7.59
4. Repeat steps 2 and 3 (but pH 6.5)	398	555.0	220,890	5.03	110.34	96.98	14.29
5. Batch adsorption (DEAE-Sephadex, pH 6.5–7.5)	293	668.0	195,490	2.40	278.33	85.82	36.05
6. Affinity chromatography (PBA-Sepharose, pH 6.5)	756	234.0	176,900	0.22	1073.39	77.67	139.04

[a] From P. S. Ong and G. M. Gaucher, *Can. J. Microbiol.* **22**, 165 (1976).
[b] Arbitrary enzyme units (pH 7.4; Cbz-gly-ONp) were used as described under Assay of Enzyme.
[c] Protein concentration was determined by the Folin–Lowry method using bovine serum albumin (Sigma) as a standard.

lysis at concentrations > 0.01 mg/ml. Activity losses are, however, limited to 20–30% due to the protective effect of autolysis peptides.[4] These peptides are readily separated from the active protease by gel filtration (Sephadex G-75 or G-200) at 4°. Gel filtration can also be used with some success as an alternative last step in the above purification. Because of autolysis and the denaturation caused by freezing or freeze-drying, the protease is best stored as an ammonium sulfate precipitate. The precipitate obtained after dialyzing a protease solution for 4 hr against 3–4 volumes of saturated ammonium sulfate at pH 7, is collected on a 0.22-μm Millipore filter, and the filter is stored in a sealed tube at −20°.

Properties of Enzyme

Type of Protease. Thermomycolin is an extracellular, fungal, alkaline serine protease. The pH optimum for the proteolysis of bovine serum albumin or casein is 8.5, with the latter optimum being somewhat broader.[5] The protease is extensively inhibited by typical serine protease inhibitors, while being unaffected by typical metallo and sulfhydryl protease inhibitors (see Table V). Furthermore stoichiometric inhibition with DFP yields a covalent serine adduct as described below. The substrate specificity while broad is compatible with that of other microbial serine proteases.

Physical Properties. Most of these characteristics (Table II) are typical of small proteins. The isoelectric point of this protease is somewhat lower than normal (7–10.5) for most alkaline serine proteases. An anomalously low molecular weight is obtained by gel filtration.[3] A survey of seven other extracellular, fungal serine proteases indicated molecular weights (gel filtration or sedimentation equilibrium) from 17,800 to 30,000. The hydrodynamic properties of thermomycolin are consistent with those expected of a typical globular protein. Thermomycolin is not as thermostable as the extracellular proteases produced by thermophilic bacteria [e.g., thermolysin, $t_{1/2}$ (80°, 0.01 mg/ml, pH 8) = 87 min (2 mM Ca^{2+})],[14] but its thermostability certainly exceeds that of most fungal proteases.[2] This thermostability is to a large extent dependent upon the conformational stability conferred upon thermomycolin by the binding of a single calcium ion.[15]

Chemical Properties. The amino acid composition of thermomycolin and other selected proteolytic enzymes is presented in Table III. The

---

[14] H. Matsubara, *in* "Molecular Mechanism of Temperature Adaptation" (C. L. Prosser, ed.), p. 283. Am. Assoc. Adv. Sci. Washington, D.C., 1967.

[15] G. Voordouw, Ph.D. Thesis, University of Calgary, 1975; and G. Voordouw and R. S. Roche, *Biochemistry* 14, 4659 (1975).

TABLE II
PHYSICAL PROPERTIES OF THERMOMYCOLIN

Property	Value	Reference
Isoelectric point (pI)	6.0	f
Extinction coefficient ($E_{1\,cm}^{1\,\%}$ at 280 nm, pH 7, 25°; ϵ at 280 nm, pH 7, 25°)	13.5 ± 0.2; 43,900 liters/mole cm	g
Partial specific volume ($\bar{v}$)	0.736	g
Molecular weight		
Sedimentation equilibrium	32,000	g
SDS gel electrophoresis	33,000	g
Amino acid analysis	32,700	h
Gel filtration	11,000–17,500[a]	i
Sedimentation coefficient, $s_{20,w}^{\circ}$	2.97 ± 0.05 S	g
Diffusion coefficient, $D_{20,w}^{\circ}$	8.4 × 10^{-7} cm^2/sec	g
Intrinsic viscosity ([η])	3.05 ± 0.05 ml/g	g
Frictional ratio (f/f_0)	1.09[b]	g
Thermostability		
$T_{1/2}$ (60 min, 0.01 mg/ml, pH 6.85)[c]	69° (1 mM Ca^{2+}) 75° (10 mM Ca^{2+})	g
$t_{1/2}$ (73°, 0.007 mg/ml, pH 7.4)[d]	7.5 min (no added Ca^{2+}) 110 min (10 mM Ca^{2+})	f
Calcium binding constant (25°, pH 7.5, $\mu = 0.1$)[e]	5.0 × 10^5 M^{-1}	j

[a] Both active or diisopropylphosphoryl-protease at various concentrations exhibited anomalously low values.

[b] Assuming a prolate ellipsoid, this value corresponds to an axial ratio of $a/b = 2.6$.

[c] $T_{1/2}$ = temperature required for 50% inactivation in a given time.

[d] $t_{1/2}$ = half-life at a given temperature.

[e] Determined by gel filtration for the binding of 1.2 ± 0.1 Ca^{2+} per mole of protease. The nonintegral value may result from an extinction coefficient that is somewhat too large.

[f] P. S. Ong and G. M. Gaucher, *Can. J. Microbiol.* **22**, 165 (1976).

[g] G. Voordouw, G. M. Gaucher, and R. S. Roche, *Can. J. Biochem.* **52**, 981 (1974).

[h] V. Dorian and K. Stevenson, *Biochem. J.*, submitted.

[i] G. Voordouw, G. M. Gaucher, and R. S. Roche, *Biochem. Biophys. Res. Commun.* **58**, 8 (1974).

[j] G. Voordouw, Ph.D. Thesis, University of Calgary, 1975; and G. Voordouw and R. S. Roche, *Biochemistry* **14**, 4659 (1975).

amino acid composition of the "serine" proteases produced by the fungi *Aspergillus flavus, A. sojae,* and *A. sulphureus* are similar[16] to that reported for Aspergillopeptidase B from *A. oryzae.* The cystine content of

[16] J. Turková, O. Mikeš, K. Hayashi, G. Danno, and L. Polgar, *Biochim. Biophys. Acta* **257**, 257 (1972).

TABLE III

AMINO ACID COMPOSITION OF THERMOMYCOLIN AND OTHER PROTEASES

Amino acid	Thermo-mycolin[a]	Aspergillo-peptidase B[b]	Aspergillo-peptidase C[c]	Thermoly-sin[d]	Subtilisin BPN[e]	α-Lytic protease[f]	Arthro-bacter[g]	Porcine elastase[h]	Moose elastase[i]
Lysine	4	13.4	12	11	11	2	9	3	9
Histidine	7	4.7	4	8	6	1	2	6	5
Arginine	14	2.7	3	10	2	12	1	12	3
Aspartic acid	39	25.2	21–22	44	28	15	24	24	23
Threonine	16	14.0	13	25	13	17	19	19	11
Serine	33	25.2	23	26	37	19	27	22	33
Glutamic acid	21	15.8	13	21	15	13	12	19	16
Proline	10	5.6	5–6	8	14	5	12	7	7
Glycine	57	24.9	21	36	33	31	29	25	26
Alanine	45	29.2	23	28	37	24	23	17	17
Half-cystine	2		0	0	0	6	4	8	10
Valine	24	15.1	16	22	30	19	12	8	22
Methionine	2	1.6	1	2	5	2	4	27	2
Isoleucine	20	11.0	10–11	18	13	8	5	10	13
Leucine	14	11.7	10	16	15	10	8	18	15
Tyrosine	8	6.7	4–5	28	10	4	15	11	11
Phenylalanine	3	5.8	6	10	3	6	10	3	3
Tryptophan	6	—	2	3	3	4	5	7	5
Total number of residues	325	~214	187–191	316	275	198	221	240	231
Molecular weight	32,700	~21,800	~19,500	34,400	26,000	20,100	23,041	25,900	24,201

[a] V. Dorian and K. Stevenson, Biochem. J. submitted.
[b] J. Turková, O. Mikeš, K. Hayashi, G. Danno, and L. Polgár, Biochim. Biophys. Acta 257, 257 (1972).
[c] A. Norwig and W. F. Jahn, Eur. J. Biochem. 3, 519 (1968).
[d] K. Titani, M. A. Hermodson, L. H. Ericsson, K. A. Walsh, and H. Neurath, Nature (London) (New Biol.) 238, 35 (1972).
[e] F. S. Markland and E. L. Smith, J. Biol. Chem. 242, 5198 (1967).
[f] J. Jurášek and D. R. Whitaker, Can. J. Biochem. 45, 917 (1967).
[g] B. V. Hofsten, H. van Kley, and D. Eaker, Biochim. Biophys. Acta 110, 585 (1965).
[h] D. M. Shotton and B. S. Hartley, Biochem. J. 131, 643 (1973).
[i] K. J. Stevenson and J. K. Voordouw, Biochim. Biophys. Acta 386, 324 (1975).

thermomycolin was quantitated as cysteic acid.[17] (1.6 residues) and also as S-carboxymethylcysteine (1.7 residues). Since no cysteine residues exist in the protease,[18] a single disulfide bridge is probably present. Diagonal peptide mapping[18,19] of thermomycolin revealed, however, only one peptide possessing cysteic acid [Lys (1.10), Arg (0.79), Cys(O_3H) (1.20), Asp (2.00), Ser (2.10), Glu (2.10), Gly (3.90), Val (1.52), Ile (2.20)]. A "mate" for this peptide could not be detected even after diagonal peptide maps were prepared under various conditions. The anomaly between the quantitation of cystine and the single peptide present on diagonal peptide maps remains unresolved.

The active-site sequence of thermomycolin was determined from peptides obtained by partial acid hydrolysis of the ^{32}P-labeled DFP-inhibited protease and is compared to other "serine" proteases in Table IV. Thermomycolin is clearly associated with the -Thr-Ser-Met- family of "serine" proteases in contrast to the -Gly-Asp-Ser-Gly- family.[20,21] Using an automatic Beckman Sequencer Model 890C the N-terminal sequence[18] of thermomycolin was determined to be

$$
\overset{+}{N}H_3\text{-Ala-Leu-Val-Thr-Gln-[Ser]-Asn-Ala-Pro-[Ser]-Trp-Gly-Leu-Gly-}
$$
$$
\text{Arg-Ile-[Ser]-Asn-Arg-Gln-Ala-Gly-Ile-Arg-Asp-Tyr-His-Tyr-}
$$

The identification of the amino acid residues in brackets is tentative, since serine and cysteine were not unambiguously resolved by gas–liquid chromatography. Serine was the more likely residue as determined by thin-layer chromatography. No obvious homology has been observed between the N-terminal sequence of thermomycolin and that of other "serine" proteases sequenced to date.

Effect of Inhibitors. A summary of the effect of various compounds on the activity of thermomycolin is presented in Table V.[5] The inactivation with ^{32}P-labeled DFP resulted in the incorporation of 1.33 moles of ^{32}P per mole of thermomycolin.[18] The ^{32}P label was associated with only a single amino acid sequence (Table IV) and thus suggests a 1:1 stoichiometry for the inhibition. A single ^{32}P-labeled DIP-labeled peptide was also isolated[18] from an extended trypsin digest [composition: Thr (0.9), Ser (2.4), Gly (1.2), Met (0.7), Leu (1.0)], sequence Leu-Ser-(Gly)-

[17] S. Moore, *J. Biol. Chem.* **238**, 235 (1963).

[18] V. Dorian and K. Stevenson, *Biochem. J.*, submitted.

[19] J. R. Brown and B. S. Hartley, *Biochem. J.* **101**, 214 (1966).

[20] G. H. Dixon, *in* "Essays in Biochemistry" (P. N. Campbell and G. D. Greville, eds.), Vol. 2, p. 147. Academic Press, New York, 1966.

[21] B. S. Hartley, *Annu. Rev. Biochem.* **29**, 45 (1960).

TABLE IV
Active-Site Sequence of Thermomycolin and Other Alkaline Proteases

Protease	Origin	Active-site sequence[a]
Thermomycolin[b]	*Malbranchea pulchella* var. *sulfurea*	Leu-Ser-(Gly)-Thr-Ser*-Met
Alkaline protease[c]	*Aspergillus oryzae*	-Thr-Ser*-Met-Ala
Alkaline protease[d]	*Aspergillus flavus*	-Gly-Thr-Ser*-Met-Ala
Subtilisin BPN'[e]	*Bacillus subtilis*	-Asn-Gly-Thr-Ser*-Met-Ala-Ser-
α-Lytic[f]	*Myxobacter sorangium* sp.	-Arg-Gly-Asp-Ser*-Gly-Gly-Ser-
Protease A[g]	*Streptomyces griseus*	-Pro-Gly-Asp-Ser*-Gly-Gly-Ser-
Alkaline protease[h]	*Arthrobacter* (strain B22)	-Ser-Ser*-Gly-

[a] Asterisk denotes the active serine residue possessing the diisopropyl phosphoryl moiety.
[b] V. Dorian and K. Stevenson, *Biochem. J.*, submitted.
[c] F. Sanger, *Proc. Chem. Soc. London* **1963**, 76 (1963).
[d] O. Mikeš, J. Turková, N. B. Toan, and F. Šorm, *Biochim. Biophys. Acta* **178**, 112 (1969).
[e] E. L. Smith, R. J. DeLange, W. H. Evans, M. Landon, and F. S. Markland, *J. Biol. Chem.* **243**, 2184 (1968).
[f] D. R. Whitaker and C. Roy, *Can. J. Biochem.* **45**, 911 (1967).
[g] P. Johnson and L. B. Smillie, *FEBS Lett.* **47**, 1 (1974).
[h] S. Wåhlby, *Biochim. Biophys. Acta* **151**, 409 (1968).

Thr-Ser-Met-. DIP-labeled proteinase is also readily reactivated at pH 3.7.[15]

In contrast to the ineffectiveness of Tos-Phe-CH₂Cl, the more apolar Cbz-Phe-CH₂Cl inactivates thermomycolin. Chloromethyl ketones which possess peptide structures, and are therefore larger than the chloromethyl ketone reagents noted in Table V, will most likely be more effective inhibitors. This situation has been observed with microbial proteases[22,23] and a mammalian protease.[24]

[22] K. Morihara and T. Oka, *Arch. Biochem. Biophys.* **138**, 526 (1970).
[23] K. Morihara, T. Oka, and H. Tsuzuki, *Arch. Biochem. Biophys.* **146**, 297 (1971).
[24] J. C. Powers and P. M. Tuhy, *Biochemistry* **12**, 4767 (1973).

Table V

INHIBITOR SPECIFICITY OF THERMOMYCOLIN

Inhibitor			Percent inhibi-tion
Type[a]	Compound	Conc. (mM)	
Divalent metal chlorides[b]	Hg, Zn, Cu	50	100
	Ni, Co, Cd	50	56, 67, 85
	Ba, Mg, Mn	50	19, 22, 26
Metalloprotease[c]	EDTA	12.5	Nil
Sulfhydryl protease[c]	p-Chloromercuribenzoate, iodoace-tate, iodoacetamide	2.5	Nil
Serine protease[c]	Diisopropyl fluorophosphate	1–2.5	100
	Phenylmethyl-, p-methylphenyl-, and p-nitrophenyl-sulfonyl fluorides	0.05	100
Toluenesulfonyl chloro-methyl ketones[c,d]	Gly, L-Ala, L-Lys, L-Phe	0.05	Nil
Cbz-halomethyl ketones[e]	L-Phe-CH$_2$Cl	0.0025	69
	L-Phe-CH$_2$Br	0.0025	100
Trypsin[c]	p-Aminobenzamidine	0.05	Nil
	Soybean, egg white, lima bean	1.3[f]	20, 37, 67

[a] Refers to the type of inhibitor or, alternatively, to the type of protease affected by the inhibitor.

[b] Dialyzed protease was incubated with inhibitor for 30 min at 30° and compared to a protease solution containing 50 mM Ca^{2+}.

[c] Protease in assay buffer was incubated with inhibitor for 1 hr at 30° and compared to a noninhibited control.

[d] Stock solutions were made up in absolute ethanol; hence protease-inhibitor solutions and control were 5 % v/v ethanol.

[e] Molar ratio of inhibitor to protease was 10; incubation was at 45°, pH 8.0, for 0.5 hr.

[f] Concentration in milligrams per milliliter.

Substrate Specificity. The specificity of thermomycolin toward small synthetic ester substrates is presented in Table VI.[5] This study indicates a preference for L-Ala > L-Tyr > L Phe ≫ Gly ≫ L-Leu > L-Trp ≫ L-Val > L-Lys, L-Pro on the carboxyl side of the bond hydrolyzed. The protease possesses elastaselike activity since the rates of hydrolysis of Ac-Ala$_3$-OMe[11] and elastin-Congo red are 22 and 8.5-fold greater, respectively, than those obtained with porcine elastase.[6] The kinetic data for the hydrolysis of Ac-Ala$_3$-OMe by thermomycolin (in 1.2 mM Tris-HCl, pH 8.5 containing 1.2 mM CaCl$_2$ and 1.2 mM KCl at 30°) are: $K_m = 1.5$ mM, $k_{cat} = 2,050$ sec^{-1} and $k_{cat}/K_m = 1,370$ mM^{-1} sec^{-1}. Hydrolysis

TABLE VI

SUBSTRATE SPECIFICITY OF THERMOMYCOLIN[a]

Substrate		Relative activity[c] (%)
Amino acid side chain	p-Nitrophenyl ester[b]	
Nil	Cbz-Gly-ONp	100
	Boc-Gly-ONp	4 (100)[d]
Aromatic	Cbz-L-Phe-ONp[e]	160
	Boc-L-Phe-ONp[f]	25 (610)
	Cbz-D-Phe-ONp[e]	27
	Cbz-L-Tyr-ONp[f]	190
	Boc-L-Tyr-ONp	19 (460)
	Cbz-L-Trp-ONp[e]	32
	Boc-L-Trp-ONp[e]	3 (80)
Branched	Cbz-L-Val-ONp[f]	9
	Boc-L-Val-ONp	0.8 (19)
	Cbz-D-Val-ONp[f]	Nil
	Cbz-L-Pro-ONp	Nil
	Cbz-L-Leu-ONp[e]	66
Unbranched	Cbz-L-Ala-ONp	213
	Boc-L-Ala-ONp	14 (340)
	Cbz-D-Ala-ONp	7
	Boc-L-Met-ONp	20 (480)
	Cbz-L-Lys-ONp[g]	Nil

[a] From P. S. Ong and G. M. Gaucher, *Can. J. Microbiol.* **22,** 165 (1976).

[b] No activity was observed with other substrates utilized by chymotrypsin (N-acetyl-L-Tyr ethyl ester); trypsin and papain (N-benzoyl L Arg ethyl ester, N-benzoyl-DL-Arg-p-nitroanilide); thermolysin, a neutral metalloprotease (furylacryloyl-Gly-L-Leu amide, N-Cbz Gly-L-Leu amide, N-Cbz-Gly-L-Phe amide); aminopeptidase (L-Leu-p-nitroanilide), and carboxypeptidase B (hippuryl-L-Arg).

[c] Assayed in 10 mM Tris buffer (pH 7.4); 0.1 mM substrate; 6.7% (v/v) acetonitrile; at 30° and calculated from units per given amount of enzyme relative to Cbz-Gly-ONp = 100%.

[d] As in c, but calculated relative to Boc-Gly-ONp = 100%.

[e] Poorly soluble in water, hence assayed as in c, but in 23.3% (v/v) acetonitrile and calculated relative to Cbz-Gly-ONp assayed in 23.3% (v/v) acetonitrile (yields only ~5% of the activity obtained at 6.7%).

[f] Same as in e but in 13.3% (v/v) acetonitrile (Cbz-Gly-ONp yields only ~50% of the activity obtained at 6.7%).

[g] Less soluble in acetonitrile, hence stock solution of substrate made up in 10% (v/v) H$_2$O-acetonitrile. Assay as in c except in 6.0% (v/v) acetonitrile. Rapid nonenzymic hydrolysis occurs.

of Cbz-Gly-ONp[7] as described under Assay of Enzyme yields a k_{cat}/K_m = 796 mM^{-1} sec^{-1}.[15] A comparative study of the esterase activity of various "serine" alkaline proteases from microorganisms, against N-

acylated peptide ester substrates, has recently appeared.[25] A "serine" protease from *A. sojae* possessed a high elastaselike activity toward Ac-Ala$_3$-OMe but was incapable of hydrolyzing elastin because its acidic nature (pI = 5.1) prevented its adsorption onto the insoluble elastin.[26]

The digestion of glucagon by thermomycolin at 25° (5 min) and 45° (40 min) is presented in Table VII and compared to a digestion with aspergillopeptidase C from *A. oryzae*. The specificity of thermomycolin toward the oxidized A- and B-chains of insulin is shown in Tables VIII and IX and is compared to other alkaline proteases of fungal, bacterial, and mammalian origins. In general, thermomycolin readily hydrolyzes the peptide bonds of sequences possessing predominantly nonpolar residues. Careful examination of the amino acids occupying positions P$_3$–P$_1$ and P$_1'$–P$_3'$ relative to the scissile bond P$_1$–P$_1'$ does not indicate any well defined pattern of hydrolysis, but within the above sequence of six amino acid residues (P$_3$–P$_3'$) at least four residues are always nonpolar. Charged amino acid residues are not, however, excluded from this sequence. This is probably because any unfavorable interaction is more than offset by the numerous favorable apolar interactions arising from neighboring residues and even from the apolar portion of the charged residue itself (i.e., Lys, Arg, and Glu).

In addition to the results with small substrates (Table VI),[5] no amino acids have been observed in digests of glucagon, insulin chains, and chymotrypsin α,[6] thereby indicating the absence of exopeptidase activity in this protease. The hydrolysis of

$$\downarrow$$
-X-Pro-

bonds by thermomycolin has not been observed. The peptides

$$\downarrow$$
Cbz-Gly-Pro-Leu-Gly-Pro

and to a lesser extent

$$\downarrow$$
Cbz-Gly-Pro-Gly-Gly-Pro-Ala

are hydrolyzed.[6] These peptides are known to be good substrates for collagenases[27,28] and subtilisin BPN'.[29] The "serine" proteases from *Streptomyces* and *Aspergillus* sp.[23,30] hydrolyzed these peptides in a

[25] K. Morihara, T. Oka and H. Tsuzuki, *Arch. Biochem. Biophys.* **65**, 72 (1974).

[26] A. Gertler and K. Hayashi, *Biochim. Biophys. Acta* **235**, 378 (1971).

[27] W. Grapmann and A. Nordwig, *Hoppe-Seyler's Z. Physiol. Chem.* **322**, 267 (1960).

[28] Y. Nagai, S. Sakakibara, H. Noda, and S. Akabori, *Biochim. Biophys. Acta* **37**, 567 (1960).

[29] K. Morihara, T. Oka, and H. Tsuzuki, *Arch. Biochem. Biophys.* **138**, 515 (1970).

[30] A. Norwig and W. F. Jahn, *Eur. J. Biochem.*, **3**, 519 (1968).

TABLE VII

Digestion of Glucagon by Thermomycolin and Aspergillopeptidase C[a]

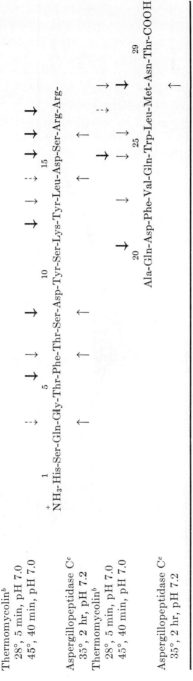

Thermomycolin[b]
28°, 5 min, pH 7.0
45°, 40 min, pH 7.0

Aspergillopeptidase C[c]
35°, 2 hr, pH 7.2

Thermomycolin[b]
28°, 5 min, pH 7.0
45°, 40 min, pH 7.0

Aspergillopeptidase C[c]
35°, 2 hr, pH 7.2

NH₃-His-Ser-Gln-Gly-Thr-Phe-Thr-Ser-Asp-Tyr-Ser-Lys-Tyr-Leu-Asp-Ser-Arg-Arg-

Ala-Gln-Asp-Phe-Val-Gln-Trp-Leu-Met-Asn-Thr-COOH

[a] The arrows indicate cleavages producing major peptides (↓), intermediate peptides (↓), and minor peptides (⋮).
[b] K. J. Stevenson and G. M. Gaucher, *Biochem. J.* **151**, 527 (1975).
[c] A. Norwig and W. F. Jahn, *Eur. J. Biochem.* **3**, 519 (1968).

TABLE VIII

DIGESTION OF THE OXIDIZED A-CHAIN OF INSULIN BY THERMOMYCOLIN AND ALKALINE PROTEASES FROM
Aspergillus oryzae AND *A. ochraceus*[a]

Aspergillopeptidase C,[b]
35°, 2 hr, pH 7.2

A. *ochraceus* protease,[c]
35°, 10 hr, pH 8.0

A. *oryzae* protease,[d]
37°, 24 hr, pH 8.0

Thermomycolin,[e]
25°, 5 min, pH 7.0

$$O_3H\ O_3H \qquad O_3H \qquad\qquad O_3H$$

$$\overset{+}{H_3}N\text{-Gly-Ile-Val-Glu-Gln-Cys-Cys-Ala-Ser-Val-Cys-Ser-Leu-Tyr-Gln-Leu-Glu-Asn-Tyr-Cys-Asn-COOH}$$

1 ... 10 ... 21

[a] Solid arrows (↓) denote major cleavage; broken arrows (⇣) represent minor cleavages.
[b] A. Norwig and W. F. Jahn, *Eur. J. Biochem.* **3,** 519 (1968).
[c] T. Kishida and S. Yoshimura, *J. Biochem. (Tokyo)* **55,** 95 (1964).
[d] F. Sanger, E. O. P. Thompson, and R. Kitai, *Biochem. J.* **59,** 509 (1955).
[e] K. J. Stevenson and G. M. Gaucher, *Biochem. J.* **151,** 527 (1975).

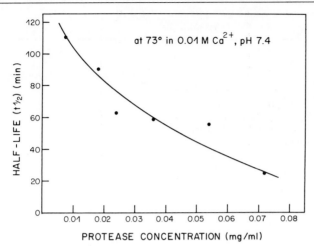

FIG. 2. The effect of protease concentration on protease stability. From P. S. Ong and G. M. Gaucher, *Can. J. Microbiol.* **22**, 165 (1976).

manner similar to thermomycolin. The latter proteases are, however, not considered to be true collagenases. Thermomycolin is similarly not a true collagenase, as it only slowly released peptides from native collagen.[6] Keratinase activity has not been observed with thermomycolin.[6]

Stability toward pH, Temperature, Autolysis, and Agitation. Thermomycolin is stable over a broad pH range of 4.5–10.5 for 2 hr, and 6.0–9.5 for 20 hr at 30°. The optimum pH for stability is ~8.5, while a rapid irreversible denaturation occurs below pH 3.5.[5] At 60°, thermal denaturation is least at pH 6.8.[15]

Thermomycolin exhibits significant thermostability especially in the presence of calcium (Table II). At 70° and 0.01 mg of protease per milliliter, maximum thermostability is achieved at a calcium ion concentration of 10 mM.[4] The first-order thermal denaturation which occurs at high temperatures (>70°) is not decreased by increasing the concentration of *purified* protease (Fig. 2), even though higher protein concentrations generally enhance thermostability.[31,32] This diminished stability at concentrations >0.01 mg/ml is due to autolysis.

Autolysis, which is the dominant inactivation process below 50°, is decreased by the presence of calcium or macromolecules. Thus $t_{1/2}$ (40°, 0.15 mg/ml, pH 7.4) equals >12 hr in the presence of 10 mM Ca^{2+}, and ~6 hr in the presence of the strong Ca^{2+} chelator, EGTA (10 mM),

[31] K. Ogasahara, K. Yutani, A. Imanishi, and T. Isemura, *J. Biochem. (Tokyo)* **67**, 83 (1970).

[32] F. C. Wedler and F. M. Hoffmann, *Biochemistry* **13**, 3215 (1974).

TABLE

HYDROLYSIS OF THE OXIDIZED B-CHAIN OF INSULIN

	O_3H		
	+ 1 5 |		
Protease	H₃N-Phe-Val-Asn-Gln-His-Leu-Cys-Gly-Ser-		

Protease	Cleavage sites (↑ major, ↑̣ minor)
Thermomycolin (25°, pH 9.0, 5 min)[a]	—
Thermomycolin (45°, pH 9.0, 5 min)[a]	↑ (His-Leu)
Aspergillopeptidase B (0°, pH 10.2, 30 min)[c]	—
Aspergillopeptidase B (25°, pH 9.2, 2 hr)[d]	↑ (His-Leu)
Aspergillopeptidase C (35°, pH 7.2, 2 hr)[e]	↑ ↑ (Gln-His, His-Leu) ... ↑
Alkaline protease from *Aspergillus flavus* (37°, pH 8.5, 1 hr)[f]	↑ ↑ (Gln-His, His-Leu) ... ↑̣
Alkaline protease from *Cephalosporium* sp. (37°, pH 10.5, 10 hr)[g]	↑ (His) ... ↑ ↑ ↑
Mammalian elastase: porcine (25°, pH 8.8, 15 min);[h] moose (37°, pH 8, 15 min)[i]	↑̣
α-Lytic protease from *Sorangium* (25°, pH 9, 1 hr)[j]	↑̣ ↑̣
Streptomyces fradia 1b (25°, pH 9, 2–3 hr)[d]	↑ ... ↑
Streptococcal proteinase[k] (25°, pH 7.6, 2.5 hr)[l]	↑ ... ↑̣ ↑̣
Alkaline protease from *Pseudomonas aeruginosa* (27°, pH 8.0, 2 hr)[m]	↑ ... ↑ ↑ ↑
Elastase from *Pseudomonas aeruginosa* (27°, pH 8.0, 5 hr)[m]	↑
Alkaline protease from marine psychrophilic *Pseudomonas* sp. No. 548 (37°, pH 8, 3 hr)[n]	↑

[a] From K. J. Stevenson and G. M. Gaucher, *Biochem. J.* **151**, 527 (1975).

[b] Solid arrows (↑) denote major cleavage, broken arrows (↑̣) represent minor cleavages.

[c] S. Spadari, A. R. Subramanian, and G. Kalnitsky, *Biochim. Biophys. Acta* **359**, 267 (1974).

[d] K. Morihara and H. Tsuzuki, *Arch. Biochem. Biophys.* **129**, 620 (1969).

[e] A. Norwig and W. F. Jahn, *Eur. J. Biochem.* **3**, 519 (1968).

[f] J. Turková and O. Mikeš, *Biochim. Biophys. Acta* **198**, 386 (1970).

[g] J. Yagi, K. Ikushima, T. Yano, H. Sakai, and M. Ajisaka, *J. Ferment. Technol.* **52**, 713 (1974).

[h] A. S. Narayanan and R. A. Anwar, *Biochem. J.* **114**, 11 (1969).

[i] K. J. Stevenson and J. K. Voordouw, *Biochim. Biophys. Acta* **386**, 324 (1975).

[j] D. R. Whitaker, C. Roy, C. S. Tsai, and L. Jurášek, *Can. J. Biochem.* **43**, 1961 (1965).

[k] Streptococcal proteinase is a thiol protease, and the substrate was the S-carboxylmethylated B-chain of insulin.

[l] B. I. Gerwin, W. H. Stein, and S. Moore, *J. Biol. Chem.* **241**, 3331 (1966).

[m] K. Morihara and H. Tsuzuki, *Arch. Biochem. Biophys.* **114**, 158 (1966).

[n] N. Kato, S. Adachi, K. Takeuchi, K. Morihara, Y. Tani, and K. Ogata, *Agr. Biol. Chem.* **38**, 103 (1972).

while the carbohydrate polymer Ficoll (Pharmacia, 5%) provided complete protection over a 12 hr period.[5] Both the second-order autoproteolysis (25°, pH 7.5, $\mu = 0.1$, 1.5 μM) and first-order thermal (70°, pH 7.0, $\mu = 0.05$, 3.0 μM), and 8 M urea (50°, pH 7.0, $\mu = 0.05$, 0.3 mM) denaturation of thermomycolase are markedly inhibited by the binding of a single Ca^{2+}.[15] Thermomycolin also possesses a significant intrinsic (i.e., Ca^{2+} independent) thermostability.[15]

Thermomycolin (28°, 0.05 mg/ml, pH 7.4, 10 mM Ca^{2+}) is also subject to surface inactivation during agitation on a reciprocal shaker (150 rpm, 3 cm stroke) in the presence of a silicone antifoam agent (0.005%).

IX

BY VARIOUS ALKALINE "SERINE" PROTEASES[a,b]

O₃H

| 10 | 15 | | 20 | 25 | 30 |

His-Leu-Val-Glu-Ala-Leu-Tyr-Leu-Val-Cys-Gly-Glu-Arg-Gly-Phe-Phe-Tyr-Thr-Pro-Lys-Ala-COOH

In the absence of this antifoam agent, or in the presence of both the antifoam agent and 5% Ficoll, no loss in activity was observed over a 12-hr
period of agitation.[5]

Acknowledgments

This project was supported by the National Research Council of Canada. The
authors also wish to acknowledge the invaluable contributions made to this study
by Poh Seng Ong, Gerrit Voordouw, Dr. Rodney S. Roche, Victor Dorian, Johanna
Voordouw, Betty Cowie, Kay Carter, Brian Ward, and Mark Sosnowski. Thanks
are also due to Dr. L. B. Smillie, Dr. G. H. Dixon, Mr. M. Carpenter, and Mr.
D. Watson for the sequencer studies.

[35] Penicillopepsin

By Theo Hofmann

Penicillopepsin is the extracellular acid protease which is produced by *Penicillium janthinellum* (NRRL 905 = C.M.I. 75589). As the partial amino acid sequences available show, it is homologous to mammalian pepsin and chymosin (see below). Similar enzymes are found in many other species of fungi.[1-5] A discussion of the structural, functional, and evolutionary relationships among acid proteases is contained in a recent review.[6]

Recently the isolation and properties of an acid protease from *Penicillium roquefortii* has also been reported.[7,8]

Cultivation of *Penicillium janthinellum* and Enzyme Production[9]

Originally penicillopepsin (then called peptidase A) was isolated from stationary surface cultures.[10] However, this procedure was time-consuming and yielded only small amounts of enzyme. Subsequently, methods were developed for growing the organism in submerged culture. Conditions have been found that give final enzyme levels about ten times as high as those of the stationary cultures and are reached in 4–8 days instead of 3–5 weeks. The use of a 10–12-liter Microferm apparatus (New Brunswick Scientific Co., New Brunswick, New Jersey) has been described in a previous volume of this series.[1] Subsequently one preparation was made in a 1500-liter fermentor at the Connaught Medical Research Laboratories, Toronto.[11] Since then we have routinely used a 250-liter Fermacell apparatus (New Brunswick Scientific Co., New Brunswick,

[1] J. Šodek and T. Hofmann, this series Vol. 19, p. 372.
[2] E. Ichishima, this series Vol. 19, p. 397.
[3] J. R. Whitaker, this series Vol. 19, p. 436.
[4] K. Arima, J. Yu and S. Iwasaki, this series Vol. 19, p. 446.
[5] M. Ottesen and W. Rickert, this series Vol. 19, p. 459 (1970).
[6] T. Hofmann, *Adv. Chem. Ser.* **136**, 146 (1974).
[7] J. C. Gripon and J. L. Bergère, *Lait* **52**, 497 (1972).
[8] C. Zevaco, J. Hermier, and J. C. Gripon, *Biochimie* **55**, 1353 (1973).
[9] Two acid carboxypeptidases (penicillocarboxypeptidases S-1 and S-2) are also produced by *Penicillium janthinellum*.[11] The procedure for cultivation and isolation of penicillopepsin described below also yields these two enzymes (see Chapter [49] of the volume).
[10] T. Hofmann and R. Shaw, *Biochim. Biophys. Acta* **92**, 543 (1964).
[11] J. Sodek and T. Hofmann, *Can. J. Biochem.* **48**, 425 (1970).

New Jersey). The procedure using this equipment is described here. It was previously reported in connection with the production of penicillo-carboxypeptidase-Sl.[12]

The medium used for growing the organism has been described previously.[10] It contains the following ingredients per liter: sucrose, 7.2 g; glucose, 3.6 g; $MgSO_4 \cdot 7H_2O$, 1.23 g; KH_2PO_4, 13.62 g; KNO_3, 2 g; $CaCl_2 \cdot 6H_2O$, 1.1 g; and the following trace elements: $ZnSO_4 \cdot 6H_2O$, 1.5 mg; $Fe(NH_4)_2(SO_4)_2 \cdot 6H_2O$, 2.8 mg; $CuSO_4 \cdot 5H_2O$, 0.32 mg; $MnCl_2 \cdot 4H_2O$, 0.14 mg; and $(NH_4)_2MoO_4$, 0.08 mg. After sterilization, separately sterilized lactic acid (10% v/v) is added to the medium (100 ml/liter) to bring the pH to approximately 3.0.

The *Penicillium janthinellum* cultures are obtainable from the Commonwealth Mycological Institute, Kew, Surrey, U.K. (strain 75589) or from the Northern Utilization Research Branch, U.S.D.A., Peoria, Illinois (strain 905). These strains are identical. Another strain (C.M.I. 75588 = NRRL 904) produces lower enzyme levels in stationary cultures.[13]

Inoculum Preparation

It has been found most convenient to prepare the inoculum in two stages. First, a spore suspension is prepared from colonies on agar slants and used to inoculate four 1-liter flasks containing 400 ml of the complete medium. The flasks are incubated on a mechanical shaker operated at 180 cycles/min at room temperature (about 22°) for 3 days until a thick suspension of mycelium is obtained. This preparation is used to seed a 12-liter Microferm (New Brunswick Scientific Co., New Brunswick, New Jersey). The medium (12 liter) and fermentor jar are sterilized by autoclaving at 115 psi for 30 min. Separately sterilized lactic acid (10% v/v, 1.2 liter) is added to the medium, followed by the mycelium suspensions from the four 1-liter flasks. A few drops of Antifoam B (Dow-Corning) are added. The Microferm is operated according to the manufacturer's instructions.

Rapid growth is obtained at 28° with continuous stirring at 180–220 rpm. Vigorous and uninterrupted aeration (4–6 liters/min) is essential for the production of penicillopepsin and the carboxypeptidases.

While the Microferm culture is growing, the 250-liter Fermacell fermentor is being prepared for use according to the manufacturer's instructions. Because of the high aeration rates required, a supply of sterile air is a necessity. A pH-control system is desirable but not essential. A

[12] S. R. Jones and T. Hofmann, *Can. J. Biochem.* **50**, 1297 (1972).
[13] T. Hofmann, *Pharm. Acta Helv.* **38**, 634 (1963).

foam-control system is not needed. The Fermacell vessel is charged with 180 liters of medium, which can be made up with tap water. It is sterilized *in situ* at 120° for 60 min. After cooling, the temperature is adjusted to 28°. Lactic acid (100%, not sterilized, 1.8 liter) is added to bring the pH to about 3.0. The mycelium suspension in the Microferm vessel, which was allowed to grow for 3–4 days, is then used as inoculum for the Fermacell. The volume is made to 200 liters with sterile water.

Immediately after inoculation, aeration is started and maintained at 6 ft³/min (about 0.85 liter/min per liter of medium). The culture is stirred continuously at 180 rpm. The pH is allowed to rise to 3.4 and subsequently maintained at that level either by an automatic pH control system or by the addition of lactic acid (100%, nonsterile) once or twice a day. The liquid level is checked daily and maintained at 200 liters by the addition of sterile water. Because of the high aeration rate evaporation losses are considerable. The penicillopepsin levels reach their maximum after 9–12 days. (In the standard trypsinogen assay described below, this gives a reading of 0.2–0.3 OD units when 0.5 μl of the medium is assayed and corresponds to approximately 1.5–2 g of penicillopepsin in the whole culture).

At this point the pH is allowed to rise until it reaches 5.4–5.6 (3–4 days). This pH is necessary for the efficient adsorption of the three enzymes from the medium by DEAE-Sephadex A-50. The pH cannot be raised by adding base because this increases the ionic strength and prevents the adsorption on the DEAE-sephadex.

Isolation and Purification of Penicillopepsin

The procedure described below differs very markedly from the procedures described previously.[1,10,11] The main reason for the changes that were introduced is that the modifications allow the simultaneous isolation in good yield of penicillopepsin and the two penicillocarboxypeptidases S-1 and S-2 which are described in chapter [49] of this volume.

Step 1. Batch Adsorption on DEAE-Sephadex A-50. When the pH of the medium has risen to 5.4–5.6, the Fermacell vessel is cooled to the temperature of the cooling water (tap water) available (10°–15°). The mycelium is removed either by a filter press or by a Sharples centrifuge. A convenient type of press is made of stainless steel with 6–10 filter plates (2 ft × 2 ft), which can be connected directly to the outlet of the Fermacell. In our laboratory a Sharples centrifuge (Model AS16) with a 10-liter stainless steel bowl is used. The Fermacell is connected to the centrifuge through pressure hose. The mycelium suspension is passed through the centrifuge at a flow rate of 10 liters/min by applying

2–3 psi to the Fermacell vessel. When 100 liters have been centrifuged, it is necessary to empty the mycelium from the centrifuge bowl. When the medium has been centrifuged, the Fermacell vessel is immediately cleaned out so that the supernatant can be returned to it for recentrifugation. Although it is not possible to obtain a clear supernatant, the second centrifugation removes a considerable amount of spores and mycelial breakdown products. The supernatant (190–200 liters) is again returned to the Fermacell vessel and kept overnight with cooling. For the following step (adsorption on DEAE-Sephadex), it is essential to work rapidly because of the presence of dextranases that digest the DEAE-Sephadex. Adsorption and elution should be completed in 24 hr or less.

The procedure is as follows. First thing in the morning, 100 g of dry DEAE-Sephadex A-50 is added to the medium. The suspension is stirred for 30 min, then centrifuged through the Sharples at a flow rate of not less than 10 liters/min. The DEAE-Sephadex slurry is removed from the centrifuge bowl and immediately cooled in an ice-bath. The supernatant is returned to the Fermacell and stirred with another batch of 100 g of DEAE-Sephadex for 30 min. The DEAE-Sephadex is again removed by centrifugation and combined with the first batch. This adsorption removes 85–90% of penicillopepsin and of the two penicillocarboxypeptidases from the medium.

The DEAE-Sephadex as removed from the centrifuge bowl is diluted with supernatant medium to a consistency where it forms an easily flowing homogeneous slurry (6–10 liters). It is transferred to a large column for elution of the enzyme. Two types of columns have been used. The fastest elution which we have obtained was with a KS370/15 sectional column (37 cm diameter $\times$ 15 cm high) (Pharmacia Fine Chemicals AB, Uppsala, Sweden) but a K100/100 column (10 cm $\times$ 60 cm, Pharmacia) was also found satisfactory. After the DEAE-Sephadex has settled in the column, the enzymes are eluted with 0.1 M Li$_2$SO$_4$ in 0.1 M sodium citrate pH 4.0. Fractions of 500 ml each are collected and assayed for activity. Over 90% of the activity is usually found in 2–4 liters. The volume of the eluate is then reduced to 500 ml in a Diaflo apparatus (Model 2000, UM-10 membrane, Amicon Corp., Lexington, Massachusetts).

Step 2. Chromatography on Bio-Gel-P-100. While the preceding steps are being carried out, a column of Bio-Gel-P100 (10 cm $\times$ 60 cm, K-100/100, Pharmacia, Uppsala) is being prepared. Fines are removed from the resin (600 g) by suspending it in 25 liters of distilled water and allowing it to settle partially; the turbid supernatant is removed several times over a period of 2 days. The settled resin is then mixed with an equal volume of 0.1 M KCl in 0.1 M pyridine–formate, pH 4.2, and poured into the column. The final flow rate should be not less than 100 ml/hr.

The concentrated enzyme (500 ml) is then applied to this column. The column is developed with 0.05 M KCl–0.05 M pyridine–formate buffer, pH 4.2, at a flow rate of 100–150 ml/hr. The column is cooled with tap water (10°–15°). Fractions of 20 ml are collected and analyzed for penicillopepsin activity, penicillocarboxypeptidase activity (for assay procedure see chapter [49]) and for absorbance at 280 nm. A typical elution pattern is shown in Fig. 1. In this latter run, fractions 151–202 (900 ml) contained the penicillopepsin. They were pooled and concentrated to 175 ml in a Diaflo apparatus (Amicon Corp., Model 400, 500-ml capacity, UM-10 membrane at 40 psi) in a cold room. The concentrate was dialyzed against 0.2 M NaCl–0.05 M sodium acetate, pH 5.0, in preparation for the next step.

Step 3. Purification of Penicillopepsin by Chromatography on DEAE-Sephadex A-50. The concentrate from the preceding step is chromatographed on a DEAE-Sephadex A-50 column (5 cm × 85 cm). The resin is prepared by swelling it in water and washing several times by sedi-

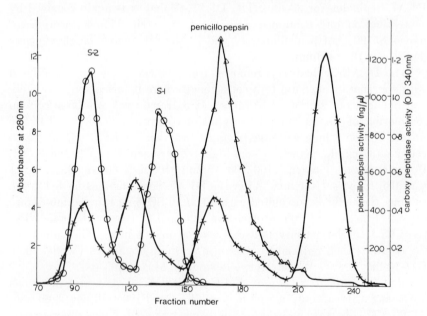

Fig. 1. Separation of penicillopepsin and penicillocarboxypeptidases S-1 and S-2 on Bio-Gel P-100. Column size, 10 cm × 60 cm; fraction volume, 20 ml; solvent, 0.05 M KCl–0.05 M pyridine acetate, pH 4.2; flow rate, 120 ml/hr. ×——×, $OD_{280\ nm}$; △——△, penicillopepsin activity, sample volume 1 μl; ○——○, carboxypeptidase activity; Sample volume, 0.5 μl.

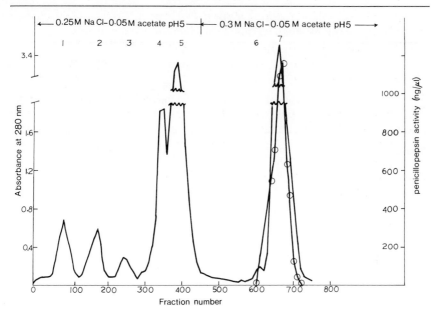

FIG. 2. Chromatography of penicillopepsin on DEAE-Sephadex-A 50. Column size, 5 cm × 85 cm; fraction volumes, 10 ml; eluent as indicated; flow rate, 40 ml/hr. ——, $OD_{280\ nm}$; ○——○, penicillopepsin activity.

mentation in order to remove fines. It is then mixed with NaCl–sodium acetate pH 5.0 buffer to a final concentration of 0.2 M NaCl–0.05 M sodium acetate, pH 5.0, degassed and poured into the column. After equilibrating it with one column volume of buffer, the protein solution is applied. The column is eluted with 0.25 M NaCl–0.05 M sodium acetate, pH 5.0 (4500 ml). In a representative run shown in Fig. 2, four inactive protein peaks were eluted with this solvent. The eluant is then changed to 0.3 M NaCl–0.05 M sodium acetate, pH 5.0. The major peak (peak 7) that elutes is penicillopepsin.

A summary of the purification is given in Table I.

Criteria of Purity. In addition to the specific activity, which can be related to the enzymic purity from the curve in Fig. 3, homogeneity of penicillopepsin is assessed by polyacrylamide gel electrophoresis in borate buffer, pH 9.5,[14] and also in SDS at pH 8.[15] A further criterion of purity is the absence of arginine and methionine from a hydrolyzate of the enzyme. The best preparations obtained have been judged to be better than 95% pure.

[14] K. Weber and M. Osborn, *J. Biol. Chem.* **244**, 4406 (1969).
[15] B. J. Davis, *Ann. N.Y. Acad. Sci.* **121**, 404 (1964).

TABLE I

SUMMARY OF PURIFICATION PROCEDURE OF PENICILLOPEPSIN

Fraction	Volume (liters)	Protein[a] (g total)	Penicillo-pepsin[b] (g total)	Specific activity (mg enzyme/mg protein)	Yield (%)
Medium	200	ND[c]	1.8	—	100
Eluate from DEAE-Sephadex (batch)	0.6	22.5	1.38	0.062	77
Bio-Gel-P 100 peak	0.5	3.42	1.28	0.38	71
Peak 7 from DEAE-Sephadex	0.3	0.868	0.8	0.92	44

[a] Estimated from the absorbance at 280 nm and assuming a specific absorbance of 1.5 (1 mg/ml) except for the last step, where the specific extinction is known to be 1.35 [T. Hofmann and R. Shaw, *Biochim. Biophys. Acta* **92**, 543 (1964)].

[b] As calculated from the calibration curve (Fig. 3).

[c] ND, not determined.

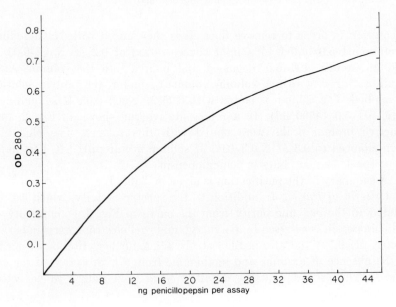

FIG. 3. Standard curve for penicillopepsin assay with trypsinogen. This curve was prepared by assaying known amounts of penicillopepsin of high purity (see text).

Assay Methods

Principle. So far no small molecular weight substrate convenient for routine assays has been discovered. For this reason and for reasons of sensitivity, the trypsinogen-activating assay originally developed by Kunitz[16] is still used routinely in our laboratory. It is based on the ability of penicillopepsin to activate trypsinogen at pH 3.4. The trypsin formed is determined by one of the standard methods at pH 8.0; for routine work, the most convenient one is that described by Kunitz[16] with casein as substrate. The complete assay given below is based on that of Kunitz[17] with the modifications given by Hofmann and Shaw.[10] The enzyme can also be assayed with bovine serum albumin at pH 2.6.

Trypsinogen Activation Assay

Reagents

ACTIVATION STEP
Sodium citrate buffer, 0.1 M, pH 3.4
KH_2PO_4, 0.1 M
HCl, 0.0025 M
Trypsinogen, $10^{-5}M$. Commercial trypsinogen, 5.0 mg (containing 50% $MgSO_4$), is dissolved in 10 ml of 0.0025 M HCl. This is stable under refrigeration for 1 day.
Enzyme dilutions, made in 0.1 M KH_2PO_4

TRYPSIN ASSAY
Sodium phosphate buffer, 0.1 M, pH 8.0
Tris (hydroxymethyl) aminomethane (Tris), 0.37 M
Trichloroacetic acid (TCA), 8.5% (w/v)
Casein, 2%, in phosphate buffer, pH 8.0. Suspend 2 g of casein (Hammarsten) in 100 ml of the phosphate buffer (pH 8.0). Heat for 15 min in a boiling water bath with occasional stirring; cool, and adjust the pH to 8.0. This solution can be kept for many months at −20°.
Trypsin, 0.05% in 0.0025 M HCl for preparing a standard curve
The concentration of active trypsin is determined by active site titration with carbobenzoxy-L-lysine p-nitrophenyl ester at pH 3.0.[18]

[16] M. Kunitz, *J. Gen. Physiol.* **21**, 601 (1938).
[17] M. Kunitz, *J. Gen. Physiol.* **30**, 291 (1967).
[18] M. L. Bender, M. L. Begue-Canton, R. L. Blakeley, L. J. Brubacher, J. Feder, C. R. Gunter, F. J. Kézdy, J. V. Killheffer, T. H. Marshall, C. G. Miller, R. W. Roeske, and J. K. Stoops, *J. Am. Chem. Soc.* **88**, 5890 (1966).

The following solutions are prepared daily: solution A: 1 volume of trypsinogen plus 2 volumes of citrate buffer, pH 3.4; solution B: 1 volume of casein plus 1 volume of Tris $(0.37\ M)$. A mixture of 3 volumes of A, 4 volumes of B, and 1 volume of KH_2PO_4 $(0.1\ M)$ should give pH 7.9–8.1. If this pH is not obtained, the concentration of the Tris can be adjusted so that the mixture will give this pH, which is essential for the trypsin assay.

Procedure. Preparation of Trypsin Standard Curve. Tris buffer $(0.37\ M$, 20 ml) is mixed with 2% casein (20 ml), citrate buffer $(0.1\ M$, pH 3.4, 20 ml), and $0.1\ M$ KH_2PO_4 (10 ml). Aliquots of 3.5 ml of this solution are pipetted into ten test tubes and warmed to 36°. At 1-min intervals, 0.5 ml of trypsin standard solutions (containing 6–140 μg of trypsin per milliliter) are added. After exactly 10 min, undigested casein is precipitated with TCA (2 ml). The tubes are left at 36° for 10 min and filtered through Whatman No. 42 (7 cm) filter paper. The extinction values at 280 nm are read against a blank prepared without trypsin. They are plotted as a function of the total amount of trypsin (in nanomoles) in each tube.

Assay. ACTIVATION STEP. Solution A (1.5 ml) is warmed to 36°. Five-tenths milliliter of enzyme solution in $0.1\ M$ KH_2PO_4 or the growth medium containing the equivalent of 5–20 ng of the pure enzyme is added to start the reaction.

TRYPSIN ASSAY. After exactly 10 min, solution B (2 ml) is added, and after a further 10 min the undigested casein is precipitated by 8.5% TCA (2 ml). After 10 min of standing at 36°, the precipitate is filtered through Whatman No. 42 (7 cm) filter paper. The extinction values at 280 nm are read against a blank containing the reagents. The number of nanomoles of trypsinogen activated is obtained from the standard curve.

Definition of Unit and Specific Activity. One unit of penicillopepsin activity is defined as the amount of enzyme required to activate 1 $\mu mole$ of trypsinogen per minute at 25°, pH 3.4, and saturating substrate concentration. Routine assays are carried out, however, at 36° with suboptimal substrate concentrations so that a conversion is required to obtain the defined units. The conversion factor is calculated from the known maximum velocities at 25° and 36° (420 and 685 $\mu moles$ of trypsinogen activated per micromole of penicillopepsin per minute) to obtain the temperature correction. The correction for nonsaturating conditions is calculated from the Michaelis–Menten equation using a value for $K_m =$ 7.6 μM. The combined conversion factor $= 0.405$. Thus the defined units $=$ (micromoles of trypsinogen activated at 36°) $\times 0.405$. The specific activity is taken as the number of units of activity per milligram of enzyme.

Figure 3 shows the relationship between the absorbance readings at 280 nm and the amount of penicillopepsin (in nanograms) in the total assay mixture. The curve, which varies slightly with different batches of Hammersten casein, was prepared by using penicillopepsin which was judged homogeneous by electrophoresis in SDS-polyacrylamide at pH 8.0, in polyacrylamide at pH 8.6, in starch gel at pH 3.0, pH 4.5, pH 6.0, and pH 8.6. It was also free of arginine and methionine ($<$0.02 residues per molecule). Several preparations gave the same curve as shown in Fig. 3 within 10%.

The standard deviations of replicate samples is about $\pm 5\%$.

Assay with Bovine Serum Albumin

  Reagents

    Bovine serum albumin (fraction V, B-grade, Calbiochem), 1% in
      McIlvaine's citric acid–phosphate buffer[19] at pH 2.6
    Trichloroacetic acid (TCA), 8.5%

Procedure. The solution of bovine serum albumin (1 ml) is warmed to 36°. The enzyme solution (50 μl) containing the equivalent of 1–25 μg of the pure enzyme is added to start the reaction. After exactly 10 min, undigested albumin is precipitated with TCA (2 ml). Ten minutes later the precipitate is filtered (Whatman No. 42, 7 cm). The extinction values at 280 nm are read against a blank prepared without enzyme. The enzyme concentration is read off a standard curve. It may be mentioned that the reading for 5 μg of the purest enzyme preparation is 0.200 OD unit.

Properties of Penicillopepsin

Stability

The stability has been determined under various conditions of pH and temperature.[11] Penicillopepsin is completely stable at pH 4.9 and room temperature for several days (S. Siu and T. Hofmann, unpublished observations). It loses activity rapidly at pH $\leq$ 2.0 and $\geq$ 7.0.[11] At pH

[19] R. G. Bates, *in* "Handbook of Biochemistry" (H. Sober, ed.), p. J-197. Chem. Rubber Publ. Co., Cleveland, Ohio, 1968.

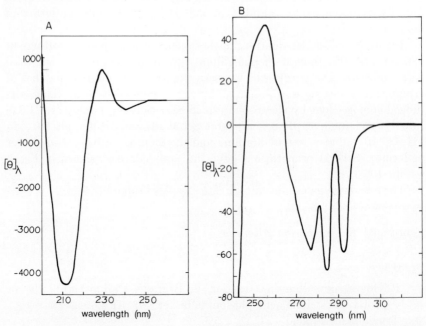

Fig. 4. Circular dichroism spectrum of penicillopepsin. Solvent: 0.02 M sodium acetate pH 4.5. (A) 200–260 nm. (B) 240–320 nm. The mean residue ellipticity is given in degrees cm² decimole⁻¹.

4.9 the enzyme is stable if kept at 55° for 1 hr, but at pH 3.0 and pH 6.1 it slowly inactivates at 37°.

Molecular Properties. The enzyme has the following physical constants:[10] molecular weight = 32,000; sedimentation constant $s_{20°,w}$ = 3.1×10^{-13} sec; diffusion constant $D_{20,w}$ = 8.4 ($\pm$0.23) $\times 10^{-7}$ cm² sec⁻¹; partial specific volume $\bar{V}$ = 0.718 (calculated from amino acid composition); electrophoretic mobility (in 0.22 M acetate, pH 4.25) is μ = -4.47 ± 0.02 cm² V⁻¹ sec⁻¹; the isoelectric point is probably below pH 3.0 as judged from behavior on isoelectric focusing; molecular extinction at 280 nm is ϵ_M = 43,200; specific extinction at 280 nm is $E^{1\,cm}_{1\,mg/ml}$ = 1.35. The circular dichroism spectrum (Fig. 4) in the far-ultraviolet shows no clear evidence for the presence of α-helix; in the near-ultraviolet region at least 5 well defined bands can be seen (Wang *et al.*[20]; S. Siu, K. J. Dorrington, and T. Hofmann, unpublished observations). These latter have been used to follow a conformational change due to secondary binding of an activator peptide.[20]

[20] T.-T. Wang, K. J. Dorrington, and T. Hofmann, *Biochem. Biophys. Res. Commun.* **57**, 865 (1974).

Amino Acid Composition.[10] Lys_5, His_3, Asp_{36}, Thr_{28}, Ser_{42}, Glu_{28}, Pro_{12}, Gly_{39}, Ala_{23}, Half-cystine$_2$, Val_{22}, Ile_{12-13}, Leu_{20}, Tyr_{14}, Phe_{19}, Trp_5, amide groups$_{43}$. It should be noted that, contrary to the earlier report,[10] penicillopepsin has one disulfide bridge. However, the absence of arginine and methionine has been confirmed.

Amino Acid Sequence. Practically all amino acids have been accounted for in 15 fragments, which are given in Tables II and III. Table II shows the alignment of the N-terminal 112 amino acids of penicillopepsin with the last four amino acids of the activation peptide of porcine pepsinogen,[21] and the first 111 amino acids of porcine pepsin,[22] the alignment of 46 residues with residues 153–197 of porcine pepsin, the diazoacetylnorleucine-reactive sites and 27 residues at the C-terminal. There is sufficient identity between the two proteins to establish unambiguously their homology. Table III shows an additional 11 fragments that cannot be aligned with certainty with corresponding parts of the porcine pepsin molecule until the complete sequence has been determined. The disulfide bridge could match with the disulfide that links Cys_{250} and Cys_{282} in porcine pepsin.[22] Manuscripts on the sequence are in preparation (A. Kurosky, L. Rao, C. I. Harris, S. R. Jones, A. Cunningham, H. M. Wang and T. Hofmann, unpublished). Some fragments have been reported.[23,24]

Table IV shows the alignment of the N-terminal sequences—as determined in a sequenator—of penicillopepsin, a protease from *Penicillium roqueforti*, *Rhizopus chinensis* pepsin, pig pepsin, and chymosin. The first 27 residues of *Rhizopus chinensis* pepsin agree with a sequence reported by Sepulveda *et al.*,[25] including the presence of approximately equal amounts of isoleucine and valine in position 15.

Tertiary Structure. The X-ray analysis of penicillopepsin is in progress. A description of the crystals has been published.[26] Crystals are obtained from ammonium sulfate (0.3 saturated) or lithium sulfate at pH between 3.5 and 6. The crystals are monoclinic, space group C2. The unit cell dimensions are a = 97.81Å, b = 46.73Å, c = 65.69Å and β = 115.55°. There are 4 molecules per unit cell. At 6 Å resolution, a three-

[21] V. B. Pedersen and B. Foltmann, *FEBS Lett.* **35**, 255 (1973).

[22] J. Tang, P. Sepulveda, J. Marciniszyn, Jr., K. Chen, W.-Y. Huang, N. Tao, D. Lin, and J. P. Lanier, *Proc. Natl. Acad. Sci. U.S.A.* **70**, 3437 (1973).

[23] C. I. Harris, A. Kurosky, L. Rao, and T. Hofmann, *Biochem. J.* **127**, 34P–35P (1972).

[24] A. Kurosky, L. Rao, and C. I. Harris, *Fed. Proc., Fed. Am. Soc. Exp. Biol.* **32**, 2031 (1973).

[25] P. Sepulveda, K. W. Jackson, and J. Tang, *Biochem. Biophys. Res. Commun.* **63**, 1106 (1975).

[26] N. Camerman, T. Hofmann, S. Jones, and S. C. Nyburg, *J. Mol. Biol.* **44**, 569 (1969).

TABLE II

SEQUENCE OF FOUR FRAGMENTS OF PENICILLOPEPSIN (PEN) AND THEIR COMPARISON WITH
CORRESPONDING SEQUENCES OF PORCINE PEPSIN (PIG)^{a,b}

N-Terminal fragment (to residue 111)

Residue No.	-4	1	5	10	15	20

Penicillopepsin: NH₂ Ala-Ala | Ser-Gly-Val-Ala-Thr-Asn-Thr-Pro-Thr-Ala- - - Asn-Asp-Glu | Glu-Tyr-Ile-Thr-Pro-Val | Thr-Ile-Gly | - | Gly-Thr | - - Thr-Leu

Porcine pepsin: ... Ala-Ala-Leu/Ile - Gly-Asp-Glu-Pro-Leu-Glu-Asn-Tyr-Leu | Asp-Thr-Glu-Tyr | Phe- - - | Gly | Thr-Ile-Gly | Ile-Gly-Thr-Pro-Ala-Gln-

	30	35	40	45	50	55

Pen: - - Asn-Leu-Asn | Phe-Asp-Thr-Gly | Ser | Ala-Asp | Leu-Trp-Val | Phe | Ser | Thr-Glu-Leu-Pro-Ala-Ser-Gln-Gln | Ser | Gly | His | Ser-Val-Tyr | Asn-Pro

Pig: Asp-Phe-Thr-Val-Ile | Phe-Asp-Thr-Gly-Ser | Ser | Asn | Leu-Trp-Val | Pro | Ser | Val-Tyr-Cys-Ser-Ser-Leu-Ala-Cys | Ser | Asp | His | Asn-Gln-Phe | Asn-Pro

	60	65	70	75	80	85	90

Pen: Ser-Ala-Thr-Gly-Lys- - - | Glu | Leu-Ser-Gly-Tyr-Thr-Trp | Ser-Ile | Ser-Tyr-Gly | Asp | Gly-Ser | Ser-Ala-Ser | Gly | Asn-Val-Phe | Thr | Asp | Ser-Val | Thr-

Pig: Asp-Asp-Ser-Ser-Thr-Phe | Glu | Ala-Thr-Ser-Gln-Glu-Leu-Ser-Ile | Thr-Tyr-Gly-Thr-Gly-Ser | Met- - - | Thr | Gly | Ile-Leu-Gly-Tyr | Asp | Thr | Val | Gln-

	95	100	105	110

Pen: Val-Gly-Gly | Val-Thr-Ala-His-Gly | Gln | Ala-Val-Gln-Ala-Gln-Ile-Ser-Ala-Gln | Phe

Pig: Val-Gly-Gly | Ile-Ser-Asp-Thr-Asn-Gln | Ile-Phe-Gly-Leu-Ser-Glu-Thr-Glu-Pro-Gly-Ser | Phe ...

Central fragment 153-197

```
       153              160                 165             170                175                180            185
Pen  /Phe-Ala-Val-Ala-Leu-Lys-His-Gln-Gln-Pro-Gly|Val|Tyr-Asp-Phe|Gly|Phe|Ile-Asp-Ser-Ser|Lys|Tyr-Thr-Gly-Ser-Leu|Tyr-Thr-Gly|Val|Asn-
Pig  ...Val-Tyr-Leu-Ser-Ser-Asn-Asp-Asp-Ser-Gly-Ser|Val|Val-Leu-Leu|Gly|Gly|Ile-Asp-Ser-Ser|Tyr|Tyr-Thr-Gly-Ser-Leu|Asn-Trp-Val-Pro|Val|- -

                  190                   197
Pen  Asn-Ser-Gln|Gly|Phe|Trp|Ser-Phe-Asn-Val|Asp-Ser-Tyr|/
Pig  Ser-Val-Glu|Gly|Tyr|Trp|Gln-Ile-Thr-Leu|Asp-Ser|Ile
```

Active site fragment 211-220

```
       211                      220
Pen  /Ser-Gly|Ile|Ala|Asp-Thr-Gly-Thr|Thr|Leu|/
Pig  ...Gln-Ala|Ile|Val|Asp-Thr-Gly-Thr|Ser|Leu|...
```

C-terminal fragment 301-327

```
       301              310                315                320              327
Pen  /Ser-Ile-Gly-Asp|Ile|Phe|Leu-Lys-Ser-Gln|Tyr-Val|Val|Phe-Asp|Ser-Asp-Gly-Pro-Gln-Leu|Gly|Phe|Ala-Pro-Gln|Ala|OH
Pig  ...Ile-Leu-Gly-Asp|Val|Phe|Ile-Arg-Gln-Tyr|Tyr-Thr|Val|Phe-Asp|Arg-Ala-Asn-Asn-Lys-Val|Gly|Leu-Ala-Pro-Val|Ala|OH
```

[a] From J. Tang, P. Sepulveda, J. Marciniszyn, Jr., K. Chen, W.-Y. Huang, N. Tao, D. Lin, and J. P. Lanier, *Proc. Natl. Acad. Sci. U.S.A.* **70**, 3437 (1973).

[b] Identical residues are enclosed in boxes.

[c] Numbering according to Tang et al.[a]

TABLE III

SEQUENCE OF ADDITIONAL FRAGMENTS OF PENICILLOPEPSIN
THAT HAVE NOT BEEN ALIGNED WITH PEPSIN

/Ser - Ser - Ile - Asn - Thr - Val - Gln - Pro - Gln - Ser - Gln - Thr - Thr - Phe/

/Phe - Asp - Thr - Val - Lys - Ser - Leu - Ala - Gln - Pro - Leu/

/Phe - Gln - Asp - Thr - Asn - Asn - Asp - Gly - Leu - Leu/

/Leu - Leu - Asx - Asp - Ser - Val - Val - Ser - Gln - Tyr/

/Thr - Ala - Gly - Ser - Gln - Ser - Gly - Asp - Gly - Phe/

/Ile - Ser - Gly - Tyr/

/Gly - Leu - Ala - Phe/

/Gly - Asp - Ser /

/ Ile - Phe - Gly - Asx - Asx - Gly /

Cystine bridge

/Tyr - Ser - Gln - Val - Ser - Gly - Ala - Gln - Gln - Asp - Ser - Asn - Ala - Gly - Gly - Tyr - Val - Phe - Asx - Asx - Ser - Thr - Asx - Leu - Pro - Cy S- Ser /

/Thr - Ala - Thr - Val - Pro - Gly - Ser - Leu - Ile - Asn - Tyr - Gly - Pro - Ser - Gly - Asn - Gly - Ser - Thr - Cy S- Leu - Gly - Gly - Ile - Gln - Ser - Asn - Ser - Gly - Ile - Gly - Phe/

TABLE IV
N-Terminal Sequences of Acid Proteases[a,b]

Penicillopepsin[c]	Ala-Ala-Ser-Gly-Val-Ala-Thr-Asn-Thr-Pro-Ala -	Asn-Asp-Glu-Glu-Tyr-Ile-Thr-Pro-Val-Thr-Ile-Gly - - -
Penicillium roquefortii[d]	Val- ? -Gly-Ser-Ala-Ile-Thr-Thr-Pro-Glu-Ala -	-Asp-Val-Glu-Tyr-Leu-Thr-Pro-Val-Thr-Ile-Gly -
Rhizopus pepsin[e]	Ala-Gly-Val-Gly-Thr-Val-Pro-Met-Thr-Asp-Tyr-Gly -	-Asn-Asp-Ile/Val-Glu-Tyr-Tyr-Gly-Gln-Val-Thr-Ile-Gly -
Pig pepsin[f]	Ala-Ala-Ala-Leu/Ile-Gly-Asp-Glu-Pro-Leu-Glu-Asn-Tyr-Leu-Asp-Thr-Glu-Tyr-Phe -	Gly-Thr-Ile-Gly-Ile -
Chymosin[g]	Gly-Phe/Gly-Glu-Val-Ala-Ser-Val-Pro-Leu-Thr-Asn-Tyr-Leu-Asp-Ser-Gln-Tyr-Phe -	Gly-Lys-Ile-Tyr-Leu-

Penicillopepsin[c]	Gly-Thr - Thr-Leu -	Asn-Leu-Asn-Phe-Asp-Thr-Gly-Ser-Ala-Asp-Leu-Trp-Val-Phe-Ser
Penicillium roquefortii[d]	(Ser-Ser) - Thr-Leu- -	Asn-Leu-Asn-Phe-Asp x Gly (Ser)/
Rhizopus pepsin[e]	- (Thr) Pro-Gly-Lys (Ser)	Phe-Asn-Leu-Asx-Phe-Asp-Thr-Gly (Ser) Thr/
Pig pepsin[f]	Gly-Thr-Pro-Ala-Gln-Asp-Phe-Thr-Val-Ile-	Phe-Asp-Thr-Gly-Ser-Ser-Asn-Leu-Trp-Val-Pro-Ser
Chymosin[g]	Gly-Thr-Pro-Gln-Glu-Phe-Thr-Val-Leu-Phe-Val-	Phe-Asp-Thr-Gly-Ser-Ser-Asp-Phe-Trp-Val-Pro-Ser

[a] The alignment given here is similar to that proposed by P. Sepulveda, K. W. Jackson, and J. Tang, *Biochem. Biophys. Res. Commun.* **63**, 1106 (1975).

[b] Symbols and abbreviations: ?, not identified; x, not identified; (Ser), not identified, but evidence suggests carbohydrate-containing residue; x, not identified by gas chromatography; (Thr), not identified by gas chromatography, only as α-aminobutyric acid by amino acid analysis; —, presumed deletion; Leu/Ile and Phe/Gly, activation site of porcine pepsin and chymosin, respectively.

[c] A. Kurosky, L. Rao, C. I. Harris, S. R. Jones, A. Cunningham, and T. Hofmann, to be published.

[d] Determined by Sequenator: J. C. Gripon, S. H. Rhee, and T. Hofmann, to be published.

[e] Residues 1-27 as determined by P. Sepulveda et al. (*Biochem. Biophys. Res. Commun.* **63**, 1106 (1975)), also by Sequenator up to 39 (S. H. Rhee and T. Hofmann, to be published) and manually up to residue 9.

[f] The first four residues are part of the activation peptide [V. B. Pedersen and B. Foltmann, *FEBS Lett.* **35**, 255 (1973), the remaining sequence is from J. Tang, P. Sepulveda, J. Marciniszyn, Jr., K. Chen, W.-Y. Hrang, N. Tao, D. Lin, and J. P. Lanier, *Proc. Natl. Acad. Sci. U.S.A.* **70**, 3437 (1973).

[g] B. Foltmann, this series Vol. 19, p. 421; V. B. Pedersen and B. Foltmann, *FEBS Lett.* **35**, 250 (1973); B. Foltmann, D. Kauffman, M. Paul and P. M. Andersen, *Neth. Milk Dairy J.* **27**, 288 (1973).

dimensional structure has been calculated from native crystals and two derivatives: $PtCl_6^{2-}$ and UO_2^{2+}.[27]

Enzymic Properties

Inhibitors. Penicillopepsin is inhibited competitively by Cbz-Glu-Tyr ($K_{I,app}$ = 50 μM), Cbz-Gln-Tyr, Cbz-Val-Tyr, and also by Cbz-Glu, but not by Cbz-Tyr and Cbz-Phe. Pepstatin is also a potent inhibitor.

Leucyl-glycyl-leucine inhibits trypsinogen activation, but activates by a factor of ten the cleavage of leucyl-tyrosyl-amide.[20]

The acid protease inhibitor diazoacetyl norleucine methyl ester and similar diazo compounds in the presence of Cu^{2+} inhibit the enzyme by reacting covalently with an aspartic acid residue in the sequence Ile-Ala-Asp-Thr-Gly-Thr-Thr-Leu,[28] a sequence which is nearly identical with that of an active site peptide from pig pepsin[29] as shown in Table II. Another inhibitor of pepsin, chymosin, and other acid proteases, 1,2, epoxy-3-p-nitrophenoxypropane, also inhibits penicillopepsin by reacting with an aspartic acid residue.[30] This has not yet been identified in penicillopepsin. In porcine pepsin it is found in the sequence -Ile-Phe-Asp-Thr-Gly-Ser-Ser-Asn-Leu-Trp-Val-[22] (residues 30–40). This sequence differs only in three residues from the corresponding sequence of penicillopepsin (Table II). Another pepsin inhibitor p-bromophenacylbromide does not inhibit. The partial inhibition by K_2PtCl_6 has helped in locating tentatively the active site region in the 6 Å structure.

Specificity. Penicillopepsin shows a specificity similar to that of pig pepsin and other acid proteinases toward the B-chain of S-sulfo-insulin,[31] but it hydrolyzes a somewhat larger number of bonds. Although the specificity cannot be defined by the amino acids involved in the scissile bond, the action on the B-chain of S-sulfo-insulin and on glucagon shows that the amino acid residue that contributes the amino group requires a large hydrophobic side chain.[31] Penicillopepsin cleaves about 15% of the bonds of native bovine serum albumin.[10] It also cleaves

$$\text{Cbz-Gly-Gly-Phe-Phe—O—CH}_2\text{—CH}_2\text{—CH}_2\text{—} \langle \text{ring} \rangle \text{N}$$

one of the best substrates of porcine pepsin.[32] The cleavage is restricted to

[27] I.-N. Hsu, A. DeJong, M. N. G. James, T. Hofmann, and S. C. Nyburg, *Abstr. Int. Congr. Cryst., 10th,* Amsterdam, 1975.

[28] J. Sodek and T. Hofmann, *Can. J. Biochem.* **48**, 1014 (1970).

[29] R. S. Bayliss, J. R. Knowles, and G. B. Wybrandt, *Biochem. J.* **113**, 377 (1969).

[30] G. Mains and T. Hofmann, *Can. J. Biochem.* **52**, 1018 (1974).

[31] G. Mains, M. Takahashi, J. Sodek, and T. Hofmann, *Can. J. Biochem.* **49**, 1134 (1971).

[32] G. P. Sachdev and J. S. Fruton, *Biochemistry* **8**, 4231 (1969).

TABLE V
MOLECULAR ACTIVITIES OF PENICILLOPEPSIN[a]

Temperature (°C)	Trypsinogen activated (μmoles/min/μmole enzyme)
0	64
10	105
15	180
20	260
25	420
30	540
35	660

[a] Conditions: 0.01 M sodium citrate, pH 3.4; trypsinogen = 10^{-4} M.

the Phe-Phe bond. Notably, however, it does not act on Cbz-His-Phe-Trp-OEt, another good pepsin substrate. For good activity, long substrate chains are required, as shown by the initial rate of bovine serum albumin cleavage (k_{cat} = 300 min^{-1})[31] and trypsinogen activation (Table V). Trypsinogens from various species are still the most specific substrates. Bovine trypsinogen is activated rapidly with the specific cleavage of the bond Lys$_6$-Ile$_7$. Chymotrypsinogen A and proelastase are not activated. K_m for trypsinogen activation is 7.6 ($\pm$2) μM at 0°, pH 3.4.[10] Molecular activities (k_{cat}) are shown in Table V. It should be noted that these measure specifically the cleavage of the single activating peptide bond.

Penicillopepsin also acts on some small substrates which have two hydrophobic amino acids at the N-terminal and a free amino group.[33,34] In addition to removing the N-terminal amino acid hydrolytically a transpeptidation reaction occurs. Thus Leu-Leu is formed from Leu-Tyr-Leu, Phe-Phe from Phe-Tyr-Thr-Pro-Lys-Ala and Met-Met from Met-Leu-Gly.[34] A covalent acyl-enzyme is involved as an intermediate.[33] Similar reactions are catalyzed by porcine pepsin.[33-35] Their interpretation has led to new concepts concerning the mechanism of action of pepsins.[34] As mentioned above, Leu-Gly-Leu, which is not a substrate, acts as efficient activator of the cleavage of Leu-Tyr-amide.[33] This activation confirms the importance of peptide chain binding in secondary sites, first described for porcine pepsin by Fruton.[36] In contrast to porcine pepsin,

[33] M. Takahashi, T.-T. Wang, and T. Hofmann, *Biochem. Biophys. Res. Commun.* **57**, 39 (1974).
[34] M. Takahashi and T. Hofmann, *Biochem. J.* **147**, 549 (1975).
[35] T.-T. Wang and T. Hofmann, *Biochem. J.* **153**, 691 (1976).
[36] J. S. Fruton, *Adv. Enzymol.* **33**, 401 (1970).

however, no evidence for a covalent amino intermediate has been found so far.

pH Optimum. Optimal trypsinogen activation is observed at pH 3.4[10]; the optimum of hydrolysis of bovine serum albumin is at pH 2.6. Hydrolysis and transpeptidation of small substrates is optimal at pH 3.6.[34]

Acknowledgments

The author is grateful to the Medical Research Council of Canada for the generous support of the work on penicillopepsin (Grants MT-1982 and MA-2438).

[36] Canine Pepsinogen and Pepsin[1]

By BEATRICE KASSELL, CHRISTINE L. WRIGHT, and PETER H. WARD

The gastric mucosa of the dog contains several pepsinogens,[2-5] one of which has been purified and characterized.[6]

Assay and Unit

A modification of the hemoglobin digestion method of Anson[7] has been described in a previous volume of this treatise.[8]

The specific activity is expressed as microgram equivalents of porcine pepsin per A_{280} unit of canine pepsinogen or pepsin.

Purification Procedure for the Pepsinogen

The procedure described here, a modification of our previous method for purifying the main pepsinogen, includes chromatography on polyly-

[1] Supported by Public Health Service Research Grant AM-09826 from the National Institute of Arthritis and Metabolic Diseases.
[2] W. B. Hanley, S. H. Boyer, and M. A. Naughton, *Nature (London)* **209**, 996 (1966).
[3] V. N. Orekhovich, L. M. Ginodman, S. M. Levitova, T. P. Levchuk, and T. A. Solov'eva, *Ionoobmen. Tekhnol. (Akad. Nauk SSSR Inst. Fiz. Khim.)* **1965**, 275 (1965).
[4] A. Z. Vafin, *Prikl. Biokhim. Mikrobiol.* **4**, 229 (1968).
[5] L. Korbová, J. Kohout, L. Krejčík, and E. Pokorny, *Abstr. Fed. Eur. Biochem. Soc. 5th Meet.*, Prague, 1968, p. 283 (1968).
[6] J. P. Marciniszyn, Jr., and B. Kassell, *J. Biol. Chem.* **246**, 6560 (1971).
[7] M. L. Anson, *J. Gen. Physiol.* **22**, 79 (1938).
[8] B. Kassell and P. A. Meitner, this series Vol. 19 [21].

sine-Sepharose 4B.[9] The use of polylysine is based on the observation of Katchalski *et al.*[10] that it is a pepsin inhibitor. The process appears to be a combination of affinity and ion-exchange chromatography.

General Precautions.[8] Pepsinogen solutions are very susceptible to accidental activation by acid, to denaturation, and to bacterial contamination. Do not freeze solutions containing NaCl. Always stir slowly without whipping. Store buffers and protein solutions under toluene (1 ml/4 liters) in the cold. Avoid long storage of protein solutions, particularly in test tubes, during chromatography steps.

Preparation of Polylysine–Sepharose 4B.[9,11] For a column volume of 200–250 ml, prepare two batches, using for each 75 g of Sepharose 4B (Pharmacia Fine Chemicals, Piscataway, New Jersey) and 375 mg of polylysine (MW 50,000–100,000, Pierce Chemical Co., Rockford, Illinois). Wash the Sepharose with water on a coarse fritted-glass funnel under suction. Suspend 75 g of washed gel in 65 ml of water at 25°. Working under a hood, add 7.5 g of CNBr dissolved in 100 ml of water to the stirred solution. Immediately adjust the pH to 11 with 5 M NaOH. Maintain the suspension at 25° and at pH 11 ± 0.3 for 8–10 min until the rapid consumption of NaOH ceases. Transfer the activated gel at once to a fritted-glass funnel and wash under suction with 2 liters of ice-cold water and 2 liters of cold 0.1 M NaHCO$_3$. Transfer the washed gel to a 250-ml screw-cap bottle, using 75 ml of the bicarbonate solution. Add the 375 mg of polylysine dissolved in an additional 75 ml. Shake gently by inversion at 4° for 24 hr. Pour the polylysine–Sepharose into a 2-cm column. Wash with 1 liter of 0.1 M NaHCO$_3$, 1 liter of 0.5 M NaCl adjusted to pH 9.2, and finally equilibrate the column with 50 mM sodium phosphate buffer, pH 6.5.

After use, wash the column with 1M NaCl and reequilibrate with starting buffer. The column gradually darkens with use, but the color does not intefere with the separation.

Step 1. Collection of the Mucosa.[6] Remove the stomachs immediately after the death of the dogs. Turn inside out and wash thoroughly in cold water to prevent activation of the zymogen by the acidic gastric juice. Cut out the fundic portion and store frozen (up to one year). Partially thaw to dissect the mucosa from the layer of muscle, trim, and grind. Carry out all subsequent operations at 4°.

Step 2. Extraction.[6] Stir the ground mucosa from 6–8 stomachs (about 350 g) with 600 ml of 0.1 M sodium phosphate buffer, pH 7.0, for 30 min.

[9] B. Nevaldine and B. Kassell, *Biochim. Biophys. Acta* **250**, 207 (1971).
[10] E. Katchalski, A. Berger, and H. Neumann, *Nature (London)* **133**, 998 (1954).
[11] P. Cuatrecasas, M. Wilchek, and C. B. Anfinsen, *Proc. Natl. Acad. Sci. U.S.A.* **61**, 636 (1968).

Add 85 g of Celite 545 and centrifuge for 90 min at 13,000 g in a refrigerated centrifuge. Resuspend the sediment in 600 ml of buffer and
centrifuge again. Filter the combined supernatant solutions through cloth
on a Büchner funnel. The crude extract has a specific potential activity
of about 30 μg per A_{280} unit and varies somewhat in pepsinogen content:
300 mg per stomach is an average amount.

Step 3. Ammonium Sulfate Fractionation.[6] Add solid ammonium sulfate gradually to the filtrate to 20% saturation, maintaining the pH at
7.0 by adding 5 M NaOH as required. Stir to avoid local excess of acidic
ammonium sulfate or of NaOH. Add 50 g of Celite; allow to stand overnight, then centrifuge at 13,000 g for 1 hr. Resuspend the sediment in
600 ml of buffer solution 20% saturated with ammonium sulfate, and
centrifuge again. Filter the combined supernatant solutions through a
Celite bed on Whatman No. 4 paper spread over cloth.

To the filtrate, add solid ammonium sulfate to 52% saturation, keeping the pH close to 7.0. Stir 30 min and allow to settle overnight. Siphon
off as much liquid as possible and collect the precipitate on a 20-cm
Büchner funnel on a Celite bed on Whatman No 4 paper over cloth.

Suspend the solid cake in 440 ml of 50 mM sodium phosphate buffer,
pH 6.5, and dialyze against 2×6 liters of the same buffer. Centrifuge
at 13,000 g for 30 min. Resuspend the Celite in 250 ml of buffer and
centrifuge again. Clarify the slightly cloudy combined supernatant solutions by centrifuging at 27,000 g for 60 min. The yield at this step is
about 70%, and the specific activity is 120 μg/A_{280} unit. All slimy material that would clog a column is removed.

Step 4. First Chromatography on Polylysine-Sepharose 4B.[9,12] Pass
the solution through the column at a flow rate of 60–70 ml/hr and elute
with the equilibrating buffer until the first peak appears (Fig. 1). Apply
a linear gradient of 900 ml of 50 mM sodium phosphate buffer, pH 6.5,
and 900 ml of 1 M NaCl in the same buffer. Peak 1 has no activity,
peaks 2 and 3 have several pepsinogens of low potential activity against
hemoglobin and peak 4 contains the main pepsinogen. Pool the fractions
of peak 4 having activity over 200 μg/A_{280} unit, concentrate on an Amicon
ultrafilter, using a UM-10 membrane, and dialyze against 2×6 liters
of starting buffer. The yield of the main pepsinogen is usually about
60% of the total activity applied to the column, but as shown previously,[6] the proportion of the different pepsinogens varies.

Step 5. Second Chromatography on Polylysine-Sepharose 4B. Repeat
the chromatography as before with the dialyzed solution of step 4, using
the same size column. A small peak of impurity is eluted first, and

[12] P. H. Ward, C. L. Wright, and B. Kassell, *Fed. Proc., Fed. Am. Soc. Exp. Biol.* 34,
534 (1975).

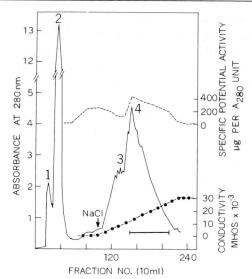

FIG. 1. Chromatography of the crude pepsinogen from step 3 on polylysine-Sepharose 4B. Column 2×77 cm; buffer, 50 mM sodium phosphate, pH 6.5; gradient, at ↓, 900 ml of buffer and 900 ml of 1 M NaCl in buffer in open interconnected bottles to give a linear gradient; fractions, 10 ml; flow rate, 60–70 ml/hr.

there is further separation of the minor pepsinogens. Pool the fractions containing potential activity over 250 $\mu g/A_{280}$ unit, about 70% of the total activity. Concentrate and dialyze as in step 4.

At this point, two batches may be combined, the first solution being kept frozen while the second is being prepared.

Step 6. Gel Filtration on Sephadex G-100.[6] Equilibrate two Sephadex G-100 columns connected in series (3.2×104 cm each) with 50mM Tris-phosphate buffer, pH 7.0. Introduce the solution in the upward direction at 6 ml/hr; wash the pump and tubing with 3×2 ml of buffer at the same flow rate. Increase the flow rate to 12 ml/hr, and collect 10-ml fractions. Several small peaks of inactive material are eluted before and after the main peak, which contains about 90% of the potential activity applied. Pool fractions of potential activity over 400 $\mu g/A_{280}$ unit. Concentrate and dialyze against the same Tris-phosphate buffer.

Step 7. Chromatography on DEAE-Cellulose.[6] Precycle the DEAE-cellulose with 0.5 M HCl and 0.5 M NaOH,[13] neutralize and equilibrate with 50 mM Tris-phosphate buffer, pH 7.0. Prepare a column 5×50 cm. Apply the sample of step 6; elute with a linear gradient of 900 ml of the equilibrating buffer and 900 ml of 0.8 M NaCl in the same buffer. Pool

[13] Whatman Data Manual No. 2000, H. Reeve Angel, Clifton, New Jersey.

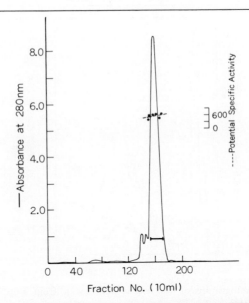

FIG. 2. Chromatography of the pepsinogen from step 6 on DEAE-cellulose. Column, 5×48 cm; linear gradient of 900 ml of 50 mM Tris-phosphate, pH 7.0, and 900 ml of 0.8 M NaCl in the same buffer; fractions, 10 ml; flow rate, 50 ml/hr.

the fractions of uniform potential activity (Fig. 2), i.e., all but a small tail portion of the main peak. Concentrate, dialyze against 1×6 liters of 50 mM sodium acetate solution, pH 7.0 (to remove phosphate) and 2×6 liters of distilled water adjusted to pH 7. Lyophilize.

The potential activity is about 600 μg/A_{280} unit with a recovery of about 90% in this step. The yield is about 500 mg of pepsinogen from 350 g of mucosa.

Preparation of the Pepsin

Step 1. Activation.[8] Dissolve 400 mg of the purified pepsinogen in water at 0° to give a concentration of 15 mg/ml. Place the solution in an ice bath on a magnetic stirrer and insert the electrode of a pH meter calibrated at 0°. Rapidly adjust the pH to 2.0 with a predetermined amount of 2.5 M HCl added from a microburette. After 1.5 min, rapidly adjust th pH to 5.3 with 2.5 M sodium acetate.

Step 2. Separation of the Activation Peptides from the Pepsin.[12,14] Apply the activation mixture to a polylysine-Sepharose 4B column $(2.6 \times 57$ cm$)$ equilibrated with 50 mM ammonium acetate buffer, pH 5.3. Elute with the same buffer until a sharp peak of peptides (Fig. 3)

¹⁴ M. Harboe, P. M. Andersen, B. Foltmann, J. Kay, and B. Kassell, *J. Biol. Chem.* **249**, 4487 (1974).

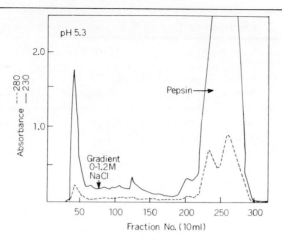

FIG. 3. Separation of the activation peptides from pepsin on polylysine-Sepharose 4B. Column 2.6 × 57 cm; buffer, 50 mM ammonium acetate, pH 5.3; gradient, 800 ml of buffer and 800 ml of 1.2 M NaCl in the same buffer; fractions, 9 ml; flow rate, 60 ml/hr.

appears (detected by A_{280}). Begin a linear gradient of 800 ml of starting buffer and 800 ml of 1 M NaCl in the same buffer. Additional peaks of activation peptides are eluted (detected with ninhydrin) and finally the pepsin. The pepsin fractions of uniform activity are pooled, concentrated, dialyzed against 2 × 6 liters of water (adjusted to pH 5.3 with acetic acid) and lyophilized. About 220 mg of pepsin is obtained, with a specific activity of 650 $\mu g/A_{280}$ unit.

Properties

Stability. Pepsinogen solutions in 50 mM Tris-phosphate buffer, pH 7.0, lose some potential activity when stored for 24 hr at 5°. The loss is concentration dependent; at 170 $\mu g/ml$ there was a 3% loss in potential activity, and at 17 $\mu g/ml$ a 15% loss occurred. Maximum stability is between pH 6 and 6.5.

Purity. The pepsinogen prepared by the present method does not differ significantly in amino acid composition from the pepsinogen already reported, which was prepared by conventional chromatography.[6] The molar ratios of amino acids present in amounts less than 10 residues per molecule are close to integral values. The potential peptic activity is almost uniformly distributed across the peak in the final step of purification. There is a single symmetrical peak in the ultracentrifuge.[6] However, disc electrophoresis at pH 8 and hydroxyapatite chromatography at pH 6.5 separates the final product into three fractions. These fractions have the same amino acid composition within experimental error and

COMPOSITION OF CANINE PEPSINOGEN AND PEPSIN

Component	Residues per molecule		Component	Residues per molecule	
	Pepsinogen[a]	Pepsin[b]		Pepsinogen	Pepsin
Lysine	6	1	Alanine	23	22
Histidine	2	1	Cystine	6	6
Arginine	6	3–4	Valine	26	21
Aspartic acid	39	35	Methionine	8	7
Threonine	20	19–20	Isoleucine	25	21
Serine	38	36–37	Leucine	25	19–20
Glutamic acid	33	29	Tyrosine	18	17
Proline	22	17	Phenylalanine	15	14
Glycine	35	34–35	Tryptophan	5	5
Amino sugar	7–8	7	Nonamino sugar	4	ND[c]

[a] J. P. Marciniszyn, Jr., and B. Kassell, *J. Biol. Chem.* **246**, 6560 (1971).
[b] C. L. Wright, P. H. Ward, and B. Kassell, unpublished observations.
[c] ND, not determined.

equal potential activity; they appear to differ in carbohydrate content.

The pepsin prepared from the purified pepsinogen is pure according to the criteria above, but also separates into three peaks of uniform activity when chromatographed on hydroxyapatite.

Physical Properties. The pepsinogen[6] has a molecular weight of 39,900 from amino acid and carbohydrate content and 41,667 by ultracentrifugation; the sedimentation coefficient, $s_{20,w}$, is 3.38 S, and the diffusion coefficient, $D_{20,w}$ is 7.37×10^{-7} cm^2 per second. The absorbance at 280 nm is 1.28 for 1 mg of protein per milliliter.

The pepsin has a molecular weight of 35,500 by equilibrium centrifugation and 35,000 by sodium dodecyl sulfate gel electrophoresis. The absorbance at 280 nm is 1.39 for 1 mg of protein per milliliter.

Activators. Like other pepsinogens, canine pepsinogen is activated by acid. Between pH 1.5 and 2.5, activation is complete in 5 min at 0°.

Specificity. The action of canine pepsin has been compared to that of porcine pepsin on several substrates. The activities of the two pepsins on hemoglobin substrate at pH 2 are approximately equal. Canine pepsin has only 1/30 of the activity of porcine pepsin in clotting milk at pH 5.3. Canine pepsin has much less activity than porcine pepsin on acetyl-Phe-Tyr at pH 2 and on Cbz-His-Phe-Trp-OEt at pH 4.

Composition. The table gives the composition of the pepsinogen and the pepsin. The most striking feature compared to other mammalian species is the large number of methionine residues.

Terminal Groups. Canine pepsinogen has amino-terminal Ala-Ile-.

[37] Proteinases from the Venom of *Agkistrodon halys blomhoffii*

By Sadaaki Iwanaga, Genichiro Oshima, and Tomoji Suzuki

Snake venoms contain several types of proteinases including endo-peptidases, nonproteolytic arginine ester hydrolases, thrombinlike enzymes and kininogenase.[1-4] The venom endopeptidase is mainly distributed in families of Crotalidae and Viperidae. A common feature of this proteinase from the Crotalidae venoms is that they all hydrolyze peptide bonds with amino groups contributed by leucine and phenylalanine residues, and require more than a hexapeptide sequence for hydrolysis of this bond. Another common characteristic point is that they are easily inactivated by EDTA and cysteine. In these properties, they differ from well known mammalian endopeptidase.

Assay Method[5]

Principle. The enzyme is assayed with 2.0% casein solution in Tris-HCl buffer, pH 8.5, at 37°. Undigested casein is precipitated with trichloroacetic acid and filtered or centrifuged. Digested casein in the supernatant is determined with the Folin–Ciocalteau reagent.

Reagents

Tris-HCl buffer, 0.4 M, pH 8.5

Casein, 2.0%: Two grams of casein (Hammarsten quality) are suspended in 100 ml of 0.4 M Tris-HCl buffer, pH 8.5 and heated for 15 min, in a boiling water bath to make complete solution.

Trichloroacetic acid (TCA), 0.44 M

Sodium carbonate, 0.4 M

Folin–Ciocalteau reagent. The commercially available reagent is diluted three times with distilled water before use.

Standard tyrosine solution: The tyrosine standard is prepared by dissolving 18.10 mg of pure, dry tyrosine in 1 liter of 0.2 M HCl.

---

[1] E. Kaiser and H. Michl, "Die Biochemie der tierischen Gifte." Deuticke, Vienna, 1958.

[2] "Venomous Animals and Their Venoms" (W. Bucherl, E. E. Buckley, and V. Deulofeu, eds.), Vol. 1. Academic Press, New York, 1968.

[3] O. B. Henriques and S. B. Henriques, "Pharmacology and Toxicology of Naturally Occurring Toxins," Vol. 1, p. 215. Pergamon, Oxford, 1971.

[4] H. F. Deutsch and C. R. Diniz, *J. Biol. Chem.* **216**, 17 (1955).

[5] Y. Murata, S. Satake, and T. Suzuki, *J. Biochem.* **53**, 431 (1963).

Procedure. Casein solution 0.5 ml, is digested for 15 min at 37° with 0.5 ml of a suitably diluted enzyme solution. The reaction is terminated by adding 1.0 ml of 0.44 M TCA; after it has stood for 30 min, the mixture is filtered through Toyo filter paper No. 2 or Whatman filter paper No. 2. To 0.5 ml of filtrate, 2.5 ml of 0.4 M Na_2CO_3, and 0.5 ml of Folin–Ciocalteau reagent are added. After the preparation has stood for 20 min, the blue color is read at 660 nm and compared against a standard tyrosine solution.

Definition of Unit and Specific Activity.[5] One unit of proteinase, expressed as proteinase units (PU) is defined as the amount of enzyme yielding an increase in color equivalent to 1.0 μg of tyrosine per minute with Folin's reagent. Specific activity is expressed as units per milligram of protein.

Purification Procedure[6-9]

The following procedures are conducted at 4°–6°.

Step 1. Chromatographic Separation of Proteinases.[7] A sample of 3 g (total absorbance at 280 nm = 3600) of lyophilized *Agkistrodon halys blomhoffii* venom is dissolved in 10 ml of 0.005 M sodium acetate and applied to a column of DEAE-cellulose (Serva, 0.68 meq/g, 2.5 × 70 cm) equilibrated with 0.005 M sodium acetate, pH 7.0. The exponential gradient elution is started with 1 liter of the equilibration buffer in the mixing vessel and 0.1 M sodium acetate, pH 7.0, in the reservoir. The latter is replaced by 0.2 M and 0.5 M sodium acetate, pH 7.0, as indicated in Fig. 1. The flow rate is adjusted to about 30 ml/hr, and the eluate is collected in 10-ml fractions. Figure 1 shows a typical elution profile. The proteinase activities are separated into three peaks, which are designated in the order of their elution from the column as proteinases a, b, and c. This pattern is very reproducible, always giving three proteolytic peaks. Proteinase a is found in the first peak in which the proteins are not adsorbed by DEAE-cellulose. Proteinase b, however, emerges occasionally a little faster or later than the associating peak of protein. Proteinase c is demonstrated always as the last peak of protein. Because the content of proteinase b is highest among the three proteinases in this venom, amounting approximately 10.4% of the lyophilized venom by weight, the following purification procedure is described on

[6] S. Satake, Y. Murata, and T. Suzuki, *J. Biochem.* 53, 438 (1963).

[7] S. Iwanaga, T. Omori, G. Oshima, and T. Suzuki, *J. Biochem.* 57, 392 (1965).

[8] G. Oshima, S. Iwanaga, and T. Suzuki, *J. Biochem.* 64, 215 (1968).

[9] G. Oshima, Y. Matsuo, S. Iwanaga, and T. Suzuki, *J. Biochem.* 64, 227 (1968).

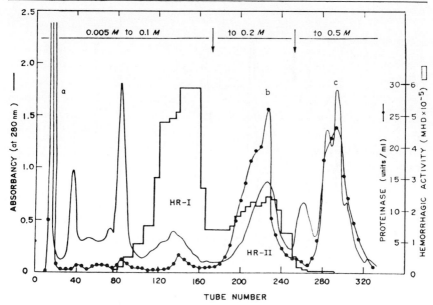

Fig. 1. Separation of three proteinases, a, b, and c, from the venom of *Agkistrodon halys blomhoffii* by column chromatography on DEAE-cellulose. Experimental conditions: column, 2.5 × 70 cm; sample volume, 3 g/10 ml; flow rate, 30 ml/hr; fraction volume, 10 ml per tube; exponential gradient, 0.005 *M* to 0.1 *M*, 0.2 *M* and 0.5 *M* sodium acetate as indicated. Reproduced from S. Iwanaga, T. Omori, G. Oshima, and T. Suzuki, *J. Biochem.* **57**, 392 (1965).

this enzyme. Further purification methods of proteinases a and c are described in the literatures.[6,9]

Step 2. Gel Filtration of Proteinase b Fraction on a Sephadex G-100 Column.[8] The fractions (tubes 190 to 250) with proteinase b activity eluted with 0.2 *M* acetate are combined and concentrated to a half volume of the pooled fraction by lyophilization. The concentrate is dialyzed against 2 liters of distilled water for 24 hr with two changes, and the dialyzed solution is then freeze-dried. The proteinase b fraction (total absorbance at 280 nm = 2530), which was prepared from 32.7 g of the unfractionated venom, is dissolved in 26.3 ml of 0.05 *M* sodium acetate, pH 7.0. The solution is applied to a column of Sephadex G-100 (6.5 × 866 cm), previously equilibrated with the same buffer and eluted with the same buffer at a flow rate of about 100 ml/hr. Fractions of 15 ml per tube are collected. Figure 2 shows the elution profile. The proteins are eluted in three peaks, and the first peak with proteolytic activity is concentrated by lyophilization. The concentrate is desalted on a column of Sephadex G-25 (6.5 × 60 cm) in distilled water and lyophilized.

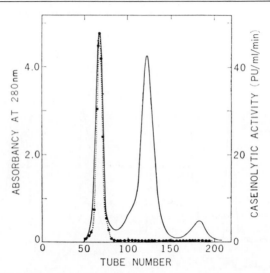

Fig. 2. Gel filtration of proteinase b fraction on a Sephadex G-100 column. Experimental conditions: column, 6.5 × 86 cm; sample volume, about 2.5 g/26.3 ml; flow rate, 100 ml/hr; fraction volume, 15 ml per tube; eluent, 0.05 M sodium acetate, pH 7.0. Reproduced from G. Oshima, S. Iwanaga, and T. Suzuki, J. Biochem. 64, 215 (1968).

Step 3. Final Purification of Proteinase b on a DEAE-Sephadex A-50 Column.[8] Part of the dried material (153 mg) obtained from step 2 is dissolved in 5.3 ml of 0.05 M sodium acetate, pH 7.0. The solution is applied to a column of DEAE-Sephadex A-50 (1.5 × 29 cm), equilibrated with the same buffer. Exponential gradient elution is carried out with 250 ml of 0.05 M sodium acetate, pH 7.0, in the mixing vessel and 0.4 M sodium acetate in the reservoir. The flow rate is adjusted to 35 ml/hr, and fractions of 3 ml per tube are collected. As shown in Fig. 3, a small amount of contaminating proteins is removed by this column, and the peak of proteolytic activity coincides with that of absorbance at 280 nm. The eluates in tubes 83 to 105 in Fig. 3 are combined and lyophilized. The specific activity of this highly purified proteinase b is 52.5, and its overall recovery is 88.5%. The purification procedures are summarized in Table I.

Properties

Homogeneity.[8] The purified proteinase b gives a symmetrical sedimentation pattern by ultracentrifugal analysis. It also gives a single band on disc polyacrylamide-gel electrophoresis when performed under

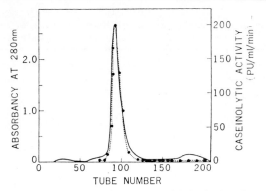

Fig. 3. Purification of proteinase b on a DEAE-Sephadex A-50 column. Experimental conditions: column, 1.5 × 29 cm; sample volume, 153 mg/5.3 ml; flow rate, 85 ml/hr; fraction volume, 3 ml per tube; exponential gradient, 0.05 M to 0.4 M sodium acetate, pH 7.0. Reproduced from G. Oshima, S. Iwanaga, and T. Suzuki, *J. Biochem.* **64**, 215 (1968).

standard conditions. Moving-boundary electrophoresis of proteinase b in several kinds of buffers with pH 3.05 to 8.51 shows a single symmetrical peak.

Physical Properties.[8,9] Physical constants of proteinase b, in addition to proteinases a and c, are listed in Table II. Sedimentation studies done on proteinase b in a protein concentration range of 2 to 7 mg/ml at pH 7.0 in sodium phosphate buffer ($\mu = 0.1$) indicate the dependence of $s_{20,w}$ on the protein concentration. Extrapolation gives a value of 5.54 S at zero concentration. Using a corrected $D^{\circ}_{20,w}$ of 5.26 × 10⁻⁷ cm² sec⁻¹ and $\bar{v}$ of 0.722 obtained from the density measurement, the molecular weight of proteinase b is calculated to be 92,000. From the values of sedimentation constant and intrinsic viscosity, the molecular weight of

TABLE I
SUMMARY OF PURIFICATION PROCEDURE OF PROTEINASE b[a]

Proteinase b fraction in the eluate of	Protein (g)	Total absorbance at 280 nm	Total proteinase units	Specific activity
DEAE-Cellulose	6.48	6238	212,700	35.5
Sephadex G-100 column	3.86	2856	197,500	51.2
DEAE-Sephadex A-50	3.42	2528	179,500	52.5

[a] The lyophilized venom (32.7 g) of *Agkistrodon halys blomhoffii* was applied on a column of DEAE-cellulose, and the proteinase b fraction obtained was further purified. Experimental conditions are described in the text.

TABLE II

PHYSICOCHEMICAL PROPERTIES OF PROTEINASES a, b, AND c
FROM THE VENOM OF *Agkistrodon halys blomhoffii*

Physical parameters	Proteinase a	Proteinase b	Proteinase c
Sedimentation coefficient, $s_{20,w}$ (S)	3.59	5.54	4.94
Diffusion coefficient, $D_{20,w}$ $(cm^2 sec^{-1})$	—[a]	5.26	5.88
Partial specific volume, $\bar{v}$	—[a]	0.722	0.695
Intrinsic viscosity $[\eta]$, (dl/g)	—[a]	0.052	0.046
Molecular weight:			
By Sephadex G-100	75,400	115,000	95,500
By Archibald procedure	50,200	95,500	72,300
Sedimentation diffusion	—[a]	92,000	67,000
Isoelectric point, pI	6.00	4.16	3.85
$A_{1\,cm}^{1\%}$ at 280 nm	9.03	7.4	10.93
Nitrogen content (%)	13.83	12.48	13.90
Polypeptide content (%)	76.5	76.5	91.2

[a] Owing to its low content in the venom, these properties of proteinase a could not be examined.

proteinase b is also computed to be 94,600 if $\beta = 2.16 \times 10^6$. Moreover, the Archibald procedure of sedimentation in the ultracentrifuge gives the average molecular weight of 92,700 and 98,200 at 0.70 and 0.38% protein concentration, respectively. The frictional ratio of proteinase b is calculated to be 1.368 from sedimentation constant, molecular weight, and intrinsic viscosity. When boundary electrophoretic mobilities of proteinase b at several pH's are plotted, the mobility calculated at pH 8.51 for the descending is -5.82×10^{-5} cm^2 $volt^{-1}$ sec^{-1}. Thus, the isoelectric point gives pH 4.18. The ultraviolet absorption spectrum of proteinase b shows an absorption curve having a maximum at 278 nm and a minimum at 252 nm at pH 7.5, and the $A_{1\,cm}^{1\%}$ value at 280 nm is estimated to be 7.4.

Chemical Properties.[8,9] The amino acid compositions of proteinases a, b, and c, determined after 24, 48, and 72 hr of acid hydrolysis, are shown in Table III. They have similar compositions; a common features include high aspartic acid and cystine contents and, in most cases, low tryptophan content. Attention is paid to other chemical compositions of proteinase b: nitrogen, 12.4% polypeptide moiety, 76.5%; neutral sugars (galactose, mannose, and trace of fucose) 8%; glucosamine, 6.5%; and sialic acid, 3%. In general, snake venom proteinase is characterized as a glycoprotein.

Stability.[6,7] Purified proteinases a, b, and c are soluble in distilled water and are not inactivated by dialysis. There is no loss of enzyme

TABLE III
Amino Acid Compositions of Snake Venom Proteinases a, b, and c

Component	Average or extrapolated value from 24-, 48-, and 72-hr hydrolysis (g amino acid residue per 100 g protein)		
	Proteinase a	Proteinase b	Proteinase c
Amino acids			
Lys	6.36	3.94	5.01
His	2.01	2.57	3.58
Ammonia	1.57	2.22	1.23
Arg	4.73	4.01	2.34
Asp	12.00	12.39	11.11
Thr	4.82	3.69	3.10
Ser	4.21	2.69	3.41
Glu	8.32	8.46	8.67
Pro	2.77	3.73	3.36
Gly	2.88	2.56	3.16
Ala	3.15	2.73	2.83
$Cys_{1/2}$	5.09	4.54	5.38
Val	3.91	4.27	3.75
Met	1.62	2.70	2.39
Ile	3.96	4.83	3.62
Leu	6.07	5.83	4.21
Tyr	4.05	3.51	6.06
Phe	3.18	2.48	2.59
Trp[a]	1.84	1.69	1.90
	82.54	78.84	77.79

	Content(%)		
Carbohydrates	Proteinase a	Proteinase b	Proteinase c
Neutral sugars	8.15	7.6	3.91
Amino sugars	4.72	6.5	3.34
Sialic acid	0.81	3.1	1.21
	13.68	17.1	8.46

[a] Estimated by the method of T. L. Goodwin and R. A. Morton, *Biochem. J.* **40**, 628 (1946).

activity when 0.1% solution of proteinase b in physiological saline is stored at —15° for one year. Moreover, the enzyme activity does not decline when freeze-thawed or lyophilized. All the proteinases are inactivated completely by heating for 10 min at 80°.

pH Profile.[6] Each of proteinases is active in the alkaline pH region, showing different pH optima; the values are 10.5, 9.8, and 8.9 for proteinases a, b, and c, respectively, when casein is used as substrate.

Metal Ion Requirement.[6] Ca^{2+} and Mg^{2+} are slightly stimulatory. All other divalent metal ions including Mn^{2+}, Cu^{2+}, Zn^{2+}, Co^{2+}, Hg^{2+}, and Cd^{2+} act as inhibitors; Cd^{2+}, particularly, has a strong inhibitory effect.

Inactivators.[6,7,10] Three proteinases are inactivated irreversibly by 10^{-2} *M* EDTA, and their activities are slightly inhibited by –SH reagents, including *p*-chloromercuribenzoate (PCMB), monoiodoacetic acid, *N*-ethylmaleimide, and 5,5′-dithiobis(2-nitrobenzoic acid). On prolonged incubation with the proteinase b, PCMB (10^{-3}) causes significant inhibition. Moreover, 10^{-2} *M* β-mercaptoethylamine, cysteine, glutathione, and β-mercaptoethanol inactivate strongly their caseinolytic activities. Soybean trypsin inhibitor, pancreatic basic trypsin inhibitor, and ϵ-aminocaproic acid have no effect on proteinase activity.

Role of Ca^{2+} and SH group in Proteinase b Activity.[10] Proteinase b contains approximately 2 g-atoms of calcium per enzyme molecule. The removal of the metals by electrodialysis or EDTA results in conformational change of the enzyme protein, as judged from the blue shift in its ultraviolet difference spectra. This conformational change of the protein with concomitant loss of proteolytic activity is irreversible, and addition of Ca^{2+} does not result in reactivation of the enzyme activity or reversal of the denaturation. In the presence of 8 *M* urea and 10^{-2} *M* EDTA, 1 mole of –SH group is titrated per enzyme molecule; in the absence of urea and EDTA, only 20% of the –SH group is measurable. It seems, therefore, that the metal is probably involved in the stabilization of the enzyme molecule, and that the –SH groups are closely related to the enzyme activity of proteinase b.

Specificity. Venom proteinase catalyzes the hydrolysis of peptide bonds of a wide variety of natural substrates, including casein[5], hemoglobin[5], gelatin, elastin,[11] collagen,[12] fibrinogen, fibrin, insulin,[13] glucagon,[13] and bradykinin.[14] In general, snake venom which contains bradykinin-releasing enzyme also includes bradykinin-destroying enzyme(s)— *Bothrops jararaca* and *Agkistrodon halys blomhoffii*, for example. In the case of the latter, the main bradykinin-destroying enzyme hydrolyzes the phenylalanyl-seryl bond of bradykinin.[15] The proteinase exhibits rather broad specificity with respect to the hydrolysis of peptide, but it

[10] G. Oshima, S. Iwanaga, and T. Suzuki, *Biochim. Biophys. Acta* **250**, 416 (1971).
[11] J. W. Simpson and A. C. Taylor, *Proc. Soc. Exp. Biol. Med.* **144**, 380 (1973).
[12] J. W. Simpson and L. J. Rider, *Proc. Soc. Exp. Biol. Med.* **137**, 893 (1973).
[13] M. Satake, T. Omori, S. Iwanaga, and T. Suzuki, *J. Biochem.* **54**, 8 (1963).
[14] T. Suzuki and S. Iwanaga, *in* "Handbook of Experimental Pharmacology" (E. G. Erdös, ed.), p. 193. Springer-Verlag, Berlin and New York, 1970.
[15] T. Suzuki, S. Iwanaga, S. Nagasawa, and T. Sato, "Hypotensive Peptides" (E. G. Erdös, N. Back, F. Sicuteri, and A. F. Wilde, ed.), p. 149. Springer-Verlag, Berlin and New York, 1966.

does not split ester and amide bonds of synthetic substrates,[16] such as N^α-toluenesulfonyl-L-arginine methyl ester, N^α-acetyltyrosine ethyl ester, leucinamide, L-leucyl-β-naphthyl amide, and N-benzoyl arginine amide. Moreover, the proteinase does not cleave the shorter peptides of di-, tri-, tetra-, and pentapeptides, and it requires at least peptide sequence more than hexapeptide in the bond split.[17] Employing the general formula, R'-NH-CHR-CO-X, cleavage of the CO–t bond is demonstrated when R represents the side chain from threonine,[13] asparagine,[17] aspartic acid,[13] tryptophan,[13] alanine,[13,17,18] histidine,[13,17,18] glycine,[13,17,18] phenylalanine,[13,18] and tyrosine.[19] The X component of the substrate is derived mainly from hydrophobic amino acids including phenylalanine,[13,16–19] leucine,[13,16–19] and valine.[18] Thus, it seems likely that venom proteinase hydrolyzes many kind of substrates with a specificity that is directed toward bonds in which amino groups are contributed by the hydrophobic residues.

Biological Activities.[20,21] Among the three proteinases of the venom of *A. halys blomhoffii*, proteinase b produces a petechial hemorrhage when it is injected intracutaneously into albino rabbit,[7] and the other proteinase c induces an edema formation on the skin of a depilated albino rabbit.[9] However, proteinase a does not show such biological actions.

Distribution.[22] The proteinases having a similar specificity are found in other Crotalidae and Viperidae venoms: *Crotalus atrox*,[19,23,24] *Trimeresurus flavoviridis*,[17] *Agkistrodon piscivorus leucostoma*,[18,25] *Bitis arietans*,[26] *Vipera russelli*,[27] *Bitis gavonica*,[28] *Causus rhombeatus*,[28] *Ophiophagus hannah*,[29] *Leptodeira annulata*,[29] and *Echis carinatus*.[30] The distribution of the proteinase and the related enzymes are shown in Table IV.

[16] T. Sato, S. Iwanaga, Y. Mizushima, and T. Suzuki, *J. Biochem.* **57**, 380 (1965).

[17] T. Takahashi and A. Ohsaka, *Biochim. Biophys. Acta* **198**, 293 (1970).

[18] A. M. Spiekerman, K. K. Fredericks, F. W. Wagner, and J. M. Prescott, *Biochim. Biophys. Acta* **293**, 464 (1973).

[19] G. Pfleiderer and A. Kraus, *Biochem. Z.* **342**, 85 (1965).

[20] T. Omori, S. Iwanaga, and T. Suzuki, *Toxicon* **2**, 1 (1964).

[21] G. Oshima, T. Omori-Sato, S. Iwanaga, and T. Suzuki, *J. Biochem.* **72**, 1483 (1972).

[22] G. Oshima, T. Sato-Ohmori, and T. Suzuki, *Toxicon* **7**, 229 (1969).

[23] R. Zwilling and G. Pfleiderer, *Hoppe-Seyler's Z. Physiol. Chem.* **348**, 519 (1967).

[24] G. Pfleiderer and G. Sumyk, *Biochim. Biophys. Acta* **51**, 482 (1961).

[25] W. Wagner, A. M. Spiekerman, and J. M. Prescott, *J. Biol. Chem.* **243**, 4486 (1968).

[26] S. J. Van Der Walt and F. J. Joubert, *Toxicon* **9**, 153 (1971).

[27] G. D. Dimitrov, *Toxicon* **9**, 33 (1971).

[28] G. R. Delpierre, *Toxicon* **5**, 232, (1968).

[29] D. Mebs, *Hoppe-Seyler's Z. Physiol. Chem.* **349**, 1115 (1968).

[30] A. Schieck, F. Kornalik, and E. Habermann, *Naunyn-Schmiedeberg's Arch. Pharmacol.* **272**, 402 (1972).

TABLE IV
Enzyme Distribution in Various Snake Venoms[a]

Snake venoms	Proteinase activities[b]	Arginine ester hydrolase activity[c]	Bradykinin-releasing activity[d]	Clotting activity[e]
Crotalidae				
Agkistrodon halys blomhoffii	24.5	1.8	25.2	−
A. piscivorus piscivorus	35.3	4.6	28.0	±
A. contortrix contortrix	49.5	11.1	72.6	+
A. contortrix mokeson	46.5	10.2	38.6	+
A. acutus	25.5	2.9	0.5	++
Crotalus adamanteus	13.8	16.8	20.4	±
C. atrox	59.0	4.2	124.3	++
C. durissus terrificus	25.7	8.9	56.5	++
C. viridis viridis	27.5	9.1	233.5	±
C. basiliscus	77.0	4.0	24.5	±
Trimeresurus flavoviridis	22.5	1.5	3.0	±
T. okinavensis	19.0	2.2	4.3	−
T. mucrosquamatus	26.8	34.1	28.6	±
T. gramineus	11.5	8.0	54.4	++
Bothrops jajararaca	7.8	2.3	10.4	++
B. atrox	17.8	3.8	5.0	++
Viperidae				
Vipera russelli	2.6	0.4	1.3	−
V. palaestinae	6.5	5.0	40.0	−
V. ammodytes	21.5	1.1	107.8	−
Echis carinatus	23.5	1.1	9.6	±
Crotalus rhombeatus	3.1	1.8	6.6	−
Bitis gavonica	11.3	2.7	215.0	±
Bitis arietans	10.0	1.0	6.57	−
Elapidae:				
Naja naja atra	0.9	—	3.92	−
Naja melanoleuca	1.5	—	ND	−
Ophiophagus hannah	6.4	—	ND	−
Naja naja samarensis	0.3	—	0.4	−
Naja nigricollis	0.5	—	0.3	±
Bungarus fasciatus	ND[f]	—	0.1	−
Bungarus multicinctus[f]	ND	—	ND	−
Dendroaspis angusticeps	1.0	—	15.6	±
Hemachatus haemachatus	2.0	—	0.2	±

[a] Activities in this table were assayed using unfractionated venoms.

[b] The activity was expressed as micrograms of tyrosine equivalent of TCA-soluble product formed per minute per milligram of protein.

[c] The activity was expressed as micromoles of α-toluenesulfonyl arginine ethyl ester hydrolyzed per minute per milligram of protein.

[d] The activity was expressed as milligrams of bradykinin released from bovine kininogen per minute per milligram of protein.

[e] The activity was determined from the time when fibrin fibers appeared: ++ within 3 min; + within 20 min; ± within 2 hr; − over 2 hr.

[f] ND, not detected.

[38] Protease from *Staphyloccus aureus*

By GABRIEL R. DRAPEAU

Staphylococci excrete extracellularly several types of proteins or enzymes, such as a variety of toxins, a nuclease, hyaluronidase, staphylokinase, lipase, and different proteases.[1] One of the proteases, which is discussed below, has been studied in some detail and is of particular interest in view of its narrow and unique substrate specificity.

Purification

The following purification procedure has been described previously[2] with some modifications. The production of protease is carried out in a 28-liter fermentor, but substantial amounts of enzyme can be obtained even from 1 or 2 liters of medium.

Organism. Strain V8 of *Staphylococcus aureus* is an excellent producer of the staphylococcal protease, but several other strains also secrete significant amounts of the enzyme. However, as previously reported, the protease from other strains may have properties different from those of the enzyme from strain V8.[3]

Cultivation. The cells are grown in 20-liter batches in a 28-liter fermentor in a medium containing (amount per liter of solution in tap water): K_2HPO_4, 2.4 g; NaH_2PO_4, 0.4 g; $CaCl_2 \cdot 2H_2O$, 0.74 g; yeast extract (General Biochemicals), 5.0 g; sodium β-glycerophosphate (Sigma, Grade III, containing approximately 50% of the α-isomer), 10 g; and casein, 10 g. Foam formation is prevented by the addition of 3–4 drops of a silicone antifoaming agent. $CaCl_2$ is sterilized separately as a 1 M solution and added before inoculation. The fermentor is inoculated with 200 ml of cells grown overnight in the same medium. Incubation is carried out at 37° for 18 hr with moderate aeration and the agitation set at 250 rpm. For production on a smaller scale, the cells are grown in 2-liter Erlenmeyer flasks containing 500 ml of the same growth medium. Each flask is inoculated with 1 ml of an overnight culture and incubated on a rotary shaker for 18 hr at 37°. At the completion of the cultivation hexadecyltrimethylammonium bromide (0.1 g/liter) is added in order to kill the cells and the incubation is continued for an additional

[1] "The Staphylococci" (J. O. Cohen, ed.). Wiley (Interscience), New York, 1972.
[2] G. R. Drapeau, Y. Boily, and J. Houmard, *J. Biol. Chem.* **247**, 6720 (1972).
[3] R. Beaudet, S. A. Saheb, and G. R. Drapeau, *J. Biol. Chem,* **249**, 6468 (1974).

hour. The suspension is then chilled to about 10° and centrifuged in a Sharples T-1 centrifuge.

Procedure. To the centrifuged culture filtrate 11 kg of ammonium sulfate are added with stirring. After stirring for 2 hr, the suspension is centrifuged in a Sharples centrifuge. The dark brown precipitate obtained is dissolved in 500 ml of distilled water. The enzyme is then precipitated by the slow but continuous addition, with vigorous stirring, of 675 ml of cold acetone (−15°). The enzyme precipitates mostly during the addition of the final 50–75 ml of acetone. The suspension is immediately centrifuged in the cold at 9000 g for 10 min. The precipitate is collected and dissolved in about 200 ml of distilled water, and insoluble material is removed by centrifugation at 16,000 g for 20 min. The enzyme solution can be frozen (−20°) at this stage, and no loss in activity is noted even after several months of storage.

The acetone fraction is applied to a 7.5 × 25 cm column of DEAE-cellulose, Whatman DE-52, which has been equilibrated with 10 mM Tris-HCl buffer, pH 7.5. The column is washed with 50 ml of the same buffer, and then a linear gradient from 0 to 0.7 M KCl made from 1.4 liter of the wash buffer and 1.4 liter of buffer containing 0.7 M KCl is applied. The flow rate is maintained at 60 ml/hr and 20-ml fractions are collected. The most active fractions are eluted in about 200 ml (fractions 35–45). The enzyme is dialyzed against distilled water in acetylated dialysis tubing[4] for 24 hr and then lyophilized. Usually 700–800 mg of pure protease can be obtained from 20 liters of culture medium.[5]

Remarks. The enzyme obtained after DEAE-cellulose chromatogra-

---

[4] Dialysis tubings are immersed in a solution containing 90 ml of pyridine and 10 ml of acetic anhydride. After standing for 20 hr at room temperature, the solution is decanted; the tubing is washed with running tap water for 1 hr, and then rinsed with distilled water [L. G. Craig, *Anal. Methods Protein Chem.* **6**, 103 (1960)].

[5] The existence of three different extracellular proteolytic enzymes produced by *S. aureus* strain V8 has been reported [S. Arvidson, T. Holme, and B. Lindholm, *Biochim. Biophys. Acta* **301**, 135 (1973); S. Arvidson, *Biochim. Biophys. Acta* **301**, 149 (1973)]. However, under the growth conditions described above, only two proteolytic enzymes could be detected. The new protease is EDTA-sensitive and appears to be unstable in the growth medium since its activity can be measured only during the log phase of the culture and practically no activity, or very little, remains after 18 hr of cultivation. It does not interfere with the purification of the staphylococcal protease because it is eluted later from the DEAE-cellulose column. Moreover, it is completely inactivated during this purification step unless Ca^{2+} is present in the buffers (S. A. Saheb and G. R. Drapeau, unpublished observations). Consequently, to avoid possible contamination with this new protease, it is important to pursue the purification through the DEAE-cellulose chromatography step.

phy usually contains some pigmented material. This pigment originates from the yeast extract present in the growth medium. It is preferable to use yeast extract which contains the least amount of colored material. In such a case, the small amount of pigmented material associated with the purified enzyme is readily removed by treatment with charcoal.

A small quantity of activated charcoal is added to the enzyme solution, and the suspension is filtered on a sintered-glass funnel containing a layer of Sephadex G-25 of 3–4 cm in thickness.

Purity. The purity of staphylococcal protease was examined by polyacrylamide gel electrophoresis, sedimentation velocity ultracentrifugation, and immunodiffusion.[2,3] These methods showed the preparation to be homogeneous.

Assay Methods

Two assay methods are described below. The first one is based on the method of Kunitz[6] and is performed at either pH 4.0 with hemoglobin or at pH 7.8 with casein as substrates. The second method is based on the esterolytic activity of the protease.

Proteolytic Activity toward Casein and Hemoglobin

Reagents

Casein solution: 1 g of casein (Hammersten quality) is dissolved with heating in 50 ml of 0.012 N NaOH. The volume is then brought to 100 ml with a 0.1 M solution of Tris-HCl buffer, pH 7.8.

Hemoglobin solution: 0.8 g of urea-denatured hemoglobin is dissolved in distilled water and the pH is adjusted to 4.0 with 1 N HCl. The volume is completed to 90 ml and then 10 ml of a 1 M solution of sodium acetate buffer, pH 4.0, is added.

Procedure. A 5-ml aliquot of casein or hemoglobin solution is added to a test tube (18 × 150 mm) and equilibrated at 37° for 5 min in a water bath. To the protein solution is added 0.1 ml of the enzyme solution in distilled water (10–40 μg of protease), and the mixture is incubated at 37° for 10 min. The reaction is stopped by the addition of 5 ml of 10% trichloroacetic acid. After standing for 15 min, the contents of the tubes are filtered through a Whatman No. 42 filter paper and the absorption is measured at 280 nm against the blank. The blank is pre-

---

[6] M. Kunitz, *J. Gen. Physiol.* **30**, 291 (1947).

pared by first mixing the casein solution with TCA and adding the enzyme solution to the casein–TCA mixture. A linear rate of increase is obtained up to an absorbance of 0.200 with both substrates.

A unit of activity is defined as the amount of enzyme which yields 0.001 OD_{280} unit of change per minute. The specific activity is calculated as the number of enzyme units per milligram of protein.

Esterase Activity toward N-CBZ-L-glutamyl-α-phenyl Ester

Reagents

N-CBZ-L-glutamyl-α-phenyl ester solution, 20 mM, in dioxane
Tris solution, 0.2 M, adjusted to pH 7.5 with phosphoric acid (chloride ions are competitive inhibitors).

Synthesis of the Substrate. N-CBZ-L-GLUTAMIC ANHYDRIDE. This compound is synthesized by the method of Harington and Mead[7] or can be obtained commercially. N-CBZ-L-glutamic acid (12.5 g) is added to 35 ml of freshly distilled acetic anhydride. The suspension is heated to 40° with occasional stirring until a clear solution is obtained. The solution is evaporated at 50° under reduced pressure. The residue is dissolved in 10 ml of chloroform. The solution is filtered, and crystallization is achieved by the addition of anhydrous ether. Crystals are collected by filtration and dried in a desiccator over P_2O_5 and paraffin. The yield is 8.55 g (75%).

N-CBZ-L-GLUTAMYL-α-PHENYL ESTER. This compound is prepared according to the method of Klieger and Gibian.[8] N-CBZ-L-glutamic anhydride (4.3 g) is dissolved in 100 ml of ether containing 3.4 ml of dicyclohexylamine. Phenol, 1.84 g (recrystallized from petroleum ether, b.p. 60°–110°) dissolved in 20 ml of ether is added. The mixture is stirred at room temperature for 18 hr. The precipitate formed is collected by filtration and dissolved while heating in a minimum volume of 100% ethanol. The crystals which form on cooling are suspended in ethyl acetate and a 2 N H_2SO_4 solution is added dropwise until the crystals are dissolved.[9] The organic phase is washed with water and then dried over $MgSO_4$. The solution is filtered and evaporated; the residue is placed in a desiccator over P_2O_5 and paraffin. The product is crystallized from isopropanol-petroleum ether (b.p. 30°–75°). The yield is 4.4 g (80%).

---

[7] C. R. Harington, and T. H. Mead, *Biochem. J.* **29**, 1602 (1935).
[8] E. Klieger and H. Gibian, *Justus Liebigs Ann. Chem.* **655**, 195 (1962).
[9] R. G. Hiskey, L. M. Beacham, and V. G. Maltl, *J. Org. Chem.* **37**, 2472 (1972).

Assay. To a 1-cm thermostatted cuvette with 50 μl of the substrate is added 2.5 ml of 0.2 M Tris-phosphate buffer, pH 7.5; after a few minutes at 35°, 50 μl (0.1–0.2 μg of protease) of the enzyme solution is added with rapid stirring. The increase in absorbance at 270 mm is recorded and used to calculate the micromoles of phenol released per minute using the molar extinction coefficient of phenol at 270 nm of 1500 M^{-1} cm^{-1}.[10] The rate of hydrolysis obtained in this way is corrected for the rate of spontaneous hydrolysis.

Properties

Enzymic Properties. The protease is active in the pH range of 3.5 to 9.5 and exhibits maximum proteolytic activity at pH 4.0 and 7.8 with hemoglobin as the substrate. An optimum pH of 7.8 is also observed with casein. Because casein tends to precipitate at pH values lower than 6.0, it is not possible to determine whether the enzyme has also a pH optimum in the vicinity of pH 4.0 for this substrate.

Stability. The enzyme is very soluble in distilled water; no evidence of autodigestion is observed, even at high enzyme concentrations. It autodigests very readily, however, at temperatures between 40° and 65°. Solutions of the enzyme in distilled water can be frozen and thawed repeatedly without loss of activity. It is also stable in the pH range of 4–10, but it slowly precipitates at pH around 4.0.

Physical Properties and Amino Acid Composition. The molecular weight of the staphylococcal protease was estimated to be 12,000 by sedimentation equilibrium studies, and 11,400 by polyacrylamide gel electrophoresis in the presence of sodium dodecyl sulfate. The sedimentation coefficient $s_{20,w}^{o}$ is 2.9 Svedberg units (S).[2] The protease, freed of pigment, has a $E_{1\,cm}^{M}$ (280 nm) of 11,500. The amino acid composition is shown in Table I. The enzyme is remarkably rich in dicarboxylic amino acids (or their amides) and proline residues. Sulfhydryl groups are absent. The amino and carboxyl terminal residues are valine and glutamic acid, respectively.

Activators and Inhibitors. The addition of Ca^{2+}, Mg^{2+}, Mn^{2+}, Zn^{2+}, EDTA, or citrate at concentrations up to 10 mM in the assay mixture, do not have any effect on substrate hydrolysis. However, the protease is strongly inhibited by diisopropyl fluorophosphate.[2]

Specificity. By contrast with most proteolytic enzymes of microbial origin, the staphylococcal protease exhibits a narrow substrate specificity. When incubated with oxidized ribonuclease in the presence of 50 mM phosphate buffer, the protease cleaves specifically peptide bonds on the

[10] A. A. Kortt and T. Y. Liu, *Biochemistry* **12**, 328 (1973).

TABLE I

AMINO ACID COMPOSITION OF PROTEASE FROM *Staphylococcus aureus* STRAIN V8[a]

Amino acid	No. of residues[b]	Amino acid	No. of residues[b]
Lysine	6	Alanine	7
Histidine	3	Cysteine	0
Arginine	1	Valine	7
Aspartic acid	29	Methionine	2[c]
Threonine	8	Isoleucine	6
Serine	4	Leucine	4
Glutamic acid	10	Tyrosine	3
Proline	11	Phenylalanine	4
Glycine	9	Tryptophan	1

[a] From G. R. Drapeau, Y. Boily, and J. Houmard, *J. Biol. Chem.* **247**, 6720 (1972).

[b] Moles of amino acid per mole of enzyme (molecular weight 12,000).

[c] The reported value of one methionine[a] is known to be an underestimate.

carboxyl-terminal side of either aspartate or glutamate. The enzyme has no activity on casein, which has all carboxyl groups blocked with glycine ethyl ester in amide linkage. This observation provides the ultimate proof of the narrow specificity of the protease and its requirement for free carboxyl groups.

The protease can be rendered even more restrictive depending on the type of buffer used during the digestion of protein substrates.[11] In 50 mM ammonium bicarbonate buffer pH 7.8, or ammonium acetate buffer, pH 4.0, only glutamyl bonds are cleaved. The protease hydrolyzes all glutamyl bonds, but those involving amino acids with bulky side chains are cleaved at a reduced rate. None of the aspartyl bonds of ribonuclease are cleaved under these conditions, while in lysosyme only an Asp—Gly bond is split at a low but detectable rate. However, this low cleavage is not observed when the digestion is carried out at pH 4.0.

Distribution. The majority of *S. aureus* strains produce extracellular proteolytic enzymes. However, on a casein agar medium, the proteolytic activities of staphylococci can be differentiated into four classes by the patterns of precipitation zones.[12] The protease isolated from a representative strain of each of these classes exhibits identical substrate specificities, but they differ significantly in electrophoretic mobilities, specific activities, and molecular weights.[3] While the protease from four different strains have a molecular weight in the vicinity of 12,000, the

[11] J. Houmard and G. R. Drapeau, *Proc. Natl. Acad. Sci. U.S.A.* **69**, 3506 (1972).

[12] F. G. Martley, A. W. Davis, D. F. Bacon, and R. C. Lawrence, *Infect. Immun.* **2**, 439 (1970).

TABLE II
ESTERASE ACTIVITY OF STAPHYLOCOCCAL PROTEASE ISOLATED FROM VARIOUS
STRAINS ON N-CBZ-GLUTAMYL-α-PHENYL ESTER

Strain	K_m $10^4\ M$	k_{cat} sec^{-1}	k_{cat}/K_m $10^4\ M^{-1}\ sec^{-1}$
V8	0.78	230	295
A	5.10	44	8.6
B	1.17	219	187
C	0.97	167	172
D	1.03	175	170

enzyme from another strain has a molecular weight of about 24,000. This enzyme does not cross-react immunologically with any of the low-molecular-weight proteases. Recently, the purification of a protease from another strain of *S. aureus* having the same substrate specificity has been reported.[13] The molecular weight of this enzyme was estimated to be 29,000. It would appear that extensive autodigestion can occur with this class of enzymes without significantly affecting their activities.

Table II shows some kinetic parameters for the protease isolated from different *S. aureus* strains.[14]

[13] A. C. Ryden, L. Ryden, and L. Philipson, *Eur. J. Biochem.* **44**, 105 (1974).
[14] J. Houmard and G. R. Drapeau, in preparation.

[39] Bromelain Enzymes

By TAKASHI MURACHI

Nomenclature. Bromelain enzymes are found in tissues of the plant family Bromeliaceae of which pineapple, *Ananas comosus* (L), is the best known. The proteolytic enzyme found in the juice of pineapple stem is called stem bromelain and the enzyme in the fruit was first described under the name of bromelin[1] and is now called fruit bromelain.[2] Systematic number EC 3.4.22.4 is given to stem bromelain, and EC 3.4.22.5 to fruit bromelain. The purpose of the present chapter is to review only recent information about these two enzymes and to focus on data accumulated since publication of Chapter 18 in Vol. 19 of this series.[3] Thus it is suggested that the reader consult that reference first.

[1] R. H. Chittenden, *Trans. Conn. Acad. Sci.* **8**, 281 (1892).
[2] R. M. Heinicke and W. A. Gortner, *Econ. Bot.* **11**, 225 (1957).
[3] T. Murachi, Vol. 19, this series [18].

Stem Bromelain

Purification

General Comments Regarding Purification. The crude stem bromelain preparation, usually obtained commercially (Dole Company, Honolulu, Hawaii), contains a number of proteolytic enzymes and nonproteolytic enzymes, including phosphatase,[2] peroxidase,[2] cellulase,[4] and other glycosidases.[5] Purification of a carboxypeptidase from commercial bromelain has recently been described.[6] The existence of at least two major components in stem bromelain having similar proteolytic activities and molecular properties except for electrophoretic mobility was first demonstrated by Murachi and Neurath.[7] El-Gharbawi and Whitaker[8] separated five proteolytically active components from a crude stem bromelain preparation by chromatography on Bio-Rex 70 resin at pH 6.1 or by zone electrophoresis in Sephadex G-75 gel. Feinstein and Whitaker[9] compared the physical and chemical properties of these five components. Further characterization of multiple components has also been reported by Silverstein and Kézdy.[10] Using SE-Sephadex C-50, Scocca and Lee[11] were able to separate two proteolytically active components that had identical carbohydrate composition. These previous reports together may imply (1) that the crude starting material contains several proteolytically active components, (2) that among these components some differ significantly in molecular size and also in electric charge, so that they can be separated by conventional chromatography or electrophoresis, and (3) that some other components, however, resemble one another so closely that they usually behave as a single component and may be fractionated only under special conditions. Therefore, the purification procedures must be such that one proteolytic enzyme could be separated from another and also from nonenzyme proteins.

Purification Procedure of Murachi et al.[12] The method consists of six steps, which have been given in detail in Vol. 19.[3] A few additional points should be noted, however. Sephadex G-50 can be used in place

---

[4] H. Suzuki, S. Imai, K. Nisizawa, and T. Murachi, *Bot. Mag. Tokyo* **84**, 389 (1971).

[5] Y. T. Li and Y. C. Lee, *J. Biol. Chem.* **247**, 3677 (1972).

[6] T. Doi, C. Ohtsuru, and T. Matoba, *J. Biochem.* (Tokyo) **75**, 1063 (1974).

[7] T. Murachi and H. Neurath, *J. Biol. Chem.* **235**, 99 (1960).

[8] M. El-Gharbawi and J. R. Whitaker, *Biochemistry* **2**, 476 (1963).

[9] G. Feinstein and J. R. Whitaker, *Biochemistry* **3**, 1050 (1964).

[10] R. M. Silverstein and F. J. Kézdy, *Fed. Proc., Fed. Am. Soc. Exp. Biol.* **29**, 929 (abstr.) (1970).

[11] J. Scocca and Y. C. Lee, *J. Biol. Chem.* **244**, 4852 (1969).

[12] T. Murachi, M. Yasui, and Y. Yasuda, *Biochemistry* **3**, 48 (1964).

of Sephadex G-100.[13] In order to minimize the hazard of autodigestion during the purification procedure, it is recommended to use sodium tetrathionate, which may reversibly block the catalytic sulfhydryl group of the enzyme protein. The concentration of sodium tetrathionate in the buffers used is 0.1 mM.[13]

Further Fractionation on SP-Sephadex.[13] The step 6 preparation can be further fractionated into two very similar components by means of chromatography on SP-Sephadex. An aqueous solution of tetrathionate-blocked fraction 6 protein (2–4 g in 100 ml) is dialyzed against 5 liters of the starting buffer, 0.05 M Tris-HCl, pH 8.0, at 4° overnight. The dialyzed solution is centrifuged and applied to a 2.5 × 100 cm column of SP-Sephadex C-50 (2.3 mEq/g), which has been equilibrated with the starting buffer containing 0.1 mM sodium tetrathionate and 0.1 mM EDTA. The column is washed with 500 ml of the starting buffer, then adsorbed enzyme proteins are eluted from the column with 2.5 liters of 0.2 M Tris-HCl, pH 8.0, containing 0.1 mM sodium tetrathionate and 0.1 mM EDTA, at a flow rate of 15 ml/hr, at 4°. This gives rise to one major and one minor component on the chromatogram with almost equal specific activity values toward casein. Yields after rechromatography of the major and minor components are 15–25% and 7.5–15% of the fraction 6 preparation on a weight basis.

Alternative Purification Procedure.[14] Five grams of commercial bromelain preparation are dissolved in 30 ml of 0.02 M sodium citrate buffer, pH 5.5, saturated with phenylmercuric acetate (less than 0.5 mM). After insoluble material has been removed by centrifugation at 7000 g for 20 min, the supernatant solution is applied to a 5 × 85 cm column of Sephadex G-75 (medium particle size), which has been equilibrated at room temperature with 0.02 M sodium citrate buffer, pH 5.5, saturated with phenylmercuric acetate. The column is washed with the same buffer at a flow rate of 40 ml/hr at room temperature. The major proteolytically active fractions are eluted between 600 and 1060 ml. The pH of the pooled fractions is adjusted to 7.2 with 0.2 M NaOH, and the solution is then applied to a 5 × 20 cm column of DEAE-cellulose (0.96 mEq/g), which has been equilibrated with 0.02 M sodium phosphate buffer, pH 7.2, containing 0.5 mM phenylmercuric acetate. The column is washed with 1 liter of the same buffer at a flow rate of 40 ml/hr and at room temperature. Most of the proteolytic activity comes off the column. The pH of the combined effluent solution is

[13] N. Takahashi, Y. Yasuda, K. Goto, T. Miyake, and T. Murachi, *J. Biochem.* (*Tokyo*) **74**, 355 (1973).
[14] S. Ota, K. Horie, F. Hagino, C. Hashimoto, and H. Date, *J. Biochem.* (*Tokyo*) **71**, 817 (1972).

adjusted to 6.0 with 0.1 M acetic acid, and the solution is then applied to a 5×35 cm column of SE-Sephadex C-25 (medium particle size, 2.0 mEq/g), which has been equilibrated with 0.03 M sodium citrate buffer, pH 6.0, containing 0.5 mM phenylmercuric acetate. The column is washed at a flow rate of 40 ml/hr and at room temperature first with 1.5 liters of the starting buffer and then with 1.5 liters of 0.3 M sodium citrate buffer, pH 6.0, containing 0.5 mM phenylmercuric acetate. The fractions that have high proteolytic activity are pooled, concentrated, and rechromatographed twice on a 5×35 cm column of SE-Sephadex under the identical conditions as above. The final preparation of the main enzyme component amounts to 10.4% of the crude enzyme. Several minor proteolytically active components can be separated during the chromatographic procedures described above with yields ranging from 0.03% to 2.2% of the starting material.[14]

Isolation of stem bromelain from an acetone powder by a single passage through a column of ϵ-aminocaproyl-D-tryptophan methyl ester coupled to Sepharose 4B has been described.[15]

Purity. The enzyme preparation purified up to step 6[12] contains at least two very closely similar components, which can be resolved only by chromatography on SP-Sephadex under specified conditions.[13] In this respect, the step 6 preparation is not pure enough. Nevertheless, since these two components are almost indistinguishable from each other and from the step 6 preparation on any other chemical, physical, and enzymic criteria,[13] the step 6 preparation can be employed as a practically pure material for most purposes. The homogeneity of the two components obtained by SP-Sephadex chromatography is verified by polyacrylamide gel electrophoresis in the presence of sodium dodecyl sulfate (SDS) and immunoelectrophoresis.[13] Enzyme preparations purified by alternative procedures are also found to be chromatographically homogeneous[14,16,17] and migrate as a single band upon electrophoresis on cellulose acetate[16] or on polyacrylamide gel.[16,18]

The two components obtained by SP-Sephadex chromatography have each single amino-terminal residue valine.[13] The main enzyme purified by Ota *et al.*[14] and the purified preparation obtained by Wharton[16] always contain significant amounts of extraneous amino end groups, particularly alanine, in addition to valine. The implication of such persistent heterogeneity of these preparations with respect to the amino end group has not been fully understood.

[15] D. Bobb, *Prep. Biochem.* **2**, 347 (1972).
[16] C. W. Wharton, *Biochem. J.* **143**, 575 (1974).
[17] S. Ota, S. Moore, and W. H. Stein, *Biochemistry* **3**, 180 (1964).
[18] L. P. Chao and I. E. Liener, *Biochem. Biophys. Res. Commun.* **27**, 100 (1967).

Properties

Physical Properties. Physical data for stem bromelain were listed in Table II of Ref. 3. The results showed the enzyme to be a more basic and larger protein than papain. The molecular weight of the step 6 preparation obtained by the method of Murachi *et al.* was calculated to be approximately 33,000 from experimentally determined sedimentation and diffusion coefficients.[12] Using a further purified preparation, reexamination of the molecular weight has been made by polyacrylamide gel electrophoresis in the presence of SDS and also by sedimentation equilibrium ultracentrifugation.[13] Measured values range from 25,600 to 28,100, and in practice it is recommended to adopt a tentative value of 28,000.[13]

Chemical Properties. Amino acid compositions of stem bromelain as reported by different investigators,[9,13,14] are given in the table. Characteristically, stem bromelain has only one cysteinyl and histidyl residue per molecule whereas papain has two histidyl residues; also bromelain contains methionyl residues that papain does not have. The amino-terminal residue is valine,[13] and the principal carboxyl terminal is glycine.[14] The enzyme is a glycoprotein having one oligosaccharide moiety per molecule, which is covalently linked to the peptide chain.[11,19] The proposed structure of the carbohydrate moiety of stem bromelain follows.[20,21]

The amino acid sequence of the peptide fragment that attaches the sugar residues has been determined.[20,22] The β-configuration of the most proximal D-mannosyl residue has been demonstrated by the use of β-D-mannosidase isolated from a crude stem bromelain preparation.[21]

[19] T. Murachi, A. Suzuki, and N. Takahashi, *Biochemistry* 6, 3730 (1967).
[20] Y. Yasuda, N. Takahashi, and T. Murachi, *Biochemistry* 9, 25 (1970).
[21] Y. C. Lee and J. R. Scocca, *J. Biol. Chem.* 247, 5753 (1972).
[22] K. Goto, T. Murachi, and N. Takahashi, *FEBS Lett.* 62, 93 (1976).

AMINO ACID COMPOSITION OF STEM AND FRUIT BROMELAINS[a]

Components	Stem bromelain				Fruit bromelain	
	1	2	3	4	5	6
Amino acids						
Lysine	14	17	17	12	5	8
Histidine	1	1	1	1	1	1
Arginine	8	8	9	6	5	9
Aspartic acid	23	23	23	16	17	32
Threonine	10	10	11	8	8	14
Serine	21	21	22	16	18	30
Glutamic acid	19	17	18	12	13	25
Proline	11	13	11	8	7	10
Glycine	27	24	26	19	18	35
Alanine	30	26	28	20	14	25
Half-cystine	9	9	7	5	6	9
Valine	16	16	18	12	11	20
Methionine	3	3	4	2	3	5
Isoleucine	17	17	18	12	9	18
Leucine	7	8	8	5	6	10
Tyrosine	14	16	16	11	13	16
Phenylalanine	8	8	7	5	4	8
Tryptophan	7	7	6	5	3	7
	245	244	250	179	161	282
Ammonia (amide)		21	28	19	24	
Glucosamine	2	2[b]	2	4	<0.2	0
Carbohydrate (%)	2.1	2.1	0.9	2.0	3.2	0

[a] Sources of values are as follows. Column 1: N. Takahashi, Y. Yasuda, K. Goto, T. Miyake, and T. Murachi, *J. Biochem. (Tokyo)* **74**, 355 (1973). Nearest integral number of residues per mole of MW 28,000 for component SB1. Column 2: Number of residues per mole of MW 33,000 reported earlier for step 6 preparation [T. Murachi, *Biochemistry* **3**, 932 (1964)] has been recalculated on the basis of an MW of 28,000. Column 3: S. Ota, K. Horie, F. Hagino, C. Hashimoto, and H. Date, *J. Biochem. (Tokyo)* **71**, 817 (1972). Number of residues per mole of MW 28,000 for component I-1. Column 4: G. Feinstein and J. R. Whitaker, *Biochemistry* **3**, 1050 (1964). For component II, taken methionine as two residues per molecule. Column 5: S. Ota, K. Horie, F. Hagino, C. Hashimoto, and H. Date, *J. Biochem. (Tokyo)* **71**, 817 (1972). Nearest integral number of residues per mole of MW 18,000 for component A. Column 6: F. Yamada, N. Takahashi, and T. Murachi, *J. Biochem. (Tokyo)* **79** in press. Nearest integral number of residues per mole of MW 31,000 for component FA2.

[b] N. Takahashi, Y. Yasuda, M. Kuzuya, and T. Murachi, *J. Biochem. (Tokyo)* **66**, 659 (1969).

Oxidative degradation of the carbohydrate residues by sodium periodate does not cause much alteration in enzymic activity toward casin as well as synthetic substrates.[23]

Stem bromelain has one reactive sulfhydryl group per molecule as determined by titration with p-chloromercuribenzoate. The sulfhydryl group is essential for catalytic activity[24] and reacts stoichiometrically with 5,5'-dithiobis(2-nitrobenzoic acid)[13] or with 2,2'-dipyridyl disulfide.[25] This cysteinyl residue (Cys) is located at position 25 from the amino-terminus.[22]

```
      1                5                  10                   15
   H-Val-Pro-Gln -Ser - Ile - Asp-Trp-Arg-Asp-Tyr-Gly -Ala-Val-Thr-Ser-

     16              20                  25
   Val-Lys-Asn-Gln -Asn-Pro-Cys-Gly -Ala -Cys-Trp-
                                |
                           - Gly -Tyr-Cys-Lys-
```

A nonapeptide sequence, which corresponds to positions 18 to 26 in the sequence illustrated, except for Asp in place of Asn at position 20, was first reported by Husain and Lowe[26] but a neutral amino acid residue Asn at this position has been confirmed.[13]

The unique histidine residue, involved in the intramolecular cross-linking of the enzyme by 1,3-dibromoacetone, has an apparent pK_a value of 6.4 as determined directly by the photooxidation reaction in the presence of methylene blue.[27,28] This value is within the *normal* range for an imidazole group.

The following is the amino acid sequence of carboxyl-terminal 33 residues[22]:

```
   -Arg-Trp-Gly-Glu-Ala-Gly-Tyr-Ile-Arg-Met-Ala-Arg-Asp-Val-Ser-
   Ser-Ser-Ser-Gly-Ile-Cys-Gly-Ile-Ala-Ile-Asp-Pro-Leu-Tyr-Pro-
   Thr-Glu-Gly-OH
```

Evidence for the structural homology among stem bromelain and other thiol proteinases, including fruit bromelain, papain, and ficin, are now being accumulated by elucidating the primary structure of the fragments[22] and by studying the immunological cross-reactions between thiol proteinases.[29-31]

[23] Y. Yasuda, N. Takahashi, and T. Murachi, *Biochemistry* **10**, 2624 (1971).

[24] T. Murachi and M. Yasui, *Biochemistry* **4**, 2275 (1965).

[25] C. W. Wharton, E. M. Crook, and K. Blocklehurst, *Eur. J. Biochem.* **6**, 565 (1968).

[26] S. S. Husain and G. Lowe, *Chem. Commun.*, p. 1387 (1968).

[27] T. Murachi and K. Okumura, *FEBS Lett.* **40**, 127 (1974).

[28] T. Murachi, T. Tsudzuki, and K. Okumura, *Biochemistry* **14**, 249 (1975).

[29] S. Iida, M. Sasaki, and S. Ota, *J. Biochem. (Tokyo)* **73**, 377 (1973).

[30] M. Sasaki, T. Kato, and S. Iida, *J. Biochem. (Tokyo)* **74**, 635 (1973).

[31] T. Kato and M. Sasaki, *J. Biochem. (Tokyo)* **76**, 1021 (1974).

Activators, Inhibitors, and Chemical Modifications. In addition to the reactions already reviewed in Vol. 19,[3] the following should be considered.

The tyrosyl residues in stem bromelain, which are in "exposed" state, and hence readily accessible to the solvent, can be acetylated with N-acetylimidazole at pH 7.5 or nitrated with tetranitromethane at pH 8.0 without change in catalytic activities.[32] Photosensitized oxidation of stem bromelain in the presence of methylene blue results in partial loss of the enzymic activity even when the essential sulfhydryl group is protected from the oxidation. The photooxidation involves histidyl, methionyl and tryptophanyl residues.[27,28]

Rabbit anti-stem bromelain antibodies inhibit the catalytic activity of stem bromelain. The inhibition occurs more strongly when casein is used as the substrate than with N^α-benzoyl-L-arginine ethyl ester (BAEE).[33] Stem bromelain cross-reacts with rabbit anti-fruit bromelain, anti-papain, and anti-ficin antibodies to varying degrees resulting in partial loss of its enzymic actvity.[29-31] Human plasma inhibits the hydrolysis of casein by stem bromelain.[34] α_2-Macroglobulin shows inhibitory effects on various proteolytic enzymes including stem bromelain, but α_1-antitrypsin does not inhibit bromelain.[35] The occurrence of several inhibitor proteins, which bind stem bromelain, in acetone powder of pineapple juice has recently been described.[36] One of these inhibitors has a molecular weight of 5600 and is composed of two peptide chains with an interchain disulfide. The amino acid sequences of these two chains have been determined.[37]

The preparation and some properties of stem bromelain covalently attached to O-carboxymethyl cellulose have been described.[25]

Specificity, Kinetic Properties, and Enzymic Mechanism. Substrate specificity and kinetic data for stem bromelain were summarized in Table IV of Chapter 18 in Vol. 19,[3] concerning the hydrolysis of amino acid esters and amides. A further significant finding is the fact that the enzyme seems to be able to interact with two or more sequential amino acids in a peptide substrate is demonstrated by using N-benzyloxycarbonyl-L-phenylalanyl-L-serine methyl ester,[16] with a $K_{m(app)}$ of 0.53 mM and k_{cat} of 3.4 sec^{-1}.

---

[32] K. Goto, N. Takahashi, and T. Murachi, *J. Biochem.* (*Tokyo*) **70**, 157 (1971).

[33] M. Sasaki, S. Iida, and T. Murachi, *J. Biochem.* (*Tokyo*) **73**, 367 (1973).

[34] M. Sasaki, H. Yamamoto, S. Iida, and S. Kimura, *Nagoya Med. J.* **17**, 49 (1972).

[35] M. Sasaki, H. Yamamoto, H. Yamamoto, and S. Iida, *J. Biochem.* (*Tokyo*) **75**, 171 (1974).

[36] S. H. Perlstein and F. J. Kézdy, *J. Supramol. Struct.* **1**, 249 (1972).

[37] M. N. Reddy, P. S. Keim, R. L. Heinrikson, and F. J. Kézdy, *J. Biol. Chem.* **250**, 1741 (1975).

With BAEE or N^α-benzoyl-L-arginine amide (BAA) as a substrate, the pH profile of $K_{m(app)}/k_{cat}$ shows a wide plateau in the range from pH 5 to pH 8.[38] This is in general agreement with the pH-profile of kinetic parameters for papain[39,40] and ficin.[41,42] The pH optimum for denatured hemoglobin is around pH 5.[43]

Evidence is given for an S-acyl-enzyme intermediate by observing an absorption band having a maximal intensity at 316 nm after admixture of bromelain and methyl thionohippurate.[44] In the acylation step, a controversial, i.e., either a proton-donating or a proton-withdrawing, function is assigned to the imidazole group of the unique histidyl residue that is spatially very close to the sulfhydryl group.[45] From the results of the photooxidation experiments, however, an entirely different mechanism has recently been proposed in which the histidyl residue may not have intimate electronic interaction with the substrate molecule during catalysis.[28]

Fruit Bromelain

Methods of assay and preparation of the acetone powder of this enzyme from the fruit juice have been described in detail in Vol. 19.[3] Here, only additional steps of purification and recently available information concerning the properties of the purified enzyme will be given.

Purification

Fractionation Procedure[14] Five grams of the acetone powder is dissolved in 30 ml of 0.02 M sodium phosphate buffer, pH 7.2, containing 0.5 mM phenylmercuric acetate. After centrifugation, the supernatant solution is applied to a 5×20 cm column of DEAE-cellulose (0.96 mEq/g), which has been equilibrated with 0.02 M sodium phosphate buffer, pH 7.2, containing 0.5 mM phenylmercuric acetate. The column is washed with 1 liter of the same buffer. The adsorbed enzyme is eluted from the column using two-convex-gradient system, 0.06 M– 0.3 M–0.5 M sodium phosphate buffer, pH 7.2, containing 0.5 mM

[38] T. Inagami and T. Murachi, *Biochemistry* 2, 1439 (1963).
[39] E. L. Smith and M. J. Parker, *J. Biol. Chem.* 233, 1387 (1958).
[40] J. R. Whitaker and M. L. Bender, *J. Am. Chem. Soc.* 87, 2728 (1965).
[41] L. A. AE. Sluyterman, *Biochim. Biophys. Acta* 85, 305 (1964).
[42] B. R. Hammond and H. Gutfreund, *Biochem. J.* 63, 61 (1956).
[43] F. Yamada, N. Takahashi, and T. Murachi, *J. Biochem. (Tokyo)* 79, in press.
[44] K. Brocklehurst, E. M. Crook, and C. W. Wharton, *Chem. Commun.* p. 1185 (1967).
[45] D. M. Blow and T. A. Steitz, *Annu. Rev. Biochem.* 39, 63 (1970).

phenylmercuric acetate. The fractions that represent the major portion of the proteolytic activity applied are pooled, concentrated, and rechromatographed on DEAE-cellulose at pH 7.2. The yield of the final product is 6.4%. Besides the main component, two minor proteolytically active components may be isolated during the course of the fractionation with yields of 0.3 and 0.5%.

Alternative Purification Procedure.[43] After the first DEAE-cellulose chromatography at pH 7.2, fruit bromelain may be further chromatographed on ECTEOLA-cellulose (0.45 mEq/g) at pH 7.2. Rechromatography on ECTEOLA-cellulose yields 260–300 mg of the purified material from 10 g of acetone powder of the juice of pineapple fruit. The product gives a single precipitation arc to anticrude fruit bromelain antibody upon immunoelectrophoresis at pH 8.6.

Properties

Unlike stem bromelain, the main fruit enzyme is an acidic protein.[17] Its isoelectric point is pH 4.6 as determined by isoelectric focusing technique.[43] The amino-terminal residue is alanine,[14] with sequence[43]:

Ala-Val-Pro-Gln-Ser-Ile-Asp-Trp-Arg-Asp-Tyr-Gly-Ala

The principal carboxyl-terminal residue is glycine.[14] The reported amino acid compositions of fruit bromelain preparations are shown in the table. One preparation is reported to contain carbohydrate that cannot be removed by purification procedures used thus far,[14] and another preparation obtained by different investigators has neither hexosamine nor neutral sugars.[43] The molecular weight of fruit bromelain is still a subject to controversy: 18,000 as determined by Sephadex G-75 gel filtration[14] and 31,000 by polyacrylamide gel electrophoresis in the presence of SDS and by Sephadex G-75 gel filtration. [43] The fruit enzyme is more active against BAA and BAEE than the stem enzyme. The following kinetic parameters are reported: for BAA[14] $K_{m(app)} = 4.0$ mM and $k_{cat} = 0.033$ sec^{-1} at pH 6.0 and 25°, and for BAEE[43] $K_{m(app)} = 0.043$ M at pH 6.0 and 25°. The pH optima for casein and denatured hemoglobin are pH 8.3 and pH 8.0, respectively.[43] Fruit bromelain catalyzes synthesis of acylamino acid anilides.[46] The enzyme is inhibited by mercurials, and activity is restored by cysteine or mercaptoethanol. The amino acid sequence around the reactive cysteinyl residue (Cys) is Asn-Glx-Asn-Pro-Cys-Gly-Ala-Cys.[43] Fruit bromelain cleaves bradykinin at either

[46] S. Ota, T. Fu, and R. Hirohata, *J. Biochem. (Tokyo)* **49**, 532 (1961).

Gly(4)-Phe(5) or Phe(5)-Ser(6) bond at comparable rates, whereas it splits angiotensin II almost exclusively at Tyr(4)-Ile(5) bond.[43] Rabbit antifruit bromelain antibodies inhibit the catalytic activity of fruit bromelain, and they also cross-react with stem bromelain.[29,31]

[40] Peptidoglutaminase (*Bacillus circulans*)[1]

By Mamoru Kikuchi and Kenji Sakaguchi

Peptidoglutaminase catalyzes the deamidation of the γ-carboxyamide of peptide-bound glutamine.[2] There are two peptidoglutaminases in *B. circulans* cells. The two enzymes, designated peptidoglutaminase I and peptidoglutaminase II, can be separated by DEAE-Sephadex chromatography.[3] The former catalyzes reaction (1) and the latter reaction (2) preferentially.[4]

$$
\begin{array}{c}
\text{CONH}_2 \\
|\\
(\text{CH}_2)_2 \\
|\\
\text{R—NH—CH—COOH}
\end{array}
+ \text{H}_2\text{O} \longrightarrow
\begin{array}{c}
\text{COOH} \\
|\\
(\text{CH}_2)_2 \\
|\\
\text{R—NH—CH—COOH}
\end{array}
+ \text{NH}_3 \quad (1)
$$

$$
\begin{array}{c}
\text{CONH}_2 \\
|\\
(\text{CH}_2)_2 \\
|\\
\text{R—NH—CH—COR}
\end{array}
+ \text{H}_2\text{O} \longrightarrow
\begin{array}{c}
\text{COOH} \\
|\\
(\text{CH}_2)_2 \\
|\\
\text{R—NH—CH—COR}
\end{array}
+ \text{NH}_3 \quad (2)
$$

Assay Method

Principle. Enzyme activity is measured by determination of ammonia formed during hydrolysis at the γ-carboxyamide of carbobenzoxy (CBZ)-L-glutamine for peptidoglutaminase I or *tert*-amyloxycarbonyl (*t*-AOC)-L-glutaminyl-L-proline for peptidoglutaminase II. The assay procedure described below is based on the direct nesslerization of ammonia.

Reagents

Sodium phosphate buffer, 0.05 M, pH 7.5

CBZ-L-glutamine or *t*-AOC-L-glutaminyl-L-proline (Peptide Insti-

[1] EC class 3.5.1.

[2] M. Kikuchi, H. Hayashida, E. Nakano, and K. Sakaguchi, *Biochemistry* **10**, 1222 (1971).

[3] M. Kikuchi and K. Sakaguchi, *Agric. Biol. Chem* **37**, 827 (1973).

[4] M. Kikuchi and K. Sakaguchi, *Agric. Biol. Chem* **37**, 1813 (1973).

tute, Protein Research Foundation, Mino-shi, Osaka, Japan), 0.2 M, adjusted to pH 6.0 with NaOH

Trichloroacetic acid (TCA), 50%

Nessler's reagent (Daiichi Pure Chemicals Co, Ltd. Tokyo)

Procedure. The reaction mixture has a total volume of 1.0 ml and contains the enzyme sample, 40 μmoles of sodium phosphate buffer and 10 μmoles of the substrate peptide. After a reaction time of 10–30 min at 30°, the reaction is stopped by the addition of 0.1 ml of TCA solution and centrifuged. The mixture is quantitatively transferred to a 10-ml graduated test tube and diluted to 9.0 ml with water, and 1.0 ml of Nessler's reagent is added. Ammonia is estimated by determining optical density at 420 nm. Enzyme and substrate blanks are included in all tests, and a standard curve is prepared with ammonium sulfate.

Definition of Unit and Specific Activity. An unit of activity is defined as the amount of enzyme that catalyzes the formation of 1 μmole of ammonia per minute under conditions of the assay. Specific activity is expressed as units per milligram of protein. Protein concentration is calculated from the biuret reaction during the early stages of purification (step 1 to step 3) and routinely by absorbance at 280 nm.

Purification Procedure[3]

Growth of Cells. Bacillus circulans ATCC 21590 is grown in a medium. composed of 1.0% polypeptone, 0.5% lactose, 0.025% $MgSO_4 \cdot 7H_2O$, 0.005% $FeSO_4 \cdot 7H_2O$, 0.025% KH_2PO_4, and 0.17% $Na_2HPO_4 \cdot 12H_2O$,[5] with the pH adjusted to 7.2. The cultures are grown in a fermentor tank of 200-liter capacity containing 150 liters of the medium at 30° for 20 hr with aeration (150 liters of air per minute) and with stirring of 150 rpm. The cells are harvested by centrifugation in a Sharples continuous centrifuge at 15,000 rpm, washed twice with 0.85% sodium chloride, and centrifuged in a Spinco Model L-2 preparative ultracentrifuge at 50,000 g for 30 min. The usual yield of cells is approximately 1.6 g (dry weight) per liter of medium.

Preparation of Cell-Free Extracts. One hundred fifty grams (dry weight) of *B. circulans* are suspended in 0.01 M phosphate buffer, pH 8.0 (buffer 1) to a final volume of 5 liters, then ruptured in a Manton-Gaulin laboratory Model 15M-8TBA homogenizer at an operating pressure of 7000 psi three times. The ruptured suspension is clarified by centrifugation at 50,000 g for 90 min. The resulting supernatant fluid

[5] M. Kikuchi and K. Sakaguchi, *Agric. Biol. Chem* **37**, 719 (1973).

(crude extract) is dialyzed overnight against 0.005 M phosphate buffer, pH 7.2 (buffer 2) at 5°. All subsequent operations are performed at 0 ~ 5°.

Step 1. Treatment with Streptomycin Sulfate. To the crude extract fraction is added 250 ml of 20% streptomycin sulfate solution (Kyowa Hakko Kogyo Co., Ltd. Tokyo). The mixture is stirred for 30 min, then centrifuged at 15,000 g for 30 min.

Step 2. Ammonium Sulfate Fractionation. Solid ammonium sulfate is added to the supernatant fraction obtained in step 1 to 0.35 saturation, the pH of the solution being maintained at 7.0 with 1 N NaOH. The solution is stirred for 30 min, then precipitated proteins are removed by centrifugation at 50,000 g using a Spinco Model L-2 ultracentrifuge. The supernatant fraction is brought to 0.5 saturation of ammonium sulfate and stirred for 30 min. The protein precipitate is collected by centrifugation at 15,000 g for 15 min, dissolved in buffer 1, and dialyzed against 10 liters of buffer 1 four times.

Step 3. DEAE-Sephadex Column Chromatography. The dialyzed solution, to which KCl is added to a final concentration of 0.1 M, is applied to a DEAE-Sephadex A-50 column (3 × 80 cm) previously equilibrated with buffer 1 containing 0.1 M KCl. The column is washed with buffer 1 containing 0.12 M KCl until the optical density at 280 nm of the effluent drops below 2.0; it is then subjected to linear gradient elution with buffer 1 containing 0.12–0.5 M KCl. Peptidoglutaminases I and II are eluted separately from this column, as shown in Fig. 1. Fractions representing each enzyme are pooled, and are futher concentrated by ammonium sulfate in the manner described in step 2. The proteins precipitating between 0.35 and 0.5 saturation are dissolved in

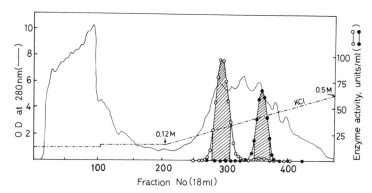

FIG. 1. Fractionation of peptidoglutaminase I and II on a DEAE-Sephadex A-50 column (3 × 80 cm). ——, protein concentration, O——O, peptidoglutaminase I activity; ●——●, peptidoglutaminase II activity.

minimum amounts of 0.01 M phosphate buffer, pH 6.0 (buffer 3) containing 0.1 M KCl.

Step 4. Gel Filtration with Sephadex G-200. Enzyme solutions I and II from step 3 are applied separately to Sephadex G-200 columns (4 × 90 cm) previously equilibrated with buffer 3 containing 0.1 M KCl and are eluted with the same buffer.

Step 5. QAE-Sephadex Column Chromatography. The active fractions from step 4 are pooled and loaded on top of a QAE-Sephadex column (2 × 40 cm) equilibrated with buffer 3 containing 0.08 M KCl. A linear gradient elution for peptidoglutaminase I is carried out using 600 ml of buffer 3 containing 0.1 M KCl in the mixing chamber and 600 ml of buffer 3 containing 0.4 M KCl in the reservoir. For peptidoglutaminase II, a linear gradient elution is performed using 500 ml of 0.2 M KCl in the mixing chamber and 500 ml of buffer 3 containing 0.5 M KCl in the reservoir. Active fractions from each enzyme are collected and dialyzed against 5 liters of 0.01 M phosphate buffer, pH 7.0 (buffer 4) with three changes.

Step 6. Hydroxyapatite Column Chromatography. The enzyme solutions obtained in step 5 are chromatographed on a hydroxyapatite column (2 × 15 cm) using linear gradient elution. The hydroxyapatite column is prepared by mixing calcium hydroxyapatite suspension (Bio-Rad Laboratories) with cellulose powder (Toyo-Roshi, 2:1, v/w). After the washing with about 50 ml of buffer 4, the enzyme solutions are applied to the column. A linear gradient is established between 0.01 (250 ml) and 0.1 M (250 ml) potassium phosphate buffer (pH 7.0). Fractions having peptidoglutaminase I activity are pooled, concentrated with a collodion bag (Sartorius Membrane filter GmbH), and dialyzed against 0.05 M Tris-glycine buffer, pH 8.9. On the other hand, the pooled eluate of peptidoglutaminase II is made 0.1 M in KCl, and the pH of this solution is adjusted to 9.5 with NaOH in order to prevent aggregation of this enzyme. Then the enzyme is concentrated in a collodion bag and dialyzed against 0.05 M Tris-glycine buffer, pH 8.9.

Step 7. Electrophoresis on Polyacrylamide Gel. The final purification for both proteins is performed by preparative electrophoresis on polyacrylamide gel with a Mitsumi Preparative disc equipment. The polyacrylamide gel column is prepared by casting a separating gel (7.5%, 5 cm height) and stacking gel (3.5%, 1 cm height). The dialyzed enzyme solution from step 6 is made 3% in sucrose and applied on the top of the stacking gel with a microtube pump. Electrophoresis is carried out for 12 hr at 50 mA for peptidoglutaminase I and at 40 mA for peptidoglutaminase II, eluting with 0.1 M Tris-HCl buffer, pH 8.1, at 0 ~ 4°. Fractions with peptidoglutaminase I or II activities are collected.

The method, summarized in Table I, results in a 714-fold overall purification with about 13% recovery of activity for peptidoglutaminase I and a 223-fold purification with about 25% recovery for peptidoglutaminase II. Both enzymes can be stored for several months at $-20°$ without significant loss of activity.

Some Criteria of Purity and Physical Properties

Both enzymes exhibit single symmetrical peaks by ultracentrifugation at 55,430 rpm, and are homogeneous by polyacrylamide disc gel electrophoresis at pH 9.4 and by immunodiffusion. Ultraviolet absorption spectra in buffer 1 containing 0.1 M KCl show a maximum absorption at 278 nm for peptidoglutaminase I and 280 nm for II. E_{280}/E_{260} ratios are calculated to be 2.04 and 1.88 for I and II, and $E_{1\text{ cm}}^{1\%}$ at 280 nm values of 7.27 and 11.62 are obtained, respectively. By gel filtration,[3] the molecular weight of peptidoglutaminase I is estimated to be approximately 90,000 and that of peptidoglutaminase II approximately 125,000.

Specificity and Kinetics.[4] Table II summarizes the available information relating to substrate specificity in terms of the $V_{\text{max}}/K_{\text{m}}$ index deduced from Michaelis–Menten kinetics. Peptidoglutaminase I specifically deamidates the γ-amide of L-glutamine residing at the carboxyl terminal of peptides and is inactive toward glutamine derivatives substituted in both α-amino and α-carboxyl groups. On the other hand, the best substrates for peptidoglutaminase II are tripeptides of the X-Gln-Y type, where X is a carbobenzoxy, t-amyloxycarbonyl-, or amino acid residue, and Y is some amino acid. Although L-glutamines in higher peptides are poor substrates for this enzyme, two L-glutamines in the oxidized insulin A chain are attacked. The high specificity of the enzyme allows the quantitative determination of peptide- or protein-bound glutamine.[6]

The enzymes show slight activity toward acetamide and proprionamide, but are inactive to crotonamide, L-citrulline, L-Leu-amide, and L-Phe-Gly-amide. A very slow liberation of hydrazine from γ-glutamylhydrazide occurs with peptidoglutaminase I. When hydroxylamine is present during reaction with glutaminylpeptides, hydroxamate formation can be detected with either enzyme, but in several systems tested[2] it has not been possible to demonstrate a real transamidase activity for the purified enzymes. Peptidoglutaminase I and II are inactive toward asparagine and its derivatives: CBZ-L-asparagine, Gly-L-Asn, CBZ-L-Ala-L-Asn, t-AOC-L-Asn-L-Pro, L-Phe-L-Asn-Gly, L-Leu-L-Asn-Gly, L-Lys-L-Asn-Gly, L-Ala-L-Asn-L-Phe and L-Glu L-Asn-L-Phe.

[6] M. Kikuchi, *Anal Biochem.* 59, 83 (1974).

TABLE I

PURIFICATION OF PEPTIDOGLUTAMINASE I AND II

Step	Peptidoglutaminase I				Peptidoglutaminase II			
	Total activity (units)	Total protein (mg)	Specific activity	Yield (%)	Total activity (units)	Total protein (mg)	Specific activity	Yield (%)
Crude extract	42,646	100,580	0.424	100	35,203	100,580	0.35	100
1. Streptomycin	41,200	80,464	0.512	96.6	34,599	80,464	0.43	98.2
2. Ammonium sulfate	34,117	40,000	0.82	80.0	33,088	40,000	0.82	93.9
3. DEAE-Sephadex	20,390	2,614	7.80	47.8	20,023	2,484	8.06	56.8
4. Sephadex G-200	15,400	811.3	18.90	36.1	13,090	341.7	38.30	37.1
5. QAE-Sephadex	10,460	72.6	144.00	24.5	12,499	227.3	54.90	34.3
6. Hydroxyapatite	8,680	34.5	251.5	20.3	8,885	113.9	78.0	25.2
7. Electrophoresis[a]	5,605	18.5	303.0	13.1	1,201[a]	15.4	78.0	

[a] About 20 mg of hydroxyapatite fraction is applied to preparative disc electrophoresis.

TABLE II

K_m, V_{max}, AND SUBSTRATE SPECIFICITY (V_{max}/K_m) FOR PEPTIDOGLUTAMINASE I AND II[a]

Substrates	Peptidoglutaminase I			Peptidoglutaminase II		
	K_m (mM)	V_{max} (rel.)	V_{max}/K_m	K_m (mM)	V_{max} (rel.)	V_{max}/K_m
CBZ-L-Gln	0.11	100	909	22.2	5	0.22
CBZ-DL-Gln	0.21	96	457	22.2	3	0.14
CBZ-D-Gln	1.56	4	2.6	—	—	—
N-Acetyl-L-Gln	0.10	100	1000	14.3	1.3	0.09
L-Tyr-L-Gln	0.16	107	669	10.0	1.5	0.15
L-Pro-L-Gln	0.19	79	416	22.2	5	0.22
Gly-L-Gln	0.33	115	348	18.2	1	0.05
L-Leu-Gly-L-Gln	0.35	130	371	23.3	1	0.04
L-Pro-L-Leu-Gly-L-Gln	0.33	119	360	—	—	—
Gly-L-Pro-L-Leu-Gly-L-Gln	0.50	120	240			
L-Gln	5.00	34	6.8	50.00	21	0.42
L-Gln-Gly		<0.1		0.66	71	108
CBZ-L-Gln-L-Pro		0		0.21	133	633
t-AOC-L-Gln-L-Pro		0		0.12	100	833
t-AOC-L-Gln-L-Leu-Gly		0		16.70	27	1.62
CBZ-L-Gln-Gly		<0.1		0.12	114	950
CBZ-L-Gln-Gly-OMe		<0.1		1.25	133	106
L-Phe-L-Gln-Gly		<0.02		0.12	77	642
L-Ser-L-Gln-Gly		<0.02		0.12	81	675
L-Lys-L-Gln-Gly		<0.02		0.09	63	700
Pyr-L-Gln-L-Ala		<0.02		0.10	65	650
L-His-L-Ser-L-Gln-Gly-L-Thr-L-Phe-L-Thr		0		50.00	11	0.22
L-Ile-L-Gln-L-Asn-CysH-L-Pro-L-Leu-Gly-NH₂ acetate		0		44.00	6	0.14

[a] V_{max} is calculated using the value of CBZ-L-glutamine for peptidoglutaminase I and t-AOC-L-glutaminyl-L-proline for peptidoglutaminase II as a control (CBZ, carbobenzoxy; t-AOC, tert-amyloxycarbonyl).

pH and Temperature Optimum. The optimum pH for deamidation of CBZ-L-glutamine by peptidoglutaminase I is about 8.0 using Tris-HCl buffer. Peptidoglutaminase II has a broad pH rate profile with an optimum pH about 8.0. Temperature optima occur at about 55° and 50° for peptidoglutaminase I and II, respectively.

Activators and Inhibitors. No appreciable activation or inactivation can be seen after adding metal salts other than $HgCl_2$ or $AgNO_3$ (10 mM). Both enzymes are completely inactivated by 1 mM N-bromosuccimide and 1 ∼ 10 mM sodium lauryl sulfate.

Section VIII

Exopeptidases

A. Aminopeptidases
Articles 41 through 46

B. Carboxypeptidases
Articles 47 through 51

[41] Human Liver Aminopeptidase

By Gwynne H. Little, Willis L. Starnes, and Francis J. Behal

Human liver aminopeptidase (HLA) catalyzes the hydrolysis of N-terminal amino acid residues from peptides, amino acid amides, or certain chromogenic synthetic substrates. The rate of hydrolysis is maximum when the residue is L-alanine, but certain other amino acids with nonpolar R groups, e.g., L-leucine, are hydrolyzed at significant rates. Since one of the first commonly available synthetic substrates for aminopeptidases was L-leucyl-β-naphthylamide, it was initially assumed by many investigators that HLA activity could be attributed to leucine aminopeptidase (LAP),[1] but it is now clear that HLA is chemically distinct from LAP. HLA catalyzes the general reaction:

$$RCH(N^+H_3)CONHR' + H_2O \rightarrow RCH(N^+H_3)COO^- + H_3N^+R'$$

$R'N^+H_3$ may be ammonium ion, an amino acid, an oligopeptide, β-naphthylamine, or p-nitroaniline. The relative rates of hydrolysis of several substrates are shown in Table I. HLA must be assayed with a variety of substrates and under a variety of conditions, depending on

TABLE I
SUBSTRATE SPECIFICITY OF HUMAN LIVER AMINOPEPTIDASE

Aminoacyl-β-naphthylamides	Relative rate (%)	Dipeptides	Relative rate (%)
L-Alanyl-β-naphthylamide	100	L-Ala-L-Trp	100
L-Phenylalanyl-β-naphthylamide	62.8	L-Ala-L-Val	47.3
L-Methionyl-β-naphthylamide	57.5	L-Ala-L-Ala	45.4
L-Leucyl-β-naphthylamide	36.4	L-Ala-L-Phe	42.8
L-Arginyl-β-naphthylamide	28.0	L-Leu-L-Leu	35.9
L-Tryptophanyl-β-naphthylamide	19.2	L-Ala-L-Leu	27.1
Glycyl-β-naphthylamide	13.7	L-Leu-L-Ala	25.6
L-Lysyl-β-naphthylamide	13.4	L-Ala-L-Ser	22.5
L-Seryl-β-naphthylamide	6.0	L-Phe-Gly	19.2
L-Threonyl-β-naphthylamide	5.7	L-Ala-L-Glu	19.0
L-Glutamyl-β-naphthylamide	3.7	L-Ala-Gly	17.8
L-Valyl-β-naphthylamide	2.8	L-Ala-L-Asp	11.4
L-Isoleucyl-β-naphthylamide	2.8	L-Leu-Gly	5.5
		L-Ile-Gly	1.5
		L-Ser-Gly	1.3

[1] See this series Vol. 19 [33].

the state of purity of the enzyme and the specific results desired. The various applicable assay methods necessary for detailed work with this enzyme are given below. These are based, first, on the substrate used, and then on the various instrumental methods available.

Assay Methods

Assay with Aminoacyl-β-naphthylamide Substrates

Colorimetric Method[2]

The release of β-naphthylamine from aminoacyl-β-naphthylamides is determined by diazotization after a discrete incubation period. This method is useful for following the progress of purification and for laboratories without a recording spectrophotometer or fluorometer.

Reagents

Phosphate buffer: 0.1 M, pH 6.8
Substrate: 3.0 mM L-alanyl-β-naphthylamide, pH 6.8
Assay mixture: 0.9 ml of buffer plus 0.5 ml of buffered substrate
β-Naphthylamine standard: 0.06 mM β-naphthylamine
Trichloroacetic acid (TCA), 2.5 M
Sodium nitrite, 15 mM
Ammonium sulfamate, 45 mM
N-(1-Naphthyl)ethylenediamine dihydrochloride, 2 mM in 95% ethanol

Procedure. Begin the reaction by adding 0.1 ml of the enzyme solution to the assay mixture at zero time and incubate with a blank (which is identical except that it lacks the enzyme solution) for 10–30 min at 37°. At the end of the incubation period add 0.5 ml of 2.5 M TCA to the sample to stop the reaction. To the blank add 0.5 ml of 2.5 M TCA solution followed immediately by 0.1 ml of enzyme.

To develop the color add 1.0 ml of sodium nitrite; after 3 min add 1.0 ml of amonium sulfamate; after 2 min add 2.0 ml of N-(1 naphthyl)-ethylenediamine. Allow the color to develop for 30 min, then read the absorbance of the blank and sample against water at 580 nm. The quantity of β-naphthylamine released is determined from the absorbance of the sample relative to that of a standard solution prepared by substituting 0.06 mM β-naphthylamine for substrate in the assay mixture.

[2] J. A. Goldbarg and A. M. Rutenburg, *Cancer* **11**, 283 (1958).

Fluorometric Method[3]

Free β-naphthylamine is intensely fluorescent whereas aminoacyl-β-naphthylamides are only slightly fluorescent. This fact permits continuous monitoring of the progress of hydrolysis by means of a recording fluorometer or spectrofluorometer. The fluorometric assay is used for kinetic studies where maximum sensitivity is desired.

Reagents

Phosphate buffer, 0.1 M, pH 6.8
Substrate: 3.0 mM L-alanyl-β-naphthylamide, pH 6.8
Assay mixture: 1.9 ml of buffer plus 1.0 ml of substrate (volumes may have to be adjusted for various-sized cuvettes)
β-Naphthylamine standard, 0.01 mM: assay mixture plus 0.1 ml of 0.3 mM β-naphthylamine

Procedure. Place the cuvette containing the assay mixture in the fluorometer and allow the temperature to equilibrate to 37°. Start the reaction, and record the rate of change in fluorescence at 410 nm relative to the β-naphthylamine standard. The excitation wavelength is 340 nm.

Spectrophotometric Method[4]

This assay is useful for kinetic work in the presence of compounds that interfere with the fluorometric assay. It lacks the sensitivity of the fluorometric assay, however.

Reagents

Phosphate buffer, 0.1 M, pH 6.8
Substrate: 3.0 mM L-alanyl-β-naphthylamide, pH 6.8
Assay mixture: 1.9 ml of buffer plus 1.0 ml of substrate

Procedure. Pipette the reaction mixture into a 3-ml quartz cuvette and allow the temperature to equilibrate to 37°. Begin the reaction by adding 0.1 ml of enzyme solution, and record the rate of increase in absorbance at 340 nm. The slope of the resulting curve is proportional to the amount of β-naphthylamine released per unit of time. The amount may be calculated from a standard curve or from the measured extinction coefficient of the difference spectrum between L-alanyl-β-naphthylamide and β-naphthylamine at 340 nm under these conditions.

[3] L. J. Greenberg, *Biochem. Biophys. Res. Commun.* 9, 430 (1962).
[4] H. Lee, J. N. LaRue, and I. B. Wilson, *Anal. Biochem.* 41, 397 (1971).

Assay with Aminoacyl-p-nitroanilide Substrates

Spectrophotometric Method[5]

This assay is more sensitive than the spectrophotometric assay for β-naphthylamides owing to the higher extinction coefficient of p-nitroaniline as compared to β-naphthylamine.

Reagents

Phosphate buffer: 0.1 M, pH 6.8
Substrate: 3.0 mM L-alanyl-p-nitroanilide, pH 6.8
Assay mixture: 1.9 ml of buffer plus 1.0 ml of substrate

Procedure. Same as for β-naphthylamides except that the rate of increase in absorbance is recorded at the absorption maximum of p-nitroaniline, 405 nm.

Assay with Dipeptide and Amino Acid Amide Substrates

Spectrophotometric Method

Reagents

Phosphate buffer: 0.1 M, pH 6.8
Substrate: nonaromatic amides or dipeptides, 3.0 mM, pH 6.8
Assay mixture: 1.9 ml of buffer plus 1.0 ml of substrate

Procedure. Begin the reaction by adding 0.1 ml of enzyme to the assay mixture in a 3-ml quartz cuvette and record the decrease in absorbance at 220 nm maintaining the temperture at 37°. The slope of the line is proportional to the amount of hydrolysis of the amide or peptide bond, and the rate in micromoles per minute may be determined from a standard curve prepared by reading the absorbance at 220 nm of mixtures of substrate and products simulating varying degrees of hydrolysis.

Colorimetric Ninhydrin Method

This method is based on the reaction with ninhydrin of α-amino groups or ammonia produced by hydrolysis of dipeptides or amino acid amides. The reaction can be carried out manually as described below or by means of an automatic amino acid analyzer, in which case the

[5] H. Tuppy, W. Wiesbauer, and E. Wintersberger, *Hoppe-Seyler's Z. Physiol. Chem.* **329**, 278 (1962).

assay mixture is set up in test tubes as in the spectrophotometric method above. After 30 min incubation at 37°, the reaction is stopped by placing the tubes in a boiling water bath for 10 min. The sample is then prepared for amino acid analysis by standard techniques. The automated procedure has the advantage of greater sensitivity since the products are separated from the substrate, which is also ninhydrin positive. The manual procedure is carried out as follows.[6]

Reagents

Phosphate buffer. 0.01 M, pH 6.8
Substrate: 3.0 mM dipeptide, pH 6.8
Assay mixture: 0.9 ml of buffer plus 0.5 ml of substrate
Picric acid, 0.04 M
Citrate-methyl Cellosolve diluent: equal volumes of 0.2 M citrate buffer, pH 5.0, and methyl Cellosolve
Ninhydrin reagent: prepared by mixing 1 volume of 0.2 M citrate buffer, pH 5.0, containing 1.6 mg of stannous chloride (SnCl$_2$· 2H$_2$O) per milliliter, with 1 volume of methyl Cellosolve, containing 40 mg of ninhydrin per milliliter
Ethanol, 60%

Procedure. Prepare assay mixture in quadruplicate for each sample to be assayed. Preincubate the tubes at 37° for 5 min, then begin the reaction in two of the tubes by adding 0.1 ml of enzyme solution. The other two tubes are blanks. Incubate all four tubes 10–30 min then stop the reactions by adding 2.0 ml of the picric acid solution. To the blanks add 2.0 ml of picric acid solution followed immediately by 0.1 ml of enzyme solution.

To develop the color take duplicate 0.25-ml aliquots from each of the four tubes and add 0.75 ml of the citrate-methyl Cellosolve diluent and 1.25 ml of the ninhydrin solution. Heat in a boiling water bath for exactly 5 min and immediately cool in ice water. Add 8.5 ml of 60% ethanol, mix thoroughly, and read the absorbance against water at 570 nm.

Micromoles of substrate hydrolyzed can be determined from a standard curve prepared by running the ninhydrin reaction on a series of mixtures of the dipeptide and its component amino acids representing varying degrees of hydrolysis.

Definition of Aminopeptidase Unit and Specific Activity. One unit of aminopeptidase activity is defined as the amount of enzyme that will liberate 1 μmole of β-naphthylamine or of p-nitroaniline or hydrolyze

[6] G. A. Fleisher, M. Pankow, and C. Warmka, *Clin. Chim. Acta* **9**, 254 (1964).

TABLE II
PURIFICATION OF HUMAN LIVER AMINOPEPTIDASE

Isolation step	Total activity[a] (units)	Re-covery (%)	Total protein (mg)	Specific activity[a]	Purifi-cation factor
Homogenization and autolysis	1990	100	103,000	0.0193	1
Ammonium sulfate fractionation	2590	130	4,720	0.521	27
Gel filtration	2560	129	578	4.43	229
Anion-exchange chromatography	1090	55	63.7	17.1	887
Hydroxyapatite chromatography	824	41	18.0	45.7	2370

[a] Units of activity and specific activity are defined in the text.

1 μmole of an amino acid amide or dipeptide per minute under the conditions described. Specific activity is defined as units per milligram of protein.

Purification of Human Liver Aminopeptidase

HLA purified by the following procedure behaves as a single species on acrylamide gels[7,8] and in sedimentation equilibrium experiments in denaturing solvents.[8] The results obtained at each step are summarized in Table II. The colorimetric assay with L-alanyl-β-naphthylamide is normally used during the purification.

Homogenization and Autolysis. Human liver is obtained at autopsy from unembalmed cadavers and frozen at $-20°$ until used. Dissect about 1000 g of liver free of fat and connective tissue, cut into small pieces, cover with phosphate buffer (0.02 M, pH 6.86), and homogenize in a Waring Blendor or similar homogenizer until thoroughly disrupted. Add enough buffer to make the final volume 4 liters, then allow the homogenate to autolyze for 24 hr at $37°$ (autolysis for periods much longer than 24 hr results in loss of activity). After autolysis, the solid residue is removed by centrifugation. Washing of the pellet provides higher yields of enzyme, but the procedure is laborious.

Ammonium Sulfate Fractionation. Add solid ammonium sulfate to a final concentration of 1.0 M and then pour onto a 10-cm bed of porous

[7] F. J. Behal, G. H. Little, and R. A. Klein, *Biochem. Biophys. Acta* **178**, 118 (1969).
[8] W. L. Starnes and F. J. Behal, *Biochemistry* **13**, 3221 (1974).

glass (Bio-Glas 200, 100–200 mesh, Bio-Rad Laboratories) in a large filter funnel (2-liter capacity, coarse sintered glass) equilibrated with 1.0 M ammonium sulfate. Elute with 1.0 M ammonium sulfate until protein ceases to appear in the effluent, and then begin elution with water and collect fractions containing aminopeptidase activity and concentrate by ultrafiltration to a volume of 50 ml or less. The efficiency of later steps can be increased if the Bio-Glas step is repeated, permitting omission of the hydroxyapatite step. The Bio-Glas is regenerated by washing with 0.1 M NaOH to remove bound protein followed by several washes with water and reequilibration with 1.0 M $(NH_4)_2SO_4$.

Gel Filtration. Apply the concentrated material from the above step to a Pharmacia K 50/100 column packed with Sephadex G-200 equilibrated with 0.1 M sodium borate buffer at pH 7.5 containing 0.5 mole of NaCl per liter. Elute the column with approximately 2000 ml of the same buffer. Pool the fractions containing aminopeptidase activity and concentrate by ultrafiltration to a volume of approximately 25 ml.

Anion-Exchange Chromatography. Dialyze the ultrafiltrate against 0.005 M Tris buffer, pH 8.6, and apply it to a 2.5×69 cm column of DEAE-cellulose equilibrated with the same buffer. Elute with 2000 ml of linearly increasing NaCl gradient at pH 7.0 with an initial NaCl concentration of 0.01 M and a limiting concentration of 0.2 M. Pool the active fractions and concentrate by ultrafiltration to a volume of about 5 ml.

Hydroxyapatite Chromatography. Dialyze the ultrafiltrate against 0.001 M phosphate buffer at pH 7.0 and apply it to a 1.0×50 cm column of hydroxyapatite (Hypatite C: Clarkson Chemical Co.) in the same buffer. Elute the column by means of a linear 500-ml gradient of phosphate buffer at pH 7.0 with the concentration increasing from 0.01 to 0.05 M. Collect 5-ml fractions and combine and concentrate those containing aminopeptidase activity.

Properties of Human Liver Aminopeptidase

HLA was earlier confused with LAP and was so designated in the literature. HLA differs markedly from LAP in several aspects: HLA is a zinc metalloenzyme which also forms a cobalt metal–enzyme complex; HLA is far more active on leucyl-β-naphthylamide than on leucylglycine, whereas LAP is maximally active on leucylglycine; HLA is maximally active on substrates with L-alanyl-N-terminal residues, whereas LAP is more active on substrates with L-leucyl residues; the HLA pH optimum is 6.8, rather than 8.8–9.0.

HLA is one of several electrophoretically distinct "isoenzymes"

TABLE III
PHYSICAL PROPERTIES OF HUMAN LIVER AMINOPEPTIDASE

Property	Method	Value
Extinction coefficient, $(E_{280}^{0.1\%}\,nm)$	From dry weight and ultraviolet spectrum	1.75 ml-mg^{-1}
Sedimentation coefficient $(s_{20,w}^0)$	Sedimentation velocity	10.1×10^{-13} sec^{-1}
Diffusion coefficient $(D_{20,w}^0)$	Synthetic boundary diffusion	4.02×10^{-7} cm^2 sec^{-1}
Partial specific volume	From amino acid content	0.730 cm^3 g^{-1}
Molecular weight of native enzyme	High speed sedimentation equilibrium	$223,000 \pm 17,000$
	Sedimentation velocity	$235,000 \pm 10,000$
	Gel filtration	$242,000 \pm 21,000$
Subunit molecular weight	Sedimentation equilibrium in guanidinium chloride	$118,000 \pm 5,000$
	Sodium dodecyl sulfate gel electrophoresis	$120,000 \pm 12,000$

which occur in human blood. These have moderately broad specificity and exhibit maximum activity when the N-terminal amino acid residue is nonpolar. In contrast, other groups of aminopeptidases present in blood, liver, and other tissues are characterized by maximum activity when the N-terminal residue is acidic or basic. The other circulating "isoenzymes" of the HLA family are derived from pancreas, kidney, and duodenum.[9]

HLA is a sialoglycoprotein containing 17.46% carbohydrate (4.32% glucosamine, 4.14% sialic acid, 9.00% hexoses). The molecular weight of the enzyme in dilute salt solutions is approximately 235,000.[8,10] In denaturing solvents, HLA dissociates to a single species with molecular weight near 118,000 as determined by sedimentation equilibrium and SDS acrylamide gel electrophoresis in the presence and in the absence of 6 M urea. Earlier evidence obtained by agarose gel chromatography in guanidinium chloride indicates that the enzyme can further dissociate into smaller subunits having a molecular weight of roughly 38,000.[10] This result has not been confirmed by means of sedimentation equilibrium experiments, but it is not yet excluded since there are unique problems inherent in the analysis of the physical properties of glycoproteins.

[9] F. J. Behal, B. Asserson, F. Dawson, and J. Hardman, Arch Biochem. Biophys. 111, 335 (1965).
[10] G. H. Little and F. J. Behal, Biochim. Biophys Acta 243, 312 (1971).

TABLE IV
CHEMICAL COMPOSITION OF HUMAN LIVER AMINOPEPTIDASE

Constituent	Residues per 118,000 daltons	Constituent	Residues per 118,000 daltons
Amino acids		Amino acids (*continued*)	
Lysine	41	Methionine	16
Histidine	19	Isoleucine[d]	35
Arginine	35	Leucine	81
Tryptophan[a]	31	Tyrosine	42
Aspartic acid[b]	105	Phenylalanine	36
Threonine	54	Carbohydrates	
Serine	57	Glucosamine	29
Glutamic acid	96	Sialic acid(s)	16
Proline	44	Hexoses	62
Glycine	43	Metals	
Alanine	58	Zinc	1
Cysteine[c]	tr		
Valine[d]	53		

[a] Corrected to zero hydrolysis time.
[b] Free acid plus amide.
[c] As cysteic acid.
[d] Corrected to infinite hydrolysis time.

HLA contains 8.3 ± 1.5 nanoatoms of tightly bound zinc per milligram of protein (i.e., one zinc ion per 118,000 dalton subunit), is activated up to 2.4-fold by the addition of Co^{2+}, which binds loosely in a noncompetitive manner[11] and is inhibited by puromycin,[12] penicillin,[13] certain amino acids,[14] peptides,[14] carboxylic acids,[15] and aliphatic amines.[15] In general, amino acids are noncompetitive inhibitors of alanyl-β-naphthylamide hydrolysis with K_i values near 1.0 mM. Dipeptides generally are competitive inhibitors with significantly lower K_i values.

Some physical properties of HLA and the method by which they were obtained are summarized in Table III. The chemical composition of HLA is presented in Table IV.

[11] C. W. Garner and F. J. Behal, *Biochemistry* 13, 3227 (1974).
[12] G. H. Little, Ph.D. Thesis, Medical College of Georgia, 1970.
[13] Unpublished observation.
[14] C. W. Garner and F. J. Behal, *Biochemistry* 14, 3208 (1975).
[15] C. W. Garner, personal communication.

[42] Crystalline Leucine Aminopeptidase from Lens[1] (α-Aminoacyl-Peptide Hydrolase; EC 3.4.11.1)

By Horst Hanson and Marlies Frohne

Leucine aminopeptidase (LAP) is a Zn-enzyme localized in the cytosol and occurring in the tissues and body fluids of all organisms. LAP, crystallized at first from bovine eye lens,[1a] shows a total or partial immunological identity in different tissues of vertebrates (liver, kidney) and also in different vertebrate species. Immunological assay can establish a phylogenetic relationship between LAP from man, cattle, and frog spanning an evolutionary time of about 280 million years.[2] LAP crystallized from bovine eye lens is nearly identical with the enzyme from the cytosol of pig kidney.[3] However, it is clearly different in its substrate specificity and in immunological behavior from particle-bound enzyme (aminopeptidase EC 3.4.11.2) of the pig kidney and the enzyme described by Himmelhoch in this series (Vol. 19 [33]).[4]

Enzyme Preparation and Crystallization[5]

Preparation from Bovine Lens

Homogenate. Calf lenses are homogenized in 0.15 M NaCl solution (100 ml per 12 g of lenses).

Zinc-Heat Precipitation. One milliliter of 0.6 M ZnSO$_4$ and 1 ml of 0.7 N NaOH are added dropwise simultaneously to 98-ml lots of the homogenate under stirring. After this, the pH is held between 7.2 and 7.4 and the precipitate is removed by centrifugation. Pooled supernatants (as much as 500 ml) are stirred for 15 min at 54°, then cooled in an ice bath for about 30 min and clarified by filtration.

Crystallization. Ammonium sulfate, 0.34 g per milliliter of the clear filtrate is slowly added with stirring. Large-scale preparations are stored

---

[1] Dedicated to Professor K. Mothes on the occasion of his 75th birthday.

[1a] D. Glässer and H. Hanson, *Naturwissenschaften* **50**, 595 (1963).

[2] D. Glässer, M. John, and H. Hanson, *Hoppe-Seyler's Z. Physiol. Chem.* **351**, 1337 (1970).

[3] H. Hanson, D. Glässer, M. Ludewig, H. G. Mannsfeldt, M. John, and H. Nesvadba, *Hoppe-Seyler's Z. Physiol. Chem.* **348**, 689 (1967).

[4] S. R. Himmelhoch and E. A. Peterson, *Biochemistry* **7**, 2085 (1968).

[5] H. Hanson, D. Glässer, and H. Kirschke, *Hoppe-Seyler's Z. Physiol. Chem.* **340**, 107 (1965).

TABLE I

ISOLATION OF LEUCINE AMINOPEPTIDASE FROM BOVINE LENS

Fraction	Volume (ml)	Protein (mg/ml)	Hydrolysis of substrate[a] (% 20 min)	Purification factor	Yield (%)
Homogenate	150	33.44	7.42	1	100
Zn-heat-preparation	122	0.406	8.61	96	95
Crystallized enzyme	7.25	0.428	8.11	1172	68

[a] Leucinamide substrate, Mn^{2+} activated enzyme, 40°, and pH 9.2.

for 3–4 days at 4° in tall vessels, during which time the crystals separate out and most of the solution can be removed by aspiration before centrifuging. The crystals are collected by centrifugation at about 15,000 g for 20 min at 4°. The crystalline sediment is washed twice with small volumes of cold water and is taken up in 0.1 M Tris-HCl buffer of pH 8.0. Insoluble material is removed by centrifugation. The enzyme solution can be stored at 4° for years without loss of activity. For recrystallization, ammonium sulfate is added until a slight turbidity is formed; the amount of ammonium sulfate needed is dictated by the concentration of LAP.

After the first crystallization, the yield is 60–70% of the LAP activity. If not otherwise stated, all steps can be carried out at room temperature.

Isolation of LAP from other Sources with the Combined Zinc-Heat Treatment Procedures

Using the zinc-heat-precipitation as described above, LAP may be purified from pig lenses (78-fold), sheep lenses (50-fold), bovine liver and kidney (7-fold)—all with a yield of 75–80%. Using 0.01 M $ZnSO_4$ and a heating period of 40 min at 54°, the LAP from bovine kidney was purified 20-fold.[6]

Macrocrystals[6]

Macrocrystals of LAP up to a magnitude of 0.35 × 0.15 mm can be prepared according to the method of Zeppezauer and co-workers[7] by

[6] D. Glässer, U. Kettmann, and H. Hanson, *Hoppe-Seyler's Z. Physiol. Chem.* **351**, 1329 (1970).
[7] M. Zeppezauer, H. Eklund, and E. S. Zeppezauer, *Arch. Biochem. Biophys.* **126**, 564 (1968).

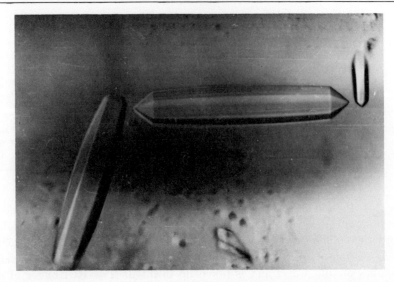

Fig. 1. Macrocrystals of leucine aminopeptidase prepared by equilibrium dialysis against ammonium sulfate solutions.

equilibrium dialysis against ammonium sulfate solutions of increasing concentrations (Fig. 1). The enzyme solution (concentration 4–10 mg/ml) is placed in a glass or X-ray capillary, which is sealed off with a diaphragm of 7% polyacrylamide. The concentration of ammonium sulfate solutions is increased continuously or discontinuously from 0.2 M to 1.0 M and pH 6.0 to 7.5.

Preparation of Apo-LAP

All buffer solutions are prepared with metal-free water, boiled, degassed, and saturated with nitrogen. Two milliliters of a solution of 6–10 mg of LAP in a 0.25 M Tris-HCl buffer, pH 7.8, containing 0.125 M NaCl are dialyzed in a collodion membrane sac for 10 hr against a 0.25 M Tris-HCl buffer, pH 7.8, containing 0.125 M NaCl, 10 μM mercaptoethanol, and 1 mM o-phenanthroline first, followed by an 8-hr dialysis against the same buffer without o-phenanthroline, while keeping the system oxygen free with bubbling of nitrogen. The enzyme solution is applied to a Sephadex G-25 column (1.0 × 20 cm) equilibrated with 0.1 M Tris-HCl buffer of pH 7.8, and the enzyme is eluted with the same. Apo-LAP must be sealed in an ampoule under nitrogen for preservation. Adding L-leucine during purification[8] did not seem to be

[8] F. H. Carpenter and J. M. Vahl, *J. Biol. Chem.* **248**, 294 (1973).

necessary. The apo-LAP can be reactivated by addition of Zn^{2+} to about full activity.[8]

Properties of Apo-LAP

Apo-LAP has no enzymic activity. Its molecular weight, disc gel electrophoretic, spectrophotometric, and immunological properties are identical to those of the native enzyme. However, attempts at crystallization were not successful.[9]

Preparation of Immobilized Leucine Aminopeptidase[10,11]

CNBr, 300–500 mg, is dissolved in 5 ml of water under magnetic stirring, and 10 ml of a Sepharose 6B suspension is admixed, and the pH is adjusted to 10.5. Activation is allowed to proceed at 15° for 6 min. After washing with water and 0.1 M sodium hydrogen carbonate on a G-3 glass filter, the activated gel is brought to a volume of 10 ml with 0.1 M of sodium bicarbonate. Two milliliters of the enzyme solution (10–20 mg of protein per milliliter) are added, and the mixture is gently shaken overnight at room temperature. The gel is filtered and washed consecutively with 0.1 M $NaHCO_3$, 1 M KCl, and H_2O; it is finally brought to a volume of 10 ml by adding 0.1 M KCl, and stored at 4°. Generally 10–60 mg of enzyme become bound per gram of dried Sepharose. The specific activity of the solid-phase enzyme suspension is 90–100% of that of the soluble enzyme.

Other methods have also been used for the insolubilization of LAP.[12]

Properties

Physical data for LAP are summarized in Table II.

Criteria of Purity

In contrast to LAP preparations from bovine kidney,[13,14] the enzyme isolated from lens contains no extraneous phosphatase[15,16] or esterase activities.

[9] U. Kettmann, unpublished results, 1971.
[10] R. Koelsch, J. Lasch, and H. Hanson, *Acta Biol. Med. Germ.* **24**, 833 (1970).
[11] R. Koelsch, *Enzymologia* **42**, 257 (1972).
[12] H. Kuthan, unpublished results, 1973.
[13] F. Binkley, *J. Biol. Chem.* **236**, 1075 (1961).
[14] G. R. J. Law, *Science* **156**, 1106 (1967).
[15] M. Ludewig, Dissertation, Martin-Luther-Universität, Halle, DDR, 1971.
[16] B. Wiederanders and H. Hanson, *Acta Biol. Med. Germ.* **26**, 1081 (1971).

TABLE II
PHYSICAL DATA FOR LEUCINE AMINOPEPTIDASE

Constants	Symbol	Value	References
Molecular weight[k]	$M_{s,D}$	$326{,}000 \pm 20{,}000$ g·mole^{-1}	a, b
Sedimentation coefficient[k] (c in g·dl^{-1})	$s_{20,w}$	$12.56 - 0.328 \cdot c \pm 0.02$ S	a
Diffusion coefficient[k] (c in g·dl^{-1})	$D_{20,w}$	$3.75 - 0.05 \cdot c \pm 0.18$ F	a
Partial specific volume (hydrated)	$\bar{v}_2$	0.751 ± 0.002 cm^3·g^{-1}	a
Partial specific volume (dry)	$\bar{v}_2$	0.739 ± 0.007 cm^3·g^{-1}	c
Partial specific volume (calculated)	$\bar{v}_2$	0.735 cm^3·g^{-1}	d
Frictional coefficient	f/f_0	1.23 ± 0.04	a
Intrinsic viscosity	$[\eta]$	3.82 cm^3·g^{-1}	d
Viscosity increment	ν'	5.19	d
Internal hydration (max)	w_{max}	0.257 g H$_2$O/g dry protein	d
Internal hydration (min)	w_{min}	0.05 g H$_2$O/g dry protein	d
Internal hydration[j]	w	0.15 g H$_2$O/g dry protein	e
Isoelectric point	pI	4.9 ± 0.2	a
Molar absorption coefficient	ϵ_{280}	$313{,}000 \pm 7{,}000$	c
Specific absorption (280 nm, pH 8.0)	$E_{1\,cm}^{1\,g/dl^{-1}}$	9.8 ± 0.2	d
pH-Dependence of specific absorption (7.4 $\leq$ pH $\leq$ 9.1; 280 nm)		$14.75 - 0.625 \cdot$ pH	f
Maximum diameter	L	13.0 nm	e
Streumassenradius[k]	R_{sc}	4.45 nm	g, h
Mean distance between two atoms	A_c	5.80 nm	h
Volume	V_c	485 nm^3	h
Correlation area	f_c	54 nm^2	e
Correlation distance	l_c	6.32 nm	e
Durchschusslänge[l]	d_c	3.51 nm	e
Surface	O	553 nm^2	e
Rotational relaxation time	ρ	339.5 nsec	i
Rotational diffusion coefficient	D_{rot}	1.47×10^6 sec^{-1}	i

[a] K. Kretschmer and H. Hanson, *Hoppe-Seyler's Z. Physiol. Chem.* **340**, 126 (1965).

[b] S. W. Melby and F. H. Carpenter, *J. Biol. Chem.* **246**, 2459 (1971).

[c] C. Pachaly, unpublished results (1973).

[d] K. Kretschmer, *Hoppe-Seyler's Z. Physiol. Chem.* **348**, 1158 (1967).

[e] H. Damaschun, G. Damaschun, R. Kröber, H.-V. Pürschel, and H. Hanson, *in* "Polymerphysik," Part I, pp. 171–172. Leipzig, 1974.

[f] M. Frohne, unpublished results (1967).

[g] K. Kretschmer and G. Kollin, *Hoppe-Seyler's Z. Physiol. Chem.* **350**, 431 (1969).

[h] G. Damaschun, H. Damaschun, H. Hanson, I. J. Müller, and H.-V. Pürschel, *Stud. Biophys.* **35**, 59 (1973).

[i] J. Lasch, *Acta Biol. Med. Germ.* **34**, 549 (1975).

[j] pH 8.0, $\mu = 0.1$.

[k] Formerly: radius of gyration.

[l] Mean value of all chords.

Some commercial LAP preparations are known to be contaminated by endopeptidases.[17] Total absence of such impurities could be shown[18] for the LAP obtained from lens.

Precipitating antibodies have been prepared against LAP from bovine eye lens and could be used in immunoelectrophoretic[3,19,20] as well as in enzyme neutralization assays.[21] The antibodies also precipitate LAP from kidney, but not the arylamidase from the same source.[3]

Disc Electrophoresis. The following modification of gel-system No. 6 of Maurer[20] is well suited for checking the purity of LAP by means of disc-gel electrophoresis. Solution I contains 10 g of acrylamide, 0.5 g of bisacrylamide, 6 g of sucrose, 12 ml of 1 N HCL, 1.65 g of Tris, and 120 μl of TEMED in 100 ml of aqueous solution, pH 7.5. Solution II consisting of 140 mg ammonium persulfate in 100 ml of water and 2.52 g of barbituric acid and 0.9 g of Tris in 1000 ml of water, pH 7.0, is used for the running buffer. Equal volumes of solutions I and II are mixed to give a 5% polyacrylamide gel. Electrophoresis is carried out at 200 V and 5 mA/gel for about 120 min, and Coomassie Brilliant Blue G 250 is used for staining.[22]

Electrophoresis of LAP under these conditions results in a sharp zone at the position of the monomer, representing more than 90% of the total protein. The rest of the protein is localized at the dimer position. Both protein bands are enzymically active.

Disc-gel electrophoresis of LAP at pHs above 9.5 (as in gel system No. 1 of Maurer[20]) leads to artifacts, tailing,[5] and degradation products.[23]

The results of Glasser *et al.* exclude the existence of isozymes.[6]

Stability

Native LAP is stable in solution (0.1 M Tris-HCl buffer pH 8.0) for years if kept in a refrigerator at 4°. At 65° the enzyme solution is stable up to 5 min, and the immobilized enzyme up to 10 min. Lyophilization of enzyme solutions leads to inactivation.

[17] L. J. Deftos and J. T. Potts, Jr., *Biochim. Biophys. Acta* **171**, 121 (1969).
[18] B. Wiederanders, J. Lasch, H. Kirschke, P. Bohley, S. Ansorge, and H. Hanson, *Eur. J. Biochem.* **36**, 504 (1973).
[19] R. J. Wieme, "Agar Gel Electrophoresis." Elsevier, Amsterdam, 1965.
[20] H. R. Maurer, "Disk-Elektrophorese, Theorie und Praxis der diskontinuierlichen Polyacrylamid-Elektrophorese," p. 43. Table 4. de Gruyter, Berlin, 1968.
[21] G. Pfleiderer, P. G. Celliers, M. Stanulovic, E. D. Wachsmuth, H. Determann, and G. Braunitzer, *Biochem. Z.* **340**, 552 (1964).
[22] W. Diezel, G. Kopperschläger, and E. Hofmann, *Anal. Biochem.* **48**, 617 (1972).
[23] B. Wiederanders, unpublished results, 1969.

TABLE III

AMINO ACID COMPOSITION OF LEUCINE AMINOPEPTIDASE OF BOVINE LENS

	Amino acid residues per mole of LAP	
Component	a	b, c
Lysine	204	190
Histidine	48	48
Arginine	126	120
Aspartic acid	282	258
Threonine[d]	150	134
Serine[d]	168	157
Glutamic acid	312	289
Proline	156	150
Glycine	252	232
Alanine	318	290
Half-cystine[e]	48	48
Valine	198	177
Methionine	66	60
Isoleucine	174	147
Leucine	234	211
Tyrosine[f]	60	48[g,h]
Phenylalanine	120	108
Tryptophan[i]	60	42[j]
Amide-NH₃[d]	210	279

[a] F. H. Carpenter and J. M. Vahl, *J. Biol. Chem.* **248**, 294 (1973).
[b] U. Kettmann, K. Kretschmer, and H. Hanson, *Hoppe-Seyler's Z. Physiol. Chem.* **394**, 1537 (1968), in some cases improved values are given, unpublished results of our laboratory.
[c] Lyophilized sample containing 9.11 % water.
[d] Extrapolated to zero time.
[e] Determined as cysteic acid.
[f] Determined from a hydrolysis performed in the presence of phenol.
[g] Determined spectrophotometrically on the intact enzyme.
[h] Determined by nitration.
[i] Determined by relating to tyrosine.
[j] Determined by solvent perturbation R. Misselwitz, D. Zirwer, M. Frohne, and H. Hanson, *FEBS Lett.* **55**, 233 (1975).

In 8 *M* urea at pH 8 the enzyme undergoes changes in both shape and enzymic activity,[15,24] whereas at pH 9 only a decrease in activity is found.[25] Removal of urea results in renaturation.[15,25]

[24] M. Frohne and H. Hanson, *Hoppe-Seyler's Z. Physiol. Chem.* **350**, 207 (1969).
[25] S. W. Melby and F. H. Carpenter, *J. Biol. Chem.* **246**, 2459 (1971).

TABLE IV
Exposure and Functional Role of Some Amino Acid Residues in Leucine Aminopeptidase

| | | | Participating residues | |
| | Total | Exposed | In Mn^{2+} | In Zn^{2+} |
Residue	number	residues	binding	binding
Trp	42	7–8[a]	—	—
Tyr	48	14 ± 2[b]	—	—
His	48	12[c]	12[c]	24[c]
Cys	36[d]	6[e]	?	?

[a] Determined by solvent and thermal perturbation [R. Misselwitz, D. Zirwer, M. Frohne, and H. Hanson, *FEBS Lett.* **55**, 233 (1976)].

[b] Determined using N-acetylimidazole (M. Frohne, unpublished results).

[c] Determined by carbethoxylation [M. Ludewig, M. Frohne, I. Marquardt, and H. Hanson *Eur. J. Biochem.* **54**, 155 (1975)].

[d] F. H. Carpenter and J. M. Vahl, *J. Biol. Chem.* **248**, 294 (1973).

[e] M. Frohne and H. Hanson, *Hoppe-Seyler's Z. Physiol. Chem.* **350**, 213 (1969).

Enzyme Structure (Tables III and IV)

Metal Content. Leucine aminopeptidase is a zinc metalloenzyme. The zinc content of the crystallized enzyme is about 6–12 g-atoms of zinc per mole of LAP (326,000).[8,26,27] Other metals could be detected in traces only.[8,27] Six gram-atoms of zinc per mole of LAP is essential for the activity. The remaining 6 binding sites also could conceivably be occupied by other metals.[8]

Secondary and Higher Structures (Tables III–V). Maximal helix content as judged by optical rotatory dispersion (ORD), exists in β-chlorethanol. Assuming $[m']_{233} = -1870$ for 0% and $[m']_{233} = -15,100$ for 100% helix, a value of $[m']_{233} = -9580$ was found corresponding to 58% helix, which agrees well with that calculated from amino acid composition.[28]

Dimer LAP.[29] The freshly prepared enzyme contains about 1% of a dimer, the amount of which may increase with time up to 6% after about 6 months of storage and even to 9% on incubation with 2 mM Mn^{2+}.

[26] U. Kettmann and H. Hanson, *FEBS Lett.* **10**, 17 (1970).

[27] H. Böttger, S. Fittkau, S. Niese, and H. Hanson, *Acta Biol. Med. Germ.* **21**, 143 (1968).

[28] U. Kettmann, K. Kretschmer, and H. Hanson, *Hoppe-Seyler's Z. Physiol. Chem.* **349**, 1537(1968).

[29] M. Ludewig and H. Hanson, *Acta Biol. Med. Germ.* **29**, 539 (1972).

TABLE V
SECONDARY STRUCTURE OF LEUCINE AMINOPEPTIDASE

Measurement method	$\% \, \alpha$	$\% \, \beta$	$\% \, \rho$	References
I. By optical rotatory dispersion measurements				
a. Moffit-Yang	21 · · · 22			a
	33 · · · 35			b
	32			c
b. Shechter-Blout	27 · · · 37			c
c. λ_c	22 · · · 23			a
	40			b
	30			c
d. $[m']_{233}$	17 · · · 20			a
	15 · · · 22			b
	17			c
II. By circular dichroism measurements				
a. $[\ominus]_{208}-[\ominus]_{222}$	37 · · · 45			c
b. Curve-fitting^e (210 · · · 240 nm)	36	15	49	c
c. 3-Point-approach	36	16	48	d

^{a} K. Kretschmer, *Hoppe-Seyler's Z. Physiol. Chem.* **349**, 846 (1968).
^{b} S. W. Melby, Thesis, University of California, Berkeley, 1970.
^{c} M. Becker, W. Schälike, and Zirwer, *Stud. Biophys.* **35**, 203 (1973).
^{d} W. Schälike, M. Becker, and D. Zirwer, *Stud. Biophys.* **35**, 221 (1973).
^{e} V. P. Saxena and D. B. Wetlaufer, *Proc. Natl. Acad. Sci. U.S.A.* **68**, 969 (1971).

The dimer can be dissociated by dilution ($\tau_{1/2} = 32$ hr) or by incubation with 0.1 M 2-thioethanol ($\tau_{1/2} = 3$ min) or with 0.1 M sodium sulfite.[15]

Subunit Structure (Table VI). Treatment of LAP with 1 M guanidinium chloride causes a partly reversible dissociation to half molecules of MW 165,000 with an enzymic activity of 100–65% of the original enzyme.[30]

In 7 M urea below pH 3, in 3.7 M guanidinium chloride below pH 8.5, or in 0.1–1% sodium dodecyl sulfate (SDS), LAP as well as its reduced and carboxyamidomethylated derivative can dissociate into subunits with or without the addition of 2-mercaptoethanol.[25]

Prolonged standing of the subunits in 1% SDS and 1% 2-mercaptoethanol results in a progressive decomposition into 5–7 components of molecular weights ranging from 55,000 to 10,000. Carboxyamidomethyla-

[30] R. Kleine and J. Lehmann, *Acta Biol. Med. Germ.* **35**, (1976).

TABLE VI
PHYSICAL PROPERTIES OF THE SUBUNITS OF LEUCINE AMINOPEPTIDASE

Parameter	Value	References
Number	6	a, b
Diameter[e]	L_{SU} 5.26	c
Molecular weight (Equilibr.)	M_{SU} 54,000 ± 4000	c
	57,000 ± 3000	b
	53,000 · · · 57,000[f]	b, d

[a] G. Damaschun, H. Damaschun, H. Hanson, I. J. Müller, and H.-V. Pürschel, *Stud. Biophys.* **35**, 59 (1973).
[b] S. W. Melby and F. H. Carpenter, *J. Biol. Chem.* **246**, 2459 (1971).
[c] H. Damaschun, G. Damaschun, R. Kröber, H.-V. Pürschel, and H. Hanson, *in* "Polymerphysik," Part I, pp. 171–172. Leipzig, 1974.
[d] J. Lehmann, unpublished results, 1973.
[e] Taken as spheres.
[f] Found by gel electrophoresis after incubation in 0.1 to 1.2% sodium dodecyl sulfate in the presence of 2-thioethanol.

tion of the subunits or standing at 37° speed up the production of the peptide chain of about MW 10,000.[30a]

The subunits are arranged in two rings around a nearly cylindrical cavity, each ring comprising three subunits.[31,32]

Activity Enhancement of LAP

Some metal ions may affect the specific activity of LAP without influence on K_m value. The activation sites of the enzyme hexamer may be filled by 6 Zn^{2+}, which in turn may be replaced by 6 Mg^{2+} or 6 Mn^{2+}, with a concomitant gain in specific activity[8] (cf. Table VII). The dissociation constants published by Carpenter and Vahl[8] will probably need some correction because of the strong complexes of Zn^{2+} with the Tris and carbonate in the buffer.

Ba^{2+} causes a slight decrease in activity and Sr^{2+} or Ca^{2+} have no influence. According to their activating ability at fixed low concentration, the relative effectiveness of the ions is: $Mn^{2+} > Mg^{2+} > Fe^{2+} > Co^{2+} \sim Ni^{2+} > [Zn^{2+} > Cu^{2+}]$, the two latter being inhibitory. In terms of being able to produce maximal activation, there seems to be no difference between Mn^{2+}, Mg^{2+}, or OH^{-15} (cf. Table VII).

[30a] R. Kleine, unpublished results, 1973.
[31] G. Damaschun, H. Damaschun, H. Hanson, I. J. Müller, and H.-V. Pürschel, *Stud. Biophys.* **35**, 59 (1973).
[32] K. Kretschmer and G. Kollin, *Hoppe-Seyler's Z. Physiol. Chem.* **350**, 431 (1969).

TABLE VII

ACTIVITY ENHANCEMENTS FOR LEUCINE AMINOPEPTIDASE (LAP)

Substrate[m]: additions	pH optima			K_D (mM) (activator)		K_m (mM) (substrate)[l]		Relative activity enhancement[n] at saturating metal ion concentrations
	Leu-NH$_2$	Leu-N$_2$H$_3$	LpNA	Leu-NH$_2$	LpNA	Leu-NH$_2$	LpNA	
None[p]	10[a]					31[a]	1.8[b] 1.9[c] 1.3[d]	1[b]
Zn^{2+} ,[p]	~9[e]						2[b]	
Mg^{2+}	9.5[e,q] 10[e,r] 7–10[e,s]			0.002 . . . 0.2[g]	3[b]	20[f] 10[g] 10[f] 5[g]	2[b] 0.2[d]	17[b]
Mn^{2+}	10[a] 9[a,h]	9[h]			2.3[b]	31[a,h] 18[i] 33[f]	2[d]	18[b]
OH⁻			9.6[j]		0.5[k]		3[c]	18[k]

[a] H. Hanson, D. Glässer, and H. Kirschke, *Hoppe-Seyler's Z. Physiol. Chem.* **340**, 107 (1965).

[b] M. Ludewig, Dissertation, Martin-Luther-Universität, Halle, DDR, 1972.

[c] M. Frohne, J. Lasch, and H. Hanson, *Acta Biol. Med. Germ.* **31**, 879 (1973).

[d] J. Lasch, W. Kudernatsch, and H. Hanson, *Eur. J. Biochem.* **34**, 53 (1973).

[e] S. W. Melby and F. H. Carpenter, *J. Biol. Chem.* **246**, 2459 (1971).

[f] F. H. Carpenter and J. M. Vahl, *J. Biol. Chem.* **248**, 294 (1973).

[g] S. W. Melby, Ph.D. Thesis, University of California, Berkeley, 1970.

[h] A. Wergin, *Naturwissenschaften* **52**, 24 (1965).

[i] S. Fittkau, U. Förster, C. Pascual, and W. H. Schunck, *Eur. J. Biochem.* **44**, 523 (1974).

[j] M. Frohne and H. Hanson, *Acta Biol. Med. Germ.* **31**, 453 (1973).

[k] M. Ludewig, unpublished results.

[l] K_m-Values are given without reference to the free substrate concentration.

[m] LAP substrates: L-leucinamide, L-leucine hydrazide, L-leucine-p-nitroanilide.

[n] Given with regard to an enzyme kept at pH 8.2.

[o] LAP of unknown Zn content at the activation site.

[p] LAP with 6 Zn^{2+} at the activation site.

[q] Activated with 10 mM magnesium sulfate at pH 8.5 and 40° for 3 hr.

[r] Activated with 10 mM magnesium sulfate at given pH and 40° for 3 hr.

[s] First exposed to 10 mM magnesium sulfate at pH 8.5 for 3 hr at 40° and subsequently kept at 26° and the given pH for 1 hr.

Complete activation at 30° is achieved in 3.5–6 hr.[15] Half-times for the activation of the enzyme by Mn^{2+}, Co^{2+}, or OH^- were estimated to be 22, 30, 32, or 25 min, respectively. Temperatures above 30°[15] as well as pH values above pH 9.5[33] should be avoided in long-time experiments because of denaturation. The activated diluted enzyme (about 0.1 mg/ ml or less) may be stored at 4° for 1 day without loss of activity.

A concentration of 2–4 mM Mn^{2+} appears to be sufficient to produce appreciable but not maximum activation at pH 8.2 Table VII).

Anions exert a synergistic activation effect without the necessity of preincubation. At 0.1 M concentration, the following order was established for anion effectiveness: $F^- < CH^- < SO_4^{2-} < Cl^- < N_3^- < Br^- < SCN^- < ClO_4^- < I^-$, maleate, sulfosalicylate.[34,35]

Specificity and Kinetic Properties

Amidase Activity. Lens LAP splits off N-terminal L-amino acids from polypeptides: aminoacyl-peptide $+ H_2O \rightarrow$ amino acid $+$ peptide. Furthermore, it hydrolyzes amino acid amides, alkylamides, arylamides, and hydrazides.

Numerous findings show that the rate of hydrolysis depends on the size and polarity of the group R′ of the substrates:

$$H_2N-\underset{R'}{CH}-C\overset{O}{\underset{\underset{H}{N}}{}}R''$$

In agreement with the concept of hydrophobic interactions between the R′ group of the substrate with similar groups of a binding site on the enzyme,[15,36] the amides, anilides, and (with some restrictions) also β-naphthylamides of leucine, phenylalanine, and the other aliphatic amino acids are cleaved best in the order of Leu > Phe > Val > Ala > Gly.[3,37] Polar side chains decrease the rate of hydrolysis (cf. Table VIII). The same relative specificity was observed for the cytoplasmic LAP from hog kidney.[3]

Investigations on the cleavage of peptide substrates show that, next to the primary substrate specificity (hydrophobicity, L-configuration of

[33] M. Frohne and H. Hanson, *Acta Biol. Med. Germ.* 31, 453 (1973).

[34] J. Lasch, W. Kudernatsch, and H. Hanson, *Eur. J. Biochem.* 34, 53 (1973).

[35] M. Ludewig, J. Lasch, U. Kettmann, M. Frohne, and H. Hanson, *Enzymologia* 41, 59 (1971).

[36] S. Fittkau, U. Förster, C. Pascual, and W.-H. Schunck, *Eur. J. Biochem.* 44, 523 (1974).

[37] S. Fittkau, D. Glässer, and H. Hanson, *Hoppe-Seyler's Z. Physiol. Chem.* 322, 101 (1961).

TABLE VIII

KINETIC DATA FOR HYDROLYSIS BY LEUCINE AMINOPEPTIDASE (pH 8.0–8.3; 25°)

Substrate	K_m (mM)	k_{cat} (min^{-1})	Activator	References
L-Leu-amide	28	110,000	None	a
	29	1,190,000	Mg^{2+}	b
D-Leu-amide	330	28	Mg^{2+}	c
L-Leu-hydrazide	23	80,000	None	a
	17.5	197,000	Mn^{2+}	d
L-Phe-hydrazide	32.9	66,300	Mn^{2+}	d
N-Methyl-L-Phe-hydrazide	196	2,050	Mn^{2+}	d
L-Leu-Gly	8.4	100,000	Mg^{2+}	e
L-Leu-Gly-Gly	3.2	70,000	Mg^{2+}	e
L-Leu-4-nitroanilide	1.1	45	None	b
L-Leu-β-naphthylamide	0.25	6	None	e
L-Leu-phenylazophenylamide	0.14	7	None	e
L-Leu-n-butylamide	5.0	—	None	f
L-Leu-hexylamide	5.0	—	None	f
L-Leu-ethanolamide	100	—	None	f

^{a} M. Frohne, unpublished results, 1969.

^{b} J. Lasch, W. Kudernatsch, and H. Hanson, *Eur. J. Biochem.* **34**, 53 (1973).

^{c} R. Koelsch, *Enzymologia* **42**, 257 (1972).

^{d} S. Fittkau, U. Förster, C. Pascual, and W.-H. Schunck, *Eur. J. Biochem.* **44**, 523 (1974).

^{e} J. Lasch, W. Kudernatsch, and R. Koelsch, unpublished results, 1971.

^{f} M. Ludewig, unpublished results, 1972.

the N-terminal amino acid and free α-amino group), the secondary substrate specificity is important to affinity and hydrolysis rate. This is determined by the nature of the side chain and the configuration of the penultimate amino acid (Table IX).

In general, the presence of a hydrophobic residue R'' increases the affinity of the enzyme toward the substrate.

The high stereospecificity of LAP activity toward N-terminal amino acids affects both binding and hydrolysis of the substrate (cf. L-Leu-amide with D-Leu-amide in Table VIII). There is also a pronounced secondary stereospecificity, insofar as L-Leu-D-Leu and related dipeptides are hardly attacked by LAP.[38] The α-amino group of the substrate has to be in the unprotonated form. Amides and peptides containing β-amino acids are not split. β-Aspartyl or γ-glutamyl bonds, N^α-acylated amino acid derivatives and peptides, and peptide bonds involving proline with its imino group (Gly-Pro, Leu-Pro, Gly-Pro-Gly) resist hydrolysis. On

[38] E. L. Smith and D. H. Spackman, *J. Biol. Chem.* **212**, 271 (1955).

TABLE IX

KINETIC DATA FOR THE HYDROLYSIS OF DIPEPTIDES OF L-AMINO ACIDS

(pH 8.3, 25°, without activator)[a]

Substrate	K_m (mM)	k_{cat} (min^{-1})
Gly-Phe	1.0	30
Ala-Phe	0.145	320
Phe-Phe	0.030	6,900
Leu-Phe	0.056	16,200
Phe-Gly	1.3	220
Trp-Phe	0.004	2,780
Phe-Trp	0.042	2,400
Trp-Trp	0.007	2,940
Phe-Tyr	0.057	6,800
Tyr-Tyr	0.038	1,470

[a] P. F. Palmberg, Ph.D. Thesis, Northwestern University, Evanston, Illinois, 1969.

the other hand, Pro-X-bonds (Pro-Gly, Pro-Gly-Pro) are cleaved with sufficient velocity.[18] Owing to sterical effects, N^α-alkylated amino acid derivatives have a lower affinity toward the enzyme and a lower rate of hydrolysis.[36]

Apart from the limitations mentioned, lens LAP has a relatively broad specificity and is capable of removing natural amino acids from randomly coiled peptides. Thus, lens LAP is useful for end group analysis and for determination of protein sequences.[18,39] It can also be used for evaluating the optical purity of synthetic peptides and the separation of racemic amino acids in their carboxyl group modified forms.[34] For this purpose, immobilized LAP is particularly useful.[10]

Esterase Activity (Table X). LAP possesses esterase activity with a pH optimum between 9.0 and 9.4, and Mn^{2+} ions, too, enhance this activity. L-Leucine and L-tryptophan esters are best; D-amino acid esters are not hydrolyzed.

None of the usual esterase inhibitors (atoxyl, quinine, NaF, diisopropyl fluorophosphonate) cause any inhibition of esterase activity.[40,41] The

[39] H. Hanson and S. Fittkau, *in* "Symposium on Modern Methods in the Investigation of Protein Structure" (F. B. Straub and F. Friedrich, eds.), pp. 9–18. Akadémiai Kiadó, Budapest, 1967.

[40] R. Kleine and H. Hanson, *Acta Biol. Med. Germ.* **9**, 606 (1962).

[41] H. Hanson, D. Glässer and R. Kleine, *Hoppe-Seyler's Z. Physiol. Chem.* **329**, 257 (1962).

TABLE X
RELATIVE RATES OF HYDROLYSIS OF L-AMINO ACID ESTERS[a,b]

Substrate		Substrate	
TrpOC$_2$H$_5$	(100)	TrpOCH$_3$	96
LeuOC$_2$H$_5$	94	TrpOC$_2$H$_5$	100
PheOC$_2$H$_5$	51	TrpOC$_3$H$_7$	58
TyrOC$_2$H$_5$	12	TrpOC$_4$H$_9$	101
GlyOC$_2$H$_5$	11		

[a] R. Kleine and H. Hanson, *Hoppe-Seyler's Z. Physiol. Chem.* **326,** 106 (1961).
[b] Mn^{2+}-activated leucine aminopeptidase at pH 9.0–9.3, 37°.

ratio of amidase activity (L-LeuNH$_2$) to esterase activity (L-LeuOC$_2$H$_5$) is 26:1 at pH 8.2.[5] (For kinetic data on the hydrolysis of the esters of Phe and Trp, see Palmberg.[42])

Transfer Activity. With amides or esters of leucine or phenylalanine as substrates, the formation of dipeptide amides or dipeptide esters is observed in the initial phase of the reaction,[43] and this transfer reaction is particularly pronounced at pH 8.5. No transamidation takes place with the D-amino acid amides.

Inhibitors. Some chloromethyl ketones and methyl ketones of L-amino acids and peptides[44] (group A in Table XI) inhibit LAP. The best inhibitors carry an α-amino group in the uncharged form and a hydrophobic side chain in the L-configuration.[36] N^{α}-Acylated halomethyl ketones (tosyl-PheCH$_2$Cl, Z-LeuCH$_2$Cl) have no inhibitory effect. D-Compounds such as D-AlaCH$_2$Cl are also ineffective.[45]

The Thr(OBu^t)-Phe peptide bond (group B) even at high enzyme concentrations is split to an extent less than 0.7% in 60 min.[46]

p-Diazo derivatives of L-PheCH$_3$ and of N-terminal L-Phe-peptides active-site-directed reagents, may be useful as active site-directed reagents for labeling leucine aminopeptidase.[47]

Assay Methods

Several methods can be used for determining the activity of leucine aminopeptidase. (A) Hydrolysis of leucinamide and determination of

[42] P. F. Palmberg, Ph.D. Thesis, Northwestern University, Evanston, Illinois, 1969.
[43] H. Hanson and J. Lasch, *Hoppe-Seyler's Z. Physiol. Chem.* **348,** 1525 (1967).
[44] S. Fittkau, *J. Prakt. Chem.* **315,** 1037 (1973).
[45] P. L. Birch, H. A. El-Obeid, and M. Akhtar, *Arch. Biochem. Biophys.* **148,** 447 (1972).
[46] R. Jost, *FEBS Lett.* **29,** 7 (1973).
[47] S. Fittkau, *FEBS Meet. 9th,* Budapest 1974, abstr. S2/j15.

TABLE XI
K_i VALUES FOR COMPETITIVE INHIBITORS

Group	Compound	K_i (mM)	References
A	LeuCH$_2$Cl	0.23	a
	PheCH$_2$Cl	0.34	a
	PheCH$_2$Br	1.2	b
	PheCH$_3$	1.98	a
	N-Methyl-PheCH$_2$Cl	23.3	a
B	N-p-Tosyl-N'-butylcarbamide	0.64	c
	N-Sulfanilyl-N'-butylcarbamide	1.45	c
	N-p-Azidobenzolsulfonyl-N'-butylcarbamide	1.8	d
	N-(3-phenylazo)sulfanilyl-N'-butylcarbamide	0.34	e
C	Thr(OBu^t)-Phe	0.025	f
	Thr(OBu^t)-Phe-Pro	0.01	f
	Thr(OBu^t)-Phe-Pro-Gln-Thr(OBu^t)-Ala-Ile-Gly	0.1	g
	Leu-Pro	13	h

[a] S. Fittkau, U. Förster, C. Pascual, and W.-H. Schunck, *Eur. J. Biochem.* **44**, 523 (1974).
[b] S. Fittkau, *Acta Biol. Med. Germ.* **35**, (1976).
[c] J. Lasch, W. Kudernatsch, and H. Hanson, *Eur. J. Biochem.* **34**, 53 (1973).
[d] J. Lasch, unpublished results, 1973.
[e] R. Koelsch, unpublished results, 1973.
[f] R. Jost, *FEBS Lett.* **29**, 7 (1973).
[g] R. Jost, A. Masson, and H. Zuber, *FEBS Lett.* **23**, 211 (1972).
[h] B. Wiederanders, J. Lasch, H. Kirschke, P. Bohley, S. Ansorge, and H. Hanson, *Eur. J Biochem.* **36**, 504 (1973).

the liberated leucine by (1) paper chromatography,[48] (2) optical absorption[43,49] or (3) potentiometric procedure.[10] (The latter method is for immobilized enzyme.) (B) Hydrolysis of leucine hydrazide[50] followed by determination of hydrazine[51] as the 4,4'-dimethylbenzilideneazine in hydrochloric acid, at 454 nm. (C) Hydrolysis of leucine-p-nitroanilide and direct spectrophotometric determination of the p-nitroanilide formed at 405 nm.[52] (D) Hydrolysis of leucine-β-naphthylamide and determination of β-naphthylamine by (1) increase of absorption[53] at 334 nm, (2)

[48] P. Bohley, *Naturwissenschaften* **49**, 326 (1962).
[49] F. Binkley and C. Torres, *Arch. Biochem. Biophys.* **86**, 201 (1960).
[50] A. Wergin, *Naturwissenschaften* **52**, 34 (1965).
[51] G. W. Watt and J. D. Crisp, *Anal. Chem.* **24**, 2006 (1952).
[52] H. Tuppy, U. Wiesbauer, and E. Wintersberger, *Hoppe-Seyler's Z. Physiol. Chem.* **329**, 278 (1962).
[53] H. J. Lee, J. N. LeRue, and I. B. Wilson, *Anal Biochem.* **41**, 397 (1971).

fluorescence,[54] or (3) coupling to Fast Garnet BGS to produce a highly colored azo dye.[55,56] The latter method is applicable to the preparation of zymograms[57] after gel electrophoresis of enzyme solutions. Several aminopeptidases distinct from leucine aminopeptidase can hydrolyze the substrates used in methods C and D. Thus their use is restricted to leucine aminopeptidase preparations known to be of high degree of purity.

Method A (1). The assay mixture contains 0.2 ml of buffer (0.5 M Tris-HCl, pH 8.5), 0.1 ml of 0.25 M substrate L-Leu-amide, adjusted to pH 8.5, 0.1 ml of enzyme (about 40 nM in 5 mM Tris-HCl of pH 8.5) and water to a final volume of 0.5 ml. The water can be replaced by a 10mM MnCl$_2$ solution for a 5- to 10-fold enhancement of activity. Simultaneously the enzyme concentration could be reduced.

After incubation for 30 min at 25°, about 10% of the substrate is hydrolyzed. Aliquots (10 μl) are spotted on a dry chromatographic paper which is then developed in a *sec*-butanol:formic acid:water mixture of 75:15:10 (v/v) for separating the liberated leucine. After a running time of about 6 hr the paper is dried, then dipped in a 0.5% ninhydrin solution in acetone; the color is allowed to develop for 5 hr at 35°. The liberated leucine is identified with the aid of reference leucine (0.05 M). The blue spots corresponding to leucine are eluted with a 0.5 M ethanolic solution of NaHCO$_3$, and absorbance is measured at 578 nm. Product concentration is calculated after correcting blank values at corresponding positions on the paper from substrate above.

If the enzyme is used without Mg^{2+} or Mn^{2+}, leucyl-leucinamide also appears as a product with a mobility faster than leucine.[43]

Method A (2). The hydrolysis of leucinamide could be followed by spectrophotometric assay. The assay mixture contains 7.5 nM enzyme in 0.2 M Tris-HCl buffer (pH 8.5) and 5–25 mM L-leucinamide. The liberated leucine is calculated from the difference in molar absorption coefficient between leucinamide and leucine at 225 nm of 104 M^{-1}cm^{-1} (leucine $\epsilon = 30\ M^{-1}$cm^{-1}).

Method A (3). The pH-static assay is performed (Radiometer TTT 1 c titrator titrigraph) at 25° with 0.1 N HCl as titrant. The reaction is started by addition of 100 μl of enzyme (ca. 0.1 μM in 0.1 M KCl) solution to 900 μl of substrate (0.05 M Leu amide in 0.1 M KCl, adjusted to pH 8.5). At a fixed pH, the substrate turnover can be calcu-

[54] L. J. Greenberg, *Biochem. Biophys. Res. Commun.* 9, 430 (1962).
[55] A. C. Bratton and E. K. Marshall, Jr., *J. Biol. Chem.* 128, 537 (1939).
[56] M. N. Green, K. C. Tsou, R. Bressler, and A. M. Seligman, *Arch. Biochem. Biophys.* 57, 458 (1955).
[57] E. Wintersberger and H. Tuppy, *Monatsh. Chem.* 91, 406 (1960).

lated using the formula

$$x = 1/\nu \cdot y$$

where x = moles of substrate transformed; y = moles of acid consumed; ν = effective proton consumption = $\rho_{Leu} - (\rho_{LA} + \alpha_{NH_3})$; ρ_{Leu} = degree of protonation of leucine ($pK_3 = 9.6$); ρ_{LA} = degree of protonation of leucine amide ($pK = 7.8$); and α_{NH_3} = degree of dissociation of ammonia ($pK = 9.25$). At pH 8.5 the correction factor (ν) has a value of 0.59.

Method B. The reaction mixture at 25° comprises 0.2 ml of 0.5 M Tris-HCl buffer (pH 8.5), 0.1 ml of enzyme (20–50 nM in 5 mM Tris-HCl, pH 8.5), 0.1 ml of 0.25 M freshly prepared Leu hydrazide, and 0.1 ml of water. Between 5 and 30 min, 10 μl of the mixture are withdrawn and added to 5 ml of fresh aldehyde solution (one part of 4-dimethylaminobenzaldehyde, 4% in ethanol, and 14 parts of 0.1 N HCl). The sample is left in the dark at room temperature for 1 hr, then absorbance is measured at 436 nm. The enzymically liberated hydrazine is calculated using a molar absorption coefficient of $\epsilon_{436} = 44,000$ $M^{-1}cm^{-1}$ for the 4,4'-dimethylbenzilideneazine.

Method C. In a thermostatted cuvette (25°), 100 μl of 0.5 M Tris-HCl buffer, 100 μl of enzyme (ca. 1 μM in 5 mM Tris-HCl, pH 8.5) and 50 μl of 0.01 M Leu-p-nitroanilide are mixed, and the change in absorbance at 405 nm is recorded. Concentration of the p-nitroaniline product is calculated using a molar absorption coefficient of $\epsilon_{405} = 9450$ $M^{-1}cm^{-1}$.

Method D (1). In a thermostatted cuvette (25°), 200 μl of 0.2–2 mM Leu-β-naphthylamide in 0.5 M Tris-HCl, pH 8.5, is mixed with 50 μl of enzyme (15–40 μM in 5 mM Tris-HCl, pH 8.5). The change in absorbance is followed at 334 nm ($\epsilon_{334} = 1500$ $M^{-1}cm^{-1}$).

Method D (3) (Zymogram Method). After electrophoresis using about 1 μl (agar gel electrophoresis) and 5 μl (polyacrylamide gel electrophoresis) of enzyme solution (10 μM in 5 mM Tris-HCl, pH 8), the gels are immersed in 10 ml of a staining solution containing 2 mg of L-leucine-β-naphthylamide, 100 mg of Fast Garnet GBS, and 40 mg of $MnCl_2 \cdot 4H_2O$ dissolved in 10 ml of 0.2 M Tris-maleate buffer of pH 6.0. After incubation for 20–60 min at room temperature, purple bands appear.

Acknowledgments

We would like to express our gratitude to the co-workers of our group (Drs. S. Fittkau, D. Gläßer, U. Kettmann, R. Kleine, R. Koelsch, K. Kretschmer, J. Lasch, M. Ludewig, I. Marquardt, B. Wiederanders) for their assistance in the preparation of this manuscript.

[43] Thermophilic Aminopeptidase I

By G. RONCARI, E. STOLL, and H. ZUBER

Various strains of the thermophilic bacterium *Bacillus stearothermophilus* produce three types of aminopeptidases: the highly thermostable aminopeptidase I (API),[1-6] two more thermolabile aminopeptidases, designated APII[1,7,8] and APIII.[1,9] The high-molecular-weight API is composed of 12 subunits, whereas the low-molecular-weight APII and APIII are composed of two identical subunits, of MN 46,000 and 47,500, respectively. Depending on the optimum growth temperature to which the *B. stearothermophilus* strains have been adapted (55° or 37°),[10] the amounts of these aminopeptidases vary greatly:[9] much API and APII is produced at 55° but little at 37°. On the other hand, in 37° cultures the amount of APIII is increased. At 37° an additional aminopeptidase (APIm) of high molecular weight (about 400,000), strongly particle bound is synthesized.[9] Like leucine aminopeptidase all aminopeptidases preferentially hydrolyze peptides containing aliphatic or aromatic amino acid residues, but show differences in their substrate specificity. APIII is an aminotripeptidase.

This article describes some new details of the isolation and properties of API (see also this series Vol. 19 [37a] and [37b]).

Assay Methods

Principle. Two methods can be used for determining the activity of API: (A) Hydrolysis of leucine *p*-nitroanilide or glutamic acid 1-(4-nitroanilide) and spectrophotometric determination of *p*-nitroanilide at 410 nm. (B) Hydrolysis of Gly-Leu-Tyr (or other tripeptides) and determination of the liberated glycine (or other amino acid) as its ninhy-

[1] G. Roncari and H. Zuber, *Int. J. Protein Res.* **1**, 45 (1969).
[2] P. Moser, G. Roncari, and H. Zuber, *Int. J. Protein Res.* **2**, 191 (1970).
[3] E. Stoll, M. A. Hermodson, L. H. Ericsson, and H. Zuber, *Biochemistry* **11**, 4731 (1972).
[4] G. Roncari, H. Zuber, and A. Wyttenbach, *Int. J. Pept. Protein Res.* **4**, 267 (1972).
[5] E. Stoll and H. Zuber, *FEBS Lett.* **40**, 210 (1974).
[6] E. Stoll, L. H. Ericsson, and H. Zuber, *Proc. Natl. Acad. Sci. U.S.A.* **70**, 3781 (1973).
[7] E. Stoll, H. G. Weder, and H. Zuber, *Biochim. Biophys. Acta* in press.
[8] P.-Å. Myrin and B. v. Hofsten, *Biochim. Biophys. Acta* **350**, 13 (1974).
[9] M. Balerna and H. Zuber, *Int. J. Pept. Protein Res.* **6**, 499 (1974).
[10] L. Jung, R. Jost, E. Stoll, and H. Zuber, *Arch. Microbiol.* **95**, 125 (1974).

drin–copper complex[11] or ninhydrin–cadmium complex[12] by a chromato-graphic method.

Method A

Reagents

$CoCl_2$, 1 mM in 0.05 M imidazole-HCl buffer, pH 7.5, or in 0.05 M Tris-HCl buffer pH 7.2 (all pH adjustments are made at room temperature)
Leucine-p-nitroanilide, 1 mM
Glutamic acid 1-(4-nitroanilide), 0.25 mM
Enzyme, in Tris-HCl buffer, pH 7.2, or imidazole-HCl buffer, pH 7.5, 1 mM $CoCl_2$

Procedure. Leucine-p-nitroanilide or glutamic acid 1-(4-nitroanilide) can be hydrolyzed either by adding the enzyme solution (1–20 μl) to 3 ml of substrate solution, followed by incubation at 30°–65°, and measurement at 410 nm in the spectrophotometer (1-cm cuvette), or directly by adding the enzyme solution to a 1-cm cuvette containing 1.5 ml of substrate solution, placed in a spectrophotometer with a cell holder heated to the appropriate temperature (incubation temperatures up to 65° can be used since the enzyme is very thermostable). The blank contains no enzyme. Calibration curve: 1–15 μm per milliliter of nitroaniline yields an absorption of 0.06–1.50. The leucine-p-nitroanilide is 10 times more sensitive to hydrolysis by API than the glutamic acid 1-(4-nitro-anilide) degradation.

Method B

Reagents

$CoCl_2$, 1 mM, in Tris-HCl buffer, pH 7.2, 0.05 M
Gly-Leu-Tyr, 0.045 M
NaOH, 0.1 N
Chromatography solvent: n-butanol:acetic acid water, 4:1:5
Color reagents as described[11,12]

Procedure. Five-tenths millimeter of the Gly-Leu-Tyr solution is mixed with the appropriate amount of enzyme solution and incubated at 30°–65°. If necessary, the pH is kept constant by adding 0.1 N NaOH

[11] F. G. Fischer and H. Dörfel, Biochem. Z. 324, 544 (1953).
[12] G. N. Atfield and J. O. R. Morris, Biochem. J. 81, 606 (1961).

(glass electrode). Aliquots of 40 μl of the incubation mixture are chromatographed together with 10–20 μl glycine (comparison solution) on the chromatography paper Wt 1. Glycine is determined as described.[11,12]

Definition of Unit and Specific Activity. One unit of enzyme is defined as the amount of enzyme that hydrolyzes 1 μmole of substrate per minute under the above conditions. Specific activity is expressed as units per milligram of protein. The protein concentration was determined either by the Lowry method[13] or by amino acid analysis (sum of the amino acid residues).

Purification Procedure

For the isolation of small amounts of API, the method already described in this series, Vol. 19 [37a] can be used (starting with disintegration of 1–10 g of bacteria by sonication). An alternative procedure, more convenient than the former for the purification of larger amounts of API (extraction up to 500 g of bacteria), is described in this chapter. The yields of activity during the purification of the enzyme and the specific activity of API (see below) isolated correspond to the values found in the earlier methods.

Preparation of Cells. Bacillus stearothermophilus (NICB 8924) cells at 55° were stored in freeze-dried state (lyophilized in a 2% aqueous solution of gelatin[14]) or were kept on brain–heart infusion agar slants (BBL) at 5°. Growth was initiated in 10 ml of 3.7% brain–heart infusion broth (BBL) at 55°.

Filter-sterilized air was bubbled through the cultivation tubes. These tube cultures were used to prepare larger-batch cultures (1, 10, and 100 liters) by repeated inoculations of 10 vol%. The following liquid media could be employed: (1) brain–heart infusion, BBL; (2) Lab Lemco medium: 3 g of Lab Lemco beef extract (Oxoid), 2 g of Bacto tryptone (Difco), 2 g of glucose, 3 g of K_2HPO_4, 1 g of KH_2PO_4 per liter, pH 7.1; (3)[15] Bacto tryptone (Difco), 2%; yeast extract (Oxoid), 1.0%; Sucrose, 1.0%; K_2SO_4, 0.13%; $Na_2HPO_4 \cdot 2H_2O$, 0.32%; $MgSO \cdot 7H_2O$, 0.027%; $MnCl_2 \cdot 4H_2O$, 0.0015%; $FeCl_3 \cdot 6H_2O$, 0.0007%; citric acid, 0.032%; pH 7.1.

The large culture medium is incubated for 15–24 hr at 55°; 10 liters per minute are aerated. At the end of the incubation the cells are cen-

[13] O. H. Lowry, N. J. Rosebrough, A. L. Farr, and R. J. Randall, *J. Biol. Chem.* **193**, 265 (1951).

[14] T. J. Sinskey and G. J. Silverman, *J. Bacteriol.* **101**, 429 (1970).

[15] K. Sargeant, D. N. East, A. R. Whitaker, and R. Elsworth, *J. Gen. Microbiol.* **65**, iii (1971).

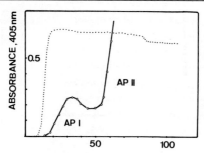

Fig. 1. Separation of aminopeptidases (AP) I and II (APIII) by Sephadex G-150 filtration Activity test: 10 μl from the fractions were assayed in 1 ml of leucine p-nitroanilide solution (1 mM). Incubation time: 6 min at 55°. (···) Protein curve (280 nm); (——) activity curve (hydrolysis of leucine-p-nitroanilide, 410 nm).

trifuged off (Sharples centrifuge). Under the conditions described, the yield is about 1 g to 10 g of bacteria (wet weight) per liter of medium.

Disintegration of the Cells. Cells, 500 g, are suspended in 1.5 liter of 0.05 M Tris-HCl buffer, pH 7.2. The cells are disrupted by pumping the cell suspension three times through a Manton–Gaulin press at 650 kg/cm². The homogenate is then centrifuged at 25,000 g for 30 min. The supernatant is collected. The following purification steps are carried out at 4°.

Step 1. Ammonium Sulfate Precipitation.[16] Solid ammonium sulfate is added to the supernatant to 47.5% saturation (295 g/liter). After 30 min, the suspension is centrifuged and the supernatant thus obtained then brought to 75% saturation with ammonium sulfate (192 g/liter). The suspension is centrifuged and the supernatant is discarded. The pellet is dissolved in 250 ml of Tris-buffer, pH 7.2.

Step 2. Sephadex G-150 Filtration. The dissolved ammonium sulfate fraction is applied to the top of a Sephadex G-150 column (10 × 100 cm), equilibrated with Tris-HCl buffer, pH 7.2. This step separates the high molecular weight API from APII, both of which hydrolyze leucine p-nitroanilide, and from APIII. Fractions of 20 ml are collected immediately after the void volume has passed (elution profile, Fig. 1).

Step 3. Heat Treatment. The API fractions from the Sephadex G-150 column are pooled. 1 M CoCl₂ solution is added with stirring to give a final concentration of 1 mM. The solution is heated to 80° for 30 min and then cooled to room temperature. The precipitate is removed by centrifugation.

Step 4. DEAE-Cellulose Chromatography. After the heat treatment the protein solution is directly applied to a DEAE-cellulose column (Cellex D, 3 × 40 cm) equilibrated with 0.05 M Tris-HCl buffer, pH 7.2,

[16] H. Hengartner, E. Stoll, and H. Zuber, *Experientia* 29, 941 (1973).

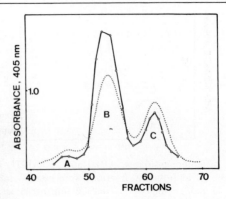

Fig. 2. Fractionation of the aminopeptidase I-hybrids by DEAE-Sephadex (A-50) chromatography. (···) Protein curve (280 nm); (——) activity curve (hydrolysis of leucine-*p*-nitroanilide, 410 nm).

containing 1 mM CoCl$_2$. The enzyme is eluted with a linear gradient of 0–0.35 M NaCl in a total volume of 1.2 liters.

Step 5. Sephadex G-150 Filtration. The active fractions from step 4 are pooled, concentrated by ultrafiltration to 20 ml and applied to a Sephadex G-150 column (5 × 100 cm), which has been equilibrated with Tris buffer (0.05 M), pH 7.2, containing 1 mM CoCl$_2$. The eluted active fractions are pooled. After this purification step the API preparation is homogeneous with respect to other proteins, but is still a mixture of the 3 API hybrids.

Step 6. DEAE-Sephadex A-50 Chromatography. The pooled fractions are further purified by chromatography on DEAE-Sephadex A-50 (2.5 × 40 cm) in Tris buffer, 0.05 M, pH 7.2, containing 1 mM CoCl$_2$. A linear gradient of 0.1–0.32 M NaCl in a total volume of 1.2 liter is applied. The enzyme was eluted in three separated activity peaks, representing three different hybrids of the aminopeptidase I (Fig. 2).

Step 7. Preparative Polyacrylamide Gel Electrophoresis. The different API hybrids after step 6 are in almost pure state. Some minor contaminants which are still present can be completely removed by preparative polyacrylamide gel electrophoresis. The electrophoresis is carried out with the Poly-Prep apparatus manufactured by Buchler Instruments (New Jersey). A 5.63% gel (80 ml) is prepared and used for separation purposes. The gel and the buffer solution are prepared as described, using the following reagents:

Preparation of the gel
  Gel buffer: 24 ml of 1 N HCl, 18.15 g of Tris, 0.25 ml of N,N,N',N'-tetramethylethylenediamine (TEMED)

Polymerization: 30 g of acrylamide and 0.8 g of bisacrylamide per 100 ml of H_2O, 0.35% ammonium persulfate
Buffers
Upper reservoir buffer: 6.32 g of Tris, 3.94 g of glycine per 1000 ml
Lower elution buffer: 12 g of Tris, 50 ml of 1 N HCl
Membrane holder buffer: 4 $\times$ concentration of the lower buffer

The ammonium persulfate is removed in a prerun. The aminopeptidase is then layered directly on top of the separating gel in 0.05 M Tris buffer, pH 7.2, containing 1 mM $CoCl_2$ and 3% sucrose. The pure fractions are pooled, concentrated by ultrafiltration and stored in 0.05 M Tris buffer, pH 7.2, containing 1 mM $CoCl_2$ or as an ammonium sulfate precipitate.

Properties[17]

Purity and Physical Properties. The purified enzyme appears to be homogeneous in the ultracentrifuge.[2] From the sedimentation constant $s^o_{20,w} = 16.5 \times 10^{-13}$ sec) and the diffusion coefficient ($D^o_{20,w} = 3.9 \times 10^{-7}$ cm² sec⁻¹), the molecular weight was calculated to be 395,000 $\pm$ 30,000 assuming a partial specific volume of 0.739 ml g⁻¹ (estimated from the amino acid composition with the aid of partial specific volumes of amino acid residues as calculated by Cohn and Edsall.[18] The molecular weight based on equilibrium centrifugation (M_{app}) is 400,000 $\pm$ 45,000. Gel filtration through Sephadex G-200 indicates a considerably higher molecular weight of about 500,000.[19] In polyacrylamide gel electrophoresis at pH 8.6[20,21] of the pure API preparation after step 5, three bands of API hybrids are found on the gel with nearly identical mobilities.[3] The rate of migration of these three API hybrids (see below) toward the anode is API C > API B > API A (API B main component). In 8 M urea and 0.01 M EDTA at pH 5.6, API dissociates into subunits ($s^o_{20,w} = 1.95 \pm 0.1 \times 10^{-13}$ sec, $D^o_{20,w} = 5.0 \pm 0.25 \times 10^{-7}$ cm² sec⁻¹, $M_{S,D} = 36,500 \pm 4,000$. Disc electrophoresis in 6 M urea at pH 4.5[3] (modified system of Reisfeld *et al.*[22]) shows the two bands of the α- and β-unit of API. On the

---

[17] Except where otherwise indicated in the following, the properties of the mixture of the API hybrids API A: 10%, API B: 70%, API C: 20%, are described.
[18] J. J. Cohn and J. T. Edsall, "Proteins, Amino Acids and Peptides," p. 375. Van Nostrand-Reinhold, Princeton, New Jersey, 1943.
[19] E. Stoll, unpublished observations.
[21] B. J. Davis, *Ann. N.Y. Acad. Sci.* **121**, 404 (1964).
[20] L. Ornstein, *Ann. N.Y. Acad. Sci.* **121**, 321 (1964).
[22] R. A. Reisfeld, V. J. Lewis, and D. E. Williams, *Nature (London)* **195**, 281 (1962).

basis of the amino acid analysis (protein concentration) and the molecular weight, the extinction coefficient was found to be 10.2 and the molar extinction coefficient $E_{280} = 4.1 \times 10^5$ (0.05 M Tris-HCl buffer, pH 7.2, 0.001 M CoCl$_2$).

Stability. The purified, native API (mixture of the three hybrids or the each single hybrid) is stable in solution (0.05 M Tris-HCl buffer, pH 7.2, 0.001 M CoCl$_2$) or as an ammonium sulfate precipitate (80% saturation) for several months if kept in a refrigerator. The enzyme (all three hybrids) is stable for 15 hr at 80° (in 0.05 M Tris-HCl buffer, pH 7.2, 0.001 M CoCl$_2$). At this temperature the activity increases by 20–30% within 30 min (increase in V_{max}, k_m is constant). In cobalt-free buffer the thermostability of API is markedly reduced (1 hr 80°:35% residual activity). The apoenzyme (below) is also very thermolabile. In 8 M urea (at pH 7–8) API is not denatured (the hydrolytic activity is only slightly reduced, if at all. After treatment with 0.3% sodium dodecyl sulfate, the activity is reduced by 8%.

Metal Chelators and Metal Content. API is a metal enzyme (complex), Co^{2+} ions conferring the highest activity upon activation (see stability) and upon recombination from apoenzyme with Co^{2+} ions. The Co^{2+} ions are strongly bound at pH 8.1; the metal enzyme complex is fairly stable up to an EDTA concentration of 0.01 M (only a slight decrease in activity occurs). At a pH of less than 6.0 the metal enzyme is converted into a stable (at room temperature), inactive apoenzyme following dialysis against 0.01 M EDTA. In polyacrylamide gel electrophoresis the apoenzyme displays a much higher migration velocity than API, which indicates possibly a dissociation of the apoenzyme into subunits or dimeric molecules at pH 9. Neutron activation analysis[4] of API isolated (mixture of hybrids) in the presence of 0.001 M Co^{2+} indicated 20.4 g-atoms of cobalt and 2.5 g-atoms of zinc bound per molecule (API$_{isol. Co,Zn}$). The apoenzyme contains 0.9 g-atom of cobalt and after recombination of the apoenzyme in the presence of 0.001 M Co^{2+} or Zn^{2+}, 17.8 g-atoms of cobalt (API$_{rec. Co}$) or 0.74 g-atom cobalt and 16.0 g-atom zinc (API$_{rec. Zn}$), respectively, are bound per molecule. The API$_{rec. Zn}$ has only 5% of the activity of the API$_{isol. Co, Zn}$, but it is more thermostable in metal-free buffer. It appears that the zinc atoms are more strongly bound in the zinc enzyme.

Enzyme Structure

API is composed of two different subunit types,[3] the α- and β-subunit. They can be separated by disc gel electrophoresis in 6 M urea at pH 4.5 and by chromatography on SE-Sephadex. The amino-terminal sequences of the α- and β-subunits (single polypeptide chains) as determined by Sequenator analysis demonstrate a remarkable degree of homology[6] (of the first 30 amino acid residues, 67% are identical or chemically similar).

Separation of Subunits by SE-Sephadex C-50 Chromatography. API (5 mg) is dissolved and dialyzed at 4° against an 8 M urea solution in 0.025 M formate buffer (pH 4.3 at room temperature) which contains 0.01 M EDTA and 5 mM cysteine or mercaptoethanol. The subunit mixture is added to an SE-Sephadex column (60 ml) which had been equilibrated at 4° with 0.02 M sodium formate buffer (pH 4.3, adjusted at room temperature). The subunits are eluted by a concentration gradient from 0 to 0.5 M NaCl in the same buffer (total volume of buffer: 300 ml).

Reactivation of API (Recombination of the Subunits). The enzyme subunits in urea can be recombined by lowering the concentration of the urea by dilution and by dialysis.[14] However, active API reactivated from a mixture of α- and β-subunits consists only of the α-type subunit, the β-subunit precipitating during the reactivation procedure. The same results are obtained by recombination of the pure α- or β-subunits.

Ratio of α- and β-Subunits (API Hybrids). The two subunits (α,β) can combine in various ratios, resulting in the hybrids $\alpha_{10}\beta_2$ (API A), $\alpha_8\beta_4$ (API B), and $\alpha_6\beta_6$ (API C).[3] In the API isolated under the condition described above (after step 5), $\alpha_8\beta_4$ is by far the main component (70%) of the mixture of the three hybrids.

The API hybrids can be separated by rechromatography on DEAE-Sephadex (see purification procedure). Since the molecular weights of the subunits are identical (SDS-disc electrophoresis) the three hybrids and the α_{12} enzyme, each composed of 12 subunits, show no difference in their molecular weights. Preliminary electron micrographs suggest that the 12 subunits are arranged in two hexagons the one stocked above the other. This arrangement would allow a certain symmetry for the proposed subunits ratios of 5:1 ($\alpha_{10}\beta_2$), 2:1 ($\alpha_8\beta_4$), and 1:1 ($\alpha_6\beta_6$) in six-membered rings. Under special conditions[5] (apoenzyme at pH 9.0) interconversion of the $\alpha_8\beta_4$ hybrid of API into α_{12} enzyme, and $\alpha_{10}\beta_2$ was observed.

Function of the Two Subunits. API consisting only of α-subunits (α_{12} enzyme) hydrolyzes leucine-p-nitroanilide and neutral peptides and, in addition slowly deformylates N-formylmethionyl peptides. However, dipeptides having amino-terminal aspartic or glutamic acid are substantially hydrolyzed only by the hybrids containing the other (β) subunit as well.[6] Since in addition Asp-Gly inhibits the hydrolysis of glutamic acid 1-(4-nitroanilide) very strongly but hardly affects the hydrolysis of leucine p-nitroanilide, it is reasonable to assume that both types of subunit have hydrolytic activity but different specificity. In this case the API hybrids can equally well be called multienzyme complexes.

Specificity. The specificity of API is similar to that of other known aminopeptidases (broad substrate specificity). It releases not only neutral (preferentially aliphatic and aromatic), but also acid and basic amino acids, including proline, from the amino end. In contrast to leucine aminopeptidase from pig kidneys[23] or eye lenses,[24] it splits leucine *p*-nitroanilide more slowly than peptides [specific activity: (a) leucine *p*-nitroanilide hydrolysis 6.6 units mg^{-1}; (b) hydrolysis of Leu-Gly, 400 units mg^{-1}; (c) hydrolysis of Gly-Leu-Tyr (splitting of Gly), 900 units mg^{-1}].

Kinetic Properties. The pH range of maximum activity for the substrate Gly-Leu-Tyr is between 9.2 and 9.4 and for the substrate leucine *p*-nitroanilide between 7.5 and 8.0. The optimum temperature for the hydrolysis of Leu-Gly (pH 7.2) is 90° (10 min). At pH 7.2 the K_m was found to be 95 mM for Leu-Gly and 8 mM (Tris-HCl buffer) for leucine *p*-nitroanilide (65°). In the case of the pure API B, the following values were found[6]: Hydrolysis of leucine *p*-nitroanilide in Tris-HCl pH 7.5: K_m, 5 mM; V_m, 7.5 units; in imidazole-HCl, pH 7.5:K_m, 0.1 mM; V_m, 0.65 unit. Hydrolysis of glutamic acid 1-(4-nitroanilide) in imidazole-HCl, pH 7.5:K_m, 0.03 mM, V_m 0.55 unit.

[23] R. Himmelhoch, this series Vol. 19 [33].
[24] H. Hanson, D. Glasser, and H. Kirschke, *Hoppe-Seyler's Z. Physiol. Chem.* 340, 107 (1965).

[44] *Aeromonas* Aminopeptidase

By John M. Prescott and Stella H. Wilkes

Aeromonas aminopeptidase is a heat-stable extracellular enzyme that can be isolated in high yield from culture filtrates of the marine bacterial species, *Aeromonas proteolytica*[1,2]; it is readily prepared in physically homogeneous form, free from traces of endopeptidase activity.[2,3] The production and isolation of this aminopeptidase are described below, whereas the procedure for isolating *Aeromonas* neutral protease, an endopeptidase also found in culture fluid of *A. proteolytica*, is given in Chapter [33] of this volume.

[1] J. M. Prescott and S. H. Wilkes, *Arch. Biochem. Biophys.* 117, 328 (1966).
[2] J. M. Prescott, S. H. Wilkes, F. W. Wagner, and K. J. Wilson, *J. Biol. Chem.* 246.
[3] S. H. Wilkes, M. E. Bayliss, and J. M. Prescott, *Eur. J. Biochem.* 34, 459 (1973).

Purification Procedure

Enzyme Production

The rapid, heavy growth, and nonfastidious nutritional requirements of *A. proteolytica* make it easy to produce large cultures of this organism in ordinary laboratory facilities. We have found laboratory-scale microbial fermentors (New Brunswick) to be efficient and convenient, although satisfactory apparatus can be improvised rather easily. Irrespective of the exact nature of equipment, the principal requirement is to provide adequate agitation and forced aeration to the culture. Of a number of media that have been tested for cell growth and enzyme production, the one that has proved best for production of the extracellular aminopeptidase consists of enzymically digested casein in diluted artificial seawater, prepared as follows.

1. *Casein Hydrolyzate.* One kilogram of casein is slowly added with stirring to 10 liters of 0.8% $NaHCO_3$ solution. After the casein has been suspended, the pH is adjusted to 8.0–8.5 by the addition of concentrated NaOH or KOH, roughly 450 mEq of base being needed. To the suspension is added 5.56 g of pancreatin (Nutritional Biochemicals), and the mixture is incubated in a cotton-plugged bottle under toluene at 37° for 40–48 hr with constant stirring; it is then stored at 5° until needed.

2. *Medium.* We have found it efficient to prepare 30 liters of medium for a production run and the subsequent isolation experiment. Medium is prepared by adding 2 liters of the casein hydrolyzate (equivalent to 200 g of casein) and 264 g of Seven Seas Marine Mix (Utility Chemical Co., Paterson, New Jersey) to each of three 14-liter fermentor jars; water is added to yield a total volume of 10 liters. This volume is the maximum that should be used in fermentor apparatus of this size because of the difficulty in controlling frothing when the containers are too full. In our earlier experiments, distilled water was used to adjust the volume, but our local tap water (pH 8.0–8.5) has proved to be equally satisfactory. The medium is autoclaved at 121° for 90 min and allowed to cool thoroughly before being moved because of the hazards inherent in handling large volumes of superheated liquids. Normally, the medium is autoclaved in fully assembled New Brunswick fermentors (Model FS314), although it can be sterilized and transferred to sterile fermentor assemblies if autoclaves large enough to accept the fully assembled apparatus are not available. When the medium has cooled to room temperature, a sterile solution of 1 g of K_2HPO_4 in 10 ml of distilled water is added; this solution is autoclaved separately to prevent precipitation of metal ions in the seawater by the phosphate. The additional phosphate has

proved to be advantageous in ensuring more uniform growth, as some lots of casein give poor yields of cells and extracellular enzyme, evidently because medium made from them is limited in phosphate. Rendering the growth medium 0.1 mM with respect to Zn^{2+} ions (also after the autoclaved medium has cooled) is an additional precaution that is sometimes used, although additional Zn^{2+} ions appear not to be so critical as the supplementary phosphate.

3. *Culture and Inoculum.* Our culture of *A. proteolytica is* a subculture of the original isolate of Merkel and Traganza,[4,5] maintained in this laboratory since 1957. A subculture from our stock has been deposited with the American Type Culture Collection (ATCC 15338). Stock cultures are maintained by monthly slant transfers on a medium consisting of Difco Bacto-Peptone (1.0 g), agar (4.0 g), $FeSO_4 \cdot 7H_2O$ (10 mg), K_2HPO_4 (10 mg), and 200 ml of seawater adjusted to pH 7.2. The ability of *A. proteolytica* to produce the enzyme is checked periodically by growing a 50-ml shake culture in the production medium described above for 18–24 hr. Normal activity is 0.1–0.4 unit/ml, assayed with L-leucyl-β-naphthylamide at 25° (see below). If activity is low, the culture should be plated and individual colonies tested for higher production, although this is rarely necessary.

Inoculum medium consisting of Seven Seas Marine Mix (20 g), Difco Bacto-Peptone (5 g), and water (500 ml) is distributed in 30-ml quantities into 125-ml Erlenmeyer flasks and autoclaved for 15 min at 121°. Loop transfers from a working slant culture are made, the inoculum is grown with vigorous shaking on a rotary shaker at room temperature for 8–12 hr, and the contents of three flasks of the rapidly dividing cells are added to each 10-liter container of autoclaved medium.

4. *Growth.* The water bath for the fermentors is set at 26°. To provide the vigorous aeration required by this organism for growth and enzyme production, filtered air is delivered at a rate of 1 liter/min per liter of medium and the culture is agitated at 400 rpm. Frothing is controlled by the addition of General Electric Antifoam 60, diluted with an equal volume of water. This silicone is unstable at sterilizing temperatures, and forms a precipitate when autoclaved at 121° for 10 min. Nevertheless, we ordinarily autoclave the silicone emulsion and minimize precipitation by cooling the solution while agitating it rapidly on a rotary shaker.

Beginning 8–10 hr after inoculation, aminopeptidase activity is assayed hourly and can be expected to reach a maximum value at approxi-

[4] J. R. Merkel and E. D. Traganza, *Bacteriol. Proc.* p. 53 (1958).
[5] J. R. Merkel, E. D. Traganza, B. B. Mukherjee, T. B. Griffin, and J. M. Prescott, *J. Bacteriol.* 87, 1227 (1964).

mately 12 hr. It is advantageous to harvest the culture filtrate at 12 hr or earlier because the production of endopeptidase increases rapidly thereafter. Cells are removed by passage through a Sharples Super Centrifuge operating at 50,000 g. The supernatant fluid is filtered through a 0.65 μm Millipore filter with a fiberglass prefilter, then through a 0.45 μm filter. These filtration steps serve to remove residual cells and debris and require little time if a large filter (293 mm) and a 20-liter pressure reservoir are used; however, they could probably be omitted without substantial effect on the isolation procedure. At this point, 22–26 liters of the original 30 liters of culture medium are recovered, the rest having been lost in transfer and by evaporation.

Enzyme Isolation

Step 1. Ammonium Sulfate Precipitation. Culture filtrate is cooled to about 5°, and 364 g of solid $(NH_4)_2SO_4$ are added to each liter. After the precipitate has been allowed to settle in the cold overnight, the supernatant liquid is removed by pumping or siphoning and the precipitate is collected by centrifugation at 13,000 g in a refrigerated (5°) centrifuge. The precipitate is dissolved in the minimal volume of 10 mM Tricine buffer, pH 8.0, and dialyzed at 4° for 18–24 hr against two changes of the same buffer. Our local tap water has proved to be as satisfactory as Tricine buffer for this dialysis step.

Step 2. Acetone Precipitation. After dialysis, the concentrated protein solution is made 0.15% with respect to NaCl and cooled to 0°, and the smallest volume of cold acetone (−5°) necessary to initiate precipitation is added. Typically about 330–370 ml of cold actone are added with stirring to 500 ml of the dialyzed enzyme solution, resulting in an acetone concentration of aproximately 40% (v/v). The precipitate, collected by centrifugation at 18,000 g, contains little aminopeptidase activity and is discarded. Sufficient acetone (840 ml) is then added to make the solution 70% (v/v) with respect to acetone, and the resulting precipitate, containing most of the aminopeptidase activity, is collected, dissolved in a minimal volume of 10 mM Tricine (pH 8.0) and dialyzed against three changes of this same buffer over a period of 36 hr.

Step 3. Heat Treatment. The thermal stability of *Aeromonas* aminopeptidase makes possible the use of a heat treatment that has proved to be highly successful in eliminating all other proteins in *A. proteolytica* culture fluids, including the neutral protease that is the most abundant extracellular protein. It is advantageous to run a pilot heating experiment on each lot of acetone-precipitated material by assaying a small sample of the dialyzed 40–70% acetone fraction, heating at 70° for 1 hr, then reassaying. Some decrease in aminopeptidase activity normally occurs, but if the loss does not exceed 25%, the rest of the preparation may be subjected to a full-scale heat treatment of several hours. If large losses in aminopeptidase activity are observed in the pilot experiment, however, additional dialysis is advisable.

The large-scale heat treatment is initiated by incubating the solution in an 80° water bath until the temperature of the enzyme solution reaches 70°, then transferring the sample to a 70° water bath. The time required to eliminate endopeptidase activity is determined by withdrawing a 50-μl sample each hour and assaying it by the gelatin-digesting test described below. Usually, 6–9 hr of treatment are required. When endopeptidase activity is no longer detectable, the sample is cooled, then dialyzed against 10 mM Tricine at pH 8.0. The heating and dialysis steps eliminate about 80% of the total protein found in the acetone-precipitated sample and result in a 4- to 5-fold increase in the specific activity of aminopeptidase.

Step 4. Chromatography. The dialyzed solution from the heat treatment typically contains about 600 mg of Lowry reactive material consisting principally of aminopeptidase with some bacterial pigments. The sample is concentrated to 30 ml by pressure filtration on an Amicon ultrafiltration membrane (UM-10), clarified by centrifugation at 10,000 g, and subjected to gel filtration on Sephadex G-75. We have found a 2.5×90 cm column to be the most practical size, although its use necessitates dividing the sample into three parts for separate chromatographic runs. The Sephadex column is equilibrated with 10 mM Tricine, pH 8.0, that is 50 μM with respect to spectrographically pure $ZnCl_2$, and a 10-ml sample is applied by ascending flow. The column is inverted and the chromatogram is developed by gravity flow elution with the same buffer at a rate of 25 ml/hr. A peak that contains opalescent material but no enzymic activity emerges, well separated from the aminopeptidase, but the amber pigment that is eluted after the enzyme peak is incompletely separated from the latter, as shown in Fig. 1. A 75% recovery of the purified aminopeptidase may be obtained, however, by judicious combination of fractions that contain no visible pigment. Electrophoresis at pH 8.5 in 7.5% polyacrylamide gel should indicate a high degree of homogeneity at this step; however, pigments are not revealed by this procedure because they migrate more rapidly than the tracking dye and, moreover, do not react with either amido black or Coomassie blue stains. Visual observation of concentrated enzyme solutions therefore provides the most reliable criterion for pigment content. The gel filtration step has also been successfully performed on Sephadex G-150 by means of ascending flow chromatography,[6] but Sephadex G-75 has provided a gen-

[6] J. M. Miller, Ph.D. Dissertation, Texas A&M University, 1972.

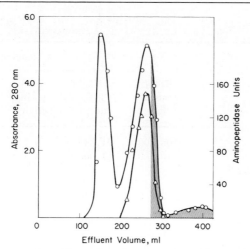

Fig. 1. Chromatography of heat-treated *Aeromonas* aminopeptidase on a 2.5 × 80 cm column of Sephadex G-75, ○, absorbance at 280 nm; △, aminopeptidase. The first peak contains opalescent material with high absorbance at 280 nm, and the shaded portion of the aminopeptidase peak indicates the presence of visible pigment; fractions containing the pigment were excluded from the enzyme preparation. The quantities of the opalescent and the pigment peaks vary somewhat with each isolation.

erally more convenient and reproducible means of separation in our hands.

If any impurities are indicated by either gel electrophoresis or visual observation of pigment, they can be removed by ion-exchange chromatography on DEAE-Sephadex A-50 in 10 mM Tricine buffer, pH 8.0, with elution by a linear gradient of NaCl from 0.2 to 0.7 M. Although seldom necessary as a purification measure, this step may be included as an additional test for homogeneity. The final product should be tested for endopeptidase activity by a 1-hr incubation of a 20-μl sample containing about 40 μg of protein, using the gelatin digesting test described below. Preparations to be used in sequence determinations may be subjected also to the oligopeptide assay for traces of endopeptidase, and should be reassayed again after several weeks, as traces of endopeptidase activity sometimes reappear upon storage of purified preparations. This probably reflects slow renaturation of some endopeptidase that has been inactivated by the heat step, but not degraded by autolysis or action of the aminopeptidase. Once detected, traces of endopeptidase activity can readily be eliminated by reheating the preparation for 1–2 hr at 70°.

Yield. Although the aminopeptidase constitutes less than 20% of the total proteins in dialyzed, concentrated culture filtrates of *A. proteo-*

TABLE I

ISOLATION OF *Aeromonas* AMINOPEPTIDASE[a]

		Aminopeptidase			
Isolation step	Total protein[b] (mg)	Total units[c]	Specific activity	Yield (%)	Purification factor
Culture filtrate	106,300	21,880	0.2	100	1
Ammonium sulfate precipitate, dialyzed	10,200	21,500	2.1	98.3	10
Acetone fractions					
0 to 43.5% (v/v) precipitate, dialyzed	2,730	940	0.3	4.3	1.5
43.5 to 70% (v/v) precipitate, dialyzed	5,350	20,300	3.8	92.8	19
Heat treated, dialyzed	565	13,320[d]	23.6	60.9	118
Sephadex G-75 chromatography	430	15,250	35.5	69.7	178

[a] The data herein have been abstracted from a previously published paper [J. M. Prescott, S. H. Wilkes, F. W. Wagner, and K. J. Wilson, *J. Biol. Chem.* **246**, 1756 (1971)].

[b] Estimated by the Lowry procedure.

[c] Determined by the colorimetric procedure of J. A. Goldbarg and A. M. Rutenburg [*Cancer* **11**, 283 (1958)] using Tricine buffer, pH 8.0, at 25°. One unit is defined as the amount of enzyme required to liberate 1 μmole of β-naphthylamine in 1 min. In this assay, the concentration of substrate in the reaction mixture is 0.34 mM, whereas the K_m of *Aeromonas* aminopeptidase has subsequently been determined to be 0.43 mM; with the spectrophotometric assay now employed, we use 0.68 mM leucyl β-naphthylamide solution. The low specific activity of the purified preparation shown here is a reflection of the difference in assay conditions.

[d] This preparation decreased in activity during the dialysis and concentration step following the heat treatment, probably because of a loss of Zn^{2+} ions, since the activity was regained upon gel filtration chromatography in the presence of 50 μM $ZnCl_2$.

lytica, we have consistently isolated 300–450 mg of purified aminopeptidase from a 30-liter culture, as is shown in Table I.

Properties

Purity. Preparations of *Aeromonas* aminopeptidase have been subjected to a variety of physical tests for heterogeneity, including electrophoresis by both gel and moving boundary techniques at pH values ranging from 4.85 to 9.5, by sedimentation velocity, immunoelectrophoresis, gel electrophoresis in the presence of dodecyl sulfate and re-chromatography on DEAE-Sephadex A-50. All these tests, as well as enzymic assays, indicate that carefully made preparations of the amino-peptidase contain a single component.

Although gel electrophoresis at pH 8.5 of unfractionated *A. proteo-lytica* culture filtrates reveals two zones with activity toward amino-peptidase substrates, the slower migrating component completely disappears during the heat treatment and the purification procedure described above always results in the isolation of the rapidly migrating aminopeptidase with no trace of the other enzyme.

Physical Properties. *Aeromonas* aminopeptidase is a metalloenzyme that contains 2 g-atoms of zinc in a single peptide with a molecular weight of 29,500, as determined by sedimentation velocity and diffusion. The value of $s_{20,w}^0$ is 3.12 S and that for $D_{20,w}$ is 9.29×10^{-7} cm^2 sec^{-1}; calculation of $\bar{v}$ from the amino acid composition gave a value of 0.724 cm^3 g^{-1}. The molecular weights obtained by sedimentation equilibrium (28,480) and gel electrophoresis in the presence of dodecyl sulfate (29,700) agree closely with that obtained from sedimentation velocity and diffusion. The enzyme is isoelectric at pH 3.0–3.5, a value consistent with its high content of acidic amino acids. The amino acid composition is half-Cys$_2$, Asp$_{30}$, Thr$_{31}$, Ser$_{26}$, Glu$_{28}$, Pro$_{13}$, Gly$_{19}$, Ala$_{29}$, Val$_{15}$, Met$_6$, Ile$_{12}$, Leu$_{16}$, Tyr$_{12}$, Phe$_9$, Lys$_8$, His$_6$, Arg$_5$, Trp$_5$. Methionine has been identified as the NH$_2$-terminal residue, and we have found the two half-cystine residues to be joined in a disulfide bridge[7]; other investigators have found that the disulfide loop consists of 29 residues or less.[8] The extinction value, $E_{1\,cm}^{1\%}$, of *Aeromonas* aminopeptidase is 14.4 at 278.5 nm, the wavelength of maximum absorbance.

Stability. *Aeromonas* aminopeptidase obviously possesses a high degree of stability, as it tolerates exposure to 70° for several hours, and it is only partially inactivated in 8 M urea. Maximum stability and activity of the enzyme both occur in the range pH 8.0–8.5. The aminopeptidase is stable to storage for several years at −20°, and it may be lyophilized and reconstituted with little loss in activity.

Specificity and Catalytic Properties. The enzyme requires substrates possessing a free α-amino group of the L-configuration, and it is also capable of releasing proline from the terminal position of certain substrates. In addition to hydrolyzing peptides, *Aeromonas* aminopeptidase cleaves amides, esters, β-naphthylamides, and *p*-nitroanilides of several amino acids. The catalytic activity of *Aeromonas* aminopeptidase

[7] S. H. Wilkes and M. E. Bayliss, unpublished results.
[8] R. A. Bradshaw and C. Coffee, personal communication, 1971.

toward the amides of all the common amino acids has been determined,[9] and the data in Table II reveal a preference by this enzyme for hydrophobic residues in the NH_2-terminal position. The penultimate residue also affects the action of this enzyme as may be seen from the differences in kinetic values for the leucyl dipeptides in Table II. Dipeptides show generally lower K_m values than amino acid amides, and those dipeptides that contain aromatic residues, methionine, or arginine in the penultimate position have higher k_{cat} values than those with hydrophobic residues.

Less specificity is evident and generally rather high hydrolytic rates are observed when the enzyme acts on oligopeptide or polypeptide substrates,[3] with enzyme-to-substrate ratios of 1:2500 having proved effective for most of the oligopeptides tested. Table III shows the residues that are liberated from several such substrates. Aspartic, glutamic, and cysteic acids are extremely resistant to removal by the aminopeptidase, whereas glycine is released slowly from oligopeptides and not at all from dipeptides. It is evident that the negative charges of glutamyl and aspartyl residues are responsible for their refractory behavior toward the enzyme, since NH_2-terminal glutaminyl and asparaginyl residues are readily liberated. Moreover, the presence of these acidic residues in the penultimate position reduces the rate of release of the NH_2-terminal residue. The ability of *Aeromonas* aminopeptidase to release proline is dependent upon the identity of the neighboring residues and evidently is enhanced by the presence of an aromatic or hydrophobic residue in the penultimate position.

Inactivation and Inhibition. *Aeromonas* aminopeptidase is inactivated by complexing agents that are capable of removing zinc from the holoenzyme. Typical conditions for inactivation are dialysis for 18 hr against 20 mM EDTA at pH 6.0, or for 72 hr against 5 mM o-phenanthroline at pH 8.0. The apoenzyme thus prepared is stable at $-20°$ and can be restored to full activity by the addition of Zn^{2+} ions. Partial activity is restored by the addition of Co^{2+} ions; Mn^{2+} ions produce a somewhat less active enzyme, and both Mg^{2+} and Ca^{2+} ions are essentially without effect.

Several compounds having the general formula $(CH_3)_2CHCH_2CH$-$(NH_2)CO$-X have been synthesized and tested for inhibition of purified *Aeromonas* aminopeptidase.[10] The structures of these compounds and their K_i values are listed in Table IV, which shows that these methyl ketone derivatives of leucine are competitive inhibitors of this enzyme.

[9] F. W. Wagner, S. H. Wilkes, and J. M. Prescott, *J. Biol. Chem.* **247**, 1208 (1971).
[10] C. M. Kettner, G. I. Glover, and J. M. Prescott, *Arch. Biochem. Biophys.* **165**, 739 (1974).

TABLE II
KINETIC VALUES FOR *Aeromonas* AMINOPEPTIDASE TOWARD SOME AMIDE, ESTER, AND DIPEPTIDE SUBSTRATES[a]

Substrate	$K_{m(app)} \pm$ SD (mM)	$k_{cat} \pm$ SD (sec^{-1})	$k_{cat}: K_{m(app)}$ (mM^{-1} sec^{-1})
Amides[b]			
Leucinamide	5.1 ± 1.0	220.0 ± 18.3	43.1
Norvalinamide	10.8 ± 2.7	66.7 ± 9.7	6.2
Norleucinamide	6.3 ± 0.5	60.0 ± 8.3	9.5
Methioninamide[c]	14.6 ± 3.4	32.0 ± 3.7	2.2
Valinamide	2.4 ± 0.9	6.5 ± 0.8	2.7
Phenylalaninamide	1.8 ± 0.1	6.2 ± 0.3	3.4
Isoleucinamide	1.0 ± 0.5	6.2 ± 1.2	6.2
Substituted amides			
Leucyl β-naphthylamide[d]	0.43 ± 0.05	29.6 ± 2.6	68.9
Leucyl-p-nitroanilide	0.019	67	3526
Esters			
Leucine methyl ester	8.5 ± 1.0	145 ± 12	17.0
Phenylalanine methyl ester	3.9 ± 0.8	41.6 ± 6.1	10.7
Leucyl dipeptides[e]			
Leu-Met	0.35 ± 0.06	53.7 ± 6.8	153.3
Leu-Arg	0.39 ± 0.07	39.3 ± 4.9	100.8
Leu-Phe	0.86 ± 0.09	72.0 ± 9.7	83.7
Leu-Trp[d]	0.96 ± 0.2	58.0 ± 9.0	60.4
Leu-Tyr[d]	1.50 ± 0.33	85.4 ± 4.8	56.9
Leu-Leu	0.18 ± 0.06	8.3 ± 0.5	46.3
Leu-Ile	0.38 ± 0.1	16.5 ± 0.33	43.4
Leu-Val	0.64 ± 0.07	15.2 ± 0.63	23.8
Leu-Ala	1.0 ± 0.08	15.4 ± 1.0	15.4
Leu-Gly	7.3 ± 2.2	31.6 ± 6.2	4.3

[a] The data shown herein are taken from F. W. Wagner, S. H. Wilkes, and J. M. Prescott, *J. Biol. Chem.* **247,** 1208 (1971).

[b] Amino acid amides that the enzyme failed to hydrolyze when tested at 20 mM concentration were alaninamide, serinamide, threoninamide, prolinamide, histidinamide, isoglutamine, isoasparagine, α-NH$_2$-butyramide, and glycinamide. Activity toward lysinamide, argininamide, tyrosinamide, and tryptophanamide was detectable by thin-layer chromatography or high-voltage electrophoresis after a 6-hr incubation at 25°.

[c] Assayed in cuvettes of 2-mm light path.

[d] Assayed in cuvettes of 1-mm light path.

[e] Leucyl dipeptides not hydrolyzed by *Aeromonas* aminopeptidase were Leu-Glu, Leu-Pro, D-Leu-L-Leu, and L-Leu-D-Leu.

TABLE III

QUALITATIVE HYDROLYSIS OF SUBSTRATES BY *Aeromonas* AMINOPEPTIDASE[a,b]

Substrate	Enzyme: substrate	Reaction time (min)	Hydrolysis
Tripeptide	1:250,000	30	Pro-Phe-Lys
Tetrapeptide	1:250,000	30	Pro-Phe-Gly-Lys
Gastrin peptide (COOH-terminal)	1:250,000	60	Trp-Met-Asp-Phe·NH_2
Pentapeptide	1:1,000	60	Phe-Asp-Ser-Ala-Val
Hexapeptide	1:25,000	60	Leu-Trp-Met-Phe-Arg-Ala
Octapeptide (COOH-terminal of glucagon)	1:1,000	30	Phe-Val-Gln-Trp-Leu-Met-Asn-Thr
Met-Lys-bradykinin	1:100	120	Met-Lys-Arg-Pro-Pro-Gly-Phe-Ser---
Adrenocorticotropic hormone	1:1,000	30	Ser-Tyr-Ser-Met-Glu-His-Phe---
Glucagon	1:2,500	140	His-Ser-Gln-Gly-Thr-Phe-Thr-Ser-Asp---
Aminoethylated B-chain insulin[c]	1:500	120	Phe-Val-Asn-Gln-His-Leu-AECys-Gly-Ser-His-Leu-Val-Glu---
Myoglobin (sperm whale)	1:150	180	Val-Leu-Ser-Glu-Gly---
Ribonuclease	1:100	120	Lys-Glu-Thr---
β-Lactoglobulin	1:200	60	Leu-Ile-Val-Thr-Thr-Met-Gln-Lys-Gly-Asp---
Lysozyme (egg white)	1:50	120	Lys-Val-Phe-Gly-Arg-Cys-Glu---
Aminoethylated lysozyme	1:300	120	Lys-Val-Phe-Gly-Arg-AECys-Glu---

[a] The data herein are taken from S. H. Wilkes, M. E. Bayliss, and J. M. Prescott, *Eur. J. Biochem.* **34**, 459 (1973).

[b] Substrate and enzyme were incubated at 37° and in 10 mM Tricine at pH 8.0. Residues liberated were identified by high-voltage electrophoresis and amino acid analysis.

[c] Reaction was carried out at pH 7.0 because of solubility of the substrate.

TABLE IV
INHIBITION CONSTANTS OF LEUCINE METHYL KETONES
FOR *Aeromonas* AMINOPEPTIDASE[a]

Abbreviated name	X[b]	$K_i{}^c$ (μM)	S.D.[d] (μM)
SALK	$-CH_2-NH-\overset{\overset{O}{\|\|}}{C}-(CH_2)_2-\overset{\overset{O}{\|\|}}{C}-OH$	6.89	1.84
SLK	$-CH_2-N$ (succinimide)	166	18.9
PLK	$-CH_2-N$ (phthalimide)	139	63.4
LeuMK	$-CH_3$	17.5	1.92
LeuCK	$-CH_2-Cl$	0.673	0.105
LeuBK	$-CH_2-Br$	0.202	0.015

[a] Reproduced with the permission of Academic Press, Inc., from C. A. Kettner, G. I. Glover, and J. M. Prescott, *Arch. Biochem. Biophys.* **165**, 739 (1974).

[b] X is substituent of compounds having the general formula

$$(CH_3)_2CHCH_2CH(NH_2)CO-X$$

[c] This value is the average of a minimum of 5 individual constants determined over a range of inhibitor concentrations from 1 to 5 K_i in 20 mM Tricine (pH 8.0) and at 25°.

[d] S.D., standard deviation of individual measurements of K_i from the mean reported in column 3.

Reversible inhibition was observed in each case, indicating that alkylation or acylation of the active site of the enzyme does not occur. One of these inhibitors, the succinamic acid derivative of leucyl bromomethyl ketone (SALK), has been coupled to aminoethyl cellulose and shown to be effective as an absorbent in affinity chromatography.[11]

[11] C. Kettner, J. Rodriguez-Absi, G. I. Glover, and J. M. Prescott, *Arch. Biochem. Biophys.* **162**, 56 (1974).

Assay Procedures

In our earlier work, aminopeptidase activity was assayed by measuring the liberation of β-naphthylamine from L-leucyl-β- naphthylamide by the method of Goldbarg and Rutenburg[12] or of Marks et al., [13] using Tricine buffer, pH 8.0, at 25°. These methods are satisfactory, but their use is restricted to discrete time values, and thus they do not permit continuous rate determinations that are possible with other methods. Activity of the *Aeromonas* aminopeptidase may be conveniently assayed by spectrophotometric recording of the rate of hydrolysis of L-leucinamide, L-leucyl-β-naphthylamide, and L-leucyl-*p*-nitroanilide. All three assays are used in this laboratory, and the latter two are described below; they have proved to be reliable on crude culture filtrates as well as with purified fractions. Protein concentrations in the crude mixture are determined by the Lowry procedure and in preparations that are relatively free of pigments, by absorbance measurements at 278.5 nm. Details of the aminopeptidase assays are described below.

1. Leucyl-β-naphthylamide Assay. The low K_m observed for leucyl-β-naphthylamide (Table II) makes this an attractive substrate, for which we have developed a convenient spectrophotometric assay. The high molar absorbance of the substrate ($E_{248nm} = 28,630$) requires a cuvette of 1-mm light path if substrate concentrations greater than K_m are to be used. For routine assays, 10 μl of a solution containing 0.03–0.1 μg of aminopeptidase are added to 0.3 ml of 0.68 mM leucyl-β-naphthylamide in 20 mM Tricine (pH 8.0), the solution is mixed by inversion, and the decrease in absorbance is measured at 25°. Complete hydrolysis of the substrate to leucine and β-naphthylamine ($E_{248nm} = 8190$) results in a decrease in absorbance of 1.39 A. One enzyme unit is the amount of enzyme that hydrolyzes 1 μmole of the substrate per minute, and purified preparations of *Aeromonas* aminopeptidase typically have specific activities of 50–60 units/mg. A stock solution (1.36 mM) of leucyl-β-naphthylamide in water is stable for several weeks at 10°, and fresh dilutions are made from this daily.

2. Leucyl-p-nitroanilide Assay. The method of Tuppy et al.[14] consists of measuring the liberation of *p*-nitroaniline by observing the increase in absorbance at 405 nm as *p*-nitroanilides are cleaved. This assay is readily adapted for use with *Aeromonas* aminopeptidase and is probably the most sensitive and convenient assay for this enzyme. A stock solution (1 mM) of the leucyl-*p*-nitroanilide is prepared by dissolving 12.57 mg

[12] J. A. Goldbarg and A. M. Rutenburg, *Cancer* **11**, 283 (1958).

[13] N. Marks, R. K. Datta, and A. Lajtha, *J. Biol. Chem.* **243**, 2882 (1968).

[14] H. Tuppy, W. Wiesbauer, and E. Wintersberger, *Hoppe-Seyler's Z. Physiol. Chem.* **329**, 278 (1962).

of substrate and 8.96 mg of Tricine in 50 ml of water. The resulting solution has a pH value of 4.5 and is stable for several weeks when stored at 10°. A 0.2 mM solution of the substrate is prepared daily by diluting the stock solution with 20 mM Tricine buffer, pH 8.0. In a typical assay, 20 μl of enzyme solution containing 0.05 μg of the aminopeptidase are added to 1.0 ml of substrate in a cuvette of 1-cm light path and 1.5-ml volume; the increase in absorbance at 405 nm is observed. One unit is defined as the amount of enzyme that will release 1 μmole of p-nitroaniline at 25° in 1 min; the liberation of this quantity results in an increase in absorbance of 10.6 Å. Specific activity of a purified preparation is about 120 units per milligram of enzyme.

3. *Gelatin-Digesting Test for Endopeptidase.* The presence of endopeptidases in preparations of aminopeptidase may be detected by the digestion of gelatin on exposed X-ray film. A small section of the film, ruled in 1 × 1-cm grids for ease of sample identification, is placed in a petri dish containing moistened cotton, and 20 μl of each enzyme solution being tested are pipetted on individual squares. The dish is covered and incubated at 37° for 30 min; the film is washed with running tap water for 1 min. The presence of endopeptidase activity is indicated by cleared zones. An estimation of the concentration of endopeptidase may be made by sequential dilutions and increased times of incubation. The method will detect 0.01 μg of *Aeromonas* endopeptidase in the 20-μl test fraction after an incubation period of 1-hr.

4. *Oligopeptide Test for Endopeptidase Activity.* Extremely sensitive tests for trace contamination with endopeptidase may be performed by incubating *Aeromonas* aminopeptidase with a pure oligopeptide of known sequence and testing for the appearance of free amino acids that cannot be accounted for by liberation from the NH$_2$-terminus. Preparations of the aminopeptidase that are contaminated with endopeptidase liberate leucine, methionine, and tyrosine from glucagon in addition to releasing the first seven NH$_2$-terminal residues of this substrate. This assay is performed by incubating enzyme and glucagon (1:2500) at 37° for 2 hr, then subjecting the mixture, along with an appropriate blank, to analysis on an amino acid analyzer. Alternatively, bovine bradykinin may be used as a substrate. This nonapeptide is not hydrolyzed by highly purified *Aeromonas* aminopeptidase, but the Gly-Phe and Pro-Phe bonds it contains are particularly susceptible to *Aeromonas* neutral protease. The enzyme and substrate are incubated in 10 mM Tricine (pH 8.0) at 37° for 2 hr in a ratio of 1:100, and the reaction mixture and blanks are analyzed in an amino acid analyzer. The detection of any amino acid, other than traces of arginine, indicates endopeptidase contamination, and the presence of free phenylalanine is particularly revealing.

[45] Extracellular Aminopeptidase from *Clostridium histolyticum*

By Efrat Kessler and Arieh Yaron

Assay Method

Principle. Clostridial aminopeptidase (CAP) is an extracellular enzyme of broad substrate specificity that has been isolated from the culture filtrate of *Clostridium histolyticum* and purified to homogeneity.[1] The enzyme cleaves efficiently various N-terminal amino acid residues, including proline, from peptides of low and of high molecular weight. The tripeptide Pro-Gly-Pro, which is resistant to most proteolytic enzymes, is used as a selective substrate for the routine assay during the purification of CAP. Proline and Gly-Pro are produced as the only products and the amount of proline formed is determined colorimetrically by the acid ninhydrin method.[2,3]

Reagents

Pro-Gly-Pro (commercially available from Miles Yeda Ltd., Rehovot, Kiryat Weizmann, Israel). A stock solution (10 mM), prepared by dissolving the peptide (2.69 mg/ml) in double-distilled water, is stored frozen at −20°.

Veronal buffer, 0.05 M, pH 8.6; sodium citrate, 0.4 M; manganous chloride, 0.1 M; and NaOH, 0.1 M. Each is passed through a Chellex-100 (analytical grade chelating resin, mesh 50–100, Bio-Rad, Richmond, California) column in Na⁺ form, preequilibrated with the respective solution, and stored at 4°. The manganous chloride solution is prepared by passing a 1 M solution through the column, and the accurate manganese concentration is determined by complexometric titration[4] with EDTA, using Eriochrome black as the indicator. The solution is diluted to 0.1 M with double-distilled water.

Mn-citrate metallobuffer is prepared fresh before the assay by mixing the above Veronal buffer, sodium citrate, MnCl₂, and NaOH (5:1:1:1, v/v).

[1] E. Kessler and A. Yaron, *Biochem. Biophys. Res. Commun.* **50**, 405 (1973).
[2] W. Troll and J. Lindsley, *J. Biol. Chem.* **215**, 655 (1955).
[3] S. Sarid, A. Berger, and E. Katchalski, *J. Biol. Chem.* **234**, 1740 (1959).
[4] F. J. Welcher, "The Analytical Uses of Ethylenediaminetetraacetic Acid." Van Nostrand-Reinhold, Princeton, New Jersey, 1958.

Ninhydrin reagent. Ninhydrin (3 g) is dissolved in a mixture of glacial acetic acid (60 ml) and 6 M phosphoric acid (40 ml) by heating to 70° with occasional shaking. The reagent is stable for at least 3 months when stored in the dark at room temperature.
Glacial acetic acid

Enzyme. The enzyme can be stored for prolonged time periods (up to one year) as a frozen solution in 0.05 M potassium phosphate buffer, pH 7.4, containing 1 mM EDTA (0.05 M PE),[5] or in 0.05 M sodium acetate, pH 5.6, 1 mM in EDTA, at an enzyme concentration of 0.1–1 mg/ml.

Procedure. The reaction solution (1 ml) is prepared by mixing 50 μl of the Pro-Gly-Pro stock solution, 200 μl of the Mn-citrate metallobuffer, and 700–750 μl of Veronal buffer. The test tubes are placed in a 40° water bath; after thermal equilibrium is reached, the reaction is started by adding 5–50 μl of enzyme solution, diluted to produce 1–10 μg of proline in 30 min. The reaction is stopped after 30 min by adding the ninhydrin reagent (2.5 ml) and glacial acetic acid (2.5 ml). The solution is heated in a boiling water bath for 30 min. After cooling, the intensity of the red color is measured by a Klett–Summerson photoelectric colorimeter, using filter No. 52. The amount of proline formed is calculated from an experimental calibration curve obtained with an equimolar mixture of Pro and Gly-Pro (in the 0.01–0.15 μmole range) under identical conditions. The color intensity given by the substrate (approximately 2% of the value obtained at 100% hydrolysis) is subtracted.

Definition of Unit and Specific Activity. One unit of activity is defined as the amount of enzyme that produces 1 μmole of proline under the above conditions. The specific activity is expressed in units per milligram of protein.

Purification Procedure

Cultivation of Bacteria and Preparation of the Culture Filtrate. Clostridium histolyticum cells (the strain was obtained from Dr. Kindler, Department of Microbiology, Tel Aviv University) are grown in the medium described by Warren and Gray[6] at 37°. Stock cultures grown for 18 hr in the same medium can be kept in stoppered test tubes at room temperature for 3–6 months.

For CAP production, the cells are grown in an 80-liter fermentor

[5] Potassium phosphate buffer, pH 7.4, containing 1 mM EDTA (dipotassium salt) is abbreviated as PE.
[6] G. H. Warren and J. Gray, Nature (London) 192, 755 (1961).

(Biotec, Sweden), completely filled with the fresh medium, and inoculated from a "starter" culture (grown for 18 hr) to an initial turbidity of 1–3 Klett Units (filter 66). This requires usually about 1% by volume of the "starter" culture. Growth at 37° is maintained for 40–44 hr, and the culture is transferred into a stainless steel container. The turbidity (Klett, filter 66) obtained is generally in the range of 170–220 Klett units. The cells are removed by continuous centrifugation, using the Szent-Györgi and Blum continuous flow system (Servall, 6–8 liter/hr, 14,000 rpm, 4°), and EDTA (1 M dipotassium salt) is added to the supernatant (1 mM). The enzyme activity in the filtrate remains constant for at least 1 week when stored at 4°. The purification procedure described below (see also Table I) is for 65 liters of culture filtrate. All the operations are performed at 4°.

Batchwise Adsorption to DEAE-Cellulose. The culture filtrate is diluted 3.5-fold with a cold solution of EDTA (1 mM). Approximately 4 liters of packed DEAE-cellulose (Whatman, DE-32), equilibrated with 0.05 M PE,[5] are added to the diluted filtrate and stirred mechanically overnight. Stirring is stopped and the adsorbent is allowed to settle for at least 2 hr. The supernatant is siphoned off, and the adsorbent is packed into a column (12 $\times$ 40 cm) under a pressure of 10 psi. The column is washed with 10 liters of 0.05 M PE,[5] and the activity is eluted with 5 liters of the above buffer, containing 0.1 M NaCl at a flow rate of 800 ml/hr. Fractions of 160 ml are collected. The enzyme activity emerges in a sharp peak (approximately 80% of the original activity is eluted in 1 liter). The main four fractions containing about 40% of the initial activity (650 ml) are pooled and concentrated by ultrafiltration (Diaflo, PM-30 membrane) to 20 ml.

Gel-Filtration on Sephadex G-150. A 10-ml aliquot is applied to a Sephadex G-150 (Pharmacia, Uppsala, Sweden) column (2.8 $\times$ 245 cm) equilibrated and eluted with 0.05 M PE[5] at a flow rate of 7–10 ml/hr, and 7-ml fractions are collected. (Collection of fractions can be started after 400 ml of effluent have emerged from the column.) Three protein peaks are obtained, the activity is found in the middle peak. The yellow pigment, originating in the growth medium, and present also after the DEAE-cellulose adsorption step, is separated from the enzyme at this stage. The main fractions of the active peak, containing about 70–80% of the applied activity are pooled (70 ml).

DEAE-Cellulose Rechromatography. The pooled fractions from the gel filtration step (70 ml) are applied to a DEAE-cellulose column (1.7 $\times$ 44 cm) equilibrated with 0.05 M PE.[5] The loaded column is washed with 200 ml of the same buffer, and the elution of activity is achieved by a linear salt gradient between the starting buffer (750 ml) and the same buffer, 0.33 M in NaCl (750 ml). The flow rate, used for application and elution, is 50 ml/hr, and 6-ml fractions are collected. The enzyme activity is eluted in a broad peak at about two-thirds of the gradient. The specific activity across the peak is constant (300–340 units/mg), and approximately 70% of the applied activity is recovered. The active fractions can be stored frozen at −20° without additional treatment or after concentration by ultrafiltration (Diaflo, PM-30 membrane). In some cases, the rechromatography on DEAE-cellulose was not required, since the enzyme obtained after the gel filtration step had a specific activity > 300 units/mg and was pure by the criteria described below.

A typical purification is summarized in Table I.

Properties

Stability. The enzyme is stable if stored frozen at −20° in 0.05 M PE[5] or in 0.05 M sodium acetate, pH 5.6, containing 1 mM EDTA. Under these conditions, it generally does not lose activity. Occasionally, however, a loss of up to 20–30% did occur. The original activity may be restored in such cases by adding dithiothreitol (2.5 mM) to the assay solution. The enzyme is very sensitive to traces of Zn^{2+}. Therefore, dialysis or ultrafiltration should be performed with EDTA-washed membranes using EDTA (1 mM) containing buffers. The enzyme loses activity upon lyophilization.

Purity and Physical Properties. The enzyme preparations obtained after the last purification step was shown to be homogeneous by polyacrylamide gel electrophoresis in the presence of sodium dodecyl sulfate (SDS)[7] and by immunodiffusion and immunoelectrophoresis, using antisera from rabbits immunized with the crude culture filtrate. Polyacrylamide gel electrophoresis at pH 8.9[8] in the absence of SDS, resulted in patterns that depended on the concentration of the enzyme solution. A single band of low mobility toward the anode is seen if the enzyme solution is stored at 4° at a concentration higher than 0.3 mg/ml, while an additional, fast moving band, generally faint and diffuse, appears if the enzyme solution was stored at a concentration lower than 0.1 mg/ml. Both components were shown to be active toward Pro-Gly-Pro and

[7] A. Shapiro, E. Viñuela, and J. V. Maizel, *Biochem. Biophys. Res. Commun.* **28**, 1815 (1967).
[8] B. J. Davis, *Ann. N.Y. Acad. Sci.* **121**, 404 (1964).

TABLE I
Purification of Clostridial Aminopeptidase

Step	Volume (ml)	Total activity[a] (units)	Total protein[b] (mg)	Recovery (%)	Specific activity (units/mg)	Purification factor
Crude culture filtrate	65,000	27,400	4,420[c]	100	6.2	1
DEAE-cellulose, ultrafiltration	20	10,850	149[c]	40	72.7	12
Sephadex G-150[d]	131	8,600	43.2	31	198	32
DEAE-cellulose rechromatography[d]	324	6,090	19.4	22	314	51

[a] Activity was determined colorimetrically by the acid ninhydrin method with Pro-Gly-Pro as the substrate (detailed in Assay Method) [see W. Troll and J. Lindsley, J. Biol. Chem. **215**, 655 (1955); S. Sarid, A. Berger, and E. Katchalski, J. Biol. Chem. **234**, 1740 (1959)].

[b] Protein concentration was determined by the method of Lowry [O. H. Lowry, M. J. Rosebrough, A. L. Farr, and R. J. Randall, J. Biol. Chem. **193**, 265 (1951)], using bovine serum albumin as the standard.

[c] Determined after precipitation with 13.3% trichloroacetic acid, and redissolving the water washed precipitate in 0.1 M NaOH.

[d] The enzyme solution obtained after the first DEAE-cellulose step was divided into two equal parts that were further processed separately. The figures given here represent combined data from the two runs.

leucine-p-nitroanilide.[9] Since the enzyme is oligomeric, the appearance of the two bands was attributed to dissociation.[10]

Absence of endopeptidase activity in the pure CAP preparation has been demonstrated by showing that no amino acids are formed on incubation of the enzyme with denatured cytochrome c.[10] In control experiments amino acids are produced with a partially purified CAP or with a mixture of pure CAP and trypsin. Gel filtration experiments with performic acid oxidized lysozyme incubated with CAP, also showed that the purified enzyme is devoid of endopeptidase activity.[1] In addition, α-N-blocked peptides, such as Z-Ala-Ala-Ala, Z-Ala-Ala-Phe-Ala, and Ac-Gly-Phe-Ala are not degraded by pure CAP.

The molecular weight of the native enzyme is 340,000 as shown by gel filtration on a calibrated Sephadex G-200 column.[11] The subunit molecular weight was found to be 56,000 by polyacrylamide gel electrophoresis of the denatured enzyme in presence of SDS after cross-linking of the enzyme with dimethyl suberimidate[12] and without cross-linking.[7] It was therefore concluded that the enzyme consists of six subunits.

Metal Ion Requirements and pH Profile. The enzyme, which is normally stored in the presence of EDTA, is not active if no metal ions are added into the assay solution. The dependence of activity on various bivalent cations was studied after removal of the phosphate and EDTA (of the buffer in which the enzyme is stored) by dialysis against 0.05 M sodium acetate, pH 5.6. Activation of the enzyme is achieved by adding Mn^{2+}, Co^{2+}, Cd^{2+}, or Ni^{2+}, the first two yielding the highest degree of activation. Mg^{2+}, Ba^{2+}, Ca^{2+}, Cu^{2+}, and Zn^{2+} do not activate the enzyme; the last two are inhibitory (see below). The dependence of activity at pH 8.6 on the Mn^{2+} and Co^{2+} concentration display typical bell-shaped curves with an optimal concentration around 50 μM for Mn^{2+} and 25 μM for Co^{2+}.

The dependence of activity on pH was studied in 0.05 M Veronal buffer (pH range 7–9) in presence of Co^{2+} or Mn^{2+}, both at their optimal concentration. A bell-shaped curve is obtained with both. The optimum pH found for the enzyme activated with Mn^{2+} is 8.6; with Co^{2+} it is 8.2. The specific activity at optimum pH is the same for both.

Inhibitors. Cu^{2+} and Zn^{2+} ions were shown to be inhibitory to the enzyme. Complete inhibition is obtained (in presence of the optimal Mn^{2+} concentration) with 1 mM Cu^{2+} or with 10 μM Zn^{2+}. The -SH reagent

[9] H. Tuppy, U. Wiesbauer, and E. Wintersberger, *Hoppe-Seyler's Z. Physiol. Chem.* **329**, 278 (1962).
[10] E. Kessler and A. Yaron, *Eur. J. Biochem.* **63**, 271 (1976).
[11] P. Andrews, *Biochem. J.* **96**, 595 (1965).
[12] G. E. Davies and G. R. Stark, *Proc. Natl. Acad. Sci. U.S.A.* **66**, 651 (1970).

p-mercuribenzoate causes complete inhibition at 1.7 mM. Diisopropyl phosphofluoridate (at a 600 M excess) has no effect on the enzyme activity.

Substrate Specificity. CAP cleaves all types of N-terminal amino acid residues, including proline and hydroxyproline, from peptides of low molecular weight (di-, tri-, and tetrapeptides) and from high-molecular-weight peptides such as poly(Pro-Gly-Pro) or reduced and carboxymethylated B chain of insulin. The free α-amino group is essential, since α-N-blocked peptides are resistant to CAP. Ionizable residues, such as Lys and Glu, some dipeptides, such as Gly-Phe, Phe-Ala, Ala-Ala, and Pro-Gly, and amino acid amides, such as alanyl amide, leucine β-naphthylamide, and leucine p-nitroanilide are hydrolyzed very slowly. Peptide bonds formed between N-terminal amino acid residues and proline are refractory to hydrolysis by CAP. This bond is known to be hydrolyzed specifically by aminopeptidase P,[13,14] and when the two enzymes are used together, stepwise degradation of polypeptides containing proline residues can be achieved. This was demonstrated with the proline-rich nonapeptide bradykinin (Arg-Pro-Pro-Gly-Phe-Ser-Pro-Phe-Arg). No hydrolysis is observed when bradykinin is incubated with CAP or with bovine lens aminopeptidase (LAP) alone. If, however, the hormone is preincubated with aminopeptidase P (6 μg/ml), and CAP (6 μg/ml) or LAP (6 μg/ml) is then added to the incubation mixture (50 mM Veronal, pH 8.6; Mn^{2+}, 2.5 mM; citrate, 10 mM; bradykinin, 0.65 mM; 40°), complete hydrolysis is achieved with CAP within 30 min. With LAP, less than 60% of serine and only about 30% of Arg$_9$ is liberated after 4 hr. This reflects the efficient action of CAP on N-terminal proline residues.

Kinetic Parameters. The kinetic parameters K_m and k_{cat} were determined for seven substrates (Table II). In these peptides, only the N-terminal residue is cleaved by CAP, since the following peptide bond either belongs to the dipeptide Gly-Gly or involves a prolyl nitrogen. Both are resistant to CAP. Affinity for a hydrophobic side chain in the N-terminal residue of the substrate is indicated by the decrease in K_m with increasing size of the hydrophobic side chain [K_m(Ala) > K_m(Val) > K_m(Phe), K_m(Leu)]. In comparison to other amino acid residues, N-terminal proline is released quite efficiently, mainly because of the relatively high k_{cat} value. The kinetic constants for Pro-Gly-Pro and Pro-Gly-Pro-Pro show that the tetrapeptide is a better substrate than the tripeptide (5-fold decrease in K_m), and since the dipeptide Pro-Gly is a very poor substrate,[1] an active site, extending over at least four amino acid residues is indicated.

---

[13] A. Yaron and D. Mlynar, *Biochem. Biophys. Res. Commun.* **32**, 658 (1968).
[14] A. Yaron and A. Berger, this series Vol. 19, p. 521.

TABLE II
KINETIC PARAMETERS

Substrate	$K_m{}^a$ (mM)	$K_i{}^b$ (mM)	$k_{cat}{}^{a,c}$ (sec^{-1})	k_{cat}/K_m (M^{-1} sec^{-1} $\times$ 10^{-4})
Leu-Gly-Gly	6.45	—	917	14.2
Phe-Gly-Gly	8.40	—	667	7.9
Val-Gly-Gly	18.5	—	276	1.5
Ala-Gly-Gly	80.0	—	1,830	2.3
Pro-Gly-Gly	40.0	—	1,360	3.4
Pro-Gly-Pro	20.0	—	561	2.8
Pro-Gly-Pro-Pro	4.44	—	337	7.6
Lys(Ant)-Phe-ONb	—	0.003^d	—	—
Lys(Ant)-Ala-ONb	—	0.03^d	—	—
Lys(Ant)-Ala-Ala-ONb	—	0.02^d	—	—
Thr(Bu^t)-Phe-Pro	—	0.5^e	—	—
Thr(Bu^t)-Phe	—	No inhibition^e	—	—

^{a} The enzymic reactions were performed under conditions described in the Assay Method. Substrate concentrations were in the range of 0.1 K_m to K_m and enzyme concentrations were 0.4–0.9 μg/ml. Initial rates were determined by following the increase of ninhydrin color according to the method described by A. T. Matheson and B. L. Tattrie [*Can. J. Biochem.* **42**, 95 (1964)], or by the acid ninhydrin method for determining proline [W. Troll and J. Lindsley, *J. Biol. Chem.* **215**, 655 (1955); S. Sarid, A. Berger, and E. Katchalski, *J. Biol. Chem.* **234**, 1740 (1959)].

^{b} Leucine *p*-nitroanilide (0.5–2.0 mM) was used as substrate; the reaction rates were followed spectrophotometrically at 405 mμ [H. Tuppy, U. Wiesbauer, and E. Wintersberger, *Hoppe-Seyler's Z. Physiol. Chem.* **329**, 278 (1962)]. Inhibition for the compounds listed is of the competitive type [E. Kessler and A. Yaron, (see Ref. 10). *Eur. J. Biochem.* **63**, 271 (1976); A. Carmel, E. Kessler, and A. Yaron, in preparation].

^{c} Assumed molecular weight 56,000.

^{d} The substrate in 0.05 M Tris-HCl, pH 7.8, MnCl$_2$ (50 μM) and dithiothreitol (2.5 mM) was incubated with clostridial aminopeptidase (CAP) (1.0–1.6 μg/ml) at 25°.

^{e} The substrate in Veronal buffer containing Mn citrate (detailed in Assay Method) was incubated with CAP (1.0–1.6 μg/ml) at 30°.

Competitive inhibition of CAP has been found with a series of substrate analogs consisting of di- and tripeptides, in which the N-terminal amino acid residue bears a bulky hydrophobic (Bu^t)[10] or aromatic (Ant)[15] group (see Table II).[16] The nitrobenzyl ester inhibitors, bearing

[15] A. Carmel, E. Kessler, and A. Yaron, manuscript in preparation.
[16] Bu^t, tertiary butyl; Ant, *o*-aminobenzoyl ("anthranyl"); –ONb, *p*-nitrobenzyloxy.

an anthranilyl fluorophore, have been originally designed as fluorogenic substrates for aminopeptidases. Leucine aminopeptidase, an enzyme that is in many respects similar to CAP, hydrolyzes the compounds quite efficiently, and the large increase in fluorescence resulting from the separation of the anthranilyl fluorophor from the quenching nitrobenzyl group makes these compounds applicable in a very sensitive quantitative method to measure the activity of leucine aminopeptidase.[15] On the other hand, the Thr(Bu^t)-peptide is known[17,18] to be a very good competitive inhibitor to leucine aminopeptidase as well.

Acknowledgment

The financial support granted by the Helena Rubinstein Foundation, Inc. is gratefully acknowledged.

[17] R. Jost, A. Masson, and H. Zuber, *FEBS Lett.* **23**, 211 (1972).
[18] R. Jost, *FEBS Lett.* **29**, 7 (1973).

[46] Acylamino Acid-Releasing Enzyme from Rat Liver

By SUSUMU TSUNASAWA and KOZO NARITA

This enzyme releases acylamino acids from acylated amino-terminal positions of a number of peptides and proteins. Hydrolysis rates depend on nature of acyl groups, terminal amino acid sequences, and tertiary structure of acyl protein substrates.[1,2] The enzyme may be useful for the removal of the N-terminal acylamino acid, from some of the N-terminally blocked peptides and proteins in amino acid sequence analysis. A similar enzyme has been found also in rabbit reticulocytes by Yoshida and Lin,[3] and acylamino acid-releasing activity exists in many animal tissues. This chapter is concerned only with studies on the enzyme isolated from rat liver.

Assay Method

Principle. The enzyme activity (*N*-acylpeptide acylaminoacylamidohydrolase) is usually assayed by the ninhydrin method according to

[1] S. Tsunazawa, K. Narita, and K. Ogata, *Proc. Jpn. Acad.* **46**, 960 (1970).
[2] S. Tsunasawa, K. Narita, and K. Ogata, *J. Biochem.* **77**, 89 (1975).
[3] A. Yoshida and H. Lin, *J. Biol. Chem.* **247**, 952 (1972).

Yemm and Cocking[4] using a synthetic substrate, N-acetyl-L-methionyl-L-threonine (Ac-L-Met-L-Thr).

Reagents

Ac-L-Met-L-Thr, 0.005 M
Sodium phosphate buffer, 0.125 M, pH 7.2
Trichloroacetic acid solution (TCA), 20%
Sodium hydroxide solution, 2.5%
Ninhydrin solution: 50 ml of 5% ninhydrin in methyl Cellosolve, 250 ml of methyl Cellosolve, and 5 ml of KCN, 0.01 M. This reagent is stored at 0°.
Acetate buffer, 4 M, pH 5.5
Aqueous ethyl alcohol, 60%

Procedure. A standard reaction mixture contains 0.1 ml of the appropriately diluted enzyme, 0.2 ml of the substrate solution, and 0.2 ml of sodium phosphate buffer. After incubation for 1 hr at 37.5°, the reaction is terminated by adding 0.5 ml of 20% TCA. The mixture is allowed to stand for about 10 min and centrifuged at 3000 g for 10 min. Then 0.5 ml of the supernatant is transferred into a test tube and it is neutralized by adding 0.5 ml of 2.5% NaOH. Next, the solution is mixed with 1.0 ml of the acetate buffer and 1.0 ml ninhydrin reagent, and heated in a boiling water bath for 15 min. After cooling, 5 ml of 60% ethyl alcohol is added as diluent and absorbance is measured at 570 nm. The number of amino groups liberated by the enzyme is calculated on the basis of a ninhydrin colorimetric standard curve for threonine. One unit of the enzyme is defined as the amount of the enzyme required for hydrolysis of 1 μmole of the substrate per hour under the above conditions. Specific activity is expressed as units per milligram of enzyme protein. Protein concentration is estimated by the method of Lowry *et al.*[5]

Remarks. In the above assay system, enzyme activity may be over-estimated by the secondary effect of contaminating aminoacylase[6,7] which would hydrolyze the acylamino acid released.

Fractionation of Enzyme

Rats (Wistar strain) weighing 130–160 g are starved overnight and decapitated; the livers are removed, weighed, and homogenized in a glass

[4] E. W. Yemm and E. C. Cocking, *Analyst* **80**, 209 (1955).
[5] O. H. Lowry, N. J. Rosebrough, A. L. Farr, and R. J. Randall, *J. Biol. Chem.* **193**, 265 (1951).
[6] J. P. Greenstein and M. Winitz, "Chemistry of the Amino Acids," Vol. 2, pp. 1755–1767. Wiley, New York, 1961.
[7] Our unpublished data.

Potter–Elvehjem homogenizer with a Teflon pestle in 2 volumes of 0.005 M sodium phosphate buffer (pH 7.2) containing 10 mM β-mercaptoethanol and 1 mM EDTA (MEP buffer). Acylamino acid-releasing enzyme is fractionated from the crude rat liver homogenates by a procedure comprising six steps, all performed at 0°–4° unless otherwise indicated. The following steps are based on the use of 100 g of liver.

Step 1. Extraction. The homogenates are centrifuged at 10,000 g for 10 min, and the supernatant is collected.

Step 2. Ammonium Sulfate Fractionation. To the extract (about 290 ml) solid $(NH_4)_2SO_4$ is added and the precipitate between 20 and 50% saturation is collected by centrifugation at 10,000 g for 30 min. The precipitate is suspended in cold 0.2 M NaCl–MEP buffer and the solution is subsequently dialyzed overnight against several changes of the same.

Step 3. Heat Treatment. The dialyzed solution is transferred into a flask and is swirled continuously in a water bath of 70° until the temperature of the solution reaches 60°. After 5 min at this temperature, it is quickly chilled in an ice bath. Precipitates are then removed by centrifugation at 10,000 g for 30 min.

Step 4. Chromatography on DEAE-Cellulose. The supernatant is applied to a DEAE-cellulose column (2.5 × 83 cm, Brown Co., 0.83 meq/g) equilibrated with 0.2 M NaCl–MEP buffer. Elution is performed

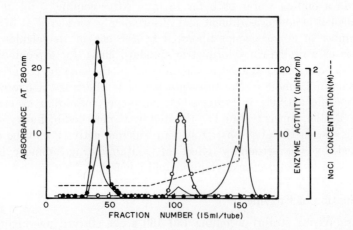

Fig. 1. Chromatography of the acylamino acid-releasing enzyme on DEAE-cellulose. The material obtained by heat treatment (total volume, 130 ml) was applied to a column of DEAE-cellulose (2.5 × 83 cm), equilibrated with 0.2 M NaCl-MEP buffer (pH 7.2). The column was washed with about 1000 ml of the equilibrating buffer, and proteins were eluted by increasing NaCl concentration linearly from 0.2 M to 0.6 M and finally with 2 M NaCl-MEP buffer (pH 7.2) as described in the text. One fraction was 15 ml. ——, Absorbance at 280 nm; ○——○, acylamino acid-releasing activity; ●——●, aminoacylase activity; - - -, NaCl concentration.

with about 1000 ml of the buffer and linearly increasing concentration of NaCl from 0.2 M to 0.6 M, and finally with 2.0 M NaCl–MEP buffer. Acylamino acid-releasing activity is clearly separated from aminoacylase activity in this step, as shown in Fig. 1.

Step 5. Chromatography on Hydroxyapatite. The pooled active fractions from DEAE-cellulose column are concentrated to a small volume (final volume, 15 ml) by ultrafiltration in a Diaflo type 52 ultrafiltration cell (Amicon Corp., membrane: XM-100). The concentrated solution is dialyzed against 0.005 M sodium phosphate buffer (pH 7.2) containing 10 mM β-mercaptoethanol (MP buffer) overnight. The dialyzed solution is applied to a hydroxyapatite column (1.5 $\times$ 10.6 cm, Bio-Rad Laboratories) which has been equilibrated with the same MP buffer. The column is washed with 3 column volumes of the equilibrating buffer. For elution, 100 ml of the same buffer is used with linearly increasing concentration of sodium phosphate from 0.02 M to 0.1 M. Finally, the column is washed 0.2 M sodium phosphate buffer (pH 7.2) containing 10 mM β-mercaptoethanol. Fractions possessing enzyme activity are collected (Fig. 2).

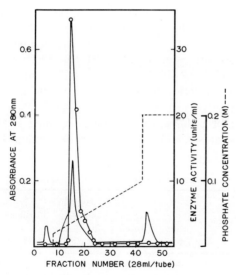

Fig. 2. Chromatography on hydroxyapatite. The enzyme fraction from the DEAE-cellulose column (total volume, 15 ml) was placed on a hydroxyaptite column (1.5 $\times$ 10.6 cm) which had been equilibrated with MP buffer (pH 7.2). After washing with three column volumes of the equilibrating buffer, the adsorbed enzyme was eluted by increasing the concentration of sodium phosphate linearly from 0.02 M to 0.1 M and finally with 0.2 M sodium phosphate buffer (pH 7.2) containing 10 mM β-mercaptoethanol as described in the text. One fraction was 2.8 ml. ——, Absorbance at 280 nm; O——O, acylamino acid-releasing activity; ---, NaCl concentration.

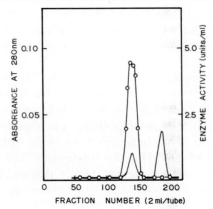

F1G. 3. Gel filtration on Sepharose 6B. The enzyme fraction from the hydroxy-apatite column (total volume, 5 ml) was applied on a Sepharose 6B column (2.5 × 94 cm), which had been equilibrated with 0.2 *M* NaCl-MEP buffer (pH 7.2) and developed with the same buffer. One fraction was 2 ml. ——, Absorbance at 280 nm; ○——○, acylamino acid-releasing activity.

Step 6. Gel Filtration on Sepharose 6B. Pooled fractions of enzyme activity are combined and dialyzed against 0.2 *M* NaCl–MEP buffer (pH 7.2) overnight, and the solution is concentrated to a small volume (final volume, 5 ml) by ultrafiltration. It is then subjected to gel filtration on a Sepharose 6B column (2.5 × 94 cm, Pharmacia), which has been equilibrated with 0.2 *M* NaCl–MEP buffer, as shown in Fig. 3. The purified enzyme fraction is concentrated to an appropriate volume by ultrafiltration.

With the above procedures, the enzyme is purified 1500-fold from the crude liver homogenates. Yields for each step are summarized in Table I.

TABLE I
PURIFICATION OF ACYLAMINO ACID-RELEASING ENZYME[a]

Step	Vol-ume (ml)	Protein (mg)	Specific activity (units/mg protein)	Purifica-tion factor	Total units	Yield (%)
Homogenate	290	21,200	0.598	1	12,600	100
Supernatant, 10,000 *g*	294	15,700	0.832	1.39	13,000	103
Precipitates at 20–50% (NH₄)₂SO₄	166	8,500	1.22	2.04	10,400	81.9
Heat treatment	130	976	7.61	12.7	7,430	58.7
DEAE-cellulose eluate	210	12.2	110	184	1,380	10.9
Hydroxyapatite eluate	37	1.57	287	480	451	3.6
Sepharose 6B eluate	2	0.15	922	1540	138	1.1

[a] Values are from 100 g of rat liver.

The final yield is very low and this may be due partly to the fact that acylamino acid-releasing activity is overestimated by the assay in crude fractions because of the secondary effect of contaminating aminoacylase.

Properties

Purity of the Enzyme. The enzyme preparation appears to be homogeneous in 7.5% cross-linked polyacrylamide gel disc electroresis with 0.05 M Tris-glycine buffer (pH 8.3) by the method of Ornstein and Davis.[8,9] In disc electrophoresis in the 5% cross-linked gel containing SDS according to Weber and Osborn,[10] it migrates as a single band.

Specificity. As shown in Table II, the enzyme yields acylamino acids and dipeptides from synthetic acyltripeptides, but it does not split acyl-amide linkages of acylamino acids, peptide linkages of N-terminal non-acylated dipeptides and pyrrolidonecarboxylamide bonds. Concerning the nature of acyl substituents, N-acetyl derivatives seem to be best. The reaction of the enzyme with N-terminally acetylated protein substrates is illustrated in Table III, also indicating that denatured ovalbumin is more susceptible to the enzyme than the native one. However, the fact that a number of N-terminal residues were identified in addition to the expected main product, suggests that the enzyme is still contaminated with an endopeptidase and cannot reliably be used in conjunction with protein substrates.

Stability. The enzyme is unstable and loses its activity completely upon freezing or lyophilization in the presence of β-mercaptoethanol. In the absence of the thiol, the enzyme loses 50% of its activity on lyophilization. Storage of the enzyme solution containing 10 mM β-mercapto-ethanol and 1 mM EDTA at 0°–4° for 1 month results in only 10% loss of activity.

pH Optimum and K_m Value. When Ac-L-Met-L-Thr is used as a substrate, the pH optimum lies between 7.2 and 7.6, and the K_m value is 0.83 mM.

Isoelectric Point. The isoelectric point of the enzyme is 4.25 as determined by the use of an isoelectric focusing apparatus (LKB Productur AB) according to the method of Vesterberg and Svensson.[11]

Molecular Weight and Subunit Structure. The approximate molecular weight of the enzyme is about 360,000 by gel filtration and about 423,000 by sucrose density gradient ultracentrifugation. SDS-polyacrylamide

[8] L. Ornstein, *Ann. N. Y. Acad. Sci.* **121**, 321 (1964).
[9] R. J. Davis, *Ann. N. Y. Acad. Sci.* **121**, 373 (1964).
[10] K. Weber and M. Osborn, *J. Biol. Chem.* **244**, 4406 (1969).
[11] O. Vesterberg and H. Svensson, *Acta Chem. Scand.* **20**, 820 (1966).

TABLE II
RELATIVE HYDROLYSIS RATES OF N-ACYLATED PEPTIDES BY ACYLAMINO
ACID-RELEASING ENZYME[a]

Substrate[b]	Relative hydrolysis rate (%)[c]	Main products
Ac-Met-Thr	100	Ac-Met + Thr
Ac-Met-Thr-OCH$_3$	28.7	Ac-Met + Thr-OCH$_3$
Ac-Met-Thr-NH$_2$	65.0	Ac-Met + Thr-NH$_2$
F-Met-Thr	28.6	F-Met + Thr
Ac-Met-Ala	94.8	Ac-Met + Ala
Ac-Met	0	None
Ac-Met-OCH$_3$	37.6	Ac-Met
Ac-Met-NH$_2$	5.3	Ac-Met
Ac-Ala-Ala	38.0	Ac-Ala + Ala
Ac-D-Ala-Ala	0	None
Ac-Leu-Ala-Gly	62.5	Ac-Leu + Ala-Gly
Ac-Phe-Gly-Phe	27.5	Ac-Phe + Gly-Phe
Ac-Trp-Ala	0	None
Ac-Ser-Tyr[d]	67.5	Ac-Ser + Tyr
Ac-Ala-Tyr[e]	79.0	Ac-Ala + Tyr
Ac-Gly-Ser[f]	13.8	Ac-Gly + Ser
Ac-Ala-Leu[g]	0	None
Ac-Gly-Asp-Val-Glu[h]	0	None
Ac-Gly	0	None
Met-Thr	0	None
⟨Glu-Gln-Trp	0	None
⟨Glu-Asn-Trp	0	None
⟨Glu-Gly-Arg-Pro-Pro-Gly	0	None
⟨Glu-Lys-Phe-Ala-Pro	0	None

[a] Rates of hydrolysis were measured after incubation with the enzyme for 1 hr using 2.0 mM substrates with the exceptions of Ac-Ser-Tyr and Ac-Ala-Tyr, which were used at 4.0 mM concentration, since serine and alanine residues involved in the last two peptides were in DL-form.

[b] Abbreviations: Ac-, acetyl; F-, formyl; -OCH$_3$, methyl ester; -NH$_2$, amide; ⟨Glu-, pyroglutamyl.

[c] The values listed are based on the rate of hydrolysis of Ac-Met-Thr selected to be 100. Ac-Met-Thr was hydrolyzed at a rate of 0.72 μmole/hr at 37.5° by the enzyme preparation (0.8 μg).

[d] N-Terminal peptide of tobacco mosaic virus coat protein [K. Narita, *Biochim. Biophys. Acta* **28**, 184 (1958)]. Serine residue involved was DL-form.

[e] N-Terminal peptide of cucumber virus 4 coat protein [K. Narita, *Biochim. Biophys. Acta* **31**, 372 (1959)]. Alanine residue involved was DL-form.

[f] N-Terminal peptide of ovalbumin [K. Narita and J. Ishii, *J. Biochem.* **52**, 367 (1962)].

[g] N-Terminal peptide of α-chain of frog hemoglobin [J. P. Chavet and R. Acher, *FEBS Lett.* **1**, 305 (1968)].

[h] N-Terminal peptide of vertebrate heart cytochromes c [G. Kreil and H. Tuppy, *Nature (London)* **192**, 1121 (1961); K. Titani, K. Narita, and K. Okunuki, *J. Biochem.* **51**, 350 (1962)].

disc-gel electrophoresis data suggest that the enzyme consists of five or six identical subunits having a subunit weight of about 75,000.

Amino Acid Composition and Terminal Residues. Amino acid composition and the N- and C-terminal amino acids of the enzyme from rat

TABLE III

N-TERMINAL RESIDUES OF THE DIGESTS OF ACETYLATED *Saccharomyces oviformis* CYTOCHROME *c* AND OVALBUMIN BY ACYLAMINO ACID-RELEASING ENZYME[a]

Acetylated yeast cytochrome *c* N-terminal (mole/mole)	Denatured ovalbumin N-terminal (mole/mole)	Native ovalbumin N-terminal (mole/mole)
Glu (0.10)	Ser (0.61)	Ser (0.31)
Ser (0.04)	Gly (0.20)	Gly (0.10)
Thr (0.03)	Glu (0.11)	—
Gly (0.02)	Thr (0.07)	Thr (0.04)

N-Terminal sequence

```
                Ac
                 |
Ac-Thr-Glu-Phe-Lys-                    Ac-Gly-Ser-Gly-Ile-Ala-
```

[a] Protein substrates (1 μmole) were digested with the enzyme (3.5 units) for 16 hr at 40° in 2 ml of 1% NH_4HCO_3. The terminal groups were characterized and estimated by the cyanate method using reported correction fractors [G. R. Stark and D. G. Smyth, *J. Biol. Chem.* **238**, 214 (1963)].

and pig livers are summarized in Tables IV and V. The appearance of C-terminal alanine and glycine and N-terminal alanine in less than stoichiometric amounts may indicate a heterogeneity of the protein. Glycine is the suggested N-terminal, and serine is the C-terminal residue of the enzyme.

Activators and Inhibitors. Heavy metal ions such as Hg^{2+} and Cu^{2+} inhibit enzyme activity completely, whereas other metal ions and EDTA neither inhibit nor stimulate. Among sulfhydryl blocking reagents, *p*-chloromercuribenzoate and $HgCl_2$ inhibit activity completely, but mono-iodoacetic acid and monoiodoactamide do not, or do so only slightly. Diisopropyl fluorophosphate also inhibits the activity, but soybean trypsin inhibitor (Kunitz) has no effect.

Distribution

In a preliminary study on the subcellular distribution of the enzyme, it has been found in the supernatant fraction by centrifugation at 105,000 *g* for 90 min. Study on tissue localization of the enzyme activity suggests that the enzyme activity exists in many tissues of animals. It is high in liver, spleen, and reticulocytes, and low in muscle, heart, and brain in rat. A similar distribution was reported by Yoshida and Lin[3] for the formylmethionine-releasing enzyme. The apparently high enzyme

TABLE IV
Amino Acid Compositions of Acylamino Acid-Releasing Enzymes from Rat and Pig Livers[a]

Amino acid	Rat enzyme residues per subunit, MW 75,000	Pig enzyme residues per subunit, MW 75,000
Asp	57.1	54.8
Thr	31.1	30.7
Ser	49.2	48.9
Glu	92.9	80.2
Pro	47.3	48.4
Gly	59.1	71.4
Ala	57.6	57.0
Half-Cys	1.3	1.3
Val	47.8	56.9
Met	12.5	13.5
Ile	21.7	20.7
Leu	83.2	74.3
Tyr	16.4	17.9
Phe	21.9	21.7
Lys	42.4	38.7
His	16.7	14.6
Arg	31.4	39.1

[a] The values listed are 24-hr hydrolyzates at 110°. No tryptophan content has been analyzed.

TABLE V
Terminal Groups of Acylamino Acid-Releasing Enzyme[a]

Amino acid	Rat liver enzyme N-terminus cyanate method[b]	Pig liver enzyme N-terminus cyanate method[b]	C-terminus Hydrazinolysis[c]	C-terminus Cpase A digestion[d]
Ser	—	—	0.78	1.01
Gly	0.88	0.85	0.31	0.54
Ala	0.26	0.24	—	0.39

[a] Number of residues per subunit.
[b] Corrected value [G. R. Stark and D. G. Smyth, J. Biol. Chem. 238, 214 (1963)].
[c] Hydrazinolysis for 6 hr at 100° (uncorrected value).
[d] Digestion for 1 hr at 37°.

activity observed in kidney may be an artifact attributable to its high content of aminoacylase.

Use of the Acylamino Acid-Releasing Enzyme for the Sequence Determination of an *N*-Acetylated Peptide

Nakamura *et al.*[12] succeeded in deducing the amino acid sequence of the acetylated octapeptide Ac-Ser-Glu-Ser-Ser-Gly-Thr-Ala-Phe by the Edman-dansyl technique after the removal of the terminal acetylserine with the present enzyme described here. The acetylated octapeptide was isolated from the tryptic and chymotryptic digests of the peptide, which had been released during activation of bovine fatcor XIII by thrombin.

The octapeptide (30 nmoles) was digested in 0.1 ml of 0.1 M NH_4HCO_3 (pH 8.0) with enzyme (1.8 units) for 24 hr at 40°. The lyophilized digest was dissolved in 0.2 ml of 10% acetic acid, and it was subjected to gel filtration on a Sephadex G-15 column (1 × 45 cm) with 10% acetic acid. The acetylserine released was eluted from the column in fractions 30–40 (1 ml/fraction), and the heptapeptide in fractions 23–26. Detection of the two components was carried out by dansylation technique on the 6 N HCl hydrolyzate of an aliquot of each fraction.

[12] S. Nakamura, S. Iwanaga, T. Suzuki, Y. Mikuni, and K. Konishi, *Biochem. Biophys. Res. Commun.* **58**, 250 (1974).

[47] Carboxypeptidase C

By H. ZUBER

Carboxypeptidases with a pH optimum around 5 are common constituents of many angiospermous plants.[1-9] These enzymes, subsequently termed acid carboxypeptidases, have the ability to remove most amino

[1] H. Zuber and P. Matile, *Z. Naturforsch. Teil B* **23**, 663 (1968).
[2] H. Zuber, *Nature (London)* **201**, 613 (1964).
[3] H. Zuber, *Hoppe-Seyler's Z. Physiol. Chem.* **349**, 1337 (1968).
[4] Y. Kubota, S. Shoji, T. Funakoshi, and H. Ueki, *J. Biochem. (Tokyo)* **74**, 757 (1973).
[5] B. Sprössler, H.-D. Heilmann, E. Grampp, and H. Uhlig, *Hoppe-Seyler's Z. Physiol. Chem.* **352**, 1524 (1971).
[6] J. R. E. Wells, *Biochem. J.* **97**, 228 (1965).
[7] W. F. Carey and J. R. E. Wells, *J. Biol. Chem.* **247**, 5573 (1972).
[8] J. N. Ihle and L. S. Dure, *J. Biol. Chem.* **247**, 5034, 5041 (1972).
[9] K. Visuri, J. Mikola, and J. M. Enari, *Eur. J. Biochem.* **7**, 193 (1969).

acid residues, including that of proline, from the COOH termini of poly-peptide chains, like those from yeast[10] and *Aspergillus*.[11]

In particular carboxypeptidases from the citrus peel (exocarp of citrus fruit)[2-4] and citrus leaves[1,5] (designated citrus carboxypeptidase or carboxypeptidase C) have been purified. In this article the purification and properties of the enzymes from citrus peel will be described. Since it is not clear whether the carboxypeptidases C from both sources are identical, a short description of the purification and properties of the carboxypeptidase from citrus leaves is included.

Assay Methods

Principle. For determining the activity of carboxypeptidase C the hydrolysis of carbobenzoxyleucyl-L-phenylalanine or carbobenzoxyglut-amyl-L-tyrosine was followed: (A) by measuring the rate of release of phenylalanine (tyrosine) after ninhydrin reaction,[12] and (B) by chro-matographic determination of the liberated phenylalaine as its ninhydrin–copper complex[13] or ninhydrin–cadmium complex.[14] With crude extracts method B is recommended.

Method A

Reagents

Substrates: 0.002 M Cbo-Leu-Phe (Cbo-Glu-Tyr), 8.24 mg Cbo-Leu-Phe dissolved in 0.3 ml of ethanol and diluted with 0.1 M sodium citrate buffer pH 5.3 to 10 ml

Ninhydrin solution: ninhydrin in dimethyl sulfoxide–lithium acetate buffer pH 5.2[12]

Procedure. Substrate solution, 1.5 ml, is incubated at 30° with 1.5 ml of suitably diluted enzyme solution. After 1 min and 21 min, 1-ml ali-quots of the reaction mixture is removed and pipetted into a test tube containing 1 ml of ninhydrin solution. The mixed solution is heated at 100° for 15 min, cooled in an ice bath, and diluted with ethanol/water (1:1, v:v) to 20 ml. The extinction (ΔE) is measured at 578 nm ($d = 1$ cm). The molar extinction coefficient (ϵ) of the reaction product at 578 nm $= 18.2$ mole^{-1} cm^{-1}.

[10] R. Hayashi, S. Moore, and W. H. Stein, *J. Biol. Chem.* **248**, 2296 (1973).
[11] E. Ichishima, *Biochim. Biophys. Acta* **258**, 274 (1972).
[12] S. Moore, *J. Biol. Chem.* **243**, 6281 (1968).
[13] F. G. Fischer and H. Dörfel, *Biochem. Z.* **324**, 544 (1953).
[14] G. N. Atfield and J. O. R. Morris, *Biochem. J.* **81**, 606 (1961).

Definition of Unit. One unit = 1 μmole of substrate hydrolyzed per minute; mU per ml enzyme solution = $\Delta E \times F$ ($F = 0.11$ for the procedure described.)

Method B

Reagents

Substrate: 0.002 M Cbo-Leu-Phe in 0.1 M sodium citrate buffer, pH 5.3

Chromatography solvent: *n*-butanol:acetic acid:water, 4:1:5.

Color reagents as described by Fischer and Dörfel[13] and by Atfield and Morris[14]

Procedure. Cbo-Leu-Phe solution, 1 ml, is incubated with 1 ml of diluted enzyme solution at 40° for 1–10 min. At the end of the reaction, the pH is adjusted to 7 by 0.1 N NaOH and 0.3-ml aliquots are applied to chromatography paper (start lines 6 cm wide, Wt 1 or Schleicher and Schuell 2043 Mgl). For comparison 10 and 20 μl of a 0.1% phenylalanine solution are pipetted onto the same sheet of paper. The spots are developed with the ninhydrin reagents[13,14] and eluted with 5 ml of methanol. The colored solution is measured at 504 nm.

Definition of Unit and Specific Activity. One unit of enzyme activity is defined as the amount that hydrolyzes 1 μmole of substrate per minute. The specific activity is expressed in units per milligram of protein.

Purification Procedure

Carboxypeptidase C from Citrus (Orange) Peel

Preparation of Homogenate and Extraction. The flavedo (outermost yellow cortical layer) of fresh oranges which should have a peel as thick as possible is scraped off with a "Bircher" grater. Two hundred-gram portions of the flavedo are homogenized (homogenizer: electric mixer, usually for 5 min) with 250 ml of 2.3% sodium chloride solution previously cooled to 0°. The homogenate is cooled in an ice bath (the temperature should not rise above 5°) and mixed with Celite (160 g per 20-kg batch of oranges) and filtered through a Büchner porcelain funnel (squeezed out with a glass stopper).

Fractionation with Ammonium Sulfate. To the filtrate (0°–5°, pH about 3.7) ammonium sulfate is added to reach 75% saturation (pH 4.1). The solution then is kept at 5° for 16 hr and is subsequently centrifuged or filtered through nylon filter cloth. The sediment is dissolved in water (400 ml) and dialyzed against water (5°) for 48 hr. To the dialyzed solution ammonium sulfate is added to bring the ammonium sulfate concentration to 30%. After standing overnight at 5° the sediment is centrifuged or filtered off (fraction E_o) and discarded. Ammonium sulfate is added to the supernatant to 70% saturation, the solution is kept at 5° overnight and the precipitate centrifuged or filtered off (Fraction E_1). This precipitate is dissolved in 100 ml water and dialyzed against water. The dialyzed solution is lyophilyzed.

Column Chromatography on CM-Sephadex. Fraction E_1 is dissolved in 50 ml of 0.025 M sodium citrate buffer/0.005 M sodium oxalate (pH 5.3) and transferred to a CM-Sephadex column (CM-Sephadex C-50, 7×13 cm). Elution is performed with a linear gradient running from 0.025 M sodium citrate buffer/0.005 M sodium oxalate solution (pH 5.3) to 0.25 M sodium citrate buffer/0.05 M sodium oxalate solution (pH 5.3, total volume 10 liters). The elution rate is 60 ml/hr (200 fractions of 20 ml). The carboxypeptidase is eluted in fractions 100–150. The combined fractions are dialyzed against 0.1 M citrate buffer (pH 5.3) and concentrated by ultrafiltration (Diaflo ultrafiltration cell, Amicon Corp., Models 401 and 10, membrane UM-20E) to 2–3 ml.

Preparation of CM-Sephadex for column chromatography: The CM-Sephadex was treated first with NaOH and then with HCl—the reverse of the usual way—because this method yielded better fractionation results. Ten grams of CM-Sephadex are soaked in water overnight. The gel is treated with 500 ml of 0.5 N NaOH and with 500 ml of 0.5 N HCl on a Büchner funnel. Between these treatments and afterward the material is washed with water until it displays a neutral pH. Treatment with HCl must not exceed 10 min. Finally, the sample is washed with 0.025 M sodium citrate buffer/0.005 M sodium oxalate (pH 5.3) until the pH of the suspension is 4.8.

Fractionation on Sephadex G-100. The concentrated enzyme solution is passed through a Sephadex G-100 column (1.3×110 cm), equilibrated, and eluted with 0.1 M sodium citrate buffer, pH 5.3. The active fractions (9×2 ml) are pooled and dialyzed against 0.01 M sodium citrate buffer and concentrated by ultrafiltration. The active material obtained is largely free of aminopeptidase activity or of protease activity. The residual acetylesterase activity can be separated by fractionation or recycling on Sephadex G-200.

Fractionation on Sephadex G-200. The concentrated solution of carboxypeptidase C from the Sephadex G-100 column is placed on a Sephadex G-200 column (3.2×53 cm, equilibrated with 0.08 M sodium citrate buffer/0.02 M sodium oxalate, pH 5.3) and fractionated by recycling

(3–6 cycles, flow rate 20 ml/hr). After the last cycle the active fractions are collected, pooled, and concentrated by ultrafiltration.

A typical purification is summarized in Table I.

Carboxypeptidase from Citrus (Orange) Leaves

Extraction and Ammonium Sulfate Fractionation. One hundred grams of freshly harvested orange leaves are homogenized in 100 ml of water in a mixer. The temperature should not exceed 5° (cooling with ice). The pH is adjusted to 5.0 with 1 N HCl, then the homogenate is stirred for 2 hr at 5°. The insoluble material is centrifuged, and the supernatant is saturated with ammonium sulfate to 30%. The precipitate is discarded by centrifugation, and the enzyme activity in the supernatant is precipitated with ammonium sulfate (90% saturation). After centrifugation the precipitate is dissolved in water, dialyzed against water (4°), and concentrated by ultrafiltration.

Gel Filtration on Sephadex G-100. The concentrated carboxypeptidase solution is passed through Sephadex G-100, equilibrated with 0.01 M sodium citrate buffer pH 4.7. The active fractions were collected and concentrated by ultrafiltration.

TABLE I

PURIFICATION OF CARBOXYPEPTIDASE C FROM CITRUS PEEL[a]

Step	Procedure	Volume (ml)	Total protein (mg)[d]	Specific activity (units/mg protein) ($\times 10^{-3}$)	Purification factor	Total units[b]	Yield in activity (%)
1	Crude extract	3000	5400	7.5[c]	—	40.5	100
2	Ammonium sulfate fractionation (E_1)	—	2800	13	1.7	37.3	92
3	CM-Sephadex	600	90	380	50.6	33.9	83.7
4	Sephadex G-100	18	13	1060	141	13.8	34.1
5	Sephadex G-200	9	7	1780	237	12.5	30.9

[a] Starting from 20 kg of oranges (*Citrus sinensis*), 10–20 days old. Weight of flavedo: 110–210 g per kilogram of oranges.

[b] One unit: hydrolysis of 1 μmole of Cbo-Leu-Phe per minute (30°).

[c] "Old" oranges 7–15 U ($\times 10^{-3}$), fresh lemons 60–100 U ($\times 10^{-3}$) from the tree [H. Zuber and P. Matile, Z. *Naturforsch.*, *Teil B* **23**, 663 (1968).]

[d] Determined by the Lowry method (O. H. Lowry, N. J. Rosebaugh, A. L. Farr, and R. J. Randall, *J. Biol. Chem.* **193**, 265 (1951)], or by amino acid analysis (sum of amino acid residues).

Adsorption on DEAE-Sephadex. DEAE-Sephadex is pretreated as per instructions (Pharmacia) and equilibrated with 0.01 M sodium citrate buffer, pH 5.0. The concentrated enzyme solution in 0.01 M sodium citrate buffer, pH 5, is loaded onto the DEAE-Sephadex and washed with several portions of the same buffer. The carboxypeptidase is eluted from the DEAE-Sephadex by washing with 0.2 M sodium citrate buffer. The enzyme solution is dialyzed against 0.01 M sodium citrate buffer pH 4.7, concentrated by ultrafiltration and lyophilized.

During this described purification procedure the specific activity of the carboxypeptidase C from orange leaves is increased from about 0.04 U per milligram of protein to about 0.8 U per milligram of protein (30°, Cbo-Leu-Phe). The total units are 175 U per kilogram of leaves.

Properties

Carboxypeptidase from Orange Peel

Purity and Physical Properties.[3] The purified enzyme did not contain protein impurities, showed only a single band after analytical disc electrophoresis, and was monodisperse in the ultracentrifuge ($s_{20,w}^{\circ} = 8.14 \times 10^{-13}$ S, $D_{20,w} = 4.60 \pm 0.05 \times 10^{-7}$ cm^2 sec^{-1}, M_{app}(c = 0.4 g/dl) = 148,700 $\pm$ 6000). The molecular weight determined by gel filtration (Sephadex G-200) was 86,000.

Stability.[3] The enzyme was stable at pH 4–6 in 0.1 M sodium citrate buffer, but was inactivated above pH 6 irreversibly. In the pure state the enzyme became inactivated upon lyophilization. However, the enzyme can be conveniently concentrated by ultrafiltration.

pH Profile.[3] The pH optimum of carboxypeptidase C is between pH 5.3 and 5.7 (degradation of Cbo-Leu-Phe and Cbo-Arg-Pro).

Specificity. The carboxypeptidase C splits off most readily aromatic and aliphatic amino acid residues at the carboxyl end, whereas other neutral, basic and acid amino acids including proline were all hydrolyzed to approximately the same extent.[3] (Table II). Glycine was split off very slowly, and in proline peptides the enzyme hydrolyzes bonds on both the carboxyl and the amino side of proline. However, in both cases the reaction rate is highly dependent upon the amino acid residues in the neighborhood of the proline residue.[15] It is especially low when the imino group of proline is linked to glycine. Dipeptides in general and the peptide bonds of C-terminal hydroxyproline are not attacked.

Kinetic Properties. The Michaelis constant (K_m) was found to be 0.5 mM (hydrolysis of Cbo-Leu-Phe) and 4 mM (hydrolysis of Cbo-

[15] A. Nordwig, *Hoppe-Seyler's Z. Physiol. Chem.* **349**, 1353 (1968).

TABLE II

RELATIVE INITIAL ACTIVITY OF CARBOXYPEPTIDASE C FROM ORANGE PEEL
AND ORANGE LEAVES TOWARD A VARIETY OF BENZOYLOXYCARBONYL DIPEPTIDES

Cbo-Dipeptides	Percent activity found	Cbo-Dipeptides	Percent activity found
Orange peel peptidase[a]			
Cbo-Leu-Phe	100	Cbo-Lys-Asp	49
Cbo-Arg-Pro	39	Cbo-Val-Lys	39
Cbo-Phe-Pro	26	Cbo-Ala-Glu(NH$_2$)·OH	58
Cbo-Pro-Phe	53	Cbo-Val-His	34
Cbo-Gly-Glu	39	Cbo-Gly-Leu	46
Orange leaves peptidase[b]			
Cbo-Leu-Phe	100	Cbo-Gly-Gly	5
Cbo-Glu-Phe	54	Cbo-Gly-His	10
Cbo-Gly-Phe	6	Cbo-Gly-Pro	0
Cbo-Pro-Phe	8	Cbo-Pro-Gly	`1
Cbo-Gly-Tyr	21	Cbo-Pro-Ala	3
Cbo-Glu-Leu	26	Cbo-Phe-Pro	20
Cbo-His-Leu	298	Cbo-Ala-Arg	20
Cbo-Leu-Gly	10		

[a] From H. Zuber, *Hoppe-Seyler's Z. Physiol. Chem.* **349**, 1337 (1968). Substrate concentration 1 mM.

[b] From B. Sprössler, H.-D. Heilmann, E. Grampp, and H. Uhlig, *Hoppe-Seyler's Z. Physiol. Chem.* **352**, 1524 (1971). Substrate concentration 1.67 mM.

Arg-Pro). The enzyme could not be inhibited by EDTA (1–10 mM) or
o-phenanthroline (5–50 mM), but was inhibited to the extent of 64%
by diisopropyl fluorophosphate (DFP) (1 mM).

It is possible that the carboxypeptidase C contains zinc in the active
site (which, however, could be only partly removed from the enzyme by
treatment with a solution containing 6 M urea, 1 mM o-phenanthroline,
and 10 mM β-mercaptoethanol in 0.1 M citrate buffer, pH 5.3).[16]

Carboxypeptidase C from Orange Leaves

The enzyme showed for the most part the same properties as carboxypeptidase C isolated from orange peel.[4]

Physical Properties. When determined by gel filtration on Sephadex
G-200, the molecular weight was 175,000 ± 10,000, which is quite different from the molecular weight of the orange peel enzyme. The isoelectric point is pH 4.5 as determined by disc electrophoresis in the pH
gradient (pH 4.0–6.0; disc electrofocusing).

[16] B. von Hofsten, G. Nässén-Puu, and J. Drevin, *FEBS Lett.* **40**, 302 (1974).

Stability. The carboxypeptidase is most stable at pH 5 (4°) and unstable below pH 4.0 and above pH 6.0. The enzyme in solution can be stored for a few days at pH 4.7–5.5 (4°) or for several months at −20°.

Specificity. Carboxypeptidase C from orange leaves has the same broad substrate specificity as the peel enzyme (Table II). Owing to their low specificity the carboxypeptidases from both sources are very suitable for sequence analysis.[4,17,18]

Inhibition. In contrast to the orange peel enzyme, carboxypeptidase C is not inhibited by 1 mM DFP solution. No inhibition was observed by EDTA or by thiol reagents.

[17] H. Zuber, *Int. Congr. Biochem., 7th,* Tokyo 1967, Abstr. III, p. 541.
[18] H. Tschesche and S. Kupfer, *Eur. J. Biochem.* **26,** 33 (1972).

[48] Carboxypeptidase Y

By RIKIMARU HAYASHI

Studies on the proteases in yeast led to the discovery of a carboxypeptidase of broad specificity.[1,2] The enzyme was originally termed yeast proteinase C,[2] but is now designated carboxypeptidase Y[3] to distinguish it from similar enzymes from other sources (see Distribution and Nomenclature). The enzyme is quite different from the pancreatic carboxypeptidases A [EC 3.4.12.2] and B [EC 3.4.12.3] in many respects; carboxypeptidase Y is free of metals and has essential serine and histidine residues. Its activity is at a maximum in the acidic region. The enzyme is very soluble even at low ionic strength. The most prominent feature of carboxypeptidase Y is its ability to release proline from peptides and proteins. These unique properties render carboxypeptidase Y[4] a useful tool in the sequence analysis and limited proteolysis of proteins and peptides in combination with pancreatic enzymes.

Assay Methods

Carboxypeptidase Y hydrolyzes both peptide substrates of the pancreatic carboxypeptidases A and B, i.e., N-acylated dipeptides, and also

[1] R. Hayashi, S. Aibara, and T. Hata, *Biochim. Biophys. Acta* **212,** 359 (1970).
[2] T. Hata, R. Hayashi, and E. Doi, *Agric. Biol. Chem.* **31,** 150 (1967).
[3] R. Hayashi, S. Moore, and W. H. Stein, *J. Biol. Chem.* **248,** 2296 (1973).
[4] The pure enyme is commercially available from the Enzyme Department of the Oriental Yeast Co., 6–10, 3 Chome, Azusawa, Itabashi-ku, Tokyo **174,** Japan.

the synthetic ester and amide substrates of chymotrypsin [EC 3.4.21.1], i.e., α,N-acetyl-L-tyrosine ethyl ester and benzoyl-L-tyrosine p-nitroanilide. Therefore, the activity of carboxypeptidase Y can be determined by a method similar to that for determining for carboxypeptidases A[5] or B[6] and chymotrypsin.[7]

In the initial stages of purification, where there are many ninhydrin-positive compounds (amino acids and peptides) and also ultraviolet-absorbing compounds (proteins and nucleic acids), a colorimetric assay for anilidase activity can be conveniently performed without interference from these compounds. The ninhydrin method for peptidase activity is convenient for analyzing large quantities of a sample, as in locating the enzyme in chromatographic eluates.

One unit of activity is defined as the amount of enzyme that produces 1 μmole of the product, or the amount of enzyme which hydrolyzes 1 μmole of substrate per minute in the respective assay.

Assay of Peptidase Activity[3,5,6,8]

Principle. This method is based on the rate of the enzymic hydrolysis of benzyloxycarbonyl-L-phenylalanyl-L-leucine (Z-Phe-Leu). The rate of the reaction can be measured either with the colorimetric ninhydrin method of Moore and Stein[9] for the estimation of liberated leucine, or spectrophotometrically by the decrease in absorbance at 224 nm.[5]

Reagents

Substrate: Z-Phe-Leu (Fluka A.G.), 0.5 mM in 0.05 M sodium phosphate buffer, pH 6.5

Enzyme. Dilute the enzyme with water or 0.01 M sodium phosphate buffer, pH 7.0, to obtain a solution containing approximately 1 μg/ml for the ninhydrin method or 300 μg/ml for the spectrophotometric method.

Procedure for the Ninhydrin Method.[3] One-tenth milliliter of enzyme solution is added to 0.9 ml of substrate solution, which has been prewarmed in a water bath at 25° in a test tube (1.5 $\times$ 17 cm). After the

[5] P. H. Pètra this series Vol. 19, p. 460.
[6] J. E. Folk, this series Vol. 19, p. 504.
[7] P. E. Wilcox, this series Vol. 19, p. 64. See also G. W. Schwert and Y. Takenaka, *Biochim. Biophys. Acta* **16**, 570 (1955).
[8] R. Hayashi, Y. Bai, and T. Hata, *J. Biochem.* **77**, 69 (1975).
[9] S. Moore and W. H. Stein, *J. Biol. Chem.* **211**, 907 (1954).

mixture has been incubated at 25° for 10 min, the reaction is terminated by the addition of ninhydrin reagent (1 ml),[9,10] and color development is performed immediately. The zero-time sample serves as a blank. Under these conditions, approximately 30% hydrolysis occurs.

For samples of unknown activity, the reaction curve should be linear up to an absorbance of about 1.0.

Procedure for the Spectrophotometric Method.[5,11] Two 1-cm cuvettes, one containing 2 ml of substrate solution and the other 2 ml of buffer, are placed in a double-beam spectrophotometer and allowed to equilibrate at 25°. The wavelength used is 224 nm. The reaction is initiated by adding 10 μl of the enzyme solution to the reaction cuvette, and the change in absorbance is recorded for 3 min. The hydrolysis of 1 μmole of substrate causes a decrease in absorbance of 0.2.

Assay of Esterase Activity[3,7,8,11]

Principle. This method is based on the titrimetric measurement of the release of protons or upon the spectrophotometric measurement of the change in ultraviolet absorbance that occurs as a result of the enzymic hydrolysis of α,N-acetyl-L-tyrosine ethyl ester (ATEE) or acetyl-L-phenylalanine ethyl ester (APEE).

A. Titrimetric method [7,8]

Reagents

Substrate: ATEE or APEE, 12.5 mM in H$_2$O
KCl, 1.25 M
Base: 0.01 N NaOH, standardized
Enzyme: 20–30 μg per milliliter of H$_2$O

Procedure. The assay is carried out in a pH stat equipped with a thermostatted vessel and a syringe containing the standardized base solution. A reaction mixture, which consists of 2 ml of substrate solution and 0.2 ml of KCl solution, is allowed to equilibrate at 25° for 5 min, then is adjusted to pH 8.0 for ATEE or pH 7.5 for APEE. The reaction is initiated by adding 50–200 μl of enzyme solution. During titration, the reaction vessel is covered with a nitrogen gas stream. The rate of the reaction is followed using the rate of consumption of the standardized alkaline solution.

[10] S. Moore, *J. Biol. Chem.* **243**, 628 (1968).
[11] Y. Bai, R. Hayashi, and T. Hata, *J. Biochem.* **77**, 81 (1975).

B. Spectrophotometric method[3,7,11]

Reagents

Substrate: ATEE, 1 mM or APEE, 5 mM, in 0.05 M sodium phosphate, pH 7.5

Enzyme: 1.2 mg per milliliter of H_2O or 0.01 M sodium phosphate buffer, pH 7.0

Procedure. Two 1-cm cuvettes, one containing 2 ml of substrate solution and the other 2 ml of buffer (0.05 M sodium phosphate, pH· 7.5) are placed in a double-beam spectrophotometer and allowed to equilibrate at 25° for 5 min. The wavelength used is 237 nm for ATEE and 230 nm for APEE. The reaction is initiated by adding 10 μl of the enzyme solution to the reaction cuvette, and the change in absorbance is recorded. The hydrolysis of 1 μmole of ATEE and APEE causes decreases in absorbance of 0.1 and 0.04, respectively.

Assay of Anilidase Activity[3,12]

Principle. This method is based on determining the amount of *p*-nitroaniline liberated during the enzyme hydrolysis of benzoyl-L-tyrosine *p*-nitroanilide (BTPNA).

Reagents

Substrate: BTPNA, 1.5 mM in dimethylformamide, stored at −20° in the dark

Buffer: Sodium phosphate, 0.1 M, pH 7.0

Inhibitor: $HgCl_2$, 0.01 M in H_2O

Enzyme: 5–30 μg per milliliter of H_2O or 0.01 M sodium phosphate buffer, pH 7.0

Procedure. The enzyme solution, aliquots of 0.2–0.8 ml, is pippetted into 15-ml tubes, then the volume in each tube is adjusted with H_2O to 0.8 ml. One milliliter of the buffer solution is added to each tube. The tubes are allowed to equilibrate at 25°, then 0.2 ml of the substrate solution is added at 30-sec intervals. After 10 min of incubation, the reaction is terminated by adding 0.5 ml of $HgCl_2$ solution. The quantity of liberated *p*-nitroaniline is estimated spectrophotometrically at 410 nm. Control tubes containing the reagents in the absence of enzyme show no self-hydrolysis. The amount of substrate hydrolyzed by the enzyme can be calculated according to the molar extinction of *p*-nitroaniline at 410 nm (ϵ = 8800).

[12] A. Aibara, R. Hayashi, and T. Hata, *Agric. Biol. Chem.* **35**, 658 (1971).

Alternatively, the liberation of p-nitroaniline is followed in a double-beam spectrophotometer by measuring the increase in absorbance at 410 nm.

Purification Procedure

General Considerations. In the autolyzate of bakers' yeast, carboxypeptidase Y is in an inactive form,[13-15] the molecular weight of which is approximately 20,000 higher than that of the active enzyme.[16] Activation is brought about by the action of yeast protease A,[17] which coexists in the autolyzate,[2,18] or by exposure to various protein denaturants.[19]

Carboxypeptidase Y is, therefore, prepared through the partial purification of the inactive form of the enzyme and protease A, followed by an activation treatment.[3,12] The procedure described here is on a laboratory scale, the procedure for large-scale production having been previously described.[3]

Step 1. Autolysis. Five pounds of fresh bakers' yeast[20] are crumbled and mixed with 500 ml of chloroform. The mixture is occasionally kneaded with a spatula until the yeast has liquefied (30–60 min). When the cooled yeast has been delivered, the container of the mixture is immersed in a water bath at 30° during kneading of the mixture. After liquefaction, the mixture is stirred for 30 min, then 1 liter of H_2O is added. The pH of the mixture is adjusted to 7.0 with 1 N NaOH under vigorous stirring, and the whole is kept at 25° for approximately 2 hr. The pH is readjusted to 7.0 with 1 N NaOH and the mixture is left at 25° for an additional 16 hr.

Step 2. Removal of Cell Debris. The autolyzed mixture is centrifuged at 4300 g for 15 min and the residue is discarded. The supernatant solutions are combined (2 liters).

Step 3. Fractionation by Ammonium Sulfate. Solid $(NH_4)_2SO_4$ (313 g/liter) is added gradually with stirring to the combined supernatant solutions, while the pH is maintained at 7.0. The suspension is stirred

[13] R. Hayashi, Y. Oka, E. Doi, and T. Hata, *Agric. Biol. Chem.* **31**, 1102 (1967).
[14] R. Hayashi, Y. Oka, E. Doi, and T. Hata, *Agric. Biol. Chem.* **32**, 359 (1967).
[15] R. Hayashi, Y. Oka, E. Doi and T. Hata, *Agric. Biol. Chem.* **32**, 367 (1967).
[16] R. Hayashi, Y. Oka, and T. Hata, *Agric. Biol. Chem.* **33**, 196 (1969).
[17] R. Hayashi and T. Hata, *Agric. Biol. Chem.* **36**, 630 (1972).
[18] T. Hata, R. Hayashi, and E. Doi, *Agric. Biol. Chem.* **31**, 357 (1967).
[19] R. Hayashi, Y. Minami, and T. Hata, *Agric. Biol. Chem.* **36**, 621 (1972).
[20] Carboxypeptidase Y from Oriental Yeast (Oriental Yeast Co.) and Fleischmann's Yeast (Standard Brands Inc.) have been found to be quite similar in all respects.

for 30 min, then the precipitate is removed by centrifugation at 8400 g for 20 min. Solid $(NH_4)_2SO_4$ (350 g/liter of the initial solution) is added to the combined supernatant solutions, while the pH is maintained as before. After the $(NH_4)_2SO_4$ is completely dissolved, the suspension is left for 2 hr, then the precipitates are collected by centrifugation at 13,000 g for 30 min.

Step 4. Activation. The precipitate (175 g in wet weight) is dissolved in 525 ml[21] of 0.05 M sodium acetate buffer, pH 5.0, and adjusted to pH 5.0 with 1 N acetic acid under stirring. To prevent the solution from forming, a few drops of n-octyl alcohol may be added. After the solution has left at 25° for 1 hr, the precipitate produced is removed by centrifugation at 4300 g for 10 min. The supernatant solution is readjusted to pH 5.0, as above, and incubated at 25° for an additional 18–19 hr with a few drops of toluene added.

The procedures for steps 1–4 may be performed at room temperature, unless otherwise specified.

Step 5. Stepwise Chromatography. All operations given hereafter are performed in a cold room at 5°.

The activated enzyme solution (660 ml) is brought to pH 7.0, and dialyzed against frequent changes of 0.01 M sodium phosphate buffer, pH 7.0. After dialysis, solid NaCl is added to make the solution 0.1 M in NaCl. The solution is then poured onto a column of DEAE-cellulose[22] (3.3 × 25 cm; volume 200 ml) equilibrated with 0.01 M sodium phosphate, pH 7.0, containing 0.1 M NaCl. After all the solution has passed through, the column is washed with 400 ml (twice the column volume) of the starting buffer. Then, the starting buffer is changed to the same buffer containing 0.3 M NaCl. The first eluate (100 ml, light yellow) is discarded, and the next eluate (100 ml, dark brown) is collected. Fractions containing the enzyme coincide with the elution of the dark eluate. In this step, nucleic acids have been completely removed and the yield of the enzyme is 80%.

In eluting the enzyme, it is recommended that one measure the enzymic activity and absorbances at 260 nm and 280 nm with each 50 ml of the eluate in order to check the separation of proteins and nucleic acids[23] and also to obtain the enzyme in high yield. A small amount of contamination may be included in order to obtain a good yield of the

[21] For 100 g of wet precipitate, 300 ml of the buffer are added to reduce the concentration of $(NH_4)_2SO_4$ to approximate 0.8 M. This salt concentration does not interfere with the activation of carboxypeptidase Y.

[22] DE-32 or DE-52 (Whatman) is preferable.

[23] Protein and nucleic acid contents of the eluate are estimated using the spectrophotometric method [D. Warburg and W. Christian, Biochem. Z. 310, 384 (1946)].

enzymic activity. The contaminating nucleic acids are removed during subsequent purification.

The enzyme fraction is collected and is precipitated by dialysis against two batches of saturated $(NH_4)_2SO_4$ (1 liter each); then it is adjusted to pH 7 at 5° over a period of 30 hr. The precipitate is collected by centrifugation at 20,000 g for 20 min.

Step 6. Chromatography on DEAE-Sephadex A-50. The precipitated enzyme is dissolved in a minimum volume of 0.01 M sodium phosphate, pH 7.0, containing 0.1 M NaCl; then it is dialyzed against this buffer with frequent changes for 30 hr. The enzyme solution is applied to a column of DEAE-Sephadex A-50 (1.5 × 10 cm) equilibrated with the same buffer. After the column has been washed with the starting buffer, elution is performed with a linear increase in the NaCl concentration from 0.1 M to 0.42 M (each chamber, 120 ml) at a flow rate of 20 ml/hr. A typical pattern of this chromatography is shown in Fig. 1. Active fractions are collected.

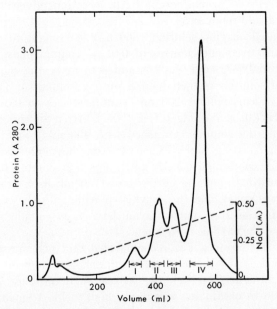

FIG. 1. Typical pattern of a first chromatography on DEAE-Sephadex A-50. Experimental conditions: column, 1.6 × 24 cm; sample volume, 10 ml (21.4 mg of protein); fraction volume, 5 ml; flow rate, 19 ml/hr; elution carried out at 5° with 0.01 M sodium phosphate, pH 7.0, with an increase of NaCl concentration from 0.1 to 0.42 M. Fractions obtained from each peak I: not identified; peaks II and III: protease A, another acid protease, and aminopeptidases; peak IV: carboxypeptidase Y. The fractions marked with horizontal arrows at peak IV were pooled for rechromatography.

Step 7. Rechromatography. Pooled fractions are dialyzed against saturated $(NH_4)_2SO_4$, as in step 5. The enzyme precipitated is collected by centrifugation at 20,000 g for 30 min. The precipitate is dissolved and dialyzed as in step 6. The enzyme solution is applied to a column of DEAE-Sephadex A-50 (1.5 × 10 cm) equilibrated as in step 6. The enzyme is eluted with a linear increase in the NaCl concentration from 0.2 to 0.42 (each chamber, 120 ml) at a flow rate of 20 ml/hr. Carboxypeptidase Y gives a single peak, and its specific activity (about 90 units/mg for the hydrolysis of Z-Phe-Leu) is constant across the peak. The pooled fraction is concentrated with $(NH_4)_2SO_4$ as described above, or lyophilized after dialysis against H_2O for 30 hr.

The degree of purification (about 450-fold) and the yield (45 mg, 30% recovery) are calculated from activities measured after an aliquot from step 2 is treated to convert the inactive form to the active enzyme (see below).

Estimation of the Potential Activity. In the extract of step 2, carboxypeptidase Y is almostly inactive.[13-15] Its potential activity is estimated after the enzyme is activated by either given method: (1) An aliquot from step 2 is adjusted to pH 5 by the addition of 1 N HCl and is allowed to stand at 25° for 19 hr with a few drops of toluene added.[13-15] (2) A 2-ml aliquot from step 2 is mixed with 1 ml of redistilled dioxane and is incubated at 25° for 15 min.[16] Activity is immediately assayed.

Discussion. The autolysis of yeast performed here is based on the modified method[3] of Lenney.[24] One hundred milliliters of chloroform per pound of yeast (half the volume used previously[2]) has been found to be enough to cause solubilization of all the potential activity of the enzyme.[3] This change has facilitated centrifugation by preventing separation of the chloroform layer and by decreasing the extraction of nucleic acids.[3]

Previously, activation was conducted by immediately incubating the solution from step 2 at pH 5 overnight[12,18] (see also Kuhn *et al.*[25]). This activation method is not recommended for obtaining the enzyme in a high yield, because chloroform dissolved in the solution renders the enzyme inactive during the activation process.

Carboxypeptidase Y is precipitated by $(NH_4)_2SO_4$ at more than 90% saturation. However, if solid $(NH_4)_2SO_4$ is added to the diluted enzyme solution, it is difficult to collect the enzyme precipitate by centrifugation because of the great dilution and the increase in the viscosity of the suspension. Dialysis against the saturated $(NH_4)_2SO_4$ used here

[24] J. F. Lenney, *J. Biol. Chem.* **221**, 919 (1956).

[25] R. W. Kuhn, K. A. Walsch, and H. Neurath, *Biochemistry* **13**, 3871 (1974).

is recommended. Ultrafiltration at 5° may be made to concentrate the enzyme without detectable inactivation.

Yeast proteinase A is an enzyme that can contaminate carboxypeptidase Y, because both enzymes have similar properties (sugar contents and isoelectric points).[18] No specific inhibitor for proteinase A has yet been found. Aminopeptidase, which are easily inactivated by EDTA, may also possibly contaminate carboxypeptidase Y.[26] Therefore, when using carboxypeptidase Y for sequence study, with the utmost care, a third chromatography on DEAE-Sephadex A-50 under conditions similar to those in step 7, or gel-filtration chromatography on Sephadex G-100 using 0.01 M sodium phosphate buffer, pH 7.0, containing 0.1 M NaCl and 1 mM EDTA as the eluent is recommended.

Properties

Purity. The purified enzyme produces a single band on disc electrophoresis both in the absence and in the presence of sodium dodecyl sulfate (SDS). No heterogeneity is evident from the usual physical analyses.[12]

Stability. A thick suspension of the enzyme in saturated $(NH_4)_2SO_4$ can be stored at —20° indefinitely. For use, a portion is dissolved in a minimum volume of H_2O and dialyzed against H_2O or 0.01 M sodium phosphate buffer, pH 7.0, to give an approximately 1% aqueous solution of the enzyme.[3] About 1-ml portions of the solution are stored in vials at —20°. The frozen solutions show no loss of activity for at least 2 years.[3] The activity of greatly diluted solutions (below 0.1 mg/ml) is quickly lost; therefore, these should be prepared just before use.[8] Repeated freezing and thawing of solutions of the enzyme, or prolonged storage at room temperature, could lead to autodigestion with the liberation of free amino acids, primarily the amino acids in the C-terminal portions of the enzyme.[3]

The salt-free pure enzyme can also be lyophilized with the loss of approximately 20% of its activity, based on the assay of activity just after the lyophilized powder is dissolved in H_2O. However, almost full activity is restored by leaving the solution overnight at 5° or —20°. Therefore, the lyophilized enzyme should be dissolved in H_2O or in 0.01 M sodium phosphate buffer, pH 7.0, one day before use.[27]

The enzyme is stable for 8 hr in the pH range 5.5–8.0 at 25° and for 2 hr in the pH range 6–8 at 37°. At higher temperatures, the enzyme is

[26] T. Masuda, R. Hayashi, and T. Hata, *Agric. Biol. Chem.* **39**, 499 (1975).
[27] R. Hayashi, Y. Bai, and T. Hata, *J. Biochem.* **77**, 1313 (1975).

TABLE I
PHYSICAL PROPERTIES OF CARBOXYPEPTIDASE Y[a]

Sedimentation constant, $s^\circ_{20,w}$	4.23 S
Diffusion constant, $D^\circ_{20,w}$	6.1×10^{-7} cm^2 sec^{-1}
Partial specific volume, $\bar{V}$	0.71 ml g^{-1}
Intrinsic viscosity, $[\eta]$	4.83×10^{-2} ml g^{-1}
Isoelectric point, pI	3.6
Absorbance, $A^{1\%}_{1\,cm}$ at 280 nm	15.0
ORD[g] parameters, λc	246 nm
$\quad a_0$	12.0
$\quad b_0$	90.5
Molar ellipticity, $[\theta]_{222}$	$-10,200$
Molecular weight	60,000,[b] 59,000[c]
	58,000,[d] 61,000[e]
	61,000[f]

[a] Data with the enzyme from Oriental Yeast [S. Aibara, R. Hayashi, and T. Hata, *Agric. Biol. Chem.* **35**, 658 (1971); R. Hayashi, and T. Hata, *Agric. Biol. Chem.* **36**, 630 (1972)].

[b] By $[\eta]$ and D.

[c] By S and D.

[d] By $[\eta]$ and S.

[e] By sedimentation equilibrium.

[f] By disc electrophoresis in the presence of sodium dodecyl sulfate [R. Hayashi, S. Moore, and W. H. Stein, *J. Biol. Chem.* **248**, 2296 (1973)].

[g] Optical rotatory dispersion.

most stable at pH 7.0. Its activity is quickly lost on incubation below pH 3 or above 60°.[25,28]

The enzyme is relatively stable in the presence of protein denaturants or certain solvents, in which the substrates of the enzyme are soluble or are in the denatured form so as to render their C-terminal available for the enzyme. About 80% of the activity is retained after incubation of the enzyme with 6 M urea at 25° for 1 hr.[3] In the presence of 10% methanol, the enzyme is completely stable at pH 5.5–8.0 for 8 hr at 25°. Even in 20% methanol, only 20% activity is lost after incubation at pH 7 for 24 hr.[29] For prolonged incubation of the enzyme, therefore, the presence of 10% methanol (or ethanol) is desirable to avoid microbial contamination. In the presence of 30% dioxane, 10% 2-chloroethanol, or 60% ethylene glycol, the enzyme is stable at pH 7.0 for at least 15 min.[19]

Physical Properties.[3,12,17] The physical constants of carboxypeptidase Y are listed in Table I. The data show that the enzyme is an acidic protein

[28] E. Doi, R. Hayashi, and T. Hata, *Agric. Biol. Chem.* **31**, 160 (1967).

[29] R. Hayashi, Y. Bai, and T. Hata, *J. Biochem.* **76**, 1355 (1974). Details will be published elsewhere.

TABLE II

AMINO ACID COMPOSITION OF THE CARBOXYPEPTIDASE Y FROM
FLEISCHMANN'S BAKERS' YEAST[a]

Amino acid	Residues per molecule[b]
Aspartic acid	65
Threonine	18
Serine	30
Glutamic acid	41
Proline	25
Glycine	34
Alanine	25
Half-cystine	11
Valine	30
Methionine	7
Isoleucine	20
Leucine	37
Tyrosine	24
Phenylalanine	27
Lysine	19
Histidine	9
Arginine	9
Tryptophan	11
Total[c]	442
Glucosamine	16

[a] R. Hayashi, S. Moore, and W. H. Stein, *J. Biol. Chem.* **248**, 2296 (1973).

[b] Calculations are based on a molecular weight of 61,000.

[c] The NH_3 content of the hydrolyzate, corrected for the 5 % and 10 % destruction of threonine and serine, respectively, is 62 amide-NH_3 groups per molecule.

with a molecular weight of approximately 61,000. The molar concentration of the enzyme is calculated using the $A_{280\,nm}^{1\,cm}$ of a 1 % solution, which is 15.0 and a molecular weight of 61,000.

Chemical Properties.[3,12,17] In Table II the amino acid composition of the carboxypeptidase Y from Fleischmann's yeast[3] is shown. Results are similar to those obtained with a preparation from Oriental Yeast.[12,29] The N-terminal residue is lysine, and the N-terminal sequence is Lys-Ile-Lys-Asp-Pro-Lys-Ile-Leu-Gly-Ile-Asp-Pro.[25] The C-terminal sequence is -Asp-Phe-Ser-Leu.[3] The enzyme is a glycoprotein having about 16 residues of glucosamine and about 15% hexose. The sugar content varies slightly with the lot of the enzyme preparation.

Carboxypeptidase Y has one SH-group per molecule, as determined by titration with *p*-hydroxymercuribenzoate and by measurement of the *S*-carboxymethylcysteine.[30] The SH-group is not available to iodoacetate,

[30] R. Hayashi, S. Moore, and W. H. Stein, *J. Biol. Chem.* **248**, 8366 (1973).

iodoacetamide, or to the Ellman reagent, 5,5'-dithiobis (2-nitrobenzoic acid) (see this series Vol. 25, [37]), unless the protein has been denatured.[30]

Activators. Unlike the pancreatic carboxypeptidases A and B, carboxypeptidase Y requires neither metals for its activity, nor any other cofactors.[27]

Inhibitors.[11,19,27,29,30] Carboxypeptidase Y is a diisopropyl-phospho-fluoridate (DFP)-sensitive enzyme. Its inactivation by [^{32}P]DFP is accompanied by the formation of 1 mole of labeled serine per mole of enzyme. A 15-residue ^{32}P-labeled (*) peptide has been isolated from a peptic digest and has been shown to have the sequence, His-Ile-Ala-Gly-Glu-Ser*-Tyr-Ala-His-Gly-Tyr-Ile-Pro-Val-Phe.[30] The enzyme is also inhibited stoichiometrically and irreversibly by phenylmethane sulfo-nylfluoride (PMSF), which also seems to react with the active serine residue.[27] The inactivation rate with DFP is much faster than that with chymotrypsin, and the rate with PMSF is one-fifteenth that for chymotrypsin (Table III). The pH dependency of the inactivation by DFP is similar to that of the enzymic hydrolysis of acetyl-L-phenylalanine ethyl ester: the most rapid inactivation occurs at pH 7.5–8.0.[27]

Site-specific reagents,[31] i.e., a chloromethyl ketone derivative of ben-zyloxycarbonyl-L-phenylalanine (ZPCK), inactivates both the peptidase and esterase activities of the enzyme in a much slower reaction than for chymotrypsin (Table III), while the D-isomer of ZPCK has no effect on either activity.[29] The optimum pH for the inactivation is pH 5.5–6.5, in accord with that for the enzymic hydrolysis of some peptide substrates. Using radioactive L-ZPCK, the reaction with the enzyme is shown to be essentially stoichiometric. Amino acid analyses of the ZPCK-inactivated enzyme, after performic acid oxidation, show the loss of one histidine residue from the native enzyme, with a yield of one residue of N$^\tau$-car-boxymethyl histidine. The DFP-inactivated enzyme fails to react with L-ZPCK, and conversely, the ZPCK-inactivated enzyme fails to react with DFP. This suggests that L-ZPCK inactivates the enzyme via a non-covalent enzyme–substrate complex by alkylating a single histidine at its N$^\tau$-position. Therefore, carboxypeptidase Y is a serine enzyme in which the charge relay system operates in the enzymic hydrolysis of peptide and ester substrates, as in chymotrypsin.

The enzyme is stoichiometrically inhibited by *p*-hydroxymercuri-benzoate (*p*-HMB), probably by its reaction with a single SH-group of the enzyme.[30] However, if 8 *M* urea is not added, the iodoacetate neither causes the loss of activity nor reacts with the SH-group after incubation

[31] E. Shaw, this series Vol. 25, p. 655.

TABLE III

RATES OF INACTIVATION OF CARBOXYPEPTIDASE Y AND CHYMOTRYPSIN BY
SOME ACTIVE SITE-DIRECTED REAGENTS[a]

Reagents[b]	$k(M^{-1} \sec^{-1})$	
	Carboxypeptidase Y	Chymotrypsin
TPCK	1.6^c	7.7^e
ZPCK	3.8^c	69^e, 11^f
ZAPCK	0^c	1.4^f
ZAGPCK	0.6^c	100^f
DFP	93.3^d	45^e
PMSF	16.1^d	248^e

[a] Data at pH 7.0 and 25°.
[b] TPCK, ZPCK, ZAPCK, and ZAGPCK, chloromethyl ketone (CK) derivatives
of toluenesulfonyl-L-Phe, carbobenzoxy-L-Phe, carbobenzoxy-L-Ala-L-Phe, and
carbobenzoxy-L-Ala-Gly-L-Phe, respectively; DFP, diisopropyl fluorophosphate;
PMSF, phenylmethane sulfonyl fluoride.
[c] R. Hayashi, Y. Bai, and T. Hata *J. Biol. Chem.*, **250**, 5221 (1975).
[d] R. Hayashi, Y. Bai, and T. Hata *J. Biochem.* **77**, 1313 (1975).
[e] E. Shaw, this series Vol. 25, p. 655.
[f] K. Morihara and T. Oka, *Arch. Biochem. Biophys.* **138**, 526 (1970). Data at 40°.

of the enzyme with 0.01 M reagent at pH 5.5, 7.0, or 8.0 for 18 hr at
room temperature. Iodoacetamide causes slight inhibition (10–20%)
after a similar incubation. The reaction of the active serine residue with
DFP is blocked by the presence of p-HMB. However, the SH-group is
not affected during the reaction with L-ZPCK: the p-HMB-inactivated
enzyme still reacts with the ZPCK in appreciable amounts. Therefore,
the single SH-group of the enzyme has no catalytic function, but may
be close to the active center of the enzyme.

Carboxypeptidase Y is sensitive to metal ions. Preincubation with a
10^{-4} M concentration of Cu^{2+}, Ag^+, or Hg^{2+} results in complete loss of
activity. Further, a 1 mM concentration of Cu^+, Mg^{2+}, Ca^{2+}, Ba^{2+}, Cr^{2+},
Mn^{2+}, Fe^{2+}, Fe^{3+}, Co^{2+}, or Ni^{2+} causes the loss of more than 50% of the
original activity.[27] Therefore, one must take the general precaution of
preventing contact with metal ions in purifying and handling carboxypeptidase Y.

EDTA and o-phenanthroline have no effect on enzymic activity.
Trypsin inhibitors from the soybean and the lima bean do not inhibit
the activity. An excess of poly-L-lysine causes a maximum decrease of
70% of the activity.[32] All the above inhibitions are parallel for both
peptidase and esterase activities.

[32] R. Hayashi and T. Hata, *Biochim. Biophys. Acta* **263**, 673 (1972).

TABLE IV

Type and Constant of Inhibition by Substrate and Product Analogs[a]

Inhibitor[b]	Peptidase		Esterase	
	Type[c]	K_i (mM)	Type[c]	K_i (mM)
L-Phenylalanine	C	3.9	C	0.4
D-Phenylalanine	M	24	—	—
L-Leu-L-Phe	C	0.9	C	2.6
CBZ-L-Phe	C	0.7	C	0.6
CBZ-D-Phe	M	1.7	M	1.3
Acetyl-L-Phe	C	20	C	25
Acetyl-D-Phe	N	260	—	—
L-Phe-OEt	M	8.3	M	9.1
D-Phe-OEt	M	5.3	M	11
CBZ-D-Phe-D-Leu	C	0.3	C	0.2
Acetyl-D-Phe-OEt	M	30	C	12
β-Phenylpropionate	N	4.5	M	1.9
β-Phenyl-L-lactate	M	5.7	M	2.6
t-Cinnamate	C	8.0	C	3.1

[a] Y. Bai, R. Hayashi, and T. Hata, *J. Biochem.*, **77**, 81 (1975).

[b] CBZ, benzyloxycarbonyl; OEt, ethyl ester.

[c] The type of inhibition is: C for the competitive; N for the noncompetitive; and M for the mixed type.

Product and substrate analogs act as reversible inhibitors of carboxypeptidase Y (Table IV). L-Amino acids and NH₂-blocked L-amino acids show the competitive type of inhibition, whereas their D-enantiomers cause less inhibition than the L-enantiomers with a noncompetitive or mixed type of inhibition. Some phenylalanine analogs, e.g., β-phenylpropionate and t-cinnamate, are also reversible inhibitors of carboxypeptidase Y. The type of, and constant (K_i) for, these inhibitors are generally parallel for both the peptidase and esterase activities.[11]

Specificity and Kinetic Properties

Peptidase Activity.[3,8,32] Carboxypeptidase Y has the ability to remove most amino acid residues, including that of proline, from the C-termini of proteins and peptides at pH 5.5–6.5.[32] Table V shows some of the kinetic parameters for substrates that vary in structure.[8,25] The absolute requirement for specificity is the presence of an L-amino acid in the terminal position. A change in the penultimate amino acids as well

TABLE V

Kinetic Parameters of Carboxypeptidase Y

Substrate[a]	Assay pH	k_{cat} (sec^{-1})	K_m (mM)	k_{cat}/K_m (mM^{-1} sec^{-1})	References
CBZ-Phe-Leu	6.5	130	0.1	1300	b
	6.0	15.1	0.24	61.8	c
CBZ-Phe-Phe	6.5	420	0.5	840	b
CBZ-Phe-Ala	6.5	120	0.56	210	b
CBZ-Phe-Pro	6.5	23	0.67	34	b
CBZ-Phe-Glu	4.5	24.5	0.36	67.5	b
	5.5	26.2	0.41	63.7	b
	6.5	96.5	16.5	5.8	b
CBZ-Phe-His	5.5	3.0	0.9	3.3	b
	6.5	11.6	3.2	3.6	b
CBZ-Phe-Gly	6.5	140	4.0	36.7	b
CBZ-Phe-βAla	6.5	66	9.0	7.4	b
CBZ-Leu-Phe	6.5	46	0.1	470	b
CBZ-Glu-Phe	4.5	15.5	0.05	290	b
	5.5	16.8	0.13	130	b
	6.5	13.3	3.3	4.0	b
CBZ-His-Phe	6.5	0.72	0.10	7.25	b
	7.0	1.82	0.23	8.09	b
	7.5	2.05	0.38	5.47	b
CBZ-Leu-Leu	6.5	6.85	0.07	100.7	b
	6.0	42.8	0.12	356.7	c
CBZ-Ala-Leu	6.0	462.0	1.28	360	c
CBZ-Val-Leu	6.0	14.1	0.10	141	c
CBZ-Ileu-Leu	6.0	15.0	0.10	150	c
CBZ-Ser-Leu	6.0	17.9	1.7	10.5	c
CBZ-His-Leu	6.0	1.4	1.8	0.78	c
CBZ-Pro-Leu	6.0	0.8	1.8	0.45	c
CBZ-Glu-Leu	6.0	29.8	8.3	3.6	c
CBZ-Nleu-Leu	6.0	43.6	0.03	777	c
CBZ-Ala-Ala	6.5	69.4	3.26	21.3	b
CBZ-Glu-Tyr	5.0	18.8	0.14	134.3	d
	5.5	19.0	0.50	39.6	d
	6.0	23.8	3.0	7.9	d
CBZ-His-Tyr	5.7	10.7	0.83	12.8	b
	6.5	8.2	1.33	6.2	b
	7.5	8.2	1.82	4.5	b
CBZ-Gly-Phe	6.5	2.2	1.7	1.3	b
	6.0	2.3	1.2	1.9	c
CBZ-Gly-Leu	6.0	6.1	2.95	2.05	c
CBZ-Gly-Arg	6.0	0.67	16.3	0.04	b,c
CBZ-Gly-Pro	6.5	~0	—	—	c
CBZ-Gly-Glu	4.5	0.32	1.15	0.28	b
	5.5	0.46	9.09	0.051	b
	6.0	0.67	24.1	0.03	c

TABLE V (*Continued*)

Substrate[a]	Assay pH	k_{cat} (sec^{-1})	K_m (mM)	k_{cat}/K_m (mM^{-1} sec^{-1})	References
CBZ-Gly-Gly	6.0, 6.5	~ 0	—	—	b, c
CBZ-Gly-Gly-Phe	6.5	2.18	1.75	1.3	b
Ac-Phe-Leu	6.5	36.1	0.74	48.8	b
Ac-Gly-Leu	6.0	1.9	45.3	0.04	c
Bz-Gly-Phe	6.5	3.8	7.7	0.49	b
Bz-Gly-Arg	6.5	1.44	10.0	0.14	b
CBZ-D-Phe-D-Leu	6.5	Not hydrolyzed			b
CBZ-L-Phe-D-Leu	6.5	~ 0			b
CBZ-D-Phe-L-Leu	6.5	~ 0			b
CBZ-Gly-D-Phe	6.0	Not hydrolyzed			c

[a] Unless otherwise specified, all amino acids had the L-configuration, except glycine. CBZ, benzyloxycarbonyl; Ac, acetyl; Bz, benzoyl.

[b] In 0.05 M buffer (sodium acetate, pH 3.5–6.0; sodium phosphate, pH 6.0–8.0) at 25° [R. Hayashi, Y. Bai, and T. Hata, *J. Biochem.* **77**, 69 (1975)].

[c] In 0.05 M 2-(N-morpholino)ethanesulfonic acid–0.15 M NaCl, pH 6.0, at 25° [R. W. Kuhn, K. A. Walsh, and H. Neurath, *Biochemistry* **13**, 3871 (1974)].

[d] By E. Doi, R. Hayashi, and T. Hata, *Symp. Enzyme Chem.* **17**, 289 (1965). In Japanese.

as the C-terminal amino acids results in large variations in the k_{cat} and K_m values. In general, when the penultimate and/or terminal residues have aromatic or aliphatic side chains, catalysis is high. When glycine is placed in the penultimate position, the release of the terminal amino acid is extremely low. The release of terminal histidine, arginine, and lysine is relatively slow. The nature of NH_2-blocking group has a decisive effect on the reaction rate of dipeptide substrates: the benzyloxycarbonyl group is the most effective followed by the benzoyl and acetyl groups. The cleavage of dipeptides is negligible.

The K_m values for peptides containing glutamic acid at the penultimate position or the C-terminal position increase markedly with a slight change in k_{cat} when the pH of the assay is altered from pH 5.5 to more than pH 6.5. Therefore, the peptide bonds involving the carboxyl and/or the amino group of glutamic acid, are well hydrolyzed at, or below, pH 5.5 rather than at pH 6.5, where the peptides composed only of neutral amino acids are most susceptible. The k_{cat} and K_m va' for peptides containing histidine slightly increase or show no change with pH.

In contrast to the carboxypeptidases A and B, terminal proline and β-alanine are rather good substrates.[8] However, the rate of splitting

TABLE VI
RELATIVE SPECIFIC ACTIVITY FOR BENZYLOXYCARBONYL DIPEPTIDES
CONTAINING PROLINE

Substrate[a]	Relative specific activity	Reference
CBZ-Glu-Tyr	100	b
CBZ-Pro-Val	8.8	b
CBZ-Pro-Tyr	1.1	b
CBZ-Gly-Pro	0	c
CBZ-Phe-Pro	74.7	c

[a] All amino acids had the L-configuration, except glycine. CBZ, benzyloxycarbonyl.
[b] For 3.3 mM substrate in 0.05 M pyridine–acetate, pH 5.5, at 25° [R. Hayashi, S. Moore, and R. B. Merrifield, *J. Biol. Chem.*, **248**, 3889 (1973)].
[c] For 1 mM substrate in 0.05 M sodium phosphate, pH 6.5, at 25° [R. Hayashi, Y. Bai, and T. Hata, *J. Biochem.* **77**, 69 (1975).

the peptide bond on either side of proline depends extensively upon the kinds of the adjacent amino acids (Table VI).[33]

The release of S-carboxymethyl cysteine and homoserine has also been shown.[34]

Broad carboxypeptidase activity, which includes the liberation of proline, has been tested on glucagon, the B chain of insulin, and reduced and carboxymethylated pancreatic ribonuclease.[3] The C-terminal sequence of the amyloid protein A has been deduced using carboxypeptidase Y.[35] The amino acid in the sequences released most slowly is glycine, or sometimes aspartic acid.

Carboxypeptidase Y rapidly hydrolyzes poly-α-L-glutamic acid at pH 4.2, liberating only glutamic acid; poly-L-lysine and poly-L-proline are never hydrolyzed. Poly-α-L-aspartic acid is slowly hydrolyzed by the enzyme.[32]

Esterase and Amidase Activities.[8,32] Carboxypeptidase Y can hydrolyze ester, amide, and *p*-nitroanilide substrates (Table VII).[32] Thus, unlike the carboxypeptidases A and B, carboxypeptidase Y does not necessarily require a free carboxyl group at the carboxyl end of the ester substrate. The kinetic parameters obtained for ATEE and APEE are compatible with those obtained for α-chymotrypsin.

The properties of the amidase action can be applied to the sequence

[33] R. Hayashi, S. Moore, and R. B. Merrifield, *J. Biol. Chem.* **248**, 3889 (1973).
[34] J. Salnikow, T-T. Liao. S. Moore, and W. H. Stein, *J. Biol. Chem.* **248**, 1480 (1973).
[35] M. A. Hermodson, R. W. Kuhn, K. A. Walsh, H. Neurath, N. Eriksen, and R Benditt, *Biochemistry* **11**, 2934 (1972).

TABLE VII
KINETIC PARAMETERS FOR THE ESTERASE AND AMIDASE
ACTIVITIES OF CARBOXYPEPTIDASE Y^a

Substrate[b]	Assay pH	k_{cat} (sec^{-1})	K_m (mM)	k_{cat}/K_m (mM^{-1}sec^{-1})
Ac-Phe-OEt	6.5	122	1.28	96.3
	8.0	120	1.20	100
Ac-Tyr-OEt	8.0	105	2.4	43.7
Hippuryl-β-phenyl lactate	6.0	21.2	0.45	47
CBZ-Phe-NH$_2$	6.0[c]	2.8	15.0	0.19
	7.0[c]	2.2	11.0	0.20
	8.0[d]	5.3	10.0	0.53
Bz-Tyr-pNA	7.0[e]	2.1	0.13	16.1

[a] R. Hayashi, Y. Bai, and T. Hata, *J. Biochem.* **77**, 69 (1975).

[b] All amino acids had the L-configuration. Ac, acetyl; Bz, benzoyl; CBZ, benzyloxycarbonyl; OEt, ethyl ester; pNA, p-nitroanilide.

[c] In 30% dimethylformamide.

[d] In 15% dimethylformamide.

[e] In 10% dimethylformamide.

analyses of peptides with amidated C-terminal groups, such as oxytocin and vasopressein.[36]

Effects of Environmental Factors on Kinetic Parameters.[8,36] The ratio of k_{cat}/K_m for carboxypeptidase Y has been shown to depend on ionizable groups with a pK_1 of 4.4 and a pK_2 of 6.5 for peptidase activity, and a pK_1 of 5.9 and a pK_2 of 8.9 for esterase activity.[8,36]

Certain solvents, i.e., ethanol and dimethylformamide, cause strong inhibition of both the peptidase and esterase activities by raising the K_m values.[8,36] The salt concentration, i.e., 0–1.5 M KCl, has a negligible effect on both activities.[8,36]

Application to Sequence Analysis and Limited Hydrolysis. The unique specificity of carboxypeptidase Y makes the enzyme applicable to sequence analysis and also to the limited hydrolysis of peptides and proteins, in combination with the carboxypeptidases A and B. Successive digestion by these enzymes is made in a volatile buffer (see this series Vol. 11, [38] and [39]), which is removable after the preceding digestion. For example, the first digestion by carboxypeptidase Y is performed in 0.1 M pyridine-acetate buffer, pH 5.5, and is stopped by making the solution 1 mM in DFP. Then, the reaction mixture is lyophilized and subjected to a second digestion by carboxypeptidase A in 0.1 M N-ethyl-

[36] Y. Bai, R. Hayashi, and T. Hata, *J. Biochem.* **78**, 617 (1975).

morpholine-acetate buffer, pH 7.6.[30] The sequence of the terminal five residues in the 15-residue ^{32}P-labeled peptide of carboxypeptidase Y has been determined by this method.[30]

Carboxypeptidase Y acts more strongly on proteins than does carboxypeptidase A, as has been shown in the hydrolyses of the amyloid A protein[35] and pancreatic deoxyribonuclease.[37] For a sequence study of proteins that are difficult to denature, digestion in 6 M urea has been used.[3]

The optimum pH for the release of the C-terminal basic amino acid is around 7, and that for the release of the C-terminal acidic amino acid is at, or below, pH 5.5. Therefore, at pH 5.5, carboxypeptidase Y hydrolyzes peptides through the acidic amino acid, but not through the basic amino acid; whereas around pH 7, the enzyme hydrolyzes them through the neutral and basic amino acids, but not through the acidic amino acid.[8,36]

Distribution and Nomenclature

DFP-sensitive carboxypeptidases have recently been isolated from plants[38-44] and molds.[45-47] These enzymes have no metals as cofacters. Catheptic enzymes from animal tissues[48,49] also share some of the same properties. Although not enough evidence has been presented to definitely define these enzymes as serine enzymes, some studies (e.g., Ihle and Dure[43] and Shaw and Wells[50]) support the hypothesis that the serine protease mechanism operates in the catalytic process of these enzymes, as it does in chymotrypsin[29] and carboxypeptidase Y.

The above carboxypeptidases display activity in the acidic region; therefore, they seem to belong to the family of "acid carboxypeptidase" originally described by Zuber and Matile.[39] However, no enzyme utiliz-

[37] T. E. Hugli, *J. Biol. Chem.* **248**, 1712 (1973).
[38] H. Zuber, *Nature (London)* **201**, 613 (1964).
[39] H. Zuber and P. Matile, *Z. Naturforsch., Teil B* **23**, 663 (1968).
[40] K. Visuri, J. Mikola, and J.-M. Enari, *Eur. J. Biochem.* **7**, 193 (1969).
[41] H. Tschesche and S. Kupfer, *Eur. J. Biochem.* **26**, 33 (1972).
[42] W. F. Carey and J. R. E. Wells, *J. Biol. Chem.* **247**, 5573 (1972).
[43] J. N. Ihle and L. S. Dure, III, *J. Biol. Chem.* **247**, 5034 and 5041 (1972).
[44] Y. Kubota, S. Shoji, T. Funakoshi, and H. Ueki, *J. Biochem.* **74**, 757 (1973).
[45] E. Ichishima, *Biochim. Biophys. Acta* **258**, 274 (1972).
[46] S. R. Jones and T. Hofmann, *Can. J. Biochem.* **50**, 1297 (1972).
[47] T. Nakadai, S. Nasuno, and N. Iguchi, *Agr. Biol. Chem.* **37**, 1237 (1973).
[48] A. I. Logunov and V. N. Orekhovich, *Biochim. Biophys. Acta* **46**, 1161 (1972).
[49] N. W. Dunn and M. T. McQuillan, *Biochim. Biophys. Acta* **235**, 149 (1971).
[50] D. C. Shaw and J. R. E. Wells, *Biochem. J.* **128**, 229 (1972).

ing the serine protease mechanism should be referred to as an "acid car-boxypeptidase," since this term is closely associated with acid protease. Acid proteases, e.g., pepsin, rennin, and a number of mold enzymes, have recently been shown to have a common catalytic center, aspartic acid.[51] According to the classification of proteases by Hartley[52] (serine, thiol, metal and acid proteases), carboxypeptidase Y should be a serine protease, not metal nor acid protease. The enzyme also differs from the thiol proteases. Thus, carboxypeptidase Y and, probably, the family of the enzymes described above should more appropriately be referred to by the generic name "serine carboxypeptidases" to distinguish them from the "metal carboxypeptidases" to which the carboxypeptidases A and B belong.

[51] J. S. Fruton, *Acc. Chem. Res.* **7**, 241 (1974).
[52] B. S. Hartley, *Annu. Rev. Biochem.* **29**, 45 (1960).

[49] Penicillocarboxypeptidases S-1 and S-2

By THEO HOFMANN

The mold *Penicillium janthinellum* secretes two carboxypeptidases of low specificity during the later stages of growth, at the time when sporulation occurs. The enzymes have been termed penicillocarboxypeptidases S-1 and S-2 (abbreviated PenCP-S-1 and PenCP-S-2). One of these enzymes has been described previously under the designation of peptidase B by Shaw.[1] On the basis of its electrophoretic mobility (cathodic migration at pH 4.2), it can now be identified with PenCP-S-2. Some of its properties have been described in a previous volume of this series.[2]

Isolation and Purification

The demonstration that there are two different enzymes and the purification of one of them (PenCP-S-1) have been described previously.[3] It involved the adsorption of the enzyme from the medium by DEAE-Sephadex, chromatography on two phenylbutylamine–Sepharose affinity columns and two preparative isoelectric focusing columns. However, the behavior of the affinity column has been erratic, and reproducible results were difficult to obtain. New isolation and purification

[1] R. Shaw, *Biochim. Biophys. Acta* **92**, 558 (1964).
[2] J. Sodek and T. Hofmann, this series Vol. 19, p. 384.
[3] S. R. Jones and T. Hofmann, *Can. J. Biochem.* **50**, 1297 (1972).

procedures have been developed that allow the purification of both carboxypeptidases and penicillopepsin from the same growth medium.

The production of the enzymes, their isolation and partial purification, is described in Chapter [35] of this volume. The enzymes are produced in a 250-liter fermentor in submerged culture. After 10–14 days the mycelium is removed and the enzymes are adsorbed on to DEAE-Sephadex A-50. After elution from the resin the three enzymes are separated on a Bio-Gel P-100 column (Chapter [35]). The two peaks containing the carboxypeptidase activities (S-1 and S-2 of Fig. 1, Chapter [35]) are then separately purified by preparative isoelectric focusing under the conditions described,[3] using the method originated by Svensson.[4]

The pooled peaks from the Bio-Gel P-100 column are separately concentrated in a Diaflo apparatus (Amicon Corp., Lexington, Massachusetts, Model 400, UM-10 membrane) to about 25 ml (PenCP-S-1) or 100 ml (PenCP-S-2), respectively, to give protein concentrations between 15 and 20 mg/ml. The concentrates are dialyzed against 0.02 M acetate, pH 4.5.

The separation is carried out on 440-ml isoelectric focusing columns (LKB Instruments, Stockholm, Sweden; column No. 8102) by the method of Vesterberg et al.[5] using a stabilizing gradient of 0 to 46% sucrose containing carrier ampholytes at a concentration of 1% in the pH range 3–5. The sucrose gradient is prepared conveniently with an ISCO Dialagrad (ISCO, Instrumentation Specialties Co., Lincoln, Nebraska; Model 380). The protein solution (5–10 ml containing not more than 100 mg of protein) is added to those sucrose solutions that will form the midregion of the gradient in order to protect the protein from exposure to either anode or cathode buffer solutions. Several runs are usually required for each enzyme to avoid overloading. The focusing column is cooled to 4° and run at a constant voltage of 600 V for 48 hr. When equilibration is reached, as indicated by a drop in the current to near zero, the column is eluted and fractions of 3 ml are collected. The pH, absorbance at 280 nm, and carboxypeptidase activity are determined. The tubes with maximum activity are pooled and combined with the solutions from several runs.

PenCP-S-1

As was shown previously[3] this enzyme exists in two forms; these appear to be identical in all properties examined except for the isoelectric

[4] H. Svensson, Arch. Biochem. Biophys., Suppl. I, 132 (1962).
[5] O. Vesterberg, T. Waldstrom, K. Vesterberg, H. Svensson, and B. Malmgren, Biochim. Biophys. Acta 133, 435 (1967).

points, which were found to be at pH 3.70 and 3.77. In a large-scale preparation the two peaks generally do not resolve, and a broadish peak of activity is obtained between pH 3.7 and 3.9. Since most of the properties of the two isoenzymes are identical, the two are not usually separated. A typical yield of PenCP-S-1 is given in Table I.

Electrophoresis in polyacrylamide gels in borate pH 9.5[6] or in phosphate-SDS pH 7.0[7] gives single bands. Two closely spaced bands are obtained by analytical isoelectrical focusing in polyacrylamide gels.[8] Sedimentation-equilibrium analysis in the ultracentrifuge indicates homogeneity.[3] The preparation should be free of penicillopepsin. When a sample of about 50 μg of PenCP-S-1 is assayed by the penicillopepsin assay described in Chapter [35], no trypsinogen activation should be observable. Since 5 ng penicillopepsin can be detected its content will be less than 0.01%. Freedom from endopeptidase activity is essential when PenCP-S-1 is used in sequencing, especially of proteins.

Another test for endopeptidase activity involves prolonged incubation with bovine serum albumin[3] (BSA). BSA (20 mg/ml) in 0.2 M acetate pH 4.7 is incubated with an equal volume of PenCP-S-1 (0.2 mg/ml) in water at 37°. A drop of toluene is added as preservative. At time intervals up to 24 hr, samples of 0.5 ml are added to 2 ml of 8.5% trichloroacetic acid. The precipitate is filtered and the absorbance at 280 nm is determined. An enzyme preparation that gave an increase in absorbance of about 0.06 was found to be devoid of endopeptidase activity when it was used to determine the C-terminal sequence of penicillopepsin (MW 32,000).[3]

PenCP-S-2

This enzyme was reported to focus at pH 4.72.[3] In recent experiments the peaks of activity were found between pH 4.57 and pH 4.7. The specific activities across the peak are constant. A typical yield of enzyme is given in Table I.

Criteria of Purity. Electrophoresis in polyacrylamide gels in borate pH 9.5[6] or in phosphate-SDS pH 7.0[7] gives single bands. In the latter case it is essential, however, to treat the protein in 1% SDS to 100° for 2 min and subsequently incubate the treated sample for 1 hr at 37° before electrophoresis. Failure to do this results in the appearance of two bands which correspond to a monomer and a dimer (see below). As with

---

[6] K. Weber and M. Osborn, *J. Biol. Chem.* **244**, 4406 (1969).

[7] B. J. Davis, *Ann. N.Y. Acad. Sci.* **121**, 404 (1964).

[8] L. Awdeh, A. R. Williamson, and B. A. Askonas, *Nature (London)* **219**, 66 (1968).

TABLE I

SUMMARY OF PURIFICATION PROCEDURE OF PENICILLOCARBOXYPEPTIDASES S-1 (PenCP-S-1) AND S-2 (PenCP-S-2)

Factor	Volume (ml)	Protein[a] (g total)	PenCP-S-1			PenCP-S-2		
			Activity (units)	Specific activity (units/mg)	Yield (%)	Activity (units)	Specific activity (units/mg)	Yield (%)
Medium	200,000	ND	48,000	—	100	70,000	—	100
Eluate from DEAE-Sephadex (batch)	600	22.5	44,000	1.95	91	61,000	2.7	87
PenCP-S-1								
Bio-Gel P-100 peak S-1	1,040	0.8	39,600	49.5	82	—	—	—
Isoelectric focusing[b]	200	0.2	16,000	80	33	—	—	—
PenCP-S-2								
Bio-Gel P-100 peak S-2	960	2.8	—	—	—	50,400	18	72
Isoelectric focusing[b]	500	1.8	—	—	—	36,000	20	51

[a] Estimated for the absorbance at 280 nm, assuming a specific absorbance of 1.5 (1 mg/ml).
[b] Combined yields from several runs.

PenCP-S-1, penicillopepsin activity is readily determined and usually found to be absent. Endopeptidase activity is also determined with bovine serum albumin as described under PenCP-S-1.

Assay Methods

Principle. Both carboxypeptidases hydrolyze Cbz-Glu-Tyr and similar peptides. The hydrolysis is followed by a ninhydrin procedure[9] or with 2,4,6-trinitrobenzene sulfonic acid[10] (TNBS). PenCP-S-1 and PenCP-S-2 can be determined separately in a mixture of the two. The method depends on the fact that PenCP-S-2 is strongly inhibited by *o*-phenanthroline, but PenCPS-S-1 is unaffected.

Assay with Ninhydrin

Reagents

Substrate: Carbobenzoxy-L-glutamyl-L-tyrosine (2 mM) in 0.02 M sodium formate, pH 4.2

Sodium acetate buffer, 4 M, pH 5.3, containing 0.2 mM NaCN (prepared fresh daily from 4.08 M sodium acetate and 10 mM NaCN)

Ninhydrin, 3%, in methyl Cellosolve

Isopropanol–water, 1:1

Procedure. Equal volumes (0.5 ml) of the enzyme containing 0.2–1.5 µg of active enzyme in water or dilute buffers and the substrate are incubated at 35° for 10 min. Ninhydrin (0.25 ml) and cyanide–acetate buffer (0.25 ml) are added. The mixture is heated to 100° for 25 min. After adding 2.5 ml of isopropanol–water the absorbance is read at 570 nm. A standard curve with tyrosine is prepared for calculations of the units.

Assay with TNBS[3]

Reagents

Substrate: Carbobenzoxy-L-glutamyl-L-tyrosine (1 mM) in 0.02 M sodium formate, pH 4.2

[9] H. Rosen, *Arch. Biochem. Biophys.* **60**, 10 (1957).
[10] K. Satake, T. Okyama, M. Ohashi, and T. Shimada, *J. Biochem.* (*Tokyo*) **47**, 654 (1960).

Borate buffer, 0.05 M, containing sufficient NaOH to give a final pH of 8.0 when equal volumes of substrate, TNBS solution (see below), and borate are mixed

2,4,6-Trinitrobenzene sulfonic acid (TNBS) 0.05% in H_2O. This reagent requires purification if a blank of the whole assay system described below gives an absorbance at 340 nm greater than 0.100. TNBS is purified as follows: 1.5 g are dissolved in water and passed through a column (2.5 cm $\times$ 20 cm) consisting of 15 g each of Norite and Celite-545 suspended in water. Fractions (10 ml) are collected, and absorbances at 280 nm are determined. The final peak, which is colorless and consists of TNBS, is pooled and freeze-dried. The purified preparation is stored cold and dry

HCl, 1 N

Procedure. To 1 ml of substrate solution at 35° are added samples of enzyme not exceeding 50 μl and containing up to 1.5 μg of enzyme. After 10 min, 1 ml of borate and 1 ml of TNBS are added. The mixtures are left for 2 hr at 40° in a water bath that is covered to exclude light. After adding 1 ml of 1 N HCl, the absorbances are read at 340 nm.

Determination of PenCP-S-1 and PenCP-S-2 in Mixture

Reagents. These are as described for TNBS assay.

o-Phenanthroline substrate: Cbz-Glu-Tyr (mM) in 0.02 M sodium acetate pH 4.2 containing 0.1 mM *o*-phenanthroline.

Procedure. The assay procedure is the same as described, but each sample is assayed twice, once with substrate alone, then with substrate containing *o*-phenanthroline. *o*-Phenanthroline (0.1 mM) causes 90% inhibition of PenCP-S-2 under these assay conditions, but it has no effect on PenCP-S-1. The absorbances at 340 nm due to each enzyme are calculated as follows:

$$A_{S2} = (A - A_{op}) \times 1.1$$
$$A_{S1} = A - A_{S2}$$

where A is the absorbance with substrate only, A_{op} is the absorbance with substrate + *o*-phenanthroline, A_{S1} is the absorbance due to PenCP-S-1, and A_{S2} that due to PénCP-S-2.

Definition of Units

One unit is defined as the amount of enzyme that liberates 1 μmole of tyrosine per assay per minute. One milliunit corresponds to ΔOD = 0.031 when measured by the TNBS assay.

Properties

Stability of Penicillocarboxypeptidases

Although no systematic studies on the stability of the two enzymes have been undertaken, experience in this laboratory shows that both are stable in the pH range 3.5–5.5 for prolonged periods. They can be kept frozen for many months, but they are completely inactivated when freeze-dried. Below pH 3 and above pH 6.5, both enzymes are slowly and irreversibly denatured.

Molecular Properties

PenCP-S-1

Molecular weight: 48,000, as determined by SDS–polyacrylamide gel electrophoresis[3] and by sedimentation- equilibrium with the meniscus depletion method.[3] The partial specific volume was calculated from the amino acid composition, $\bar{V} = 0.716$. Specific absorbance $E_{1\,cm}^{1\,mg/ml} = 2.6$. The two isoenzymes[3] have isoelectric points of 3.7 and 3.77.

Amino Acid Composition. The composition of PenCP-S-1 is given in Table II. The most remarkable features are the presence of one cysteine (determined as carboxymethylcysteine after reaction with [14C]iodo-acetic acid) and three glucosamine residues. The latter show that the protein is a glycoprotein. The carbohydrate content, assuming two hexoses per glucosamine[11] is about 3%.[3]

PenCP-S-2[12]

This enzyme is usually found as a dimer, molecular weight 128,000–140,000, as determined by sedimentation equilibrium in the ultracentrifuge with the meniscus depletion method. At very low protein concentrations (less than 50 μg/ml) there is slight deviation from linearity indicating the possibility of dissociation into monomers. The elution position from the Bio-Gel P-100 column during purification (see Fig. 1, Chapter [35]) is compatible with a molecular weight over 100,000. On the other hand, when PenCP-S-2 is electrophoresed in SDS–polyacryl-amide gels *without* mercaptoethanol, a single band with molecular weight around 65,000 is obtained (as mentioned above, rigorous treatment with SDS is required to dissociate the dimer completely). Since no mercapto-

[11] R. G. Spiro, *Annu. Rev. Biochem.* **39**, 599 (1970).
[12] The results reported in this section are as yet unpublished (S. R. Jones and T. Hofmann).

TABLE II

Amino Acid Compositions of Penicillocarboxypeptidases S-1 (PenCP-S-1) and S-2 (PenCP-S-2)

Amino acid	PenCP-S-1[a] (moles per 48,400 daltons)	PenCP-S-2[b] (moles per 64,000 daltons)
Lysine	13	23
Histidine	4	14
Arginine	6	13
Aspartic acid	61	63
Threonine	23	34
Serine	39	29
Glutamic acid	46	61
Proline	24	38
Glycine	35	37
Alanine	33	31
Half-cystine as cystine	6	6
Cysteine	1	1
Valine	27	33
Methionine	3–4	19
Isoleucine	18	28
Leucine	20	43
Tyrosine	26	34
Phenylalanine	23	34
Tryptophan	16	8–12
Glucosamine	3	9
Hexoses (assumed)	6	18

[a] S. R. Jones and T. Hofmann, *Can. J. Biochem.* **50,** 1297 (1972). The compositions of the isoenzymes are the same within experimental error.

[b] The analysis was carried out as described for PenCP-S-1. S. R. Jones and T. Hofmann, to be published.

ethanol is required, the dimer formation is noncovalent and does not involve the cysteine residues, of which there is one per monomer (see Table II). Ichishima[13] found evidence for a monomer–dimer interaction with an acid carboxypeptidase from *Aspergillus saitoi.*

Partial specific volume, 0.724 (calculated from the amino acid composition). Specific absorbance, $E_{1\,cm}^{1\,mg/ml} = 1.85$; isoelectric point, 4.6–4.7.

Amino Acid Composition. The composition of PenCP-S-2 is shown in Table II. A comparison with that of PenCP-S-1 shows clearly that the two carboxypeptidases are quite different proteins, as also shown by the different isoelectric points and molecular weights. Especially notable are the differences in histidine and methionine. Like PenCP-S-1, how-

[13] E. Ichishima, *Biochim. Biophys. Acta* **258,** 274 (1972).

ever, PenCP-S-2 also has only one cysteine residue per monomer, and it too is a glycoprotein. The carbohydrate content, assuming two hexoses per glucosamine[11] is about 7%.

Enzymic Properties

Inhibitors. Both PenCP-S-1 and PenCP-S-2 are inhibited by diisopropyl fluorophosphate (DFP) and phenylmethylsulfonyl chloride, two typical serine-protease inhibitors. Inhibition is virtually complete (<0.05% residual activity). In contrast to other serine proteases, both enzymes are also inhibited by p-hydroxymercuribenzoic (PHMB), p-hydroxymercuribenzene sulfonate, phenylmercuric acetate, $HgCl_2$, and $AgNO_3$, all typical reagents for sulfhydryl groups.[3] The rate of inactivation is relatively slow,[3] but inactivation is virtually complete (<0.1% residual activity). The inhibition by PHMB is almost fully reversible by dithioerythritol. The PHMB-inhibited enzymes still react with DFP, and after reaction can no longer be reactivated.[14] DFP therefore reacts with a group other than the -SH. The evidence suggests that enzymic action probably depends on a cysteine and a serine residue. Furthermore photooxidation of PenCP-S-2 in the presence of methylene blue also causes inactivation[15]; the SH-group is not oxidized. The pH dependence of the rate of loss of activity has an inflection around pH 5.5 and suggests a possible histidine involvement. However, several chloromethyl ketone derivatives tried failed to inhibit either of the two carboxypeptidases, although the related enzymes, carboxypeptidase Y[16] and an *Aspergillus* carboxypeptidase[17] are inhibited by chloromethyl ketones.

As mentioned above, o-phenanthroline inhibits PenCP-S-2, but not PenCP-S-1. The inhibition is not due to metal chelation because addition of various metal ions does not restore activity and the o-phenanthroline can be removed from the inhibited enzyme by dialysis against 1 mM EDTA with complete recovery of activity.

Both enzymes, but especially PenCP-S-2, are sensitive to carboxylic acids.[1] The inhibition by aliphatic monocarboxylic acids increases with chain length and approaches 100% with octanoic acid (10 mM).[1] This creates a problem when kinetic studies are carried out, because there are no effective buffers without carboxyl groups in the pH range 3–5.

[14] T. Hofmann and S. R. Jones, *Fed. Proc., Fed. Am. Soc. Exp. Biol.* **32**, 577 (1973).
[15] T. Hofmann, H. Lee, S. R. Jones, and T. Murachi, *Fed. Proc., Fed. Am. Soc. Exp. Biol.* **33**, 1307 (1974).
[16] R. Hayashi, Y. Bai, and T. Hata, *J. Biochem. (Tokyo)* **76**, 1355 (1974).
[17] E. Ichishima, S. Sonoki, K. Hirai, Y. Torii, and S. Yokoyama, *J. Biochem. (Tokyo)* **72**, 1045 (1972).

TABLE III
KINETIC PARAMETERS FOR SOME PENICILLOCARBOXYPEPTIDASE S-2
(PENCP-S-2) SUBSTRATES[a]

Substrates	k_{cat} max (min^{-1})	pH
"Carboxypeptidase" *substrates*		
Ac-Tyr-Phe	8200	5.0
Cbz-Ala-Ala	>540	4.2
Ac-Tyr-Gly	>180	4.2
	350	5.5
Bz-Gly-Phe	840	5.5
"Esterase" substrates		
Bz-Arg-OEt	1900	6.0
Ac-Tyr-OMe	3000	6.0
Ac-Ala$_3$-OEt	>380	6.0
"Amidase" substrates		
Bz-Arg-NH$_2$	>120	6.0
Endopeptidase		
Various peptides	<0.02	4.7

[a] S. R. Jones and T. Hofmann, unpublished.

Formate ions up to 0.02 M show no inhibition, but acetate ions are inhibitory at that concentration. In current kinetic studies in this laboratory, the pH is usually controlled in a pH stat.

In spite of the differences in molecular properties, the studies with the inhibitors show that these enzymes are closely related as far as their mechanism of action is concerned.

Specificity of Penicillocarboxypeptidases

As far as peptide and protein substrates are concerned, both enzymes are specific exopeptidases without detectable endopeptidase activity.[3,18] However, they also act on N-substituted amino acid esters and amides. For PenCP-S-2 k_{cat} values have been determined for a variety of substrates. These are listed in Table III. PenCP-S-1 also acts on these and other substrates, but accurate kinetic constants have not been measured as yet.

[18] S. R. Jones and T. Hofmann, to be published.

Both enzymes degrade glucagon and the B chain of insulin sequentially,[3,18] and cleave all amino acids including lysine, arginine, and proline. They have been used in the determination of the C-terminal sequence of porcine intestinal calcium-binding protein.[19]

The relative rates of hydrolysis of small peptides differ for the two enzymes, but a systematic comparative study has not been carried out.

Use of PenCP-S-1 in Sequencing[20]

Table IV summarizes the semiquantitative assessment of the relative cleavage rates of various amino acids by PenCP-S-1.[20] The results were obtained with over 40 peptides isolated as part of the determination of the amino acid sequence of penicillopepsin.[20] Table IV shows that a number of amino acids, especially those with bulky side chains, are readily cleaved. These include lysine and arginine and, rather surprisingly, proline, which can be released very rapidly from a number of peptides.[20] Large differences, however, are found in the rate of release of the smaller amino acids—glycine, alanine, serine, and threonine. Although there is as yet insufficient information to warrant many general conclusions, it can be said that, if two or more of the four small amino acids listed above are in sequence, their release is slow. In our hands PenCP-S-1 was found useful not only because it has a wider specificity than pancreatic carboxypeptidases, but also because it has a considerably higher catalytic efficiency[20] and lower background, even after prolonged incubation.[20]

pH Dependencies

The "pH optima" for Cbz-Glu-Tyr are as follows PenCP-S-1: pH 4.3[3]; PenCP-S-2: pH 3.9[18]; the pH optimum for Cbz-Ala-Ala with PenCP-S-1: pH 4.7.[3] Esterase and amidase activities are optimal in the pH range 5.5–6.5 (Table III). The falloff in the activity on "carboxypeptidase" substrates on the alkaline side is controlled by a loss of binding as shown by a large increase in K_m above pH 5.[21]

[19] K. J. Dorrington, A. Hui, T. Hofmann, A. J. W. Hitchman, and J. E. Harrison, J. Biol. Chem. 249, 199 (1974).

[20] A. Hui, L. Rao, A. Kurosky, S. R. Jones, G. Mains, J. W. Dixon, A. Szewczuk, and T. Hofmann, Arch. Biochem. Biophys. 160, 577 (1974).

[21] H. Lee and T. Hofmann, to be published.

TABLE IV
APPROXIMATE RELATIVE RATES OF RELEASE OF AMINO ACIDS FROM PEPTIDES BY PENICILLOCARBOXYPEPTIDASE S-1[a]

Amino acid	Approximate rate[b]	Number of peptides examined
Alanine[c]	Medium rapid	4
	Slow	1
Arginine	Rapid	2
Asparagine	Medium rapid	3
Aspartic acid	Slow	3
Glutamic acid	Medium to rapid	6
Glutamine	Rapid	5
Glycine	Rapid	2
	Medium rapid	1
	Slow	4
	Not released	1
Histidine	Medium rapid	1
Isoleucine	Medium to rapid	1
Leucine	Rapid	5
	Medium rapid	2
Lysine	Rapid	2
	Medium rapid	3
Methionine	Medium to rapid	1
Phenylalanine	Rapid	1
	Medium rapid	2
Proline	Rapid	2
	Medium rapid	3
Serine	Rapid	4
	Medium rapid	4
	Slow	1
Threonine	Medium rapid	4
	Not released	2
Tryptophan	Medium rapid	1
Tyrosine[d]	Rapid	2
	Medium rapid	2
Valine	Medium rapid	4

[a] From A. Hui, L. Rao, A. Kuroski, S. R. Jones, G. Mains, J. W. Dixon, A. Szewczuk, and T. Hofmann, *Arch. Biochem. Biophys.* **160**, 577 (1974).
[b] Rates are based on results obtained with over 40 peptides, B chain of insulin, glucagon, porcine Ca-binding proteins, and a light chain of immunoglobulin. They have been roughly estimated on the following basis: Amino acids at the C-terminal of a peptide were considered rapidly released if over 50% was cleaved in 5 min at molar ratios of enzyme to peptide of 1:100. Amino acids were considered slowly released when less than 50% was cleaved after 2 hr. Amino acids were considered medium rapidly released if they fell between these limits. The release of amino acids not at the terminal position of the peptide was roughly estimated from a comparison of the release of the two neighboring residues.
[c] Alanine was rapidly cleaved from carbobenzoxy-alanyl-alanine [R. Shaw, *Biochim. Biophys. Acta* **92**, 558 (1964)].
[d] Tyrosine was rapidly cleaved from carbobenzoxy-glutamyl-tyrosine.

Distribution of Similar Enzymes

DFP-inhibited carboxypeptidase have been isolated from a variety of sources: citrus peel,[22] citrus leaves,[23] yeast,[24] *Aspuergillus saitoi*,[13] *Aspergillus oryzae*,[25] French bean leaves (phaseolain),[26] germinating barley,[27] germinating cotton seedlings,[28] tomatoes,[29] watermelons,[30] and bromelain powder.[31] Furthermore, Doi[32] has suggested that cathepsin A from pig kidney is identical with catheptic carboxypeptidase and possibly related to the plant and fungal carboxypeptidases.

[22] H. Zuber, *Nature (London)* **201**, 613 (1964).
[23] B. Sprossler, H.-D. Heilmann, E. Grampp, and H. Uhlig, *Hoppe-Seyler's Z. Physiol. Chem.* **352**, 1524 (1971).
[24] T. Hata, R. Hayashi, and E. Doi, *Agric. Biol. Chem.* **31**, 357 (1967).
[25] T. Nakadai, S. Nasumo, and N. Iguchi, *Agric. Biol. Chem.* **36**, 1343, 1473, 1481 (1972).
[26] W. F. Carey and J. R. E. Wells, *J. Biol. Chem.* **247**, 5573 (1972).
[27] K. Visuri, J. Mikola, and T.-M. Enari, *Eur. J. Biochem.* **7**, 193 (1969).
[28] J. N. Ihle and L. S. Dure, *J. Biol. Chem.* **247**, 5034, 5041 (1972).
[29] T. Matoba and E. Doi, *Agric. Biol. Chem.* **38**, 1901 (1974).
[30] T. Matoba and E. Doi, *Agric. Biol. Chem.* **38**, 1891 (1974).
[31] E. Doi, C. Ohtsuru, and T. Matoba, *J. Biochem. (Tokyo)* **75**, 1063 (1974).
[32] E. Doi, *J. Biochem.* **75**, 881 (1974).

[50] Dipeptidyl Carboxypeptidase from *Escherichia coli*

By A. YARON

Assay Methods

Principle. Dipeptidyl carboxypeptidase (DCP) from *E. coli* is an intracellular enzyme, which hydrolytically removes dipeptides from the carboxyl end of low- and high-molecular-weight peptides.[1] Two methods for measuring the enzymic activity are described.

Method 1: The assay used for monitoring the enzyme purification is based on the release of glycylproline from the carboxyl end of the sequential polypeptide $\mathrm{Dnp(Pro\text{-}Gly\text{-}Pro)}_n$, using the colorimetric ninhydrin method for quantitative determination of the dipeptide.

$$\mathrm{Dnp(Pro\text{-}Gly\text{-}Pro)}_{n-1}\text{-}Pro\text{-}Gly\text{-}Pro \rightarrow \mathrm{Dnp(Pro\text{-}Gly\text{-}Pro)}_{n-1}\text{-}Pro + Gly\text{-}Pro \quad (1)$$

where n is the average number of the repeating units Pro-Gly-Pro per molecule.

[1] A. Yaron, D. Mlynar, and A. Berger, *Biochem. Biophys. Res. Commun.* **47**, 897 (1972).

Only one molecule of Gly-Pro is formed from each polymer chain, since the release of the next dipeptide would require the cleavage of a secondary amide, to which a proline residue donates the nitrogen. Such bonds are resistant to hydrolysis by DCP. The sequential polymer is rich in proline and has no free amino groups. It is therefore resistant to most peptidases. The only known enzyme capable of hydrolyzing it is collagenase.[2] No interference by this enzyme was experienced with DCP preparations purified beyond the ammonium sulfate precipitation step of the purification procedure described below. Aminopeptidase P[3] and a dipeptidase activity present in the extract of *E. coli*, both are capable to hydrolyze glycylproline to glycine and proline if incubated under the conditions used for the assay of DCP in the presence of Mn^{2+} ions. As the assay for DCP does not require the adding of cations, the presence of these enzymes does not prevent the quantitative determination of the activity of DCP by analysis of the Gly-Pro formed. However, if crude enzyme preparations are used, they should be dialyzed and/or EDTA (0.1 mM) should be added before the assay. In the purification procedure described below, aminopeptidase P is separated from DCP in the acetone fractionation step, and the dipeptidase activity by hydroxyapatite chromatography.

Method 2: In the absence of interfering peptidases benzyloxycarbonyltetraalanine can be used as the substrate and the hydrolysis by DCP (Eq. 2) may be followed either by the colorimetric ninhydrin method (not described here) or by the potentiometric method, using a recording pH stat.

$$Z-Ala_4 \rightarrow Z-Ala_2 + Ala_2 \tag{2}$$

At pH 8.1, which is the pK of the amino group of alanylalanine,[4] half a mole of base is consumed per mole of Z-Ala$_4$ hydrolyzed.

The Colorimetric Ninhydrin Method Using Dinitrophenyl-Poly (Prolyl-Glycyl-Prolyl) as the Substrate

Preparation of Dnp(Pro-Gly-Pro)$_n$. This compound is prepared by dinitrophenylation of poly (Pro-Gly-Pro)[5] as follows: Fluorodinitrobenzene (300 mg) is dissolved in ethanol (20 ml) and the solution is added to an aqueous solution (10 ml) containing poly (Pro-Gly-Pro) (500 mg,

[2] E. Harper, A. Berger, and E. Katchalski, *Biopolymers* **11**, 1607 (1972).

[3] A. Yaron and A. Berger, this series Vol. 19, p. 521.

[4] E. Ellenbogen, *J. Am. Chem. Soc.* **74**, 5198 (1952).

[5] J. Engel, J. Kurtz, E. Katchalski, and A. Berger, *J. Mol. Biol.* **17**, 255 (1966). The polymer is commercially available from Miles-Yeda Ltd., Kiriat Weizmann, Rehovot, Israel.

average molecular weight 1300) and sodium bicarbonate (252 mg). The solution, protected against light, is left at room temperature overnight, extracted with three 50-ml portions of ether and concentrated to a volume of 2.5 ml by ultrafiltration (UM-05 membrane, Diaflo ultrafiltration cell, Amicon Corp.). This solution is applied to a column (2.2 × 100 cm) of Sephadex G-15 and eluted with water at a flow rate of 30 ml/hr. The effluent is monitored at 360 nm and collected in 7-ml fractions. Most of the material appears as a major peak beginning immediately after the void volume and emerging in about 80 ml. Smaller peaks that follow are discarded, and fractions containing material from the major peak are pooled, lyophilized, and dried *in vacuo* over sulfuric acid. The yield is about 300 mg. The number average chain length (n) of the polymer used by us was 4.9. This was established by quantitative spectrophotometric determination of the Dnp end groups ($\epsilon_{360\ nm} = 16,800$, phosphate buffer pH 7.0). Complete substitution of N-terminal amines by Dnp was demonstrated by showing that the polymer is resistant to hydrolysis by clostridial aminopeptidase.[6] This enzyme releases one proline residue from the amino end of every unsubstituted chain of (Pro-Gly-Pro)$_n$.

Reagents

    Substrate stock solution is prepared by dissolving Dnp(Pro-Gly-Pro)$_n$ in water (7.2 mM, concentration is determined spectrophotometrically, $\epsilon_{360\ nm} = 16,800\ M^{-1}\ cm^{-1}$)

    Veronal buffer, 0.05 M, pH 8.15

    Sodium acetate buffer, 4 M, pH 5.3, containing 0.2 mM sodium cyanide, freshly prepared before the assay from 4.08 M sodium acetate and 10 mM sodium cyanide

    Ninhydrin, 3% in methyl Cellosolve (ethyleneglycol monomethyl ether)

    Isopropanol in water, 50%

    Enzyme, the enzyme solution is suitably diluted with the Veronal buffer to obtain a solution containing 0.2–1.0 unit/ml

Procedure. The substrate stock solution (50 μl) is mixed with Veronal buffer (0.95 ml) and placed in a water bath at 40°; the enzyme solution (10 μl) is added. After 15 min the incubation mixture is analyzed for ninhydrin color.[7] To each tube, containing 1 ml of incubation mixture, the ninhydrin solution (0.5 ml) and the acetate-cyanide buffer (0.5 ml) are added. The tubes are heated in a boiling water bath for

[6] E. Kessler and A. Yaron, *Biochem. Biophys. Res. Commun.* **32**, 658 (1968); see also *Eur. J. Biochem.* (1976), in press, and this volume [45].

[7] H. Rosen, *Arch. Biochem. Biophys.* **60**, 10 (1957).

20 min and cooled with tap water. A mixture of isopropanol and water (1:1 v/v) is added and the mixture is well mixed. The optical density at 570 nm is determined and the amount of glycylproline is calculated from a calibration curve constructed with known amounts of the dipeptide. The absorption of Dnp at 570 nm is small and is corrected by subtracting the substrate blank.

Definition of Unit and Specific Activity. One enzyme unit is defined as the amount of enzyme which produces 1 μmole of glycylproline per minute under the conditions of the above described assay. This is 10.3 larger than the previously defined unit.[1] The specific activity is expressed in units per milligram of protein.

The rate of hydrolysis is proportional to enzyme concentration in the range from zero to about 17 mU/ml, but deviates from linearity at higher concentrations.

The Potentiometric Assay

Reagents

> Substrate stock solution: a 10 mM solution of Z-Ala$_4$ in 0.1 M KCl is prepared. A sufficient amount of 1 M KOH in 0.1 M KCl is added to bring the pH to 8.1
> Base, 2 mM KOH

Procedure. Assays are performed in a pH stat. We used the Radiometer, Copenhagen assembly consisting of a pH meter Model 26 with a glass electrode, connected through an agar bridge to a calomel electrode. The titrant is delivered from a 0.25-ml burette of an automatic pipette ABU-12 controlled with a Titrator 11 and coupled to a recorder (Servograph REC 51). The reaction vessel, thermostatted at 40° is equipped with a magnetic stirrer, and the titrated solutions are kept under a constant stream of CO_2-free argon.

The substrate stock solution (0.5 ml) and 0.1 M KCl (0.5 ml) are placed into the reaction vessel and the pH is adjusted to 8.1. When thermal equilibrium is reached, there is no uptake of base. The enzyme solution is mixed in, and the uptake of base is recorded as a function of time. Initial rates are calculated from records with a time scale usually between 10 sec/cm and 60 sec/cm. Enzyme concentration is chosen so that the line recorded has an inclination of 15°–60°.

To obtain the number of micromoles of substrate hydrolyzed per minute, the number of micromoles of base titrant consumed per minute is multiplied by 2, since 0.5 mole of base is consumed per mole of substrate hydrolyzed. One unit of DCP as defined by the ninhydrin colorimetric method was found to correspond to a hydrolysis rate of $Z-Ala_4$ of 3.24 μmoles/min. The potentiometric assay can be used only in the absence of other peptidases.

Purification Procedure

The enzyme is isolated from *E. coli* B wild type (American type culture collection catalog No. 23226). The following procedure (see also Table I) results in a 1200-fold purification. Purification starts from 10 kg of packed cells. The amount of pure enzyme obtained in a single run is limited by the capacity of the preparative polyacrylamide gel electrophoresis, which is the last purification step. All operations are carried out at 4°.

Preparation of Cell-Free Extract. Mass cultures of *Escherichia coli*, strain B are grown for 5 hr in a 500-liter fermentor (Biotec, Sweden) with aeration, on Davis media, tryptone (0.6%) and glucose (0.2%). The cells are harvested in a Sharples centrifuge yielding 4 g of wet cells per liter. The wet cells are stored frozen at −20°. The frozen mass (10 kg) is dispersed at 4° in a solution of 0.9% KCl (38 liters) by occasional shaking during 2 days. DNase (2.5 mg) is added and the cells are ruptured by passing the cold suspension through a Manton–Gaulin homogenizer (Model 15 M, Everett, Massachusetts) at 5000 psi at a rate of 46 liters/hr. Cell debris is removed from the extract by centrifugation, using a Westphalia Separator AG, Model KDD 605 at 10,000 rpm and a flow rate of 12 liters/hr After each 10 liters of suspension that pass, the sediment filling the head of the centrifuge is removed. The total weight of packed sediment is 7 kg. The effluent emerges at 20°–23° when running tap water is used to cool the centrifuge. The collected solution (40 liters) is stored at 4° for 15 hr.

Heat Treatment. The cell-free extract is rapidly warmed to 50° by passing it through a spiral glass tubing of 0.5 cm internal diameter and 125-ml capacity that is immersed in a water bath (50 liters) kept at 51° with two B. Brown, Thermomix II immersion thermostats. A finger pump (Sigma Motor Model T8) is used to keep a flow rate of 200 ml/hr. The solution emerges at 50° and is kept at that temperature for about 15 min by continuing the flow through a 30-meter Tygon tubing of 1.12 cm internal diameter immersed in a 50° water bath. The temperature of the suspension is now lowered by passing it through additional 7 meters of Tygon tubing immersed in an ice bath. The effluent emerging at 28° is collected and stored at 4° for about 9 hr. For determination of enzymic activity, the supernatant is first dialyzed against 0.9% KCl.

TABLE I

Purification of Dipeptidyl Carboxypeptidase from *Escherichia coli* B[a]

Step	Volume (ml)	Activity[b] (U/ml)	Total units	Protein[c] (mg/ml)	Specific activity (U/ml)	Activity yield (%)	Purification factor
Cell-free extract	40,000	0.0845[d]	3380[d]	31	0.00273	—	1.0
Heat treatment	40,000	0.0932	3730	10	0.00932	100	1.2
Ammonium sulfate, 0.4–0.6 saturation	7,000	0.176	1230	15.1	0.0116	33.0	8.2
Acetone fractionation (44–58%), dialysis	1,130	0.534	603	6.93	0.0770	16.1	79
DEAE-cellulose, ultrafiltration	20	20.4	408	27.6	0.739	10.9	177
Gel filtration	113	2.92	330	1.77	1.65	8.85	660
Hydroxyapatite, ultrafiltration	56	4.12	230	0.67	6.15	6.16	1230
Electrophoresis, ultrafiltration	48	3.88	186	0.34	11.4	5.0	

[a] Figures are given for 10 kg of wet cells.

[b] The activity was determined with Dnp-poly(Pro-Gly-Pro) as the substrate by the ninhydrin colorimetric method. One unit of activity is defined as the amount of enzyme that produces 1 μmole of Gly-Pro per minute under the assay conditions.

[c] Protein concentrations were determined by the Lowry method [O. H. Lowry, N. J. Rosebrough, A. L. Farr, and R. J. Randall, *J. Biol. Chem.* **193**, 265 (1951)], using bovine serum albumin as standard.

[d] The figures obtained with the crude extracts are approximate values, since interfering activities prevented accurate measurements.

The same results are obtained for activity with nondialyzed solutions, but blanks are much higher.

Ammonium Sulfate Fractionation. The suspension from the previous step is divided into two equal parts, and each is treated separately. Saturated ammonium sulfate solution[8] (13.3 liters) is added at 4° to the suspension (20 liters) under mechanical stirring. After 90 min at 4°, the solid material (1.5 kg of packed precipitate) is removed by centrifugation, using the Westphalia separator at 10,000 rpm, at a flow rate of 15 liters/hr. To each liter of the stirred supernatant, saturated ammonium sulfate solution (0.5 liter) is added, stirring is continued for 2 hr at 4°, and the precipitate formed is collected by centrifugation (Sharples Super Centrifuge 1A, open-type turbine, 40,000 rpm, 15 liters/hr). The precipitate obtained from the two runs is dissolved in 0.05 M sodium acetate, pH 5.6 (7 liters), and subjected to acetone precipitation without delay. For determination of protein and activity of the enzyme a sample is dialyzed against 0.05 M sodium acetate, pH 5.6.

Acetone Fractionation. Cold acetone (kept at −20° prior to use, 5.6 liters) is added slowly (over 1 hr) to the solution obtained in the previous step (7 liters). The mixture is kept at 8°–11° by cooling with an external ice-salt bath. Stirring is continued at 8°–11° for 20 min, and the precipitate (about 250 g) is removed with a continuous centrifuge (Westphalia Separator, 10,000 rpm, 20 liters/hr).

Cold acetone (4.2 liters) is added to the supernatant (12.5 liters) during about 20 min, keeping the temperature of the well mixed solution at 8°–11°. Stirring is continued for an additional hour, and the precipitate formed is collected by centrifugation (Sharples Super Centrifuge 1A, open-type turbine, 40,000 rpm, 30 liters/hr). About 70–80 g of packed precipitate is obtained. This is dissolved in 0.05 M sodium acetate, pH 5.6 (1 liter), the yellow solution obtained is dialyzed against 0.005 M phosphate buffer, pH 6.0 (final volume is 1130 ml), and stored at −20°. At this stage the enzyme is quite stable, about 9% of activity being lost in 1 month.

Ion-Exchange Chromatography on DEAE-Cellulose. The dialyzed solution from the previous step (1100 ml) is applied under gravity (90 ml/hr) to a 11.5 × 28.5 cm bed of DEAE-cellulose (Whatman DE 23) equilibrated with 0.01 M sodium phosphate, pH 6.0. The column is then washed with 4 liters of 0.01 M sodium phosphate, pH 6.0, 0.05 M in KCl, and eluted with a 36-liter linear gradient to 0.24 M KCl in 0.01 M sodium phosphate, pH 6.0. Fractions of 200 ml are collected at 1200 ml/hr, and the effluent is monitored at 280 nm. Fractions between 15,090 and

---

[8] The saturated ammonium sulfate solution was prepared as described previously by Yaron and Berger.[3]

18,660 ml, containing the activity, are combined, concentrated to 20 ml by ultrafiltration (Amicon PM-30 membrane), and stored frozen at —20°. No loss of activity was observed after 20 days of storage.

Gel Filtration on Sephadex G-150. The above concentrated solution (19 ml) is applied to a 2.8 × 245 cm column of Sephadex G-150 (Pharmacia) preequilibrated with 0.1 M sodium phosphate, pH 6.0. The same buffer is used for elution, 6.7-ml fractions being collected at 13.5 ml/hr. The active fractions are combined (113 ml) and stored at —20°. A 20% loss of activity was observed after 5 months of storage.

Hydroxyapatite Column Chromatography. An aliquot of the above solution (24 ml) is applied under gravity to a hydroxyapatite[9] column (1.3 × 92 cm) equilibrated with 0.01 M sodium phosphate, pH 6.0. A 2-liter linear gradient from 0.01 M to 0.08 M sodium phosphate, pH 6.0, is used for elution. Fractions of 11 ml are collected at a flow rate of 24 ml/hr. The active fractions are concentrated to 12 ml by ultrafiltration (PM-30 membranes). The solution is kept frozen at —20°.

Polyacrylamide Gel Electrophoresis. Samples from the previous purification step (150 μl) containing about 100 μg of protein and 20% sucrose, are applied to 7% gels (6.5 × 50 mm). The gels are prepared and the electrophoresis is performed according to Davis,[10] omitting the spacer gel, and using the Shandon disc electrophoresis appartus. Prior to application of the sample a current of 3 mA per gel is applied for 25 min. Separation is achieved by applying 3 mA per gel for 90 min at 4°, six gels being used at a time. The location of proteins in the gel is detected spectrophotometrically at 280 nm, using a Gilford Model 2410-S linear transport attachment. Two bands appear, the band from the cathode side is cut out as a 2 mm slice and several such slices are mixed with glass beads (Superbrite glass beads, type 130-5005, Reflective Product Division 3 M Co., Minnesota; about one quarter of the volume of the slices is used) and ground with a glass rod. Phosphate buffer, 0.01 M, pH 6.0, is then added; the suspension is well mixed and filtered through a fine sintered-glass filter, and more of the buffer is used for washing (total volume of 8 ml per six slices). The filtrate is dialyzed against 0.05 M borate buffer, pH 8.17 (three changes of 50 ml at 4°; the dialysis bags pretreated by heating them at 100° for 20 min in 0.01 M EDTA, are washed with double-distilled water and stored at 4°) and concentrated by ultrafiltration. One run with six gels yields 0.26 mg[11] of the pure enzyme.

[9] A. Tiselius, S. Hjerten, and O. Levin, *Arch. Biochem. Biophys.* **65**, 132 (1956).

[10] B. J. Davis, *Ann. N.Y. Acad. Sci.* **121**, 404 (1964).

[11] O. H. Lowry, N. J. Rosebrough, A. L. Farr, and R. J. Randall, *J. Biol. Chem.* **193**, 265 (1951).

Stability. When stored at —20° this preparation retained 70% of the original activity after 5 weeks. It is advantageous to keep stored the preparations obtained by hydroxyapatite chromatography in the previous step. Even when loss of activity occurs, pure preparations, having the highest specific activity (110–120 units/mg) can be recovered by electrophoresis, but with an accordingly lower yield.

Purity and Properties

Polyacrylamide Gel Electrophoresis. The enzyme migrates as a single component when subjected to polyacrylamide gel electrophoresis in 7.5% gels at pH 8.3 and 7.0 by the method of Ornstein and Davis,[12] in gels of graded porosity,[13] and when the denatured enzyme is subjected to gel electrophoresis in presence of sodium dodecyl sulfate.[14]

Immunodiffusion and Immunoelectrophoresis. Rabbits were immunized with a partially purified enzyme, and the serum was used to establish the homogeneity of DCP by immunodiffusion and by immunoelectrophoresis. A single precipitin arc is observed with the pure DCP, whereas a number of arcs are seen with a DCP preparation obtained after the gel-filtration step.

Molecular Weight. The molecular weight of the native DCP was estimated as 97,000 by electrophoresis through polyacrylamide gels of graded porosity,[15] using ovalbumin (MW 45,000) and bovine serum albumin (MW 68,000 for monomer, 138,000 for dimer) as the markers.

pH Profile. The dependence of DCP activity on pH is represented by a bell-shaped curve with a maximum at pII 8.2. This was shown with $Dnp(Pro-Gly-Pro)_n$ as the substrate, in 0.05 M Veronal buffer, in the pH range from 7.0 to 9.0.

Metal Ion Requirement. The enzyme, as obtained, is active without the addition of metal ions. Exhaustive dialysis against 0.1 mM EDTA does not affect the specific activity. Addition of Mg^{2+}, Mn^{2+}, Zn^{2+}, Cd^{2+}, and Ni^{2+} has a very small effect, but the addition of Co^{2+} (1 μM to 1 mM) increases the activity about 5–8 times, with a maximal effect at 50 μM. $Z-Ala_3$ was used in these experiments as the substrate, and the release of alanylalanine was followed by the colorimetric ninhydrin method of Rosen.[7]

[12] L. Ornstein and B. J. Davis, *Ann. N.Y. Acad. Sci.* **121**, 428 (1964).
[13] L. Kamm and J. Mes, *J. Chromatogr.* **62**, 383 (1971).
[14] A. L. Shapira, E. Viñuela, and J. V. Maizel, *Biochem. Biophys. Res. Commun.* **28**, 1815 (1967).
[15] L. O. Anderson, H. Borg, and M. Mikaelsson, *FEBS Lett.* **20**, 199 (1972).

Substrate Specificity. DCP hydrolyzes the penultimate peptide bond of α-*N*-blocked tripeptides, free tetrapeptides, and higher peptides. The following compounds are typical examples:

$$\text{Z-Ala-}\!\overset{\downarrow}{\text{Ala}}\text{-Ala,} \quad \text{Ala-Ala-}\!\overset{\downarrow}{\text{Ala}}\text{-Ala,} \quad \text{Lys-Lys-}\!\overset{\downarrow}{\text{Lys}}\text{-Lys,}$$

$$\text{Gly-Gly-}\!\overset{\downarrow}{\text{Phe}}\text{-Ala,} \quad \text{Boc-Ala-Glu-}\!\overset{\downarrow}{\text{Ala}}\text{-Ala,} \quad \text{Ala-Ala-Ala-}\!\overset{\downarrow}{\text{Lys}}\text{-Phe,}$$

$$\text{Lys-Ala-}\!\overset{\downarrow}{\text{Ala}}\text{-Lys-Ala-Ala,} \quad \text{Arg-Pro-Pro-}\!\overset{\downarrow}{\text{Gly}}\text{-Phe-}\!\overset{\downarrow}{\text{Ser}}\text{-Pro-}\!\overset{\downarrow}{\text{Phe}}\text{-Arg}$$

(0.05 *M* borate buffer, pH 8.1, 75 μ*M* CoSO$_4$, 40°, 2 μg/ml enzyme). Free tripeptides are not hydrolyzed, therefore N-protected tripeptides are the smallest substrates for DCP. A free carboxyl group is required, since no hydrolysis occurs with tetraalanine amide. Secondary peptide bonds to which the nitrogen is donated by a proline residue, as in Z-Phe-Pro-Ala or polyproline, are not cleaved. Neither is a peptide consisting of a chain of several glycine residues, such as Z-Gly-Gly-Gly-Gly. Peptides with a C-terminal D-residue are resistant to DCP.

Since DCP acts on the carboxyl end of polypeptides by hydrolyzing successively every second peptide bond, one may expect to find it useful for application in studies of protein structure.

Kinetic Parameters

Dependence of Specific Activity on Enzyme Concentration. Linear increase of activity with increasing enzyme concentration was found up to a DCP concentration of 1.5 μg/ml with Boc-Ala$_3$ as the substrate (5m*M* in 50 m*M* borate buffer, pH 8.15, at 40°). At higher concentration, the dependence deviated from linearity, the specific activity becoming smaller with increasing enzyme concentration. Thus, while the rate of hydrolysis per 1 μg of enzyme per milliliter was 8 nmoles min^{-1} in the linear range, it was 7 nmoles min^{-1} at an enzyme concentration of 2 μg/ml and only 6 nmoles min^{-1} at an enzyme concentration of 3 μg/ml.

Michaelis–Menten Parameters. The kinetic parameters, K_m and k_{cat}, for several tri- and tetrapeptides are given in Table II. Initial rates were obtained by the pH stat method or spectrophotometrically by measuring the absorption at 225 nm. The kinetic constants were calculated from Lineweaver–Burk plots that were linear in the substrate concentration range of 0.1–7.5 m*M*. The enzyme shows high affinity for basic residues as can be seen by the substrate inhibition observed for Lys$_4$ and the pronounced product inhibition with lysyl and arginyl dipeptides.

TABLE II
KINETIC PARAMETERS FOR THE HYDROLYSIS OF PEPTIDES BY
DIPEPTIDYL CARBOXYPEPTIDASE

Substrate or competitive inhibitor	$K_m{}^a$ (mM)	$K_i{}^c$ (μM)	k_{cat} (sec^{-1})	Substrate conc. range (mM)
Boc-Ala$_3$	0.71	—	34	0.8–5.0
Ala$_4$	0.44	—	139	0.6–5.0
S-Ala$_4$	0.41	—	131	0.5–5.0
Ala-Ala-Phe-Ala	1.27	—	225	1.0–5.0
Gly-Ala-Phe-Ala	1.39	—	194	1.0–5.0
Ala-Gly-Phe-Ala	0.61	—	156	1.0–5.0
Gly-Gly-Phe-Ala	6.06	—	116	1.0–5.0
Lys$_4$	~0.10^b	—	~30^b	0.08–1.0
Ala$_2$	—	396	—	—
Lys-Ala	—	19.3	—	—
Ala-Lys	—	9.0	—	—
Phe-Arg	—	0.92	—	—

^{a} K_m values were determined spectrophotometrically. (For Z-Ala$_4$ the pH-stat method was also used, resulting in identical K_m value.) The composition of the reaction mixture was: substrate in 0.05 M borate buffer, pH 8.15 (no metal added) at 40°, enzyme (1.0 to 6 nM). The difference in molar extinction coefficients, $\Delta\epsilon_{225}$ accompanying the hydrolysis was determined with authentic mixtures of the product peptides and the substrates. In the pH-stat method, the substrate and enzyme in 0.1 M KCl were kept at pH 8.15 with 0.002 M KOH, at 40°. The reaction mixture volume was 1 ml.
^{b} Substrate inhibition.
^{c} Competitive inhibition, pH-stat method used with Z-Ala$_4$ as the substrate (A. Yaron and D. Mlynar, unpublished results).

Occurrence

In its specificity requirements DCP resembles the "angiotensin I-converting enzyme" present in mammalian tissue. This "converting enzyme," first discovered by Skeggs et al.[16] in horse plasma was shown to be a dipeptidyl carboxypeptidase or peptidyldipeptide hydrolase (EC 3.4.15.1), previously also termed kininase II.[17] It thus appears that dipeptidyl carboxypeptidase belongs to a family of enzymes occuring both in bacterial and mammalian organisms. It is of interest that during the various purification steps, DCP was accompanied by a dipeptidase, an enzyme that specifically hydrolyzes dipeptides, which are the prod-

[16] L. T. Skeggs, W. H. Marsh, J. R. Kahn, and N. P. Shumway, J. Exp. Med. 99, 275 (1954).
[17] H. Y. T. Yang, E. G. Erdös, and Y. Levin, Biochim. Biophys. Acta 214, 374 (1970).

ucts of DCP action. The two enzymes were not separated by ion-exchange column chromatography on DEAE-cellulose, by gel filtration, or by polyacrylamide gel electrophoresis. Even the activity–pH curves were found to be very similar. Separation of the two enzymes was eventually achieved by chromatography through hydroxyapatite columns. The complementary action of the two enzymes on the carboxyl ends of polypeptide chains represents an alternative mechanism for carboxypeptidase action by which a polypeptide chain is successively degraded from the carboxyl end to amino acids. An interesting regulatory mechanism is indicated by the finding that the dipeptides formed by DCP act as competitive inhibitors of this enzyme and in turn serve as substrates for dipeptidase, the activity of which depends on Mn ions.

Acknowledgment

The financial support granted by the Helena Rubinstein Foundation Inc. is gratefully acknowledged.

[51] Exocellular DD-Carboxypeptidases-Transpeptidases from *Streptomyces*

By JEAN-MARIE FRÉRE, MÉLINA LEYH-BOUILLE, JEAN-MARIE GHUYSEN, MANUEL NIETO, and HAROLD R. PERKINS

Strains and Culture

Strains. Strains R39 and R61 are soil isolates.[1,2] Their designations are arbitrary. In strain R39, the cross-link between the peptide units of the wall peptidoglycan extends from the C-terminal D-alanine of one unit to the amino group at the D-center of *meso*-diaminopimelic acid of another unit[3] (peptidoglycan of chemotype I).[4] The interpeptide bond is in α-position to a free carboxyl group. In strain R61, the cross-link extends from a C-terminal D-alanine of a peptide unit to a glycine residue attached to the ε-amino group of LL-diaminopimelic acid of another peptide unit[5] (peptidoglycan of chemotype II).[4] The exocellular DD-

[1] M. Welsch, *Rev. Belge Pathol. Med. Exp.* 18, Suppl. 2, 1 (1947).
[2] M. Welsch and A. Rutten-Pinckaers, *Bull. Soc. R. Sci. Liege,* 3–4, 374 (1963).
[3] J.-M. Ghuysen, M. Leyh-Bouille, J. N. Compbell, R. Moreno, J.-M. Frère, C. Duez, M. Nieto, and H. R. Perkins, *Biochemistry* 12, 1243 (1973).
[4] J.-M. Ghuysen, *Bacteriol. Rev.* 32, 425 (1968).
[5] M. Leyh-Bouille, R. Bonaly, J.-M. Ghuysen, R. Tinelli, and D. J. Tipper, *Biochemistry* 9, 2944 (1970).

carboxypeptidases-transpeptidases produced by both strains (1) catalyze hydrolysis and transfer reactions according to the general equations R-D-Ala-D-Ala + H_2O → D-Ala + R-D-Ala (DD-carboxypeptidase activity) and R-D-Ala-D-Ala + NH_2-R′ → D-Ala + R-D-Ala-R′ (transpeptidase activity) and (2) react with β-lactam antibiotics (i.e., penicillins and cephalosporins) to form equimolar and inactive antibiotic–enzyme complexes of various half-lives. Both R39 strain and, to a much lesser extent, strain R61 also produce an exocelular β-lactamase (penicillinase, EC 3.5.2.6) which hydrolyzes penicillin to penicilloic acid.

Culture Media. Peptone oxoid medium contains 1% peptone oxoid, 0.1% K_2HPO_4, 0.1% $MgSO_4 \cdot 7H_2O$, 0.2% $NaNO_3$, and 0.005% KCl in water.

Glycerol-casein medium contains per liter of final volume: 20 g of glycerol, 40 ml of 10% casein solution (w/v), and 50 ml of salt suspension. The salt suspension contains per liter: 5 g of NaCl, 1 g of $CaCO_3$, 1 g of $MgSO_4 \cdot 7H_2O$, 10 g of K_2HPO_4, and 1 g of $FeSO_4 \cdot 7H_2O$. The casein is previously dissolved at 70° with the help of KOH (0.04 g per gram casein). The solution is then cooled and neutralized.

Agar-APG medium contains per liter of final volume: agar 20 g; asparagine, 0.5 g; peptone oxoid, 0.5 g; glucose, 10 g; and K_2HPO_4, 0.5 g.

Agar-KC medium contains per liter of final volume: agar, 20 g; partially hydrolyzed keratin from white hens' feathers, 2.5 g; partially hydrolyzed casein, 2.5 g; NaCl, 0.5 g; $CaCO_3$, 0.1 g; $MgSO_4 \cdot 7H_2O$, 0.1 g; K_2HPO_4, 1 g; and $FeSO_4 \cdot 7H_2O$, 0.1 g. The final pH is 7.5. Hydrolyzed keratin and casein are prepared as follows: 100 g of dried, white hens' feathers are treated for 1 hr at 100° with 1 liter of 0.125 N KOH. After centrifugation, casein (100 g) is added to the supernatant fraction and dissolved by heating at 70°. The pH of the mixture (final volume, 1 liter) is adjusted to 7.5–8.0; 25 ml contain 2.5 g of both partially hydrolyzed keratin and casein.

Maintenance of Strains. Strain R39 is grown at 28° on slants of agar-KC, and strain of R61 on slants of agar-APG. Abundant sporulation occurs after 4–5 days. The strains are then maintained at 4°.

Assay Methods for DD-Carboxypeptidase Activity

Standard Reaction

$$Ac_2\text{-L-Lys-D-Ala-D-Ala} + H_2O \rightarrow \text{D-Ala} + Ac_2\text{-L-Lys-D-Ala} \tag{1}$$

Unit. One unit of DD-carboxypeptidase catalyzes the hydrolysis of 1 μmole of N^α,N^ε-diacetyl-L-lysyl-D-alanyl-D-alanine into D-alanine and

N^α,N^ε-diacetyl-L-lysyl-D-alanine per minute at $37°$ under conditions of enzyme saturation by the substrate.

Substrate. Tripeptide Ac₂-L-Lys-D-Ala-D-Ala is prepared as described by Nieto and Perkins.[6]

Standard Incubation Conditions. Streptomyces R39 enzyme: Ac₂-L-Lys-D-Ala-D-Ala ($0.25~\mu$mole), is incubated with the enzyme at $37°$ in $30~\mu$l (final volume) of $0.03~M$ Tris-HCl buffer, pH 7.5, supplemented with $3~mM$ MgCl₂. Final substrate concentration is $8~mM$, i.e., $10 \times$ the K_m value ($0.8~mM$).

Streptomyces R61 enzyme: Ac₂-L-Lys-D-Ala-D-Ala ($0.36~\mu$mole) is incubated with the enzyme at $37°$ in $30~\mu$l of either 5–$10~mM$ sodium phosphate buffer pH 7.5 or $10~mM$ Tris-HCl buffer pH 7.5. Final substrate concentration is $12~mM$, i.e., equivalent to the K_m value. For this enzyme, assays are thus carried out at half the maximum velocity.

The D-Ala released during the reaction is estimated by using one of the following techniques.

Chemical Estimation of Free Alanine

The following is a modification of the technique of Ghuysen *et al.*[7]

FDNB Reagent. Fluorodinitrobenzene, $130~\mu$l in 10 ml of 100% ethanol.

Procedure. Samples containing 10–50 nmoles of alanine are mixed with 10% K₂B₄O₇ and water to give a total volume of $100~\mu$l of 1% K₂B₄O₇. FDNB reagent ($10~\mu$l) is added. The solutions are mixed and incubated at $60°$ for 30 min. After acidification with $50~\mu$l of $12~N$ HCl, the DNP-alanine is extracted three times with $200~\mu$l of ether. The ether extracts are evaporated in a stream of hot air and dried *in vacuo.* The residues are dissolved in methanol and chromatographed at room temperature on thin-layer plates of silica gel G in chloroform:methanol:acetic acid (220:25:5, v/v/v). DNP-alanine moves faster than DNP-Tris. After drying, the DNP-alanine spots are transferred to 1-ml tubes and eluted by vigorous mixing with $500~\mu$l of water:ethanol:25% (specific gravity 0.91) ammonia (100:100:0.54, v/v/v). After centrifugation, the optical density of the supernatant fractions is measured at 360 nm. The molar extinction coefficient for DNP-alanine is about $15,000$.

Enzymic Estimation of 4-Alanine

The following modification of the technique of Ghuysen *et al.*[7] permits many simultaneous tests to be carried out in a very short time.

[6] M. Nieto and H. R. Perkins, *Biochem. J.* **123,** 789 (1971).
[7] J.-M. Ghuysen, D. J. Tipper, and J. L. Strominger, this series Vol. 8, p. 685.

Reagents

o-Dianisidine (Merck, analytical grade): 10 mg/ml in methanol (freshly prepared)

K pyrophosphate buffer, 0.1 M, pH 8.3

FAD (monosodium; Boehringer), 0.3 mg/ml in pyrophosphate buffer

Peroxidase (Boehringer, Reinheitsgrad 1, für analytische Zwecke; suspension 10 mg/ml) to be diluted to 10 μg/ml in H_2O

D-Amino acid oxidase (Boehringer, für analytische Zweke; Kristall-suspension, 5 mg/ml)

Enzymes and coenzyme mixture (freshly prepared): pyrophosphate buffer:FAD solution:diluted peroxidase solution:D-amino acid oxidase suspension (20:10:5:1, v/v/v/v)

Procedure. Samples (30 μl) containing 5–40 nomoles of D-alanine (from the hydrolyzed substrate) are mixed with 5μl of o-dianisidine solution and 70 μl of the enzymes–coenzyme mixture. To such solutions incubated for 5 min at 37°, is added 400 μl of methanol–water (v/v). After an additional 2-min incubation at 37°, the optical density at 460 nm is imediately measured. (Coloration of the solution is silghtly labilized after addition of the methanol–water solution.) Blanks consist of the same mixtures as above lacking Ac$_2$-L-Lys-D-Ala-D-Ala tripeptide. Standards are blanks containing known amounts of D-alanine.

Assay Method for β-Lactamase

DD-Carboxypeptidases-transpeptidases and β-lactamases react with penicillin. The reaction mechanisms and the reaction products are, however, different (see below). A complete separation between these two classes of enzymes is thus one major goal of the purification procedure of the DD-carboxypeptidases-transpeptidases.

Standard Reaction for β-Lactamase

$$\text{Benzylpenicillin} + H_2O \rightarrow \text{benzylpenicilloic acid} \qquad (2)$$

β-Lactamase Unit. One unit of β-lactamase catalyzes the hydrolysis of 1 μmole of benzylpenicillin per minute at 30° and under conditions of enzyme saturation by the substrate. The K_m value of the β-lactamase from *Streptomyces* R39 is 70 μM.[8] That of the β-lactamase from *Streptomyces* R61 is not known.

---

[8] K. Johnson, J. Dusart, J. N. Campbell, and J.-M. Ghuysen, *Antimicrob. Agents Chemother.* 3, 289 (1973).

Reagents

Sodium acetate buffer, 1 M pH 3.6

Color reagent: equal volumes of a water-soluble starch solution (0.8%, w/v) and of a 240 μM I_2 + 4.8 mM KI solution

Procedure. The following is a microscale adaptation of the technique of Novick and Dubnau.[9] Benzylpenicillin (0.3 μmole) is incubated with the enzyme preparation in 30 μl (final volume) of 0.025 M sodium phosphate buffer pH 7.0. The benzylpenicillin concentration in the mixture is 10 mM; hence, any β-lactamase with a K_m value for this antibiotic equal to or lower than 1 mM is saturated. After 10–30 min, 200 μl of 1 M acetate buffer and then 200 μl of color reagent are added in sequence to the reaction mixture. After 10 min at 25°, the optical density of the solution is measured at 620 nm. A control consisting of the same mixture lacking enzyme is incubated as above. Acetate buffer, the same amount of enzyme as used in the test, and finally the color reagent are added, and the optical density at 620 nm is determined. A decrease of the optical density of 0.1 corresponds to about 0.37 nmole of benzylpenicilloic acid produced.

Excretion of DD-Carboxypeptidase-Transpeptidase and β-Lactamase by *Streptomyces* R39[10]

Streptomyces R39 is grown for 24 hr at 28° with shaking in peptone medium. After two successive subcultures of increasing size, 100 liters of culture in exponential phase are used to inoculate 400 liters of the same peptone medium contained in a 500-liter tank. This culture is grown at 28° for about 90–100 hr with mechanical stirring (120 rpm), and an air-flow rate of 100 liters/min at an air pressure of 1.5 × 10⁵ Pa. Silicone A emulsion (Dow Corning Co., Midland, Michigan; 20 ml) is used as antifoam. After centrifugation, the DD-carboxypeptidase activity in the culture fluid is about 2.5 munits/ml or 1.5 munits/mg of protein, and β-lactamase activity is 50 munits/ml. Similar results are obtained on a smaller scale by growing *Streptomyces* R39 in a New Brunswick Shaker incubator, in 1-liter flasks containing 400 ml of peptone medium. Under both conditions, maximal DD-carboxypeptidase activity and maximal β-lactamase activity occur almost simultaneously. Both activities then disappear progressively and are negligible after 6 days. With time, the two enzyme activities increase and decrease independently of each other.

[9] R. P. Novick, *J. Gen. Microbiol.* **33**, 121 (1963).

[10] J.-M. Frère, R. Moreno, J.-M. Ghuysen, H. R. Perkins, L. Dierickx, and L. Delcambe, *Biochem. J.* **143**, 233 (1974).

Excretion of DD-Carboxypeptidase–Transpeptidase and β-Lactamase
by *Streptomyces* R61[11]

Streptomyces R61 excretes larger amounts of its DD-carboxypeptidase–
transpeptidase in glycerol–casein medium than in peptone medium. The
strain is grown at 28° in a New Brunswick Shaker incubator in 1-liter
flasks containing 500 ml of glycerol–casein medium. Often, the cultures
exhibit two peaks of DD-carboxypeptidase activity, the first one occurring
after about 100 hr (about 11 mU/ml or 13 mU/mg protein) and the
second one after about 300 hr (about 18 mU/ml). No corresponding
increase in mycelium production is observed. Large amounts of free amino
acids are present in the 100–200 hr cultures, and they subsequently disappear, suggesting that part of the bacterial population resumes growth
from autolysis products. Under the above growth conditions, *Streptomyces* R61 is a poor producer of exocellular β-lactamase. Its β-lactamase
has not been studied.

Purification of the DD-Carboxypeptidase–Transpeptidase from
Streptomyces R39 (for 500 Liters of Culture Fluid)[10]

Step 1. The enzyme is adsorbed from 500 liters of culture fluid on
3.7 kg (wet weight) of DEAE-cellulose (MN 2100 DEAE, Macherey,
Nagel and Co., D-156 Düren, Germany) equilibrated against 0.1 M
Tris-HCl, pH 7.5. All subsequent steps are performed at 4°. The enzyme
is eluted from the DEAE-cellulose by two subsequent treatments with
5 liters of 0.1 M Tris-HCl buffer, pH 7.5, containing 1 mM $MgCl_2$ and
0.4 M NaCl. The solution is concentrated to 1.5 liter on Carbowax 4000
and solid ammonium sulfate is added to 50% saturation. The precipitate
is discarded, and the $(NH_4)_2SO_4$ concentration in the supernatant is increased to 90% saturation. The precipitate is collected by centrifugation,
dissolved in 240 ml of 0.1 M Tris-HCl buffer, pH 7.7, containing 1 mM
$MgCl_2$ and the solution is dialyzed against the same buffer.
Step 2. After step 1, the enzyme solution is applied to a 600 ml column
of DEAE-cellulose (4 x 48 cm) equilibrated against 0.1 M Tris-HCl buffer, pH 7.7, containing 1 mM $MgCl_2$ and 0.1 M NaCl. Some enzymically
inactive proteins are eliminated by washing the column with the same
buffer, and others are eluted by increasing the NaCl concentration in the
buffer to 0.19 M. The resin is then treated with an increasing convex gradient of NaCl (mixing flask: 970 ml of 0.1 M Tris-HCl buffer + 1 mM

[11] M. Leyh-Bouille, J. Coyette, J.-M. Ghuysen, J. Idczak, H. R. Perkins, and M. Nieto,
Biochemistry **10**, 2163 (1971).

MgCl$_2$ + 0.19 M NaCl; upper flask: 0.1 M Tris-HCl buffer + 1 mM MgCl$_2$ + 0.28 M NaCl). The enzyme is eluted at a NaCl concentration of about 0.24 M. The active fractions are pooled and concentrated to 20 ml by ultrafiltration through Amicon UM-10 membranes.

Step 3. After step 2, the concentrated solution is filtered through a 400-ml column of Sephadex G-100 (3 × 45 cm) previously equilibrated against 0.05 M cacodylate-HCl buffer, pH 6.0, containing 1 mM MgCl$_2$ + 0.3 M NaCl. The enzyme is eluted at a K_D value of 0.21 (and well separated from two yellow-brown pigments presenting K_D values of 0–0.02 and 0.72, respectively). The active fractions are pooled (90 ml).

Step 4. After step 3, the solution is applied to a 30-ml (2 × 10 cm) column of DEAE-Sephadex A-50, previously equilibrated against 0.05 M cacodylate-HCl buffer, pH 6.0, containing 1 mM MgCl$_2$ + 0.3 M NaCl. Enzymically inactive proteins are eliminated by washing the column first with 300 ml of the same cacodylate-HCl-MgCl$_2$ buffer containing 0.4 M NaCl and then with 300 ml of the same buffer containing 0.47 M NaCl. The enzyme is then eluted with a convex gradient of NaCl (mixing flask 500 ml of cacodylate-HCl buffer + 1 mM MgCl$_2$ + 0.47 M NaCl; upper flask: cacodylate-HCl buffer + 1 mM MgCl$_2$ + 0.5 M NaCl). A single peak of protein is obtained, which closely correlates with the activity. The active fractions are pooled and concentrated to 20 ml by ultrafiltration. The concentrated solution is filtered through the 400-ml column of Sephadex G-100, in 0.05 M cacodylate-HCl buffer, pH 6.0, containing 1 mM MgCl$_2$ and 0.4 M NaCl. The most active fractions of the eluted peak exhibit the same high specific activity. They are pooled, concentrated by ultrafiltration and dialyzed against 0.1 M Tris-HCl buffer, pH 7.7, containing 0.2 M NaCl and 0.05 M MgCl$_2$ (fraction A, Table I, specific activity: 17.1 units/mg protein). The other fractions of the eluted peak have lower specific activities. They are pooled, concentrated, and dialyzed (fraction B, Table I). Fraction B (specific activity: 13 units/mg protein) may be stored as it is and used for experiments that do not require an enzyme of absolute purity. Fraction A behaves as a homogeneous protein in several analytical tests (see below). However, benzylpenicillin binding occurs at a ratio of 0.9 mole of benzylpenicillin per mole of enzyme, showing that some impurities are still present in the preparation (see below, Fig. 4).

Step 5. After step 4, to fraction A (1 ml, 1.6 mg of protein) is added 1 ml of acetone previously cooled to −20°. The mixture is stirred for 30 min at −20°; the precipitate is collected by centrifugation at −5° and redissolved in 1 ml of 0.1 M Tris-HCl buffer, pH 7.7, containing 0.2 M NaCl and 0.05 M MgCl$_2$. At this stage, benzylpenicillin binding occurs at a ratio of 1.04 ± 0.03 mole of benzylpenicillin per mole of enzyme (see Fig. 4). When compared to fraction A, the final preparation has a

TABLE I

PURIFICATION OF THE DD-CARBOXYPEPTIDASES-TRANSPEPTIDASES FROM *Streptomyces* R39 AND R61[a]

Enzyme	Step[b]	Total protein[c] (mg)	Activity (total units)	Yield (%)	Specific activity (units/mg of protein)	Enrichment
R39	Culture supernatant	8.4×10^5	1250	100	0.0015	1
	1	2.9×10^3	1050[d]	84	0.36	240
	2	166	830	66	5.0	3300
	3	65	720	58	11.0	6700
	4A	21.8	375	30	17.1	11400
	4B	11.4	147	12	13.0	8700
	5[e]	1.4			19.8	12700
R61	Culture supernatant	350×10^3	4200	100	0.012	1
	1	7×10^3	2500	60	0.36	30
	2	59	1700	40	29	2400
	3A	11.2	970	23	86	7200
	3B	8	400	9.5	50	4150

[a] From J.-M. Frère, R. Moreno, J.-M. Ghuysen, H. R. Perkins, L. Diericks, and L. Delcambe, *Biochem. J.* **143**, 233 (1974); and J.-M. Frère, J.-M. Ghuysen, H. R. Perkins, and M. Nieto, *Biochem. J.* **135**, 463 (1973).

[b] The R39 enzyme preparations after steps 4 and 5 and the R61 enzyme preparation after step 3A are devoid of β-lactamase activity.

[c] The protein concentration was determined either by measuring the extinction at 280 and 260 nm and using the formula C (mg/ml) = $1.54\ E_{280} - 0.76\ E_{260}$ or (for the final preparations) by measuring the amount of total amino groups available to fluorodinitrobenzene after 6 M HCl hydrolysis (100° 20 hr; standard: bovine serum albumin).

[d] The total activity obtained after $(NH_4)_2SO_4$ fractionation by adding up the total units of all fractions (0–50%, 50–90%, and >90%) was always equal to about 130% of the total activity of the original solution. It is possible that an inhibitor was eliminated during this step.

[e] Step 5 was carried out on a fraction of the preparation obtained after step 4A.

specific activity increased by about 11%. The enzyme recovery is about 90%. Table I gives the total recoveries and enrichments in specific activity after each step of the purification procedure.

Purification of the DD-Carboxypeptidase–Transpeptidase from *Streptomyces* R61 (for 400 Liters of Culture Fluid)

The following is an adaptation of the technique of Frère *et al.*[12]

Step 1. The enzyme is adsorbed from 400 liters of culture fluid on 10 kg of Amberlite XE64 H$^+$ or CG50 H$^+$ by adjusting the pH to 4.0 with acetic acid. The Amberlite-adsorbed enzyme complex is suspended in 20 liters of cold 0.1 M K$_2$HPO$_4$, and the pH of the suspension is brought to 8.0 by dropwise addition of concentrated ammonia. The resin is removed by filtration, and the filtrate is clarified by centrifugation. The adsorption of the enzyme on the resin and its elution as well as all subsequent steps are performed at 4°. Solid (NH$_4$)$_2$SO$_4$ is added to the eluted enzyme solution and the precipitate obtained at 40% saturation is discarded. Protein (128 g, wet weight) precipitated when the (NH$_4$)$_2$SO$_4$ concentration is raised to 60% saturation is redissolved in 1 liter of 0.01 M Tris-HCl buffer and dialyzed twice against 50 liters of the same buffer. The dialyzed solution is stirred with 500 g (wet weight) of DEAE-cellulose previously equilibrated against 0.01 M Tris-HCl buffer pH 8.0. The enzyme is eluted from the DEAE-cellulose by treating the resin, batchwise, with 1 liter of the same buffer containing 0.1 M NaCl. The eluate is dialyzed against water.

Step 2. After step 1, the enzyme is adsorbed on a 600 ml column of DEAE-cellulose (4 × 40 cm) equilibrated against 0.01 M Tris-HCl buffer pH 8.0. The column is treated with an increasing convex gradient of NaCl (mixing flask, at constant volume: 800 ml of 0.01 M Tris-HCl buffer; solution added: same buffer + 0.13 M NaCl). The active fractions are pooled, concentrated to 20 ml by ultrafiltration and filtered through a 425-ml column (3 × 60 cm) of Sephadex G-75 equilibrated against 0.01 M cacodylate-HCl buffer pH 6.0. The column is washed with the same cacodylate-HCl buffer and the enzyme is eluted just before the main peak of protein. The active fractions are pooled. At this stage, the solution is still faintly yellow.

Step 3. After step 2, the enzyme is adsorbed on a 30-ml column (2 × 10 cm) of DEAE-Sephadex A-50 previously equilibrated against 0.01 M

---

[12] J.-M. Frère, J.-M. Ghuysen, H. R. Perkins, and M. Nieto, *Biochem. J.* **135**, 463 (1973).

cacodylate-HCl buffer pH 6.0. The enzyme is eluted with an increasing linear gradient of NaCl (0 to 0.15 M) in the same buffer. The yellow pigment remains fixed on the top of the column. The active fractions in the center of the eluted peak of protein exhibit a constant high specific activity. They are pooled, concentrated to 10 ml by ultrafiltration, and extensively dialyzed against water (preparation 3A). The other fractions are treated likewise (preparation 3B). Table I gives the total recoveries and enrichments in specific activity after each step of the purification procedure.

Physicochemical Properties of DD-Carboxypeptidases–Transpeptidases from *Streptomyces* R39 and R61

The properties described in the present paragraph (together with the techniques of titration of the *Streptomyces* enzymes by β-lactam antibiotics: see below) constitute the best available criteria of purity. For more details, see Frère *et al.*[10, 12]

Diffusion Constant $(D_{20,w})$, Molecular Weight (MW), and Frictional Ratio (f/f_o)

Buffers. The following buffers are used: 0.1 M Tris-HCl, pH 7.7 + 3 mM $MgCl_2$ + 0.5 M NaCl (density, 1.02) for the R39 enzyme; 0.01 M Tris-HCl pH 8.0 + 0.09 M NaCl (density, 1.002) for the R61 enzyme.

Procedure. The enzyme solutions (dialyzed against the relevant buffer) are analyzed by equilibrium sedimentation at a speed of 13,000 rpm for 22 hr at 20° and initial protein concentration of 2–3 mg/ml. The rates of diffusion are measured by plotting $A^2/(H^2F^2)$ against time (A = area and H = maximum height of the peak; F = total enlargement used). The apparent molecular weights (M_{app}) at any point x of the solution column (about 2 mm long) are calculated according to the equation

$$M_{app} = [RT/(1 - \bar{v}\rho)\omega^2](1/C_x)(dc/dx)$$

where $\bar{v}$ = partial specific volume (supposed to be equal to 0.75 cm³ g⁻¹), ρ = density of the solvent (see above); C = protein concentration at point x. The molecular weights are obtained by plotting $1/M_{app}$ against concentration at 0.1-mm intervals of the column. The results are given in Table II.

TABLE II

MOLECULAR WEIGHT (MW), DIFFUSION CONSTANT $(D_{20,w})$, FRICTIONAL RATIO
(f/f_0), AND MOLAR ACTIVITY OF DD-CARBOXYPEPTIDASES–TRANSPEPTIDASES
FROM *Streptomyces* R39 AND R61[a]

Enzyme	MW	$D_{20,w}$ $(\times 10^{-3}$ cm^2 sec$^{-1})$	f/f_0	Molar activity[b] (min^{-1})
R39	53,500	7.88	1.07	1050
R61	37,000	8.45	1.12	3300

[a] From J.-M. Frère, R. Moreno, J. M. Ghuysen, H. R. Perkins, L. Dierickx, and
L. Delcambe, *Biochem. J.* **143**, 233 (1974); and J.-M. Frère, J.-M. Ghuysen,
H. R. Perkins, and M. Nieto, *Biochem. J.* **135**, 463 (1973).
[b] On Ac$_2$-L-Lys-D-Ala-D-Ala.

Remarks

1. With a homogeneous protein, the plot of $1/M_{app}$ vs C_x during sedimentation equilibrium gives rise to a line parallel to the abscissa. Figures 1A and 1B show the results obtained with the R61 enzyme after various steps of purification, and Fig. 1C with the purified R39 enzyme.

2. When equilibrium sedimentation of the purified R39 enzyme is carried out in the same Tris–HCl–MgCl$_2$ buffer as above except that the NaCl concentration is 0.2 M instead of 0.5 M, the plot of $1/M_{app}$ vs concentration gives a line with a slope of -1.1×10^{-4} (with the same coordinates as in Fig. 1). This is because the R39 protein aggregates at low ionic strength.

Sephadex Filtrations. Filtrations are carried out on 1.5 × 65 cm columns of Sephadex G-100, using dextran blue, ovalbumin, myoglobin, bovine serum albumin, and chymotrypsinogen as molecular weight standards. In 0.01 M Tris-HCl buffer, pH 8.0, the apparent molecular weight of the R61 enzyme is about 38,000. The apparent molecular weight of the R39 enzyme decreases as the ionic strength of the buffer increases. Ionic strengths equal to or higher than 0.07 M and ionic strengths lower than 0.07 M are obtained with 0.1 M and 0.01 M Tris-HCl buffers, pH 7.7, respectively, supplemented with the appropriate amount of NaCl. The apparent molecular weight is over 100,000 at $I = 0.008$ M, 86,000 at $I = 0.025$ M, 70,000 at $I = 0.070$ M, and about 60,000 at $I > 0.20$ M.

Analytical Polyacrylamide Gel Electrophoresis

In the Absence of Sodium Dodecyl Sulfate (SDS). The gels (6 × 71 mm) are prepared with 7% acrylamide and 0.18% N,N'-methylenebisa-

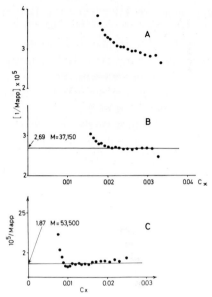

FIG. 1. Reciprocal of the apparent molecular weight $(1/M_{app})$ of *Streptomyces* DD-carboxypeptidases-transpeptidases as a function of the concentration in the column (C_x) during sedimentation equilibrium. (A) R61 enzyme (84% pure); (B) R61 enzyme after step 3A of purification; (C) R39 enzyme after step 5 of purification. Experiments are carried out in 0.01 M Tris-HCl buffer, pH 8.0, with ionic strength adjusted to 0.1 with NaCl for the R61 enzyme and in 0.1 M Tris-HCl buffer pH 7.7 + 3 mM MgCl$_2$ and 0.5 M NaCl for the R39 enzyme. For other conditions, see text and the following articles, from which Fig. 1 is reprinted by courtesy of the Biochemical Society. J.-M. Frère, R. Moreno, J.-M. Ghuysen, H. R. Perkins, L. Dierickx, and L. Delcambe, *Biochem. J.* **143**, 233 (1974); J.-M. Frère, J.-M. Ghuysen, H. R. Perkins, and M. Nieto, *Biochem. J.* **135**, 463 (1973).

crylamide in 0.375 M Tris-HCl buffer, pH 8.4, and polymerized in the presence of 0.08% ammonium persulfate and 0.03% N,N,N',N'-tetramethylethylenediamine. The electrolyte is 0.025 M (in Tris) Tris-glycine buffer pH 8.4. In some cases, gels are submitted to prior electrophoresis for 3 hr at 4 mA/gel (preruns). Electrophoreses of the samples (25 μg of protein) are performed for 135 min at room temperature and at 3 mA/gel. Quartz tubes are used so that the same gel can be scanned at 280 nm and then removed from the tube and sliced into sections (which are eluted and assayed for DD-carboxypeptidase activity). Parallel gels are stained with Coomassie blue as usual. Irrespective of the procedure used for detection, the enzymes give a single band. On gels that are not prerun, migrations toward the anode are 3.5 cm for the R61 enzyme, 4.4 cm for the R39 enzyme, and 4.4 cm for bromophenol blue. On gels that are prerun, migrations toward the anode are 4.3 cm for the R39 enzyme and 6.4 for bromophenol blue.

In the Presence of SDS. Electrophoreses are performed at room temperature in 0.01 M sodium phosphate, pH 7.1, in the presence of 0.1% (w/v) SDS (Weber and Osborn,[13]) with prior incubations of the proteins for 15 hr at 37° in the same phosphate buffer containing 1% (w/v) SDS, with or without the addition of 2-mercaptoethanol (1%, final volume). Each enzyme gives rise to a single band and the mobilities (when compared to that of similarly treated standard proteins) indicate a molecular weight of 56,000 ± 3000 (7 determinations) for the R39 enzyme and of 39,000 ± 1600 (4 determinations) for the R61 enzyme. These values are unaltered when mercaptoethanol is omitted during pretreatment of the proteins with SDS. Each of the R39 and R61 enzymes thus consists of one single polypeptide chain.

Electrofocusing. The gels (6 × 0.5 cm) contain 2.5% carrier ampholytes (pH 3–6), 7% acrylamide, 0.18% N,N'-methylenebisacrylamide, and 0.08% ammonium persulfate. For other conditions, see Frère *et al.*[12] The R61 enzyme has an isoelectric point of 4.8 ± 0.14 (4 determinations). Because of the high tendency of the R39 enzyme to aggregate at low ionic strength, its isoelectric point cannot be determined by this technique. In normal gel electrophoresis at pH 8.4, the R39 enzyme is more anionic than the R61 enzyme (see above).

Spectra. All spectra are determined at 25° in 0.1 M Tris-HCl buffer, pH 7.7, + 0.05 M MgCl$_2$ + 0.2 M NaCl for the R39 enzyme and in 1 mM Tris-HCl buffer, pH 7.4, for the R61 enzyme.

Ultraviolet Absorption. $E_{1\,cm}^{1\%}$ values at 280 nm: 9.7 for the R39 enzyme and 10.0 for the R61 enzyme.

Fluorescence Emission. Excitation at 285 nm produces an emission maximum at 340 nm with the R39 enzyme[10] and at 320 nm with the R61 enzyme.[14]

Circular Dichroism. Circular dichroism spectrum of the R61 enzyme is given by Nieto *et al.*[14]

Amino Acid Compositions. The compositions are shown in Table III. By summing the residue mole percentages of the hydrophilic amino acids plus one half of the mole percentages of the intermediate class,[15] the polarity index then found is 39.7 for the R39 enzyme and 40.5 for the R61 enzyme.

[13] K. Weber and M. Osborn, *J. Biol. Chem.* **244**, 4406 (1969).
[14] M. Nieto, H. R. Perkins, J.-M. Frère, and J.-M. Ghuysen, *Biochem. J.* **135**, 493 (1973).
[15] R. A. Capaldi and G. Vanderkooi, *Proc. Natl. Acad. Sci. U.S.A.* **69**, 930 (1972).

TABLE III

AMINO ACID COMPOSITION OF THE DD-CARBOXYPEPTIDASES-TRANSPEPTIDASES
FROM *Streptomyces* R39 AND R61

Residue	Strain R39 Residues/ enzyme molecule (MW = 53,300)	Total mass	% (in number)	Strain R61 Residues/ enzyme molecule (MW = 38,000)	Total mass	% (in number)
Asp	50	5750	9.5	38	4370	10.9
Thr	34	3434	6.4	38	3838	10.9
Ser	38	3306	7.2	29	2523	8.3
Glu	57	7353	10.8	28	3612	8.0
Pro	25	2425	4.7	11	1067	3.1
Gly	66	3762	12.5	32	1824	9.1
Ala	82	5822	15.5	34	2414	9.7
Cys[a]	2	204	0.37	3	306	0.86
Val	51	5049	9.7	30	2970	8.6
Met	3	393	0.57	6	786	1.71
Ile	10	1130	1.9	9	1017	2.6
Leu	50	5650	9.5	33	3729	9.4
Tyr	10	1630	1.9	13	2119	3.7
Phe	11	1617	2.1	12	1764	3.4
His	9	1233	1.7	8	1096	2.3
Lys	5	640	0.95	8	1024	2.3
Arg	19	2945	3.6	14	2170	4.0
Trp[a]	6	1122	1.1	4	748	1.14
	528	53483[b]		350	37305[b]	

[a] Half-cystine is measured as cysteic acid after performic acid oxidation of 200 μg of protein. Tryptophan is measured from the UV spectrum in alkali.
[b] Taking into account a mass of 18 for H_2O.

Interaction between DD-Carboxypeptidases–Transpeptidases from *Streptomyces* R39 and R61 and β-Lactam Antibiotics[16–18]

The reactions and properties that are described below are relevant to experimental conditions where the enzymes, by themselves, are perfectly stable, i.e., in 0.1 M Tris-HCl buffer, pH 7.7, + 0.05 M $MgCl_2$ +

[16] J.-M. Frère, J.-M. Ghuysen, P. E. Reynolds, R. Moreno, and H. R. Perkins, *Biochem. J.* **143**, 241 (1974).
[17] J.-M. Frère, M. Leyh-Bouille, J.-M. Ghuysen, and H. R. Perkins, *Eur. J. Biochem.* **50**, 203 (1974).
[18] J.-M. Frère, J.-M. Ghuysen, and M. Iwatsubo, *Eur. J. Biochem.* **57**, 343 (1975).

0.2 M NaCl for the R39 enzyme and in 5–10 mM sodium phosphate buffer, pH 7.0, for the R61 enzyme.

Reaction. At room temperature, the R39 and R61 enzymes react readily with β-lactam antibiotics to form equimolar and inactive antibiotic–enzyme complexes. When maintained at 37°, these complexes undergo spontaneous breakdown during which the enzyme is reactivated whereas the antibiotic molecule is released in a chemically altered form. The reactions occur according one of the two following possible mechanisms:

$$\text{E} + \text{P} \underset{}{\overset{K}{\rightleftharpoons}} \text{EP} \overset{k_3}{\rightarrow} \text{EP}^* \overset{k_4}{\rightarrow} \text{E} + \text{X} \tag{3}$$

or

$$\text{E} + \text{P} \underset{k_2}{\overset{k_1}{\rightleftharpoons}} \text{EP} \overset{k_3}{\rightarrow} \text{EP}^* \overset{k_4}{\rightarrow} \text{E} + \text{X} \tag{4}$$

E = active enzyme; P = intact antibiotic molecule; EP = inactive antibiotic–enzyme complex; EP* = inactive complex after isomerization of the antibiotic molecule; X = released and chemically altered antibiotic molecule. Mechanism (3) is the simplest one that accounts for all the experimental facts so far accumulated. Table IV gives the values of the various constants (K = dissociation constant of complex EP) for the reactions between the R61 enzyme and various β-lactam antibiotics. Table IV also gives the half-lives of various EP* complexes formed with both R61 and R39 enzymes. For more details, see Frère *et al.*[18]

Effects on β-Lactam Antibiotics. Reaction of benzylpenicillin with both R39 and R61 enzymes[16,17] yields complexes that subsequently release a compound X that is neither benzylpenicillin nor benzylpenicilloic acid (nor a product arising from a spontaneous degradation of free benzylpenicillin). Unlike benzylpenicilloic acid, compound X cannot be titrated with the iodine reagent (see assay method for β-lactamase).

Reaction of cephaloridine with the R39 enzyme[16] causes a 70% decrease of the absorbance at 250 nm, as observed when the β-lactam ring of cephaloridine is hydrolyzed by β-lactamase. Reaction of cephalosporin 87-312 (see below) with the R39 enzyme[16] causes a shift of the absorption maximum of the antibiotic molecule from 386 to 482 nm. The absorption spectrum of the R39 enzyme–cephalosporin 87-312 complex is identical to that of cephalosporin 87-312 hydrolyzed by β-lactamase (Fig. 2). In both cases, the ratio $\epsilon_{482}/\epsilon_{386}$ is equal to 2.40.

The R61 enzyme–cephalosporin 87-312 complex, once formed, has a ratio $\epsilon_{482}/\epsilon_{386}$ equal to 1.20.[17] This value is considerably lower than normally expected if the β-lactam ring were hydrolyzed as with β-lactamase. The difference spectrum between the R61 enzyme–cephalosporin 87-312

TABLE IV

VALUES OF THE CONSTANTS FOR THE REACTION BETWEEN THE R61 ENZYME AND VARIOUS β-LACTAM ANTIBIOTICS

$$(E + P \underset{}{\overset{K}{\rightleftharpoons}} EP \xrightarrow{k_3} EP* \xrightarrow{k_4} E + X)$$

AND k_4 VALUES FOR THE BREAKDOWN OF EP* COMPLEXES FORMED WITH THE R39 ENZYME[a]

| Antibiotic | R61 | | | | R39 |
	K (mM)	k_3 (sec^{-1})	k_3/K (M^{-1} sec^{-1})	k_4 (sec^{-1})	k_4 (sec^{-1})
Benzylpenicillin	13 (25°)	179 (25°)	1.20×10^4 (25°)	0.21×10^{-4} (25°)	3×10^{-6}
				1.4×10^{-4}	
Carbenicillin	0.109	0.091	830	1.4×10^{-4}	5×10^{-6}
Ampicillin	7.2	0.77	107	1.4×10^{-4}	$>4.4 \times 10^{-6}$
Penicillin V	>1	>1	1,500	2.8×10^{-4}	ND
Cephalosporin C	>1	>1	1,150	1×10^{-6}	$>0.3 \times 10^{-6}$
Cephaloglycine	0.4	8.5×10^{-3}	21	3×10^{-6}	$>0.8 \times 10^{-6}$
Cephalosporin 87/312	0.2(10°)	>0.1(10°)	460(10°)	3×10^{-4}	1.5×10^{-6}
Cephaloridin	ND	ND	ND	ND	0.6×10^{-6}

[a] At 37° unless otherwise indicated; ND, not done.

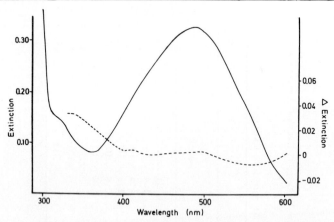

Fig. 2. Absorption spectrum of cephalosporin 87-312 after treatment with the *Streptomyces* R39 enzyme and with β-lactamase. ——, Solution containing 15.5 nmoles of *Streptomyces* R39 protein in 0.4 ml of Tris–NaCl–MgCl₂ buffer was mixed with 0.2 ml of a 67.7 μM solution of cephalosporin 87-312. The mixture was incubated for 5 min at room temperature before the absorption spectrum was determined. ---, Difference spectrum between cephalosporin treated with *Streptomyces* R39 enzyme and cephalosporin treated with β-lactamase. Note the change of scale. Solution containing 6 units of β-lactamase in 0.4 ml of Tris–NaCl–MgCl₂ buffer was mixed with 0.2 ml of the same solution of cephalosporin 87-312 as above. The mixture was incubated at 37° until the absorbance at 482 nm was a maximum (less than 5 min). Reprinted by courtesy of the Biochemical Society from J.-M. Frère, J.-M. Ghuysen, P. E. Reynolds, R. Moreno, and H. R. Perkins, *Biochem. J.* **143**, 241 (1974).

complex, once formed, and the same concentration of cephalosporin 87-312 hydrolyzed by β-lactamase shows extrema at 390 and 525 nm (Fig. 3). Incubation at 37° of the R61 enzyme–cephalosporin 87-312 complex results in the reactivation of the enzyme and in a parallel decrease of the two extrema in the difference spectrum (Fig. 3).

Molar Activity on Benzylpenicillin. On the basis of the k_4 values for the breakdown of the EP* complexes (Table IV), the specific activity (in equivalents of β-lactamase unit and at 37°) of the R61 enzyme is 2.2×10^{-4} unit per milligram of protein and that of the R39 enzyme is 0.35×10^{-5} unit per milligram of protein. Molar activity (in min^{-1}) is 8.4×10^{-3} for the R61 enzyme and 2.1×10^{-4} for the R39 enzyme. These values are to be compared with molar activities on Ac₂-L-Lys-D-Ala of 3300 and 1050, respectively.

Effects on Circular Dichroism (CD) and Fluorescence of Streptomyces Enzymes. The circular dichroism and fluorescence spectra of the R39 enzyme are not modified by benzylpenicillin.

The near UV CD spectrum of the R61 enzyme is affected by benzylpenicillin whereas the peptide region of the CD spectrum in the far UV

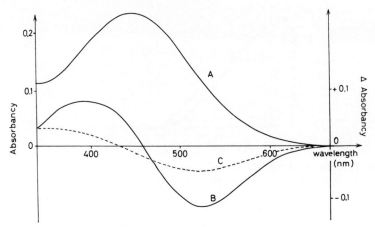

Fig. 3. Absorption spectrum of cephalosporin 87-312 after treatment with the *Streptomyces* R61 enzyme and with β-lactamase. The spectra were recorded with 0.5 ml cuvettes and an optical pathway of 1.0 cm. Buffer: 5 mM sodium phosphate pH 7.0. For other conditions, see text. Curve A: Absorption spectrum of a freshly prepared R61 enzyme–cephalosporin 87-312 complex. Concentration of the complex, 15 μM. Curve B: Difference spectrum between the freshly prepared R61 enzyme–cephalosporin 87-312 complex and an equivalent amount of cephalosporin 87-312 hydrolyzed by penicillinase. The solutions contained 7.5 nanoequivalents of cephalosporin 87-312 (either combined with the enzyme or hydrolyzed by penicillinase) in a final volume of 500 μl. Curve C: Difference spectrum between the same complex as in B, maintained for 60 min at 37°, and an equivalent amount of cephalosporin 87-312 hydrolyzed by penicillinase. Note that after 60 min at 37°, 65% of the initially inhibited activity had recovered and, parallel to this, the intensities of the two extrema in the difference spectrum had decreased to 40% of the original values. Reprinted by courtesy of the Federation of European Biochemical Societies from J.-M. Frère, M. Leyh-Bouille, J.-M. Ghuysen, and H. R. Perkins, *Eur. J. Biochem.* **50**, 203 (1974).

remains unchanged.[14] The overall change in CD is too small, and the concentration of enzyme too high, to be convenient for quantitative work.

Saturating concentrations of benzylpenicillin decrease the fluorescence emission at 320 nm of the R61 enzyme by 25–30%.[14] Quenching of the fluorescence of the enzyme (concentration, 0.76 μM) in the presence of an equimolar amount of benzylpenicillin is not immediate and is maximal after 5–10 min.[14]

Titration of DD-Carboxypeptidases–Transpeptidases from *Streptomyces* R39 and R61 by β-Lactam Antibiotics

Buffers. Tris-HCl buffer, 0.1 M, pH 7.7 + 0.2 M NaCl + 0.05 M MgCl$_2$ (Tris-NaCl-MgCl$_2$ buffer) is used for the R39 enzyme, and 5–10 mM sodium phosphate buffer is used for the R61 enzyme.

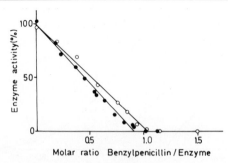

Fig. 4. Titration of the *Streptomyces* R39 protein with benzylpenicillin, based on the inhibition of enzyme activity. ●——●, Enzyme after step 4A of purification; ○——○, enzyme after step 5 of purification. For conditions, see text. Reprinted by courtesy of the Biochemical Society from J.-M. Frère, J.-M. Ghuysen, P. E. Reynolds, R. Moreno, and H. R. Perkins, *Biochem. J.* **143**, 241 (1974).

Antibiotics. Benzylpenicillin and cephaloridine solutions are made fresh, by weight, in the appropriate buffer.

Cephalosporin 87-312 [i.e., 3-(2,4-dinitrostyryl)-(6R-7R)-7-(2-thienyl-acetamido)-ceph-3-em-4-carboxylic acid, E-isomer][19] solutions (usually 0.1 mM) are made by dissolving the antibiotic in 1 ml of dimethyl-formamide, and the volume of the solution is adjusted to 250 ml with the appropriate buffer. The final concentration is estimated by measuring the extinction at 386 nm and by using a molar coefficient $\epsilon_{386}^{1\,cm}$ of 17,500.

Titration of the R39 and R61 Enzymes with Benzylpenicillin Based on the Disappearance of Enzyme Activity.[16] Samples of a 5 μM benzyl-penicillin solution (either in Tris-NaCl-MgCl$_2$ or phosphate buffer) are added stepwise to 100 μl of a solution containing 0.3 nmole of either R39 or R61 enzyme in the relevant buffer. After each addition, the solution is mixed and maintained at room temperature for 5 min; a sample (2–10 μl) is removed and used for the measurement of the residual DD-carboxy-peptidase activity (by incubating the sample with Ac$_2$-L-Lys-D-Ala-D-Ala for 10 min at 37°). After correction for the decrease in the amount of enzyme owing to the removal of samples and, in the case of the R61 enzyme, after correction of the estimated residual enzyme activity (see below), the end points of the titration occur at a molar ratio of benzyl-penicillin to enzyme of 1:1 (Fig. 4).

Because of the short half-life of the R61 enzyme–benzylpenicillin complex (Table IV), breakdown of the complex and reactivation of the

[19] C. O'Callaghan, A. Morris, S. A. Kirby, and A. H. Shingler, *Antimicrob. Agents Chemother.* **1**, 283 (1972).

enzyme occur during estimation of residual enzyme activity at 37°. The exact residual activity[17] (as percent, after addition of x moles of antibiotic) is equal to

$$\frac{(A_m/A_0) + [(1 - e^{-k_4t})/k_4t] - 1}{(1 - e^{-k_4t})/k_4t}$$

where A_m is the amount of hydrolyzed tripeptide measured in each case, A_0 the amount of hydrolyzed tripeptide obtained with the same amount of uninhibited enzyme, and t the duration of the incubation with the tripeptide (i.e., 600 sec). For the k_4 value, see Table IV.

Breakdown of the R61 enzyme–antibiotic complex formed at the beginning of the titration also occurs during the subsequent steps of the titration. Since the titration is carried out at 22°, however, the error thus introduced is negligible.

Titration of the R61 Enzyme with Benzylpenicillin Based on Fluoresence Quenching.[14] Binding of benzylpenicillin to the R61 enzyme can be followed by the fluorescence quenching at 320 nm (ΔF_{320}). Samples of a 0.232 mM solution of benzylpenicillin in 10 mM sodium phosphate buffer pH 7.0 are added stepwise at 25° to 2 ml of a solution containing 1.4 nmoles of R61 enzyme in the same buffer. Readings are taken 10 min after each addition to allow time for completion of the reaction. Excitation is at 273 nm. Extrapolation of the two lines of the curve intersect at a point where the molar ratio of benzylpenicillin to enzyme is 1:1 (Fig. 5).

Titration of the R39 Enzyme with Cephaloridine.[16] Samples of 0.5 mM cephaloridine in Tris-NaCl-MgCl$_2$ buffer are added stepwise, at room temperature, to 0.4 ml of a solution containing 6.5 nmoles of R39 enzyme in the same buffer. After each addition, the extinction of the solution is measured at 250 nm and the DD-carboxypeptidase is measured in a 2-μl sample. A plot of the increased extinction of the solution at 250 nm as a function of the amount of cephaloridine added yields two lines intersecting at a point. On the basis of these data and the determination of enzyme activity, the end points of the reaction occur at a molar ratio of cephaloridine to enzyme of 1:1. (For illustration, see Frère *et al.*[16])

Titration of the R39 and R61 Enzymes with Cephalosporin 87-312. Samples of a solution of cephalosporin 87-312 in either Tris-NaCl-MgCl$_2$ or phosphate buffer, are added stepwise at room temperature to 0.4 ml of a solution of either R39 or R61 enzyme in the relevant buffer. (For concentration of enzymes and cephalosporin 87-312, see legend of Fig. 6.) After each addition, the mixture is maintained at room temperature for 5 min, after which time the absorbances of the solutions are measured both at 386 nm (intact cephalosporin) and at 482 nm (modified cephalo-

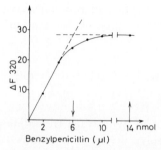

FIG. 5. Titration of the *Streptomyces* R61 protein with benzylpenicillin, based on fluorescence quenching at 320 nm (ΔF_{320}). R61 enzyme (2 ml of 26.5 μg/ml) is dissolved in 10 mM sodium phosphate buffer, pH 7.0, at 25°. Excitation is at 273 nm and the fluorimeter settings are: sensitivity 20, meter multiplier 0.01. Fluorescence intensity (F) is expressed in arbitrary units, the maximum emission being taken as 100. The sodium benzylpenicillin solution (0.232 mM) is in 10 mM sodium phosphate buffer, pH 7.0. Readings are taken 10 min after each addition to allow time for equilibration. After 10 μl of penicillin solution has been added, 2.32 mM penicillin (5 μl) is added to ensure saturation (↑). The arrow ↓ indicates the titration end point. Reprinted by courtesy of the Biochemical Society from M. Nieto, H. R. Perkins, J.-M. Frère. and J.-M. Ghuysen, *Biochem. J.* **135**, 493 (1973).

sporin), and the DD-carboxypeptidase activity is measured in 2-μl samples. With the R61 enzyme, the residual enzyme activity after each addition of x moles of cephalosporin 87-312 is corrected as indicated above (see titration with benzylpenicillin). As revealed by the three procedures, the end points of the titration of both enzymes occur at a molar ratio of cephalosporin 87-312 to enzyme of 1.25:1 (Fig. 6). This high ratio might be due to the fact that the amount of cephalosporin 87-312 used is estimated on the basis of a ϵ_{386}^{cm} value of 17,500. An underestimation of this coefficient would result in an overestimation of the amount of antibiotic required to estimate the enzyme.

In the case of the R39 enzyme, the extinction values at 386 and 482 nm can be transformed into concentrations, and from these into nanomoles, of intact and modified cephalosporin 87-312 from the following equations:[16]

$$\epsilon_{386} = 17,500 \text{ [intact cephalosporin]} + 7700 \text{ [modified cephalosporin]}$$
$$\epsilon_{482} = 2600 \text{ [intact cephalosporin]} + 16,700 \text{ [modified cephalosporin]}$$

where 17,500 and 2600 = molar extinction coefficients of intact cephalosporin and 7700 and 16700 = molar extinction coefficients of cephalosporin hydrolyzed by β-lactamase, at the relevant wavelengths.

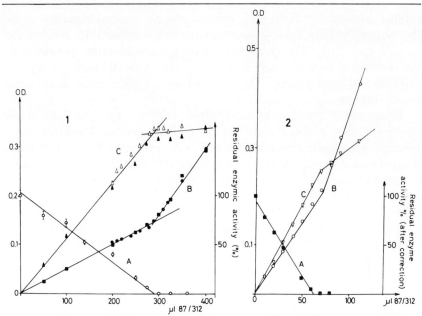

Fig. 6. Titrations of *Streptomyces* R39 enzyme (1) and *Streptomyces* R61 enzyme (2) with cephalosporin 87-312 based (A) on the inhibition of enzyme activity, (B) on the increase of the absorbance of the solution at 386 nm, and (C) on the increase of the absorbance of the solution at 482 nm. The R39 enzyme (13.3 nanomoles) is titrated with a 58.8 M cephalosporin 87-312 solution. The R61 enzyme (7.8 nanomoles) is titrated with a 0.15 mM cephalosporin 87-312 solution. After each addition of cephalosporin 87-312, the optical densities of the solutions are normalized for final volumes of 800 μl (R39 enzyme) and 510 μl (R61 enzyme). For other conditions, see text. Reprinted by courtesy of the Federation of European Biochemical Societies from J.-M. Frère, M. Leyh-Bouille, J.-M. Ghuysen, and H. R. Perkins, *Eur. J. Biochem.* **50,** 203 (1974).

Hydrolysis Reactions Catalyzed by the DD-Carboxypeptidases–Transpeptidases from *Streptomyces* R39 and R61

General Reaction

$$\text{X-L-R}_3\text{-R}_2\text{-R}_1\text{(OH)} + \text{H}_2\text{O} \rightarrow \text{X-L-R}_3\text{-R}_2\text{(OH)} + \text{R}_1 \qquad (5)$$

Standard reaction (1) is an example.

Procedure. The liberated R_1 residue can be estimated by the fluorodinitrobenzene as indicated above for alanine. Samples containing known amounts of R_1 residue are treated similarly.

Substrate Requirements. These were studied with peptides presenting the above general sequence by measuring the amount of C-terminal R_1

residue released.[11,20] With each enzyme, the C-terminal sequence giving the highest activity is D-Ala-D-Ala, but the preceding L-R$_3$ residue also has a large effect, both enzymes exhibiting a considerable specificity for the occurrence of a long aliphatic side chain at the R$_3$ position. Typical Michaelis–Menten kinetics are observed over a wide range of substrate concentrations.

Effects of Peptide Analogs. Peptides that are close analogs of the substrate donor inhibit the activity of the R61 enzyme.[21] Ac-D-Ala-D-Asp behaves as a competitive inhibitor with a K_i value of 3.2 mM (substrate: Ac$_2$-L-Lys-D-Ala-D-Ala). It has been suggested that the two D-Ala-D-Ala C-terminal residues of the substrates and inhibitors are mainly responsible for the initial binding whereas the side chain of the L-R$_3$ residue is critical in inducing catalytic activity.

None of the peptide inhibitors of the R61 enzyme have any effect on the carboxypeptidase activity of the R39 enzyme. Some of them are good substrates for this enzyme.[21]

Concomitant Hydrolysis and Transfer Reactions Involving Distinct Donor and Acceptor Peptides, Catalyzed by the DD-Carboxy-peptidases–Transpeptidases from *Streptomyces* R39 and R61

General Reaction

$$[^{14}C]Ac_2\text{-L-Lys-D-Ala-D-Ala} \begin{cases} H_2O \\ \\ NH_2\text{-R} \end{cases} \begin{cases} [^{14}C]Ac_2\text{-L-Lys-D-Ala} \quad\quad \text{(a)} \\ D\text{-Ala} \\ [^{14}C]Ac_2\text{-L-Lys-D-Ala-R} \quad \text{(b)} \end{cases} \quad (6)$$

(a) = hydrolysis pathway; (b) = transfer pathway. D-Ala is the reaction product common to both pathways.

Donor Substrate. The tripeptide Ac$_2$-L-Lys-D-Ala-D-Ala radioactively labeled with ^{14}C in both acetyl groups (specific activity, 10,000 dpm/nanomole) is prepared as described by Perkins *et al.*[22]

Assay Method (with meso-Diaminopimelic Acid as Acceptor). R39 enzyme: [^{14}C]Ac$_2$-L-Lys-D-Ala-D-Ala (5 mM) and *meso*-diaminopimelic

[20] M. Leyh-Bouille, M. Nakel, J.-M. Frère, K. Johnson, J.-M. Ghuysen, M. Nieto, and H. R. Perkins, *Biochemistry* **11**, 1290 (1972).

[21] M. Nieto, H. R. Perkins, M. Leyh-Bouille, J.-M. Frère, and J.-M. Ghuysen, *Biochem. J.* **131**, 163 (1973).

[22] H. R. Perkins, M. Nieto, J.-M. Frère, M. Leyh-Bouille, and J.-M. Ghuysen, *Biochem. J.* **131**, 707 (1973).

acid (8 mM) are incubated at 37° with 2.2 pmoles (0.12 μg) of enzyme in 30 μl of 0.03 M Tris-HCl buffer, pH 7.5, containing 0.1 M NaCl and 0.02 M MgCl$_2$. After 30 min, about 18 and 12.5% of the tripeptide donor are converted into [^{14}C]Ac$_2$-L-Lys-D-Ala and [^{14}C]Ac$_2$-L-Lys-D-Ala-(D)-*meso*-diaminopimelic acid, respectively.

R61 enzyme: [^{14}C]Ac$_2$-L-Lys-D-Ala-D-Ala (5.5 mM) and *meso*-diaminopimelic acid (8 mM) are incubated at 37° with 1.35 pmoles (0.05 μg) of enzyme in 30 μl of 5 mM sodium phosphate pH 7.0. After 60 min, about 10% of the tripeptide donor is converted into [^{14}C]Ac$_2$-L-Lys-D-Ala and the same amount into [^{14}C]Ac$_2$-L-Lys-D-Ala-(D)-*meso*-diaminopimelic acid.

Estimation of [^{14}C]Ac$_2$-L-*Lys*-D-*Ala*-D-*Ala* (Residual Donor), [^{14}C]Ac$_2$-L-*Lys*-D-*Ala* (Hydrolysis Product), *and* [^{14}C]Ac$_2$-L-*Lys*-D-*Ala*-R (Transpeptidation Product)

Reagents

Collidine:acetic acid:water (9.1:2.65:1000, v/v/v), buffer, pH 6.4
Liquid scintillation: 2,2-p-phenylenebis (5-phenyloxazole) (POPOP), 100 mg; 2,5-diphenyloxazole (PPO), 4 g; toluene, 1 liter

Procedure. Samples of the reaction mixtures (30 μl) containing 10,000–20,000 dpm are diluted with 40 μl of water. These are spotted as bands, 30 cm from the cathode on 4 cm × 1.5 m strips of Whatman 3 MM paper and subjected to electrophoresis at pH 6.4 for 4 hr at 60 V/cm, under a Sol.T Shell. A Gilson high voltage, 10,000 V, Electrophoretor Model DW equipped with a cooling device is used as power source. Residual [^{14}C]Ac$_2$-L-Lys-D-Ala-D-Ala and the hydrolysis product [^{14}C]Ac$_2$-L-Lys-D-Ala move about 65 and 75 cm, respectively, toward the anode. Transpeptidation products [^{14}C]Ac$_2$-L-Lys-D-Ala-R move differently depending upon the nature of the R residue (about 55 cm where R = *meso*-diaminopimelic acid). The radioactive compounds are located on the dried strips with a Packard Radiochromatogram Scanner Model 7201. Cuts of the radioactive spots (10 mm section) are placed in vials, to each of which is added 0.75 ml of the scintillation liquid. Counting is performed in a Packard Tri-Carb liquid scintillation spectrometer.

Specificity Profiles for Acceptors. The range of substrates that function as acceptors reflects the type of cross-linking that exists in the peptidoglycan of the organism which produces the exocellular enzyme (see the above section: Strains). Correspondingly, glycine and many peptides with an N-terminal glycine residue act as acceptors in transpeptidation reactions catalyzed by the R61 enzyme (although other amino compounds also function).[22] In marked contrast, suitable transpeptidation acceptors for the R39 enzyme must have an amino group in α-position to the carboxyl group of a D-amino acid (or glycine).[3,22,23]

Kinetics. The kinetics are necessarily complex. For a theoretical analysis, see Frère.[24] For application to the R61 enzyme (with either Gly-L-Ala or *meso*-diaminopimelic acid as acceptor) see Frère *et al.*[25]

The proportion of the enzyme activity channeled in either transpeptidation or hydrolysis depends upon the environmental conditions.[3,23,25] Transpeptidation is increased and hydrolysis decreased by raising the pH of the reaction mixture and the concentration of acceptor within it (this latter behaves as a noncompetitive inhibitor of the hydrolysis pathway). Replacement of part of the water of the reaction mixture by a solvent of low polarity preferentially decreases the hydrolytic activity of the enzyme so that transpeptidation then largely supersedes hydrolysis. In the case of the R39 enzyme, transpeptidase activity is increased at high ionic strengths.[3]

With some peptide acceptors, transpeptidation itself is inhibited at· high acceptor concentrations.[3,23,25] For instance, when increasing concentrations of tetrapeptide L-Ala-D-αGln-(L)-*meso*-A$_2$pm-(L)-D-Ala are provided as acceptor to the R39 enzyme (with 0.27 mM Ac$_2$-L-Lys-D-Ala-D-Ala as donor), transpeptidation rises to a maximum at an acceptor concentration of about 0.8 mM, and at higher concentrations both transpeptidation and hydrolysis reactions are progressively inhibited until eventually the tripeptide donor remains unused. In this example, this phenomenon is dependent on the α-amido group on the D-glutamic acid residue of the tetrapeptide acceptor.[3]

For extensive studies on the functioning of the R39 enzyme as a transpeptidase in relation to the degree of saturation of its donor site, see Ghuysen *et al.*[23] Under conditions where the donor site of the enzyme is saturated, the rate of the total reaction (hydrolysis + transpeptidation) is the same as the rate of hydrolysis alone when no acceptor is added, i.e., the enzyme has the same turnover number for D-alanine release whether or not acceptor is present.

---

[23] J.-M. Ghuysen, P. E. Reynolds, H. R. Perkins, J.-M. Frère, and R. Moreno, *Biochemistry* **13**, 2539 (1974).

[24] J.-M. Frère, *Biochem. J.* **135**, 469 (1973).

[25] J.-M. Frère, J.-M. Ghuysen, H. R. Perkins, and M. Nieto, *Biochem. J.* **135**, 483 (1973).

Concomitant Hydrolysis and Transfer Reactions Catalyzed by the
DD-Carboxypeptidases–Transpeptidases from *Streptomyces*
R39 and R61 and in Which the Same Peptide Acts as Donor
and Acceptor

General Reaction

$$
\left. X\text{-L-R}_3\text{-D-Ala-D-Ala} \atop (H) \right\}
\left\{
\begin{array}{l}
H_2O \searrow \quad \left\{ \begin{array}{l} X\text{-L-R}_3\text{-D-Ala} \\ \quad | \\ \quad (H) \\ \text{D-Ala} \end{array} \right. \quad\quad (a) \\[2em]
\\
X\text{-L-R}_3\text{-D-Ala-D-Ala} \nearrow \quad \begin{array}{l} X\text{-L-R}_3\text{-D-Ala(D-Ala)} \\ \searrow X\text{-L-R}_3\text{-D-Ala}\text{---}\!\!\rfloor \quad (b) \end{array}
\end{array}
\right.
\tag{7}
$$

(a) hydrolysis pathway; (b) transfer pathway.

The C-terminal D-Ala-D-Ala sequence of the dimer formed can also be hydrolyzed through the carboxypeptidase activity of the enzyme.

Kinetics. Examples of such studies can be found in Ghuysen *et al.*[23] for the R39 enzyme substrate:

$$
\text{L-Ala-D-Glu---(L)-}meso\text{-A}_2\text{pm-(L)-D-[}^{14}\text{C]Ala-D-[}^{14}\text{C]Ala}
$$

in which the amino group on the D center of *meso*-diaminopimelic acid functions as acceptor, and in Zeiger *et al.*[26] for the R61 enzyme substrate:

$$
\text{[}^{14}\text{C]Ac-L-Lys-D-Ala-D-Ala} \atop \text{Gly---}\!\!\rfloor
$$

Inhibition of DD-Carboxypeptidases–Transpeptidases from
Streptomyces R39 and R61 by β-Lactam Antibiotics in the
Presence of Substrates

Procedure. Ideally, the experiments are carried out under conditions where the enzyme concentration ($10-20$ nM) is considerably lower than the concentrations of antibiotics (0.3 μM and higher). Substrate(s), antibiotic (when present), enzyme and buffer are precooled and mixed together at $0°$. The solutions are incubated at $37°$ and after a given time (usually 60 min) the reaction products are estimated. For more details, see Frère *et al.*[25]

[26] A. R. Zeiger, J.-M. Frère, J.-M. Ghuysen, and H. R. Perkins, *FEBS Lett.* **52**, 221 (1975).

Lineweaver–Burk and Dixon Plots. Inhibition by benzylpenicillin, carbenicillin, and penicillin V of the hydrolysis reaction (substrate: Ac$_2$-L-Lys-D-Ala-D-Ala) catalyzed by the R61 enzyme in the absence of acceptor is competitive. Apparent K_i values are 40 nM for benzylpenicillin, 0.33 μM for penicillin V, and 1.05 μM for carbenicillin.

Inhibition by penicillin V of concomitant hydrolysis and transfer reactions catalyzed by the R61 enzyme (substrates: Ac$_2$-L-Lys-D-Ala-D-Ala + *meso*-diaminopimelic acid) is competitive with regard to the donor both in the hydrolysis pathway (apparent K_i value = 0.77 μM) and in the transpeptidation pathway (apparent K_i value = 0.62 μM) and is noncompetitive with regard to the acceptor.

The R61 enzyme–antibiotic complexes have half-lives of 40/80 min; hence the competitive inhibition theory, which is based on a rapid equilibrium process, is not applicable. For a theoretical analysis of these results and the exact physical meaning of the apparent K_i values, see Frère *et al.*[27]

Unusual Dixon plots were reported for the inhibition of the R39 enzyme by benzylpenicillin.[20] The benzylpenicillin concentrations used in these studies (made at a time when the molecular weight of the enzyme was not known) were very close to that of the enzyme itself. The results actually reflect the titration of the enzyme by the antibiotic.

Note Added in Proof.

As a result of its interaction with the R61 enzyme, benzylpenicillin is degraded into phenylacetylglycine[28] and *N*-formyl-D-penicillamine.[29] Phenylacetylglycine is also produced by the interaction of benzylpenicillin with the R39 enzyme.

The models (3) or (4) also apply to the interaction between the R39 enzyme and β-lactam antibiotic.[30]

[27] J.-M. Frère, J.-M. Ghuysen, and H. R. Perkins, *Eur J. Biochem.* **57**, 353 (1975).
[28] J. M. Frère, J.-M. Ghuysen, J. Degelaen, A. Loffet, and H. R. Perkins, *Nature (London)* **258**, 168 (1975).
[29] J.-M. Frère, J.-M. Ghuysen, H. Vanderhaeghe, P. Adriaens, J. Degelaen, and J. De Graeve, *Nature (London)*, **260**, 451 (1976).
[30] N. Fuad, J.-M. Frère, J.-M. Ghuysen, C. Duez, and M. Iwatsubo, *Biochem. J.* **155**, in press.

Section IX
Naturally Occurring Protease Inhibitors

A. Specific Inhibitors of Clotting and Lysis in Blood
Articles 52 through 54

B. Inhibitors from Bacteria
Article 55

C. Inhibitors from Plants
Articles 56 through 64

D. Protease Inhibitors from Various Sources
Articles 65 through 79

[52] Human α_2-Macroglobulin

By PETER C. HARPEL

The α_2-macroglobulin (α_2M) was first isolated from human serum by Schultze *et al.*,[1] and shown to be separate from IgM by Müller-Eberhard *et al.*[2] and Wallenius *et al.*[3] Haverback *et al.*[4] demonstrated that trypsin and chymotrypsin formed a complex with a serum protein possessing an α_2 electrophoretic mobility. This enzyme–protein complex retained enzymic activity; however, the susceptibility of the trypsin in the complex to soybean trypsin inhibitor was markedly decreased. Mehl *et al.*[5] isolated α_2M and demonstrated its trypsin-binding activity. Recent studies suggest that α_2M may form complexes with all proteolytic enzymes,[6] thereby serving as a mechanism for the removal of activated proteases from the circulation. The α_2M differs from other proteolytic enzyme inhibitors in that the active site of the enzyme in the α_2M–enzyme complex retains its catalytic potential for small-molecular-weight substrates, whereas the hydrolysis of naturally occurring protein substrates appears to be sterically hindered.[7] A number of recent reviews of the physiology and biochemistry of α_2M have appeared.[6,8-10]

Serum Concentration

The serum concentration of α_2M in healthy individuals was reported by Adham *et al.*[11] to be 215 ± 11 mg per 100 ml of serum, a value in

[1] H. E. Schultze, I. Göllner, K. Heide, M. Schönenberger, and H. G. Schwick, *Z. Naturforsch. Teil B* **10**, 463 (1955).

[2] H. J. Müller-Eberhard, H. G. Kunkel, and E. C. Franklin, *Proc. Soc. Exp. Biol. Med.* **93**, 146 (1956).

[3] G. Wallenius, R. Trautman, H. G. Kunkel, and E. C. Franklin, *J. Biol. Chem.* **225**, 253 (1957).

[4] B. J. Haverback, B. Dyce, H. F. Bundy, S. K. Wirtschafter, and H. A. Edmondson, *J. Clin. Invest.* **41**, 972 (1962).

[5] J. W. Mehl, W. O'Connell, and J. DeGroot, *Science* **145**, 821 (1964).

[6] A. J. Barrett and P. M. Starkey, *Biochem. J.* **133**, 709 (1973).

[7] P. C. Harpel and M. W. Mosesson, *J. Clin. Invest.* **52**, 2175 (1973).

[8] P. C. Harpel, M. W. Mosesson, and N. R. Cooper, *in* "Proteases and Biological Control" (E. Reich, D. Rifkin, and E. Shaw, eds.), p. 387. Cold Spring Harbor Laboratory, New York, 1975.

[9] R. Bourrillon and E. Razafimahaleo, *in* "Glycoproteins. Their Composition Structure and Function" (A. Gottschalk, ed.), Vol. 5, p. 699. Elsevier, Amsterdam, 1972.

[10] M. Steinbuch and R. Audran, *Proteinase Inhibitors, Proc. Int. Res. Conf., 2nd (Bayer Symp. V)*, Grosse Ledder, 1973, p. 78. Springer-Verlag, Berlin and New York, 1974.

[11] N. F. Adham, P. Wilding, J. Mehl, and B. J. Haverback, *J. Lab. Clin. Med.* **71**, 271 (1968).

agreement with several other studies.[12,13] James *et al.*[14] found a higher concentration, 265 ± 55 mg per 100 ml, in agreement with earlier studies.[15,16] Although Adham[11] found no difference in $\alpha_2 M$ in men and women, other investigators[12,14,17] have found higher concentrations in females. $\alpha_2 M$ increases significantly in pregnancy[18,19] and in association with oral contraceptives[18]; however, estrogen administration was not found to elicit an increase of $\alpha_2 M$ in females.[11] Infants possess 2.5 times as much $\alpha_2 M$ as do adults.[12]

The $\alpha_2 M$ does not appear to be an acute-phase protein reactant, as it does not increase in patients with rheumatoid arthritis who show high haptoglobin levels and C-reactive protein,[17] nor does it increase following surgery in patients showing the acute-phase protein reaction.[20] Greatly elevated levels of $\alpha_2 M$ are found in ataxia telangiectasia, atopic dermatitis, mongolism,[14] and the nephrotic syndrome.[17,21-25] Increased levels have been reported in pulmonary disorders, diabetes, mellitus, and agammaglobulinemia[14] and in patients with liver disease.[17,26] Decreased $\alpha_2 M$ levels are reported in multiple myeloma,[14] in females with rheumatoid arthritis,[17] and following the intravenous infusion of the plasminogen activators streptokinase or urokinase.[27-30] In preeclampsia without associated proteinuria, $\alpha_2 M$ levels fall, but when preeclampsia is accompanied by proteinuria, $\alpha_2 M$ levels increase.[31]

[12] P.-O. Ganrot and B. Scherstén, *Clin. Chim. Acta* **15**, 113 (1967).

[13] P. Wilding, N. F. Adham, J. W. Mehl, and B. J. Haverback, *Nature* (*London*) **214**, 1226 (1967).

[14] K. James, G. Johnson, and H. H. Fudenberg, *Clin. Chim. Acta* **14**, 207 (1966).

[15] W. H. Hitzig, *in* "Die Plasmaprotein in der klinischen Medizin," Vol. III, p. 110. Springer-Verlag, Berlin and New York, 1963.

[16] W. A. Weiss, *Klin. Wochenschr.* **43**, 273 (1965).

[17] J. Housley, *J. Clin. Pathol.* **21**, 27 (1968).

[18] C. H. W. Horne, P. W. Howie, R. J. Weir, and R. B. Goudie, *Lancet* **1**, 49 (1970).

[19] P.-O. Ganrot and B. Bjerre, *Acta Obstet. Gynaecol. Scand.* **46**, 126 (1967).

[20] R. A. Crockson, C. J. Payne, A. P. Ratcliff, and J. F. Soothill, *Clin. Chim. Acta* **14**, 435 (1966).

[21] A. Peterkofsky, L. Levine, and R. K. Brown, *J. Immunol.* **76**, 237 (1956).

[22] L. Hartman, G. Langrue, and J. Moretti, *Rev. Fr. Etud. Clin. Biol.* **3**, 1052 (1958).

[23] H. E. Schultze and H. G. Schwick, *Clin. Chim. Acta* **4**, 15 (1959).

[24] W. J. Steines and J. W. Mehl, *J. Lab. Clin. Med.* **67**, 559 (1966).

[25] R. Kluthe, V. Hagemann, and N. Kleine, *Vox Sang.* **12**, 308 (1967).

[26] P. S. Damus and G. A. Wallace, *Thromb. Res.* **6**, 27 (1975).

[27] F. Spöttl and F. Holzknecht, *Thromb. Diath. Haemorrh.* **24**, 100 (1970).

[28] M. Fischer, *in* "Thrombolytic Therapy: Transaction of the 19th Annual Symposium on Blood," p. 153. Schattauer, Stuttgart, 1971.

[29] J. E. Niléhn and P.-O. Ganrot, *Scand. J. Clin. Lab. Invest.* **20**, 113 (1967).

[30] H. Arnesen and M. K. Fagerhol, *Scand. J. Clin. Lab. Invest.* **29**, 259 (1972).

[31] C. H. W. Horne, P. W. Howie, and R. B. Goudie, *J. Clin. Pathol.* **23**, 514 (1970).

Assay

The concentration of α_2M antigen is measured by radial immunodiffusion using commercially available immunoplates (Hyland, Division Travenol Laboratories, Inc., Yonkers, New York; Behring Diagnostics, Inc., Somerville, New Jersey).

The trypsin-binding activity of α_2M is determined using N-α-benzoyl-DL-arginine-p-nitroanilide (BAPA) as substrate for the α_2M–trypsin complex following the addition of soybean trypsin inhibitor to inhibit unbound trypsin as detailed by Ganrot.[32,33]

The esterolytic activity of the purified α_2M reflecting the proteases that are bound to the α_2M is measured using toluenesulfonyl arginine methyl ester (TAME) as a substrate, as previously described.[7]

Purification Procedure

Reagents. All buffers contain 0.02% sodium azide to retard bacterial growth.

Phosphate buffer: monobasic potassium phosphate, 20 mM, containing 0.1 M NaCl is added to dibasic sodium phosphate, 20 mM, containing 0.1 M NaCl to pH 7.4

Potassium bromide solution: 10 g of KBr in 100 ml of Tris-HCl, 25 mM, pH 8.0, containing 25 mg crude soybean trypsin inhibitor (Worthington Biochemical Corporation, Freehold, New Jersey).

Tris buffer 1: Tris-HCl, 50 mM Tris, pH 8.0, containing 20 mM NaCl. The conductivity is 4.0 mmho/cm at 27° using a type CDM 2e conductivity meter with a CDC 104 probe (Radiometer Co., Copenhagen, Denmark)

Tris buffer 2: 50 mM Tris, pH 8.0, containing 0.4 M NaCl. Conductivity is 37 mmho/cm

Tris-citrate-NaCl buffer: Tris, 50 mM, pH 8.0, containing 5.4 mM sodium citrate and 0.15 M NaCl

Barbital buffer: 71.84 g of sodium barbital are dissolved in 7 liters of distilled H$_2$O, then 51.6 ml of 1 N HCl is added to pH 8.6. Volume is brought to 8 liters. Conductivity is 2.8 mmho/cm.

Phosphate buffer for electrophoresis: 224 ml of 1 M Na$_2$HPO$_4$, 128 ml of 1 M NaH$_2$PO$_4$ brough to 8 liters with distilled H$_2$O; pH 7.0. Conductivity is 5.0 mmho/cm

---

[32] P.-O. Ganrot, *Clin. Chim. Acta* **14**, 493 (1966).
[33] P.-O. Ganrot, *Acta Chem. Scand.* **21**, 602 (1967).

Procedure

It has been found that the α_2-macroglobulin purified using glass containers has a 50- to 100-fold increase in the TAMe esterase activity of the final preparation as compared to α_2M prepared utilizing nonwettable surfaces and in the presence of soybean trypsin inhibitor.[34] This esterolytic activity is presumably due to plasma proteases, which are activated through the glass-induced activation of Hageman factor-dependent pathways.[35,36] Thus, all purification procedures are performed using plastic or siliconized glass surfaces.

Venous blood (1000 ml) is collected in standard blood donor plastic bags containing an acid-citrate dextrose anticoagulant. The fresh plasma is harvested immediately by centrifugation. Crude soybean trypsin inhibitor, 50 mg dissolved in 30 ml phosphate buffer 1 is added to the plasma.

Blood coagulation factors II, VII, IX, X are then removed by adsorption to eliminate factor II (prothrombin) activation and subsequent complex formation with α_2M. Barium chloride (1 M in H_2O) is added to the plasma to a final concentration of 0.1 M and mixed by magnetic stirring for 10 min at room temperature. The precipitate is removed by centrifugation (2000 rpm for 15 min in an International Refrigerated Centrifuge, Model PR-2). Barium sulfate powder (USP for X-ray diagnosis, Mallinckrodt Chemical Works, St. Louis, Missouri), 50 mg per milliliter of plasma is added and mixed for 10 min at room temperature. The barium sulfate is removed by centrifugation. This barium sulfate adsorption step is repeated a second time.

Polyethylene Glycol (PEG) Precipitation. To the plasma supernatant, add 2 volumes of phosphate buffer. Add 50% polyethylene glycol (w/w in H_2O) (Sentry Carbowax, 4000, USP, flakes, Union Carbide, Clifton, New Jersey), to a final concentration of 4%. Mix for 30 min at room temperature. The precipitate, which contains the bulk of the plasma fibrinogen, is removed by centrifugation in a Sorvall RC2-B Centrifuge at 9000 rpm for 30 min at 4°. To the 4% PEG supernatant add 50% PEG to achieve a 12% concentration. Let stand at 4° for 18 hr. The precipitate is harvested as above and contains the α_2-macroglobulin. The 12% supernatant contains C$\bar{1}$ inactivator and is used as the starting material to purify this plasma inhibitor as detailed in this volume, [65].

Ultracentrifugation in KBr. The 4–12% PEG precipitate from 500

---

[34] P. C. Harpel, *J. Exp. Med.* **132**, 329 (1970).

[35] O. D. Ratnoff, *Adv. Immunol.* **10**, 145 (1969).

[36] C. G. Cochrane, S. D. Revak, and K. D. Wuepper, *J. Exp. Med.* **138**, 1564 (1973).

ml of plasma is dissolved in 90 ml of KBr solution. Centrifuge and discard the precipitate which forms. The KBr-α_2M solution is then centrifuged in a Beckman fixed-angle type 50 Ti Rotor at 40,000 rpm, 5° for 20 hr in a Beckman L2-65B ultracentrifuge. After centrifugation the tubes are sliced at their midpoint and the low-density, lipoprotein-rich top fraction is discarded. The viscous material in the bottom of each tube is carefully resuspended and pooled. Two volumes of Tris-HCl, 50 mM, pH 8.0, are added, and the α_2-macroglobulin fraction is precipitated by adding 50% PEG to a final concentration of 15%. After mixing 30 min, the precipitate is recovered by centrifugation at 9000 rpm, for 45 min in the Sorvall Centrifuge and dissolved in Tris buffer 1.

DEAE-Cellulose Chromatography. All chromatography procedures are performed at 4°. The KBr purified fraction is then applied to a 5.0 × 50 cm column of DE-52 (Reeve Angel and Co. Inc., Clifton, New Jersey) prepared as detailed in the Manufacturer's instructions, and equilibrated with Tris buffer 1. After the sample is on the gel, the column is washed with buffer until $A_{280\ nm}$ is below 0.02, then a linear NaCl gradient is started. The gradient is formed by 2 liters of Tris buffer 2, running into 2 liters of Tris buffer 1, both contained in 2-liter Erlenmeyer flasks. The α_2M is identified by double-diffusion analysis on agar plates[37] using rabbit antihuman α_2M serum (Behring Diagnostics, American Hoechst Corp., Somerville, New Jersey) and is found as indicated in Fig. 1, at a conductivity of 7.5–11 mmho/cm at 27°.

Gel Filtration Chromatography. The fractions containing α_2M are concentrated using an ultrafiltration apparatus (Amicon Corp., Lexington, Massachusetts) and a UM-10 membrane. Precipitation with PEG at this stage of purification resulted in decreased yields. The DEAE-purified α_2M is concentrated to 10.0 ml and applied to a 5.0 × 80 cm column of Bio-Gel A-5m agarose (Bio-Rad Laboratory, Richmond, California) in Tris–citrate–saline buffer. The fractions containing immunologically identifiable α_2M are pooled (Fig. 2) and concentrated by ultrafiltration to 6.0 ml.

Pevicon Block Electrophoresis. Preparative block electrophoresis with Pevicon-C870 (Mercer Chemical Corp., New York) was performed as described.[38] After washing the Pevicon in H_2O 10–12 times with decanting to remove the fines, the material is added to barbital buffer. The poured Pevicon should be 0.5 cm in depth when it has settled on the electrophoresis plate. Two sets of vessels are used on either side of the Pevicon plate. The proximal vessel contains barbital buffer, and the dis-

---

[37] G. Mancini, J.-P. Vaerman, A. O. Carbonaro, and J. F. Heremans, *Protides Biol. Fluids, Proc. Colloq.* **15**, 370. (1963).

[38] H. J. Müller-Eberhard, *Scand. J. Clin. Lab. Invest.* **12**, 33 (1960).

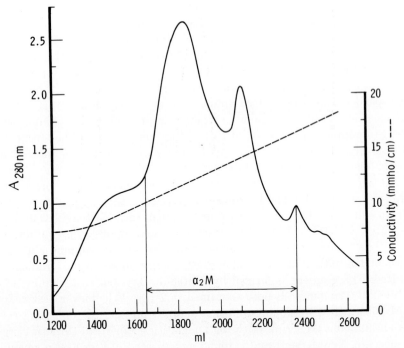

Fig. 1. DEAE-cellulose chromatography of α_2-macroglobulin. The 12% polyethylene glycol precipitate, after ultracentrifugation in potassium bromide as described in the text, is applied to a 5.0 × 50-cm column of DE-52. A linear NaCl gradient is started after wash-through of the frontal peak with the starting buffer. The fractions containing immunologically identified α_2M are indicated.

tal vessel phosphate buffer for electrophoresis. The buffer vessels are connected by glass bridges. The concentrated α_2M (6.0 ml) is applied in a 6-inch narrow V slit approximately 3 inches from the cathodal end of the electrophoresis plate, and the slit is then closed with a spatula. Electrophoresis is at 300 V for 18 hr at 4°. The milliamperage should not exceed 90. Segments (0.5 inch) of Pevicon are removed and eluted with 6.0 ml of Tris–citrate–NaCl buffer. The fractions containing α_2M, as identified by double-diffusion analysis, are pooled, concentrated by ultrafiltration, and frozen at —70° in small portions. The yield is approximately 10–15% of the concentration in the starting plasma.

Properties

Stability. The α_2M is both heat and acid sensitive. Loss of trypsin-binding activity is partly reversible after acid treatment, depending upon the pH. Activity is irreversibly lost at pH 3.0, whereas about 25% of the

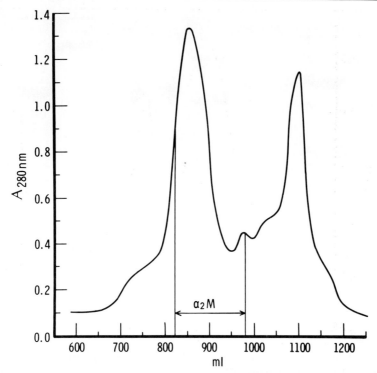

FIG. 2. Gel filtration chromatography of α_2-macroglobulin. The α_2M containing fractions obtained following DE-52 chromatography (Fig. 1) were pooled, concentrated, and applied to a 5.0 × 100-cm column of Bio-Gel A-5m agarose in Tris-citrate–saline buffer. The fractions containing immunologically identified α_2M are indicated.

activity remained at pH 4.0.[39] Ammonium sulfate has been found to destroy the trypsin-binding activity of α_2M, thus making this agent unsuitable for use in its purification.[5] Aliphatic amines also inactivate α_2M. Pretreatment of α_2M with 0.25 M methylamine at pH 7.4 reversibly induced a loss of plasmin-binding activity. In contrast, 0.25 M hydrazine with 1% α_2M at pH 7.5 irreversibly destroyed its plasmin-binding activity.[40] Whereas quick freezing in Dry Ice and acetone and storage at −70° preserves the enzyme-binding activity for at least 1 year, repeated freezing and thawing at warmer temperatures causes a loss in functional activity.

[39] S. M. Howard, Studies of trypsin-binding α_2-macroglobulin of human plasma. Ph.D. Dissertation, University of Southern California, 1966.
[40] M. Steinbuch, L. Pejaudier, M. Quentin, and V. Martin, *Biochim. Biophys. Acta* **154**, 228 (1968).

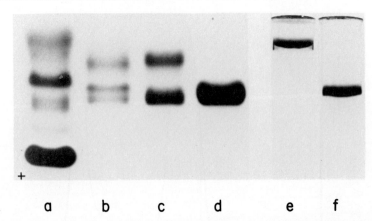

FIG. 3. Electrophoretic analysis of α_2-macroglobulin fractions. The following preparations were analyzed by cellulose acetate electrophoresis: a, serum; b, α_2M after DE-52 chromatography; c, α_2M after Bio-Gel A-5m agarose chromatography; d, α_2M-final product following Pevicon block electrophoresis. The α_2M was analyzed by sodium dodecyl sulfate–acrylamide (5% gel) electrophoresis: e, unreduced α_2M; f, α_2M reduced with dithiothreitol. The direction of electrophoresis is from the top to the bottom of the figure. The anode is indicated (+).

Purity. The purified inhibitor forms a single precipitin arc with a mobility of an α_2-globulin following immunoelectrophoresis using rabbit anti-whole human serum. Electrophoretic analysis on cellulose acetate of fractions of α_2M obtained during several different steps in the purification procedure are illustrated in Fig. 3. The final product, as analyzed in acrylamide gels containing sodium dodecyl sulfate (SDS), is shown in Fig. 3. Unreduced α_2M (Fig. 3e) demonstrates a single major band with a slower mobility than the band observed after disulfide bond cleavage with dithiothreitol (Fig. 3f). This subunit chain has an apparent molecular weight of 185,000.[41]

Physical Properties. The molecular weight of α_2M has been variously reported to be 820,000,[42] to 650,000.[43] The most recent estimates by sedimentation equilibrium indicate a weight of 725,000.[44,45] The sedimentation constant $(s^0_{20,w})$ is 19.6[42] to 17.3.[39] The partial specific volume has

[41] P. C. Harpel, *J. Exp. Med.* **138**, 508 (1973).

[42] M. Schönenberger, R. Schmidtberger, and H. E. Schultze, *Z. Naturforsch. Teil B* **13**, 761 (1958).

[43] R. Saunders, B. J. Dyce, W. E. Vannier, and B. J. Haverback, *J. Clin. Invest.* **50**, 2376 (1971).

[44] J. M. Jones, J. M. Creeth, and R. A. Kekwick, *Biochem. J.* **127**, 187 (1972).

[45] R. C. Roberts, W. A. Riesen, and P. K. Hall, *Proteinase Inhibitors Proc. Int. Res. Conf., 2nd (Bayer Symp. V), Grosse Ledder, 1973*, p. 63. Springer-Verlag, Berlin and New York, 1974.

been calculated to be 0.735.[42-45] The frictional ratio (f/f_{min}) is 1.43.[42] The absorbance $(A_{1\,cm}^{1\%}$ 280 nm) is 8.1[46] or 8.7.[39] The inhibitor is precipitable by 25% ethanol, 1.2–1.8 M ammonium sulfate, 0.0065 M Rivanol, 0.6 M perchloric acid, and 0.15 M trichloroacetic acid.[47] α_2M is very soluble in distilled water at neutral pH, but at acid pH (4.5–5.0) the solubility is greatly diminished.[9] The isoelectric point is 5.4.[42]

Subunit Structure. Although a number of studies have investigated the structure of the α_2M molecule, there is no consensus as to its subunit structure.[41,42,44,45,48-53] Under acid conditions (pH 2.5–4.5), or by treatment with urea, the α_2M dissociates into 11 S components.[44,49] This suggests that the molecule possesses a dimeric structure and that the two half-molecules are joined by noncovalent bonds.[49] Reduction of the molecule with thiols yields subunits whose estimated molecular weight is 196,000 by density-gradient equilibrium ultracentrifugation[49] and 185,000 by SDS-gel electrophoresis.[41,45,53] After complex formation with either bovine trypsin, human plasmin, thrombin, or plasma kallikrein, the reduced α_2M preparation demonstrates a new band with an apparent molecular weight of 85,000.[41] These results suggest that each half of the molecule consists of two subunit chains (MW 185,000) joined by disulfide bonds, and that a region at or near the center of the subunit chain is cleaved by the proteases, which are inhibited by the α_2M.[41,54] Derivatives with molecular weights of 80,000–85,000 were also obtained by reduction in guanidine-HCl and by performic oxidation of α_2M preparations not incubated with proteases.[45,53] The methods of purification used, however, did not eliminate the possibility of protease generation during the preparation of α_2M. Ultrafiltrates of human sera were found to contain α_2M-related antigens with sedimentation rates of 5–10 S.[55] The relation of these proteins to the subunit structure of α_2M has not been clarified.

Specificity. Human plasma α_2M may bind all proteinases.[6] The follow-

[46] M. Schönenberger, *Z. Naturforsch Teil B* **10**, 474 (1955).

[47] H. E. Schultze and J. F. Heremans, "Nature and Metabolism of Extracellular Proteins," Vol. 1, p. 204. Elsevier, Amsterdam, 1966.

[48] E. Razafimahaleo, J.-P. Frénoy, and R. Bourrillon, *C. R. Acad. Sci. Ser. D* **269**, 1567 (1969).

[49] C. Gentou, *C. R. Acad. Sci. Ser. D* **266**, 2358 (1968).

[50] H. E. Schultze, H. Haupt, K. Heide, G. Möschlin, R. Schmidtberger, and G. Schwick, *Z. Naturforsch Teil B* **17**, 313 (1962).

[51] M. D. Poulik, *Biochim. Biophys. Acta* **44**, 390 (1960).

[52] T. Isliker, *Helv. Med. Acta* **25**, 41 (1958).

[53] J.-P. Frénoy, E. Razafimahaleo, and R. Bourrillon, *Biochim. Biophys. Acta* **257**, 111 (1972).

[54] Z. Werb, M. C. Burleigh, A. J. Barrett, and P. M. Starkey, *Biochem. J.* **139**, 359 (1974).

[55] C. J. van Oss and P. M. Bronson, *Prep. Biochem.* **2**, 93 (1972).

ing serine proteinases have been shown to either be bound or inhibited by α_2M: bovine trypsin,[5,32,56] human cationic trypsin,[57] acetylated bovine trypsin,[58] bovine α-chymotrypsin,[4,43,59] human chymotrypsin,[57] human plasmin,[7,29,60-62] bovine plasmin,[63] human thrombin,[64,65] human plasma kallikrein,[34,66] human plasminogen activator,[67] human urokinase,[68] human polymorphonuclear leukocyte elastase,[69,70] porcine elastase,[71] human and rabbit polymorphonuclear leukocyte neutral proteinase[72,73]; the neutral proteinases from *Staphylococcus aureus*, *Proteus vulgaris*,[74] *Trichophyton mentagrophytes*,[75] and ovine testis[6]; subtilopeptidase A and B.[76,77]

α_2M reacts with the following thiol proteinases: human cathepsin B1,[78] papain, ficin, and bromelain[6]; the carboxyl proteinases human cathepsin D[6] and the acid proteinase from *Periplaneta* sp.[79] It inhibits the following

---

[56] S. Nagasawa, B. H. Han, H. Sugihara, and T. Suzuki, *J. Biochem.* (*Tokyo*) **67**, 821 (1970).

[57] J. Bieth, M. Aubry, and J. Travis, *Proteinase Inhibitors, Proc. Int. Res. Conf., 2nd* (*Bayer Symp. V*), Grosse Ledder, 1973, p. 53. Springer-Verlag, Berlin and New York, 1974.

[58] Y. Jacquot-Armand and G. Krebs, *Biochem. Biophys. Res. Commun.* **36**, 815 (1969).

[59] M. Iwamoto and Y. Abiko, *Biochim. Biophys. Acta* **214**, 402 (1970).

[60] H. E. Schultze, N. Heimburger, K. Heide, H. Haupt, K. Störiko, and H. G. Schwick, *Proc. 9th Congr. Eur. Soc. Haematol, Lisbon.* p. 1315 (abstr.) (1963).

[61] M. Steinbuch, M. Quentin, and L. Pejaudier, *Protides Biol. Fluids, Proc. 13th Colloq.* p. 375 (1965).

[62] P. O. Ganrot, *Clin. Chim. Acta* **16**, 328 (1967).

[63] H. Sugihara, S. Nagasawa, and T. Suzuki, *J. Biochem.* (*Tokyo*) **70**, 649 (1971).

[64] G. F. Lanchantin, M. L. Plesset, J. A. Friedmann, and D. W. Hart, *Proc. Soc. Exp. Biol. Med.* **121**, 444 (1966).

[65] M. Steinbuch, C. Blatrix, and F. Josso, *Rev. Fr. Etud. Clin. Biol.* **13**, 179 (1968).

[66] P. C. Harpel, *J. Clin. Invest.* **50**, 2084 (1971).

[67] A. D. Schreiber, A. P. Kaplan, and K. F. Austen, *J. Clin. Invest.*, **52**, 1394 (1973).

[68] D. Ogston, B. Bennett, R. J. Herbert, and A. S. Douglas, *Clin. Sci.* **44**, 73 (1973).

[69] K. Ohlsson and I. Ohlsson, *Scand. J. Clin. Lab. Invest.* **34**, 349 (1974).

[70] K. Ohlsson, *Proteinase Inhibitors, Proc. Int. Res. Conf., 2nd* (*Bayer Symp. V*), Grosse Ledder, 1973, p. 96. Springer-Verlag, Berlin and New York, 1974.

[71] J. S. Baumstark, *Biochim. Biophys. Acta* **207**, 318 (1970).

[72] A. Koj, J. Chudzik, W. Pajdak, and A. Dubin, *Biochim. Biophys. Acta* **268**, 199 (1972).

[73] S. F. Brown, C. W. Hook, and M. P. Tragakis, *Invest. Ophthalmol.* **11**, 149 (1972).

[74] F. Kueppers and A. G. Bearn, *Proc. Soc. Exp. Biol. Med.* **121**, 1207 (1966).

[75] R. F. Yu, F. Grappel, and F. Blank, *Experientia* **28**, 886 (1972).

[76] J. Dolovich and V. Wicher, *J. Lab. Clin. Med.* **77**, 951 (1971).

[77] V. Wicher and J. Dolovich, *Immunochemistry* **10**, 239 (1973).

[78] P. M. Starkey and A. J. Barrett, *Biochem. J.* **131**, 823 (1973).

[79] J. C. Landureau and M. Steinbuch, *Z. Naturforsch. Teil B* **25**, 231 (1970).

metal proteinases: thermolysin (*Bacillus thermolyticus*), clostridiopepti-
dase A,[54,80] and collagenases from human skin,[81] human and rabbit cor-
neal tissue,[82,83] rabbit synovial cell, rabbit skin,[6,54] tadpole tail,[84] and
human polymorphonuclear leukocytes.[54,70] Arvin, the coagulating enzyme
from the Malayan pitviper (*Ankistrodon rhodostoma*)[85] and the cationic
form of aspartate aminotransferase[86,87] also bind with α_2M. The α_2M has
been identified as the zinc-binding protein of human plasma.[88]

Inactivated forms of proteinases or their zymogens fail to form a
complex with α_2M, nor do the hydrolases hyaluronidase, β-acetylglucosa-
minidase, ribonuclease, β-glucuronidase or the exopeptiase aminopeptidase M bind to α_2M.[6]

Enzyme Binding Properties. There is no consensus as to the moles of
trypsin or of plasmin bound to α_2M. Human α_2M has been found to bind
2 mole equivalents of trypsin[24,33] and of plasmin,[67] whereas other investigators have reported equimolar binding ratios for trypsin,[6,89,90] and for
plasmin.[33,91] Formation of the trypsin-α_2-macroglobulin complex results
in retention of the esterase or amidase activity of the enzyme, however,
in comparison with free trypsin, Rinderknecht *et al.*[90] have documented
marked alterations of the Michaelis–Menten constant, pH optimum, and
sensitivity to ionic strength using N-carbobenzoxyglycylglycyl-L-arginine-
2-naphthylamide as substrate. The α_2M is significantly more effective as
an inhibitor of the proteolytic activity of trypsin and plasmin toward native protein substrates. The binding of trypsin and plasmin to α_2M, for
example, results in reduction, but not total absence, of caseinolytic activity.[62,90] The thrombin-α_2M complex retained clotting activity.[92] Further, α_2M purified from plasma possessed the ability to degrade native
fibrinogen.[7] This activity was not inhibited by protease inhibitors indicating that the putative α_2M-plasmin complex was responsible for the

[80] Z. Werb, *Biochem. Soc. Trans.* **1**, 382 (1973).
[81] A. Z. Eisen, K. J. Block, and T. Sakai, *J. Lab. Clin. Invest.* **75**, 258 (1970).
[82] M. B. Berman, J. C. Barber, R. C. Talamo, and C. E. Langley, *Invest. Ophthalmol.* **12**, 759 (1973).
[83] M. Berman, J. Gordon, L. A. Garcia, and T. Gage, *Exp. Eye Res.* **20**, 231 (1975).
[84] S. Abe and Y. Nagai, *Biochim. Biophys. Acta* **278**, 125 (1972).
[85] W. R. Pitney and E. Regoeczi, *B. J. Haematol.* **19**, 67 (1970).
[86] T. R. C. Boyde and I. F. Pryme, *Clin. Chim. Acta* **21**, 9 (1968).
[87] T. R. C. Boyde, *Biochem. J.* **111**, 59 (1969).
[88] A. F. Parisi and B. L. Vallee, *Biochemistry* **9**, 2421 (1970).
[89] V. Hamberg, P. Stelwagen, and H.-S. Ervast, *Eur. J. Biochem.* **40**, 439 (1973).
[90] H. Rinderknecht, R. M. Fleming, and M. C. Geokas, *Biochim Biophys. Acta* **377**, 158 (1975).
[91] P. C. Harpel, *Fed. Proc. Fed. Am. Soc. Exp. Biol.* **34**, 344 (abstr.) (1975).
[92] H. Rinderknecht and M. C. Geokas, *Biochim. Biophys. Acta* **295**, 233 (1973).

activity rather than free enzyme dissociating from the inhibitor as has been suggested previously.[93]

Microheterogeneity. Utilizing a variety of separation methods, the electrophoretic heterogeneity of α_2M derived from serum has been demonstrated.[43,94,95] Saunders *et al.*[43] demonstrated five α_2M bands by preparative polyacrylamide gel electrophoresis. These species possessed amidase activity of the same specific activity, however, the slower migrating fractions were capable of binding trypsin, whereas the fast-moving component was not. This suggested that the heterogeneity observed was due to α_2M protease interactions. Isoelectric focusing in a pH 4–6 gradient demonstrated four molecular forms of α_2M.[95] The amino acid composition of the fractions of α_2M was found to be similar. In contrast, as the pI of the fractions increased a decrease in N-acetylneuraminic acid content and an increase in the mannose, galactose ratio was observed (see Table II). These data support the concept that the microheterogeneity of α_2M in human serum is due in part to variation in carbohydrate content.

Genetic Polymorphism. Three apparently different genetic polymorphisms of α_2M in man have been described. Berg and Bearn[96] first detected a sex-linked α_2M antigen designated the Xm system. Leikola *et al.*[97] described the AL-M system, independently of Xm and transmitted as a dominant autosomal trait. Gallango and Castillo[98] have detected immunoelectrophoretic variations of α_2M in the sera of a Venezuelan Mestizo population. The three major variants, A, B, and C, are distinguished by their differing electrophoretic mobilities. The inheritance appears to be of Mendelian autosomal codominant type. Allotypes of rabbit α_2-macroglobulins have also been described.[99–101]

Distribution. α_2M has been identified in a variety of inflammatory effusions being found in pleural fluid, ascites,[102,103] and synovial fluid.[104–106]

[93] C. M. Ambrus and G. Markus, *Am. J. Physiol.* **199,** 491 (1960).

[94] P.-O. Ganrot and C. B. Laurell, *Clin. Chim. Acta* **14,** 137 (1966).

[95] J.-P. Frénoy and R. Bourrillon, *Biochim. Biophys. Acta* **371,** 168 (1974).

[96] K. Berg and A. G. Bearn, *J. Exp. Med.* **123,** 379 (1966).

[97] J. Leikola, H. H. Fudenberg, R. K. Kasukawa, and F. Milgrom, *Am. J. Hum. Genet.* **24,** 134 (1972).

[98] M. L. Gallango and O. Castillo, *J. Immunogenetics* **1,** 147 (1974).

[99] K. L. Knight and S. Dray, *Biochemistry* **7,** 1165 (1968).

[100] K. L. Knight and S. Dray, *Biochemistry* **7,** 3830 (1968).

[101] B. H. Berne, S. Dray, and K. L. Knight, *J. Immunol.* **105,** 856 (1970).

[102] J. Bieth and P. Metais, *Clin. Chim. Acta* **22,** 639 (1968).

[103] J. Bieth, J.-G. Barth, and P. Metais, *Clin. Chim. Acta* **30,** 621 (1970).

[104] S. Abe and Y. Nagai, *J. Biochem. (Tokyo)* **71,** 919 (1972).

[105] S. Abe and Y. Nagai, *J. Biochem. (Tokyo)* **73,** 897 (1973).

[106] G. Shtacher, R. Maayan, and G. Feinstein, *Biochim. Biophys. Acta* **303,** 138 (1973).

TABLE I
AMINO ACID COMPOSITION OF HUMAN α_2-MACROGLOBULIN (g/100 ml)

Amino acid	Heimburger[a]	Dunn[b]	Bourrillon[c]
Lysine	5.34	6.47	4.82
Histidine	2.7	3.03	2.55
Arginine	3.82	4.27	3.58
Aspartic acid	7.16	7.24	7.92
Threonine	5.35	5.47	5.11
Serine	5.5	5.35	5.77
Glutamic acid	12.3	12.6	13.5
Proline	4.07	4.21	2.75
Glycine	2.75	2.77	3.05
Alanine	3.44	3.49	3.94
Half-cystine	1.14	1.58	2.59
Valine	6.8	6.98	6.52
Methionine	1.53	1.71	1.49
Isoleucine	3.09	3.26	3.04
Leucine	7.76	8.09	7.87
Tyrosine	4.84	4.76	5.3
Phenylalanine	5.37	4.96	5.12
Tryptophan	1.3	1.71	—
Carbohydrate	8.4	8.64	10.1

[a] N. Heimburger, K. Heide, H. Haupt, and H. E. Schultze, *Clin. Chim. Acta* **10**, 293 (1964).

[b] J. T. Dunn and R. G. Spiro, *J. Biol. Chem.* **242**, 5549 (1967).

[c] R. Bourrillon and E. Razafimahaleo, *in* "Glycoproteins. Their Composition, Structure and Function" (A. Gottschalk, ed.), Vol. 5, p. 699. Elsevier, Amsterdam, 1972.

It occurs in human semen[107] and has also been identified as a surface antigen of human and mice B lymphocytes.[108,109] The α_2M is associated with the membrane and granule fractions of human platelets.[110] Preliminary studies suggest that human lymphocytes may synthesize α_2M.[111] Rabbit antisera against human α_2M reacts with trypsin-binding proteins in the monkey, goat, sheep, cow, donkey, zebra, dog, cat, pig, and guinea pig.[112] α_2M has been shown to be synthesized by hepatic tissue from 29-day-old human embryos.[113]

[107] A. Mroveh and N. F. Adham, *Clin. Chim. Acta* **28**, 259 (1970).

[108] J. N. McCormick, D. Nelson, A. M. Tunstall, and K. James, *Nature (London) New Biol.* **246**, 78 (1973).

[109] K. James, A. M. Tunstall, A. C. Parker, and J. N. McCormick, *Clin. Exp. Immunol.* **19**, 237 (1975).

[110] R. Nachman and P. C. Harpel, unpublished data.

[111] A. M. Tunstall and K. James, *Clin. Exp. Immunol.* **17**, 697 (1974).

[112] K. James, *Immunology* **8**, 55 (1965).

[113] D. Gitlin and A. Biasucci, *J. Clin. Invest.* **48**, 1433 (1969).

TABLE II

CARBOHYDRATE COMPOSITION OF HUMAN α_2-MACROGLOBULIN (g/100 ml)

Component	Heim-burger[a]	Dunn[b]	Bour-rillon[c]	Isoelectric focusing fractions[d]			
				A	B	C	D
Galactose	1.8	1.22	2.0	2.79	1.72	1.45	1.28
Mannose	1.8	1.86	2.5	3.04	3.14	3.01	2.99
N-Acetyl-glucosamine	2.9	3.62	3.95	3.57	3.20	3.39	3.36
N-Acetyl-neuraminic acid	1.8	1.7	1.83	2.74	2.33	2.23	1.62
Fucose	0.1	0.24	0.71	—	—	—	—

[a] N. Heimburger, K. Heide, H. Haupt, and H. E. Schultze, *Clin. Chim. Acta* **10**, 293 (1964).

[b] J. T. Dunn, and R. G. Spiro, *J. Biol. Chem.* **242**, 5549 (1967).

[c] R. Bourrillon, and E. Razafimahaleo, *in* "Glycoproteins. Their Composition, Structure and Function" (A. Gottschalk, ed.), Vol. 5, p. 699. Elsevier, Amsterdam, 1972.

[d] J.-P. Frénoy and R. Bourrillon, *Biochim. Biophys. Acta* **371**, 168 (1974).

Composition. The amino acid composition of α_2M has been determined in several studies,[9,89,114-116] and representative data are shown in Table I. A variable number of terminal NH_2 groups have been described for human α_2M, perhaps reflecting the binding of proteases to α_2M when purified from serum. Schultze[117] has identified amino-terminal serine, valine, and aspartic acid, and Razafimahaleo, *et al.*[48] have found these amino acids plus glutamic acid, alanine, and glycine. The carbohydrate composition of α_2M and of four α_2M fractions obtained by isoelectric focusing is indicated in Table II.

Analysis of the glycopeptides associated with α_2M has indicated considerable variation in size and composition of the individual carbohydrate units.[118] It has been postulated that this variability is a function of the degree of completion of the oligosaccharide chains. The basic carbohydrate unit contains an internal portion consisting of three residues of mannose and two of glucosamine to which more external sugars are bound in varying amounts.[118] The linkage of the carbohydrate moieties to the asparagine of the protein is of the glycosylamine type.

[114] N. Heimburger, K. Heide, H. Haupt, and H. E. Schultze, *Clin. Chim. Acta* **10**, 293 (1964).

[115] J. T. Dunn and R. G. Spiro, *J. Biol. Chem.* **242**, 5549 (1967).

[116] J. Demaille, J. Broussal, C. Colette, F. Guilleux, and M. B. Magnan de Bornier, *C. R. Acad. Sci. Ser. D* **270**, 2133 (1970).

[117] H. E. Schultze, *Verh. Dtsch. Ges. Inn. Med.* **66**, 225 (1960).

[118] J. T. Dunn and R. G. Spiro, *J. Biol. Chem.* **242**, 5556 (1967).

[53] Antithrombin–Heparin Cofactor

By PAUL S. DAMUS and ROBERT D. ROSENBERG

At the beginning of this century, Contejean,[1] Morawitz,[2] and others recognized that thrombin gradually lost activity when added to defibrinated plasma or serum. On the basis of these data, they surmised that a specific inactivator of the proteolytic enzyme antithrombin must be present under physiological conditions. In 1916, McLean[3] isolated heparin from the liver, as well as the heart, and demonstrated its potent anticoagulant properties. Confusion about its inhibitory effect on purified procoagulants was resolved in 1939 by Brinkhous *et al.*,[4] who showed that heparin was effective as an anticoagulant only in the presence of a plasma component, which they termed heparin cofactor. In the 1950's, the kinetic studies of Waugh and Fitzgerald[5] and Monkhouse *et al.*[6] indicated that plasma antithrombin activity and plasma heparin cofactor activity are intimately related. These investigators suggested that heparin acts to accelerate the rate at which antithrombin neutralizes thrombin. In 1968, Abildgaard[7] was able to isolate small amounts of an α_2-globulin from human plasma, which functions in both capacities. In 1973, we provided the first reproducible means for isolating large quantities of this human inhibitor and advanced convincing physiochemical evidence that plasma antithrombin activity as well as plasma heparin cofactor activity reside in the same molecular species.[8] Furthermore, we have demonstrated that antithrombin–heparin cofactor functions in a manner analogous to other protease inhibitors, except that heparin uniquely accelerates its action in apparent allosteric fashion.[8] The results of this mechanistic analysis led us to examine the interactions of this inhibitor and its mucopolysaccharide cofator with other serine proteases of the coagulation–fibrinolytic mechanism.[9–12] Based upon the data obtained from these

[1] C. Contejean, *Arch. Physiol. Norm. Pathol.* **7**, 45 (1895).
[2] P. Morawitz *in* "The Chemistry of Blood Coagulation." Thomas, Springfield, Illinois, 1968.
[3] J. McLean, *Am. J. Physiol.* **41**, 250 (1916).
[4] K. Brinkhous, H. P. Smith, E. D. Warner, and W. H. Seegers, *Am. J. Physiol.* **125**, 683 (1939).
[5] D. F. Waugh anl M. A. Fitzgerald, *Am. J. Physiol.* **184**, 627 (1956).
[6] F. C. Monkhouse, E. S. Framer, and W. H Seegers, *Circ. Res.* **3**, 397 (1955).
[7] U. Abildgaard, *Scand. J. Clin. Lab. Invest.* **21**, 89 (1968).
[8] R. D. Rosenberg and P. S. Damus, *J. Biol. Chem.* **248**, 6490 (1973).
[9] P. S. Damus, M. Hicks, and R. D. Rosenberg, *Nature (London)* **246**, 355 (1973).
[10] R. F. Highsmith and R. D. Rosenberg, *J. Biol. Chem.* **249**, 4335 (1974).
[11] J. S. Rosenberg, P. McKenna, and R. D. Rosenberg,
[12] N. W. Stead, A. P. Kaplan, and R. D. Rosenberg, unpublished data.

studies, we have established that virtually all these enzymes are slowly but progressively neutralized by antithrombin–heparin cofactor in the absence of heparin, but are instantaneously inactivated by the inhibitor in the presence of this acidic mucopolysaccharide. Thus, we have shown that this inhibitor is a critical component in modulating the activity of the hemostatic mechanism.

The present chapter surveys current methods of assaying for antithrombin–heparin cofactor and provides a variety of techniques for isolating this inhibitor from human, rabbit, and canine plasma. In addition, this communication describes, in some detail, the specificities and properties of this plasma protease inhibitor.

Assay Methods—General Considerations

The potency of an antithrombin-heparin cofactor preparation can be estimated by incubating this plasma inhibitor with a serine protease in the presence or the absence of heparin and measuring the time-dependent neutralization of enzymic activity. Thus to define an assay procedure capable of quantitating the inhibitory activity of this plasma component, it is necesssary to select an appropriate serine protease for incubation with antithrombin-heparin cofactor, to consider whether heparin should be included in or omitted from reaction mixtures, to determine the number of timed observations needed to describe the neutralization process, and to choose a proper substrate for the measurement of enzymic activity.

Either thrombin or factor X_a have been customarily employed to assay for antithrombin–heparin cofactor. The utilization of these enzymes rather than trypsin, plasmin, or other serine proteases clearly increases assay specificity when crude plasma fractions are examined. Other protease inhibitors contribute in a relatively minor way to the plasma's capacity for inactivating either of these coagulation system serine proteases. However, this selectivity of inhibitor action is wholly dependent upon employing low concentrations of either enzyme. If higher levels are used, the neutralization of these enzymes by other plasma inhibitors grow in relative importance.

Human or bovine thrombin can readily be obtained in pure and stable form by a variety of chromatographic techniques.[13-15] Furthermore, an interlaboratory activity standard is available from the National Institutes of Health. Human and bovine factor X_a have only recently been isolated

[13] J. R. Rosenberg, D. L. Beeler, and R. D. Rosenberg, *J. Biol. Chem.* **250**, 1607 (1975).
[14] R. L. Lundblad, *Biochemistry* **10**, 2501 (1971).
[15] R. D. Rosenberg and D. F. Waugh, *J. Biol. Chem.* **245**, 5049 (1970).

in a physically homogeneous state.[13,16] Furthermore, only small quantities of these enzymes can be prepared with these purification techniques. The procedures themselves are more complex and prolonged than those utilized for the isolation of thrombin. Therefore, relatively impure preparations of bovine factor X_a have been employed in the past to assay for human, bovine, or rabbit antithrombin–heparin cofactors.[17,18] In addition, no interlaboratory activity standard exists for this enzyme, and recourse to a plasma titration curve is necessary.

The assay procedure can be conducted in the presence or in the absence of heparin. If this acidic mucopolysaccharide is utilized, the time required for incubation of thrombin or factor X_a with the sample is minimal. In addition, the use of heparin in conjunction with a short incubation period further augments assay specificity. This is due to the relatively slow action of other plasma protease inhibitors and their inability to be accelerated by heparin. If this acidic mucopolysaccharide is omitted from the assay procedure, the incubation of thrombin or factor X_a with inhibitor must be prolonged (3 min to 10 min) to attain any significant degree of enzyme inactivation. In principle, the precision of the assay is optimal when enzyme neutralization is quantitated at multiple time points.[19] In practice, this is an onerous procedure and inactivation of the enzyme by inhibitor is measured after the shortest possible incubation period. The utilization of this abbreviated technique is essential during purification of antithrombin–heparin cofactor, where many samples must be rapidly analyzed for inhibitor activity.

The neutralization of thrombin by inhibitor is usually quantitated by adding an aliquot of the incubation mixture to a solution of fibrinogen and recording the clotting time of this substrate.[20,21] On occasion, other techniques have been employed to monitor enzyme inactivation, such as a reduction in the rate of hydrolysis of toluenesulfonyl (tosyl) lysine methyl ester (TLME) or tosyl arginine methyl ester (TAME). These small synthetic substrates of thrombin are preferable to fibrinogen, since they are easily characterized and can be prepared in a highly reproducible fashion. However, assay procedures utilizing either TAME or TLME are relatively insensitive when compared to the fibrinogen clotting assay. Furthermore, they require high concentrations of reactants. Thus, there is a reduction in assay specificity when these techniques are employed to

[16] K. Fujikawa, M. E. Legaz, and E. W. Davie, *Biochemistry*. **11**, 4882 (1972).
[17] E. T. Yin, R. Wessler, and J. B. Butler, *J. Lab. Clin. Med.* **81**, 289 (1973).
[18] S. N. Gitel and S. Wessler, *Thromb. Res.* in press.
[19] F. Monkhouse, F. C., this series Vol. 19, p. 915.
[20] M. Gerandas, *Thromb. Diath. Haemorrh.* **4**, 50 (1960).
[21] U. Abildgaard, *Scand. J. Clin. Lab. Invest.* **19**, 190 (1967).

estimate inhibitor activity in crude plasma fractions and relatively low concentrations of antithrombin–heparin cofactor cannot be measured with these assay techniques. However, the recent introduction of the N-benzoyl-L-phenylalanyl-L-valyl-L-arginine-p-nitroanilide may alter previous conclusions about the usefulness of small synthetic substrates. This tripeptide permits the spectrophotometric measurement of low concentrations of thrombin, and it may replace fibrinogen as the substrate of choice for quantitating antithrombin or heparin cofactor activity.[22] At this time, there is relatively little experience in utilizing it for this purpose and significant methodological difficulties exist when ths substrate is employed in conjunction with heparin.

The inactivation of factor X_a by inhibitor is usually assayed by admixing an aliquot of these reactants with plasma and recording the clotting time of this substrate after phospholipid addition and recalcification.[17,18] This technique has considerable interlaboratory variability due to the complexity of the substrate and the lack of a readily available enzyme standard. However, it has been successfully utilized by Yin, Wessler, and co-workers in their studies of the purification and properties of rabbit antithrombin-heparin cofactor.[23,24]

The subsequent section describes assays of antithrombin-heparin cofactor activity which are based upon the neutralization of thrombin. Techniques that measure inhibitor activity by quantitating the inactivation of factor Z_a have been summarized elsewhere.[18] A recent analysis of assay methodology indicates that procedures that employ either thrombin or factor X_a give similar, if not identical, results.[18] We have chosen to present procedures that utilize thrombin because of the ease in obtaining purified preparations of this protein, the greater reproducibility of substrate required for the assay of enzyme activity, and the existence of an interlaboratory enzyme reference standard.

Assay Method

Principle. A sample is incubated with thrombin in the presence or in the absence of heparin for a specified time. Inhibition of the activity of thrombin is quantitated by measuring the time required for residual enzyme to clot a fibrinogen solution.

Reagents

Buffer A: 0.1 M sodium chloride in 0.1 M Tris-HCl (pH 8.3)
Buffer B: 0.15 M sodium chloride in 0.01 M Tris-HCl (pH 7.5)

[22] U. Abildgaard.
[23] E. T. Yin, S. Wessler, and P. J. Stoll, *J. Biol. Chem.* **246**, 3712 (1971).
[24] E. T. Yin and S. Wessler, *Biochem. Biophys. Acta* **201**, 387 (1970).

Substrate: Dissolve 50 mg of human fibrinogen (KABI, Stockholm, Sweden) in 7 ml of buffer A or buffer B at 37°; add 3 ml of 15% (w/v) solution of gum acacia (Howe and French, Salem, Massachusetts). Incubate at 37° for at least 1 hr prior to use. At this temperature, the substrate solution is stable for approximately 3 hr.

Standard enzyme solution: Purified preparations of either bovine or human thrombin are utilized.[13,15]

Heparin, diluted to 25 units/ml in buffer B

Defibrinated pooled human plasma: Human plasma samples are pooled from 10 or more normal donors using 0.38% (w/v) sodium citrate as anticoagulant. Samples are defibrinated in aliquots of 15 ml by placing them in a 54° water bath for 3 min. The samples are placed on ice for 5 min, centrifuged at 200 g for 15 min, and the supernatants are stored at −70°.

Progressive Antithrombin Assay (No Heparin in the Incubation Mixtures).[20] Place a series of glass tubes (10 × 75 mm) in a water bath at 37° and add 0.3 ml of substrate dissolved in buffer A. Dilute thrombin extensively with buffer A in an albuminized plastic tube to a concentration of approximately 0.010 absorbance unit and place this solution on ice. Add 0.1 ml of this enzyme solution to 0.4 ml of buffer A and incubate at 37° in an albuminized plastic tube for 3 minutes. Remove several 0.1 ml aliquots, add each sample to a randomly selected glass tube containing substrate, and record clotting time. A buffer time average (θ buffer) of approximately 18 sec is optimal. If θ buffer deviates markedly from this value, the initial thrombin dilution must be altered appropriately. This sequence of steps is repeated with a sample substituted for buffer A in the incubation mixture. The sample must be either extensively diluted with buffer A or dialyzed against this solution. Significant alterations in ionic strength will invalidate the assay results. In addition, a series of dilutions (1:10–1:40) of defibrinated pooled plasma in buffer A are similarly analyzed. All determinations are randomized and performed in triplicate. The progressive antithrombin activity is obtained from this relationship:

$$\text{Log } \theta/\theta \text{ (buffer)} = \text{constant} \times \text{antithrombin concentration}$$

Heparin Cofactor Assay (Heparin in the Incubation Mixtures).[21] Place a series of glass tubes (10 × 75 mm) in a water bath at 37° and add 0.2 ml of substrate dissolved in buffer B. Dilute the thrombin preparation extensively with buffer B in an albuminized plastic tube to a concentration of approximately 0.002 absorbance unit per milliliter and place this solution in an ice bath. Randomly select several glass tubes containing substrate and add to each tube first, 0.1 ml of buffer B and then 0.1 ml of heparin solution. Incubate for 1 min at 37°, add 0.1 ml of the enzyme solution, and record the clotting time. A buffer time average (θ buffer) of approximately 20 sec is optimal. If θ buffer deviates markedly from this value, the initial thrombin dilution must be altered appropriately. The substitution of buffer B for the heparin solution in the initial incubation mixture should yield a θ identical to θ buffer. The sequence of steps is repeated with a sample substituted for buffer B in the initial incubation mixture. The sample must either be extensively diluted with buffer B or previously dialyzed against this buffer. Significant deviations in ionic strength or pH will invalidate the assay results. A series of dilutions of defibrinated plasma (1:20–1:60) in buffer B are similarly analyzed. All determinations are randomized and performed in triplicate. Heparin cofactor activity is obtained from a relationship that is functionally identical to that used for determining progressive antithrombin activity (see above).

Definition of Unit and Specific Activity. The potency of antithrombin–heparin cofactor preparations have been defined by comparing them with the thrombin neutralizing activity of heat-defibrinated pooled plasma. The activity of this reference standard is arbitrarily set at 100 units/ml. The specific activity of a preparation is equivalent to its inhibitory unit content per milligram of protein.

In the future, the activity of antithrombin–heparin cofactor products will probably be defined in terms of the amount of purified thrombin neutralized instantaneously (assay in the presence of heparin) or per unit time (assay in the absence of heparin). The unitage of this enzyme preparation ought to be calibrated with active-site titrants[25] or by titration against the available National Institutes of Health thrombin reference standard. This alteration in the determination of inhibitor activity will permit interlaboratory comparisons of purified products.

Purification Procedures

Two basic types of purification procedures have been devised in order to isolate antithrombin–heparin cofactor. The first approach utilizes classical adsorption, chromatographic, and electrophoretic methodology. Human and rabbit antithrombin–heparin cofactor have been obtained in homogeneous form by this type of purification procedure, and these techniques are summarized in the next section. These procedures demand that approximately 2.5 weeks be devoted to the isolation of this plasma protein and final yields average 15–20%. The second approach employs the affinity

[25] K. J. Kezdy, L. Lorand, K. M. Miller, *Biochemistry* **4**, 2302 (1965).

matrix heparin-Sepharose for the purification of inhibitor. For some animal plasmas the use of this technique is sufficient in order to isolate homogeneous antithrombin–heparin cofactor. For plasmas of other species of animals, one or two additional chromatographic steps are required if the inhibitor is to be obtained in pure form. Approximately 2–7 days are required to complete the preparation of antithrombin–heparin cofactor by this methodology, and final yields range from 35 to 95%. These procedures are also summarized in the next section. Both types of purification techniques have been presented, since the affinity chromatographic procedures may have inherent limitations. For example, the adsorption–desorption of the inhibitor from heparin-Sepharose may subtly alter the conformation of antithrombin–heparin cofactor. This might effect the *in vivo* survival of this protein or the kinetics of its inactivation of coagulation–fibrinolytic system serine proteases. Comparative studies with preparations from both types of procedures will be required to settle this issue.

Classical Purifications

Human Antithrombin-Heparin Cofactor[8]

Starting Material. Human plasma is obtained by plasmaphoresis of normal donors using ACD* or 0.38% sodium citrate as anticoagulant. The use of fresh plasma is mandatory for optimal yield and maximal specific activity. If frozen plasma is employed, a 25–50% reduction in these parameters is observed.

Step 1. Preparation of a Crude Concentrate by Adsorption-Elution from Aluminum Hydroxide. Barium carbonate is aded to 1000–1500 ml of plasma at a concentration of 50 mg/ml. The resulting supernatant solution is defibrinated by pipetting 15-ml aliquots into glass tubes and placing them in a water bath at 54° for 3 min. The samples are placed on ice for 5 min and centrifuged at 2000 g for 15 min. To the combined supernatant solutions, 20% (w/v) aluminum hydroxide, prepared according to Loeb,[26] is added. The resulting creamy suspension is stirred continuously for 15 min at 24° and centrifuged at 2000 g for 15 min at 4°. The supernatant is discarded, and the aluminum hydroxide gel is washed with 500 ml of 0.15 M sodium chloride. To elute inhibitor activity, 0.36 M ammonium phosphate (pH 8.1) is added to the aluminum hydroxide gel in a volume equivalent to 25% of the initial heat-defibrinated supernatant solution, and the suspension is gently agitated for 15 min at 24°. The cloudy solution is centrifuged at 2000 g for 15 min at 4°. The supernatant

[26] J. Loeb, *Arch. Sci. Physiol.* **10**, 129 (1956).

solution is clarified by centrifugation at 27,000 g for 30 min at 4°. The ammonium phosphate elution of antithrombin–heparin cofactor from the aluminum hydroxide is repeated two additional times. All three aluminum hydroxide eluates are combined and concentrated from approximately 540 ml to about 35 ml. This concentration step requires 12–24 hr and employs an ultrafiltration apparatus equipped with a PM-30 membrane. The recovery of antithrombin averages 55% for step 1. The specific activity of this step 1 product averages 10 thrombin inhibitory units per milligram of protein (average specific activity of plasma = 2.0 thrombin inhibitory units per milligram of protein). Step 1 preparations are stable at 4° for at least 4 weeks.

Step 2. Sephadex Gel Filtration. A Sephadex G-200 column (5.0 × 100 cm) is equilibrated at 4° with 1.0 M sodium chloride in 0.05 M Tris-HCl (pH 8.3). The gel matrix is conditioned to prevent adsorption of antithrombin-heparin cofactor. This is accomplished by an initial filtration of approximately 80 ml of defibrinated plasma. After passage of one column volume, the step 1 product ∼5500 mg of protein) is applied, and flow rates of 60–70 ml/hr are maintained. The great bulk of antithrombin–heparin cofactor activity emerges in the final protein peak. The recovery of inhibitor activity for step 2 averages 65% and the specific activity of this product averages 13 thrombin inhibitory units per milligram of protein. Step 2 preparations are stable for many weeks at 4°, but lose significant activity if frozen at −20 to −70°. Prior to the initiation of step 3, a step 2 preparation is dialyzed for 8–12 hr against 10–15 liters of 0.1 M Tris-HCl (pH 9.0).

Step 3. DEAE-Cellulose A-50 Chromatography. The dialyzed step 2 products (∼2700 mg of protein) is applied to a DEAE-cellulose A-50 column (5.0 × 30.0 cm) equilibrated with 0.1 M Tris-HCl (pH 9.0). A linear salt gradient is employed for this fractionation with the reservoir containing 1200 ml of 0.25 M sodium chloride in 0.1 M Tris-HCl (pH 9.0) and the mixing chamber containing 1200 ml of 0.1 M Tris-HCl (pH 9.0). Chromatography is conducted at 4° with flow rates maintained at approximately 57 ml/hr. The inhibitor elutes over a narrow range of sodium chloride concentration with the peak of biological activity centered at an added ionic strength of 0.11 M sodium chloride. With different lots of DEAE-Sephadex A-50, the added sodium chloride concentration required to elute maximal amounts of the inhibitor varies from 0.11 M sodium chloride to 0.14 M sodium chloride. However, the purity and yield of the step 3 product is independent of the DEAE-Sephadex A-50 lot number. This step has an average yield of 73% and the step 3 product has an average specific activity of 134 thrombin inhibitory units per milligram of protein. Step 3 products stored at 4° are stable for 4–6 weeks.

Prior to step 4, step 3 preparations are dialyzed for 8–12 hr against 4 liters of 0.1 M Tris-HCl (pH 8.3).

Step 4. DEAE-Cellulose Chromatography. The dialyzed step 3 product ($\sim$190 mg of protein) is applied to a DEAE-cellulose (DEAE-52, Whatman) column (2.5 $\times$ 40 cm) at 4° equilibrated with 0.1 M Tris-HCl buffer (pH 8.3). A linear salt gradient is employed to elute the inhibitor with the reservoir containing 950 ml of 0.1 M sodium chloride in 0.1 M Tris-HCl (pH 8.3) and the mixing chamber containing 950 ml of 0.1 M Tris-HCl (pH 8.3). Chromatography is conducted at 4° with flow rates maintained at 40 ml/hr. The inhibitor emerges over a narrow ionic strength range with the peak of biological activity centered at an added ionic strength of 0.068 M NaCl. Chromatographic parameters have been remarkably constant for several lots of DEAE-cellulose. Fractions with specific activities of 378 thrombin inhibitory units per milligram of protein are pooled. The recovery of inhibitor activity averages 66% for step 4, and the specific activity of this product averages 494 thrombin inhibitory units per milligram protein. If stored at 4°, step 4 products are stable for several weeks. Prior to step 5, the step 4 preparation is concentrated from 100 ml to approximately 4 ml by ultrafiltration with a PM-30 membrane and dialyzed for 2 hr against 0.5% (w/v) ampholytes in water.

Step 5. Isoelectric Focusing in Sucrose Density Gradients. The dialyzed step 4 preparation ($\sim$34 mg of protein) is fractionated in an LKB 8102 column utilizing the standard isofocusing technique described by Vesterberg and Svensson.[27]

The sucrose density gradient is prepared manually with 1.7% concentration of pH 4–6 carrier ampholytes The cathode solution of monoethanolamine is placed on the bottom of the column. The protein sample is applied in mid-column after being mixed with an equal volume of dense solution. The sulfuric acid anode solution is applied at the top of the column. During isofocusing, the column is maintained at 6°. An initial potential of 250 V is established and is gradually increased to 750 V over the next 12 hr. The total time of isofocusing is 38 hr. The contents of the column are removed at 2.5ml/min under a constant hydrostatic head. The inhibitor is distributed over 0.15 pH units with the peak of biological activity centered at pH 5.11. The recovery of antithrombin–heparin cofactor for step 5 averages 65%, and the specific activity averages 945 thrombin inhibitory units per milligram of protein. Thus, the overall yield for the entire five-step preparative technique averages 11%.

Ampholytes are removed from the final products by extensive dialysis and filtration at 4° through a conditioned Sephadex G-50 column (1.25 $\times$ 60 cm) equilibrated with 1.0 sodium chloride in 0.05 M Tris-HCl (pH 8.3).

[27] O. Vesterberg and H. Svensson, *Acta Chem. Scand.* **20**, 820 (1966).

Rabbit Antithrombin–Heparin Cofactor[23]

Starting Material. Plasma is obtained from rabbits using 0.01 *M* sodium citrate as anticoagulant and quickly frozen. It may also be purchased in 1-liter quantities from Pel-Freez Biochemicals, Rogers, Arkansas.

Step 1. Preparation of Plasma Concentrate. One-liter batches of freshly frozen plasma are thawed overnight at 4°. The plasma is centrifuged at 2000 *g* for 15 min at 4°. It is treated with barium sulfate at a concentration of 100 mg/ml at room temperature with continuous stirring for 30 min. The barium sulfate is removed by centrifugation at 2000 *g* for 10 min at 4°, the plasma is decanted and the barium sulfate adsorption is repeated once again. Finally, the plasma is rapidly passed through an asbestos filter pad (Seitz filter pad, grade 7) in a Seitz filtration unit under nitrogen at room temperature and lypholyzed.

Step 2. Gel Filtration on Sephadex G-200. Lyophilized material equivalent to 1 liter of rabbit plasma is reconstituted in 250 ml of deionized water and dialyzed against two changes of 4 liters of 0.04 *M* sodium chloride for 24 hr. A fatty layer forms and is removed by careful decanting and filtration through Whatman No. 1 filter paper. The material (60 g of protein) is diluted to 400 ml in 0.145 *M* sodium chloride and chromatographed at room temperature on a Sephadex G-200 column (10 × 100 cm) equilibrated with 0.145 sodium chloride. Flow rates are maintained at approximately 150 ml/hr. The bulk of inhibitor activity emerges in the final protein peak. The step 2 recovery of antithrombin–heparin cofactor, as judged by factor X_a inhibitory activity determination, averages 97%. The specific activity of these products ranges from 0.21 to 0.27 factor X_a inhibitory units per milligram of protein (average specific activity of plasma = 0.042 factor X_a inhibitory units per milligram of protein). Prior to the initiation of step 3, the step 2 preparation is extensively dialyzed against 0.04 *M* sodium chloride at 4° for 24 hr and subsequently lyophilized.

Step 3. Chromatography on DEAE-Sephadex A-50. Lyophilized material from two separate step 2 preparations are pooled and reconstituted in 80 ml of 0.05 *M* sodium chloride in 0.1 *M* Tris-HCl (pH 8.32).This material is extensively dialyzed against the same solvent at 4° for 48 hr. The entire fraction (4000–6000 mg of protein) is filtered through a DEAE-Sephadex A-50 column (4 × 50 cm) at room temperature, which has been previously equilibrated with 0.05 *M* sodium chloride in 0.1 Tris-HCl (pH 8.32). To remove contaminating proteins, the chromatographic column is washed with approximately 550 ml of 0.1 *M* sodium chloride in 0.1 *M* Tris-HCl (pH 8.32). The inhibitory activity is eluted with 600 ml of 0.125 *M* sodium chloride in 0.14 *M* Tris-HCl (pH 8.32).

It emerges as a narrow peak of biological activity during the next column volume. The recovery of antithrombin–heparin cofactor, as judged by factor X_a inhibitory activity determinations, averages 61% for step 3. The specific activity of this product ranges from 4.5 to 6.0 factor X_a inhibitory units per milligram of protein. Prior to the initiation of step 4, the step 3 preparation is dialyzed for 3 hr at room temperature against several changes of deionized water and then lyophilized.

Step 4. Chromatography on DEAE-Cellulose. Lyophilized material from a single step 3 preparation is reconstituted in 5 ml of deionized water and dialyzed extensively for 24 hr at 4° against 0.02 M sodium chloride in 0.1 M Tris-HCl (pH 8.32). The dialyzed fraction is centrifuged at 34,000 g for 1 hr. Approximately 2–3 ml (35–70 mg of protein) are applied to a DEAE-cellulose column (1 $\times$ 35 cm, Selectacel, 1.17 meq per gram of exchange capacity, Carl Schleicher and Schuell Company, Keene, New Hampshire) previously equilibrated with 0.02 M sodium chloride in 0.1 M Tris-HCl (pH 8.32) at 4°. A linear salt gradient is employed for this fractionation with 150 ml of 0.1 M sodium chloride in 0.1 M Tris-HCl (pH 8.32) in the reservoir and 150 ml of 0.02 M sodium chloride in 0.1 M Tris-HCl (pH 8.32) in the mixing chamber. The flow rate is maintained at 7–10 ml/hr. The peak of inhibitor activity elutes at an added ionic strength of approximately 0.04 M sodium chloride. The chromatographic patterns of inhibitor activity vary considerably and may exhibit one to three distinct peaks depending upon the step 3 preparations utilized as well as the amount of protein fractionated. However, each of these peaks of activity appears to be of equivalent electrophoretic homogeneity and factor X_a inhibitory specific activity. Thus, all fractions containing inhibitor activity are pooled. The recovery of antithrombin–heparin cofactor, as judged by factor X_a inhibitory activity determinations, averages 50% for step 4. The specific activity of this preparation ranges from 7.8 to 15.6 factor X_a inhibitory units per milligram of protein. Prior to the initiation of step 5, the step 4 product is concentrated by dialysis at 4° against 40% poly-N-vinylpyrrolidone dissolved in water and adjusted to pH 7.0 with 0.01 M NaOH.

Step 5. Rechromatography on Sephadex G-200. An entire step 4 preparation concentrated to approximately 3 ml is applied to a Sephadex G-200 column (2.5 $\times$ 100 cm), previously equilibrated with 0.1 M sodium chloride in 0.05 M Tris-HCl (pH 8.32) at 4°. The flow rate is maintained at approximately 9 ml/hr. The bulk of inhibitor activity is eluted with a distribution coefficient equivalent to that of serum albumin. The recovery of antithrombin–heparin cofactor, as judged by factor X_a inhibitory activity, averages 57% for step 5. The specific activity of these preparations range from 7.8 to 14.4 factor X_a inhibitory units per milligram of protein. This final step in the purification procedure is necessary to remove minor contaminants present in the step 4 product. The overall yield for the entire five-step preparative technique averages 20%.

Affinity Chromatographic Procedures

Heparin-Sepharose has recently been utilized in the purification of antithrombin–heparin cofactor.[28-31] A variety of methods have been employed to prepare this affinity matrix. In all these instances, cyanogen bromide activation of Sepharose beads has been used to bind the acidic mucopolysaccharide to this matrix. However, various concentrations and differing types of heparin preparations have been employed in forming the affinity resin. We have utilized heparin preparations with covalently linked peptide as originally described by Iverius.[32] The acidic mucopolysaccharide is bound to the Sepharose via this peptide spacer and is, therefore, at some distance from the surface of the matrix. This may facilitate specific interactions of antithrombin–heparin cofactor with the affinity ligand. Preparation of this chromatographic matrix has been carried on for some time in our laboratory, and the appropriate methodology is presented below. Other investigators have utilized heparin preparations without covalently attached peptide to produce this affinity matrix.[28,30] Higher concentrations of these acidic mucopolysaccharides are required to bind reasonable quantities of heparin to the Sepharose beads. However, these chromatographic matrices have also been reported to be useful in the purification of antithrombin–heparin cofactor.[28,30]

Preparation of Heparin–Sepharose.[32] Heparin utilized in preparing the affinity matrix was obtained in crude form (stage 14) from the Wilson Chemical Corporation, Chicago, Illinois. It was further purified by precipitation from 1.2 M sodium chloride with cetylpyridium chloride according to the method described by Lindahl *et al.*[33] These preparations average 175 USP units/mg.

Sepharose 4B is activated at 4° with cyanogen bromide.[32] This is accomplished by adding 2.25 g of this reagent per 75 ml of "settled" gel and maintaining the pH of the suspension between 11.0 and 11.5 for 5 min with NaOH. The Sepharose matrix is then extensively washed by

[28] M. Miller-Andersson, H. Borg, and L. O. Anderson, *Thromb. Res.* **5**, 439 (1974).
[29] P. S. Damus and G. A. Wallace, *Biochem. Biophys. Res. Commun.* **61**, 1147 (1974).
[30] E. T. Yin, L. Isenkramer, and J. V. Butler, *in* "Heparin: Structure, Function, and Clinical Implications" (R. E. Bradshaw and S. Wessler, eds.), p. 239. Plenum, New York, 1975.
[31] R. D. Rosenberg, unpublished observations.
[32] P.-H. Iverius, *Biochemistry* **124**, 677 (1971).
[33] U. Lindahl, L. A. Cifonelli, B. Lindahl, and L. Rodén, *J. Biol. Chem.* **240**, 2817 (1965).

suction on a sintered-glass funnel with two hundred times its volume of ice-cold distilled water and suspended in twice its volume of 0.1 M NaHCO$_3$ (pH 8.5). Purified heparin is added to this matrix at a concentration of 2.67 mg of acidic mucopolysaccharide per milliliter of "settled" gel, and the suspension is stirred for 16 hr at 4°. Ethanolamine (1.2 ml per 100 ml of gel) is then added and stirring continued for 2 hr when the gel is to be used to purify canine heparin cofactor. The gel matrix is then extensively washed by suction on a sintered-glass funnel with 200 times its volume of 1 M glycine (pH 8.5), distilled water, 1.0 M sodium chloride in 0.01 M Tris-HCl (pH 7.5), and 0.15 M sodium chloride in 0.01 M Tris-HCl (pH 7.5). The heparin–Sepharose can be reutilized 10 or more times in the purification of antithrombin–heparin cofactor, if it is immediately reequilibrated with 0.15 M sodium chloride in 0.01 M Tris-HCl (pH 7.5) and stored at 4° in 0.1% sodium azide.

Chromatographic Procedures. Approximately 300–500 ml of heparin–Sepharose is added to plasma anticoagulated with ACD or 0.38% sodium citrate, and the gel suspension is gently stirred for 30 min at 24°. If the affinity matrix has been prepared as described above, it will adsorb all the antithrombin–heparin cofactor present in 5–10 times its volume of plasma. If the chromatographic resin has been formed by other techniques, it may only bind one-third to one-fifth as much inhibitor. The gel matrix is gently washed by adding two-thirds its volume of 0.15 M sodium chloride in 0.01 M Tris-HCl (pH 7.5), centrifuging the suspension at 1000 g for several minutes, and discarding the supernatant.

Two different approaches can be utilized to differentially elute the inhibitor and free it of contaminating proteins. On the one hand, the gel suspension can be placed in a glass column (5 × 15 cm) and washed at a flow rate of 7–10 ml/min with 0.4 M sodium chloride in 0.01 M Tris-HCl (pH 7.5) at 4°. The inhibitor can then be eluted at the same flow rate with 2.0 M sodium chloride in 0.01 M Tris-HCl (pH 7.5) at 4°. On the other hand, the gel suspension can be placed on a glass column of dimensions similar to that described above, and the chromatographic matrix can be slowly washed with a linear ionic strength gradient. In the case of the human and rabbit inhibitor, the linear ionic strength gradient ranges from 0.15 M sodium chloride in 0.01 M Tris-HCl (pH 7.5) to 1.5 M sodium chloride in 0.01 M Tris-HCl (pH 7.5) with the total volume of the gradient approximately equivalent to that of the chromatographic matrix. Both inhibitors elute an ionic strength of approximately 1.0. In the case of canine inhibitor, the linear ionic strength gradient ranges from 0.3 M sodium chloride in 0.01 M Tris-HCl (pH 7.5) to 1.2 M sodium chloride in 0.01 M Tris-HCl (pH 7.5) with the total volume of the gradient equivalent to 4–6 times that of the chromatographic matrix. The canine inhibitor elutes at approximately 0.7 to 0.9 ionic strength.

At the present time, canine antithrombin–heparin cofactor can be obtained in homogeneous form by elution from heparin–Sepharose with linear ionic strength gradients.[29] The overall yield for this one-step isolation technique is 95%.

Unfortunately, neither human nor rabbit antithrombin–heparin cofactor can be prepared in pure form by affinity methodology. Batch elution of either inhibitor from heparin–Sepharose yields products with specific activities of ~500 units/mg. A more time-consuming but selective fractionation of these proteins with ionic strength gradients can result in preparations whose specific activities are 50% higher than those attainable by the more rapid batch elution procedure. However, rabbit or human inhibitor isolated with either type of elution schedule requires fractionation by two additional classical chromatographic techniques, if physically homogeneous final products are desired. In most instances, investigators have utilized DEAE-Sephadex or DEAE-cellulose ion-exchange chromatography at pH 8.0–8.3 and Sephadex G-200 gel filtration under conditions similar or identical to those previously outlined in the *Classical Purifications* section of this communication. With respect to the purification of human antithrombin–heparin cofactor, we have observed that rapid batch adsorption-elution of plasma from heparin–Sepharose, followed by DEAE-cellulose chromatography (step IV) and Sephadex G-200 gel filtration, is an efficient means for obtaining electrophoretically homogeneous inhibitor.[31] The overall yield for isolating human and rabbit antithrombin–heparin cofactor by several different three-step affinity and classical isolation techniques is approximately 30%.[31,30]

Properties

Stability. Canine, rabbit, and human antithrombin–heparin cofactor are relatively stable between pH 7.0 and 11.5, but lose considerable activity in acidic solutions. The human inhibitor is somewhat unstable at ionic strengths below 0.15. This does not appear to be true of the rabbit and canine protein. All forms of antithrombin–heparin cofactor are stable below 60°, but each loses considerable potency above this temperature.

Purity. Antithrombin–heparin cofactor has been isolated as a single component from dog, rabbit, and human plasma. The criteria utilized to demonstrate homogeneity include disc gel electrophoresis, isoelectric focusing in sucrose density gradients, sodium dodecyl sulfate (SDS) gel electrophoresis in the presence and in the absence of reducing agents, and immunoelectrophoresis.[7,8,23,29–31]

Physical Properties. The molecular weight of human and rabbit antithrombin–heparin cofactor is between 62,000 and 67,000 by SDS gel electrophoresis, gel filtration on Sephadex G-200, and analytic ultracentrifugation.[8,17,28,29] Furthermore, both inhibitors exist as single polypeptide chains.[8] Canine antithrombin–heparin cofactor has a somewhat higher molecular weight of 77,000–80,000 as determined by SDS gel electrophoresis and gel filtration on Sephadex G-200.[29] The human plasma protease inhibitor has been found to have an isoelectric point of 5.11 as measured by isofocusing electrophoresis in sucrose density gradients.[8]

Specificity and Reactive Sites. When purified human antithrombin–heparin cofactor is incubated with purified human thrombin, a slow progressive neutralization of thrombin's activity is observable by coagulation or esterolytic assay.[8] When the inhibitor is preincubated with heparin, prior to the admixture of thrombin, an instantaneous neutralization of the enzyme's activity occurs.[8] Using results obtained with both the esterolytic assay as well as the fibrin end-point assay, it is possible to demonstrate that the equivalence point of the reaction occurs at a thrombin to antithrombin–heparin cofactor molar ratio of approximately 1:1. Furthermore, the admixtures of thrombin and inhibitor results in a new molecular species that cannot be disrupted by any combination of denaturing and reducing agents. Therefore, the inactivation process can be monitored by SDS gel electrophoresis. This technique demonstrates that interaction of antithrombin–heparin cofactor (MW $\sim$63,000) and thrombin (MW $\sim$34,000) results in the gradual formation of the thrombin–antithrombin complex (MW $\sim$89,000). In the presence of heparin, the formation of complex is instantaneous.[8] The molecular weight of the complex when compared to those of the enzyme and inhibitor demonstrates that the stoichiometry of the reaction is 1:1. The stoichiometry is identical in the presence and in the absence of heparin.[8]

The activity of thrombin is neutralized by complex formation with the inhibitor via an active center (serine)-reactive site (arginine) interaction. Heparin binds to lysyl residues of antithrombin and accelerates the inactivation. This dramatic increase in the rate of complex formation appears to be due to a heparin-induced conformational alteration of the inhibitor, which renders the reactive site arginine residue more accessible to the active serine center of thrombin.[8]

Previous studies had shown that factor X_a was inhibited by antithrombin–heparin cofactor and that heparin dramatically accelerated the neutralization process.[23,34,35] Based upon this finding and a knowledge of the mechanism of action of this inhibitor, it was predicted that all serine

[34] W. H. Seegers, E. R. Cole, C. R. Harmison, and F. C. Monkhouse, *Can. J. Biochem.* **42**, 359 (1964).
[35] R. Biggs, K. W. E. Denson, N. Akman, R. Borrett, and M. Hadden, *Br. J. Hematol.* **19**, 283 (1970).

proteases produced within the coagulation–fibrinolytic mechanism would be inactivated by antithrombin–heparin cofactor and heparin would accelerate each of these interactions.[8,9] It was subsequently demonstrated that partially purified human factor XI_a is inactivated by antithrombin–heparin cofactor in a slow, progressive fashion when heparin is absent, but neutralized instantaneously when heparin is present.[9] Human factor IX_a and human factor XII_a are also inhibited by antithrombin–heparin cofactor,[11,12] and the rate of their neutralization is accelerated by heparin. In each of these cases, undissociable enzyme–inhibitor complexes are observable, and their molecular weights indicate that the molar stoichiometry of the reactions are $1:1$.[9,11,12] Although the data have not yet been published, we have observed similar phenomena for human factor X_a–antithrombin. Thus, it is now established that all the known serine proteases of the intrinsic coagulation cascade (factors IX_a, X_a, XI_a, XII_a, and thrombin) are neutralized by antithrombin–heparin cofactor via complex formation with the inhibitor and that heparin accelerates the rate of each of these interactions.[8,9,11,12]

In the allied fibrinolytic system, it has been recently demonstrated that plasmin is neutralized by antithrombin–heparin cofactor and that heparin accelerates this neutralization in a fashion similar to that observed for the serine proteases of the intrinsic coagulation cascade.[10] As expected, a $1:1$ stoichiometric complex of plasmin and inhibitor is apparen in the presence and in the absence of heparin.[10] It has been assumed that α_2-macroglobulin is the major plasma inhibitor of the fibrinolytic mechanism. However, recent experiments have indicated that heparin-activated antithrombin will complex with all the available plasmin, even if a large excess of α_2-macroglobulin is present. Thus, heparin can act via antithrombin–heparin cofactor to inhibit the fibrinolytic mechanism as well as the intrinsic coagulation cascade.

As a logical extension of this pattern of inhibitor specificity, the ability of antithrombin–heparin cofactor to inactivate serine proteases of the complement- and kinin-generating systems has also been examined. It has been shown that this inhibitor does not neutralize the esterolytic activity of C^1_s in the presence or in the absence of heparin. However, the mucopolysaccharide itself can prevent the proteolysis of subsequent components of the complement system by C^1_s.[36] With respect to the kinin-generating system, it has been demonstrated that antithrombin–heparin cofactor in the absence of heparin slowly neutralizes the esterolytic and proteolytic action of the serine protease human kallikrein.[37] In the pre-

[36] H. Struck, H. Colton, and F. Rosen, unpublished observations.
[37] B. Lahiri, R. D. Rosenberg, A. Bagdasarian, B. Mitchell, R. C. Talmo, and R. W. Colman, *Fed. Proc., Fed. Am. Soc. Exp. Biol.* 33, 642 (1974).

sence of this acidic mucopolysaccharide, a modest 1.5- to 2-fold acceleration of the inhibitory process occurs. This is a unique example of an
interaction in which a soluble serine protease is neutralized by antithrombin–heparin cofactor, but where heparin does not dramatically accelerate
complex formation. It would appear that additional specificity is required
on the part of a serine protease, if heparin is to catalyze its instantaneous
inactivation by antithrombin–heparin cofactor.

Antithrombin–heparin cofactor may also be an important modulator
of other physiological systems. For example, this plasma component
is capable of greatly increasing the potency of macrophage inhibitory
factor (MIF) by neutralizing the activity of a unique serine protease
bound to the macrophage surface. However, if heparin is added to this
system, the inactivation of the enzyme by antithrombin–heparin cofactor
is prevented. This phenomenon appears to be due to the electrostatic
repulsion of the highly negative heparin–antithrombin complex from the
similarly charged macrophage surface.[38] These types of interactions
may explain the known effects of heparin upon delayed hypersensitivity
reactions. Antithrombin–heparin cofactor is also known to be an inhibitor of trypsin and chymotrypsin, although the physiological significance
of these neutralizations remains unclear.[7]

[38] H. G. Remold and R. D. Rosenberg, *J. Biol. Chem.* in press (1975).

[54] Hirudin

By Dániel Bagdy, Éva Barabas, László Gráf, Torben Ellebaek
Petersen, and Staffan Magnusson

Chemically pure hirudin was first prepared by Markwardt and co-
workers, who also characterized its reaction with thrombin.[1] Data were
also given on the antithrombin effect of hirudin *in vivo*, including its
resorption, elimination, and toxicity.[2,3] Chemical composition and certain physical constants of hirudin were measured by de la Llosa *et al.*[4]
and Tertrin *et al.*[5]

[1] F. Markwardt, this series Vol. 19, p. 924.
[2] F. Markwardt and H. Landmann, *in* "Handbuch der Experimentelle Pharma-
kologie," Vol. 27, pp. 105–115. Springer-Verlag, Berlin and New York, 1971.
[3] P. G. Barton and E. Thye Yin, *in* "Metabolic Inhibitors," Vol. 4, (R. M. Hoch-
ster, M. Kates, and J. H. Quastel, eds.), pp. 246–248. Academic Press, New York,
1973.
[4] P. de la Llosa, C. Tertrin, and M. Jutisz, *Bull. Soc. Chim. Biol.* **45**, 69 (1963).
[5] C. Tertrin, P. de la Llosa, and M. Jutisz, *Biochim. Biophys. Acta* **124**, 380 (1966).

Preparation of Hirudin

The preparation starts from whole leeches, *Hirudo medicinalis*, rather than from the head part of the animals as was the case in the early work.[6,7] Leeches 30–60 mm weighing 1–2 g were starved for 2–3 weeks while kept at 4°–10°. The starved animals were then frozen and stored at −10°. It must be emphasized that starvation is an essential condition for obtaining a high yield of crude hirudin.

Extraction of Whole Leeches with Water-Miscible Organic Solvents.[8] Two kilograms of starved leeches are thawed and minced at 0°, and 6 liters of 80% (v/v) of cold aqueous acetone are added with stirring at 0°–5° for 10 min. Hirudin is extracted by the addition of sodium chloride to a final concentration between 0.2 and 0.5 M, and by trichloroacetic acid (TCA) to a final concentration of 0.1–0.4 M with constant stirring for 30–60 min, with a pH range of 2.5–3.5. The insoluble residue is removed by decanting and filtering, and the extraction step is repeated with 2 liters of the 80% cold aqueous acetone containing NaCl and TCA. Hirudin is precipitated from the combined extracts by adding 2 volumes of cold acetone at −10°. The precipitate is centrifuged, washed with cold acetone, and then taken up in acetate buffer solution of $\mu = 0.1$, pH 5–6, at 0–5°. After separating and discarding the insoluble residue, crude hirudin is precipitated with cold ethanol or acetone at −5°, washed, and dried. The yield amounts to about 80,000–140,000 National Institute of Health (NIH) Anti-thrombin Units (AT-U) per kilogram of leeches, and the specific activity of the preparation is 200–400 NIH AT-U per milligram of protein.

Extraction of Whole Leeches in Aqueous Medium and Preparation of Crude Hirudin.[9] Waldschmitz-Leitz et al.[10] found that low yields of hirudin were obtained by purely aqueous extractions at acidic or alkaline pHs. Thus until 1973 purification methods focused on the use of water-miscible organic solvents. By reinvestigating the problem, it was found that hirudin can be extracted efficiently from whole leeches in aqueous media under appropriate conditions,[9] so that the yield of hirudin in the aqueous extracts equals that obtained with water-miscible organic solvents. The aqueous extracts, however, contain much less pigment.

Ground leeches (5 kg) are stirred with 15 liters of 0.5 M NaCl solution for 30–60 min at room temperature. After acidification with 3 N HCl

[6] F. Markwardt, *Naturwissenschaften* **42**, 537 (1955).
[7] F. Markwardt, *Hoppe-Seyler's Z. Physiol. Chem.* **308**, 147 (1957).
[8] D. Bagdy and S. Török, Hung. Patent No. 150, 600 (1962).
[9] D. Bagdy, I. Bihari, and S. Török, Hung. Patent No. 150, 833 (1962).
[10] F. Waldschmitz-Leitz, P. Stadler, and F. Steigerwaldt, *Hoppe-Seyler's Z. Physiol. Chem.* **183**, 39 (1929).

to pH 2.5–3.0, the temperature is raised and kept at 70° for 15 min, with constant stirring. The mixture is cooled to room temperature, and the insoluble residue is removed by centrifugation at 2500–3000 g for 30 min. The supernatant is adjusted to pH 6.5–7.0, and the precipitating material is separated and discarded. By addition of 1.5 volume of ethanol to 1 volume of extract, some further purities are precipitated and removed by centrifugation. The supernatant is then concentrated to 1.5–3.0 liters *in vacuo* at 30°–40°.

A further purification is carried out by fractionation with acetone. To the concentrate, an equal volume of acetone is added, and the precipitate is separated and discarded. Hirudin is then precipitated from the supernatant by addition of 3–4 volumes of cold acetone at 0°–5°. The precipitate is dissolved in 0.5–0.8 liter of cold 0.1 M TCA, the insoluble residue is removed by centrifugation, and the solution is diluted with 4–5 volumes of cold water. Then 2 g of activated bentonite is added per 100 ml of solution with constant stirring. Adsorption of hirudin to bentonite at room temperature is complete within 1 hr.

After repeatedly washing the adsorbent with water, elution of hirudin is accomplished in two steps, using for each, 1 liter of 45% (v/v) acetone at pH 8.0–8.5, room temperature. The combined eluates are concentrated *in vacuo* at 20°–30° to about 0.5 liter and adjusted to pH 4.5–5.0. The precipitate is discarded, and the supernatant containing hirudin is filtered and lyophilized or is precipitated with 4 volumes of cold acetone and dried with acetone and ether.

The average yield is 0.2 g of crude hirudin per kilogram of leeches, and the specific activity varies from 200 to 700 NIH AT-U per milligram of protein. A flow chart is given in Fig. 1.

Crude hirudin preparations obtained by either extraction process still contain a considerable amount of protease inhibitors[11] and inert proteins. Hirudin may be purified as follows.

Chromatography of Crude Hirudin on ECTEOLA Cellulose.[12] Crude hirudin can be completely adsorbed to anionic cellulose derivatives, especially ECTEOLA cellulose in a mild acidic medium at low ionic strength. Elution may be carried out by increasing the ionic strength stepwise or gradually. Stepwise elution results in a 3- to 5-fold increase in specific activity. Gradient elution, however, is more effective, with an 8- to 10-fold increase in purity. Typical results are shown in Fig. 2.

Crude hirudin, 400 mg, 260 AT-U/mg, is chromatographed on a 1.5 ×

[11] H. Fritz, K.-H. Oppitz, M. Gebhardt, I. Oppitz, and E. Werle, *Hoppe-Seyler's Z. Physiol. Chem.* **350**, 91 (1969).
[12] D. Bagdy, É. Barabás, G. Jécsay, and E. F. Kazi, Hung. Patent No. 152, 836 (1965).

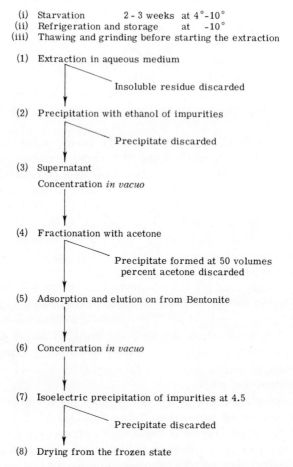

    (i)   Starvation   2 - 3 weeks at 4°-10°
    (ii) Refrigeration and storage   at   -10°
   (iii) Thawing and grinding before starting the extraction

    (1) Extraction in aqueous medium

                    Insoluble residue discarded

    (2) Precipitation with ethanol of impurities

                    Precipitate discarded

    (3) Supernatant
         Concentration *in vacuo*

    (4) Fractionation with acetone

                    Precipitate formed at 50 volumes
                      percent acetone discarded

    (5) Adsorption and elution on from Bentonite

    (6) Concentration *in vacuo*

    (7) Isoelectric precipitation of impurities at 4.5

                    Precipitate discarded

    (8) Drying from the frozen state

Fig. 1. Flow chart of the large-scale preparation of crude hirudin from whole leeches (*Hirudo medicinalis*). Yield: 80,000–140,000 NIH antithrombin units (AT-U) per kilogram of leeches. Specific activity: 200–700 AT-U per milligram of protein.

23 cm column of ECTEOLA cellulose equilibrated with 0.02 M acetate buffer, pH 5.0. Elution is carried out with a gradient of 0 to 0.5 M NaCl at a 30 ml/hr flow rate at room temperature; 5-ml fractions are collected. Antithrombin activity is located sharply in the $\mu = 0.19$–0.22 region. Fractions 23–31, containing hirudin, are concentrated *in vacuo* at 30° and desalted by dialysis against water. Hirudin is precipitated from the eluate with 4 volumes of cold acetone at —5°, with a yield of 44 mg and a specific activity of 2000 AT-U/mg. The yield in relation to the starting material is 85–90%. In spite of the definite increase in specific

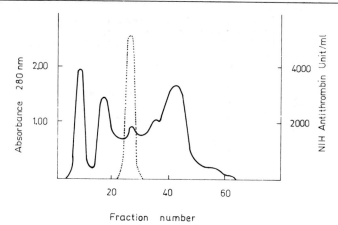

Fraction number

FIG. 2. Typical results of chromatography of crude hirudin on ECTEOLA cellulose.

activity, the sample still contains a significant amount of brownish yellow pigment.

Chromatography of Partially Purified Hirudin on Sephadex-CM C-25.[13] After ECTEOLA cellulose chromatography, hirudin was subjected to chromatography on Sephadex CM C-25. Optimal purification can be obtained at pH 5.0, $\mu = 0.02$. The contaminating pigments remain on the column, and practically colorless antithrombin fractions can be eluted by increasing the ionic strength.

Hirudin, 260 mg, 2000 AT-U/mg, dissolved in 5 ml of acetate buffer ($\mu = 0.02$, pH 5.0) is applied to a 0.8×25 cm Sephadex CM C 25 column previously equilibrated with 0.02 M acetate buffer of pH 5.0. Elution is carried out with a gradient of 0 to 0.5 M NaCl at a flow rate of 20 ml/hr at room temperature; 5-ml fractions are collected. Hirudin appears in fractions 19–23. These pooled fractions are concentrated by dialysis against water first, and then precipitated with 4 volumes of cold acetone ($-5°$) and dried. The yield is 66 mg of protein and a specific antithrombin activity of 5800 AT-U/mg (Fig. 3).

Gel Filtration of Hirudin on Sephadex G-75.[13] After the ECTEOLA cellulose and Sephadex CM C-25 steps, hirudin is applied to Sephadex G-75. Hirudin 42 mg, 5800 AT-U/mg, is dissolved in 1 ml of 0.1 M NaCl and placed on a 1.2×52 cm Sephadex G-75 column previously equilibrated with 0.1 M NaCl, at flow rate 15 ml/hr, and fraction volume 5 ml. The active fractions are combined, concentrated *in vacuo*, desalted by dialysis against water, and lyophilized. The gel filtration profile

[13] D. Bagdy, É. Barabás, and L. Gráf, *Thromb. Res.* **2**, 229 (1973).

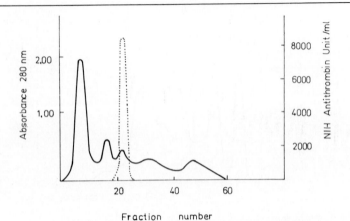

FIG. 3. Chromatography of partially purified hirudin on Sephadex-CM C-25.

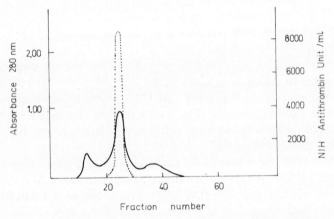

FIG. 4. Gel filtration profile of hirudin on Sephadex G-75.

shows three components (see Fig. 4), of which only the second contains antithrombin activity. Rechromatography on Sephadex G-75 results in a single peak containing pure hirudin. This final product has a specific activity of 8000–10,000 AT-U per milligram of protein.

Although the sequence of steps of isolation and purification may be varied, the order presented above proved to be the most advantageous and could be used on an industrial production scale.

Properties and Purity

Criteria of Purity. The single *N-terminal amino acid residue* of hirudin was recently characterized as valine[14] instead of the previously

[14] L. Gráf, A. Patthy, É. Barabás, and D. Bagdy, *Biochim. Biophys. Acta* **310,** 416 (1973).

reported isoleucine. In addition, the dansyl method after Edman degradation indicated valine as the second amino acid residue. Thus, the N-terminal sequence of hirudin appears to be Val-Val-.

The amino acid sequence from the C-terminal end was reported to be: -Ala-Gly-Ser-Gln-Leu.[15] Assuming that the molecular weight of hirudin is 12,200, the amino acid composition (number or residues per mole) is as follows[13]: 14 Asp, 6 Thr, 6 Ser, 16 Glu, 6 Pro, 13 Gly, 3 Ala, 6 ½Cys, 7 Val, 3 Ile, 5 Leu, 3 Tyr, 2 Phe, 5 Lys, 2 His. This composition appears to be close to the data of de la Llosa *et al.*[4] except for two differences. One pertains to methionine, which is absent from our preparation and also from the preparations of Markwardt and Walsmann.[16] The second difference is that, compared to the analysis of de la Llosa *et al.*,[4] alanine is present in higher concentration in our preparations.

Amino Acid Sequence. At the Department of Molecular Biology, University of Aarhus, Denmark, two batches of hirudin with specific activities of 6700 per 54 mg and 4000 AT-U/mg per 66 mg, respectively, have been used to investigate the amino acid sequence of hirudin. The following preliminary data have been obtained:

```
  1   2   3   4   5   6   7   8   9  10  11  12  13  14  15  16  17  18  19
Val-Val-Tyr-Thr-Asp-Cys-Thr-Glu-Ser-Gly-Gln-Asn-Leu-Cys-Leu-Cys-Glu-Gly-Ser-

 20  21  22  23  24  25  26  27  28  29  30  31  32  33  34  35  36  37  38
Asn-Val-Cys-Gly-Gln-Gly-Asn-Lys-Cys-Ile-Leu-Gly-Ser-Asp-Gly-Glu-Lys-Asn-Gln-

 39  40  41  42  43  44  45  46  47  48  49  50  51  52  53  54  55  56
Cys-Val-Thr-Gly-Glu-Gly-Thr-Pro-Lys-Pro-Gln-(Ser,His,Asx,Asx,Asx,Gly)-Phe-

                                 57  58  59  60  61  62      63      64  65
                               Glu-Glu-Ile-Pro-Glu-Glu-TyrSO₃⁻-Leu-Gln
```

There is a possibility that the Asp-Gly- sequence 33-34 has arisen from an Asn-Gly sequence during purification of hirudin by heat treatment at low pH. This may also explain why the region 50-55 was obtained in low yield. The position of the disulfide bridges has not yet been determined. The N-terminal sequence Val-Val- confirms earlier results,[14] and our C-terminal sequence of -Leu-Gln would not be in conflict with the experimental data of de la Llosa *et al.*,[15] which, however, they interpreted as representing -Gln-Leu. We have found no evidence for carbohydrate in hirudin. Tyr-63 is substituted with an *O*-sulfate ester group and free tyrosine-*O*-sulfate was obtained from the peptic peptide Tyr SO_3^--Leu-Gln after degradation with carboxypeptidase A. The function of the Tyr SO_3^- residue is not known.

An interesting similarity between hirudin and prothrombin, with

[15] P. de la Llosa, C. Tertrin, and M. Jutisz, *Biochim. Biophys. Acta* **93**, 40 (1964).
[16] F. Markwardt and P. Walsmann, *Hoppe-Seyler's Z. Physiol. Chem.* **348**, 1381 (1967).

TABLE I
REDUCTION AND ALKYLATION OF HIRUDIN

DTT/disulfide bond[a] (molar ratio)	Urea (moles)	Alkylating agent	CM-Cys[b] (residues/ 7000 g)	[14C]CM[c] (residues/ 7000 g)	Anti-thrombin effect (%)
5	0	ICH$_2$CONH$_2$	1.8	—	55
5	0	ICH$_2$[14]COOH	1.9	1.8	30
20	0	ICH$_2$[14]COOH	3.4	3.6	20
20	8	ICH$_2$CONH$_2$	6.2	—	10
20	8	ICH$_2$[14]COOH	6.3	6.6	3

[a] Dithiothreitol (DTT)/protein disulfide bond.
[b] S-Carboxymethylcysteine content per mole of hirudin as determined in amino acid analyzer and calculated on the basis of 13 glutamic acid residues, as found in the sequence, instead of 16 as given by D. Bagdy, É. Barabás, and L. Gráf, *Thromb. Res.* **2**, 229 (1973).
[c] 14C label covalently bound to the protein.

possible relevance to the specific binding of thrombin, is that hirudin contains the sequence of -Val-Thr-Gly-Glu-Gly-Thr-Pro-Lys-Pro- in residues 40–48 and that prothrombin has a very similar sequence, -Val-Thr-Val-Glu-Val-Ile-Pro-Arg-Ser-, in residues 149–157.[17,18] This similarity is particularly interesting because the only peptide bond in prothrombin attacked by thrombin is the Arg-Ser bond between residues 156 and 157.

Structure–Activity Relationships. Some preliminary data on the effects of side-chain modifications of hirudin have been published by Tertrin et al.[19] In order to further explore structure–function relationships in hirudin, the biological importance of the disulfide bonds was investigated. Performic acid oxidation of hirudin resulted in complete loss of biological activity. Table I summarizes studies on the reduction and alkylation of the disulfide bonds in hirudin. It may be seen that even partial reduction and alkylation, affecting only two disulfides in hirudin, carried out with 5–20 m excess of dithiothreitol over protein disulfide content in the absence of denaturants, was accompanied by a

[17] S. Magnusson, L. Sottrup-Jensen, T. E. Petersen, and H. Claeys, *in* "Prothrombin and Related Coagulation Factors" (H. C. Hemker and J. Veltkamp, eds.), pp. 22–46. Universitaire Press, Leiden, Holland, 1975.
[18] S. Magnusson, T. E. Petersen, L. Sottrup-Jensen, and H. Claeys, *in* "Proteases and Biological Control" (E. Reich et al., eds.), Chapter 9. Cold Spring Harbor Laboratory, Cold Spring Harbor, New York (1975).
[19] C. Tertrin, P. de la Llosa, and M. Jutisz, *Bull. Soc. Chim. Biol.* **49**, 1837 (1967).

TABLE II

PHYSICAL CONSTANTS OF HIRUDIN

Constant	Value	References[a]
Isoelectric point (pH)	4.0	5, 6
	3.8	7
	3.9	2, 15
Molecular weight	16,000	7
	13,000	4
	15,000	5
	9,060	20
	10,800	15
	12,200	13
Sedimentation constant, $s_{22,w}^{\circ}$ (S)	0.98	1, 2, 20
	1.37	5
Diffusion constant (cm^2 sec^{-1})	10.8×10^{-7}	1, 2, 20
	11.5×10^{-7}	5
Partial specific volume (V)	0.741	20
	0.71	5

[a] Numbers refer to text footnotes.

considerable decrease in biological activity. More disulfide bonds could be cleaved in the presence of 8 M urea leading to the complete reduction and alkylation of the disulfide bonds in hirudin, and also to an additional decrease in the biological activity. The antithrombin potency of hirudin, however, was not completely abolished even after such extensive modifications. Thus the possibility cannot be excluded that smaller fragments of hirudin might eventually be isolated with antithrombin activity.[16] It may also be noted from Table I that the reduced and carbamidomethylated derivatives of hirudin appear to be more active than the corresponding reduced and carboxymethylated ones.

Physical Constants of Hirudin. Certain physical data for hirudin are summarized in Table II.[1,2,4-7,13,15,20] As can be seen, there is great variation in apparent molecular weights as obtained by different methods. The preliminary sequence data suggest, however, a molecular weight of less than 10,000.

Assay Method

The assay is that of Markwardt[1]; it is based on the specific and stoichiometric reaction between hirudin and thrombin. Only a few gen-

[20] H. Triebel and P. Walsmann, *Biochim. Biophys. Acta* **120**, 137 (1966).

eral remarks will be added. For consistent results, it is desirable to use not only a standardized thrombin preparation, but also a standardized bovine fibrinogen that is free of prothrombin and plasminogen. Commercially available fibrinogens generally contain a considerable quantity of impurities and are rather unstable. A standard bovine fibrinogen free of prothrombin and plasminogen can be prepared according to Barabás *et al.*[21] The stock solution containing 3% (w/v) fibrinogen in 0.3 M saline, stored in suitable aliquots at $-10°$, is stable for at least 6 months. After thawing it is completely soluble and can be diluted and buffered as prescribed by Markwardt.[1]

Special attention must be paid to the thrombin preparation used in hirudin assay. Activity labeling of commercial thrombin preparations is often inaccurate and confusing.

Specificity. In addition to inhibiting thrombin, the hirudin prepared by the method described shows very substantial inactivation of bovine factor IX_a.[22] This observation is of special interest from both theoretical and practical points of view. In the course of purification of crude hirudin prepared from whole leeches, new proteases have been detected and separated. Two of these inhibit trypsin and plasmin,[11] and a third shows a marked elastase-inhibiting effect.[23]

[21] É. Barabás, D. Bagdy, and J. Pintér, *Bibl. Haematol.* (*Basel*) **38**, 498 (1971).
[22] E. W. Davie, personal communication, 1975.
[23] P. Tolnay, D. Bagdy, and É. Barabás, in press.

[55] Structures and Activities of Protease Inhibitors of Microbial Origin

By HAMAO UMEZAWA

The study of enzyme inhibitors produced by microorganisms was initiated by H. Umezawa.[1] Compared with inhibitors obtained from animal and plant tissues, which are mostly of macromolecular nature, the inhibitors found in microbial culture filtrates are relatively small, and protease inhibitors are widely distributed in various species of actinomycetes which also produce a number of proteases. These inhibitors are generally resistant to being metabolized by animals.

Assay methods of testing activities of various proteases can be applied to search for inhibitors in culture filtrates which are heated (100°) for 3 min prior to testing. The assay methods employed at the

[1] H. Umezawa, "Enzyme Inhibitors of Microbial Origin." Univ. of Tokyo Press, Tokyo, 1972.

time of discovery of each inhibitor, special fermentation conditions, and methods for extraction and purification of the inhibitors are described in this chapter.

Leupeptins—Inhibiting Plasmin, Trypsin, Papain, and Cathepsin B

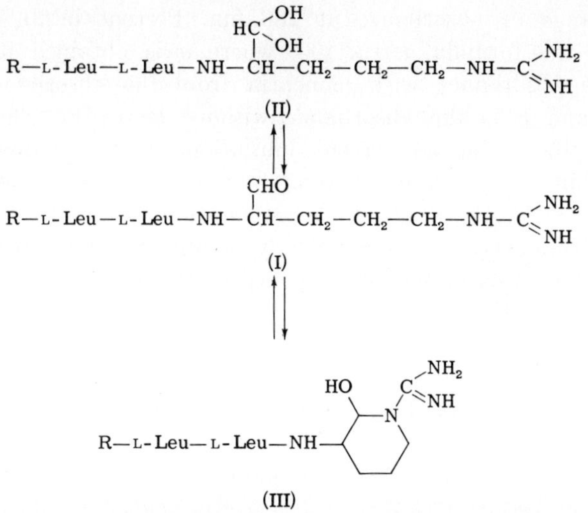

$$R—\text{L-Leu}—\text{L-Leu}—NH—CH—CH_2—CH_2—CH_2—NH—C\underset{NH}{\overset{NH_2}{\diagdown}}$$

with $HC\overset{OH}{\underset{OH}{\diagup}}$ on the CH

(II)

$$R—\text{L-Leu}—\text{L-Leu}—NH—CH—CH_2—CH_2—CH_2—NH—C\underset{NH}{\overset{NH_2}{\diagdown}}$$

with CHO on the CH

(I)

(III)

$$R = CH_3CO \text{ or } CH_3CH_2CO$$

In a search for plasmin-inhibiting activity, leupeptins were found in *Streptomyces roseus, Streptomyces roseochromogenes, Streptomyces chartreusis, Streptomyces albireticuli, Streptomyces thioluteus, Streptomyces lavendulae, Streptomyces noboritoensis,* and some other unclassified species,[2] and the inhibitors were found to have the structures of propionyl or acetyl-L-leucyl-L-leucyl-L-argininal or their analogs in which each leucine is replaced with L-isoleucine or L-valine.[3,4] The di-*n*-butyl acetals of leupeptins are employed in the purification of leupeptins. During extraction, racemization in the assymmetric center of the argininal residue occurs. Only the L-argininal form is active,[5,6] and, in aqueous solution,

[2] T. Aoyagi, S. Miyata, M. Nanbo, F. Kojima, M. Matsuzaki, M. Ishizuka, T. Takeuchi, and H. Umezawa, *J. Antibiot.* **22,** 558 (1969).

[3] S. Kondo, K. Kawamura, J. Iwanaga, M. Hamada, T. Aoyagi, K. Maeda, T. Takeuchi, and H. Umezawa, *Chem. Pharm. Bull.* **17,** 1896 (1969).

[4] K. Kawamura, S. Kondo, K. Maeda, and H. Umezawa, *Chem. Pharm. Bull.* **17,** 1902 (1969).

[5] B. Shimizu, A. Saito, A. Ito, K. Tokawa, K. Maeda, and H. Umezawa, *J. Antibiot.* **25,** 515 (1972).

[6] H. Sakeki, Y. Shimada, N. Kawakita, B. Shimizu, E. Ohki, K. Maeda, and H. Umezawa, *Chem. Pharm. Bull.* **21,** 163 (1973).

leupeptins are present in forms II and III, with only traces of form (I) present.[7]

Assay

Principle. Inhibition of plasmin hydrolysis of fibrinogen is measured by the change of absorbance at 280 nm. Percent inhibition is calculated by the formula $a/b \times 100$, where a is obtained by subtraction of the absorbance with leupeptin from the absorbance without leupeptin, and b is the absorbance without leupeptin. The data are plotted in the probit as percent inhibition on the ordinate and in the logarithm of the value corresponding to the concentration of the inhibitor on the abscissa. The concentration of inhibitor required for 50% inhibition (ID_{50}) is graphically evaluated. Leupeptin activity directed against other proteases are tested by the same principle.

Reagents

Buffer: phosphate buffer saline, 0.05 M at pH 7.2

Substrate: Dissolve 2 g of fibrinogen (Armour Pharm. Co.) with warming (37°) in 100 ml of buffer.

Plasminogen: Dilute 100 ml of human serum in 2 liters of water, and adjust pH to 5.2 with 2 N acetic acid. After 30 min, collect the sediment by centrifugation at 3000 rpm for 10 min and suspend in 100 ml of water. Again collect the sediment by centrifugation and dissolve with 100 ml of buffer. This solution is stable for several weeks at 4°.

Streptokinase (Varidase, Lederle Lab.): Prepare a stock solution containing 10,000 units/ml in water. This solution is stable for several weeks at 4°. Before use, dilute with buffer to contain 200 units per 0.1 ml.

Inhibitor solution: Dilute with water to give about 50% inhibition of the enzyme.

Procedure. Pipette 0.5 ml of plasminogen solution and 0.4 ml of buffer with or without an inhibitor into a series of 15 $\times$ 100 mm test tubes at 37°. Add 0.1 ml of streptokinase and incubate for 5 min. The reaction is initiated by rapidly mixing in 2.0 ml of substrate. Exactly 20 min later, the reaction is stopped by adding 1.5 ml of 1.7 M perchloric acid. After 1 hr at room temperature and centrifugation, absorbance of the supernatant is measured at 280 nm.

---

[7] K. Maeda, K. Kawamura, S. Kondo, T. Aoyagi, T. Takeuchi, and H. Umezawa, *J. Antibiot.* **24**, 402 (1971).

Purification Procedure[2,3]

Step 1. Fermentation and Extraction. *Streptomyces roseus* is grown in a medium containing 1% glucose, 1% starch, 2% peptone (Daigo Eiyo Co.), and 0.5% sodium chloride (pH 7.0) for 71 hr at 27°. Centrifuge the cultured broth (10 liters, $ID_{50} = 0.015$ ml/ml) at 3000 rpm for 10 min. To the supernatant (9 liters, pH 7.3), add carbon (135 g). After stirring for 30 min, separate the carbon by filtration, wash with water (6 liters), and elute with acidic 80% methanol (pH 2, 1500 ml and 1200 ml). Neutralize the active eluate with Amberlite IR-45 (OH form) and concentrate to dryness (22 g of a brownish powder, $ID_{50} = 95$ μg/ml). Dissolve the brownish powder in water (500 ml, pH 7.0) and pour over a column of carbon (110 g). Wash the column with water (2.0 liters) and 0.02 N hydrochloric acid (2.0 liters), and elute with 0.02 N hydrochloric acid in 80% methanol. Collect the active eluate, neutralize with Amberlite IR-45 (OH form) and concentrate to dryness (8.9 g of yellowish powder, $ID_{50} = 53$ μg/ml). Dissolve the yellowish powder (7.5 g) in 50 ml of methanol, pour over a column of alumina (750 g), and develop with methanol. Concentrate to dryness (a crude leupeptin hydrochloride, 2.3 g, $ID_{50} = 12$ μg/ml).

Step 2. Separation of Leupeptin Containing Propionyl from Leupeptin Containing Acetyl as Their Di-n-Butyl Acetals. Reflux the crude leupeptin hydrochloride (19.2 g, $ID_{50} = 15$ μg/ml) in butanol (270 ml) for 2 hr. Wash the butanol solution with water (270 ml) and concentrate to dryness (23.4 g). Dissolve (14.0 g) in butanol–butyl acetate–acetic acid–water (4:8:1:1) (20 ml) and subject to a column chromatography of silicic acid (Mallinckrodt, 900 g) developed with the same solvent. Concentrate the first eluate (150 ml, positive Sakaguchi) to dryness (a white powder, 1.5 g of hydrochloride of di-n-butyl acetal of leupeptin containing propionyl group). Concentrate the second eluate (500 ml, positive Sakaguchi) to dryness (5.3 g, a mixture of hydrochlorides of di-n-butyl acetals of leupeptins containing propional and acetyl). Concentrate the third eluate (1400 ml) to dryness (a white powder, 3.4 g of hydrochloride of di-n-butyl acetal of leupeptin containing acetyl).

Step 3. Isolation of Hydrochloride of Leupeptin Containing Propionyl. Dissolve the di-n-butyl acetal hydrochloride (93 mg) in 0.01 N hydrochloric acid (5.0 ml) and heat at 60° for 3 hr. Neutralize the solution to pH 6.0 with Amberlite IR-45 (OH form) and concentrate to dryness (72 mg).

Isolation of Hydrochloride of Leupeptin Containing Acetyl. Dissolve the di-n-butyl acetal hydrochloride (1.0 g) in 0.01 N hydrochloric acid (50 ml) and subject to the same processes described above to obtain pure hydrochloride of leupeptin containing acetyl (800 mg).

Properties

Physicochemical Properties.[2] Leupeptin is very stable and, as generally used, it is a mixture of the hydrochlorides of propionyl- and acetyl-L-leucyl-L-leucyl-DL-argininal in a 3:1 ratio of propionylleupeptin to acetylleupeptin. Hydrochlorides of leupeptins are soluble in water, methanol, ethanol, butanol, acetic acid, and dimethyl formamide, but are poorly soluble in ethyl acetate, acetone, chloroform, carbon tetrachloride, ethyl ether, or n-hexane. Negative ninhydrin and positive Rydon-Smith, red tetrazolium, 2,4-dinitrophenylhydrazine, Sakaguchi, and diacetyl reactions are obtained. In thin-layer chromatography using silica gel G, propionylleupeptin and acetylleupeptin move with a R_f 0.6–0.7 in n-propanol–water (7:3) and R_f 0.3–0.4 in ethyl acetate–acetic acid–water (60:17:17). In thin-layer chromatography with butanol–butyl acetate–acetic acid–water (4:2:1:1), propionylleupeptin gives two spots with R_f 0.45–0.50 and R_f 0.35–0.45, and acetylleupeptin also shows two spots with R_f 0.35–0.45 and R_f 0.30–0.35. In this thin-layer chromatography procedure, propionylleupeptin di-n-butyl acetal shows a single spot of R_f 0.65 and acetylleupeptin di-n-butyl acetal shows a single spot of R_f 0.60.

Biological Properties and Activity.[2,8] The mixture of propionylleupeptin and acetylleupeptin inhibits a number of enzyme reactions: plasmin hydrolysis of fibrinogen and fibrin; trypsin hydrolysis of casein and of BAA (α-N-benzoyl-L-arginine amide HCl); papain hydrolysis of casein and of BAA; cathepsin B hydrolysis of BAA. Leupeptin delays coagulation of human and rabbit blood for up to 30 min. It does not inhibit coagulation of mouse, rat, and dog blood. It inhibits boar acrosin, all kallikreins prepared from porcine pancreas, submandibular gland, and urine,[9] and cathepsin B.[10] Leupeptin also inhibits the stimulation of ^{3}H-labeled thymidine incorporation by phytohemagglutinin in guinea pig peripheral blood lymphocytes when added prior or together with the stimulant[11]; at 10–50 μg/ml, it inhibits the growth of normal and polyoma-transformed baby hamster kidney cells in culture with a stronger effect against the latter.[12] Oral administration of leupeptin inhibits chemi-

---

[8] T. Aoyagi and H. Umezawa, *Cold Spring Harbor Conf. Cell Proliferation* **2**, 429 (1975).

[9] H. Fritz, B. Fölg-Brey, and H. Umezawa, *Hoppe-Seyler's Z. Physiol. Chem.* **354**, 1304 (1973).

[10] H. Ikezawa, T. Aoyagi, T. Takeuchi, and H. Umezawa, *J. Antibiot.* **24**, 488 (1971).

[11] M. Saito, T. Hagiwara, T. Aoyagi, and Y. Nagai, *Jpn. J. Exp. Med.* **42**, 509 (1972).

[12] A. McIlhinney and B. L. M. Hogan, *Biochem. Biophys. Res. Commun.* **60**, 348 (1974).

cal tumorigenesis in mouse skin[13] and in colon,[14] and vascular metastasis of hepatoma to lung in rats. It also inhibits chemically induced malignant transformations *in vitro*. Given to rabbits orally, the compound is well absorbed, and about 25% is excreted in urine. It inhibits carrageenin edema and leupeptin ointment applied to a burn immediately suppresses pain and blister formation.

Kinetic Properties. Leupeptin shows a competitive inhibition of the hydrolysis of TAME (*N*-*p*-toluenesulfonyl-L-arginine methyl ester) and BAEE (N^{α}-benzoyl-L-arginine methyl ester) by crystalline trypsin.[1] Its dissociation constant (K_i) is 0.34 μM with TAME and 0.13 μM with BAEE. But the inhibitory mechanism of leupeptin is a noncompetitive type for the hydrolysis of BAPA (N^{α}-benzoyl-L-arginine *p*-nitroanilide) with almost the same dissociation constant.[1]

Antipain—Inhibiting Papain, Trypsin, Cathepsins A and B

Antipain,[15] [(*S*)-1-carboxy-2-phenylethyl]carbamoyl-L-arginyl-L-valyl-argininal,[16] is produced by *Streptomyces michigaensis, Streptomyces yokosukaensis,*[1] and other species of actinomycetes. Although not yet confirmed, the L-argininal form is thought to be active and it has been shown that the C-terminal aldehyde group is the critical functional group for inhibition of papain.[17]

Assay

Principles. Antipapain activity is measured by the change of absorbance at 280 nm during the hydrolysis of protein substrates. ID_{50} is obtained by the type of calculation described for measuring leupeptin activity.

[13] M. Hozumi, M. Ogawa, T. Sugimura, T. Takeuchi, and H. Umezawa, *Cancer Res.* **32,** 1725 (1972).

[14] T. Matsushima, R. S. Yamamoto, K. Hara, T. Sugimura, T. Takeuchi, and H. Umezawa, *Igaku no Ayumi* (Progress in Medicine) **88,** 710 (1974).

[15] H. Suda, T. Aoyagi, M. Hamada, T. Takeuchi, and H. Umezawa, *J. Antibiot.* **25,** 263 (1972).

[16] S. Umezawa, K. Tatsuta, K. Fujimoto, T. Tsuchiya, H. Umezawa, and H. Naganawa, *J. Antibiot.* **25,** 267 (1972).

[17] J. O. Westerich and R. Wolfenden, *J. Biol. Chem.* **247,** 8195 (1972).

Purification Procedure[15,16]

Step 1. Fermentation and Extraction. Streptomyces michigaensis is grown in a medium containing 2.0% glucose, 1.0% N–Z amine, 0.2% yeast extract, 0.3% NaCl, 0.1% $MgSO_4 \cdot 7H_2O$, 0.1% K_2HPO_4, 0.0007% $CuSO_4 \cdot 5H_2O$, 0.0001% $FeSO_4 \cdot 7H_2O$, 0.0008% $MnCl_2 \cdot 4H_2O$, 0.0002% $ZnSO_4 \cdot 7H_2O$ (pH 7.2) for 2–3 days in shake culture or for 22–28 hr in tank fermentation at 27°. Pass the culture filtrate (4.0 liters) through a column (4 × 30 cm) of active carbon. Wash the column with distilled water (4.0 liters) and elute antipain with 0.5 N HCl–80% methanol. Adjust pH of the active eluate (1.0 liter) to 7.0 and concentrate to dryness under reduced pressure at 40° (6.64 g, $ID_{50} = 3.75$ $\mu g/ml$).

Step 2. Chromatography on CM-Sephadex C-25. Prepare a 3 × 30 cm column of CM-Sephadex C-25 equilibrated with 0.01 M ammonium formate. Apply the step 1 powder dissolved at 10 mg/ml in 0.01 M ammonium formate to the column. Elute antipain with a gradient of ammonium formate from 0.01 M to 1.0 M. Collect the antipain fraction eluted with 0.4 M ammonium formate concentration. Charge the active eluate on a carbon column (10 ml) and wash with distilled water (200 ml). Elute antipain with 0.05 N HCl–80% methanol; concentrate to dryness (antipain dihydrochloride monohydrate, 96 mg, $ID_{50} = 0.06$ $\mu g/ml$).

Reagents

Buffer: borate buffer, 0.05 M at pH 7.4, containing 0.05 M NaCl
Substrate: Dissolve 2 g of purified casein with warming in 100 ml water adjusted to pH 7.4.
Papain (E. Merck, Germany): Prepare a stock solution containing 2 mg/ml in L-cysteine HCl (4 mg/ml) solution. Before use, dilute with cysteine HCl solution to contain 200 μg per 0.2 ml.
Inhibitor solution: Dilute with water to give about 50% inhibition.

Procedure. 1.0 ml of substrate and 0.8 ml of buffer with or without the inhibitor are pipetted into test tubes at 37°. After 3 min, 0.2 ml of enzyme is admixed; 20 min later the reaction is stopped by adding 2.0 ml of 1.7 M perchloric acid and the absorbance of the supernatant is read at 280 nm.

Properties

Physicochemical Properties.[15,16] Antipain dihydrochloride monohydrate is a white amorphous powder: m.p. 170–177° (dec.), $[\alpha]_D^{20} -10°$ (c 1, H_2O); maxima at 247 nm ($E_{1cm}^{1\%}$ 7.09), 252 nm ($E_{1cm}^{1\%}$ 6.83), 257 nm ($E_{1cm}^{1\%}$ 6.29),

263 nm ($E_{1cm}^{1\%}$ 4.64), 267 nm ($E_{1cm}^{1\%}$ 3.48) in H_2O. It is soluble in water, methanol, and dimethyl sulfoxide; slightly soluble in ethanol; and insoluble in benzene, hexane, petroleum ether, ethyl ether, and chloroform. Positive Rydon-Smith, permanganate, and Sakaguchi reactions and negative ninhydrin reactions are obtained. R_f values in thin-layer chromatography: 0.4 with n-butanol–butyl acetate–acetic acid–water (4:2:1:1) on silica gel G; 0.55 with n-butanol–ethanol–water (4:1:2) on cellulose (Avicel). Antipain dipicrate monohydrate: m.p. 149–144°, $[\alpha]_D^{23}$ −0.5° (c 0.5, H_2O).

Biological Properties and Activity.[15] Antipain inhibits papain (ID_{50} = 0.16 μg/ml, on casein substrate), trypsin (ID_{50} = 0.26 μg/ml, on casein), thrombokinase (ID_{50} = 20 μg/ml), cathepsin A (ID_{50} = 1.19 μg/ml, on N^α-carbobenzoxy-L-glutamyl-L-tyrosine[18]), cathepsin B (ID_{50} = 0.595 μg/ml, on BAA[18]) and plasmin (ID_{50} = 93 μg/ml, on fibrinogen). It inhibits acrosin (strongly, on BAEE and BAPA), and pancreatic submandibular and urinary kallikreins (weakly, on BAEE[9]). It inhibits carrageenin edema and blood coagulation in man and rabbit. The LD_{50} for mice is >1.0 g/kg intraperitoneal; >125 mg/kg intravenous.

Chymostatins—Inhibiting Chymotrypsins

Chymostatins[19] were found in *Streptomyces hygroscopicus*, *Streptomyces lavendulae*,[1] and other species of actinomycetes. Chymostatin A is N-{[(S)-1-carboxy-2-phenylethyl]carbamoyl}-α-[2-iminohexahydro-4(S)-pyrimidyl]-L-glycyl-L-leucyl-phenylalaninal,[20] and in chymostatins B and C, L-leucine is replaced with L-valine or L-isoleucine, respectively.[20] During extraction, partial racemization of the phenylalanine residue occurs. L-Phenylalaninal is thought to be an essential functionality in chymostatins.

[18] H. Ikezawa, K. Yamada, T. Aoyagi, T. Takeuchi, and H. Umezawa, *J. Antibiot.* **25**, 738 (1972).

[19] H. Umezawa, T. Aoyagi, H. Morishima, S. Kunimoto, M. Matsuzaki, M. Hamada, and T. Takeuchi, *J. Antibiot.* **23**, 425 (1970).

[20] K. Tatsuta, N. Mikami, K. Fujimoto, S. Umezawa, H. Umezawa, and T. Aoyagi, *J. Antibiot.* **26**, 625 (1973).

Assay

Principle. Antichymotrypsin activity is measured by the change of absorbance at 280 nm during the hydrolysis of a protein substrate. ID_{50} is obtained by the type of calculation used to measure inhibition of leupeptin.

Reagents

Buffer: borate buffer, 0.05 M at pH 7.4, containing 0.05 M NaCl

Substrate: Dissolve 2 g of purified casein (prepared by Norman's method) with warming in 100 ml of water adjusted to pH 7.4. Chymotrypsin (Sigma Chemical Co.): a stock solution containing 1 mg/ml in 0.001 M HCl. Before use, dilute with buffer containing 0.02 M $CaCl_2$ to contain 4 μg/0.2 ml.

Inhibitor solution: Dilute with water to give about 50% inhibition.

Procedure. The procedure is similar to that described for measuring antipain activity.

Purification

Fermentation, Extraction, and Purification. Streptomyces hygroscopicus is grown in a medium that contains 2.5% glycerol, 0.5% meat extract, 0.5% peptone, 1.0% yeast extract, 0.2% NaCl, 0.05% $MgSO_4$ · $7H_2O$, 0.05% K_2HPO_4, 0.32% $CaCO_3$ (pH 7.0) at 27° for 24–27 hr in shake culture. Inoculate 1.5 liters to 140 liters of the medium in a 200-liter fermentor and continue the fermentation at 27° under aeration and stirring for 24–36 hr. Extract the culture filtrate (ID_{50} = 0.0007 ml/ml) with 80 and 60 liters of *n*-butanol and concentrate the butanol extract under reduced pressure to a syrup (200 g). Wash the syrup with 2 liters of ethyl acetate and dry (40 g of a crude powder, ID_{50} = 1.0 μg/ml). Dissolve the powder (1.2 g) in butanol–water, 9:1, and subject to silica gel chromatography, using the same solvent. Pour the active eluate on a column of Dowex 1 in the chloride form, and elute with distilled water. Evaporate the active effluent to give pure chymostatin (0.15 g, ID_{50} = 0.15 μg/ml), which is crystallized from methanol.

Properties

Physicochemical Properties.[19,20] White crystals, m.p. 205°–207° (dec.), $[\alpha]_D^{22}$ +9°, (c 0.25, acetic acid), consist of chymostatins A and B with a trace of C. They are soluble in acetic acid and dimethyl sulfoxide; slightly soluble in water, methanol, ethanol, propanol, butanol, and ethylene glycol; and insoluble in ethyl acetate, butyl acetate, ether, hexane, petroleum ether, chloroform, and benzene. They give a positive Rydon-Smith and a negative ninhydrin test. R_f = 0.45 in thin-layer chromatography on silica gel with butanol–methanol–water (4:1:2).

Biological Properties and Activity.[19] Chymostatin inhibits hydrolysis of casein by α, β, γ, δ chymotrypsins rather effectively (ID_{50} = 0.15 μg/ml) and that by papain less efficiently (ID_{50} = 7.5 μg/ml). It does not inhibit plasmin, trypsin, and kallikrein, but it inhibits cathepsins: ID_{50} = 62.5 μg/ml against A, 2.6 μg/ml against B, 49.0 μg/ml against D.[10] Intraperitoneal injection of 1.5–100 mg/kg inhibits carrageenin edema in the hind paw of the rat (1.5 mg/kg causes 30% inhibition).

Elastatinal—Inhibiting Elastase

Elastatinal,[21] *N*-[(*S*)-1-carboxyisopentyl]carbamoyl-α-[2-iminohexahydro-4(*S*)-pyrimidyl]-(*S*)-glycyl-(*S*)-glutaminyl-(*S*)-alaninal,[22] was found in various species of actinomycetes, and one of them is closely related to *Streptomyces griseoruber*.

Assay

Principle. Antielastase activity is measured by the change of absorbance at 492 nm during the hydrolysis of congo red-labeled elastin. ID_{50} is obtained by calculation, as described in the section on leupeptin.

Reagents

Buffer 1: Tris chloride, 0.2 M, pH 8.8
Buffer 2: phosphate buffer, 0.5 M, pH 6.0
Substrate: Dissolve 200 mg of elastin–Congo red (Boehringer, Mannheim, Germany) with stirring in 100 ml of buffer.
Elastase (Boehringer, Mannheim, Germany): Prepare a stock solu-

[21] H. Umezawa, T. Aoyagi, A. Okura, H. Morishima, T. Takeuchi, and Y. Okami, *J. Antibiot.* **26**, 787 (1973).
[22] A. Okura, H. Morishima, T. Takita, T. Aoyagi, T. Takeuchi, and H. Umezawa, *J. Antibiot.* **28**, 337 (1975).

tion containing 1 mg/ml in buffer. Before use, dilute with buffer to contain 10 μg/0.2 ml.

Inhibitor solution: Prepare an inhibitor solution as described in the section of leupeptin.

Procedure. Pipette 1.0 ml of the substrate and 0.8 ml of buffer 1 with or without an inhibitor into a series of test tubes at 37°. After 3 min, add 0.2 ml of the enzyme solution and mix well. Thirty minutes later, the reaction is terminated by adding 2.0 ml of buffer 2, and absorbance of the supernatant is measured at 492 nm.

Purification Procedure[21]

Step 1. Fermentation and Extraction. The elastatinal-producing streptomyces is grown in a medium containing 3.0% glucose, 2.0% soybean meal, 0.3% NaCl, 0.25% NH$_4$Cl, 0.6% CaCO$_3$ for 48–66 hr in the shaking culture or for 38–48 hr in the tank fermentation at 27°. Pass the culture filtrate (26,360 ml, ID$_{50}$ = 9 μl/ml) through a column of Amberlite XAD-4 (1.5 liters, 7.4 × 41 cm). Wash with distilled water (10.5 liters) and elute with 30% acetone. Pool the active fraction and concentrate under reduced pressure at 40° (500 ml, ID$_{50}$ = 0.225 μl/ml). Adjust pH to 3.1 by formic acid and remove the precipitate by centrifugation.

Step 2. Chromatography on Dowex 50-X8. Prepare a 4.5 × 82 cm (1.3 liters) column and equilibrate with 50 mM pyridine–formic acid buffer (pH 3.1). Apply the supernatant of step 1, and elute with a gradient of 75 mM pyridine–formic acid buffer (pH 4.0) to 100 mM pyridine–formic acid buffer (pH 5.0) (3722 ml, ID$_{50}$ = 2.15 μl/ml). Pool the active fraction and pass through a column of Amberlite XAD-4 (300 ml, 2.9 × 45 cm). Wash with water (4.8 liters) and 3% acetone (1 liter), and elute with 10% acetone (1357 ml, ID$_{50}$ = 1.13 μl/ml). Pool the active fraction, remove acetone, pass through 2 × 17 cm column of Dowex 1-X2 (acetate form), and lyophilize (a white powder, 1145 mg, ID$_{50}$ = 0.95 μg/ml).

Properties[21]

Physicochemical Properties. Elastatinal is a white powder, m.p. 196°–204° (dec.), $[\alpha]_D^{25}$ +2° (c 1.0, H$_2$O); pK_a' 3.7, >11.0. It is soluble in water, methanol, pyridine and dimethyl sulfoxide; slightly soluble in ethanol, n-propanol, acetone, and chloroform; and insoluble in n-butanol, ethyl acetate, hexane, toluene benzene, and ethyl ether. The purified material gives positive Rydon-Smith, Folin, triphenyltetrazolium chloride,

2,4-dinitrophenyl hydrazine, nitroprusside-ferricyanide, and silver nitrate-sodium hydroxide tests, and negative ninhydrin and Sakaguchi tests. The R_f value in silica gel thin-layer chromatography is 0.31 with n-butanol–acetic acid–water (4:1:1) and 0.60 with methanol–pyridine–water (20:1:5).

Biological Activity. It inhibits elastase, but not the other proteases. Intravenous injection of 250 mg/kg to mice does not cause death.

Kinetic Properties. Inhibition of elastase by elastatinal is competitive with the substrate. K_i is 0.24 μM with acetyl-alanyl-alanyl-alanine p-nitroanilide and 0.21 μM with acetyl-alanyl-alanyl-alanine methyl ester as substrates.

Pepstatin—Inhibiting Acid Proteases

Pepstatin,[23] isovaleryl-L-valyl-L-valyl-[(3S,4S)-4-amino-3-hydroxy-6-methyl]heptanoyl-L-alanyl-[(3S,4S)-4-amino-3-hydroxy-6-methyl] heptanoic acid,[24,25] was found in *Streptomyces testaceus* n.sp., *Streptomyces argenteolus* var. *toyokaensis*, and various other species of actinomycetes. Stereochemistry of the new amino acid (3S,4S)-4-amino-3-hydroxy-6-methylheptanoic acid (AHMHA) was determined by X-ray analysis[26] and chemical synthesis.[27] Pepstatin-producing strains also produce other pepstatins which contain an n-caproyl or an isocaproyl group instead of the isovaleryl group.[28] Production of individual pepstatins depends on fermentation conditions; pepstatin containing the n-caproyl group is the main component formed in a casein medium, whereas pepstatin containing isovaleryl group is the main component in a peptone medium. Pepstatins containing other fatty acids are also found as minor components. Pepstatin-producing strains also produce pepstanones in which the terminal AHMHA in pepstatin is replaced with 3-

[23] H. Umezawa, T. Aoyagi, H. Morishima, M. Matsuzaki, M. Hamada, and T. Takeuchi, *J. Antibiot.* **23**, 259 (1970).
[24] H. Morishima, T. Takita, T. Aoyagi, T. Takeuchi, and H. Umezawa, *J. Antibiot.* **23**, 263 (1970).
[25] H. Morishima, T. Takita, and H. Umezawa, *J. Antibiot.* **25**, 551 (1972).
[26] H. Nakamura, H. Morishima, T. Takita, H. Umezawa, and Y. Iitaka, *J. Antibiot.* **26**, 255 (1973).
[27] H. Morishima, T. Takita, and H. Umezawa, *J. Antibiot.* **26**, 115 (1973).
[28] T. Miyano, M. Tomiyasu, H. Iizuka, S. Tomisaka, T. Takita, T. Aoyagi, and H. Umezawa, *J. Antibiot.* **25**, 489 (1972).

amino-5-methylhexanone-2. Another pepstatin-producing strain, which was classified as *Streptomyces parvisporogenes*, produces yet another family of pepstatins, in which isovaleryl is replaced by an acetyl, butyl, or propionyl group,[29] and *Streptomyces naniwaensis* produces pepstatin blocked by an acetyl group.[30] Hydroxypepstatin, in which L-alanine is replaced with L-serine, is produced by *Streptomyces testaceus*.[31]

Among the pepstatins, pepstanones, and hydroxypepstatin, biological activity has been studied mostly on the pepstatin which has an isovaleryl group.

Assay

Principle. Antipepsin activity is measured by the change of absorbance at 280 nm during casein hydrolysis. ID_{50} is obtained by a calculation such as described for leupeptin.

Reagents

Buffer: KCl-HCl, 0.02 M at pH 2.0
Substrate: Dissolve 600 mg of purified casein with warming in 100 ml of 0.75% lactic acid.
Porcine pepsin (Sigma Chemical Co., U.S.A.): Prepare a stock solution containing 1 mg/ml in 0.01 N HCl. This solution is stable for several weeks at 4°. Before use, dilute with buffer to contain 4 μg/0.1 ml.
Inhibitor solution: Dilute with water to give about 50% inhibition.

Procedure. Pipette 1.0 ml of substrate and 0.9 ml of buffer with or without an inhibitor into a series of test tubes at 37°. After 3 min, add 0.1 ml of enzyme and mix well; the reaction is terminated 30 min later by adding 2.0 ml of 1.7 M perchloric acid. Absorbance of the protein-free supernatant is measured at 280 nm.

Purification Procedure

Step 1. *Streptomyces testaceus* is grown in a medium containing 1.0% glucose, 1.0% starch, 0.75% peptone, 0.75% meat extract, 0.3% NaCl,

[29] T. Aoyagi, Y. Yagisawa, M. Kumagai, M. Hamada, H. Morishima, T. Takeuchi, and H. Umezawa, *J. Antibiot.* 26, 539 (1973).
[30] M. Fukumura, S. Satoi, N. Kuwana, and S. Murao, *Agric. Biol. Chem.* 35, 1310 (1971).
[31] H. Umezawa, T. Miyano, T. Murakami, T. Takita, T. Aoyagi, T. Takeuchi, H. Naganawa, and H. Morishimi, *J. Antibiot.* 26, 615 (1973).

0.1% $MgSO_4 \cdot 7H_2O$, 0.1% K_2HPO_4, 0.0007% $CuSO_4 \cdot 5H_2O$, 0.0001% $FeSO_4 \cdot 7H_2O$, 0.0008% $MnCl_2 \cdot 4H_2O$, 0.0002% $ZnSO_4 \cdot 7H_2O$ for 3–5 days in shake culture or for 60–70 hr in a tank fermentation at 27°. Extract the culture filtrate (4.5 liters) with n-butanol (2 liters) and concentrate to dryness under reduced pressure (4.0 g, ID_{50} = 0.07 $\mu g/ml$). Apply the crude powder (2.1 g) dissolved in water (30 ml) to the column (1.5 × 41 cm) chromatography of active carbon. Wash the column with distilled water (400 ml) and 40% methanol (30 ml), elute the active material with 80% methanol, collect the active eluate, and concentrate to dryness under reduced pressure (a white powder 0.7 g, ID_{50} = 0.027 $\mu g/ml$). Crystallize from methanol (colorless needles of pepstatin A; ID_{50} = 0.01 $\mu g/ml$).

Properties

Physicochemical Properties.[23,24] Pepstatin (isovaleryl-L-Val-L-Val-AHMHA-L-Ala-AHMHA) is obtained as colorless needles; m.p. 228–229° (dec.), $[\alpha]_D^{27}$ −90° (c 0.288, methanol). It is soluble in methanol, ethanol, acetic acid, pyridine and dimethyl sulfoxide; and slightly soluble or insoluble in ethyl acetate, ether, benzene, chloroform, and water. The compound gives positive Rydon-Smith and permanganate tests and a negative ninhydrin reaction. It readily yields sodium, magnesium, and calcium salts and methyl and ethyl esters and amide. The R_f value for isovaleryl pepstatin in silica-gel thin-layer chromatography is 0.76 with n-butanol–butyl acetate–acetic acid–water (4:4:1:1) and 0.15 with chloroform–methanol–acetic acid (92.5:6:1.5).

Biological Properties and Activity.[32] Strong inhibition is observed against the following acid protease preparations: pepsin, proctase B, *Trametes sanguinea*, *Aspergillus saitoi*, *Xylaria* sp. Pepstanones, hydroxypepstatin, and esters and amides show inhibitory activities similar those of pepstatin against acid proteases.[33] Pepstatin inhibits gastricsin, and this activity is about 100 times lower than that against pepsin. It does not inhibit proctase A.[32] Pepstatin exerts strong inhibition against cathepsin D prepared from swine liver,[33] rabbit liver,[34] human liver,[34] beef lung, rabbit lung,[35] rabbit alveolar macrophage,[35] oil-induced rabbit

---

[32] T. Aoyagi, S. Kunimoto, H. Morishima, T. Takeuchi, and H. Umezawa, *J. Antibiot.* **24**, 687 (1971).

[33] T. Aoyagi, H. Morishima, R. Nishizawa, S. Kunimoto, T. Takeuchi, H. Umezawa, and H. Ikezawa, *J. Antibiot.* **25**, 689 (1972).

[34] A. J. Barrett and J. T. Dingle, *Biochem. J.* **127**, 439 (1972).

[35] M. H. McAdoo, A. M. Dannenberg, C. J. Hayes, S. P. James, and J. H. Sanner, *Infect. Immunol.* **7**, 655 (1973).

peritoneal macrophage,[35] glycogen-induced rat peritoneal macrophage,[36] and pig brain.[37] Pepstatin inhibits renin in vitro[38,39] and also probably in vivo[40]; in affinity column it is utilized for renin purification.[41,42] It also inhibits the acid protease prepared from the erythrocytic stage of Plasmodium beghei.[43]

Pepstatin has a low toxicity: LD_{50} by intraperitoneal injection is 1090 mg/kg for mice, 820 mg/kg for rabbits and 450 mg/kg for dogs; LD_{50} by oral administration, more than 2000 mg/kg.[23] Daily oral administration of 800 mg/kg to monkeys does not cause any sign of toxicity, and oral administration does not lead to absorption of the compound. Intraperitoneal injection of 50 mg/kg to rats or dogs results in 2–10 μg/ml of blood level (maintained for 24 hr) and can be detected in urine for a period of 72 hr.[23] Oral administration shows strong preventive effect of gastric ulcer in Shay rats.[23] The intraperitoneal injection prevents carrageenin edema.[23] Pepstatin inhibits human cartilage degradation[44] and focus formation of YH-7 mouse cells by murine sarcoma virus.[45] Addition of pepstatin to a medium increases the protein content of rhodotorula.[46]

Kinetic Properties. Pepstatin binds with pepsin, yielding a stoichiometric complex demonstrable by chromatography.[47] Binding of pepstatin with the active site on pepsin has been suggested.[48] K_i of pepstatin against pepsin hydrolysis of N-acetyl-L-phenylalanyl-L-diiodotyrosine is less than 1 nM and against Phe·Gly·His·Phe(NO₂)·Phe·Ala·PheOMe it is 97 pM.[49] The binding is strong enough so that pepstatin can be used

[36] T. Kato, K. Kojima, and T. Murachi, Biochim. Biophys. Acta 289, 187 (1972).
[37] N. Marks, Science 181, 949 (1973).
[38] F. G. J. Lazar and H. Orth, Science 175, 656 (1972).
[39] M. Overturf, M. Lenard, and W. M. Kirkendall, Biochem. Pharmacol. 23, 671 (1974).
[40] R. P. Miller, C. J. Poper, C. W. Wilson, and E. DeVito, Biochem. Pharmacol. 21, 2941 (1972).
[41] P. Corvol, C. Devaux, and J. Menard, FEBS Lett. 34, 189 (1973).
[42] K. Murakami, T. Inagami, A. M. Michelakis, and S. Cohen, Biochem. Biophys. Res. Commun. 54, 482 (1973).
[43] M. R. Levy and S. C. Chou, Biochim. Biophys. Acta 334, 423 (1974).
[44] J. T. Dingle, A. J. Barrett, and A. P. Poole, Biochem. J. 127, 443 (1972).
[45] Y. Yuasa, H. Shimojo, T. Aoyagi, and H. Umezawa, J. Natl. Cancer Inst. 54, 1255 (1975).
[46] S. Murao, M. Arai, K. Nakahara, and M. Tsuchiya, Agric. Biol. Chem. 36, 2041 (1973).
[47] S. Kunimoto, T. Aoyagi, H. Morishima, T. Takeuchi, and H. Umezawa, J. Antibiot. 25, 251 (1972).
[48] G. P. Sachdev, A. D. Brownstein, and J. S. Fruton, J. Biol. Chem. 248, 6292 (1973).
[49] S. Kunimoto, T. Aoyagi, R. Nishizawa, T. Komai, T. Takeuchi, and H. Umezawa, J. Antibiot. 27, 413 (1974).

as a titrant for pepsin.[49] Pepstatin also binds to cathepsin D in an equimolar ratio.[50]

Phosphoramidon—Inhibiting Thermolysin

Phosphoramidon, N-(α-L-rhamnopyranosyloxyhydroxyphosphinyl)-L-leucly-L-tryptophan,[51] was found in *Streptomyces tanashiensis* and various other species of actinomycetes.[52] Phosphoryl-L-leucyl-L-tryptophan, which is obtained by a mild hydrolysis of phosphoramidon, is the most active.

Assay

Principle. Antihermolysin activity is measured by the change of absorbance at 280 nm during the hydrolysis of casein. ID_{50} is calculated in the manner described for leupeptin.

Reagents

Buffer: Tris-chloride, 0.1 M, pH 7.5, containing 0.02 M $CaCl_2$, 0.06 M NaCl

Substrate: Dissolve 2 g of purified casein with warming in 100 ml of water adjusted to pH 7.4.

Thermolysin (Nakarai Chem. Co., Japan): Prepare a stock solution containing 1 mg/ml in buffer. This solution is stable for several weeks at 4°. Before use, dilute with the same buffer to 0.75 μg/0.1 ml.

Inhibitor solution: Dilute with water to give about 50% inhibition.

[50] J. F. Woessner, *Biochem. Biophys. Res. Commun.* **47**, 965 (1972).
[51] S. Umezawa, K. Tatsuta, O. Izawa, T. Tsuchiya, and H. Umezawa, *Tetrahedron Lett.* **97**, (1972).
[52] H. Suda, T. Aoyagi, T. Takeuchi, and H. Umezawa, *J. Antibiot.* **26**, 621 (1973).

Procedure. The procedure is similar to antipain measurement, except that 0.1 ml of enzyme is used.

Purification Procedure

Step 1. Fermentation and Extraction. Streptomyces tanashiensis is grown in a medium containing 0.25% glycerol, 0.5% meat extract, 0.5% polypeptone, 1.0% yeast extract, 0.2% NaCl, 0.05% $MgSO_4 \cdot 7H_2O$, 0.05% KH_2PO_4, and 0.32% $CaCO_3$ (adjust to pH 7.4) for 2–4 days in shaking culture. Add 150 g of activated carbon powder to the culture filtrate (7700 ml, $ID_{50} = 2.8$ $\mu l/ml$), collect the carbon by filtration and wash with 5 liters of water. Suspend the carbon in 5 liters of methanol, elute the active material at pH 8.0 with 2 N NH_4OH, and concentrate to dryness under reduced pressure at 40° (11.4 g, $ID_{50} = 19.5$ $\mu g/ml$).

Step 2. Chromatography of DEAE-Sephadex A-25. Prepare a 3 × 20 cm column of the DEAE-Sephadex A-25 equibrated with 1 M acetic acid. Apply the active material of step 1 at a concentration of 20 mg/ml in 1 M acetic acid. Elute phosphoramidon with a gradient of sodium chloride from 0.0 M to 1.0 M in 1 M acetic acid. Evaporate the active eluate under reduced pressure, extract the active residue with ethanol (50 ml), and concentrate to dryness under reduced pressure (354.2 mg, $ID_{50} = 0.85$ $\mu g/ml$).

Step 3. Chromatography of Sephadex LH-20. Prepare a 2.5 × 120 cm column of Sephadex LH-20 washed with methanol. Apply the active material of step 2 dissolved in 1.5 ml of methanol. Elute phosphoramidon with methanol. Evaporate the active fraction under reduced pressure (135 mg, $ID_{50} = 0.4$ $\mu g/ml$).

Properties

Physicochemical Properties. Phosphoramidon is crystallized as the sodium, ammonium, cyclohexylammonium, or dicyclohexylammonium salt. For sodium salt of phosphoramidon: m.p. 173–178° (dec.), $[\alpha]_D^{20}$ −33.6° (c 1.0, H_2O); maxima at 221 nm ($E_{1\,cm}^{1\%}$ 480), 275 nm (sh.)($E_{1\,cm}^{1\%}$ 76), 282 nm ($E_{1\,cm}^{1\%}$ 81), 289.5 nm ($E_{1\,cm}^{1\%}$ 69.5) in H_2O. The salt is soluble in water, methanol, and dimethyl sulfoxide; less soluble in ethanol and ethyl acetate; and insoluble in benzene, hexane, petroleum ether, ethyl ether, and chloroform. The compound gives positive Ehrlich, ammonium molybdate-perchloric acid, and Rydon-Smith tests and a negative ninhydrin reaction. R_f values in thin-layer chromatography are 0.32 with *n*-butanol–acetic acid–water (4:1:1) on silica gel; 0.67 with *n*-butanol-acetone–acetic acid–5% ammonia–water (35:25:15:15:10) on cellulose (Avicel).

Biological Activity. Phosphoramidon inhibits thermolysin and the related enzyme; very specifically. The intravenous injection of 1.0 g/kg does not cause death in mice.

Kinetic Properties.[8] Inhibition of thermolysin by phosphoramidon is competitive with substrate; K_i for carbobenzoxy-glycyl-L-leucine amide is 28 nM. It is of interest that phosphoryl-L-leucyl-L-tryptophan (a mild hydrolysis product of phosphoramidon) shows a stronger inhibitory activity: $K_i = 2.0$ nM.

[56] Proteinase Inhibitors from Plant Sources

By YEHUDITH BIRK

Distribution and Occurrence

Proteinlike proteinase inhibitors are widely distributed in the plant kingdom. Most of the inhibitors are present in the seeds of the various plants, but they are not necessarily restricted to this part of the plant. Certain storage organs, such as seeds from the Leguminosae family and tubers from the Solanaceae family, are excellent sources of proteinase inhibitors. The inhibitors are diverse in number and in specificity toward various proteolytic enzymes. Several different kinds of inhibitors can be present in a single tissue—for example, barley grains, soybeans, and potato tuber. The multiplicity of plant proteinase inhibitors may partly be ascribed to the self- and mixed-association of a few monomers in each plant as well as to partial proteolysis of the inhibitors during purification, especially when the inhibitors are purified by affinity chromatography. The frequent presence of several proteinase inhibitors in the same source tissue, and the finding that the same inhibitor often inhibits more than one enzyme, are partly responsible for the difficulty in establishing their nomenclature.

The physiological significance of plant proteinase inhibitors has been questioned for a long time. The fact that certain seeds, such as soybeans and wheat grains, contain inhibitors of growth and of larval gut proteases of the insects *Tribolium* and *Tenebrio*,[1] suggest the possibility that these inhibitors may have evolved as a defense mechanism against predatory

---

[1] Y. Birk, *Proteinase Inhibitors, Proc. Int. Res. Conf., 2nd (Bayer Symp. V)*, Grosse Ledder, 1973, p. 355. Springer-Verlag, Berlin and New York, 1974.

insects.[2] This working hypothesis is further supported by the finding that wounding of the leaves of potato or tomato plants by insects induces a rapid accumulation of chymotrypsin inhibitor I.[3] This process is associated with the release and translocation of a hormone that induces rapid accumulation of proteinase inhibitors,[4] and it thus demonstrates that insect behavior can influence the protein composition of plant leaves.

Characterization

The molecular weights of plant inhibitors are mainly in the range from 3000 to 25,000. Many of them have been isolated in pure form and characterized with respect to amino acid composition, partial or full amino acid sequence, chemistry of the reactive sites, and the nature of the complexes formed with the respective proteinases. The inhibitors are frequently multiheaded as a consequence of gene elongation (via gene duplication or multiplication); this feature is being used for the study of evolution of specificity proteinase inhibitors.[5] It is generally assumed that inhibition of several enzymes by the same inhibitor is done either at separate nonoverlapping reactive sites or at separate but overlapping reactive sites, or at the same reactive site. Several mechanisms have been proposed[6,7] for enzyme–inhibitor interactions.

General Methods of Assay

Inhibitory activity and inhibitor concentration measurements are based on formation of enzyme–inhibitor complex and determine the decrease in the enzymic hydrolysis of natural or synthetic substrates. Inhibitor concentration can be determined in crude extracts provided the association constant of enzyme–inhibitor is high and the exact concentration of the active enzyme is known. For determination of specific inhibitory activity, assays should be performed at 50% inhibition of the enzyme, provided that inhibition is stoichiometric in this range. Suf-

[2] S. W. Applebaum and Y. Birk, in "Insect and Mite Nutrition" (J. G. Rodriguez, ed.), p. 629. North-Holland Publ., Amsterdam, 1972.
[3] T. R. Green and C. A. Ryan, Science 175, 776 (1972).
[4] T. R. Green and C. A. Ryan, Plant Physiol. 51, 19 (1972).
[5] M. Laskowski, Jr., I. Kato, T. R. Leary, J. Schrode, and R. Sealock, Proteinase Inhibitors, Proc. Int. Res. Conf., 2nd (Bayer Symp. V), Grosse Ledder, 1973, p. 597. Springer-Verlag, Berlin and New York, 1974.
[6] W. R. Finkenstadt, M. A. Hamid, J. A. Mattis, J. Schrode, R. Sealock, D. Wang, and M. Laskowski, Jr., Proteinase Inhibitors, Proc. Int. Res. Conf., 2nd, (Bayer Symp. V), Grosse Ledder, 1973, p. 389. Springer-Verlag, Berlin and New York, 1974.
[7] H. Ako, R. J. Foster, and C. A. Ryan, Biochemistry 13, 132 (1974).

ficient preincubation of enzyme and inhibitor should be allowed to reach inhibition equilibrium.

Chapters [56]–[63] deal with the most characterized inhibitors from plant sources. As stated already by Kassell[8] in this series, it has not been possible to express the units of inhibiting activity in a uniform manner; the assays and units are those used by the individual investigators whose isolation procedures are described. The chapter by Kassell,[8] the reviews by Liener and Kakade[9] and by Laskowski, Jr. and Sealock,[10] and the Proceedings of the First and Second International Research Conferences on Proteinase Inhibitors[11,12] should be consulted for earlier as well as for further information.

Acknowledgment

The author is grateful to Dr. R. Hofstein for her most valuable assistance in preparing Chapters [56]–[63].

[8] B. Kassell, this series Vol. 19 [66].
[9] I. E. Liener and M. L. Kakade, in "Toxic Constituents of Plant Foodstuffs" (I. E. Liener, ed.), p. 8. Academic Press, New York, 1969.
[10] M. Laskowski, Jr., and R. W. Sealock, in "The Enzymes" (P. D. Boyer, ed.), 3rd ed., Vol. 3, p. 375. Academic Press, New York, 1971.
[11] Proteinase Inhibitors Proc. First Int. Res. Conf. 1st, Munich, 1970, de Gruyter, Berlin, 1971.
[12] Proteinase Inhibitors, Proc., Int. Res. Conf., 2nd (Bayer Symp. V), Grosse Ledder, 1973. Springer-Verlag, Berlin and New York, 1974.

[57] Proteinase Inhibitors from Legume Seeds

By YEHUDITH BIRK

The presence of proteinlike trypsin inhibitors in legume seeds is by now a well established fact. Different legume seeds contain one or more inhibitors. Most of them have a molecular weight of about 8000 with a high concentration of cystine ($\sim 20\%$) and no free SH groups. The abundance of S—S bonds accounts for the considerable resistance to overall denaturation and to proteolytic digestion. Establishment of the amino acid sequence of the Bowman–Birk soybean inhibitor[1] and of the lima bean inhibitor IV[2] shows the existence of two homologous regions in these proteins. The region located in the first half of the molecule

[1] S. Odani and T. Ikenaka, J. Biochem. (Tokyo) 71, 839 (1972).
[2] F. C. Stevens, S. Wuerz, and J. Krahn, Proteinase Inhibitors, Proc. Int. Res. Conf., 2nd (Bayer Symp. V), Grosse Ledder, 1973, p. 344. Springer-Verlag, Berlin and New York, 1974.

BB   Asp-Asp-Glu-Ser-Ser-Lys-Pro-Cys-

                           10

LB   Ser-Gly-His-His-Glu-His-Ser-Thr-Asp-Glx-Pro-Ser -Glx-Ser-Ser-Lys-Pro-Cys-

---

          10   20

BB   Cys-Asp-Gln-Cys -Ala -Cys-Thr-Lys-Ser-Asn-Pro-Pro-Gln-Cys-Arg-Cys -Ser-Asp-

                                    30

GB   (Val, Cys) Thr-Ala-Ser-Ile -Pro-Pro-Gln (Cys, Ile, Cys, Thr, Asx,

      20   30

LB   Cys-Asn-His-Cys-Ala-Cys-Thr-Lys-Ser-Ile -Pro-Pro-Gln-Cys-Arg-Cys–Thr–Asp-

                 Leu   Ser

---

          30   40

BB   Met-Arg-Leu-Asn-Ser-Cys-His-Ser-Ala-Cys-Lys- Ser-Cys-Ile -Cys-Ala -Leu-Ser-

                40   50

GB   Val) Arg-Leu-Asx-Ser-Cys-His-Ser-Ala-Cys-Lys-Ser-Cys-Met-Cys-Thr-Arg-Ser-

                40   50

LB   Leu -Arg-Leu-Asp-Ser-Cys-His-Ser-Ala-Cys-Lys-Ser-Cys-Ile -Cys-Thr-Leu-Ser-

      Phe

---

           50   60

BB   Tyr -Pro-Ala-Gln-Cys-Phe-Cys-Val-Asp-Ile -Thr-Asp-Phe-Cys-Tyr-Glu-Pro-Cys-

                    60   70

GB   Met-Pro-Gly-Lys -Cys-Arg-Cys-Leu-Asx-Thr-Thr-Asx-Tyr-Cys-Tyr-Lys-Ser-Cys-

                    60   70

LB   Ile -Pro-Ala-Gln-Cys- Val (Cys, Thr, Asx) Ile-Asx-Asp-Phe-Cys-Tyr-Glu-Pro-Cys-

---

                70

BB   Lys-Pro-Ser -Glu- - - - -Asp-Asp-Lys-Glu-Asn

                80

GB   Lys-Ser -Asx-Ser -Gly-Glx -Asx-Asx

                80

LB   Lys-Ser -Ser -His-Ser -Asp-Asp-Asp-Asn-Asn-Asn

contains the trypsin inhibitory site, and the one located in the second half of the molecule contains the chymotrypsin inhibitory site. The striking similarity between the Bowman–Birk inhibitor and the lima bean inhibitor is also expressed in their antigenicity, since specific antibodies to the former cross-react with the latter and vice versa.[3] A very high degree of homology exists between these two inhibitors and garden bean inhibitor II.[4] This can be seen from the accompanying comparison (where BB, GB, and LB stand for the Bowman–Birk soybean inhibitor, garden bean inhibitor II, and lima bean inhibitor IV, respectively). However, in the

---

[3] Y. Birk, *Proteinase Inhibitors, Proc. Int. Res. Conf., 2nd (Bayer Symp. V)*, Grosse Ledder, 1973, p. 355. Springer-Verlag, Berlin and New York, 1974.

[4] K. A. Wilson and M. Laskowski, Sr., *Proteinase Inhibitors Proc. Int. Res. Conf., 2nd (Bayer Symp. V)*, Grosse Ledder, 1973, p. 286. Springer-Verlag, Berlin and New York, 1974.

COMPARATIVE EFFECTS OF ENZYMIC TREATMENTS OF THE BOWMAN–BIRK INHIBITOR (AA) FROM SOYBEANS, THE GROUNDNUT INHIBITOR (GI), AND THE CHICK PEA INHIBITOR (CI) ON THEIR INHIBITORY ACTIVITIES[a]

| Treatment of inhibitor | Percent residual inhibitory activity of | | | | | | Newly formed COOH terminal amino acid (mole/mole) removed from | | |
| | AA against | | GI against | | CI against | | | | |
	Trypsin	Chymo-trypsin	Trypsin	Chymo-trypsin	Trypsin	Chymo-trypsin	AA	GI	CI
Native	100	100	100	100	100	100			
Incubation with trypsin at pH 3.75	100	100	90	10	15	40			
Incubation with trypsin as above followed by CPB at pH 8	30	100	90	10	15	40	Lys	Arg	Lys
Incubation with chymotrypsin at pH 3.75	100	15	90	90	30	0			
Incubation with chymotrypsin as above followed by CPA at pH 8	100	15	90	90	30	0	Leu	None	Tyr

[a] From Y. Birk, *Proteinase Inhibitors, Proc. Int. Res. Conf. 2nd (Bayer Symp. V)*, Grosse Ledder, 1973, p. 355. Springer-Verlag, Berlin and New York, 1974.

garden bean inhibitor II the trypsin reactive site is located in the second half of the molecule, at a position homologous to the chymotrypsin-reactive sites of the Bowman-Birk and lima bean inhibitors. Comparison of the reactive sites of the Bowman–Birk soybean inhibitor (AA) with those of the groundnut inhibitor (GI) and the chick pea inhibitor (CI)[3] indicate that, although all the three are double-headed inhibitors, the reactive sites in CI are overlapping and the trypsin-reactive site in GI is not the trypsin-inhibitory site (see the table).

[58] Trypsin and Chymotrypsin Inhibitors from Soybeans

By YEHUDITH BIRK

The two prevalent trypsin inhibitors in soybeans are the Kunitz soybean inhibitor (SBTI) and the Bowman–Birk inhibitor (inhibitor AA). They differ markedly from each other in size, amino acid composition, structure, and properties. A detailed description of the preparation of these two inhibitors, as well as their physical and kinetic properties and specificity are outlined by Kassell in this series, Vol. 19 [66c].

Only certain selected features of the inhibitors, such as their amino acid sequence, their susceptibility to modifications, and their biological activities, will be mentioned here.

Kunitz Soybean Trypsin Inhibitor (SBTI)

Properties

Kinetic Properties.[1-3] The kinetics of interaction of the inhibitor with bovine β-trypsin have been studied extensively by Laskowski, and

---

[1] M. Laskowski, Jr., R. W. Duran, W. R. Finkenstadt, S. Herbert, H. F. Hixson, D. Kowalski, J. A. Luthy, J. A. Mattis, R. E. McKee, and C. W. Niekamp, *Proteinase Inhibitors, Proc. Int. Res. Conf., 1st,* Munich, 1970, p. 117. de Gruyter, Berlin, 1971.

[2] J. A. Luthy, M. Praissman, W. R. Finkenstadt, and M. Laskowski, Jr., *J. Biol. Chem.* **248**, 1760 (1973).

[3] W. R. Finkenstadt, M. A. Hamid, J. A. Mattis, J. Schrode, R. W. Sealock, D. Wang, and M. Laskowski, Jr., *Proteinase Inhibitors, Proc. Int. Res. Conf., 2nd (Bayer Symp. V),* Grosse Ledder, 1973, p. 389. Springer-Verlag, Berlin and New York, 1974.

associates. The rate of formation of the stable trypsin–inhibitor complex from bovine β-trypsin and either virgin inhibitor (with reactive site peptide bond intact) or modified inhibitor (with reactive site peptide bond hydrolyzed) is monitored by following the change in absorbance at 260 nm after mixing of equimolar reactants in a stopped-flow apparatus. At low reactant concentrations both reactions are second order with respect to concentration and to time. As the concentration of reactants is raised, the reaction order changes gradually to first order. This observed behavior—that at least two processes, a second-order process followed by a first-order one, are sequentially involved in stable complex formation—has been formulated in the following mechanism of interaction:

$$ T + I \underset{k_{-1}}{\overset{k_1}{\rightleftharpoons}} L \underset{k_{-2}}{\overset{k_2}{\rightleftharpoons}} C \underset{k_{-3}}{\overset{k_3}{\rightleftharpoons}} L^* \underset{k_{-4}}{\overset{k_4}{\rightleftharpoons}} T + I^* $$

where T is the free trypsin, I and I* are the virgin and modified inhibitors, respectively, C is the stable complex, and L and L* are loose, noncovalent complexes T-I and T-I*, respectively. The rate constants k_2 and k_3 increase with increasing pH in the pH range of 4.5–8.0 and reach plateau values near pH 8.0. The dissociation rate of the stable complex is very fast near pH 2 and declines sharply as the pH is raised.

Amino Acid Sequence.[4-7] The complete amino acid sequence of the inhibitor, which consists of 181 amino acid residues and two disulfide bridges, is shown in Fig. 1.

Alterations of Amino Acid Residues in the Reactive Site.[8,9] The effects of replacements, insertions, and modifications of amino acid residues in the reactive site of the inhibitor on enzyme–inhibitor interaction have been studied extensively by Laskowski, Jr., and associates (references to experimental details are listed by Kowalski *et al.*[9]). The primary structure of the reactive site of the inhibitor modified with catalytic amounts of trypsin at acid pH is

60   61   62   63   64   65   66   67

-Pro-Ser-Tyr-Arg-OH   H-Ile-Arg-Phe-Ile-

A schematic presentation of the so-called enzymic mutation of the in-

[4] T. Koide and T. Ikenaka, *Eur. J. Biochem.* 32, 401 (1973).

[5] T. Koide, S. Tsunazawa, and T. Ikenaka, *Eur. J. Biochem.* 32, 408 (1973).

[6] T. Koide and T. Ikenaka, *Eur. J. Biochem.* 32, 417, (1973).

[7] T. Ikenaka, S. Odani, and T. Koide, *Proteinase Inhibitors, Proc. Int. Res. Conf., 2nd (Bayer Symp. V)*, Grosse Ledder, 1973, p. 325, Springer-Verlag, Berlin and New York, 1974.

[8] D. Kowalski and M. Laskowski, Jr., *Biochemistry* 11, 3451 (1972).

[9] D. Kowalski, T. R. Leary, R. E. McKee, R. W. Sealock, D. Wang, and M. Laskowski, Jr., *Proteinase Inhibitors, Proc. Int. Res. Conf., 2nd (Bayer Symp. V)*, Grosse Ledder, 1973, p. 311. Springer-Verlag, Berlin and New York, 1974.

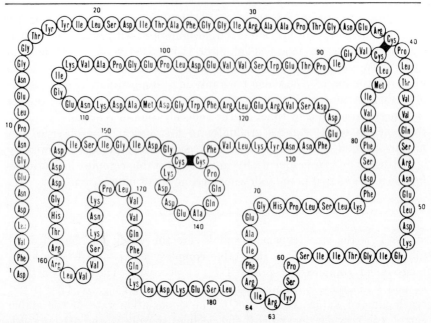

FIG. 1. Complete amino acid sequence of Kunitz soybean trypsin inhibitor. Reproduced from T. Koide and T. Ikenaka, *Eur. J. Biochem.* **32**, 417 (1973).

hibitor is shown in Fig. 2. Replacement of Arg_{63} by Lys causes the enzyme–inhibitor complex to dissociate more slowly. Replacement of Arg_{63} by Trp converts a strong trypsin inhibitor into a strong chymotrypsin inhibitor, which has no measurable association with trypsin. Substitution of Phe for Arg_{63} does not change significantly the kinetic properties of the inhibitor toward trypsin or chymotrypsin.

Insertion of Glu, Ala, or Ile by chemical methods in the reactive site on the α-amino group of $H\text{-}Ile_{64}$ of the protected modified inhibitor causes inactivation of the inhibitor, but it does not prevent trypsin-catalyzed bond synthesis between $ArgOH_{63}$ and the new amino acid.

Removal, by chemical methods, of $H\text{-}Ile_{64}$ inactivates the inhibitor. Replacement of $H\text{-}Ile_{64}$ by Ala, Leu, Gly, and Ile restore strong inhibitory activity.

Nitration of Tyr_{62} interferes with trypsin–inhibitor complex formation, but does not prevent it.

Reconstitution of an Active Fragment from Two Inactive Fragments of the Inhibitor.[7,10,11] Cyanogen bromide treatment of the inhibitor yields

[10] I. Kato and N. Tominaga, *FEBS Lett.* **10**, 313 (1970).
[11] T. Koide, T. Ikenaka, K. Ikeda, and K. Hamaguchi, *J. Biochem. (Tokyo)* **75**, 805 (1974).

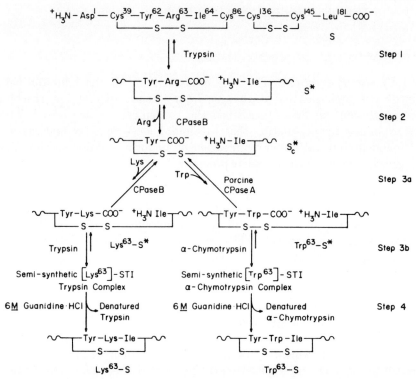

FIG. 2. The sequence of reactions employed in the enzymic mutations of Kunitz soybean trypsin inhibitor. The relative lengths of the arrows qualitatively indicate the equilibrium positions of the reactions. Reproduced from D. Kowalski, T. R. Leary, R. E. McKee, D. W. Sealock, D. Wang, and M. Laskowski, Jr., *Proteinase Inhibitors, Proc. Int. Res. Conf., 2nd (Bayer Symp. V)*, Grosse Ledder, 1973, p. 311. Springer-Verlag, Berlin and New York, 1974.

two inactive fragments ABC and D, respectively. At neutral pH the fragments associate to give an active complex, designated STI-C, that possesses more than 80% of the native inhibitory activity. The CD spectrum of STI-C is similar to that of the native inhibitor, but a large change is observed in the CD spectra between, before, and after mixing of the two fragments. The native inhibitor contains two tryptophan residues at positions 93 and 117. The former is in fragment ABC and the latter in D. Tryptophan-93 plays an important role in the inhibitory activity of the inhibitor and also somewhat in the formation of the tertiary structure of STI-C. Tryptophan-117 plays a minor role in the inhibitory activity of the inhibitor and does not participate in the association of STI-C.

Bowman–Birk Inhibitor from Soybeans (Inhibitor AA)

Properties

Kinetic Properties.[12] The rate of association of the inhibitor with trypsin is biphasic, and it changes from second order at low reactant concentrations to first order when the concentrations are raised. This is explained by postulating that the initial loose complex undergoes a subsequent intramolecular rearrangement to form a stable species. The suggested model for the complete process is

$$T + I \underset{k_{-1}}{\overset{k_1}{\rightleftharpoons}} C' \underset{k_{-2}}{\overset{k_2}{\rightleftharpoons}} C$$

where T, I, C′, and C represent trypsin, inhibitor, initial complex, and final complex, respectively. The rate constant for dissociation of the complex increases at acid pH. The dissociation has a high activation energy.

The combination of trypsin and chymotrypsin with the inhibitor results in characteristic changes in absorption and fluorescence emission spectra, indicating an alteration of the microenvironments of the enzyme chromophores as a consequence of the interaction.

The inhibitor, which has been modified by prolonged incubation with catalytic amounts of trypsin at acid pH, forms a complex with trypsin which has the same rate constant and activation energy of dissociation as the complex formed by the native inhibitor.

Amino Acid Sequence.[7,13–15] The complete covalent structure of the Bowman–Birk inhibitor has been fully established by Ikenaka and associates and is given in Fig. 3. The inhibitor consists of two large polypeptide moieties having almost identical sizes and structures. One is the trypsin-inhibitory region, Cys_8 through Cys_{24} plus Cys_{58} through Cys_{62}, and the other is the chymotrypsin inhibitory region Cys_{32} through Cys_{51}. They are linked to each other by two polypeptide bridges of 7 and 6 residues, Ser_{25} to Ser_{31} and Val_{52} to Phe_{57}, respectively. The trypsin-inhibitory region has the two additional peptide chains of the amino- and carboxy-terminal parts of the protein. Each reactive site abides in a nonapeptide loop formed by a single disulfide bridge. The striking similarity in the amino acid sequence around the two reactive sites is expressed in

---

[12] R. F. Steiner, *Eur. J. Biochem.* **27**, 87 (1972).
[13] S. Odani, T. Koide, and T. Ikenaka, *Proc. Jpn. Acad.* **47**, 621 (1971).
[14] S. Odani, T. Koide, and T. Ikenaka, *J. Biochem.* (*Tokyo*) **71**, 831 (1972).
[15] S. Odani and T. Ikenaka, *J. Biochem.* (*Tokyo*) **74**, 697 (1973).

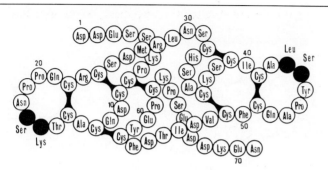

FIG. 3. Complete covalent structure of Bowman–Birk inhibitor. Residues at the two reactive sites are shown as solid black circles. Reproduced from S. Odani and T. Ikenaka, *J. Biochem. (Tokyo)* **74**, 697 (1973).

the identical distribution of five half-cystine residues around the trypsin- and chymotrypsin-inhibitory sites, as well as in the homology in the amino acid sequences around these sites:

8   16 17   26
Cys-*Cys*-Asp-Glu-*Cys*-Ala-*Cys*-Thr-*Lys*-*Ser*-Asn-Pro-Pro-Gln-*Cys*-Arg-*Cys*-Ser-Asp
35   43 44   53
Ala-*Cys*-Lys-Ser-*Cys*-Ile-*Cys*-Ala-*Leu*-Ser-Tyr-Pro-Ala-Gln-*Cys*-Phe-*Cys*-Val-Asp

Chemical Modifications. Guanidation of the inhibitor does not affect its inhibitory activity. Subsequent incubation with trypsin at acid pH results in splitting of the homoarginine–serine bond in the active site of the guanidated inhibitor.[16] Maleylation, succinylation, and treatment with nitrous acid affect mainly and selectively the trypsin-inhibitory site. Treatment with CNBr affects considerably both activities. Disruption of the S—S bonds by oxidation with performic acid causes a complete loss of the inhibitory activities, both against trypsin and chymotrypsin.[17,18]

Partial reduction of the inhibitor with sodium borohydride[19] results in a linear decrease in activity toward trypsin and chymotrypsin, which reaches zero at the reduction of an average of 4 disulfide bonds. It is possible to regenerate the activity in very dilute solution of reduced inhibitor, by using a disulfide–thiol redox buffer.

Scission of the inhibitor into two small fragments having either

[16] D. S. Seidl and I. E. Liener, *Biochem. Biophys. Res. Commun.* **42**, 1101 (1971).
[17] Z. Madar, Y. Birk, and A. Gertler, *Comp. Biochem. Physiol.* **48B**, 251 (1974).
[18] Y. Birk, *Proteinase Inhibitors, Proc. Int. Res. Conf., 2nd (Bayer Symp. V)*, Grosse Ledder, 1973, p. 355. Springer-Verlag, Berlin and New York, 1974.
[19] J. M. Hogle and I. E. Liener, *Can. J. Biochem.* **51**, 1014 (1973).

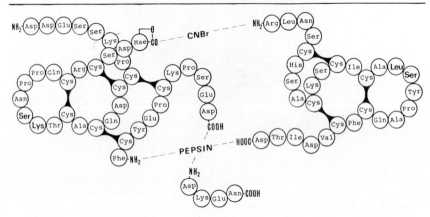

Fig. 4. Probable sites of peptide bond cleavage and the structure of the Bowman–Birk inhibitor fragments. The two proteinase reactive sites are shown by boldface type. Reproduced from T. Ikenaka, S. Odani, and T. Koide, *Proteinase Inhibitors, Proc. Int. Res. Conf., 2nd (Bayer Symp. V)*, Grosse Ledder, 1973, p. 325. Springer-Verlag, Berlin and New York, 1974.

trypsin- or chymotrypsin-inhibitory activity was achieved by Ikenaka and associates.[7,20] In order to split the methionyl at the sole methionine residue 27, the inhibitor is allowed to react with CNBr in 70% formic acid for 20 hr at 2°. The reaction mixture is diluted with water and lyophilized. The lyophilized material is digested with pepsin for 24 hr at pH 2.5 and 40°. After acidification of the reaction mixture with formic acid, the mixture is gel-filtered through a Bio-Gel P-10 column (1.5 × 200 cm) previously equilibrated with 0.1 M formic acid. The elution is monitored by measuring the absorbance of the effluent at 230 nm. Four fractions are obtained. Fraction I has both inhibitory activities. Fraction II inhibits trypsin but not chymotrypsin. Fraction III inhibits chymotrypsin but not trypsin. Fraction IV is a tetrapeptide derived from the carboxy-terminal part of the native inhibitor. The cleavage of the "double headed" inhibitor into two "single-headed" inhibitor fragments is illustrated in Fig. 4. One fragment (fraction II) with 38 amino acid residues includes residues 1–27 plus 57–67 within which the trypsin inhibitory site Lys_{16}-Ser_{17} is located. These two peptides are held together with 4 disulfide bridges. The other fragment (fraction III) consists of 29 amino acid residues comprising of residues 28–56, which include the chymotrypsin inhibitory site Leu_{43}-Ser_{44}. The inhibitory activities are 84% against trypsin and 16% against chymotrypsin, respectively, as compared with the intact inhibitor. The dissociation constants of "frag-

[20] S. Odani and T. Ikenaka, *J. Biochem. (Tokyo)* **74**, 857 (1973).

ment-enzyme complexes" are much higher than those of "native inhibitor-enzyme complexes."

Biological Activities. Ingestion of inhibitor (at ~0.5% level) causes pancreatic enlargement in rats, chicks, and quails. The enlargement of the pancreas is characterized by a general increase in biosynthesis of pancreatic proteins and particularly of proteolytic enzymes.[17,21,22] When the native inhibitor in the diet of Japanese quails is replaced by succinylated inhibitor, or by inhibitor treated with nitrous acid or by a 1:1 complex of the inhibitor with trypsin, no pancreatic enlargement occurs and no significant increase of pancreatic proteinases is noticed, indicating that the trypsin-inhibitory site is involved in the effect on the pancreas.[17,18]

Interaction of the inhibitor with its specific antibodies results in about 90% inactivation of the trypsin-inhibitory activity. When specific antibodies to the native inhibitor are preincubated with inhibitor that has been inactivated at the trypsin inhibitory site (by incubation with catalytic amounts of trypsin at acid pH and subsequent removal of Lys_{16} by carboxypeptidase B), cross-reaction of these antibodies with the native inhibitor is prevented, indicating that the trypsin inhibitory site of the inhibitor is not identical with its antigenic determinant.[18]

[21] A. M. Konijn, Y. Birk, and K. Guggenheim, *J. Nutr.* **100**, 361 (1970).
[22] A. M. Konijn, Y. Birk, and K. Guggenheim, *Am. J. Physiol.* **218**, 1113 (1970).

[59] Lima Bean Trypsin Inhibitors

By YEHUDITH BIRK

Lima bean (*Phaseolus lunatus*) trypsin inhibitors comprise at least four biologically active variants.[1] The assay methods, purification procedure, several physical and kinetic properties, specificity, and amino acid composition of trypsin inhibitors from lima beans have been described in an earlier treatise in this series.[2] This chapter includes information on the amino acid sequence of the inhibitors, on their reactive sites, and on their interactions and complexes with trypsin and chymotrypsin.

Physical Properties. The monomer has an apparent molecular weight of about 9000, but it undergoes a concentration-dependent dimeriza-

[1] F. C. Stevens, *Proteinase Inhibitors, Proc. Int. Res. Conf., 1st,* Munich, 1970, p. 149. de Gruyter, Berlin, 1971.
[2] B. Kassell, this series Vol. 19 [66d].

tion.[3] High speed sedimentation equilibrium studies show that variants I and II do not self-associate at moderate pH and protein concentration (<4 g/liter). Variants III and IV display pH-dependent self-association properties. Variant IV exists as a monomer at pH 2.0, dimerizes at pH 7.0, and associates to at least a trimer at the intermediate pH of 4.65.[4]

Kinetic Properties. The complex of the inhibitor with either trypsin or chymotrypsin has no inhibitory activity against the enzyme building it, but has full activity against the other enzyme.[1,5] The ternary complex between inhibitor, trypsin, and chymotrypsin has no inhibitory activity against either of the two enzymes. All these complexes have a 1:1 molar stoichiometry between the inhibitor and each enzyme.[3]

The inhibitory activities of the four variants toward trypsin are essentially identical. The extent of modification by catalytic amounts of trypsin at acid pH of variants I–IV are 93%, 90%, 93%, and 82%, respectively. The kinetics of complex formation between trypsin and the four variants, either in their native state or modified, is the same. The variants differ with regard to chymotrypsin-inhibitory activity: while variant IV is a strong chymotrypsin inhibitor, variant I does not achieve more than 60% inhibition, and variants II and III appear to have intermediate inhibitory activity. The extent of modification of the four variants by incubation with catalytic amounts of chymotrypsin at acid pH is identical—approximately 90% in every case. The difference in affinity for chymotrypsin of the modified variants parallels their inhibitory activity.[6]

Amino Acid Sequence. Sequence determinations on variants I and IV showed that neither of the two variants represent a single homogeneous molecular species. Although there is great variability at the amino-terminal end of the molecule, the variability ceases starting with residue 13. The complete amino acid sequence of variants IV and I, as established by Stevens and associates,[7] is

```
                                    10
IV        Ser-Gly-His-His-Glu-His-Ser-Thr-Asp-Glx-Pro-Ser-Glx-Ser-Ser-Lys-
I                                      Asp-Glx-Pro-Ser-Glx-Ser-Ser-Lys-


              20                                            30
IV    Pro-Cys(Cys,Asx)His(Cys{ Leu  Cys)Thr-Lys-Ser-Ile-Pro-Pro-Gln-Cys-
                               Ala,
I     Pro-Cys(Cys,Asx)His(Cys, Ala,Cys)Thr-Lys-Ser-Ile-Pro-Pro-Gln-Cys-
```

[3] J. Krahn and F. C. Stevens, *FEBS Lett.* **13**, 339 (1971).
[4] J. D. Sakura and S. N. Timasheff, *Arch. Biochem. Biophys.* **159**, 123 (1973).
[5] J. Krahn and F. C. Stevens, *Biochemistry* **9**, 2646 (1970).
[6] J. Krahn and F. C. Stevens, *FEBS Lett.* **28**, 313 (1972).
[7] F. C. Stevens, S. Wuerz, and J. Krahn, *Proteinase Inhibitors, Proc. Int. Res. Conf., 2nd (Bayer Symp. V)*, Grosse Ledder, 1973, p. 344. Springer-Verlag, Berlin and New York, 1974.

IV   Arg-Cys-$\frac{\text{Thr}}{\text{Ser}}$-Asp-$\frac{\overset{40}{\text{Leu}}}{\text{Phe}}$-Arg-Leu-Asp-Ser-Cys-His-Ser-Ala-Cys-Lys-Ser-

I   Arg-Cys-Thr-Asp-Leu-Arg-Leu-Asp-Ser-Cys-His-Ser-Ala-Cys-Lys-Ser-

IV $\overset{50}{\text{Cys}}$-Ile-Cys-Thr-Leu-Ser-Ile-Pro-Ala-$\overset{60}{\text{Gln}}$-Cys-Val(Cys,$\frac{\text{Thr}}{\text{Asx}}$,Asx)Ile-

I   Cys-Ile-Cys-Thr-Leu-Ser-Ile-Pro-Ala-Gln-Cys-Val(Cys,Asx,Asx)Ile-

IV $\frac{\text{Asx}}{\text{Thr}}$-Asp-Phe-Cys-Tyr-$\overset{70}{\text{Glu}}$-Pro-Cys-Lys-Ser-Ser-His-Ser-Asp-Asp-$\overset{80}{\text{Asp}}$-

I   Asx-Asp-Phe-Cys-Tyr-Glu-Pro-Cys-Lys(Ser,Ser,His,Ser,Asx, Asx, Asx,

IV   Asn-Asn-Asn

I   Asx)

Reactive Sites and Chemical Modifications. Variant IV is a "double-headed" inhibitor that contains two independent reactive sites, for trypsin and chymotrypsin.[1,5] The trypsin-inhibiting site is the Lys_{26}–Ser_{27} peptide bond[7,8] and the chymotrypsin-inhibiting site is either a Leu_{53}–Ser_{54} or a Phe_{53}–Ser_{54} peptide bond. The two active sites lie in two homologous regions of the molecule that are comprised of residues 23–34 for the trypsin-inhibiting site and residues 50–61 for the chymotrypsin-inhibiting site.[1]

Guanidation of the inhibitor results in 95% conversion of the lysine groups to homoarginine. Conversion of the critical Lys-Ser peptide bond does not block the interaction of the inhibitor with trypsin since guanidated inhibitor shows an activity similar to that of the native inhibitor. Guanidation of the inhibitor after it has been modified with trypsin at acid pH results in loss of 80% of the activity, indicating that the conversion of the new C-terminal lysine to homoarginine prevents complex formation with trypsin.[9]

Complete reduction of the disulfide bonds of the inhibitor results in loss of biological activity. All the disulfide bonds are equally accessible to the reducing reagents. After air oxidation, up to 50% of the trypsin-inhibitory activity and full chymotrypsin-inhibitory activity is regained. Both inhibitory activities are equally sensitive to reduction and are lost as a linear function of the average number of disulfide bonds reduced and carboxymethylated. The disulfide bonds are stabilized when the inhibitor is in the form of a molar complex with trypsin. Under these conditions only one out of a possible eight disulfide bonds can be reduced in the inhibitor with up to a 10-fold molar excess of the reducing agent dithioerythritol. The modified inhibitor obtained after dissociation of reduced and carboxymethylated trypsin–inhibitor complex is fully active against both trypsin and chymotrypsin.[10]

[8] J. Krahn and F. C. Stevens, *Biochemistry* 11, 1804 (1972).
[9] R. F. Steiner, C. Horan, and A. Lunasin, *FEBS Lett.* 38, 106 (1973).
[10] F. C. Stevens and E. Doskoch, *Can. J. Biochem.* 51, 1021 (1973).

[60] Trypsin Isoinhibitors from Garden Beans (*Phaseolus vulgaris*)

By Yehudith Birk

Assay Methods[1]

Trypsin inhibitory activity is assayed spectrophotometrically by measuring the inhibition of the tryptic hydrolysis of benzoyl-L-arginine ethyl ester (BAEE) as described in an earlier volume of this treatise.[2]

Chymotrypsin-inhibitory activity is determined according to a modification of the method of Hummel[3] as described by Kress *et al.*,[4] using benzoyl-L-tyrosine ethyl ester (BTEE) as substrate.

Reagents

Buffer: Tris-chloride 50 mM, pH 8.0, containing 10 mM CaCl$_2$

Substrate: BTEE (Mann), 1.07 mM in 50% (w/w) aqueous methanol

Enzyme: α-Chymotrypsin (Worthington), 1 mg/ml in 1 mM HCl

Procedure. Assays are performed in 10-mm quartz cuvettes of 3.5-ml capacity. The cuvettes are held at 25° in a thermostatted compartment of a spectrophotometer, preferably with a recording mechanism. Prepare the reference solution in a cuvette by mixing 1.5 ml of substrate solution with 1.5 ml of buffer. Place 1.5 ml of substrate solution and 1.4 ml of buffer in the assay cuvette. Allow the solutions to come to thermal equilibrium in the cell compartment and zero the instrument at 256 nm. In a small test tube, mix 30 μl of the enzyme solution with the desired amount of inhibitor in a final volume of 1 ml of the buffer. Preincubate for 5 min at 25°. Start the reaction by addition of 100 μl of the preincubation mixture to the assay cuvette. Record the increase in absorbance for about 5 min. To determine the activity of the chymotrypsin without inhibitor perform the same assay, but preincubate chymotrypsin alone.

Definition of Unit.[1] A unit of inhibitor is defined as the amount of inhibitor that inhibits the enzymic activity of 1 mg of active enzyme (either trypsin or chymotrypsin) in the assay.

[1] K. A. Wilson and M. Lakowski, Sr., *J. Biol. Chem.* **248**, 756(1973).

[2] P. J. Burk, this series Vol. 19 [67].

[3] B. C. W. Hummel, *Can. J. Biochem. Physiol.* **37**, 1393 (1959).

[4] L. F. Kress, K. A. Wilson, and M. Laskowski, Sr., *J. Biol. Chem.* **243**, 1758 (1968).

Purification Procedure[1]

Step 1. Extraction, Ammonium Sulfate Fractionation, and Trichloroacetic Acid Precipitation. The starting material is 3.5 kg of garden beans, *Phaseolus vulgaris* var. 'Great Northern.' Grind the beans to a fine meal in a blender. Extract with 14 liters of 0.05 M HCl at room temperature with stirring. After 2 hr pass the mixture through several layers of cheesecloth on a Büchner funnel to remove the larger particles. Add a few milliliters of toluene and hold the filtrate overnight at 4°. Clarify the extract by centrifugation at 9500 rpm at 0° for 30 min in the GSA rotor of a Sorvall RC-2B centrifuge (the same conditions apply also to other centrifugations below). Adjust the clarified extract (approximately 9 liters) to pH 5.0 with 10 M NaOH and add solid ammonium sulfate to attain 65% saturation at 0° (407 g/liter). Let the suspension stand overnight at 4°, and then centrifuge and discard the supernatant. Dissolve the precipitate in a minimal amount of water (approximately 1 liter). Add a sufficient amount of 25% (w/v) trichloroacetic acid to attain a final concentration of 5%. After 5 min at room temperature remove the inactive precipitate by centrifugation. Adjust the supernatant immediately to pH 5.0 with 10 M NaOH and repeat the precipitation step with 65% saturation ammonium sulfate. Recover the resulting precipitate by centrifugation and dissolve in a minimal volume of water. Dialyze against three 8-liter changes of water and remove the resulting precipitate by centrifugation. Lyophilize the salt-free, crude, inhibitor solution and store as a powder at −20°. The initial purification procedure is summarized in Table I.

Step 2. Gel Filtration on a Sephadex G-50 Column. Prepare a 21.5 × 82 cm column of Sephadex G-50 equilibrated with 0.01 M HCl. Dissolve

TABLE I
SUMMARY OF THE INITIAL PURIFICATION PROCEDURE[a,b]

Steps	Protein (A_{280} units)	Inhibitor units	Specific activity (units/ A_{280})	Recovery (%)	Purification factor
Crude extract	76,000	3,520	0.046	100	1
65% Ammonium sulfate	21,170	2,730	0.128	77.5	2.8
2.5% Trichloroacetic acid	10,200	2,190	0.215	62.1	4.7
65% Ammonium sulfate	5,420	2,040	0.375	57.8	8.2
Sephadex G-50	436	1,925	4.42	54.7	96

[a] K. A. Wilson and M. Laskowski, Sr., *J. Biol. Chem.* **248**, 756 (1973).
[b] Starting material was 3.5 kg of bean meal.

7 g of the crude inhibitor from step 1 in a small volume of 0.01 M HCl and apply to the column. Elute with the same solvent at a rate of 300 ml/hr. After discarding the first 6.4 liters of effluent, collect 20-ml fractions. Pool the active fraction (approximately tubes 450–650).

Step 3. Chromatography on a DEAE-Cellulose Column. Prepare a 2.5 × 92 cm column of DEAE-cellulose equilibrated with 0.05 M sodium phosphate buffer, pH 7.0—the starting buffer. Apply half of the active solution from step 2 to the column, and elute first the inactive proteins with 270 ml of the starting buffer. Then apply a linear gradient of NaCl, 0 to 0.2 M, in the starting buffer (1500 ml per bottle). Collect 6-ml fractions at a rate of 60 ml/hr. Pool the 4 active fractions I–IV as indicated in Fig. 1.

Step 4. Further Purification of Pool II from Step 3 on SP-Sephadex. Prepare a 2.5 × 93 column of SP-Sephadex equilibrated with 0.05 M sodium formate, pH 3.8, containing 0.12 M NaCl. Apply the material recovered in pool II from step 3 to the SP-Sephadex column. Elute with a linear gradient of NaCl 0.12 to 0.50 M, in 0.05 M sodium formate pH 3.8, 1500 ml per bottle. Collect 6-ml fractions at a rate of 42 ml per hour. From the several active peaks (see Fig. 2), pool the one eluting at 0.2 M NaCl (approximately tubes 150–200). Rechromatograph under the same conditions, and pool again the main inhibitor fraction. Subject it to equi-

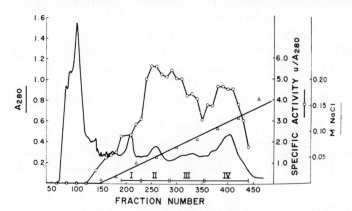

Fig. 1. Ion-exchange chromatography of the active solution from step 2 on DEAE-cellulose. One-half of the material from the gel filtration step was charged on a column of DEAE-cellulose (2.5 × 92 cm) in 0.05 M sodium phosphate, pH 7.0. After eluting with 270 ml of the same buffer, a linear gradient of NaCl in 0.05 M sodium phosphate, pH 7.0, 0 to 0.2 M (1500 ml per bottle), was begun. Elution was at 60 ml/hr, with 6-ml fractions collected. Pools were made as indicated. ——, absorbance at 280 nm; ○——○, specific activity; △——△, NaCl concentration. Reproduced from K. A. Wilson and M. Laskowski, Sr., *J. Biol. Chem.* **248,** 756 (1973).

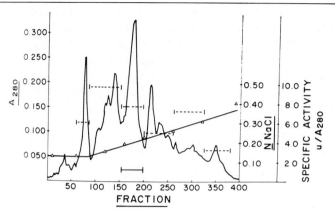

FIG. 2. Ion-exchange chromatography of pool II (Fig. 1) on SP-Sephadex. Pool II from the DEAE-cellulose chromatography was charged on a column of SP-Sephadex (2.5 × 93 cm) in 0.05 M sodium formate containing 0.12 M NaCl, pH 3.80. The column was eluted at 42 ml/hr with a linear gradient of NaCl, 0.12 to 0.50 M, in 0.05 M sodium formate, pH 3.80, 1500 ml per bottle. Fractions of 6 ml were collected. ——, Absorbance at 280 nm; ----, specific activity of pooled fractions; △——△, NaCl concentrations. Reproduced from K. A. Wilson and M. Laskowski, Sr., *J. Biol. Chem.* **248**, 756 (1973).

librium chromatography on a 2.5 × 89 cm column of SP-Sephadex in 0.05 M sodium formate containing 0.12 M NaCl, pH 3.8. Elute with the same buffer, collecting 6-ml fractions at a rate of 40 ml/hr. The protein starts emerging after about 240 fractions. Pool the active fraction that elutes with constant specific activity (approximately tubes 335–390). This fraction is called isoinhibitor I.

Step 4a. Further Purification of Pool IV from Step 3 on SP-Sephadex. Prepare a 2.5 × 83 cm column of SP-Sephadex equilibrated with 0.05 M sodium formate containing 0.10 M NaCl, pH 3.75—the starting buffer. Apply one sixth of the material recovered in pool IV from step 3 to the SP-Sephadex column. Elute first with 150 ml of the starting buffer. Then apply a linear gradient of NaCl, 0.1–0.5 M, in 0.05 M sodium formate pH 3.75 (1200 ml per bottle). Collect 6-ml fractions at a rate of 60 ml/hr. Pool the two active peaks (Fig. 3) eluting at 0.24 M NaCl (peak A) and at 0.3 M NaCl (peak B). To further purify peak A, subject half of it to equilibrium chromatography on a SP-Sephadex column (2.5 × 93 cm) in 0.05 M sodium formate containing 0.19 M NaCl pH 3.75. Elute with the same buffer collecting 4 ml fractions at a rate of 48 ml/hr. Pool the active fraction (approximately tubes 225–275). This fraction is called isoinhibitor II.

To further purify peak B, submit it to equilibrium chromatography

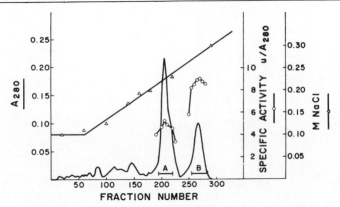

FIG. 3. Ion-exchange chromatography of pool IV (Fig. 1) on SP-Sephadex. One-sixth of pool IV (220 mg) was charged on a column of SP-Sephadex (2.5 × 83 cm) in 0.05 M sodium formate containing 0.10 M NaCl, pH 3.75. The column was first eluted with 150 ml of the same buffer. A linear gradient of NaCl, 0.10 to 0.50 M in 0.05 M sodium formate, pH 3.75, 1200 ml per bottle was then started. Elution was at 60 ml/hr, with 6-ml fractions collected. Pools were made as shown and marked A and B; ——, Absorbance at 280 nm; ○——○, specific activity; △——△, NaCl concentration. Reproduced from K. A. Wilson and M. Laskowski, Sr., *J. Biol. Chem.*, **248**, 756 (1973).

on a SP-Sephadex column (1.5 × 85 cm) in 0.05 M sodium formate containing 4 M urea and 0.21 M NaCl, pH 3.75. Elute with the same buffer collecting 4-ml fractions at a rate of 35 ml/hr. The material is resolved into two partly overlapping peaks. Pool them separately and rechromatograph under identical conditions. Pool the main portions of the peaks. These materials are isoinhibitors IIIa and IIIb.

Properties

Purity.[1] Isoinhibitors I and II have constant specific activities across their peaks in equilibrium chromatography, and they exhibit single bands in polyacrylamide gel electrophoresis at pH 2.3 and 9.5. Isoinhibitor IIIb shows a single band in electrophoresis at pH 2.3 but a second, minor band at pH 9.5. Isoinhibitor IIIa is highly heterogeneous.

Physical Properties.[1] The molecular weights of the inhibitors are 8100–9000 by calculation from amino acid analyses.

The optical factors at pH 8 of isoinhibitors I, II, and IIIb are 3.54, 2.31, and 3.40 mg per A_{280} unit, at pH 8.0, and 3.69, 2.38, and 3.40 at pH 2.0, respectively, as calculated from the molar recovery of amino acids.

Inhibitory Activities.[1,5] Trypsin isoinhibitors I, II, and IIIb strongly

[5] K. A. Wilson and M. Laskowski, Sr., *Proteinase Inhibitors, Proc. Int. Res. Conf., 2nd (Boyer Symp. V)*, Grosse Ledder, 1973, p. 286. Springer-Verlag, Berlin and New York, 1974.

inhibit bovine trypsin (2.57, 2.65, and 2.87 units/mg, respectively), but they vary in the character of inhibition. Inhibitor II is relatively a much weaker inhibitor than either I or IIIb, and it is significantly displaced from the trypsin–inhibitor complex by the 1 mM substrate, whereas inhibitors I and IIIb are not. The dissociation constant of inhibitor II with bovine β-trypsin is two to three orders of magnitude greater than those of inhibitor I or IIIb at any pH above 4.5. Thus at pH 5.0 the complexes of trypsin and inhibitor I and IIIb are only slightly dissociated, while that of inhibitor II is over 50% dissociated. Even at pH 8.0 the complex of trypsin with inhibitor II is approximately 10% dissociated, whereas the complexes of trypsin with either inhibitor I or IIIb are less than 1% dissociated at pH values above 6.

The isoinhibitors also vary in their activity toward α-chymotrypsin. Inhibitors I and II only weakly inhibit chymotrypsin, each with a specific activity of 0.20 units/mg whereas inhibitor IIIb is a potent α-chymotrypsin inhibitor with a specific activity of 2.84 units/mg. This inhibitor has two independent sites for trypsin and chymotrypsin. The complex of the inhibitor with trypsin still inhibits chymotrypsin, and vice versa, the complex with chymotrypsin inhibits trypsin.

TABLE II

AMINO ACID COMPOSITIONS OF *Phaseolus vulgaris* INHIBITORS[a,b]

Amino acid	Isoinhibitor I	Isoinhibitor II	Isoinhibitor IIIb
Aspartic acid	11	10	11
Threonine	3	5	5
Serine	11	13	12
Glutamic acid	6	5	7
Proline	6	6	6
Glycine	1	2	2
Alanine	3	2	3
Half-cystine	14	14	14
Valine	1	2	0
Methionine	0	2	0
Isoleucine	3	3	4
Leucine	3	2	2
Tyrosine	1	2	1
Phenylalanine	1	0	2
Lysine	4	4	4
Histidine	3	3	5
Arginine	3	3	3
	74	78	81
Molecular weight:	8086	8371	8884

[a] K. A. Wilson and M. Laskowski, Sr., *J. Biol. Chem.* **248**, 756 (1973).

[b] Results are expressed as residues per molecule.

Inhibitor II is a strong inhibitor of elastase when assayed with Congo red–elastin system. It inhibits both trypsin and elastase simultaneously and independently. Inhibitors I and IIIb also inhibit elastase, but with specific activities less than one-quarter that of inhibitor II.

Reactive Site Residues.[1,5] Isoinhibitor I has lysine in the trypsin-reactive site while isoinhibitor II has arginine. The location of the trypsin reactive site in isoinhibitor II is at Arg_{53}–Ser_{54}, and that of the elastase reactive site is at Ala_{26}–Ser_{27}.

Amino Acid Composition.[1] The amino acid composition of isoinhibitors I, II, and IIIb is shown in Table II. The inhibitors contain no free —SH groups and no carbohydrates.

Peptide Maps.[1] The peptide maps of inhibitors I and IIIb are very similar; much less similarity is apparent between inhibitor II and either I or IIIb. Isoinhibitors I and II have no peptides in common.

Partial Sequence.[5] The partial amino acid sequence of isoinhibitor II is

```
                          30
-(Val,Cys)Thr-Ala-Ser-Ile-Pro-Pro-Gln(Cys,Ile,Cys,Thr,Asx,Val)
         40                                    50
    Arg-Leu-Asx-Ser-Cys-His-Ser-Ala-Cys-Lys-Ser-Cys-Met-Cys-Thr-Arg-Ser-
              60                                    70
    Met-Pro-Gly-Lys-Cys-Arg-Cys-Leu-Asx-Thr-Thr-Asx-Tyr-Cys-Tyr-Lys-Ser-
    Cys-Lys-Ser-Asx-Ser-Gly-Glx-Asx-Asx
```

[61] A Trypsin and Chymotrypsin Inhibitor from Groundnuts (*Arachis hypogaea*)

By YEHUDITH BIRK

Assay Methods[1]

Inhibition of Proteolysis

Principle. Inhibition of proteolysis is determined by Kunitz's casein digestion method,[2] which has been described in a previous volume of this treatise.[3]

[1] A. Tur-Sinai, Y. Birk, A. Gertler, and M. Rigbi, *Biochim. Biophys. Acta* **263**, 666 (1972).
[2] M. Kunitz, *J. Gen. Physiol.* **30**, 291 (1947).
[3] M. Laskowski, Sr., this series Vol. 2 [3].

Definition of Unit. Kunitz[2] defined 1 tryptic or chymotryptic unit as the enzyme activity which causes an increase of 1 unit of absorbance at 280 nm per minute of digestion of casein, under the standard conditions. The specific activity is expressed as units per microgram of enzyme protein. Enzyme-inhibiting activity is expressed as units of enzyme inhibited, and specific activity as units of enzyme inhibited per absorbance unit, at 280 nm, of inhibitor.

Inhibition of Esterolysis

Principle. Inhibition of the esterolytic activity of trypsin is determined on the substrate p-toluenesulfonyl-L-arginine methyl ester (TAME) either spectrophotometrically or by a potentiometric method.

Inhibition of the esterolytic activity of chymotrypsin is determined on the substrate acetyl-L-tyrosine methyl ester (ATEE) in a similar manner.

Spectrophotometric Assay

Reagents

Buffer: Tris-chloride, 46 mM, pH 8.1, containing 11.5 mM CaCl$_2$
Substrate: Dissolve 37.9 mg of TAME (Sigma) in 10 ml of water. Mix 3 ml of this solution with 26 ml of the buffer.
Enzyme: Trypsin (Worthington), 1 mg/ml in 1 mM HCl

Procedure. The reaction is performed in a Gilford 2400 spectrophotometer, at constant temperature (25°). Preincubate for 5 min 25 μl of enzyme solution with the inhibitor (up to 150 μl) in 1 ml of the buffer, at room temperature. In each of two 1-cm quartz cuvettes, pipette 3 ml of substrate and place one of them as a blank into the spectrophotometer. Set the wavelength to 247 nm. Start the reaction by adding 100 μl of the enzyme–inhibitor mixture (after 5 min of preincubation) to the second cuvette. Mix thoroughly and record change in absorbance for about 4 min.

Definition of Units. One trypsin unit is defined as 1 μmole of substrate hydrolyzed per minute of reaction. One inhibition unit is defined as unit of enzyme inhibited.

Specific activity is defined as inhibition units per absorbance unit, at 280 nm, of the inhibitor.

Chymotrypsin. Inhibition of the esterolytic activity of chymotrypsin is determined on ATEE in the same way as described above for trypsin except for the following changes.

Substrate: Dissolve 24.5 mg of ATEE (Sigma) in 10 ml of water.
Mix 6 ml of this solution with 24 ml of the buffer
Enzyme: Chymotrypsin (Worthington), 1 mg/ml in 1 mM HCl

The preincubation mixture contains 40 μl of the enzyme with up to
150 μl of the inhibitor in 1 ml of the buffer. The change in absorbance is
recorded at 237 nm.

Potentiometric Assay

Reagents

Buffer: Tris-chloride 3 mM, pH 8.0, containing 45 mM KCl and
150 mM CaCl$_2$
Substrate: TAME (Sigma) 56.85 mg in 10 ml of water
Enzyme: Trypsin (Worthington), 250 μg/ml in 1 mM HCl

Procedure. Assays are performed in a Radiometer TTT1c pH stat at
a constant temperature (25°). The syringe is filled with standardized
0.1 M NaOH, and the titrator is set to raise the pH to 8.0. In the reaction
vessel preincubate for 2 min 50 μl of the enzyme with up to 200 μl of
the inhibitor, in 1 ml of the buffer. Start the reaction by adding 1 ml of
substrate. Record the rate of NaOH uptake for 2 min.

The slope of the linear relationship between NaOH consumed at the
time of reaction, gives a direct measure of micromoles of substrate consumed per minute.

Definition of Unit. The units are the same as for the spectrophotometric method.

Chymotrypsin. Inhibition of esterolytic activity of chymotrypsin is
determined on ATEE in the same way as described above for trypsin,
except for the following changes:

Substrate: Dissolve 48.56 mg of ATEE (Sigma) in 10 ml of water
Enzyme: Chymotrypsin (Worthington) 200 μg/ml in HCl, 1 mM

Purification Procedures

Method A[1]

The starting material is defatted groundnut flour prepared by two
consecutive ether extractions, in a Soxhlet aparatus, of coarsely ground
groundnuts, Virginia Sihit Meshubahat variety. The defatted meal is
then finely ground to flour. All operations are performed at room temperature. The purification is summarized in Table I.

TABLE I

YIELD AND POTENCY OF VARIOUS FRACTIONS OBTAINED DURING THE PREPARATION OF THE BASIC TRYPSIN- AND CHYMOTRYPSIN-INHIBITOR FROM GROUNDNUTS[a]

Fraction	Vol. (ml)	A_{280}	Absorbance units	Specific activity TIU/AU[b]	Specific activity ChIU/AU[c]	Total activity TIU	Total activity ChIU	Yield (%) TIU	Yield (%) ChIU	TIU/ChIU
Acidic extract	2840	22.50	63900	0.164	0.075	10480	4793	100	100	2.2
Aqueous solution of ammonium sulfate precipitate	180	60.00	10800	0.585	0.292	6318	3153	60	66	2.0
Dialyzate	236	16.00	3776	1.540	0.732	5815	2764	56	58	2.1
Active fraction from DEAE-cellulose column	360	0.80	288	10.000	4.920	2880	1417	27	29	2.0
Active fraction from Ca-phosphate column	350	0.32	112	20.000	9.333	2240	1044	22	22	2.1
Active fraction from CM-cellulose column	150	0.09	13	113.140	53.333	1471	693	14	14	2.1

[a] A. Tur-Sinai, Y. Birk, A. Gertler, and M. Rigbi, *Biochim. Biophys. Acta* **263**, 666 (1972).
[b] TIU/AU = trypsin-inhibiting units/absorbance unit.
[c] ChIU/AU = chymotrypsin-inhibiting units/absorbance unit.

Step 1. Extraction and Ammonium Sulfate Fractionation. Suspend 370 g of the starting material in 3.7 liters of 0.02 M HCl, yielding a pH of 4.8. Stir for 1 hr. Remove the undissolved residue by centrifugation for 20 min at 1000 g. Bring the supernatant to pH 7.0 with 1 M NaOH and clarify by filtration. Add solid ammonium-sulfate to 70% saturation, collect the resulting precipitate by centrifugation, and dissolve in 180 ml of distilled water. Dialyze swiftly against distilled water (12 changes, every 45 min, against 5 liters of distilled water, each).

Step 2. Chromatography on a DEAE-Cellulose Column. Prepare a DEAE-cellulose column (35 × 3.6 cm) equilibrated with 3 mM ammonium acetate buffer pH 7.0—the starting buffer. Centrifuge the solution from step 1 and treat the yellowish active supernatant (about 240 ml) with ammonium acetate, pH 7.0, to 3 mM concentration. Apply to the column and elute first with about 800 ml of the starting buffer and then with ammonium acetate buffer, 0.06 M, pH 7.0. Collect 13-ml fractions at a flow rate of 260 ml/hr. Pool the first active fraction that emerges with the hold-up volume of the starting buffer (approximately tubes 14–45), and dialyze swiftly against 1 mM phosphate buffer pH 6.8.

Step 3. Chromatography on a Calcium Phosphate Column. Prepare a calcium phosphate column (16 × 3.3 cm) according to the procedure described in a previous volume of this treatise.[4] Apply the dialyzate from step 2 to the column. Elute primarily with 300 ml of 1 mM phosphate buffer, pH 6.8, and then with 400 ml of 0.1 M of the same buffer. Collect 13-ml fractions at a flow rate of 300 ml/hr. Pool the active fraction (approximately tubes, 11–35) and dialyze swiftly against distilled water.

Step 4. Chromatography on a CM-Cellulose Column. Prepare a CM-cellulose column (40 × 1.8 cm) equilibrated with 30 mM ammonium acetate buffer pH 5.0. Treat the dialyzate from step 3 with ammonium acetate to 30 mM and apply to the column. Elute with ammonium acetate buffer pH 5.0, consecutively, as following: 450 ml of 0.03 M, 650 ml of 0.12 M, and 455 ml of 0.2 M. Collect 13-ml fractions at a flow rate of 150 ml/hr. Pool the active fractions (approximately tubes 95–106) and dialyze for 5 hr against distilled water. Lyophilize and store in a desiccator.

Method B[5]

The starting material is the dialyzed solution from step 1 (Method A). All the operations are carried out at 4°. Prepare a chymotrypsin-

[4] O. Levin, this series Vol. 5 [2].
[5] Y. Birk, S. Khalef, and M. Rotman, personal communication.

Sepharose column[6] (3.6 × 16 cm). Equilibrate with Tris-chloride, 50 mM, pH 8.0, containing 20 mM CaCl$_2$ and 200 mM KCl—the starting buffer. Centrifuge the solution from step 1 for 20 min at 1000 g. Bring the supernatant to pH 8.0 and to the molarity of the starting buffer and apply to the column. Elute inactive proteins first with the starting buffer until no more protein appears in the effluent and then elute with 0.2 M KCl pH 2.0. Collect 5-ml fractions at a flow rate of 100 ml/hr maintained with a peristaltic pump. Pool the active fraction and dialyze against distilled water. Lyophilize and store at room temperature.

Properties

Stability.[1] The inhibitor is stable when kept in acetate buffer solutions of pH 2–11 for 24 hr at 5°, or when boiled for 15 min in Trischloride buffer, pH 8.0. It is stable when incubated with trypsin or chymotrypsin at pH 8.0 for 36 hr at 5° and at room temperature.

Purity[1,5] The inhibitor prepared by Method A is a pure homogeneous protein according to polyacrylamide gel electrophoresis at pH 4.5 and pH 8.9, and by isoelectric focusing in polyacrylamide gels in the pH range of 3–10.

The inhibitor prepared by affinity chromatography (Method B) gave 4 active bands—presumably isoinhibitors—in microzone electrophoresis on cellulose acetate strips at pH 6.9.

Physical Properties.[1] The molecular weight is 7450–7700 by sedimentation and diffusion measurements, sedimentation equilibrium, gel chromatography and amino acid composition. The $s_{20,w}$ is $1.4 × 10^{-13}$, the $D_{20,w}$ is $14.07 × 10^{-7}$, and the partial specific volume, as calculated from the amino acid composition, is 0.69. The specific extinction at 280 nm is $E_{1\,cm}^{1\%} = 2.5$. The inhibitor has an isoelectric pH of 8–9, estimated by isoelectric focusing in polyacrylamide gels.

Specifity.[1,7] The inhibitor inhibits both trypsin and chymotrypsin. The enzyme–inhibitor complex is formed by 1 mole of inhibitor with 1 mole of trypsin or chymotrypsin. Full inhibition of proteolysis and nearly full inhibition of esterolysis is attained by this ratio.

The trypsin–inhibitor complex formed at pH 7.6 is devoid of tryptic activity as well as of trypsin- and chymotrypsin-inhibiting activities. Similarly the complex formed between chymotrypsin and the inhibitor lacks chymotryptic activity and chymotrypsin- and trypsin-inhibiting activities. Incubation of the inhibitor with trypsin at pH 3.75 splits a

[6] S. C. March, I. Parikh, and P. Cuatrecasas, *Anal. Biochem.* **60**, 149 (1974).
[7] Y. Birk, *Proteinase Inhibitors, Proc. Int. Res. Conf., 2nd (Bayer Symp. V)*, Grosse Ledder, 1973, p. 355. Springer-Verlag, Berlin and New York, 1974.

single Arg-X bond, which results in loss of chymotrypsin-inhibiting activity, whereas the ability to inhibit trypsin is not affected even upon removal of the newly formed carboxy-terminal Arg by carboxypeptidase B. Incubation of the inhibitor with chymotrypsin at pH 3.75 did not open any peptide bond in the inhibitor and did not affect its trypsin- or chymotrypsin-inhibiting activity.

All four isoinhibitors prepared by method B form distinct complexes with trypsin and with chymotrypsin when submitted to electrophoresis on paper or on cellulose acetate strips at pH 6.9.

Amino acid composition.[1] The amino acid composition of the inhibitor prepared by method A is given in Table II. No free -SH groups or carbohydrates could be detected.

TABLE II

AMINO ACID COMPOSITION OF TRYPSIN- AND CHYMOTRYPSIN-
INHIBITOR FROM GROUNDNUTS[a,b]

Amino acid	Number of residues per molecule
Aspartic acid	8
Glutamic acid	6
Glycine	4
Alanine	3
Valine	5
Leucine	1
Isoleucine	0
Serine	5
Threonine	7
Half-cystine	14
Methionine	0
Proline	7
Phenylalanine	2
Tyrosine	1
Histidine	2
Lysine	2
Arginine	7
Tryptophan	0
	74

[a] A. Tur-Sinai, Y. Birk, A. Gertler, and M. Rigbi, *Biochim. Biophys. Acta* **263**, 666 (1972).

[b] Values are given in molar ratios; the molecular weight of the inhibitor was taken as 7580.

[62] Proteinase Inhibitors from Cereal Grains

By YEHUDITH BIRK

Trypsin inhibitors are widely distributed among the cereal grains including[1] wheat, corn, rye, oats, buckwheat, barley, rice, and sorghum.[2]
As stated earlier,[3] they have been isolated from whole wheat flour, corn
seed, rye and wheat germ, barley, and oats and recently also from rice[4]
and rye seeds.[5] The amino acid sequence of the corn seed inhibitor[6] and
the partial sequences of the rye and wheat germ inhibitors[7] have been
reported. The preparation and characterization of the trypsin inhibitor
from barley, as reported by Mikola and Suolinna,[8] was described in detail in this series.[3] Mikola and Kirsi[9] later showed that trypsin inhibitors
are present both in the embryos and endosperms of grains of barley,
wheat, and rye, the inhibitors in the two respective organs being dis-
similar to each other.

The presence, in barley, of inhibitors of *Aspergillus oryzae* proteinases has been known since 1955.[10] They were further studied by
Mikola and Suolinna,[11] who have purified and characterized a group of
isoinhibitors from barley that inhibit the alkaline proteinases of *Aspergillus oryzae, Bacillus subtilis, Streptomyces griseus*, and *Alternaria
tenuissima* and chymotrypsin. These inhibitors have been selected for
detailed description herein.

Studies have also been carried out on formation of trypsin inhibitors
and of *Aspergillus*-proteinase inhibitors in developing barley grain[12] and
in germinating barley embryos.[13] Recently, the wheat inhibitor of

---

[1] I. E. Liener and M. L. Kakade, *in* "Toxic Constituents of Plant Foodstuffs" (I. E.
Liener, ed.), p. 8. Academic Press, New York, 1969.
[2] J. Xavier-Filho, *J. Food Sci.* **39**, 422 (1974).
[3] B. Kassell, this series Vol. 19 [66a].
[4] T. Horiguchi and K. Kitagishi, *Plant Cell Physiol.* **12**, 907 (1971).
[5] A. Polanowski, *Acta Soc. Bot. Pol.* **43**, 27 (1974).
[6] K. Hochstrasser, K. Illchmann, and E. Werle, *Hoppe-Seyler's Z. Physiol. Chem.*
351, 721 (1970).
[7] K. Hochstrasser, M. Muss, and E. Werle, *Hoppe-Seyler's Z. Physiol. Chem.* **350**,
249 (1969).
[8] J. Mikola and E.-M. Suolinna, *Eur. J. Biochem.* **9**, 555 (1969).
[9] J. Mikola and M. Kirsi, *Acta Chem. Scand.* **26**, 787 (1972).
[10] K. Matsushima, *J. Agric. Chem. Soc. Jpn.* **29**, 405 (1955).
[11] J. Mikola and E.-M. Suolinna, *Arch. Biochem. Biophys.* **144**, 566 (1971).
[12] M. Kirsi, *Physiol. Plant.* **29**, 141 (1973).
[13] M. Kirsi, *Physiol. Plant.* **32**, 89 (1974).

Tribolium castaneum larval proteinases[14] has been purified and characterized.[15]

Aspergillus Proteinase Inhibitor from Barley

Assay Method[11]

Principle. The inhibition of hydrolysis of casein by the alkaline proteinase of *Aspergillus oryzae* is measured by the change in absorbance at 280 nm of a trichloroacetic acid supernatant of the reaction mixture.

Reagents

Buffer 1: sodium glycinate, 50 mM, pH 10.3
Substrate: casein, 1%, in buffer 1
Buffer 2: buffer 1 containing 20 mM EDTA.
Enzyme solution: *Aspergillus oryzae* proteinase (crude, type II Sigma Chemical Corp., St. Louis, Missouri) 600 μg per milliliter of buffer 2
Inhibitor solution: Dilute with buffer 2 to a concentration to give about 50% inhibition of the enzyme.
Trichloroacetic acid: 5% in water (w/v)

Procedure. Pipette 1 ml of substrate into a series of test tubes in a 35° bath. In a second series of small test tubes, place 2 ml of enzyme solution plus 2 ml of diluted inhibitor for unknowns. In the same way pipette 2 ml of enzyme solution and 2 ml of buffer for control containing no inhibitor. Allow to stand at room temperature for 10 min. Timing with a stopwatch, add 1 ml of the enzyme and inhibitor mixture (or enzyme and buffer mixture) to the substrate and mix well. Incubate at 35°. Exactly 30 min later, stop the reaction by the addition of 3 ml of 5% trichloroacetic acid. Allow the tubes to stand at room temperature for 30 min for precipitation. Then clarify by centrifugation. Measure the absorbance at 280 nm against a blank without enzyme or inhibitor. Subtract the absorbance of the unknowns from the absorbance of the enzyme standard.

Definition of Unit. One unit of enzyme is the amount that causes an increase of 1 unit of absorbance at 280 nm per minute of digestion of casein under the standard conditions. Enzyme-inhibiting activity is expressed as units of enzyme inhibited. This is based on the casein digestion method of Kunitz,[16] originally devised for the determination of

[14] S. W. Applebaum and A. M. Konijn, *J. Insect Physiol.* **12**, 665 (1966).
[15] R. Lippmann, Y. Birk, and S. W. Applebaum, personal communication.
[16] M. Kunitz, *J. Gen. Physiol.* **30**, 291 (1947).

TABLE I
PURIFICATION OF *Aspergillus* PROTEINASE
INHIBITOR FROM PIRKKA BARLEY[a]

Fraction	Volume (ml)	Total activity (units)	Specific activity[b] (units/ E_{280})	Re-covery (%)
Extract	2160	268,000	19[c]	
Ammonium sulfate precipitate, 20–40 %	232	196,000		100
Same after dialysis	280	180,000	32	92
DEAE-cellulose peak	342	123,000	102	64
Diethylether precipitate	60	108,000	370	55
Sephadex G-75 peak, freeze-dried	190 mg	98,000	580	50

[a] J. Mikola and E.-M. Suolinna, *Arch. Biochem. Biophys.* **144**, 566 (1971).

[b] Ratio of activity (units/ml) to the E_{280} of the solution (1.0-cm cuvette).

[c] Determined separately after removal of small-molecular compounds by gel filtration.

trypsin and later applied to the determination of trypsin inhibitors and chymotrypsin. The original method is described in detail in a previous volume of this series.[17]

Purification Procedure[11]

The procedure is summarized in Table I. All operations are effected at $+5°$, unless otherwise stated. Between the different stages of purification the inhibitor solutions should be frozen rapidly and then stored at $-18°$.

Step 1. Extraction. The starting material is Pirkka barley (a Finnish six-row variety). Mix thoroughly 1600 g of finely ground barley with 3200 ml of water. Incubate the thick slurry (pH 5.9) at $5°$ for 2 hr without agitation. Then separate the extract by centrifugation and allow to stand overnight a $5°$.

Step 2. Ammonium Sulfate Precipitation. Add solid ammonium sulfate to 20% saturation, stir the suspension for half an hour, and remove the inactive precipitate by centrifugation. To precipitate the active material, add to the supernatant fraction ammonium sulfate to 40% saturation and stir for 1 hr. Collect the active precipitate by centrifugation and dissolve in one-tenth the original volume of water. Remove the small amount of insoluble material by centrifugation. Dialyze the active solution extensively against 35 mM Tris chloride buffer, pH 7.5.

[17] M. Laskowski, Sr., this series Vol. 2 [3].

Step 3. Chromatography on DEAE-Cellulose. Prepare a 5 × 14 cm column of DEAE-cellulose, equilibrated with 35 mM Tris chloride buffer, pH 7.5—the starting buffer. Apply half of the solution of step 2 (about 140 ml) to the column. Elute inert proteins first with 100 ml of the starting buffer and then elute the active material with 125 mM Tris chloride buffer, pH 7.5. Treat the other half of the solution similarly, and combine the active eluates.

Step 4. Precipitation by Ethanol and Diethyl Ether. All the operations should be effected at −15°, and the organic solvents should be chilled to −18° before use. Mix 1 volume of the active fraction from step 3 with 3 volumes of 94% ethanol. After 30 min, remove the large, inactive precipitate by centrifugation. To the supernatant fraction add 2 volumes of diethyl ether. After 30 min collect the active precipitate by centrifugation. Dissolve in one-fifth the original volume of 25 mM acetic acid and lyophilize.

Step 5. Gel Filtration on Sephadex G-75. Prepare a 2.5 × 95 cm column of Sephadex G-75, equilibrated with sodium acetate buffer pH 4.9, ionic strength 0.1. Dissolve the active lyophilization residue from step 4 in a small volume of the same buffer and apply to the column. Elute with the same buffer, collecting 6.7-ml fractions at the rate of 30 ml/hr at a temperature of 25°. The active peak appears approximately in tubes 200–300. Pool the active peak and dialyze against 9 volumes of water for 18 hr. Adjust the dialyzed solution to pH 7.5 with 1 N NaOH and precipitate the inhibitors with ethanol and ether exactly as in step 4. Dissolve the precipitate in 25 mM acetic acid and lyophilize.

Properties[11]

Stability. The inhibitor is heat labile. Incubation at 100° for 15 min at pH 5.4 and ionic strength of 0.05 results in complete inactivation.

Purity. As found from polyacrylamide gel electrophoresis and isoelectric focusing, the purified inhibitor consists of 4–5 isoinhibitors, similar in molecular size but with different isoelectric points, which were already present at the beginning of the purification procedure. The inhibitor preparation is relatively free from contaminating proteins.

Physical Properties. The molecular weight is 25,000 as figured from the elution volume of the active peak from the Sephadex G-75 column. A solution of 1 mg of inhibitor per milliliter has an absorbance of 0.882 at 280 nm.

Specificity. The inhibitor (preparation from step 5) inhibits the alkaline proteinases of *Aspergillus oryzae, Streptomyces griseus* (casein substrate, pH 10.3), *Alternaria tenuissima* (casein substrate pH 9.5), and

TABLE II
AMINO ACID COMPOSITION OF BARLEY *Aspergillus*
PROTEINASE INHIBITOR[a]

Amino acid	Residues/MW 25,000 in		
	20-hr hy- drolyzate	72-hr hy- drolyzate	Mean
Aspartic acid	15.1	14.1	14.6
Glutamic acid	21.3	21.2	21.3
Glycine	12.7	13.3	13.0
Alanine	18.8	19.8	19.3
Valine	19.5	25.1	25.1[b]
Leucine	8.1	9.0	8.6
Isoleucine	7.6	11.3	11.3[b]
Serine	14.0	12.5	14.4[c]
Threonine	8.9	8.4	9.1[c]
Half-cystine	0.4	0.4	0.4
Methionine	6.2	6.1	6.2
Proline	20.3	21.3	20.9
Phenylalanine	4.2	4.9	4.6
Tyrosine	2.6	2.4	2.5
Histidine	1.6	1.7	1.7
Lysine	11.8	12.0	11.9
Arginine	8.7	8.9	8.8
Amide-NH_2	12.0	—	11.1[d]
Tryptophan	ND[e]	ND	ND

[a] J. Mikola and E.-M. Suolinna, *Arch. Biochem. Biophys.* **144**, 566 (1971).
[b] 72-hr value.
[c] Extrapolated to zero time.
[d] Corrected for destruction of serine and threonine.
[e] Not determined.

Bacillus subtilis (casein substrate, pH 10.3) and also chymotrypsin (casein substrate, pH 7.6, and glutaryl-L-phenylalanine *p*-nitroanilide substrate, pH 7.6). It does not inhibit trypsin (BAPA substrate, pH 8.2), the endopeptidases of germinating barley (gelatin substrate, pH 5.4), pepsin (hemoglobin substrate, pH 2.0), and papain (casein substrate, pH 7.6). It also has no effect on the neutral proteinases of *Aspergillus oryzae*, *Streptomyces griseus*, and *Bacillus subtilis* (casein substrates, pH 6.5).

Kinetic Properties. The inhibition of *Aspergillus* proteinase is linear up to 65% and reaches its maximum at about 95% inhibition. That of *Bacillus subtilis* is linear up to 30%, with a maximum of about 75% inhibition. Inhibition of *Streptomyces griseus* and *Alternaria tenuissima* proteinases has no linear section, and it reaches maximum inhibition at

75% and 80%, respectively. Inhibition of chymotryptic digestion of casein is linear up to 20% with a maximum of 75%, and that of glutaryl-L-phenylalanine p-nitroanilide up to 70% with a maximum of 95%.

Amino Acid Composition. The amino acid composition expressed as residues per molecular weight of 25,000 is shown in Table II. The inhibitor contains no carbohydrates.

[63] Proteinase Inhibitors from Potatoes

By YEHUDITH BIRK

A large number of proteinase inhibitors have been separated and isolated from potatoes. In spite of a certain confusion in the nomenclature used by different groups of investigators, the potato proteinase inhibitors seem to belong to four major groups.

The chymotrypsin inhibitor I, described in an earlier volume in this series,[1] has been found to be similar to potato proteinase inhibitor I.[2,3] Chymotrypsin Inhibitor I is composed of four subunit components,[4,5] which will be described in detail in the present section. The second group comprises proteinase inhibitors IIa and IIb, studied extensively by Iwasaki and associates.[6-10] Both of them inhibit stoichiometrically chymotrypsin and the bacterial proteinase nagarse, but inhibitor IIa inhibits also trypsin. Inhibitor IIa is probably identical with inhibitor A7b described by Belitz and associates as one of thirteen inhibitors of trypsin and chymotrypsin isolated by isoelectric focusing.[11-13] To the

[1] C. A. Ryan and B. Kassell, this series Vol. 19 [66f].
[2] T. Kiyohara, T. Iwasaki, and M. Yoshikawa, *J. Biochem. (Tokyo)* **73**, 89 (1973).
[3] T. Kiyohara, M. Fujii, T. Iwasaki, and M. Yoshikawa, *J. Biochem. (Tokyo)* **74**, 675 (1973).
[4] J. C. Melville and C. A. Ryan, *Arch. Biochem. Biophys.* **138**, 700 (1970).
[5] J. C. Melville and C. A. Ryan, *J. Biol. Chem.* **247**, 3445 (1972).
[6] T. Iwasaki, T. Kiyohara, and M. Yoshikawa, *J. Biochem. (Tokyo)* **70**, 817 (1971).
[7] T. Iwasaki, T. Kiyohara, and M. Yoshikawa, *J. Biochem. (Tokyo)* **72**, 1029 (1972).
[8] T. Iwasaki, T. Kiyohara, and M. Yoshikawa, *J. Biochem. (Tokyo)* **73**, 1039 (1973).
[9] T. Iwasaki, T. Kiyohara, and M. Yoshikawa, *J. Biochem. (Tokyo)* **74**, 335 (1973).
[10] T. Iwasaki, T. Kiyohara, and M. Yoshikawa, *J. Biochem. (Tokyo)* **75**, 843 (1974).
[11] H. D. Belitz, K. P. Kaiser, and K. Santarius, *Biochem. Biophys. Res. Commun.* **42**, 420 (1971).
[12] K. Santarius and H. D. Belitz, *J. Biochem. (Tokyo)* **154**, 206 (1974).
[13] K. P. Kaiser, L. C. Bruhn, and H. D. Belitz, *Z. Lebensm.-Unters. Forsch.* **154**, 339 (1974).

third group belong the two kallikrein inhibitors PKI-56 and PKI-64, which are potent inhibitors of human plasma kallikrein and of rat serum kallikrein, but inhibit only weakly trypsin, chymotrypsin, pepsin, plasmin, nagarse, and the proteinase of *Serratia* and *Streptomyces griseus*.[14-17] The fourth group includes the low-molecular-weight inhibitor of pancreatic carboxypeptidase A and B,[18-21] which will be described in detail herein.

Chymotrypsin Inhibitor I from Potatoes and Its Subunit Components

Assay Methods[5]

Chymotrypsin inhibitor I (henceforth inhibitor I) is assayed by a number of techniques. The assays of chymotrypsin- and trypsin-inhibiting activities (Method 1) are based on the spectrophotometric method of Hummel[22] for determination of the esterase activities of chymotrypsin and trypsin and have already been described in Vol. 19 of this series.[23,24] The second is an immunological method for measurement of inhibitor concentration,[25] and will be described herein.

Method 1. The esterase activity of chymotrypsin on *N*-benzoyl-L-tyrosine ethyl ester (BTEE), in the absence and in the presence of the inhibitor, is determined with a recording spectrophotometer. The reaction is initiated by adding the solution containing enzyme and inhibitor. The activity of the chymotrypsin is calculated from the slope of the reaction curve without inhibitor, and the inhibition is calculated from the difference between this value and the slope of the reaction curve containing inhibitor. One unit of esterase is the amount of enzyme that hydrolyzes 1 μmole of substrate per minute. One unit of inhibition is the difference in units between the reaction mixture containing enzyme without inhibitor

[14] Y. Hojima, H. Moriya, and C. Moriwaki, *J. Biochem. (Tokyo)* **69**, 1019 (1971).
[15] Y. Hojima, H. Moriya, and C. Moriwaki, *J. Biochem. (Tokyo)* **69**, 1027 (1971).
[16] Y. Hojima, C. Moriwaki, and H. Moriya, *J. Biochem. (Tokyo)* **73**, 923 (1973).
[17] Y. Hojima, C. Moriwaki, and H. Moriya, *J. Biochem. (Tokyo)* **73**, 933 (1973).
[18] J. M. Rancour and C. A. Ryan, *Arch. Biochem. Biophys.* **125**, 380 (1968).
[19] C. A. Ryan, *Biochim. Biophys. Res. Commun.* **44**, 1265 (1971).
[20] C. A. Ryan, G. M. Hass, and R. W. Kuhn, *J. Biol. Chem.* **249**, 5495 (1974).
[21] C. A. Ryan, G. M. Hass, R. W. Kuhn, and H. Neurath, *Proteinase Inhibitors, Proc. Int. Res. Conf. 2nd (Bayer Symp. V)*, Grosse Ledder, 1973, p. 565. Springer-Verlag, Berlin and New York, 1974.
[22] B. C. W. Hummel, *Can. J. Biochem. Physiol.* **37**, 1393 (1959).
[23] K. A. Walsh and P. E. Wilcox, this series Vol. 19 [3].
[24] K. A. Walsh, this series Vol. 19 [4].
[25] C. A. Ryan, *Anal. Biochem.* **19**, 434 (1967).

and the reaction mixture containing enzyme and inhibitor. The reaction between chymotrypsin and inhibitor I is stoichiometric, and therefore the amount of chymotrypsin neutralized can be used to calculate the amount of inhibitor in each preparation.

The esterase activity of trypsin in the absence and in the presence of inhibitor is determined in a similar manner on the substrate p-toluene-sulfonyl-L-arginine methyl ester (TAME).

Method 2. The immunological determination of inhibitor concentration is performed by measuring the diameter of radial precipitin zones formed by diffusion of the antigen (inhibitor I) through agar gels containing anti-inhibitor I antibodies.

Reagents

> Antiserum: Prepare rabbit anti-inhibitor I serum by injecting young rabbits weighing 4–5 pounds subcutaneously with 1 mg of four times crystallized inhibitor I emulsified in complete Freund's adjuvant, twice weekly for 3 weeks. Rest the rabbits for approximately 3 months and inject again twice weekly for 2 weeks. After 10 days bleed the rabbits through an ear vein. Collect 30 ml of blood twice weekly for 3 weeks. Immediately after collection chill the blood and store overnight in a refrigerator. Centrifuge at $2000g$ for 15 min, draw off the clear serum, and freeze. The frozen serum is fully active after several months of storage. The dilution of antiserum in agar is determined by preliminary tests with serial dilutions ranging from 1:25 to 1:200.

> Agar gels: Use 2% Noble agar (Difco) in 0.1 M sodium Veronal, 0.9% NaCl, pH 8.2, containing 1:10,000 thimerosal. Dissolve the agar in buffer over a steam bath and cool to 45°–50°. Add a drop of trypan blue to enhance the dark-field illumination of the precipitin zone. Add the antiinhibitor serum and mix thoroughly. Pour the agar–antibody solution into plastic 100 × 15 mm petri dishes to a depth of about 2 mm (15 ml of agar per plate) and allow to cool and solidify. Using a thin-walled hollow cylinder (4.3 mm o.d.), punch 6–8 wells in the gels at 2-cm intervals. Remove the center agar of the wells by gentle suction using a Pasteur pipette.

Procedure. Fill the wells in the agar with the test solutions. Cover the petri dishes, and incubate in a humid atmosphere at room temperature. After 40 hr measure the diameter of each radial precipitin zone carefully under dark-field illumination. Use a vernier caliper precise to 0.1 mm, aided by a 1.5 × magnifier 12-cm in diameter. Consider parallax carefully during measurements.

Definition of Units. The diameter of the rings is a log–log function of antigen concentration. To obtain a standard curve, use four times crystallized inhibitor I as a standard.

Purification Procedure[5]

The preparation of inhibitor I and its individual subunits comprises of 4 stages, three of which have been fully described in Vol. 9[1] (steps 1 through 6) and are summarized in Table I. In the fourth and last stage, four-subunit components are obtained from the gel-filtered inhibitor (step 5[1]) by chromatography on sulfoethylcellulose using the dissociating solvent of 0.2 M formic acid in 8 M urea, pH 2.8.

Chromatography of Gel-Filtered Inhibitor on Sulfoethyl Cellulose. Equilibrate sulfoethyl cellulose with water for at least 1 day prior to use and remove the fines by decantation. Wash the ion-exchanger successively with approximately 20 volumes of 0.1 N NaOH, 0.1 N hydrochloride, distilled water, and then with 0.2 M formic acid in 8 M urea pH 2.8—the starting buffer. Add the washed exchanger to about 5 volumes of the starting buffer and pour into the column (1×10 cm) as a thick slurry. To equilibrate, allow at least 10 void volumes of starting buffer to flow through the column. Dissolve 100 mg of gel-filtered inhibitor (the lyophilized desalted preparation from step 5[1]) in the starting buffer at least 0.5 hr prior to application to the column. Apply the solution to the column and elute inhibitor I fractions with a linear salt gradient of

TABLE I
SUMMARY OF PURIFICATION OF INHIBITOR I[a]

Step	Total protein (mg)	Total inhibitor I[b] (mg)	Purification factor	Recovery (%)
1. Original juice, filtered at pH 3.0	8940	212	0	100
2. Ammonium sulfate precipitate, 0–70%	5975	156	1.8	73
3. Heat, 80°, 5 min, filter	771	149	17.9	70
4. Sephadex G-75 chromatography, Tris buffer, pH 8.2				
5. Sephadex G-75 chromatography, guanidine hydrochloride, pH 8.0	100	100	39.6	17

[a] J. C. Melville and C. A. Ryan, *J. Biol. Chem.* **247**, 3445 (1972).
[b] Radial diffusion [C. A. Ryan, *Anal. Biochem.* **19**, 434 (1967)] with four times crystallized inhibitor I as a standard.

0 to 0.4 M KCl in the starting buffer. Collect fractions of 1.5 ml at a flow rate of about 20 ml/hr. Figure 1 shows the elution of four subunit components (protomers) of inhibitor I. Isolate each peak, desalt, and lyophilize, and rechromatograph separately under the same conditions as described above. The rechromatographed peaks should elute at the same salt concentration as in the original chromatography.

Properties

Those properties of inhibitor I that have been covered in an earlier article in this series,[1] will not be listed herein.

Stability.[5] Gel-filtered inhibitor dissociates at pH 3 into dimers with a consequent change in molecular weight from 39,000 to about 20,000. The unfractionated inhibitor is susceptible to digestion with pepsin at pH 3 but its four isolated variants A, B, C, and D are not equally digested. Thus the immunological reactivity of species D was rapidly destroyed by pepsin whereas species A, B, and C were more stable.

Purity.[5] Electrophoresis on cellulose acetate strips in 8 M urea of the four individual monomers (henceforth protomers) as compared to gel-filtered inhibitor I shows that the four protomers represent the four electrophoretically identifiable components of the inhibitor. Only the

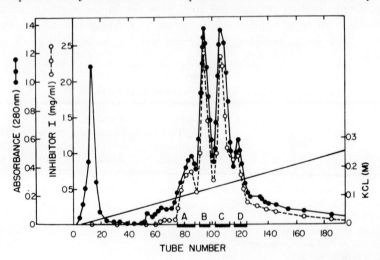

FIG. 1. Chromatography of gel-filtered inhibitor on a sulfoethylcellulose column (1 × 10 cm) equilibrated with 8 M urea, 0.2 M formic acid, pH 2.8, and eluted with a linear KCl gradient, represented by the solid line (——). The column was charged with 102 mg of the preparation. ○, Inhibitor I as determined by radial diffusion assays[25]; ●, absorbance at 280 nm. Fractions of 1.4 ml were collected at 3-min intervals. Reproduced from J. C. Melville and C. A. Ryan, *J. Biol. Chem.* 247, 3445 (1972).

protomer from peak C (Fig. 1) is similar to crystalline inhibitor I in its electrophoretic mobility. The ability of the protomers to hybridize with each other to form stable tetrameric isoinhibitors suggests that native inhibitor is a heterogeneous mixture of subunits whose properties reflect those of the individual proteins that comprise it. The immuno-electrophoretic data suggest that partially purified inhibitor I does exist as a mixture of heterogeneous tetrameric species.

Physical Properties.[5] A summary of the information concerning the molecular weights of inhibitor I, the protomers, and the complex of inhibitor I with chymotrypsin, is given in Table II.

Specificity and Kinetic Properties.[5] The inhibitory activities of tetramers reconstituted from the four protomer types are compared in Table III. Molar combining ratios of the most effective inhibitors, B and C, based on a molecular weight of 39,000 for inhibitor and 25,000 for chymotrypsin, show 4 moles of chymotrypsin inhibited per mole of inhibitor.

Immunological Properties.[5] The cross-reactivity among the tetrameric species A, B, C, and D is identical; this shows that the protomers possess considerable regions that are immunologically very similar.

Amino Acid Composition.[5] The amino acid composition of inhibitor tetramers reconstituted from the individual protomers A, B, C, and D

TABLE II

SUMMARY OF MOLECULAR WEIGHT DETERMINATIONS OF INHIBITOR I AND ITS COMPLEX WITH CHYMOTRYPSIN[a]

Protein	Method	Molecular weight or equivalent weight
Inhibitor I	Ultracentrifuge, Archibald	$38,000 \pm 2400$
Four times crystalized	Ultracentrifuge, velocity	$37,500 \pm 2800$
	Sephadex G-75, Tris buffer, pH 8.2	$39,000 \pm 1000$
Dissociation-purified inhibitor I	Ultracentrifuge, equilibrium	$39,900 \pm 1000$
	Sephadex G-75, Tris buffer, pH 8.2	$39,000 \pm 1100$
Monomers	Sephadex G-75, 4 M guanidine, pH 8.5	$9,800 \pm 200$
	Sephadex G-75, 8 M urea, pH 2.8	$10,000 \pm 500$
	Electrophoresis in 0.1 % sodium dodecyl sulfate	$9,300 \pm 1100$
	α-Chymotrypsin inhibition (synthetic substrate)	$9,500 \pm 500$
Inhibitor I–chymotrypsin complex	Sephadex G-200	$140,000 \pm 4600$

[a] J. C. Melville and C. A. Ryan, *J. Biol. Chem.* **247**, 3445 (1972).

TABLE III
Inhibition of Chymotrypsin and Trypsin Esterase Activities by Reassociated Tetramers Isolated from Inhibitor I[a,b]

| | Enzyme inhibited[c] (mg/mg inhibitors) | |
Tetrameric species	Chymotrypsin	Trypsin
A	0.72	0.49
B	2.81	1.23
C	2.81	0.22
D	1.67	0.22

[a] J. C. Melville and C. A. Ryan, *J. Biol. Chem.* **247**, 3445 (1972).

[b] The substrate for chymotrypsin was BTEE, and for trypsin, TAME, using the assay system of B. C. W. Hummel [*Can. J. Biochem. Physiol.* **37**, 1393 (1959)]. BTEE, N-benzoyl-L-tyrosine ethyl ester; TAME, p-toluenesulfonyl-L-arginine methyl ester.

[c] Calculated at 50 % inhibition.

TABLE IV
Amino Acid Composition of Inhibitor I Subunits and Gel-Filtered Inhibitor I[a]

| | Amino acid residues per molecule[b] | | | | |
| | Subunits | | | | Gel-filtered inhibitor I |
Amino acid	A	B	C	D	
Lysine	5.2	5.0	5.9	5.8	5.2
Histidine	0.4	0.4	0.7	1.0	0.6
Arginine	3.9	3.6	4.6	3.9	3.6
Aspartic acid	11.6	10.4	10.5	8.4	9.1
Threonine	3.6	2.8	3.1	3.3	2.9
Serine	5.1	4.0	4.2	3.4	4.2
Glutamic acid	11.1	9.3	8.6	8.0	9.2
Proline	6.6	7.1	5.4	4.3	5.4
Glycine	6.9	5.5	7.1	8.1	6.3
Alanine	6.5	2.3	2.4	2.5	4.7
Half-cystine	2.5	2.1	2.6	2.1	2.1
Valine	9.6	8.0	8.7	6.1	7.6
Methionine	1.1	0.6	0.6	1.4	0.8
Isoleucine	7.7	6.3	7.7	5.6	5.9
Leucine	8.0	9.8	0.4	6.9	7.7
Tyrosine	2.3	0.6	1.0	1.3	1.3
Phenylalanine	3.1	3.2	2.3	1.9	2.4

[a] J. C. Melville and C. A. Ryan, *J. Biol. Chem.* **247**, 3445 (1972).

[b] Calculated on the basis of a molecular weight of 9500 for all species of proteins.

and of gel-filtered inhibitor I, from which they were prepared, is shown in Table IV. All four of the protomer peaks have an NH$_2$-terminal glutamic acid, as determined by the dansylation procedure.

Sequence.[26,27] The complete amino acid sequences of the variants isolated as promoters A, B, C and D are as follows.

```
                                 (Ser) 10
A       Glu Phe Glu Cys Asp Gly Lys Leu Gln Trp Pro Glu Leu Ile Gly Val Pro
B       Glu Phe Glu Cys Lys Gly Lys Leu Gln Trp Pro Glu Leu Ile Gly Val Pro
C   Lys Glu Phe Glu Cys Lys Gly Lys Leu Gln Trp Pro Glu Leu Ile Gly Val Pro
D   Lys Glu Phe Glu Cys Asn Gly Lys Leu Gln Trp Pro Glu Leu Ile Gly Val Pro


               20                    30
A       Thr Lys Leu Ala Lys Glu Ile Ile Glu Lys Gln Asn Ser Leu Ile Ser Asn
B       Thr Lys Leu Ala Lys Gly Ile Ile Glu Lys Gln Asn Ser Leu Ile Thr Asn
                                                                        Ser
C       Thr Lys Leu Ala Lys Gly Ile Ile Glu Lys Gln Asn Ser Leu Ile Thr Asn
                                                                        Ser
D       Thr Lys Leu Ala Lys Gly Ile Ile Glu Lys Gln Asn Ser Leu Ile Thr Asn
                                                                        Ser


               40                    50
A       Val His Ile Leu Leu Asn Gly Ser Pro Val Thr Leu Asp Ile Leu Gly Asp
B       Val His Ile Leu Leu Asn Gly Ser Pro Val Thr Leu Asp Ile Leu Gly Asp
C       Val Gln Ile Leu Leu Asn Gly Ser Pro Val Thr Leu Asp Ile Leu Gly Asp
D       Val Gln Ile Leu(Leu)Asn Gly Ser Pro Val Thr Leu Asp Ile Leu Gly Asn
                        Lys


          (Met)              60
A       Val Val Gln Leu Pro Val Val Gly Met  Asp Phe Arg Cys Asp Arg Val Arg
          (Met)                       (Phe)
B       Val Val Asp Ile Pro Val Val Gly Met  Asp Tyr Arg Cys Asp Arg Val Arg
C       Val Val Asp Ile Pro Val Val Gly Met  Aps Tyr Arg Cys Asn Arg Val Arg
D       Val Val Asp Ile Pro Val Val Gly Met  Asp Tyr Arg Cys Asn Arg Val Arg


          70                   80
A       Leu Phe Asp Asp Ile Leu Gly Ser Val Val Gln Ile Pro Arg Val Ala
B       Leu Phe Asp Asp Ile Leu Gly Tyr Val Val Gln Ile Pro Arg Val Ala
C       Leu Phe Asp Asn Ile Leu Gly Asn Val Val Gln Ile Pro Arg Val Ala
                                Tyr
D       Leu Phe Asp Asn Ile Leu Gly Asn Val Val Gln Ile Pro Arg Val Ala
                                Ser
```

[26] M. Richardson, *Biochem. J.* **137**, 101 (1974).
[27] M. Richardson and L. Cossins, *FEBS Lett.* **45**, 11 (1974).

Carboxypeptidase Inhibitor from Potatoes

Assay Methods[20,28]

Principle. The method is based upon measurement of change in ultraviolet absorption at 254 nm that results from enzymic hydrolysis of *N*-benzoyl-glycyl-L-phenylalanine (hippuryl-L-phenylalanine) to *N*-hippuric acid and phenylalanine, in the absence and in the presence of inhibitor.

Reagents

Buffer: Tris chloride, 20 mM, pH 7.5, in 0.5 M sodium chloride
Substrate: Hippuryl-L-phenylalanine, 1 mM, in the buffer
Enzyme: Dilute stock bovine carboxypeptidase A (Worthington) to a concentration of 10–25 μg/ml in 10% LiCl, immediately before use. Absorbance index at 278 nm = 1.94/cm/mg/ml.
Inhibitor: Dilute with the buffer to a concentration that gives 50% inhibition of the enzyme.

Procedure. Place two 1-cm quartz cuvettes, containing 3 ml of substrate, in a double-beam spectrophotometer and allow to equilibrate at 25°. Start the reaction by adding 0–50 μl of a solution of enzyme or of enzyme mixed with inhibitor. Reset the absorbance at 254 nm immediately to zero and record increasing absorbance.

Definition of Specific Activity.[20] The specific activity of the inhibitor is expressed as micrograms of carboxypeptidase A inhibited by 1 μg of inhibitor at 50% inhibition.

Purification Procedure[1,5,20,21]

The starting material is crude inhibitor I, prepared from Russet Burbank potato tuber juice through a series of steps including homogenization and extraction at pH 3, fractionation with ammonium sulfate (0 to 40%), heat treatment, desalting and lyophilization, according to the procedure of Melville and Ryan,[5] described in Vol. 19.[1]

Step 1. Gel Filtration on Sephadex G-75. Dissolve 3 g of crude inhibitor I from step 4[1] in 50 ml of 0.1 M KCl in 0.05 M Tris-chloride buffer, pH 8.2, and clarify by centrifugation at 15,000 *g* for 10 min. Apply to a 10 × 100 cm column of Sephadex G-75 equilibrated with the same buffer. Apply the sample to the column from the bottom, and use an upward flow of about 500 ml/hr for elution. Collect fractions of 50 ml. Figure 2 shows the elution pattern. Pool the active fractions (peak III), lyophilize, and store at 0°.

[28] J. E. Folk and E. W. Schirmer, *J. Biol. Chem.* **238**, 3884 (1963).

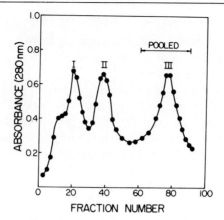

Fig. 2. Chromatography of crude inhibitor I on Sephadex G-75. A column, 10 × 100 cm, was eluted with 0.1 *M* KCl, 0.05 *M* Tris-HCl buffer, pH 8.2. Three grams of crude inhibitor, dissolved in 50 ml of buffer and clarified by centrifugation at 15,000 *g* for 10 min, was applied to the column. An upward flow rate of 500 ml/hr was employed, and 52-ml fractions were collected. Reproduced from C. A. Ryan, G. M. Hass, and R. W. Kuhn, *J. Biol. Chem.* 249, 5495 (1974).

Step 2. Chromatography on Phosphocellulose. Dissolve 50 mg of the lyophilized material from step 1 in 0.01 *M* sodium citrate buffer pH 4.3. Apply to a 1.4 × 13 cm column of phosphocellulose equilibrated with the same buffer. After elution of the breakthrough peak employ a linear gradient of sodium chloride (0 to 0.25 *M*) in the equilibration buffer. Collect 5.4-ml fractions with a flow rate of 60 ml/hr. Figure 3 shows the elution pattern. Pool the active fractions (eluted by 0.18–0.19 *M* sodium chloride), desalt by passing through a 1.5 × 50 cm column of Bio-Gel P-2, lyophilize, and store at 0°. A summary of the purification of the inhibitor from 1000 g of potatoes is shown in Table V.

TABLE V

SUMMARY OF PURIFICATION OF CARBOXYPEPTIDASE INHIBITOR FROM 1000 g
OF RUSSET BURBANK POTATOES[a]

Step	Carboxypeptidase			
	Protein (mg)	Inhibitor (mg)	Recovery (%)	Specific activity
Original juice filtered at pH 3	8940	30	100	0.003
Crude inhibitor I	771	18	60	0.024
Sephadex G-75	104	18	60	0.173
Phosphocellulose	14	14	48	10.000

[a] C. A. Ryan, G. M. Hass, and R. W. Kuhn, *J. Biol. Chem.* **249,** 5495 (1974).

Properties[19,20]

Stability. The inhibitor is stable upon exposure to 80° for 5 min. It is also stable after 50 min of incubation at pH 7.8 with either trypsin or chymotrypsin at room temperature.[19]

Purity. The inhibitor elutes symmetrically from phosphocellulose with constant specific inhibitory activity across the peak (Fig. 3). It migrates as a single component in polyacrylamide gel electrophoresis at pH 4.3, and elutes from a Sephadex G-50 column as a symmetrical peak.[20]

Physical Properties. The molecular weight is estimated to be 3100 ± 300 by gel filtration,[20] 3600 ± 200 by light scattering,[19] and 3000 by titration of carboxypeptidase A activity.[19]

Specificity. The inhibitor inhibits carboxypeptidases A and B.[20] It has no inhibitory activity toward trypsin and chymotrypsin.[19] The inhibitor, when preincubated for 2 min at pH 7.65 with increasing quantities of carboxypeptidase A or carboxypeptidase B, becomes inactive toward the other enzyme. This suggests that if there are two sites on the inhibitor,

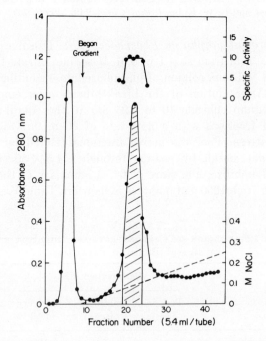

Fig. 3. Chromatography of 48 mg of Sephadex G-75 peak III on phosphocellulose. A column, 1.4 × 13 cm, was equilibrated with 0.01 *M* sodium citrate buffer, pH 4.3. A linear gradient of sodium chloride was employed to elute carboxypeptidase inhibitor. The flow rate was 60 ml/hr. Reproduced from C. A. Ryan, G. M. Hass, and R. W. Kuhn, *J. Biol. Chem.* **249**, 5495 (1974).

TABLE VI

AMINO ACID COMPOSITION OF CARBOXYPEPTIDASE INHIBITOR[a]

Amino acid	Number of amino acid residues per molecule
Aspartic acid	5
Threonine	2
Serine	2
Glutamic acid	2–3
Proline	3
Glycine	3
Alanine	4
Valine	1
Isoleucine	1
Tyrosine	1
Phenylalanine	1
Lysine	2
Histidine	2
Arginine	1
Tryptophan	1–2
Half-cystine	6
	37–39

[a] C. A. Ryan, G. M. Hass, and R. W. Kuhn, *J. Biol. Chem.* **249**, 5495 (1974).

one for each enzyme, the smallness of the inhibitor and the close proximity of the sites prevents the enzymes from reacting simultaneously.[19]

Kinetic Properties. Inhibition of carboxypeptidase A is linear up to about 75% inhibition. The K_i value for the inhibition of this enzyme is $5 \times$ nM. Inhibition of carboxypeptidase B is linear up to about 70% of inhibition, and the K_i value is 50 nM.

Amino Acid Composition. The amino acid composition of the carboxypeptidase inhibitor is presented in Table VI. No free thiol groups are present. The inhibitor has a blocked NH$_2$-terminal residue and contains glycine in the COOH-terminal position.

Amino Acid Sequence.[21] The partial amino acid sequence of the inhibitor is

5   10

Glu-(Glx,His,Ala)-Asp-Pro-Ile-Cys-Asn-Lys

20

Pro-Cys-Lys-Thr-His-Asp-Asp-Cys-Ser-Gly

30

Ala-Trp-Phe-Cys-Gln-Ala-Cys-Trp-Asn-Ser

Ala-Arg-Thr-Cys-Gly-Pro-Tyr-Val-Gly

[64] Acidic Cysteine Protease Inhibitors from Pineapple Stem[1]

By ROBERT L. HEINRIKSON and FERENC J. KÉZDY

Commercial pineapple stem acetone powder is a rich source of an unusually complex mixture of proteases (bromelains) and their poly-peptide inhibitors. Because of the great multiplicity both of proteolytic enzymes and of their peptide inhibitors in the same biological source, the purification and characterization of individual components are chal-lenging undertakings even by present-day technology. The existence of bromelain inhibitors in the pineapple stem acetone powders was un-detected until very recently, and their subsequent purification revealed the presence of at least seven inhibitor fractions. Structural analysis of one of these fractions (VII), indicated extensive microheterogeneity. The multiplicity of these proteins has been shown to be a consequence of genetic substitutions in the polypeptide chains, limited proteolysis, partial deamidation, variable content of carbohydrate, and partial cyclization of an amino terminal residue.

The present article describes procedures for the purification and assay of the bromelain inhibitors and for the isolation of an enzymologically pure bromelain required for assay of the inhibitors. Also included are the results of recent covalent structural studies of inhibitor VII in which the chemical basis of the microheterogeneity is documented.

Assay Procedures

Assay of Bromelain[2]

The esterolytic activity of bromelain solutions is assayed spectro-photometrically at pH 4.60, using the chromogenic substrate p-nitrophenyl N^{α}-benzyloxycarbonyl-L-lysinate (CLN).[3] The assay is carried out in the presence of 1 mM L-cysteine, and thus the phenylmercury complex of bromelain can be assayed as well as active bromelain. Several other convenient assay procedures have been described for assaying bromelain

[1] Supported in part by grants GM-13885 from the National Institutes of Health, and BMS-75-03147 and GB-29098 from the National Science Foundation.
[2] R. M. Silverstein, *Anal. Biochem.* **62**, 478 (1974).
[3] M. L. Bender, F. J. Kézdy, and J. Feder, *J. Am. Chem. Soc.* **84**, 4953 (1965).

at near neutral pH,[4] but these assays would not detect the inhibitor, which has a pH optimum in the strongly acidic range.

Reagents

CLN, 15 mM, in acetonitrile containing 20% v/v water. The substrate is suspended in acetonitrile, the required amount of water is added, and the volume is adjusted with acetonitrile. The resulting solution is stable at room temperature for not more than a few days.

Buffer. Sodium acetate–acetic acid, 10 mM, containing 0.1 M KCl, 1.0 mM L-cysteine, pH 4.60, freshly prepared.

Bromelain. The concentration of the enzyme (or of its phenylmercury derivative) in the solution to be assayed should be of the order of 0.2 to 2.0 mg/ml.

Procedure. Three milliliters of buffer, in a 1-cm path length glass or quartz cuvette, is placed in the cell compartment of a spectrophotometer and thermostatted at $25 \pm 0.5°$. An appropriate aliquot of the enzyme solution (2–100 μl) is added and thoroughly mixed with the flattened tip of a glass rod 2.5 mm in diameter. The solution is incubated for 1 min to allow activation of the enzyme, and then 50 μl of CLN solution is added with stirring. The increase in absorbance at 340 nm is measured for a few minutes; a linear increase should occur at least up to 0.4 absorbance unit (AU). The slow spontaneous hydrolysis of the substrate should also be determined in a blank reaction, where 50 μl of buffer is added instead of the enzyme solution.

Calculation of Bromelain Concentration and Units. Under the conditions of the assay, the change in molar absorptivity of the substrate is equal to 6320 M^{-1} cm^{-1}. The kinetic parameters for the bromelain A-catalyzed reaction are[2]: $k_{cat} = 7.4$ sec^{-1} and $K_m = 57$ μM. If V is the rate of the reaction expressed in units of ΔAU_{340}/min and V_0 the rate of spontaneous hydrolysis, then for an aliquot of Y μl of enzyme solution used in the assay, the molarity of the enzyme solution is given by the equation

$$E_A = \frac{(V - V_0)[3.05 + (Y \times 10^{-3})]}{Y \times 2.28 \times 10^3} M$$

For bromelain B the kinetic parameters are $k_{cat} = 9.8$ sec^{-1} and $K_m = 82$ μM, and thus

$$E_B = \frac{(V - V_0)[3.05 + (Y \times 10^{-3})]}{Y \times 2.77 \times 10^3} M$$

[4] T. Murachi, this series Vol. 19 [18].

Based upon a molecular weight of 35,000 for the enzyme,[4] one calculates that 1 international unit is equal to 93 μg of enzyme A and 76.5 μg of enzyme B under the conditions of the assay.

For routine assays, a convenient relative enzyme unit can be defined as the amount of enzyme that, when added to the reaction mixture, will yield a rate of 1 AU_{340}/min. With these units and a protein concentration expressed in units of AU_{280}, the specific activity of pure bromelain A is 12.9 units per A_{280} per milliliter.

Assay of Bromelain Inhibitor

The bromelain-catalyzed hydrolysis of CLN is inhibited competitively by the pineapple stem protease inhibitor. The assay is performed under conditions where $S_o \ll K_m$, i.e., where the apparent first-order rate constant of the enzymic reaction is proportional to the concentration of free enzyme, and the amount of enzyme bound to the substrate is negligible with respect to the total enzyme. The reaction produces a spectral change of the order of 0.2 AU and, for this reason, a recording spectrophotometer of good sensitivity is desirable for the measurement.

Reagents

CLN, 1.25 mM, in acetonitrile containing 10% (v/v) water. The solution, when stored at 4°, is stable for several days.

Buffer: sodium acetate–acetic acid (20 mM) containing 0.1 M KCl, 1 mM L-cysteine, pH 4.60, freshly prepared.

Bromelain A. The enzyme, or its phenylmercury complex, should be enzymically pure, i.e., free of noninhibitable esterolytic activity, which accompanies bromelain A and B in trace amounts in the pineapple stem acetone powder.[5] The concentration of the enzyme solution should be approximately 30 μM, in a 10 mM acetic acid–sodium acetate buffer, 0.1 M KCl, pH 4.60.

Inhibitor. The inhibitor stock solution should be in the concentration range of 10–25 μM, i.e., A_{280} = 0.06–0.15 AU/cm.

Procedure. Three milliliters of the buffer, 50 μl of the bromelain A solution, and 50 μl of the inhibitor solution are mixed thoroughly in a 1-cm path length quartz cuvette and placed in the thermostatted cell compartment of a recording spectrophotometer. The temperature of the solution should be 25 ± 0.5°. After at least 2 min of preincubation, 50 μl of the CLN solution is added on the tip of a flattened glass rod; the solution is stirred for 2 sec, and the absorbance of the solution is recorded

[5] R. M. Silverstein and F. J. Kézdy, *Arch. Biochem. Biophys.* **167**, 678 (1975).

as a function of time at 317 nm. The measurement should begin not later than 15–20 sec after mixing. Under the conditions specified, the half-life of the first-order reaction is about 10–20 sec. The reaction is monitored for seven half-lives, and the absorbance of the solution at the end of the reaction (A_∞) noted. Five to ten points are selected on the reaction curve, their absorbance (A) is read as a function of the time (t) in seconds, and the apparent first-order rate constant (k) is calculated from the slope of a plot of $\ln (A_\infty - A)$ vs t.

The rate constant in the absence of inhibitor (k_0) is determined in a similar manner, using 50 μl of buffer instead of the inhibitor solution.

With the parameters $k_{cat}/K_m = 1.30 \times 10^5 \ M^{-1} \ sec^{-1}$ and $K_i = 81.5$ nM,[6] the first-order rate constants expressed in units of sec^{-1} allow one to calculate the molarity of the inhibitor solution by the equation

$$[I] = (k_0 - k)[4.85 \times 10^{-4} + (5.13 \times 10^{-6}/k)]M$$

If bromelain B is used for the assay $(k_{cat}/K_m = 1.19 \times 10^5 \ M^{-1} \ sec^{-1};$ $K_i = 0.1 \ \mu M$ the molarity of the inhibitor solution is given by the equation:

$$[I] = (k_0 - k)[5.28 \times 10^{-4} + (6.30 \times 10^{-6}/k)]M$$

Purification of Bromelain

Several methods of purification of the enzymes have been described previously.[4] The following procedure is preferred by the authors, since it yields an enzymically pure and well characterized product in the form of the phenylmercury complex, which can be stored indefinitely, even in solution.[5,6]

Starting Material. The acetone powder from the juice of the pineapple stem is commercially available under the name bromelain (e.g., Sigma Chemical Co., Nutritional Biochemicals Corp.). All steps are carried out at room temperature, and all buffers are saturated with phenylmercuric acetate.

Step 1. Extraction. Acetone powder, 2.5 g, is suspended in 10 ml of 0.1 M potassium phosphate buffer pH 6.10 and stirred for 10 min. After centrifugation for 10 min in a desk-top centrifuge (3500 rpm), the supernatant is subjected to gel permeation chromatography.

Step 2. Gel Filtration. The supernatant from step 1 is applied to a column (2.5 × 75 cm) of Sephadex G-75 equilibrated with 0.1 M potassium phosphate buffer, pH 6.10, and eluted with the same buffer. Fractions of 14.5 ml are collected, and the A_{280} of the eluate is measured (Fig. 1). The enzyme is eluted in fractions 12–18, followed by the inhibi-

[6] S. H. Perlstein and F. J. Kézdy, *J. Supramol. Struct.* 1, 249 (1973).

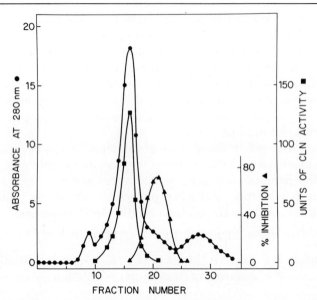

Fig. 1. Gel filtration of crude extract on Sephadex G-75. Commercial acetone powder, 2.5 g, was extracted by 10 ml of pH 6.10, 0.10 M phosphate buffer saturated with phenylmercury acetate and centrifuged. The supernatant was applied to a column (2.5 × 75 cm) of Sephadex G-75 and eluted with the buffer that was used for extraction; 14.5-ml fractions were collected. Reproduced from S. H. Perlstein and F. J. Kézdy, *J. Supramol. Struct.* **1**, 249 (1973).

tor in fractions 19–24. The enzyme fractions are assayed and pooled; the resulting solution is purified further by ion-exchange chromatography. The pooled inhibitor fractions yield the seven inhibitor species as described below.

Step 3. Ion-Exchange Chromatography. The bromelain preparation from step 2 is applied directly to a column (1.5 × 30 cm) of Bio-Rex 70 equilibrated with 0.1 M potassium phosphate buffer pH 6.10. It is of utmost importance that the ion-exchange column should be equilibrated with the buffer: the difference between the starting buffer and the eluate at the end of the equilibrium should not exceed 0.01 pH unit and 1000 mho of conductance. After application of the enzyme solution to the column, the enzyme is eluted with a linear gradient produced from 225 ml of 0.1 M phosphate buffer, pH 6.10, 0.1 M KCl, and 225 ml of 0.1 M phosphate buffer, pH 6.10, 0.6 M KCl. Fractions of 3 ml are collected at a flow rate of 50 ml/hr. Two major protein peaks are eluted, as measured by absorbance at 280 nm; bromelain A appears at 0.2 M KCl, and bromelain B at around 0.26 M KCl. After enzymic assay, the appropriate fractions are pooled and stored at 4°.

As compared with the crude extract, the specific activity of the pure enzyme is increased 15-fold when measured by the CLN assay. The yields (corrected for the presence of inhibitors in the crude extract) are of the order of 50% for the total process. Enzymes A and B appear to be present in approximately equal amounts in the acetone powders tested.

Bromelains A and B are homogeneous proteins by the following criteria: disc gel electrophoresis at two pH's, ion-exchange chromatography at two pHs, sedimentation in the analytical ultracentrifuge, and end group analysis. They are also pure by enzymic criteria, such as titration of the active-site sulfhydryl groups, constant specific activity upon rechromatography, and kinetic properties.

The substrate specificities of bromelains A and B are nearly identical. Toward p-nitrophenyl esters of Z-L-amino acids, the relative specificity of enzyme A is in the order L-Lys (550) > L-Ala (280) > L-Tyr (140) > Gly (110) > L-Asn (51) > L-Val, L-Leu, L-Ile, L-Trp (<1).

Purification of Bromelain Inhibitor

The nearly 4-fold increase in total bromelain activity obtained by gel filtration of pineapple stem acetone powder extracts (Fig. 1) was the first indication that these crude bromelain preparations contain an enzyme inhibitor that is removed during this stage of the purification. Accordingly, gel filtration on Sephadex G-75 is a first step common to the purification of both bromelain and its inhibitor. As may be seen in Fig. 1, a shoulder is resolved on the trailing edge of the bromelain peak, and material eluted in this region is inhibitory when added back to the enzyme. Proteolytic digestion of the polypeptide inhibitor is minimized by developing the column with buffers containing phenylmercury acetate, which inhibits bromelain by reacting with the active site sulfhydryl group. A second passage of bromelain inhibitor through the same column results in complete removal of bromelain.

The inhibitor preparation isolated by gel filtration contains numerous components, nearly all of which display inhibitory activity. These inhibitors may be resolved by polyacrylamide gel electrophoresis or, on a preparative scale, by ion-exchange chromatography on columns of DEAE-Sephadex.[6] In the latter procedure, the inhibitor isolated by gel filtration is desalted on a column (2.5 × 95 cm) of Sephadex G-25 (fine), equilibrated with 1% acetic acid, and lyophilized. This material (30–40 mg) is dissolved in 5 ml of 0.1 M Tris-HCl, pH 7.55, and the solution is applied to a column (2 × 45 cm) of DEAE-Sephadex A-25 previously equilibrated with the same buffer. Elution of protein is carried out at room temperature first with 400 ml of the same buffer, followed by a linear gradient of increasing NaCl concentration from 0 to 0.2 M (250 ml each of initial and final buffer solution). Fractions of 10 ml are collected at a flow rate of 50 ml/hr; a typical elution profile is shown in Fig. 2. Appropriate fractions are pooled, desalted by gel filtration on Sephadex G-25, lyophilized, and stored at 4°. This procedure results in the separation of seven polypeptide inhibitor fractions, all of which possess approximately the same specific inhibitory activity toward bromelain A. Each inhibitor component, labeled I–VII in Fig. 2, appears as a single band following polyacrylamide gel electrophoresis and staining with Coomassie Brilliant Blue R-250. Minor components, if present, always appear in positions characteristic of the adjacent inhibitor fractions. Gel electrophoresis in the presence of sodium dodecyl sulfate and 2-mercaptoethanol also reveals, in each case, a single band, the mobility of which corresponds to a polypeptide of molecular weight less than 10,000.[6]

Although most criteria of homogeneity applied to the seven inhibitor preparations are consistent with their being pure, the results obtained from compositional and amino-terminal analysis are, in most cases, suggestive of inhomogeneity. Amino acid analytical data reveal great compositional similarities among the seven inhibitors. The polypeptides each contain about 52 amino acids, and the molecular weight calculated on this basis is about 5600. All the fractions are lacking in methionine,

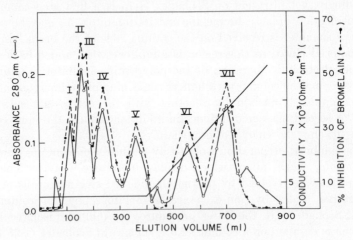

FIG. 2. Purification of bromelain inhibitors by ion-exchange chromatography on DEAE-Sephadex. The sample applied to the column contained 30–40 mg of inhibitor isolated from 2.5 g of acetone powder by gel filtration (Fig. 1). The column (2 × 45 cm) was developed with Tris buffer, pH 7.55, containing a linear gradient of increasing NaCl concentration from 0 to 0.2 M. Reproduced from M. N. Reddy, P. S. Keim, R. L. Heinrikson, and F. J. Kézdy, *J. Biol. Chem.* 250, 1741 (1975).

TABLE I
COMPOSITIONAL ANALYSES OF BROMELAIN INHIBITOR VII AND COMPONENTS
ISOLATED THEREFROM AFTER PERFORMIC ACID OXIDATION

| | Nearest integral number of residues | | | | |
Amino acid	Inhibitor VII	A-1	A-2	B-1	B-2
Cysteic acid	10	7	7	3	3
Aspartate + asparagine	5–6	6	5	0	0
Threonine	2–3	1	2	1	1
Serine	3	2	2	1	1
Glutamate + glutamine	4	2	3	1	2
Proline	3	2	2	1	1
Glycine	2	2	2	0	0
Alanine	2–3	2	1	1	1
Valine	2	1	1	1	1
Isoleucine	2	2	2	0	0
Leucine	3	2	2	1	1
Tyrosine	4	4	4	0	0
Phenylalanine	2	2	2	0	0
Lysine	6	6	6	0	0
Arginine	0–1	0	0	1	0
	50–54	41	41	11	11

histidine, and tryptophan, and inhibitor VII has considerably less than the one full residue of arginine present in fractions I–VI. However, although most of the amino acids may be easily quantified in terms of a well defined number of residues, it is not possible to assign a specific residue number for aspartic acid, threonine, alanine, and arginine. A typical analysis is shown in Table I for performic acid-oxidized inhibitor VII. Except for the diminished arginine content of this fraction, it is essentially indistinguishable from inhibitors I–VI. Further evidence of inhomogeneity is derived from automated Edman degradation of the desalted inhibitor fractions.[7,8] Six cycles of this procedure revealed multiple amino acids at nearly every stage. The amino-terminal residues most often encountered were threonine, alanine, and aspartic and glutamic acids. Residue three was predominantly tyrosine. These preliminary indications of heterogeneity among the bromelain inhibitor fractions prompted the undertaking of covalent structural studies. The results thus obtained for two of the fractions, IV and VII, provide a basis for explaining in part the origins and types of structural diversity en-

[7] P. Edman and G. Begg, Eur. J. Biochem. 1, 80 (1967).
[8] R. L. Heinrikson and K. J. Kramer, Prog. Bioorg. Chem. 3, 141 (1974).

countered among the various inhibitors (see section on Protein Chemistry).

Properties of the Inhibitors

Stability. The inhibitors appear to be remarkably stable even under extreme conditions: no loss of activity was observed when they were kept at pH 3.0 for several weeks, or when kept at 90° for 10 min at pH 7.0. With the lyophilized powder, no loss of activity was observed at 4° after 1 year.

Physical Properties. The molecular weight of inhibitor VII is calculated to be 5651 from the amino acid composition. Owing to the low value of MW, no accurate determination was possible by SDS-polyacrylamide disc gel electrophoresis. Sedimentation velocity measurements with the analytical ultracentrifuge yielded a molecular weight value in agreement with that from the amino acid composition. The specific absorptivity of the inhibitors was found to be $A^{1\%}_{280} = 9.7$ AU at pH 4.60. The isoelectric points of inhibitors I–VII range from 4.5 to 4.9, as determined by isoelectric focusing.

Kinetic Properties. The inhibitors inhibit competitively the bromelain-catalyzed hydrolysis of CLN. In this reaction, 1 mole of bromelain reacts with 5500 ± 650 g of inhibitor to yield an inactive complex. With inhibitor VII at pH 4.60, $K_i^A = 82$ nM for bromelain A and $K_i^B = 0.1$ μM for bromelain B.

Study of the pH dependence of the inhibition revealed that the protonated forms of an acid of p$K = 4.45$ and one of p$K = 5.03$ are required for inhibitory action. Thus, inhibition is maximal between pH 3 and pH 4; at higher pH values it decreases with the square of the proton concentration. At pH 8, inhibitory activity is virtually undetectable.

Specificity. Papain and ficin are inhibited competitively by inhibitor VII at pH 4.6, with $K_i \simeq 10$ μM. At the same pH, a slight inhibition of the bovine trypsin-catalyzed hydrolysis of CLN was also observed.

Protein Chemistry.[9] In view of the fact that the bromelain inhibitors not only constitute a complex array of structural entities with distinct electrophoretic and chromatographic properties, but also vary within any given fraction, the ensuing discussion of their covalent structure will be confined, by and large, to inhibitor VII (Fig. 2), the only preparation studied thus far in detail. Amino-terminal analysis of VII gave aspartic acid, threonine, and alanine in major yields with a small amount of glutamic acid. This, coupled with its compositional analysis (Table I)

[9] M. N. Reddy, P. S. Keim, R. L. Heinrikson, and F. J. Kézdy, *J. Biol. Chem.* **250**, 1741 (1975).

TABLE II
AMINO ACID SEQUENCES OF THE PEPTIDE COMPONENTS OF INHIBITOR VII

Peptide	Sequence
A-1	Asp⎤Glu-Tyr-Lys-Cys-Tyr-Cys⎤Ala⎤Asp-Thr-Tyr-Ser-Asp-Cys-
	1 5 10
A-2	PCA⎤Glu-Tyr-Lys-Cys-Tyr-Cys⎤Thr⎤Asp-Thr-Tyr-Ser-Asp-Cys-
A-1 and A-2	-Pro-Gly-Phe-Cys-Lys-Lys-Cys-Lys-Ala-Glu-Phe-Gly-Lys-Tyr-
	15 20 25
A-1 and A-2	-Ile-Cys-Leu-Asp-Leu-Ile-Ser-Pro-Asn-Asp-Cys-Val-Lys-COOH
	30 35 40
B-1	(Thr)Ala-Cys-Ser-Glu-Cys-Val-Cys-Pro-Leu⎤Arg⎤COOH
	1 5 10
B-2	(Thr)Ala-Cys-Ser-Glu-Cys-Val-Cys-Pro-Leu⎤Gln⎤COOH
	1 5 10

prompted attempts at further purification by column chromatography, isoelectric focusing, and high-voltage paper electrophoresis at pH 2 and pH 6.5. In all these systems VII migrated as a single component.

Inhibitor VII has five disulfide bonds in a total of 52 residues, and therefore its tertiary structure is highly constrained. Performic acid oxidation followed by gel filtration on Sephadex G-25 yields two well resolved peaks, one of which (peak A) contains peptides 41 residues in length; the other (peak B) is composed of undeca- and decapeptides. High-voltage paper electrophoresis of peaks A and B reveals that each comprises two major components, A-1 and A-2 and B-1 and B-2. The compositions of these components are listed in Table I. Clearly, the 41-residue peak A peptides are very similar in composition, as are the B peptides. Moreover, these compositional analyses, at last, yield integral values for all the amino acids, thus indicating that the peptides are pure.

The covalent structures of the inhibitor VII components have been determined by automated Edman degradation and conventional sequence analysis, and the results are summarized in Table II. The bromelain inhibitors are composed of two chains linked covalently by disulfide bonds. A search for the putative single-chain precursor in carefully prepared extracts of fresh pineapple stem has been unsuccessful; the inhibitors thus obtained appear to be less diverse than those isolated from commercial powders, but are still composed of two chains.[10] On the

[10] M. N. Reddy and F. J. Kézdy, unpublished observations.

basis of their charge characteristics and yields, the best pairing would involve A-1 and B-2 in one molecule and A-2 and B-1 in another. The two major inhibitors differ only at three loci; position 1 (PCA or Asp) and position 8 (Thr or Ala) in the A chains, and the COOH-terminus of the B chains (Arg or Gln). The amino-terminal threonine residues in the B chains are enclosed within parentheses because about 20% of both B-1 and B-2 consists of decapeptides lacking threonine and having alanine as the NH_2-terminal residue.

The structural analysis explains the nonintegral values for aspartic acid, threonine, alanine, and arginine obtained in the analysis of inhibitor VII (Table I). Moreover, the amino-terminal analyses are in accord with the existence of major components beginning with aspartic acid (A-1), threonine, and alanine (B chains).

Of special interest in regard to the sequence analysis is the astonishing degree of structural variation within the components of a preparation (VII) which defied further purification. The microheterogeneity among the peak VII isoinhibitors arises from a number of different contributing factors. First, the two major isoinhibitors [A-1:B-2] and [A-2:B-2] are obviously the products of two similar but different structural genes with the result that the molecules are identical except for residues 1 and 8 in the A chains and the COOH-terminal residue of the B chains. A second indication of microheterogeneity is in the variability in the NH_2-terminal residue of the B chains, apparently the cause of partial and limited proteolysis. If the B chains represent the NH_2-terminal portion of a single-chain precursor, this might indicate partial exopeptidase digestion. This explanation is probably more likely than our earlier proposal[9] that the B-chain is COOH-terminal and results from the enzymic excision of a "bridge" peptide in the single-chain inhibitor. A third factor, which appears to confer variability on the population of isoinhibitors, is that of deamidation. Our preliminary studies of inhibitor fraction IV indicate that the major component (80–90%) is [A-2:B-1],[10] and yet the charge properties of this molecule are consistent with its having less free carboxyl groups than the [A-2:B-1] in fraction VII. In fact, our studies with freshly prepared acetone powders reveal considerably less heterogeneity in the inhibitor components than observed in preparations from commercial material, suggestive of the possibility that the array of inhibitors I–VII (Fig. 2) is generated primarily by deamidation. Finally, the correlation of carbohydrate content with arginine[6] may suggest that Thr-8, which is in the A chain associated with the B peptide terminating in arginine, can undergo glycosylation.

Having discussed in some detail the structural features of the bromelain inhibitors, it is appropriate to close with some remarks about their function. Unfortunately, little is known about how these polypeptides inhibit bromelain. In an earlier report[9] we proposed that the pineapple stem bromelain inhibitors are synthesized as single-chain proteins in which the A-peptides are NH_2-terminal, the B-peptides are COOH-terminal, and the two are bridged by a missing link of 4–6 residues. With this alignment, the half-cystine residues show a high degree of homology with those in numerous trypsin inhibitors. This hypothesis poses one major problem, however, in that the active sites of the trypsin inhibitors at residues 15 and 16[11] correspond to a Pro-Gly sequence in the brome-lain inhibitor, a structure unlikely to be involved in formation of a co-valent intermediate. If a thioester intermediate is formed during the course of bromelain inhibition, it might be expected on the basis of the enzyme specificity that the carbonyl functions of arginine, lysine, or glutamine might be involved.[5] In a recent series of experiments we have shown that carboxypeptidase B cleaves only one bond in the native in-hibitors [A-1:B-2] and [A-2:B-1] and that leads in both cases to the removal of the COOH-terminal residue in the B chain, i.e., either arginine in B-1 or glutamine in B-2. The resulting polypeptides are no longer inhibitory toward bromelain. In view of these findings, we are led to surmise that the B chains are amino-terminal in the presumptive sin-gle chain inhibitor precursor and that the active sites are at position 11. Cleavage of [B-1:A-2] would then leave a Glx residue at the NH_2-terminus of A-2 which could then undergo cyclization to PCA. Similar cleavage of [B-2:A-1] would yield NH_2-terminal aspartic acid. Since R-COOH is a better electrophile at the low pH of optimum inhibition than is R—C(=O)—NH—R′, it may be that the two-chain polypeptide is much more inhibitory than the single-chain species and that a search for the latter will be in vain.

In any event, these recent findings offer many possibilities for ex-perimentation in regard to the mechanism of action of the bromelain in-hibitors. In addition, the polypeptides themselves may provide an in-teresting model for the study of microheterogeneity in plant systems.

[11] B. Kassell, this series Vol. 19 [66].

[65] CĪ Inactivator

By PETER C. HARPEL

Normal human serum contains an inhibitory activity directed against the proteolytic and esterolytic activity of CĪ, the activated first com-ponent of human complement.[1,2] Pensky et al.[3] achieved partial purifica-

[1] O. D. Ratnoff and I. H. Lepow, J. Exp. Med. 106, 327 (1957).
[2] L. R. Levy and I. H. Lepow, Proc. Soc. Exp. Biol. Med. 101, 608 (1959).
[3] J. Pensky, L. R. Levy, and I. H. Lepow, J. Biol. Chem. 236, 1674 (1961).

tion of the protein responsible for this inhibitory activity. The inhibitor, $C\bar{1}$ inactivator ($C\bar{1}$ In), was found to be identical to the α_2-neuramino-glycoprotein isolated by Schultze et al.,[4] a protein whose function had not previously been determined.[5] Donaldson and Evans[6] showed that the serum of patients with hereditary angioneurotic edema lacked $C\bar{1}$ inhibitory activity. Landerman et al.[7] demonstrated that these sera also failed to inhibit plasma kallikrein and serum globulin permeability factor. The inhibitory defect in these patients was shown to be associated with a deficiency in $C\bar{1}$ In.[8,9] The normal concentration of $C\bar{1}$ In as determined by immunodiffusion is 18 ± 5 mg per 100 ml of serum.[10]

Assay[11]

Principle. $C\bar{1}$ In inhibits the hydrolysis of N-α-acetyl-L-lysine methyl ester (ALME) by $C\bar{1}s$, the activated subunit of the first component of complement.

$C\bar{1}s$ Purification. $C\bar{1}s$ is prepared from human serum according to a modification of the method of Haines and Lepow[12] as described previously in detail.[11]

Procedure. The $C\bar{1}s$ is diluted so that 0.1 ml of enzyme solution releases approximately 4.5 μmoles of methanol per hour from the substrate, ALME.[11] Purified $C\bar{1}$ inactivator is diluted to a concentration ranging from 1 to 5 μg/ml, and 0.2 ml of inhibitor is incubated with 0.1 ml $C\bar{1}s$ for 10 min at 37°, after which 0.3 ml of crude soybean trypsin inhibitor (2 mg/ml; Worthington Biochemical Corp., Freehold, New Jersey) and 1.4 ml of 0.1 M phosphate buffer, pH 7.5, is added. ALME (19.1 mg per milliliter of phosphate buffer; Cyclo Chemical, Division Travenol Laboratories, Inc., Los Angeles, California), 0.3 ml, is added to achieve a final substrate concentration of 10 mM, and the mixture is incubated at 37°.

One-milliliter samples are withdrawn after 1 and 61 min of incubation at 37° and are added to tubes containing 0.5 ml of 0.75 M perchloric acid (32.2 ml of 70% perchloric acid made to 500 ml with distilled water) in

[4] H. E. Schultze, K. Heide, and H. Haupt, *Naturwissenschaften* **49**, 133 (1962).
[5] J. Pensky and H. G. Schwick, *Science* **163**, 698 (1969).
[6] V. H. Donaldson and R. R. Evans, *Am. J. Med.* **35**, 37 (1963).
[7] N. S. Landerman, M. E. Webster, E. L. Becker, and H. E. Ratcliffe, *J. Allergy* **33**, 330 (1962).
[8] F. S. Rosen, P. Charache, J. Pensky, and V. Donaldson, *Science* **148**, 957 (1965).
[9] I. Gigli, S. Ruddy, and K. F. Austen, *J. Immunol.* **100**, 1154 (1968).
[10] F. S. Rosen, C. A. Alper, J. Pensky, M. R. Klemperer, and V. H. Donaldson, *J. Clin. Invest.* **50**, 2143 (1971).
[11] P. C. Harpel, *J. Immunol.* **104**, 1024 (1970).
[12] A. L. Haines and I. H. Lepow, *J. Immunol.* **92**, 456 (1964).

an ice bath. The precipitate is removed by centrifugation and 1 ml is added to thick-walled boiling tubes; 0.1 ml of a 2% solution of potassium permanganate is added. One minute later, 0.1 ml of a 10% solution of freshly prepared sodium sulfite is added and the solution immediately decolorized by shaking. Four milliliters of a chromotropic acid solution are added and mixed thoroughly for 30 sec on a Vortex mixer. The chromotropic acid is made by adding 10 ml of 2% 4,5-dihydroxy-2,7-naphthalenedisulfonic acid disodium salt, practical (Distillation Products Industries, Rochester, New York) in distilled water to 90 ml of a 29 N solution of reagent grade sulfuric acid (Merck and Co., Inc. Rahway, New Jersey). The sulfuric acid solution is made by diluting 260 ml of concentrated sulfuric acid with 64 ml of water. The final volume is 5.2 ml. The tubes are heated in a dry heating block at 100° for 30 min, after which they are cooled in tap water. Optical densities are read in a spectrophotometer in a 1-cm cuvette at 580 nm. The 1-min sample serves as the zero time blank.

The reagent blank read against distilled water gives $A_{580 \, nm}$ readings below 0.030. All optical density readings are corrected for spontaneous hydrolysis of the substrate during the 60-min incubation period. This varies between 0.015 and 0.030. The standard consists of adding 1.0 ml of 0.75 μM methanol in phosphate buffer to 0.5 ml of 0.75 M perchloric acid and treated as in the test procedure. The final methanol concentration is therefore 0.5 μM. Optical density values for 10 methanol standards prepared in duplicate were 0.197 ± 0.009 as read against a reagent blank.

Definition and Calculation of $C\overline{1}$ Inactivator Units.[11] One unit of $C\overline{1}$ inactivator activity is defined as the amount of inhibitor that inhibits 10 units of $C\overline{1}s$.[2] A unit of $C\overline{1}s$ is that volume of enzyme which causes the release of 1 μmole of methanol per hour from the substrate, ALME. The concentration of inhibitor units per milliliter of purified $C\overline{1}$ inactivator using the ALME esterase assay is calculated by the following formula:

$$\frac{(1.725)(C\overline{1} \text{ In dilution})[A_{580} \, C\overline{1}s - (A_{580} \, C\overline{1}s + C\overline{1} \text{ In})]}{(A_{580}CH_3OH \text{ standard})(ml \, C\overline{1} \text{ In})(10)}$$

Inhibition of $C\overline{1}s$ by $C\overline{1}$ In is linear in the range of 20–50% inhibition.

Purification Procedure[13]

The initial procedures for the purification of $C\overline{1}$ In from freshly obtained venous blood are identical to those described in Chapter [52][14] for the purification of α_2-macroglobulin.

[13] P. C. Harpel and N. R. Cooper, *J. Clin. Invest.* **55**, 503 (1975).
[14] P. C. Harpel, this volume [52].

DEAE-Cellulose Chromatography. All chromatographic procedures are performed at 4°. The 12% polyethylene glycol plasma supernatant prepared by the methods detailed in Chapter [52] is directly applied to a 5.0 × 50 cm column of DE-52 (Reeve Angel and Co., Inc., Clifton, New Jersey) prewashed with Tris buffer 1 (cf. Chapter [52]) containing 50 μg of Polybrene (hexadimethrine bromide) per milliliter. After the sample is on the gel, the column is washed with Tris buffer 1, until $A_{280\,nm}$ is below 0.02, then a linear NaCl gradient is started using Tris buffer 2 as the limiting buffer. The $C\bar{1}$ In is identified by double diffusion analysis on agar plates using rabbit anti-human $C\bar{1}$ inactivator serum (Behring Diagnostics, American Hoechst Corp., Somerville, New Jersey) and is found as indicated in Fig. 1, at a conductivity of 13.5–15.5 mmho/cm (27°).

Gel Filtration Chromatography. The fractions containing $C\bar{1}$ In are pooled and concentrated by ultrafiltration to 10.0 ml and applied to a 5.0 × 80 cm column of Bio-Gel A-5m agarose (Bio-Rad Laboratory,

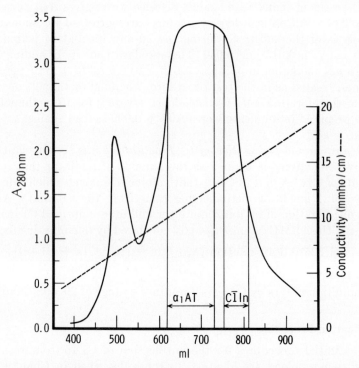

FIG. 1. DEAE-cellulose chromatography of $C\bar{1}$ inactivator. The 12% polyethylene glycol supernatant prepared as described in the text, is applied to a 5.0 × 50 cm column of DE-52. Following wash-through of the frontal peak with starting buffer, a linear NaCl gradient is started. The fractions containing immunologically identifiable α_1-antitrypsin and $C\bar{1}$ In are indicated.

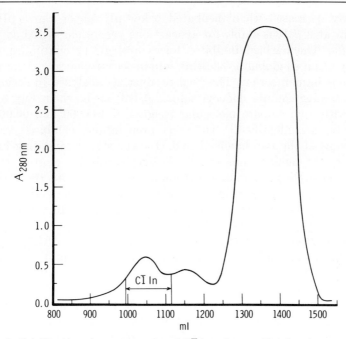

FIG. 2. Gel filtration chromatography of C$\overline{\text{I}}$ inactivator. The fractions containing C$\overline{\text{I}}$ In following DE-52 chromatography (Fig. 1) were pooled, concentrated and applied to a 5.0 × 80 cm column of Bio-Gel A-5m aragose in Tris-citrate-saline buffer. The fractions containing immunologically identified C$\overline{\text{I}}$ In are indicated.

Richmond, California) in Tris–citrate–saline buffer. The fractions containing C$\overline{\text{I}}$ In (Fig. 2) are then pooled, concentrated, and subjected to preparative electrophoresis in Pevicon as described for α_2-macroglobulin.[14]

Insoluble Concanavalin A Chromatography. After elution from the Pevicon block, the C$\overline{\text{I}}$ In preparation is passed through a 20-ml column of concanavalin A–Sepharose (Pharmacia Fine Chemicals, Inc., Piscataway, New Jersey) to remove the contaminating albumin dimer. The inhibitor is eluted from the affinity column with 10% methyl-α-D-mannopyranoside in 50 mM Tris-HCl buffer, pH 8.0, containing 0.5 M EDTA.[15] The fractions containing C$\overline{\text{I}}$ In are pooled, dialyzed against Tris buffer 1, concentrated by ultrafiltration, and frozen at $-70°$ in small portions. The average yield of C$\overline{\text{I}}$ inactivator by this method is 10–15% of the starting plasma concentration.

Properties

Stability. C$\overline{\text{I}}$ In is unstable at temperatures higher than 48° and its inhibitory activity is abolished at 60°.[3] The inhibitory activity also pro-

[15] A. P. Kaplan, personal communication.

gressively decreased when incubated below pH 5.5 or above pH 10.5 for 18 hr at 2°.[3] It is stable for at least one year when stored at −70°.

Purity. The purified inhibitor forms a single precipitin arc with a mobility of an α_2-globulin following immunoelectrophoresis using rabbit anti-whole human serum. The final product, as analyzed in acrylamide gels containing sodium dodecyl sulfate (SDS) is shown in Fig. 3. Two bands with approximate molecular weights of 105,000 and 96,000, respectively, are identified.[13] Different preparations exhibited variable proportions of the two bands (I and II). The slower migrating band is generally the major component. Antigen–antibody crossed electrophoresis as well as immunoelectrophoresis demonstrate the immunological identity of bands I and II (Fig. 3). The specific activity of the C$\bar{1}$ In preparation measured with ALME as a substrate for C$\bar{1}$s, is 217 inhibitor units per milligram of C$\bar{1}$ In.[13]

Physical Properties. The molecular weight of C$\bar{1}$ In is 104,000 by sedimentation equilibrium centrifugation.[16] The sedimentation constant $(s_{20,w})$ is 3.7.[4,16] The adsorbance $(A_{1cm}^{1\%}$ 280 nm) is 4.5–5.0.[4,16] The isoelectric point is 2.7–2.8.[16] The inhibitor is precipitable by 2.0–2.7 M ammonium sulfate at pH 7.0, by 0.0065 M Rivanol and is soluble in 0.6 M perchloric acid, 0.2 M trichloroacetic acid, and 40% ethanol.[16] No subunits of C$\bar{1}$ In have been found after the use of disulfide bond cleaving agents,[13,16] indicating that the inhibitor molecule consists of a single polypeptide chain.

Specificity. C$\bar{1}$ In inhibits enzymes of the complement, kinin-forming, coagulation, and fibrinolytic systems. These include C$\bar{1}$ and its subcomponents C$\bar{1}$s and C$\bar{1}$r,[2,3,9,17,18] plasma kallikrein, the permeability factor of dilution,[7,18–20] plasmin,[13,16,18,21] Hageman factor[22] and its active fragments,[23] and plasma thromboplastin antecedent (PTA).[22,24] It has a relatively weak affinity for trypsin and chymotrypsin.[3,16] It fails to inhibit thrombin.[3]

Enzyme Binding Properties. C$\bar{1}$ In binds and inhibits both C$\bar{1}$s and plasmin in a 1:1 molar ratio.[9,13] Both the proteolytic and esterolytic

[16] H. Haupt, N. Heimburger, T. Krantz, and H. G. Schwick, *Eur. J. Biochem.* **17**, 254 (1970).
[17] I. H. Lepow and M. A. Leon, *Immunology* **5**, 222 (1962).
[18] O. D. Ratnoff, J. Pensky, D. Ogston, and G. B. Naff, *J. Exp. Med.* **129**, 315 (1969).
[19] L. J. Kagen, *Br. J. Exp. Pathol.* **45**, 604 (1964).
[20] I. Gigli, J. W. Mason, R. W. Colman, and K. F. Austen, *J. Immunol.* **104**, 574 (1970).
[21] A. D. Schreiber, A. P. Kaplan, and K. F. Austen, *J. Clin. Invest.* **52**, 1394 (1973).
[22] C. D. Forbes, J. Pensky, and O. D. Ratnoff, *J. Lab. Clin. Med.* **76**, 809 (1970).
[23] A. D. Schreiber, A. P. Kaplan, and K. F. Austen, *J. Clin. Invest.* **52**, 1402 (1973).
[24] P. C. Harpel, *J. Clin. Invest.* **50**, 2084 (1971).

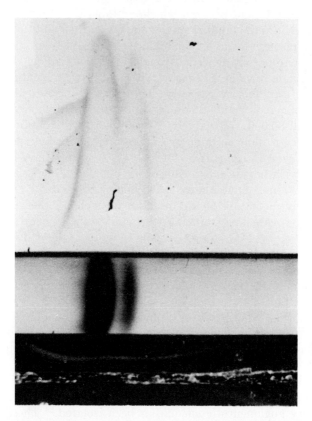

FIG. 3. Antigen-antibody crossed electrophoresis and immunoelectrophoresis of purified human plasma C1̄ inactivator. After electrophoresis of a C1̄ inactivator preparation on an acrylamide gel (5 %) containing sodium dodecyl sulfate (SDS) (electrophoresis in stained gel was from left to right), the gel was sliced longitudinally and embedded in agarose for immunodiffusion using rabbit anti-C1̄ inactivator antibody (beneath stained gel). Other gel slices were overlaid with agarose containing the antibody and electrophoresed at right angles to the gel (top panel). The two peaks that appeared after crossed electrophoresis and corresponded to the two C1̄ inactivator bands are designated I and II. From P. C. Harpel and N. R. Cooper, *J. Clin. Invest.* **55,** 593 (1975).

activities of these enzymes are fully inhibited by C$\overline{1}$ In. The interaction of C$\overline{1}$ In with C$\overline{1}$s, plasmin or kallikrein results in a higher molecular weight complex whose apparent molecular weight is consistent with equimolar binding ratios.[13,25] These enzyme-inhibitor complexes do not dissociate when incubated in a 1% SDS–5M urea solution. The light chain of both C$\overline{1}$s and plasmin which contains the active site of these enzymes, provides the binding site for the C$\overline{1}$ In.[13] In addition to binding with C$\overline{1}$ In, plasmin degrades the inhibitor producing identifiable derivatives with an apparent molecular weight of 96,000. One of these derivatives is active in forming a complex with plasmin.[13] Plasmin also inactivates the C$\overline{1}$s inhibitory capacity of C$\overline{1}$ In.[13,26] Trypsin degrades the inhibitor producing two nonfunctional derivatives with molecular weights of 96,000 and 86,000, respectively. No complex formation between trypsin and C$\overline{1}$ In is demonstrable by SDS-acrylamide gel electrophoresis.[13]

C$\overline{1}$ Inactivator Variants. Two forms of C$\overline{1}$ deficiency have been defined in patients with hereditary angioneurotic edema.[8,9] The sera of 85% of affected individuals have a diminished ability to inhibit C$\overline{1}$ as well as a low serum concentration of the inhibitor antigen. In the variant form, 15% of the patients have normal or elevated serum concentrations of the C$\overline{1}$ In antigen. This C$\overline{1}$ In protein fails to inhibit C$\overline{1}$s.[8,10] Eight kindred with nonfunctional inhibitor protein demonstrated a genetic heterogeneity characterized by differences in the ability to bind C$\overline{1}$s, in electrophoretic mobility in agarose, and in the ability to inhibit the esterolytic activity of C$\overline{1}$s.[10] The sera of some affected individuals possess several times the normal concentration of antigenically intact but functionally deficient C$\overline{1}$ In protein. Their C$\overline{1}$ In displays two different electrophoretic mobilities as demonstrated by immunoelectrophoresis in agar gels. These results are apparently explained by the noncovalent binding of some of the inhibitor molecules to albumin.[10,27]

The C$\overline{1}$ In protein from a mother and daughter with the variant form of hereditary angioneurotic edema has been isolated.[28] These abnormal inhibitors shared immunologic identity with normal C$\overline{1}$ In protein, however, they were inactive in inhibiting the functional activity of C$\overline{1}$s. Analysis of the abnormal inhibitors by SDS-acrylamide gel electrophoresis suggested that each consisted of a single polypeptide chain, the mobility of which was slower than that of normal C$\overline{1}$ In. The apparent molecular

[25] P. C. Harpel, M. W. Mosesson, and N. R. Cooper, *in* "Proteases and Biological Control" (E. Reich, D. Rifkin, and E. Shaw, eds.), p. 387. Cold Spring Harbor Laboratory, Cold Spring Harbor, New York, 1975.

[26] P. C. Harpel, *J. Clin. Invest.* **49**, 568 (1970).

[27] C.-B. Laurell, J. Lindegren, I. Malmros, and H. Martensson, *Scand. J. Clin. Lab. Invest.* **24**, 221 (1969).

[28] P. C. Harpel, T. E. Hugli, and N. R. Cooper, *J. Clin. Invest.* **55**, 605 (1975).

weight of the patients' inhibitors was 109,000. The structural abnormality appeared to be a property of the core molecule, as indicated by studies of proteolytic derivatives produced by trypsin and plasmin. No major differences in amino acid composition of the abnormal as compared to the normal CĪ In were demonstrated; nor were differences in susceptibility to *Vibrio cholerae* neuraminidase identified.

Composition. The amino acid composition of CĪ In as determined in two studies[16,28] is shown in Table I. Comparison of the compositions of the isolated abnormal inhibitors from two patients in a kindred with the variant form of hereditary angioneurotic edema demonstrated slightly higher values for the acidic amino acids and lower phenylalanine content than found in the normal CĪ In.[28]

TABLE I
AMINO ACID COMPOSITION OF CĪ INACTIVATOR (% AMINO ACID)

Component	Haupt[a]	Harpel[b]
Lysine	4.43	4.08
Histidine	1.59	1.53
Arginine	2.72	1.55
Aspartic acid	5.92	6.19
Threonine[c]	5.88	6.42
Serine[c]	4.34	5.94
Glutamic acid	7.64	6.93
Proline	3.53	3.97
Glycine	0.93	2.26
Alanine	2.49	4.08
Half-cystine[d]	0.55	0.68
Valine	3.59	4.52
Methionine	1.79	1.21
Isoleucine	2.26	2.28
Leucine	7.09	7.46
Tyrosine	2.64	1.15
Phenylalanine	4.14	3.38
Tryptophan	1.31	1.31[e]
Carbohydrates	34.70	34.70[e]
	100.00	99.6

[a] H. Haupt, N. Heimburger, T. Kranz, and H. G. Schwick, *Eur. J. Biochem.* **17** 254 (1970).

[b] P. C. Harpel, T. E. Hugli, and N. R. Cooper, *J. Clin. Invest.* **55,** 606 (1975).

[c] Not corrected for partial destruction during hydrolysis.

[d] Half-cystine was determined as cysteic acid.

[e] Tryptophan and carbohydrate values reported by Haupt *et al.* were applied for comparison in this study.

The carbohydrate composition of C$\overline{1}$ In is indicated in Table II. It is rich in carbohydrates and in sialic acid.[4,16]

TABLE II

CARBOHYDRATE COMPOSITION (%) OF HUMAN C$\overline{1}$ INACTIVATOR

Component	Schultze[a]	Haupt[b]
Hexose (galactose:mannose)	12	10.8
	(1:2)	
Acetylhexosamine	13	9.2
N-Acetylneuraminic acid	17	14.3
Fucose	0.6	0.4
Total carbohydrate	42.6	34.7

[a] H. E. Schultze, K. Heide, and H. Haupt, *Naturwissenschaften* **49**, 133 (1962).
[b] H. Haupt, N. Heimburger, T. Kranz, and H. G. Schwick, *Eur. J. Biochem.* **17**, 254 (1970).

[66] The Inter-α-Trypsin Inhibitor

By M. STEINBUCH

An α-glycoprotein, called protein π (the greek letter π standing for Paris), was isolated from human plasma by Steinbuch and Loeb.[1]

Heimburger and Schwick[2] showed that a protein migrating in the inter-α-region (between α$_1$- and α$_2$-globulins), had an inhibitory activity against trypsin. Heide *et al.*[3] purified the active component and called it inter-α-trypsin inhibtor (IαI). This protein proved later to be identical with protein π.

[1] M. Steinbuch and J. Loeb, *Nature* (*London*) **192**, 1196 (1961).
[2] N. Heimburger and H. G. Schwick, *Thromb. Diath. Haemorrh.* **7**, 432 (1962).
[3] K. Heide, N. Heimburger, and H. Haupt, *Clin. Chim. Acta* **11**, 82 (1965).

Assay Methods

A. In Whole Serum

1. Semiquantitative Methods, such as Fibrin–Agar Electrophoresis[2,4]

Serum is separated by electrophoresis on agar containing fibrin. A known amount of trypsin is allowed to diffuse into the supporting medium. In general, trypsin causes clearing of the opaque gel, but the zones containing a trypsin inhibitor will remain unchanged. In combination with immunoelectrophoresis and standardization of the digested areas with the use of controlled amounts of enzyme and of other known inhibitors, it is possible to evaluate the approximate amount of the inhibitors.

Reagents

Agarose (or agar)

Human or bovine fibrinogen: We prefer to use human fibrinogen devoid of plasminogen. This is prepared by washing fraction I in NaCl containing lysine and ethanol, followed by absorption of the remaining plasminogen onto $Ca_3(PO_4)_2$.

Barbital buffer, pH 8.2: 19.42 g of sodium acetate and 29.42 g barbital-sodium are dissolved in 1 liter of water, and 8 volumes of this stock solution are mixed with 9.6 volumes of water and brought to pH 8.2 with 0.1 M HCl. A 1:3 dilution gives an ionic strength of 0.05.

$CaCl_2$, 0.025 M

Antisera against the serum proteins to be examined

Trypsin solution, 0.175 mg/ml

Procedure. A 1.5% solution of agarose dissolved in barbital buffer, pH 8.2, is heated to 50°–55°; 9 volumes of this are added to 2 volumes of a 0.4% solution of fibrinogen dissolved in barbital buffer and 1 volume of similarly warmed (50°–55°) 0.025 M $CaCl_2$ is added; 3-ml portions of the mixture are pipetted onto microscopic slides. Solidification is achieved in about 5 min. With bovine fibrinogen, the slides must be heated for one hour at 80° in a moist chamber.

Electrophoretic seperation is carried out at 6 V/cm for 45 min. The trypsin solution and/or antiserum are filled in troughs cut parallel to the

[4] N. Heimburger, H. Haupt, and H. G. Schwick, Proteinase Inhibitors, Proc. Int. Res. Conf. 1st, Munich 1970, p. 1. de Gruyter, Berlin, 1971.

migration direction. If a "sandwich" technique is used, the serum (or inhibitor solution) is first subjected to electrophoretic seperation in starch gel or polyacrylamide (PAA), the use of a sieving gel facilitating the identification of the inhibitor. The agarose-fibrin slides are then pressed for 24 hr against the surface of these gels. The inhibitory activity of samples containing different molecular varieties of a purified inhibitor may be studied by electrophoresis on polyacrylamide containing 0.4% fibrinogen. The unstained gel is covered for 24 hr with a paper strip which had previously been dipped into a trypsin solution (0.1%).

2. Quantitative Methods

Any of the techniques described by K. A. Walsh and P. E. Wilcox for serine proteases in this series Vol. 19 [3] might be used. The only variation will be the addition of the inhibitor (serum) to the enzyme solution, incubation time before starting the assay, and pretreatment of the serum sample.

All plasma antiproteinases, with the exception of α_2-macroglobulin (α_2M), inhibit the esterase activity of trypsin and of related enzymes. α_1-Antitrypsin (α_1AT) is the main trypsin inhibitor, and it can be inactivated by heating the serum (56°, 10 min). α_2M can be inactivated by aliphatic amines, such as methylamine (0.5 M) or hydroxylamine (0.25 M). C$\bar{1}$ inactivator, if present, may be detected by the inhibition of the hydrolysis of a substrate for C$\bar{1}$ esterase. The approximate quantity of each inhibitor can thus be evaluated.

Immunological methods can be used for measuring IαI protein, but they provide no information about its activity. The methods especially recommended are passive hemagglutination/inhibition tests[5] and Laurell's crossed immunoelectrodiffusion.[6] The former needs only few specific antibodies and permits checking of many sera in a simple way. The second gives precise information regarding the amount and molecular forms of the inhibitor in the sample under consideration. However, it requires a larger volume of the specific antiserum.

B. Techniques for Assay of Purified IαI

This may be done by measuring (1) inhibition of esterase activity of trypsin using tosyl-L-arginine methyl ester (TAME), benzoyl-L-

[5] H. Haupt, N. Heimburger, T. Krantz, and H. G. Schwick, *Eur. J. Biochem.* **17**, 254 (1970).
[6] C. B. Laurell, *Scand. J. Lab. Invest.* **29**, Suppl. 124 (1972).

arginine ethyl ester (BAEE), or benzoyl-DL-arginine p-nitroanilide (BApNa) as substrates; (2) inhibition of the proteolytic activity of trypsin, using casein, fibrin, or hemoglobin as substrates.

IαI being an instantaneous inhibitor, the assay procedure can be started immediately after the addition of IαI to the trypsin solution. Semipurified IαI preparations must be free of interferring inhibition.

Isolation Techniques

IαI is an unusually labile protein, and aggregation as well as degradation are observed during purification. There is some evidence that the degradation is due to proteolytic modification. Among many substances studied only EDTA and known inhibitors of $C\overline{1}$-esterase provided some protection to IαI. A homogeneous material is more easily obtained from fraction III as starting material than with fresh plasma.

Protein π was originally isolated by chromatography of plasma on DEAE-cellulose at pH 5, whereas IαI was obtained by differential precipitation involving Rivanol, ammonium sulfate, and ethanol.[3]

Two purification techniques are given below, one (A) using freshly recalcified plasma, (B) the other Cohn fraction III as starting material.

Procedure A: Heimburger et al.[4,5]

1. Fresh plasma is recalcified by adding 12 ml of 1.3 M CaCl$_2$ to 1 liter of starting material. The fibrin is removed by filtration (paper).

2. The serum is diluted with an equal volume of water.

3. DE-32 cellulose, 30 g (wet), are added for each liter of diluted serum. After adsorption, the cellulose is collected on a Büchner filter, thoroughly washed with phosphate buffer, 0.02 M and then poured into a column. The proteins are eluted with a linear gradient of a sodium phosphate buffer of pH 6.9, from 0.02 M to 0.4 M (Fig. 1).

Ion-exchange chromatography allows the separation of $C\overline{1}$ inactivator from IαI.

4. Fractions containing IαI are pooled and ammonium sulfate is added to a final concentration of 1.8 M.

5. The precipitate is dissolved in minimal volume of 0.15 M NaCl and is subjected to gel filtration on a Sephadex G-150 column.

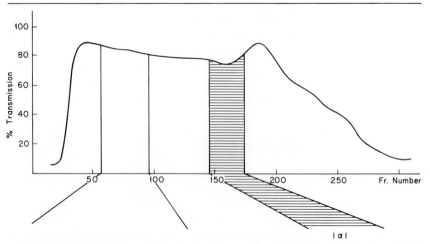

FIG. 1. Elution profile of IαI obtained by a linear gradient of phosphate buffer 0.02–0.4 *M*, pH 6.9. Wet DE32, 1.2 kg, was used for the batch adsorption of 20 liters of recalcified serum. Elution is achieved on a column (10 × 40 cm); 25-ml fractions are collected.

Procedure B: Steinbuch *et al.*

This method may be used either with fraction III as starting material or with ACD-plasma. In the latter case 1.5 g of EDTA are added per liter of plasma, the pH is brought to 5 and the protein concentration to 2.5% by dilution with 0.1 *M* acetate buffer, pH 5. Fraction III is a product obtained by a slightly modified Nitschmann technique,[7] as shown in Fig. 2. Fresh fraction III should be used for the isolation of IαI. As the precipitate of fraction III contains ethanol, it must be kept cold and handled without delay.

1. Fraction III (1 kg) is dissolved in 4.6 liters of 0.1 *M* acetate buffer containing EDTA (1.5 g/liter), to yield a solution of about 25 g of protein per liter. Insoluble material is removed by centrifugation.

2. DEAE-cellulose is equilibrated with 0.1 *M* acetate buffer, pH 5, and 1.5 g of the wet material is added to each gram of protein. Batch adsorption is carried out for a period of 20 min with stirring. The cellulose is removed by centrifugation, washed twice in 0.2 *M* acetate buffer, pH 5, and then poured into a column (5 × 60 cm) and eluted with 0.4 *M* acetate buffer containing EDTA (1.5 g/liter).

3. The IαI in the eluate is precipitated by 30% ethanol with lowering of the temperature to −5°.

[7] J. Nitschmann, P. Kistler, and W. Lergier, *Helv. Chim. Acta* **37,** 886 (1954).

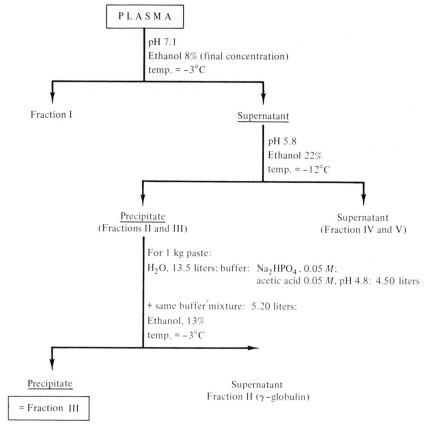

FIG. 2. Flow sheet for preparation of fraction III by a modified Nitschmann technique [J. Nitschmann, P. Kistler, and W. Lergier, *Helv. Chim. Acta* **37**, 886 (1954).

4. The sediment obtained by centrifugation in the cold is immediately dissolved in 0.15 *M* NaCl at pH 7, dialyzed against 0.2 *M* acetate, pH 6.5, containing EDTA (1.5 g/liter) and subjected to gel filtration on a Sephadex G-150 column (2.6 × 100 cm) (Fig. 3).

The first peak contains homogeneous IαI (see Fig. 4) whereas modified IαI showing several bands of more anodic mobility is found in the second peak. Nevertheless, the two peak materials show immunological identities. The average yield of crude IαI is about 1.3 g from 1 kg of paste and approximately 300–400 mg of the pure protein are obtained after gel filtration on Sephadex G-150.

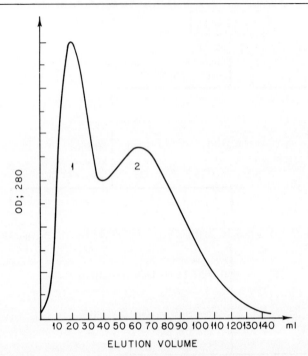

FIG. 3. Gel filtration of crude IαI on a Sephadex G-150 column (2.6 × 100 cm) with 0.2 M acetate buffer, pH 6.5 containing EDTA 1.5 g/liter. Flow rate 30 ml/hr.

Properties

Stability and Other Properties. IαI is a very unstable protein and one of the few plasma proteins denatured and aggregated by heating at 56°.[8] The aggregated protein sediments at 19 S in the ultracentrifuge, and its mobility in zone electrophoresis corresponds approximately to that of α_2M, even though the solution remains clear and inhibitory activity remains practically unchanged. However, the heated protein reacts only poorly with the corresponding antiserum, whereas antisera raised against the heated protein also react with the native protein. Some aggregated material is frequently observed in purified preparations (Fig. 4), and a small percentage of it also forms after storage of IαI in the frozen state. IαI isolated from fresh plasma often shows several bands in PAA and PAA/agarose electrophoresis (Fig. 5). There is evidence that the heterogeneity is due to proteolysis, but its exact cause is uncertain. When a small amount of plasmin is added to a homogeneous preparation of IαI, a multiband pattern similar to the one shown in Fig. 5 for peak 2 is seen. In SDS-PAA electrophoresis, IαI yields a single

[8] M. Steinbuch and R. Audran, *C. R. Acad. Sci.* **260**, 7058 (1965).

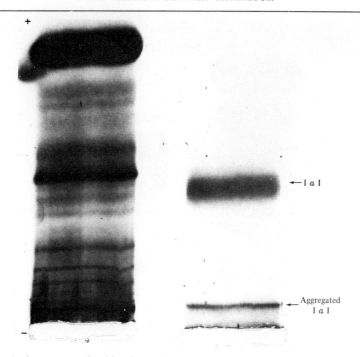

Fig. 4. Agarose/acrylamide electrophoresis of IαI (right) corresponding to peak 1 of Fig. 3. The small additional band near the origin corresponds to aggregated IαI as shown by immunodiffusion. Serum profile is shown on left.

190,000 MW component even in the presence of reducing agents. IαI can be kept at 4° under sterile conditions for 2–3 weeks. Freeze-drying is possible without loss of activity although it is most frequently accompanied by minor molecular alterations as shown by zone electrophoresis.

Crude IαI obtained after chromatography on DEAE-cellulose may appear homogeneous in cellulose acetate (or paper) electrophoresis and even in immunoelectrophoresis (Fig. 6), but sieving gel electrophoresis or analytical ultracentrifugation reveals the heterogeneity (7 S and 4 S components). Gel filtration on Sephadex G-150 (Fig. 4) achieves separation of the slow (native?) from the more anodic (breakdown?) component.

Physical and Chemical Properties. The molecular weight was initially evaluated by ultracentrifugation as 180–200,000.[1] More recently Heimburger published a value of 160,000[4] whereas 190–200,000 were found by SDS-polyacrylamide electrophoresis in our laboratory.[8a]

[8a] M. Steinbuch, R. Audran, P. Lambin and J. M. Fine, "23rd Colloq. on Protides of the Biological Fluids" (H. Peters ed.), p. 115–118. Elsevier, Amsterdam, 1975.

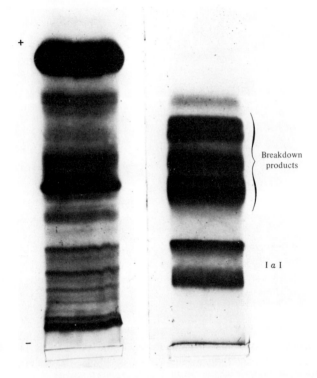

FIG. 5. Agarose/acrylamide electrophoresis of an IαI preparation obtained from ACD-plasma without addition of any stabilizing substance. Right: Peak 2 material from Fig. 3. The anodic bands show immunological identity with the main band (intact IαI) and represent active trypsin inhibitors in fibrin/agarose electrophoresis. Left: Serum control.

IαI has very characteristic electrophoretic properties. In paper and cellulose acetate electrophoresis it takes (barbital buffer; pH 8.2–8.6) an inter-α position, which is the reason for the name given to the inhibitor (Fig. 6). In polyacrylamide, acrylamide/agarose and starch gel IαI behaves as a posttransferrin, taking a position between $\beta_1 C$ and $\beta_1 A$-globulin (Fig. 5). In immunoelectrophoresis IαI is identified by a precipitation-line in the α_1-region (Fig. 6).

The absorbance ($E_{1\,cm}^{1\%}$ 280 nm) is 7.1.[3] IαI is a glycoprotein containing 3.4% of hexose; 3.5% (4.3%) hexosamine, and 2% sialic acid. The amino acid composition as measured by Heimburger et al.[4] is shown in the table.

Arginine is the N-terminal residue. Zn^{2+} was found in both the early preparation of protein π and in IαI isolated under the conditions described under procedure B. Atomic absorption analysis of several IαI

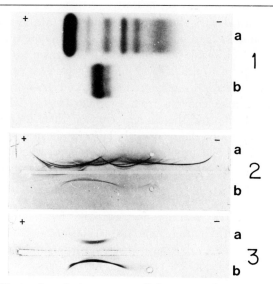

Fig. 6. (1) Electrophoresis in acetate cellulose, control serum in 1a and IαI in 1b and 3. (2) Immunoelectrophoresis. 2a and 3a, NHS; 2b and 3b, IαI. Whole antiserum was used in 2 and anti-IαI-serum in 3.

AMINO ACID COMPOSITION OF INTER-α-TRYPSIN INHIBITOR[a]

Amino acid	%	Amino acid	%
Lysine	5.63	Alanine	3.37
Histidine	2.20	Half-cystine	1.69
Ammonia	1.44	Valine	5.86
Arginine	5.18	Methionine	2.00
Aspartic acid	9.50	Isoleucine	3.91
Threonine	4.64	Leucine	7.40
Serine	4.87	Tyrosine	4.01
Glutamic acid	11.97	Phenylalanine	5.00
Proline	3.69	Tryptophan	1.69
Glycine	2.90	Carbohydrates	8.40

Total = 95.35

[a] Values (in percent of amino acid residue) published by N. Heimburger, H. Haupt, and H. G. Schwick, *Proteinase Inhibitors, Proc. Int. Res. Conf. 1st*, Munich, 1970, p. 1. de Grugter, Berlin, 1971.

preparations showed a Zn^{2+} content of 0.6–0.95 g-atom per molecular weight of 190,000.

Specificity. IαI inhibits mainly trypsin, having a higher affinity for bovine than for the human enzyme; trypsin I of Figarella,[9] which

[9] C. Figarella, G. A. Negri, and O. Guy, *Proteinase Inhibitors, Proc. Int. Res. Conf. 2nd (Bayer Symp. V)*, Grosse Ledder 1973, p. 213. Springer-Verlag, Berlin and New York, 1974.

seems to be identical with the cationic trypsin of Mallory and Travis,[10] is much less inhibited than human trypsin 2 (= anionic trypsin). Bovine chymotrypsin is inhibited less than bovine trypsin[11] but more than human chymotrypsin.[11] Inhibition of human trypsin by various inhibitors was studied by Travis[12] and by Feeney and co-workers.[13] Human acrosin (a trypsinlike enzyme from spermatozoa) is inhibited by IαI with a higher affinity for the human enzyme than for boar acrosin.[14] Only a weak inhibition is seen with human plasmin. Affinity of the inhibitor for papain and bromelain is rather poor, whereas a somewhat better reaction was observed for an insect protease, probably related to cocoonase.[15]

IαI belongs to the class of trypsin inhibitors having an arginine in its active center. The activity is destroyed by modification of the guanidino group with butanedione.[16] Acetylation and succinylation of lysines are without effect, but as little as 30% modification of arginines abolishes activity. The latter treatment also diminishes immunological reactivity of IαI.[4]

Kinetic Properties. Inhibition of bovine trypsin appears to be stoichiometric. Most stable complexes are those formed with the bovine enzyme, where $K_i = 21$ pM; $K_i = 9$ nM was found for the complex formed between IαI and human cationic trypsin.[11] Complexes formed with human and bovine chymotrypsins are less stable. Trypsin bound to IαI dissociates promptly in the presence of α_2M to form a stronger complex with this antiprotease.[11] The α_2M–trypsin complex has esterase activity whereas the IαI–trypsin complex has none.

Distribution. INTACT MOLECULE. The normal concentration of IαI is between 300 and 500 mg/l of plasma.[17] This inhibitor has also been found in human cervical mucus together with other inhibitors, namely,

[10] P. A. Mallory and J. Travis, *Biochemistry* 12, 2847 (1973).
[11] J. Bieth, M. Aubry, and J. Travis, *Proteinase Inhibitors, Proc. Int. Res. Conf., 2nd (Bayer Symp. V)*, Grosse Ledder, 1973, p. 53. Springer-Verlag, Berlin and New York, 1974.
[12] J. Travis, *Biochem. Biophys. Res. Commun.* 44, (1971).
[13] R. E. Feeney, G. E. Means, and J. C. Bigler, *J. Biol. Chem.* 244, 1957 (1969).
[14] H. Fritz, N. Heimburger, M. Meier, M. Arnhold, L. Zaneveld, and G. Schumacher, *Hoppe-Seyler's Z. Physiol. Chem.* 353, 1953 (1972).
[15] J. C. Landureau and M. Steinbuch, *Z. Naturforsch. Tail B* 25, 231 (1970).
[16] H. Fritz, B. Brey, M. Müller, and M. Gebhardt, *Proteinase Inhibitors, Proc. Int. Res. Conf. 1st*, Munich, 1970, p. 28. de Gruyter, Berlin, 1971.
[17] N. Heimburger, *Proc. Int. Res. Conf., 2nd Proteinase Inhibitors, 2nd (Bayer Symp. V.)*, Grosse Ledder, 1973, p. 14. Springer-Verlag, Berlin and New York, 1974.

α_1AT,[18,19] levels being higher during the nonfertile phases of the ovulatory cycles. Preliminary observations indicate that acrosin inhibitors, such as IαI, impair[19] sperm penetration through cervical mucus.

The IαI of *Gorilla gorilla* and *Panroglodytes* show complete immunological identity with the human inhibitor, which gives a partial identity with that of *Pongo pygmaeus*.[20]

RELATION TO OTHER INHIBITORS. Hochstrasser and co-workers[21-23] found considerable antiproteolytic activity in secretions of the upper respiratory tract. Of this activiy, 80% belongs to relatively low-molecular-weight, acid-stable inhibitors (MW 14,000 and 20,000). These inhibitors cross-react with the monospecific antiserum against IαI. Also, their active sites are similar in the sense that blocking of arginine destroys their activities.[23a] These substances inhibit leukocyte proteases involved in the inflammatory process.[23]

Human urine contains an acid-stable inhibitor (MW 16,000) first described by Astrup and co-workers,[24] called Mingin,[25] which is also immunologically related to IαI.[26]

Other acid-stable proteinase inhibitors were described 20 years ago.[27] Hochstrasser's group recently showed the existence of inhibitors (MW 22,000 and 40,000) in the perchloric acid supernatant of human plasma. Both are immunologically related to IαI and to the inhibitors found in urine.[28] Furthermore, the plasma and urine inhibitors have identical amino acid compositions and inhibitory properties.

[18] O. Wallner and H. Fritz, *Hoppe-Seyler's Z. Physiol. Chem.* **355**, 709 (1974).

[19] G. F. B. Schumacher and L. J. D. Zaneveld, *Proteinase Inhibitors, Proc. Int. Res. Conf., 2nd (Bayer Symp. V)*, Grosse Ledder, 1973, p. 178. Springer-Verlag, Berlin and New York, 1974.

[20] K. Bauer, *Humangenetik* **8**, 357 (1970).

[21] K. Hochstrasser, R. Reichert, S. Schwarz, and E. Werle, *Hoppe-Seyler's Z. Physiol. Chem.* **353**, 221 (1972).

[22] K. Hochstrasser, R. Reichert, S. Schwarz, and E. Werle, *Hoppe-Seyler's Z. Physiol. Chem.* **354**, 923 (1973).

[23] K. Hochstrasser, R. Reichert, M. Matzner, and E. Werle, *Z. Klin. Chem. Klin. Biochem.* **10**, 104 (1972).

[23a] K. Hochstrasser, R. Reichert, and N. Heimburger, *Hoppe-Seyler's Z. Physiol. Chem.* **354**, 587 (1973).

[24] T. Astrup and J. Sterndorf, *Scand. J. Clin. Lab. Invest.* **7**, 239 (1955).

[25] T. Astrup and U. Nissen, *Nature (London)* **203**, 255 (1964).

[26] G. J. Proksch and J. I. Routh, *J. Lab. Clin. Med.* **79**, 491 (1972).

[27] R. J. Peanasky and M. Laskowski, *J. Biol. Chem.* **204**, 153 (1953).

[28] K. Hochstrasser, H. Feuth, and O. Steiner, *Hoppe-Seyler's Z. Physiol. Chem.* **354**, 927 (1973).

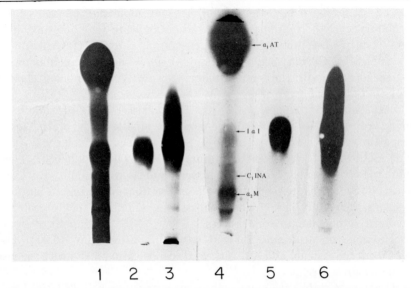

FIG. 7. Electrophoresis of IαI (samples 2 and 5) and of split products (samples 3 and 6) obtained after the addition of plasmin; 5% polyacrylamide gels; barbital buffer as described under Reagents. Samples 1, 2, and 3 were stained with Amido black. Samples 4, 5, and 6 were separated on polyacrylamide containing 0.4% fibrin. The latter gels were covered for 24 hr with a paperstrip which had previously been dipped into a trypsin solution and were stained with Amido black only later. Samples 1 and 4 contain serum. The undigested fibrin appears clearly after tryptic digestion. These zones show the inhibitory activity of IαI and its split products.

The origin of these IαI-related, but smaller, inhibitors is unknown.[28] As shown in Fig. 7, degradation products produced by plasmin are still active trypsin inhibitors. They are more potent inhibitors than the native protein.

[67] Trypsin-Kallikrein Inhibitors from Snails (*Helix pomatia*)

By HARALD TSCHESCHE and THOMAS DIETL

Proteinase inhibitors have long been known in the species of Annelida. Investigations on another androgyne species from the pedigree of Mollusca led to the discovery of proteinase inhibitors in snails (*Helix pomatia*)[1], in cuttle fish (*Loligo vulgaris*)[2] and in mussels (*Mytilus edulis*).[3]

[1] H. Tschesche, T. Dietl, R. Marx, and H. Fritz, *Hoppe-Seyler's Z. Physiol. Chem.* **353**, 483 (1972).

[2] H. Tschesche and A. v. Rücker, *Hoppe-Seyler's Z. Physiol. Chem.* **354**, 1447 (1973).

[3] The inhibitor activity of mussels (*Mytilus edulis*) against trypsin is rather low

The snail (*Helix pomatia*) produces two inhibitory proteins differing in molecular size and properties.[1,4,5] A group of low-molecular-weight, acid- and heat-stable trypsin-kallikrein isoinhibitors of broad specificity is synthesized from the epidermal parts of the body walls and is excreted into the mucus.[4,6] An acid- and heat-labile inhibitor of higher molecular weight is produced and secreted from the albumin gland.[4-10] Easy separation of both inhibitor proteins obtained from extracts of undissected snails can be achieved by a single step of ammonium sulfate fractionation.[1,4,5]

The epidermal (or mucous) inhibitors are related in structure and properties to the bovine trypsin-kallikrein inhibitor (Kunitz),[11-17] the cow colostrum inhibitor,[18-19] the turtle egg white inhibitor,[20] the Russell's viper venom inhibitor,[21] and the toxins I and K from black mamba venom.[22]

with 25.5 mIU (BAPA)/mg wet live weight (without shell). The activity is present in the foot [23 mIU (BAPA)/mg], in the gill [18 mIU (BAPA)/mg], and in other body parts [27 mIU (BAPA)/mg]. It is stable against 3% perchloric acid treatment at 75° for 1 min. H. Tschesche and A. V. Rücker, unpublished results.

[4] H. Tschesche and T. Dietl, *Protides Biol. Fluids Proc. Colloq.* **20**, 437 (1973).

[5] H. Tschesche and T. Dietl, *Eur. J. Biochem.* **30**, 560 (1972).

[6] H. Tschesche and T. Dietl, *Hoppe-Seyler's Z. Physiol. Chem.* **353**, 1189 (1972).

[7] G. Uhlenbruck, I. Sprenger, and I. Ishiyama, *Z. Klin. Chem. Klin. Biochem.* **9**, 361 (1971).

[8] G. Uhlenbruck, I. Sprenger, and G. Hermann, *Z. Klin. Chem. Klin. Biochem.* **9**, 494 (1971).

[9] I. Springer, G. Uhlenbruck, and G. Hermann, *Enzymologia* **432**, 83 (1972).

[10] G. Hermann, G. Uhlenbruck, I. Sprenger, and H. Franke *Protides Biol. Fluids Proc. Colloq.* **20**, 433 (1973).

[11] M. Kunitz and J. H. Northrop, *J. Gen. Physiol.* **19**, 991 (1936).

[12] B. Kassell, M. Radicevic, S. Berlow, R. J. Peanasky, and M. Laskowski, Sr., *J. Biol. Chem.* **238**, 3274 (1963).

[13] B. Kassell and M. Laskowski, Sr., *Biochem. Biophys. Res. Commun.* **20**, 463 (1965).

[14] F. A. Anderer, *Z. Naturforsch. Teil B* **20**, 462 (1965).

[15] T. Dietl and H. Tschesche, *Proteinase Inhibitors, Proc. Int. Res. Conf., 2nd* (*Bayer Symp. V*), Grosse Ledder, 1973, pp. 437–445.

[16] T. Dietl and H. Tschesche, *Eur. J. Biochem.* **58**, 453 (1975).

[17] H. Tschesche and T. Dietl, *Eur. J. Biochem.* **58**, 439 (1975).

[18] D. Pospíšilová, V. Dlouhá, and F. Šorm, *Fed. Eur. Biochem. Soc. 5th Meet. Abstr.* No. 971 (1968).

[19] D. Čechová-Pospíšilová, V. Svestková, and F. Šorm, *Fed. Eur. Biochem. Soc. 6th Meet. Abstr.* No. 225 (1969).

[20] M. Laskowski, Jr., I. Kato, T. R. Leary, J. Schrode, and R. Sealock, *Proteinase Inhibitors, Proc. Int. Res. Conf. 2nd* (*Bayer Symp. V*), Grosse Ledder, 1973, pp. 597–611.

[21] H. Takahashi, S. Jwanoga, T. Kitagawa, and T. Suzuki *in Proteinase Inhibitors, Proc. Int. Res. Conf. (Bayer Symp. V),* Grosse Ledder, 1973, pp. 265 276.

[22] D. J. Strydom, *Nature (London) New Biol.* **243**, 88 (1973).

A comparison of these structures is included in this chapter. The group of isoinhibitors isolated from the cuttle fish (*Loligo vulgaris*) seems to be of a different class of inhibitor proteins not structurally related to the snail inhibitors.[2]

Assay

Determination of the inhibitors exhibiting broad specificity is based on the inhibition of trypsin.

Principle.[23] Inhibition of the tryptic hydrolysis of benzoyl-DL-arginine *p*-nitroanilide (BAPA) is measured by the decrease in absorbance at 405 nm per minute, as has already been described in this series.[24]

Definition of Unit and Specific Activity. One enzyme unit corresponds to the hydrolysis of 1 μmole of substrate per minute (ΔA_{405}/min cm = 3.32 for 3 ml). One inhibitor unit decreases the activity of two enzyme units by 50%, thus decreasing the absorbance by 3.32 per minute and cm for 3 ml.[23] The specific activity equals inhibitor units (BAPA) per milligram protein.

Trypsin-Kallikrein Inhibitor (*Helix pomatia*) (HPI)

Purification Procedure

Purification after aqueous tissue extraction involves separation from the albumin gland inhibitor by ammonium sulfate fractionation. It is then based on the selective adsorption of the inhibitor from the redissolved ammonium-sulfate precipitate onto polymer-bound trypsin, trypsin resin. The inhibitor attached to the matrix-bound trypsin is then washed free of contaminating material. Thereafter the trypsin–inhibitor complex is dissociated under acidic conditions, and the inhibitor is removed. The procedure is that of Fritz *et al.*[25] Since the snail inhibitors are extremely resistant to enzymic hydrolysis, as is the bovine tissue inhibitor (Kunitz),[26] this method is especially suited for the isolation of these inhibitors in their native form.

[23] H. Fritz, G. Hartwich, and E. Werle, *Hoppe-Seyler's Z. Physiol. Chem.* **345**, 150 (1966).

[24] B. Kassell, this series Vol. 19, p. 845.

[25] H. Fritz, M. Gebhardt, R. Meister, K. Illchmann, and K. Hochstrasser, *Hoppe-Seyler's Z. Physiol. Chem.* **351**, 571 (1970).

[26] B. Kassell and T. W. Wang, *Proteinase Inhibitors Proc. Int. Res. Conf., 1st,* Munich, 1970, pp. 89–94.

If isolation of the albumin gland inhibitor is not desired, it can be irreversibly denatured by heating the aqueous extract for 1 min with 3% perchloric acid to 65°. The step of ammonium-sulfate fractionation may then be omitted before selective affinity binding of the acid-stable inhibitor onto the trypsin resin; however, it facilitates the further purification.

Trypsin resin. Preparation and assay of trypsin resin (EMA-copolymer-trypsin resin) has already been described in this series.[24] Because of its anionic ion exchange and less hydrophilic properties, this resin is less suitable than the cellulose-bound trypsin,[27] which can be obtained commercially (bovine trypsin bound to CM-cellulose, Merck AG, Darmstadt, Germany).

The trypsin resin should be subjected to excessive washings with the basic and acidic buffers prior to use in the affinity step in order to remove all labile and extractable trypsin.

Step 1. Crude Extract.[1,5] Homogenize 2 kg of frozen snails (*Helix pomatia*) previously freed from shells with 4 liters of ice-water at 0°–4° for approximately 10 min. The solution is centrifuged for 1 hr at 24,000 g in a refrigerated centrifuge, and 4.1 liters of clear supernatant solution are collected from the sediment. The amount of inhibitor in the solution is determined and should approximate 1400 inhibitor units (BAPA) against trypsin.

Separation from Albumin Gland Inhibitor

Step 2. Ammonium Sulfate Precipitation.[1,5] Dissolve 1839 g of solid ammonium sulfate (447 g/liter, 63% saturation) at 3° in the supernatant solution. A precipitate containing the low-molecular-weight trypsin-kallikrein inhibitors is formed, while the albumin gland inhibitor remains in the ammonium sulfate solution. The ammonium sulfate solution may then be stored at 0°–4° for several months (over 2 years in the author's laboratory) without deterioration until work-up of the albumin gland inhibitor is desired. It contains about 600 (BAPA) units of inhibitory activity measured against trypsin. This amounts to about 43% of the total inhibitory activity of the crude extract.

The ammonium sulfate precipitate is separated at 0°–4° by centrifugation at 24,000 g for 30 min and is immediately redissolved in 1.9 liters of water. Small amounts of insoluble material are removed by a second centrifugation at 24,000 g for 30 min. A clear supernatant solution

[27] H. Fritz, M. Gebhardt, R. Meister, K. Illchmann, and K. Hochstrasser, *Hoppe-Seyler's Z. Physiol. Chem.* **351**, 571 (1970).

containing the low-molecular-weight inhibitors with about 800 (BAPA) units against trypsin is obtained. This amounts to 57% of the total inhibitory activity of the crude extract.

Step 3. Affinity Chromatography.[1,5,25] CM-cellulose bound trypsin is equilibrated overnight with 0.1 *M* triethanolamine/HCl buffer, pH 7.8, containing 0.4 *M* NaCl and 0.01 *M* CaCl$_2$ and then used in batch operation. The resin (5 g) is stirred at 0°–4° for 1 hr with 1 liter of the inhibitor solution. The final pH being adjusted to pH 7.0. The inhibitory activity is measured in the supernatant. Only traces should be detectable. The suspension is then cleared by centrifugation.

The cake is washed six times with 0.1 *M* triethanolamine/HCl buffer, pH 7.8. The inhibitors are desorbed from the resin by repeated extraction for 10 min with 0.4 *M* KCl/HCl of pH 1.7 and removed from the resin by centrifugation. The pH of the suspension should be checked. This procedure is repeated until the supernatant is free from inhibitor. The acid solution (2 liters) is adjusted to pH 5–6 immediately by addition of 1 *N* KOH and concentrated by rotary evaporation at 35°. The yields of inhibitor in this step are generally 80–100% but depend on the particular resin and its preconditioning. After reequilibration with pH 7.8 buffer, the resin can be reused for another batch operation.

Step 4. Sephadex G-50 Filtration and Desalting.[1,5] The concentrated inhibitor solution is halved. Precipitated salt is removed by centrifugation or filtration, and the precipitate is washed with a small amount of ice cold 50% water–ethanol mixture. Each half is passed through a column of Sephadex G-50 fine, 60 × 740 mm, equilibrated with 0.1 *N* acetic acid. Some high-molecular-weight impurities and salts are thus removed.[5] Tests should be made with AgNO$_3$/HNO$_3$ in order to find the breakthrough peak of chloride ions. The salt-free inhibitor solution is collected and lyophilized.

At this stage, the inhibitor is 85–95% pure with 85% yield. This mixture of isoinhibitors is designated inhibitor preparation I[1,5] and has a specific activity about 3500 mIU (BAPA)/mg. The final purification may be carried out by a second passage through the trypsin resin.

Separation and Purification of Individual Isoinhibitors

Step 5. I. Ion Equilibrium Chromatography.[5,16] The inhibitor preparation I (70 mg) is dissolved in 2 ml of 0.05 *M* ammonium acetate pH 4.5, and placed on top of a column, 1.5 × 140 cm, of SE-Sephadex C-25, 40–120 μm, previously equilibrated with 0.05 *M* ammonium acetate, pH 4.9. The less basic isoinhibitors A–J are successively eluted by a pH gradient (Fig. 1) prepared by mixing 0.05 *M* ammonium acetate,

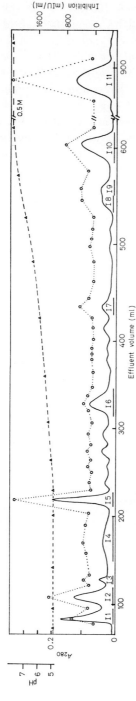

Fig. 1. Elution pattern of snail inhibitor preparation I after equilibrium chromatography on SE-Sephadex C-25. Sample: 70 mg of lyophilized inhibitor in 2 ml of 0.05 M ammonium acetate, pH 4.5. The column (1.5 × 140 cm) was equilibrated with 0.05 M ammonium acetate, pH 4.9–8.0, up to 650 ml of effluent volume. Then 0.5 M ammonium acetate, pH 8.0, was applied for final elution. Flow rate: 14 ml/hr.——, Absorbance at 280 nm; ○ ··· ○, inhibitor activity in mIU (BAPA)/ml; △--- △, pH. The isoinhibitors B, E, G, H, and K are purified from fractions I2, I5, I7, I8, and I.11, respectively.

pH 8.0, from a 2-liter reservoir to 100 ml of the equilibrating buffer. The flow rate is adjusted to 14 ml/hr. About 45% of all the isoinhibitors are resolved and eluted at this buffer molarity when the effluent pH has reached pH 7 (Fig. 1). Elution of the more basic isoinhibitors (mainly isoinhibitor K) is then achieved by applying a 0.5 M ammonium acetate buffer, pH 8.0, to the top of the column. Individual fractions are collected, as shown by horizontal bars, in Fig. 1, and concentrated by rotary evaporation at 35°. They are desalted by passage through columns of Bio-Gel P-2 equilibrated with 0.1 N acetic acids and then are lyophilized. The final purification of the isoinhibitors A–J is done by rechromatography under identical or slightly modified conditions.[1,5]

Step 5. II. Isolation of Isoinhibitor K.[16] If en bloc separation of isoinhibitor fractions A–J from the basic ones is desired the inhibitor preparation I (370 mg) is placed on top of a column, 4.5 × 30 cm of SE-Sephadex C-25 equilibrated and eluted with 0.05 M ammonium acetate, pH 8.0 (Fig. 2). At a flow rate of 24 ml/hr, fractions of 3 ml are collected. Isoinhibitor fractions A–J are obtained after 510 ml of effluent volume. The basic fractions containing isoinhibitor K are then eluted by a linear gradient prepared by mixing 0.5 M ammonium acetate to 100 ml of starting buffer in a mixing chamber (Fig. 2). The major fraction (85%) corresponds to isoinhibitor K and equals 40% of the total inhibitor preparation I.

Pure isoinhibitor K is obtained after rechromatography of the de-salted major fraction on a column, 4.5 × 30 cm, of SE-Sephadex C-25, equilibrated with 0.1 M ammonium acetate, pH 8.0, and a gradient up to 0.3 M buffer.[16]

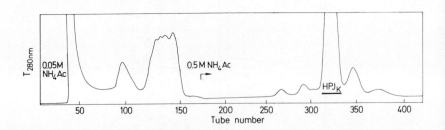

FIG. 2. Ion equilibrium chromatography of snail inhibitor preparation I on SE-Sephadex C-25 for quick separation of isoinhibitor K. Sample: 370 mg of lyophilized inhibitor in 10 ml of 0.05 M ammonium acetate, pH 8.0. The column (4.5 × 30 cm) was equilibrated and eluted with 0.05 M ammonium acetate, pH 8.0, up to 510 ml effluent volume. Then 0.5 M ammonium acetate, pH 8.0, was mixed to 100 ml of the equilibrating buffer. Flow rate: 25 ml/hr. Fraction volume 3 ml/tube. ——, Absorbance at 280 nm.

Properties

Stability. The isoinhibitors (mixture and individual components) show remarkable stability against heat, acid and alkali denaturation, and/or enzymic degradation. They may be heated in dilute acid or at neutral pH at 100° or in 3% perchloric or trichloroacetic acid at 75° without loss of activity.[1,5] The dry lyophilized material has practically indefinite stability if bacterial breakdown is prevented. The isoinhibitors are not inactivated by trypsin α or β,[5] chymotrypsin α[5] plasmin,[5] porcine pancreatic kallikrein,[5] porcine pancreatic elastase,[28] carboxypeptidases A or B,[28] carboxypeptidase C from orange leaves.[28] However, it is digested by thermolysin at 60°–80°.[28,45] Isoinhibitor K is less stable than the bovine inhibitor (Kunitz) against proteolytic inactivation by thermolysin or against thermal denaturation at pH 8.0.[45]

Purity. The isoinhibitors B, E, G, H, and K were purified by similar rechromatographic procedures on SE-Sephadex C-25 as described.[1,5,15,16] They were homogeneous by cellulose acetate and polyacrylamide gel electrophoresis, by SDS-polyacrylamide gel electrophoresis, and amino acid analysis. Isoinhibitor K purified as described herein gives the correct amino acid analysis after step 5.

Physical Properties. The molecular weight by SDS-polyacrylamide gel electrophoresis[16] is 6500 and by amino acid analysis[15,16] is 6463 for isoinhibitor B, 6431 for E, 6591 for G, 6575 for H, and 6463 for K (see the table). From 1:1 stoichiometric titration with trypsin a molecular weight of 6500 is calculated.[5]

The high resolution ^{1}H NMR spectrum at 270 MHz[29] gives a spectrum typical for the structure of the trypsin-kallikrein inhibitor (Kunitz).[30] However, deuterium exchange of some labile protons seems to be more rapid in the snail isoinhibitor K when compared to the bovine inhibitor[31] and the cow colostrum inhibitor.[29,32]

Physiological Properties. No immune cross-reaction of the isoinhibitor preparation I with antibodies against the bovine inhibitor (Kunitz)[33] or the snail albumin gland inhibitor[34] could be detected.

---

[28] T. Dietl and H. Tschesche, *Eur. J. Biochem.* **58**, 453 (1975), and *Protides Biol. Fluids Proc. Colloq.* **23**, 271 (1976). We are grateful to Dr. H. Fritz for gifts of boar acrosin and porcine serum kallikrein.

[29] G. Wagner, K. Wüthrich, and H. Tschesche, *FEBS Lett.*

[30] A. Masson and K. Wüthrich, *FEBS Lett.* **31**, 114 (1973).

[31] K. Wüthrich, personal communication, 1974.

[32] K. Wüthrich, G. Wager, and H. Tschesche, *Protides Biol. Fluids Proc. Colloq.* **23**, 201 (1976).

[33] E. Truscheit, personal communication, 1972.

[34] G. Uhlenbruck, personal communication, 1972.

In contrast to the very basic bovine inhibitor (Kunitz), which is totally trapped in rat kidneys after intraveneous injection,[35] the less basic snail isoinhibitor preparation I is almost completely excreted into the rat urine after 48 hr.[36,37]

The inhibitor preparation I exhibits the same antiplasmin activity in fibrinolysis, but only about one-tenth of the anticoagulation activity in the blood clotting test as the bovine inhibitor (Kunitz).[38] The anti-streptokinase activity seems to be blood dependent and thus ranges from equal to ten times higher activity than the bovine inhibitor (Kunitz).[38] There is almost no activity observed in the blood platelet tests.[38]

The snail isoinhibitor preparation I is almost twice as effective than the trypsin-kallikrein inhibitor (Kunitz) in diminishing the lethal rate of rats with experimental burn shocks.[38a]

Specificity. Inhibitor preparation I inhibits the following enzymes: trypsins of cow[1,5] and pig,[28] the trypsinlike proteinase of Pronase E from *Streptomyces griseus*,[28] boar acrosin[28] (the latter only slightly), porcine plasmin,[1,5] porcine pancreatic kallikrein,[1,5] kallikrein from porcine urine,[28] porcine serum kallikrein[1] (the latter only slightly), bovine α-chymotrypsin,[1,5] the chymotrypsin-like proteinase of Pronase E from *Streptomyces griseus*[28] (the latter only slightly), fungi proteinase K from *Tritirachium album* Limber (Merck, Darmstadt, Germany),[28] *Aspergillus* proteinase P (Röhm, Darmstadt, Germany),[28] alkaline bacillus proteinase 2231 (Röhm, Darmstadt, Germany),[28] insulin-degrading proteinase from rat kidney, muscle, or liver,[39,40] and an insulin degrading activity from human erythrocytes.[41]

Numerous other proteinases have been tested for inhibition, but are not inhibited: pepsin,[28] subtilisin,[28] porcine pancreatic elastase,[16,28] cathepsin B,[42] bacterial proteinase N (M) (Röhm, Darmstadt, Ger-

[35] H. Fritz, K.-H. Oppitz, D. Meckel, B. Kemkes, H. Haendle, H. Schult, and E. Werle, *Hoppe-Seyler's Z Physiol. Chem.* **350**, 1541 (1969).

[36] H. Tschesche, H. Jering, G. Schorp, and T. Dietl, *Proteinase Inhibitors, Proc. Int. Res. Conf. 2nd (Bayer Symp. V)*, Grosse Ledder, 1973, pp. 362–377. Springer-Verlag, Berlin and New York, 1974.

[37] H. Jering, G. Schorp, and H. Tschesche, *Hoppe-Seyler's Z. Physiol. Chem.* **355**, 1129 (1974).

[38] R. Marx, personal communications, 1972 and 1974.

[38a] G. Kwiecinsky, PhD thesis, Faculty of Medicine, University Munich, 1976.

[39] J. S. Brush and H. Tschesche, *Proteinase Inhibitors, Proc. Int. Res. Con., 2nd (Bayer Symp. V)*, Grosse Ledder, 1973, pp. 581–585. Springer-Verlag, Berlin and New York, 1974.

[40] J. S. Brush and H. Tschesche, unpublished results, 1973.

[41] H. Tschesche, T. Dietl, H. J. Kolb, and E. Standl, *Proteinase Inhibitors, Proc. Int. Res. Conf., 2nd (Bayer Symp. V)*, Grosse Ledder, 1973, pp. 586–593. Springer-Verlag, Berlin and New York, 1974.

[42] A. J. Barrett, personal communication, 1973.

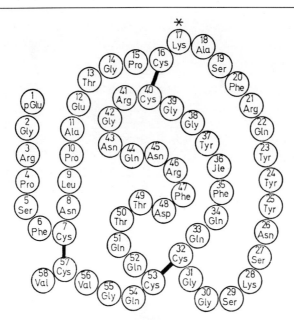

FIG. 3. A schematic representation of the covalent structure of the isoinhibitor K of snails (*Helix pomatia*) based on data from references 15–17 and 45.

many),[28] bacterial collagenase,[28] fungi proteinase P (Röhm, Darmstadt, Germany),[28] *Aspergillus oryzae* protease,[28] clostripain,[28] papain,[28] trypsinlike proteinase from wheat,[41] plasminogen activator,[43] pancreatic carboxypeptidases A[28] and B,[28] or citrus carboxypeptidase C.[17,28]

The individual isoinhibitors differ in association constant and specificity; e.g., isoinhibitor H inhibits porcine pancreatic kallikrein more effectively than isoinhibitor K[15,28] whereas isoinhibitor K is more effective against porcine plasmin than are isoinhibitor B or E.[28] The isoinhibitors B, E, and G all inhibit porcine pancreatic kallikrein, but in contrast to the entire mixture, inhibitor preparation I, they do not inhibit porcine serum kallikrein.[1,5] The insulin-degrading activity from rat liver is more effectively inhibited by the fraction of isoinhibitors A–D or the isoinhibitors F, G, or K.[39] The natural mixture of all isoinhibitors is more effective than isoinhibitor K in the inhibition of insulin degradation by hemolyzed human erythrocytes.[41]

Kinetic Properties. The inhibition of trypsin follows a 1:1 stoichiometry.[5] The dissociation constants (K_i) of the complex of trypsin and isoinhibitors B, H, E, K, and the total mixture are not significantly different from that of the bovine trypsin–kallikrein inhibitor

---

[43] E. Reich, personal communication, 1973.

Positions are marked at 10, 20, 30; the reactive site region is boxed.

	Sequence (→)
Snail isoinhibitor K	pGlu Gly Arg Pro Ser Phe Cys Asn Leu Pro Ala Glu Thr Gly Pro Cys Lys Ala Ser Phe Arg Gln Tyr Tyr Asn Ser Lys Ser Gly Gly Cys Gln Gly Phe Ile Tyr Gly Gly
Pancreatic (Kunitz)	Arg Pro Asp Phe Cys Leu Glu Pro Pro Tyr Thr Gly Pro Cys Lys Ala Arg Ile Ile Arg Tyr Phe Tyr Asn Ala Lys Ala Gly Leu Cys Gln Thr Phe Val Tyr Gly Gly
Bovine colostrum	Phe Gln Thr Pro Asp Leu Cys Gln Leu Pro Gln Ala Arg Gly Pro Cys Lys Ala Ala Leu Leu Arg Tyr Phe Tyr Asn Ser Thr Ser Asn Ala Cys Glu Pro Phe Thr Tyr Gly Gly
Turtle egg white	Lys Gln Asn Gly Arg Asp Ile Cys Arg Leu Pro Glu Gln Gly Pro Cys Lys Gly Arg Ile Pro Arg Tyr Phe Tyr Asn Pro Ala Ser Arg Met Cys Glu Ser Phe Ile Tyr Gly Gly
Russell's viper	His Asp Arg Pro Thr Phe Cys Asn Leu Ala Pro Glu Ser Gly Arg Cys Arg Gly His Leu Arg Arg Ile Tyr Tyr Asn Leu Glu Ser Asn Lys Cys Lys Val Phe Phe Tyr Gly Gly
Toxin K	Ala Ala Lys Tyr Cys Lys Leu Pro Leu Arg Ile Gly Pro Cys Lys Arg Lys Ile Pro Ser Phe Tyr Tyr Lys Trp Lys Ala Lys Gln Cys Leu Pro Phe Asp Tyr Ser Gly
Toxin I	Gln Pro Leu Arg Lys Leu Cys Ile Leu His Arg Asn Pro Gly Arg Cys Tyr Gln Lys Ile Pro Ala Phe Tyr Tyr Asn Gly Lys Lys Lys Gln Cys Glu Gly Phe Thr Trp Ser Gly

	40											50								
Snail isoinhibitor K	Cys	Arg	Gly	Asn	Gln	Asn	Arg	Phe	Asp	Thr	Gln		Gln	Cys	Gln	Gly	Val	Cys	Val	
Pancreatic (Kunitz)	Cys	Arg	Ala	Lys	Arg	Asn	Asn	Phe	Lys	Ser	Ala	Glu	Asp	Cys	Met	Arg	Thr	Cys	Gly	Gly
Bovine colostrum	Cys	Gln	Gly	Asn	Asn	Asn	Asn	Phe	Glu	Thr	Thr	Glu	Met	Cys	Leu	Arg	Ile	Cys	Glu	Pro
Turtle egg white	Cys	Lys	Gly	Asn	Lys	Asn	Asn	Phe	Lys	Thr	Lys	Ala	Glu	Cys	Val	Arg	Ala	Cys	Arg	Pro
Russell's viper	Cys	Gly	Gly	Asn	Ala	Asn	Asn	Phe	Glu	Thr	Arg	Asp	Glu	Cys	Arg	Glu	Thr	Cys	Gly	Gly
Toxin K	Cys	Gly	Gly	Asn	Ala	Asn	Arg	Phe	Lys	Thr	Ile	Glu	Glu	Cys	Arg	Arg	Thr	Cys	Val	Gly
Toxin I	Cys	Gly	Gly	Asn	Ser	Asn	Arg	Phe	Lys	Thr	Ile	Glu	Glu	Cys	Arg	Arg	Thr	Cys	Ile	Arg

Snail isoinhibitor K						
Pancreatic (Kunitz)	Ala					
Bovine colostrum	Pro	Gln	Gln	Thr	Asp	Lys	Ser
Turtle egg white	Pro	Glu	Pro	Gly	Val	Cys ... 43 more residues
Russell's viper	Lys					
Toxin K	Lys					
Toxin I	Lys					

Fig. 4. Comparison of the amino acid sequences of the snail isoinhibitor K,[11-14] the bovine pancreatic trypsin-kallikrein inhibitor (Kunitz)[15,17] [see also J. Chauvet, G. Nouvel, and R. Acher, *Biochim. Biophys. Acta* **92**, 200 (1964); V. Dlouhá, P. Pospíšilová, B. Meloun, and F. Šorm, *Collect. Czech. Chem. Commun.* **30**, 1311 (1965)], the cow colostrum inhibitor,[18,19] the turtle egg white inhibitor,[20] the Russell's viper venom inhibitor,[21] and the toxins K and I from black mamba venom.[22] Invariant residues are framed. Superscript numbers refer to text footnotes that cite sources of data on which sequences are based.

AMINO ACID COMPOSITION OF ISOINHIBITORS FROM THE SNAIL (*Helix pomatia*)

Amino acid	Isoinhibitors (moles amino acid per mole protein)				
	B	E	G	H	K
Asparatic acid	7	6	7	6	5
Threonine	2	2	2	2	3
Serine	5	5	3	4	4
Glutamic acid	6	7	8	9	9
Proline	5	5	3	3	3
Glycine	6	7	6	6	8
Alanine	3	3	2	2	2
Valine	0	0	3	3	2
Half-cystine	6	6	6	6	6
Methionine	1	1	0	0	0
Isoleucine	2	2	2	2	1
Leucine	2	1	2	1	1
Tyrosine	4	4	4	4	4
Phenylalanine	4	4	4	4	4
Lysine	1	1	2	2	2
Arginine	4	4	4	4	4
	58	58	58	58	58
Molecular weight	6463	6431	6591	6575	6463
Reactive site residue	Arg	Arg		Lys	Lys

(Kunitz).[5,15,28] From titrations according to Green and Work[44] values (K_i) between 10^{-9} M and 10^{-11} M at pH 7.8[5,15,28] can be estimated which are probably the lower limit. However, there are significant differences between individual isoinhibitors and the dissociation of the complexes with plasmin,[28] pancreatic kallikrein,[15,28] or insulin-degrading proteinases.[39,40,41] Isoinhibitor F is a competitive inhibitor of insulin degradation by the rat liver enzyme.[39]

The reaction of purified isoinhibitors B, H, E, G, or K and of the mixture, preparation I, is quick but not instantaneous. The complex formation with trypsin, chymotrypsin, plasmin, and pancreatic kallikrein is almost completed after 1 min at concentrations of 10^{-7} mole/liter at pH 7.8 and 25°.[5,28]

Distribution. The isoinhibitors are very likely synthesized in the subepithelial glands of the skin. They are found in the foot, mantle, and mucus of the snail, but not in the inner organs, i.e., liver, kidney, stomach, intestines, or sex organs.[6] There was found no significant change in the inhibitor content of snails during active life or hibernation.[6]

[44] N. M. Green and E. Work, *Biochem. J.* **54**, 347 (1953).

Amino Acid Compositions. All isoinhibitors B, H, E, G, or K contain the same total number of 58 amino acid residues but differ by substitution of single amino acid residues. All are devoid of carbohydrate content. The compositions are shown in the table.[1,4,5,15,16]

Structure and Homologies. The amino acid sequence[15,17] and the positioning of the three disulfide bridges[45] of the snail isoinhibitor K is given in Fig. 3. The N-terminus is blocked by pyroglutamic acid.[15-17] The exposed disulfide bridge Cys_{16}–Cys_{40} is accessible to selective sodium borohydride reduction,[45] as is the corresponding bridge in the bovine inhibitor (Kunitz).[46]

The homologies between isoinhibitor K and the bovine trypsin–kallikrein inhibitor (Kunitz),[12,13] the cow colostrum inhibitor,[18,19] the turtle egg white inhibitor,[20] the Russell's viper venom inhibitor,[21] and the toxins K and I from black mamba venom[22] are presented in Fig. 4. There are 28 amino acid residues of the snail inhibitor identical with that of the bovine inhibitor (Kunitz). This corresponds to a higher degree of homology between the snail isoinhibitor K and the bovine inhibitor (Kunitz) than between the latter and the cow colostrum inhibitor with 21 identical residues.

[45] T. Dietl and H. Tschesche, *Hoppe-Seyler's Z. Physiol. Chem.* **357**, 139 (1976).
[46] L. F. Kress and M. Laskowski, Sr., *J. Biol. Chem.* **242**, 4925 (1967).

[68] Albumin Gland Inhibitor from Snails (*Helix pomatia*)

By THOMAS DIETL and HARALD TSCHESCHE

The resistance of anti-A-agglutinins from the albumin gland of snails (*Helix pomatia*) toward enzymic proteolysis led to the discovery of the albumin gland proteinase inhibitor.[1,2] This inhibitor is different in molecular size and inhibitory and other properties[3-6] from the epidermal trypsin–kallikrein inhibitors (this volume [67]) and is present only in

[1] G. Uhlenbruck, I. Sprenger, and I. Ishiyama, *Z. Klin. Chem. Klin. Biochem.* **9**, 361 (1971).
[2] G. Uhlenbruck, I. Sprenger, and G. Hermann, *Z. Klin. Chem. Klin. Biochem.* **9**, 491 (1971).
[3] H. Tschesche, T. Dietl, R. Marx, and H. Fritz, *Hoppe-Seyler's Z. Physiol. Chem.* **535**, 483 (1972).
[4] H. Tschesche and T. Dietl, *Eur. J. Biochem.* **30**, 560 (1972).
[5] H. Tschesche and T. Dietl, *Protides Biol. Fluids, Proc. Colloq.* **20**, 437 (1973).
[6] T. Dietl and H. Tschesche, *Hoppe-Seyler's Z. Physiol. Chem.* in press (1976).

the albumin gland.[7] It has not yet been studied as intensively as the low-molecular-weight inhibitors.

Assay[8]

Principle. Inhibition of the tryptic hydrolysis of benzoyl-DL-arginine *p*-nitroanilide (BAPA) is measured by the change in absorbance at 405 nm as described in a previous volume of this treatise.[9,9a]

Definition of Unit and Specific Activity. One trypsin unit corresponds to the hydrolysis of 1 μmole of substrate per minute (ΔA_{405}/min cm = 3.32 for 3 ml). One inhibitor unit decreases the activity of two enzyme units by 50%, thus decreasing the absorbance by 3.32/minute and cm for 3 ml.

The specific activity corresponds to inhibitor units (BAPA) per milligram of protein.

Purification Procedure

Steps I, II, and III. These steps in the isolation are identical with the steps in the purification procedure of the epidermal, heat- and acid-stable trypsin-kallikrein inhibitors described in another chapter of this volume [67]. The separation of the epidermal and the albumin gland inhibitor is based on their different solubilities in ammonium sulfate.[3-5] The albumin gland inhibitor is not precipitated by ammonium sulfate at 63% saturation and is thus obtained in solution designated as inhibitor preparation II.[4,5]

Step IV. Dialysis of Inhibitor Preparation II. The ammonium sulfate solution containing the albumin gland inhibitor has a specific activity of about 9.5 mIU (BAPA)/(ml $\cdot A_{280\,nm}^{1\,cm}$). It is dialyzed for 16 hr at 0° in an Amicon dialyzer in H 1 D $\times$ 50 tubings against deionized water and is then lyophilized. The yield is 90% of the starting activity.

Step V. Sephadex G-100 Filtration. The inhibitor (208 mg) is dissolved in 0.1 M ammonium sulfate and gel filtered through a column, 4.4 $\times$ 40 cm, of Sephadex G-100 fine, previously equilibrated with 0.1 M ammonium sulfate. The flow rate is adjusted to 20 ml/hr. Some impurities of higher molecular weight are eluted prior to the inhibitor. The fractions containing active inhibitor are combined, concentrated by rotary evapora-

[7] H. Tschesche and T. Dietl, *Hoppe-Seyler's Z. Physiol. Chem.* **353**, 1189 (1972).
[8] H. Fritz, G. Hartwich, and E. Werle, *Hoppe-Seyler's Z. Physiol. Chem.* **345**, 150 (1966).
[9] B. Kassell, this series Vol. 19, p. 845.
[9a] H. Fritz, I. Trautschold, and E. Werle, *in* "Methoden der Enzymatischen Analyse" (H. U. Bergmeyer, ed.), p. 1021. Verlag Chemie, Weinheim, 1970.

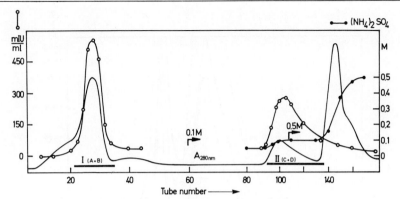

Fig. 1. Elution pattern of snail albumin gland inhibitor after ion-exchange chromatography on DEAE-Sephadex A-50. Sample: 166 mg in 0.01 M ammonium sulfate. The column, 1.5 × 20 cm, was equilibrated with 0.01 M ammonium sulfate, pH 5.5, and eluted with the equilibrating solvent up to tube 60 in order to obtain peak I (inhibitors A and B). Then 0.1 M ammonium sulfate, pH 5.5, was applied for elution of peak II (inhibitors C and D). Inactive contaminations were eluted by 0.5 M ammonium sulfate. Flow rate: 5 ml/hr. ——, Absorbance at 280 nm; O——O, inhibitor activity in mIU (benzoyl-DL-arginine p-nitroanilide)/ml; ●——●, molarity in ammonium sulfate.

tion, and dialyzed against 0.01 M ammonium sulfate. The inhibitor solution has a specific activity of 147 mIU (BAPA)/(ml · $A_{280\,nm}^{1\,cm}$) and is kept frozen. The yield is 80% of the starting activity in this step.

Step VI. DEAE-Sephadex Chromatography. The inhibitor solution is placed on top of a column, 1.5 × 20 cm, of DEAE-Sephadex A-50 previously equilibrated with 0.01 M ammonium sulfate, pH 5.5. The column is eluted at a flow rate of 5 ml/hr. With the equilibrating solvent, a peak I containing isoinhibitors A and B is displaced from the column, yielding 39% of the starting activity in this step. With 0.1 M ammonium sulfate a second peak, II, containing isoinhibitors C and D representing 42% of the starting activity is eluted (Fig. 1). Both inhibitor fractions I and II are concentrated and dialyzed against 0.01 M ammonium sulfate and stored frozen.

Step VII. Separation of Isoinhibitors A and B. The inhibitor fraction I from step VI is placed on a column, 0.9 × 60 cm, of DEAE-Sephadex A-50 equilibrated with 0.1 M ammonium sulfate, pH 9. Elution of the column is carried out at a flow rate of 5 ml/hr with 0.1 M ammonium sulfate, pH 8.0. Two inhibitory active peaks are successively eluted (Fig. 2). The first inhibitor, designated isoinhibitor A, corresponds to 17% of the starting activity in this step. Its specific activity is 413 mIU (BAPA)/(ml $A_{280\,nm}^{1\,cm}$). The second peak, designated inhibitor B, Corresponds to 53% of the starting activity. Its specific activity is 480 mIU

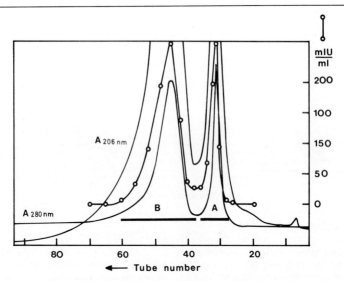

FIG. 2. Elution profile of snail albumin gland inhibitor peak I (Fig. 1) after equilibrium chromatography on DEAE-Sephadex A-50. The column, 0.9 × 60 cm, was equilibrated with 0.1 M ammonium sulfate, pH 9, and eluted with 0.1 M ammonium sulfate, pH 8.0. Flow rate: 5 ml/hr. ——, Absorbance at 280 nm and 206 nm; ○——○, inhibitor activity in mIU (benzoyl-DL-arginine p-nitroanilide)/ml.

(BAPA)/(ml $A_{280\,nm}^{1\,cm}$). Both inhibitors were dialyzed against 0.01 M ammonium sulfate, concentrated, and stored frozen.

Step VIII. Separation of Isoinhibitors C and D. The inhibitor fraction II from step VI is subjected to chromatography on a column, 0.9 × 60 cm, of DEAE-Sephadex A-50 equilibrated and eluted with 0.075 M ammonium sulfate, pH 5.2 up to tube 80. Elution of the column is continued with 0.1 M ammonium sulfate, pH 5.2, at a flow rate of 5 ml/hr. Two inhibitory active peaks are eluted (Fig. 3). The first inhibitor is designated inhibitor C and corresponds to 52% of the starting activity in this step. Its specific activity is 438 mIU (BAPA)/(mg $A_{280\,nm}^{1\,cm}$). The second peak containing isoinhibitor D represents 58% of the starting activity and has a specific activity of 455 mIU (BAPA)/(mg $A_{280\,nm}^{1\,cm}$). Both inhibitors were dialyzed against 0.01 M ammonium sulfate, concentrated, and stored frozen.

Step VIIIa. Rechromatography of Inhibitors C and D. Isoinhibitor C was rechromatographed on a column, 0.9 × 45 cm, of DEAE-Sephadex A-50 equilibrated with 0.05 M ammonium sulfate, pH 9.2, at a flow rate of 5 ml/hr. A single symmetric peak was obtained that was dialyzed against deionized water and lyophilized.

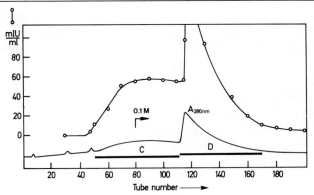

Fig. 3. Elution pattern of snail albumin gland inhibitor peak II (Fig. 1) after ion-exchange chromatography on DEAE-Sephadex A-50. The column, 0.9 × 60 cm, was equilibrated and eluted with 0.075 M ammonium sulfate, pH 5.2, up to tube 80 and then with 0.1 M ammonium sulfate, pH 5.2. Flow rate: 5 ml/hr. ——, Absorbance at 280 nm; ○——○, inhibitor activity in mIU (benzoyl-DL-arginine p-nitroanilide)/ml.

Isoinhibitor D was rechromatographed on a column, 0.9 × 60 cm, of DEAE-Sephadex A-50 equilibrated with 0.1 M ammonium sulfate, pH 9.3, at a flow rate of 5 ml/hr. The inhibitor was eluted as a major peak with a small inactive contamination preceding this peak. It was dialyzed against deionized water and lyophilized.

The overall yield of the separated isoinhibitors A–D were 4% of inhibitor A, 10% of B, 8% of C, and 11% of D when calculated for the initial activity present in the ammonium sulfate solution.

Properties

Stability. The isoinhibitors (mixture and individual components) are labile to acid and heat denaturation. Their activity is lowered by heating above 50° and is destroyed within 1 min upon heating at 100° at neutral pH[10] or in 3% perchloric or trichloroacetic acid at 65°.[5] The dry lyophilized material is stable at room temperature and can be kept for several months under refrigeration. The inhibition of bovine trypsin remains unchanged even with excess of trypsin (17-hr test) indicating permanent inhibition.[10]

Purity. The isoinhibitors A, B, C, and D obtained by DEAE-Sephadex A-50 ion-exchange chromatography are about 90% homogeneous by polyacrylamide gel electrophoresis at pH 9.5.[11]

[10] T. Dietl and H. Tschesche, *Protides Biol. Fluids Proc. Colloq.* **23,** 271 (1976).
[11] T. Dietl and H. Tschesche, in preparation.

Physical Properties. The molecular weight of the albumin gland isoinhibitor C by its gel retention behavior on Sephadex G-100 was 73,000 ∓ 7000,[10] that calculated from the complex with bovine trypsin was 70,000 ± 7000.[10] A molecular weight in this region is in accord with the stoichiometry of the titration of bovine trypsin with the inhibitor.

The inhibitor is a glycoprotein. The carbohydrates amount to about 35% by weight. The molecular weight calculated from the amino acid analysis (see the table) and the carbohydrate composition (see the table) is 74,000 and in good agreement to the other values.[10]

Specificity. The isoinhibitors all inhibit bovine and porcine trypsin[10,11] and porcine plasmin.[11] They do not inhibit bovine α-chymotrypsin even after 1 hr of preincubation at pH 7.8 and 25°,[10,11] nor porcine pancreatic kallikrein,[5] thermolysin,[11] papain,[11] or subtilisin.[10] Somewhat contradictory observations on a surprisingly broad specificity have been reported.[1,2,12]

Kinetic Properties. The inhibitors form 1:1 stoichiometric complexes with bovine trypsin and porcine plasmin. The inhibition is not instantaneous but progresses 15–20 min at pH 7.8 and 25° under assay conditions.[10,11] The reaction with porcine plasmin progresses even much more slowly. The association constant with bovine trypsin was determined from titrations according to Green and Work[10,11] to be $K_{ass} = 1.5 \times 10^9$ liters/mole.

Immunogenic Properties. The albumin gland inhibitors are good immunogens.[1,2,12,13] Immunsera prepared in New Zealand rabbits have been used in immunoelectrophoretic studies to characterize four isoinhibitors by their different electrophoretic mobilities.[2] A cross-reactivity of the *Helix pomatia* antiserum with *Helix aspersa* antigens was observed.[12,13] Several distinct inhibitory components of albumin gland inhibitors from other animal species have been detected by fibrin-agar-electrophoresis and immunoelectrophoresis.[12–14] However, no cross-reactivity of *Helix pomatia* antiserum with the epidermal inhibitors was detectable.[15]

Reactive Site. Isoinhibitors A–D are inactivated upon reaction with 1,2-butanedione in borate buffer and thus are arginine inhibitors.[15a]

Distribution. The occurrence of albumin gland inhibitors appears to

[12] G. Hermann, U. Uhlenbruck, I. Sprenger, A. C. Vales, and H. Franke, *Naturwissenschaften* **59**, 125 (1972).

[13] G. Hermann, G. Uhlenbruck, I. Sprenger, and H. Franke, *Protides Biol. Fluids, Proc. Colloq.* **20**, 433 (1973).

[14] I. Sprenger, G. Uhlenbruck, and G. Hermann, *Enzymologia* **43**, 83.

[15] G. Uhlenbruck, personal communication, 1972.

[15a] T. Dietl and H. Tschesche, *Hoppe-Seyler's Z. Physiol. Chem.* **357**, in press (1976).

Amino Acid and Carbohydrate Composition[a] of Snail Isoinhibitors
A–D and Human Plasma α_1-Proteinase Inhibitor[b]

Component	Moles amino acid per mole protein				
	A[c]	B[c]	C[c]	D[c]	α_1-Proteinase inhibitor
Amino acids					
Asparagine	43	51	50	46	47
Threonine	27	28	25	21	26
Serine	39	45	55	40	23
Glutamic acid	55	60	60	60	55
Proline	20	19	19	17	23
Glycine	24[d]	29[d]	28[d]	23[d]	23
Alanine	39[d]	47[d]	37[d]	37[d]	25
Valine	23	27	23	23	25
Half-cystine	4	6	2	0	4
Methionine	8	9	9	9	6
Isoleucine	23	26	26	22	20
Leucine	31	34	34	33	48
Tyrosine	11	12	15	8	6
Phenylalanine	24	30	27	27	27
Lysine	31	30	29	36	41
Histidine	9	9	9	9	13
Arginine	16	19	17	16	7
Tryptophan	5				1
Carbohydrates					
Fucose		5			
Mannose		36			21[e]
Galactose		54			
Glucose		9			
Galactosamine		5			17[f]
Glucosamine		39			
Sialic acid		2			2

[a] T. Dietl and H. Tschesche, *Protides Biol. Fluids, Proc. Colloq.* **23**, 271 (1976).
[b] R. Panell, D. Johnson, and J. Travis, *Biochemistry* **13**, 5439 (1974).
[c] Hydrolysis time 16 hr.
[d] Peak included amino acid and amino sugar.
[e] Total number of hexose units.
[f] Total number of hexosamine units.

be common to all snail species tested and has been confirmed for the species *Helix pomatia*,[1-5] *Helix aspersa*,[12,13] *Ampullarius australis*,[13] *Zebrina dentrita* (Müller),[16] *Cyclostoma elegans* (Muller),[16] *Zoonites algirus*,[16] and *Galba truncatula*.[16] In contrast to the species *Helix*, the

[16] G. Hermann, *Med. Welt* **24**, 2001 (1973).

cyclostomes are not pulmonary animals and hermaphrodites. As has been demonstrated for the species *Helix pomatia*, the labile high-molecular-weight inhibitors are exclusively produced in the albumin gland.[7] A report on heat-labile inhibitors in the blood of snails (*Helix pomatia*)[17] could not be confirmed and obviously is due to cross contamination by the albumin gland inhibitor and salt effects.

Composition. The amino acid compositions of the isoinhibitors A–D are presented in the table.[10,11] A close similarity to the composition of the human-α_1-proteinase inhibitor[18] is striking. The high content of proline is unusual. Free cysteinyl-SH is absent.

The carbohydrate content of isoinhibitor B was investigated[10,19] and revealed the presence of the monosaccharide units fucose, mannose, galactose, glucose, galactosamine, glucosamine, and sialic acid. The maximal numbers of monosaccharide units was determined after kinetic studies on the acid hydrolysis[19] and are given in the table as calculated for a total molecular weight of 70,000.

[17] E. Werle, W. Appel, and E. Happ, *Naunyn-Schmiedeberg's Arch. Exp. Pathol. Pharmakol.* **234**, 364 (1958).
[18] R. Panell, D. Johnson, and J. Travis, *Biochemistry* **13**, 5439 (1974).
[19] R. Klauser, T. Dietl, and H. Tschesche, *Hoppe-Seyler's Z. Physiol. Chem.*, in preparation.

[69] Trypsin–Kallikrein Inhibitors from Cuttlefish (*Loligo vulgaris*)

By HARALD TSCHESCHE

The isolation of proteinase inhibitors from snails stimulated the investigation of other members of the Class Mollusca. This article summarizes the information[1,2] available about the inhibitors from the edible cuttlefish (*Loligo vulgaris*) captured in the California area of Monterey and San Pedro.

Assay

The inhibitors having a broad specificity are assayed by their inhibition of trypsin.

[1] H. Tschesche and A. v. Rücker, *Hoppe-Seyler's Z. Physiol. Chem.* **354**, 1447 (1973).
[2] H. Tschesche and A. v. Rücker, *Hoppe-Seyler's Z. Physiol. Chem.* **354**, 1510 (1973).

Principle.[3] Inhibition of tryptic hydrolysis of benzoyl-DL-arginine p-nitroanilide (BAPA) is measured by the decrease in absorbance at 405 nm/min as has already been described in a previous volume of this treatise.[4]

Definition of Unit and Specific Activity. One trypsin unit corresponds to the hydrolysis of 1 μmole of substrate per minute (ΔA_{405}/min cm = 3.32 for 3 ml). One inhibitor unit decreases the activity of two enzyme units by 50%, thus causing a decrease in absorbance of 3.32/min and cm for 3 ml.[3] The specific activity is expressed in units (BAPA) per milligram protein.

Purification Procedure

The main step of purification is based on the selective adsorption of the inhibitors from precleaned extracts by insoluble trypsin (cellulose trypsin-resin) as described in this volume [67, 73–75].

Step 1. Crude Extract.[1] The starting material is the commercially available frozen cuttlefish. A portion of 200 g is homogenized with 400 ml of water in a Waring Blendor. The cloudy, yellow extract is cleared by centrifugation. The inhibitor content equals 53 mIU (BAPA) per gram wet weight of cuttlefish.

Step 2. Deproteinizing. The clear extract is deproteinized by addition of an equal volume 6% perchloric acid and heated at 65° for 1 min. The suspension is rapidly cooled and neutralized by 2 N KOH. The precipitates formed are removed by centrifugation, and the solution is concentrated ten to twelve times in a rotary evaporator at 35°.

Step 3. Extraction of Lipids. All lipids are extracted from the deproteinized solution by addition of the same volume of chloroform or methylene chloride under vigorous shaking in a separating funnel. Separation of the two layers of the emulsion thus formed is facilitated by dissolving 7–10% of ammonium sulfate in the aqueous phase and using centrifugation for 30 min at 10,000 g, 4°. The aqueous phase contains 94–98% of the inhibitory activity against trypsin found in the crude extract.

Step 4. Ammonium Sulfate Precipitation. By addition of solid ammonium sulfate to the inhibitor solution this is brought to 75% saturation. The inhibitors are then precipitated and removed by filtration. The ammonium sulfate precipitates can be stored as a suspension at 4° for several months under the mother liquor without loss of activity. The filter

[3] H. Fritz, G. Hartwich, and E. Werle, *Hoppe-Seyler's Z. Physiol. Chem.* **345**, 150 (1966).
[4] B. Kassell, this series Vol. 19, p. 845.

cake is redissolved in 10–20 times its volume in water. The yield is 95–98%.

Step 5. Affinity Chromatography. Trypsin bound to CM-cellulose is preconditioned and freed from soluble trypsin as described in this volume [67, 73–75]. The procedure is that described by Fritz *et al.*[5] The trypsin resin (5 g), equilibrated overnight in 0.1 M triethanolamine-HCl buffer, pH 7.8, containing 0.4 M NaCl and 0.01 M CaCl$_2$, is added to 1 liter of the inhibitor solution. The suspension is stirred for 6 hr at 4°–5° and aliquots of the supernatant are withdrawn at time intervals and tested for residual inhibitor activity. When the adsorption of the inhibitor is finished the resin is filtered on a glass frit and washed 8 times by 250 ml of 0.1 M triethanolamine-HCl buffer, pH 7.8, 0.4 M in NaCl and 0.01 M in CaCl$_2$. The loaded resin is transferred into 250 ml of 0.4 M KCl/HCl, pH 1.7. The inhibitor is desorbed under stirring the suspension for 14 min. The pH of the solution should be checked and readjusted to pH 1.7 if necessary. This procedure is repeated until all inhibitor is extracted. The acidic solutions are combined and neutralized immediately and stored at 4°.

Step 6. Gel filtration. The inhibitor solution is concentrated in a rotary evaporator at 35° and the precipitated NaCl is removed by filtration. The concentrated solution is freed from the major portion of salt by passage through a column of Bio-Gel P-2 equilibrated in 0.1 M acetic acid. The fractions containing inhibitor are combined and again concentrated by rotary evaporation. The concentrate is placed on top of a column (1.5 × 120 cm) of Sephadex G-50 fine equilibrated in 0.1 M acetic acid and eluted at a flow rate of 9 ml/hr. The peak containing the inhibitor is collected and lyophilized. The ratios of the inhibitory activities against the enzymes trypsin, chymotrypsin, plasmin, and kallikrein remains unchanged from step 1 to step 6.

The inhibitor preparation has a high specific activity of 3580 mIU (BAPA)/mg against trypsin. The preparation is 90–95% pure inhibitor with 80–85% yield, but contains more than a dozen isoinhibitors.[1]

Step 7. Isolation of Individual Isoinhibitors. The inhibitor preparation can be resolved into various isoinhibitors by pH- and concentration-gradient elution chromatography on SE-Sephadex C-25 as shown in Fig. 1. The homogeneous isoinhibitors A, B, E, and L (L almost pure) can be purified from individual fractions by four more steps of ion-equilibrium chromatography on SE-Sephadex C-25 with buffers of different but constant pH.[1]

[5] H. Fritz, H. Schult, M. Hutzel, M. Wiedemann, and E. Werle, *Hoppe-Seyler's Z. Physiol. Chem.* 348, 308 (1967).

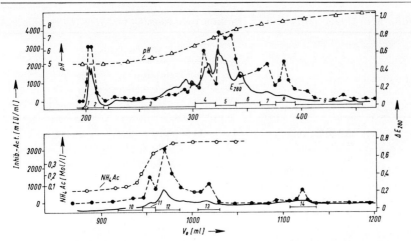

FIG. 1. Elution pattern of isoinhibitors from cuttlefish (*Loligo vulgaris*) on SE-Sephadex C-25. Sample: 90 mg of lyophilized inhibitor preparation in 2.5 ml of 0.05 M ammonium acetate, pH 4.5. A column, 1.5 × 140 cm, equilibrated with 0.05 M ammonium acetate, pH 5.0, was eluted with a pH gradient from pH 5 to 9 up to 780 ml of effluent volume (upper panel). Then a buffer of pH 9.0 with the molarity increasing from 0.05 M to 0.5 M was applied (lower panel). Flow rate 11 ml/hr. ——, Absorbance at 280 nm; ●---●; inhibition of trypsin in mIU (benzoyl-DL-arginine *p*-nitroanilide)/ml; △---△, pH of the effluent; ○---○, molarity in ammonium acetate. Fractions 1, 2, 5, and 12 correspond to isoinhibitors A, B, E, and L.

Properties

Stability. The inhibitors are stable in 3% perchloric acid and can be heated in this medium to 65° for 1 min without significant loss of activity.

Purity. The preparation after step 6 is at least 90–95% pure inhibitor as calculated on the basis of specific activity, 1:1 molar association with trypsin, and an average molecular weight of the inhibitor of 6700. It can be resolved into numerous isoinhibitors by chromatography on SE-Sephadex C-25.[1]

Physical Properties. The molecular weights calculated from titrations with trypsin are 6800 for isoinhibitors A and B, and 7200 for isoinhibitor E, from the amino acid compositions are 6630, 6686, and 6925 for isoinhibitors, A, B, and E, respectively.[1]

Specificity. The inhibitor preparation as the mixture of all isoinhibitors inhibits the enzymes: bovine α- and β-trypsin,[1] bovine α-chymotrypsin,[1] porcine pancreatic kallikrein,[1] and porcine plasmin.[1] Isoinhibitor E

AMINO ACID COMPOSITION OF THE ISOINHIBITORS FROM CUTTLEFISH (*Loligo vulgaris*)

	Residues per molecule			
	Isoinhibitors			
Amino acid	A	B	E	L
Aspartic acid	11	11	11	11
Threonine	5	5	2	4
Serine	3	3	2	1
Glutamic acid	6	5	4	5
Proline	2	2	—	1
Glycine	7	7	6	7
Alanine	3	3	3	3
Valine	2	1	3	2
Half-cystine	8	8	8	8
Methionine	2	2	2	3
Isoleucine	2	2	2	0–1 $\left.\right\}\Sigma = 2^a$
Leucine	3	3	2	1–2
Tyrosine	1	1	1	1
Phenylalanine	3	3	3	3
Lysine	3	4	8	8
Histidine	—	—	3	3
Arginine	1	2	2	5
	62	62	62	(67)

[a] The values add up to two residues independent from the time of acid hydrolysis [H. Tschesche and A. v. Rücker, *Hoppe-Seyler's Z. Physiol. Chem.* **354**, 1447 (1973).

is a better inhibitor of porcine pancreatic kallikrein than isoinhibitors A, B, or L.[1,6] The inhibitors do not inhibit porcine serum kallikrein.[7]

Kinetic Properties. Titrations of isoinhibitors A, B, E, and L with trypsin indicate a 1:1 stoichiometric association. The dissociation constants estimated according to Green and Work[8] are 16 nM for isoinhibitors A and B, 0.64 nM for isoinhibitors E, and 0.22 nM for isoinhibitor L.[1]

The reactions of the isoinhibitors with the enzymes are not instantaneous and differ in velocity. Inhibition equilibrium is attained in 3–5 min with trypsin (isoinhibitors A and B react more slowly than E and L),[1] in 5 min with chymotrypsin,[1] and in 5–8 min with porcine pancreatic

[6] Porcine pancreatic kallikrein was kindly provided by Dr. Schmidt-Kastner, Bayer AG, Wuppertal, Germany.
[7] We are indebted to Dr. H. Fritz, University of Munich, Germany, for a gift of this enzyme.
[8] N. M. Green and E. Work, *Biochem. J.* **54**, 347–352 (1953).

kallikrein (isoinhibitor E reacts more rapid than A, B, and L)[1] at pH 7.8 and 25°. The inhibition of trypsin is permanent.

Distribution.[2] The acid-stable inhibitors occur in body wall, tentacles, head, intestines, stomach, gill, and sex organs.

Amino Acid Composition.[1] The compositions of isoinhibitors A, B, E, and L are given in the table. The absence of proline from isoinhibitor E is unusual. All isoinhibitors contain four disulfide bridges.

Reactive Site Residues.[1] Isoinhibitor E contains lysine, isoinhibitor L arginine as the primary reactive site residue P_1.[9]

[9] The notation is that of A. Berger and I. Schechter, *Biochem. Biophys. Res. Commun.* **27**, 157 (1967).

[70] Trypsin-Plasmin Inhibitors (Bdellins) from Leeches

By HANS FRITZ and KAJ KREJCI

The occurrence of the thrombin-specific inhibitor hirudin in salivary glands of the leech *Hirudo medicinalis* is well known.[1,2] Commercially available samples of hirudin contain, however, in addition to the thrombin-specific inhibitor, large amounts of trypsin–plasmin inhibitors[3,4] and an appreciable amount of a chymotrypsin inhibitor.[4] For example, 1 g of the starting material from Medimpex Budapest (270 ATE[5]/mg) contains about 230 mg of the bdellins and only about 27 mg of the thrombin inhibitor hirudin, and 1 g of sterile "Hirudin" samples from Serva Heidelberg (3400 ATE/mg) contain, besides 300 mg of hirudin, about 75 mg of the bdellins. The purest hirudin preparations (>10 000 ATE/mg) isolated by Markwardt *et al.*[1] are free of bdellins.[3] The presence of considerable amounts of strong plasmin inhibitors in commercially available hirudin may explain the conflicting results in blood clotting studies,[6] as a consequence of which, common use of hirudin in medical therapy is hindered. Bdellins occur in all parts of the body of the leech, but the

[1] F. Markwardt, this series, Vol. **19**, p. 924; further references are given there.

[2] P. de la Llosa, C. Tertrin, and M. Justisz, *Biochim. Biophys. Acta* **93**, 40 (1964).

[3] H. Fritz, K.-H. Oppitz, M. Gebhardt, I. Oppitz, and E. Werle, *Hoppe-Seyler's Z. Physiol. Chem.* **350**, 91 (1969).

[4] H. Fritz, M. Gebhardt, R. Meister, and E. Fink, *Proteinase Inhibitors, Proc. Int. Res. Conf., 1st,* Munich, 1970, p. 271. de Gruyter, Berlin, 1971.

[5] Antithrombin units.

[6] A superposition of thrombin and plasmin inhibition using such samples was originally observed by R. Marx, Munich, who thus stimulated our investigations in this direction.

highest concentration is found in the region of the outer sexual organs.[7] This indicates perhaps a functional relationship to the trypsin or trypsin–plasmin inhibitors occurring in seminal vesicles[8] or seminal plasma,[9,10] which, like the bdellins,[11] are strong inhibitors of the sperm acrosomal proteinase acrosin.[12] In view of a possible therapeutic use of the bdellins, it is of interest to note that they are excreted into the urine after intravenous injection.[13]

Assay Methods

Known methods were used to estimate the inhibition of the enzymes employed: bovine trypsin and α-chymotrypsin, porcine (Novo Industri A/S) and human (Deutsche Kabi) plasmin, boar[12] and human[14] sperm acrosin, porcine pancreatic kallikrein (Bayer AG), porcine and human plasma kallikrein[15] and subtilisin Novo (Novo Industri A/S) with the following substrates: N^{α}-benzoyl-DL-arginine p-nitroanilide (BAPA) (trypsin,[16,17] porcine plasmin,[4,16] acrosin[18]), N^{α}-benzoyl-L-arginine ethyl ester (BAEE) (human plasmin,[19] pancreatic[16,20] and plasma[15,20] kallikrein), N^{α}-succinyl-L-phenylalanine p-nitroanilide (SUPHEPA) (chymotrypsin[9,16]), and azocasein (subtilisin[16,21]).

Definition of Units. One inhibitor unit corresponds to the reduction of the enzyme-catalyzed hydrolysis of the substrate by 1 μmole/min.

---

[7] R. Marx, unpublished data.
[8] E. Fink and H. Fritz, this volume [73].
[9] H. Schiessler, E. Fink, and H. Fritz, this volume [75].
[10] H. Fritz, H. Tschesche, and E. Fink, this volume [74].
[11] H. Fritz, B. Förg-Brey, H. Schiessler, M. Arnhold, and E. Fink, *Hoppe-Seyler's Z. Physiol. Chem.* **353**, 1010 (1972).
[12] W.-D. Schleuning and H. Fritz, this volume [27].
[13] H. Fritz, K.-H. Oppitz, D. Meckl, B. Kemkes, H. Haendle, H. Schult, and E. Werle, *Hoppe-Seyler's Z. Physiol. Chem.* **350**, 1541 (1969).
[14] H. Fritz, B. Förg-Brey, E. Fink, M. Meier, H. Schiessler, and C. Schirren, *Hoppe-Seyler's Z. Physiol. Chem.* **353**, 1943 (1972).
[15] G. Wunderer, K. Kummer, and H. Fritz, *Hoppe-Seyler's Z. Physiol. Chem.* **353**, 1646 (1972).
[16] H. Fritz, I. Trautschold, and E. Werle, *in* "Methoden der Enzymatischen Analyse" (H. U. Bergmeyer, ed.), 3rd ed., Vol. I, p. 1105. Verlag Chemie, Weinheim, 1974.
[17] B. Kassell, this series Vol. 19, p. 845.
[18] H. Fritz, B. Förg-Brey, E. Fink, H. Schiessler, E. Jaumann, and M. Arnhold, *Hoppe-Seyler's Z. Physiol. Chem.* **353**, 1007 (1972); cf. the method described in this volume [27].
[19] The direct BAEE assay described in this volume [27] was used.
[20] The combined BAEE/ADH assay described in this volume [27] was used.
[21] H. Fritz and K. Hochstrasser, this volume [76].

The inhibitor units given throughout refer to the inhibition of bovine trypsin, using BAPA as substrate.[16,17]

Purification Procedure

Starting Material. We used commercially available "Hirudin" from Medimpex Budapest, Hungary, characterized by its antithrombin activity of 270 ATE/mg or trypsin inhibitory activity of 0.9–1.0 IU/mg. This material corresponds to the "raw hirudin" (500 AT-U/mg) described by Markwardt.[1] Any other fraction, e.g., acetone powder produced from whole leeches fasted for about 3 months, may be employed either in step 1 or 2.

Step 1. Separation of the Bdellins from Hirudin. A 4.2 × 44 cm DEAE-cellulose column (acetate form) is equilibrated with 0.2 M sodium acetate, pH 6.0. The solution of 4.5 g of Medimpex "Hirudin" in 6 ml of the equilibration buffer is applied to the column, which is developed with the same buffer at a rate of 60 ml/hr. About 95% of the trypsin inhibitory activity applied is eluted in the inhibitor(bdellins)-containing fractions in a total buffer volume of 1.0–1.2 liters.

Subsequently, the thrombin inhibitor, which is completely retained on the DEAE-cellulose, is eluted from the column, employing a linear gradient formed from 2 liters each of the starting buffer and 0.6 M sodium acetate, pH 6.0. About 75% of the thrombin inhibitory activity applied is recovered.

The combined bdellin fractions are concentrated by ultrafiltration using an Amicon UM-05 membrane. The desired inhibitor concentration of about 25 IU/ml may be also attained by evaporation *in vacuo* (water bath temperature, 30°) and dialysis (4–5 hr) to remove excess salts.[4] Alternatively, desalting on Sephadex G-25, Bio-Gel P-2 or Merck gel PGM-2000 equilibrated and developed with 0.1–0.01 M acetic acid followed by lyophilization is possible, too.

Step 2. Affinity Chromatography. In principle, any water-insoluble trypsin derivative may be used for this step.[22,23] In our hands, the trypsin-cellulose ("bovine trypsin polymer bound to CM-cellulose, 7–10 U/mg") from E. Merck, Darmstadt, proved to be especially suitable under the conditions employed. Working temperature is 4° throughout.

The adsorbent (10 g) is prepared by suspending it repeatedly in 0.5 M NaCl, 0.05 M TRA-HCl pH 7.8, and 1.0 M KCl-HCl, pH 2.5,

[22] H. Fritz, B. Brey, M. Müller, and M. Gebhardt, *Proteinase Inhibitors, Proc. Int. Res. Conf. 1st.* Munich, 1970, p. 28. de Gruyter, Berlin, 1971.
[23] G. Wunderer, H. Fritz, W. Brümmer, N. Hennrich, and H.-D. Orth, *Biochem. Soc. Trans.* 2, 1324 (1974).

until no trypsin activity is detectable in the supernatant solutions. After reequilibration in the pH 7.8 buffer, the trypsin-cellulose is suspended in the bdellin solutions from step 1 (about 400 ml containing approximately 1000 IU) previously adjusted to pH 7.8, and approximately 0.4 M Na⁺. This suspension is gently shaken or stirred for 20 min. Subsequently, non-inhibitory contaminants are removed by suspending the loaded adsorbent five times in the pH 7.8 buffer and once in a 1:10 dilution of it, each time for about 3 min. The wash solutions are removed by short centrifugation.

Dissociation of the inhibitors from trypsin and thus their elution is achieved by suspending the adsorbent three or four times in 300 ml of the pH 2.5 buffer, each time for 10 min. (If the first extract contains only negligible amounts of inhibitors, the pH of the suspension must be adjusted to 2.5 prior to the second extraction.) Depending on the trypsin-cellulose employed, 80–95% of the inhibitors bound are recovered in the combined acidic extracts. The acidic inhibitor solution is neutralized and desalted by repeated ultrafiltration using Amicon UM-05 membranes or by gel filtration (cf. step 1). A white powder with a specific activity of 3.5–4.2 IU/mg is obtained after lyophilization and stored at 4°.

Step 3. Separation of the Bdellins into Two Fractions A and B. A 1.6 × 60 cm DEAE-cellulose (OH⁻ form) column is equilibrated and developed with 0.1 M sodium acetate pH 6.0, at a rate of 65 ml/hr. The solution of up to 500 mg of bdellin A, B mixture in 3 ml of the equilibration buffer is applied to the column. About 55% of the inhibitory activity is eluted with the equilibration buffer in a total volume of 500–600 ml. These unretarded inhibitor forms are named bdellins A.

The inhibitors retarded (about 45% of the inhibitory activity applied) on the DEAE-cellulose column, named bdellins B, are subsequently eluted with 0.1 M sodium acetate, 0.4 M NaCl, pH 6.0, in a total volume of 170–240 ml.

The inhibitor solutions are concentrated by ultrafiltration (Amicon UM-05 membrane) and desalted by repeated ultrafiltration or gel filtration on appropriate columns (cf. step 1) equilibrated and developed with 5% (v/v) acetic acid. About 15% of the total inhibitory activity is lost during this procedure. Specific activities of 3.3–3.5 IU/mg for bdellins A and 4.2–4.8 IU/mg for bdellins B are estimated after lyophilization.

Step 4. Equilibrium Chromatography of Bdellins A. The 1.2 × 24 cm DEAE-cellulose column is equilibrated and developed with 0.05 M ammonium acetate, pH 6.3, at a rate of 6.5 ml/hr, 2.7 ml/tube. A solution of 30 mg bdellins A from step 3 in 2 ml of the pH 6.3 buffer is applied to the column. The distribution of the inhibitor activity among the eluted fractions (cf. Fig. 1) is given in Table I. The antitryptic activity profile shows also three major inhibitor peaks corresponding to the fractions 1,

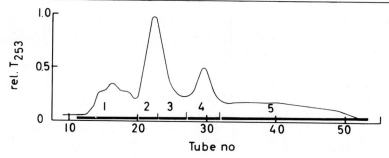

FIG. 1. Separation of the multiple forms of bdellin A by equilibrium chromatography. For experimental details, see step 4. Tubes were combined to yield the fractions as indicated. ——, relative transmission at 253 nm (rel. T_{253}).

$2 + 3$ (main peak) and 4 indicated in Fig. 1. The main fractions $2 + 3$ and 4 are desalted by repeated ultrafiltration (Amicon UM-05 membrane) and lyophilized. For complete separation of the inhibitor fractions, equilibrium chromatography may be repeated under identical conditions. The specific activity of fractions $2 + 3$ and 4 were estimated to 3.8 IU/mg.

Step 5. Equilibrium Chromatography of Bdellins B. The 1.2 × 55 cm DEAE-cellulose column is equilibrated and developed with 0.1 M sodium acetate pH 5.5 at a rate of 12 ml/hr, 4 ml/tube. A solution of 50 mg of

TABLE I

PURIFICATION OF THE BDELLINS

Step	Separation of	Specific activity (IU/mg)	Yield[a] (%)			
	Commercial samples	0.9–1.0				
1	Bdellins from hirudin		95			
2	Impurities (affinity chromatography)	3.5–4.2	75			
3	Bdellins A	3.3–3.5	55⎫			
	Bdellins B	4.2–4.8	45⎭ 100			
4	Bellins A in multiple forms	3.8[b]	80			
	fraction		1	2,3	4	5
	distribution[c]		14	37	11	14
5	Bdellins B in multiple forms	4.8–5.1[d]	83			
	fraction		1	2	3	4,5,6
	distribution[c]		7	18	34	12

[a] Mean values of several purifications.
[b] Fractions 2, 3, and 4.
[c] Of the inhibitory activity among the fractions.
[d] Fractions 2 and 3.

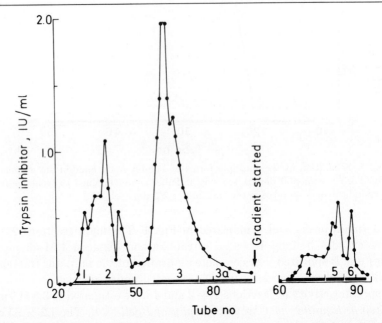

FIG. 2. Separation of the multiple forms of bdellin B by equilibrium chromatography. For experimental details, see step 4. Tubes were combined to yield the fractions as indicated. ———, relative transmission at 253 nm.

bdellins B from step 3 in 2 ml of the pH 5.5 buffer is applied to the column. After elution of fraction B-3 (cf. Fig. 2), the residual inhibitors retained are eluted with 0.1 M sodium acetate, 0.4 M NaCl, pH 5.5, or further separated by gradient elution chromatography.[4] The distribution of the inhibitor activity among the eluted fractions is given in Table I. The main fraction B-3 may be rechromatographed under identical conditions after desalting be repeated ultrafiltration and lyophilization. The specific activity of bdellin B-3 was estimated as 4.8–5.1 IU/mg.

Comments on the Purification Procedure. Step 1, separation of the thrombin inhibitor hirudin from the bdellins, was chosen to avoid a possible limited degradation of hirudin during affinity chromatography. This step can therefore be omitted. Degradation or limited proteolysis of the bdellins during affinity chromatography occurs, according to our experience, only to a very small extent, if at all. Hence, prolonged incubation times for adsorption and elution of the inhibitors may be used. If ammonium acetate buffers were employed in steps 3 and 5, desalting could be achieved directly by lyophilization.

As indicated by the elution profile in Fig. 2, bdellin B-3 seems to be

TABLE II
INHIBITION SPECIFICITY OF THE TRYPSIN-PLASMIN
INHIBITORS "BDELLINS" FROM LEECHES

Enzyme	Substrate[a]	Inhibition by bdellin	
		A[b]	B[b]
Trypsin, bovine	BAPA	+ +	+ +
Plasmin, porcine	BAPA	+ +	+ +
human	BAEE	+ +	+
Acrosin, boar	BAPA	+	+ +
human	BAPA	+	+ +
Chymotrypsin, bovine	SUPHEPA	−	−
Kallikrein, pancreatic plasma	BAEE	−	−
	BAEE	−	−
Subtilisin	Azocasein	−	−

[a] BAPA, N^α-benzoyl-DL-arginine p-nitroanilide; BAEE, N^α-benzoyl-L-arginine ethyl ester; SUPHEPA, N^α-succinyl-L-phenylalanine p-nitroanilide.

[b] In most cases, mixtures of multiple forms were employed. If inhibition was observed, the purified fractions were applied, too. Strong, + +; weaker, +; no inhibition, −.

a mixture of two strongly related inhibitors the preparative separation of which has still to be achieved; separation into two fractions by electrophoresis is possible.[24]

Properties

Inhibition Specificity. The bdellins are strong inhibitors of trypsin, plasmin, and sperm acrosin (Table II). The high affinity of bdellin B-3 to boar[11] and human[25] acrosin and of bdellins A to human plasmin[26] is especially remarkable. These inhibitors may be used, therefore, to discriminate between plasmin- and plasma kallikrein-catalyzed BAEE hydrolysis (cf. Table II). In the examples mentioned, the inhibitors form equimolar complexes with the appropriate enzymes (K_i values between 10 and 0.1 nM which can be dissociated by acidification to pH 2–3).

The bdellin mixtures A and B still contain small amounts (2–5% by weight) of a chymotrypsin inhibitor which is separated during equilibrium chromatography.

[24] H. Fritz and R. Biekarg, unpublished data, 1974.
[25] H. Fritz, B. Förg-Brey, M. Meier, M. Arnhold, and H. Tschesche, *Hoppe-Seyler's Z. Physiol. Chem.* 353, 1950 (1972).
[26] W. Graf von Kalckreuth, M.D. Thesis (Dissertation). Medical Faculty of the University of Munich, 1972.

Amino Acid Compositions and Molecular Weights. The amino acid compositions of the bdellin mixtures A and B and of some purified inhibitor forms are given in Table III. Obviously, bdellins A and B differ significantly in their amino acid compositions, especially in the number of disulfide bridges connecting the polypeptide chain. Bdellin B-3 is the natural inhibitor of a protein nature with the lowest molecular weight (4830) found so far. In view of the high cystine content, the molecular weights of the bdellins A are also relatively low (cf. Table III).

On the basis of the amino acid compositions thus far elucidated[4] it may be concluded that the various forms of the bdellins A and B, respectively, are multiple forms of an original bdellin A or B molecule.

Reactive Sites. Lysine residues are located in the reactive sites of both bdellin groups A and B.[4] During affinity chromatography cleavage of the

TABLE III
AMINO ACID COMPOSITIONS (MOLES/MOLE) AND
MOLECULAR WEIGHTS OF BDELLINS[a]

Residue	Bdellin					
	A	A-2,3	A-4	B	B-2	B-3
Asp	7.19	8	7	5.21	5	5
Thr	3.01	3	3	3.39	4	4
Ser	2.97	3	3	1.93	3	2
Glu	6.45	5	5	5.81	6	6
Pro	2.80	3	3	0.60	1	0
Gly	4.73	4	4	3.71	4	4
Ala	3.96	4	4	3.42	4	4
$Cys_{1/2}$	9.41	10	10	4.32	6	6
Val	4.19	5	5	4.08	4	4
Met	0.42	1	1	0.19	0	0
Ile	1.01	1	1	0.19	0	0
Leu	1.58	1	1	1.69	2	2
Tyr	1.21	1	1	0.95	1	1
Phe	1.95	2	2	0.35	0	0
Lys	4.82	5	5	1.75	2	1
His	2.80	3	3	6.15	5	5[b]
Arg	1.05	0	0	1.01	1	1
Trp	0.12	0	0	0.12	0	0
Total		59	58		48	45
Molecular weight		6339				4830

[a] Experimental values and the compositions of further fractions are given by H. Fritz, M. Gebhardt, R. Meister, and E. Fink, *Proteinase Inhibitors, Proc. Int. Res. Conf., 1st*, Munich, 1970, p. 271.

[b] In stored fractions; in freshly prepared fractions 6 residues are found.

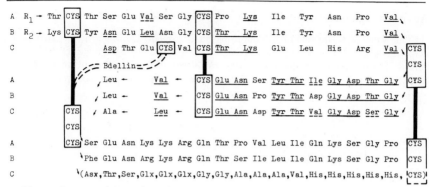

FIG. 3. Structural homology between pancreatic secretory trypsin inhibitors and bdellin B-3 from leeches. A, porcine pancreatic inhibitor [H. Tschesche and E. Wachter, *Eur. J. Biochem.* **16**, 187 (1970); cf. H. Tschesche, *Ang. Chem. Int. Ed.* **13**, 10 (1974); and D. C. Bartelt and L. J. Greene, *J. Biol. Chem.* **246**, 2218 (1971)]; B, human pancreatic inhibitor [L. J. Greene and D. C. Bartelt, *in* "Protides of the Biological Fluids—23rd Colloquium" (H. Peeters, ed.) Pergamon, Oxford, 1976.]; C, bdellin B-3. N-Terminal sequences:

R1   H₂N-Thr-Ser-Pro-Gln-Arg-Glu-Ala→
R2   H₂N-Asp-Ser-Leu-Gly-Arg-Glu-Ala→

reactive-site peptide bond seems to occur in about 15–20% of the material applied.[4]

Structure and Homology of Bdellins B. The structure of bdellin B-3 so far elucidated is shown in Fig. 3. The structural investigations were complicated by the finding that the preparation is composed of two very closely related inhibitor forms still to be separated. In automatic Edman degradation,[27] an overlap of identical sequences occurs after position 11 or 12 indicating the lack of one amino acid in one of the fractions or only partial degradation in this position due to a structural particularity.

The comparison of the sequences of the pancreatic secretory trypsin inhibitor (Kazal type), the seminal acrosin inhibitor and bdellin B-3 shows clearly that the structure of bdellin B-3 is homologous to those of the other two inhibitors. Hence, the Kazal type inhibitor structure seems to occur more universally in nature as originally assumed.[28] In view of the positions of the subsites on the inhibitor molecules involved in complex formation,[29] the very small first disulfide loop in bdellin B-3 seems to be especially interesting.

[27] P. Fietzek and K. Krejci, unpublished data, 1974.
[28] M. Laskowski, Jr., and R. W. Sealock, *in* "The Enzymes" (P. D. Boyer, ed.), 3rd ed., Vol. III, p. 375. Academic Press, New York, 1971.
[29] A. Rühlmann, D. Kukla, P. Schwager, K. Bartels, and R. Huber, *J. Mol. Biol.* **77**, 417 (1973).

Acknowledgments

This work was supported by Sonderforschungsbereich-51, München. The structural investigations were mainly done in the Institut für Physiologische Chemie und Physikalische Biochemie der Universität München. We are grateful to Prof. Dr. Th. Bücher, Dr. W. Machleidt, and Dr. J. Otto for their support and help in these studies.

[71] Trypsin Inhibitor from Cow Colostrum

By Dana Čechová

Colostrum of mammals contains various amounts of antitryptic activity. Laskowski and Laskowski[1,2] observed a trypsin inhibitor in cow, hog, and human colostrum. They isolated and crystallized a low-molecular-weight inhibitor from bovine[2] colostrum and, later with their co-workers, one from hog colostrum.[3-6] The protein obtained was electrophoretically heterogeneous. Several new methods of isolation of the inhibitor from cow[7-9] and hog[10] colostrum were developed. The new data showed that the trypsin inhibitors from both hog and cow colostrum were glycoproteins[9-11] occurring in several forms.[9,10] The presence of multiple forms seems to be due both to microheterogeneity in amino acid composition and to differences in sugar moiety. The latter could result from the use of improper isolation methods. The primary structure of the cow colostrum proteinase inhibitor has been determined[8,12] and was found to be similar to the covalent structure of the basic pancreatic

[1] M. Laskowski, Jr., and M. Laskowski, *Fed. Proc., Fed. Am. Soc. Exp. Biol.* **9**, 194 (1950).

[2] M. Laskowski, Jr., and M. Laskowski, *J. Biol. Chem.* **190**, 563 (1951).

[3] M. Laskowski, B. Kassell, and G. Hagerty, *Biochim. Biophys. Acta* **24**, 300 (1957).

[4] M. Laskowski and M. Laskowski, Jr., *Adv. Protein Chem.* **9**, 203 (1954).

[5] M. Laskowski, this series Vol. 2 [3].

[6] R. Vogel, I. Trautschold, and E. Werle, "Natural Proteinase Inhibitors." Academic Press, New York, 1968.

[7] D. Čechová, V. Jonáková-Švestková, and F. Šorm, *Collect. Czech. Chem. Commun.* **35**, 3085 (1970).

[8] D. Čechová and E. Ber, *Collect. Czech. Chem. Commun.* **39**, 680 (1974).

[9] D. Čechová, *Collect. Czech. Chem. Commun.* **39**, 647 (1974).

[10] L. F. Kress, S. R. Martin, and M. Laskowski, Sr., *Biochim. Biophys. Acta* **229**, 836 (1971).

[11] D. Čechová-Pospíšilová, V. Švestková, and F. Šorm, *Fed. Eur. Biochem. Soc., 6th Meet.*, Abstr. No. 222 (1969).

[12] D. Čechová, V. Jonáková, and F. Šorm, *Collect. Czech. Chem. Commun.* **36**, 3343 (1971).

polyvalent inhibitor (Kunitz),[12] the trypsin inhibitor from *Helix pomatia*,[13] and the structures of other inhibitors and snake toxins.[14]

The inhibitor mentioned above is not the only one present in bovine colostrum[15]; there are two other inhibitors of higher molecular weights, which are both acid and heat labile.

Assay Method[16]

Principle. Inhibition of tryptic hydrolysis of N^α-benzoyl-DL-arginine *p*-nitroanilide (BAPA) is measured in terms of the change in absorbance at 405 nm.

Reagents

Buffer: Tris-chloride, 0.1 M, pH 7.8, containing 0.025 M CaCl$_2$
Substrate: Dissolve 100 mg of BAPA with heating in 100 ml of water
Trypsin: Dissolve 10 mg in 100 ml of 0.02 M HCl
Dilute inhibitor solution in buffer: 25 μg of inhibitor per 1 ml of solution

Procedure. The trypsin solution (0.03 ml), the inhibitor (0.02 ml) and 1.95 ml of the buffer are incubated at 25° for 10 min in a cuvette (1 cm light path) before starting the enzymic reaction by adding 1 ml of substrate. The increase in absorbance at 405 nm is recorded for 5 min. One enzyme unit corresponds to the hydrolysis of 1 μmole of substrate per minute ($\Delta A_{405}/\text{min} = 3.32$ for 3 ml of mixture). One inhibitor unit decreases the activity of two enzyme units by 50%.

The specific activity is expressed in enzyme units inhibited by 1 ml of the protein solution ($A_{280} = 1.0$).

Purification Procedure[9]

The procedure, summarized in Table I, is based on the selective adsorption of the inhibitor from a crude extract to trypsin-Sepharose in the manner described for the isolation of the basic pancreatic trypsin inhibi-

[13] T. Dietl and H. Tschesche, *Proteinase Inhibitors, Proc. Int. Res. Conf., 2nd (Bayer Symp. V)*, Grosse Ledder, 1973, p. 254.
[14] Comparison of the known structures is presented by M. Laskowski, Jr., *Proteinase Inhibitors, Proc. Int. Res. Conf., 2nd (Bayer Symp. V)*, Grosse Ledder, 1973, p. 597.
[15] V. B. Pederson, V. Keil-Dlouhá, and B. Keil, *FEBS Lett.* **17**, 23 (1971).
[16] A modification of the method of H. Fritz, G. Hartwich, and E. Werle, *Hoppe-Seyler's Z. Physiol. Chem.* **345**, 150 (1966).

TABLE I

ISOLATION OF INHIBITOR FROM COW COLOSTRUM[a]

Fraction	Recovery (%)	Specific activity (units/A_{280})
Colostrum	100	—
Filtrate after precipitation of colostrum	83	0.02
Material after purification by affinity chromatography	70	3.23
Material after purification on DEAE-cellulose[b]	43	4.36

[a] D. Čechová, *Collect. Czech. Chem. Commun.* **39,** 647 (1974).

[b] Average specific activity of material from peaks 2 through 7 (Fig. 1).

tor.[17] The inhibitor is then liberated by dissociating the complex in acid media.

Preparation of Trypsin-Sepharose

The method of Kassell and Maciniszyn[17] was used, with all operations carried out at 0°. Sepharose 4 B (70 g) is placed on a glass filter and is washed several times with distilled water, after which the gel is suspended in 50 ml of distilled water. A solution of cyanogen bromide (9 g/100 ml of water) is added to the Sepharose with stirring and maintenance of pH at 11 by 5.0 M NaOH for 6–7 min. The Sepharose was subsequently washed with 4 liters of cold distilled water and with 4 liters of cold 0.05 M sodium borate of pH 9, on a glass filter. Trypsin (2 g) is dissolved in 200 ml of cold borate buffer at pH 9 just prior to adding it to the activated Sepharose. After stirring overnight in the cold room, a column (diameter 3.5 cm) is poured and the gel is washed consecutively with 4 liters of cold buffer (0.05 M sodium borate, 0.01 M CaCl$_2$, pH 9), with 2 liters of 0.5 M NaCl containing 0.01 M CaCl$_2$ at a rate of 20 ml/hr with 0.1 M sodium acetate (pH 4) containing 0.3 M NaCl, and, finally, with 0.01 M CaCl$_2$ until the pH of the effluent becomes constant.

Inhibitor Purification[9]

Step 1. Crude Extract. Colostrum is obtained during the first hours after parturition and it can be stored frozen (—20°). Before use, the thawed colostrum is centrifuged at 1200 g for 30 min, left for 2 hr in a refrigerator, when a lipid layer forming on top is removed. One milliliter of colostrum contains approximately an amount of inhibitor which would

[17] B. Kassell and M. B. Marciniszyn, *Proteinase Inhibitors, Proc. Int. Res. Conf., 1st,* Munich, 1970, p. 43.

neutralize 500 μg of active trypsin. One liter of colostrum is diluted with 2 liters of distilled water and is allowed to stand at room temperature until it curdles. The precipitate is separated by centrifugation and the clear supernatant is kept separately. The precipitate is extracted with 1 liter of water and the mixture is centrifuged again. The two supernatants are pooled and the pH of the solution is adjusted to 7.5 whereupon some contamination proteins precipitate. These are removed either by centrifugation or by filtration over an asbestos filter coated with a thin kieselguhr layer. The filtrate is used for affinity chromatography.

Step 2. Affinity Chromatography. The inhibitor solution is passed through the column at a rate of 10 ml/min, collecting 200-ml fractions, until excess inhibitor appears in the effluent. The column is washed with pH 4 buffer until absorbance at 280 nm drops below 0.01 (usually 2 liters). The inhibitor is eluted with 0.01 M HCl, containing 0.3 M NaCl and 0.01 M CaCl$_2$ at a flow rate of 2 ml/min collecting 20-ml fractions (the column is reequilibrated with the pH 4 buffer). Inhibitor-containing fractions are pooled and adjusted to pH 6.0, desalted by passage through a column of Sephadex G-25 in 0.01 M ammonium carbonate and then lyophilized.

Step 3. Chromatography on DEAE-Cellulose. The lyophilized sample (50 mg) is taken up in 10 ml of distilled water and its conductivity and pH is adjusted to equal those of the equilibrating buffer. The solution is applied to a column of DEAE-cellulose (1.9 × 28.5 cm), equilibrated with 0.03 M Tris-HCl buffer at pH 7.3. As seen in Fig. 1, several forms of the inhibitor are eluted with a linear gradient to 0.2 M NaCl in 0.03 M Tris-HCl buffer at pH 7.3 (1200 ml).

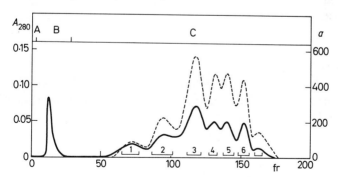

Fig 1. Chromatography of on DEAE-cellulose at pH 7.3. ——, absorbance at 280 nm; ---, relative activity (*a*) expressed as micrograms of trypsin inhibited by 1 ml of sample. A, application of sample; B, washing with 0.03 M Tris-HCl buffer; C, elution by linear gradient to 0.02 M NaCl in 0.03 M Tris-HCl buffer at a flow rate of 4.5 ml/15 min (per fraction); fr, number of fractions; 1–7, pooled fractions. From D. Čechová, *Collect. Czech. Chem. Commun.* **39,** 647 (1974); reproduced with permission.

Other Methods of Purification[8]

A procedure involving precipitation of contaminating proteins from colostrum by 2.5% trichloroacetic acid rather than natural curdling can be used for the initial separation of the inhibitor from colostrum. The inhibitor can then be precipitated from the solution by ammonium sulfate (80% saturation) and desalted on a Sephadex G-25 column in 0.2% formic acid and lyophilized. Affinity chromatography on a column of trypsin-Sepharose is used next. After elution and desalting, the inhibitor (400 mg) is placed on a column of CM-Sephadex C-25 (2.6 × 27 cm) equilibrated with 0.05 M potassium formate (pH 3.5) and is eluted by a linear gradient of increasing ionic strength (3 liters, to 0.2 M KCl), collecting 15-ml fractions at 30-min intervals. The center portion of the active peak represents a mixture of inhibitor forms, but it is free of other contaminating proteins. Trichloroacetic acid cannot be used if the inhibitor is to be obtained in intact form because sialic acid can be split off by this acid treatment.

Properties

Stability. The inhibitors from cow and hog colostrum are remarkably stable at acid pHs as well as at high temperatures. They can be heated in 2.5% trichloroacetic acid at 80° for 5 min without loss of activity.[3-5] By contrast, activity is lost after 24-hr treatment with 0.1 M NaOH.[4] Both inhibitors are relatively stable toward the action of proteolytic enzymes. They are not hydrolyzed in their native states by either trypsin or chymotrypsin. However, the inhibitor from cow colostrum in which one disulfide bond has been selectively cleaved by reduction-alkylation is readily digested by trypsin.[8] The cow colostrum inhibitor becomes inactivated after 24-hr incubation in a 0.1% pepsin solution (pH 1.5, 37°), whereas the hog colostrum material retains 30% of its activity after 8 days of incubation in the same medium.[3]

Purity. The purified isoinhibitors are homogeneous by disc-gel electrophoresis[9] at pH 4.5 and at pH 8.3. Their amino acid composition is shown in Table II. None of the inhibitors contain histidine or valine, and when isolated with omission of the trichloroacetic acid precipitation step, no free amino-terminal groups are detectable.[18] Otherwise, phenylalanine is found as the only amino-terminal amino acid.[7]

Physical Properties. The molecular weight of the inhibitor calculated from the stoichiometry of the reaction with trypsin[19] is 10,500. A value of

---

[18] D. Čechová and V. Jonáková, unpublished observations.
[19] M. Laskowski, Jr, P. H. Mars, and M. Laskowski, *J. Biol. Chem.* **198**, 745 (1952).

TABLE II
AMINO ACID COMPOSITION OF TRYPSIN INHIBITOR FROM COW COLOSTRUM[a]

Amino acid	Residues/mole	Amino acid	Residues/mole
Lysine	2	Alanine	4
Histidine	0	Half-cystine	6
Arginine	3	Valine	0
Aspartic acid	8	Methionine	1
Threonine	6	Isoleucine	1
Serine	3	Leucine	5
Glutamic acid	10	Tyrosine	3
Proline	7	Phenylalanine	4
Glycine	4	Tryptophan	0

[a] D. Čechová, V. Jonáková-Švestoková, and F. Šorm, *Collect. Czech. Chem. Commun.* **35**, 3085 (1970). The main form was isolated by the trichloroacetic acid procedure.

11,000 is obtained[20] from amino acid and carbohydrate analysis. A molecular weight of 20,000 is obtained from gel filtration experiments.[15] This anomalous value may be due to the presence of a sugar moiety. The molecular weight of the complex of cow colostrum trypsin inhibitor with trypsin by ultracentrifugation[11] gave a value of $35,000 \pm 1800$. The isoelectric point of the inhibitor[19] is pH 4.2. A conversion factor of 2.00 and 1.9 can be applied to calculate the milligram quantities of cow colostrum[19] and hog colostrum[3] trypsin inhibitors, respectively, from absorbancy measurements at 280 nm. Absorption maxima are at 277 nm.[3,19]

Specificity. The following enzymes are inhibited: bovine trypsin[1] and chymotrypsin,[1] plasmin from porcine organs and urine,[21] boar[22,23] and bull[23a] acrosins. Thrombin and hog kallikreins from urine and submandibular glands were also tested,[21] but no inhibition could be seen.

Kinetic Properties. The reaction with trypsin is stoichiometric, with a bimolecular rate constant[24] (25° and pH 7.8) of 3×10^5 liters/mole per second and a dissociation constant (K_i) of 4 pM. Both constants are

[20] H. Tschesche, K. Rainer, D. Čechová, and V. Jonáková, *Hoppe-Seyler's Z. Physiol. Chem.* **356**, 1759 (1975).

[21] D. Čechova and H. Fritz, to be published.

[22] E. Fink, H. Fritz, E. Jaumann, H. Schiessler, B. Förg-Brey, and E. Werle, *Protides Biol. Fluids, Proc. Colloq.* **20**, 425 (1973).

[23] H. Fritz, B. Förg-Brey, H. Schiessler, M. Arnhold, and E. Fink, *Hoppe-Seyler's Z. Physiol. Chem.* **353**, 1010 (1972).

[23a] D. Čechová and H. Fritz, *Hoppe-Seyler's Z. Physiol. Chem.* in press.

[24] D. Čechová, N. F. Kazanskaya, N. I. Larionova, and I. V. Berezin, *Biokhimiya* **39**, (1975).

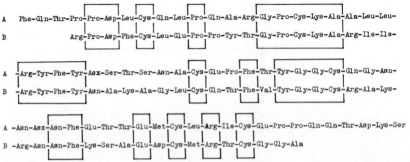

FIG. 2. Comparison of the amino acid sequences of the main component of the cow colostrum inihibitor (A) and of the basic pancreatic inhibitor (B). Residues occupying identical positions in the two proteins are boxed. From D. Čechová, V. Jonáková, and F. Šorm, *Collect. Czech. Chem. Commun.* **36**, 3343 (1971); reproduced with permission.

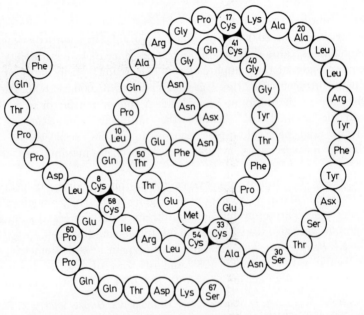

FIG. 3. Schematic representation of the structure of the cow colostrum trypsin inhibitor. From D. Čechová and E. Ber, *Collect. Czech. Chem. Commun.* **39**, 680 (1974); reproduced with permission.

close to the values measured for the basic pancreatic trypsin inhibitor (Kunitz).[25] Inhibitions of chymotrypsin and plasmin are competitive, with respective K_i values of 0.11 μM and 32 nM.[21]

Composition. Cow colostrum trypsin inhibitor is a glycoprotein con-

[25] B. Kassell, this series Vol. 19 [66b].

taining 67 amino acid residues[12] (Table II) and comprising a sugar moiety of more than 30% of its molecular weight.[20] The sugar is composed of fucose, mannose, galactose, glucose, galactosamine, glucosamine, and sialic acid.[20] Hog colostrum trypsin inhibitor has a similar composition.[10] As far as the multiple forms of both the cow and hog inhibitors are concerned, some variations could be shown in the lysine[7,26,27] and threonine[7] contents, and different forms containing 1, 2, or 3 lysine residues could be isolated. Differences in content of other amino acids[18,27] are also likely. The primary structures of two inhibitor forms have already been determined,[12,28] differing only by two amino acids: replacement of threonine in position 3 by lysine.[28] There is a 40% homology in the primary structure of the colostrum inhibitor with that of the basic pancreatic trypsin inhibitor,[12] and a comparison of the two structures is given in Fig. 2. The disulfide bonds in the molecule of cow colostrum trypsin inhibitor are shown in Fig. 3; their positions are identical with the disulfides in the basic pancreatic trypsin inhibitor.[8]

[26] U. Kucich, Ph.D. Thesis, State University of New York at Buffalo, 1972.
[27] M. Laskowski, Sr., personal communication.
[28] V. Jonáková and D. Čechová, Collect. Czech. Chem. Commun., in press.

[72] Human Pancreatic Secretory Trypsin Inhibitor[1]

By Lewis J. Greene, Merton H. Pubols,[2] and Diana C. Bartelt

Pancreatic juice, the exocrine secretion of pancreas, contains a polypeptide trypsin inhibitor[3-8] in addition to hydrolytic enzymes and inactive enzyme precursors (zymogens). The pancreatic secretory trypsin inhibitor (PSTI) prevents the premature trypsin-catalyzed activation of zymogens within the pancreas and the pancreatic duct. In the small in-

[1] Research carried out at Brookhaven National Laboratory under the auspices of the U.S. Atomic Energy Commission.
[2] Visiting Biochemist at Brookhaven National Laboratory 1968 to 1969, supported by the National Institutes of Health Special Fellowship GM-40285-01.
[3] M. H. Kalser and M. Grossman, Gastroenterology 29, 35 (1955)
[4] B. J. Haverback, B. Dyce, H. Bundy, and H. A. Edmondson, Am. J. Med. 29, 424 (1960).
[5] L. J. Greene, M. Rigbi, and D. S. Fackre, J. Biol. Chem. 241, 5610 (1966).
[6] H. Fritz, F. Woitinas, and E. Werle, Hoppe-Seyler's Z. Physiol. Chem. 345, 168 (1966).
[7] L. J. Greene, J. J. DiCarlo, A. J. Sussman, and D. C. Bartelt, J. Biol. Chem. 243, 1804 (1968).
[8] M. H. Pubols, D. C. Bartelt, and L. J. Greene, J. Biol. Chem. 249, 2235 (1974).

testine, enterokinase initiates the activation of trypsinogen, which is responsible for the activation of all other zymogens. Enterokinase is not inhibited by PSTI.[9]

Kazal et al.[10] isolated the first pancreatic secretory trypsin inhibitor[11] from bovine pancreas in 1948. It could be distinguished from the bovine trypsin–kallikrein inhibitor (Kunitz inhibitor)[12] on the basis of its inability to inhibit bovine chymotrypsin or porcine pancreatic kallikrein and because the inhibition is temporary in the presence of excess trypsin. Bovine and porcine PSTI,[13] as well as the bovine trypsin–kallikrein inhibitor,[14] have been described in this treatise. More recent information may be found in review articles[15,16] and reports of conferences.[17,18]

The trypsin inhibitor activity present in human pancreas and pancreatic juice was first demonstrated by Haverback et al.[4] Partially purified inhibitor was later identified as Kazal-type on the basis of its inhibitory properties.[6,19,20] This assignment is independently indicated by the amino acid sequence of the human inhibitor, which is homologous to the pancreatic secretory trypsin inhibitors of bovine, porcine, and ovine origin. A comparison of these structures is included in this article.

The procedure[8] detailed here for the isolation of multiple chromatographic forms of human PSTI utilizes inactive pancreatic juice or tissue as the starting material. The inhibitor is not exposed to proteolytic enzymes and is isolated by gel filtration and ion exchange chromatography in the free form (not complexed with trypsin). The use of trypsin-affinity chromatography for the isolation of inhibitor from autolyzed human pancreas has recently been reported.[21]

---

[9] S. Maroux, J. Baratti, and P. Desnuelle, *J. Biol. Chem.* **246**, 5031 (1971).

[10] L. A. Kazal, D. S. Spicer, and R. A. Brahinsky, *J. Am. Chem. Soc.* **70**, 3034 (1948).

[11] Pancreatic secretory trypsin inhibitors are also referred to in the literature as "Kazal-type" inhibitors or specific inhibitors from pancreatic juice.

[12] M. Kunitz and J. H. Northrop, *J. Gen. Physiol.* **19**, 991 (1936).

[13] P. J. Burck, this series Vol. 19 [67].

[14] B. Kassell, this series Vol. 19 [66b].

[15] M. Laskowski, Jr., and R. W Sealock, *in* "The Enzymes" (P. D. Boyer, ed.), 3rd ed., Vol. III, p. 375. Academic Press, New York, 1971.

[16] H. Tschesche, *Angew. Chem.* **13**, 10 (1974).

[17] "Proteinase Inhibitors, Proceedings of the First International Research Conference (Bayer Symposium V) Munich, 1970. de Gruyter, Berlin, 1971.

[18] *Proteinase Inhibitors, Proc. Int. Res. Conf., 2nd (Bayer Symp. V)*, Grosse Ledder, 1973. Springer-Verlag, Berlin and New York, 1974.

[19] H. Fritz, I. Hüller, M. Wiedemann, and E. Werle, *Hoppe-Seyler's Z. Physiol. Chem.* **348**, 405 (1967).

[20] P. J. Keller and B. J. Allan, *J. Biol. Chem.* **242**, 281 (1967).

[21] G. Feinstein, R. Hofstein, J Koifmann, and M. Sokolovsky, *Eur. J. Biochem.* **43**, 569 (1974).

Assay Method

The inhibition of the trypsin hydrolysis of p-toluenesulfonyl-L-arginine methyl ester (TAME) is used as the quantitative assay for the trypsin inhibitor. The pH-stat method is recommended because it can be used to assay trypsin inhibitor in tissue extracts where nucleotide absorption interferes with the spectrophotometric determination. The assay is carried out with a final substrate concentration of 0.008 M TAME in 0.005 M Tris-HCl buffer, 0.1 M KCl, 0.02 M CaCl$_2$ at pH 7.8, 25°.[5] Crystalline bovine trypsin (Worthington Biochemical Co., Freehold, New Jersey) is used. A description of the procedure used in our laboratory has appeared in a previous volume of this treatise.[13]

Definition of Unit and Specific Activity. One inhibition unit is the amount of inhibition that has caused the reduction of TAME hydrolysis by 1 μmole/min. Specific activity is defined as inhibitor units per A_{280}^{1cm}.

Preparation of Human PSTI[8]

Individual samples of pancreatic juice and tissue are initially screened for trypsin and trypsin inhibitor activity to ensure that no tryptic activation has occurred. This precaution is taken so that the inhibitor can be separated in the "free" form from trypsin, trypsinogen, or inhibitor-trypsin complex by gel filtration on Sephadex G-75. The inhibitor is then purified and separated into five chromatographic forms by gradient elution chromatography on DEAE-cellulose and SP-Sephadex. Pancreatic tissue is subjected to a preliminary fractionation by ammonium sulfate precipitation and then processed in the same manner as the pancreatic juice.

Step 1a. Collection of Pancreatic Juice. Human pancreatic juice is collected by catheterization of the pancreatic duct after related surgical procedures.[22,23] The pancreatic juice is frozen as soon as possible after collection. Trypsin activity is determined by the pH-stat method on an aliquot of juice containing 250–500 μg protein. Samples that do not have demonstrable trypsin activity and contain inhibitor activity are used. Pancreatic juice was stored at $-22°$ after lyophilization.

Step 1b. Extraction of PSTI from Pancreas. Postmortem human pancreas from approximately 100 individuals who did not have diagnosed pancreatic disease were stored at $-22°$ for up to 12 months. Minced tissue in 50–100 g portions is suspended in 5 volumes (w/v) of chilled 0.1 mM diisopropyl phosphofluoridate (DFP) and homogenized in a

[22] A. Morgan, L. A. Robinson, and T. T. White, *Am. J. Surg.* **115**, 131 (1968).
[23] C. Figarella and T. Robeiro, *Scand. J. Gastroenterol.* **6**, 133 (1971).

Waring blender for 2 min at maximum speed. The suspension is adjusted to pH 4.5 with 6 N HClO$_4$ and centrifuged at 13,000 g for 45 min at 5° in a Sorvall GSA rotor. If the pH of the tissue suspension is brought to pH 3 or less, large losses of trypsin inhibitor activity occur. The sediment is reextracted, and the combined supernatants are brought to 70% saturation by the slow addition of 472 g of solid ammonium sulfate per liter of supernatant at 5°. After centrifugation at 13,000 g, 5°, for 45 min the supernatant is discarded. The precipitate is resuspended in 3 volumes of 0.1 mM DFP and stored at −22°. Each preparation is assayed for trypsin and inhibitor activity before being combined with others for gel filtration on Sephadex G-75. The DFP present does not inhibit the trypsin under the conditions of the assay.

Step 2. Gel Filtration on Sephadex G-75. Two 7.6 × 180 cm columns are prepared from 1.5 kg of Sephadex G-75 which has been swelled overnight in 50 liters of 50% acetic acid. The fines are removed by suction provided by a water pump. The acetic acid is replaced with distilled water and then by the eluting buffer (0.5 M KCl, 0.01 M Tris-HCl, pH 8.1, and 0.1 mM DFP) by settling, followed by decantation with suction. Equilibration of the gel is completed by storage at 4° for 4–5 days. Two centimeters of glass beads are overlaid on top of the sintered-glass disks fitted at the bottom of each column. The gel suspension is deaerated at 4° with a water pump before the columns are poured in sections. Thirty liters of buffer are passed through each column before the columns are connected in series. The flow rate is limited to 120 ml/hr by a peristaltic pump attached to the outlet of the second column. Effluent is collected in 60-ml fractions. The columns are operated in a cold room at 2°–4°.

Lyophilized human pancreatic juice is suspended in 200–400 ml of cold distilled water at 4° to give a final protein concentration of 1–2%. One milliliter is removed for the determination of inhibitor activity later to be used for the calculation of inhibitor recovery. The solution is then made 1 mM with respect to DFP by the addition of 0.1 M DFP (in isopropyl alcohol) and held at 4° for 2 hr with stirring before gel filtration on Sephadex G-75 (Fig. 1). The inhibitor activity, corresponding to a molecular weight of ∼6000, is retarded relative to most of the secretory protein. Amylase activity is anomalously eluted after the inhibitor in the effluent corresponding to fractions 240–270. Effluent with inhibitor-specific activity greater than 12 (fractions 193–203, indicated by the solid bar) is combined and lyophilized. This fraction is denoted "low molecular weight fraction."

Step 3. Gel Filtration on Sephadex G-25. Sephadex G-25 is swelled and degassed in 10 volumes of distilled water by placing the beaker in a boiling water bath for 2 hr. After settling, the supernatant and fines are

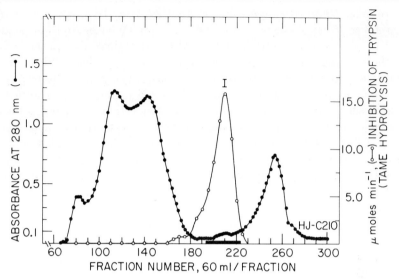

FIG. 1. Gel filtration of human pancreatic juice on Sephadex G-75. Sample, 34 $A_{280}^{1\,cm} \times 190$ ml. ●——●, Absorbance at 280 nm; ○——○, trypsin inhibition. TAME, p-toluenesulfonyl-L-arginine methyl ester. From M. H. Pubols, D. C. Bartelt, and L. J. Greene, *J. Biol. Chem.* **249**, 2235 (1974).

decanted by suction. The eluting buffer (0.05 M ammonium bicarbonate, pH 8.1) is then added to the swelled gel at 4 or 5 times the settled gel volume. The Sephadex is gently stirred and allowed to settle; and fines are removed by suction. Buffer addition, gel settling, and decantation are repeated three times. The equilibrated gel is kept at 2°–4° overnight before packing the columns as described for Sephadex G-75.

The lyophilized low-molecular-weight fraction from the Sephadex G-75 column (Fig. 1) is dissolved in 400 ml of 0.1 mM DFP and desalted on a Sephadex G-25 (medium) column, 7.6 × 72 cm, equilibrated and developed with 0.05 M ammonium bicarbonate buffer, pH 8.1, containing 0.1 mM DFP at 200 ml/hr, 4°. Fractions of 30 ml are collected. The effluent is monitored by measurement of absorbance at 280 nm, trypsin-inhibitor activity, and effluent conductivity. Fractions with inhibitor specific activity greater than 20 are pooled and lyophilized. Because the inhibitor activity is partially included within the matrix of Sephadex G-25, this preparation is then resubmitted to gel filtration on two columns (1.8 × 160 cm) of Sephadex G-25 (Superfine) connected in series. The columns are equilibrated and developed with 0.05 M ammonium bicarbonate, pH 8.1, at 15 ml/hr. Effluent with a specific activity greater than 90 is combined and lyophilized twice to remove the volatile buffer.

Step 4. Chromatography on DEAE-Cellulose. To prepare resin for two 1.8 × 45 cm columns, 120 g DE-52 resin (Whatman) is suspended in 2 liters of 0.280 Tris-HCl buffer, pH 9.0 (ten times the concentration of the starting elution buffer). After the suspension has settled for 10 min, the supernatant is removed by suction. The resin is resuspended with one additional liter of buffer, and the pH of the slurry is adjusted, if necessary. The resin suspension is degassed with a water pump for 30 min and stored at 2°–4°. The column is poured in the cold room at 2°–4° using a final slurry volume which is 1.5 times of the wet settled volume. The column is then equilibrated with the starting elution buffer (0.028 M Tris-HCl, pH 9.0) by passing 2 liters or more of buffer through the column, until the effluent pH and conductivity are exactly the same as that of the starting elution buffer.

The lyophilized fraction containing inhibitor activity derived from the Sephadex G-25 column is suspended in 25 ml of 0.014 M Tris-HCl buffer, pH 9.0. The pH is corrected to pH 9.0, and the conductivity is reduced, if necessary, by dilution to a value below that of the starting elution buffer. After sample application, the column is operated with starting elution buffer at 40 ml/hr with a piston pump until 1320 ml have passed through the column (tube 66, indicated by the arrow in Fig. 2). A linear gradient is then applied to the column. It is prepared from 2 liters each of equilibrating buffer and 0.028 M Tris-HCl buffer, 0.2 M potassium chloride, pH 9.0, using mixing chambers of the same cross sectional area.[24] The inhibitor activity from both pancreatic juice (Fig. 2, top) and tissue (Fig. 2, bottom) are each resolved into two chromatographic components, A and B, which are eluted with the same effluent conductivity and present in similar proportions. The effluent corresponding to each peak of inhibitor activity is combined, lyophilized, desalted by gel filtration on Sephadex G-25 (Superfine), 1.8 by 72 cm, and developed with 0.05 M ammonium bicarbonate buffer, pH 8.1, at 4°. After gel filtration, the active material, located by inhibitor activity and absorbance at 280 nm, is pooled and lyophilized.

Step 5. Gradient Elution Chromatography on SP-Sephadex. To prepare resin for two 0.9 × 70 cm columns, 15 g of SP-Sephadex G-25 are suspended in 1 liter of 0.5 M acetic acid overnight, and fine particles are removed by settling several times in distilled water using suction to remove the supernatant. The resin is equilibrated by suspending it in 1 liter of 0.1 M ammonium acetate, pH 4.5 (0.1 M in acetic acid) overnight. The supernatant is removed and the equilibration process is repeated three times. The column is poured at 4° with glass beads overlaying the sintered-glass disks at the bottom of the column.

[24] R. M. Bock and N.-S. Ling, *Anal. Chem.* **26**, 1546 (1954).

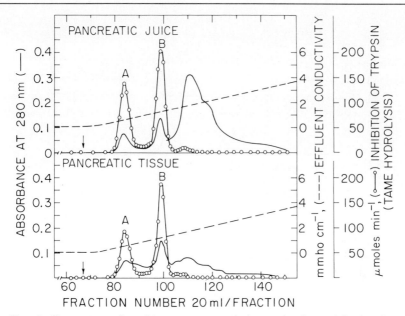

FIG. 2. Chromatography of human pancreatic low-molecular weight fraction on DEAE-cellulose. *Top:* Pancreatic juice low-molecular weight fraction, 4.3 $A_{280}^{1\,cm}$ × 3 ml. *Bottom:* Pancreatic tissue low-molecular-weight fraction, 2.4 $A_{280}^{1\,cm}$ × 35 ml. ——, Absorbance at 280 nm; O——O, trypsin inhibition. TAME, *p*-toluenesulfonyl-L-arginine methyl ester. From M. H. Pubols, D. C. Bartelt, and L. J. Greene, *J. Biol. Chem.* **249**, 2235 (1974).

The sample, fractions A or B, derived from either pancreatic juice or tissue, after desalting and lyophilization is dissolved in 2–5 ml of 0.05 M ammonium acetate buffer, pH 4.3, and is applied to the column (0.9 × 70 cm), which is equilibrated in 0.1 M ammonium acetate buffer, pH 4.5. The pH and conductivity of the sample should be less than that of the equilibrating buffer. After sample application, the column is developed with a linear gradient prepared from 400 ml each of equilibrating buffer and 0.1 M ammonium acetate buffer, pH 7.0, at 10 ml/hr, 4°, with a piston pump. Effluent is collected in 5-ml fractions. The effluent is monitored by measurement of inhibitor activity absorbance at 280 nm and effluent pH. The active fractions of each inhibitor are pooled and lyophilized. Figure 3 shows a comparison of SP-Sephadex chromatography elution profiles of the multiple chromatographic forms of human pancreatic secretory trypsin inhibitor from pancreatic juice and tissue.

The table summarizes the purification of human pancreatic secretory trypsin inhibitor from pancreatic juice and the ammonium sulfate fraction derived from postmortem pancreatic tissue. The values given in

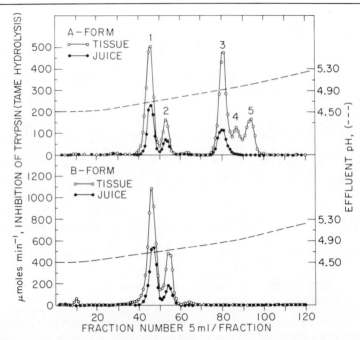

Fig. 3 Comparison of SP-Sephadex chromatography elution profiles of the multiple chromatographic forms of human pancreatic secretory trypsin inhibitor isolated from pancreatic juice and tissue. Forms A and B of inhibitor were prepared by chromatography on DEAE-cellulose (cf. Fig. 2). O——O, Trypsin inhibitor derived from tissues; ●——●, trypsin inhibitor derived from pancreatic juice; - - -, effluent pH. TAME, *p*-toluenesulfonyl-L-arginine methyl ester. From M. H. Pubols, D. C. Bartelt, and L. J. Greene, *J. Biol. Chem.* **249**, 2235 (1974).

the table are the averages for several preparations of human PSTI, each requiring one or more chromatographic columns for each step. The overall yield for the preparation of the multiple chromatographic forms of human PSTI is 55% from pancreatic juice and 38% from the ammonium sulfate fraction derived from postmortem pancreases. Occasionally, during the early stages of the preparation from tissue, trypsin activation occurs, thereby reducing the yield of inhibitor.

The low concentration of inhibitor in the starting material necessitates the use of large columns and thus large elution volumes at the beginning of the procedure, as well as the repetition of several underloaded analytical columns to achieve the separation of the closely chemically related multiple chromatographic forms. As has been noted, the inhibitor is isolated in the free form and is not exposed to active trypsin or acidic conditions below pH 4.5. Recently, a procedure starting with autolyzed human pancreas, utilizing trypsin affinity chromatography with elution by 0.01 N NHCl and isoelectric focusing, has been described for the isolation of several forms of human PSTI from pancreas.[20] In the absence of documentation of the homogeneity of these materials, it is not possible to compare the efficacy of the isolation procedure nor the chemical properties of the inhibitors with those described here.

Trypsin Inhibitor Content of Human Pancreatic Juice and Tissue

The average amount of trypsin inhibitor activity in more than 50 samples of pancreatic juice from 10 individuals was 0.3 mg/100 mg of protein with a range of 0.1–0.6 mg. Unactivated human pancreas contained, on the average, 3 mg inhibitor/100 g wet tissue (range 1–6 mg/100 g). Approximately 40% of the 10 kg of postmortem human pancreases, examined in 50–100 g portions, contained active trypsin.

Properties

Multiple Chromatographic Forms. Pancreatic juice and tissue contain the same multiple chromatographic forms of human PSTI (cf. Fig. 3). They could be distinguished on the basis of chromatographic behavior on ion-exchange resins and by acrylamide electrophores at pH 8.3 and 4.5. The multiple chromatographic forms had the same amino acid composition after acid hydrolysis, specific activity for the inhibition of bovine trypsin, and mobility in sodium dodecyl sulfate (SDS) acrylamide electrophoresis. The three major forms, A_3, B_1, and B_2 had Asx-Ser as their amino-terminal residues but could be distinguished on the basis of asparagine/aspartic acid content and susceptibility to enzymic hydrolysis. The available chemical and physical evidence indicates that the multiple chromatographic forms differ only in asparagine content.

Criteria of Homogeneity. All forms of human PSTI are homogeneous by acrylamide electrophoresis at pH 4.5 and 8.3, SDS acrylamide electrophoresis, amino acid analysis, and on the basis of the stoichiometry of their interaction with trypsin. Forms A_3, B_1, and B_2 had Asx as the only demonstrable amino-terminal residue. The only form examined by high-speed equilibrium centrifugation, B_1, behaved as a homogeneous solute.

Physical Properties. The ultraviolet spectra of forms B_1 and B_2 were identical with the molar absorptivity, $\epsilon_{2755}^{1cm} = 5950$. The protein absorbance index, A_{280}^{1cm} (10 mg/ml) is 8.4. The spectra are consistent with the absence of tryptophan and similar to those published for bovine[5] and porcine[7] PSTI.

Equilibrium sedimentation studies have been carried out with the B_1 form of inhibitor at 20°, 56,000 rpm in 0.1 M KCl, 0.01 M Tris-HCl, pH 7.8, with 3-mm column height. It behaved as a homogeneous, ideal solute giving a measured molecular weight of 6300 ± 200. The specific volume used, 0.71, was calculated from the amino acid composition.

Electrophoretic examination of human PSTI is conveniently carried out in a vertical slab apparatus using Tris-glycine buffer, pH 8.3, and 20% acrylamide in the presence and absence of SDS.[25] The β-alanine–acetic acid, pH 4.5, buffer system of Reisfeld et al.[26] is also used in 20% acrylamide gels. Inhibitor 2–5 μg, was detected by staining at 40° for 1 hr with Coomassie blue, 0.25% in methanol–acetic acid–water (5:1:4). Gels were distained by diffusion.

Amino Acid Composition and Sequence. Human PSTI exists as multiple chromatographic forms having identical amino acid compositions but differing in asparagine content. All forms contain 56 amino acid residues/molecule, which are arranged in a single linear polypeptide chain. The major, most highly amidated form, A_3, has the following amino acid composition: Asp_3, Asn_5, Thr_4, Ser_3, Glu_4, Gln_2, Pro_3, Gly_5, Ala_1, Cys_6, Val_2, Ile_3, Leu_4, Tyr_3, Phe_1, Lys_4, and Arg_3. It does not contain methionine, tryptophan, histidine, glucosamine, or galactosamine. The calculated minimal chemical weight, 6242, corresponds to that obtained by equilibrium ultracentrifugation, SDS acrylamide gel electrophoresis and the stoichiometry of the interaction with trypsin assuming a 1:1 molar ratio.

The amino acid sequence determination was carried out on the mixture of chromatographic forms. Asparagine or aspartic acid were assigned on the basis of the most highly amidated form A_3 when both amino acids were recovered in enzymic hydrolyzates of peptides. The amino acid sequence of human PSTI[27] is compared with inhibitor from porcine,[28,29] bovine,[30] and ovine[31,32] pancreas in Fig. 4. Amino acids identical in all four structures are indicated by the boxes. The structural homology of the series is apparent. The reactive site residue P_1 (residue 18) is either

[25] F. W. Studier, *J. Mol. Biol.* **79**, 237 (1973).

[26] R. A. Reisfeld, V. J. Lewis, and D. E. Williams, *Nature (London)* **195**, 281 (1962).

[27] D. C. Bartelt and L. J. Greene, *J. Biol. Chem.* submitted for publication.

[28] H. Tschesche and E. Wachter, *Eur. J. Biochem.* **16**, 187 (1970).

[29] D. C. Bartelt and L. J. Greene, *J. Biol. Chem.* **246**, 2218 (1971).

[30] L. J. Greene and D. C. Bartelt, *J. Biol. Chem.* **244**, 2646 (1969).

[31] K. Hochstrasser, W. Schramm, H. Fritz, S. Schwarz, and E. Werle, *Hoppe-Seyler's Z. Physiol. Chem.* **350**, 893 (1969).

[32] H. Tschesche, E. Wachter, S. Kupfer, R. Obermeier, E. Reidel, E Haenisch, and M. Schneider, *Proteinase Inhibitors, Proc. Int. Res. Conf. 1st* Munich, 1970, p. 207. de Gruyter, Berlin, 1971.

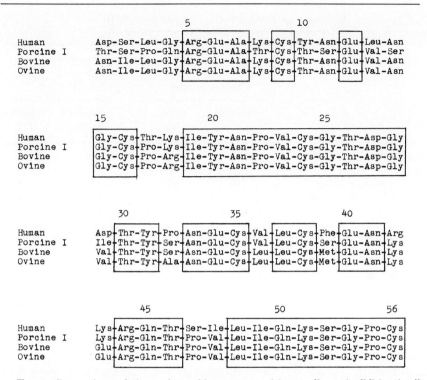

FIG. 4. Comparison of the amino acid sequences of human,[27] porcine[28,29] bovine,[30] and ovine[31,32] pancreatic secretory trypsin inhibitors. Superscript numbers refer to text footnotes that cite appropriate references.

lysine or arginine. Other amino acids near the reactive site (residues 15–28) are identical, with the exception of the P_2 position (residue 17) where threonine occurs in human PSTI and proline in the other species.

Inhibitor Specific Activity (Bovine Trypsin). The calculated value for the inhibition of bovine trypsin by human PSTI is 1790 μmoles (TAME hydrolysis)/min/A_{280}. It is based on the specific activity of trypsin (410 units/mg),[5] the protein absorbance index of the inhibitor and the molecular weights of the components, on the basis of a 1:1 molar complex. The specific activity of the inhibitor in the SP-Sephadex column effluents was 1500–1800 (cf. the table). A decrease of 10–20 in specific activity was sometimes observed after lyophilization of ammonium bicarbonate buffer and ammonium acetate buffer solutions of human PSTI.

Specificity. The initial studies of the specificity of partially purified preparations of inhibitor showed that it inhibited bovine trypsin but not α-chymotrypsin, pancreatic kallikrein, plasma kallikrein, plasmin, throm-

SUMMARY OF PURIFICATION PROCEDURE FOR HUMAN
PANCREATIC SECRETORY TRYPSIN INHIBITOR

Fraction	Pancreatic juice		Pancreatic tissue, ammonium sulfate fraction	
	Specific activity[a]	Yield (%)	Specific activity[a]	Yield (%)
Step 1. Juice (tissue)	3	(100)	—[b]	(100)
Step 2. Sephadex G-75	49	87	—[b]	70
Step 3. Sephadex G-25	200	87	—[b]	70
Step 4. DEAE-cellulose	1350–1600[c]	74	550–900[c]	60
Step 5. SP-Sephadex	1500–1800[c]	55	1500–1800[c]	38

[a] Reduction of p-toluenesulfonyl-L-arginine methyl ester hydrolysis (bovine trypsin) by 1 mole/min/A_{280}.

[b] Not reported because of interference by UV-absorbing materials in these fractions.

[c] Range of values found for all forms of inhibitor.

bin, or papain.[6,19,20] Temporary inhibition in the presence of excess trypsin was also demonstrated.[19,20]

Recent studies with homogeneous preparations of human PSTI show that it effectively inhibits bovine, porcine, and both human trypsins on a 1:1 molar basis.[8,33,34] The inhibition of α-chymotrypsin is much less effective, requiring a 4-fold molar excess to obtain 30% inhibition.[23] Inhibitors prepared from autolyzed pancreas by trypsin-affinity chromatography have been reported to inhibit bovine and both human trypsins, to give partial inhibition of bovine α-chymotrypsin, and to have no inhibitory activity toward two human chymotrypsins.[35]

Most studies of the specificity of trypsin inhibitors have used enzyme turnover assays to determine free enzyme concentration. This method can detect interactions with large association constants but is not suitable for measuring weaker interactions (cf. Laskowski and co-workers[15,36]). Since most of the reports in the literature which describe either weak or no interactions between human PSTI and other serine

[33] C. Figarella, G. A. Negri, and O. Guy, *Proteinase Inhibitors, Proc. Int. Res. Conf., 2nd (Bayer Symp. V)*, Grosse Ledder, 1973, p. 213. Springer-Verlag, Berlin and New York, 1974.

[34] L. J. Greene, D. E. Roark, and D. C. Bartelt, *Proteinase Inhibitors, Proc. Int. Res. Conf., 2nd (Bayer Symp. V)*, Grosse Ledder, 1973, p. 188. Springer-Verlag, Berlin and New York, 1974.

[35] G. Feinstein, R. Hoffstein, and M. Sokolovsky, *Proteinase Inhibitors, Proc. Int. Res. Conf., 2nd (Bayer Symp. V)*, Grosse Ledder, 1973, p. 199. Springer-Verlag, Berlin and New York, 1974.

proteinases have used turnover assays, these data should be considered as tentative and deserve to be examined by burst titrant[36] and other appropriate methods. Some of the differences in specificity reported between bovine and porcine PSTI (see Table III of Burck[13]) and human PSTI probably reflect differences in the type of assay used or in the assay conditions.

A discussion of the evolution of the specificity of protein-proteinase inhibitors is given by Laskowski et al.[36]

[35] M. Laskowski, Jr., I. Kato, T. R. Leary, J. Schrode, and R. W. Sealock, Proteinase Inhibitors, Proc. Int. Res. Conf., 2nd (Bayer Symp. V), Grosse Ledder, 1973, p. 597. Springer-Verlag, Berlin and New York, 1974.

[73] Proteinase Inhibitors from Guinea Pig Seminal Vesicles

By EDWIN FINK and HANS FRITZ

Studies on a trypsin inhibitor in guinea pig seminal vesicles were initiated by Haendle et al.[1] Furthermore investigation of inhibitor preparations obtained by affinity chromatography on trypsin resin revealed the presence of two inhibitors, distinguishable on the basis of inhibition characteristics and physical and chemical properties:[2-4] the one was characterized as a trypsin inhibitor, TI; the other as a trypsin-plasmin inhibitor, TPI. Their physiological role was unknown.

In 1968 Stambaugh and Buckley[5] described a proteinase, localized in the acrosome of mammalian spermatozoa, which is responsible for penetration of the sperm through the zona pellucida of the egg. Following the suggestion of Zaneveld, this enzyme was named acrosin.[6] Both inhibitors from guinea pig seminal vesicles turned out to be potent inhibitors of acrosins from various species.[6-8] This finding, which suggested an impor-

[1] H. Haendle, H. Fritz, I. Trautschold, and E. Werle, Hoppe-Seyler's Z. Physiol. Chem. 343, 185 (1965).
[2] E. Fink, Ph.D. Thesis, Faculty of Science, University of Munich, 1970.
[3] H. Fritz, E. Fink, R. Meister, and G. Klein, Hoppe-Seyler's Z. Physiol. Chem. 351, 1344 (1970).
[4] E. Fink, G. Klein, F. Hammer, G. Müller-Bardorff, and H. Fritz, Proteinase Inhibitors, Proc. Int. Res. Conf., 1st, Munich, 1970, p. 223. de Gruyter, Berlin, 1971.
[5] R. Stambaugh and J. Buckley, Science 161, 585 (1968).
[6] L. J. D. Zaneveld, K. L. Polakoski, R. T. Robertson, and W. L. Williams, Proteinase Inhibitors, Proc. Int. Res. Conf., 1st, Munich, 1970, p. 236. de Gruyter, Berlin, 1971.
[7] H. Haendle, H. Ingrisch, and E. Werle, Klin. Wochenschr. 48, 824 (1970).
[8] R. L. Stambaugh, in "Biology of Mammalian Fertilization and Implantation" (K. S. Moghissi and E. S. E. Hafez, eds.), p. 185. Thomas, Springfield, Illinois, 1972.

tant function of proteinase inhibitors in fertilization, stimulated investigations of trypsin inhibitors from the male reproductive system of other species.[4,9-14]

Assay Methods

Principle. Both trypsin and plasmin hydrolyze N^{α}-benzoyl-DL-arginine p-nitroanilide (DL-BAPA). The enzyme activity is measured by following the change in absorbance at 405 nm. Inhibitory activity is measured by the reduction of enzymic activity.

Inhibition of Trypsin

This has been described elsewhere.[10,15,16]

Inhibition of Plasmin[15]

Reagents

Substrate: 100 mg of DL-BAPA dissolved by heating and stirring in 100 ml of water
Buffer: 0.2 M triethanolamine-HCl, pH 7.8, 0.05 M lysine
Plasmin: Novo Industri, Copenhagen; 250 mg are dissolved in 100 ml of 0.001 M HCl. The solution is stored at 4°.
Inhibitor solutions are prepared in buffer.

Procedure. Incubate 0.1 ml of plasmin solution, 1.8 ml of prewarmed (25°) buffer, and 0.1 ml of inhibitor solution for 3 min in 3-ml cuvettes (1 cm light path) placed in a spectrophotometer connected to a circulating water bath, 25°. The enzymic reaction is started by adding 1 ml of

[9] E. Fink, E. Jaumann, H. Fritz, H. Ingrisch, and E. Werle, *Hoppe-Seyler's Z. Physiol. Chem.* **352**, 1591 (1971).
[10] H. Schiessler, E. Fink, and H. Fritz, this volume [75].
[11] H. Fritz, H. Tschesche, and E. Fink, this volume [74].
[12] H. Fritz, B. Förg-Brey, E. Fink, H. Schliessler, E. Jaumann, and M. Arnhold, *Hoppe-Seyler's Z. Physiol. Chem.* **353**, 1007 (1972).
[13] H. Schiessler, M. Arnhold, and H. Fritz, *Proteinase Inhibitors, Proc. Int. Res. Conf., 2nd (Bayer Symp. V)*, Grosse Ledder, 1973, p. 147. Springer-Verlag, Berlin and New York, 1974.
[14] For further references see this volume [74] and [75].
[15] H. Fritz, I. Trautschold, and E. Werle, *in* "Methoden der enzymatischen Analyse" (H. U. Bergmeyer, ed.), p. 1103. Verlag Chemie, Weinheim, 1974.
[16] B. Kassell, this series Vol. 19, p. 844.

prewarmed (25°) substrate solution. After mixing with a plastic spatula, the absorbance at 405 nm is read every minute for 10 min. In the enzyme reference sample, inhibitor solution is replaced by buffer.

Definition of Units and Specific Activity. One enzyme unit corresponds to the hydrolysis of 1 μmole of substrate per minute (ΔA_{405}/min = 3.32 for 3 ml). One inhibitor unit corresponds to the reduction of one enzyme unit. The specific activity is expressed in inhibitor units per milligram of protein.

Isolation

Two procedures were developed for the isolation of inhibitors from the aqueous extract of guinea pig seminal vesicles.[2-4] The first employs affinity chromatography on insolubilized trypsin; the second procedure is based on conventional methods, such as gel filtration and ion-exchange chromatography.

Desalting. At several stages in both isolation methods, the preparations have to be desalted. This is achieved by gel filtration on Sephadex G-25 fine columns with a bed height of 80 cm and a bed volume of 10-fold the applied solution volume. The columns are eluted with 0.5% (v/v) acetic acid at a flow rate of 6 ml/cm^2/hr; fractions of 10 ml are collected.

Preparation of the Crude Extract. Adult male guinea pigs are anesthetized and exsanguinated. The abdomen is opened, and the seminal vesicles are removed. The glands are kept frozen until use.

Thawed seminal vesicles, 500 g, are homogenized in 1 liter of water (0°–4°), and the resulting homogenate is centrifuged for 30 min at 1200 g. The precipitate is rehomogenized in 500 ml of water and centrifuged. This extraction is repeated five to eight times until the trypsin-inhibiting capacity of the supernatant is approximately one-tenth that of the first supernatant. All aqueous extracts are combined; their total trypsin-inhibiting capacity is usually between 1000 and 1500 IU.

Another 250–300 IU of inhibitory activity are obtained by continuing the extraction with 3% sodium chloride solution instead of water. Thus a total of up to 1750 IU can be extracted from 500 g of seminal vesicles.

Method I

Step 1. Deproteinizing of the Crude Extract. To the crude extract an equal volume of 6% (w/v) perchloric acid is added. After centrifugation (30 min, 1200 g), the precipitate is extracted three times with 3% perchloric acid. The combined acidic solutions are neutralized by careful

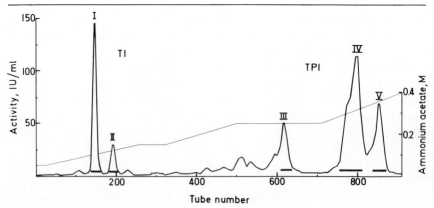

FIG. 1. Fractionation of the inhibitor preparation obtained by affinity chromatography on SP-Sephadex C-25 as described in Method I, step 3. TI, trypsin inhibitor; TPI, trypsin–plasmin inhibitor.

addition of solid potassium hydrogen carbonate. After standing overnight at 4°, the precipitate of potassium perchlorate is filtered off.

Step 2. Affinity Chromatography. Potassium chloride and triethanolamine hydrochloride are added to the neutralized, deproteinized extract to a final concentration of 0.15 M and 0.05 M, respectively, and the pH is adjusted to 7.8 by addition of 2 N NaOH. Then sufficient trypsin resin[10,17] is added to bind at least 90% of the inhibitor (any type of resin-bound trypsin may be used). The pH is readjusted to 7.8, then the suspension is stirred for 30 min and centrifuged. The supernatant, containing less than 10% of the total inhibitory activity, is discarded. The trypsin resin (with the inhibitor bound) is packed into a column of suitable size (diameter to height ratio 1:5) and washed (10 ml/cm² per hour) with 0.05 M triethanolamine-HCl, pH 7.8, 0.15 M KCl, 0.01 M CaCl₂, until readings of absorbance show the eluate to be protein-free.

The inhibitor is dissociated from the insolubilized trypsin and eluted from the column with 0.2 M KCl solution, adjusted to pH 2.0 with HCl at a flow rate of 10 ml/cm² per hour; fractions of 5–10 ml are collected.

The fractions containing inhibitory activity are pooled, desalted as described above, and lyophilized. The recovery is 70%; the specific activity, 2.9 IU/mg.

Step 3. Chromatography on SP-Sephadex[18] (Fig. 1). Of the inhibitor preparation, 1100 IU (380 mg) are applied to a SP-Sephadex C-25

---

[17] H. Fritz, M. Gebhardt, E. Fink, W. Schramm, and E. Werle, *Hoppe-Seyler's Z. Physiol. Chem.* **350**, 129 (1969).

[18] The SE-Sephadex C-25 used in the original work can be replaced by SP-Sephadex C-25 without any changes in the elution conditions.

column, 1.8 $\times$ 120 cm, equilibrated with 0.05 M ammonium acetate, pH 5.4, 0.02% NaN$_3$. The column is developed at a flow rate of 10 ml/cm^2 per hour and 10-ml fractions are collected.

The elution is performed in the following way. Step a: A linear concentration gradient is produced from 1.5 liters each of 0.05 M and 0.15 M ammonium acetate buffer, pH 5.4, 0.02% NaN$_3$. Elution is continued with 350 ml of the 0.15 M buffer. Step b: Elution is done with a linear concentration gradient produced from 1 liter each of 0.15 M and 0.25 M ammonium acetate buffer, pH 5.4, 0.02% NaN$_3$, followed by 2 liters of the second buffer. Step c: Elution with a linear concentration gradient produced from 1.5 liters each of 0.25 M and 0.4 M ammonium acetate buffer, pH 5.4, 0.02% NaN$_3$.

In step a, the TI is eluted, mainly in two peaks (I and II, Fig. 1); in steps b and c, the TPI appears in several small and three large peaks (III, IV, and V, Fig. 1). The fractions representing the peak areas are pooled, desalted as described above, and lyophilized. The total recovery in step 3 is 90%, the specific activity of the main fractions is about 3.2 IU/mg.

Method II

Step 1. Gel Filtration of the Crude Extract on Sephadex G-50. The crude extract is concentrated using a rotary evaporator to a final concentration of trypsin inhibitor activity of 2.5 IU/ml, any precipitate is removed by centrifugation. Then 50–100 ml of the concentrate are applied to a Sephadex G-50 medium column, 3.5 $\times$ 100 cm, equilibrated with 0.5% (v/v) acetic acid. The column is developed at a flow rate of 40 ml/hr, and fractions of 5 ml are collected. The fractions are tested for chloride, and the salt-free inhibitory active fractions are pooled and lyophilized. A yellow-to-brown powder with a specific activity of about 0.06 IU/mg is obtained.

Step 2. Separation of TI and TPI by Chromatography on SP-Sephadex. The crude inhibitor preparation, 1750 IU, 30 g, is dissolved in 75 ml of 0.02 M ammonium acetate, pH 5.4, 0.02% NaN$_3$ (the pH of the solution is corrected to 5.4), and applied to a SP-Sephadex C-25 column, 4.5 $\times$ 30 cm, equilibrated with 0.05 M ammonium acetate, pH 5.4, 0.02% NaN$_3$. The column is eluted at a flow rate of 20 ml/cm^2 per hour; 10-ml fractions are collected.

The elution is performed in the following way: Step a: Elution is done with 0.05 M ammonium acetate, pH 5.4, 0.02% NaN$_3$, until the inhibitory activity in the eluate reaches a constant, low level. Approximately 15% of the applied inhibitory activity is eluted along with a high amount of impurities. Step b: Elution is done with a linear concentration gradient produced from 500 ml each of ammonium acetate 0.05 M and 0.2 M, pH 5.4, 0.02% NaN$_3$. Elution with the 0.2 M buffer is continued until the inhibitory activity reaches a constant, low level. During this step about 55% of the applied activity is eluted, representing the TI. Step c: Elution is done with a linear gradient produced from 500 ml each of 0.2 M and 0.6 M ammonium acetate pH 5.4, 0.02% NaN$_3$. Elution is continued with the 0.6 M buffer until no more inhibitory activity is detectable in the eluate. During this step about 30% of the applied activity is eluted, representing the TPI.

The inhibitor fractions of steps b and c are pooled separately; the two pools are desalted as described above and lyophilized. The specific activity of each of the two preparations is about 0.25 IU/mg.

Step 3. Separation of the TI into Multiple Forms by Chromatography on SP-Sephadex. Of the TI preparation, 6.2 g, 1900 IU are dissolved in 20 ml of starting buffer and applied on top of a SP-Sephadex C-25 column, 1.8 × 120 cm, equilibrated with the starting buffer, 0.05 M ammonium acetate, pH 5.4, 0.02% NaN$_3$. The column is washed with 750 ml of starting buffer, then a linear concentration gradient produced from 800 ml each of 0.05 M and 0.2 M ammonium acetate, pH 5.4, 0.02% NaN$_3$, is applied; the elution is continued with the second buffer until no more inhibitory activity is detectable in the eluate. The fractions representing the peak areas are pooled, desalted as described above, and lyophilized; recovery is 90%, and the specific activity is about 3 IU/mg.

The TI preparation from step 2 is separated into five distinct peaks, the first peak starting to elute at an ammonium acetate concentration of 0.1 M; the two main peaks elute at ammonium acetate concentrations of 0.12 and 0.2 M and account for 19% and 62%, respectively, of the total activity eluted.

Step 4. Rechromatography of the Multiple TI Forms on SP-Sephadex. The TI fractions separated in step 3 are rechromatographed under equilibrium conditions on a SP-Sephadex C-25 column, 0.9 × 100 cm, equilibrated and developed with ammonium acetate, pH 5.4, 0.02% NaN$_3$, at a flow rate of 8 ml/cm^2 per hour. The ammonium acetate concentration chosen is 10% lower than the concentration at which the respective fraction began to appear in the eluate in step 3. The active fractions are pooled, desalted as described above and lyophilized.

Step 5. Separation of the TPI into Multiple Forms by Chromatography on SP-Sephadex. Of the TPI preparation from step 2, 5.5 g, 1400 IU are dissolved in 30 ml of 0.1 M ammonium acetate, pH 5.4, and applied on top of a SP-Sephadex C-25 column, 3.7 × 100 cm, equilibrated with 0.2 M ammonium acetate, pH 5.4, 0.02% NaN$_3$. The column is developed at a flow rate of 10 ml/cm^2 per hour with a linear concentration gradient produced from 750 ml each of 0.2 M and 0.35 M ammonium acetate, pH 5.4, 0.02% NaN$_3$, followed by elution with the second buffer until no more inhibitor appears in the eluate; fractions of 8 ml are collected. The fractions of each peak are pooled, desalted as described above, and lyophilized; the recovery is 95%; the specific activity about 1 IU/mg.

The TPI preparation from step 2 is separated into more than six unsymmetrical peaks. The main fraction, which appears after elution with about two bed volumes of the second buffer, accounts for 50% of the total activity eluted.

Step 6. Rechromatography of the TPI Fractions from Step 5 on SP-Sephadex. Fractions from step 5 are rechromatographed under starting conditions on a SP-Sephadex C-25 column, 0.6 × 115 cm, equilibrated and developed at a flow rate of 5 ml/hr with a potassium phosphate buffer of ionic strength 0.1.[19] Depending on the elution position of the respective peak in step 5, the pH of the phosphate buffer is chosen between 7.1 and 7.6. Fractions of 2 ml are collected. The elution is continued until no more inhibitor is detectable in the eluate. Elution with up to 20 bed volumes may be necessary. The recovery is about 80%; the specific activity up to 3.4 IU/mg.

Properties

Stability. Lyophilized preparations were stored at −20° over several years without any detectable loss of activity. Lyophilized material has to be incubated in salt-containing solutions prior to the assay; if deionized water is used as solvent, no inhibitor activity is detected by DL-BAPA assay. Both inhibitors are stable in solutions between pH 2 and 9 at room temperature.

Molecular Weight. The molecular weights of both inhibitors estimated by gel filtration to 6600–6800 are in good agreement with the values calculated from the amino acid compositions (Table I).

Amino Acid Composition. Table I shows the amino acid compositions of inhibitor fractions isolated by method I. TI fractions I and II (Fig. 1) have identical compositions, but in TI fraction I the peptide bond Arg-X of the reactive site is split.[2-4] This modification occurs during affinity chromatography. Both the TPI fractions IV and V (Fig. 1) consist of two inhibitors, IVa, IVb and Va, Vb. The five TPI fractions differ in their amino acid composition. TPI fraction III (Fig. 1) is the fraction with the lowest number of amino acids. Fractions IVa, IVb, Va, and Vb

[19] A. A. Green, *J. Am. Chem. Soc.* **55**, 2331 (1933).

TABLE I

AMINO ACID COMPOSITION OF PROTEINASE INHIBITORS
FROM GUINEA PIG SEMINAL VESICLES

Amino acid	TI	TPI[b]				
	I, II	III	Va	Vb	IVa	IVb
Aspartic acid	6	6				
Threonine	1	4				
Serine	2	5	6	6	6	
Glutamic acid	10	4				
Proline	5	2			3	3
Glycine	6	5				
Alanine	1	0			1	1
Half-cystine	6	6				
Valine	3	3				
Methionine	0	1				
Isoleucine	4	1				
Leucine	5	3				
Tyrosine	2	4				
Phenylalanine	0	3				4
Lysine	1	4	5	5	5	5
Histidine	2	3				
Arginine	6	4				
Tryptophan	0	0				
	60	58	59	60	62	63
Molecular weight	6772	6687	6815	6902	7071	7218

[a] The roman numerals correspond to the fractions in Fig. 1. TI, Trypsin inhibitor; TPI, trypsin–plasmin inhibitor.

[b] Amino acids are quoted for TPI fractions IVa, IVb, Va, and Vb where the composition differs from that of fraction III.

differ from fraction III in the peptide chain, being extended from the N-terminal valine residue by 1–5 amino acids (see Table II). None of the inhibitors contains tryptophan or amino sugars.

The multiple forms of both inhibitors isolated by method II are not yet completely characterized. According to amino acid analyses, at least some of the isolated forms of TPI are identical with those isolated by method I.

Inhibition Properties. Neither of the two inhibitors is effective on porcine pancreatic kallikrein, bovine chymotrypsin, or porcine pepsin; both inhibit bovine trypsin and acrosins from boar,[20] rabbit,[6,8] hamster,[21]

[20] H. Fritz, B. Förg-Brey, E. Fink, H. Schiessler, E. Jaumann, and M. Arnhold, *Hoppe-Seyler's Z. Physiol. Chem.* **353,** 1007 (1972)

[21] E. Fink, H. D. Notdurft, and H. Fritz, unpublished results.

TABLE II
N-Terminal Amino Acid Sequences of Five Forms of the
Trypsin-Plasmin Inhibitor Isolated by Method I

N-terminal sequence	Inhibitor fraction[a]
Val	III
Lys-Val	Va
Ser-Lys-Val	Vb
Ala-Pro-Ser-Lys-Val	IVa
Phe-Ala-Pro-Ser-Lys-Val	IVb

[a] The roman numerals correspond to the fractions in Fig. 1.

and human.[22] Only the TPI inhibits porcine plasmin. Inhibition constants (K_i) were calculated by the method of Green and Work[23]; for bovine trypsin $K_i(TI) = 2 \times 10^{-9}\ M$ and $K_i(TPI) = 3 \times 10^{-9}\ M$, and for porcine plasmin $K_i(TPI) = 3 \times 10^{-9}\ M$.

Reactive Site. Treatment with excess maleic anhydride resulted in the loss of the inhibitory activity of TPI against both trypsin and plasmin, indicating the presence of a lysine residue in the reactive center of TPI. If TI isolated by method II or TI fraction II (Fig. 1) was treated with excess maleic anhydride, no loss of activity was observed, but incubation of the maleylated TI with the 2,3-butanedione reagent of Grossberg and Pressman[24] caused rapid loss of the trypsin-inhibiting activity. This result implicates arginine in the reactive center. On the other hand, treatment of TI fraction I (Fig. 1) with maleic acid anhydride causes loss of the inhibitory activity. But in this fraction the Arg-X bond of the reactive center is split, and loss of inhibitory activity results from acylation of the free α-NH$_2$ group of the residue X. Owing to this cleavage of the reactive site peptide bond, Tschesche and Obermeier[25] identified residue X as isoleucine.

Covalent Structure. The partial sequence of TPI is shown in this volume in [74], Fig. 2.

Acknowledgment

This work was supported by Sonderforschungsbereich 51, Munich (Grant B-3).

[22] H. Fritz, B. Förg-Brey, M. Meier, M. Arnhold, and H. Tschesche, *Hoppe-Seyler's Z. Physiol. Chem.* **353,** 1950 (1972).

[23] N. M. Green and E. Work, *Biochem. J.* **54,** 347 (1953).

[24] A. L. Grossberg and D. Pressman, *Biochemistry* **7,** 272 (1968).

[25] H. Tschesche and R. Obermeier, *Proteinase Inhibitors, Proc. Int. Res. Conf., 1st,* Munich, 1970, p. 135. de Gruyter, Berlin, 1971.

[74] Proteinase Inhibitors from Boar Seminal Plasma

By Hans Fritz, Harald Tschesche, and Edwin Fink

During investigations on the distribution and possible function of proteinase inhibitors in various animals, Haendle *et al.* found high antitryptic activity in seminal vesicles and seminal plasma of various species including the boar.[1] The antiproteolytic activity derived from acid-stable, low-molecular-weight proteins. These were also found in lower concentrations in extracts of boar testes and epididymis.[2] Later, a mixture of related inhibitory active glycoproteins was isolated by Fink *et al.* employing affinity chromatography on water-insoluble trypsin-resin. These were characterized by the average molecular weights, estimated by gel filtration, and some molecular properties and their strong inhibition of bovine trypsin and porcine plasmin.[3] Independently, Zaneveld *et al.* purified also an acid-stable, low-molecular-weight trypsin inhibitor from boar seminal plasma and calculated its molecular weight as 6781 on the basis of the amino acid composition.[4] The purification of the inhibitors from boar seminal plasma and spermatozoa to homogeneity was reported by Tschesche *et al.*[5] The boar seminal inhibitors described by Polakoski *et al.*[6] are most probably identical with the inhibitors presented in this chapter (cf. the discussion on this subject by Fritz *et al.*[7] and Tschesche and Kupfer[8]).

Trypsin inhibitors from boar seminal plasma and spermatozoa are strong inhibitors of boar and human acrosins.[9,10] This inhibitory speci-

---

[1] H. Haendle, H. Fritz, I. Trautschold, and E. Werle, *Hoppe Seyler's Z. Physiol. Chem.* **343**, 185 (1965).

[2] H. Haendle, Dissertation (M.D. Thesis), Medical Faculty of the University of Munich, 1969.

[3] E. Fink, G. Klein, F. Hammer, G. Müller-Bardorff, and H. Fritz, *Proteinase Inhibitors, Proc. Int. Res. Conf., 1st, Munich, 1970*, p. 225.

[4] L. J. D. Zaneveld, K. L. Polakoski, R. I. Robertson, and W. L. Williams, *Proteinase Inhibitors, Proc. Int. Res. Conf., 1st*, p. 236.

[5] H. Tschesche, S. Kupfer, O. Lengel, R. Klauser, M. Meier, and H. Fritz, *Proteinase Inhibitors, Proc. Int. Res. Conf., 2nd (Bayer Symp. V), Grosse Ledder, 1973*.

[6] K. L. Polakoski and W. L. Williams, *Proteinase Inhibitors, Proc. Int. Res. Conf., 2nd (Bayer Symp. V), Grosse Ledder, 1973*, p. 156.

[7] H. Fritz, H. Schiessler, W.-B. Schill, H. Tschesche, N. Heimburger, and O. Wallner, *in* "Proteases and Biological Control" (E. Reich, D. Rifkin, and E. Shaw, eds.) p. 737. Cold Spring Harbor Laboratory, Cold Spring Harbor, New York, 1975.

[8] H. Tschesche and S. Kupfer, unpublished results, 1975.

[9] H. Fritz, B. Förg-Brey, E. Fink, H. Schiessler, E. Jaumann, and M. Arnhold, *Hoppe-Seyler's Z. Physiol. Chem.* **353**, 1007 (1972).

[10] H. Fritz, B. Förg-Brey, M. Meier, M. Arnhold, and H. Tschesche, *Hoppe-Seyler's Z. Physiol. Chem.* **353**, 1950 (1972).

ficity, the unusual glycoprotein nature and the unexpected homology between the pancreatic secretory trypsin inhibitors (Kazal type) and the boar seminal acrosin inhibitors reported recently by Tschesche et al.[11] make these inhibitors an interesting subject for studies in the areas of fertilization, structure–function relationships, and inhibitor evolution.[7,8]

Assay Methods

Principles. The substrate N^α-benzoyl-DL-arginine p-nitroanilide (BAPA) was used for measuring proteolytic activity and the inhibition of bovine trypsin,[12-14] porcine plasmin,[12,15] and boar and human acrosin[9,10,14] by the change in absorbance at 405 nm.

Known methods were employed to assay the inhibition of bovine chymotrypsin,[12,14] porcine pancreatic kallikrein,[12] bovine thrombin,[16] and human neutral leukocytic proteinases.[14]

Definition of Unit and Specific Activity. One enzyme unit corresponds to the hydrolysis of 1 μmole of substrate (BAPA) per minute (ΔA_{405}/min cm $= 3.32$ for 3 ml). One inhibitor unit reduces the activity of two enzyme units by 50%, thus decreasing the absorbance by 3.32/minute and cm for 3 ml. The specific activity corresponds to inhibitor units (BAPA) per milligram of protein.[12,13]

Purification Procedure

Different procedures may be used to isolate proteinase inhibitors from seminal plasma and spermatozoa, depending on the starting material and the desired objectives, e.g., isolation on a large scale or in high yield and of material with high specific activity or of apparent homogeneity with respect to the protein moiety. Seminal plasma contains various glycosidases and proteinases[17] suspected of causing degradation of the inhibitors starting immediately after ejaculation. As long as effective inactivation of these enzymes cannot be achieved, acidification of the

[11] H. Tschesche, S. Kupfer, R. Klauser, E. Fink, and H. Fritz, in "Protides of the Biological Fluids, 23rd Colloquium" (H. Peeters, ed.), p. 255. Pergamon, Oxford and New York, 1976.
[12] H. Fritz, I. Trautschold, and E Werle, in "Methoden der enzymatischen Analyse" (H. U. Bergmeyer, ed.), p. 1103. Verlag Chemie, Weinheim, 1974.
[13] B. Kassell, this series Vol. 19, p 844.
[14] H. Schiessler, E. Fink, and H. Fritz, this volume [75].
[15] E. Fink and H. Fritz, this volume [73].
[16] The kallikrein estimation procedure was used.[12]
[17] L. J. D. Zaneveld, K. L. Polakoski, and G. F. B. Schumacher, in "Proteases and Biological Control" (E. Reich, D. Rifkin, and E. Shaw, eds.) p. 683. Cold Spring Harbor Laboratory, Cold Spring Harbor, New York, 1975.

seminal plasma is the method of choice in order to prevent hydrolytic degradation. Cleavage of carbohydrate residues, e.g., of sialic acids and fucose, and hydrolysis of acid-labile peptide (Asp-Pro bonds[11]) and carboxamide bonds, is an inevitable disadvantage. Nevertheless, sufficient amounts of active, and in the protein backbone unaffected, inhibitors have been obtained under the conditions described below, using deproteinization of the seminal plasma in 3% perchloric acid or acidification to pH 1–2, followed by affinity chromatography on CM-cellulose trypsin.

Step 1. Collection of Seminal Plasma and Spermatozoa. Boar semen is collected using an artificial vagina, filtered through gauze, and centrifuged at 600 *g* for 10 min. The supernatant is centrifuged for 30 min at 1000 *g* and 4°, and treated according to step 2 or stored at −40°. The sperm pellet is twice resuspended in saline and centrifuged in order to remove unspecifically adsorbed substances. The pellet is then processed according to step 2 (Section II) or frozen at −196° and stored at −40°.[18]

I. Purification of Seminal Plasma Inhibitors

Step 2. Acid Deproteinization. A prepurification of the inhibitors and denaturation of the seminal plasma enzymes is possible by one of the following procedures.

METHOD 1: DEPROTEINIZATION. The seminal plasma is mixed with an equal volume of 6% (w/v) perchloric acid (see method I, step 1[15]). The clear inhibitor solution thus obtained may be directly applied to affinity chromatography on a trypsin resin (step 3) after neutralization (pH 7.8) and addition of solid sodium chloride to a final concentration of about 0.4 *M*.

METHOD 2: ACIDIFICATION. Seminal plasma is adjusted to pH 1 by addition of 3% (w/v) perchloric acid or 2 *N* HCl, which inactivates the enzymes and increases the yield of free inhibitor by dissociation of the acrosin–inhibitor complex. The acidified solution is stored for 15 hr at 4° and then centrifuged (30 min, 1200 *g*). The supernatant inhibitor solution is readjusted to pH 7.8 by addition of triethanolamine and made up to a final concentration of 0.4 *M* sodium chloride.

Step 2 may be omitted and trypsin resin be directly applied in a batch operation to the native seminal plasma at 4°.

Step 3. Affinity Chromatography. Deproteinized inhibitor solutions of low viscosity may be applied to any water-insoluble trypsin derivative available. A suitable flow rate in the affinity column is essential. Native seminal plasma or turbid inhibitor solutions are best applied to such

---

[18] W.-D. Schleuning and H. Fritz, this volume [27].

trypsin derivatives, which can be easily removed by centrifugation. Trypsin cellulose (bovine trypsin bound to CM-cellulose, 7–10 U/mg, E. Merck, Darmstadt, West Germany) is especially suitable. Working temperature is 4° throughout. Details of the procedure are described in this volume[14,15,19] and by Fritz et al.[7]

The affinity adsorbent (10 g) is preconditioned by suspending it repeatedly in 0.4 M NaCl, 0.1 M triethanolamine–HCl, pH 7.8, and in 1.0 M KCl–HCl, pH 2.0, until the supernatant is free of tryptic activity. After reequilibration in pH 7.8 buffer, the trypsin cellulose is added to native seminal plasma or the inhibitor solution from step 2 and gently stirred or shaken for 30–60 min. About 100 ml of wet trypsin cellulose is recommended per liter of seminal plasma containing 700–1000 IU (trypsin inhibition). Subsequently, the suspension is centrifuged (1000 g, 5 min) and the inhibitor-loaded resin (precipitate) is washed five times with the pH 7.8 buffer and once with a 1:10 dilution of the same buffer.

The inhibitors are dissociated from trypsin at acid pH. The adsorbent (100 ml wet volume) is suspended four to five times in 300 ml of 0.4 M KCl–HCl, pH 2.0, and then filtered or centrifuged. The first extract containing only low amounts of inhibitors is discarded. The pH of the second suspension has to be readjusted to 2.0 with 2 N HCl. The major portion of inhibitors is found in the second extract.

The acidic extracts are combined, neutralized, and desalted by repeated ultrafiltration using the Diaflo membranes UM-05 or UM-2. After lyophilization, inhibitor preparations with specific activities of 1.9–2.2 IU/mg are obtained. The yields depend on the quality of the starting material: If deproteinized inhibitor solutions are applied, between 82 and 93% of the inhibitors bound to the trypsin cellulose are recovered. Native seminal plasma contains besides the acid-stable inhibitors, also high-molecular-weight (above 45,000), acid-labile inhibitors, probably of plasmatic origin. These inhibitors are also bound—at least partially—to the resin; however, they are denatured during the acidic extraction at pH 2.0. Depending on the varying amounts of these inhibitors in native seminal plasma, between 74 and 90% of the inhibitory activity bound to the affinity adsorbent is thus recovered in the acidic extracts. The purity of this inhibitor preparation is 70–80%. However, the inhibitor material is heterogeneous and, besides the homologous isoinhibitors A, A₁, and (or) B, contains at least 15–20 multiple chromatographic forms due to microheterogeneity in the carbohydrates.

Step 3a. Gel Filtration. The mixture of inhibitors obtained in step 3 is separated into two major fractions by gel filtration on Sephadex G-75

---

[19] H. Tschesche, this volume [69].

either in slightly alkaline (pH 7.6) or acidic (pH 2.6) salt-buffer solutions. The main portion of the inhibitors is eluted in the molecular weight range 5500–6800 and 10,500–12,800 (pH 7.6) or even higher. In a typical experiment, 3 ml of the inhibitor solution containing 13 IU are applied to a 1.6 × 122 cm column of Sephadex G-75 equilibrated and developed with 0.4 M NaCl, 0.1 M triethanolamine-HCl, pH 7.6, at a flow rate of 12 ml/hr; fractions of 4 ml are collected.[7] The two major inhibitor fractions eluted in the molecular weight range mentioned above were desalted by repeated ultrafiltration and lyophilized (cf. step 3). The material used for immunization experiments was subjected to a second gel filtration under the same conditions. The fractions were designated BSTI-II (MW about 6000) or BSTI-I (MW about 12,000), respectively.

Native seminal plasma when fractionated by gel filtration under the same conditions separates into three major acrosin–trypsin inhibitor fractions, a high-molecular-weight, acid-labile one, and the acid-stable inhibitor fractions I and II.[7,20] Both inhibitor fractions I and II can be purified independently by affinity chromatography as described in step 3. Inhibitor preparations with specific activities of 2.0 IU/mg for BSTI-I and 2.8 IU/mg for BSTI-II are thus obtained. The difference between the inhibitor fractions I and II seems to refer only to the carbohydrate portions. The same general fractionation patterns have been obtained applying inhibitor material from step 3 and from the fraction II (step 3a) to chromatography on SE-Sephadex (step 4). Both preparations, from step 3 and fractions I or II from step 3a, seem to be equally well suited for further fractionation.

Step 4. Chromatography on SE-Sephadex.[5] SE-Sephadex C-25 is equilibrated with 0.025 M ammonium acetate, pH 6.8, transferred to a chromatography column, 4.5 × 40 cm, which is equilibrated with the 0.025 M buffer until the pH of the effluent is 6.8. Inhibitor, 200 mg (1.95 IU/mg) is dissolved in the equilibration buffer and applied to the column. Elution is carried out with a four-step ammonium acetate gradient of increasing molarity at pH 6.8 and room temperature. The gradient is prepared by mixing buffers of the following concentrations: 0.05 M to 0.025 M (100 ml) until tube 200 (a), 0.15 M to 0.05 M (200 ml) until tube 400 (b), 0.75 M to 0.15 M (200 ml) until tube 600 (c). Finally, 0.75 M buffer is applied to the column (d and e in Fig. 1). The flow rate is adjusted to 40 ml/hr and 4-ml fractions are collected. The column effluent is monitored by its absorption at 280 nm and by its trypsin-inhibiting activity. The active material is distributed among four major peaks (I, II, III, IV in Fig. 1) and several smaller ones. The elution

[20] H. Fritz, W.-D. Schleuning, and W.-B. Schill, *Proteinase Inhibitors, Proc. Int. Res. Conf., 2nd (Bayer Symp. V), Grosse Ledder, 1973,* p. 118.

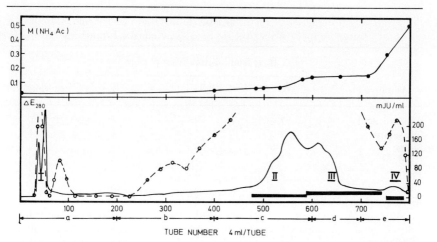

Fig. 1. Gradient elution chromatography of crude boar seminal plasma inhibitor on SE-Sephadex C-25. For experimental conditions, see step I, 4. *Top:* ●——●, ammonium acetate gradient molarity. *Bottom:* ——, absorbance at 280 nm; ○---○, trypsin inhibition.

pattern varies considerably with inhibitor materials derived from seminal plasma of different animals and is strongly dependent on the age and pretreatment of the seminal plasma. Each peak is pooled separately as indicated by the bars. The pools are concentrated by rotary evaporation, desalted by gel filtration on Bio-Gel P-2 columns, 1.5×80 cm, equilibrated with 0.1 M acetic acid, and lyophilized.

Peak I contains the most acidic inhibitor which still contains one residue of sialic acid per molecule.[21] The preparation is contaminated with some of the inactive material which is mainly eluted in the subsequent fraction containing the cleaved acidic and neutral sugars (sialic acid, fucose, mannose, galactose, galactosamine, glucose, and glucosamine). The specific activity of this inhibitor is about 1 IU/mg after rechromatography under the same conditions. The main portion of the inhibitory activity is recovered in fractions II and III. The preparations obtained from these fractions are devoid of sialic acid. The specific activity of peak II is about 2.3 IU, and of peak III about 2.8 IU/mg. Peak IV was present in some, but not all, batches of seminal plasma inhibitors investigated. Its specific activity is about 2.8 IU/mg. Isoinhibitor B (67 amino acid residues; see Table I) was isolated from this peak after rechromatography.

Amino acid analysis and cellulose acetate electrophoresis at pH 7.8 reveal that none of the fractions is homogeneous. However, preparations

[21] G. Decker, Diplomarbeit, Technical University of Munich, 1974.

TABLE I

AMINO ACID COMPOSITION OF SEMINAL ACROSIN INHIBITORS[a]

Amino acid	Boar isoinhibitors from plasma			Boar spermatozoa
	A	A₁	B	
Aspartic acid	7	7	7	7
Threonine	4	4	4	4
Serine	5	5	5	5
Glutamic acid	5	5	5	5
Proline	3	3	3	3
Glycine	5	5	5	5
Alanine	2	2	2	2
Valine	1	1	1	1
Half-cystine	6	6	6	6
Methionine	1	1	1	1
Isoleucine	2	2	2	2
Leucine	2	2	2	2
Tyrosine	3	3	3	3
Phenylalanine	5	5	5	5
Lysine	5	4	6	4
Histidine	3	3	3	3
Arginine	5	6	6	5
Tryptophan	1	1	1	(1)
	65	65	67	65
Carbohydrates	+	+	+	+
Molecular weight	~11500	~11500	~12000	~11500

[a] H. Tschesche, S. Kupfer, O. Lengel R. Klauser, M. Meier, and H. Fritz, *Proteinase Inhibitors, Proc. Int. Res. Conf. 2nd (Bayer Symp. V)*, Grosse Ledder 1973, p. 156.

II to IV represent seminal inhibitors of about 95% purity with high specific activities depending on the respective carbohydrate content.

Step 5. Rechromatography on SE-Sephadex. The main fraction II from step 4 is rechromatographed on a column, 1.5×100 cm, of SE-Sephadex C-25. The resin is pretreated as in step 4 and equilibrated with 0.05 M ammonium acetate, pH 6.6. The inhibitor is dissolved in 2–3 ml of 0.05 M ammonium acetate, pH 6.0, and applied to the column, which is eluted at room temperature with the equilibrium buffer. The flow rate is 10 ml/hr, and 2-ml fractions are collected. The activity appears in one or two broad peaks II-1, II-2 or III-1, III-2 (see below), respectively. The fractions pooled from each peak are concentrated by rotary evaporation, desalted by gel filtration on Bio-Gel P-2 in 0.1 M acetic acid, and lyophilized.

The fractions II-1, and II-2, contain isoinhibitor A. The peaks are not homogeneous as revealed by cellulose acetate electrophoresis, but represent multiple chromatographic forms identical in amino acid composition but different in carbohydrate composition (see Table II). If the amino acid analysis is not satisfactory, rechromatography should be repeated. The specific activity of this preparation of isoinhibitor A is 2.4 IU/mg, depending on the amount of carbohydrates preserved. Fraction III from step 4 is rechromatographed under the same conditions, but repeated rechromatography may be necessary to obtain material with constant amino acid composition representing one of the isoinhibitors A, A_1, or B. The specific activity is higher than that of peak II preparations and varies between 2.8 and 3.1 IU/mg, depending on the amount of carbohydrates present.

II. Purification of Inhibitors from Spermatozoa

Step 2. Acidic Extraction.[18] The frozen sperm pellet (from step 1) is thawed at room temperature and suspended in twice its volume of cold (4°) 2% (v/v) acetic acid adjusted to pH 2.0 with 2 N HCl. The suspension is gently stirred for 5 min and then centrifuged at 1000 g for 5 min at 4°. The extraction is repeated five times. The extracts are combined, centrifuged at 17,000 g for 100 min, and concentrated to one-tenth the original volume by ultrafiltration, using Amicon UM-2 membranes at 4°. Alternatively, lyophilization of the acidic sperm extract is also possible.

Step 3. Separation of Acrosin and Inhibitors.[18] The acidic sperm extract is subjected to gel filtration. A column, 5 × 130 cm, of Sephadex G-75 is equilibrated with 2% (v/v) acetic acid previously adjusted to pH 2.0 by addition of 2 N HCl. The column is suitable for handling 15–20 ml of the concentrated solution (or 170 mg of lyophilized extract) and is developed at a flow rate of 30 ml/hr with 7.5-ml fractions per tube. Acrosin is completely separated (first peak, tubes 60 to 80) from the inhibitory activity contained in a single peak. The inhibitor fractions are combined, concentrated by rotary evaporation, and lyophilized. The inhibitor is eluted from the column in the molecular weight range of about 6000.

Step 4. Gel Filtration at pH 7.8. The preparation obtained in step 3 contains two inhibitor fractions distinguishable by their solubility in neutral and acidic media. One inhibitor fraction can be dissolved by extraction of the preparation from step 3 with 0.4 M NaCl, 0.1 M triethanolamine-HCl, pH 7.6. This extract is passed through a Sephadex G-75 column, 1.6 × 220 cm, equilibrated with the same buffer. The inhibitory activity is eluted in a symmetrical peak; the pooled fractions are concentrated and desalted by ultrafiltration and lyophilized. This material is processed further in step 5. The other inhibitor fraction of the material from step 3 is soluble only in acidic solutions (below pH 4). It is inhibitory-active against trypsin and acrosin; any further characterization has not yet been done.

Step 5. Chromatography on SE-Sephadex.[5] In principle the same conditions for ion-exchange chromatography can be used as described for the seminal plasma inhibitors (see Section I, step 4). A column, 1.5 × 100 cm, of SE-Sephadex C-25 preconditioned as described above (see Section I, step 4) is equilibrated with 0.015 M ammonium acetate, pH 6.8, and developed with an ammonium acetate gradient at a rate of 7 ml/hr. Fractions of 1 ml per tube are collected. After tube 20, a gradient is applied to the column by mixing 0.1 M to 100 ml of 0.015 M buffer; after tube 150, 0.3 M buffer is mixed with 100 ml of 0.1 M buffer. The absorbance at 280 nm and the inhibitory activity against trypsin is monitored. The main inhibitor fractions, tubes 200 to 225, are combined, concentrated, desalted on Bio-Gel P-2 in 0.1 M acetic acid, and lyophilized.

Step 6. Rechromatography on SE-Sephadex. The inhibitor is subjected to rechromatography under slightly modified conditions. A smaller column, 0.9 × 60 cm, of SE-Sephadex C-25 equilibrated with 0.05 M ammonium acetate, pH 6.8, is eluted with an ammonium actate gradient formed from 0.05 M to 0.3 M buffer. Inhibitor of 95% purity is collected from the main fraction, its amino acid composition is presented in Table I.

Properties

Stability. The isoinhibitors A, A_1, or B are stable in acid and neutral solution. There is no loss of activity when the seminal plasma inhibitors are heated to 100° in neutral buffers or to 75° in 3% perchloric acid or if heated for 23 hr at 40° in the dry state. The inhibitors are digested by thermolysin at 25°.[8]

Physical Properties. The boar seminal plasma inhibitor is a glycoprotein.[3,5,8,11] It exists in at least two forms with respect to its protein part. Isoinhibitor A and A_1 contain 65 amino acid residues; isoinhibitor B contains two additional amino acid residues.[5] Minimal molecular weights of 7599 and 7856 are calculated on the basis of the amino acid composition for the protein portions of the inhibitor molecules. Including the carbohydrate residues (Table II), however, molecular weights from 11,000 up to 12,000 are calculated for some of the multiple chromato-

TABLE II
CARBOHYDRATE COMPOSITION OF BOAR SEMINAL PLASMA ISOINHIBITORS[a]

Mono-saccharides[b]	Isoinhibitors				
	A	A[c]	A₁	B	B[c]
Fucose	0.05	0.91	0.74	0.05	0.86[e]
Mannose	1.84	1.67	2.92	1.83	1.29
Galactose	1.98	2.96	5.92	1.55	1.38
Glucose	4.14[d]	0.66	0.96	1.28	1.34
Galactosamine	2.27	2.38	5.16	2.49	2.44
Glucosamine	3.40	2.80	7.96	1.35	2.87
Sialic acid	0.05	0.09[f]	0.41	0.05	0.05
	13–15	13	25	9–10	12–13

[a] Calculated for total MW 11,000.
[b] Determined after HCl-methanolysis, acetylation, and silylation by gas chromatography.
[c] Other peak from SE-Sephadex separation of isoinhibitors.
[d] Sample dialyzed for 60 hr against 25 liters of distilled water prior to carbohydrate determination.
[e] Determined by cysteine-H₂SO₄ method according to Dische and Shettles.
[f] Determined by resorcinol method according to Svennerholm.

graphic forms.[5,11] The molar UV absorbance of isoinhibitor A₁ of MW 12,000 as calculated from the amino acid and carbohydrate composition is found at pH 6.0 to $\epsilon^{1\,cm}_{278\,nm}$ = 11,700.[8] From titrations of bovine trypsin[3] and boar acrosin[5] and the specific inhibitor activity, a molecular weight of approximately 13,000 is calculated for the high-molecular-weight form.

The inhibitors isolated by the procedure described in this article differ strongly in their carbohydrate content. This might explain the differences in the molecular weights of the boar seminal plasma inhibitors reported earlier (about 13,500 in acidic solutions[3,6] and 12,000 in neutral solution[3]). The occurrence of forms with lower molecular weights (7000–5800 and even 1600)[4,6,7,20] might be due to a more extensive removal of carbohydrate residues and perhaps also C- and N-terminal amino acid residues either by the hydrolytic effect of enzymes or acid-induced cleavages.

Purity. Isoinhibitor A, A₁, or B are microheterogeneous in their carbohydrates and therefore are resolved in cellulose acetate electrophoresis at pH 7.8 into several bands having the same amino acid compositions.[5] At least one to two dozen multiple chromatographic forms could be distinguished in the crude preparation from step 3 by ion equilibrium chromatography on SE-Sephadex C-25.[5]

Specificity. Isoinhibitors A, A_1, or B and all unidentified intermediate chromatographic forms have equal effects on the tryptic digestion of casein, gelatin, BAEE, and BAPA. They inhibit acrosin of boar[5-7,9] and man,[10] trypsins of cow[3,4,5-7] and pig,[3,5] and human plasmin.[22] α-Chymotrypsin is inhibited to some extent by excess of any of the inhibitors.[7] Human thrombin, human neutral leukocytic proteinases and porcine pancreatic kallikrein are not inhibited, though seminal plasma contains very small amounts of inhibitors for the latter enzyme.[7]

Kinetic Properties. The inhibition of acrosin, trypsin, and plasmin by inhibitors from seminal plasma and spermatozoa is instantaneous and stoichiometric.[3,5-7,9-11] The inhibition constants derived from titrations (with BAPA as substrate and against NPGB-titrated enzymes) are as follows: for boar acrosin, $K_i = 1.4$ nM; for bovine β-trypsin, $K_i = 3.2$ nM, and for porcine plasmin, $K_i = 0.12$ μM.[5]

When the virgin inhibitor reacts with catalytic amounts of boar acrosin at pH 3.75, the reactive site peptide bond Arg_{19}-Gln_{20} is hydrolyzed, and modified inhibitor is formed.[5] The modified inhibitor retains its activity against acrosin, trypsin, and plasmin. A thermodynamic equilibrium between $76 \mp 2\%$ virgin and $24 \mp 2\%$ modified inhibitor is attained, giving a K_{hyd} of 0.33.[5,8,11]

The inhibition of trypsin is of a temporary mode.[5,11,21] The inhibitor even in 0.5 M excess over trypsin is completely inactivated after 20 hr at pH 7.8 (0.02 M in $CaCl_2$) and 25°.[11] The inhibition of boar acrosin is pseudopermanent; if acrosin is incubated with a 0.7 M excess of inhibitor at pH 7.8 (0.02 M in $CaCl_2$) at 22°, the inhibitor is almost completely inactivated after 20 days.[11,21] At pH 3.75 and 0.5 M excess, the inhibitor is inactivated after 4.5 days at 22°.[21]

Amino Acid Composition. The compositions of the isoinhibitors A, A_1, and B are shown in Table I. All inhibitors have identical compositions except for the lysine and arginine residues. All contain one residue each of methionine and tryptophan.

Carbohydrate Composition. The carbohydrate composition of several chromatographic forms obtained after Sephadex G-75 gel filtration followed by ion equilibrium chromatography on SE-Sephadex C-25 using the system described in step 4 and 5 are given in Table II. The different compositions reveal a microheterogeneity. All inhibitors contain the monosaccharide units galactose, glucose, mannose, galactosamine, and

---

[22] W. Kalckreuth, Dissertion (M.D. thesis), Medical Faculty, University of Munich, 1972.

```
                       Thr Arg Lys Gln Pro Asn Cys  -   -   -   -  Asn Val Tyr Arg Ser His Leu
Boar seminal
Guinea pig seminal     Phe Ala Pro Ser Lys Val Asx Ser Cys  -  Arg  -  Pro Asx Ser Asx  -  Arg Tyr Val Glx
Porcine pancreatic I   Thr Ser Pro Gln Arg Glu Ala Thr Cys  -   -   -  Thr Ser Glu Val Ser  -  Gly
Bovine pancreatic      Asn Ile Leu Gly Arg Glu Ala Lys Cys  -   -   -  Thr Asn Glu Val Asn  -  Gly
```

```
Boar seminal            -   -  Phe Phe Cys Thr Arg Gln Met Asp Pro Ile Cys Gly Thr Asn Gly Lys Ser Tyr Ala Asn Pro Cys
Guinea pig seminal     Arg  -   -  Tyr Met Cys Thr Lys (Glx Leu Asx Pro Val [Cys, Gly, Thr, Asx, Gly, His, Thr] Tyr)
Porcine pancreatic I   His Leu  -   -  Cys Pro Lys Ile Tyr Asn Pro Val Cys Gly Thr Asp Gly Ile Thr Tyr Ser Asn Glu Cys
Bovine pancreatic       -   -  Cys Pro Arg Ile Tyr Asn Pro Val Cys Gly Thr Asp Gly Val Thr Tyr Ser Asn Glu Cys
```

```
Boar seminal           Ile Phe Cys Ser Glu Lys Gly Leu Arg Asn Gln Lys Phe Asp Phe Gly His Trp Gly His Cys Arg Glu Tyr Thr Ser Ala Arg Ser
Guinea pig seminal                                                  (Phe Thr Phe Ser His Tyr Gly Arg)
Porcine pancreatic I   Val Leu Cys Ser Glu Asn Lys Lys Arg Gln Thr Pro Val Leu Ile Gln Lys Ser Gly Pro Cys
Bovine pancreatic      Leu Leu Cys Met Glu Asn Lys Glu Arg Gln Thr Pro Val Leu Ile Gln Lys Ser Gly Pro Cys
```

Fig. 2. Alignment of the homologous amino acid sequences of the boar seminal plasma acrosin inhibitor (top line) [H. Tschesche, S. Kupfer, R. Klauser, E. Fink, and H. Fritz, in "Protides of the Biological Fluids, 23rd Colloquium" (H. Peeters, eds.), p. 225, Pergamon, Oxford and New York, 1976.] and the partial sequence of the guinea pig seminal vesicles trypsin-plasmin inhibitor (a) with the sequences of the porcine (b, c, d) and bovine (e) pancreatic secretory trypsin inhibitors. For further structurally homologous inhibitors, see H. Tschesche, S. Kupfer, R. Klauser, E. Fink, and H. Fritz, in "Protides of the Biological Fluids, 23rd Colloquium" (H. Peeters, ed.), p. 225, Pergamon, Oxford and New York, 1976; this volume [70], [73], and [76]. I. Kato, J. Schrode, K. A. Wilson, and M. Laskowski, J., in "Protides of the Biological Fluids, 23rd Colloquium" (H. Peeters, ed.), Pergamon, Oxford and New York, 1976; L. J. Greene and D. C. Bartelt, in "Protides of the Biological Fluids, 23rd Colloquium" (H. Peeters, ed.), Pergamon, Oxford and New York, 1976.; (a) E. Fink, Dissertation, Ph.D. Thesis, Faculty of Natural Science, University of Munich, 1970; and unpublished results. (b) H. Tschesche, E. Wachter, S. Kupfer, and K. Niedermeier, Hoppe-Seyler's Z. Physiol. Chem. 350, 1247 (1969). (c) H. Tschesche and E. Wachter, Eur. J. Biochem. 16, 187 (1970). (d) D. C. Bartelt and L. J. Greene, J. Biol. Chem. 246, 2218 (1971). (e) L. J. Greene and D. C. Bartelt, J. Biol. Chem. 244, 2646 (1969).

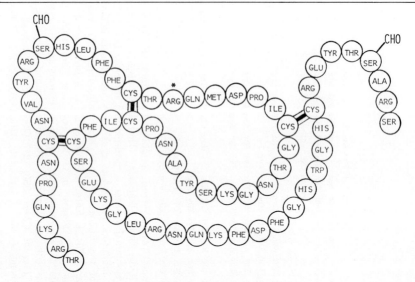

Fig. 3. Schematic diagram of the covalent structure of the boar seminal plasma acrosin inhibitor A₁ [H. Tschesche, S. Kupfer, R. Klauser, E. Fink, and H. Fritz, *in* "Protides of the Biological Fluids, 23rd Colloquium" (H. Peeters, ed.), p. 225, Pergamon, Oxford and New York, 1976; H. Tschesche and S. Kupfer, unpublished results, 1975.

glucosamine; some in addition contain fucose and sialic acid. The sialic acid content is due to N-acetylneuraminic acid and glycolylneuraminic acid.[23] The carbohydrates are attached by O-glycosidic bonds to Ser_{12} and Ser_{63}[8,11] forming two carbohydrate moieties.

Covalent Structure. The amino acid sequence of inhibitor A₁ is given in Fig. 2 and compared to those of other members of this family of proteinase inhibitors (Kazal type) thus far elucidated.[11] The positions of the disulfide bridges[8] and the probable attachment sites of the carbohydrate moieties are shown in Fig. 3.

Biological Aspects[7]

Antibodies against BSTI-I, BSTI-II (see Section I, Step 3a) and the mixture of both inhibitors have been raised in rabbits. BSTI-I-directed antibodies cross-react with BSTI-II inhibitor and BSTI-II-directed antibodies with BSTI-I inhibitor. Using the indirect immunofluorescence method, the inhibitors have been localized in the epithelial cells of the tail of the epididymis, of the ductus deferens, of the seminal vesicles, of the prostate (smaller cell aggregates only), and of the urethra by intense

[23] R. Klauser and H. Tschesche, unpublished results, 1975.

fluorescence. No antibody response is found in the midpiece and head of the epididymis and testes tissue. The seminal plasma inhibitors are mainly produced in the seminal vesicles. No immunological cross-reaction was observed with the inhibitors present in the epididymis and testes.

The indirect immunofluorescence antibody technique revealed also that ejaculated spermatozoa carry seminal inhibitors, probably at the outer acrosomal region, that the inhibitors are removed during residence (capacitation) in the uterine fluid (6 hr), and that uterine spermatozoa can be recharged with seminal inhibitors.

The inhibition of *in vivo* fertilization by treatment of capacitated spermatozoa by boar seminal inhibitors has been reported.[4,6]

Acknowledgments

This work was supported by Sonderforschungsbereich 51, Munich, and WHO Geneva (Grants Nos. 2837 and 74,053). We wish to thank S. Kupfer and M. Meier for skillful technical assistance as far as unpublished work is concerned.

[75] Acid-Stable Proteinase Inhibitors from Human Seminal Plasma

By Hans Schiessler, Edwin Fink, and Hans Fritz

Antitryptic activity in human seminal plasma was first detected by Rasmussen and Albrechtson.[1] Haendle *et al.*[2] described the occurrence of acid-stable trypsin inhibitors in testes, epididymis, and seminal vesicles as well as in the seminal plasma of many mammals, including man.

Fink *et al.*[3] and Suominen and Niemi[4] showed that the antitryptic activity in human seminal plasma is due to two different trypsin inhibitors, the human seminal plasma inhibitors (HUSI) I and II. HUSI-I, a trypsin–chymotrypsin inhibitor, and HUSI-II, a trypsin-acrosin inhibitor, were further characterized in our laboratory.[5,6] α_1-Antitrypsin, an

[1] J. Rasmussen and O. K. Albrechtson, Fertil. *Steril.* **11**, 264 (1960).

[2] H. Haendle, H. Fritz, I. Trautschold, and E. Werle, *Hoppe-Seyler's Z. Physiol. Chem.* **343**, 185 (1965).

[3] E. Fink, E. Jaumann, H. Fritz, H. Ingrisch, and E. Werle, *Hoppe-Seyler's Z. Physiol. Chem.* **352**, 1591 (1971).

[4] J. Suominen and M. Niemi, *J. Reprod. Fertil.* **29**, 163 (1972).

[5] H. Schiessler, M. Arnhold, and H. Fritz, *Proteinase Inhibitors, Proc. Int. Res, Conf., 2nd, (Bayer Symp. V)*, Grosse Ledder, 1973 p. 147. Springer-Verlag, Berlin and New York, 1974.

[6] H. Schiessler, M. Arnhold, and H. Tschesche, *Abstr. Annu. Meet. Soc. Study Reprod.*, 1974, p. 79.

acid-unstable glycoprotein, is also present in human seminal plasma.[7]

Trypsin inhibitors isolated from human spermatozoa[5,6,8–10] show very similar characteristics to those of HUSI-I and HUSI-II.

Assay Methods

Trypsin Inhibition

Principle. The inhibition of the tryptic hydrolysis of N^α-benzoyl-DL-arginine p-nitroanilide (DL-BAPA) is measured by following the change in absorbance at 405 nm. This standard assay for the determination of the inhibitory activity has been described by Fritz et al.[11] and also by Kassell[12] in this series.

However, N^α-benzoyl-L-arginine p-nitroanilide (L-BAPA) is a better substrate than DL-BAPA in this test system,[13] since the D-isomer is a competitive inhibitor of trypsin. L-BAPA has the further advantage of greater solubility and a higher turnover number with trypsin.

Chymotrypsin Inhibition[11]

Principle. The inhibition of the chymotrypsin-catalyzed release of p-nitroaniline from N^α-succinyl-L-phenylalanine p-nitroanilide (SUPHEPA) is measured by following the change in absorbance at 405 nm.

Reagents

Substrate: 250 mg of SUPHEPA dissolved in 50 ml of buffer
Buffer: 0.2 M triethanolamine (TRA), pH 7.8, containing 0.02 M CaCl$_2$
Enzyme: crystalline N^α-p-toluenesulfonyl-L-lysyl chloromethyl ketone (TLCK)-treated chymotrypsin is dissolved in 0.0025 N HCl at a concentration of 2 mg/10 ml. The solution is stored at 4°.

[7] G. F. B. Schumacher, *Proteinase Inhibitors, Proc. Int. Res. Conf., 1st,* Munich, 1970 p. 245. de Gruyter, Berlin, 1971.
[8] H. Fritz, B. Förg-Brey, E. Fink, M. Meier, H. Schiessler, and C. Schirren, *Hoppe-Seyler's Z. Physiol. Chem.* 353, 1943 (1972).
[9] L. J. D. Zaneveld, B. M. Dragoje, and G. F. B. Schumacher, *Science* 177, 702 (1972).
[10] F. N. Syner and R. Kuras, *Abstr. Annu. Meet. Soc. Study Reprod.,* 1974, p. 108.
[11] H. Fritz, I. Trautschold, and E. Werle, in "Methoden der Enzymatischen Analyse" (H. U. Bergmeyer, ed.), 2nd ed., Vol. I, p. 1021. Verlag Chemie, Weinheim, 1970.
[12] B. Kassell, this series Vol. 19, p. 845.
[13] H. Naketa and S. Ishi, *J. Biochem.* 72, 281 (1972).

Procedure. The mixture of 0.1 ml of the enzyme solution (corresponding to 6.7 μg of chymotrypsin per milliliter), the inhibitor solution and the buffer (final volume 2.0 ml) is preincubated for 5 min. Preincubation and the kinetic test are performed at a temperature of 25°. The enzymic reaction is started by addition of 1.0 ml of the substrate solution to the preincubation mixture. The increase in absorption at 405 nm is followed for 20 min. The chymotrypsin activity of the inhibitor-free control serves as a reference value for the calculation of the inhibitory activity.

For the determination of very low inhibitor concentrations or kinetic data, the following substrates are recommended: N^α-benzoyl-L-arginine ethyl ester (BAEE) and N^α-acetyl-L-tyrosine ethyl ester (ATEE). The conditions to be used for the trypsin-catalyzed hydrolysis of BAEE and the chymotrypsin-catalyzed hydrolysis of ATEE are given by Schwert and Takenaka.[14] The assay of trypsin or chymotrypsin inhibition with these substrates is performed in identical test systems. The trypsin-inhibition test with BAEE is extensively described by Burck[15] in this series.

Acrosin Inhibition

Principle. The substrate specificities of acrosin and trypsin are very similar.[16] Spectrophotometric measurement of the rate of acrosin-catalyzed hydrolysis of BAEE, DL-BAPA, L-BAPA, or N^α-benzoyl-L-lysine *p*-nitroanilide (L-BLNA) is suitable for quantitative determination of acrosin.[17] Therefore, the acrosin-inhibition assay may be performed using test systems employed for trypsin inhibition.[18-21] However, purified acrosin is strongly adsorbed and inactivated on glass surfaces; the use of plastic cuvettes is recommended to minimize this effect.[22]

[14] G. W. Schwert and Y. Takenaka, *Biochim. Biophys. Acta* **16**, 570 (1955).
[15] P. J. Burck, this series Vol. 19, p. 907.
[16] H. Schiessler, W.-D. Schleuning, and H. Fritz, *Hoppe-Seyler's Z. Physiol. Chem.* **356**, 1931 (1975).
[17] W.-D. Schleuning and H. Fritz, this volume [27].
[18] H. Fritz, B. Förg-Brey, E. Fink, H. Schiessler, E. Jaumann, and M. Arnhold, *Hoppe-Seyler's Z. Physiol. Chem.* **353**, 1007 (1972).
[19] H. Schiessler, H. Fritz, M. Arnhold, E. Fink, and H. Tschesche, *Hoppe-Seyler's Z. Physiol. Chem.* **353**, 1638 (1972).
[20] H. Fritz, B. Förg-Brey, M. Meier, M. Arnhold, and H. Tschesche, *Hoppe-Seyler's Z. Physiol. Chem.* **353**, 1950 (1972).
[21] L. J. D. Zaneveld, G. F. B. Schumacher, H. Fritz, E. Fink, and E. Jaumann, *J. Reprod. Fertil.* **32**, (1973).
[22] E. Fink, H. Schiessler, M. Arnhold, and H. Fritz, *Hoppe-Seyler's Z. Physiol. Chem.* **353**, 1633 (1972).

Definition of Units and Specific Activity. One enzyme unit corresponds to the hydrolysis of 1 μmole of substrate per minute under the given conditions. The change in absorbance due to activity $A_{405}^{1cm}/\text{min} = 3.32$ for the substrates BAPA, BLNA, and SUPHEPA and $A_{253}^{1cm}/\text{min} = 0.385$ for BAEE. One inhibitor unit causes the reduction of the enzyme activity by one enzyme unit. The specific activity is expressed in inhibitor units per milligram of protein.

Inhibition of Leukocytic Proteinases

Principle. Azocasein is digested by neutral proteinases from leukocytes to products that are soluble in trichloroacetic acid (TCA). The activity of the proteinases is measured by following the increase in absorbance at 366 nm in the TCA solution (due to the split products). Inhibition is calculated from the difference between the absorbance of the inhibitor-containing samples and the control.

Reagents

Substrate: 2 g of azocasein dissolved in 100 ml of warm buffer
Buffer: 0.1 M Na$_2$HPO$_4$, pH 7.65
Enzyme: Proteinases from human leukocytes. The enzyme preparation is extracted from leukocytes of maxillopharyngeal abscesses according to Hochstrasser *et al.*[23] Of the enzyme-containing powder, 75 mg are dissolved in 5 ml of buffer. Insoluble material is removed by centrifugation at 4°. In the test system used, 1.12–4.5 mg of the enzyme preparation caused an increase in absorbance A_{366}^{1cm} from 0.28×10^{-3} to 1.2×10^{-3} per minute.
Inhibitor solution: About 0.1 mg of inhibitor is dissolved in 1.0 ml of buffer.
TCA: 5 g of TCA are dissolved in 5 ml of distilled water.

Procedure. Constant amounts of the leukocytic enzyme preparation, 4.5 mg in 0.3 ml buffer, are incubated at 25° for 1 hr with increasing amounts of inhibitor in the phosphate buffer, pH 7.65, in a final volume of 1 ml. The enzyme reaction is started by addition of 2.0 ml substrate solution to each of the preincubation mixtures. After 3 hr reaction time at 37°, proteolysis is stopped by addition of 3.0 ml of TCA solution. The absorbance of the centrifuged and filtered supernatant is measured at 366 nm.

The values obtained were corrected by subtraction of the absorbance of a reference blank containing all reagents but incubated for 0 min.

[23] K. Hochstrasser, R. Reichert, S. Schwartz, and E. Werle, *Hoppe-Seyler's Z. Physiol. Chem.* **353**, 221 (1972).

TABLE I
SUMMARY OF THE PURIFICATION PROCEDURE

Step and fraction	Total inhibitory activity[a] (IU)	Yield (%)	Specific activity[a] (IU/mg)
Sperm plasma, 1 liter	120	(100)	0.00009
1 SP-Sephadex C-50	108	90	0.017
2 Trypsin cellulose	92	78	1.2
3 Sephadex G-75	82.5	69	
HUSI[c]-I	38.3	32	1.37
HUSI-II	44.2	37	1.3
4 SP-Sephadex C-25			
HUSI-I	34.5	90[b]	1.6
HUSI-II	40	90[b]	2.22

[a] Trypsin inhibition, substrate: N^α-benzoyl-DL-arginine p-nitroanilide.
[b] Referred to the corresponding preparation from step 3.
[c] HUSI, human seminal plasma inhibitor.

Purification Procedure

The procedure is summarized in Table I.

Starting Material

A trypsin inhibitor activity of 150–330 mIU/ml is detectable in freshly ejaculated human seminal plasma.[24] This corresponds to the inhibition of about 0.1–0.3 mg of bovine trypsin. We received most of our material from andrology clinics. These ejaculates had been frozen and stored up to one year. The inhibitor concentration in several of these batches was up to 50% less than in freshly collected ejaculates.[25]

Degradation of the inhibitors by proteinases of seminal plasma and spermatozoa may be responsible of this reduction. The average inhibitor concentrations of abnormal ejaculates (oligospermia, azoospermia, aspermia, hypocinetic spermatozoa) and normal ejaculates are not significantly different. The seminal plasma of such ejaculates is also suitable for inhibitor isolation.

Storage Conditions

In order to prevent extensive degradation, it is recommended that the spermatozoa be separated from the seminal plasma by centrifugation

[24] H. Haendle, Dissertation, Medical Faculty, University of Munich, 1969.
[25] C. Buck, *Andrologie* **5**, 23 (1973).

(600 *g*, 20 min, 4°) immediately after liquefaction of the ejaculates. Spermatozoa-free plasma should be frozen preferably below —70°. After thawing, the inhibitors should be separated from the residual seminal fluid without delay.

Step 1. Chromatography on SP-Sephadex C-50

This step should be performed rapidly and under constant cooling (4°) after thawing the frozen ejaculates or seminal plasma.

Reagents

Ammonium acetate buffer, 0.05 *M*, pH 5.4; 15 liters containing 0.02% (w/v) NaN$_3$

SP-Sephadex C-50, 30 g; i.e., 1 liter of resin, swollen and equilibrated with buffer

NaCl, 292 g

Procedure. One liter of a frozen ejaculate pool is thawed (4°) and centrifuged (600 *g*) for 2 hr at 4°. The supernatant is removed and tested for inhibitory activity. The sediment serves for the isolation of the inhibitors from spermatozoa and of acrosin.

The high-molecular-weight, acid-labile proteins have to be denatured by acidification of the seminal plasma before the concentration of the acid-stable inhibitors can be estimated.

Seminal plasma, 0.5 ml, is diluted with the same volume of perchloric acid (6% w/v). Denatured proteins are removed by centrifugation (6000 *g*, 15 min, 4°). The clear supernatant is neutralized by addition of solid potassium hydrogen carbonate. If solutions of high ionic strength are applied to the preincubation mixture (enzyme plus inhibitor plus buffer), inhibitor-free controls of the same ionic strength should be employed in order to avoid misleading results.

The seminal plasma is diluted with the same volume of buffer. The mixture is adjusted to pH 5.4 by addition of acetic acid. Thirty grams of SP-Sephadex C-50, equilibrated with the buffer solution, are added and the suspension is stirred slowly for 2 hr at 4°. After centrifugation (600 *g*, 15 min, 4°) the resin-free supernatant is decanted and the adsorbent is washed three times with 1 liter of buffer solution. The adsorbent is then packed into a 7 × 50 cm column. The column is washed with the buffer solution at a rate of 140 ml/hr until the transmission of the eluate has dropped to that of the buffer solution. Then, the column is developed with a linear sodium ion gradient formed from 2 liters of starting buffer (0.05 M ammonium acetate buffer pH 5.4) and 2 liters of the same buffer containing 1 M sodium chloride.

The inhibitors are eluted in a single peak between 0.15 M and 0.5 M sodium ion concentration. The combined inhibitor fractions are neutralized with 2 N NaOH and concentrated by evaporation using a rotary evaporator (water bath temperature, 30°). Desalting is performed by passing the concentrated inhibitor solution (max. 25 ml) through a Sephadex G-25 column (120 × 3.6 cm) equilibrated and developed with 2% (v/v) acetic acid. The salt-free inhibitor fractions are combined and lyophilized. Specific activities between 19 and 25 mIU/mg (trypsin inhibition; substrate, DL-BAPA) were estimated for the material thus obtained.

Step 2. *Affinity Chromatography*[26,27]

Reagents

A. "Trypsin, polymer bound on CM-Cellulose" purchased from E. Merck, Darmstadt (article No. 24,582, capacity 7–10 U/mg). One gram of the trypsin cellulose is equilibrated with buffer B. Fine particles are removed. The trypsin cellulose is poured into a cooled (0°) column (1.2 × 10 cm) and washed with buffer B until trypsin and other proteins are no longer detectable in the eluate. Used trypsin cellulose may be stored in acidic buffer (C) at 4°.

B. Washing buffer: 0.1 M triethanolamine-HCl, 0.4 M NaCl, pH 7.8

C. Acidic buffer: 0.4 M KCl-HCl, pH 1.8

Procedure. The dry material from step 1 is dissolved in 50 ml of buffer B. The inhibitor solution is applied to the column, running at a flow rate of 16 ml/hr. The effluent is collected until excess inhibitor appears in the eluate. Then the column is washed with buffer B until the effluent is entirely protein free. Dissociation of the inhibitors from the trypsin–inhibitor complex, and thus their elution from the column, is achieved by applying buffer C. Elution is stopped when the effluent is free of inhibitor. After this procedure the trypsin-cellulose column has to be reequilibrated with buffer B.

Batchwise operation is also possible. The trypsin–cellulose is carefully suspended in the inhibitor solution for 1 hr (4°). The adsorbent is

[26] B. Kassell, this series Vol. 19, p. 846.
[27] H. Fritz, B. Brey, M. Müller, and M. Gebhardt, *Proteinase Inhibitors, Proc. Int. Res. Conf. 1st*, Munich, 1970 p. 28. de Gruyter, Berlin, 1971.

then washed 6 times each with 50 ml of buffer B by repeated suspension (for 5 min) and centrifugation. Elution of the inhibitors is achieved by repeated (6 times) suspension of the adsorbent in 50 ml of buffer C each time. The degree of purification and yields are comparable for both methods.

The acidic inhibitor fractions are combined and concentrated by ultrafiltration at 4° to a final volume of 10 ml in an Amicon cell equipped with an UM-05 membrane.

Step 3. Separation of HUSI-I and HUSI-II by Gel Filtration on Sephadex G-75

Separation of the antitryptic activity into two fractions, a trypsin–chymotrypsin inhibitor, HUSI-I, and a trypsin–acrosin inhibitor, HUSI-II, is achieved by gel filtration on Sephadex G-75.

Concentrated inhibitor solution from step 2, 5 ml, is applied to a water-cooled (10°) Sephadex G-75 column (128 × 1.8 cm) equilibrated and developed with 2% (v/v) acetic acid at a flow rate of 16 ml/hr. Fractions of 5 ml are collected. After elution of impurities at the void volume, the trypsin-inhibiting activity appears separated into two fractions. The first fraction (v/v_o = 1.84) corresponding to HUSI-I is followed by the HUSI-II fraction (v/v_o = 2.23). The inhibitory activity should be estimated in each tube between 40 and 80. The contents of the tubes containing HUSI-I and HUSI-II are pooled separately and lyophilized. Tubes in the overlap region (Nos. 57–61) containing both inhibitors have to be rechromatographed in the same system.

If the column is calibrated with reference proteins, estimation of the molecular weights of the inhibitors is possible.

Step 4. Chromatography on SP-Sephadex C-25

Both HUSI-I and HUSI-II from step 3 are submitted to gradient elution chromatography using identical conditions.

Reagents

SP-Sephadex C-25 (Pharmacia Fine Chemicals)
Buffer I: 0.05 M Na$_2$HPO$_4$, pH 6.12
Buffer II: 0.05 M Na$_2$HPO$_4$, pH 6.12, 0.4 M NaCl (Na$^+$ = 0.5 M)

Procedure. HUSI-I or HUSI-II, 50 mg, is dissolved in 2–3 ml of buffer I. The clear solution is applied to the SP-Sephadex C-25 column (120 × 1 cm) equilibrated with buffer I. The column is developed with a linear sodium ion gradient formed from 0.8 liter each of buffer I and buffer II at a flow rate of 7 ml/hr; 3 ml fractions are collected.

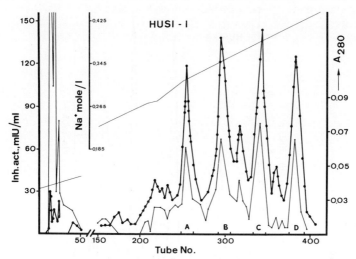

Fig. 1. Fractionation of human seminal plasma inhibitor (HUSI) I by gradient elution chromatography on SP-Sephadex C-25 HUSI-I separated from HUSI-II by gel filtration (step 3) was employed. The column (120 × 1 cm) was equilibrated with sodium phosphate buffer (Na$^+$ = 0.1 M), pH 6.12, and developed with a linear sodium ion gradient formed from 0.8 liter each of starting buffer and 0.4 M NaCl. Flow rate: 7 ml/hr, 3 ml/fraction. See text, step 4.

Chromatography of the inhibitors HUSI-I and HUSI-II results in the separation of each inhibitor fraction into 4 multiple forms (cf. Figs. 1 and 2). The sodium ion concentration, by which the varous inhibitor fractions appear in the eluate, as well as the distribution of the inhibitory activity among the fractions, are given in Table II. Tubes containing the same inhibitor fraction are combined, concentrated by evaporation, and desalted by gel filtration on a Sephadex G-25 column (70 × 2 cm) equilibrated and developed with 2% (v/v) acetic acid. After lyophilization of the salt-free inhibitor fractions, each individual form may be subjected to equilibrium chromatography.

Step 5. Final Purification by Equilibrium Chromatography

The final purification of each individual fraction may be carried out by equilibrium chromatography on a SP-Sephadex C-25 column (120 × 1 cm; flow rate 7 ml/hr; 3 ml/fraction) equilibrated and developed with a phosphate buffer of constant sodium ion concentration. The exact concentration used corresponds to that at which the applied inhibitor is eluted during the fractionation in step 4 (cf. Table II). Desalting is performed as in step 4.

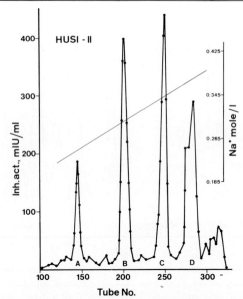

Fɪɢ. 2. Fractionation of human seminal plasma inhibitor (HUSI) II by gradient elution chromatography on SP-Sephadex C-25. HUSI-II obtained by gel filtration on Sephadex G-75 (step 3) was employed. The conditions of this chromatographic step are identical to those mentioned in Fig. 1. See text, step 4.

TABLE II

Sᴇᴘᴀʀᴀᴛɪᴏɴ ᴏғ ᴛʜᴇ Mᴜʟᴛɪᴘʟᴇ Fᴏʀᴍѕ ᴏғ Hᴜᴍᴀɴ Sᴇᴍɪɴᴀʟ Pʟᴀѕᴍᴀ Iɴʜɪʙɪᴛᴏʀ (HUSI) I ᴀɴᴅ II ғʀᴏᴍ Sᴛᴇᴘ 3 ʙʏ Gʀᴀᴅɪᴇɴᴛ Eʟᴜᴛɪᴏɴ Cʜʀᴏᴍᴀᴛᴏɢʀᴀᴘʜʏ ᴏɴ SP-Sᴇᴘʜᴀᴅᴇᴋ C-25[a]

	HUSI-I fraction			
	A	B	C	D
Na⁺ (moles/liter)	0.305	0.34	0.38	0.41
Inhibitory activity (%)[b]	15	30	24	31
	HUSI-II fraction			
	A	B	C	D
Na⁺ (moles/liter)	0.21	0.28	0.335	0.365
Inhibitory activity (%)[b]	9	26	24	41

[a] The multiple forms A to D of both inhibitors HUSI-I and HUSI-II appear in the eluate at sodium ion concentrations shown in the table. Rechromatography of each fraction is performed using these same sodium ion concentrations.

[b] Related to the total inhibitory activity eluted.

Properties

Stability. Both inhibitors are stable in solutions of pH 2–9 at room temperature. Some lyophilized batches of HUSI-II (step 3) showed a reduction of the specific activity after prolonged storage at 4°.

Molecular Weight. Molecular weight values between 11,000 and 14,500 were found for HUSI-I and 4000 to 6500 for HUSI-II by gel filtration experiments.[3-5] From the amino acid compositions shown in Table III, a molecular weight near 10,500 is calculated for HUSI-I fraction C and 6217 for HUSI-II fraction D.

Amino Acid Composition. Table III shows the amino acid compositions of HUSI-I fraction D and C and HUSI-II fraction D. Compared

TABLE III

AMINO ACID COMPOSITION OF ACID-STABLE PROTEINASE INHIBITORS FROM HUMAN SEMINAL PLASMA[a]

	Amino acid residues/molecule		
	HUSI[b]-I		HUSI-II
Amino acid	D	C	D
Aspartic acid	8	8	6
Threonine	4	4	3
Serine	6	6	4
Glutamic acid	7	8	3
Proline	12	12	5
Glycine	9	9	5
Alanine	3	3	1
Half-cystine	12	12	6
Valine	5	5	1
Methionine	3	3	1
Isoleucine	1	1	3
Leucine	4	5	2
Tyrosine	2	2	3
Phenylalanine	2	2	1
Lysine	12	12	3
Histidine	0	1	2
Arginine	4	4	5
Tryptophan	0	0	0
	94	97	54
Molecular weight	10,130	10,510	6,217

[a] The molecular weights given were calculated from the amino acid compositions. Amino sugars were not detectable.

[b] HUSI, human seminal plasma inhibitor.

TABLE IV

INHIBITION SPECIFICITY OF HUMAN SEMINAL PLASMA INHIBITOR (HUSI) I AND II

Proteinases[a]	Substrate[b]	HUSI-I	HUSI-II
Trypsin,			
Human, cationic[c]	TAME	+	+
Human, anionic[c]	TAME	+	+
Bovine[3–5]	BAEE, DL-BAPA	+	+ +
Acrosin,			
Human[5,10,20,21]	BAEE, DL-BAPA	−	+ +
Boar[5,18]	DL-BAPA	−	+ +
Chymotrypsin, bovine[3–5]	ATEE, SUPHEPA	+ +	−
Leukocytic proteinases,	Azocasein	+ +	−
human, neutral[5]			
Elastase,			
Human, granulocytic[d]	Elastine	+	−
Porcine, pancreatic	Elastine	−	−
Chymotrypsin-like proteinase	ATEE, azocasein	−	−
from human sperm plasma[e]			

[a] Superscript numbers refer to text footnotes.

[b] TAME, N^α-toluenesulfonyl-L arginine methyl ester; BAEE, N^α-benzoyl-L-arginine ethyl ester; DL-BAPA, N^α-benzoyl-DL-arginine p-nitroanilide; ATEE, N^α-acetyl-L-tyrosine ethyl ester; SUPHEPA, N^α-succinyl-L-phenylalanine p-nitroanilide.

[c] C. Figarella, G. A. Negri, and O. Guy, Proteinase Inhibitors, Proc. Int. Res. Conf., 2nd, (Bayer Symp. V), Grosse Ledder, 1973 p. 213. Springer-Verlag, Berlin and New York, 1974.

[d] K. Ohlsson, personal communications.

[e] H. Fritz, M. Arnhold, B. Förg-Brey, L. J. D. Zaneveld, and G. F. B. Schumacher, Hoppe-Seyler's Z. Physiol. Chem. 353, 1651 (1972).

to HUSI-I fraction D, HUSI-I fraction C contains three additional amino acid residues, whereas the compositions of fraction A and B are identical with that of fraction D. Both inhibitors contain neither tryptophan nor amino sugar residues.

The amino acid compositions of inhibitors isolated from freshly collected ejaculates may show differences from the compositions given in Table III. Peptidases present in the ejaculate may cause loss of amino acids or peptides by partial proteolytic degradation.

N-Terminal Residues. The only N-terminal residue for HUSI-I fraction D found by substractive Edman degradation[28] was tyrosine followed by leucine.

Inhibition Specificity (cf. Table IV). Neither of the two inhibitors

[28] M. Percy and B. Buchwald, Anal. Biochem. 45, 60 (1972).

has any effect on the enzyme activity of human plasmin (Deutsche Kabi GmbH) and porcine plasmin (Novo Industri A/S), porcine pancreatic kallikrein,[29] subtilisin (Serva AG), *Aspergillus oryzae* proteinases (Röhm GmbH), and Pronase (Merck AG, Darmstadt).

Reactive Site. Trypsin inhibition activity of HUSI-I fraction D was not diminished by treatment with excess maleic anhydride and 2,3-butanedione reagent which are used for the identification of lysine or arginine residues in biologically active regions of proteins.[30] This indicates that neither a lysine nor an arginine residue is present in the reactive site of this inhibitor.

Kinetic Properties. The following dissociation constants (K_i) for the respective complexes were calculated according to Green and Work.[31]

$$\text{HUSI-I}_D: \text{complex with} \begin{cases} \text{bovine } \alpha\text{-chymotrysin} & 5 \times 10^{-10} \ M \\ \text{bovine trypsin} & 4 \times 10^{-9} \ M \end{cases}$$

$$\text{HUSI-II: complex with} \begin{cases} \text{bovine trypsin} & 1 \times 10^{-9} \ M \\ \text{human acrosin} & 9 \times 10^{-10} \ M \end{cases}$$

Immunology. Antibodies produced by immunization of rabbits with HUSI-I (step 4, fractions A–D) cross-react with the trypsin–chymotrypsin inhibitor obtained from human spermatozoa and with the acid-stable trypsin–chymotrypsin inhibitor from cervical mucus,[32,33] but not with HUSI-II fractions A–D.

Possible Biological Function.[32] We assume, that the human seminal plasma inhibitor HUSI-I belongs to a special class of trypsin–chymotrypsin inhibitors including the inhibitors from respiratory tract secretions[23,34] and the inhibitor from cervical mucus.[33] These inhibitors form strong complexes with neutral proteinases from leukocytes and may protect mucous membranes against the hydrolytic action of these enzymes.

On the basis of the inhibition properties it may be concluded that HUSI-II, the trypsin–acrosin inhibitor in human seminal plasma, is the natural antagonist of the sperm acrosin.

---

[29] H. Fritz, I. Eckert, and E. Werle, *Hoppe-Seyler's Z. Physiol. Chem.* **348**, 1120 (1967).

[30] H. Fritz, E. Fink, M. Gebhardt, K. Hochstrasser, and E. Werle, *Hoppe-Seyler's Z. Physiol. Chem.* **350**, 933 (1969).

[31] N. M. Green and E. Work, *Biochem. J.* **54**, 347 (1953).

[32] H. Fritz, H. Schiessler, W.-B. Schill, H. Tschesche, N. Heimburger, and O. Wallner *in* "Proteases and Biological Control" (E. Reich, D. B. Rifkin, E. Shaw, eds.), p. 737, Cold Spring Harbor Laboratory, 1975.

[33] O. Wallner and H. Fritz, *Hoppe-Seyler's Z. Physiol. Chem.* **355**, 709 (1974).

[34] K. Hochstrasser, R. Reichert, M. Matzner, and E. Werle, *Z. Klin. Chem. Klin. Biochem.* **10**, 1 (1972).

[76] Proteinase (Elastase) Inhibitors from Dog Submandibular Glands

By Hans Fritz and Karl Hochstrasser

The highest concentration of a proteinase inhibitor found so far in animal tissues occurs in the submandibular glands of the dog.[1] Depending on the secretory state of the gland, 1 g of fresh tissue contains up to 5 mg of inhibitor. The inhibitor is excreted into the saliva.[2] Outstanding properties of this inhibitory protein are its double-headed nature[3]—the two reactive sites are independent from each other—the strong inhibition of pancreatic elastase and subtilisin, besides other trypsin- and chymotrypsin-like enzymes,[3] and its structural homology to the pancreatic secretory trypsin inhibitors (cf. below).[4]

Assay Methods

The following enzymes were employed: bovine trypsin and α-chymotrypsin; porcine plasmin (2.68 Novo units/mg) and subtilisin (crystalline bacterial proteinase, 22.0 Anson trypsin units/g) from Novo Industri A/S; Pronase E (70,000 PUK/g from *Streptomyces griseus*) and porcine pancreatic elastase (cryst., suspension, 15 E/mg) from E. Merck, Darmstadt; alkaline *Aspergillus oryzae* protease [3500 PU (pH)/mg protein] from Röhm & Haas GmbH, Darmstadt.

Known methods were used to estimate the inhibition of trypsin (substrate: N^α-benzoyl-DL-arginine *p*-nitroanilide, BAPA),[5,6] chymotrypsin (substrate: N^α-succinyl-L-phenylalanine *p*-nitroanilide, SUPHEPA),[5,7] and plasmin (substrate: BAPA).[3,5]

[1] E. Werle, I. Trautschold, H. Haendle, and H. Fritz, *Ann. N. Y. Acad. Sci.* **146**, 464 (1968).

[2] H. Haendle, M.D. Thesis (Dissertation), Medical Faculty of the University of Munich, Munich, 1969.

[3] H. Fritz, E. Jaumann, R. Meister, P. Pasqay, K. Hochstrasser, and E. Fink, *Proteinase Inhibitors, Proc. Int. Res. Conf., 1st,* Munich, 1970, p. 257.

[4] K. Hochstrasser and H. Fritz. *Hoppe-Seyler's Z. Physiol. Chem.,* **356**, 1659 and 1859 (1975). K. Hochstrasser, G. Bretzel, E. Wachter, and S. Heindl, *Hoppe-Seyler's Z. Physiol. Chem.* **356**, 1865 (1975).

[5] H. Fritz, I. Trautschold, and E. Werle, *in* "Methoden der Enzymatischen Analyse" (H. U. Bergmeyer, ed.), 3rd ed., Vol. I, p. 1105. Verlag Chemie, Weinheim, 1974.

[6] B. Kassell, this series Vol. 19, p. 845.

[7] H. Schiessler, E. Fink, and H. Fritz, this volume [75].

Inhibition of Subtilisin, Aspergillus oryzae Protease, and Pronase. Proteinase activity and enzyme inhibition are measured with azocasein (Pentex-PP from Fluka AG) as substrate. Constant amounts of the enzymes (25 μg of subtilisin, 25 μg of *A. oryzae* protease, 50 μg of pronase) are incubated with increasing amounts of inhibitor (0–15 μg) in 1.0 ml of 0.1 *M* sodium potassium phosphate, pH 7.6, for 5 min at 30°. Subsequently, 2.0 ml of azocasein solution (2%, w/v) in the same buffer are added and the mixture is incubated for 10 min at 30°. The enzymic reaction is stopped by the addition of 3.0 ml of aqueous trichloroacetic acid (5%, w/v). After 30 min at room temperature, the extinction at 366 nm of the supernatant is read against a blank sample. The assay method has been described in detail.[5]

Inhibition of Elastase. The activity of elastase is measured according to Sachar *et al.*[8] with elastin-orcein (E. Merck Darmstadt) as substrate. Elastase inhibition is determined in the following manner: A mixture of 0.15 ml of the elastase suspension (containing about 0.75 mg of elastase in 0.2 *M* Tris-HCl, pH 8.8) and the inhibitor (0–15 μg) solution is filled up to 1.50 ml with 0.2 *M* Tris-HCl, pH 8.8. This incubation mixture is briefly (5 min) shaken and mixed with 20 mg of elastin-orcein. The test sample is vigorously shaken for 30 min at room temperature. The enzymic reaction is stopped by the addition of 2.0 ml of 0.5 *M* phosphate buffer, pH 6.0. After centrifugation, the extinction at 578 nm of the supernatant is read against a blank sample.

One *inhibitor unit* corresponds to the reduction of the enzyme-catalyzed hydrolysis of the substrate by 1 μmole/min. The inhibitor units given throughout refer to the inhibition of bovine trypsin, applying BAPA as substrate.[5] The titration curves with the other enzymes are given by Fritz *et al.*[3]

Purification Procedure

Dog submandibular glands are extracted either with water or with diluted perchloric acid to minimize proteolytic degradation. Working temperature during the extractions (step 1a and 1b) is 0°–4° throughout.

Step 1a. Extraction with Water. Dog submandibular glands (Pel Freeze Biologicals, USA) containing 10–14 IU per gram of tissue, are thawed and homogenized in deionized water, 2 liters per 100 g of tissue. After centrifugation the supernatant is adjusted to pH 6.0–6.5 and mixed with 100 g of CM-cellulose (H⁺ form), the suspension being stirred for 2 hr. The CM-cellulose adsorbent is washed three times, each with 500

[8] L. A. Sachar, K. K. Winter, M. Sicher, and S. Frankel, *Proc. Soc. Exp. Biol. Med.* **90**, 323 (1955).

ml of 0.01 M sodium acetate, pH 5.0. Subsequently, the inhibitors are eluted by suspending the adsorbent in 0.01 M triethanolamine (TRA)-HCl, 5% (w/v) NaCl, pH 8.0, for 10 min, several times. Of the inhibitory activity found in the homogenate, 90–95% is recovered in the combined eluates.

The eluent is desalted by dialysis against deionized water, 4 hr. Concentration either by ultrafiltration (Amicon UM-2 membrane) or evaporation *in vacuo* is followed by gel filtration on Sephadex G-50 columns equilibrated and developed with 5% (v/v) acetic acid. Lyophilization of the salt-free inhibitor fractions yields a white powder with a specific activity of 1.4–1.8 IU/mg. About 18% of the inhibitory activity is lost during these procedures.

Step 1b. Extraction with Perchloric or Formic Acid. The glands are either homogenized directly in 3% (w/v) perchloric acid or the aqueous extract (cf. step 1a) is diluted with the same volume of 6% perchloric acid. The precipitate is removed by centrifugation and extracted three times with 3% perchloric acid. The combined acidic extracts are neutralized by the addition of crystalline K_2CO_3 (or 5 M K_2CO_3 solution). Precipitated $KClO_4$ is removed by filtration. The inhibitor solution thus obtained is diluted with five times its volume of water before CM-cellulose is added. Continue as described in step 1a. If less than 80% of the inhibitory activity is adsorbed to CM-cellulose, the salt concentration of the inhibitor solution has to be reduced either by dialysis or ultrafiltration.

Adsorption of the inhibitors to precipitated proteins may be minimized if the glands are homogenized in 3% (v/v) formic acid. After centrifugation, the supernatant is concentrated by evaporation *in vacuo* or ultrafiltration and fractionated on Sephadex G-75 in 3% (v/v) formic acid. The inhibitor-containing fractions of the eluate are lyophilized. The material is directly applied to step 2a "equilibrium run."

Step 2a. Chromatography on CM-Cellulose; Inhibitors from Aqueous Extracts. EQUILIBRIUM RUN. A 2 × 30 cm CM-cellulose column is equilibrated and developed with 0.01 M TRA-HCl, 0.05 M NaCl, pH 8.5, at a rate of 12 ml/hr, 3.5 ml/tube. The solution of 100–150 mg inhibitor from step 1a in 5 ml of the elution buffer is applied to the column. The inhibitor activity appears in 3 fractions in the eluate (cf. Fig. 1) in the following distribution: 14 + 50 = 64% in fractions A_1 + A_2, 7% in fraction B and 28% in fraction C; 99% in total.

GRADIENT RUN. If the inhibitors from step 1a are separated by gradient elution chromatography as described in step 2b, but with a modified gradient (0.05 M NaCl in the starting buffer and 0.5 M NaCl in the second buffer), the inhibitor activity is eluted in two major fractions, about

50% in fraction I and 30% in fraction II (cf. Fig. 2), the residual activity in intermediate fractions; total, 97%.

The inhibitor fractions are desalted by repeated ultrafiltration (Amicon UM-2 membrane) or gel filtration on Bio-Gel P-2 in 0.01 M acetic acid followed by lyophilization.

Step 3a. Rechromatography. INHIBITORS A_2 AND I FROM STEP 2a. For complete separation of impurities from inhibitor A_2 (cf. Fig. 1), fraction A_2 from the equilibrium run or fraction I from the gradient run are rechromatographed on a 2×65 cm CM-cellulose column equilibrated and developed with 0.01 M TRA-HCl, 0.05 M NaCl, pH 8.5 at a rate of 6 ml/hr, a 2 ml/tube. About 93% of the inhibitory activity applied (230 IU) is thus eluted in a single fraction A_2 (symmetrical absorption and activity peak) shortly after the contamination which contains the residual activity (cf. fraction A_1 in Fig. 1). Part of the inhibitor activity present in fraction A_1 from step 2a can be separated also from the contamination under the same conditions.

Inhibitor fraction A_2 is desalted by repeated ultrafiltration (Amicon UM-2 membrane), preferably by gel filtration on Sephadex G-50 equilibrated and developed with 5% (v/v) acetic acid followed by lyophilization. A specific activity of 2.3 IU/mg is estimated for inhibitor A_2.

INHIBITOR II FROM STEP 2a. Inhibitor II (gradient run) from step 2a is rechromatographed under the conditions of the equilibrium run of step 2a. From 255 IU applied, 8% of the inhibitory activity eluted in frac-

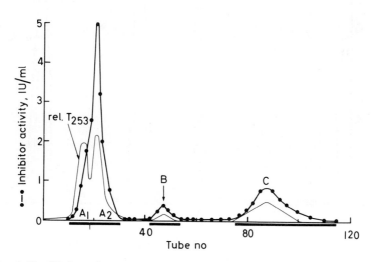

FIG. 1. Equilibrium chromatography of the dog submandibular inhibitors on CM-cellulose (step 2a, equilibrium run). Tubes were combined to yield the fractions as indicated. —, relative transmission at 253 nm (rel. T_{253}).

tion A_2; in fraction B, 27%; and in fraction C, 54%; in total, 95%. The inhibitor fractions are desalted as described above and lyophilized. Specific activities of 2.2 IU/mg are estimated for these inhibitors, too.

Step 2b. Gradient Elution Chromatography; Inhibitors from Acidic Extracts. A 1.6 × 30 cm CM-cellulose column is equilibrated and developed with 0.01 M TRA-HCl, pH 8.0, at a rate of 10.5 ml/hr, 5 ml/tube. The solution of 100–140 mg of inhibitor from step 1b in 2 ml of the starting buffer is applied to the column. As soon as the protein content and the inhibitor activity in the eluate decrease (cf. Fig. 2), the column is developed with a linear gradient formed from 0.5 liters each of the starting buffer and 0.01 M TRA-HCl, 0.3 M NaCl, pH 8.0. The inhibitor activity is eluted in two major fractions—about 35% in fraction I*, 27% in fraction II*, and the residual activity in intermediate fractions; in total, 98%.

Inhibitor fractions I* and II* are desalted by repeated ultrafiltration or gel filtration, cf. step 2a.

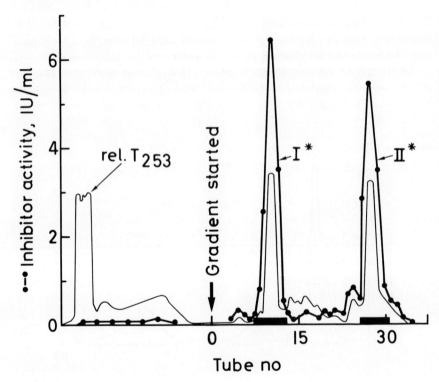

FIG. 2. Gradient chromatography of the dog submanibular inhibitors on CM-cellulose (step 2b or 2a, gradient run). Tubes were combined to yield the fractions as indicated. —, relative transmission at 253 nm. (rel. T_{253}).

Step 3b. Rechromatography. Inhibitors I* and II* from step 2b are rechromatographed under the same conditions except that the slope of the NaCl-gradient is only two-thirds the one used in step 2b. After desalting and lyophilization, the specific activity of both inhibitors I* and II* is estimated as 2.3 IU/mg.

Comments on the Purification Procedure. Removing acid-labile proteins with perchloric acid simplifies the isolation and prevents enzymic degradation or limited cleavage of the inhibitor molecules. However, owing to strong adsorption of the inhibitors to the precipitated proteins, the extraction has to be repeated several times. Adsorption is minimized if formic acid is used for deproteinization (extraction). Partial deamidation of the inhibitors in the acidic solutions or cleavage of acid-labile bonds cannot be excluded. Therefore the purification procedure should be selected in view of the desired application of the inhibitors.

Properties

Amino Acid Compositions and Molecular Weights. The amino acid compositions of the inhibitors obtained from acidic extracts, I* and II*, and from aqueous extracts, A_2 and C, are given in Table I. Although inhibitors I* and A_2 have identical compositions, they may differ in the degree of amidation and the number of internal peptide bond cleavages.

Compared to inhibitor I*, a glutamine or glutamic acid residue is exchanged by a lysine residue in inhibitor II*. Hence, the assumption of the synthesis of one of these inhibitors by a mutated gene seems logical. Inhibitor C is very probably derived from inhibitor II*; glycine, the N-terminal residue, and proline (cf. Fig. 3 below) may be split off by exopeptidases during the first purification steps.

Remarkably, the molecular weights calculated for these inhibitors (cf. Table I) are about twice as high as those of the pancreatic secretory trypsin inhibitors (Kazal type)[9,10] and that of the trypsin–kallikrein inhibitor from bovine organs.[6] For inhibitor A_2 a molecular weight of approximately 12,000 was also estimated by gel filtration, and of about 11,900 from sedimentation studies in the ultracentrifuge.[3] Similar values were reported by other authors.[11]

Inhibition Specificity. The inhibitors from dog submandibular glands (DSI) have an unusually broad inhibition spectrum (Table II). The most striking feature is the strong inhibition of enzymes with chymo-

---

[9] H. Tschesche, *Angew. Chem.* **86**, 21 (1974); *Angew. Chem. Int. Ed.* **13**, 10 (1974).

[10] L. J. Greene, D. E. Roark, and D. C. Bartelt, *Proteinase Inhibitors, Proc. Int. Res. Conf., 2nd (Bayer Symp. V)*, Grosse Ledder, 1973, p. 188.

[11] For references see Fritz *et al.*[3]

TABLE I
AMINO ACID COMPOSITIONS (MOLES/MOLE) AND MOLECULAR WEIGHTS OF DOG
SUBMANDIBULAR INHIBITORS (DSI)[a]

Residue	DSI		
	I* and A₂	II*	C
Asp	13	13	13
Thr	7	7	7
Ser	8	8	8
Glu	9	8	8
Pro	6	6	5
Gly	9	9	8
Ala	6	6	6
Cys₁/₂	12	12	12
Val	4	4	4
Met	3	3	3
Ile	5	5	5
Leu	6	6	6
Tyr	5	5	5
Phe	4	4	4
Lys	10	11	11
His	3	3	3
Arg	5	5	5
Trp	0	0	
	115	115	113
Molecular weight	12,750	12,750	12,595

[a] Experimental values are given by H. Fritz, E. Jaumann, R. Meister, P. Pasqay, K. Hochstrasser, and E. Fink, *Proteinase Inhibitors, Proc. Int. Res. Conf., 1st,* Munich, 1970, p. 257.

trypsin-like substrate specificity: equimolar complexes with chymotrypsin, pancreatic elastase, and subtilisin are formed almost to the equivalence point in the highly diluted solutions of the test systems[3] indicating K_i values near or below 1×10^{-10} mole/liter. Similar strong inhibition of the *Aspergillus oryzae* protease and of a casein-splitting protease present in the Pronase employed is observed.[3] These inhibitors may be used, therefore, to calculate the number of active enzyme molecules present in enzyme preparations on the basis of the titration curves (cf. Table 5 of Fritz *et al.*[3]).

For comparison, the inhibition spectrum of the trypsin–kallikrein inhibitor from bovine organs is also shown in Table II. The specificity of this inhibitor is directed mainly against trypsin-like proteinases.

Reactive Sites. Two independent reactive sites are located on the surface of the dog submandibular inhibitor (DSI) molecules at such a distance that the formation of ternary complexes, e.g., with trypsin *and*

TABLE II

Inhibition Specificity of the Dog Submandibular Inhibitor (DSI) and the Trypsin–Kallikrein Inhibitor from Bovine Organs (TKI)[a]

Enzyme	Substrate[b]	Inhibition by	
		DSI[c]	TKI[d]
Trypsin, bovine	BAPA	++	++
Plasmin, porcine	BAPA	+	++
Kallikrein, pancreatic	BAEE	−	++
Kallikrein, plasma	BAEE	−	+
Chymotrypsin, bovine	SUPHEPA	++	++
Elastase, pancreatic	Elastin	++	−
Subtilisin, Novo	Azocasein	++	−
Aspergillus oryzae protease	Azocasein	++	−
Pronase	Azocasein	++[e]	+[f]
Collagenase	Synthetic[g]	−	−

[a] Strong, ++; weaker, +; −, no inhibition.
[b] BAPA, N^α-benzoyl-DL-arginine p-nitroanilide; BAEE, N^α-benzoyl-L-arginine ethyl ester; SUPHEPA, N^α-succinyl-L-phenylalanine p-nitroanilide.
[c] Cf. H. Fritz, E. Jaumann, R. Meister, P. Pasqay, K. Hochstrasser, and E. Fink, *Proteinase Inhibitors, Proc. Int. Res. Conf., 1st,* Munich, 1970.
[d] See B. Kassell, this series Vol. 19, p. 845.
[e] Probably a chymotrypsin-like component of Pronase.
[f] A trypsinlike component of Pronase (see B. Kassell, this series Vol. 19, p. 845.
[g] p-Phenylazobenzyloxycarbonyl-L-Pro-L-Pro-L-Leu-Gly-L-Pro-D-Arg-OH [E. Wünsch and H.-G. Heidrich, *Hoppe-Seyler's Z. Physiol. Chem.* **333**, 149 (1963).

chymotrypsin or subtilisin, is possible[3]; the inhibitors are "double-headed" with "not overlapping" reactive centers.[12] If the arginine residue at the antitryptic center is modified, only the trypsin-inhibiting property is lost. Exhaustive maleylation does not decrease the inhibitory effect against trypsin and chymotrypsin; this means that the reactive site peptide bonds are not cleaved in the isolated inhibitors.[13]

Inhibitors with reactive centers that do not overlap and similar inhibition specificities are also present in egg white (ovomucoids, ovoinhibitors),[12,14] soybeans,[15] and lima beans.[16]

[12] R. E. Feeney and R. G. Allison (eds.) "Evolutionary Biochemistry of Proteins," p. 199f. Wiley (Interscience), New York, 1969.
[13] D. Kowalski, T. R. Leary, R. E. McKee, R. W. Sealock, D. Wang, and M. Laskowski, Jr., *Proteinase Inhibitors, Proc. Int. Res. Conf., 2nd (Bayer Symp. V),* Grosse Ledder, 1973, p. 311.
[14] R. E. Feeney, *Proteinase Inhibitors, Proc. Int. Res. Conf., 1st,* Munich, 1970, p. 189.
[15] Y. Birk and A. Gertler, *Proteinase Inhibitors, Proc. Int. Res. Conf., 1st,* Munich, 1970, p. 142.
[16] F. C. Stevens, *Proteinase Inhibitors, Proc. Int. Res. Conf., 1st,* Munich, 1970, p. 149.

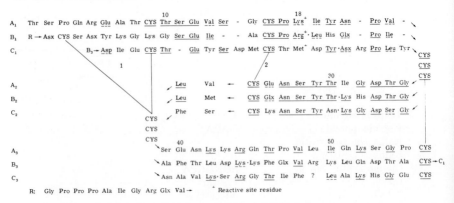

                                    10   18
A₁   Thr Ser Pro Gln Arg Glu Ala Thr CYS Thr Ser Glu Val Ser - Gly CYS Pro Lys⁺ Ile Tyr Asn - Pro Val -
B₁   R → Asx CYS Ser Asx Tyr Lys Gly Lys Gly Ser Glu Ile - - Ala CYS Pro Arg⁺·Leu His Glx - Pro Ile -
C₁   B₃→Asp Ile Glu CYS Thr - Glu Tyr Ser Asp Met CYS Thr Met⁺ Asp Tyr·Asx Arg Pro Leu Tyr
                                                                                                                CYS
                1   2   CYS
                                                                        30   CYS
A₂   Leu   Val ← CYS Glu Asn Ser Tyr Thr Ile Gly Asp Thr Gly
B₂   Leu   Met ← CYS Glx Asn Ser Tyr Thr·Lys His Asp Thr Gly
C₂   Phe   Ser ← CYS Lys Asn Ser Tyr Asn·Lys Gly Asp Ser Gly
                       CYS
                       CYS
                       CYS   40   50
A₃   Ser Glu Asn Lys Lys Arg Gln Thr Pro Val Leu Ile Gln Lys Ser Gly Pro CYS
B₃   Ala Phe Thr Leu Asp Lys·Lys Phe Glx Val Arg Lys Leu Gln Asp Thr Ala CYS→C₁
C₃   Asn Ala Val Lys·Ser Arg Gly Thr Ile Phe ? Leu Ala Lys His Gly Glu CYS

       R: Gly Pro Pro Pro Ala Ile Gly Arg Glx Val→   * Reactive site residue

FIG. 3. Comparison of the structures of the inhibitors from porcine pancreas and dog submandibular glands. Details are discussed in the text; text footnote 20. A, porcine pancreatic secretory trypsin inhibitor; B, antitryptic half of the dog submandibular inhibitor; C, antichymotryptic half of the dog submandibular inhibitor.

On the basis of the formation of the appropriate ternary complexes (trypsin and chymotrypsin *or* subtilisin)[3,14] and the specificity requirements of the proteinases,[17,18] it can be concluded that the chymotrypsin- and the subtilisin-reactive site on the DSI molecule are identical; this same center might be also responsible for the inhibition of elastase and *Aspergillus oryzae* protease. Therefore, the broad inhibition spectra of this inhibitor (DSI) and of the inhibitors from egg white[14] should be mainly caused by the antichymotryptic centers; the inhibition of chymotrypsin by the trypsin–kallikrein inhibitor from bovine organs (cf. Table II) is a special case.

Structural Homology. The pancreatic secretory trypsin inhibitors (Kazal type) belong to a homologous class of proteins which are structurally strongly related to each other.[9,19] The amino acid sequence and the location of the disulfide bridges of the porcine pancreatic trypsin inhibitor[9] are shown in Fig. 3. The structural studies on the dog submandibular inhibitors are still in progress (K. Hochstrasser, E. Wachter, and G. Bretzel); however, on the basis of the results thus far available,[20] the peptides could be clearly aligned with the known structure of the pancreatic trypsin inhibitor (cf. Fig. 3). Obviously, the DSI molecules are

[17] D. Shotton, *Proteinase Inhibitors, Proc. Int. Res. Conf., 1st,* Munich, 1970, p. 47.
[18] C. S. Wright, *J. Mol. Biol.* **67,** 151 (1972).
[19] L. J. Greene and D. C. Bartelt, *in* "Protides of the Biological Fluids—23rd Colloquium," (H. Peeters, ed.), Pergamon, Oxford, 1976.
[20] The positions of some of the residues have still to be confirmed so that some minor changes are possible later on. A dot between two residues indicates that this sequence has thus far not been confirmed by the analysis of overlapping peptides.

composed of two covalently linked halves, the antitryptic part B and the antichymotryptic part C, which are structurally related to each other and to the pancreatic inhibitors. The homology in the amino acid sequences of the peptide chains A_2, B_2, and C_2 and in the location of the disulfide bridges 2 and 3 and the reactive site residues (arginine for trypsin inhibition and perhaps methionine for chymotrypsin and subtilisin inhibition) is especially remarkable. Details of the structural studies and further relations to the structures of other known inhibitors will be discussed elsewhere.[4]

Acknowledgment

This work was supported by Sonderforschungsbereich 51, Munich (B-3, B-8).

[77] Proteinase (Elastase) Inhibitors from the Ciliated Membranes of the Human Respiratory Tract

By KARL HOCHSTRASSER

From the ciliated membranes of the human respiratory tract, two acid-stable proteinase inhibitors are excreted into the mucus.[1-3] The inhibitors inactivate trypsin, chymotrypsin, and Pronase. The outstanding physiologically interesting properties of the inhibitors are the strong inhibition of the neutral proteinases from leukocytes and its double-headed nature. This means that the reactive sites for trypsin and chymotrypsin are independent from each other, similarly to the dog submandibular inhibitor (see this volume [76]).

The inhibitors represent 80% of the total antiproteolytic activity of the secretions (20% is caused by the humoral inhibitor α_1-antitrypsin). During infections the inhibitors are complexed in a high yield by leukocytic proteinases.[4,5] From these findings the physiological function of the inhibitors can be decided. Obviously the inhibitors protect the mucous

[1] K. Hochstrasser, H. Haendle, R. Reichert, and E. Werle, *Hoppe-Seyler's Z. Physiol. Chem.* **352**, 954 (1971).

[2] K. Hochstrasser, R. Reichert, S. Schwarz, and E. Werle, *Hoppe-Seyler's Physiol. Chem.* **353**, 221 (1972).

[3] K. Hochstrasser, R. Reichert, S. Schwarz, and E. Werle, *Hoppe-Seyler's Z. Physiol. Chem.* **354**, 923 (1973).

[4] R. Reichert, K. Hochstrasser, and G. Conradi, *Pneumonologie* **147**, 13 (1972).

[5] K. Hochstrasser, K. Schorn, B. Rasche, and C. Raffelt, *Pneumonologie* **152**, 15 (1975).

membranes from destruction by proteinases liberated from disintegrating leukocytes, as α_1-antitrypsin protects the alveoli of the lung.[6]

Assay Methods

The following enzymes were employed: bovine trypsin and chymotrypsin; Pronase E (70,000 PKU/g from *Streptomyces griseus*), and leukocytic proteinase. Cytosols of human leukocytes were used for preparing the alkaline proteinases. The cytosols were filtered on Sephadex G-100. The fractions hydrolyzing azocasein were used for inhibitor measurement. By disc electrophoresis and negative staining,[7] it was shown that the main proteolytic activity in the preparations is caused by the elastolytic enzymes, as described by Ohlsson[8] and Janoff.[9] By affinity chromatography on insoluble elastin, according to the methods of Ohlsson[8] and Legrand and co-workers,[10] the specific elastolytic activity was isolated from the cytosols. The caseinolytic activity of these preparations was inhibited in a similar ratio as the total caseinolytic activity of the crude enzyme preparation.

Inhibition of Trypsin and Chymotrypsin

Known methods were used to estimate the inhibition of trypsin (substrate: N-α-benzoyl-DL-arginine p-nitroanilide)[11,12] and chymotrypsin (substrate: N-α-succinyl-L-phenylalanine p-nitroanilide).[11,13] One inhibitor milliunit (mIU) corresponds to the reduction of the trypsin-catalyzed hydrolysis of the substrate by 1 μmole/min.[11]

Inhibition of Pronase and Leukocytic Elastase

Proteinase activity and enzyme inhibition are measured with azocasein [the hydrolysis of the specific elastase substrate succinyltrialanine-4-nitroanilide by leukocytic elastases is inhibited very strongly also

[6] C. B. Laurell, *J. Clin. Lab. Invest.* **28**, 1 (1971).
[7] K. Hochstrasser, and K. Schorn, *Hoppe-Seyler's Z. Physiol. Chem.* **355**, 640 (1974).
[8] K. Ohlsson, and I. Olsson, *Eur. J. Biochem.* **42**, 519 (1974).
[9] A. Janoff, *Lab. Invest.* **29**, 458 (1973).
[10] Y. Legrand, G. Pignaus, J. P. Caen, B. Robert, and L. Robert, *Biochem. Biophys. Res. Commun.* **63**, 224 (1975).
[11] H. Fritz, I. Trautschold, and E. Werle, *in* "Methoden der Enzymatischen Analyse" (H. U. Bergmeyer, ed.) 3rd ed., Vol. I, p. 1105. Verlag Chemie, Weinheim, 1974.
[12] B. Kassell, this series Vol. 19, p. 845.
[13] H. Schiessler, E. Fink, and H. Fritz, this volume [75].

(personal communication from Dr. Ohlsson)] as substrate (Pentax-PP from Fluka). Constant amounts of the enzymes (50 μg of Pronase, 125 μg of leukocytic elastase) are incubated with increasing amounts of inhibitor (0–50 μg = 0.85 mIU) in 1 ml of triethanolamine-HCl buffer 0.2 M, pH 7.8, for 5 min at 30°. Subsequently, 2 ml of azocasein solution (2% w/v) in the same buffer are added, and the mixture is incubated for 30 min at 30°. The enzymic reaction is stopped by the addition of 3 ml of aqueous trichloracetic acid (5% w/v). Precipitated azocasein is removed by centrifugation, and the extinction at 366 nm of the supernatant is read against a blank sample. The titration curve of leukocytic elastase by the mucus inhibitor is given in Fig. 1. For titration curves of the other enzymes see Hochstrasser et al.[1-3]

Purification Procedure

Step 1. Crude Inhibitor Concentrate. Diluted bronchial mucus is obtained from intubated patients from intensive care wards. The mucus-containing solutions are lyophilized. In the material about 80% of the

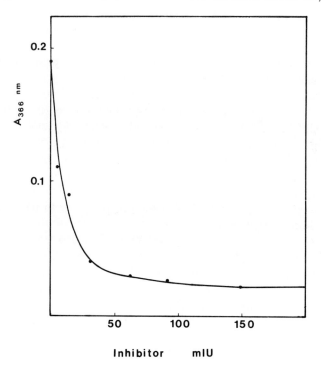

Fig. 1. Titration curve of caseinolytic activity of leukocytic elastase by bronchial mucus inhibitor. Substrate azocasein.

inhibitors are complexed with leukocytic proteinases.[4] By perchloric acid treatment the inhibitors are liberated from the complex. Therefore, lyophilized mucus is dissolved in 3% perchloric acid to a final concentration of 15% (w/v). Precipitated material is removed by centrifugation. The supernatant is neutralized by 5 N KOH and separated from precipitated $KClO_4$. The inhibitor-containing solution is desalted by ultrafiltration. From the lyophilized mucus, the yield is about 15–30 mIU of inhibitor per milligram. The solutions are very opalescent and turbid. By twice freezing and thawing it, this material is coagulated and can be removed by ultracentrifugation. For the isolation procedure, solutions are prepared containing about 150 mIU/ml.

Step 2. Isolation of Inhibitors by Reversible Binding to Insolubilized Trypsin. Insoluble trypsin was prepared by covalent binding of trypsin to Bio-Gel[14] or CM-cellulose.[15] Similar results are obtained with both trypsin resins. The crude solution, adjusted to pH 7.8, is stirred with a small sample of trypsin resin at 0°. After 10 min the decrease of inhibitor activity in a filtered sample is estimated. From the result of this measurement the amount of trypsin resin needed for complete absorption of the inhibitors can be calculated. In consequence, adding trypsin resin is continued until the inhibitor activity is decreased to about 15 mIU/ml. The amount of trypsin resin necessary to absorb the total inhibitor activity cannot be given exactly. The binding capacity of the resin preparations is very variable.

The solution nearly free of inhibitor activity is removed from the resin by centrifugation. The resin is suspended in cold (2°C) triethanolamine-HCl buffer, 0.2 M, pH 7.8, containing 0.2 M NaCl, then centrifuged. This procedure is repeated until the supernatant is colorless.

The adsorbed inhibitors are eluted from the resin by acidifying the suspension in 0.3 M KCl to pH 2.0 with hydrochloric acid. After removal of the trypsin resin, the inhibitor-containing solution is neutralized and concentrated by ultrafiltration. By this procedure up to 95% of the absorbed inhibitor activity is recovered.

Step 3. Gel Filtration. About 500,000 mIU($= 350$ mg of protein)/50 ml from step 2 are applied to a column (3×220 cm) filled with Sephadex G-75 and equilibrated with 0.05 M sodium tetraborate buffer, pH 8.0 containing 0.1 M NaCl. The inhibitor activity is eluted in two fractions. From the elution volume, molecular weights are determined for the minor inhibitor fraction, about 20,000; and for the main inhibitor fraction, 14,000. The inhibitor activities corresponding to both inhibitors are in a

---

[14] I. K. Inman, and H. M. Dintzis, *Biochemistry* **8**, 4074 (1969).
[15] W. Brünner, N. Hennrich, M. Klockow, H. Lang, and H.-D. Orth, *Eur. J. Biochem.* **25**, 129 (1972).

ratio of 15% : 85%. The inhibitors are isolated from the solutions by lyophilization following desalting by ultrafiltration.

Comments on the Purity of the Inhibitors. Provided that the inhibitors react with trypsin in a 1:1 molar ratio, a theoretical specific activity can be predicted from the molecular weights. Specific activity = (mw trypsin)/(mw inhibitor) $\times$ 1000. By this equation, specific activities for the higher and lower molecular weight inhibitors of 1200 and 1700, respectively, can be calculated. The specific activities of the isolated inhibitors are 1150 and 1750 per milligram of protein. This indicates the high purity of the material. Ion-exchange chromatography as described in our earlier papers[1-3] is not necessary.

The inhibitors show electrophoretic polymorphism.[2] This phenomenon is caused by splitting of peptide bonds *in situ* by proteolytic enzymes. As shown in many cases, proteinase inhibitors can be modified by cleavage of peptide bonds without loss of inhibitor activity. Not only peptide bonds in the active sites of the inhibitors are hydrolyzed,[16-18] but also other sensitive bonds.[19]

Properties

Both inhibitors are glycoproteins with a glucosamine content of 5.6 and 8.2, respectively. Probably the inhibitors are derivatives from the humoral proteinase inhibitor inter-α-trypsin inhibitor. The inhibitors can be precipitated with antibodies specific for the humoral inhibitor.[20] In plasma there are acid-stable proteinase inhibitors with an immunological relationship to the inter-α-trypsin inhibitor.[21] The inhibitors are not identical with the mucus inhibitors because they differ in molecular weights and in peptide mapping.[22] We assume that the mucus inhibitors are "secretory pieces" of the humoral inhibitor liberated by a specific converting system in the ciliated membranes.

[16] W. R. Winkenstadt and M. Laskowski Jr., *J. Biol. Chem.* **242,** 771 (1967).
[17] H. Tschesche, E. Wachter, and G. Kallup, *Hoppe-Seyler's Z. Physiol. Chem.* **350,** 1662 (1969).
[18] K. Hochstrasser, K. Illchmann, and E. Werle, *Hoppe-Seyler's Z. Physiol. Chem.* **351,** 721 (1970).
[19] K. Hochstrasser, and G. Bretzel, *Protides Biol. Fluids, Proc. Colloq.* 23 (1975).
[20] K. Hochstrasser, R. Reichert, and N. Heimburger, *Hoppe-Seyler's Z. Physiol. Chem.* **354,** 587 (1973).
[21] K. Hochstrasser, H. Feuth, and O. Steiner, *Hoppe-Seyler's Z. Physiol. Chem.* **354,** 972 (1973).
[22] K. Hochstrasser, H. Feuth, and K. Hochgesand, *Proteinase Inhibitors, Proc. Int. Res. Conf., 2nd (Bayer Symp. V)*, Grosse Ledder, 1973, p. 111. Springer-Verlag, Berlin and New York, 1974.

AMINO ACID COMPOSITIONS OF THE BRONCHIAL MUCUS INHIBITORS (BSI)[a]

Residue	BSI-I (moles/mole)	BSI-II (%)
Asp	16	13.9
Thr	9	6.3
Ser	15	4.1
Glu	12	8.4
Pro	12	9.4
Gly	6	4.8
Ala	5	3.8
$Cys_{1/2}$	12	8.1
Val	2	3.4
Met	2	5.8
Ile	6	3.8
Leu	6	6.2
Tyr	3	5.3
Phe	3	3.9
Lys	8	6.3
His	4	2.1
Arg	4	4.2

[a] The results are calculated from 24, 48, and 72 hr hydrolysis.

The amino acid compositions of the inhibitors are given in the table.

Acknowledgment

This work was supported by Sonderforschungsbereich 51, Munich, of the Deutsche Forschungsgemeinschaft.

[78] Proteinase Inhibitors from the Venom of Russell's Viper

By SADAAKI IWANAGA, HIDENOBU TAKAHASHI, and TOMOJI SUZUKI

Snake venom contains a number of physiologically active polypeptides and peptides, including neurotoxin, cardiotoxin, cytotoxin, and kinin potentiator and others, and the chemical structures of some of them have been established.[1-3] Recently, a new class of polypeptides,

[1] A. T. Tu, *Annu. Rev. Biochem.* **42**, 235 (1973).
[2] E. Karlsson, *Experientia* **29**, 1319 (1973).
[3] L. Ryden, D. Cabel, and D. Eaker, *Int. J. Peptide Protein Res.* **5**, 261 (1973).

which inhibits proteolytic activities of kallikrein, plasmin, trypsin, and α-chymotrypsin, has been found in several snake venoms.[4] These proteinase inhibitors are mainly distributed in the venoms of members of the Viperidae and Elapidae,[5] and the inhibitors isolated from the venom of Russell's viper (RVV), *Hemachatus haemachatus* (HHV), and *Naja nivea* (NNV) are a basic polypeptide with a molecular weight of about 6500.[6-8]

Assay Method

Principle. The hydrolysis of N^α-toluenesulfonyl-L-arginine methyl ester (TAME) by trypsin is measured by the hydroxamate method of Robert.[9]

Reagents

Tris-HCl buffer, 0.4 M, pH 8.5

Substrate solution: 0.1 M TAME (378.9 mg of TAME·HCl are dissolved in 10 ml of distilled water and stored at 4°)

Enzyme. Three times recrystallized bovine trypsin (Worthington Biochemical Corp., Freehold, New Jersey) is dissolved in 0.001 M HCl to give a final concentration of 2 mg/ml and stored at 0°. The enzyme solution is diluted to five times with the same hydrochloric acid before use.

Alkaline hydroxylamine solution: 2 M (13.9%) aqueous hydroxylamine hydrochloride is mixed with an equal volume of 3.5 M NaOH just before use. Store 2 M hydroxylamine at 4°.

Trichloroacetic acid (TCA) solution: 6 g of TCA dissolved in 100 ml of 3 M HCl

Ferric chloride solution: 14.87 g of $FeCl_3 \cdot 6H_2O$ dissolved in 500 ml of 0.04 N HCl and stored in a dark bottle

Snake venom: Lyophilized or desiccated venoms were dissolved in 0.15 M NaCl to give a concentration of 1.0 mg/ml, and the solutions were heated at 80° for 10 min to inactivate proteolytic enzymes contained in the venom. Precipitates appearing after heat treatment were removed by centrifugation at 5000 rpm for 10 min.

[4] H. Takahashi, S. Iwanaga, and T. Suzuki, *FEBS Lett.* **27**, 207 (1972).

[5] H. Takahashi, S. Iwanaga, and T. Suzuki, *Toxicon* **12**, 193 (1974).

[6] H. Takahashi, S. Iwanaga, and T. Suzuki, *J. Biochem.* **76**, 709 (1974).

[7] H. Takahashi, S. Iwanaga, Y. Hokama, T. Suzuki, and T. Kitagawa, *FEBS Lett.* **38**, 217 (1974).

[8] H. Takahashi, S Iwanaga, T. Kitagawa, Y. Hokama, and T. Suzuki, *J. Biochem.* **76**, 721 (1974).

[9] P. S. Robert, *J. Biol. Chem.* **232**, 285 (1958).

Procedure. For the determination of inhibitory activity, a mixture of 0.1 ml of trypsin solution (4 μg) and unfractionated venom or the fraction containing venom proteinase inhibitor in 0.9 ml of 0.4 M Tris-HCl buffer, pH 8.5, is incubated for 10 min; then 0.1 ml of 0.1 M TAME is added. After 30 min, the reaction is terminated by the addition of 0.5 ml of 6% TCA, and then 1.0 ml of alkaline hydroxylamine solution is added. (If a precipitate appears after addition of TCA, the preparation is allowed to stand at room temperature for at least 30 min, and the precipitate is removed by filtration or centrifugation; the filtrate or supernatant is mixed with the hydroxylamine solution.) After the mixture has stood at room temperature for 30 min, 2.0 ml of ferric chloride solution is added; the color developed is measured in 30 min at 500 nm. Formation of gas bubbles in the colorimeter cell is largely prevented if the solution is read against a reagent blank prepared by substituting buffer for enzyme and TAME. A blank of TAME should also be included, which is prepared by substituting buffer for enzyme.

Definition of Unit. One unit is defined as the amount causing reduction of TAME hydrolysis by 1 μmole/min.[6]

Purification Procedure[4,6]

Step 1. Gel filtration on Sephadex G-75. Lyophilized venom of Russell's viper (100 mg, lot S62B-206, obtained from Sigma Chemical Co., St. Louis, Missouri) is dissolved in 5 ml of 0.04 M Tris-HCl buffer, pH 8.5, containing 0.1 M NaCl, and the solution is applied to a column of Sephadex G-75 (2.0 × 128 cm) equilibrated with the same buffer. The column is eluted with the equilibration buffer at 4° and fractions of 5 ml are collected at a flow rate of 50 ml/hr. Figure 1 shows a typical elution pattern. The E peak contains proteinase inhibitors, and the fractions indicated by a solid bar are combined and lyophilized. The dried material (about 32 mg) is dissolved in 2 ml of distilled water and applied to a column (2.5 × 94 cm) of Sephadex G-50 equilibrated with 0.1 M (NH$_4$)HCO$_3$, pH 8.2, to remove Tris-HCl buffer. The resulting unvolatile buffer-free fractions containing the inhibitor are combined, lyophilized, and sublimated at 40°.

Step 2. Chromatography on SE-Sephadex C-25. The dried material (15 mg) of step 1 is dissolved in 2 ml of 0.04 ammonium formate buffer, pH 4.0, and applied to a column of SE-Sephadex C-25 (2.0 × 30 cm) equilibrated with the same buffer. Linear gradient elution is started with 500 ml of each of the equilibration buffer in the mixing vessel and 0.2 M ammonium acetate, pH 9.5, in the reservoir. Then, the column is eluted with a gradient, formed with 0.2 M buffer to 0.5 M, pH 9.5, to elute all

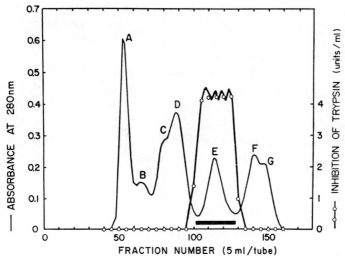

Fig. 1. Gel filtration of Russell's viper venom on a Sephadex G-75 column (2.5 × 128 cm). Experimental conditions: sample volume, 5 ml per 100 mg of crude venom; elution buffer, 0.04 M Tris-HCl buffer, pH 8.5, containing 0.1 M NaCl; flow rate, 50 ml/hr; fraction volume, 5 ml/tube. The fractions indicated by the solid bar were collected and lyophilized. Reproduced from H. Takahashi, S. Iwanaga, and T. Suzuki, *J. Biochem.* **76**, 711 (1974)

the adsorbed materials. Fractions of 5 ml are collected at 4° at a flow rate of 30 ml/hr. Figure 2 shows the elution pattern. Inhibitory activities were found in two peaks, and these fractions were combined and lyophilized (Fig. 2). Ammonium acetate is removed from the dried material *in vacuo* (0.005 mm Hg) at 40°. The overall yields of RVV inhibitor I and II are about 1 and 5 mg, respectively, from 100 mg of crude venom. The above procedures are reproducible and are applicable also for the isolation of proteinase inhibitors from the venoms of *Hemachatus haemachatus*[10] and *Naja nivea*.[11]

Properties[4,6]

Purity. Both RVV inhibitor I and II give a single band on disc polyacrylamide-gel electrophoresis in the presence and in the absence of sodium dodecyl sulfate (SDS).

Physical Properties. The molecular weights of both inhibitors I and II are about 7200 (±1000), as estimated with SDS gel (7.5%) electro-

[10] H. Takahashi, S. Iwanaga, T. Kitagawa, Y. Hokama, and T. Suzuki, *Proteinase Inhibitor, Proc. Int. Res. Conf., 2nd (Bayer Symp. V)*, Grosse Ledder, 1973, p. 265.
[11] T. Tatsuki, Y. Hokama, S. Iwanaga, and T. Suzuki, *Seikagaku* **46**, 704 (1974). In Japanese.

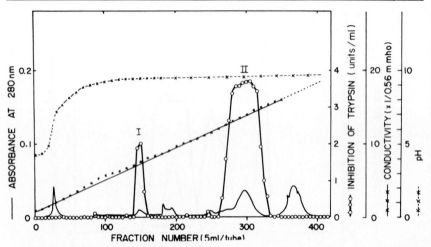

FIG. 2. Purification of the Russell's viper venom proteinase inhibitor on a SE-Sephadex column (2.0 × 30 cm). Experimental conditions: sample, the pooled fraction (15 mg) of Fig. 1; flow rate, 35 ml/hr; fraction volume, 5 ml/tube. A linear gradient elution starts with 500 ml each of 0.04 M ammonium formate buffer, pH 4.0, and 0.2 M ammonium acetate buffer, pH 9.5, and then a gradient of 0.2 M to 0.5 M buffer, pH 9.5, is used to elute all adsorbed material. Reproduced from H. Takahashi, S. Iwanaga, and T. Suzuki, *FEBS Lett.* 27, 207 (1972).

phoresis. The ultraviolet absorption spectra of these inhibitors are identical and show an absorption maximum at 278 nm and minimum at 256 nm at pH 9.0. The extinction coefficients of 1.0% polypeptide solutions of inhibitors I and II at 280 nm are 5.3 and 5.5, respectively.

Stability. RVV inhibitors I and II are stable in the frozen and lyophilized state for at least one year and no loss of inhibitory activity is observed when inhibitor II is heated at pH 2 to 7.5 at 90° for 15 min. However, above pH 8.0 it is denatured.

Specificity. The following enzymes are strongly inhibited by RVV inhibitors I and II: plasma and pancreatic kallikreins, trypsin, α-chymotrypsin, and plasmin from bovine source and human plasmin. The inhibitor II does not inhibit human and bovine thrombins, snake venom kininogenase, Reptilase, bromelain, ficin, papain, subtilisin BPN′, thermolysin, and carboxypeptidases A and B.

Kinetic Properties. RVV inhibitors I and II inactivate bovine trypsin, probably by formation an enzyme-inhibitor complex in a molar ratio of 1:1. The K_i values of inhibitor II, measured using synthetic ester substrate (TAME), are 0.76 nM for bovine trypsin, 0.14 nM (N^α-acetyltyrosine ethyl ester) for bovine α-chymotrypsin; 0.29 nM for bovine plasma kallikrein, and 1.0 nM for bovine plasmin.

TABLE I

AMINO ACID COMPOSITIONS[a] OF PROTEINASE INHIBITORS ISOLATED FROM THE
VENOMS OF RUSSELL'S VIPER, *Hemachatus haemachatus*, AND *Naja nivea*

Amino acid	Russell's viper, II	Russell's viper, I	H. haemachatus II	N. nivea II	N. nivea Ia	N. nivea Ib
Lys	3	2	2	2	2	4
His	2	2	1	2	2	1
Ammonia	(12)	(11)	(7)	(6)	(7)	(6)
Arg	7	5	5	7	6	4
Asp	8	7	5	4	4	5
Thr	3	2	3	3	3	3
Ser	2	2	1	1	1	2
Glu	5	5	6	6	7	4
Pro	2	5	2	2	2	3
Gly	8	7	6	6	6	3
Ala	2	3	5	5	5	4
$Cys_{1/2}$	6	4	6	6	6	6
Val	1	1	1	1	1	3
Met	0	0	0	0	0	1
Ile	1	2	3	3	3	3
Leu	3	1	4	3	3	3
Tyr	3	1	3	2	2	2
Phe	4	3	4	4	4	3
Trp	0	0	0	—	—	—
	60	52	57	57	57	54

[a] The compositions were calculated (in residues per mole) by extrapolation or from the average values estimated on samples after 24-, 48-, and 72-hr hydrolyzates.

Amino Acid Composition.[6,10] The compositions of several inhibitors isolated from the venoms of Russell's viper, *Hemachatus haemachatus*, and *Naja nivea* are shown in Table I. Common features include relatively high cystine content and few or no methionine and tryptophan residues; no sugar residue is detected in any inhibitor.

Amino Acid Sequence.[7,8] The sequences of RVV inhibitor II and HHV inhibitor II are shown in Fig. 3, in comparison with those of bovine pancreatic trypsin inhibitor[12] and snail isoinhibitor K.[13] The overall primary structure of snake venom proteinase inhibitors is quite similar to those of the Kunitz-type inhibitors, indicating 50–60% sequence homologies. Moreover, the 6 half-cystine residues of these four inhibitors are in the same positions along the polypeptides (Fig. 4). The reactive sites of

[12] B. Kassell, M. Radicevic, M. J. Anfield, and M. Laskowski, Sr., *Biochem. Biophys. Res. Commun.* **18**, 255 (1965).

[13] H. Tschesche, *Angew. Chem.* **86**, 21 (1974).

```
RVV II    His-Asp-Arg-Pro-Thr-Phe-Cys-Asn-Leu-Ala-Pro-Glu-Ser-Gly-Arg-Cys-Arg-Gly-His-Leu-
HHV II          Arg-Pro-Asp-Phe-Cys-Glu-Leu-Pro-Ala-Glu-Thr-Gly-Leu-Cys-Lys-Ala-Tyr-Ile-
SNAIL IK  Pyr-Gly-Arg-Pro-Asp-Phe-Cys-Glu-Leu-Pro-Ala-Glu-Thr-Gly-Pro-Cys-Lys-Ala-Ser-Phe-
BPTI            Arg-Pro-Asp-Phe-Cys-Leu-Glu-Pro-Pro-Tyr-Thr-Gly-Pro-Cys-Lys-Ala-Arg-Ile-
```

```
          Arg-Arg-Ile-Tyr-Tyr-Asn-Leu-Glu-Ser-Asn-Lys-Cys-Lys-Val-Phe-Phe-Tyr-Gly-Gly-Cys-
          Arg-Ser-Phe-His-Tyr-Asn-Leu-Ala-Ala-Gln-Lys-Cys-Leu-Glu-Phe-Ile-Tyr-Gly-Gly-Cys-
          Arg-Gln-Tyr-Tyr-Tyr-Asn-Ser-Lys-Ser-Gly-Gly-Cys-Gln-Gln-Phe-Ile-Tyr-Gly-Gly-Cys-
          Ile-Arg-Tyr-Phe-Tyr-Asn-Ala-Lys-Ala-Gly-Leu-Cys-Gln-Thr-Phe-Val-Tyr-Gly-Gly-Cys-
```

```
          Gly-Gly-Asn-Ala-Asn-Asn-Phe-Glu-Thr-Arg-Asp-Glu-Cys-Arg-Glu-Thr-Cys-Gly-Gly-Lys
          Gly-Gly-Asn-Ala-Asn-Arg-Phe-Lys-Thr-Ile-Asp-Glu-Cys-Arg-Arg-Thr-Cys-Val-Gly
          Arg-Gly-Asn-Gln-Asn-Arg-Phe-Asp-Thr-Thr-Gln-Gln-Ile-Cys-Gln-Gly-Val-Cys-Val
          Arg-Ala-Lys-Arg-Asn-Asn-Phe-Lys-Ser-Ala-Glu-Asp-Cys-Met-Arg-Thr-Cys-Gly-Gly-Ala
```

Fig. 3. Comparison of the amino acid sequences of Russell's viper venom (RVV) inhibitor II, *Hemachatus haemachatus* (HHV) inhibitor II, snail isoinhibitor K, and bovine pancreatic trypsin inhibitor (Kunitz type). Based on H. Takahashi, S. Iwanaga, Y. Hokama, T. Suzuki, and T. Kitagawa, *FEBS Lett.* **38**, 217 (1974). B. Kassell, M. Radicevic, M. J. Anfield, and M. Laskowski, Sr., *Biochem. Biophys. Res. Commun.* **18**, 255 (1965); and H. Tschesche, *Angew. Chem.* **86**, 21 (1974).

RVV inhibitor II and HHV inhibitor II are still unknown, but the most probable site for interaction with trypsin is the peptide linkage Arg_{17}–Gly_{18} in the former, and Lys_{15}–Ala_{16} in the latter, because the homologies of all the inhibitors around these portions are extremely high. These results are very interesting in relation to the evolution of these inhibitors, since evolutionary histories of cows, snakes, and snails are quite different.

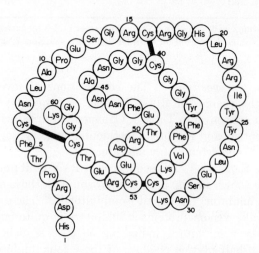

Fig. 4. A schematic representation of the structure of the proteinase inhibitor isolated from the venom of Russell's viper. Based on T. Takahashi, S. Iwanaga, Y. Hokama, T. Suzuki, and T. Kitagawa, *FEBS Lett.* **38**, 217 (1974); and H. Takahashi, S. Iwanaga, T. Kitagawa, Y. Hokama, and T. Suzuki, *J. Biochem.* **76**, 721 (1974).

TABLE II
DISTRIBUTION OF PROTEINASE INHIBITORS IN VARIOUS SNAKE VENOMS[a,b]

Family	Bovine plasma kallikrein	Bovine plasma plasmin	Bovine pancreatic trypsin	Bovine pancreatic α-chymotrypsin
Viperidae				
V. russelli	12	18	5	5
V. ammodytes	40	91	22	77
B. arietans	88	110	74	217
Elapidae				
N. hannah	100	62	55	80
N. nivea	100	12	3	15
N. haje	100	42	60	20
H. haemachatus	4	20	22	13
D. angusticeps	125	34	18	34
D. polylepis	215	43	28	105

[a] From H. Takahashi, S. Iwanaga, and T. Suzuki, *Toxicon* **12**, 193 (1974).

[b] The values were expressed as the quantities (μg) to show 50% inhibition of the activity of the proteinases. The following venoms showed no inhibitory activity when 100 μg of each was used: *V. palestinae, E. carinatus, B. gabonica, N. naja atra, N. melanoleuca, N. nigricollis, N. naja samarensis, N. naja, B. fasciatus, B. multicinctus, L. semifasciata, L. laticaudata, A. halys blomhoffii, A. acutus, A. rhodostoma, A. piscivorus leucostoma, A. contortrix contortrix, A. piscivorus piscivorus, A. contortrix mokeson, C. atrox, C. adamanteus, C. viridis viridis, C. basiliscus, C. durissus terrificus, B. atrox, B. jararaca, T. flavoviridis, T. gramineus, T. okinavensis, T. mucrosquamatus, Causus rhombeatus.*

Distribution.[5,14] Proteinase inhibitor of the same type as described here is demonstrated also in the venoms of several snakes of the Elapidae and Viperidae (Table II). However, it is not found in Crotalidae and Hydrophiidae venoms.

[14] D. J. Strydom, *Nature (London) New Biol.* **243**, 88 (1973).

[79] Broad-Specificity Inhibitors from Sea Anemones

By GERT WUNDERER, LÁSZLO BÉRESS, WERNER MACHLEIDT, and HANS FRITZ

In the purification of neurotoxins from *Anemonia sulcata*, fractions with antitryptic activity were observed.[1] This inhibitory activity could be ascribed to a variety of basic polypeptides.[2,3] The broad specificity

[1] L. Béress and R. Béress, *Kiel. Meeresforsch.* **27**, 117 (1971).

[2] H. Fritz, B. Brey, and L. Béress, *Hoppe-Seyler's Z. Physiol. Chem.* **353**, 19 (1972).

[3] G. Wunderer, K. Kummer, H. Fritz, L. Béress, and W. Machleidt, *Proteinase Inhibitors, Proc. Int. Res. Conf., 2nd (Bayer Symp. V)*, Grosse Ledder, 1973, p. 277.

of these inhibitors for trypsin, chymotrypsin, plasmin, and kallikreins closely resembles that of the basic pancreatic trypsin inhibitor (BPTI).[2]

The present article describes our revised procedure for the isolation of the proteinase inhibitors from *A. sulcata*.

Assay Method

Trypsin inhibition by the sea anemones inhibitors is tested with N^α-benzoyl-DL-arginine *p*-nitroanilide (BAPA) as substrate, as described by Kassell.[4]

Definition of Units. One tryptic unit is defined as the enzyme activity that cleaves 1 μmole of DL-BAPA per minute under the described conditions. One inhibitor unit reduces the trypsin-catalyzed hydrolysis of DL-BAPA by 1 μmole/min.

Purification Procedure

Step 1. Preparation of the Sea Anemones. Anemonia sulcata were collected in the bay of Naples. The fresh sea anemones are washed and centrifuged for 10 min at 1500 *g*. The supernatant is discarded because it contains only negligible amounts of inhibitors (and toxins). The remaining sediment is deep-frozen and stored at −20°. After this procedure the dry weight of the material amounts to 13% of its wet weight.

Step 2. Extraction. Two kilograms of prepared sea anemones are homogenized in 2 liters of 96% ethanol with a Cenco Waring blender in several portions. The homogenate is heated transiently to 60°. After cooling to room temperature, the precipitated proteins and cell fragments are spun down (15 min at 2000 *g*). The brown supernatant is collected. The sediment is homogenized for a second time in 1 liter of 50% ethanol, heated to 50°, cooled, and centrifuged. This time the residue is discarded. The combined supernatants are dialyzed for 16 hr against 40 liters of deionized water and filtered through a paper disk (Seitz K2) on a 30-cm Büchner funnel.

Step 3. Batchwise Adsorption of the Inhibitors on CM-Cellulose. Batch I: The extracts obtained in step 2 are diluted with deionized water to a conductivity of 4 mS × cm⁻¹, about 16 liters total, and adjusted to pH 6.5 with acetic acid. Then 60 g of dry CM-cellulose (Serva; capacity: 0.55 mEq/g) are added; after readjustment of the pH to 6.5, the suspension is stirred for 30 min. The suspension is filtered through a fritted glass filter, and the loaded cellulose exchanger is washed with deionized water. The CM-cellulose, batch I, contains most of the inhibitors

---

[4] B. Kassell, this series, Vol. 19, p. 845.

(60.5%) and a small amount of toxin II (9% of total toxicity, cf. Béress et al.[5,6]).

Batch II: The filtrate of batch I is adjusted to pH 5.0 with acetic acid and diluted to a conductivity of 2 mS $\times$ cm^{-1}. Then 60 g of dry CM-cellulose are added. After correction of pH, the suspension is stirred for 30 min and filtered; the cellulose is washed thoroughly with deionized water. This CM-cellulose, batch II, contains the remaining inhibitors and another part of toxin II (20% of total toxicity).

For the isolation of the toxins along with the inhibitors, a third batch (SP-Sephadex C-25, pH 3.2, 2 mS $\times$ cm^{-1}) is processed. This batch III does not contain inhibitors.

For the isolation of the inhibitors, the proteins adsorbed in batches I and II are eluted from CM-cellulose with about 1.5 liters of buffer (0.05 M Tris-HCl, pH 8.0, 1 M NaCl) each. The elutes are concentrated by ultrafiltration with an Amicon UM-05 or UM-2 filter to final volumes of about 70 ml.

Step 4. Gel Filtration on Sephadex G-50. The concentrated solutions obtained in step 3 are each chromatographed on a 7 $\times$ 120 cm Sephadex G-50 medium column equilibrated and developed with 2% acetic acid (v/v) at a flow rate of 3 ml/min. The inhibitors are eluted in 3 to 4 times the void volume separated from accompanying proteins and colored products. The inhibitor-containing fractions are collected and concentrated in a rotary vacuum evaporator. As the second half of the inhibitor peak still contains salts, it is desalted again on Sephadex G-25 (column: 5.6 $\times$ 120 cm, equilibrated and developed with 2% acetic acid (v/v), 120 ml/hr) and finally lyophilized. The yields and purification factors calculated from steps 2–4 are summarized in Table I.

Step 5. Ion-Exchange Chromatography on SP-Sephadex. The inhibitor fraction from step 4 can be separated into 10 multiple forms by ion-exchange chromatography on SP-Sephadex. A microanalytical system was used to find optimal conditions for the separation.[7] This system consists of a 0.1 $\times$ 100 cm microbore column filled with SP-Sephadex C-25, equilibrated with 25 mM phosphate buffer, pH 6.13 (50 mM Na$^+$) and developed with a concave gradient of sodium chloride in 25 mM phosphate buffer. Continuous flow-cell photometry at 224, 253, and 280 nm permits a detection of less than 1 nmole (about 7 μg) of inhibitor in the eluates as well as the characterization of peaks by their different UV absorption. From the reproducible results obtained with this system we were able to select the most suitable sodium ion concentration range for

[5] L. Béress, R. Béress, and G. Wunderer, *FEBS Lett.* **50**, 311 (1975).
[6] L. Béress, R. Béress, and G. Wunderer, *Toxicon* **13**, 359 (1975).
[7] W. Machleidt, W. Kerner, and J. Otto, *Z. Anal. Chem.* **252**, 151 (1970).

TABLE I

YIELDS AND PURIFICATION FACTORS CALCULATED FOR STEPS 2, 3, AND 4
IN THE ISOLATION PROCEDURE

Step	Procedure	Volume (ml)	IU[a]	$(IU \times ml)/A_{280}{}^{b}$	Purification factor	Yield (%)
2	Combined extracts	3750	9700	0.065	1	100
	dialysis	6500	8400	0.072	1.1	86.5
3	Eluate from batch I	1570	6550	3.92	60.5	67.5
	Eluate from batch II	1380	1820	1.41	21.7	18.8
4	Inhibitor fraction after					
	Sephadex G 50: Batch I	1570	6500	5.64	86.8	66
	Batch II	910	776	2.30	35.4	16.9

[a] Trypsin inhibitory activity, substrate: N^{α}-benzoyl-DL-arginine p-nitroanilide; 1 IU is equivalent to the inhibition of about 1 mg of bovine trypsin.
[b] Absorption at 280 nm.

the preparative separation of the inhibitors (Fig. 1). The same system was used to check the purity of the fractions isolated on a preparative scale.

The preparative 1.5 × 100 cm SP-Sephadex C-25 column is equilibrated with a 50 mM phosphate buffer (100 mM Na$^+$), pH 6.13, and developed with a linear gradient of NaCl in the starting buffer up to 400 mM Na$^+$ and in a second step from 400 to 1000 mM Na$^+$ (flow rate 70 ml/hr; Fig. 2). The 9 inhibitor fractions obtained in this chromatography are each desalted on a 3.5 × 120 cm Sephadex G-50 superfine column (50 mM NH$_4$HCO$_3$ pH 8.1, flow rate 50 ml/hr) and lyophilized.

Step 6. Rechromatography of the Inhibitors on SP-Sephadex. Inhib-

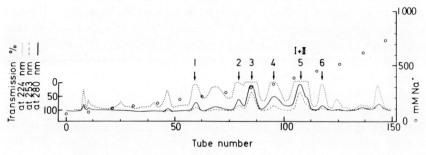

FIG. 1. Separation of proteinase inhibitors from *Anemonia sulcata*. Analytical scale: 0.1 × 100 cm SP-Sephadex C-25 column equilibrated with 25 mM phosphate buffer, pH 6.13, 50 mM Na$^+$, and developed with a concave NaCl gradient (up to 1 M Na$^+$), 0.385 ml/hr, 2 tubes/hr, 25°. 250 nmoles of the inhibitor mixture were applied. Transmissions were registered continuously at 224 (·····), 253 (----), and 280 (——) nm.

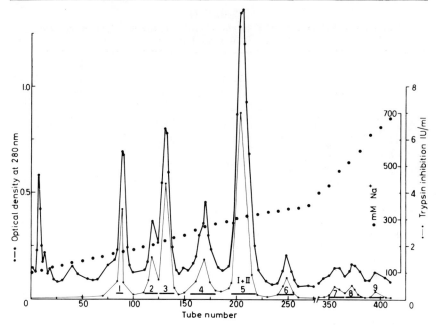

FIG. 2. Preparative separation of proteinase inhibitors on a 1.5 × 100 cm SP-Sephadex C-25 colum equilibrated with 50 mM phosphate buffer, pH 6.13, 100 mM Na⁺, and developed first with a linear NaCl gradient up to 400 mM Na⁺ followed by a second linear NaCl gradient from 400 to 1000 mM Na⁺, flow rate 70 ml/hr, 6 tubes/hr, 25°. Up to 1 g of the inhibitor mixture was applied. Upper boldface line: absorption at 280 nm– thin line: inhibitor activity in IU.

itor fractions 2, 3, 5I, 5II, 6, 7, and 8 (cf. Fig. 2) are further purified by chromatography on a SP-Sephadex C-25 column (1 × 100 cm, flow rate 50 ml/hr). For this purpose the column is equilibrated with 50 mM phosphate buffer pH 6.13 and developed in equilibrium with selected Na⁺ concentrations. The yields of the inhibitors obtained in step 5 and the conditions for rechromatography (step 6) are given in Table II. The rechromatographed inhibitors are desalted on Sephadex G-50 as described in step 5 and lyophilized. Inhibitor I is further purified by affinity chromatography on trypsin covalently bound to CM-cellulose as described earlier.[2,8]

Properties

Stability. Like other low-molecular-weight trypsin inhibitors, the inhibitors from *Anemonia sulcata* are very stable. They do not loose activ-

[8] G. Wunderer, H. Fritz, W. Brümmer, N. Hennrich, and H.-D. Orth, *Biochem. Soc. Trans.* **2**, 1324 (1974).

TABLE II
RELATIVE YIELDS IN SP-SEPHADEX CHROMATOGRAPHY (STEP 5) AND
CONDITIONS FOR RECHROMATOGRAPHY (STEP 6)

		Inhibitors								
		1	2	3	4	5	6	7	8	9
Relative yields	Batch I	2.4	5.2	14.0	18.5	24.4	4.8	3.2	2.8	0.5
(%) in step 5	Batch II	2.3	5.2	14.0	21.0	29.0	3.0	1.6	0.4	0.2
(cf. Fig. 2)										
Na$^+$ concentration,		—^{a}	205	205	—^{b}	330	350	470	510	—^{b}
used in rechroma-										
tography (mM)										
(Step 6)										

^{a} Enriched by affinity chromatography [H. Fritz, B. Brey, and L. Béress, *Hoppe-Seyler's Z. Physiol. Chem.* **353**, 19 (1971); G. Wunderer, H. Fritz, M. Brümmer, N. Hennrich, and H.-D. Orth, *Biochem. Soc. Trans.* **2**, 1324 (1974)].
^{b} Not rechromatographed.

ity when incubated at room temperature in the pH range 1.5 to 10 for several hours. Lyophilized material stored at —20° retains its activity for at least 3 years.

Molecular Weight. As estimated by gel filtration on calibrated Sephadex G-50 columns, the isolated inhibitors have molecular weights between 5500 and 7000. These values are confirmed by SDS gel electrophoresis of inhibitor 5II in its reduced form. They are also in agreement with those values calculated from the specific trypsin inhibitory activity.

Amino Acid Compositions. Table III gives the amino acid compositions of inhibitors 1 to 7 as calculated for molecular weights between 5500 and 7000. Inhibitor 5 II (the major component) is composed of 60 residues and has a calculated molecular weight of 6799.8. Its molecular absorption coefficient is $A_{1\,cm}^{1\%}$ at 280 nm = 7.24.

N-Terminal Residues. With the dansylation technique, the N-terminal amino acid residues of inhibitors 1–5 II were found to be isoleucine, whereas for inhibitors 6–9 an end group could not be detected. In manual Edman degradation of inhibitor 5 II, isoleucine was found as a single N-terminal residue, in 81% yield, followed by asparagine and glycine.

Specificity. Each of the isolated inhibitors inhibits strongly bovine trypsin and chymotrypsin, porcine plasmin, and porcine pancreatic kallikrein.[2,3] Human and porcine serum kallikreins,[9] human plasmin,[10] as

[9] G. Wunderer, K. Kummer, and H. Fritz, *Hoppe-Seyler's Z. Physiol. Chem.* **353**, 1646 (1972).
[10] W. Kalckreuth, Dissertation, Faculty of Medicine, University of Munich, 1972.

TABLE III
AMINO ACID ANALYSIS DATA[a] FOR INHIBITORS 1 TO 7 (MOLES PER MOLE)

Residue	1	2	3	4	5 I	5 II	6	7
Asp	6.00	6.00	6.00	6.00	6.14	6.00	6.34	6.00
Thr	1.91	1.75	1.00	1.79	1.03	0.99	1.40	3.45
Ser	4.71	3.80	4.32	4.11	3.55	3.94	3.83	3.62
Glu	4.07	4.95	5.06	4.89	5.01	5.00	4.06	2.24
Pro	3.72	2.88	2.75	2.89	2.64	2.18	2.11	2.90
Gly	8.27	6.73	6.11	7.21	7.00	6.95	7.00	6.29
Ala	2.15	2.05	1.59	2.75	2.06	1.75	2.67	6.98
Cys[b]	5.98	5.05	5.82	2.82	5.79	5.95	5.31	5.30
Val	2.09	2.14	3.17	3.05	2.39	3.92	2.82	2.63
Met	—	—	—	0.30	—	—	—	0.85
Ile	1.84	0.92	1.62	1.69	1.47	1.82	1.00	0.48
Leu	2.87	1.89	2.13	2.53	2.11	2.86	2.83	1.22
Tyr	2.16	3.43	3.17	2.60	2.36	3.94	2.07	2.07
Phe	1.39	2.37	2.08	2.64	2.98	2.99	2.46	1.79
His	1.98	0.99	1.78	0.83	1.07	1.00	1.56	1.07
Lys	4.46	3.70	4.00	4.10	3.72	3.76	3.87	6.85
Arg	4.98	4.67	5.72	4.66	6.39	6.31	5.90	2.00

[a] The values given were obtained from 20-hr hydrolyzate and are not corrected.
[b] Determined as cysteic acid after performic acid oxidation.

well as boar sperm acrosin,[11] are also inhibited by the mixture of the multiple inhibitor forms.

Kinetic Properties. Trypsin is inhibited stoichiometrically (molar ratio 1:1) by inhibitor 5 II. The dissociation constant of the complex of inhibitor 5 II with bovine trypsin is determined to approximately 3×10^{-10} moles per liter according to Green and Work.[12] The affinities of the *Anemonia sulcata* inhibitors and of BPTI to trypsin, plasmin, and chymotrypsin are comparable[2]; the affinities to porcine pancreatic kallikrein, however, increase with the basicity of the inhibitors.[3] In contrast to our results published earlier,[3] we could neither degrade nor modify inhibitor 5II with TPCK-treated trypsin from E. Merck, Darmstadt, under various acidic conditions (4 mole/100 moles trypsin; pH 3.0, 3.75, 4.0, and 5.0). This discrepancy may be explained by the action of accompanying proteinases in the trypsin preparation used formerly.

Discussion

The procedure described permits the isolation of 10 proteinase inhibitors from *Anemonia sulcata*. It proved to be reproducible in at least 20

[11] H. Fritz, B. Förg-Brey, H. Schiessler, M. Arhnold, and E. Fink, *Hoppe-Seyler's Z. Physiol. Chem.* **353**, 1010 (1972).
[12] N. M. Green and E. Work, *Biochem. J.* **54**, 347 (1953).

independent preparations. The procedure can be used as a general method for the separation of basic and neutral polypeptides from tissues. For example, using this procedure 3 neurotoxins from *A. sulcata*[5,6] and 4 neurotoxins from *Condylactis aurantiaca*[13] could be separated from the more basic inhibitors. For the isolation of all basic and neutral polypeptides, we recommend that the complete procedure, including 3 batches, be followed. If only the proteinase inhibitors from *A. sulcata* are to be isolated, a single batch II is sufficient.

As judged from chromatographic data, amino acid compositions, and N-terminal end-group determinations, at least inhibitors 2, 3, 5 I, 5 II, and 6 are pure polypeptides. Inhibitor 5 II was also shown to be pure using SDS gel electrophoresis. The amino acid compositions found for the other inhibitors suggest the presence of minor impurities. A further purification of these fractions was not attempted. A comparison of the amino acid compositions shows that the inhibitors from *Anemonia sulcata* differ only in minor amino acid exchanges, i.e., they are very probably isoinhibitors.

The main inhibitor fraction 5 II was chosen for structural investigations.[14,15] From its amino acid composition, a homology with the known sequences of BPTI,[16] CTI,[17] and inhibitors from snails,[18] and vipers[19] is to be expected. This is supported by preliminary results of the sequence determination.[14,15] The N-terminal sequence of inhibitor 5 II, Ile-Asn-Gly-Asp-Cys-Glu-Leu-Pro-Lys-Val-Val-Gly-Pro-Cys-Arg-Ala-Arg-Phe-Pro-Arg-Tyr-Tyr-Tyr-Asn-Ser-Ser-Ser-Lys-Arg-Cys-Glx-Lys-Phe-Ile-Tyr-Gly-Gly-Cys-Arg- is homologous with the sequence in positions 5–39 (which includes the active center) of BPTI, and there is a further sequence Lys-Val-Cys-Gly-Val-Arg-Ser in inhibitor 5 II which is partially homologous with positions 53–56 of BPTI. Therefore we confirm that the inhibitor 5 II from sea anemones contains Arg in its reactive center as postulated earlier.[2]

[13] R. Béress, L. Béress, and G. Wunderer, *Hoppe Seyler's Z. Physiol. Chem.* **357**, 409 (1976).

[14] G. Wunderer, L. Béress, W. Machleidt, and H. Fritz *in* "Protides of the Biological Fluids, 23rd Colloquium" (H. Peeters, ed.), p. 285. Pergamon, Oxford and New York, 1976.

[15] G. Wunderer, Dissertation, Technical University of Munich, 1975.

[16] B. Kassell and M. Laskowski, Sr., *Biochem. Biophys. Res. Commun.* **18**, 255 (1965).

[17] D. Čechová, V. Jonáková, and F. Šorm, *Proteinase Inhibitors, Proc. Int. Res. Conf., 1st,* Munich, 1970, p. 105. de Gruyter, Berlin, 1971.

[18] T. Dietl and H. Tschesche, *Proteinase Inhibitors, Proc. Int. Res. Conf., 2nd (Bayer Symp. V)* Grosse Ledder, 1973, p. 254. Springer-Verlag, Berlin and New York, 1974.

[19] H. Takahashi, S. Iwanaga, Y. Hokama, T. Suzuki, and T. Kitagawa, *FEBS Lett.* **38**, 217 (1974).

Author Index

Numbers in parentheses are footnote reference numbers and indicate that an author's work is referred to although his name is not cited in the text.

Frère, J.-M., 610, 613, 615(10), 617, 618, 619(10, 12), 620, 621, 622 (10, 12), 623, 624(16, 17, 18), 626, 627(14), 628(16), 629(14, 16, 17), 630, 631, 632, 634(3, 22), 635(23, 25), 636(20)

Frey, E. K., 289, 303(2)

Friedmani, J. A., 166, 172(46)

Friedmann, J. A., 127, 133, 136(25), 137, 156, 168, 171, 172(80), 174, 648

Fritz, H., 36, 289, 291, 296(6), 297, 300, 303 (6), 306, 326, 327, 331, 332, 334(20, 41), 335(20, 27, 41), 336, 338(28), 339 (28, 32, 38, 49), 340 (28a, 32), 341, 342 (28a, 32, 41), 671, 678(11), 682, 770, 772, 773(1), 774, 775(1), 776(1, 25), 778(1), 779(1), 780(1), 781(1), 785 (1), 786(3), 791(3), 793, 794, 796, 797, 798(4), 799(4, 16), 802(4), 803 (11), 804(4), 805(4), 807, 811, 812(21), 813, 814(6), 822, 823(31), 824(6, 19), 825, 826(4), 827(3, 4), 828(10), 831 (3, 4), 832, 833, 834, 835(7, 9, 10), 836(11, 15), 837(7, 14, 15), 838(5, 7), 840, 841 (18), 842(3, 5, 11), 843 (3, 5, 7, 11, 20), 844 (3, 5, 7, 9, 10, 11), 845, 846 (7, 11), 847, 848(5), 849, 853, 857(3, 5), 858, 859, 860, 861(3, 5), 865(3), 866(3), 867(3), 868(3), 869(4), 870, 881, 882(2), 885(2), 886(2, 3), 887(2, 3), 888(2)

Frolme, M., 508, 510, 511, 514, 515, 516, 518(35)

Fruton, J. S., 450, 451, 587, 692

Fu, T., 484

Fuad, N., 636

Fudenberg, H. H., 640, 650

Fujii, M., 728

Fujii, S., 314

Fujikawa, K., 31, 32, 33, 34(8), 35(8, 9), 36(16a), 53, 54, 55, 65, 66, 74, 76, 81, 82(1, 3, 5), 83, 89, 90(4), 92, 93, 94, 95, 96, 97(9), 98, 99(14), 102(16), 106(19, 22), 191, 194(5), 202(5), 655

Fujimoto, K., 683, 684(16), 685, 686(20)

Fukumura, M., 690

Fukushima, H., 289, 293(8), 297

Funakoshi, T., 561, 562(4), 567(4), 568(4), 586

Furie, B., 191, 196, 197(6), 198(10, 11), 199, 202(6), 204(11, 20), 205

Furie, B. C., 191, 192, 194(9), 195, 196, 197 (6), 198(10, 11), 199, 202(6), 203(9), 204(11), 205

Furlanetto, R. W., 4, 5(7), 7(8), 8(8), 181

G

Gaddum, P., 330

Gage, T., 649

Gallango, M. L., 650

Galvanek, E. G., 317

Ganrot, P.-O., 124, 151, 152, 640, 641, 648 (29, 32), 649(33, 62), 650

Garattini, S., 222

Garcia, L. A., 649

Garner, C. W., 503

Garner, D. L., 331, 339(31), 340

Gatmaitan, J. S., 377, 379(2), 381, 385(2, 3), 391, 392(7)

Gaucher, G. M., 415, 416(2, 5), 417(5), 418(4, 5), 419(1, 2, 5, 6), 420, 421(2, 3, 4, 5), 422, 424(5), 426(5, 6), 427, 428 (5, 6), 429, 430, 431(4, 5, 6), 432(5), 433(5)

Gebhardt, M., 671, 678(11), 770, 774, 775, 776(25), 797, 798(4), 799(4), 802(4), 804(4), 805(4), 828, 853, 859

Geokas, M. C., 649

Gerandas, M., 655, 657(20)

Geratz, J. D., 173

Gertler, A., 419, 426(11), 428, 705, 707(17), 716, 718(1), 719, 721(1), 722(1), 867

Gerwin, B. I., 432

Ghuysen, J.-M., 610, 612, 613, 614, 615 (10), 617, 618, 619(10, 12), 620, 621, 622(10, 12), 623, 624(16, 17, 18), 626, 627(14), 628(16), 629(14, 16, 17), 630, 631, 632(11), 634(3, 22), 635(23, 25), 636(20)

Gibian, H., 472

Giddings, J. C., 120, 121(45)

Gigli, I., 314, 317(35), 752, 756(9), 758(9)

Giglio, J. R., 166, 173(51)

Gilboa, E., 331, 339(33), 341

Ginodman, L. M., 452

Girey, G. J. D., 315, 317

Girol, D., 27

Girolami, G., 135

Gitel, S. N., 116, 153, 655, 656(18)

H

Subject Index

A

SEP 1 2 1994

14 DAY

14 DAY   14 DAY   14 DAY

FEB 2 6 1992   OCT 2 6 1993   DEC 19 1994

FEB 2 6 1992   RETURNED

                RETURNED   DEC 2
14 DAY   OCT 2 5 1993

JUL 1 5 1993   14 DAY

                AUG 3 1 1994
14 DAY

OCT 1 8 1993   14 DAY

RETURNED   SEP 2 1 1994

OCT - 5 1993   Series 4128